Malcolm Griffiths
January 1997

STP 1295

Zirconium in the Nuclear Industry: Eleventh International Symposium

E. Ross Bradley and George P. Sabol, editors

ASTM Publication Code Number (PCN):
04-012950-04

ASTM
100 Barr Harbor Drive
West Conshohocken, PA 19428-2959

ISBN: 0-8031-2406-6
PCN: 04-012950-04
ISSN: 1050-7558

Copyright © 1996 AMERICAN SOCIETY FOR TESTING AND MATERIALS, West Conshohocken, PA. All rights reserved. This material may not be reproduced or copied, in whole or in part, in any printed, mechanical, electronic, film, or other distribution and storage media, without the written consent of the publisher.

Photocopy Rights

Authorization to photocopy items for internal, personal, or educational classroom use, or the internal, personal, or educational classroom use of specific clients, is granted by the American Society for Testing and Materials (ASTM) provided that the appropriate fee is paid to the Copyright Clearance Center, 222 Rosewood Drive, Danvers, MA 01923; Tel: 508-750-8400; online: http://www.copyright.com/.

Peer Review Policy

Each paper published in this volume was evaluated by three peer reviewers. The authors addressed all of the reviewers' comments to the satisfaction of both the technical editor(s) and the ASTM Committee on Publications.

The quality of the papers in this publication reflects not only the obvious efforts of the authors and the technical editor(s), but also the work of these peer reviewers. The ASTM Committee on Publications acknowledges with appreciation their dedication and contribution to time and effort on behalf of ASTM.

Printed in Ann Arbor, MI
November 1996

Foreword

The Eleventh International Symposium on Zirconium in the Nuclear Industry was held in Garmisch-Partenkirchen, Germany on 11–14 Sept. 1995. The sponsor of the event was ASTM Committee B-10 on Reactive and Refractory Metals and Alloys.

The symposium chairman was E. Ross Bradley, Sandvik Special Metals Corporation, the symposium co-chairman was Erich Tenckhoff, Siemens AG, and the editorial chairman was George P. Sabol, Westinghouse Electric Corporation. Serving as editors of this publication were E. Ross Bradley and George P. Sabol.

Contents

Overview   xi

KROLL AWARD PAPERS

Learning from History: A Case Study in Nuclear Fuel—J. A. L. ROBERTSON   3

Behavior and Properties of Zircaloys in Power Reactors: A Short Review of Pertinent Aspects in LWR Fuel—F. GARZAROLLI, H. STEHLE, AND E. STEINBERG   12

CORROSION MECHANISMS—I

Microstructure of Oxides on Zircaloy-4, 1.0Nb Zircaloy-4, and Zircaloy-2 Formed in 10.3-MPa Steam at 673 K—H. ANADA AND K. TAKEDA   35
Discussion   54

The Importance of Oxide Morphology for the Oxidation Rate of Zirconium Alloys—G. WIKMARK, P. RUDLING, B. LEHTINEN, B. HUTCHINSON, A. OSCARSSON, AND E. AHLBERG   55
Discussion   73

Effect of Annealing Temperature on Corrosion Behavior and ZrO_2 Microstructure of Zircaloy-4 Cladding Tube—H. ANADA, B. J. HERB, K. NOMOTO, S. HAGI, R. A. GRAHAM, AND T. KURODA   74
Discussion   92

Microstructure of Oxide Films Formed during the Waterside Corrosion of the Zircaloy-4 Cladding in Lithiated Environment—D. PÊCHEUR, J. GODLEWSKI, P. BILLOT, AND J. THOMAZET   94
Discussion   113

Mechanisms of LiOH Degradation and H_3BO_3 Repair of ZrO_2 Film—B. COX, M. UNGURELU, Y.-M. WONG, AND C. WU   114
Discussion   136

PWR Zircaloy Fuel Cladding Corrosion Performance, Mechanisms, and Modeling—B. CHENG, P. M. GILMORE, AND H. H. KLEPFER   137
Discussion   159

Corrosion Mechanisms—II

Correlation Between Electrochemical Properties and Corrosion Resistance of Zirconium Alloys—Y. ITO AND T. FURUYA — 163
Discussion — 180

Long-Term In Situ Corrosion Investigation of Zr Alloys in Simulated PWR Environment by Electrochemical Measurements—H. GÖHR, J. SCHALLER, H. RUHMANN, AND F. GARZAROLLI — 181
Discussion — 202

Anodic Protection Provided by Precipitates in Aqueous Corrosion in Zircaloy—T. ISOBE, T. MURAI, AND Y. MAE — 203
Discussion — 217

Investigation of In-Pile Grown Corrosion Films on Zirconium-Based Alloys—O. GEBHARDT, A. HERMANN, G. BART, H. BLANK, F. GARZAROLLI, AND I. L. F. RAY — 218
Discussion — 241

Microstructure Evolutions and Iron Redistribution in Zircaloy Oxide Layers: Comparative Effects of Neutron Irradiation Flux and Irradiation Damages—X. ILTIS, F. LEFEBVRE, AND C. LEMAIGNAN — 242
Discussion — 263

Oxide Characteristics and Corrosion and Hydrogen Uptake in Zr-2.5 Nb CANDU Pressure Tubes—B. D. WARR, P. A. W. VAN DER HEIDE, AND M. A. MAGUIRE — 265
Discussion — 291

Corrosion and Hydrogen Effects

The Effect of the Trace Impurity Uranium on PWR Aqueous Corrosion of Zircaloy-4—H. R. PETERS AND J. L. HARLOW — 295
Discussion — 317

Detrimental Role of Hydrogen on the Corrosion Rate of Zirconium Alloys—M. BLAT AND D. NOEL — 319
Discussion — 336

Hydrogen Pickup and Redistribution in Alpha-Annealed Zircaloy-4—B. G. KAMMENZIND, D. G. FRANKLIN, H. R. PETERS, AND W. J. DUFFIN — 338
Discussion — 370

Ex-Reactor Mechanical Behavior

A Unified Model to Describe the Anisotropic Viscoplastic Behavior of Zircaloy Cladding Tubes—P. DELOBELLE, P. ROBINET, P. BOUFFIOUX, P. GEYER, AND I. LE PICHON — 373
Discussion — 393

The Influence of Temperature and Yield Strength on Delayed Hydride Cracking in
    Hydrided Zircaloy-2—P. EFSING AND K. PETTERSSON 394
Discussion 404

IN-REACTOR MECHANICAL BEHAVIOR

Effects of Hydride Precipitate Localization and Neutron Fluence on the Ductility of
    Irradiated Zircaloy-4—A. M. GARDE, G. P. SMITH, AND R. C. PIREK 407
Discussion 429

Fracture Toughness of Zircaloy Cladding Tubes—V. GRIGORIEV, B. JOSEFSSON, AND
    B. ROSBORG 431
Discussion 447

A Model for Analysis of the Effect of Final Annealing on the In- and Out-of-Reactor
    Creep Behavior of Zircaloy Cladding—M. LIMBÄCK AND T. ANDERSSON 448
Discussion 468

Properties of an Irradiated Heat-Treated Zr-2.5 Nb Pressure Tube Removed from the
    NPD Reactor—C. K. CHOW, C. E. COLEMAN, M. H. KOIKE, A. R. CAUSEY, C. E. ELLS,
    R. R. HOSBONS, S. SAGAT, V. F. URBANIC, AND D. K. RODGERS 469
Discussion 491

Link Between Results of Small- and Large-Scale Toughness Tests on Irradiated Zr-
    2.5Nb Pressure Tube Material—P. H. DAVIES AND R. S. W. SHEWFELT 492
Discussion 517

Modeling In-Reactor Deformation of Zr-2.5Nb Pressure Tubes in CANDU Power
    Reactors—N. CHRISTODOULOU, A. R. CAUSEY, R. A. HOLT, C. N. TOMÉ, N. BADIE,
    R. J. KLASSEN, R. SAUVÉ, AND C. H. WOO 518
Discussion 537

EFFECT OF IRRADIATION ON MICROSTRUCTURE

Effect of In-PWR Irradiation on Size, Structure, and Composition of Intermetallic
    Precipitates of Zr Alloys—F. GARZAROLLI, W. GOLL, A. SEIBOLD, AND
    I. RAY 541
Discussion 555

In Situ Studies of Phase Transformations in Zirconium Alloys and Compounds Under
    Irradiation—A. T. MOTTA, J. A. FALDOWSKI, L. M. HOWE, AND P. R. OKAMOTO 557
Discussion 579

Evolution of Microstructure in Zirconium Alloys During Irradiation—M. GRIFFITHS,
    J. F. MECKE, AND J. E. WINEGAR 580
Discussion 602

Influence of Neutron Irradiation on Dislocation Structure and Phase Composition of Zr-Base Alloys—V. N. SHISHOV, A. V. NIKULINA, V. A. MARKELOV, M. M. PEREGUD, A. V. KOZLOV, S. A. AVERIN, S. A. KOLBENKOV, AND A. E. NOVOSELOV 603
Discussion 622

Non-Linear Irradiation Growth of Cold-Worked Zircaloy-2—R. A. HOLT, A. R. CAUSEY, N. CHRISTODOULOU, M. GRIFFITHS, E. T. C. HO, AND C. H. WOO 623
Discussion 637

Influence of Iron in the Nucleation of ⟨c⟩ Component Dislocation Loops in Irradiated Zircaloy-4—Y. DE CARLAN, C. REGNARD, M. GRIFFITHS, D. GILBON, AND C. LEMAIGNAN 638
Discussion 652

EFFECTS OF PROCESSING ON STRUCTURES AND PROPERTIES

Effects of Extrusion-Billet Preheating on the Microstructure and Properties of Zr-2.5Nb Pressure Tube Materials—R. CHOUBEY, S. A. ALDRIDGE, J. R. THEAKER, C. D. CANN, AND C. E. COLEMAN 657

Zircaloy-2 Lined Zirconium Barrier Fuel Cladding—C. D. WILLIAMS, M. O. MARLOWE, R. B. ADAMSON, S. B. WISNER, R. A. RAND, AND J. S. ARMIJO 676
Discussion 692

Effects of Microstructure on Ductility and Fracture Resistance of Zr-1.2Sn-1Nb-0.4Fe Alloy—S. A. NIKULIN, V. I. GONCHAROV, V. A. MARKELOV, AND V. N. SHISHOV 695
Discussion 709

Influence of Processing Variables and Alloy Chemistry on the Corrosion Behavior of ZIRLO Nuclear Fuel Cladding—R. J. COMSTOCK, G. SCHOENBERGER, AND G. P. SABOL 710
Discussion 724

Effects of Thermomechanical Processing on In-Reactor Corrosion and Post-Irradiation Mechanical Properties of Zircaloy-2—P. Y. HUANG, S. T. MAHMOOD, AND R. B. ADAMSON 726
Discussion 756

Embrittlement of Reactor Core Materials—P. H. KREYNS, W. F. BOURGEOIS, C. J. WHITE, P. L. CHARPENTIER, B. F. KAMMENZIND, AND D. G. FRANKLIN 758
Discussion 782

IN-REACTOR BEHAVIOR

Zirconium Alloy E635 as a Material for Fuel Rod Cladding and Other Components of VVER and RBMK Cores—A. V. NIKULINA, V. A. MARKELOV, M. M. PEREGUD, Y. K. BIBILASHVILI, V. A. KOTREKHOV, A. F. LOSITSKY, N. V. KUZMENKO, Y. P. SHEVNIN, V. K. SHAMARDIN, G. P. KOBYLYANSKY, AND A. E. NOVOSELOV 785
Discussion 804

Corrosion Behavior of Duplex and Reference Cladding in NPP Grohnde—O. A. BESCH, S. K. YAGNIK, K. N. WOODS, C. M. EUCKEN, AND E. R. BRADLEY — 805
   Discussion — 824

Development of New Zirconium Alloys for a BWR—Y. ETOH, S. SHIMADA, T. YASUDA, T. IKEDA, R. B. ADAMSON, J.-S. F. CHEN, Y. ISHII, AND K. TAKEI — 825
   Discussion — 849

Comparison of the Long-Time Corrosion Behavior of Certain Zr Alloys in PWR, BWR, and Laboratory Tests—F. GARZAROLLI, Y. BROY, AND R. A. BUSCH — 850
   Discussion — 863

In-BWR and Out-of-Pile Nodular Corrosion Behavior of Zry-2/4 Type Melts with Varying Fe, Cr, and Ni Content and Varying Process History—H. RUHMANN, R. MANZEL, H.-J. SELL, AND D. CHARQUET — 865
   Discussion — 883

Development of Pressure Tubes with Service Life Greater Than 30 Years—C. E. COLEMAN, B. A. CHEADLE, C. D. CANN, AND J. R. THEAKER — 884

Indexes — 899

Overview

From its inaugural meeting in Philadelphia in 1968, the ASTM Symposium on Zirconium in the Nuclear Industry has been the premiere vehicle for discussion and documentation of the scientific and technological bases for the utilization of zirconium-based alloys in water-cooled reactors. The eleventh conference in this symposium series, held in September 1995 in Garmisch-Partenkirchen, Germany, continued this tradition of excellence. Attendees to this conference numbered 209, representing 16 countries. After careful peer review and editing, forty-one technical papers presented at this conference are published in this book. The highlights of the oral discussions have also been captured and appear at the end of each paper. This publication also includes two papers that are significant contributions to zirconium technology which served as the basis for the authors receiving the W. J. Kroll awards for 1993 and 1994. In Garmisch-Partenkirchen the awards for these years were presented to J. A. L. Robertson and to the team comprised of Friedrich Garzarolli, Heinz Stehle, and Eckard Steinberg, respectively. These Kroll Award papers represent historical as well as technical significance for the use of zirconium alloys in the nuclear industry.

Since their development in the 1950s and introduction into commercial nuclear power plants in the 1960s, the zirconium-based alloys Zircaloy-2 and -4, Zr-1Nb, and Zr-2.5Nb are the alloys currently used in the world's reactors. However, with increasing fuel duty, the margins displayed by these alloys have eroded, and considerable research has been conducted to improve these materials and also to develop more advanced alloys. Optimization of alloying constituents and processing parameters, coupled with a more basic understanding of performance-limiting phenomena, are the primary themes of most of the papers contained herein. Fully half of the papers are directly concerned with the corrosion of Zr-based alloys, and several trends are developing, which include: (1) uniform corrosion resistance is favored when oxide grains are columnar, rather than equiaxed in shape; (2) lithium-accelerated corrosion of Zircaloys corresponds to the formation of equiaxed grains in the oxide in preference to the usually occurring columnar grains; (3) in-reactor acceleration of corrosion observed in PWRs may be partially due to a lithium enhancement even at the low lithium concentrations used in PWRs, and (4) an effect due to hydrogen pickup and accumulation at the metal oxide interface may provide a significant contribution to the in-reactor acceleration of corrosion.

The detailed characterization of the effects of irradiation on the microstructure of irradiated Zircaloys has confirmed the loss of iron from second phase particles, with or without amorphization of the particles. Also, the correspondence between iron in solution in the matrix, the formation of $\langle c \rangle$-type dislocations, and the onset of accelerated irradiation induced growth has been verified. Unfortunately, the role of dissolved iron in the matrix in the nucleation of $\langle c \rangle$-type dislocations has not been established. One observation that has been verified, however, is the low irradiation growth in Zr-Nb-Sn-Fe alloys, also presumably due to suppression of $\langle c \rangle$ dislocation formation.

Several papers in the symposium focused on fuel clad modeling, and although a schism still exists between fundamental material properties and fuel performance predictive codes, the modeling papers presented are attempts to link component response to the quantifiable material behavior in a manner consistent with qualitative structural observations.

In summary, the data, analyses, hypotheses, and theories presented in this book represent the current state of zirconium technology as applied to nuclear power reactors. These contributions add strength to the foundation of our knowledge in this important and challenging technology.

E. Ross Bradley

Sandvik Special Metals Corporation, Kennewick, Washington; symposium chairman and STP editor

George P. Sabol

Westinghouse Electric Corporation, Pittsburgh, Pennsylvania; editorial chairman and STP editor

Kroll Award Papers

J. A. L. Robertson[1]

Learning from History: A Case Study in Nuclear Fuel

REFERENCE: Robertson, J. A. L., **"Learning from History: A Case Study in Nuclear Fuel,"** *Zirconium in the Nuclear Industry: Eleventh International Symposium, ASTM STP 1295,* E. R. Bradley and G. P. Sabol, Eds., American Society for Testing and Materials, 1996, pp. 3–11.

ABSTRACT: The award of the 1993 W. J. Kroll Zirconium Medal recognized the value of cooperative, multidisciplinary, applied research in tackling practical problems. This paper suggests that several other lessons relevant to the current debate on science-and-technology (S&T) policy can be drawn from our experience a quarter of a century ago. It outlines how close cooperation among those involved with the fuel for the Canadian CANDU heavy-water reactors identified a problem, then proceeded to solve it expeditiously. This capability for a rapid response to an unforeseen problem was no accident, but arose out of the conditions that existed at the Chalk River Laboratory of Atomic Energy of Canada Limited (AECL) and a deliberate policy to maintain this capability even when the utility's power reactors were demonstrating excellent performance.

KEYWORDS: nuclear fuel, history, science policy

The 1993 Kroll Medal was awarded for "contributions to the application of zirconium alloys to thin-walled fuel cladding through management of the program that resulted in mitigation of pellet-cladding-interaction (PCI) by graphite coating of the inner surface of the fuel cladding. This highly successful program has almost eliminated fuel failures from PCI in CANDU fuel." The award thereby recognizes the value of cooperative, multidisciplinary research in tackling practical problems. As the team member who reported the results of the program and thus gets to accept the award on behalf of all those involved at the Chalk River Laboratory of Atomic Energy of Canada Limited (AECL), I believe that the resulting history teaches several other lessons still worth learning today, when there is much talk of science-and-technology (S&T) policy, but little actual policy.

History—The Problem

First, something about the Canadian CANDU reactors and their fuel for those more familiar with light water reactors (LWRs). While both types are water moderated and cooled, in CANDU reactors the water is heavy water and the moderator is separated from the coolant by having the fuel in horizontal *pressure tubes* that pass through a tank of moderator, the *calandria* [1]. Both types use the same materials for their fuel, sintered uranium-dioxide pellets in sheaths of a zirconium alloy (Zircaloy), but their geometries differ considerably. The fuel for current CANDU reactors is 50 cm long and 10 cm in diameter, a cylindrical bundle of 28 or 37 elements held together by welded end-plates and held apart by brazed spacers—that for the prototype Douglas Point reactor consisted of only 19 elements with a diameter of 8 cm. For refueling, fresh CANDU fuel bundles are loaded, with the reactor at power, into individual pressure tubes

---

[1] P.O. Box 2047, Deep River, Ontario, K0J 1P0.

that carry the heavy-water coolant. They therefore require none of the extraneous channel components of the much longer LWR fuel, such as flow tubes and assembly hardware.

Because of the use of heavy water, which is a much better moderator than light water, CANDU fuel is able to use natural uranium. As a result, the burnup required of the fuel before discharge is only about 160 MWh/kg U [1], much less than for LWRs. The smaller bundles and the lower burnup in combination mean that, although there are fewer CANDU reactors than LWRs, there is already experience from the performance of about one million CANDU bundles. The large-volume production of a stable design has yielded benefits in quality assurance and costs.

From the start, the performance of CANDU fuel has been excellent. In the early 1970s well under 1% of all bundles loaded into the Douglas Point reactor and Ontario Hydro's first four-unit station at Pickering had failed, including those suspected of having failed but in which no defect could be positively identified. The failure rate for elements was about an order of magnitude lower. These statistics were determined by the CANDU designer and developer, AECL, collaborating closely with the utility, Ontario Hydro, to monitor performance and to examine any failures under water at the station and in shielded facilities at the AECL laboratories.

However, the program has not always been without its problems. In the Spring of 1970, this close monitoring revealed an increase in the failure rate, still within the 1% figure [2]. Thanks to the CANDU design's ability to locate and replace failed fuel without having to shut down the reactor, the economic penalty was unimportant, but fuel management was disrupted and the release of fission products into the coolant impeded maintenance. Therefore, we decided that the cause of the failures should be identified and, if possible, eliminated.

Now, a quarter of a century later, the puzzlement at the time is difficult to imagine, even to remember. An initial reaction was to look for manufacturing defects to explain the failures, either as undetected cracks in the thin (0.4-mm) Zircaloy sheath or as leaks in the element's end-welds. However, no supporting evidence could be found. The sheaths of failed elements were heavily hydrided, and we had previously demonstrated that the deliberate introduction of hydrogenous impurities, moisture or oil, with the fuel pellets could lead to such failures [3]. However, the fuel had been well dried before loading, and some apparently intact elements contained not only the expected amount of gaseous fission products but also deuterium that must have come from the coolant. Therefore, it was concluded that the hydriding was a consequence, not a cause, of the failures. It was known that gross overpower would strain the sheath through thermal expansion of the uranium dioxide [4], but reactor records showed no power surges, and elements adjacent to the failed ones exhibited no anomalous strain. Other potential mechanisms were examined and eliminated, at least tentatively.

In CANDU fuel management, bundles are moved progressively through the reactor core until they have achieved their design burnup. Thus, a bundle's normal progression involves an increase in power to its maximum value after a prolonged period at relatively low power and thereafter a move into a position of lower power before being discharged. Detailed analysis of reactor records showed that many of the failed bundles had undergone such power increases shortly before exhibiting evidence of failure; others had apparently failed after the reactor power had been raised to 100% of the design value after several months at 75% [2].

Since fresh bundles had survived irradiation in developmental tests at powers considerably higher than the value resulting in failures, the prior exposure at low power must somehow be making them susceptible to failure. The failure probability correlated equally well with the magnitude of the power increase and with the final power attained, so that it could not be determined which variable was the more important. Also, many bundles passed through the region of maximum power without failing, suggesting that some delayed failure mechanism at the high power was involved.

This vague and tentative understanding of the cause of the failures was confirmed and sharpened by controlled experiments in the NRU test reactor [2,5]. Bundles were irradiated for several months at low power under conditions simulating those of a power reactor, examined to confirm their integrity, then returned to positions of higher power. Failures were detected by fission-product releases, from a few minutes to a few days later. The failure probability was found to be insensitive to the rate at which power is increased and to the sheath thickness, so that the thick-walled sheaths of LWR fuel would not be immune. Subsequent examination confirmed the failures. Thus, the power increase probably caused a sheath defect that allowed coolant ingress; internal corrosion resulted in the hydriding and mechanical degradation of the sheath observed in bundles discharged from the power reactors.

At this stage, a statistical analysis was performed to determine which combinations of prior burnup, power increase, and final power resulted in failure. The results were mapped in diagrams termed "Fuelograms" [6]. This allowed the operators to modify their fuel-management procedures to avoid failures, albeit at the expense of some unwanted restrictions.

History—The Solution

This ended the first phase of the program, finding an immediate solution to the failures, and started the second, providing fuel with improved performance to relieve the restrictions on fuel management. Any researcher likely to make a useful contribution was pressed into service.

According to conventional ideas of science, one would first diagnose the mechanistic cause of the failures and then, from that, devise countermeasures. In practice, there was no clear understanding of the mechanism. One favored theory was simply that the sheath ductility had been sufficiently reduced by neutron bombardment to cause failure when it was stressed by thermal expansion of the fuel on a power increase. This would be inconsistent with the already established macroscopic properties of irradiated Zircaloy [7], but thermal cracking of the uranium-dioxide pellets was believed to concentrate the stress and hence the strain. Laboratory simulations measured the strain concentration in both unirradiated and irradiated sheaths [8].

In one reactor experiment, fresh uranium-dioxide pellets were loaded into an already irradiated sheath and the element put in the reactor for the standard power-ramp test [9]. The fact that it did not fail showed that something in the irradiated fuel was essential to the failure mechanism.

Another broad theory focused on chemical mechanisms. It was well known, on the one hand, that volatile iodine (a fission product) is released from the hot center of the fuel and, on the other hand, that iodine attacks zirconium. After all, this is the basis of the "other" (non-Kroll) process to produce crystal-bar zirconium! Perhaps we should have been trying to explain why irradiation in a reactor does not turn a fuel element inside out. The answer is that the sheath would normally be protected by a thin surface layer of zirconium oxide, but this would presumably be cracked by the sheath strain, exposing bare Zircaloy to iodine attack. The iodine would be expected to be chemically combined with fission-product cesium and therefore innocuous, but perhaps bombardment by neutrons or fission fragments would release elemental iodine. And so on. . . .

By the Summer of 1970, all experiments and arguments to elucidate the mechanism had proved inconclusive. All participants were therefore urged to propose potential solutions *assuming* that their favorite mechanism was responsible. Some aimed to reduce the sheath strain, others to improve the sheath ductility, others to reduce the stress concentration over fuel cracks, and yet others to protect against the iodine. All 17 proposed modifications were then fabricated in experimental elements for comparison in the same in-reactor test that had been used to show the effects of power ramps.

Before all the results were in, one modification proved significantly superior to the then-standard CANDU fuel elements in surviving power increases after a prolonged period of low power. It consisted of a thin layer of graphite on the inner surface of the sheath, but with no other change in the specifications. Since this solution involved negligible parasitic neutron absorption and would add little to the fabrication cost, it was immediately adopted as the reference design, and the two Canadian fuel fabricators (Canadian General Electric and Canadian Westinghouse) initiated programs to develop production processes. Discussions were held with Ontario Hydro to agree what further tests were necessary to ensure no unforeseen adverse effects of the graphite before the modified fuel bundles could be accepted into its CANDU reactors. These were:

1. Post-irradiation metallography to demonstrate no reactions between the various materials.
2. Irradiation of a modified element with a deliberate hole in the sheath to demonstrate no more rapid deterioration than for a standard element.
3. In-reactor measurement of the gas pressure in a modified element to demonstrate no appreciable pressure rise due to the formation of carbon oxides.

As a result of the successful outcome of all these tests, the graphite layer was adopted as an integral component of all CANDU fuel and still is to this day.

In retrospect, stress-corrosion cracking seems a simple explanation of the failures, with stress concentrations over the fuel cracks providing high local stresses and iodine providing the corrosive agent [10]. The role of the graphite may be twofold. Those researchers backing mechanical causes proposed the graphite as a lubricant to reduce the stress concentrations, while those backing chemical causes proposed it for its known adsorption of iodine, a property exploited in charcoal filters for removing iodine from gaseous effluents. The modified CANDU fuel incorporating the graphite layer was designated "CANLUB" fuel, apparently emphasizing the lubricating role. However, fractographic evidence suggested that while both roles contribute, the chemical one predominates [11].

As the result of intensive development programs by the fuel fabricators, with full cooperation from the utility, the first CANLUB fuel was loaded into the Pickering reactors only two years from first detection of the problem.

Prehistory

This capability for a rapid response to an unforeseen problem was no accident but arose out of the conditions that existed at AECL's Chalk River Laboratory and a deliberate policy to maintain this capability even when the Pickering reactors were demonstrating excellent performance. The favorable conditions consisted of at least two components: equipment and people.

The equipment included a laboratory for fabricating experimental fuel elements, the NRX and NRU test reactors, the in-reactor test loops that simulate a power-reactor environment, a device that allows fuel bundles to be moved within a loop while the reactor is at steady power, bundles with demountable elements so that the potential solutions could be tested under identical conditions and failed elements could be removed to allow the others to continue their testing, shielded cells for post-irradiation examination, and all the ancilliary services.

The people included scientists and engineers covering virtually every discipline, so that relevant expertise was readily available. Those scientists who were part of the international scientific subculture were able to maintain an awareness of developments elsewhere. Just as important as the researchers were all the skilled trades and other occupations that together helped to operate a large enterprise.

The CANLUB development drew upon these resources together with the understanding of relevant properties and phenomena that these had already yielded. A small group had been doing applied research on the behavior of uranium dioxide and zirconium alloys at Chalk River since the late 1950s. The numbers involved directly, about 30 professionals and technical support, were small compared with those engaged in comparable research elsewhere, but they were backed up by a magnificent service organization and excellent facilities.

Neutron economy [12] provided the incentive for thin-walled sheaths, but implementation depended on laboratory tests to define tolerances on the fuel/sheath clearance to avoid collapse of the sheath into longitudinal ridges. In-reactor tests showed that, if these ridges formed, failure would soon result, presumably by flexing during power cycles [13]. The specification for the metallurgical condition of the sheath to permit the use of thin walls depended on previous studies on the mechanical and corrosion properties of Zircaloy, with and without radiation [7,14].

Neutron economy also provided the incentive to extract as much power as possible from each fuel element. In setting an upper limit to the permissible power per unit length, an international controversy developed over the temperature distribution in the fuel. Work at Chalk River demonstrated how the metallographic cross section of irradiated fuel elements, interpreted elsewhere as showing central melting, resulted from solid-state diffusion of pores at temperatures roughly 1000°C below the melting point [15,16].

Laboratory experiments elucidated how porosity and stoichiometry affect the thermal conductivity of unirradiated uranium dioxide [17]. An in-reactor experiment disproved the claim of another laboratory that the large grains that develop in the center of the uranium dioxide in service had a much higher thermal conductivity than the as-sintered material [18]. Other laboratory experiments provided the means for estimating the heat-transfer coefficient for the fuel/sheath interface and hence the effect of the filling gas on the fuel temperatures [19].

This research gave the designers of CANDU fuel a reasonable understanding of the temperature distribution in the fuel elements and hence reassurance that high powers could be extracted without causing central melting. However, the Canadian designers were unique internationally in arguing that knowledge of the uranium dioxide's thermal conductivity was unnecessary and could be misleading as long as the values were so controversial. Instead, any potential failure mechanism, e.g., excessive sheath strain or excessive internal pressure due to fission-product releases, could be characterized empirically in terms of a function (the integral with respect to temperature of the thermal conductivity from surface to center) that could, in turn, be related to the power per unit length [20]. The ability to use this empirical approach to the design of CANDU fuel was possible because of the wealth of experimental results from tests in the NRX reactor. The work related to temperature distribution was reviewed in 1962 [21].

A similar review of work related to sheath strain was published in 1964 [4]. The results were interpreted in terms of a simple physical model consisting of a plastic core surrounded by an annulus of elastic fragments of uranium dioxide held in place by a weak envelope, the sheath. The plastic core meant that dishing of the end faces of the uranium-dioxide pellets could be exploited to control longitudinal and diametral strain in the sheath [22]. Other research had shown how cracks in the fuel heal under irradiation [23] and had discovered irradiation-enhanced elimination of fine porosity [24]. While the resulting in-service densification of the uranium dioxide resulted in large gaps developing within the sheaths of LWR fuel, CANDU designers remained unconcerned, largely because of our higher as-sintered fuel density, our shorter fuel bundles, and our horizontal orientation, but also because we had not encountered any such problem in our extensive post-irradiation examinations.

The behavior of microscopic porosity in the fuel played an important role in the release of fission-product gases, which affected the internal pressure on the sheaths. Initially, the pressure was calculated from fractional releases measured in post-irradiation examinations; later, through an improved understanding of the relevant phenomena [25], a computerized physical model was developed to provide the predictions [26].

Before the CANLUB development, the same group had elucidated other failure mechanisms. One mechanism, sheath hydriding through the inclusion of moisture or oil in the fuel, has already been mentioned. Another one, involving impurities in the fuel, had been encountered in LWR fuel development. Experiments in the NRX reactor demonstrated that high levels of fluoride impurities, possible in the enriched fuel of LWRs, would lead to rapid deterioration of the sheath if any defect allowed coolant ingress [27]. CANDU fuel is not susceptible because the fabrication of natural uranium-dioxide does not involve the uranium hexafluoride process.

In 1975, all known fuel failure mechanisms were reviewed, along with possible remedies [28]. It is a measure of the maturity of the nuclear fuel industry that the compilation is still largely valid.

Lessons

The award of the Kroll Medal highlights a major lesson to be learned from this experience, the importance of applied research. However, I hope that the account has added an essential qualification, that the applied research must be based on a sound understanding of the underlying science.

When the approach to fuel development at Chalk River is compared with that elsewhere in the same period, it can be seen that we relied more on experimental results and empirical correlations and less on highly sophisticated computer models based on first principles. This can be partly attributed to our good fortune in having available such excellent facilities, but a principle is also involved. With a system as complex as a fuel element under irradiation, many assumptions and simplifications have to be made to obtain a mathematical solution: more confidence can be placed on an empirical correlation interpreted with recognition of the significant phenomena. Over the years, the two approaches have converged.

Many of the other lessons that can be learned from this experience will seem no more than common sense to those who practice science and technology (S&T) but some may seem heretical to those who write about S&T policy.

Innovation is supposed to proceed linearly from some discovery in pure science through applied science to result in a new technology. In practice, many technological advances are made empirically with science coming along later to explain why they are successful. In the present example, science provided useful guidance in seeking a solution, but it was empirical tests that led to the selection of CANLUB fuel.

Our experience disproved the common belief that patents are necessarily good indicators of innovation. AECL could not obtain a patent on CANLUB fuel, apparently because the lubricating and gettering properties of graphite were already well known. This may only reflect our own inadequacies in a quasi-scientific, quasi-legal area, but experience with other aspects of CANDU development suggests a greater significance. Despite AECL's strength in basic research, most of the basic discoveries leading to the CANDU reactor were made elsewhere. AECL was weak in basic patents but strong in adopting and adapting others' discoveries. This was achieved through an informed awareness of international developments and by application of the knowledge to Canadian circumstances and objectives through applied research.

At first sight, this may seem to favor importing technology rather than developing it indigenously. While this may be valid in some instances, conventional wisdom greatly underestimates the support needed to nurture an imported technology. In AECL's experience, importing heavy-water production technology under very different conditions from its source and with little technical support resulted in severe difficulties, but importing Zircaloy-sheathed uranium-dioxide technology into an environment that already had strong technical and industrial support was highly effective. The "Prehistory" section of this account demonstrates the magnitude and range of technical effort needed for ongoing support of a technology that was imported originally.

A related topic is technology transfer. AECL had already in the 1960s developed techniques to achieve this, including:

1. A conscious effort to maintain awareness of developments elsewhere through open publication in journals and attendance at professional meetings.
2. The letting of development contracts with commercial fabricators, resulting in transfer of the technology to them and also transfer of the fabrication know-how to the researchers and designers.
3. The temporary attachment to the laboratories of staff from the fabricators and utilities to participate in the development program. When they returned to their parent organizations, they not only transferred the technology but also provided personal contacts for continuing cooperation.
4. Regular meetings for the exchange of information between staff from the developers, the fabricators, and the utilities.

Common themes for these four techniques are that technology transfer is a two-way process and that it is achieved better by people than by paper. The speed with which the CANLUB solution was developed and implemented testifies to their effectiveness.

Even where a successful commercial innovation can be traced back to some esoteric scientific discovery—and most can if one tries hard enough—the intervening period is usually long compared with that of patent protection. Pure research, while vital for the advancement of mankind in the long term, rarely benefits directly those who fund it. European research of the 19th century provided the basis for U.S. industrialization in the 20th century, while Japanese industry is now benefiting from U.S. science earlier this century. Any affluent country has an obligation to support curiosity-driven pure research in proportion to its relative wealth as an altruistic activity.

Part of the trouble is the habit of talking of "S&T" and "R&D" (research and development) as if these were homogeneous topics. Even when one can trace a continuous path from pure scientific research to the final product, the nature of the activities and their needs vary greatly along the way. The greatest distinction along the path is determined by the motivation for the work, between pure, or curiosity-driven, research on the one hand and mission-oriented programs on the other. For the latter, the mission should be defined by industrial and social policies so that the objectives and beneficiaries are clearly identified. The R&D activities are a necessary part of the program, along with all the other activities involved in the technology, such as quality control, marketing, and trouble shooting. Thus, the mission manager should determine the R&D funding out of the program budget, which should contain an appropriate allowance. R&D is only one means, not the end.

In this and other respects, the CANLUB development was a microcosm for the whole CANDU development. The entire R&D program was directed at achieving the explicit objective, and the program components were scheduled to be completed at compatible dates. This mission-oriented R&D contrasts with the more laissez-faire model for S&T policies that are often supported by governments. In the latter, unsolicited proposals for R&D are funded if they are in general areas of policy priorities and receive endorsement in peer reviews. The implicit assumption is that if the outcome is successful, someone will develop it commercially. Unfortunately, there is a shortage of scientific entrepreneurs, and nobody seems to know how to develop them. Until this can be achieved, directed, mission-oriented R&D is the more efficient way to employ limited R&D resources. Using the metaphor of a rope, market pull is more effective than research push.

The distinction between mission-oriented and curiosity-driven R&D should not be interpreted as preventing those responsible for missions from performing R&D at a very basic level where this may be beneficial to the mission. Several of the proposals for potential solutions to the CANDU fuel problem resulted from the researchers' familiarity with the basic mechanisms of radiation damage and corrosion in zirconium alloys.

The CANLUB development also illustrated the fact that most modern technologies require large interdisciplinary programs with expensive, efficient, and helpful support services, combining: underlying and applied science; engineering development and design; safety, economic, and market analysis; and the operation of test facilities as well as pilot and prototype plants.

This experience also illustrates another truth, one that may be unpalatable to policymakers. Failures are an integral and inevitable part of research. Seventeen potential solutions were pursued for one to be adopted. If the results of a research program could be predicted, there would be no point in doing the research. In this respect, research can be compared to geological exploration, where hundreds of holes are drilled before one profitable mine or oil well can be developed. Unless research budgets acknowledge this truth, researchers will be blamed for inevitable cost overruns.

Several of these lessons, derived here from experience with fuel development, could have been derived from other aspects of AECL's development program for CANDU reactors. All lessons from the overall program, together with the evidence for them, will be available in a technical history of AECL soon to be published jointly by McGill (Montreal, Quebec) and Queen's (Kingston, Ontario) University Presses.

References

(AECL reports are published by Atomic Energy of Canada Limited, Chalk River, Ontario, K0J 1J0.)

[1] Robertson, J. A. L., "The CANDU Reactor System: An Appropriate Technology," *Science*, Vol. 199, No. 10, February 1978, pp. 657–664.
[2] Robertson, J. A. L., "Improved Performance for UO_2 Fuel," *Canadian Engineering Journal*, November/December 1973.
[3] Bain, A. S. in *Transactions of ANS*, Vol. 12, No. 1, 1969, p. 99.
[4] Notley, M. J. F., Bain, A. S., and Robertson, J. A. L., "The Longitudinal and Diametral Expansions of UO_2 Fuel Elements," Report AECL-2143, 1964.
[5] Bain, A. S., Wood, J. C., and Coleman, C. E., "Fuel Designs to Eliminate Defects on Power Increases," *Proceedings*, International Conference on Nuclear Fuel Performance, London, England, British Nuclear Energy Society, October 1973, pp. 56.1–56.5.
[6] Penn, W. J., Lo, R. K., and Wood, J. C., "CANDU Fuel—Power Ramp Performance Criteria," *Nuclear Technology*, Vol. 34, 1977, pp. 249–268.
[7] Hardy, D. G., "The Effects of Neutron Irradiation on the Mechanical Properties of Zirconium Alloy Fuel Cladding in Uniaxial and Biaxial Tests," *Irradiation Effects on Structural Alloys for Nuclear Reactor Applications, ASTM STP 484*, American Society for Testing and Material, West Conshohocken, PA, 1970, pp. 215–258.
[8] Coleman, C. E., "Simulation of Interaction between Cracked UO_2 Fuel and Zircaloy Cladding," International Conference in Berkeley, Gloucester, UK, The Metals Society, London, September 1973, pp. 302–307.
[9] MacDonald, R. D., Hardy, D. G., and Hunt, C. E. L., "Unirradiated UO_2 in Irradiated Zirconium Alloy Sheathing," *Transactions of ANS*, Vol. 17, No. 2, 1973, p. 216.
[10] Cox, B. and Wood, J. C., "Iodine Induced Cracking of Zircaloy Fuel Cladding—A Review," *Proceedings*, Conference on Corrosion Problems in Energy Conversion and Generation, 146th Meeting of the Electrochemical Society, New York, NY, October 1974, pp. 275–321.
[11] Wood, J. C., Surette, B. A., Aitchison, I., and Clendening, W. R., "Pellet Cladding Interaction—

Evaluation of Lubrication by Graphite," *Journal of Nuclear Materials,* Vol. 88, No. 1, 1980, pp. 81–94.
[12] Lewis, W. B., "Designing Heavy Water Reactors for Neutron Economy and Thermal Efficiency," Report AECL-1163, 1961.
[13] MacDonald, R. D. and Bain, A. S., "The Irradiation of Sintered UO_2 in 0.1 mm Stainless Steel Sheaths," Report AECL-1159, 1960.
[14] Cox, B., "Effects of Irradiation on the Oxidation of Zirconium Alloys in High Temperature Aqueous Environments," *Journal of Nuclear Materials,* Vol. 28, No. 1, 1968, pp. 1–47.
[15] Bain, A. S., "The Heat Rating to Produce Central Melting in Various UO_2 Fuels," *Symposium on Radiation Effects in Refractory Fuel Compounds, ASTM STP 306,* American Society for Testing and Materials, West Conshohocken, PA, 1961, pp. 30–46.
[16] Lawson, V. B. and MacEwan, J. R., "Thermal Simulation Experiments with a UO_2 Fuel Rod Assembly," Report AECL-994, 1962.
[17] Ross, A. M., "The Dependence of the Thermal Conductivity of Uranium Dioxide on Density, Microstructure, Stoichiometry and Thermal-Neutron Irradiation," Report AECL-1096, 1960.
[18] Notley, M. J. F., "The Thermal Conductivity of Columnar Grains in Irradiated UO_2 Fuel Elements," Report AECL-1822, 1963.
[19] Ross, A. M. and Stoute, R. L., "Heat Transfer Coefficient between UO_2 and Zircaloy-2," Report AECL-1552, 1962.
[20] Mooradian, A. J. and Robertson, J. A. L., "CANDU Fuelling Costs—Breaking the 1-Mill/kWh Barrier," *Nucleonics,* Vol. 18, No. 10, October 1960, pp. 60–65.
[21] Robertson, J. A. L., Ross, A. M., Notley, M. J. F., and MacEwan, J. R., "Temperature Distribution in UO_2 Fuel Elements," *Journal of Nuclear Materials,* Vol. 7, No. 3, 1962, pp. 225–262.
[22] Bain, A. S., Robertson, J. A. L., and Ridal, A., "UO_2 Irradiations of Short Duration. Part II," Report AECL-1192, 1961.
[23] Bain, A. S., "Cracking and Bulk Movement in Irradiated Uranium Oxide Fuel Elements," Report AECL-1827, 1963.
[24] Ross, A. M., "Irradiation Behaviour of Fission-Gas Bubbles and Sintering Pores in UO_2," *Journal of Nuclear Materials,* Vol. 30, No. 2, 1969, pp. 134–142.
[25] MacEwan, J. R. and Morel, P. A., "Migration of Xenon through a UO_2 Matrix Containing Trapping Sites," *Nuclear Applications,* Vol. 2, 1960, pp. 158–170.
[26] Notley, M. J. F., "Calculation of Fission-Product Gas Pressures in Operating UO_2 Fuel Elements," *Nuclear Applications,* Vol. 3, 1967, pp. 334–342.
[27] Notley, M. J. F. and Robertson, J. A. L., "Zircaloy-UO_2 Failures Tied to Fluorides," *Nucleonics,* Vol. 19, No. 3, Mar. 1961, pp. 77–79.
[28] Robertson, J. A. L., "Nuclear Fuel Failures: Their Causes and Remedies," *Proceedings of the ANS/CNA Topical Meeting in Commercial Nuclear Fuel Technology Tomorrow,* preprinted volume for ANS/CNA Meeting, Toronto, Canada, April 1975, pp. 2.1–2.14.

F. Garzarolli,[1] H. Stehle,[1] and E. Steinberg[1]

Behavior and Properties of Zircaloys in Power Reactors: A Short Review of Pertinent Aspects in LWR Fuel

REFERENCE: Garzarolli, F., Stehle, H., and Steinberg, E., **"Behavior and Properties of Zircaloys in Power Reactors: A Short Review of Pertinent Aspects in LWR Fuel,"** *Zirconium in the Nuclear Industry: Eleventh International Symposium, ASTM STP 1295,* E. R. Bradley and G. P. Sabol, Eds., American Society for Testing and Materials, 1996, pp. 12–32.

ABSTRACT: Zircaloy-2 and -4, developed mainly in the United States, have been used in Germany for fuel rod claddings and in-core structural components from the beginning of reactor technology. Extensive studies of the material properties of the Zircaloys have been performed in Siemens laboratories since 1957. Irradiation testing was done in several test reactors. However, the combined effects of irradiation and real environmental conditions were determined through many experimental irradiations in existing power reactors. Elaborate examinations in the reactor pools of such experimental materials and of many standard fuel rods and assemblies after intermediate, full, and intentionally extended exposures were the main source of information and hard data. These programs were supported by several utilities and in certain areas carried out in cooperation with others.

Zircaloy-2 and -4 turned out to be very reliable materials that fulfilled all requirements for normal operation and likewise the requirements for postulated accidental conditions and for intermediate storage for many years. Optimization of Zircaloy-2 and -4 during recent years includes both optimization of microstructure and of chemical composition. BWRs and PWRs need differently optimized materials. Today's more demanding operation conditions and discharge burnups required a further optimization of the Zircaloys and for "hot" PWRs even the development of more corrosion-resistant Zr alloys. A significant improvement of PWR corrosion behavior can be achieved with Zr alloys using the alloying elements of Zircaloy with somewhat modified concentrations. Sn should be below or at least in the lower range of the ASTM specification range for Zircaloy-4, Fe and Cr should be somewhat higher, and Si should be specified as an alloying element rather than as an impurity.

KEYWORDS: zirconium alloys, Zircaloy-2/4, irradiation growth, creep, corrosion, in-reactor creep, in-reactor corrosion, precipitates, iodine stress corrosion, neutron irradiation, radiation effects, nuclear application, operating conditions in LWRs

The low-alloyed Zirconium-base alloys Zircaloy-2 and -4 were developed for fuel rod claddings and for fuel assembly and other in-core structural components in the late 1950s and early 1960s, mainly in the United States (Bettis Atomic Power Laboratories) [1]. The incentives came not only from the outstanding high neutron transparency of zirconium (after purification from the high cross-section hafnium companion in the natural ores) but also from the fact that the austenitic stainless steels suffered from irradiation-assisted stress corrosion cracking in LWR environments. Zirconium alloys formed from the early beginning a sound basis for any heavy water/natural or low-enriched uranium power reactor strategy, being quite superior to aluminum alloys. Therefore, important work for the Zircaloys was also done in Canada.

In Germany, the first domestic water-cooled power reactor project—the multi-purpose re-

---

[1] Siemens AG, Power Generation Group (KWU), 91050 Erlangen, Germany.

search reactor, MZFR, Karlsruhe, designed and built by Siemens and also the Experimental Power Station, VAK, Kahl—used Zircaloy fuel cladding and Zircaloy structural components. Soon, the German electric power industry decided to proceed with the LWR concept as created in the United States. Based on the experience gained with the MZFR and VAK fuel, there was no skepticism about using Zircaloy cladding even for the first cores of the first commercial power reactors Kernkraftwerk Gundremmingen and Obrigheim. Later on, step by step, the control rod guide tubes and spacer grids were also made from Zircaloy, leading finally to "full zirconium fuel assemblies." However, it is worthwhile to mention that our first important information on the irradiation-induced length growth of Zircaloy tubing came from pool-site measurements of MZFR fuel bundles. Also, we recognized at that time the detrimental effect of residual moisture inside a fuel rod.

Extensive studies of the material properties of importance for the in-pile behavior of the Zircaloys were performed in the laboratories since 1957. Irradiation testing was done in test reactors in Europe, Canada, and United States. However, to determine the combined effects of irradiation and real environmental conditions on a broader scale, many "pathfinder" irradiations, often in specially designed carrier fuel assemblies, were performed in existing power reactors over the years. Elaborate examinations in the reactor pools of such experimental materials and of many standard fuel rods and assemblies after intermediate, full, and intentionally extended exposures were the main source of information and relevant data. These programs were supported by several utilities and in certain areas carried out in cooperation with Combustion Engineering and General Electric in the United States, as well as Nuclear Fuel Industries in Japan. Sophisticated techniques and special equipment [2] were developed for nondestructive pool-site inspection and measurements, especially for dimensional changes (for instance, axial growth and creepdown, ovalization, and ridge formation along the rods) and for waterside corrosion (for instance, oxide layer thickness, formation of oxide nodules, potential oxide spalling, fretting marks, and so on). The technical progress in the field of the Zircaloys together with the progress in the UO_2 area were the key to the increase of discharge burnups from about 10 MWd/kgU (BWR) and 20 MWd/kgU (PWR) to about 50 MWd/kgU, nowadays, and the reduction of the average fuel rod failure rates from about 1E-2 to about 1E-5 per cycle.

Altogether, the Zircaloys turned out to be very reliable materials that fulfilled all requirements for normal operation and also the requirements for the behavior under postulated accidental conditions and for long-term intermediate storage after discharge. The community of nuclear fuel engineers should be grateful to those who invented these alloys. Only later, when the operational conditions became more demanding (e.g., unrestricted load-follow operation in BWRs and increased average coolant outlet temperature in PWRs) and the pressure to minimize the amount of spent fuel called for markedly increased discharge burnups, a further optimization of the materials became necessary. This has led to improved fabrication and quality control methods and to the concept of Zr liner and duplex tubing. Very recently, more corrosion-resistant Zircaloy versions with slightly modified composition were examined and developed, whereby the other good behavioral qualities were not impaired. In the following, a short review is given on the most important aspects with regard to the properties and the behavior of the Zircaloys in LWRs, especially for its use as cladding material.

Operating Conditions

Structural materials inside the core of a water-cooled power reactor are affected by (multiaxial) mechanical stress, fast neutron radiation, and chemical attack from the coolant, respectively, in case of the cladding inner surface by the fission products and undesired contaminants (i.e., moisture) of the UO_2 pellets. In addition, same "free" oxygen becomes available inside a fuel rod because the average valency of the fission products is somewhat less than that of the fissioned uranium atoms. Gamma radiation effects the chemical nature of the environment. In many cases, only the combination of two or even all of these environmental loads may lead to the defection of a fuel rod or an assembly (see, for instance, Ref 3). Irradiation effects add to the complexity of the problem.

Mechanical stresses, irradiation field, discharge burnups, and, in general, the whole area of "fuel rod internal chemistry" [4] are very similar in the two LWR systems. This is not true for the coolant chemistry and the cladding temperature. The concentration of free oxygen in the coolant as a consequence of radiolytic decomposition of the water and oxygen/hydrogen recombination differs strongly between pressurized and boiling systems, as shown in Table 1. The operating temperature of the cladding is essentially given by its surface temperature, that is, dictated by the system and the temperature step across a possibly existing oxide layer; the temperature step within the metallic body is typically in the range of 10 to 40°C. The figures given in the table should be regarded as approximate lower and upper extremes.

In pressurized systems, hydrogen addition in a concentration of 2 to 4 ppm suppresses the free oxygen to < 1 ppb. In boiling systems, the oxygen concentration is around 300 ppb. In several U.S. and Swedish boiling water plants, hydrogenation is being done to reduce the propensity to stress corrosion cracking of pipings. Typical concentrations of the additives are listed in Table 1. In PWRs, boric acid is used for reactivity control and LiOH for pH control. Sometimes Zn is added to BWR coolant to reduce plant activation and to minimize stress corrosion cracking of steam generator tubes. Nominally, the cladding temperature seems not to be very different in the two systems. Nevertheless, the difference is significant because of the strong temperature dependence of the corrosion kinetics. The higher oxygen level of BWR coolants is counterbalanced by its lower temperature and vice versa for the PWR coolants. Therefore, corrosion is an important issue in both systems.

Dimensional Behavior

In the early fuel element designs, irradiation-induced or at least irradiation-enhanced dimensional changes such as growth and creep had not been considered. Repair or premature discharge were necessary in some cases. As a consequence, length and diameter changes of fuel rods, structural components, and material samples were studied quite extensively [5–7]. The current database extends up to burnups of about 75 MWd/kgU ($\approx$1.5E22, cm^{-2}) for fuel rods, up to burnups of about 100 MWd/kgU ($\approx$2E22, cm^{-2}) for structural components (BWR channels), and up to a fast fluence of about 3E22 (cm^{-2}) for material samples (neutron energies > 1 MeV).

The growth of fuel rods is the result of three contributions: irradiation growth, anisotropic

TABLE 1—*Typical operational conditions for fuel rods in LWRs.*

		BWR	PWR
Discharge burnup	MWd/kgU	55	60
Exposure time	days	1800	1500
Fast neutron fluence	cm^{-2}, E>1 MeV	1E22	1E22
System pressure	bar	70	158
Cladding hoop stress	N/mm^2	−60	−100
Cladding temperature	°C	280–320	290–400
Coolant chemistry	O$_2$ ppb	300	<1
	H$_2$ ppm	0.003	3
	B ppm	...	1500–0
	Li ppm	...	2

creepdown, and mechanical interaction between fuel and cladding. Irradiation growth as well as the additional increase in length resulting from cladding creepdown under the external overpressure depend on the texture of the cold-worked cladding tubes. The texture formation is due to the hexagonal crystal structure and the fact that slip occurs primarily on the prism planes and that any deformation in basal planes requires twinning at higher stress levels. The details depend on the final fabrication (deformation) steps at room temperature [8]. The texture of pilgered tubes and rolled strips is characterized by peak concentration of the (0002)-diffraction poles in the radial (to tangential) direction. The Kearns factors ($fa + fr + ft = 1$) describe the effective volume fraction of the (0002)-diffraction poles of the individual grains in the three principal directions: axial, radial, and tangential. In cladding tubes, the Kearns factors are typically about $fa = 0.05$, $fr = 0.6$ to 0.8, and $ft = 0.2$ to 0.4. The contribution of the anisotropic creepdown to the total growth depends on the ratio of fr to ft. If fr is larger than ft, creepdown of the cladding tube causes an additional growth, whereas in the opposite case a shrinkage does occur. For usual texture of cladding tubes, the length increase due to anisotropic creepdown is about 10% of the total diameter decrease. Irradiation growth itself depends only on fa as far as texture is concerned; it is proportional to a factor $(1-3fa)$. Another parameter that influences irradiation growth is the material condition. Figure 1 shows the effect of the fast fluence and the degree of recrystallization on irradiation growth. Whereas cold-worked stress-relieved material grows fast already at fluences below 1E22 (cm^{-2}, >1 MeV), recrystallized material exhibits high growth rates only at fluences above about 1E22 (cm^{-2}, >1 MeV). The behavior of partially recrystallized Zircaloys is in between. Low growth of recrystallized material is probably due to preferred precipitation of irradiation-induced vacancies and/or interstitials and can be recovered by annealing. High growth rates are believed to be due to an interaction of irradiation-induced vacancies and interstitials with dislocations, and this type of growth cannot

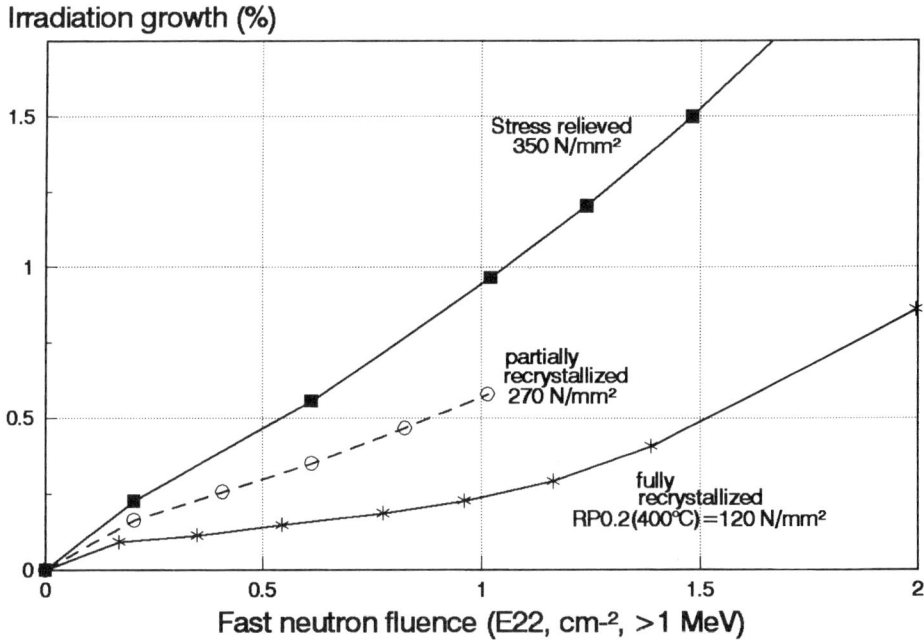

FIG. 1—*Irradiation growth of Zircaloy samples at 300°C.*

TABLE 2—*Material parameters influencing irradiation growth.*

Material Parameter	Effect	Remark
Texture: (fa up)	Strong (down)	
Grain size: (down)	Strong (down)	only below 2 μm
Cold work and degree of recrystallization: (up)	Strong (down)	in the 0–60% range
Tin content: (down)	Strong (down)	
Carbon content: (down)	Moderate (down)	at >100 ppm

be recovered. In cold-worked stress-relieved material, the dislocations result from the cold deformation. In fully recrystallized material c-type dislocations are known to form at high fluences. This was confirmed by transmission electron microscopy (TEM) at a fluence of 1.5E22 (cm^{-2}, >1 MeV) [7]. Other material parameters that influence growth are grain size, tin content, and carbon content as shown in Table 2. In the Tables 2, 3, and 4, the direction of property change with the direction of change in the variables is indicated by "up" and "down" (i.e., increasing fa leads to decreasing irradiation growth and so on).

Creep under the membrane or other stresses is the other mechanism that affects the dimensions. There are two important processes that can contribute to total creep deformation: thermally activated creep and irradiation-induced creep. Thermal creep rate decreases with time (has a primary and a steady-state part) and depends on temperature according to an Arrhenius law. The stress dependency of steady-state creep can be expressed by the so-called strain rate sensitivity $n = dln$(creep rate)/dln(stress). n was expected to be a constant, at least over significant ranges of stress, but instead was found to increase continuously with stress level [9]. Texture leads only to a moderate creep anisotropy. However, creep differs if tested under tension and compression (the so-called creep strength differential). Furthermore, thermal creep depends on several material parameters [9–11], as summarized in Table 3.

To study irradiation-induced creep, many measurements have been performed in the reactor pools. Diameter changes were determined as a function of fluence, respectively burnup, for fuel rods with standard and experimental claddings with many material variations, as well as on experimental creep samples (tubular specimens with different internal pressures) irradiated in the core of power reactors. Irradiation-induced creep was found to depend mostly on fast neutron flux and stress as shown in Fig. 2. The influence of temperature is very small [5,6]. From Fig. 2 it can be deduced that creep is faster under tensile than under compressive conditions as in the case of thermal creep. Furthermore, it can be seen that there is a stress-free diameter decrease due to superimposed irradiation growth, that is, shrinkage perpendicular to the tube axis (the volume of the material stays constant). Figure 3 shows the state of knowledge in 1978 with regard to the temperature dependency [5]. Later it was found that the effect of

TABLE 3—*Material parameters influencing thermal creep.*

Material Parameter	Effect	Remark
Texture:	Moderate	
Grain size: (up)	Strong (down)	only below 2 μm
Cold work and degree of recrystallization:	Strong	complex, in the 0–60% recryst. range
Tin content: (down)	Strong (up)	in the 0–2% range
Niobium content: (down)	Moderate (up)	in the 0–0.5% range
Oxygen content: (down)	Weak (up)	in the 500–2000 ppm range

TABLE 4—*Material parameters influencing in-reactor creep.*

Material Parameter	Effect	Remark
Grain size: (up)	Strong (down)	only below 2 μm at >320°C
cold work and degree of recrystallization: (down)	Strong (up)	in the 0–60% recrystallization range
Tin content (down)	Strong (up)	
Niobium content (down)	Moderate (up)	in the 0–0.5% range
Oxygen content (down)	Weak (up)	

temperature on in-reactor creep is more complex, at least in the temperature range of PWR fuel rod operation. Figure 4 gives the diameter decrease versus the reciprocal midwall temperature of experimental PWR fuel rods for claddings with different Sn content measured after an exposure of about 300 days. For normalization of the data taken after 298 to 316 days, a linear dependency of flux and time has been used. The figure reveals the large influence of the Sn content on in-reactor creep. In addition, it indicates that the creep rate increases with temperature in the temperature range of PWR fuel rod operation only if Sn is low. Otherwise, it even decreases slightly with increasing temperature. Probably Sn influences the irradiation-induced defect structure or even forms very fine precipitates. Fine precipitates have been seen by TEM in irradiated Zircaloy [7], although it is not clear so far what type they are. The material parameters influencing in-reactor creep are listed in Table 4. The influence of the size of the intermetallic precipitates was deduced to be of minor importance.

From a comparison of Table 4 and Table 3, it can be concluded that almost the same material

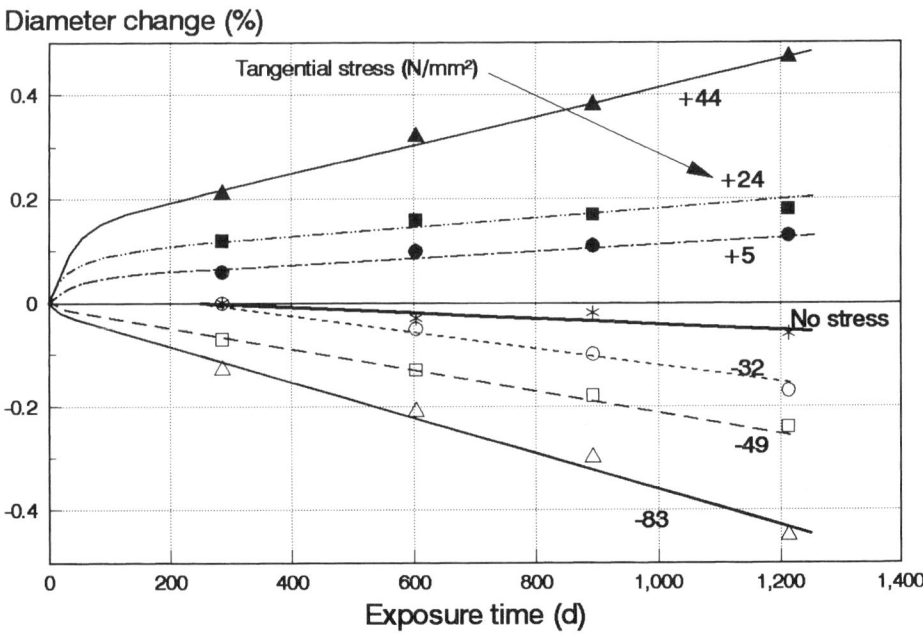

FIG. 2—*In-reactor diameter change of tubular Zircaloy samples with different membrane stresses.*

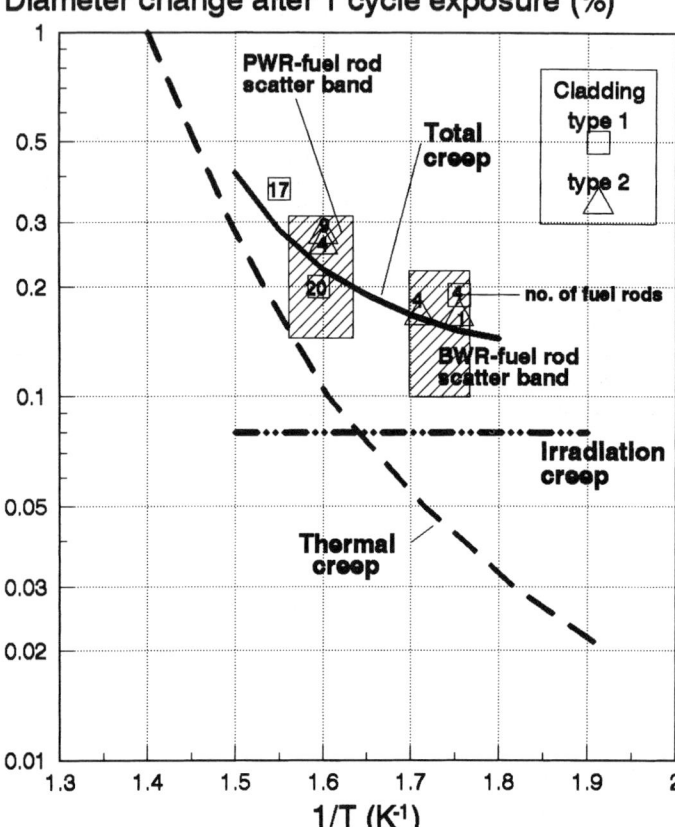

FIG. 3—*Arrhenius plot of in-reactor creep of stress-relieved cladding (fraction of thermal and irradiation creep)* [5].

parameters affect in-reactor creep and thermal creep. However, some of these parameters may have only a small effect on in-reactor creep at low temperatures (in the temperature range of BWR fuel rod operation). At least, this was observed for the grain size.

The results of many measurements of dimensional changes of PWR fuel rods and assemblies up to 1984, also including other aspects not discussed in details here, were presented at an IAEA Specialists' Meeting [12].

For dry storage of spent fuel, it is also important to know how irradiation to high neutron fluences affects thermal post-pile creep. Experiments with pre-irradiated tubular samples (up to four cycles) were performed at temperatures of 350 to 400°C and membrane stresses of 0.50 and 70 N/mm² up to 8000 h [13]. These tests revealed that a small fraction of the creep deformation occurring during irradiation is recovered if annealed at 350 to 400°C. Besides this recovery effect, post-pile creep was found to be substantially lower than thermal creep of unirradiated material.

For analyzing hypothetical accident conditions, it is necessary to know the creep behavior at high temperatures (600 to 900°C). Therefore, single and multiple rod investigations have been performed to study the behavior of Zircaloy cladding tubes under simulated loss of coolant

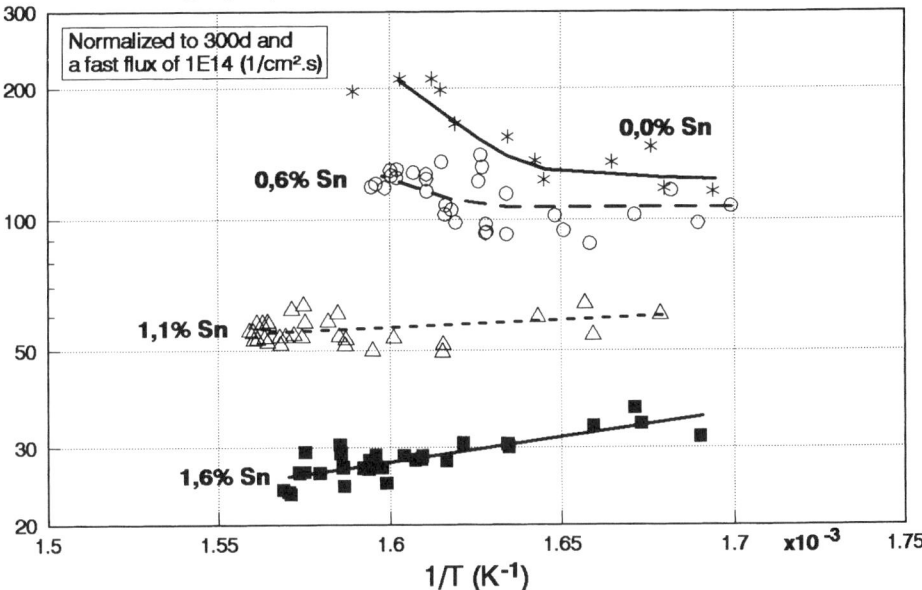

FIG. 4—*Diameter decrease of experimental claddings with different Sn contents.*

accident (LOCA) conditions [*14*]. It was found that the deformation and bursting behavior of tubes with different metallurgical conditions (i.e., cold-worked, stress-relieved, or fully recrystallized) and chemical composition can vary quite significantly under such tests. Therefore, single-rod LOCA tests were also included in the material optimization programs to ascertain that the modified cladding tubes behave like classical tubes [*15*].

Corrosion

General

Waterside zirconium corrosion results in the formation of an adherent protective oxide layer on the exposed surface whereby a certain part of the corrosion hydrogen is absorbed by the metal. The oxygen concentration profile shows a sharp step across the oxide/metal interface that only becomes smeared at much higher temperatures than under discussion here. Laboratory studies on waterside corrosion are done in water or steam autoclaves, and the corrosion kinetics are determined by measuring the weight gain as a function of insertion time. In pool-site investigation, for instance on fuel rods, the oxide thickness is measured directly using an eddy-current distance probe [*2*] based on the fact that the ZrO_2 layer has good electrical insulating properties.

In high-temperature water or steam, normally a rather uniform oxide layer is formed. Up to an oxide thickness of about 2 to 3 μm, the corrosion kinetics slows down according to a time-dependence proportional to $t^{1/3}$. Then a transition occurs to an almost linear time dependence. The temperature dependence of the linear post-transition corrosion rate as well as the time to transition can be well described by an Arrhenius law, with an activation temperature, Q/R, of 1.42 E4 K.

In-reactor, the corrosion kinetics are enhanced compared to the laboratory results. In a BWR environment, the enhancement starts from the beginning but decreases with increasing layer thickness. In spite of this enhancement, the oxide layer thickness remains normally low, even on BWR fuel rods [17]. In a PWR environment, the corrosion rate is enhanced at first when an oxide thickness of about 5 μm is reached, which is after the classical transition to the time-linear regime (second transition is in Fig. 5). This enhancement is independent of temperature and amounts to a constant factor of about 4 [19]. Because of the delayed start of the irradiation effect, it was believed in early days that the corrosion rates were rather unaffected in PWRs [16].

It is clear from these general remarks that a good surface finish and a proper final surface treatment are very important for a reliable corrosion behavior. Furthermore, continued damage to the protective oxide layer, for instance by undue vibration at the contact spots between the spacer grids and the fuel rods, can lead to a rapid local corrosion effect known as *fretting corrosion*. Debris from the coolant circuit, which can become caught between the fuel rods or on the grids, are equally harmful. Therefore, avoiding fretting corrosion is an important task of the fuel designers.

Corrosion Under Heat Flow Conditions in PWRs

Corrosion rate is controlled by the temperature at the innermost part of the oxide layer very close to the oxide/metal interface. The bulk of the oxide has no protective function. However, under heat flow conditions this non-protective part of the layer increases the temperature at the oxide/metal interface, thereby leading to a self-acceleration of the corrosion through the positive temperature feedback.

The linear corrosion kinetics after start of the irradiation enhancement can be written as:

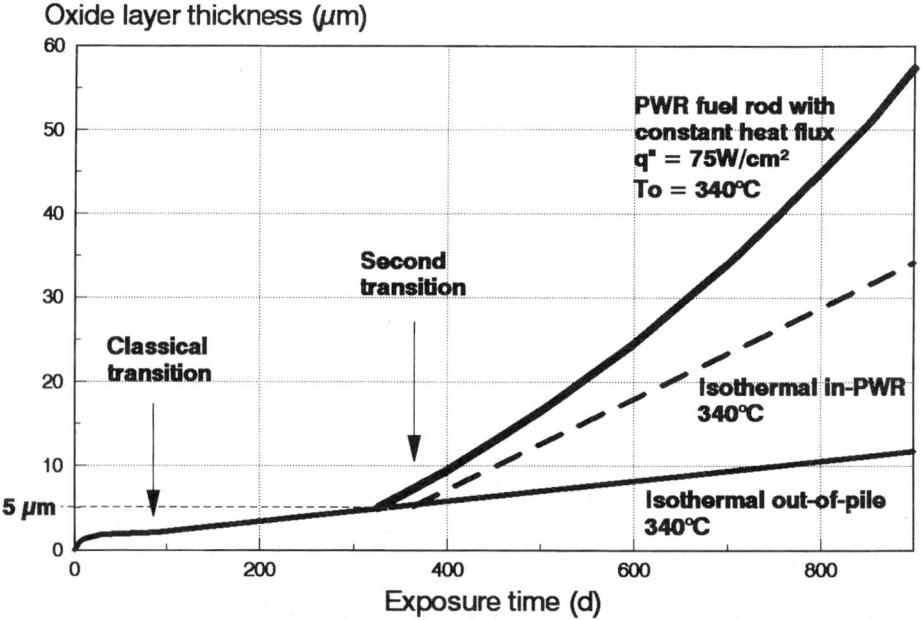

FIG. 5—*Oxide layer thickness as function of time (schematically)* [19].

$$\frac{ds}{dt} = F_R \cdot A \cdot \exp\left(-\frac{Q}{RT(s)}\right),$$

with s = oxide layer thickness and F_R = irradiation enhancement factor; A (the rate constant, [cm/d]) and Q/R (Q = activation energy) have the usual meaning and can easily be determined in autoclave tests. $T(s)$ depends on heat flux, q'', and thermal conductivity of the oxide, λ, according to

$$T(s) = T_0 + s \cdot \frac{q''}{\lambda},$$

where T_0 is the surface temperature of the fuel rod. The above differential equation can be approximately integrated leading to the equation:

$$s = s_{trans} + \frac{\lambda}{q''} \cdot \frac{RT_0^2}{Q} \left[\exp\left(\frac{q''}{\lambda} \cdot \frac{Q}{RT_0^2} \cdot s^{(0)}\right) - 1\right]$$

where s_{trans} is the oxide thickness at the second transition ($\sim 5~\mu m$) and

$$s^{(0)} = F_R \cdot A \cdot \exp\left(-\frac{Q}{RT_0}\right) \cdot (t - t_{trans})$$

represents the linear isothermal in-pile behavior after the second transition. This crude solution shows the important influence of the heat flow and the oxide thermal conductivity and reveals that the linear time dependence changes to an exponential one. For a thorough application of the model, numerical methods are necessary.

The above outlined concept for the modeling of the self-enhancement under heat flow conditions was published in 1975 [16]. To overcome some skepticism and to establish the involved parameters more precisely, an extensive research program was performed in cooperation with CE during the years 1979 to 1982, which was sponsored by EPRI (see for instance Ref 18 and several other reports edited by EPRI). In the course of this program, the thermal conductivity and the density of the ZrO_2 in the oxide layer and its microstructure also were investigated.

The effects of irradiation and of heat flow on the corrosion of fuel rods are shown schematically in Fig. 5 under the assumption of fixed parameters. Figure 6 presents a map of the experience with fuel rods inserted up to thick oxide layers [20]. It seems that the performance limit in such a map is given by a locus curve that represents a certain rather high corrosion rate, leading to defection. This locus curve, calculated for this special case, approximately represents a hyperbola with $sq''/\lambda = \Delta T_{oxid} \approx 50°C$. On the other hand, a certain thinning of the wall should not be exceeded even if the rods would not become defective. Altogether, corrosion is a life-limiting phenomenon for PWR fuel rods, and design, layout, and insertion planning must be done with great care.

Nodular Corrosion in BWRs

An anomaly in the corrosion behavior of the Zircaloys can occur in BWRs (oxygenated environment) and out-of-reactor in high-pressure steam above 400°C, preferentially above 500°C. This anomaly, called "nodular corrosion," is characterized by the formation of white pustules, appearing in metallographic cross sections as lense-shaped oxide with significantly higher thickness than the uniform oxide. Under continued exposure, these pustules grow to-

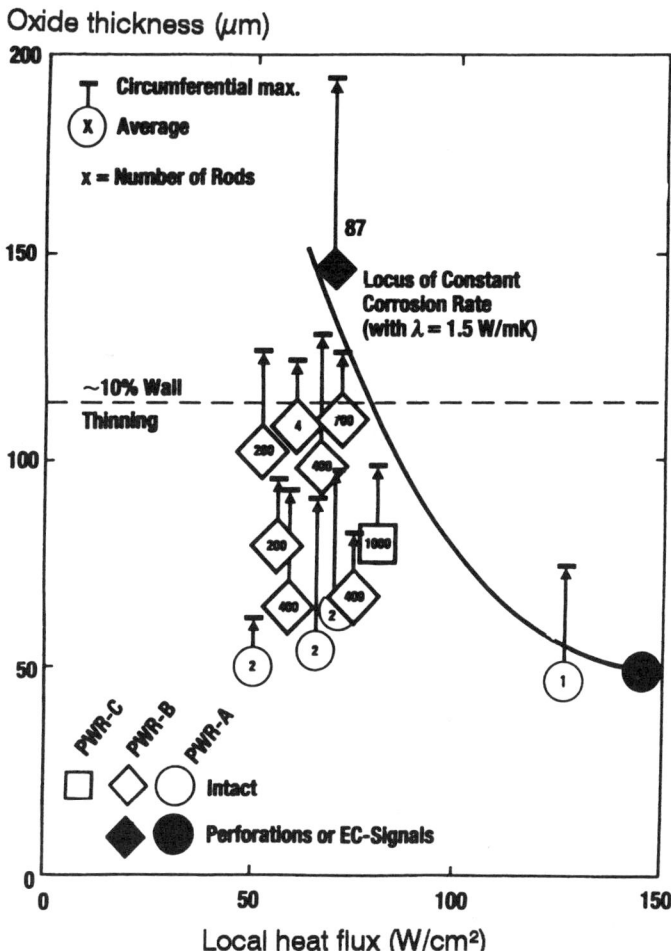

FIG. 6—*Map of experience with thick oxide layers in PWR [20].*

gether to an almost uniform very thick oxide layer. Experience from various BWRs has shown that such corrosion can vary considerably depending on water chemistry and cladding condition. However, nodular corrosion does not depend on temperature and normally does not limit fuel rod life. This has been validated by operating experience from fuel rods with oxide layer thicknesses of up to 250 μm. As far as the water chemistry is concerned, it is known today that nitrogen in the coolant (which can be introduced with the cooling water supply) increases the rate of nodular corrosion and that especially conductivity transients during the first month of fuel operation can initiate heavy nodular corrosion in high power regions of the core [17,20].

It has been well established that the size distribution of the Zr(FeCrNi) intermetallics has an important influence on corrosion behavior [20-22,24]. Figure 7 is an updated diagram, first presented in 1986 [20], showing in-pile and out-of-pile corrosion rates versus the mean size of these intermetallics. The corrosion rates are higher (except for the 500°C steam autoclave tests) when the size of the intermetallics is very small (especially smaller as usual). However, in BWRs this effect can only be seen at burnups in excess of about 20 MWd/kgU [24]. Coarse

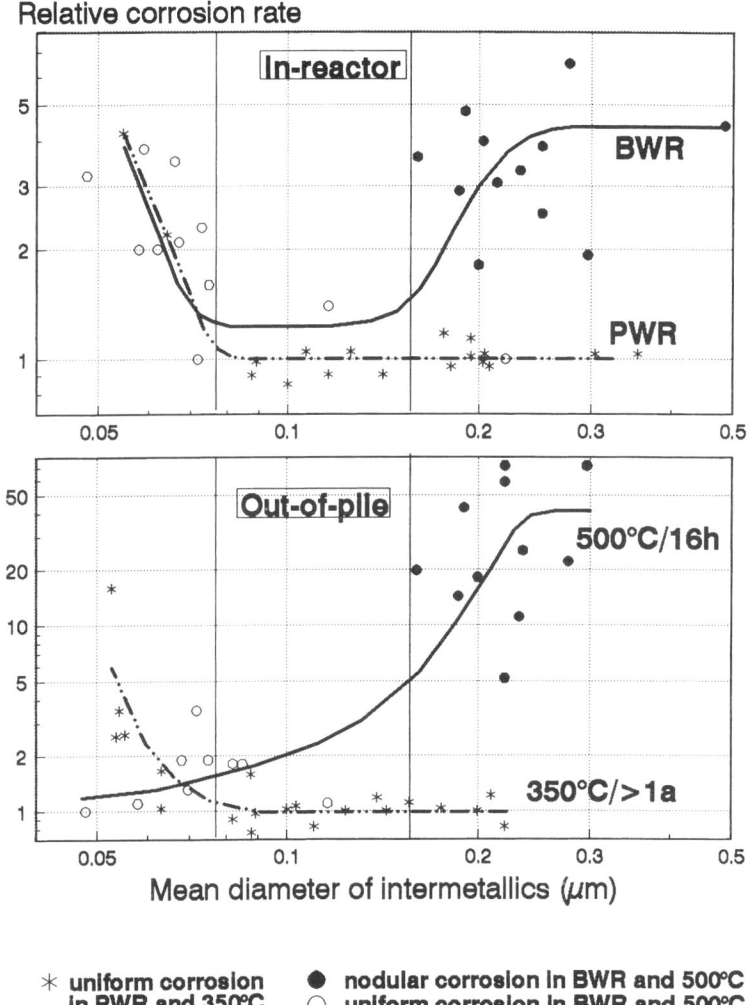

FIG. 7—*Corrosion of Zircaloy versus size of intermetallics.*

intermetallics, on the other hand, lead to nodular corrosion in BWRs as well as in 500°C steam (closed dots in the figure), but are harmless in PWRs.

Size of Intermetallics

The size of the intermetallics depends on the quenching rate from the β-phase (>950°C), where Fe, Cr, and Ni are in solid solution, and the subsequent annealing temperatures and times in the α-phase (<800°C) during fabrication [26]. It has to be noted that the determination of the size distribution of intermetallic precipitates is time consuming, and the results depend on the resolution of the electron microscope used. Furthermore, the size distribution varies within the cladding tube wall. The size of precipitates is smallest at the outer surface and increases to the inner surface. Most of the reported data of Fig. 7 are from a midwall position. In any case, a comparison of different data needs a lot of care. Therefore, an empirical process-controlling method was developed [27–30] based on the addition of the effectiveness of all heat treatments involved after the last β-quenching. The effectiveness of a single step at (constant) temperature T_i (K) during a timespan t_i (h) is given through $k \cdot t_i \cdot \exp(-T_A/T_i)$ with T_A = 3.2E4 K. The "rate factor" k is arbitrarily chosen as $k = 1E14$ (h^{-1}). An accumulated particle growth parameter PGP after i annealing steps is then defined by

$$PGP = \Sigma_i k \cdot t_i \cdot \exp(-T_A/T_i)$$

The effectiveness of the initial quenching process from β- to α-phase can be approximately considered by adding the quotient 0.45/CDR to this sum, where CDR means the cooling down rate in the α-phase in K/s. It has been proven in many experiments that this empirical PG-parameter can well be used to correlate the increase of the mean size in the logarithmic gaussian distribution of the intermetallics with time and temperature of consecutive heat treatments. Results, including β-quenched and α-annealed specimens as well as samples from production tubing, which have shown that an "accumulated annealing parameter" can be used, were published in 1989 [22].

For in-reactor corrosion, it has been shown that the size distribution of the intermetallics can be influenced by the fast neutron flux through the acceleration of dissolution processes as well as through the acceleration of foreign atom diffusion within the Zr lattice. At BWR operating temperature, the dissolution of initially fine intermetallics may be the reason for the increase of the corrosion rate at burnups in excess of about 20 MWd/kgU. At higher temperatures, where lattice diffusion is increased, a coarsening of the intermetallics may occur.

Modified New Alloys; DUPLEX Tubes

Our experience on the influence of the main alloy constituents and impurities is outlined tentatively in Table 5. The given valuation is understood relative to the behavior of the standard alloy composition. The table shows that the results partly conflict with respect to different types of exposure. Most important are the results listed up in the columns "In-PWR" and "Out-of-Pile Water," showing the unfavorable effect of Sn and the beneficial effect of Fe. Figure 8 gives quantitatively the influence of the Fe content in 370 and 400°C autoclave tests, relevant also for the behavior in-PWR. The plot clearly recommends to increase the Fe concentration beyond the range of 0.18 to 0.24% specified in ASTM Zircaloy-4. With respect to Sn, the results suggest a reduction within the ASTM range [32] or even better below the ASTM range (1.2 to 1.7%) for PWR application, whereas for BWRs Sn should remain within the ASTM

TABLE 5—*Effect of alloying elements on corrosion resistance.*

Alloying Element	In-BWR	In-PWR	Out-of-Pile Water	In $\geq$ 500°C Steam
Sn	Favorable	Unfavorable	Unfavorable	Unfavorable
Fe	Weak beneficial	Beneficial	Beneficial	Beneficial
Cr	Weak beneficial	Beneficial	Indifferent	Beneficial
O	Unfavorable	Indifferent	Indifferent	Weak beneficial
Si	Beneficial	Weak beneficial	Weak beneficial	Indifferent
C	Indifferent	Weak unfavorable	Weak unfavorable	Indifferent

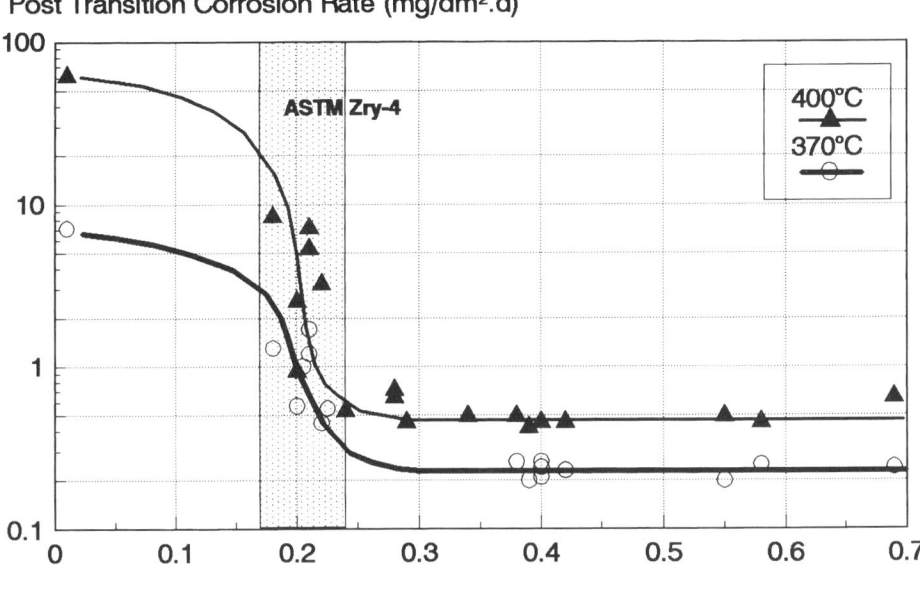

FIG. 8—*Influence of Fe on corrosion in laboratory autoclave testing* [31].

range. Si is considered to be an important constituent and should not be regarded as an impurity. Altogether, it seems worthwhile to expand the ASTM list of Zr alloys for nuclear application.

For plants with high-coolant outlet temperatures using optimized low-leakage loading patterns, the corrosion resistance of even the best performing Zircaloy-4 is not sufficient to reach the current burnup targets. For this goal, alternative alloys with compositions outside the ASTM specification for Zircaloy-4 as mentioned before have been studied in out-of-pile programs. In-reactor testing was performed in the Gösgen and Grohnde nuclear plants, which provide realistic conditions for modern fuel cycle managements. These programs were performed jointly with the utilities and in close cooperation with the prematerial suppliers CEZUS and TWCA [23]. The initial exploratory program covered a variety of alloy systems. In total, about 120 different alloys have been fabricated into strip material and cladding tubes to study the out-of-pile corrosion behavior in steam and water with and without lithium. Over two thirds of these alloys were further investigated through insertion in water rods in a PWR for up to six years. Pathfinder fuel rods with these alloys have been irradiated for up to seven cycles and a maximum rod average burnup of 80 MWd/kgU. From this program, the DUPLEX ELS tube was selected as cladding for the modern PWR fuel generation for high burnups in "hot" PWRs [23,25,28]. This tube has an outer layer with a reduced Sn content below the ASTM specification range for Zircaloy and a Fe and Cr content above the ASTM specification range for Zircaloy-4; the designation ELS means extra low Sn.

The DUPLEX cladding combines high corrosion resistance and the well known properties of Zircaloy with regard to mechanical strength, growth, and creep behavior for normal operating conditions and for the postulated accidents [35–37]. Meanwhile, quite a large experience exists with this cladding (about 125 000 fuel rods are under irradiation in several different plants, and burnups up to >60 MWd/kgU have been reached). Figure 9 shows that this new cladding has

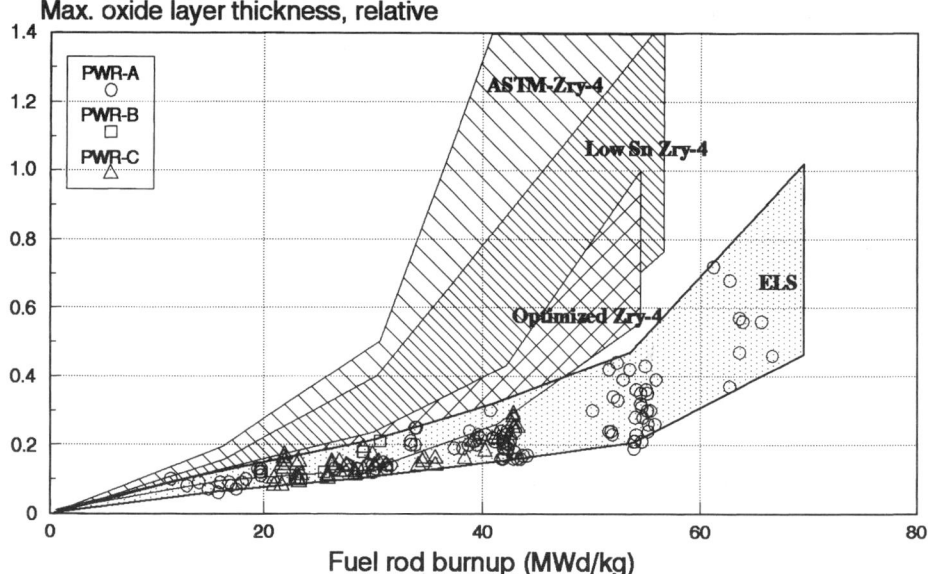

FIG. 9—*Corrosion behavior of ELS cladding compared with different Zircaloy-4 cladding materials* [25].

a higher corrosion resistance than the best Zircaloy, especially at high burnups. In addition to the DUPLEX tube, which is used for high burnups in "hot" PWRs, a new optimized Zircaloy with increased Fe and Cr content (above the ASTM specification range for Zircaloy-4) was selected for moderate burnups, respectively, for high burnups in "moderate hot" PWRs [31]. This cladding material has a higher corrosion resistance and a lower hydrogen pick-up fraction than Zircaloy-4.

Besides the corrosion under normal operational conditions, off-normal conditions have also to be studied to explore the allowable operational range and to analyze the behavior under hypothetical accident conditions. Therefore, studies on the influence of various impurities in the water on corrosion of Zircaloy [38] and the behavior at temperatures of interest for LOCA analysis have been performed. The latter tests had the primary goal to confirm that the new Zr alloy does not behave worse than Zircaloy-2/4 under these conditions [39].

Hydrogen Pickup

Some of the corrosion hydrogen is generally picked up by the underlying metal. The pickup fraction was found to be rather low in an oxygenated environment (out-of-pile and in-BWR). In hydrogenated environments, higher pick-up fractions have been found [1,17,20,33]. Hydrogen is in solid solution at operation temperatures up to about 100 to 150 ppm (corresponding to an oxide layer thickness of 10 to 50 μm at the usual cladding wall thickness). At higher concentrations, precipitation of hydride platelets occurs. Because of the lower solubility at the colder outside region and due to diffusion in the temperature gradient, precipitation of hydrides occurs mainly at the outer 10 to 40% of the wall. There is now a quite large database including fuel rods with peak oxide layer thickness in excess of 100 μm and average hydrogen contents above 500 ppm, which assures that such concentrations in the cladding do not affect the fuel rod behavior. Also, local variations of the highly mobile hydrogen have no adverse effects on the operational behavior, although these variations can be quite high. For fuel assembly structural components (e.g., spacer, guide tubes), it was shown [*34*] that room-temperature test conditions are most critical. Component tests simulating at room temperature all important loads that may arise during operation and handling revealed nothing detrimental even at hydrogen concentrations of 2000 ppm.

Oxide Morphology

To support the above-outlined empirical correlations, some work was also directed to a better understanding of the mechanism of the corrosion process [*40–42*]. This work was performed in cooperation with the prematerial suppliers, several universities, and the Siemens Research Center. These studies confirmed that the outer part of the oxide layer is porous and that the corrosion rate is controlled by the thickness and quality of a thin barrier layer at the metal/oxide interface. It could well be that the unknown chemistry in the pore system is the key to understanding the radiation enhancement. The intermetallics are embedded in the barrier layer in metallic form at the beginning. It was supposed that the mode of oxide crystallization at the metal/oxide interface plays an important role. If nucleation of new grains is suppressed, crystal growth leads to columnar grains that behave as an efficient barrier layer. If frequent nucleation of new oxide grains at the metal/oxide interface occurs, an equiaxed grain structure is formed. The latter structure is sensitive to grain boundary cracking and becomes porous easily. It is assumed that the electrical conductivity of the barrier layer, which depends on size and frequency of large intermetallics, influences proton migration. Hydrogen at the metal/oxide interface may influence the mode of crystallization. This hypothesis has still to be proven experimentally, but it turned out that this is not a simple issue [*42*].

Iodine Stress Corrosion Cracking (I-SCC)

Post-irradiation examinations, especially of defective fuel rods, have shown that pellet/cladding interaction of the UO_2 fuel pellets with the Zircaloy cladding tubes, caused by power ramping, can limit the performance of LWR fuel rods. In the course of a fast power increase, for instance after an extended period of a part-load operation, the cladding wall is stressed through the thermal expansion of the UO_2 pellets. At the same time, a surplus of fission product iodine—being a prerequisite for this defection mechanism—may be released. In contrast, this PCI defection mechanism is not observed during power cycling operation since then the time span at low power is too short for an appreciable creepdown of the cladding onto the pellets. As a consequence, several power ramping programs were performed in the research reactors at Jülich and at Petten in cooperation with different partners (as CE, NFI, and others) and in international cooperation [*43–46*]. These tests revealed the fuel performance limits for practical applications and confirmed the low-deformation SCC mode of cracking.

Laboratory tests with iodine [*46–49*] resulting in cracks with similar fractographic characteristics as in-pile-generated defects can be used for studying the SCC behavior of Zircaloy. I-SCC has been found to occur in smooth specimens only at an iodine load of $>$1E-6 g/cm^2 [*47*]. Figure 10 summarizes measurements of crack length within pressurized tubular samples with different iodine loads. Diametrical creep deformation as a function of test time and the strain when perforation occurs also are given. Initial cracks up to a depth of ~3 μm are formed during the first 0.1 h. The growth rate increases with increasing iodine concentration, but is constant for a given I-concentration over quite a long period. Perforation of samples occurs when the depth of the fastest crack penetrates the wall. These results suggest that the mechanism of I-SCC consist of at least two steps: crack initiation and crack propagation. Crack initiation occurs within a short time after loading as concluded from acoustic emission. It arises probably from

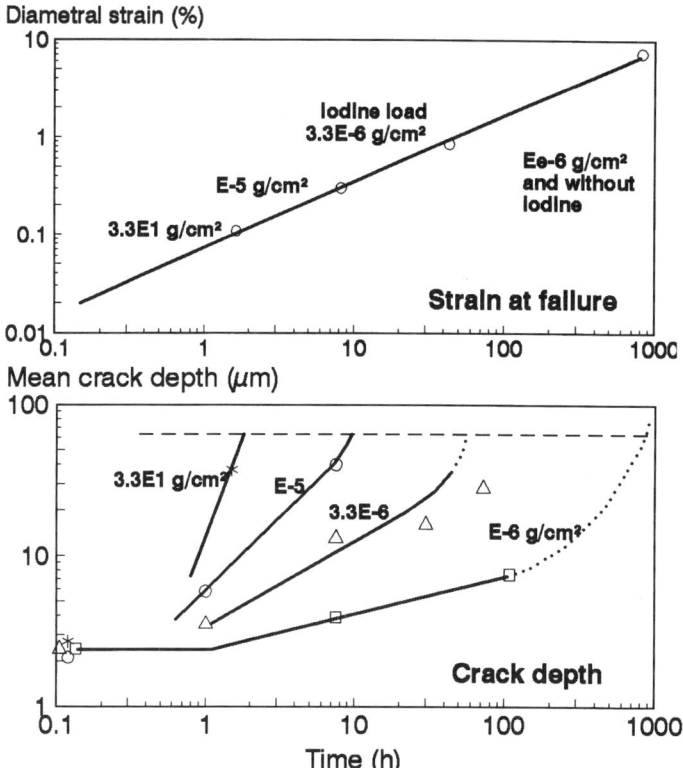

FIG. 10—*Creep deformation and crack growth of tubular samples at 225 N/mm² under various loads* [48].

a mechanical/chemical attack of the grain boundaries or at surface imperfections. Crack propagation can operate via several mechanisms; at small crack sizes, crack growth is due mainly to the interconnection of aligned cracks, whereas at larger sizes the growth of individual cracks is the dominating mechanism [47–49]. Irradiation was found not to alter I-SCC failure stress. However, strain to failure is reduced due to irradiation hardening [46]. Extensive tests have been done to study the material properties influencing I-SCC. Besides the yield strength, which is high after irradiation in any case due to the irradiation hardening, the most important parameters are texture and grain size. A fine grain size was found to be beneficial. Figure 11 indicates that the cracks occur mainly in planes that are tilted about 30° to the basal planes [47]. In this experiment, the test sample was cut out of a thick Zircaloy plate with a perpendicular orientation of the basal poles, as indicated in the figure. As a consequence of these facts, tubes with strongly radially oriented basal poles have a high resistance against I-SCC.

Much attention was given to remedies that either reduce the mechanical interaction at pellet/pellet interfaces and across radial cracks in the pellets or hinder the fission products from coming into contact with the Zircaloy. Finally, the barrier cladding concept of GE turned out as the best solution [50]. The barrier cladding has an inner layer of ductile pure Zr that relaxes more easily the local stresses. However, the performance in case of some cladding defects was rather unsatisfactory. Therefore, a modification of the liner material was necessary. New concepts use a material with improved corrosion resistance by additions of Fe [29,51,52]. Fur-

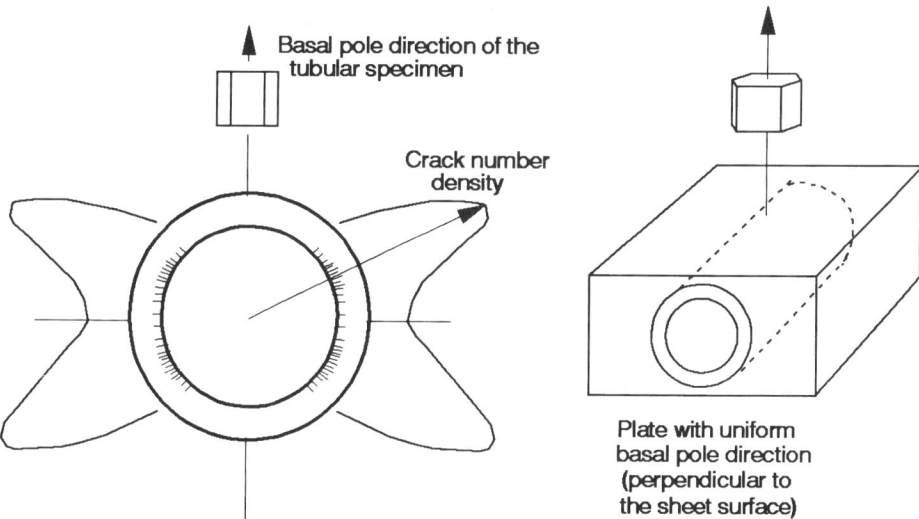

FIG. 11—*Influence of texture on crack development in iodine stress corrosion testing in laboratory* [44].

thermore, based on the above-mentioned results on texture and grain size, a highly texturized and very-fine-grained through-wall Zircaloy-2 cladding tube version, the LTP-2, was developed. The designation LTP stands for low temperature processing [52]. This tube version combines resistance against nodular corrosion and against I-SCC and has low uniform corrosion.

Conclusions

1. Zircaloy-2 and -4 turned out to be very reliable materials that fulfilled all requirements for normal operation, for postulated accidental conditions, and for long-time intermediate storage.

2. To determine the behavior and the properties of Zr alloys in power reactors, detailed examination programs—especially in the reactor pools—are very important. The support by our customer utilities was the prerequisite for achieving the status of today's knowledge.

3. Optimization of Zircaloy-2 and -4 includes both optimization of microstructure and of chemical composition. BWRs and PWRs need differently optimized materials because of their different coolant chemistry and their slightly different temperature levels. The cooperation with the pre-material suppliers was very important.

4. Current more-demanding operating conditions and discharge burnups required a further optimization of the Zircaloys and for "hot" PWRs even the development of more corrosion-resistant Zr alloys.

5. A significant improvement of PWR corrosion behavior can be achieved with Zr alloys using the alloying elements of Zircaloy with somewhat modified concentrations. Sn should be below or at least in the lower range of the ASTM specification range for Zircaloy-4, Fe and Cr should be somewhat higher, and Si should be specified as an alloying element rather than as an impurity. It is recommended that ASTM considers the new needs and expands the list of Zr alloys for nuclear application.

6. Careful evaluation of all operationally induced effects is important and is further recommended to meet future technical requirements.

References

[1] Kass, S., "The Development of the Zircaloys," *Proceedings*, USAEC Symposium on Zirconium Alloy Development, Castlewood, Pleasanton, CA, November 1962, p. 1 (USAEC-Report GEAP 4089).
[2] Knaab, H. and Knecht, K., "Pool-Site Fuel Inspection and Examination Techniques Applied by the Kraftwerk Union AG Fuel Service," *Proceedings*, 26th ANS Conference on Remote Systems Technology, 1978.
[3] Garzarolli, F., Stehle, H., and v. Jan, R., "The Main Causes of Fuel Element Failures in Water Cooled Power Reactors," *Atomic Energy Review*, Vol. 17, No. 1, IAEA, Vienna, 1979.
[4] Stehle, H. and Assmann, H., "Chemie in Reaktorbrennstäben mit oxidischen Kernbrennstoffen," *Chemiker-Zeitung*, Vol. 108, 1984, pp. 65–82.
[5] Garzarolli, F., Manzel, R., Reschke, S., and Tenckhoff, E., "Review of Corrosion and Dimensional Behavior of Zircaloy Under Water Reactor Conditions," *Zirconium in the Nuclear Industry: Fourth International Symposium, ASTM STP 681*, American Society for Testing and Materials, West Conshohocken, PA, 1979, p. 91.
[6] Garde, A. M., Smerd, P. G., Garzarolli, F., and Manzel, R., "Influence of Metallurgical Conditions on the In-Reactor Dimensional Changes of Zircaloy Fuel Rods," *Zirconium in the Nuclear Industry: Sixth International Symposium, ASTM STP 824*, American Society for Testing and Materials, West Conshohocken, PA, 1984, p. 289.
[7] Garzarolli, F., Dewes, P., Maussner, G., and Basso, H. H., "Effects of High Neutron Fluences on Microstructure and Growth of Zircaloy-4," *Zirconium in the Nuclear Industry: Eight International Symposium, ASTM STP 1032*, American Society for Testing and Materials, West Conshohocken, PA, 1989, p. 641.
[8] Tenckhoff, E., "Verformungsmechanismen Textur und Anisotropie in Zirconium und Zircaloy," *Metallkundlich-Technische Reihe*, Vol. 5, Gebrüder Borntrager Belin, Stuttgart, 1980.
[9] Stehle, H., "Progress in Zircaloy-4 Canning Technology for PWR-Fuel," *Proceedings*, Fourth United Nations International Conference on the Peaceful Uses of Atomic Energy, Geneva, Switzerland, 6–16 Sept. 1971.
[10] Stehle, H., Steinberg, E., and Tenckhoff, E., "Mechanical Properties, Anisotropy, and Microstructure of Zircaloy Canning Tubes," *Zirconium in the Nuclear Industry: Third International Symposium, ASTM STP 633*, American Society for Testing and Materials, West Conshohocken, PA, 1977, p. 486.
[11] Garzarolli, F., Manzel, R., Schönfeld, H., and Steinberg, E., "Influence of Final Annealing on Mechanical Properties of Zircaloy Before and After Irradiation," Sixth SMIRT Conference, Paris, 17–21 Aug. 1981, Vol. C2, p. C2/1.
[12] Manzel, R., Knaab, H., and Stehle, H., "The Dimensional Behavior of LWR-Fuel," *Proceedings*, IAEA Specialists' Meeting, Tokyo, 26–30 Nov. 1984.
[13] Kaspar, G., Peehs, M., and Steinberg, E., "Experimental Investigations of Post-Pile Creep of Zircaloy Cladding Tube," *Proceedings*, Eighth SMIRT Conference, Brussels, Belgium, 19–23 Aug. 1985.
[14] Cheliotis, G., Ortlieb, E., and Weidinger, H. G., "Single and Multi Rod Investigations for the Experimental and Theoretical Verification of LOCA Fuel Behavior," *Proceedings*, ANS/ENS Topical Meeting on Reactor Safety Aspects of Fuel Behavior, Sun Valley, Idaho, 1981, Vol. 2, p. 306.
[15] Eberle, R., Cheliotis, G., Fuchs, H. P., and Garzarolli, F., "The DUPLEX Conception: Making Corrosion Resistant Cladding Behave in a LOCA Like Classic Tubing," *Proceedings*, IAEA Technical Committee Meeting on Behavior of Core Materials and Fission Product Release in Accident Conditions in LWRs, Aix-en-Provance, France, 16–19 March 1992, p. 17.
[16] Stehle, H., Kaden, W., and Manzel, R., "External Corrosion of Cladding in PWR," *Nuclear Engineering and Design*, Vol. 33, 1975, p. 155.
[17] Garzarolli, F. and Manzel, R., "Corrosion Resistance of Zircaloy in KWU Power Reactors," Deutsches Atomforum e.V., *Proceedings*, Reaktortagung 1977, Mannheim, Germany, 29 March–1 April 1977.
[18] Garzarolli, F. et al., "Water Corrosion of Zircaloy Fuel Rods," Electric Power Research Institute Report EPRI NP 2789, 1982 (Final Report).
[19] Garzarolli, F., Bodmer, R. P., Stehle, H., and Trapp-Pritsching, D., "Progress in Understanding PWR Fuel Rod Waterside Corrosion," *Proceedings*, ANS Topical Meeting on LWR Fuel Performance, Orlando, Vol. I, 1985, pp. 3–55.
[20] Garzarolli, F. and Stehle, H., "Behavior of Structural Materials for Fuel and Control Elements in Light Water Cooled Power Reactors," *Proceedings*, IAEA International Symposium on Improvements in Water Reactor Fuel Technology and Utilization, Stockholm, Sweden, 15–19 Sept. 1986, p. 387.
[21] Garzarolli, F., Stehle, H., Steinberg, E., and Weidinger, H. G., "Progress in the Knowledge of

Nodular Corrosion," *Zirconium in the Nuclear Industry: Seventh International Symposium, ASTM STP 939*, American Society for Testing and Materials, West Conshohocken, PA, 1987, pp. 364–386.

[22] Garzarolli, F., Steinberg, E., and Weidinger, H. G., "Microstructure and Corrosion Studies for Optimized PWR and BWR Zircaloy Cladding," *Zirconium in the Nuclear Industry: Eighth International Symposium, ASTM STP 1023*, American Society for Testing and Materials, West Conshohocken, PA, 1989, p. 202.

[23] Fuchs, H. P. et al., "Cladding and Structural Material Development for the Advanced Siemens PWR Fuel FOCUS," *Proceedings*, ANS-ENS International Topical Meeting on LWR Fuel Performance, Avignon, France, 1991, p. 682.

[24] Garzarolli, F., Schumann, R., and Steinberg, E., "Corrosion Optimized Zircaloy for BWR Fuel Elements," *Zirconium in the Nuclear Industry: Tenth International Symposium, ASTM STP 1245*, American Society for Testing and Materials, West Conshohocken, PA, 1994, p. 709.

[25] van Swam, L. F., Garzarolli, F., and Steinberg, E., "Advanced PWR Cladding," *Proceedings*, ANS Internat. Topical Meeting on LWR Fuel Performance, West Palm Beach, FL, 1994, p. 303.

[26] Maussner, G., Steinberg, E., and Tenckhoff, E., "Nucleation and Growth of Intermetallic Precipitates in Zircaloy-2 and Zircaloy-4 and Correlation to Nodular Behavior," *Zirconium in the Nuclear Industry: Seventh International Symposium, ASTM STP 939*, American Society for Testing and Materials, West Conshohocken, PA, 1987, p. 202.

[27] Charquet, D., Steinberg, E., and Millet, Y., "Influence of Variations in Early Fabrication Steps on Corrosion, Mechanical Properties, and Structure of Zircaloy-4 Products," *Zirconium in the Nuclear Industry: Seventh International Symposium, ASTM STP 939*, American Society for Testing and Materials, West Conshohocken, PA, 1987, pp. 431–447.

[28] Steinberg, E., Weidinger, G., and Schaa, A., "Analytical Approaches and Experimental Verification to Describe the Influence of Cold Work and Heat Treatment on the Mechanical Properties of Zircaloy Cladding Tubes," *Zirconium in the Nuclear Industry: Sixth International Symposium, ASTM STP 824*, American Society for Testing and Materials, West Conshohocken, PA, 1984, p. 106.

[29] Siemens, U.S. Patent No. 5,245,645, 1991.

[30] Seibold, A. and Woods, K., "BWR Advanced Material," *Proceedings*, ANS International Topical Meeting on LWR Fuel Performance, West Palm Beach, FL, 1994, p. 633.

[31] Seibold, A., Garzarolli, F., and Steinberg, E., "Optimized Zry-4 with Enhanced Fe and Cr Content and DUPLEX Cladding: The Answer to Corrosion on PWR," *Proceedings*, International KTG]ENS Topical Meeting on Nuclear Fuel, TOPFUEL 95, Würzburg, Germany, 12–15 March 1995, Vol. II, p. 117.

[32] Weidinger, H. G. and Lettau, H., "Advanced Material and Fabrication Technology for LWR Fuel," *Proceedings*, IAEA International Symposium on Improvements in Water Reactor Fuel Technology and Utilization, Stockholm, Sweden, 15–19 Sept. 1986, p. 451.

[33] Garzarolli, F. and Holzer, R., "Waterside Corrosion Performance of Light Water Power Reactor Fuel," *Nuclear Energy*, Vol. 31, No. 1, 1992, pp. 65–86.

[34] Jahreis, W., Manzel, R., and Ortlieb, E., "Effect of the Hydrogen Content and of Irradiation on the Mechanical Behavior of Fuel-Structural-Components from Zircaloy," *Proceedings*, Jahrestagung Kerntechnik '93, Köln, Germany, 25–27 May 1983, p. 303.

[35] Siemens, U.S. Patent 4,735,768, 1985.

[36] Siemens, U.S. Patent 4,963,316, 1987.

[37] Siemens, U.S. Patent Specification 92 P 3128P, 1992.

[38] Garzarolli, F., Pohlmeyer, J., Trapp-Pritsching, S., and Weidinger, H. G., "Influence of Various Additions to Water on Zircaloy Corrosion in Autoclave Tests at 350°C," *Proceedings*, IAEA Technical Committee Meeting on Fundamental Aspects of Corrosion of Zirconium Base Alloys in Water Reactor Environments, Portland, Oregon, 11–15 Sept. 1989.

[39] Eberle, R., Perkins, R. A., and Steinberg, E., "Deformation and Oxidation of Alternative PWR Claddings at LOCA Temperatures," this publication.

[40] Garzarolli, F. et al., "Oxide Growth Mechanism on Zirconium Alloys," *Zirconium in the Nuclear Industry: Ninth International Symposium, ASTM STP 1132*, American Society for Testing and Materials, West Conshohocken, PA, 1991, p. 395.

[41] Weidinger, H. G. et al., "Corrosion-Electrochemical Properties of Zirconium," *Zirconium in the Nuclear Industry: Ninth International Symposium, ASTM STP 1132*, American Society for Testing and Materials, West Conshohocken, PA, 1991, p. 499.

[42] Beie, H. J. et al., "Examinations of the Corrosion Mechanism of Zirconium Alloys," *Zirconium in the Nuclear Industry: Tenth International Symposium, ASTM STP 1245*, American Society for Testing and Materials, West Conshohocken, PA, 1994, p. 615.

[43] Stehle, H., von Jan, R., and Knaab, H., "LWR Fuel Behavior During Operational and Overpower Transients," *Proceedings*, ANS/ENS Topical Meeting, San Valley, Idaho, 1981.

[44] Knaab, H., Lang, P. M., and Mogard, H., "Overview on International Experimental Programs on Power Ramping and Fission Gas Release," *Res Mechanica*, Vol. 4, 1985, p. 87.
[45] Vogl, W., Ruyter, I., and Markgraf, J., "The Petten Ramp Test Program of KWU/KFA During the Years 1976 to 1981," *Proceedings*, IAEA Specialists' Meeting on Power Ramping and Cycling Behavior of Water Reactor Fuel, Petten, 8–9 Sept. 1982.
[46] Garzarolli, F., Manzel, R., Peehs, M., and Stehle, H., "Observations and Hypothesis on Pellet-Clad Interaction Failures," *KERNTECHNIK*, Vol. 20, 1978, p. 27.
[47] Peehs, M., Stehle, H., and Steinberg, E., "Out-of-Pile Testing of I-SCC in Zircaloy in Relation to the PCI-Phenomen," *Zirconium in the Nuclear Industry: Fourth International Symposium, ASTM STP 681*, American Society for Testing and Materials, West Conshohocken, PA, 1979, p. 244.
[48] Peehs, M., Jung, W., Stehle, H., and Steinberg, E., "Experiments to Settle an I-SCC Hypothesis for Zry Tubing," *Proceedings*, IAEA Specialist Meeting on PCI in Water Reactors, Risö, September 1980, p. 170.
[49] Steinberg, E., Peehs, M., and Stehle, H., "Development of the Crack Pattern During Stress Corrosion in Zircaloy-Tubes," *Journal of Nuclear Materials*, Vol. 118, 1983, pp. 286–293.
[50] Armijo, J. S., Coffin, L. F., and Rosenbaum, H. S., "Development of Zirconium-Barrier Fuel Cladding," *Zirconium in the Nuclear Industry: Tenth International Symposium, ASTM STP 1245*, American Society for Testing and Materials, West Conshohocken, PA, 1994, p. 3.
[51] Steinberg, E. and Manzel, R., "Iron-Enhanced Zirconium Liner as an Answer to BWR Cladding Failures," *Proceedings*, International KTG/ENS Topical Meeting on Nuclear Fuel, TOPFUEL 95, Würzburg, Germany, 12–15 March 1995, Vol. II, p. 66.
[52] Seibold, A. and Ortlieb, E., "Überblick über die Entwicklungen auf dem Gebiet der Hüllrohr- und Strukturwerkstoffe für SIEMES DWR- und SWR-Brennelemente," *Proceedings*, Fachtagung der KTG-Fachgruppe "Brennelemente," Kernforschungszentrum Karlsruhe, 29–30 Nov. 1993.

Corrosion Mechanisms—I

Hiroyuki Anada[1] *and Kiyoko Takeda*[1]

Microstructure of Oxides on Zircaloy-4, 1.0Nb Zircaloy-4, and Zircaloy-2 Formed in 10.3-MPa Steam at 673 K

REFERENCE: Anada, H. and Takeda, K., "**Microstructure of Oxides on Zircaloy-4, 1.0Nb Zircaloy-4, and Zircaloy-2 Formed in 10.3-MPa Steam at 673 K,**" *Zirconium in the Nuclear Industry: Eleventh International Symposium, ASTM STP 1295*, E. R. Bradley and G. P. Sabol, Eds., American Society for Testing and Materials, 1996, pp. 35–54.

ABSTRACT: The microstructure of ZrO_2 formed on sheet materials of Zircaloy-2 (Zr2), Zircaloy-4 (Zr4), and an alloy of 1.0% Nb added to Zircaloy-4 (1Nb-Zr4) was analyzed using HRTEM (high-resolution transmission electron microscopy). The relationship between the corrosion behavior of the alloys and the microstructure is discussed. Stress-relieved sheet specimens of the three alloys were prepared and corrosion tested under static conditions in steam at 673 K and 10.3 MPa for a total of 220 days. The order of corrosion resistance in 673-K steam was Zr2, 1Nb-Zr4, and Zr4. Several transitions were observed in the corrosion kinetic curve of 1Nb-Zr4 and Zr2. However, only the first transition was observed in the curve of Zr4. Oxide structure in the pre-transition region on Zr4 was analyzed to be in the following order from the outside surface: columnar m-ZrO_2, t-ZrO_2 layer, substoichiometric Zr oxide layer, and α-Zr matrix. The t-ZrO_2 layer was approximately 50 to 80 nm thick, and the substoichiometric Zr oxide layer was approximately 100 to 200 nm. These layers were absent in the microstructure of the oxide in the post-transition region. The substoichiometric Zr oxide layer consisted of m-ZrO_2 grains that were less than 10 nm in diameter and some as yet unidentified grains that had lattice parameters similar to distorted and significantly oriented α-Zr. However, the t-ZrO_2 layered structure and the substoichiometric Zr oxide layer structure were observed in the post-transition oxides on Zr2 and 1Nb-Zr4. It was also observed that transformation of columnar grains to fine equiaxed grains had occurred near the lateral cracks and the incorporated intermetallic precipitates in post-transition oxides. It is implied from these results that the t-ZrO_2 layer and the substoichiometric Zr oxide layer structures play an important role as a barrier layer in controlling the occurrence of kinetic transitions.

KEYWORD: zirconium alloy, TEM, corrosion resistance, oxide film, morphology, crystal structure

Zirconium alloys are used as cladding materials in light water reactors. To support the extension of fuel burnup in light water reactors, corrosion resistance of Zircaloy cladding needs to be improved. Therefore, corrosion behavior of Zircaloy cladding material is undergoing extensive investigation from various points of view, such as chemical composition [1] and manufacturing conditions [2,3]. In addition, studies are aimed toward a mechanistic understanding of the corrosion phenomenon. Oxide microstructure, examined using transmission electron microscopy (TEM), could provide important information such as crystal structure, grain size, and grain shape for the oxide film, which would be expected to shed significant light on the corrosion behavior of Zr alloys [4], for example: grain shape and grain growth of oxide on Zr4

---

[1] Research engineer, Research and Development Center, Sumitomo Metal Industries, Ltd., 1-8 Fuso-cho, Amagasaki, Japan.

has been correlated with corrosion behavior [5]; differences in microstructure related to nodular and uniform corrosion has been discussed [6]; micropores were shown to play an important role in corrosion behavior [7,8]; and oxidation and microstructure of the intermetallic precipitates has been shown to influence the microstructure of Zirconium oxide [9,10].

The work reported herein was based on studies of the microstructure of oxide formed on Zircaloy-4, the 1% Nb added to a Zircaloy-4 alloy, and Zircaloy-2 at 673 K in 10.3 MPa steam using HR-TEM (high resolution transmission electron microscopy) to characterize microstructure, especially that associated with transitions occurring during corrosion.

Experimental Procedure

Material and Corrosion Test

Sheet specimens of Zircaloy-2 (Zr2), Zircaloy-4 (Zr4), and 1% Nb added to Zircaloy-4 (1Nb-Zr4) were prepared. Table 1 shows the chemical compositions, and Fig. 1 shows manufacturing conditions of these alloys. The small ingots, approximately 0.2 kg in weight, were β-solution treated at 1323 K for 1 h in vacuo. They were hot rolled at 923 K and cold rolled two additional times to a final thickness of 1.0 mm and intermediate annealed at 923 K for 2 h. The cold-rolled Zr4 sheet finally was stress relieved at 723 K for 2 h. The annealing parameter [$\Sigma Ai = \Sigma ti \times \exp(-40\,000/T)$ (h)] was calculated to be 8×10^{-19} (h).

Rectangular specimens, 1 by 40 by 40 mm in size, were cut from the sheets. The specimens were ground using SiC paper and were degreased in acetone. Pickling was not conducted. The corrosion tests were conducted at 673 K in 10.3 MPa steam for various times up to a total of 220 days under static autoclave conditions.

Microstructure of ZrO_2

TEM examinations were performed on the specimens that had been corrosion tested for 10, 30, and 160 days. In the case of Zr2 and 1Nb-Zr4, specimens oxidized for 30 days were selected. An apparent transition after 20 days was observed in the weight gain versus time curve as described later. Therefore the oxide microstructure characterization concentrated on differences between the pre-transition and the post-transition regions.

The oxide films were examined in cross section using JEOL:200CX TEM, JEOL:JEM-3010, and JEOL:JEM-4000EX HR-TEM instruments. Selected area diffraction (SAD) patterns and lattice images were obtained and analyzed to determine the crystal structures. The beam spot diameter for the SAD pattern was 200 nm. Figure 2 depicts the specimen preparation procedure for the TEM examination of the oxide in cross section. Corroded specimens were glued face to face along the external oxide surfaces to improve handling. The specimen was cut, ground, and polished to a thin foil specimen 50 μm in thickness. Ar gun ion thinning was performed to produce the foil.

TABLE 1—*Chemical compositions of Zr2, Zr4, and 1Nb-Zr4 (wt%).*

	Sn	Fe	Cr	Ni	Nb
Zr2	1.49	0.23	0.084	0.05	<0.001
Zr4	1.45	0.23	0.098	<0.001	<0.001
1Nb-Zr4	1.44	0.20	0.086	<0.001	1.0

β-Heat Treatment : 1323K, WQ
↓
Hot rolling: 923K, 2hr
↓
Intermediate annealing:923K, 2hr
↓
Cold rolling: Reduction=60%
↓
Intermediate annealing: 923K, 2hr
↓
Cold rolling:Reduction=70%
↓
Stress relieving:723K, 2hr

$[\Sigma A=8\times10^{-19} : \Sigma A=\sum t\times exp(-40000/T)]$

FIG. 1—*Manufacturing conditions of the sheet specimen.*

Results

Weight Gain

The weight gain versus time data for the three alloys are shown in Fig. 3. Corrosion resistance of Zr4 was poor compared to that of 1Nb-Zr4. Weight gain for Zr4 was approximately seven times greater than that for 1Nb-Zr4 after the 220-day exposure. The most corrosion-resistant alloy was the Zr2, for which the weight gain was half of that of 1Nb-Zr4 after the 220-day exposure.

A transition in the weight gain curve for Zr4 occurred at approximately 70 mg/dm² following 20-day exposure. The corrosion rate was approximately cubic before the transition, becoming linear after the transition. No additional transition was apparent after the first transition, and weight gain increased significantly.

Two transitions for 1Nb-Zr4 were observed at approximately 32 and 100 mg/dm² after 15 and 50-day exposures. However, the second transition was not as distinct as that for Zr2; for the latter, a third transition was also observed. The first transition in the weight gain curve of Zr2 also occurred at 32 mg/dm² and 20 days. In addition to the first transition, the other two transitions occurred at 85 and 130 mg/dm² after 70 and 130-day exposures. The corrosion rate was approximately cubic before the first transition, which was the same as that of Zr4. The weight gain for Zr2 was less than half of that of Zr4 even before the first transition. The corrosion rates in the early stage are different for the three alloys.

The transition behavior was different for the three alloys. In order to look at possible correlations to the microstructure of the oxide films, TEM examinations were conducted for oxide corresponding to the pre-transition region (A), the post-transition region (B), and the accelerated region (C) in the corrosion curve of Zr4 as identified in Fig. 3. The post-transition regions D and E were selected for 1Nb-Zr4 and Zr2, respectively.

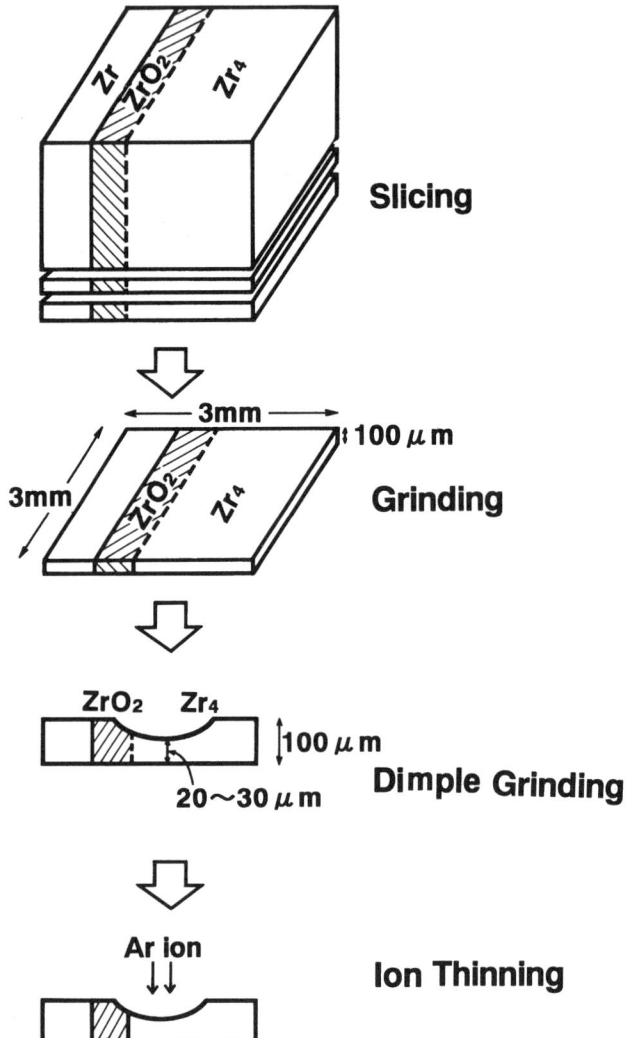

FIG. 2—*Thin foil preparation procedure for cross-sectioned specimen of Zr/ZrO_2.*

Oxide Microstructure and Corrosion of Zr4

Pre-Transition Region—Figure 4 shows a TEM image of the oxide-to-Zr4 interface region in cross section after the 10-day exposure corresponding to a weight gain of 60 mg/dm². The interface has a small bulged pattern. The oxide microstructure in the pre-transition region was analyzed to be in the following order from the outside surface: columnar grains, a substoichiometric Zr oxide layer, and an alloy matrix. The columnar grains grew perpendicular to the interface. The layered structure underneath the columnar grains and sandwiched to the α-Zr matrix was 100 to 200 nm in width and existed continuously and horizontally along the interface.

The lattice image can provide crystal structure, grain shape, and grain size of the ZrO_2 crystallites in the oxide film. Figure 5 shows the lattice image of Area 1 in Fig. 4. The growth

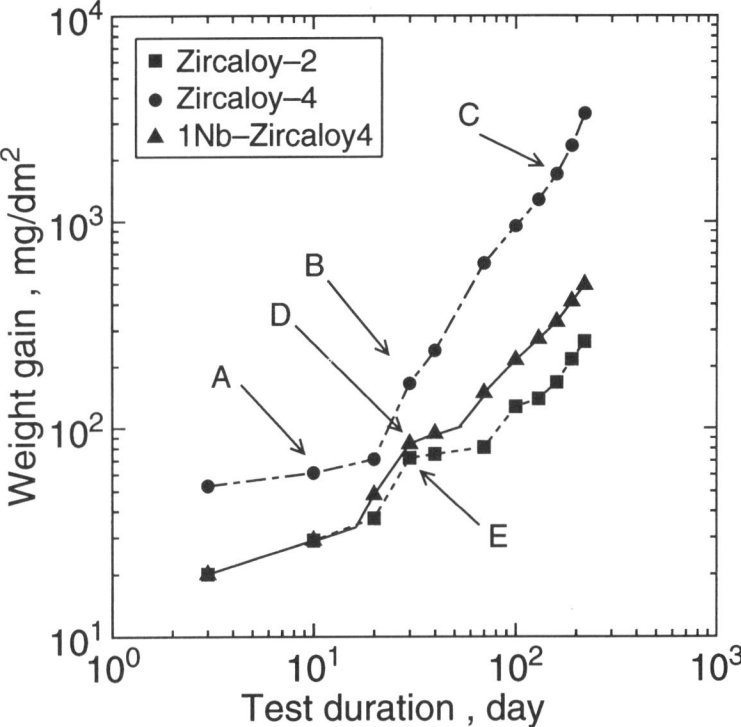

FIG. 3—*Weight gain for Zr2, Zr4, and 1Nb-Zr4 in 10.3-MPa steam at 673 K.*

fronts of the columnar ZrO_2 and the layered structure are clearly delineated. The layered structure consisted of equiaxed monoclinic ZrO_2 (m-ZrO_2) and some unidentified grains with a grain size of less than 10 nm. The lattice of m-ZrO_2 in the layered structure was distorted and elongated slightly from previously reported values (JCPDS). Some of the lattice spacings of the unidentified grains were close to that of α-Zr, and others were close to that of m-ZrO_2. This layer is defined herein as a substoichiometric Zr oxide layer.

Tetragonal ZrO_2 (t-ZrO_2) grains, which are indicated as "t" in Fig. 5, were adjacent to the outside surface of the substoichiometric layer. The t-ZrO_2 grains were approximately 80 nm wide and 50 nm thick. The t-ZrO_2 grains appeared to form a layer approximately several grains thick. However, the t-ZrO_2 layer was not entirely continuous. This was because some fine monoclinic ZrO_2 grains and unidentified Zr grains were interspersed with t-ZrO_2 crystallites, which are marked "m" and "u" in Fig. 5. The lattice spacing of such material was close to that of m-ZrO_2. However, the lattice angle was different from the literature values of m-ZrO_2.

In summary, two layer structures are identified, i.e., the continuous substoichiometric ZrO_2 layer structure on the alloy matrix and the t-ZrO_2 layer on top of the substoichiometric ZrO_2 layer. Both layered structures contained the unidentified structures that were very similar to a distorted m-ZrO_2 or α-Zr crystal structure.

The columnar grains appeared to grow on the t-ZrO_2 layer structure. All columnar grains were identified as m-ZrO_2. The size of the columnar grain was approximately 10 nm wide and more than 30 nm long. The growth of the columnar grain was apparently oriented to produce a specific orientation. The (100) plane in m-ZrO_2 appeared to grow perpendicular to the ZrO_2/

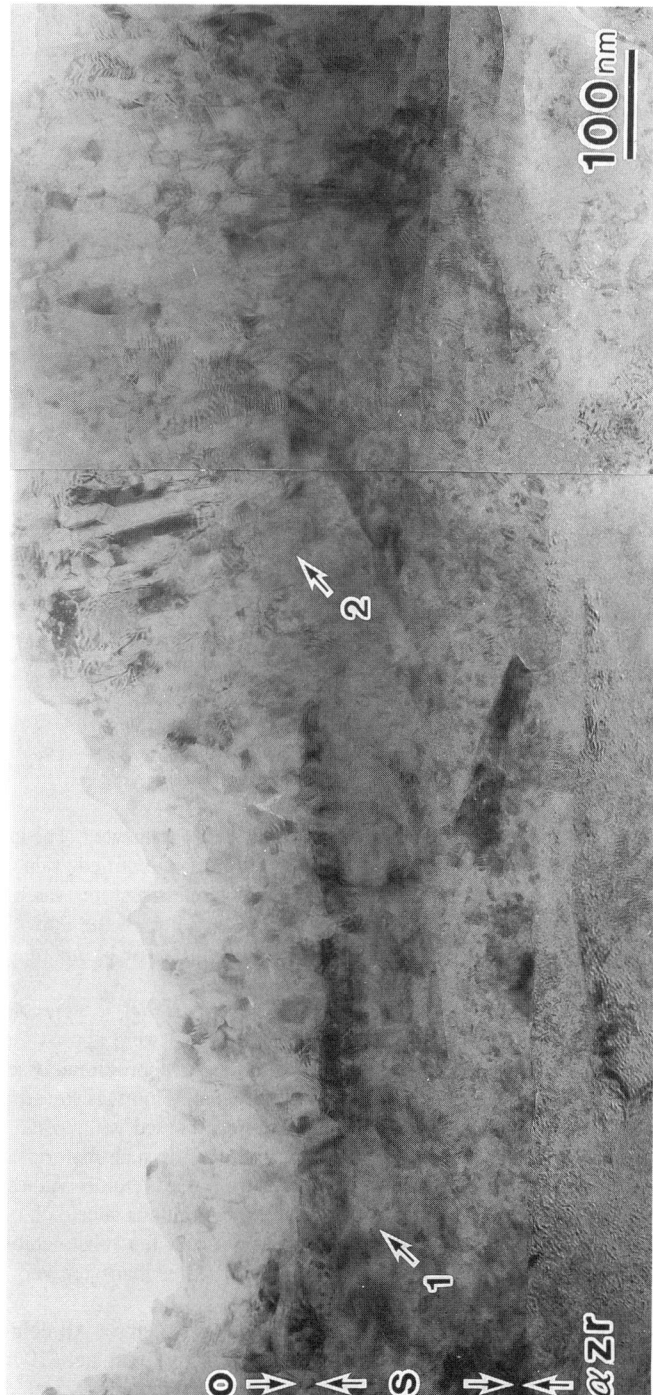

FIG. 4—*TEM image for $ZrO_2/Zr4$ interface region in cross section before the first transition (weight gain: 60 mg/dm^2); S and O denote the substoichiometric layer and Zr oxide, respectively.*

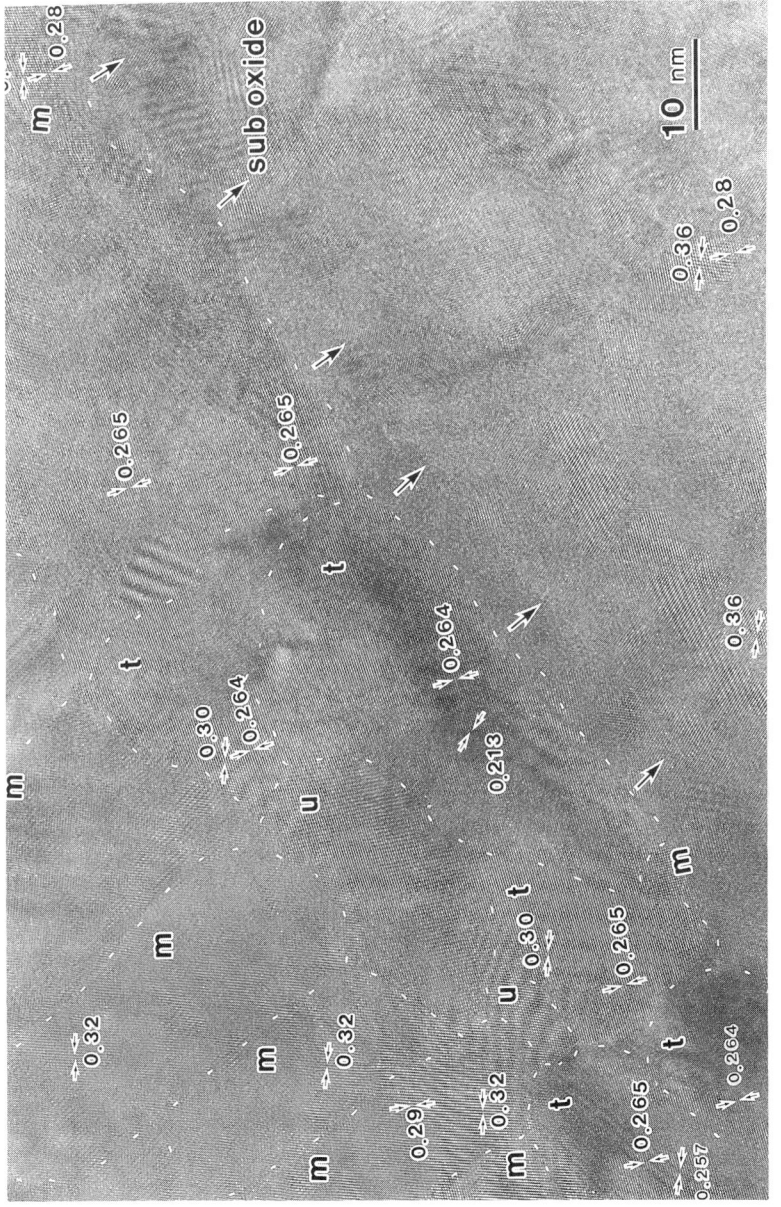

FIG. 5—Lattice image of Area 1 in Fig. 4 showing the growth fronts of columnar grain and t-ZrO$_2$ and substoichiometric layers; m, t, and u denote m-ZrO$_2$, t-ZrO$_2$, and unidentified Zr oxide.

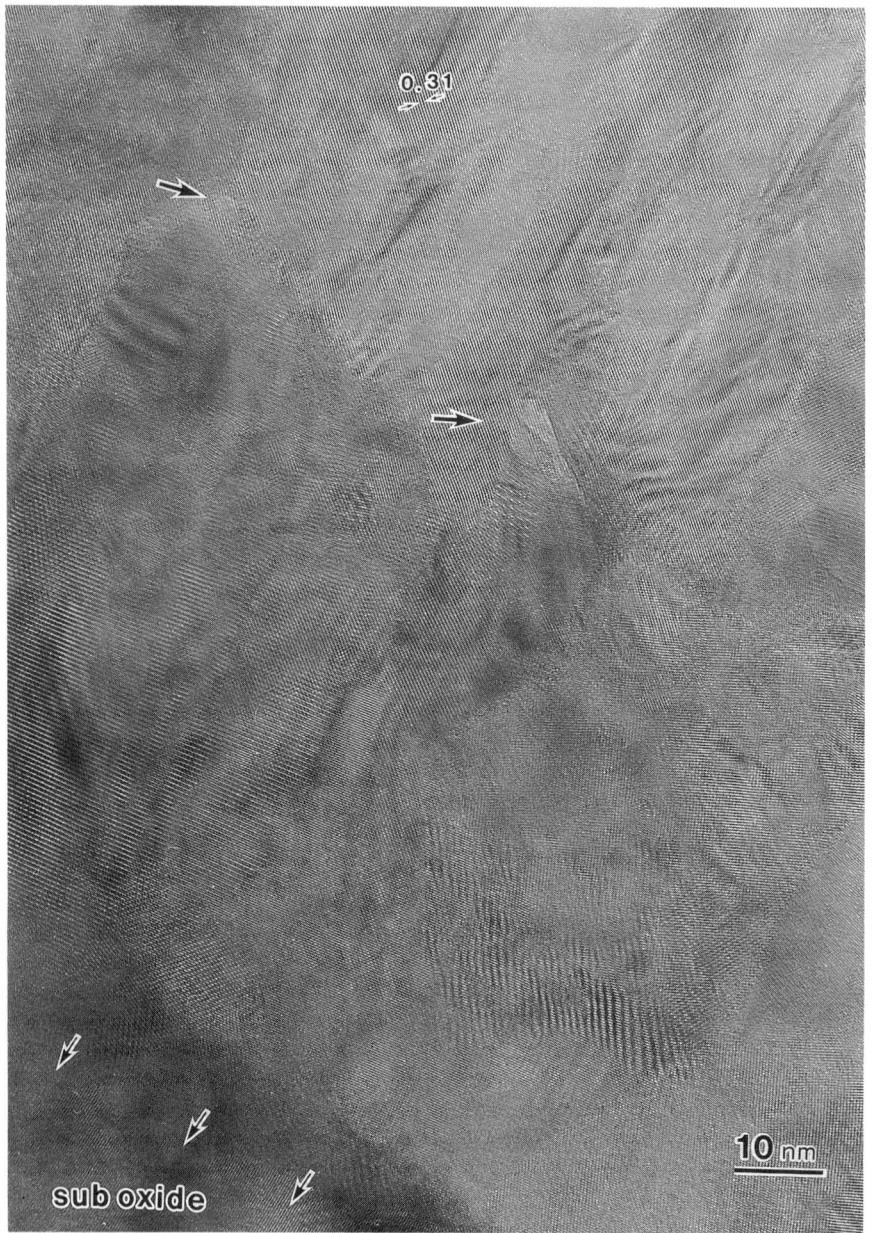

FIG. 6—*Lattice image of Area 2 in Fig. 4 showing the growing columnar grain and substoichiometric layer.*

Zr4 interface. The (101) plane and (002) plane of the t-ZrO_2 layer, including many dislocations, would appear to be expanded to become the (-111) plane and (111) plane of m-ZrO_2, respectively, according to the correlation of the lattice spacings and the lattice angles of t-ZrO_2 and m-ZrO_2.

Figure 6 shows the lattice images of the oxide crystallites in Area 2 in Fig. 4. The columnar grain was 10 to 15 nm wide and 30 to 50 nm in length. The columnar grains were ripened at the outer side away from the oxide-to-metal interface. It was obvious that the columnar grains were orientated. Micropores, indicated by arrows in Fig. 6, were observed between the columnar grain at 70 nm from the growth fronts. The size of micropore was from a few nm to less than 2 nm. Micropores were not observed at the oxide-to-metal interface and did not appear to be interconnected, either.

Post-Transition Region—Figure 7 shows a TEM image of the ZrO_2/Zr4 cross section obtained for the post-transition Region B corresponding to a weight gain of 150 mg/dm^2 in the corrosion curve in Fig. 3. Lateral cracks in the bulk oxide, indicated by arrows in Fig. 7, were observed spaced at intervals of approximately 200 nm. They appeared to exist intrinsically because of the regular intervals. The cracks, however, might have been artificially generated during the sample preparation.

Unlike in the pre-transition region oxide, columnar grain growth directly on the alloy was also observed in the vicinity of the oxide-to-alloy interface in the post-transition region as shown in Fig. 8a. The transition had not influenced the shape of the ZrO_2 grains. Another

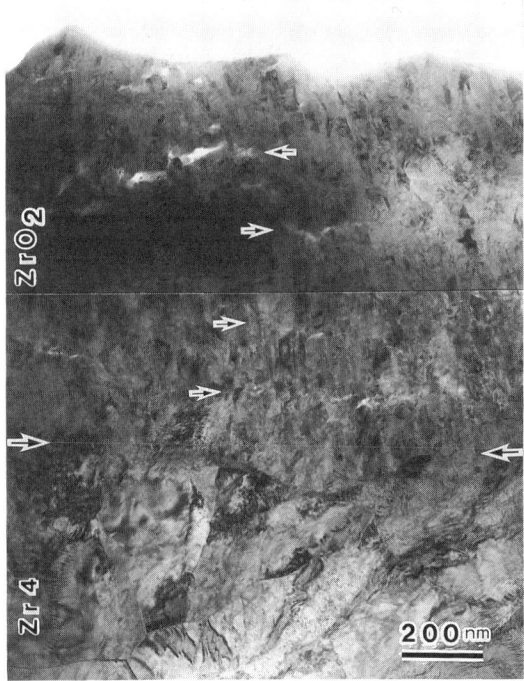

FIG. 7—*Cross-sectional view of the ZrO_2 and Zr4 in the post-transition region (weight gain: 160 mg/dm^2); large arrow and small arrow identify the interface and the lateral cracks in the oxide, respectively.*

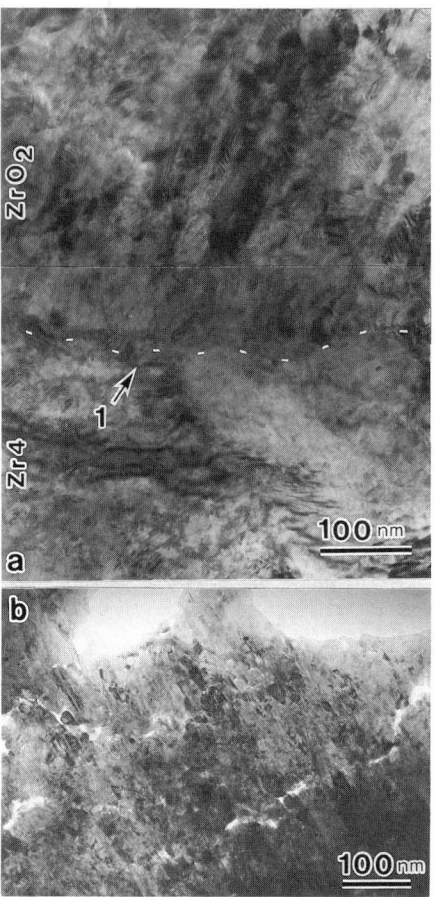

FIG. 8—*TEM micrograph of cross-sectioned post-transition oxide:* (a) *columnar grain growth directly at the $ZrO_2/Zr4$ interface and substoichiometric layer disappeared;* (b) *equiaxed grain 4 μm from the interface.*

feature of the post-transition oxide was the presence of equiaxed grains at the lateral cracks. It appeared that the columnar grains had a tendency to transform to the fine equiaxed grains around lateral cracks. The extent of transformation to equiaxed grains increased with distance away from the oxide-to-alloy interface as shown in Fig. 8b. The columnar grains appeared to have ripened slightly when compared to their size near the oxide-to-alloy interface.

Figure 9 shows a lattice image of the interface region 1 in Fig. 8a. It is noteworthy that only m-ZrO_2 was identified and that t-ZrO_2 was not found even in the vicinity of the oxide-to-alloy interface. Evidence for the presence of t-ZrO_2 and substoichiometric ZrO_2, similar to that observed in pre-transition oxides, was not found in this post-transition oxide.

To summarize, the TEM micrograph showed in the post-transition oxides the presence of columnar m-ZrO_2 grains and equiaxed grains, which were predominantly located at lateral cracks in the oxide. t-ZrO_2 and substoichiometric ZrO_2 were either absent or were too thin to be detectable.

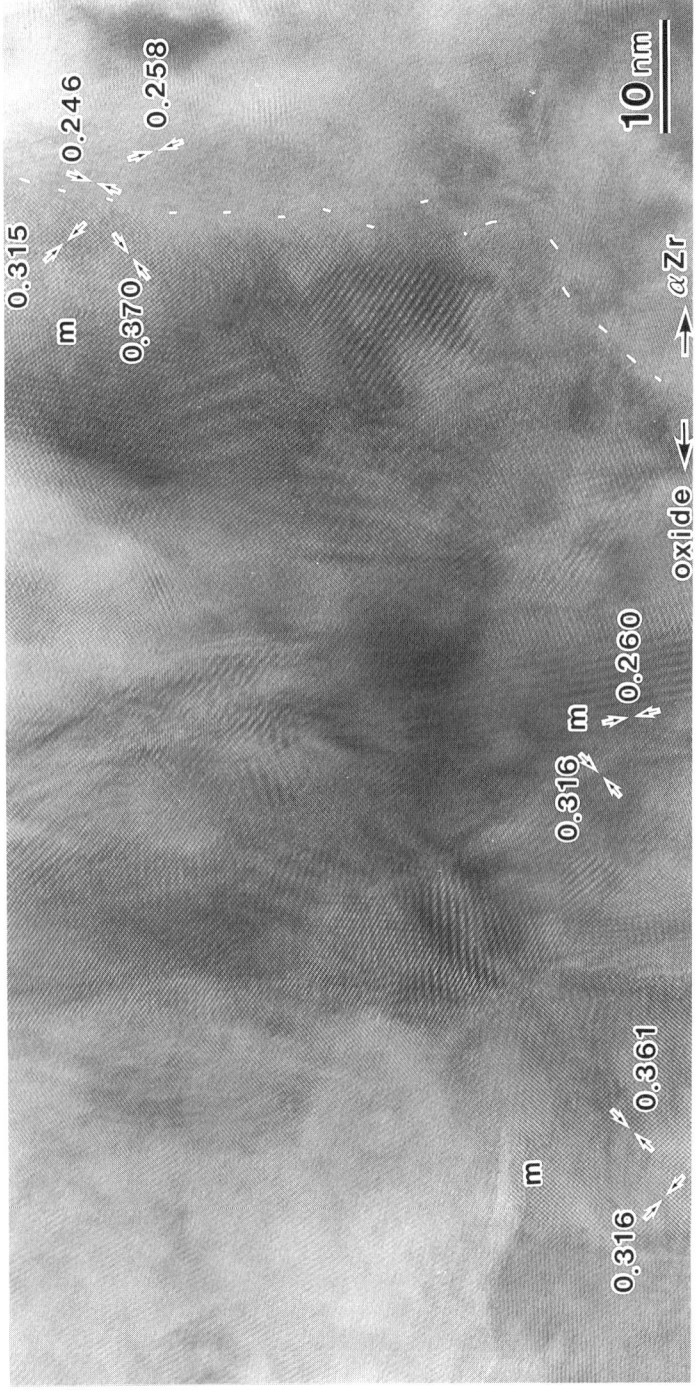

FIG. 9—*Lattice image of columnar grain of m-ZrO_2 and α-Zr matrix in the vicinity of the oxide-to-alloy interface; post-transition oxide on Zr4.*

Accelerated Corrosion Region—Figure 10 shows a typical lateral crack and an incorporated intermetallic precipitate in a cross-sectioned oxide on Zr4 after a 160-day exposure (1500 mg/dm^2 weight gain). Equiaxed grains and intermetallic precipitates were observed in the vicinity of the lateral cracks.

The ZrO_2 Microstructure on 1Nb-Zr4 and on Zr2 after the First Transition Region

The microstructure of ZrO_2 formed on the 1Nb-Zr4 alloy was observed in cross section. A representative bright field image and a set of SAD patterns were shown in Figs. 11a and 11b, respectively. The SAD patterns were taken by moving a microspot beam of 200 nm in diameter from the Zr4 matrix to the oxide surface step by step as shown in Fig. 11. Locations 1 and 2 in Fig. 11a were identified as α-Zr. The SAD patterns from Location 3 showed a dominant α-Zr pattern and a weak t-ZrO_2 pattern. t-ZrO_2 existed at Locations 4 and 5 in the oxide adjacent to the interface. Other locations adjacent to the α-Zr matrix were also found to be t-ZrO_2. Hence, it was concluded that t-ZrO_2 existed as a layer along the interface similar to that observed for the pre-transition oxide growth on Zr4. The beam diameter was 200 nm for the SAD examinations. Therefore, the width of the t-ZrO layer was equal to or less than 200 nm. At Locations 6 to 9, the oxide was m-ZrO_2. The substoichiometric ZrO_2 layer was also observed underneath the t-ZrO_2 layer, and its width was approximately 100 nm.

Figure 12 shows representative cross-sectional views of the oxide film formed on Zr2 after

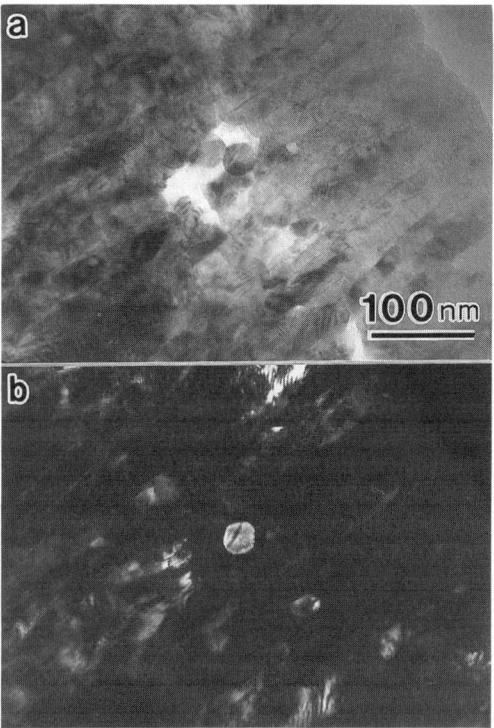

FIG. 10—*TEM micrograph of oxide cross section and incorporated intermetallic precipitate for oxide growth in accelerated corrosion region, identified "c" in Fig. 3.*

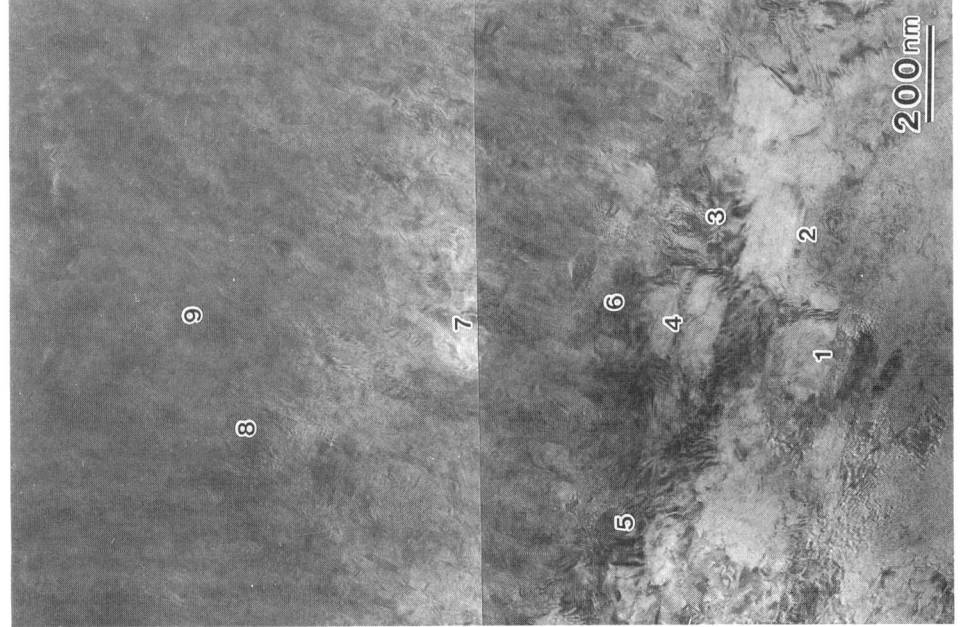

FIG. 11—TEM micrograph and SAD patterns for oxide on 1Nb-Zr4: (a) bright field image, (b) SAD patterns for selected positions (weight gain: 80 mg/dm²).

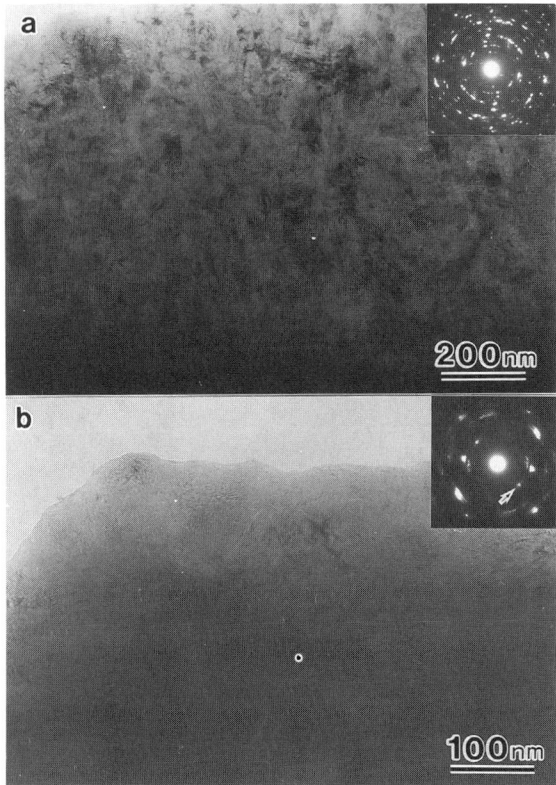

FIG. 12—*TEM micrograph and SAD patterns for oxide on Zr2: (a) bright field image; (b) SAD pattern (weight gain: 70 mg/dm^2).*

a total exposure of 30 days, corresponding to a weight gain of 60 mg/dm^2. The columnar growth of m-ZrO$_2$ grains was readily recognized as shown in Fig. 12a. The SAD patterns indicated that oxide was predominantly m-ZrO$_2$ with weak contributions from t-ZrO$_2$. Figure 12b shows the oxide cross section in the vicinity of the interface. The SAD pattern was identified as that of t-ZrO$_2$ (arrow in the figure) and α-Zr. The contributions to the SAD pattern from m-ZrO$_2$ was very weak. It was concluded that the oxide adjacent to the alloy matrix is a t-ZrO$_2$ layered structure similar to that observed on 1Nb-Zr4 in the post-transition region. The t-ZrO$_2$ layer is estimated to be approximately ≤200 nm in width.

Discussion

The TEM examination in this study shows the layered structure of ZrO$_2$ formed in steam. Figure 13 illustrates the layered structures of ZrO$_2$ on Zr4 in pre- and post-transition and the microstructure transformation occurring at the transition. In pre-transition, the t-ZrO$_2$ layer and the substoichiometric layer form on Zr4. The columnar m-ZrO$_2$ develops on the t-ZrO$_2$ layer. The micropores were 200 nm from the oxide-to-metal interface. Tetragonal ZrO$_2$ is stabilized and forms the t-ZrO$_2$ layered structure adjacent to the substoichiometric Zr oxide layer in pre-

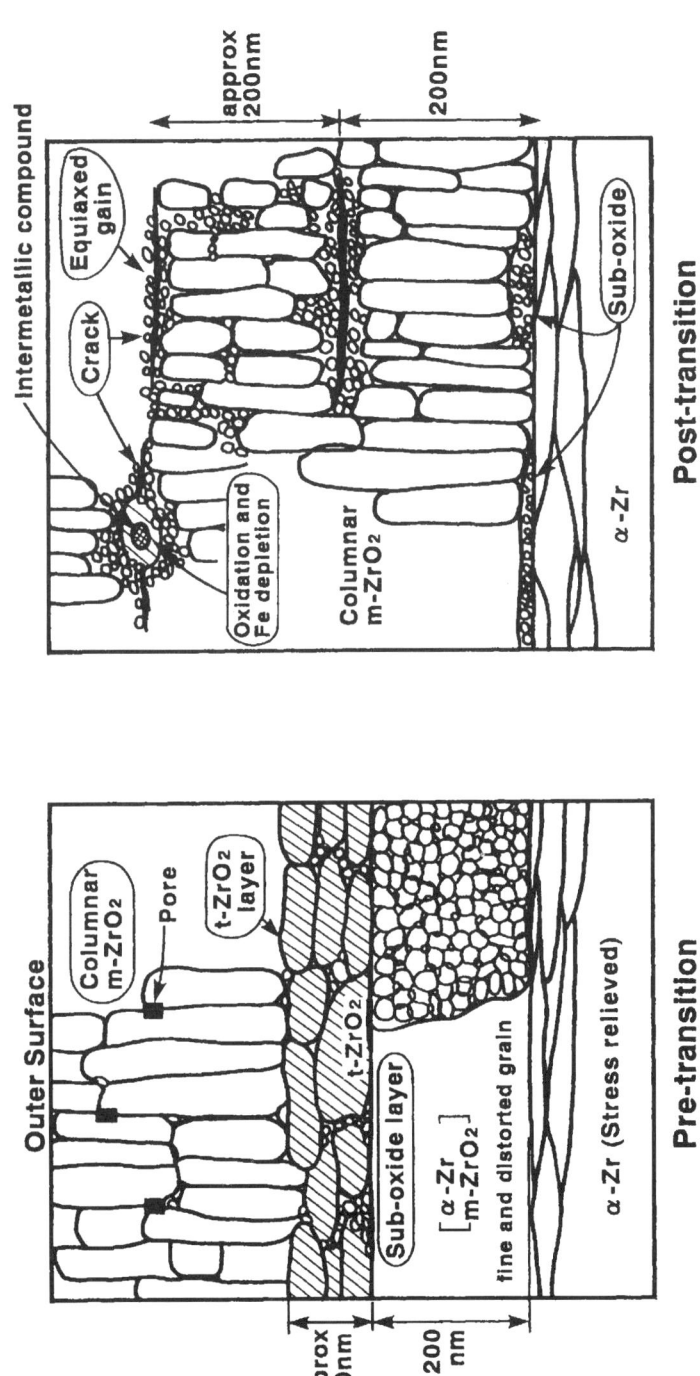

FIG. 13—*Illustration for microstructure of Zr oxide and matrix alloy for Zr4 in pre- and post-transition.*

transition. The substoichiometric Zr oxide layer consists of m-ZrO_2 and unidentified grain less than 10 nm in diameter. The t-ZrO_2 layer and the substoichiometric Zr oxide layer disappear in post-transition for Zr4. In post-transition for 1Nb-Zr4 and Zr2, however, both the t-ZrO_2 layer and the substoichiometric layer remain and were thicker than those in pre-transition for Zr4. Takeda et al. [11] indicate that the t-ZrO_2 layer structure was stabilized at the interface region before the first transition in corrosion kinetic curve for Zr4. Godlewski et al. [12] show using laser Raman spectroscopy that t-ZrO_2 was stabilized at the interface of the Zr4 metal where the Zr oxide consisted of a mixture of m-ZrO_2, t-ZrO_2, and t-ZrO_2 transformed to m-ZrO_2 when the transition occurred. The thickness of the t-ZrO_2 layer was less than 200 nm in this study. The laser beam diameter was too large to determine the distribution of the crystal structure in each grain. Wadman et al. [5] show tetragonal ZrO_2, which was a columnar grain grown directly on the α-Zr matrix. This study shows that t-ZrO_2 was transformed from the substoichiometric Zr oxide grain. There was a high dislocation density in the substoichiometric and in t-ZrO_2, which suggested a stress induced by the volume expansion by oxidation of Zr [13]. And the t-ZrO_2 layer observed in this study would be stabilized by the induced stress.

The surfaces of the specimen for the corrosion test were not pickled but grounded in this study. In general, characteristics of the grounded surface are different from those of the pickled surface, for example, residual stress induced by grinding, roughness, the intermetallic compound distribution on the surface, and so on. However, the characterized oxide in this study was more than 4 μm, which would be thick enough compared with the thickness that influenced the surface treatment. In addition, the surface treatment was the same among the three alloys. Therefore, the authors believe that damage to the surface does not affect the results in this study.

The transition behavior corresponded to the transformation from t-ZrO_2 to m-ZrO_2. It is considered that transformation of t-ZrO_2 to m-ZrO_2 had an important role in the transition behavior. The transition corresponded to the transformation of ZrO_2 [12]. This study shows that the t-ZrO_2 layer structure adjacent to the substoichiometric layer in the pre-transition region disappeared in the post-transition region. When, in particular for Zr4, the t-ZrO_2 layer disappeared at the transition, weight gain increased sharply. After the transition, the t-ZrO_2 layer structure was stabilized again in the corrosion-resistant alloy such as in the Zr2 and 1Nb-Zr4 as corrosion proceeded. In the corrosion kinetic curve for these alloys, there were several transitions. Therefore, the t-ZrO_2 layer acts as a barrier layer against transportation of oxygen ions. The results from this study would support "the barrier effect" of the oxide film adjacent to the interface [6].

Thickness of the substoichiometric Zr oxide and the t-ZrO_2 layers is important for the transition kinetics. The substoichiometric Zr oxide and t-ZrO_2 layers formed on Zr2 and on 1Nb-Zr4 in the post-transition were approximately 200 nm. They were thicker than those on Zr4 in pre-transition. Several transitions were observed in the corrosion kinetic curve for Zr2 and for 1Nb-Zr4. The thick t-ZrO_2 and thick substoichiometric Zr oxide layers would act as a barrier layer. As a result, the additional transitions would occur in the corrosion kinetic curve.

In addition, the t-ZrO_2 layer formed on Zr4 even before the transition was thin, which would influence weight gain in the early stage for Zr4. Weight gain in pre-transition for Zr4 was greater. This is because the t-ZrO_2 layer was not satisfactorily thick as the barrier layer.

The authors considered that stability of the t-ZrO_2 barrier would be important in the corrosion resistance of Zr alloy. The t-ZrO_2 barrier stability is influenced by several factors such as a micropore diffusion of oxygen ions in the t-ZrO_2 barrier and in the columnar m-ZrO_2 layer, mechanical properties of t-ZrO_2, and characteristics of the matrix.

The induced micropores caused the transition in the corrosion curve of Zircaloy [14]. In particular, the accelerated corrosion behavior correlated with the induced micropore [7]. However, if the transition corresponded to the transformation of the oxide, any micropores were not observed at the oxide surface adjacent to the oxide-to-alloy interface. Some isolating micropores were observed at 200 nm from the interface. Therefore, the micropores appear to accelerate diffusion of oxygen ions locally and to determine grain growth and grain shape. However, the total corrosion rate would not be controlled by the induced micropore. The authors consider that the micropores were induced by volume change owing to the transformation from t-ZrO_2 to m-ZrO_2.

The substoichiometric layer formed on the Zr-Sn-Nb alloy did not form on the recrystallized Zircaloy-4 [5]. In this study, the substoichiometric layer formed apparently on the Zr4 in the pre-transition region and almost disappeared in the post-transition region. The sub-oxide layer was recognized even after the first transition in the 1Nb-Zr4 and the Zr2. The transition repeatedly occurred in the corrosion curve of 1Nb-Zr4 and Zr2; however, the only first transition occurred in that of Zr4. The substoichiometric layer consisted of the fine m-ZrO_2 grain and the unidentified grain. The unidentified grain has a lattice spacing very close to that of α-Zr; however, the plane angle was distorted. The stress induced by the volume expansion of ZrO_2 [13] might cause the lattice distortion of α-Zr. As a result, the substoichiometric layer was recognized as corresponding to the transition behavior.

Grain size of the columnar m-ZrO_2 had ripened, and the columnar grain had transformed to the fine equiaxed grain as corrosion proceeded. Grain boundary diffusion of oxygen ions governed the corrosion rate of Zr [15,16]. The increase of the fine equiaxed grains increased the diffusion path and accelerated the corrosion rate. The columnar grains were dominantly observed and the transformed equiaxed grain was submissive, in particular in the 1Nb-Zr4 and the Zr2 oxides.

It was not concluded from this study that Nb and Ni influence the grain size and crystal structure of ZrO_2. Nb and Ni would influence the diffusion of oxygen ions and the mechanical properties of ZrO_2. Niobium is in solid solution in ZrO_2 [17]. A stable Nb oxide is Nb_2O_3, and the valency of the Nb ion is $+5$ in the oxide. If the Nb ion was solid-soluted as the $+5$ ion and replaced with the Zr ion ($+4$) in the ZrO_2 lattice, the density of oxygen vacancies would be reduced. In contrast, if Ni was in solid solution in the ZrO_2 matrix, the density of oxygen vacancies would increase. The reduction of oxygen vacancies would possibly stabilize t-ZrO_2 and suppress mobility of oxygen ions.

In addition, nickel has an effect on electronic conductivity in ZrO_2 [18,19]. Semi-conductor characteristics of ZrO_2 would be important to migration of oxygen ions in the ZrO_2 layer. The oxygen potential deviation would possibly influence the semiconductor characteristics of ZrO_2 [20]. It would influence the diffusion process of oxygen ions in the oxide whether the Zr oxide is an n-type or a p-type semiconductor.

The micropores, as mentioned above, would also possibly affect the oxygen potential locally. Hence, the oxygen potential deviation would influence the oxidation of the incorporated intermetallic precipitate [9]. This study shows that the incorporated intermetallic precipitate existed at lateral cracks and columnar grains transformed to the fine equiaxed grains around the precipitates. The intermetallic precipitate among the alloys used in this study was well known. The oxidation behavior of the intermetallic precipitates would be distinguished, which influences the ZrO_2 microstructure, i.e., the grain shape, grain size, and migration of oxygen ions during corrosion.

Conclusions

The microstructure of ZrO_2 grown in steam on Zr4, 1Nb-Zr4, and Zr2 was investigated by HR-TEM, and the following conclusions were drawn:

1. Zr2, 1Nb-Zr4, and Zr4 are in the order of corrosion-resistance in the 673-K steam condition. Several transitions were observed in the corrosion kinetics of Zr2 and 1Nb-Zr4; however, only the one transition was observed for Zr4.

2. Oxide microstructure in the pre-transition region for all three alloys consisted of: columnar m-ZrO_2/t-ZrO_2 layer/substoichiometric Zr oxide layer/α-Zr matrix. In the pre-transition region for Zr4, the t-ZrO_2 layer structure was approximately 50 to 80 nm thick, and the substoichiometric Zr oxide layer was approximately 100 to 200 nm thick. The substoichiometric Zr oxide layer consisted of m-ZrO_2 grain, less than 10 nm in diameter, and unidentified grains that were similar to the distorted and significantly oriented α-Zr.

3. In the post-transition region, the t-ZrO_2 layer structure and the substoichiometric layer were observed in the case of Zr2 and 1Nb-Zr4 but not in the case of Zr4.

4. The columnar grains of m-ZrO_2 transformed to fine equiaxed grain near lateral cracks and the incorporated intermetallic precipitates in post-transition oxide.

5. The presence of the t-ZrO_2 layer and the substoichiometric Zr oxide layer is correlated to the occurrence of kinetic transitions.

Acknowledgments

The authors thank Mr. K. Hanafusa, Mr. Y. Kohzuki and Mr. H. Nakamura for the TEM examination and corrosion test.

References

[1] Eucken, C. M., Finden, P. T., Trapp-Pritsching, S., and Weidinger, H. G., "Influence of Chemical Composition on Uniform Corrosion of Zirconium-Base Alloys in Autoclave Tests," *Zirconium in the Nuclear Industry, (Eighth Symposium) ASTM STP 1023*, L. F. P. Van Swam and C. M. Eucken, Eds., American Society for Testing and Materials, West Conshohocken, PA, 1989, pp. 113–127

[2] Shemel, J. H., Charquet, D., and Wadier, J.-F., "Influence of the Manufacturing Process on the Corrosion Resistance of Zircaloy-4 Fuel Cladding," *Zirconium in the Nuclear Industry (Eighth Symposium), ASTM STP 1023*, L. F. P. Van Swam and C. M. Eucken, Eds., American Society for Testing and Materials, West Conshohocken, PA, 1989, pp. 141–152.

[3] Anada, H., Herb, B., Nomoto, K., Hagi, S., Graham, R., and Kuroda, T., this publication.

[4] Bradley, E. R. and Perkins, R. A., "Characterization of Zircaloy Corrosion Films by Analytical Transmission Electron Microscopy," *Proceedings*, IAEA Technical Committee Meeting on Fundamental Aspects of Corrosion of Zr-Base Alloys in Water Reactor Environments, 1989, Portland, OR.

[5] Wadman, B., Lai, Z., Andren, H.-O., Nystrom, A.-L., Rudling, P., and Pettersson, H., "Microstructure of Oxide Layers Formed During Autoclave Testing of Zirconium Alloys," *Zirconium in the Nuclear Industry (Tenth Volume), ASTM STP 1245*, A. M. Garde and E. R. Bradley, Eds., American Society for Testing and Materials, West Conshohocken, PA, 1994, pp. 579–598.

[6] Garzarolli, F., Seidel, H., Tricot, R., and Gros, J. P., "Oxide Growth Mechanism on Zirconium Alloys," *Zirconium in the Nuclear Industry (Ninth Volume), ASTM STP 1132*, C. M. Eucken and A. M. Garde, Eds., American Society for Testing and Materials, West Conshohocken, PA, 1991, pp. 395–415.

[7] Cox, B. and Yomaguchi, Y., "The Development of Porosity in Thick Zirconia Films," *Journal of Nuclear Materials*, Vol. 210, 1994, pp. 303–317.

[8] Zhou, B.-X., "Electron Microscopy Study of Oxide Films Formed on Zircaloy-2 in Superheated Steam," *Zirconium in the Nuclear Industry (Eighth Symposium), ASTM STP 1023*, L. F. P. Van Swam and C. M. Eucken, Eds., American Society for Testing and Materials, West Conshohocken, PA, 1989, pp. 360–373.

[9] Percheur, D., Lefebvre, F., Motta, A. T., Lemaingnan, C., and Charquet, D., "Oxidation of Intermetallic Precipitates in Zircaloy-4: Impact of Irradiation," *Zirconium in the Nuclear Industry (Tenth Volume), ASTM STP 1245*, A. M. Garde and E. R. Bradley, Eds., American Society for Testing and Materials, West Conshohocken, PA, 1994, pp. 687–708.

[10] Anada, H., Nomoto, K., and Shida, Y., "Corrosion Behavior of Zircaloy-4 Sheets Produced Under Various Hot-rolling and Annealing Conditions," *Zirconium in the Nuclear Industry (Tenth Volume), ASTM STP 1245*, A. M. Garde and E. R. Bradley, Eds., American Society for Testing and Materials, West Conshohocken, PA, 1994, pp. 307–327

[11] Takeda, K., Anada, H., and Kajimura, H., to be presented to *Journal of the Japanese Institute of Metallurgy*.
[12] Godlewski, J., Gros, J. P., Lambertin, M., Wadier, J. F., and Weidinger, H., "Raman Spectroscopy Study of the Tetragonal-to-Monoclinic Transition in Zirconium Oxide Scales and Determination of Overall Oxygen Diffusion by Nuclear Microanalysis of O^{18}," *Zirconium in the Nuclear Industry, (Ninth Volume) ASTM STP 1132*, C. M. Eucken and A. M. Garde, Eds., American Society for Testing and Materials, West Conshohocken, PA, 1991, pp. 416–436.
[13] Godlewski, J., "How the Tetragonal Zirconia is Stabilized in the Oxide Scale that is Formed on a Zirconium Alloy Corroded at 400°C in Steam," *Zirconium in the Nuclear Industry, (Tenth Volume), ASTM STP 1245*, A. M. Garde and E. R. Bradley, Eds., American Society for Testing and Materials, West Conshohocken, PA, 1994, pp. 663–683.
[14] Cox, B., "Oxidation of Zirconium and Its Alloy," *Advances in Corrosion Science and Technology*, M. G. Fontana and R. W. Staele, Eds., Plenum Press, New York, Vol. 5, 1976.
[15] Cox, B. and Pemsler, J. P., "Diffusion of Oxygen in Growing Zirconia Films," *Journal of Nuclear Materials*, Vol. 28, 1968, pp. 73–78.
[16] Lightstone, J. B. and Pemsler, J. P., *Proceedings*, Symposium on Mass Transport in Oxides, National Bureau of Standard, October 1967, NBS Special Publication 296, August 1968.
[17] Ortali, P. and Jehn, H., *Metall. Ital.*, Vol. 66, 1974, p. 263.
[18] Isobe, Y., Fuse, M., and Kobayashi, K., "Additive Element Effects on Electronic Conductivity of Zirconium Oxide Film," *Journal of Nuclear Science and Technology*, Vol. 31, 1994, pp. 546–551.
[19] Inagaki, M., Kan-no, M., and Maki, H., "Effect of Alloying Elements in Zircaloy on Photo-Electrochemical Characteristics of Zirconium Oxide Film," *Zirconium in the Nuclear Industry (Ninth Volume), ASTM STP 1132*, C. M. Eucken and A. M. Garde, Eds., American Society for Testing and Materials, West Conshohocken, PA, 1991, pp. 437–459.
[20] Kofstad, P., *Nonstoichiometry, Diffusion, and Electrical Conductivity in Binary Metal Oxides*, Wiley-Interscience, New York, 1972, p. 159.

DISCUSSION

H. M. Chung[1] (written discussion)—I believe the stability of the tetragonal structure (of oxide) is well known to be sensitive to minor impurities and stress. Could you comment on the possible effects of reactor operating conditions and irradiation on the phase structure you have observed under simulated conditions? Were the alloys fabricated from the same raw material or Zr?

H. Anada (author's closure)—It is well known that the intermetallic precipitates are amorphousized by irradiation and that there is a depletion of Fe in the intermetallic precipitates. There is a possibility that these effects influence the microstructure of the oxide.

B. Cox[2] (written discussion)—Would you care to offer an explanation for the poor corrosion behavior of the Zircaloy-4 sheet, compared with Zircaloy-2, in 400°C steam. Normally, Zircaloy-2 and Zircaloy-4 fabricated in the same way would be expected to give very similar weight gain curves for at least the first 200 days or so.

H. Anada (author's closure)—I think the difference between the corrosion behavior of Zr-2 and Zr-4 was caused by the difference of the optimized value for ΣAi value between them. ΣAi for Zr-4 in this study was 8.9×10^{-19} and was the same as that for Zr-2. The optimized ΣAi value for Zr-4 is known to be higher than this value. That for Zr-2 is not clear now but will be lower than that for Zr-4.

A. T. Motta[3] (written discussion)—You suggest that the corrosion of intermetallic precipitates incorporated into the oxide could cause or influence the transformation of columnar grains into equiaxed grains. What mechanism do you propose for this phenomenon?

H. Anada (author's closure)—I think that Fe depletion from the incorporated precipitates is important for the transformation from columnar grains to equiaxed grains. Chemical composition in the ZrO_2 would change locally after depletion of Fe, and when depleted Fe is oxidized. Stress distribution in the ZrO_2 would be changed near the precipitate.

G. P. Sabol[4] (written discussion)—(1) 650°C is close to the $\alpha/\alpha + \beta$ temperature for Zr-4 + 1% Nb. What phases were present in the metal as a result of the heat treatment prior to oxidation? (2) The corrosion rate for Zr-4 seems high, probably as a result of the low ΣA-time. Are these results representative of properly heat treated Zr-4?

H. Anada (author's closure)—(1) α-Zr matrix and β-Nb precipitates were observed prior to oxidation. (2) Yes. The Zr-4 material used in the present study was very sensitive to the heat treatment. We had confirmed that the corrosion rate for the high ΣAi Zr-4 material was low and comparable.

---

[1] Avgonne National Laboratory, Avgonne, IL.

[2] University of Toronto, Toronto, Ontario, Canada.

[3] Pennsylvania State University, University Park, PA.

[4] Westinghouse NMD, Energy Center, Pittsburgh, PA.

Gunnar Wikmark,[1] Peter Rudling,[1] Börje Lehtinen,[2] Bevis Hutchinson,[2] Anders Oscarsson,[2] and Elisabet Ahlberg[3]

The Importance of Oxide Morphology for the Oxidation Rate of Zirconium Alloys

REFERENCE: Wikmark, G., Rudling, P., Lehtinen, B., Hutchinson, B., Oscarsson, A., and Ahlberg, E., "**The Importance of Oxide Morphology for the Oxidation Rate of Zirconium Alloys,**" *Zirconium in the Nuclear Industry: Eleventh International Symposium, ASTM STP 1295*, E. R. Bradley and G. P. Sabol, Eds., American Society for Testing and Materials, 1996, pp. 55–73.

ABSTRACT: The oxide growth rate of zirconium alloys, e.g., Zircaloy-2 and Zircaloy-4, has been proposed to be controlled mainly by the transformation of the zirconium oxide from tetragonal to monoclinic structure at some distance from the metal-oxide interface, leading to cracking. This oxide growth rate model is inconsistent with our results. Zirconium alloys of varying chemical composition but with identical manufacturing process had markedly different oxide growth and hydriding properties in autoclave testing (400°C steam). The materials were characterized by several methods, e.g., electron microscopy (SEM), X-ray diffraction (XRD), and electrochemical impedance (EIS). The SEM and some of the XRD investigations of the oxide were performed on the metal-oxide interface after dissolution of the metal. The oxide growth developed through three different stages with an altered oxide morphology at the metal-oxide interface at each stage. The developments of the stages were correlated with the oxide growth rate. Impedance measurements suggested that the oxide film had three layers, the outermost being extensively porous. Relaxation of the oxide film stress showed that the compressive stress in the oxide was not essential for retention of a significant amount of the tetragonal phase.

KEYWORDS: zirconium alloys, corrosion, tetragonal phase, film stress, porosity, SEM, impedance spectroscopy, X-ray diffraction

The fundamental processes involved in corrosion and hydriding of zirconium alloys have been investigated for more than 40 years. Several models have been proposed correlating corrosion and hydriding with the microstructure of the alloy or oxide [1]. The models can, however, in general explain only a fraction of the results presented in the literature. There is thus today no available single model that can explain the bulk of observations obtained from autoclave and in-pile tests. One reason for this is a lack of consistent microstructural data.

Previous studies have indicated that second-phase particle distributions and matrix chemical composition may have a large impact on corrosion and hydriding of zirconium-base alloys. However, the relation between these factors and the microstructure has not been fully understood.

To elucidate the correlation between the corrosion and hydriding properties to the microstructure and porosity of the oxide, three materials with alloy compositions slightly outside the

---

[1] Specialists, Fuel Division, ABB Atom, S-721 63 Västerås, Sweden.

[2] Research scientist, professor, and research scientist, respectively, Institute of Materials Research, Stockholm.

[3] Associate professor, Department of Inorganic Chemistry, Gothenburg University, Gothenburg, Sweden.

ASTM specification for Zircaloy-2 were investigated. Autoclave tests showed that the corrosion properties were markedly different for the three alloys, although they had been manufactured identically with only the alloy composition differing.

Materials

Three different zirconium-base materials were investigated. All three materials had a composition slightly deviating from that of Zircaloy-2, with one alloy also containing 0.1% niobium. The chemical compositions are shown in Table 1.

The materials were manufactured into 0.5-mm-thick sheets. The total heat input during the manufacturing process resulted in an accumulated annealing parameter log $A = -15.9$, calculated by use of an activation energy, Q, of 263 kJ/mol and the annealing times in hours. The manufacturing process of the materials and results from corrosion tests in 400 and 520°C by steam autoclaving of samples with an approximate size of 25 by 30 mm have been reported in a previous communication [2]. Some results from the 400°C test are listed in Table 2. All the samples with an oxide thinner than 2 μm were at the pretransition oxidation stage [2].

All three materials had a slightly different surface appearance on the two sides after autoclaving. The reason for this difference was not investigated but could be due to the use of laboratory cold rolling. However, this difference resulted only in different corrosion rates for Samples A83 and A91, where one side had a notably thicker oxide. The side with the thinner oxide was normally investigated unless otherwise noted in the text.

The bulk alloy of each of the samples was investigated regarding grain size and second-phase particle size and chemical composition by use of SEM (scanning electron microscope) and AEM (analytical electron microscope) [2]. Results of the grain size, SPP size, and chemistry determinations are shown in Table 3.

The textures of the alloys and of the oxides of the samples were determined by X-ray diffraction by use of pole figures. All three materials had an ordinary sheet texture for the metal. The tetragonal oxide exhibited a fiber texture with the $\langle 100 \rangle$ direction parallel to the sheet normal direction. The fiber texture in the monoclinic oxide phase was tilted 10 to 20° from the sheet normal direction.

Experimental Procedure

SEM Investigations

Micrographs of finely polished surfaces of the samples were obtained using an SEM (JEOL 6400) equipped with a detector for back-scattered electrons (LINK Tetra) in order to perform size measurements of grains and second-phase particles (SPP). By optimizing the microscope condition, it was possible to simultaneously image the SPPs (Z-contrast) and the grain structure (channeling contrast).

TABLE 1—*Chemical composition of materials.*

Material	Sn, %	Fe, %	Cr, %	Ni, %	Si, ppm	O, %	Nb, ppm	Al, ppm
A-7	1.48	0.20	0.10	0.15	105	0.127	930	47
A-8	1.39	0.09	0.27	0.04	59	0.210	<50	88
A-9	1.55	0.13	0.11	0.06	460	0.146	<50	100

TABLE 2—*Results from autoclaving in 400°C steam at 10.4 MPa.*

Material	Sample	Exposure Days	Appearance after Autoclaving	ΔW, mg/dm^2	$\Delta t_{\Delta W}{}^a$, μm	$\Delta t_{SEM}{}^b$, μm	$[H]_{mean}{}'$, ppm
A-7	A-72	0.5	Black	9	0.6	0.5	ND^c
A-7	A-73	3	Black	16	1.1	1.2	ND
A-7	A-70	14	Black	25	1.7	1.5	33
A-7	A-71	101	Black	69	4.6	5	202
A-8	A-84	0.5	Black	9	0.6	0.6	ND
A-8	A-85	3	Black	17	1.1	0.9	ND
A-8	A-82	14	Black	26	1.7	1.6	53
A-8	A-83	101	One side black, other non-uniform	273	18^d	6 to 12^e	498
A-9	A-92	0.5	Black	10	0.7	0.6	ND
A-9	A-93	3	Black	14	0.9	1.1	ND
A-9	A-90	14	Black	27	1.8	1.9	54
A-9	A-91	101	One side black, other non-uniform	142	9.4^d	7	555

^{a} Estimated from weight gain, ΔW, assuming that 15 mg · dm^{-2} corresponds to 1 μm of oxide.
^{b} Determined by SEM.
^{c} ND means not determined.
^{d} Mean value for both sides but one side was considerably thicker at some parts.
^{e} Uneven thickness on the side with thicker oxide.

The sample surfaces were prepared by electrolytic polishing in a solution consisting of 10% perchloric acid and 90% ethanol at −30°C, applying a voltage of 20 V. A platinum wire was used as cathode.

The grain and particle structures were copied manually from Polaroid photographs to a transparent plastic film prior to quantification in an image analyser (IBAS 2000). The computer program for measuring the diameter of area equivalent circles was used for both grains and particles.

SEM studies of the oxide structure at the metal oxide interface were performed on specimens where the metal phase had been removed. This was accomplished by grinding and polishing,

TABLE 3—*Grain size and SPP size and Chemistry.*

Material	Grain Mean Size, μm	Standard Deviation, μm	SPP Mean Size, nm	Standard Deviation, nm	Mean Ratio, [Fe]/[Cr]^a	Mean Ratio, [Fe]/[Ni]^b	Recrystallized Area, %
A-7	2.2	1.4	170	75	0.8	0.9	69
A-8	2.3	1.4	220	140	0.4	...^{c}	91
A-9	2.5	1.4	210	120	0.9	0.9	96

^{a} The ratio determined for the SPP with Zr, Fe, and Cr, but no Ni.
^{b} The ratio determined for the SPP with Zr, Fe, and Ni, but no Cr.
^{c} No such particles were found.

from one side, to reduce the sample thickness to 0.1 mm, followed by dissolution of the remaining metal by use of 15% bromine in methanol. Oxide flakes a few mm wide were fixed on specimen holders and provided with a thin layer of carbon by sublimation in order to avoid charging of the oxide in the microscope.

X-ray Diffraction Investigations

A Siemens D5000 with an open Eulerian cradle was used in all X-ray measurements. Cr Kα radiation was employed except for a few measurements, where Co Kα was used because of its greater penetration in the zirconium oxide.

In the residual stress measurements of the monoclinic oxide, the $(-1,1,1)$ peak had to be used as this is the only useful peak free from disturbing neighboring peaks, including peaks from tetragonal oxide phase and zirconium metal. This peak has a low Bragg angle ($2\Theta \approx 42.4°$), and consequently extreme care had to be taken to calibrate the goniometer. Powder of monoclinic zirconium oxide was used to check the exact position of the peak at different tilting angles, and the position of the peak was compensated accordingly. Six different tilting angles between 21 and 60° were used. The angle 0° was omitted since this peak was rather weak and less reliable. Peak positions were determined as mean chord positions at 50% of top height.

The strain in the oxide was measured by determining the variations in the d-spacing at different sample tilt angles, Ψ. The changes in the apparent d-spacing with the angle Ψ were recalculated into strain and plotted against $\sin^2 \Psi$. Assuming a equibiaxial stress condition, the slope of the curve is equal to $\dfrac{\sigma (1 + \nu)}{E}$, where σ is the stress, E, is Young's modulus, and ν is Poissons's ratio. A value of 170 GPa for ZrO_2 was used for Young's modulus. No value for Poisson's ratio was found in the literature but was assumed to be 0.3. The average standard deviation for all stress measurements was 30 MPa.

Residual stresses in the monoclinic oxide were measured directly on the received sample surfaces without further sample preparation. For two of the samples, stresses were also measured on identical samples at the metal-oxide interface after removal of the metal. The samples were first glued with a thermoplastic cement on a glass support plate to carry the stresses in the oxide layer. Gentle grinding reduced the sample thickness to 0.2 mm before the remaining zirconium alloy was removed by chemical dissolution by 15% bromine in methanol. Stress relaxation was accomplished by heating the cemented samples on a hot plate. Relaxation by heating was repeated twice, the second leading to melting of the cement, with subsequent stress measurements after each relaxation.

The relative amount of tetragonal oxide was determined semi-quantitatively by comparing the area intensities of the $(1,1,1)$ peak for the tetragonal phase to the $(-1,1,1)$ peak for the monoclinic phase. The area intensities were determined at 45° tilting angle where the tetragonal peak was strongest as a result of its texture.

Pole figures for the monoclinic and tetragonal oxide phases, as well as the zirconium alloy, were determined using the Schulz reflection method. Pole figures were compensated for background and defocusing.

Electrochemical Investigations

A PAR 273A potentiostat coupled with a Solartron 1250 frequency response analyzer were used for the impedance measurements. The frequency range covered was 100 kHz to 10 mHz with an ac amplitude of 5 mV rms.

The impedance measurements were made in 0.5 M sulphuric acid (p.a.) at the open circuit potential. All potential values refer to the sodium chloride saturated calomel electrode (SSCC), with $E = +236$ mV versus normal hydrogen electrode (NHE). The samples were cleaned ultrasonically in ethanol and dried before immersion in the electrolyte.

The cell consisted of a large concentric counter electrode and a stationary working electrode with the area of 1 cm^2. For the two samples with different oxide thicknesses on the two sides, measurements were made on both sides, one at the time.

The value of the dielectric constant for ZrO_2 applied in the evaluation of the thickness of the oxide reported in the literature varies between 20 to 31.5 [1,3–10]. One reason for the variation is the uncertainty in the surface roughness of the samples [3–5,11]. In this work, a value of 22 was adopted for the dielectric constant.

Results

Oxide Morphology at Varying Oxide Thicknesses

The morphology of the oxide at the metal-oxide interface, after removal of the metal, was investigated by SEM for twelve samples autoclaved for 0.5 to 101 days (see Table 2). For these experiments, the electron beam was normal to the metal-oxide interface plane. The morphology of the oxide from the samples after 14 days of autoclaving appeared to be alike at lower magnifications, showing a grainy structure (Fig. 1). At higher magnification, Samples A82 and A90 exhibited a well-developed needle structure (Fig. 2), which was not evident in Sample A70 (Fig. 3), or in any of the samples autoclaved for 0.5 or 3 days. The needles had a thickness of approximately 50 nm and a length of less than 1000 nm. The angles between the needles were often found to be 120°, which is equal to the angle between the prism planes in the hexagonal zirconium crystal. This needle structure has previously been found using SEM on oxidized samples of Zircaloy-2 and Zircaloy-4 when the metal was dissolved by electropolishing [12].

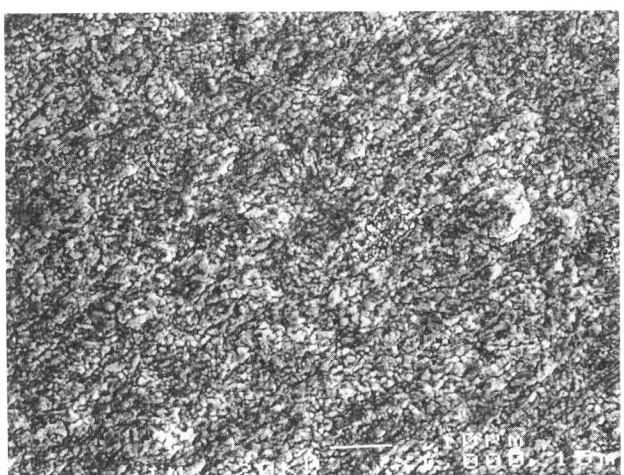

FIG. 1—*The common grainy microstructure of the metal-oxide interface oxide of the samples seen in SEM at low magnification (X1000) after 14 days of autoclaving. (Example taken from Sample A82.)*

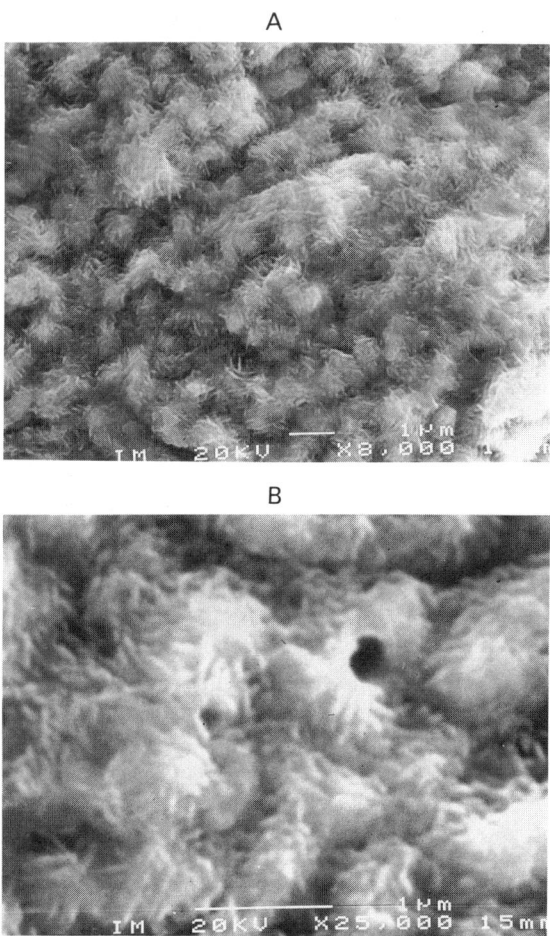

FIG. 2—*The needle microstructure of the metal-oxide interface oxide of Samples A82 and A90 seen in SEM at higher magnification of* (A) *X8000 (example from Sample A82) and* (B) *X25 000 (example from Sample A90).*

The needle structure had disappeared completely after 101 days of autoclaving in Sample A83 and was replaced by the typical cauliflower structure as shown in Fig. 4. In Sample 91, a mixed structure was seen, with some scattered cauliflower structures in a finer needle structure (Fig. 5). Sample A71, also exposed for 101 days in the autoclave, had at this stage developed the needle structure similar to that for the other two materials after 14 days (Fig. 6).

The development from a diffuse structure, via a needle structure, to the cauliflower structure appears to be common of all samples (Fig. 7) despite the difference in alloy composition and results from autoclaving. The difference in the development of the microstructure is rather a matter of the time needed to reach each stage of morphology. This time is well correlated with the corrosion resistance of the alloys, with the most-resistant alloys being slower in developing the inner oxide into a cauliflower structure, which has previously been reported for thick oxides [13].

The oxide thickness of Sample A71 was larger than 4 μm and still exhibited needle structure,

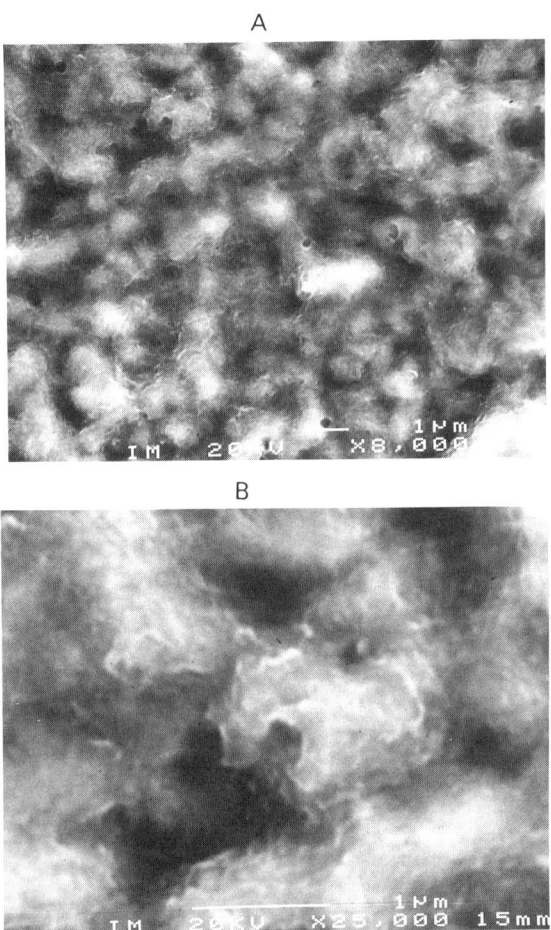

FIG. 3—*The diffuse phase without needle microstructure of the metal-oxide interface oxide of Sample A70 seen in SEM at higher magnification of* (A) *X8000 and* (B) *X25 000.*

indicating that this structure was not correlated directly to the transition from cubic to linear oxide thickness growth. The size of the agglomerates of needles, as shown in Figs. 2 and 6, had a size close to the grain size of the metal.

Stress Levels and Presence of Tetragonal Phase

The stress levels in the oxide films were determined for the autoclaved samples by X-ray diffraction without further treatment. For Samples A70 and A83, the stress levels were also determined on the oxide at the metal-oxide interface by removal of the metal. The results from the stress determinations are shown in Table 4.

When using X-ray diffraction to determine the amount of tetragonal phase from the outer side of the oxide, it is important to compare different samples having similar oxide thicknesses since the amount of tetragonal phase is much less at the outmost part of thick oxides [14,15].

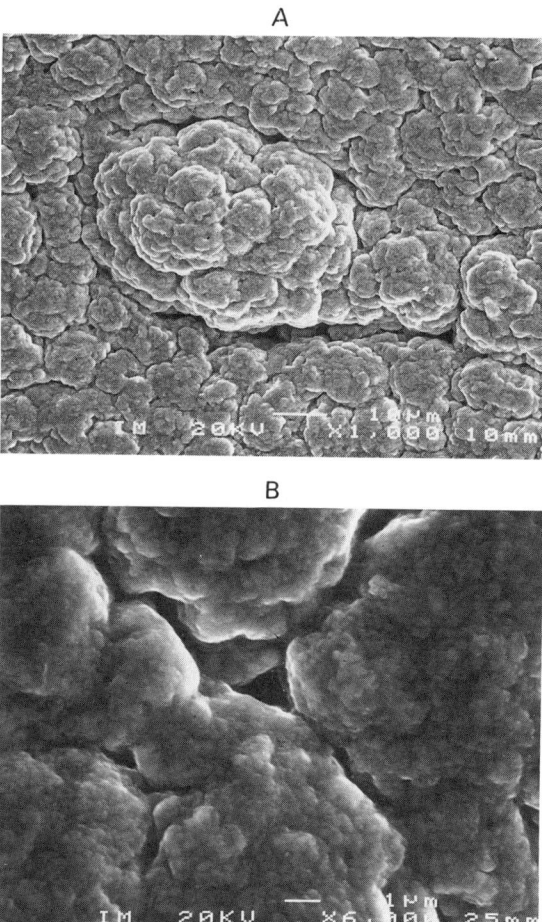

FIG. 4—*The cauliflower structure of the metal-oxide interface oxide of Sample A83 seen in SEM at* (A) *X1000 and* (B) *X6000 magnification.*

Comparison of the almost equally thick oxides on Samples A70, A82, and A91 resulted in a relative concentration of tetragonal phase in samples decreasing in the order A70 > A90 > A82 when using Co-Kα radiation. The relative concentration ratio was approximately 3:2:1, respectively, when the intensities have been compensated for texture effects. For the thicker oxides, the difference in relative amount of tetragonal phase determined was even more pronounced. This order is the same as found for the corrosion resistance (see Table 2). A higher concentration of tetragonal phase did not correlate with a higher stress level for the studied materials (see Table 4).

Cr-Kα has a penetration depth yielding 95% of measured intensity response from a 3-μm-thick layer. Measurements from the outside of the oxide could reveal the (111) reflection of the tetragonal phase in the thicker oxides of Samples A71, A83, and A91. No response was, however, detected for zirconium metal in Samples A83 and A91 at tilting angles giving maximum response according to texture. The same measurements were repeated with Co-Kα ra-

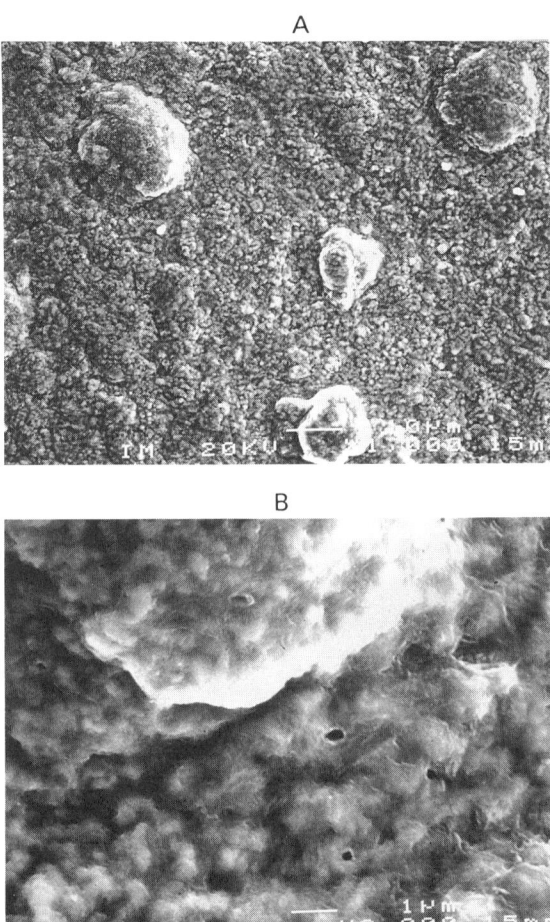

FIG. 5—*The mixed needle and cauliflower structure of the metal-oxide interface oxide of Sample A91 seen in SEM at (A) X1000 (B) X8000 magnification.*

diation, which has a 50% greater penetration depth. The response for the tetragonal phase was increased, but still the response from zirconium was absent for Sample A83 and barely detectable in A91. This demonstrates that tetragonal phase was present several μm from the metal-oxide interface and not only very close to the interface. Since the exact distribution of tetragonal phase within the film was not known, it was not possible to make reliable quantitative measurements by X-ray diffraction.

The results in Table 4 show that it is possible to remove the oxide onto a support and still retain a considerable stress in the oxide film in Samples A70 and A83. Relaxation of stress was produced by warming the supported sample to allow the cement to soften and melt. The relative amount of tetragonal phase was determined from the metal-oxide interface for both samples before relaxation and at subsequent stages of the relaxation. The results from individual measurements are depicted in Fig. 8.

The tetragonal phase was found to persist to a significant extent also when the stresses in

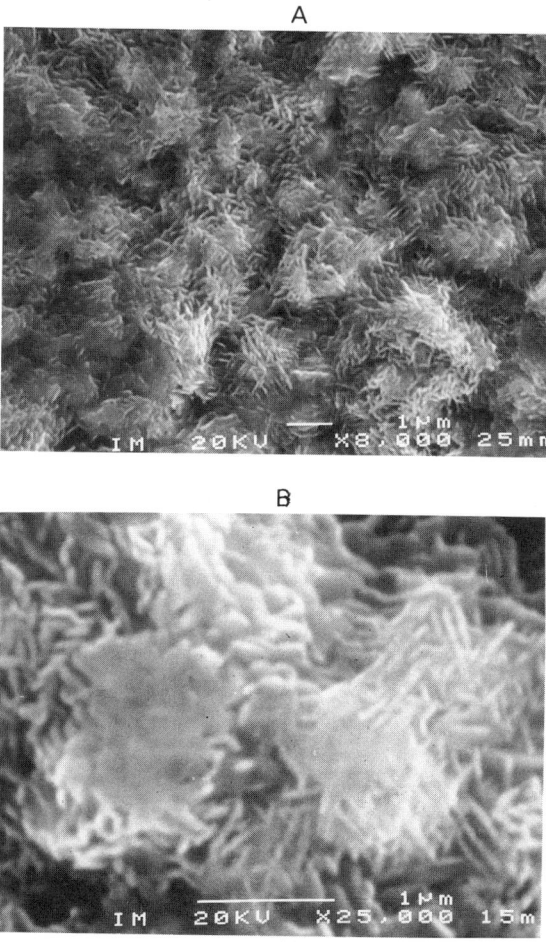

FIG. 6—*The needle structure of the metal-oxide interface oxide of Sample A71 seen in SEM at (A) X8000 and (B) X25 000 magnification.*

the oxide were far lower than the 3 GPa generally reported to be required to stabilize this phase at ambient temperature [16,17]. This existence of tetragonal phase with virtually no supporting stress does thus not apply only to isolated small crystals, but also to the oxide film on a macroscopic scale.

The present investigation could not inherently reveal any stress gradient in the film. The data in Table 4 show that the determined stress levels were 50 to 100% higher when measured from the outside with the metal still intact than those measured from the metal-oxide interface with the metal dissolved. Hence, it can be inferred that the stress level decreased due to dissolution of the metal, i.e., that the applied method could not retain the full stress at the metal-oxide interface before metal dissolution.

It was not possible to determine the exact amount of tetragonal oxide due to the presence of texture in the film. Another complicating factor was the fact that the tetragonal and monoclinic phases are mixed in the oxide. However, considering the intensity of the respective peaks in

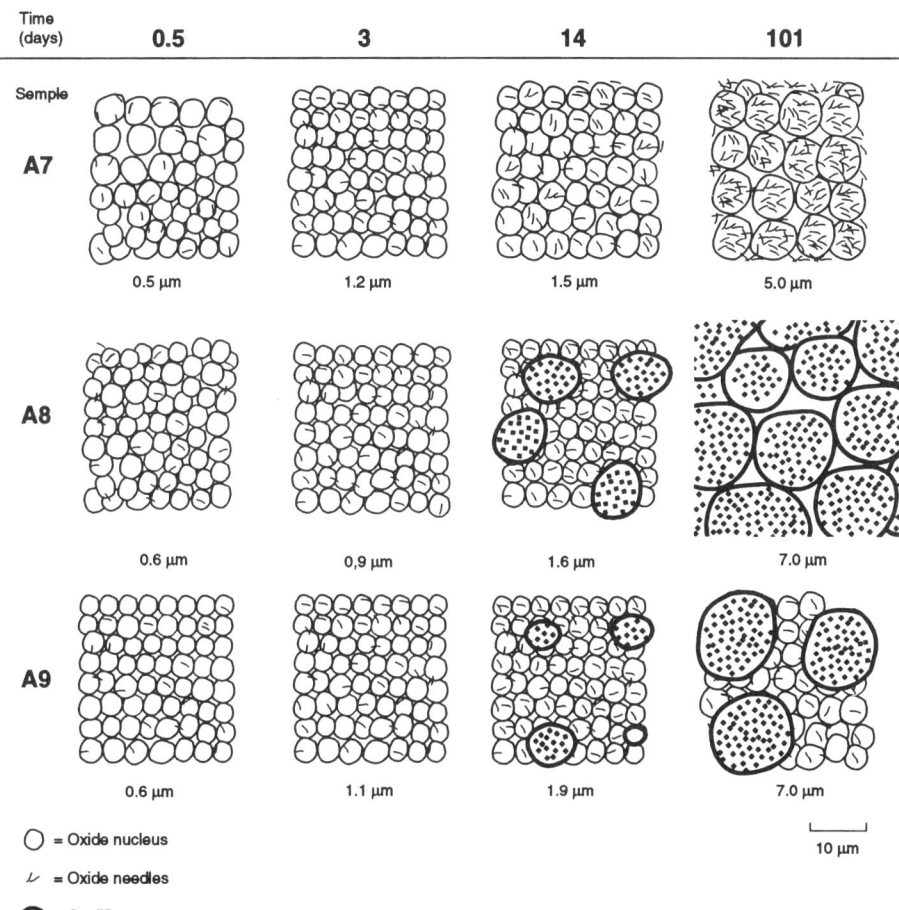

FIG. 7—*Schematic development of oxide structures at various autoclaving exposures. The oxide film thickness determined by SEM is given below each picture.*

TABLE 4—*Results from measurements of the compressive oxide stress[a] by X-ray diffraction.*

Material	Sample	$\Delta t_{SEM'}$, μm	Stress, MPa	Stress, 2nd Determination, MPa	Stress, Metal Removed, MPa	Stress, Metal Removed and Oxide Relaxed, MPa
A-7	A-70	1.5	710	740	370	70
A-7	A-71	5	700	...	...	...
A-8	A-82	1.6	770	...	...	...
A-8	A-83	6 to 12	550	580	410	190
A-9	A-90	1.9	680	...	...	...
A-9	A-91	7	610	...	...	...

[a] Values in table are averages of 2 to 3 measurements.

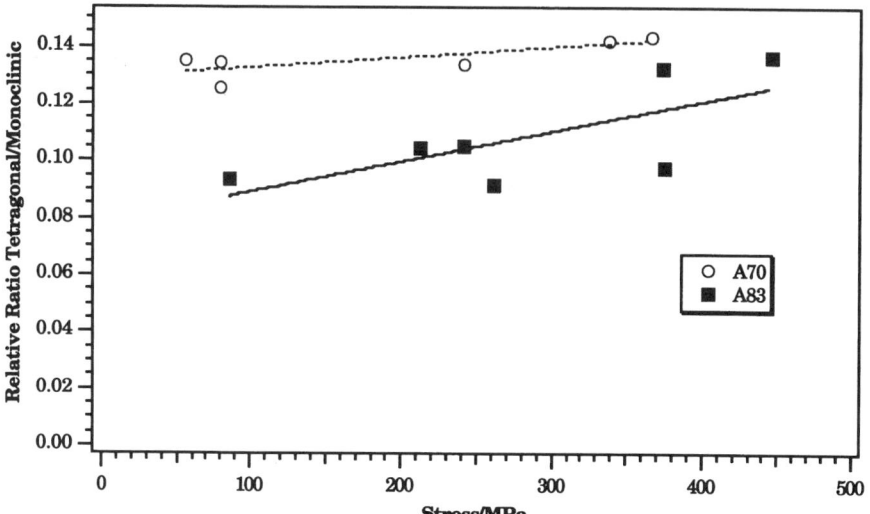

FIG. 8—*The relative amount of tetragonal phase in the oxide from the metal-oxide interface side as a function of the stress. The straight lines are least-squares fit to Samples A70 (---) and A83 (——).*

comparison with the evaluation made by Godlewski [14], it can be implied that the amount of tetragonal phase was more than 5%. This is considerable less than the 40% reported from Raman measurements for the oxide closest to the metal-oxide interface [14], but could be explained by the rather large penetration depth of the X-rays, recording contributions for a rather thick layer, as well as the loss due to stress relaxation, as discussed above.

Figure 8 also shows that the remaining amount of tetragonal phase was rather similar in Samples A70 and A83 despite the large difference in oxide thickness, the former being still in the pretransition stage.

Electrochemical Response and Oxide Porosity

Measurements with electrochemical impedance spectroscopy (EIS) can provide information about the porosity of the oxide and the presence of oxide splitting into different layers within the oxide. In the present study, it was found that electrochemical models such as shown in Fig. 9 were adequate in describing the behavior of the oxide. The CPE (constant phase element) was in this context interpreted as a distorted capacitance, i.e., the impedance of the CPE, Z_{CPE}, was given by

$$Z_{CPE} = \frac{(j\omega)^\alpha}{C}$$

where j is $\sqrt{-1}$, ω is the angular frequency, C the capacitance, and α the dispersion factor. For an ideal capacitor, α would be -1. The dispersion could be attributed to different phenomena, such as surface roughness, dielectric relaxation, porosity, etc. [18–20]. The dispersion is often marked for thick porous oxide films, and different models have been proposed for the interpretation of impedance data [21–23].

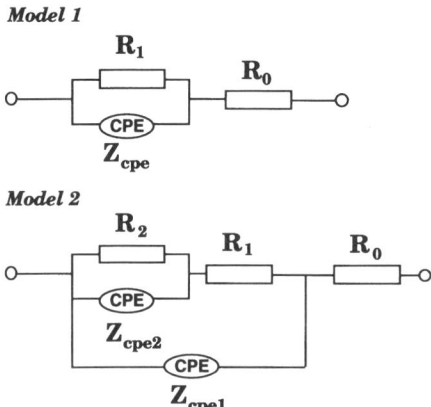

FIG. 9—*Electrochemical models used for oxide film with one* (Model 1) *and two* (Model 2) *layers detectable in the present study.*

The oxide thickness of layers within the oxide film was determined from the capacitive behavior of the film by use of the relation

$$C = \frac{\varepsilon_0 \varepsilon_r A}{d}$$

with ε_0 being the dielectric constant in vacuum, ε_r the relative dielectric constant, A the area of the electrode, and d the thickness of the layer forming the capacitance.

The EIS measurements were conducted repeatedly with six of the autoclaved samples immersed in the sulphuric acid electrolyte for up to three weeks. The values for α and C where determined by fitting parameter values in either of the models in Fig. 9 to the recorded data in the complex plane and Bode impedance and phase representations simultaneously. Generally, it was possible to determine the thickness and dispersion factor rather accurately for the first time constant but not for the second at lower frequency due to the limited frequency range investigated. The measurements thus produced a response for two layers, but only the thickness of the sum of those two layers could be evaluated. This thickness is referred to as the inner layer, although it consists of two layers, the barrier layer and an outer (actually middle) layer.

It was found that the dispersion factor of the inner layer was always in the interval $-0.8 < \alpha < -0.98$ (see Fig. 10); i.e., none of the films behaved as an ideal capacitor. The dispersion factor showed only a slight drift during the immersion except for Samples A83 and A71, where a more pronounced change was noted.

The inner oxide layers determined by EIS were always considerably thinner than the layers determined from the weight gain or SEM. One experiment was performed to ascertain that the difference in thickness was not an artefact. This was performed on a pickled and autoclaved specimen of Material A9. The oxide thickness estimated from the weight gain was 2.4 μm. Determination in the sulphate electrolyte resulted in a thickness of 150 nm after 28 days of immersion. Exchanging the electrolyte for a highly resistant and viscous electrolyte resulted in a recorded thickness of 850 nm. Measurements on a 72-nm non-porous anodic film formed on Material A9 showed no difference in oxide thickness determined by EIS with the low- or high-resistive electrolyte. This demonstrated that the low-resistive 0.5 M sulphate electrolyte will penetrate the oxide film through the pores, thereby short-circuiting the current transport. Even

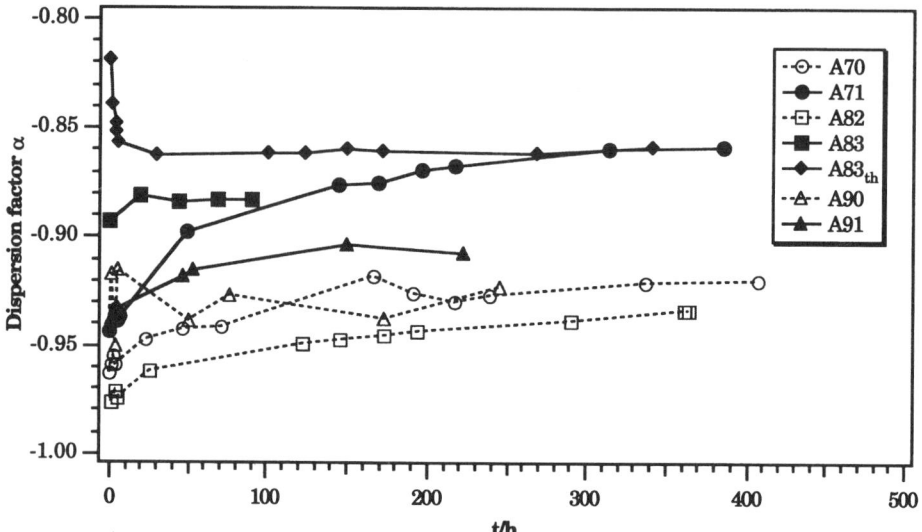

FIG. 10—*Dispersion factor of the CPE during immersion in the electrolyte. Filled markers show the post-transition films. $A83_{th}$ indicates measurements on the thicker side of Sample A83.*

the highly resistive electrolyte penetrated the oxide film to a substantial extent. Thus, the porosity of the outer oxide film is gradually decreasing from the surface into the film. It seems, therefore, that the outer part of the oxide has no influence at all on the transport and corrosion processes until a sufficiently low porosity is attained at some specific depth of the film where the inner layer interface is reached.

The determined capacitance of the oxide film increased with time for all samples, which corresponded to a decreasing film thickness with time, as shown in Fig. 11. The change in the

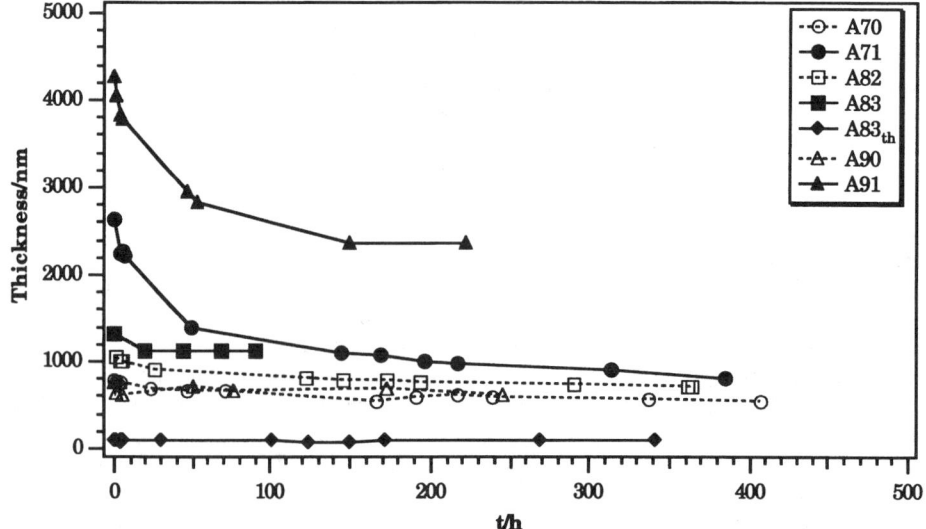

FIG. 11—*Thickness of inner layer during immersion in the electrolyte. Filled markers show the post-transition films. $A83_{th}$ indicates measurements on the thicker side of Sample A83.*

determined thickness of the inner layer was most pronounced for those samples having the thickest inner layers at the time of immersion. For the extremely thick film of A83, the inner layer was extremely thin, evidently due to the fact that almost the whole thickness, except for a thin 100-nm layer, was porous enough to allow significant electrolyte penetration.

The dispersion factors, α, for the constant phase elements, according to the applied electrochemical models, attained with time a common value of approximately -0.93 for the inner layer in the pre-transition oxide films. For the post-transition films, α came close to -0.86 despite significantly different oxide thicknesses in these cases. This observation was not applicable to Sample A91, however.

The open circuit potential of the samples in electrolyte was found to reach a fairly stable value after 200 h of immersion in the electrolyte (Fig. 12). All immersed samples with pre-transition films attained a positive potential versus the reference electrode. For the thicker films, the potential was lower except for Sample A83. The low potentials recorded correlated with a thick oxide film, a relatively thick inner layer, and a high hydrogen content in the alloy. Sample A83 did not attain a low potential even for the side with a thick inner layer. However, the open circuit potential is also given by the secondary phase particles [24].

Discussion

The oxide thickness determined by the impedance measurements was in all cases less than that determined by weight gain and SEM (see Table 2). This observation is explained by a three-layer model for the oxide film. The inner-most layer is compact and impervious to electrolyte. The middle layer has micro-pores, allowing a limited penetration of electrolyte, while the third, outermost, layer must be porous to allow considerable charge transport. The frequency range and experimental setup used in the impedance measurements limited the detectability of resistance to approximately $100 \ \Omega \cdot cm^2$. By use of calculations by Wanklyn and Silvester [3], it can be estimated that only a fraction of 10^{-6} of the outermost film has to be void from the

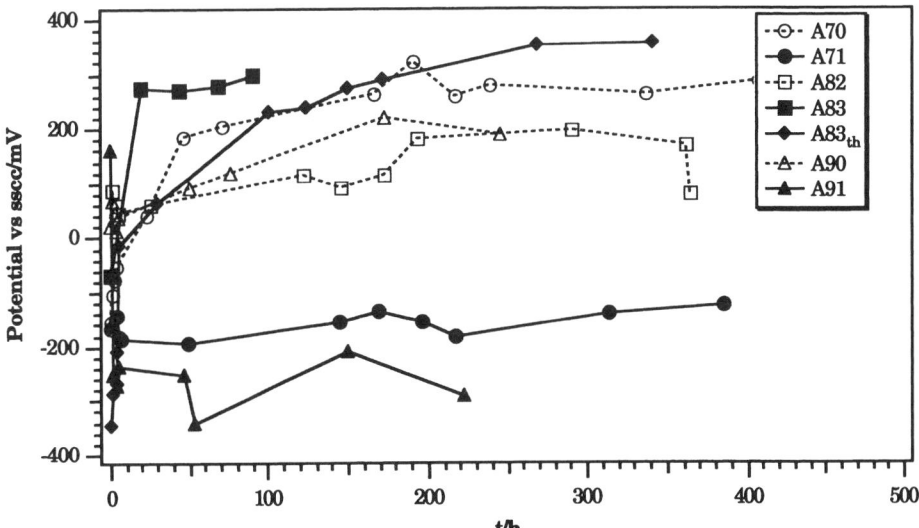

FIG. 12—*Open circuit potential during immersion in the electrolyte. Filled markers show the post-transition films. $A83_{th}$ indicates measurements on the thicker side of Sample A83.*

outer surface to the middle layer and still provide an ohmic resistance less than 100 $\Omega \cdot cm^2$ with the used electrolyte. A total porosity of 1% of the film formed on Zircaloy was reported by Urquhart and Vermilyea [11]. They also estimated the number of pores of 1.8 to 2.8-nm radius penetrating the film as $6 \cdot 10^9$ cm^{-2}, giving approximately 0.1% void fraction. It is hence quite reasonable to have an outer layer not discernible by impedance measurements. Due to the electrochemical nature of the corrosion process, the outer layer will not in any aspect be rate limiting for the Zircaloy corrosion process in water. This further implies that there can never be substantial porosity at the metal-oxide interface, as for instance reported by Cox and Yamaguchi [10], that can be filled with electrolyte since this would make the oxide film not discernible by impedance measurements at similar conditions as used here. The apparent pores in the cauliflower structure depicted in Fig. 4 has previously been shown to be filled with metal [25] before the metal was dissolved.

The shift in the dispersion factor α to a lower absolute value for the inner layer constant phase element in the post-transition films compared to the value for the pre-transition films indicates an increase in the extent or nature of the porosity of the inner layer. The porosity of the inner layer was, however, still very limited.

The tetragonal phase determined at the metal-film interface was shown to be capable of existing without the support of stress. This has been noted on the microscopic level previously [17], but it was here made evident that the tetragonal phase is also present on a macroscopic level in an oxide film. The reason for an almost complete absence of tetragonal phase in the outer parts of an oxide film [14] is thus not the relaxation of stress during outward growth of the oxide leading to a condition where phase transformation occurs spontaneously, producing cracks in the oxide. Instead, our results support the hypothesis presented by Hutchinson and Lehtinen [25] that the loss of tetragonal phase is due rather to the presence of cracking and the cracks being stopped by the martensitic transition of the tetragonal phase to the monoclinic. The three different layers of oxide found by the impedance measurements on the investigated materials have a decreasing degree of porosity with increasing depth towards the metal-oxide interface. This correlates with the increasing fraction of tetragonal phase found with depth [14]. Cracks growing into the barrier layer would be detrimental to the protective effect of this layer by increasing the porosity of the layer. A better resistance to cracking and formation of pores would be obtained with a thicker barrier layer and a larger fraction of tetragonal phase present in the layer. This is in good agreement with experience from other investigations of Zircaloy materials [26].

The three-layer model of the film presented here also conforms with the results reported for zirconium alloys by Godlewski et al. [14]. Their reported two layers closest to the metal-oxide interface, containing significant amounts of tetragonal phase, correspond to our inner layer. The cumulative thickness of their two layers was 0.7 to 3.0 μm, being abruptly thinner at cyclic transition points. In this investigation, the thicknesses of the inner layers that were demonstrated to contain tetragonal phase were 0.6 to 2.5 μm after a longer immersion time for all but the sample with the thickest oxide. For extremely thick oxides, there appears to be another not completely studied mechanism, further diminishing the inner layer thickness. This thinner inner layer was found for the sample with fully developed cauliflower morphology at the metal-oxide interface.

Conclusions

Although the materials had very different chemistries and exhibited markedly different corrosion behavior in the autoclave, the development through three stages of oxide morphology at the metal-oxide interface was common to all materials.

The oxide films had three layers. The barrier layer and the middle layer could not be fully separated. The outer layer, which could form most of the film for post-transition films, was porous to the extent that it was not discernible by impedance spectroscopy.

The tetragonal phase formed on the studied zirconium alloys can be stabilized at room temperature in the absence of high compressive stress. This implies that the loss of the tetragonal phase is associated with the cracking in the oxide rather than the opposite.

References

[1] "Corrosion of Zirconium Alloys in Nuclear Power Plants," IAEA-TECDOC-684, IAEA, Vienna, 1994.
[2] Rudling, P., Mikes-Lindbäck, M., Lehtinen, B., Andrén, H.-O., and Stiller, K., "Corrosion Performance of New Zircaloy-2-Based Alloys," *Zirconium in the Nuclear Industry: Tenth International Symposium, ASTM STP 1245*, A. M. Garde and E. R. Bradley, Eds., ASTM, West Conshohocken, PA, 1994, pp. 599–614.
[3] Wanklyn, J. N. and Silvester, D. R., "A Study of Corrosion Films on Zirconium and Its Alloys by Impedance Spectroscopy," *Journal of Electrochemical Society*, Vol. 105, No. 11, 1958, pp. 647–654.
[4] Meisterjahn, P., Hoppe, H. W., and Schultze, J. W., "Electrochemical and XPS Measurements on Thin Oxide Films on Zirconium," *Journal of Electroanalytical Chemistry*, Vol. 217, 1987, pp. 159–185.
[5] Bardwell, J. A. and McKubre, C. H., "AC Impedance Spectroscopy of the Anodic Film on Zirconium in Neutral Solution," *Electrochimica Acta*, Vol. 36, Nos. 3/4, 1991, pp. 647–653.
[6] Rosecrans, P. M., "Application of Alternating Current Impedance to Characterize Zirconium-Alloy Oxidation Films," Knolls Atomic Power Laboratory Report, KAPL-4149, 1982.
[7] Khalil, N., Bowen, A., and Leach, J. S., "The Anodic Oxidation of Valve Metals—II. The Influence of Anodizing Conditions on the Transport Processes During the Anodic Oxidation of Zirconium," *Electrochimica Acta*, Vol. 33, No. 12, 1988, pp. 1721–1727.
[8] Patrito, E. M., Torresi, R. M., Leiva, E. P. M., and Macagno, V. A., "Potentiodynamic and AC Impedance Investigation of Anodic Zirconium Oxide Films," *Journal of Electrochemical Society*, Vol. 137, No. 2, 1990, pp. 524–530.
[9] Patrito, E. M. and Macagno, V. A., "Ellipsometric Investigations of Anodic Zirconium Oxide Films," *Journal of Electrochemical Society*, Vol. 140, No. 6, 1993, pp. 1576–1585.
[10] Cox, B. and Yamaguchi, Y., *Journal of Nuclear Materials*, Vol. 210, 1994, pp. 303–317.
[11] Urquhart, A. W. and Vermilyea, D. A., "Characterization of Zircaloy Oxidation Films," *Zirconium in Nuclear Applications, ASTM STP 551*, 1974.
[12] Ekbom, L. R., Studsvik Report NF (P) -82/13, 1982.
[13] Ding, Y. and Northwood, D., *Corrosion Science*, Vol. 36, No. 2, 1994, pp. 259–282.
[14] Godlewski, J., Gros, P., Lambertin, M., Wadier, J. F., and Weidinger, H., "Raman Spectroscopy Study of the Tetragonal-to- Monoclinic Transition in Zirconium Oxide Scales and Determination of Overall Oxygen Diffusion by Nuclear Microanalysis of O^{18}," *Zirconium in the Nuclear Industry: Ninth International Symposium, ASTM STP 1132*, J. M. Euken and A. M. Garde, Eds., ASTM, West Conshohocken, PA, 1991, pp. 416–436.
[15] Godlewski, J., "How the Tetragonal Zirconia is Stabilized in the Oxide Scale that is Formed on a Zirconium Alloy Corroded at 400°C in Steam," *Zirconium in the Nuclear Industry: Tenth International Symposium, ASTM STP 1245*, A. M. Garde and E. R. Bradley, Eds., ASTM, West Conshohocken, PA, 1994, pp. 663–686.
[16] Arashi, K., Yagi, T., Akimoto, S., and Kudoh, Y., *Physical Reviews*, Vol. B41, 1990, p. 4309.
[17] Beie, H.-J., Mitwalsky, A., Garzarolli, F., Ruhmann, H., and Sell, H.-J., "Examinations of the Corrosion Mechanism of Zirconium Alloys," *Zirconium in the Nuclear Industry: Tenth International Symposium, ASTM STP 1245*, A. M. Garde and E. R. Bradley, Eds., ASTM, West Conshohocken, PA, 1994, pp. 615–643.
[18] Macdonald, R., *Impedance Spectroscopy*, Wiley, New York, 1992.
[19] Jacquelin, J., "Theoretical Impedance of Rough Electrodes with Smooth Shapes of Roughness," *Electrochimica Acta*, Vol. 39, 1994, p. 2673.
[20] Paasch, G., Micka, K., and Gersdorf, P., "Theory of the Electrochemical Impedance of Macroinhomogeneous Porous Electrodes," *Electrochimica Acta*, Vol. 38, 1993, p. 2653 and references cited therein.
[21] Bataillon, C. and Brunet, S., "Electrochemical Impedance Spectroscopy on Oxide Formed on Zircaloy in High Temperature Water," *Electrochimica Acta*, Vol. 39, Vol. 3, 1994, pp. 455–465.

[22] Gebhart, O., Schaller, J., et al., "A Phase Reference Procedure for Interpretation of Impedance Spectroscopy Experiments," *Electrochimica Acta,* Vol. 38, No. 5, 1993, pp. 633–641.
[23] Göhr, H., Schaller, J., et al., "Impedance Studies of the Oxide Layer on Zircaloy After Previous Oxidation in Water Vapour at 400°C," *Electrochimica Acta,* Vol. 38, No. 14, 1993, pp. 1961–1964.
[24] Weidinger, H. G., Ruhmann, H., Cheliotis, G., Maguire, M., and Yau, T.-L., "Corrosion-Electrochemical Properties of Zirconium Intermetallics," *Zirconium in the Nuclear Industry: Ninth International Symposium, ASTM STP 1132,* J. M. Euken and A. M. Garde, Eds., ASTM, West Conshohocken, PA, 1991, pp. 499–535.
[25] Hutchinson, B. and Lehtinen, B., *Journal of Nuclear Materials,* Vol. 217, 1994, pp. 243–249.
[26] Garzarolli, F., Seidel, H., Tricot, R., and Gros, J. P., "Oxide Growth Mechanism on Zirconium Alloys," *Zirconium in the Nuclear Industry: Ninth International Symposium, ASTM STP 1132,* J. M. Euken and A. M. Garde, Eds., ASTM, West Conshohocken, PA, 1991, pp. 395–415.

DISCUSSION

Vincent Urbanic[1] (written discussion)—How confident are you that the needle-like structure you observe in your SEM examinations of the oxide-metal interface is not the result of the precipitation of zirconium compounds from the metal dissolution process during sample preparation?

G. Wikmark et al. (authors' closure)—We are quite convinced for several reasons. One is because the needle structure appearing in some, but not all, of the samples logically follows the oxide film growth. The needle structure has previously been found for electropolished samples, i.e., a metal removal process quite different from the methanol/bromine used in our work, as cited in the paper. A similar needle structure, in that case called platelets, was also reported after metal dissolution in various media by Gebhard et al. at this conference.

Brian Cox[2] (written discussion)—Following up on Vince Urbanic's comment on your needle structure, I, too, think it is an artefact of the oxide stripping technique. The needles (if they are ZrO_2 and not ZrH_2) probably come from oxygen dissolved in the metal during oxidation, which is not very soluble in bromine/methanol. Thus, the probability of observing needles should be inversely proportional to the corrosion rate.

G. Wikmark et al. (authors' closure)—We have not proven that the needles consist of stoichiometric zirconia, but rather that it is not metal since they are not dissolved in the methanol/bromine used for metal dissolution. The oxygen diffusion front into the metal has been reported to be of the order of a few tens of nm, which is much less than the needle extension into the former metal. We do, however, also believe that the occurrence of needles correlates with the corrosion rate as described in the paper.

---

[1] Atomic Energy of Canada, Ltd., Chalk River, Ontario, Canada.
[2] University of Toronto, Toronto, Ontario, Canada.

Hiroyuki Anada,[1] *Brett J. Herb,*[2] *Ken-ichi Nomoto,*[1] *Shigeki Hagi,*[1] *Ronald A. Graham,*[2] *and Takahiro Kuroda*[1]

Effect of Annealing Temperature on Corrosion Behavior and ZrO_2 Microstructure of Zircaloy-4 Cladding Tube

REFERENCE: Anada, H., Herb, B. J., Nomoto, K., Hagi, S., Graham, R. A., and Kuroda, T., "**Effect of Annealing Temperature on Corrosion Behavior and ZrO_2 Microstructure of Zircaloy-4 Cladding Tube,**" *Zirconium in the Nuclear Industry: Eleventh International Symposium, ASTM STP 1295*, E. R. Bradley and G. P. Sabol, Eds., American Society for Testing and Materials, 1996, pp. 74–93.

ABSTRACT: This paper describes the corrosion behavior and the ZrO_2 microstructure of Zircaloy-4 (Zry-4) cladding tubes that were intermediate annealed at various temperatures. The corrosion behavior of the cladding tubes was studied by autoclave tests performed under 633 K water condition and 673 K steam condition. A TEM examination shows that the microstructure of ZrO_2 formed on the Zry-4 matrix consisted of both the columnar structure and the equiaxed grain. The grain size of the columnar grain was approximately 30 by 200 nm, while that of the equiaxed grain was less than 20 nm. The equiaxed grain was dominantly observed near lateral cracks and around intermetallic compounds that were incorporated into the ZrO_2 film. An analysis of the HR-SEM images indicated that the equiaxed grain to columnar grain volume ratio increased with increasing weight gain, especially after the first transition. The equiaxed grain to the columnar grain volume fraction decreased with increasing annealing temperature, which corresponded to decreasing weight gain. It was suggested that grain boundary diffusion of oxygen ions was accelerated by grain-size change of the oxide owing to the ZrO_2 microstructure transformation from the large columnar grains to the fine equiaxed grains. The ZrO_2 microstructure transformation might be caused dominantly by the oxidation of the intermetallic precipitates. The intermetallic precipitates were fine and uniformly distributed in the low-temperature TREX annealed Zry-4. This resulted in high-temperature TREX annealing being beneficial for improving corrosion resistance of the Zry-4 tube in PWR environments.

KEYWORDS: Zircaloy-4, corrosion resistance, oxide morphology, annealing temperature, β-solution treatment

Zircaloy has been employed as a fuel cladding tube in light water reactors because of a low cross section for the thermal neutron absorption and good corrosion resistance. To extend the fuel burnup of the light water reactor, an issue is to improve corrosion resistance of the Zircaloy cladding tube. The annealing procedure in a manufacturing process of the Zircaloy-4 cladding tube has been well known to affect corrosion behavior significantly. This paper describes the effect of annealing temperature on the corrosion behavior and the ZrO_2 microstructure of the Zircaloy-4 (Zry-4) cladding tube.

The corrosion resistance of the Zircaloy-4 cladding tube has been well known to depend on both the chemical composition and the manufacturing conditions. Sn and Si, for example, are

---

[1] Sumitomo Metal Industries, Ltd., 1–8 Fuso-cho, Amagasaki, Japan.
[2] Teledyne Wah Chang, 1600 N.E. Old Salem Road, Albany, OR.

important elements to control the corrosion resistance of Zry-4 [1,2]. The heat treatment condition in the manufacturing process is another of the important factors for corrosion resistance [3]. Hence, the heat treatment is sometimes controlled using the cumulative annealing parameter (ΣAi) [4]. The heat treatments, especially the β-solution treatment and intermediate annealing, influence the size distribution of the intermetallic precipitates in the Zry-4 matrix, which affect corrosion behavior [5,6]. However, a correlative effect of the β-solution treatment and the intermediate annealing on the precipitation and on the corrosion resistance have not been well understood yet [3]. The intermetallic precipitates in Zry-4, $Zr(Cr,Fe)_2$, which are incorporated into the oxide on Zry-4 during the corrosion process, are oxidized gradually according to oxygen partial pressure [7]. The precipitate chemical composition varies due to diffusion of Fe into the surrounding oxide [6,7]. The role of the intermetallic precipitate has been thought to be an important factor in the corrosion mechanism.

It is also important to analyze the microstructure of ZrO_2 to understand the corrosion behavior. This is because oxide character adjacent to the oxide-to-metal interface controls the corrosion rate [8]. Many researchers have characterized the oxide [9–11], and TEM examinations were often employed [12–14].

This paper describes the effect of the β-quenching rate and the TREX annealing temperatures on the microstructure of ZrO_2 and corrosion resistance of the Zry-4 cladding tube.

Experimental Procedure

Sample Preparation

Table 1 shows the chemical composition of a low-Sn Zircaloy-4 (Zry-4) ingot that was manufactured to cladding tubes 9.5 mm in outside diameter. In the manufacturing process, the quenching method of the β-solution treatment and temperatures in a TREX annealing procedure were modified as shown in Table 2. The cumulative annealing parameters [4] were also summarized in Table 2.

An 8-mT ingot was vacuum melted and forged to billet sizes of approximately 190 mm in diameter. One billet was water quenched and the other was polyethyleneglycol water quenched immediately after heating at 1323 K. The cooling rate was measured with a thermocouple in each billet at 3 mm inside from the billet outer surface. The cooling rate from 1323 to 1073 K was calculated to be more than 200 K/s for the water-quenched billet (WQ) and to be 20 K/s for the polyethyleneglycol water-quenched billet (PWQ). The WQ Zry-4 and the PWQ Zry-4 denote the cladding tubes manufactured from the water-quenched billet and the polyethyleneglycol water-quenched billet, respectively. The billets were hot extruded at 923 K, cold rolled, and annealed at temperatures as shown in Table 2. The TREX tubes were cold reduced four additional times to a final size of 9.5 mm OD and intermediate annealed three times at 923 K for 7.2 ks. The 9.5-mm OD cladding tubes received a final stress relieving at 723 K for ks.

The intermetallic precipitates in the Zry-4 matrix were observed by TEM (transmission electron microscopy: JEOL-200CX) to determine their crystal structure. SEM (scanning electron microscopy: JEOL S-510) examinations were performed to obtain the size distribution of the precipitates in the Zry-4 matrix.

TABLE 1—*Chemical composition of Zircaloy-4 ingot (wt%).*

Sn	Fe	Cr	Si	C	Zr
1.29	0.22	0.11	0.0089	0.0139	Bal.

TABLE 2—β-quenching and TREX annealing conditions and cumulative annealing parameter $[\Sigma Ai = \Sigma ti \times exp(-40\,000/T)]$.

ID	β-Quenching Coolant	TREX Annealing	ΣAi
A	Water	922 K, 7.2 ks (2 h)	1.1×10^{-18}
B	Water	973 K, 7.2 ks (2 h)	3.6×10^{-18}
C	Water	1005 K, 7.2 ks (2 h)	1.1×10^{-17}
D	Water	1023 K, 7.2 ks (2 h)	2.2×10^{-17}
E	Water	1061 K, 7.2 ks (2 h)	8.6×10^{-17}
F	Polyethyleneglycol water	922 K, 7.2 ks (2 h)	1.1×10^{-18}
G	Polyethyleneglycol water	1005 K, 7.2 ks (2 h)	1.1×10^{-17}
H	Polyethyleneglycol water	1061 K, 7.2 ks (2 h)	8.6×10^{-17}

Corrosion Tests

Static water autoclave tests and static steam autoclave tests were conducted at 633 K for 400 days and 673 k for 300 days, respectively. Specimens 40 mm in length for the water test and 50 mm in length for the steam test were cut from the cladding tubes and rinsed in acetone. Weight gain values were obtained during the corrosion tests.

Observation of Oxide Film

TEM and HR-SEM (high resolution scanning electron microscopy equipped with a field emission gun: Hitachi S-4100) were used to characterize the oxide morphology formed on the outer surface of the Zry-4 tube. The TEM examination was conducted especially for the observation of the microstructure of ZrO_2 surrounding the intermetallic compounds that were incorporated into the oxide. The HR-SEM examinations were conducted to determine the microstructure of fractured surface ZrO_2, in particular, adjacent to the interface of the Zry-4 tube and the ZrO_2.

Cross-sectional samples for the TEM observation were prepared by a previously described procedure [15]. In the sample preparation for the HR-SEM observation of the fractured surface, the oxide layer was removed from the Zry-4 matrix by immersing in nitric hydrofluoric acid solution. The fractured surface in the range of 1 μm from the oxide-to-metal interface was observed, and the images were analyzed to obtain the volume fraction of the columnar grains and the equiaxed grains of ZrO_2.

Result

Corrosion Behavior and TREX Annealing Temperature

Figures 1a and 1b show the correlation between the weight gain and the corrosion time of the WQ Zry-4 tubes in 633 K water and in 673 K steam, respectively. Weight gain showed several transitions in the weight gain–time curves of the WQ Zry-4 tubes, which were TREX annealed above 1005 K. On the other hand, after the first transition observed for the 922 K annealed material, the weight gains increased significantly as the test time proceeded, no transition was recognized, and weight gain increased linearly.

Figures 2a and 2b exhibit weight gain–corrosion time dependence curves of the PWQ Zry-4 tubes. The controlled cooling rate of the PWQ Zry-4 billet was one tenth of the cooling rate of the WQ billet. The corrosion rate was almost the same among the PWQ Zry-4 tubes. An effect of the TREX annealing temperature on the corrosion behavior as it was recognized in the WQ Zry-4 tube was not apparent in the PWQ Zry-4 tubes.

Figure 3a demonstrates the effect of the TREX annealing temperature on weight gain after

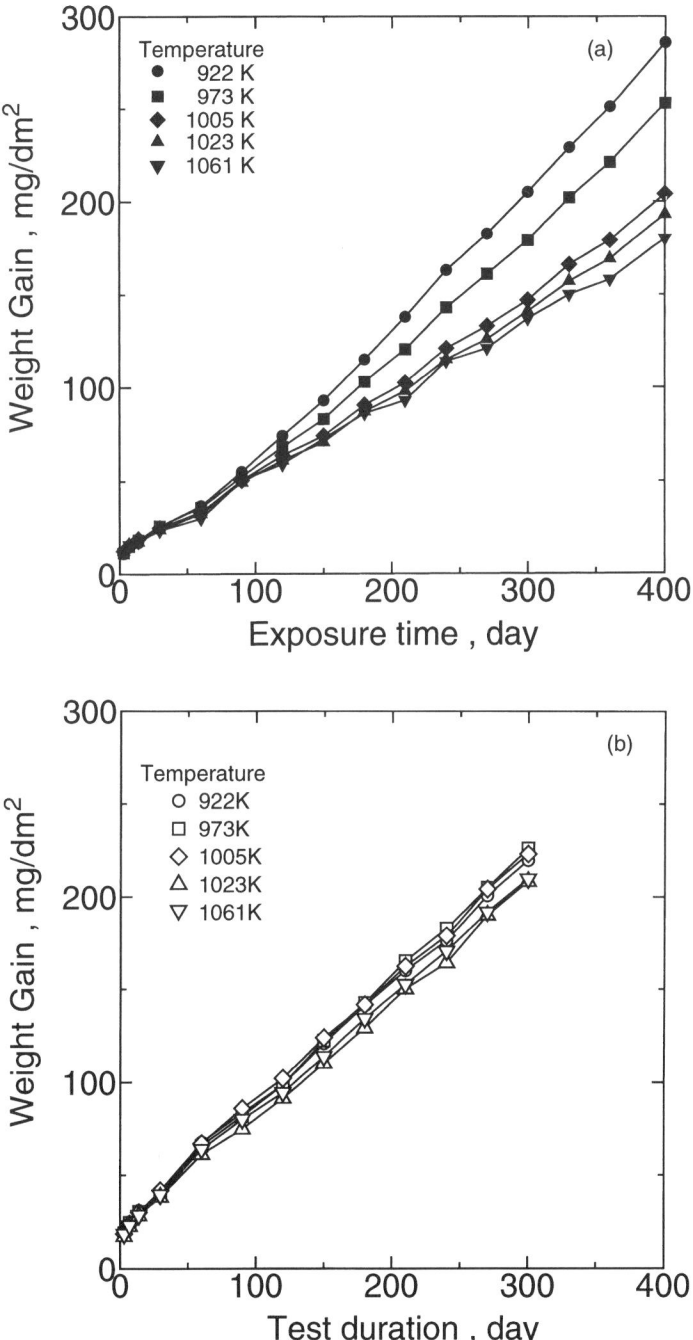

FIG. 1—*Weight gain and test time dependency of the WQ Zry-4 tube:* (a) *633 K water test;* (b) *673 K steam test.*

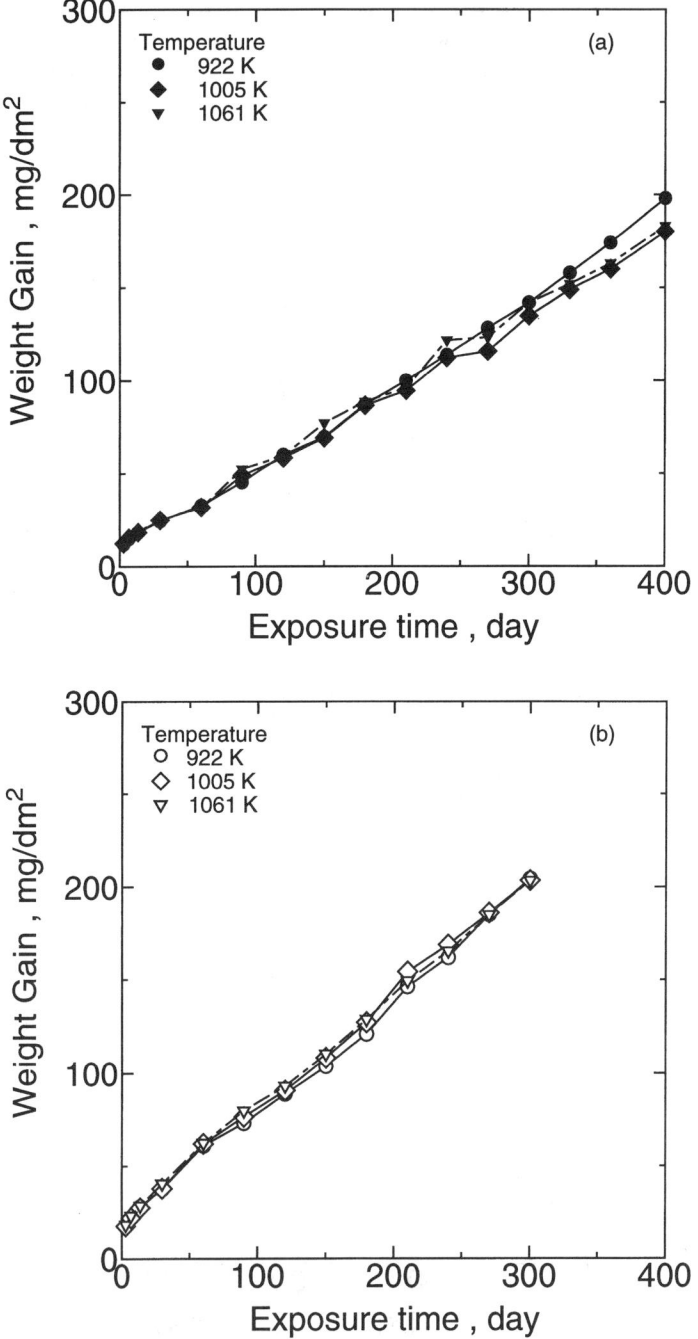

FIG. 2—*Weight gain and test time dependency of the PWQ Zry-4 tube:* (a) *633 K water test;* (b) *673 K steam test.*

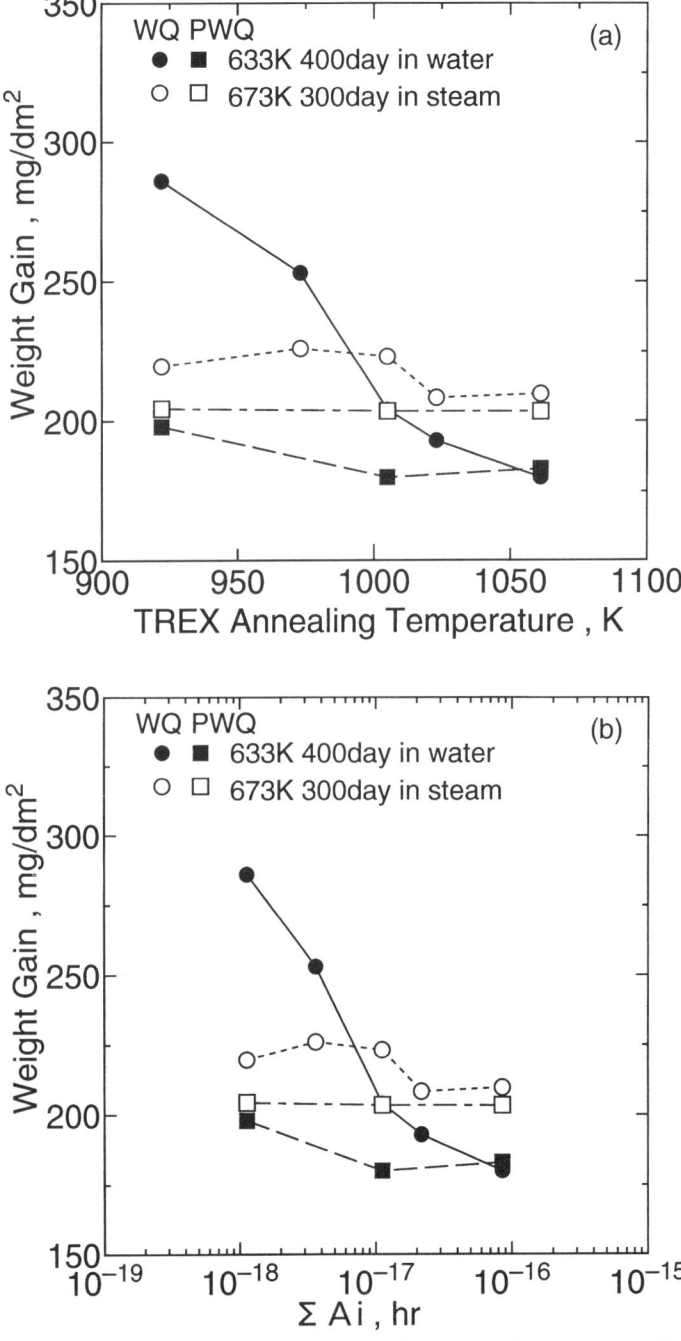

FIG. 3—*Effect of TREX annealing condition on weight gain in 633 K water and in 673 K steam:* (a) *TREX temperature,* (b) *cumulative annealing parameter* (ΣAi).

400 days exposure in 633 K water and 300 days exposure in 673 K steam. The relationship between the weight gain values and the cumulative annealing parameter was plotted in Fig. 3b. The TREX annealing temperature influences the corrosion behavior of the WQ Zry-4 tubes significantly compared with that of the PWQ Zry-4 tubes. Weight gain of the WQ Zry-4 tubes decreased with increasing TREX annealing temperature. However, weight gains of the PWQ Zry-4 tubes were almost the same among the specimens. This tendency was more apparent in the 633 K water test than in the steam corrosion test.

Observation of Precipitates in the Zry-4 Matrix

The precipitates in the Zry-4 matrix were observed by TEM, and the typical micrographs are exhibited in Fig. 4. The precipitates were identified dominantly as the $ZrCr_2$ (hcp) type intermetallic compound, with significantly fewer of the $ZrFe_2$ type intermetallic compounds by selected area diffraction patterns. The chemical compositions of the intermetallic compounds were spot analyzed by EDX (energy disperse X-ray analyzer). Table 3 shows calculated Fe/Cr ratios that were obtained by analyzing ten precipitates on each sample. Crystal structure and the Fe/Cr ratio of the intermetallic compounds were independent of the β-solution treatment and the TREX annealing conditions.

Size distributions of the intermetallic compounds were apparently sensitive to the cooling rate in the β-solution treatment and the TREX annealing temperature. Figures 5a and 5b show a typical histogram of the size distribution of the precipitates. An average diameter was calculated from the size distribution and correlated with TREX annealing temperature as shown

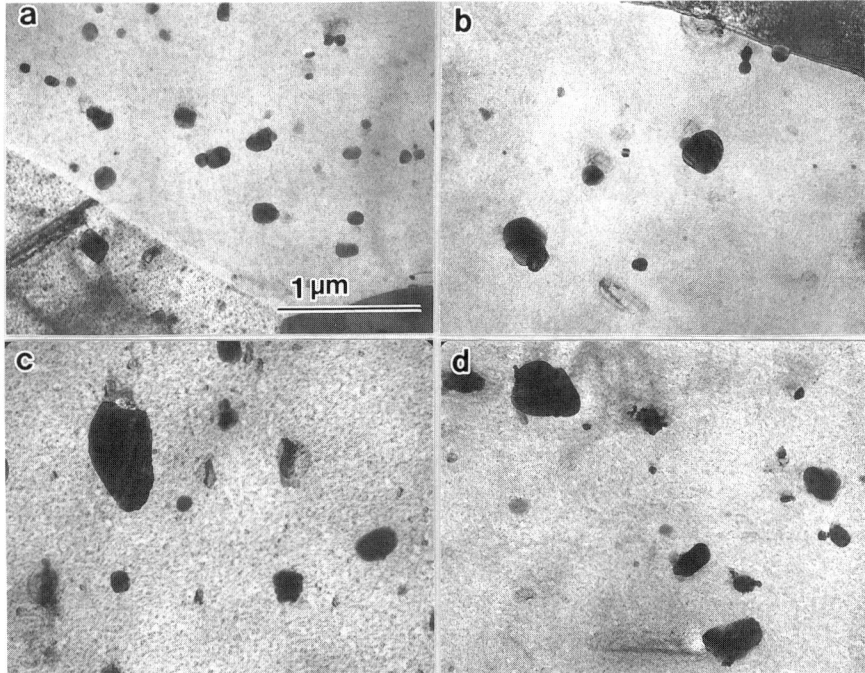

FIG. 4—*TEM micrographs of the intermetallic compound precipitated in Specimen A, Specimen C, Specimen E, and Specimen G.*

TABLE 3—*The Fe/Cr ratios in the intermetallic precipitates ($ZrCr_2$ type).*

Specimen ID	Fe/Cr Ratio		
	Minimum	Maximum	Average
A	1.44	2.00	1.71
C	1.36	1.75	1.61
E	1.48	1.99	1.67
G	1.49	2.01	1.80

in Fig. 5c. For both the WQ Zry-4 tubes and the PWQ Zry-4 tubes, the sizes of the precipitates increased by increasing the TREX annealing temperature. As expected, the size of the precipitates in the PWQ Zry-4 increased more than that in the WQ Zry-4.

Observation of Oxide Film

Figure 6 shows a representative TEM micrograph of the cross-sectional oxide on the WQ Zry-4 tube, Specimen E, exposed 60 days in water. A precipitate, showing P in Fig. 6, is observed to be incorporated into the ZrO_2 film. The columnar ZrO_2 grains grew perpendicularly to the matrix. The grain size was approximately 30 nm and elongated perpendicularly more than 200 nm. Equiaxed grains were observed around the precipitate. Their grain size, on the other hand, was less than 20 nm and was significantly finer than that of the columnar grains.

This precipitate was incorporated into the oxide and was located at a distance about 200 nm from the interface. This precipitate was identified as the $ZrCr_2$ (hcp) type intermetallic compound. As a result of EDX analysis, the Fe/Cr ratio was found to be 1.9 in the center of the precipitate and 0.9 near the edge. The decrease of the Fe/Cr ratio indicated that Fe diffused out to surrounding oxide. In addition, a portion of the outer surface of the intermetallic compound might be oxidized.

Figure 7 shows HR-SEM images of the fractured surface of the oxide on the WQ Zry-4 tubes corroded for 3 days and 120 days in 633 K water. The first transition in the corrosion curve was observed at 60 days as shown in Fig. 1a. Hence, we can compare the microstructure before and after the first transition. The columnar grains were dominantly observed on Specimen A, which was corroded for three days as shown in Fig. 7a. The columnar grains transformed to many equiaxed grains between 3 days and 120 days exposure as shown in Fig. 7b. The transformation had occurred apparently after the first transition in Specimen A, the corrosion resistance of which was inferior to other specimens. On the other hand, Fig. 7c shows that columnar grains were observed dominantly in the oxide on Specimen E, which was corroded for 120 days and was well past transition. It is suggested that the transformation from columnar grains to equiaxed grain correlated with the corrosion behavior.

The equiaxed grain to columnar grain volume fraction was obtained by image processing of the fracture surface, where the area analyzed was 1 μm from the oxide-to-metal interface. The analyzed samples were the WQ Zry-4 tubes in which the effect of TREX annealing temperature was apparent on the corrosion resistance. Figure 8a shows the relationship between the fraction of oxide that is equiaxed and the TREX annealing temperature. The relationship to the weight gain values also is shown. The equiaxed grain fractions vary as do the weight gain values, decreasing with increasing TREX annealing temperature.

The equiaxed grain volume fraction in the 922 K annealed Zry-4 tube increased as corrosion time proceeded as shown in Fig. 8b. For this specimen, the weight gain rate increased linearly

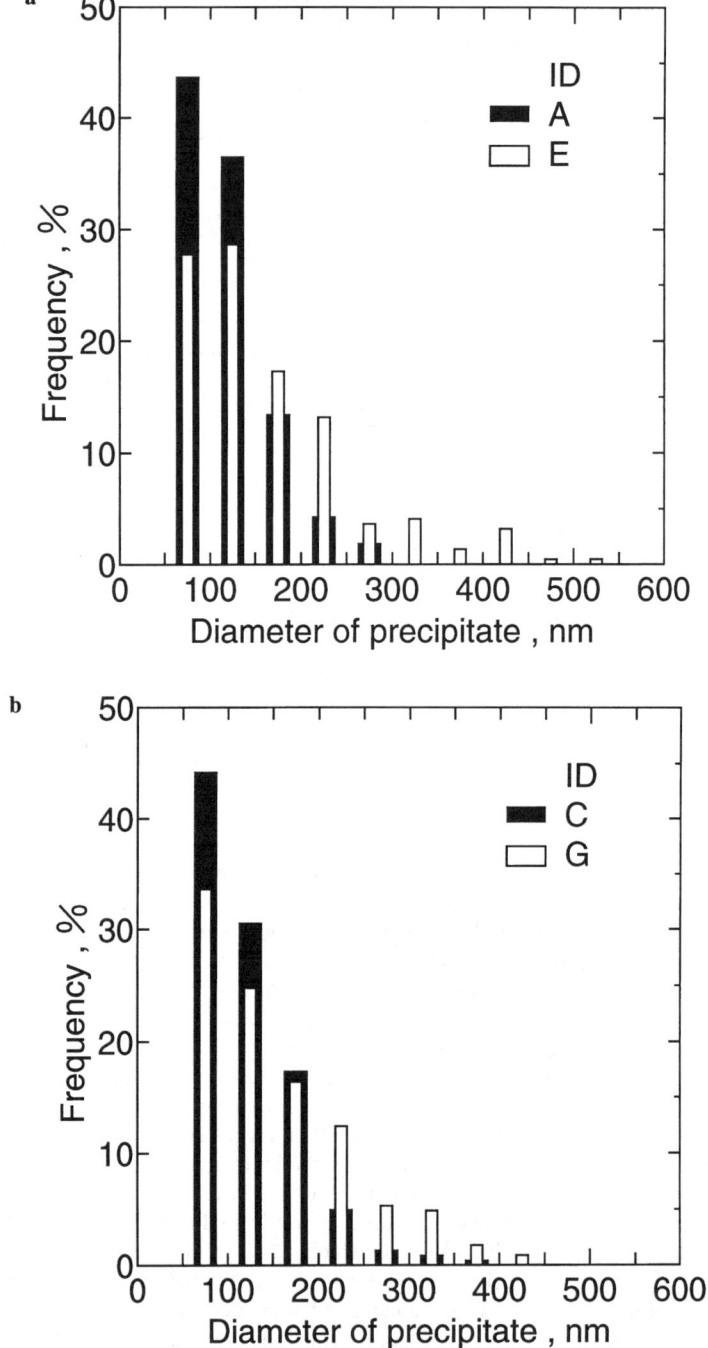

FIG. 5—*Size distribution of the intermetallic compound obtained by SEM image analyzing:* (a) *effect of TREX annealing temperature;* (b) *effect of quenching rate; and* (c) *average diameter.*

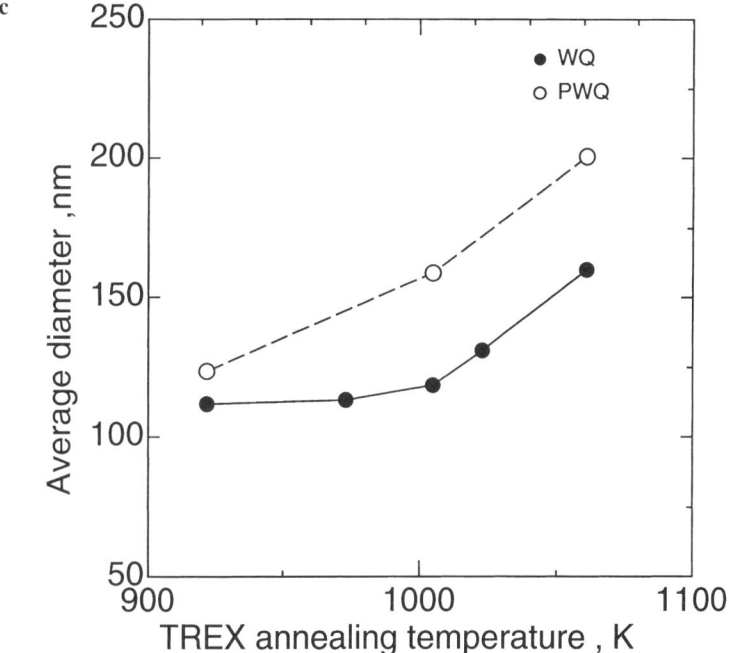

FIG. 5—*Continued*

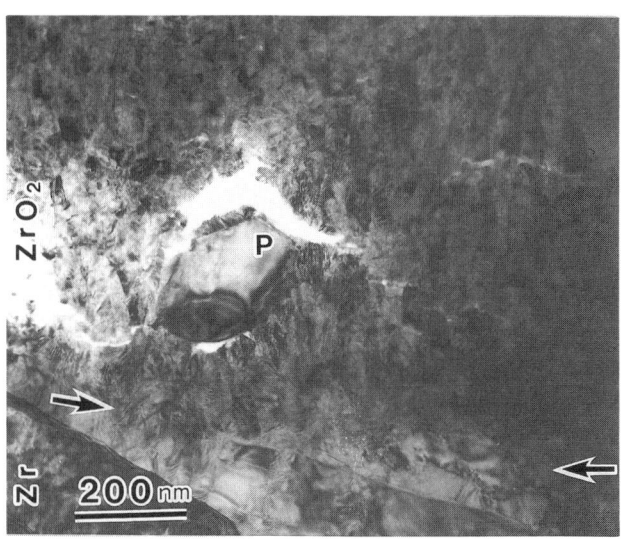

FIG. 6—*TEM micrograph of typical ZrO_2 and incorporated precipitate in the vicinity of ZrO_2/Zry-4 interface.*

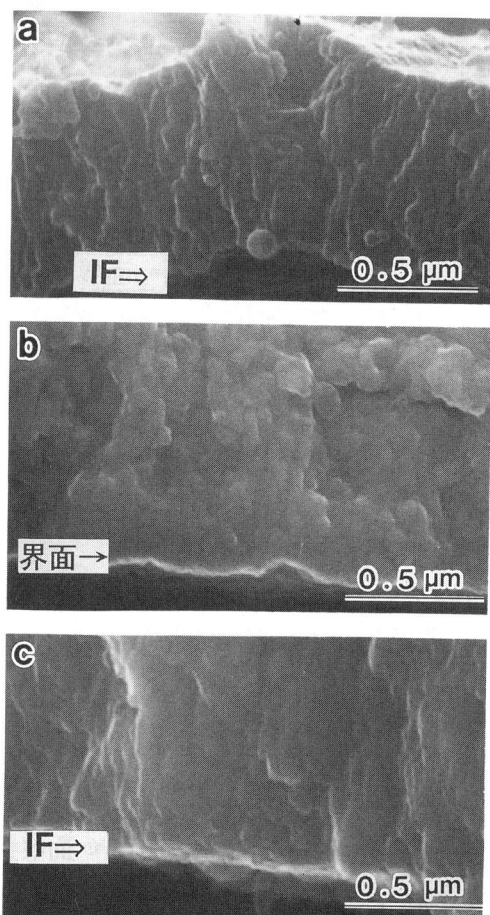

FIG. 7—*Fractured surface of the oxide formed in 633-K water observed by HR-SEM. "IF" in the figure indicates the ZrO_2-to-Zry4 interface:* (a) *Specimen A (3 days),* (b) *Specimen A (120 days), and* (c) *Specimen E (240 days).*

after the first transition occurred. On the other hand, the equiaxed grain volume fraction was approximately constant in the 1005 and 1061 K annealed Zry-4 tube. Several transitions were observed in the corrosion curves of these specimens. It is considered that the corrosion rate and transition behaviors could be correlated with the ZrO_2 grain transformation, i.e., the change in grain size. It seems important to understand how the TREX annealing temperature would affect the microstructure of the Zry-4 matrix and ZrO_2 oxide microstructure.

Discussion

Microstructure of ZrO_2 and Corrosion Resistance

This paper describes the effect of the cooling rate from the β-solution treatment and the TREX annealing temperatures on the corrosion behavior of Zry-4 cladding tubes. As a result,

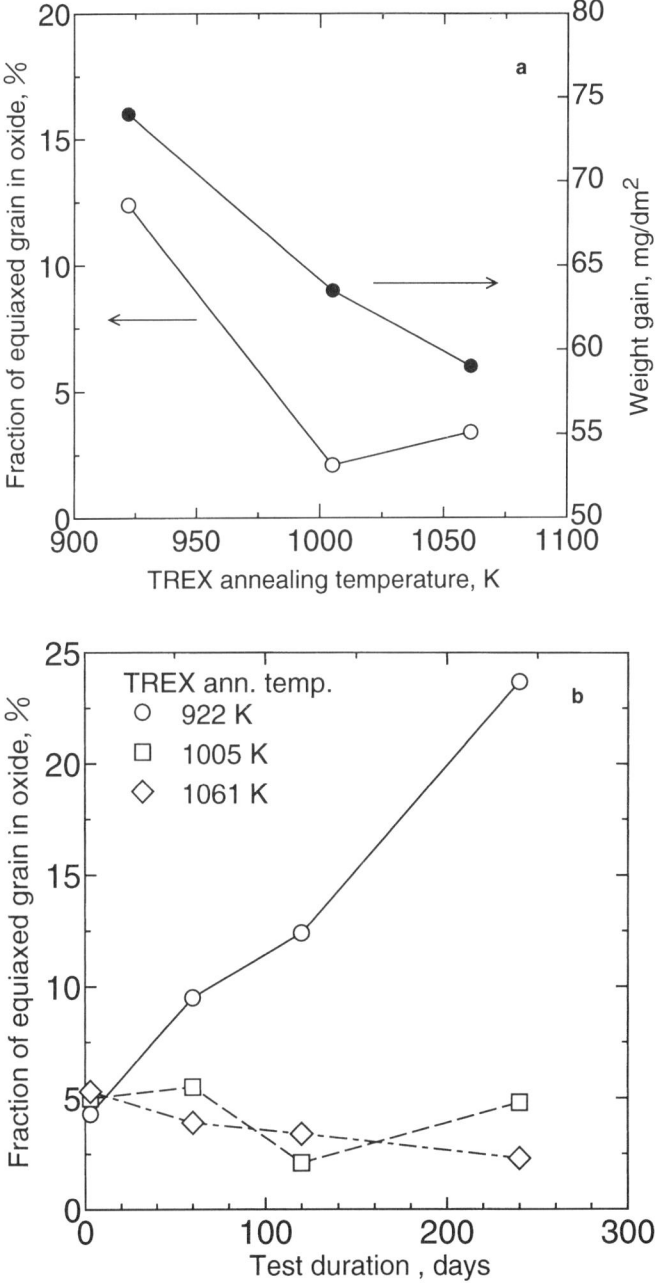

FIG. 8—*Relationship between the equiaxed grain volume fraction and corrosion behavior, showing effectiveness of:* (a) *TREX annealing temperature and weight gain, and* (b) *test time and TREX temperature.*

an accelerated corrosion behavior was recognized in the low-temperature TREX-annealed WQ Zry-4. The high-temperature annealing was beneficial, i.e., it improved the corrosion resistance of the Zry-4. The small cooling rate in the β-solution treatment weakened the enhancement of the corrosion rate in the low-temperature TREX-annealed Zry-4. This is now discussed on the basis of the microstructure transformation of ZrO_2 and the role of the intermetallic precipitates.

It is considered in this study that corrosion resistance of the Zry-4 tubes is controlled by the microstructure of ZrO_2 in the vicinity of the ZrO_2-to-Zry-4 interface. In particular, the decreasing grain size due to the transformation from the columnar to equiaxed grains would be in agreement with the accelerated corrosion behavior in the Zry-4 tubes that were TREX annealed at relatively low temperatures. The equiaxed grains of ZrO_2 are observed in the early corrosion stage of the Zircaloy-4 sheet, which did not exhibit corrosion resistance [12]. Equiaxed grains are also found in the case of the nodular corrosion, which resulted in an accelerated corrosion process [9].

There are two possibilities that the microstructure transformation could cause the accelerated corrosion to the Zry-4. One is a change of diffusion path of oxygen and the other is the destruction of the barrier ZrO_2 layer by the creation of a micropore and/or cracks.

The microstructure transformation would influence significantly the diffusion of oxygen ions in the Zr oxide, which is the diffusion species in the corrosion process [16]. It is reported that oxygen ions diffuse in the Zr oxide more dominantly through short circuit paths or grain boundaries of ZrO_2 than the lattice of ZrO_2 [8,17–20]. It is easily shown that the microstructure transformation of ZrO_2 from large columnar grains to fine equiaxed grains increases the grain boundary area and/or the short circuit diffusion paths in the ZrO_2, which would accelerate the oxygen ion diffusion. Comparing the diffusion coefficient of oxygen ions in ZrO_2, the diffusion coefficient in grain boundaries is smaller than that of bulk [21] at 677 K. This would appear to be a cause of the accelerated corrosion of the Zry-4 that was annealed at low temperature.

The other possibility also important for the corrosion resistance is that the growing oxide film of the columnar grain would be a protective oxide layer. In general, the corrosion rate of Zry-4 increased after the transition that corresponds to the transformation of the crystal structure of ZrO_2 [8,22]. The stability of the protective layer would be important.

The columnar grains would grow under compressive stresses, which are loaded perpendicularly to a long axis of the columnar grains [23]. The transformation from the columnar grain to the equiaxed grain would suggest that the compression stress would be released or would not be loaded any longer. Induced micropore from the oxide transformation might accelerate the corrosion [24,25]. The microstructure transformation is often associated with the lateral crackings in the oxide [15]. When the micropore and the crack would be induced, the diffusion species might be altered from an oxygen ion oxidizing species in solution in the water. The cracks and the micropores would cause further acceleration of corrosion [26].

Microstructure of ZrO_2 and Intermetallic Compound Distribution

The microstructure of ZrO_2 would influence corrosion behavior, as described above. The microstructure transformations occurred frequently around the intermetallic precipitates and near a lateral crack [10,15], hence, the microstructure transformation correlated with intermetallic precipitates being incorporated into the oxide.

Figure 9 illustrated the transforming of oxide grains and the role of the intermetallic precipitate during corrosion. The intermetallic precipitates are incorporated into the oxide (A in Fig. 9) and oxidized gradually in the oxide [6,7,27,28]. Thus, Fe, Cr, and Zr in the intermetallic compound would be oxidized in the ZrO_2 layer. Some of the Fe and Cr diffuse into the sur-

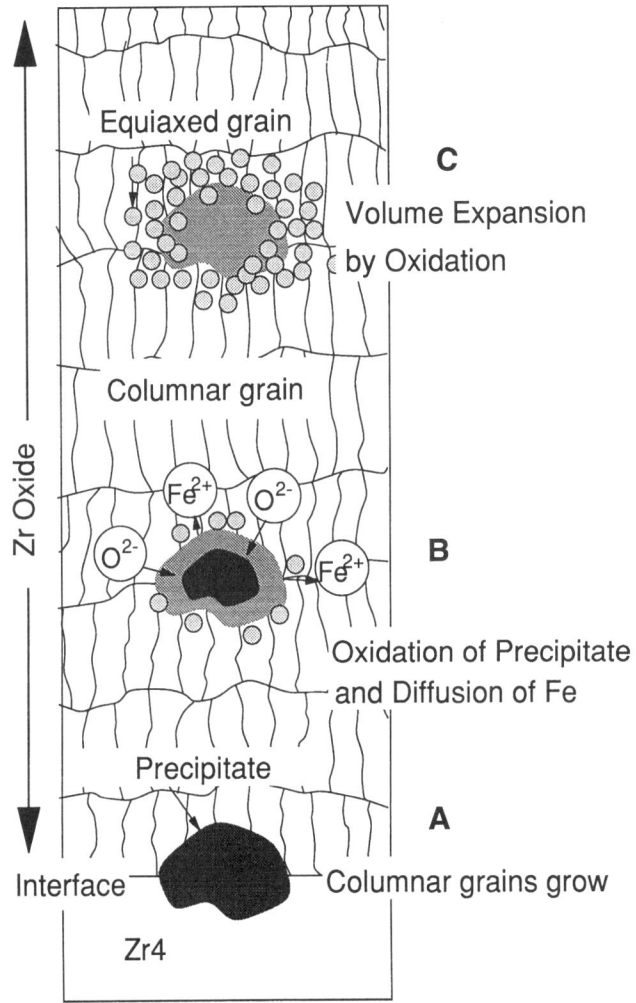

FIG. 9—*Schematic illustration for the role of the incorporated precipitate in the ZrO_2.*

rounding ZrO_2 and are oxidized to Fe oxide and Cr oxide (B in Fig. 9). The volume of these elements would expand, obeying the Pilling-Bedworth ratio [29] as shown in Table 4. When the volume expansion would occur in the ZrO_2, the surrounding columnar ZrO_2 layer would be destroyed by the high local stresses, which would reduce the compressive stress heterogeneously and the protectiveness to the inward diffusion of the oxygen ion. When the columnar grain was destroyed, the transformation to the fine equiaxed grain would have occurred (C in Fig. 9).

The intermetallic precipitate incorporated into the ZrO_2 would have the possibility to be a trigger to diminish the stable and protective ZrO_2 layer as described above. A large number of the fine intermetallic precipitates distributed uniformly in the Zr matrix would degrade the corrosion resistance of the Zry-4 when they were incorporated into oxide and oxidized. The

TABLE 4—*Pilling-Bedworth ratio (PBR) of Zr, Fe, and Cr.*

Metal	Oxide	PBR
Zr	ZrO_2	1.56
Fe	FeO	1.68
	Fe_2O_3	2.14
	Fe_3O_4	2.10
Cr	Cr_2O_3	2.07

low TREX annealing temperature and high cooling rate in the solution treatment refined the intermetallic compound, which would degrade the corrosion resistance.

Figure 10 shows the relationship between the average diameter of the precipitates and weight gain. Weight gain decreased rapidly with increasing diameter from 100 up to 150 nm and saturated above 150 nm. This result shows that both the cooling rate and the TREX annealing temperature correlated significantly with the corrosion resistance. This tendency was more obvious in the water test result than in the steam test result. This seems to correlate with test temperature, which would affect the release of the induced compressive stress in the ZrO_2.

In this study, both the cooling rate and the TREX annealing temperature correlated with corrosion resistance. When the cooling rate following the β-solution treatment was less than that for water quenching, the corrosion resistance of the Zry-4 was not sensitive to the TREX annealing temperature. The relationship between the cooling rate and the uniform corrosion resistance of the TREX tube shell of Zry-4 was reported for a suitable value of approximately

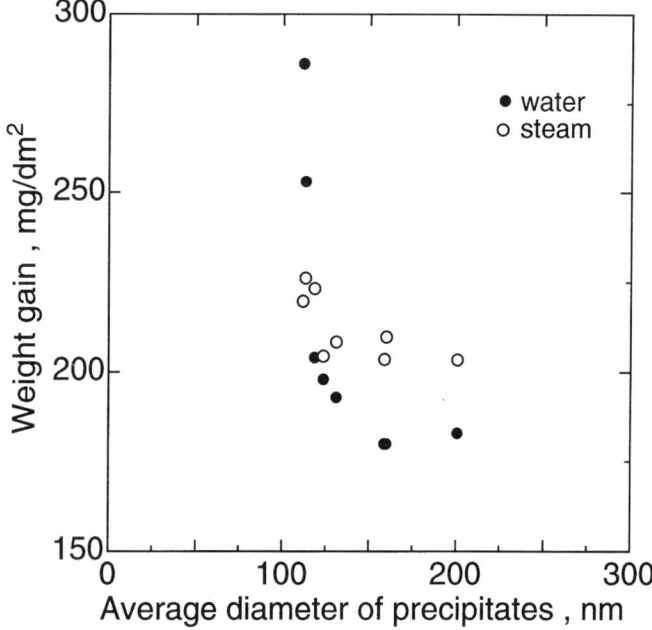

FIG. 10—*Relationship between the average diameter of the precipitates and weight gain.*

30 K/s [*3*]. It is confirmed that corrosion resistance of the slow cooled Zry-4 cladding following the β-solution treatment does not show the accelerated corrosion.

Conclusions

The effect of the cooling rate in the β-solution treatment and the TREX annealing temperature on the corrosion behavior and the microstructure of ZrO_2 was studied, and the following conclusions were drawn.

1. The equiaxed grain volume fraction of the ZrO_2 at a distance of 100 nm from the ZrO_2/Zry-4 interface increased with increasing weight gain, especially after the first transition. The volume fraction of oxide equiaxed grain decreased with increasing annealing temperature, which then corresponds to decreasing weight gain.
2. The transformation from the columnar grain to the equiaxed grain was observed dominantly around the oxidized intermetallic precipitates that were incorporated into the ZrO_2.
3. The corrosion resistance of the Zry-4 cladding tube that was water quenched from the β temperature at a rate of 200 K/s was improved with increasing TREX annealing temperature. The corrosion resistance of cladding that was cooled at 20 K/s from the β annealing temperature was not sensitive to the TREX annealing temperature.
4. The corrosion resistance was apparently degraded with decreasing size of the precipitated intermetallic compounds. It would suggest that the oxidation of the intermetallic compound was a cause for an accelerated corrosion.

Acknowledgments

The authors appreciate the advice of Dr. Y. Shida and would also like to thank Mr. K. Nakamura and Mr. Y. Kohzuki for the experimental work.

References

[1] Garde, A. M., Pati, S. R., Krammen, M. A., Smith, G. P., and Endter, R. K., "Corrosion Behavior of Zircaloy-4 Cladding with Varying Tin Content in High-Temperature Pressurized Water Reactors," *Zirconium in the Nuclear Industry (Tenth Volume), ASTM STP 1245*, A. M. Garde and E. R. Bradley, Eds., American Society for Testing and Materials, West Conshohocken, PA, 1994, pp. 760–778
[2] Eucken, C. M., Finden, P. T., Trapp-Pritsching, S., and Weidinger, H. G., "Influence of Chemical Composition on Uniform Corrosion of Zirconium-Base Alloys in Autoclave Tests," *Zirconium in the Nuclear Industry (Eighth Symposium), ASTM STP 1023*, L. F. P. Van Swam and C. M. Eucken, Eds., American Society for Testing and Materials, West Conshohocken, PA, 1989, pp. 113–127.
[3] Shemel, J. H., Charquet, D., and Wadier, J.-F., "Influence of the Manufacturing Process on the Corrosion Resistance of Zircaloy-4 Fuel Cladding," *Zirconium in the Nuclear Industry (Eighth Symposium), ASTM STP 1023*, L. F. P. Van Swam and C. M. Eucken, Eds., American Society for Testing and Materials, West Conshohocken, PA, 1989, pp. 141–152.
[4] Garzarolli, F., Steinberg, E., and Weidinger, H. G., "Microstructure and Corrosion Studies for Optimized PWR and BWR Zircaloy Cladding," *Zirconium in the Nuclear Industry (Eighth Symposium), ASTM STP 1023*, L. F. P. Van Swam and C. M. Eucken, Eds., American Society for Testing and Materials, West Conshohocken, PA, 1989, pp. 202–212.
[5] Garzarolli, F., Schumann, R., and Steinberg, E., "Corrosion Optimized Zircaloy for Boiling Water Reactor (BWR) Fuel Elements," *Zirconium in the Nuclear Industry (Tenth Volume), ASTM STP 1245*, A. M. Garde and E. R. Bradley, Eds., American Society for Testing and Materials, West Conshohocken, PA, 1994, pp. 709–723.
[6] Anada, H., Nomoto, K., and Shida, Y., "Corrosion Behavior of Zircaloy-4 Sheets Produced Under Various Hot-rolling and Annealing Conditions," *Zirconium in the Nuclear Industry (Tenth Volume)*,

ASTM STP 1245, A. M. Garde and E. R. Bradley Eds., American Society for Testing and Materials, West Conshohocken, PA, 1994, pp. 307–327.

[7] Pecheur, D., Lefebvre, F., Motta, A. T., Lemaignan, C., and Charquet, D., "Oxidation of Intermetallic Precipitates in Zircaloy-4: Impact of Irradiation," *Zirconium in the Nuclear Industry (Tenth Volume), ASTM STP 1245*, A. M. Garde and E. R. Bradley, Eds., American Society for Testing and Materials, West Conshohocken, PA, 1994, pp. 687–708.

[8] Godlewski, J., Gros, J. P., Lambertin, M., Wadier, J. F., and Weidinger, H., "Raman Spectroscopy Study of the Tetragonal-to-Monoclinic Transition in Zirconium Oxide Scales and Determination of Overall Oxygen Diffusion by Nuclear Microanalysis of O^{18}," *Zirconium in the Nuclear Industry (Ninth Volume), ASTM STP 1132*, C. M. Eucken and A. M. Garde, Eds., American Society for Testing and Materials, West Conshohocken, PA, 1991, pp. 416–436.

[9] Garzarolli, F., Seidel, H., Tricot, R., and Gros, J. P, "Oxide Growth Mechanism on Zirconium Alloys," *Zirconium in the Nuclear Industry (Ninth Volume), ASTM STP 1132*, C. M. Eucken and A. M. Garde, Eds., American Society for Testing and Materials, West Conshohocken, PA, 1991, pp. 395–415.

[10] Wadman, B. and Andren, H.-O., "Microanalysis of the Matrix and the Oxide-Metal Interface of Uniformly Corroded Zircaloy," *Zirconium in the Nuclear Industry (Ninth Volume), ASTM STP 1132*, C. M. Eucken and A. M. Garde, Eds., American Society for Testing and Materials, West Conshohocken, PA, 1991, pp. 461–475.

[11] Kubo, T. and Uno, M., "Precipitate Behavior in Zircaloy-2 Oxide Films and Its Relevance to Corrosion Resistance," *Zirconium in the Nuclear Industry (Ninth Volume), ASTM STP 1132*, C. M. Eucken and A. M. Garde, Eds., American Society for Testing and Materials, West Conshohocken, PA, 1991, pp. 476–498.

[12] Wadman, B., Lai, Z., Andren, H.-O., Nystrom, A.-L. Rudling, P., and Pettersson, H., "Microstructure of Oxide Layers Formed During Autoclave Testing of Zirconium Alloys," *Zirconium in the Nuclear Industry (Tenth Volume), ASTM STP 1245*, A. M. Garde and E. R. Bradley, Eds., American Society for Testing and Materials, West Conshohocken, PA, 1994, pp. 579–598.

[13] Dobler, U., Knop, A, Ruhmann, H., and Beie, H.-J., "On the Initial Corrosion Mechanism of Zirconium Alloy: Interaction of Oxygen and Water with Zircaloy at Room Temperature and 450°C Evaluated by X-ray Absorption Spectroscopy and Photoelectron Spectroscopy," *Zirconium in the Nuclear Industry (Tenth Volume), ASTM STP 1245*, A. M. Garde and E. R. Bradley, Eds., American Society for Testing and Materials, West Conshohocken, PA, 1994, pp. 644–662.

[14] Bradley, E. R. and Perkins, R. A., "Characterization of Zircaloy Corrosion Films by Analytical Transmission Electron Microscopy," *Proceedings*, IAEA Technical Committee Meeting on *Fundamental Aspects of Corrosion of Zr-Base Alloys in Water Reactor Environments*, 1989, Portland, OR.

[15] Anada, H. and Takeda, K., this publication.

[16] Cox, B., *Oxidation of Zirconium and its Alloy, Advances in Corrosion Science and Technology*, M. G. Fontana and R. W. Staele, Eds., Plenum Press, New York, Vol. 5, 1976.

[17] Cox, B. and Pemsler, J. P., "Diffusion of Oxygen in Growing Zirconia Films," *Journal of Nuclear Materials*, Vol. 28, 1968, pp. 73–78.

[18] Lightstone, J. B. and Pemsler, J. P., *Proceedings*, Symposium on Mass Transport in Oxides, National Bureau of Standards, October 1967, NBS Special Publication 296, August 1968.

[19] Matuda, K., Anada, H., and Bishop, H. E., "^{18}O Tracer Study of the Oxidation of Zircaloy-4 in Steam," *Surface and Interface Analysis*, Vol. 21, 1994, pp. 349–355.

[20] Sabol, J. P. and Dalgaard, S. B., *Journal of the Electrochemical Society*, Vol. 112, 1975, pp. 316–317.

[21] Kofstad, P., *Nonstoichiometry, Diffusion, and Electrical Conductivity in Binary Metal Oxides*, Wiley-Interscience, New York, 1972, p. 159.

[22] Takeda, K., Anada, H., and Kajimura, H., "Relationship Between Uniform Corrosion Resistance and Crystal Structure of Oxide on Zircaloy-4," to be presented to *Journal of the Japan Journal of Metals*.

[23] Godlewski, J., "How the Tetragonal Zirconia is Stabilized in the Oxide Scale that is Formed on a Zirconium Alloy Corroded at 400°C in Steam," *Zirconium in the Nuclear Industry (Tenth Symposium), ASTM STP 1245*, A. M. Garde and E. R. Bradley, Eds., American Society for Testing and Materials, West Conshohocken, PA, 1994, pp. 663–683.

[24] Zhou, B.-X., "Electron Microscopy Study of Oxide Films Formed on Zircaloy-2 in Superheated Steam," *Zirconium in the Nuclear Industry (Eighth Symposium), ASTM STP 1023*, L. F. P. Van Swam and C. M. Eucken, Eds., American Society for Testing and Materials, West Conshohocken, PA, 1989, pp. 360–373.

[25] Cox, B. and Yomaguchi, Y., "The Development of Porosity in Thick Zirconia Films," *Journal of Nuclear Materials,* Vol. 210, 1994, pp. 303–317.

[26] Hutchinson, B. and Lehtinen, B., "A Theory of the Resistance of Zircaloy to Uniform Corrosion," *Journal of Nuclear Materials,* Vol. 217, 1994, pp. 243–249.

[27] Pecheur, D., Lefebvre, F., Motta, A. T., and Lemaignan, C. in *Journal of Nuclear Materials,* Vol. 189, 1992, pp. 318–332.

[28] Harada, M., Kimpara, M., and Abe, K., "Effect of Alloying Elements on Uniform Corrosion Resistance of Zirconium-Based Alloys in 360°C Water and 400°C Steam," *Zirconium in the Nuclear Industry (Ninth Volume), ASTM STP 1132,* C. M. Eucken and A. M. Garde, Eds., American Society for Testing and Materials, West Conshohocken, PA, 1991, pp. 368–391.

[29] Kofstad, P., *High Temperature Oxidation,* Elsevier Applied Science, London, 1988, p. 244.

DISCUSSION

A. T. Motta[1] (written discussion)—(1) Did you see any evidence of metallic Fe in the early stages of corrosion as we saw in our previous work? (2) Could it be that the reason the equiaxed grains form around the precipitates is that they are tetragonal ZrO_2 grains stabilized by alloying elements (Fe, Cr)? In our previous work we observed that in the oxidized precipitates, tetragonal oxide formed regardless of the depth the precipitate was found in the oxide (e.g., even at 1 µm).

Anada et al. (authors' closure)—In this study, we've never found metallic Fe nor tetragonal oxide in the oxidized precipitates. Therefore, we could not make a comment on tetragonal stabilization around the oxidized precipitates. There is a possibility for Fe or Cr to stabilize tetragonal ZrO_2 grains.

Y. Etoh[2] (written discussion)—(1) What is the difference of the precipitate size effect between the water and steam test? (2) Which is better to predict in-pile corrosion?

Anada et al. (authors' closure)—The test temperature could affect the result in this study. In the previous study, the water test would predict the in-pile corrosion performance in the PWR environment.

B. Warr[3] (written discussion)—I have a question regarding the presence of artifacts introduced during sample preparation. How important is this? For example, do you believe the equiaxed grains observed between the columnar grains and the metal-oxide interface could be due to stress relaxation during foil preparation?

Anada et al. (authors' closure)—I think that stress relaxation is one reason for the transformation from columnar grain to equiaxed grain. There is a possibility of artifacts during sample preparation. However, I believe that the tendency of columnar-to-equiaxed grain volume ratio was not influenced qualitatively because the same tendency was confirmed by a TEM examination.

Bo Cheng[4] (written discussion)—Your results show that the fast quench rate by water quenching results in higher corrosion rates (as compared with PWR) at the same ΣAi value. Do your results support a need to prevent a high quench rate for PWR application?

Anada et al. (authors' closure)—Yes. Low quench rate reduces susceptibility of ΣAi to corrosion resistance of Zry-4 in PWR application.

Vincent Urbanic[5] (written discussion)—Do you seen any influence of intermetallic precipitate size in the oxide on the transformation of columnar oxide to the equiaxed oxide?

Anada et al. (authors' closure)—I did not directly examine the size effects of the intermetallic precipitates in oxides. I estimate the size effect in the oxide from information of the size distribution of the intermetallic precipitates in the alloy.

A. Strasser[6] (written discussion)—Did you notice or evaluate the effect of the base metal

---

[1] Pennsylvania State University, University Park, PA.
[2] Nippon Nuclear Fuel Development Co., Ibaraki-Ken, Japan.
[3] Ontario Hydro, Toronto Canada.
[4] EPRI, Palo Alto, CA.
[5] Atomic Energy of Canada, Ltd., Chalk River, Ontario, Canada.
[6] S. M. Stiller Corp., Pleasantville, NY.

texture differences on oxide structure and morphology? Particularly between the sheet specimens of your first paper and the tube specimens of your second paper.

Anada et al. (authors' closure)—I did not evaluate the effect of the base metal texture in this study. I think from my experience that the difference in the texture between the sheet specimens and the tube specimens could be small.

Dominique Pêcheur,[1] *Joël Godlewski,*[1] *Philippe Billot,*[1] *and Joël Thomazet*[2]

Microstructure of Oxide Films Formed during the Waterside Corrosion of the Zircaloy-4 Cladding in Lithiated Environment

REFERENCE: Pêcheur, D., Godlewski, J., Billot, P., and Thomazet, J., "**Microstructure of Oxide Films Formed during the Waterside Corrosion of the Zircaloy-4 Cladding in Lithiated Environment,**" *Zirconium in the Nuclear Industry: Eleventh International Symposium, ASTM STP 1295,* E. R. Bradley and G. P. Sabol, Eds., American Society for Testing and Materials, 1996, pp. 94–113.

ABSTRACT: Zircaloy-4 cladding materials have been oxidized in a lithiated environment in autoclave and in out-of-pile loop tests. In such oxidation tests, a strong enhancement of the oxidation rate can occur depending on the water chemistry conditions and on the oxidation time. In this work, in order to improve our understanding of the detrimental effect of lithium on the corrosion behavior of the Zircaloy-4 cladding, the microstructure of oxide films has been characterized by TEM. Simultaneously, the lithium profiles and concentrations in the oxide layers have been determined using the SIMS technique, special attention being paid to the metal-oxide interface.

These analyses revealed the existence of a correlation between the oxidation rate, the oxide microstructure, and the lithium profiles in the oxide films. Before the occurrence of the strong acceleration of the oxidation rate, the whole oxide layer is mainly composed of columnar grains and the lithium has no access to the metal-oxide interface. On the contrary, after the occurrence of such a strong accelerated corrosion rate, the inner part of the oxide layer is composed mainly of equiaxed grains (up to the metal-oxide interface), lithium then having access to the metal-oxide interface.

These experimental results are described extensively and analyzed, and their contribution to the understanding of the influence of lithium on the oxidation process of Zircaloy-4 cladding material is discussed as a function of an existing thin inner barrier layer at the metal-oxide interface.

KEY WORDS: Zircaloy-4, lithium, oxidation, oxide layer, zirconia, microstructure

In pressurized water reactors, lithium is added to the primary coolant to maintain the pH_{300} of the coolant between the recommended values of 7.2 and 7.4 so as to reduce circuit activity levels [*1*]. Until now, the recommended lithium content in the primary coolant did not exceed 2.2 ppm for 12-month cycles ([B] ≤ 1200 ppm) and 3.5 ppm for 18-month cycles ([B] ≤ 1400 ppm). However, local boiling regimes at the fuel rod surface in the hottest channels of the core can lead to a local enrichment of lithium (and boron) on the surface of the cladding [*2,3*]. Moreover, high concentrations of lithium can be present in thick porous oxide films or within crud layers deposited on fuel cladding.

Therefore, the impact of lithium on the oxidation rate of zirconium alloys has been studied

---

[1] Research engineers and head of laboratory, respectively, CEA-Cadarache, DRN/DEC/SECA, 13108 Saint Paul Lez Durance Cedex, France.
[2] Consulting engineer, FRAMATOME NUCLEAR FUEL, 10 rue Juliette Récamier, 69456 Lyon Cedex 06, France.

for lithium levels in the coolant between 2.2 and 7000 ppm for many years [4–12]. In low-lithium-level conditions (a few ppm), no significant accelerated corrosion is observed, even for high exposure times (>>100 days). However, high lithium levels (a few ten ppm) are known to increase drastically the oxidation rate of zirconium alloys after a given time, which depends on the lithium content and on the exposure temperature. For instance, in the case of Zircaloy-4 materials oxidized in autoclave at 633 K with 70 ppm lithium in the water, a strong enhancement of the oxidation rate occurs after 100 days (at least), while it occurs after only a few days of oxidation with more than 700 ppm lithium. Little or no pre-transition region is, moreover, observed with such high lithium concentrations.

Up to now, several hypotheses have been proposed to account for the enhancement of oxidation rates due to lithium. They are based on either:

1. An increase in anion vacancies in the oxide caused by the substitution of zirconium by lithium ions in the zirconia lattice [13].
2. The generation of pores caused by preferential dissolution of cubic ZrO_2 in zirconia films exposed to more than 700 ppm lithium [14–15] or by the continuous formation and dissolution of Li_2O in oxide film [16,17].
3. A modified crystal growth mechanism induced by LiOH [18] and especially caused by the formation of surface OLi groups, which impede the diametral and columnar growth of the oxide crystallites [9].

In this work, to improve our understanding of the lithium influence on the oxidation rate, quantitative lithium profiles in the oxide layer and oxide microstructure have been studied using secondary ion mass spectrometry and transmission electron microscopy, respectively. These two types of analyses were performed on the same samples across the oxide layer from the metal/oxide interface to the water/oxide interface according to varied oxide thicknesses, exposure times, oxidation rates, and oxidation conditions. Based on the results of these analyses, the incorporation of lithium in the oxide layer and its impact on the oxide morphology are discussed in order to interpret the strong lithium enhancement of the oxidation rate.

Experimental Procedure

Materials

The materials tested in the experimental program were standard Zircaloy-4 tubing [19]. Their chemical compositions were Zr-1.2 to 1.5% Sn-0.19 to 0.24% Fe-0.09 to 0.13% Cr-0.108 to 0.135% O.

Oxidation Tests

Oxidation tests were performed on as-received tubes (surface preparation: belt polishing + polishing wheel [19]), without mechanical or chemical polishing being performed prior to the oxidation tests. Two types of oxidation facilities were used: autoclave and out-of-pile loop.

1. Autoclave tests were conducted in water at 633 K and 18 MPa under two types of chemical conditions: Li = 1.5 ppm (B = 650 ppm), which is close to PWR's water chemistry conditions, and Li = 70 ppm (B = 0 ppm), which is a very high lithium content.
2. An out-of-pile loop test was conducted in the REGGAE loop in water (636 K, 19 MPa)

with 10 ppm lithium and 650 ppm boron. This type of facility simulates the PWRs operating conditions except for the neutron flux, the rods corroded in the loop being electrically heated (heat flux equal to 100 W/cm^2).

Oxidation kinetics obtained after autoclave and out-of-pile loop tests are plotted in Fig. 1, oxide thicknesses being determined by metallographic examinations. The main observations are the following:

1. For autoclave tests, the post-transition oxidation rate is higher with 70 ppm Li than with 1.5 ppm Li and 650 ppm B. Moreover, after about 150 days with 70 ppm Li, a strong enhancement of the corrosion rate occurs, whereas no such accelerated oxidation rate is observed after 900 days with 1.5 ppm lithium and 650 ppm boron.
2. For the out-of-pile loop test performed with 10 ppm Li and 650 ppm B, the post-transition oxidation rate is similar to the one obtained in autoclave with 70 ppm Li during the major part of the test, up to 150 days of oxidation. However, drastic accelerated oxidation rate is not observed after 243 days of oxidation.

Oxide Characterization

The microstructure of the oxide layers was characterized using a Philips EM 420 120-kV for scanning transmission electron microscopy, and the chemical composition data were obtained by energy dispersive X-ray spectroscopy. TEM observations were carried out on cross-sectional thin foils (i.e., the observation plane being perpendicular to the oxide-metal interface), which were prepared using a Balzers ion milling machine (8-keV Ar ions, 200 μA).

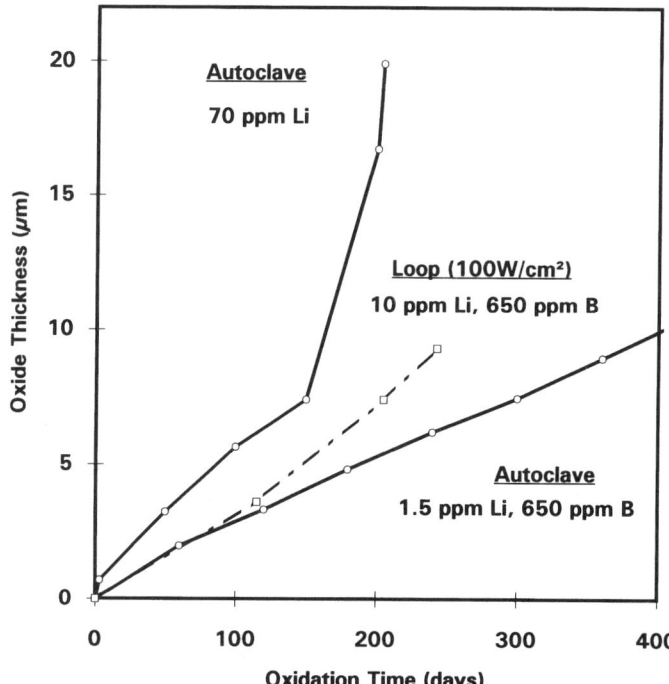

FIG. 1—*Oxidation kinetics of Zircaloy-4 tubing obtained in different lithiated environments.*

Chemical analyses were performed using secondary ion mass spectrometry (SIMSLAB apparatus of V G Instrument), which is a sensitive technique for obtaining the lithium concentration through the oxide thickness. Two types of analysis were conducted: frontal analysis and cross-sectional analysis.

1. Frontal analysis consists of sputtering several atomic layers of the sample surface with a Ga^+ primary ion beam scanning over an area 50 by 50-μm square (Fig. 2a). The positive secondary ions (Li^+, Zr^+, ZrO^+) are extracted from this region by electric fields, energy and mass being analyzed using a mass spectrometer. For this type of analysis, the ion beam and the direction of analysis are perpendicular to the metal-oxide interface (Fig. 2a). The depth resolution is equal to a few nanometers. However, since the metal/oxide interface roughness is rather significant, the lithium profile close to the metal/oxide interface is difficult to estimate. For this reason, frontal analysis is not used to study the lithium distribution close to the metal-oxide interface. It is used mainly for quantitative analyses using a zirconia layer implanted with lithium ions.
2. Concerning cross-sectional analysis, it is performed by scanning the primary ion beam along a line parallel to the surface of the sample (Fig. 2b). In this case, the ion beam is parallel to the metal/oxide interface. The lateral resolution is close to 1 μm, which corresponds to the diameter of the ion beam. To improve the lateral resolution, taper cross sections of the oxide scale were prepared. This special sample preparation consists of performing a mechanical polishing of the surface of the tube with a low angle (Fig. 2c). This type of preparation is used to increase the apparent oxide thickness. With an angle of 4°, 1 μm of real oxide thickness corresponds to 15 μm of apparent oxide thickness. Thus, with an ion beam diameter close to 1 μm, one micron of oxide is described by 15 points of analysis (Fig. 2d). Taper cross sections are used mainly to study the distribution of lithium close to the metal-oxide interface, which is located halfway down the ZrO^+ signal decrease (Fig. 2d).

Experimental Results

Lithium Profile in the Oxide Layer

In order to understand the impact of the water chemistry on the oxidation kinetics and, especially, the strong enhancement of the oxidation rate due to lithium, the lithium distribution in the oxide layers was analyzed by SIMS on one oxide film formed in autoclave with 1.5 ppm Li and 650 ppm B ($\approx$9 μm) and on the following five oxide films formed in autoclave with 70 ppm Li:

1. One "thin" oxide layer (0.7 μm) formed during the pre-transition period (three days).
2. Three "medium" oxide layers (3 to 7 μm) formed during the post-transition period, before the large enhancement of the oxidation rate (50, 100, and 150 days).
3. One thick oxide layer (20 μm) formed during the strong enhancement of the oxidation rate (204 days).

The lithium profiles carried out on each oxide film using taper cross sections and the lithium contents measured using frontal analyses are presented in Fig. 3 and in Table 1, respectively. The main results are the following:

1. Before the large enhancement of the oxidation rate:
   a. In the outer part of the oxide films, the lithium profile is rather flat. The lithium concentration on this plateau depends on exposure time and water chemistry. For the test

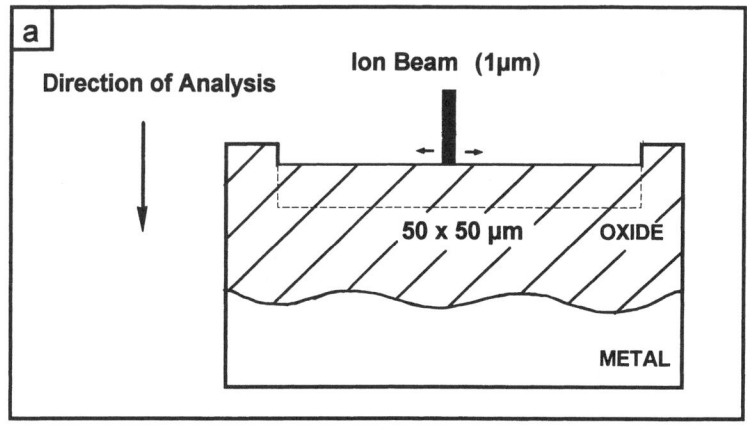

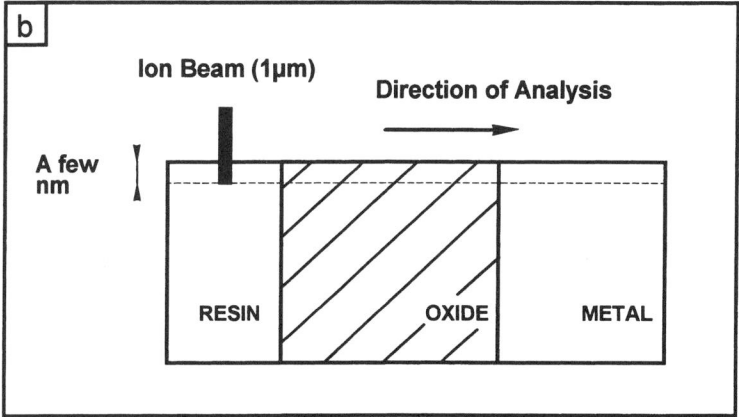

FIG. 2—*Sampling methods used for SIMS analyses:* (a) *frontal analysis,* (b) *cross-sectional analysis,* (c) *taper cross-section analysis,* (d) *type of profile obtained on a taper cross section by scanning the primary ion beam along a line perpendicular to the metal-oxide interface.*

performed with 70 ppm Li, the lithium concentration increases from 120 ppm before the transition time to 240 ppm after the transition time (Table 1). For a similar oxide thickness (7 to 9 μm), the oxide film formed with 1.5 ppm Li (B = 650 ppm) contains about 20 ppm lithium (Fig. 3a), while the oxide film formed with 70 ppm Li (B = 0) contains about 240 ppm Li (Fig. 3b).

b. In the innermost part of the oxide films, the lithium content decreases rapidly (Fig. 3). Close to the metal-oxide interface, the lithium concentrations are below 20 ppm. The barrier layer where the lithium does not penetrate significantly is close to 1 μm for post-transition oxide films. It is, however, reduced to 0.5 μm just before the significant enhancement of the oxidation rate after 150 days of oxidation (Fig. 3b, Table 1).

2. After the large enhancement of the oxidation rate (204 days):

a. The lithium concentration on the plateau increases very rapidly ([Li] $\approx$ 600 ppm) compared to less than 240 ppm before the large enhancement of the oxidation rate (Table 1, Fig. 3b).

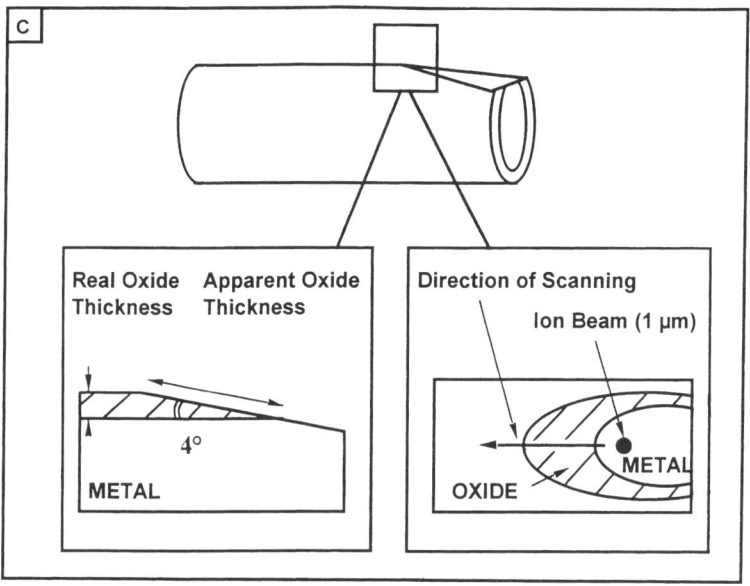

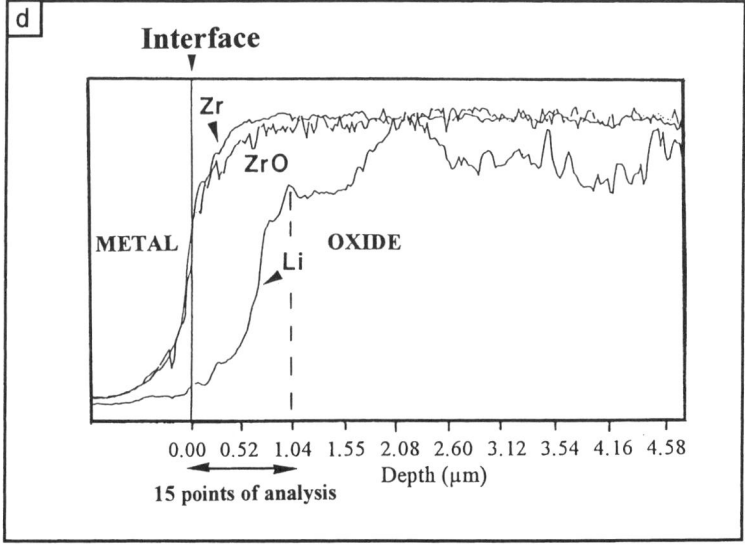

FIG. 2—*Continued*

b. The lithium reaches significantly the metal/oxide interface ([Li] ≈ 240 ppm) (Fig. 3).

The same lithium analyses were also performed on three post-transition oxide films formed in the out-of-pile loop test with 10 ppm Li and 650 ppm B (Table 1). As presented in Fig. 4, the shape of the lithium profiles is similar to that observed on oxide films formed in autoclave (before the large accelerated oxidation rate) with, especially, the existence of a thin inner barrier that limits the lithium ingress towards the metal-oxide interface.

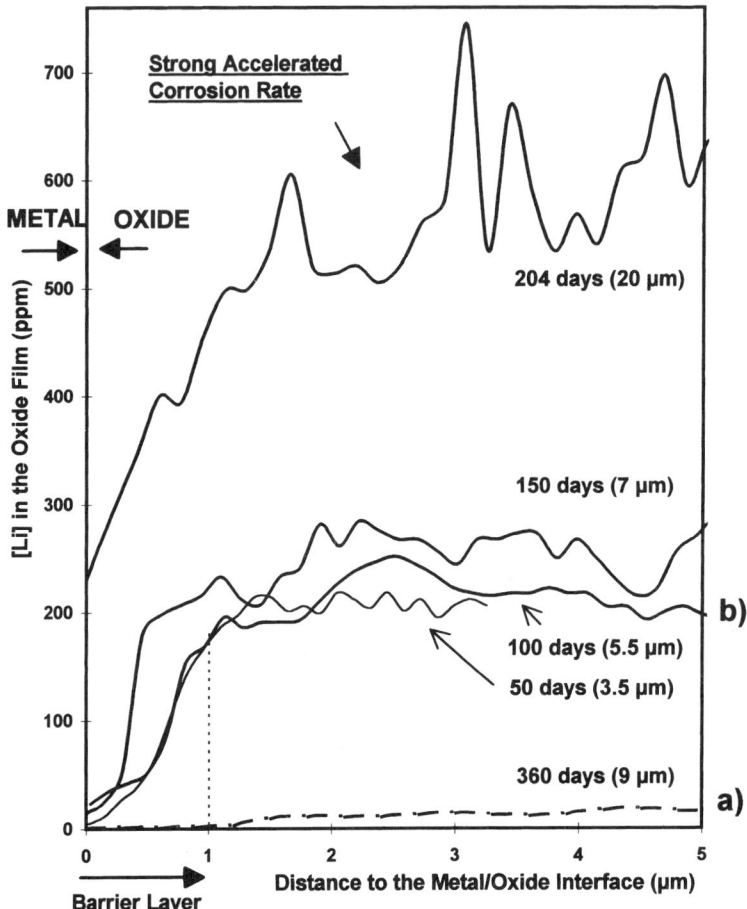

FIG. 3—*Evolution of the lithium profiles in oxide films formed on Zircaloy-4 cladding in autoclave:* (a) *with 1.5 ppm Li and 650 ppm B*, (b) *with 70 ppm Li and 0 B.*

Microstructure of Oxide Layers

To study the origin of the large lithium enhancement of the oxidation rate, two types of oxide layers were analyzed by TEM: oxide films formed before the large accelerated corrosion rate and oxide films grown after the strong enhancement of the oxidation rate.

Oxide Layer Formed Before the Large Enhancement of the Oxidation Rate—The oxide films analyzed were formed in autoclave (Li = 70 ppm, 100 days, 5 µm) and in out-of-pile loop (Li = 10 ppm, B = 650 ppm, 115 and 205 days, 4 and 7 µm, respectively).

Regarding the microstructure of such oxide layers, no significant differences were observed between the oxide films formed in autoclave and those formed in loop. They are both mainly composed of columnar grains of textured monoclinic and quadratic (or cubic)[1] zirconia except

---

[1] Because of the slight difference in the crystal parameter existing between quadratic and cubic zirconia, it was not possible to differentiate these two types of structure by electron diffraction.

TABLE 1—*Characteristics of the lithium distribution and content measured in oxide films formed under various oxidation conditions.*

Oxidation Test	Duration, days	Oxide Thickness, μm	[Li] in Plateau, ppm	Thickness Where the Li Does Not Penetrate Significantly, μm	[Li] Near the Metal/Oxide Interface, ppm
1.5 ppm Li, 650 ppm B, Autoclave test, 633 K	360	9	20	≈1.3	5
70 ppm Li, 0 B, autoclave test, 633 K	3	0.7	120	≈0.6	10
	50	3.5	200	≈1.0	10
	100	5.5	220	≈1.0	20
	150	7	240	≈0.5	20
	204	≈20	600	0	240
10 ppm Li, 650 ppm B, out-of-pile loop test, 636 K	115	3.5	140	≈1	10
	205	7.5	200	≈1	30
	243	10	240	≈1	50

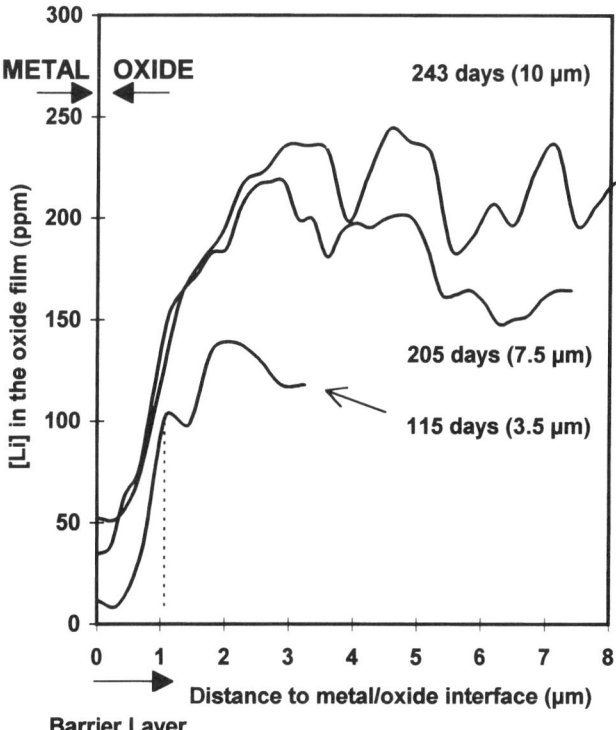

FIG. 4—*Evolution of the lithium profiles in oxide films formed on Zircaloy-4 cladding in out-of-pile loop with 10 ppm Li and 650 ppm B.*

at the outermost part of the oxide film, where the grains are equiaxed (Fig. 5). These columnar grains are perpendicular to the metal-oxide interface (Fig. 5a). Their width is between 20 and 50 nm, and their length is between 100 and 300 nm, the latter decreasing when going towards the oxide-water interface (Fig. 5b). Concerning the equiaxed grains, they are present occasionally on the extreme outer surface of the oxide film over a width below 200 nm (Fig. 5c). Their mean diameter is close to 20 nm, and their intergranular cohesion is weak. Such equiaxed grains also are observed occasionally close to circumferential cracks in the oxide layer.

Oxide Layer Formed after the Large Enhancement of the Oxidation Rate—The oxide layer analyzed was formed in autoclave with 70 ppm lithium during 204 days.

1. The oxide grains formed before the strong accelerated corrosion rate (i.e., located in the outer part of the oxide film) are not modified. They are mainly columnar except at the outermost part of the oxide film where they are equiaxed (Fig. 6a).
2. On the contrary, the new oxide grains formed after the strong accelerated corrosion rate (i.e., located in the inner part of the oxide layer at more than 15 μm from the water-oxide interface) are mainly equiaxed (Fig. 6b). These equiaxed grains are smaller than the columnar grains, their mean length being close to 50 nm (Fig. 7a). Moreover, their intergranular cohesion is weak, open grain boundaries being observed around the oxide grains. This type of oxide grains is observed up to the metal-oxide interface (Fig. 7b).

A summarize of the results obtained by SIMS and TEM is presented in Fig. 8.

Discussion

Incorporation of Lithium in the Oxide Layer

Before the strong enhancement of the oxidation rate, the lithium concentration close to the metal-oxide interface is very low ($\leq$20 ppm) (Fig. 3 and Fig. 4). During the oxide growth, the lithium ingress towards the metal-oxide interface is thus impeded by an inner oxide layer. Its thickness is close to 1 μm (Table 1). Interestingly, it is similar to the thickness of the barrier layer measured at the metal-oxide interface by Cox [15] on samples oxidized with 0.01 to 1 M LiOH at 573 K and close to the thickness of a high-conductivity region at the metal-oxide interface [20]. Moreover, at the beginning of the strong accelerated corrosion rate observed with 70 ppm Li, this thickness decreases from 1 to 0.5 μm (Table 1). Based on these observations, it can be assumed that the inner layer, where lithium does not penetrate significantly, corresponds to the dense barrier layer at the metal-oxide interface, which limits the oxidation rate. Thus, this dense inner layer, which is known to be a protective barrier against corroding species, could also act as a protective barrier towards lithium incorporation.

Regarding the lithium content in the outer part of the oxide film, it depends on exposure time, on water chemistry conditions, and on the oxidation rate (Table 1).

Based on earlier unpublished results, two forms of lithium appear to exist in the oxide layer: a form incorporated in the zirconia network and another contained in the porosity. This observation is supported by other works [9], which show that diluted acid nitric leach of a thick oxide film formed with 3500 ppm Li and 600 ppm B induces a decrease of its lithium content from 820 ppm to 100 to 120 ppm. In addition, the fraction of the leachable form (in the acid nitric solution) appears to depend on the type of the oxide film since the same acid leach performed on thin pre-transition oxide films (i.e., low porosity level) does not affect the lithium concentration profiles [9].

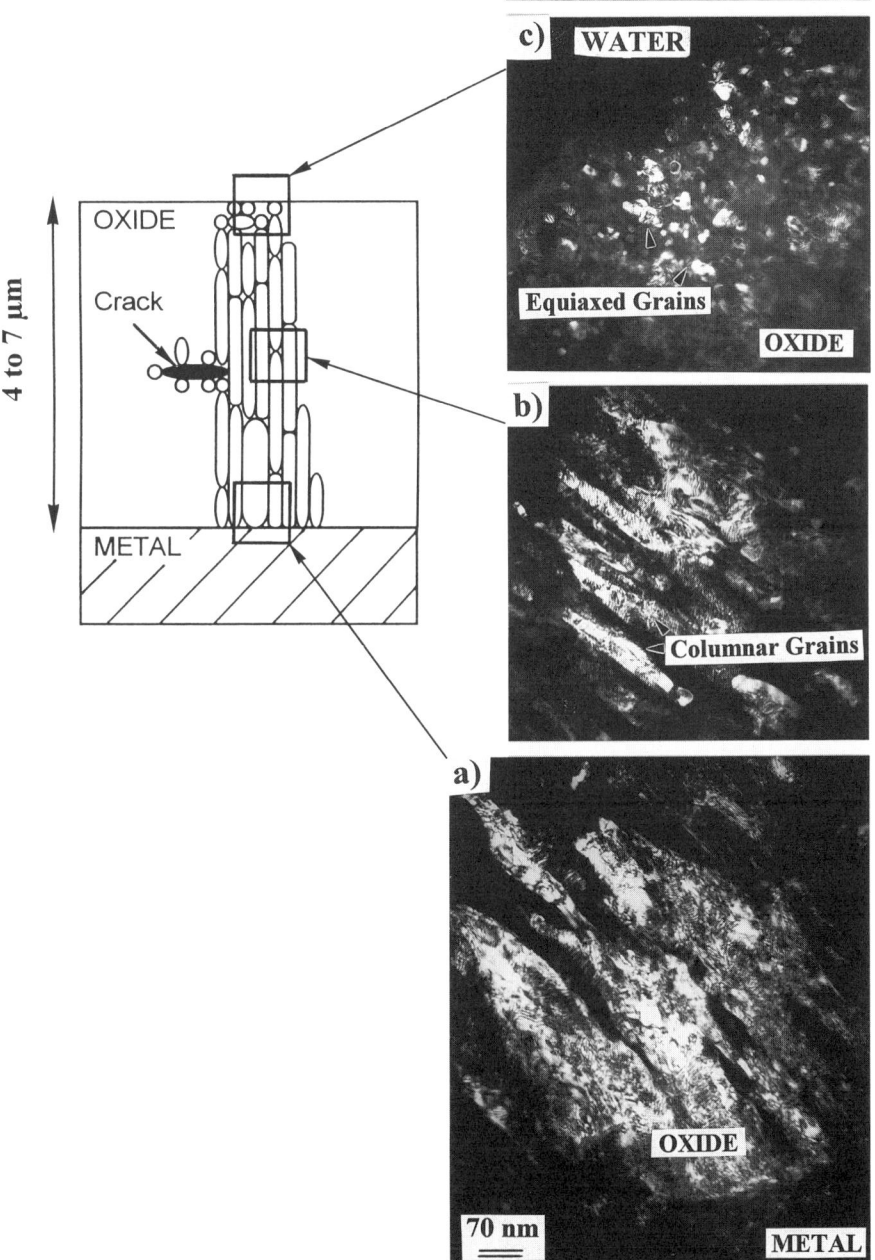

FIG. 5—*Oxide layer formed before the strong enhancement of the oxidation rate after 100 days of oxidation in autoclave with 70 ppm Li (TEM examination in dark field):* (a) *columnar grains observed at the metal-oxide interface,* (b) *columnar grains observed in the bulk of the oxide film,* (c) *fine-equiaxed grains observed occasionally on the extreme outer surface.*

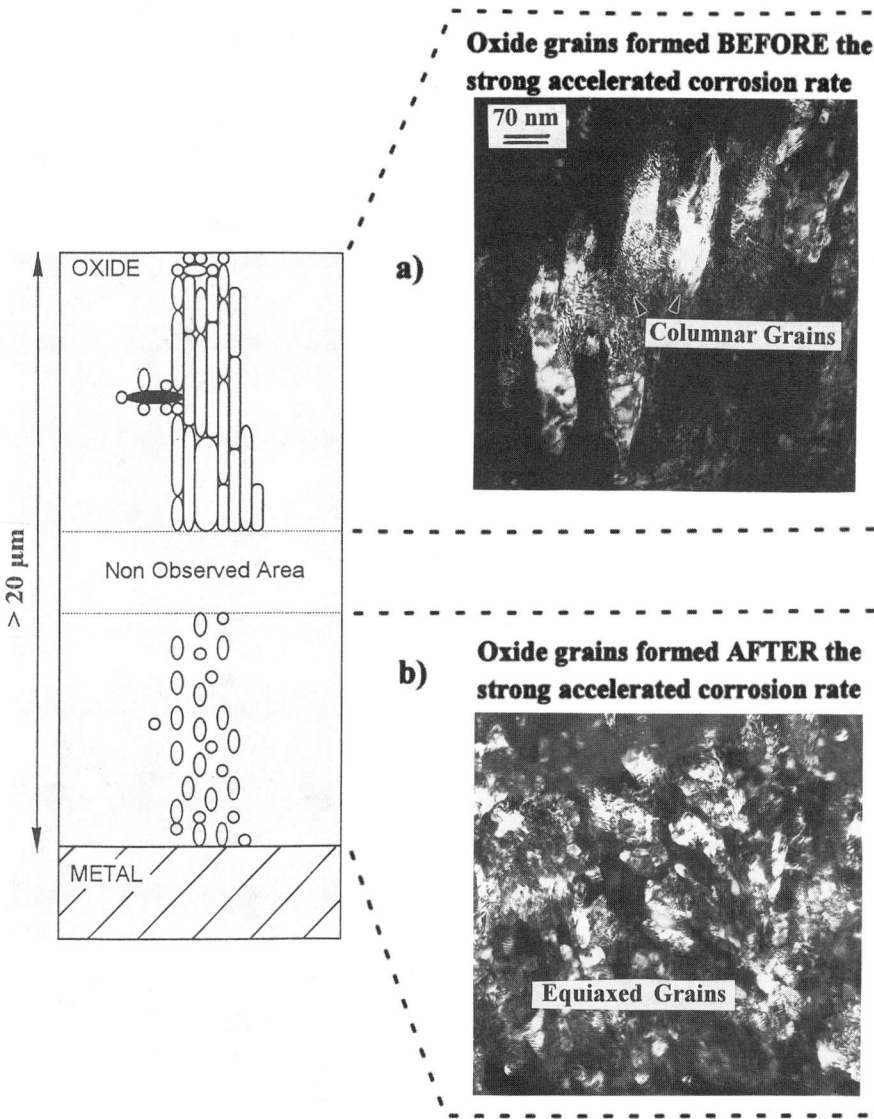

FIG. 6—*Oxide layer formed after the strong enhancement of the oxidation rate after 204 days of oxidation in autoclave with 70 ppm Li (TEM examination in dark field):* (a) *columnar grains observed on the outer part of the oxide layer up to 9 μm from the water-oxide interface at least,* (b) *equiaxed grains observed in the inner part of the oxide layer up to the metal-oxide interface.*

In the present work, leaching in dilute nitric acid has not been performed systematically yet. However, it is clear that before the strong accelerated corrosion rate, most of the lithium has a strong interaction with the oxide grains since after leaching of a post-transition oxide film (7 μm, 200 ppm Li) in dilute acid nitric, no large fraction of lithium is leached out (only 10%). On the contrary, after the strong accelerated corrosion rate, a large fraction of lithium has a

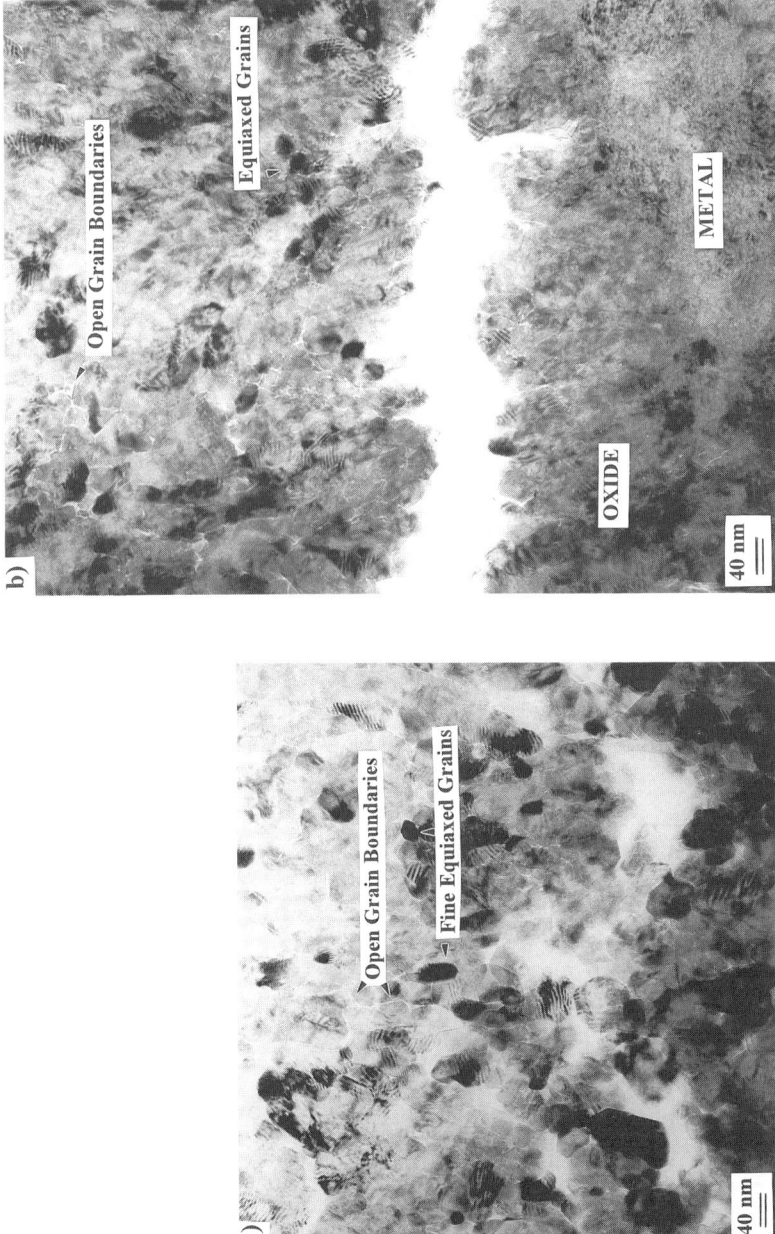

FIG. 7—*Equiaxed grains with open grain boundaries observed by TEM (bright field): (a) at about 2 µm from the metal-oxide interface, (b) at the metal-oxide interface. The oxide film analyzed was formed in autoclave with 70 ppm Li after strong enhancement of the oxidation rate (204 days).*

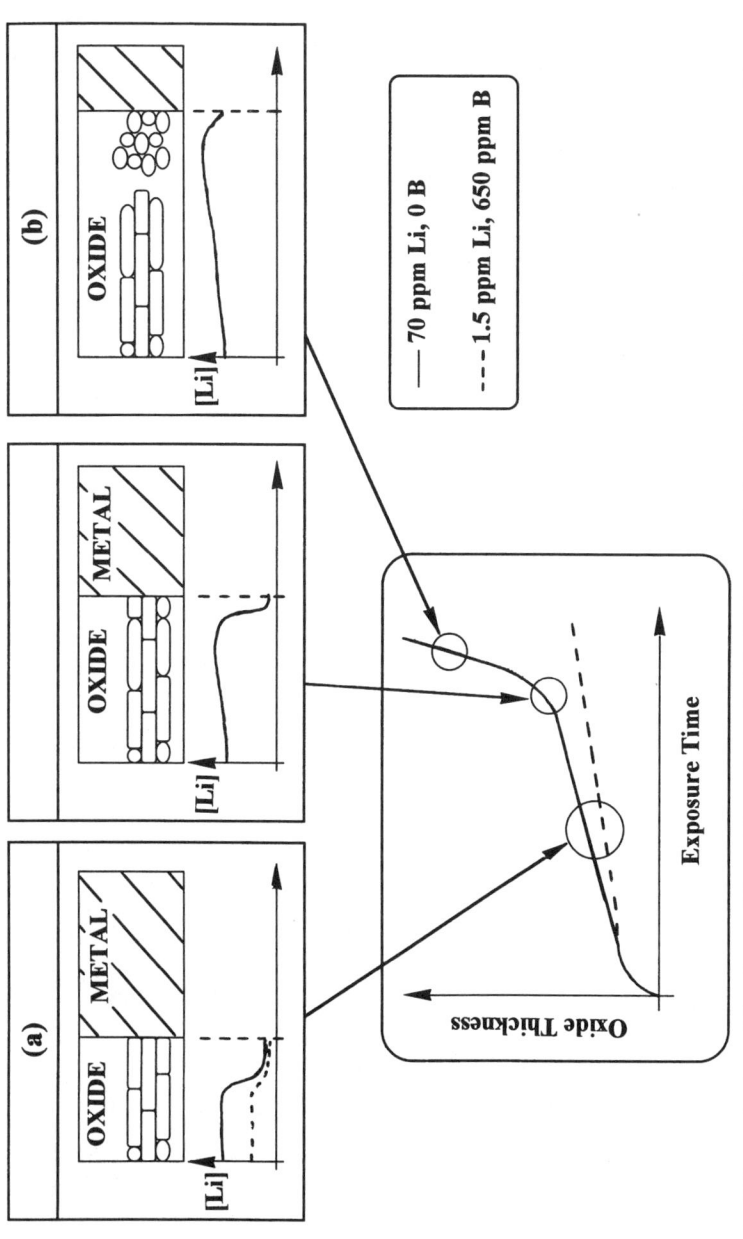

FIG. 8—*Schematic view of the shape of the oxide grains and of the lithium profile in oxide layers formed during the same oxidation test: (a) before the strong enhancement of the oxidation rate, (b) after such an enhancement.*

weak interaction with the oxide grains since after leaching of a thick oxide film in dilute acid nitric (20 μm, 600 ppm Li), a large fraction of lithium is easily leached out (>60%). This large fraction of leachable lithium can be correlated to the existence of a new type of oxide grains formed in this thick oxide film: fine-equiaxed grains with open grain boundaries.

Impact of LiOH on the Oxide Morphology

Before the occurrence of a strong accelerated oxidation rate, the oxide film formed in the lithiated environment is composed mainly of columnar grains similar to those observed in oxide films formed in steam or in pure water [20–22]. However, after such an occurrence, equiaxed grains with a weak intergranular cohesion are formed in the bulk of the oxide layer up to the metal-oxide interface (see Fig. 6). This result is in good agreement with the observations of Beie [20], who noticed a large fraction of equiaxed grains with open grain boundaries in an oxide layer formed on Zircaloy-4 material with 70 ppm Li after 294 days (623 K, 16.5 MPa). As presented in Fig. 3, additional chemical analyses show that the presence of such equiaxed grains is correlated to a high lithium content at the metal-oxide interface (240 ppm). On the other hand, such equiaxed grains are observed only after large enhancement of the oxidation rate. Therefore, the appearance of equiaxed grains in the oxide layer can be derived from two interpretations: the high oxidation rate or a direct effect of the lithium present at the metal-oxide interface.

1. The first interpretation is based on a correlation noticed between oxidation rate and oxide morphology. It is the result of the usual observation of oxide films formed with columnar grains during unaccelerated growth kinetics and oxide films formed with equiaxed grains during accelerated growth kinetics. Such equiaxed grains have been observed in oxide layers formed on materials with low corrosion resistance [21,22] or in materials resulting from severe oxidation conditions such as high oxygen levels or nodular corrosion [18]. Moreover, they have been also observed occasionally on the extreme outer surface of oxide film (Fig. 5c) growing during the first steps of the pre-transition period at a rapid oxidation rate. Therefore, the appearance of equiaxed grains in the inner part of the oxide film formed after 204 days of oxidation with 70 ppm Li could be the consequence of the high oxidation rate. The presence of equiaxed grains with a weak intergranular cohesion would allow the lithium to reach the metal-oxide interface, the access of lithium to the metal-oxide interface being the consequence of the existence of equiaxed grains.
2. The second interpretation is based on a direct interaction between lithium and the oxide grains. According to Ramasubramanian [9], the diametral and columnar growth of oxide crystallites could be impeded by OLi groups, resulting in an increase of the diffusional area of oxidizing species and in accelerated corrosion rates. Based on this second hypothesis, the access of lithium to the metal-oxide interface would then be at the origin of the equiaxed grains.

In addition, as the surface oxide thickens to more than 20 μm, the oxide will absorb not only lithium but also hydrogen. Thus, if the hydrogen content located at the metal-oxide interface is high, hydrogen could also be involved in the modification of the oxide grain morphology. However, the results of the measurements of the hydrogen content performed before and after the strong accelerated corrosion rate to confirm this hypothesis are not known at the present time.

Concerning the impact of LiOH on the oxide morphology, it is worth noting that, in the outer part of the oxide film formed after the strong accelerated corrosion rate, the oxide grains are not modified. They are still columnar as before the large enhancement of the oxidation rate (Fig. 6a). This means that, contrary to what is observed at 573 K with more than 700 ppm Li in the coolant [14,15], after 204 days of oxidation at 633 K, no dissolution of the zirconia crystallites has occurred on the surface of the oxide film formed with 70 ppm Li.

Thus, the impact of LiOH on the microstructure of the oxide layer can either be a dissolution of zirconia at the water-oxide interface or a modification of the morphology of the new oxide grains formed at the metal-oxide interface. These two mechanisms clearly depend on the water chemistry conditions and on the exposure time. The dissolution process appears to be activated for lithium levels in the coolant above 700 ppm within a few days (at 573 K), whereas the new oxide nucleation (or growth) process can be observed for lithium contents less than 70 ppm after a given incubation time (at 633 K).

It can be added that the dissolution process can, however, probably occur even for lithium content in the water below 700 ppm depending on the lithium level in the oxide layer. This assumption is based on the analysis of oxidized precipitates incorporated in an oxide film formed in the out-of-pile loop test (20 ppm Li, high oxidation rate). In this oxide film composed mainly of fine-equiaxed grains with a weak intergranular cohesion, ten oxidized precipitates were observed at a few microns from the metal-oxide interface. Surprisingly, on the one hand, the zirconium content of five precipitates was very low, a few percents with respect to a few ten percents for precipitates located in the metallic matrix and for the other oxidized precipitates (Fig. 9). On the other hand, the intergranular cohesion of the nanocrystallites present in these oxidized precipitates was very weak, and no cubic zirconia was identified. These observations suggest that a preferential dissolution of cubic zirconia occurred in these precipitates. Since the lithium content in such oxidized precipitates is sharply greater than in the surrounded matrix [3], we can suppose that this dissolution of cubic zirconia is due to very high lithium levels present in the precipitate (>1000 ppm). This result concerning the dissolution of cubic zirconia in precipitates could then be extended to the cubic zirconia present in the oxide matrix provided that the lithium content in the oxide layer is high enough.

Impact of LiOH on the Oxidation Kinetics

As presented in Fig. 1, before the strong accelerated corrosion rate, the post-transition corrosion rate is slightly increased by 70 ppm Li in the coolant with respect to 1.5 and 650 ppm B. This trend is consistent with the observations of Mac Donald [11], who noticed a slight difference in the post-transition oxidation rate between samples oxidized with 70 ppm lithium and samples oxidized in pure water.

This slight effect of the lithium (70 ppm Li) on the post-transition rate is not only due to a pH impact since the tests performed with KOH instead of LiOH (at equivalent coolant pH) lead to no significant evolution of the oxidation kinetics [6,23]. Based on TEM analyses, it is not correlated to a significant evolution of the oxide morphology. Moreover, it is not due to a significant increase of the anionic vacancies in the dense inner layer since, based on SIMS analyses, the lithium content in this inner barrier layer is very low ($\leq$20 ppm, see Fig. 3). In fact, it has to be correlated to a high lithium concentration in the plateau (200 ppm), which could modify the characteristics of the inner barrier layer. The incorporation of cationic species from the coolant to the oxide layer appears to have a significant influence on the oxidation process. This assumption is supported, on the one hand, by the absence of any significant acceleration of the oxidation kinetics in the KOH environment linked to the difficult ingress of potassium in the oxide layer [23] and, on the other hand, by other works dealing with the influence of alkali metal hydroxides on the corrosion of Zr-based alloys [6].

During the autoclave test with 70 ppm Li, a strong enhancement of the oxidation occurs after

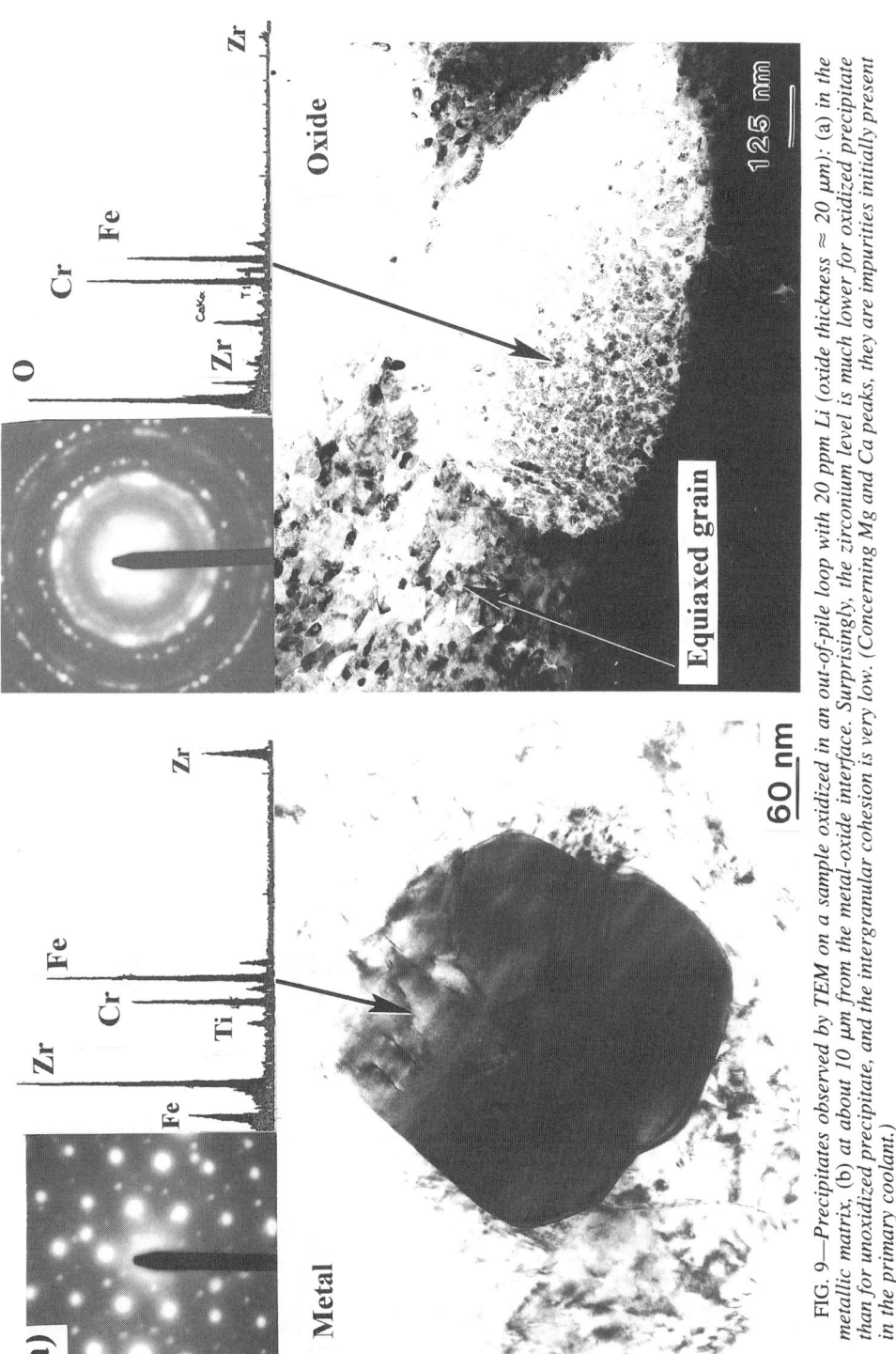

FIG. 9—*Precipitates observed by TEM on a sample oxidized in an out-of-pile loop with 20 ppm Li (oxide thickness ≈ 20 μm): (a) in the metallic matrix, (b) at about 10 μm from the metal-oxide interface. Surprisingly, the zirconium level is much lower for oxidized precipitate than for unoxidized precipitate, and the intergranular cohesion is very low. (Concerning Mg and Ca peaks, they are impurities initially present in the primary coolant.)*

a given incubation time (150 to 200 days, see Fig. 1). At this time, lithium and equiaxed grains with a weak intergranular cohesion are present at the metal-oxide interface. According to these observations, it is clear that the dense barrier layer, present at the metal-oxide interface before the large enhancement of the oxidation rate, is disappearing. And this is the reason why the oxidation rate is so high.

Although the origin of the disappearance of this dense inner barrier layer is not known, it is clear that it is not due to a dissolution of the outer part of the oxide film (see previous section). Since the thickness of the barrier, where lithium does not penetrate, has decreased just before the large enhancement of the oxidation rate (from 1 to 0.5 μm, see Table 1), it can be assumed that the protective inner barrier decreases gradually before disappearing.

In summary, based on the present work, it is clear that the strong accelerated corrosion rate caused by LiOH is due to the disappearance of the inner barrier layer. However, to improve our knowledge of the lithium effect on the oxidation kinetics, we now need to study the origin of the disappearance of this inner barrier layer and to learn, especially: (1) if it is a gradual phenomena as proposed or a threshold effect, (2) if it is a chemical or a mechanical effect, and (3) which types of species are involved (LiOH, Li^+, OH^-, H).

Conclusions

In this work, Zircaloy-4 cladding materials have been oxidized in different types of lithiated environments. Depending on water chemistry conditions and exposure time, a strong accelerated corrosion rate can occur. The systematic SIMS and TEM analyses performed on the same oxide films formed after different exposure times have brought the following results:

1. Before the strong enhancement of the oxidation rate:
   a. A thin inner barrier layer impedes the lithium ingress towards the metal-oxide interface.
   b. The lithium content in the oxide film depends on the water chemistry conditions.
   c. The oxide films are composed mainly of long columnar grains with a good intergranular cohesion.

2. After the strong enhancement of the oxidation rate:
   a. No inner barrier layer limits the lithium ingress towards the metal-oxide interface.
   b. The lithium content in the oxide film has strongly increased, but only a small fraction of lithium has a strong interaction with the oxide grains.
   c. The columnar oxide grains located in the outer part of the oxide film (i.e., those formed before the strong accelerated corrosion) are not modified. On the contrary, the new oxide grains formed at the metal-oxide interface are equiaxed and smaller. Moreover, the intergranular cohesion is weak.

Based on these results, it is suggested that:

1. The dense inner layer, which controls the oxidation rate, impedes the lithium ingress towards the metal-oxide interface.
2. The enhancement of the oxidation rate observed in lithiated environment is due to a gradual degradation of the dense inner layer as a consequence of the incorporation of lithium in the oxide film. After a given time, depending on water chemistry conditions, this inner barrier layer is disappearing, leading to strong oxidation rates related to a new oxide grain morphology at the metal-oxide interface (equiaxed grains with open grain boundaries).

Acknowledgments

The authors would like to thank Science et Surface Society (Lyon—France) for SIMS and TEM analyses, particularly F. Bossut for his expert thin foils preparation, and A. Blanc (CEA-DRN/DEC/SECA Cadarache-France) for his contribution to the out-of-pile loop test.

References

[1] Anthoni, S., Ridoux, P., Menet, O., and Weber, C., "Water Chemistry of Nuclear Reactor Systems," British Nuclear Energy Society (BNES), Bournemouth, UK, 1992, p. 9.
[2] Peybernès, J., Doctoral thesis, Université de Provence (Aix-Marseille I), Marseille, France, July 1994.
[3] Billot, P., Robin, J. C., Giordano, A., Peybernès, J., Thomazet, J., and Amanrich, H. in *Zirconium in the Nuclear Industry, Tenth International Symposium, ASTM STP 1245*, A. M. Garde and E. R. Bradley, Eds., American Society for Testing and Materials, West Conshohocken, PA, 1994, p. 351.
[4] Bramwell, I. L., Parsons, P. D., and Tice, D. R. in *Zirconium in the Nuclear Industry, Ninth International Symposium, ASTM STP 1132*, C. M. Eucken and A. M. Garde, Eds., American Society for Testing and Materials, West Conshohocken, PA, 1990, p. 628.
[5] Dumont, A., Thomazet, J., Charquet, D., and Sevenat, J., *Proceedings,* 3ème Colloque International sur la Contribution des Expertises des Matériaux à la Résolution des Problèmes Rencontrés dans les REP, Société France d'Energie Nucléaire (SFEN), Fontevraud, France, 12–16 Septembre 1994.
[6] Jeong, Y. H., Ruhmann, H., and Garzarolli, F., *Proceedings,* AIEA Technical Meeting on Influence of Water Chemistry on Fuel Cladding Behaviour, Rez, Czech Republic, 4–8 Oct. 1993.
[7] Perkins, R. A. and Busch, R. A. in *Zirconium in the Nuclear Industry, Ninth International Symposium, ASTM STP 1132*, C. M. Eucken and A. M. Garde, Eds., American Society for Testing and Materials, West Conshohocken, PA, 1990, p. 595.
[8] Cox, B. and Wong, Y. M. in *Zirconium in the Nuclear Industry, Ninth International Symposium, ASTM STP 1132*, C. M. Eucken and A. M. Garde, Eds., American Society for Testing and Materials, West Conshohocken, PA, 1990, p. 643.
[9] Ramasubramanian, N. and Balakrishnan, P. V. in *Zirconium in the Nuclear Industry, Tenth International Symposium, ASTM STP 1245*, A. M. Garde and E. R. Bradley, Eds., American Society for Testing and Materials, West Conshohocken, PA, 1994, p. 378.
[10] Han, J. H. and Rheem, K. S., *Journal of Nuclear Materials,* Vol. 217, 1994, p. 197.
[11] Mac Donald, S. G., Sabol, G. P., and Sheppard, K. D. in *Zirconium in the Nuclear Industry, Sixth International Symposium, ASTM STP 824*, D. G. Franklin and R. B. Adamson, Eds., American Society for Testing and Materials, West Conshohocken, PA, 1984, p. 519.
[12] Sabol, G. P., Kilp, G. R., Balfour, M. G., and Roberts, E. in *Zirconium in the Nuclear Industry, Eighth International Symposium, ASTM STP 1023*, L. F. P. Van Swam and C. M. Eucken, Eds., American Society for Testing and Materials, West Conshohocken, PA, 1989, p. 227.
[13] Hillner, E. and Chirigos, J. N., WAPD-TM6307, Westinghouse Electric Corp., Bettis Atomic Laboratory, Pittsburgh, PA, August 1962.
[14] Cox, B. and Wu, C., *Journal of Nuclear Materials,* Vol. 199, 1993, p. 272.
[15] Cox, B. and Wu, C., *Journal of Nuclear Materials,* Vol. 224, 1995, p. 169.
[16] Ramasubramanian, N., Precoanin, N., and Ling, V. C. in *Zirconium in the Nuclear Industry, Eighth International Symposium, ASTM STP 1023*, L. F. P. Van Swam and C. M. Eucken, Eds., American Society for Testing and Materials, West Conshohocken, PA, 1989, p. 187.
[17] Ramasubramanian, N. in *Zirconium in the Nuclear Industry, Ninth International Symposium, ASTM STP 1132*, C. M. Eucken and A. M. Garde, Eds., American Society for Testing and Materials, West Conshohocken, PA, 1990, p. 613.
[18] Garzarolli, F., Seidel, H., Tricot, R., and Gros, J. P. in *Zirconium in the Nuclear Industry, Ninth International Symposium, ASTM STP 1132*, C. M. Eucken and A. M. Garde, Eds., American Society for Testing and Materials, West Conshohocken, PA, 1990, p. 395.
[19] Mardon, J. P., Charquet, D., and Senevat, J. in *Zirconium in the Nuclear Industry, Tenth International Symposium, ASTM STP 1245*, A. M. Garde and E. R. Bradley, Eds., American Society for Testing and Materials, West Conshohocken, PA, 1994, p. 328.
[20] Beie, H. J., Mitwalsky, A., Garzarolli, F., Ruhmann, H., and Sell, H. J. in *Zirconium in the Nuclear Industry, Tenth International Symposium, ASTM STP 1245*, A. M. Garde and E. R. Bradley, Eds., American Society for Testing and Materials, West Conshohocken, PA, 1994, p. 615.
[21] Warr, B. D., Elmoselhi, M. B., Newcomb, S. B., Mac Intyre, N. S., Brennenstuhl, A. M., and

Lichtenberger, P. C. in *Zirconium in the Nuclear Industry, Ninth International Symposium, ASTM STP 1132,* C. M. Eucken and A. M. Garde, Eds., American Society for Testing and Materials, West Conshohocken, PA, 1990, p. 740.

[22] Wadman, B., Laï, Z., Andrén, H. O., Nyström, A. L., Rudling, P., and Pettersson, H. in *Zirconium in the Nuclear Industry, Tenth International Symposium, ASTM STP 1245,* A. M. Garde and E. R. Bradley, Eds., American Society for Testing and Materials, West Conshohocken, PA, 1994, p. 579.

[23] Pêcheur, D., Giordano, A., Picard, E., and Billot, P., *Proceedings,* AIEA Technical Meeting on "Influence of Water Chemistry on Fuel Cladding Behaviour," 4–8 Oct. 1993, Rez, Czech Republic.

DISCUSSION

H. M. Chung[1] (written discussion)—Breakaway oxidation of Zircaloy-4 is well known to occur in environments not containing Li. The addition of some elements such as Nb is also known to suppress breakaway oxidation. Your implication of Li in promoting formation of the equiaxed oxide (at the metal/oxide boundary) and breakaway oxidation does not seem consistent with these observations.

D. Pecheur et al. (authors' closure)—This work does not deal with the implication of Li in promoting the breakaway oxidation that occurs for Zircaloy-4 material for 2 to 3-µm-thick oxide. It deals with the implication of lithium in the strong accelerated corrosion rate that occurs in the post-transition period for oxide film thicker than 7 µm.

B. Warr[2] (written discussion)—Could a "matrix" effect due to a changing oxide structure have changed the observed Li concentration in the oxide? In a paper in the next session we will show the presence of a matrix effect in Zr-2.5 Nb pressure tube material.

D. Pecheur et al. (authors' closure)—To take the matrix effect into account, we have normalized Li concentration with respect to the Zr signal, assuming that the matrix effect, which is due to changing oxide structure, will affect the Zr signal, too. Since the Li/Zr ratio is higher in the inner oxide scale with equiaxed grains than in the outer oxide scale with columnar grains, we can suppose that the Li concentration is different for these two types of oxide layers.

Dr. Young S. Kim[3] (written discussion)—Your results showed that the outer oxide scale had columnar structure, while the inner oxide scale consists mainly of equiaxed grains after the enhanced corrosion occurred. My question is: What is the percentage of columnar grains on the outer oxide scale formed after enhanced corrosion?

D. Pecheur et al. (authors' closure)—No statistics study was performed concerning the percentage of columnar grains in the oxide film analyzed by TEM. However, based on TEM observations, it appears that the quantity of columnar grains in the outer oxide scale formed after the enhanced corrosion rate is similar to that in the oxide films formed before the accelerated oxidation rate.

Brian Cox[4] (written discussion)—Your overheads gave the impressions that you have comparative data in pure water. Do you find that post-transition corrosion rates in LiOH (1.5 ppm Li) and H_3BO_3(650 ppm B) solutions are higher, the same as, or lower than those in pure water?

D. Pecheur et al. (authors' closure)—No specific oxidation tests were performed to compare the oxidation kinetics in pure water and in water chemistry conditions close to PWRs. However, based on data reported in the literature and on unpublished results, it can be assumed that the oxidation rate in pure water is slightly higher than with 1.5 ppm Li and 650 ppm B. Typically, the oxidation rate of stress-relieved annealed Zircaloy-4 tubing (1.5% Sn) is close to 0.55 mg/dm^2/day in pure water at 633 K and close to 0.35 mg/dm^2/day with 1.5 ppm Li and 650 ppm B.

---

[1] Avgonne National Laboratory, Avgonne, IL.
[2] Ontario Hydro, Toronto, Ontario, Canada.
[3] Korea Atomic Energy Research Institute, Daejom, Korea.
[4] University of Toronto, Toronto, Canada.

Brian Cox,[1] Mihaela Ungurelu,[1] Yin-Mei Wong,[1] and Chenguang Wu[1]

Mechanisms of LiOH Degradation and H_3BO_3 Repair of ZrO_2 Films

REFERENCE: Cox, B., Ungurela, M., Wong, Y.-M., and Wu, C., "**Mechanisms of LiOH Degradation and H_3BO_3 Repair of ZrO_2 Films,**" *Zirconium in the Nuclear Industry: Eleventh International Symposium, ASTM STP 1295*, E. R. Bradley and G. P. Sabol, Eds., American Society for Testing and Materials, 1996, pp. 114–136.

ABSTRACT: During a program to elucidate the mechanisms by which LiOH accelerates the corrosion of zirconium alloys and boric acid inhibits this effect, specimens were exposed to 0.01, 0.1, and 1.0 M LiOH solutions at 300°C (573 K) or 360°C (673 K) with and without the addition of boric acid. Results showed that local dissolution of the ZrO_2 films formed pores whose depth was a function of the LiOH concentration and probably also of the temperature, alloy composition, and structure. Below a critical LiOH concentration, only superficial porosity was developed in short experiments. Above this critical concentration (which lies between 0.1 and 1.0 M LiOH for Zircaloy-2 at 300°C) porosity develops throughout the initially impervious oxide and no pretransition corrosion kinetics are observed. Below this critical concentration, no effect of LiOH is observed on the pretransition oxidation kinetics until pores and cracks start to develop in the oxide prior to the oxidation rate transition. At this point (1.5 to 1.8 μm) LiOH can concentrate in the freshly developed pores by chemical extraction of water from the solution in the pores to form new ZrO_2. Once the critical LiOH concentration is reached in the pores, enlargement or extension of the pores can occur by dissolution. This process should occur beneath the relatively untouched pretransition oxide when the bulk solution is not concentrated enough to attack the oxide surface. Hydrothermal redeposition of much of the dissolved ZrO_2 occurs on specimen surfaces or within the porous oxide. Boric acid has no effect on ZrO_2 dissolution by LiOH. It is considered that in concentrated solutions the solubility product of some complex lithium zirconate borate can be exceeded and this can plug the pores. In dilute solutions, therefore, boric acid can only operate inside the porous oxide film, where the chemical concentration mechanism should be equally effective for both LiOH and H_3BO_3. Any Li or B found subsequently in the film will be there as a consequence and not a cause of the corrosion process. It would be expected to occur in at least two forms, Li or B within pores or adsorbed on pore walls, and Li or B that is incorporated in hydrothermally deposited oxide. These two forms of doping would be "leachable" and "non-leachable," respectively.

KEYWORDS: zirconium alloys, corrosion, high-temperature water, LiOH, boric acid

Lithium hydroxide is one of only two chemicals that seriously degrade the corrosion resistance of zirconium alloys when present at moderate concentrations in high-temperature water [*1–9*]; the other is fluoride [*10,11*]. Whereas all fluorides apparently have serious effects, other alkalis have only minor and decreasing effects in the order LiOH > NaOH > KOH > NH_4OH [*1*]. Furthermore, no other salt of lithium apparently has any effect on zirconium alloy corrosion. Thus, the phenomenon is neither an effect of pH, [OH^-], nor of Li^+ but is a property of LiOH.

---

[1] Centre for Nuclear Engineering, University of Toronto, 184 College Street, Toronto, Ontario, M5S 3E4, Canada. Brian Cox is professor of nuclear engineering and chairman of the Centre.

Furthermore, it is a property of LiOH solutions in water since no effects are observed in fused LiOH [*12*].

Attempts to explain the phenomenon have concentrated on the evidence for an apparent incorporation of Li^+ in oxide films on material undergoing enhanced corrosion in LiOH solutions [*2,6,13–17*]. However, none of these studies have demonstrated that this Li^+ incorporation is the cause rather than the effect of the enhanced corrosion. Some have now shown that the correlation between Li^+ concentration in the oxide and corrosion rate is poor [*14,15*]. A simple inspection of corrosion curves shows no effect of LiOH on pretransition corrosion [*7*]; this is sufficient to show that Li^+ in the oxide cannot be affecting oxygen diffusion.

Enhanced corrosion in LiOH solutions is observed to be inhibited by the presence of boric acid [*18–20*], commonly added to the PWR primary coolant to control reactivity during each fuel cycle. Boric acid contents are high at the start of a cycle and are progressively reduced to zero at the cycle end [*21*]. The corrosion of fuel cladding alloys in reactors with coolant chemistry control using LiOH accelerates as the burnup increases and may require removal of fuel before the end of its reactivity capability if maximum oxide thicknesses exceed some selected value such as 100 μm [*22,23*]. By careful selection of the thermal hydraulic parameters used in modeling these effects, it is possible to explain this accelerating corrosion based on the rise in the temperature at the oxide/metal interface [*24*].

In reactor operating cycles where elevated LiOH concentrations (3.5 ppm) have been used [*25–29*] and in loop tests at similar or higher LiOH concentrations [*30*], no definite evidence for an effect of LiOH concentration has been visible, although in several instances it has been argued to be present. Calculations of the elevation of the boiling point of water by LiOH [*31*] show that, in the presence of thick oxide films, the temperature rise across the oxide film is more than enough to concentrate LiOH to more than 1 m at the bottoms of pores in such an oxide. Reaching equilibrium concentrations such as this presupposes minimal mixing between the solution in the pores in the oxide and the bulk solution, and at present the extent to which such mixing prevents the attainment of the theoretically high LiOH concentrations in thick porous oxides is not known. Alternatively, perhaps the interface temperatures are high enough to produce almost anhydrous LiOH at the interface, and perhaps this would not cause enhanced corrosion [*12*]. It has been argued that the failures to clearly observe any effects of increasing the LiOH concentration in the primary coolant is evidence for the absence of a lithium hydroxide concentration effect. However, the theoretical LiOH concentration attainable at the metal/oxide interface is a function only of the elevation of the boiling point constant and the temperature at the interface (if the porosity extends there). It is not dependent on the initial LiOH concentration in the primary coolant.

For the past six years at the Centre for Nuclear Engineering, a program aimed at elucidating the mechanism of enhanced corrosion in LiOH and its inhibition by boric acid has been pursued [*32–35*]. With the recent measurements of the solubility of ZrO_2 in 300°C (573 K) LiOH solutions, this work has been de-emphasized, so it appears useful at this time to assess what has been learned, and to try to reach some conclusions on whether or not this changes the picture of what is happening in reactor and results in any suggestions for improved water chemistry.

Experimental

Specimens were approximately 2 by 3 cm parallelepipeds cut from a batch of Zircaloy-2 sheet or short (1-cm) lengths of standard Zircaloy-4 fuel cladding of PWR size from several batches. Analyses have been given previously [*32*]. Specimens were pickled initially in mixed nitric/hydrofluoric acids. The batches of Zircaloy-4 tubing represent three different metallurgical conditions: (1) a batch given a stress-relief anneal (B); (2) a batch given a recrystallization anneal (K); and (3) a batch given a late β-quench (F). Both Batches B and F were annealed in vacuo at 600°C (873 K) for 1 h and then tested in this condition. The anneal resulted in equiaxed grain structures in both batches, although the grain size remained small in Batch F. It increased the mean precipitate size significantly in Batch F, but only resulted in a minor increase in the mean precipitate size of Batch B. Specimens were corroded in triplicate.

The experimental techniques applied in this work have been described in detail in previous publications [32–35]. During this work, two 2.0 L static autoclaves have been used for the 300°C tests that were new at the start of the program. One has been used only for corrosion tests in water or steam; the other has been used exclusively for experiments in LiOH solutions. Since all the specimens exposed in these autoclaves have been examined at high magnification by TEM and SEM without observing any evidence for oxide spallation, it was concluded that any zirconium found in these autoclaves had to have been transferred by dissolution. Typically about 1 L of solution was added to the autoclaves before each test. Other autoclaves used for long-term tests at 360°C were older and had seen a variety of solutions used in them.

No evidence of zirconium had been found previously [32] in the cold solutions after autoclave tests, so it was inferred that any zirconium dissolved at high temperature that was not hydrothermally deposited on the specimen surfaces was being redeposited on the autoclave walls. Typically, the wetted area of the autoclave walls was ~700 cm^2 compared with a specimen surface area of 50 cm^2. Accordingly, the autoclave walls were abraded with garnet paper and all the rubbings, filter papers, and abrasive papers used were calcined at either 650°C (923 K) or 850°C (1123 K). The samples obtained were irradiated for 16 h at 5 kW in the University of Toronto SLOWPOKE reactor. The iron background counts were too high to give adequate sensitivity for counting the 17-h ^{97}Zr isotope at 747 keV, so the samples were allowed to decay for two months to allow the background to decrease, when the 65.5-day ^{95}Zr isotope was counted at 724 and 756 keV. Initially, the autoclave walls were abraded until no more ^{95}Zr could be detected. This procedure was then repeated after subsequent corrosion tests. These short (one-day) experiments were performed in 1 M LiOH with freshly pickled specimens, and the autoclave walls were re-abraded to measure any additional zirconium now present.

The specimens from these tests were examined by the usual techniques: SEM, TEM replicas using a two-stage Formvar-carbon technique, FTIR, a-c impedance, and, finally, the oxide from some areas was stripped using a bromine/ethyl acetate solution and examined in transmission in the Hitachi H-800 TEM. The a-c impedance of the specimens was measured after each exposure at a fixed frequency of 10^3 Hz during immersion in a 1.0 M NH$_4$NO$_3$ solution at room temperature. The change in impedance was followed until the measurements reached a steady state value when an impedance spectrum from 10 to 2 × 10^5 Hz was measured. Finally, several platinum spots were evaporated onto the specimen surfaces (after a brief exposure to 300°C air to decompose the NH$_4$NO$_3$) and the impedance of the total oxide was measured.

Results and Discussion

Dissolution of the Oxide Film

Exposure of specimens bearing oxide films grown in 300°C (573 K) pH 7 water for three days in 0.1 M LiOH solution showed that large pores developed at the sites of residual intermetallics remaining in the specimen surfaces after only two days exposure to the LiOH solution (Fig. 1). After four days exposure, the whole surface was covered with an array of fine pores (Fig. 2). These pores became more distinct, but did not appear to increase in number, and by the time the exposures in 0.1 M LiOH had reached 25 days, it was becoming progressively more difficult to extract the plastic replicas from the pores, suggesting that material might be being redeposited in the mouths of the pores or on the surface of the oxide, which had now

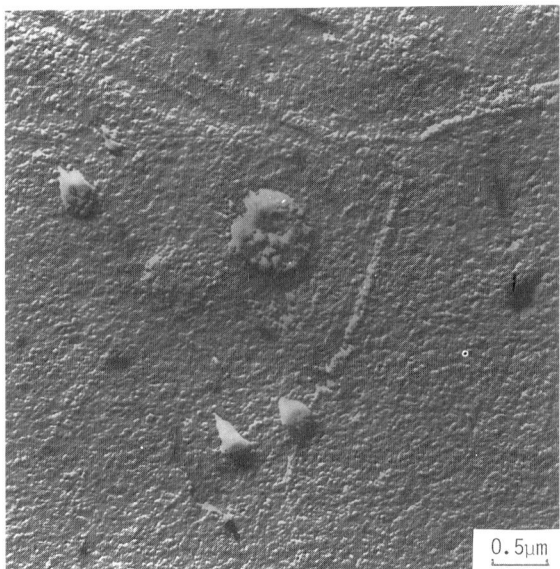

FIG. 1—*Earliest evidence of attack by 0.1 M LiOH on an oxide grown in 300°C (573 K) water occurred after two days in 300°C (573 K) 0.1 M LiOH at intermetallic sites. (TEM replica.)*

developed a characteristic roughness that was not there initially. Previous work has shown that there is usually no difficulty in getting Formvar to enter the pores in the oxide; difficulties in replication usually arise from difficulties in extracting it again in order to make the carbon copy of the plastic [36,37].

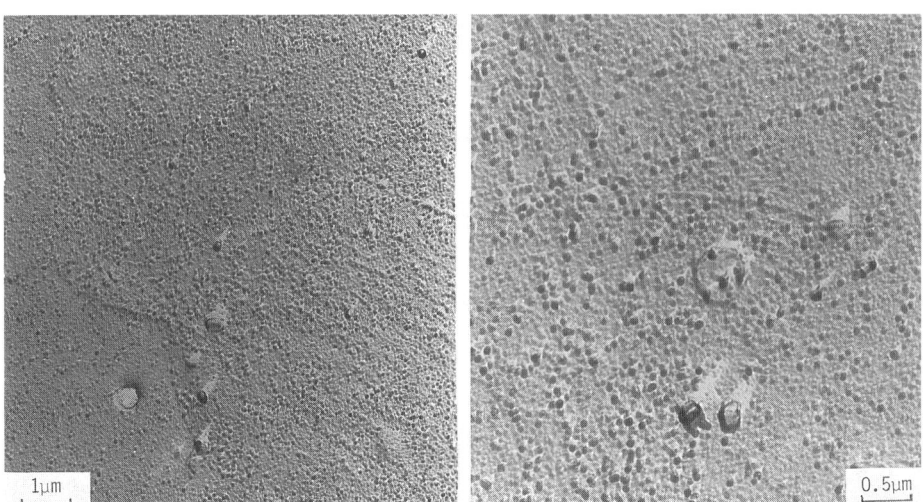

FIG. 2—*Widespread development of fine porosity in an oxide formed in pH 7 water at 300°C (573 K) after twelve days exposure in 0.1 M LiOH at 300°C (573 K). (TEM replicas.)*

A-c impedance measurements during immersion in 1 M NH_4NO_3 solution (interpreted as a simple parallel capacitance and resistance) showed that the porosity that developed in 0.1 M LiOH was essentially superficial and did not reduce the thickness of the barrier film to any great extent (Fig. 3). Whereas, when specimens were exposed to 1.0 M LiOH for one day, similar tests showed that the porosity now penetrated essentially right through the oxide film. If the impedance data are interpreted using the effective area of the pores (and not the whole area of the sample) for the films when saturated with NH_4NO_3, the effective thickness of any oxide at the bottoms of the pores is equal to that of the air-formed oxide. TEM of stripped oxides suggested that this is the correct interpretation. No large increase in oxidation was observed during such a transient, but, following a return to 0.1 M LiOH, regrowth of the barrier film gave an oxidation transient (Fig. 4) similar to the initial growth of the pretransition oxide film in 300°C (573 K) water. Raman spectroscopy on similar oxide films [34] suggested that this development of porosity occurred by the preferential dissolution of the tetragonal/cubic oxide phase. This is in agreement with observations under hydrothermal dissolution conditions [38] where cubic ZrO_2 dissolved in preference to monoclinic ZrO_2 and crystals redeposited hydrothermally were always monoclinic ZrO_2.

This would also explain why the oxide on intermetallic particles was attacked earlier and more severely by LiOH solutions (Fig. 1). Evidence suggests that Fe and Ni are only soluble to a small extent in ZrO_2, but that this solid solution is sufficient to stabilize the tetragonal or cubic forms of ZrO_2 [39,40]. Thus, the probability is that in pretransition oxides, the film on any $Zr(Fe/Cr)_2$ or Zr_2Fe/Ni precipitates left in the initial specimen surface will contain a much higher proportion of tetragonal ZrO_2 than the oxide on the zirconium matrix [39]. These oxides on the intermetallics would, therefore, be expected to dissolve more rapidly than those on the matrix when exposed to LiOH. Since there can only be one equilibrium solubility for ZrO_2 in

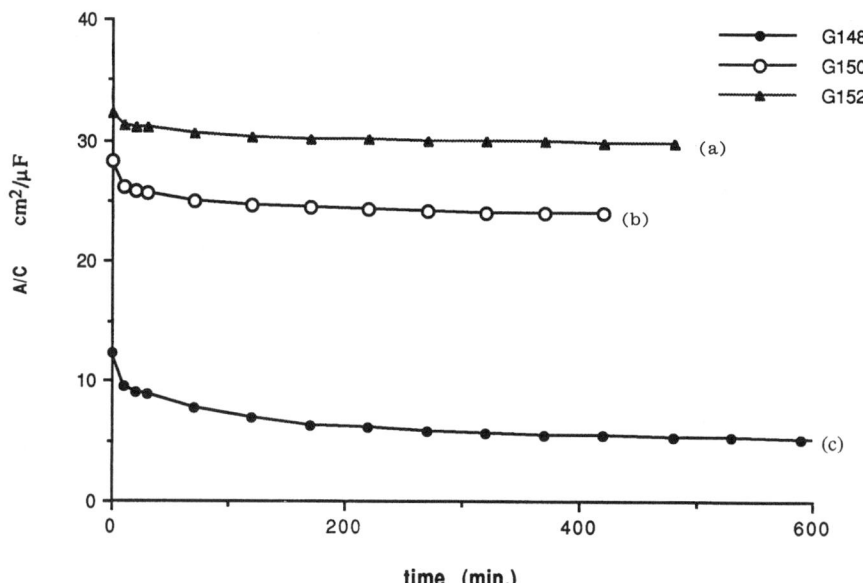

FIG. 3—*A-c impedance (expressed as an inverse parallel capacitance) as a function of time of immersion in 1.0 M NH_4NO_3 electrolyte for oxides formed in pH 7 water for 42 days (a), in pH 7 water for 3 days followed by 25 days in 0.1 M LiOH (b), and after 25 days in 0.1 M LiOH + 1 day in 1.0 M LiOH (c). All exposures at 300°C (573 K).*

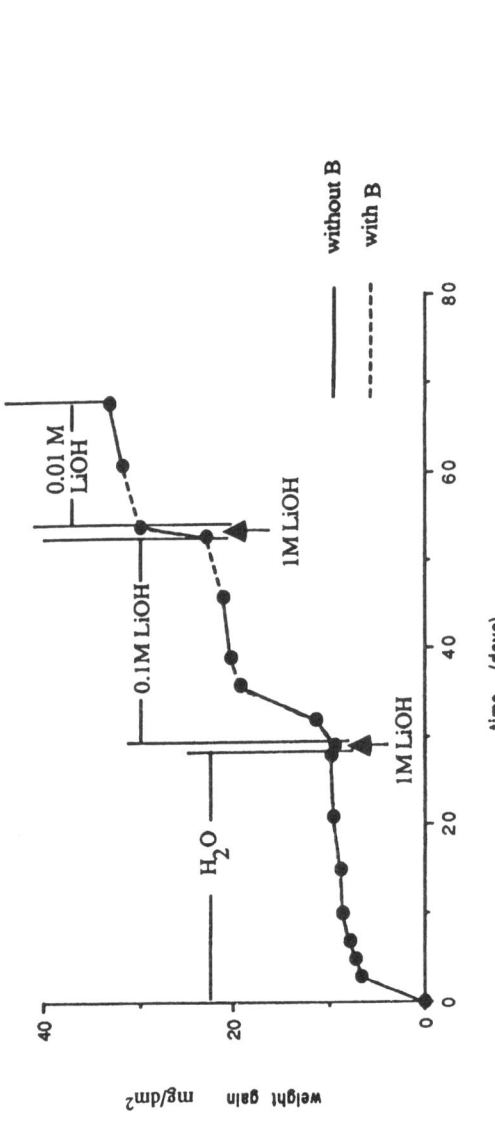

FIG. 4—*Kinetic transients observed at 300°C (573 K) after one day of exposure to 1.0 M LiOH and a return to 0.1 or 0.01 M LiOH.*

high-temperature water, observations of preferential dissolution of tetragonal or cubic ZrO_2 [34,38] probably indicate differences in the speed of dissolution of the more isotropic phases compared with monoclinic ZrO_2. A slightly higher solubility for t-ZrO_2 than for m-ZrO_2, which would never be attained, however, could result in a continuous transport process.

Solubility of ZrO_2

Measurements of the solubility of ZrO_2 in 1 M LiOH by neutron activation analysis give a value of ~1.1 ± 0.1 μg/L at 300°C. The results also showed that zirconium had accumulated in the autoclave during the tests in LiOH solutions to at least three times this level after a long series of tests at different LiOH concentrations. Thus, the redeposited monoclinic ZrO_2 on the autoclave walls does not appear to redissolve as rapidly as tetragonal ZrO_2 from the specimens the next time the autoclave is heated up. Perhaps this is partially because iron oxide gets deposited on it and covers it up. Instead, further tetragonal ZrO_2 dissolves from the specimens and redeposits on the specimens and the autoclave walls. Severe corrosion of the autoclave and crud deposition on the specimens was observed after all tests in 1.0 M LiOH. In these tests the surface area of the specimens was small compared with the surface area of the autoclave, so that, unless there was a strong preference for ZrO_2 to redeposit on the specimen surfaces, this redeposition would not materially affect the calculated solubility. If this is the case, then for short tests (one day), in which little accumulation of ZrO_2 on the autoclave walls should occur, the technique should give a reasonable value for the equilibrium solubility of ZrO_2. However, accumulation of ZrO_2 on the autoclave walls and evidence for preferential redeposition on specimen surfaces could introduce significant errors in addition to those resulting from the counting statistics. However, the numbers quoted here should represent lower-bound values for the solubility. The further question of the extent to which the "solubility" represents merely a transport process, with dissolution in one area of the autoclave and redeposition elsewhere, must also remain open at present.

The observations that steady weight losses were not observed in all these tests [34] and that weight gains in static autoclave tests do not depend significantly on the number of specimens in the autoclave suggest that there is a strong preference for redeposition on zirconia surfaces. The rapid disappearance of surface scratches (Fig. 5) during exposures in concentrated LiOH solutions confirms that ZrO_2 is being redeposited from solution on the specimen surfaces. Such behavior would be typical of crystal growth under hydrothermal conditions [38] and might be expected to lead to large equiaxed crystallites in oxides where a significant fraction of the oxide film had been redeposited hydrothermally from solution. The observation of such large equiaxed crystallites in ZrO_2 films might then be expected for oxides formed in concentrated LiOH solutions or in reactors with primary coolant containing LiOH, where a similar radiation-enhanced dissolution of primary knock on damage tracks has been postulated [41,42].

During this work, evidence for a small solubility of zirconia in pH 7 water at 300°C (573 K) was found. The result was within the detection limits of the technique and gave a value of 10 ± 10 ng/L. Bearing in mind that these solubilities are probably only lower-bound values because of the strong apparent partitioning in favor of redeposition on zirconia surfaces, it should be recognized that some effects of ZrO_2 dissolution may be present even in pH 7 water in the laboratory. This seems to be corroborated by the observation of porosity developing in the oxides on intermetallics and along a few grain boundaries during oxidation in water (Fig. 6).

Effect of LiOH on the Electrical Properties of ZrO_2 Films

In an earlier paper [34], it was reported that the electrical polarization curves of thin pre-transition oxide films formed in LiOH solutions differed significantly from those of similar

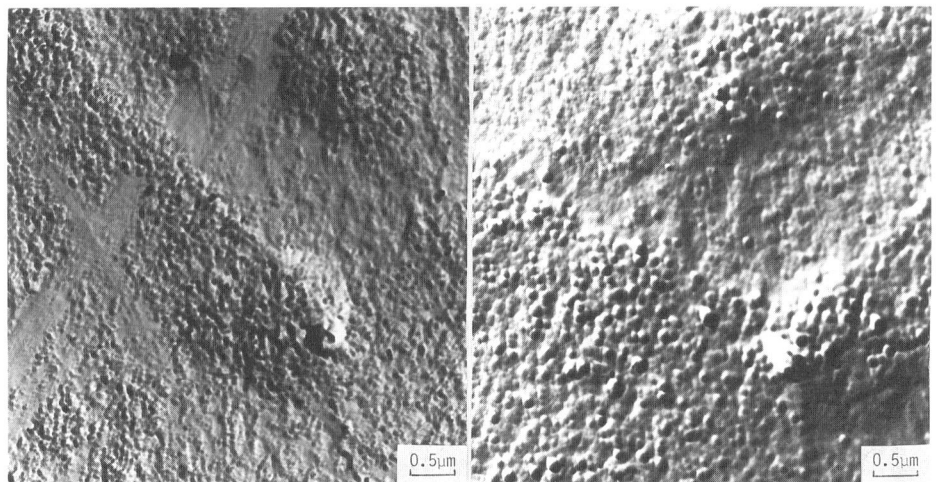

FIG. 5—*Disappearance of surface scratches after three days exposure to 0.1 M LiOH at 300°C. (TEM replicas.)*

oxide films formed in water when measured following immersion in a fused salt at 300°C (573 K). The new results on ZrO_2 dissolution now allow a reinterpretation of these results. These effects on the electrical conductivities of thin pretransition oxide films probably reflect the observations of the more rapid attack of the LiOH-containing solutions on the oxides formed on surface intermetallics. It has been established that in such films the electronic currents flow primarily at the intermetallic particles [43], so that even a small degradation of the oxide formed on these particles by LiOH-containing solutions could have a major effect on the overall conductivity of the oxide.

Effect of LiOH on the Kinetic Transition During Zircaloy Oxidation

LiOH solutions that are 0.1 M or less in concentration appear capable, at most, only of generating superficial pores in pretransition oxide films; very dilute solutions probably have no effect on ZrO_2 surface topography. By contrast, 1.0 M solutions of LiOH can generate porosity that passes essentially right through a pretransition oxide within one day. In the last situation, a period of pretransition corrosion should not be evident in the kinetic curves, which should be linear from time zero. In such a regime the corrosion rate should be equal to that of a bare metal surface in the particular solution and should be essentially independent of LiOH concentration at very high concentrations. A small dependence on concentration may be evident at the start of the regime since the impedance data cannot distinguish between the presence of a small remaining barrier oxide and its complete absence. That such an effect occurs at high LiOH concentrations is seen when the inverse corrosion rate is plotted against the inverse of the LiOH concentration [24] where a tendency to reach a limiting rate at very high LiOH concentrations is seen. The results reported here show that the pretransition corrosion period should disappear for solutions between 0.1 and 1.0 M LiOH. This is in agreement with the kinetics presented by Hillner and Chirigos [2] who showed that the pretransition corrosion kinetics disappeared between 0.1 and 0.5 M LiOH. However, it would be expected that the precise value of the critical LiOH concentration where pretransition kinetics disappear would depend on temperature and the alloy involved.

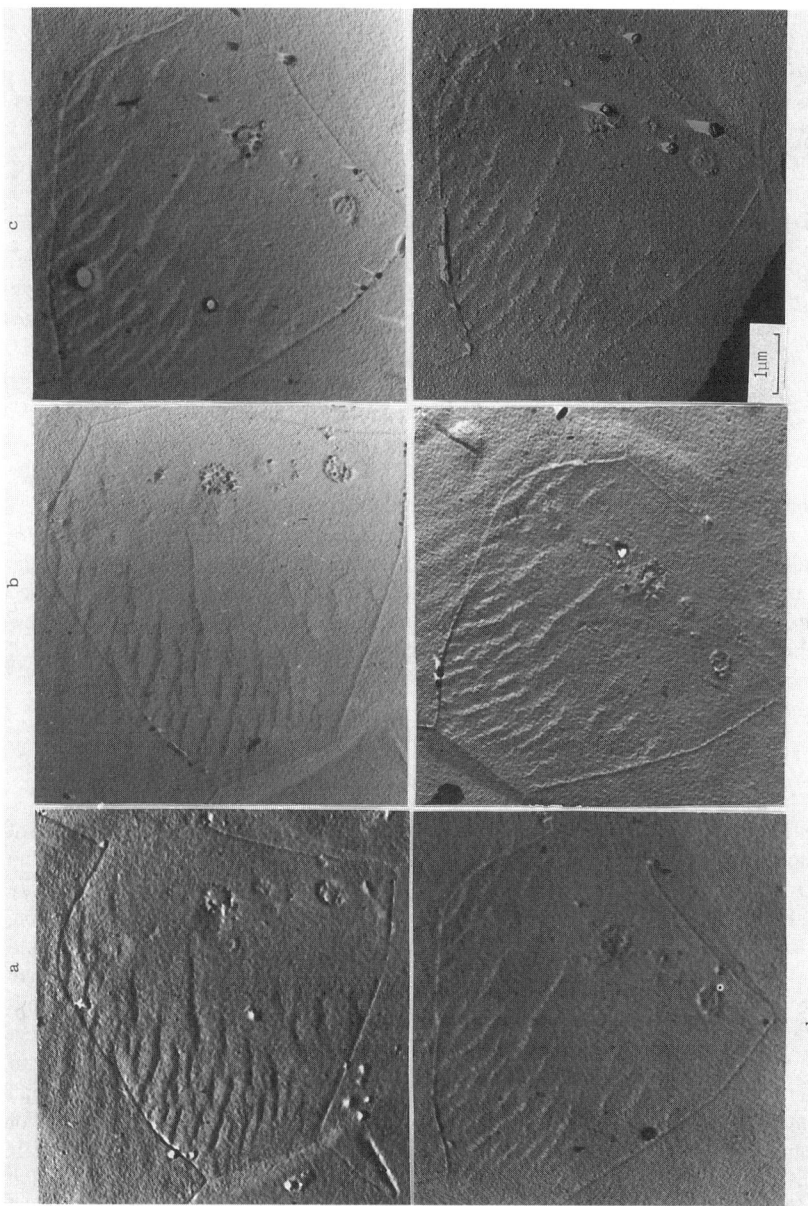

FIG. 6—*Successive TEM replicas of the same area of a specimen surface after successive exposures in 300°C (573 K) pH 7 H_2O for totals of: (a) 5, (b) 7, (c) 10, (d) 15, (e) 21, and (f) 42 days. Note the porosity developing at intermetallics and some grain boundaries (top L and bottom center).*

In solutions of much lower concentration than this critical value, effects on the pretransition kinetics will be small until porosity starts to develop in the film as a precursor to the normal rate transition. This process occurs even in the absence of either LiOH or water and, therefore, cannot be associated with an oxide dissolution process. Early work on the onset of transition [44] showed that porosity or cracking did become evident in oxide films 1.3 to 1.7 μm thick, a long time before the rate transition. At the time, these were ascribed to cracks forming at specimen edges. However, it may be that small pores had developed at intermetallic particles even on smooth surfaces at about the same time. In this work, we have seen that in pH 7 water such porosity develops at intermetallics and a few grain boundaries (Fig. 6). Once pores of this nature are present in the surface, LiOH solutions would be able to concentrate in them because the oxidation process at the bottom of a pore essentially removes water from the solution in the pore to form fresh oxide [45]. This process does not require a heat flux to drive it, although it will obviously occur more rapidly in the presence of a heat flux. If the pores are fine enough that mixing of the concentrated liquid in the pores with the bulk solution is minimal, such a mechanism would explain accelerated corrosion in dilute LiOH solutions only after the rate transition and an earlier onset of the rate transition in such cases.

This accelerated corrosion would occur by the enlarging and spreading of porosity at the bottoms of the pores to produce a porous inner film that left the outer pretransition oxide relatively unaffected because LiOH concentrations in the pores in this outer oxide will be too low for any significant dissolution of it to occur. Some dissolved Zr from the inner porous layer may still escape through the initial flaws in the pretransition film and hydrothermally deposit

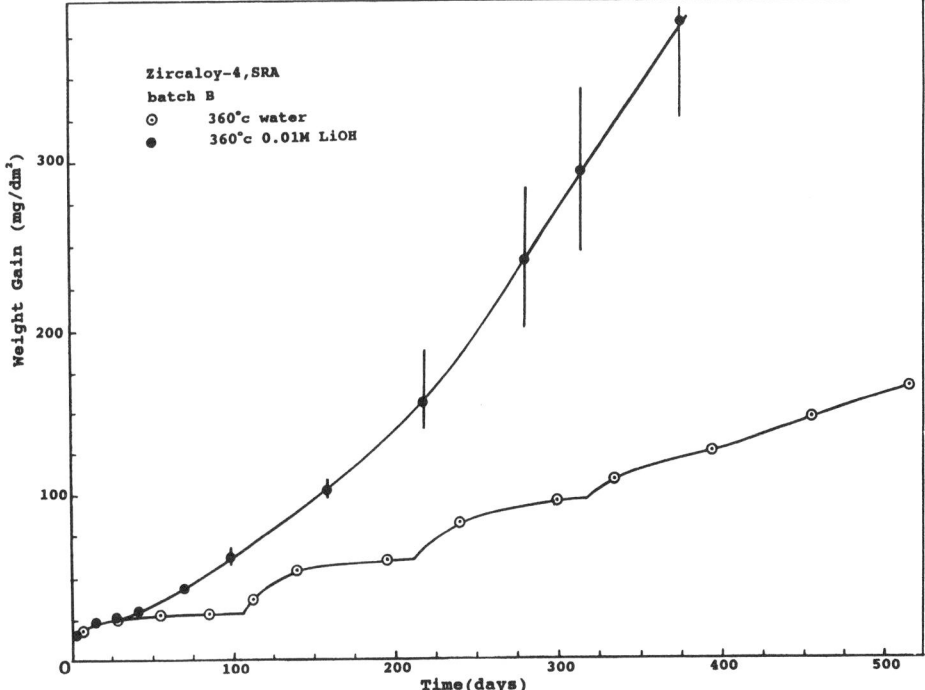

FIG. 7—*Comparison of the corrosion kinetics of an SRA batch of Zircaloy-4 cladding (B) in pH 7 water and 0.01 M LiOH at 360°C (633 K).*

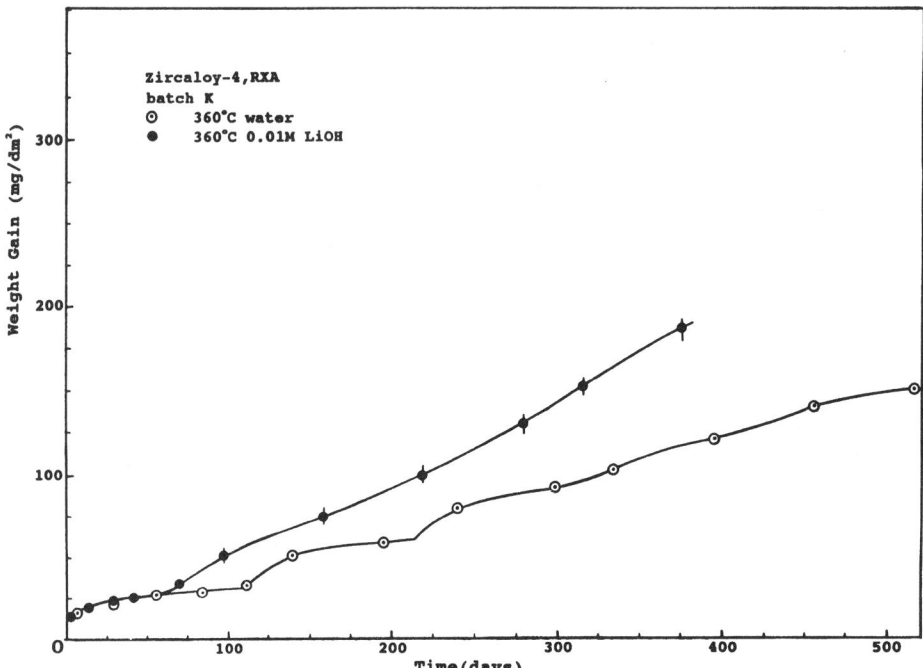

FIG. 8—*Comparison of the corrosion kinetics of an RXA batch of Zircaloy-4 cladding (K) in pH 7 water and 0.01 M LiOH at 360°C (633 K).*

on the outside of the oxide. Thus, we have the possibility of a "sandwich" type of oxide morphology, with hydrothermally deposited monoclinic ZrO_2 on the outside, a 1.0 to 1.5-μm layer of the original pretransition oxide next, and an inner porous monoclinic film with high LiOH concentrations present.

Results for comparison tests on Zircaloy-4 cladding samples in pH 7 water and 0.01 M LiOH at 360°C (633 K) are presented in Figs. 7–11. Scatter bands are shown if they were larger than the spot size on the figure. It can be seen that in general the corrosion kinetics satisfy the expectations of the hypothesis presented above. Examination of Figs. 7–11 shows that, except for the late β-quenched batch (F), the oxidation curves in 0.01 M LiOH departed from those in pH 7 water at oxide thicknesses of 1.5 to 1.8 μm, precisely the region where the mercury porosimeter studies showed the first appearance of a few pores or cracks [44] and where we have seen porosity in specimens oxidized in water here. The late β-quenched material (F) showed an earlier separation of the two curves, but also much scatter in the pretransition weight gains in the LiOH solution, an unusual observation (Fig. 9). This suggests that porosity may start to develop much earlier in these oxide films. The effect of annealing at 873 K was minor for Batch F in LiOH (Fig. 11), although a significant improvement occurred for batch B (Fig. 10).

All the kinetic curves in pH 7 water showed well-developed cycles of post-transition oxidation [46]. These cycles were much less well developed for the kinetic curves in 0.01 M LiOH and were completely absent for both SRA (B) and β-quenched (F) material. The post-transition kinetics for both these materials showed a steady acceleration in the oxidation rates that tended to a constant value at long times (Figs. 7 and 9). If the cyclic post-transition curves are indicative

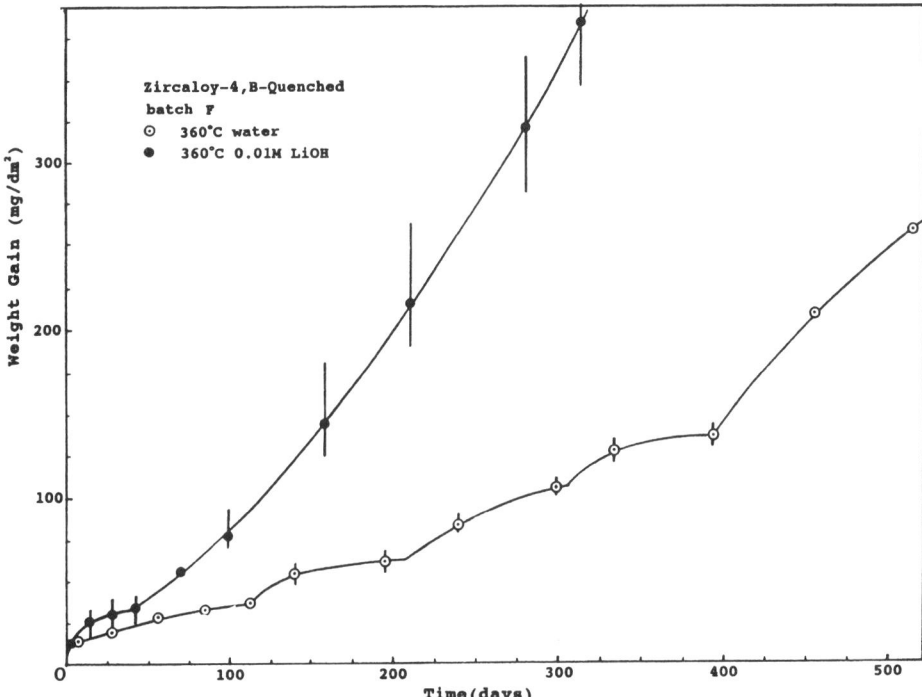

FIG. 9—*Comparison of the corrosion kinetics of a late β-quenched batch of Zircaloy-4 cladding (F) in pH 7 water and 0.01 M LiOH at 360°C (633 K).*

of cycles of growth of oxide crystallites with different morphologies, leading to layering in the oxide film [36,37], the absence of such cycles in the kinetic curves may signify the absence of distinct layers in the post-transition oxide films. The accelerating nature of the curves may be interpreted to indicate an increasing concentration of LiOH at the bottoms of the pores and generally finer porosity in such films than in the oxide on the RXA material (K) where such an accelerating post-transition curve is not observed (Fig. 8).

The annealing of the SRA material (B) resulted in oxidation curves (Fig. 10) that were slightly better than for the RXA material (K) in both pH 7 water and 0.01 M LiOH. However, annealing the late β-quenched material (F) merely delayed the accelerating post-transition curve in 0.01 M LiOH, but did not materially change its characteristics (Fig. 11). The maximum post-transition rate observed in both pH 7 water and 0.01 M LiOH was little affected by the anneal, suggesting little change in pore-size distribution. Since the mean precipitate size in this material was increased from <0.1 to ~0.25 μm by the anneal, it would appear that precipitate size is not a controlling factor in determining pore size in post-transition oxides. Nor does it appear that residual strain in the metal lattice was an important factor for the β-quenched material (F), although it may have been for the SRA material (B) where the change in precipitate size on annealing was minor, but the lattice strain was largely eliminated (metal grains extinguished uniformly in polarized light on rotating the specimen). The only distinguishing factor for the β-quenched material (F) after annealing was its small grain size, and this could only be important if the prior metal grain boundaries were important sites for the development of porosity in the oxide. Some evidence for this was obtained both in earlier work [44] and here (Fig. 6),

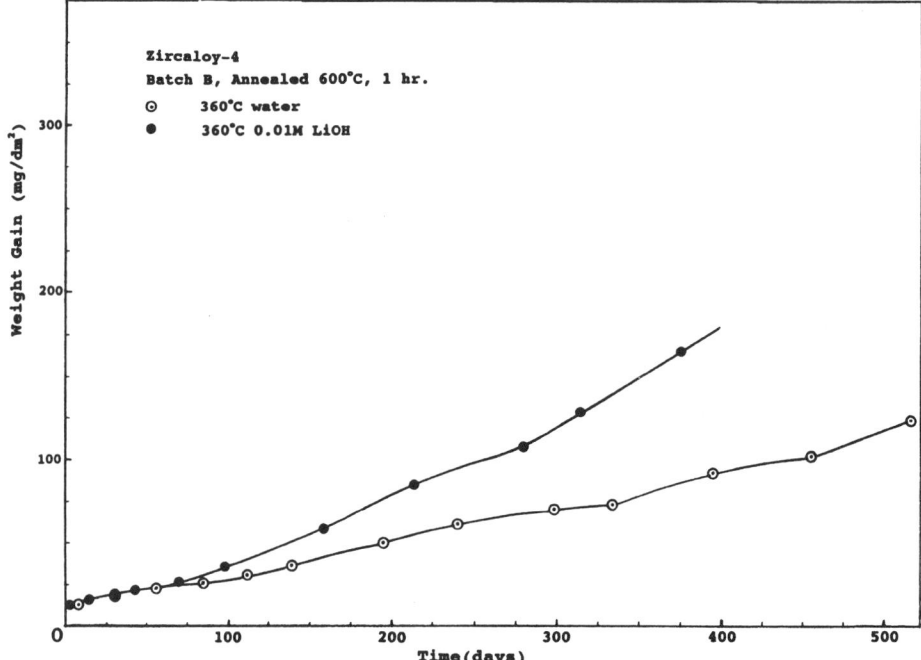

FIG. 10—*Effect of a 1-h 600°C (873 K) vacuum anneal on the corrosion kinetics of Batch B cladding in pH 7 water and 0.01 M LiOH at 360°C (633 K).*

where formation of pores at grain boundaries was observed in water, and where the only other site for porosity to develop in pretransition oxides was at intermetallics. However, the opportunity to examine these specimens for such evidence vanished when subsequent severe degradation of these oxide films occurred during later autoclave exposures.

Although there has been some evidence in the past for greater sensitivity of some batches of Zircaloy-4 to accelerated corrosion in LiOH solutions [6] and for greater sensitivity of niobium alloys to these solutions than for the Zircaloys [8], further studies of the factors involved would seem to be worthwhile, especially considering the relatively poor behavior of the SRA material. The important factor to examine would appear to be the extent of mixing between the concentrated LiOH in the pores in a thick post-transition oxide and the bulk solution. This would be the primary process limiting the [LiOH] attainable at or near the oxide-metal interface. To a first approximation, the pore diameter should be the most important factor, and, if the largest pores in an oxide are comparable in size to the precipitates, small precipitates (giving small pores) should lead to the least mixing of the solution in the pores and hence to the highest corrosion rates in LiOH.

Synergistic Effects of LiOH and H_3BO_3

The inhibiting effects of boric acid on the accelerated corrosion of Zircaloys in LiOH solutions are well established experimentally [18–20]. When specimens preoxidized in pH 7 water at 300°C (573 K) were transferred to 0.1 M LiOH + 1000 ppm (wt) B (as H_3BO_3), no difference could be detected in the development of surface porosity (Fig. 12) when compared with pores developed in the absence of H_3BO_3 (see Fig. 2). Nor did the presence of boron have any effect

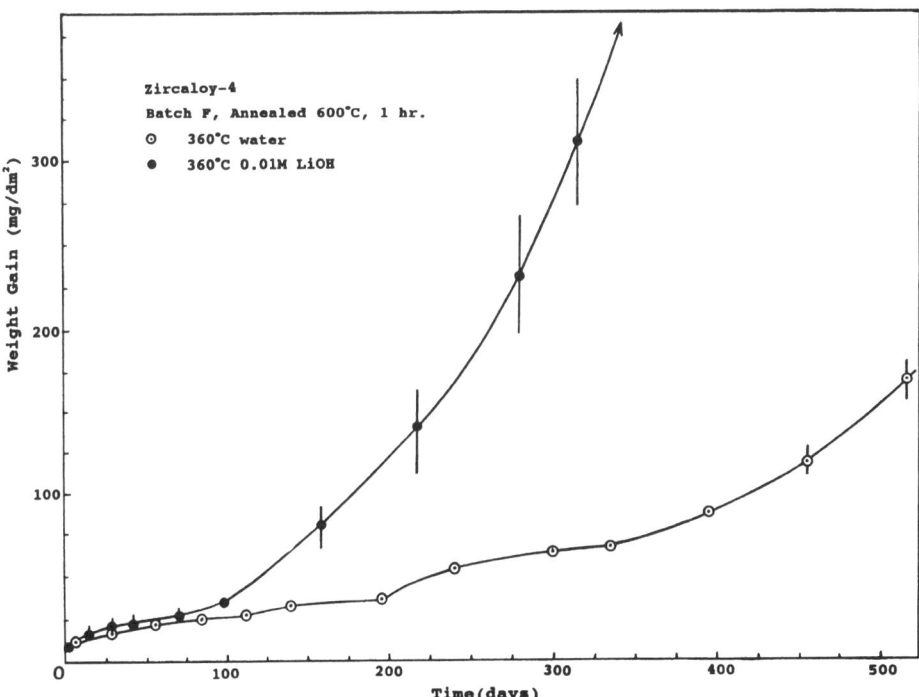

FIG. 11—*Effect of a 1-h 600°C (873 K) vacuum anneal on the corrosion kinetics of Batch F cladding in pH 7 water and 0.01 M LiOH at 360°C (633 K).*

on the recovery of the oxide film after a transient (one-day) exposure to 1.0 M LiOH at 300°C (573 K). Only when the boric acid was present in the 1.0 M LiOH solution during a transient exposure did it stop the development of porosity that penetrated through the oxide (Fig. 13) and thereby eliminated the subsequent high corrosion rate transient (Fig. 14). The enhanced redeposition of oxide under these conditions was shown further by apparently larger hemispherical crystallites forming on specimen surfaces in boric acid containing solutions (Fig. 15) and the more rapid disappearance of pre-existing scratches on specimen surfaces (see Fig. 5). The precise morphology of the surface crystallites appeared to change after each autoclave exposure when the same area was examined, suggesting that the dissolution and redeposition processes are occurring continuously.

Since the additions of boric acid obviously work in practice [18–20], the combined evidence supports a hypothesis that they function only in pores in the oxide and that high concentrations of both LiOH and H_3BO_3 must exist in these regions for the boric acid to be effective. The "chemical concentration" process (which extracts water from the solution in the pores to form additional ZrO_2) will obviously be equally effective at concentrating both LiOH and boric acid in the pores. It is not clear at present whether or not heat fluxes are equally effective at concentrating LiOH and H_3BO_3 in thick porous oxide films.

The necessity for high concentrations of both LiOH and H_3BO_3 in porous films suggests that the plugging of pores occurs and that it is necessary to exceed the solubility product of some bulky complex salt of $Li_xZr_yB_zO_n$ to effectively prevent the development of deep pores in the oxide. This appears to be the most probable mechanism for the effect of boric acid since it appears to have no effect on the dissolution of ZrO_2 by LiOH (see Fig. 12). Furthermore,

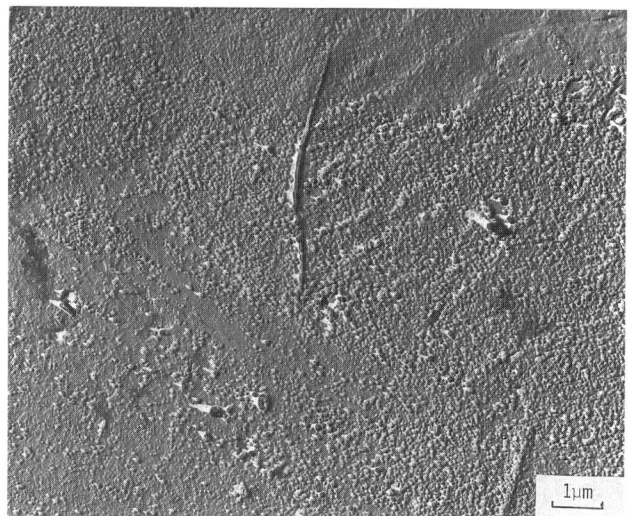

FIG. 12—*Fine porosity developed in an oxide grown in pH 7 water at 300°C (573 K) for 42 days after 7 days exposure to 0.1 M LiOH + 1000 ppm B as H_3BO_3. Compare with Fig. 2a. (TEM replica.)*

this is the mechanism that has been postulated for the inhibiting effects of boric acid on the growth of thick porous oxide films in broach holes and crevices on the secondary side of PWR steam generators [47–49] and is analogous to the "sealing" of porous anodic alumina films in order to increase its general corrosion resistance [50]. Insufficient boric acid to completely plug the porosity might then result in restricted mixing of the solution in the pores with the bulk solution and worse corrosion than in the complete absence of boric acid, as observed by Bramwell et al. [19].

Incorporation of Li^+ in ZrO_2 Films

The hypothesis we have proposed here for accelerated corrosion in concentrated LiOH solutions does not require the prior incorporation of any lithium ions in the ZrO_2 films. It can function as immediately as a ZrO_2 surface is exposed to a sufficiently concentrated LiOH solution at a sufficiently high temperature. Below a certain critical LiOH concentration, the preferential dissolution of tetragonal or cubic ZrO_2 crystallites exposed to the solution is incapable of producing pores that penetrate right through the oxide, and the observation of accelerated corrosion must await the development of deeper pores as a precursor to the normal oxidation rate transition. Once these are present, LiOH can concentrate in these pores and enlarge or extend them. This concentration process can occur chemically by extraction of water to form ZrO_2 or by application of a heat flux.

None of these steps requires the presence of any Li^+ in the oxide film, and in fact the absence of any difference in kinetics in 0.01 M LiOH solutions prior to an oxide thickness of ~1.7 μm shows that no incorporation of Li^+ takes place at this stage. The observation of enhanced electronic conductivity in the oxides during this period of growth has now been explained as an effect of preferential attack by LiOH containing solutions on the oxide formed on the intermetallic particles, and no implied incorporation of Li^+ in the oxide lattice is necessary.

All the data point to Li^+ incorporation as being an effect of any enhanced corrosion that

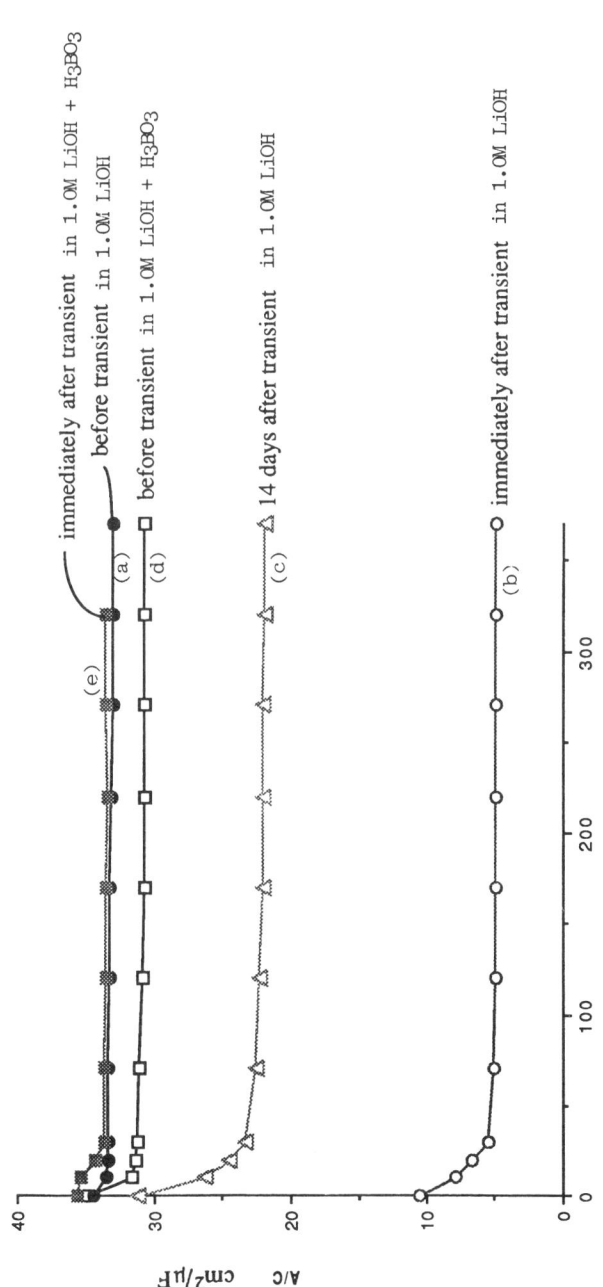

FIG. 13—A-c impedance curves (interpreted as an inverse parallel capacitance) versus immersion time in 1.0 M NH_4NO_3, showing the degradation and regrowth of the protective oxide film after a one-day exposure to 1.0 M LiOH (a-c) and the lack of any effect resulting from a one-day exposure to 1.0 M LiOH + 8000 ppm B as boric acid (d,e).

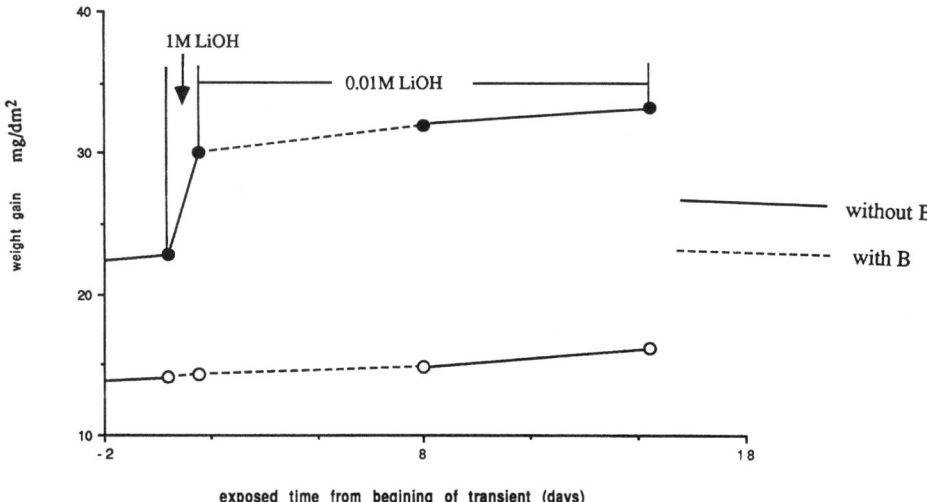

FIG. 14—*Effect of boric acid addition (8000 ppm B) on the transient effects of a one-day exposure to 1.0 M LiOH.*

occurs and not its cause. The necessity to concentrate the solution in the porous oxide to get sufficient oxide dissolution to cause accelerated oxidation ensures that adsorbed but leachable LiOH will be found in porous oxide films. The redeposition of hydrothermal monoclinic ZrO_2 on the outer surfaces of the oxide films and within the oxide when a steep LiOH concentration gradient is established through the oxide will carry some Li^+ with it that will be incorporated in the redeposited ZrO_2 in a non-leachable form. The precise concentrations of lithium in the two forms will bear only a very general relationship to the extent of accelerated oxidation, as has been demonstrated already [14,15,20].

Consequences for Reactor Operation

A LiOH concentration process may be in operation in all high-power, high-burnup fuel cladding already. Increasing the external LiOH concentration would not then affect the concentration achieved at the metal/oxide interface provided that the extent of mixing between the solution in the porous oxide and in the bulk coolant is not affected by changes in the oxide pore structure. The LiOH concentration in the bulk water might be expected to affect the rate at which equilibrium is reached at the bottoms of pores; however, the rate of increase of pore depth (approximately equal to the rate of oxide thickening) is probably slow enough that equilibrium concentrations should always be attained if mixing in the oxide is restricted. However, in practice the effects of increasing the LiOH concentration would not be evident for most of each fuel cycle because there would be enough boric acid present in the oxide to inhibit such effects until near the end of the fuel cycle. Once the boric acid concentration fell into the range where Bramwell et al. [19] observed worse corrosion than in LiOH alone, the corrosion rate could increase rapidly. Thus, the end of each fuel cycle, where little or no boric acid is present, and thermal hydraulic conditions, where the oxide/metal interface temperature exceeds the saturation temperature of the coolant by at least 5°, would appear to be a very vulnerable situation for pressurized water reactors operating under current water chemistry guidelines.

That such LiOH concentration effects are already present may be indicated by a comparison of the oxide thicknesses formed on fuel cladding in Russian and Western reactors where sat-

FIG. 15—*Surface crystallites formed on ZrO_2 surfaces by hydrothermal recrystallization:* (a) *oxide surface formed in H_2O,* (b) *oxide surface after seven days in 0.1 M LiOH,* (c) *oxide surface after 7 days in 0.1 M LiOH + 1000 ppm B (as H_3BO_3).*

uration temperatures and heat fluxes are comparable. Russian water chemistry uses ammonia and potassium hydroxide and does not add LiOH to the coolant [51,52]. Li production from fast neutron reactions in boric acid ($^{10}B \xrightarrow{n\alpha} {}^{7}Li$) is small and is controlled by the water purification system to limit the Li concentration. However, it is probably unreasonable to compare the corrosion rate of the Zr-1%Nb cladding used in VVERs in this way with the corrosion of Zircaloy-4 cladding. For this reason, it will be interesting to see how a typical Western fuel cladding performs in VVER water chemistry when the first Czech reactor at Temelin begins operating.

Conclusions

From the work presented here, it is concluded that:

1. Values of the solubility in 1.0 M LiOH at 300°C (573 K) are ~1.1 + 0.1 µg/kg. A very small solubility may be present in pH 7 water at 300°C (573 K). The value of ~10 ± 10 ng/kg was not outside the range of the counting errors indicated.

2. Below some critical LiOH concentration, which is a function of alloy and temperature, only superficial pores can be generated. Above the critical concentration, pores are produced right through the oxide film and no pretransition oxidation kinetics are observed.

3. For Zircaloy-2 at 300°C (573 K), this critical concentration lies between 0.1 and 1.0 M.

4. In solutions of significantly lower concentration than the critical concentration no effect of LiOH is seen on the pretransition corrosion kinetics until porosity begins to develop prior to the oxidation rate transition. Only for the late β-quenched batch did this appear not true, and in this case the scatter between specimens was uncommonly large and included one specimen showing no effect.

5. LiOH may be concentrated in such pores by a chemical concentration process involving the extraction of water from the solution in the pores to form ZrO_2. If so, once the solution in the pores is concentrated enough, preferential dissolution of cubic/tetragonal ZrO_2 at the bottoms of the pores can enlarge or extend them and cause enhanced corrosion by the formation of a porous inner layer of oxide. The extent of this enhancement will depend on the degree of LiOH concentration that can be achieved, which will depend on the initial pore size through an effect of this on the mixing of the solution in the pores and the bulk water. If the largest pores generated are related to the size of the intermetallic precipitates, this could relate small precipitate size to poor corrosion resistance in LiOH solutions.

6. SRA and late β-quenched Zircaloy-4 appear to be particularly susceptible to corrosion acceleration by this mechanism, suggesting that pores formed in their oxides are smaller diameter than those formed in RXA material.

7. Boric acid does not affect the dissolution of ZrO_2 in LiOH, but in concentrated (1.0 M) solutions appears to plug pores by precipitating a complex salt thought to be a lithium zirconate borate.

8. All evidence of lithium ion concentration in zirconia films is a consequence of the above processes and not a cause. In particular, hydrothermal redeposition of ZrO_2 on specimen surfaces and within porous oxides will contain a variable concentration of Li^+ in a non-leachable form.

Acknowledgments

The authors are indebted to the CANDU Owner's Group and to the Natural Sciences and Engineering Council of Canada for the funding that has allowed this program to proceed and to the Nuclear Fuel Industry Research program for permission to include some results obtained under their auspices.

References

[1] Coriou, H., Grall, L., Meunier, J., Pelras, M., and Willermoz, H., *Journal of Nuclear Materials*, Vol. 7, 1962, p. 665.
[2] Hillner, E. and Chirigos, J. N., "The Effect of Lithium Hydroxide and Related Solutions on the Corrosion Rate of Zircaloy in 680°F Water," Report WAPD-TM-307, Bettis Atomic Power Laboratory, West Mifflin, PA, 1962.
[3] Kass, S., *Corrosion*, Vol. 25, 1969, p. 30.
[4] Murgatroyd, R. A. and Winton, J., *Journal of Nuclear Materials*, Vol. 23, 1967, p. 249.
[5] Urbanic, V. F., "Chemical Control of Fuel Channel Crevice Corrosion—LiOH versus KOH," Report CRNL-2220, Atomic Energy of Canada Ltd., Chalk River, 1983.
[6] McDonald, S. G., Sabol, G. P., and Shepard, K. D., "The Effect of Lithium Hydroxide on the Corrosion Behaviour of Zircaloy-4," *Zirconium in the Nuclear Industry: Sixth International Symposium, ASTM STP 824*, American Society for Testing and Materials, West Conshohocken, PA, 1984, p. 519.
[7] Garzarolli, F., Pohlmeyer, J., Trapp-Pritsching, S., and Weidinger, H. G., "Influence of Various

Additions to Water on Zircaloy-4 Corrosion in Autoclave Tests at 350°C," *Proceedings,* IAEA Technical Committee Meeting on Fundamental Aspects of Corrosion of Zirconium-Base Alloys in Water Reactor Environments, Portland, OR, September 1989, (IWGFPT/34), pp. 65–72.

[8] Sabol, G. P., Kilp, G. R., Balfour, M. G., and Roberts, E., "Development of a Cladding Alloy for High Burnup," *Zirconium in the Nuclear Industry: Eighth International Symposium, ASTM STP 1023,* American Society for Testing and Materials, West Conshohocken, PA, 1989, pp. 227–244.

[9] Perkins, R. A. and Busch, R. A., "Corrosion of Zircaloy in the Presence of LiOH," *Zirconium in the Nuclear Industry: Ninth International Symposium, ASTM STP 1132,* American Society for Testing and Materials, West Conshohocken, PA, 1991, pp. 595–612.

[10] Kass, S., *Corrosion,* Vol. 16, 1960 (137t) and Vol. 17, 1961 (566t).

[11] Berry, W. E., "Effect of Fluoride Ions on the Aqueous Corrosion of Zirconium Alloys," *Corrosion of Zirconium Alloys, ASTM STP 368,* American Society for Testing and Materials, West Conshohocken, PA, 1964, pp. 28–40.

[12] Ramasubramanian, N., *Journal of the Electrochemical Society,* Vol. 127, 1980, pp. 2566–2572.

[13] Billot, P., Beslu, P., Giordano, A., and Thomazet, J., "Development of a Mechanistic Model to Assess the External Corrosion of the Zircaloy Claddings in PWRs," *Zirconium in the Nuclear Industry: Eighth International Symposium, ASTM STP 1023,* American Society for Testing and Materials, West Conshohocken, PA, 1989, pp. 165–184.

[14] Ramasubramanian, N., Preocanin, N., and Ling, V. C., "Lithium Uptake and the Accelerated Corrosion of Zirconium Alloys," *Zirconium in the Nuclear Industry: Eighth International Symposium, ASTM STP 1023,* American Society for Testing and Materials, West Conshohocken, PA, 1989, pp. 187–201.

[15] Ramasubramanian, N., "Lithium Uptake and the Corrosion of Zirconium Alloys in Aqueous Lithium Hydroxide Solutions," *Zirconium in the Nuclear Industry: Ninth International Symposium, ASTM STP 1132,* American Society for Testing and Materials, West Conshohocken, PA, 1991, pp. 613–627.

[16] Billot, P., Beslu, P., and Robin, J. C., "Consequences of Lithium Incorporation in Oxide Films Due to Irradiation Effect," *Proceedings,* International Topical Meeting on LWR Fuel Performance—Fuel for the 90's, Avignon, France, April 1991, American and European Nuclear Societies, Vol. 2, pp. 757–769.

[17] Billot, P., Robin, J.-C. C., Giordano, A., Peybernes, J., Thomazet, J., and Amanrich, H., "Experimental and Theoretical Studies of Parameters that Influence Corrosion of Zircaloy-4," *Zirconium in the Nuclear Industry: Tenth International Symposium, ASTM STP 1245,* American Society for Testing and Materials, West Conshohocken, PA, 1994, pp. 351–377.

[18] Tice, D. R., Huddart, G., and Bramwell, I. L., "Corrosion of Zircaloy-4 Fuel Cladding in High Concentration Lithium and Boron Conditions Simulating Extended Burnup in a PWR," *Proceedings,* British Nuclear Energy Society Conference on Materials for Nuclear Reactor Core Applications, Bristol, UK, October 1987, p. 57.

[19] Bramwell, I. L., Parsons, P. D., and Tice, D. R., "Corrosion of Zircaloy-4 PWR Fuel Cladding in Lithiated and Borated Water Environments," *Zirconium in the Nuclear Industry: Ninth International Symposium, ASTM STP 1132,* American Society for Testing and Materials, West Conshohocken, PA, 1991, pp. 628–642.

[20] Ramasubramanian, N. and Balakrishnan, P. V., "Aqueous Chemistry of Lithium Hydroxide and Boric Acid and Corrosion of Zircaloy-4 and Zr-2.5Nb Alloys," *Zirconium in the Nuclear Industry: Tenth International Symposium, ASTM STP 1245,* American Society for Testing and Materials, West Conshohocken, PA, 1994, pp. 378–399.

[21] Mathur, P. K. and Narasimhan, S. V., "Chemistry of Primary Coolant in Water Cooled Nuclear Reactors," *Coolant Technology of Water-Cooled Reactors,* Vol. 1: *Chemistry of Primary Coolant in Water Cooled Reactors,* IAEA-TECDOC-667, International Atomic Energy Agency, Vienna, 1992, pp. 98–129.

[22] Van Swam, L. F. P. and Shann, S. H., "The Corrosion of Zircaloy-4 Fuel Cladding in Pressurised Water Reactors," *Zirconium in the Nuclear Industry: Ninth International Symposium, ASTM STP 1132,* American Society for Testing and Materials, West Conshohocken, PA, 1991, pp. 758–781.

[23] Baur, K., Lisdat, R., and Puschel, H., "Fuel Behaviour in High Performance PWRs," *Proceedings,* 1994 International Topical Meeting on Light Water Reactor Fuel Performance, W. Palm Beach, FL, April 1994, American Nuclear Society, pp. 22–30.

[24] Cox, B., "Modelling the Corrosion of Zirconium Alloys in Nuclear Reactors Cooled by High Temperature Water," *Modelling Aqueous Corrosion from Individual Pits to System Management,* Trethewey, K. R. and Roberge, P. R., Eds., NATO ASI Series E: Applied Sciences, Vol. 266, Kluwer Academic Publishers, Dordrecht, Netherlands, 1994, pp. 183–200.

[25] Kilp, G. R., Balfour, M. G., Stanutz, R. N., McAtee, K. R., Miller, R. S., Boman, L. H., Wolfhope,

N. P., Ozer, O., and Yang, R. L., "Corrosion Experience with Zircaloy and ZIRLO in Operating PWRs," *Proceedings,* International Topical Meeting on LWR Fuel Performance—Fuel for the 90's, Avignon, France, April 1991, American and European Nuclear Societies, Vol. 2, pp. 730–741.

[26] Shann, S. H., Van Swam, L. F. P., and Martin, L. A., "Effects of Coolant Li Concentration on LWR Fuel Performance," *Proceedings,* International Topical Meeting on LWR Fuel Performance—Fuel for the 90's, Avignon, France, April 1991, American and European Nuclear Societies, Vol. 2, pp. 742–756.

[27] Swan, T. and Polley, M. V., "Evaluation of Zircaloy Fuel Clad Oxidation at Millstone 3 PWR," *Proceedings,* IAEA Technical Committee Meeting on Influence of Water Chemistry on Fuel Cladding Behaviour," Rez, Czech Republic, October 1993, IAEA-IWGFPT, to be published.

[28] Sabol, G. P., Weiner, R. A., McAtee, K. R., Leech, W. J., Correal-Pulver, O. A., Stanutz, R. N., Comstock, R. J., and Miller, R. S., "In-Reactor Corrosion Performance of ZIRLO and Zircaloy-4 and Related Corrosion Modelling," Presented at 1994 International Topical Meeting on Light Water Reactor Fuel Performance, W. Palm Beach, FL, April 1994, American Nuclear Society.

[29] Willse, J. T., Mitchell, D. B., and Williams, P. F., "Recent Results from the B&W Fuel Company's Fuel Performance Improvement Program," *Proceedings,* 1994 International Topical Meeting on Light Water Reactor Fuel Performance, W. Palm Beach, FL, April 1994, American Nuclear Society, pp. 45–52.

[30] Karlsen, E. and Vitanza, C., "Effects of Pressurised Water Reactor (PWR) Coolant Chemistry on Zircaloy Corrosion Behaviour," *Zirconium in the Nuclear Industry: Tenth International Symposium, ASTM STP 1245,* American Society for Testing and Materials, West Conshohocken, PA, 1994, pp. 779–789.

[31] Balakrishnan, P. V., unpublished results, Atomic Energy of Canada Ltd., Chalk River, 1994.

[32] Cox, B. and Wong, Y-M., "Effects of LiOH on Pretransition Zirconium Oxide Films," *Zirconium in the Nuclear Industry: Ninth International Symposium, ASTM STP 1132,* American Society for Testing and Materials, West Conshohocken, PA, 1991, pp. 643–662.

[33] Cox, B. and Wu, C., "Degradation of Zirconium Oxide Films in LiOH," *Oxide Films on Metals and Alloys,* MacDougall, B. R., Alwitt, R. S., and Ramanarayanan, T. A., Eds., *Proceedings,* Vol. 92–22, Electrochemical Society, Pennington, NY, 1992, pp. 265–279.

[34] Cox, B. and Wu, C., *Journal of Nuclear Materials,* Vol. 199, 1993, p. 272.

[35] Cox, B. and Wu, C., *Journal of Nuclear Materials,* Vol. 224, 1995, p. 169.

[36] Cox, B., *Journal of Nuclear Materials,* Vol. 41, 1971, pp. 96–100.

[37] Cox, B. and Donner, A., *Journal of Nuclear Materials,* Vol. 47, 1973, pp. 72–78.

[38] Somiya, S., "Hydrothermal Reactions for Materials Science and Engineering," Elsevier Applied Science, Barking, Essex, UK, 1989.

[39] Pecheur, D., Lefebvre, F., Motta, A. T., Lemaignan, C., and Charquet, D., "Oxidation of Intermetallic Precipitates in Zircaloy-4: Impact of Irradiation," *Zirconium in the Nuclear Industry: Tenth International Symposium, ASTM STP 1245,* American Society for Testing and Materials, West Conshohocken, PA, 1994, pp. 687–705.

[40] Valigi, M., Gazzoli, D., Dragone, R., Gherandi, M., and Minelli, G., *Journal of Materials Chemistry,* Vol. 5, 1995, pp. 183–189.

[41] Cox, B. and Fidleris, V., "Enhanced Low-Temperature Oxidation of Zirconium Alloys Under Irradiation," *Zirconium in the Nuclear Industry: Eighth International Symposium, ASTM STP 1023,* American Society for Testing and Materials, West Conshohocken, PA, 1989, pp. 245–265.

[42] Cox, B., "A New Model for the In-Reactor Corrosion of Zirconium Alloys," *Proceedings,* IAEA Technical Committee Meeting on Influence of Water Chemistry on Fuel Cladding Behaviour, Rez, Czech Republic, October 1993, International Atomic Energy Agency, Vienna, IWGFPT, to be published.

[43] Cox, B., "Oxidation of Zirconium and its Alloys," *Advances in Corrosion Science and Technology,* Vol. 5, M. G. Fontana and R. W. Staehle, Eds. Plenum Press, NY, 1976, pp. 173–391.

[44] Cox, B., *Journal of Nuclear Materials,* Vol. 27, 1968, pp. 1–11.

[45] Cox, B., "What is Wrong with Current Models for In-Reactor Corrosion," *Proceedings,* IAEA Technical Committee Meeting on Fundamental Aspects of Corrosion on Zirconium Base Alloys in Water Reactor Environments, Portland, OR, September 1989, International Atomic Energy Agency, Vienna, IWGFPT/34, pp. 167–173.

[46] Peters, H. R., "Improved Characterisation of Aqueous Corrosion Kinetics of Zircaloy-4," *Zirconium in the Nuclear Industry: Sixth International Symposium, ASTM STP 824,* American Society for Testing and Materials, West Conshohocken, PA, 1984, pp. 507–518.

[47] Saint Paul, P. and Slama, G., "Steam Generator Materials Degradation," *Proceedings,* Fifth International Symposium on Environmental Degradation of Materials in Nuclear Power Systems-Water Reactors, Monterey, CA, August 1991, American Nuclear Society, La Grange Park, IL, pp. 39–49.

[48] Hermer, R. E., "Boric Acid Application Guidelines for Intergranular Corrosion Inhibition," Report EPRI-NP-5558, Electric Power Research Institute, Palo Alto, CA, December 1987.
[49] Weres, O. and Tsao, L., "Phase Relations and Fluid Compositions in Steam Generator Crevices," Report EPRI-NP-5138, Electric Power Research Institute, Palo Alto, CA, April 1987.
[50] Evans, U. R., *The Corrosion and Oxidation of Metals: Scientific Principles and Practical Applications,* 1st ed., Edward Arnold, London, 1960, pp. 241–252.
[51] Bibilashvili, Y. K., Dubrovin, K. P., and Smirnov, A. V., "Corrosion and Hydriding of WWER Fuel Rod Cladding," *Proceedings,* IAEA Technical Committee Meeting on Influence of Water Chemistry on Fuel Cladding Behaviour, Rez, Czech Republic, October 1993, IAEA-IWGFPT, to be published.
[52] Dragunov, Y. G., Markov, Y. V., Rybalchenko, I. L., Ryazantsev, I. L., and Chabak, A. F., "Water Chemistry in Soviet Nuclear Power Plants," *Coolant Technology of Water Cooled Reactors,* Vol. 1: *Chemistry of Primary Coolant in Water Cooled Reactors,* IAEA-TECDOC-667, International Atomic Energy Agency, Vienna, 1992, pp. 108–114.

DISCUSSION

G. D. Moan[1] *(written discussion)*—In many of the micrographs, the surface texture adjacent to the grain boundaries appeared different from that in the grain centers. If the effect is real, does it provide any information about the mechanisms?

B. Cox et al. (authors' closure)—The replication of small pores in ZrO_2 films is near the limit of the capabilities of this technique because the small plastic filaments in these pores tend to break off at the surface rather than pulling out of the pore. Thus, the uniform replication of small pores over a large surface area is difficult, and variations from area to area are common. This variability appears to become more common as exposures to LiOH increase, perhaps because redeposition of ZrO_2 partially blocks the pores near the surface. We have endeavored to overcome this by repeated replication (both times) of the same area and comparison of the successive replicas. In general, when this is done no evidence for the presence of a G.B. zone free of pores is seen. Only in the case of a specimen exposed in 0.1 M LiOH + H_3BO_3 (1000 ppm B) was anything like this seen. In this instance, the solution may have been close to the critical concentration of H_3BO_3 needed to plug pores, and if pores near to G.B.s were smaller, these might have plugged first.

J. Piippo[2] *(written discussion)*—At what temperature did you perform the electrochemical impedance spectroscopy measurements?

B. Cox et al. (authors' closure)—Room temperature.

C. Lemaignan[3] *(written discussion)*—In all the experiments where the sample is placed back in the autoclave, we have, at the same time, oxide growth and oxide dissolution by the solutions of LiOH. Have you tried to analyze the behavior of the zirconia only, for instance by putting ZrO_2 with removed base metal in the Li-bearing autoclave?

B. Cox et al. (authors' closure)—No, we have not yet done any studies on ZrO_2 alone. At present, we have plans to look at samples of bulk ZrO_2 with various stabilizing elements to see if we can see any differences in ZrO_2 solubility. The oxides on our specimens are too thin to do tests such as you suggest. We have stripped them for TEM studies, but only get small pieces to examine that would be unsuitable for autoclaving.

P. Billot[4] *(written discussion)*—You mentioned in your presentation that above a critical concentration between 70 and 700 ppm, the development of porosity throughout the oxide film occurs with no pretransition phase. When corrosion conditions imposed are close to 1.5 and 2 ppm of lithium, how can we obtain this critical value of lithium into the growing oxide films?

B. Cox et al. (authors' closure)—There are two ways to achieve high LiOH concentrations in ZrO_2 films from an initially low LiOH external environment. The first is heat flux. If the temperature at the oxide/metal interface exceeds the saturation temperature of the system by about 5°C degrees, the elevation of the boiling point of water by LiOH will result in an equilibrium LiOH concentration at the O/M interface of 1.0 M. In the absence of heat flux, if the pores in the oxide are small enough to restrict mixing of the solution in the pores with the bulk solution, the oxidation reaction that forms fresh ZrO_2 removes water from the LiOH solution and concentrates it. This can be an efficient process as shown by enhanced concentration in crevices in low LiOH concentration solutions.

---

[1] Atomic Energy of Canada, Ltd., Chalk River, Ontario, Canada.
[2] VTT Manufacturing Technology, Finland.
[3] CEA, Grenoble, France.
[4] Commissariat a l'Energie Atomique, St. Paul lez Durance, France.

B. Cheng,[1] P. M. Gilmore,[2] and H. H. Klepfer[2]

PWR Zircaloy Fuel Cladding Corrosion Performance, Mechanisms, and Modeling

REFERENCE: Cheng, B., Gilmore, P. M., and Klepfer, H. H., **"PWR Zircaloy Fuel Cladding Corrosion Performance, Mechanisms, and Modeling,"** *Zirconium in the Nuclear Industry: Eleventh International Symposium, ASTM STP 1295*, E. R. Bradley and G. P. Sabol, Eds., American Society for Testing and Materials, 1996, pp. 137–160.

ABSTRACT: Oxide axial profiles for Zircaloy-4 cladding measured for 17 fuel rods irradiated from 50 to 60 GWD/t in three different pressurized water reactors (PWRs) were studied. Computer simulation of the magnitude and shape of the various profiles involved five contributing interrelated effects: (1) thermal feedback, (2) radial thermal redistribution of corrosion hydrogen, (3) metallurgical variables (tin content and radiation-induced changes in intermetallic particle size distribution), (4) lithium hydroxide exposure history, and (5) radiation effects in the oxide film. Published research results supporting each separate effect are reviewed, and the selection of a predictive simulation for each phenomenon and the effect of varying its relative weight are reported. The effect on cladding corrosion of hydrogen thermal redistribution to the oxide-metal interface, coupled with thermal feedback, was found to be a predominant factor in predicting the magnitude and shape of cladding oxide profiles at high burnups for PWR fuel rods. Tin content, coupled with thermal feedback, was also found to have a very important effect on the predicted magnitude of oxide profiles at all burnups. Radiation-induced dissolution of fine $Zr(Cr,Fe)_2$ intermetallic particles in the Zircaloy-4 cladding primarily affected the predicted shape of the oxide profiles at the colder inlet region of the fuel rods. Lithium hydroxide (LiOH) exposure history and radiation effects in the oxide film appeared to play non-negligible secondary roles in determining the oxide profiles at all burnups.

KEYWORDS: PWR Zircaloy-4 cladding, high burnup fuel rods, in-reactor corrosion behavior, corrosion modeling, oxide thickness, LiOH, tin content, effect of hydrogen thermal redistribution on corrosion, neutron damage on oxide films, intermetallic particles

Improving fuel cycle economy will require extending fuel burnup by means of longer cycle operation, longer total residence time, and higher heat rating. The increased duty on fuel cladding has dictated improving cladding corrosion resistance to assure fuel reliability. Improved cladding corrosion resistance is being observed due to improvements in Zircaloy-4 composition, especially lower tin content, and due to improved cladding process control. It can also be attributed to new commercial products such as Zirlo and extra-low tin (ELS) duplex cladding [1–5]. Based on various considerations, including oxide spalling, wall thinning, and cladding hydriding, maintaining the oxide thickness below 90 to 120 μm is highly desirable to maintain fuel reliability. There is a need to ascertain that the zirconium oxide thickness on cladding will not exceed either licensing or reliability limits at high burnups. Meeting this need demands a well-benchmarked and qualified model or code for zirconium oxide thickness calculations.

Several PWR fuel rod corrosion models have been developed for interpreting Zircaloy-4 cladding oxide data and predicting cladding oxide thickness for fuel reload design applications.

---

[1] Manager, Light Water Reactor Fuel Reliability, Electric Power Research Institute (EPRI), 3412 Hillview Avenue, P.O. Box 10412, Palo Alto, CA 94304.

[2] Consulting engineer and vice president, respectively, S. Levy Incorporated, 3425 S. Bascom Avenue, Campbell, CA 95008.

The EPRI/KWU/CE model developed by Garzarolli et al. [6] took a simple approach of attributing all in-reactor corrosion to an effect of fast neutron flux on the protectiveness of the zirconium oxide. This approach provides a useful tool for interpreting some fuel rod oxide data, but is not accurate enough, particularly in interpreting the axial oxide profiles [7]. The COCHISE model developed by Billot et al. [8,9] and a Westinghouse model [4] both incorporate a lithium (Li) term to interpret the effect of LiOH in the primary coolant on the oxide growth rate. In both cases, the observed corrosion rate acceleration at high burnups is attributed mainly to a corrosion enhancement effect of lithium. But such an approach cannot be supported fully by laboratory data [10,11] and in-core loop simulation tests [12]. On the contrary, research results published by Garde et al. [13] and by Kim et al. [14] show that oxide growth kinetics are increased by the hydrides accumulated at the metal-oxide interface. Recent laboratory studies by Kido [16] indicate a significant effect of massive hydrides on Zircaloy-4 corrosion. Recent results by Cheng et al. [15] suggest that the uniform oxide growth rate can be increased by neutron-induced amorphization and dissolution of precipitates, $Zr(Cr,Fe)_2$, with an associated redistribution of iron (Fe) to the Zircaloy matrix. The degree of both the hydride accumulation at the metal-oxide interface and the neutron-induced precipitate dissolution are expected to increase with increasing burnup. Therefore, both must be considered, in addition to LiOH, in interpreting high burnup oxide data.

With proper interpretation of high burnup oxide data, it is anticipated that high burnup cladding corrosion behavior can be more accurately predicted and better corrosion reduction strategies can be devised. The objective of this paper is to evaluate all variables reported to be contributing to the uniform oxide growth kinetics of Zircaloy-4 cladding in PWRs, particularly at high burnups. The relative effects of these variables, including material chemistry and cladding processing, irradiation damages, coolant chemistry, and hydrides, are treated mathematically and incorporated into a new parametric corrosion code named EPRI PWR Fuel Cladding Corrosion (PFCC).

Approaches

Thermal history normalization is required to account for the effects of temperature variation on corrosion behavior. Coolant inlet and outlet temperatures differ from plant to plant. Power histories vary from cycle to cycle in the same plant and also from rod to rod within an operating cycle. A PC-based single channel thermal hydraulic model was developed under EPRI sponsorship to calculate the fuel rod surface temperature along the rod length [17]. This thermal hydraulic model was benchmarked against more elaborate models. The thermal hydraulic model has been coupled with various corrosion models to assess their relative predictive value [7].

The second issue in predicting cladding oxide thickness is accounting for the affect of separate effects described in published research results. Each separate effect introduces a feedback to all other single effects contributing to the aggregate prediction. This feedback occurs because, for the oxidation of Zircaloy cladding under the influence of an inside-to-outside heat flux, any contribution to oxide thickening increases the rate-controlling oxide-metal interface temperature. An iterative calculational loop is necessary. A mathematical consequence is that the combination of modest single effects produces large aggregate effects, especially at high oxide thicknesses.

Large aggregate effects from the separate in-reactor variables can be seen in the oxide data in Fig. 1, where the rod peak oxide thickness data of 17 fuel rods from three PWRs are compared. Some characteristics of the three different PWRs and the different fuel rod dimensions of the fuel rods included in this study are given in Table 1. Some rods were measured for oxide thickness at two consecutive outages. After 500 to 2500 days, the observed corrosion rates vary by a factor of 3 and the rod peak oxide thicknesses were up to five times the value

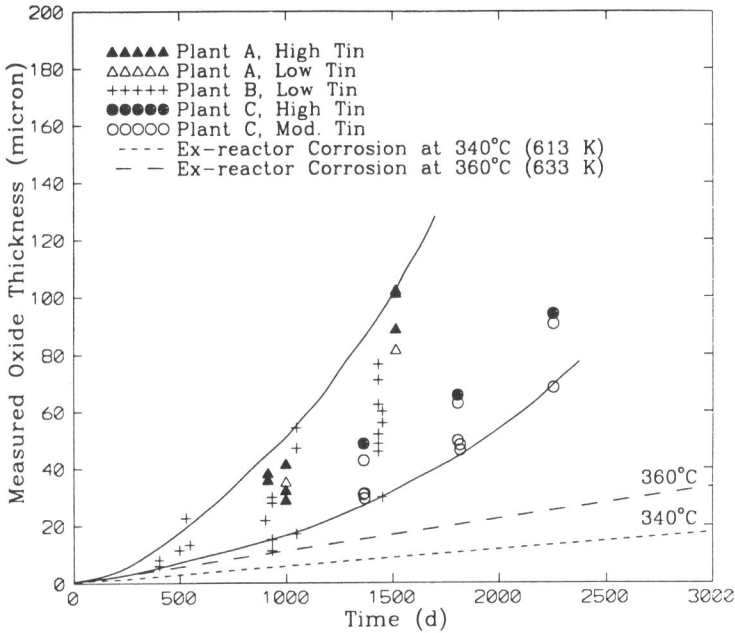

FIG. 1—*Variations in measured maximum oxide growth rates for 17 different fuel rods irradiated in three different reactors and compared to ex-reactor rates.*

predicted from ex-reactor tests conducted at 360°C (633 K). Since the peak metal-oxide interface temperatures in these 17 rods have been calculated in the range of 310 to 360°C (583 to 633 K), the added in-reactor corrosion can be attributed to the different aggregate effects of separate in-reactor variables, including in-reactor changes in material variables.

The range of different oxide profile shapes for the 17 rods studied is illustrated in Fig. 2. It shows the measured axial oxide profiles for two rods with comparable burnups and tin contents. Both a highly top-peaked profile and a flatter axial profile must be explained. In part to address the effect of each parameter, emphasis focused on the cladding oxide thickness profiles from each single fuel rod under study. This emphasis imposes additional constraints on over-weighting the importance of any individual separate effect in the oxide thickness growth simulations.

In the course of this work, the published EPRI oxide growth model [6] was improved by incorporating separate effects for cladding alloy tin content and iron content, for burnup-de-

TABLE 1—*Characteristics of Plants A, B and C shown in Fig. 1.*

	Plant A	Plant B	Plant C
Cladding diameter, cm	0.95	0.95	1.12
Wall thickness, cm	0.11	0.11	0.14
System pressure, mPa	15.51	15.51	15.51
Coolant inlet temperature, °C	290	293	285
Coolant inlet flow rate, kg/m^2-s	3581	3852	3675
Average power density, kW/m	18.7	17.8	20.6
Cycles length, days	319;281;399;516	530;404;499	438;454;472;442;449

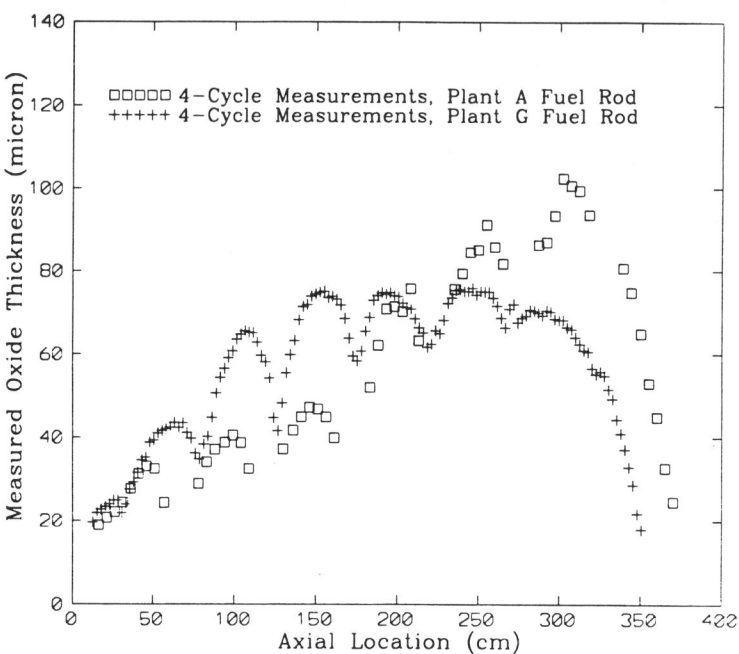

FIG. 2—*Illustration of different axial oxide profiles for high-burnup PWR fuel rods: Plant A—54 GWD/t, Plant G—46 GWD/t, comparable tin content, different power histories.*

pendent, radiation-induced changes in intermetallic particles, for time-dependent buildup of hydrides at the metal-oxide interface, and for LiOH history. The process by which the aggregate corrosion rate equations and the separate effects formulations were developed involved: (1) derivation or construction of provisional equations and formulations employing published technology; (2) adjusting provisional separate effects formulations for consistency with published experimental data; (3) iterative comparisons of the resulting aggregate corrosion code predictions against measured oxide profiles for the 17 benchmark fuel rods employed; and (4) repeating this entire process cycle three times. Over 100 trial formulations (each with predicted oxide profiles compared to 6 to 17 measured benchmark rod oxide profiles) were involved for each of the three process cycles. The predictions for each 17 rod profile trial were compared at each axial measurement point. The benchmark rods provided a total of 1369 axial point-to-point comparisons for each trial.

Results and Discussion

The corrosion of the Zircaloys is known to involve two distinct stages: the cubic growth of thin oxide (<2 μm) and a transition to linear growth of thicker oxides. For Zircaloy-4 PWR fuel cladding, transition to linear corrosion occurs early, after 100 days or less of a 1500 to 2200-day exposure.

Aggregate Effects of In-Reactor Variables

The starting point for the in-reactor pre-transition cubic equations in the EPRI PFCC code were those suggested by Billot and co-workers [8,9]. Their equations reflected essentially no difference in pre-transition oxidation rate from those observed in ex-reactor heated loop tests. However, 117-day exposure experiments in the Halden test reactor [18] were noted to result in oxide thicknesses greater than the 2-μm oxide thickness associated with transition. Ex-reactor times to transition reported by Hillner [19] and by others were found to be very much longer than 117 days at the corrosion temperatures involved in the Halden tests. The unavoidable conclusion from these experimental results is that there is an effect of in-reactor environment on time to transition.

Each separate effect was considered as a possible contributor to in-reactor reduction in time to transition. By the process of elimination, it was decided to use the Billot pre-transition expression to reach 2 μm thickness in a time inversely proportional to a lithium hydroxide factor (F_{Li1}). In addition, to cover cases of very low cumulative annealing parameter (ΣA_i) cladding, ultrafine particle size, and lower-than-Halden cladding temperatures, the time-to-transition expression was also made inversely proportional to the post-transition intermetallic particle dissolution factor, F_{Fe} (described later).

The resulting PFCC in-reactor pre-transition expressions are

$$\frac{ds^3}{dt} = C_1 \cdot \exp\left(\frac{-Q_1}{RT_{mo}}\right) \cdot F_{Li1} \cdot [1 + F_{Fe}] \qquad (1)$$

and

$$t_{trans} = \frac{s_{trans}^3}{\left(\dfrac{ds^3}{dt}\right)} \qquad (2)$$

Equations 1 and 2, when applied with the PFCC post-transition expression, were benchmarked against short-term Halden data and the long-term benchmark rod plenum data points available. Agreement was achieved within the rather large uncertainty in measuring very thin oxides. On the other hand, assuming either no time to transition (no cubic corrosion) or assuming the ex-reactor times to transition gave very much poorer agreement.

The following mathematical expression in the EPRI-PFCC code accounts for the separate effects on the post-transition linear kinetics:

$$\frac{ds}{dt} = C_2 \cdot \exp(-Q_2/RT_{mo}) \cdot F_{Li2} \cdot F_{Sn} \cdot F_{QQ} \cdot [1 + F_{H2} + F_{Fe} + F_\phi] \qquad (3)$$

where the symbols in Eqs 1, 2, and 3 are:

s = oxide thickness, μm,
t = time, days,
C_1 = pre-transition rate constant, μm^3/days,
Q_1/R = pre-transition activation energy, K,
F_{Li1} = exp $[(0.17[Li] - 20.4[Li]/T) \cdot f]$ as proposed by Billot,
[Li] = lithium concentration, ppm,
f = fitting constant,
C_2 = post-transition rate constant, μm/days,
F_{Li2} = post-transition lithium hydroxide enhancement factor,
F_{Sn} = tin content enhancement factor,
F_{QQ} = heat flux (QQ) enhancement,
F_{H2} = hydrogen (H2) radial redistribution enhancement factor,

F_{Fe} = precipitate irradiation effect enhancement factor,
F_ϕ = radiation effect enhancement factor,
Q_2/R = activation energy, K, and
T_{mo} = metal-oxide interface temperature, K.

and where the metal-oxide interface temperature, T_{mo}, is calculated from the single channel model-calculated cladding wall temperature using 1.5 W/m-K for the thermal conductivity of zirconium oxide (ZrO_2). Each of the separate effects (F_{xx}) are discussed in the next section.

In Eq 3, the first two factors on the right-hand side, $C_2 \cdot \exp(-Q_2/RT_{mo})$, can be viewed as describing the post-transition, ex-reactor corrosion rate of Zircaloy-4 in isothermal corrosion tests. In fact, the value of Q_2 is exactly that determined by Hillner from such tests and C_2 is within 7.6% of Hillner's value. The 7.6% higher pre-exponential constant C_2 was derived by fitting to Billot's ex-reactor heat transfer loop test data [8,9].

In Eq 3, all of the factors outside the brackets could be viewed as the expression for oxide growth rate in relatively short-term exposures of Zircaloy-4 cladding tubes in ex-reactor heat transfer loops with various lithium/boron chemistries. In fact, the factors outside the brackets were constrained in this investigation to be consistent with the ex-reactor loop test results reported by Billot and co-workers [8,9]. For relatively short-term exposures, radial hydrogen redistribution effects on corrosion are not expected to be significant and therefore F_{H2} need not be included outside the brackets. The terms inside the brackets in Eq 3 could be viewed as describing incremental contributions to oxide growth rate present (F_{Fe} and F_ϕ) or become important (F_{H2}) only in-reactor after long times (times greater than a year, for example). Alternative expressions, including one with the F_{Li2} factor as a term inside the bracket and one with F_{H2} outside the bracket, were tested; but the expression given as Eq 3 produced predictions in better agreement for the oxide profiles for all 17 benchmark rods.

Figures 3a and 3b present the measured versus predicted axial profile for a low tin (1.3 to 1.4 wt%) rod with a rod average burnup of 52 GWD/t. The rod was particularly well characterized prior to service; the ΣA_i was about 8×10^{-18} h. The oxide profile predicted by the aggregate effects of the separate variables in Eq 3 is consistent with the measured values along the entire rod except for the three badly overpredicted points at the last span (exit) region. Overpredictions for the exit span were not uncommon for rods in the PFCC studies; the accuracy of the coupled nuclear-thermal hydraulic inputs and the possible effect of axial hydrogen redistribution in this region are under study.

When rods less well characterized for cladding tin content and metallurgical variables are employed, the agreement between prediction and measured oxide thickness understandably is not as good for well-characterized cladding cases. Figure 4 presents the comparison of EPRI-PFCC prediction to measured oxide thickness data for the 17 benchmark rods shown in Fig. 1. These 17 rods were from three different PWRs (six from Plant A, six from Plant B, and five from Plant C). The power histories, rod average burnup, and hot operating residence times were unique to each rod. The average burnups ranged from 50 to 60 GWD/t. The peak lithium concentrations in the three plants differed, ranging from 1.2 to 3.5 ppm. These rods came from two different fuel fabrication vendors and were irradiated in two different fuel assembly designs. Both normal and purposely low-tin cladding are represented. All the prediction-to-measurement points along the profiles are plotted, resulting in a total of 1369 comparisons.

Figure 4 presents three sets of dotted lines. Those closest to the solid, exact prediction-equals-measurement line reflect the ± 5 μm uncertainty reported for measuring oxide thickness. The next-closest lines reflect a preliminary code input value uncertainty effect estimated at ± 10 μm. The dotted lines farthest from the perfect agreement line represent the maximum observation of ± 20 μm values for side-by-side rods. Only the six rods from Plant A had complete measurements for all metallurgical variables used in the PFCC code. The excellent comparisons

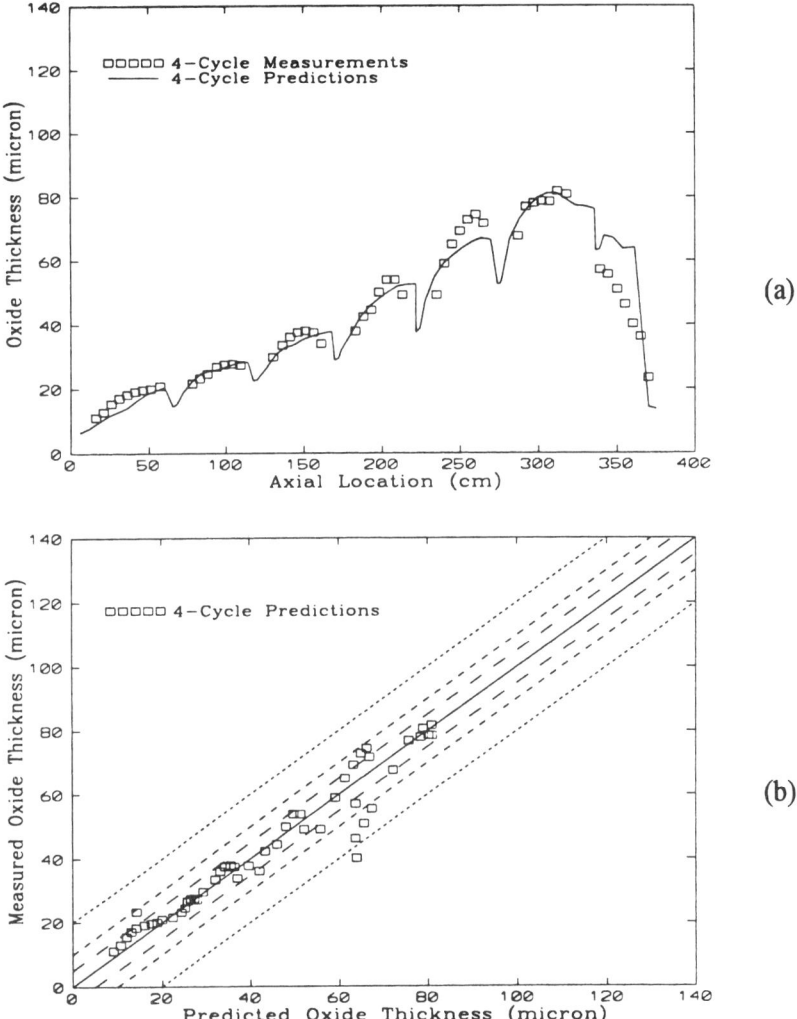

FIG. 3—*Measured and PFCC-predicted oxide profiles for a Plant A fuel rod.*

for these rods are represented by triangles in Fig. 4. For the other rods, estimated or default values had to be used for the predictions. Most of the data from Plant B with deviations greater than +20 µm were from a single fuel batch, and the large deviation is believed to result mostly from unspecified material variability.

Separate Effects Mechanisms

The separate effects on Zircaloy cladding corrosion incorporated in Eq 3 are discussed in the following order in this section:

1. Thermal feedback, $\exp(-Q_2/RT_{mo})$.
2. Radiation effects in oxide films, F_ϕ.

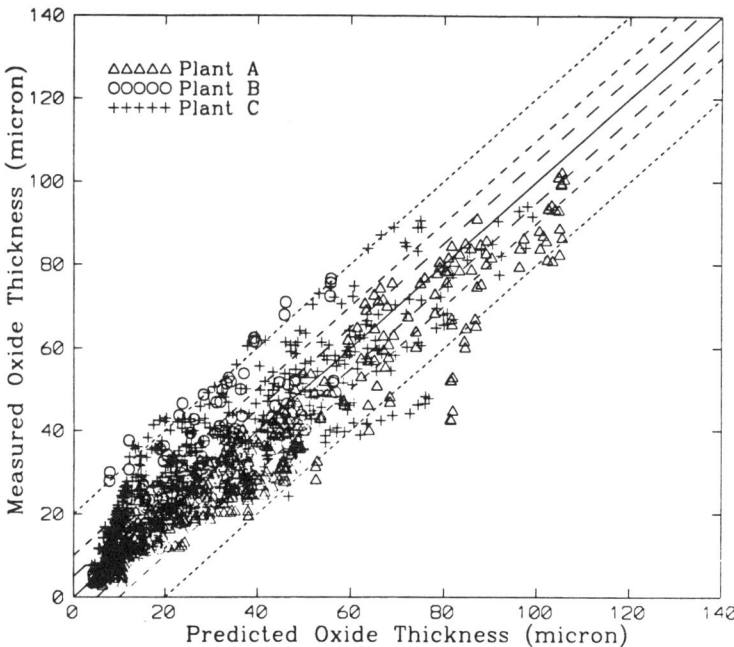

FIG. 4—*Measured versus PFCC-predicted oxide thickness at 1369 comparison points on 17 PWR fuel rods irradiated to high burnup in three different plants.*

3. Corrosion hydrogen radial thermal redistribution, F_{H2}.
4. Metallurgical variables: Sn, and precipitate irradiation effects, F_{Sn} and F_{Fe}.
5. Lithium hydroxide (boron) exposure, F_{Li}.

For each, the published research results supporting the effect are mentioned, the selection of a predictive simulation is discussed, the effect of varying the relative weighting of that effect on predicted fuel cladding oxide profile is noted, and for the last four effects the consequence of not including this effect (and its contribution to thermal feedback) is presented to illustrate the relative importance of each separate effect. The heat flux factor (F_{QQ}) is a normalizing or fitting factor that was, in any event, found to be very small (less than 1.09) and is not discussed below. The more important heat flux effects are imbedded in the coolant surface and metal-oxide temperature calculations, in the redistribution of hydrogen calculations, and in the intermetallic particle dissolution calculations. Discussion of these matters is included below.

Thermal Feedback

Ex-reactor, linear post-transition (>2 μm oxide thickness) corrosion data for Zircaloy-4, taken at various constant temperatures of interest for PWR applications and at very long times, were reported by Hillner [*19*] to follow the familiar Arrhenius expression

$$\frac{ds}{dt} = k \exp(-Q_2/RT)$$

where k is an invariant frequency factor (μm/days), Q_2 is an invariant activation energy having a value of 24 825 cal/mol, R is the gas constant (cal/mol-K), and T is the corrosion temperature in kelvin.

Some of the early in-reactor corrosion correlations modified the frequency factor by an invariant "in-reactor corrosion acceleration" factor having a value of 2 to 3 and defined T as the metal-oxide interface temperature calculated under the appropriate heat flux [20]. Later, the frequency factor invariance was abandoned and proposed to depend on in-reactor variables, for example neutron flux [6], metallurgical variables [21], and lithium concentration [8,9]. The k value in the present PFCC code is represented by the various single effects shown in Eq 3.

The literature expresses no doubt that in-reactor corrosion of Zircaloy cladding involves an exponential thermal feedback. But the quantitative value of the activation energy has been assigned different values by different investigators. The activation energy was changed from the ex-reactor invariant value to a different invariant value that better correlated in-reactor data. Billot proposed abandoning an invariant activation energy for one dependent on lithium content [8].

The EPRI PFCC studies also explore a variable-dependent activation energy to improve predictive capability. The variables affecting activation energy are discussed for the appropriate single effects elsewhere in this section. However, in the studies reported here, the activation energy variation (from that established by Hillner) is constrained to that consistent with the temperature effect along a single rod, i.e., the oxide profiles constrain acceptable activation energy values.

The effect on the predicted oxide profile of varying the activation energy value from that found by Hillner was studied for the PFCC formulation. Increasing the activation energy has the expected effect of predicting an oxide profile more peaked at the next-to-the-exit (hottest) span in the fuel assembly. The predicted profiles are flatter for lower activation energies as expected. The best predicted profile fits for all 17 rods (whose oxide growths are shown in Fig. 1) were produced using the Hillner value (modified by metal-oxide interface hydrogen content and coolant Li concentration, as discussed below).

Radiation Effects in Oxide Films

The fast neutron flux imposed on PWR cladding is on the order 10^{13} to 10^{14} n/cm^2-s. Garzarolli [6] reviewed evidence that high neutron flux affects either or both (1) the structure of the oxide films on Zircaloy cladding and (2) the chemistry within pores in the oxide. Either change in the oxide was expected to accelerate the Zircaloy corrosion by an amount depending on the intensity of the neutron flux (ϕ). Based on correlating available data (largely lower burnup), Garzarolli proposed a corrosion predictor in which the frequency factor depended on fast neutron flux taken to the 0.24 power.

In the studies reported here, this fast flux predictor was tested against the Plant A high burnup rod oxide profiles. The fast flux predictor consistently under-predicted the rod inlet peak oxide thicknesses, while the maximum oxide peaks were over-predicted. However, because the evidence for radiation damage in the oxide is expected and experimentally observed [6], a flux dependent term, $c \cdot \phi^{0.24}$ (where c is a fitting constant), was retained in the PFCC formulation. When the relative importance (the value of c) was increased in the PFCC formulation, it exaggerated the peaking in the predicted profile with little effect on the predicted inlet oxide thicknesses.

If the oxide damage term is omitted from the PFCC formulation, the effect on the predicted oxide profile displayed in Fig. 3a is modified as shown in Fig. 5. The radiation effects in the

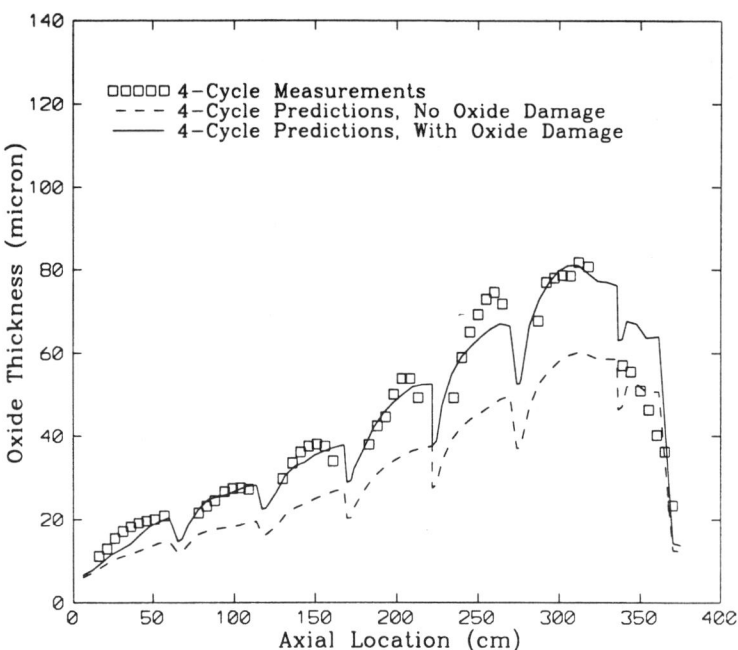

FIG. 5—*Illustration of PFCC prediction of separate effect of radiation damage in oxide factor.*

oxide film, while important for predicting axial oxide profiles, do not dominate in PFCC predictions for peak oxide thickness.

Corrosion Hydrogen Thermal Redistribution

Earlier hot cell examination results reported by Garde [*13*] and Kim [*14*] clearly show a correlation of increased oxide thickness with increased hydride content at the metal-oxide interface. Recent hot cell studies (in programs sponsored by EPRI [*22*], EPRI and Empire State Electric Energy Research Corporation [*23*], and the United States Department of Energy (DOE) [*24*]) have produced further data on the corrosion enhancement due to hydrides. Re-examination of these data allowed quantifying the effects of hydrides. Local oxide thickness enhancements of greater than 78% were found to be associated with local regions of high hydride precipitation density. Since these comparisons were made between locations with a hydride rim and adjacent locations with various amounts of hydrides, the real effect of hydrides alone should be greater than 78%. Laboratory tests performed at 360°C (633 K) by Kido [*16*] show a 170% increase in oxide thickness when the specimen was precharged with hydrogen to about 2200 ppm. The results in Ref *16* taken at 360°C (633 K) are the only available data systematically describing the dependence of Zircaloy-4 corrosion on hydrogen content. Clearly additional data at other temperatures would be useful. Blat and Noel [*25*] reported at this symposium on some data taken at 400°C (673 K). Their data not only confirm a significant increase in Zircaloy-4 corrosion rate when a hydride rim is present at the metal/oxide interface, but appear to be in good quantitative agreement with Kido's data [*16*]. The results reported in Ref *16* show that the corrosion of Zircaloy-4 depends (above the solubility limit) on the logarithm of the hydrogen content at the corroding metal-oxide interface.

The present PFCC code incorporates the thermal redistribution of hydrides in fuel cladding and its effect on cladding corrosion as the F_{H2} factor in Eq 3. The mathematical simulation for hydrogen radial thermal redistribution employed is that of White, Sawatsky, and Woo [26] modified to add the influx of corrosion hydrogen at the cold outer surface. The complex differential equations for radial thermal redistribution in PFCC require very long computation times using personal computers. A simplified approach is to generate sets of curves for predicted coldside hydrogen content versus oxide thickness such as that shown as Fig. 6 using Plant A rod conditions at four heat flux levels. Curves were generated employing different instantaneous hydrogen pickup fractions. The curves shown in Fig. 6 are for a constant instantaneous pickup fraction of 13%. PFCC interpolates for each calculational time step the added hydrogen redistribution employing mathematical tables derived from Fig. 6. The result is an estimated value of the hydrogen content at the metal-oxide interface at each time step. This hydrogen concentration is then used in the Kido expression for the effect of hydrogen on corrosion to give an incremental corrosion rate addition due to hydrogen redistribution.

Better oxide profiles resulted when the instantaneous hydrogen pickup fraction was set to decrease with oxide thickness. Better oxide profiles also resulted when the activation energy for corrosion of Zircaloy/zirconium hydride (ZrHx) mixtures was diminished from the Hillner value in proportion to the metal-oxide interface hydrogen content as given by the expression employed in PFCC:

$$Q_2 = (Q_H - Q_{Hillner}) \cdot \frac{F_{H2}}{F_{Rim}} + Q_{Hillner}, \text{ cal/mol}$$

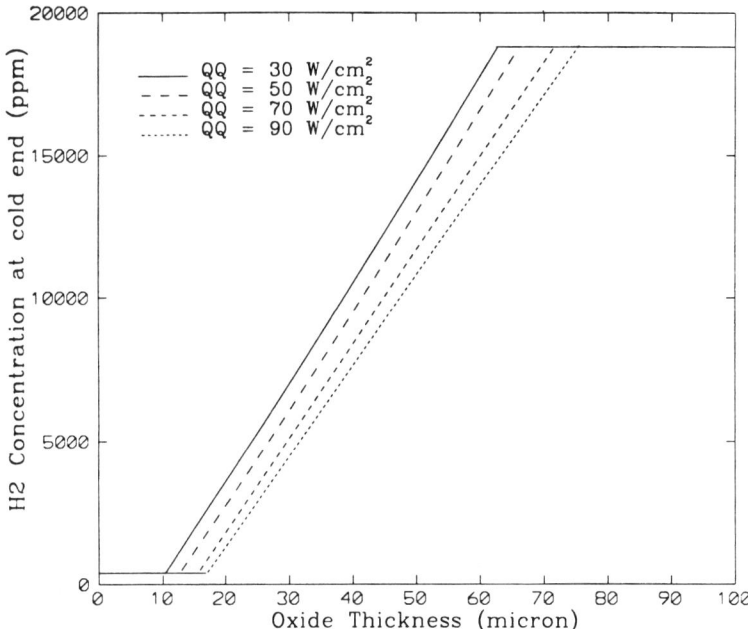

FIG. 6—*Calculated relationship between oxide thickness and cold end hydrogen concentration due to hydrogen thermal redistribution at various heat fluxes and a constant pickup fraction of 13%.*

where

$F_{H2} = 0$ for metal-oxide interface hydrogen content [H$_2$] less than 400 ppm,
0.699 log ([H2]/400) for 400 ppm $\leq$ [H$_2$] $\leq$ 18 811.25 ppm, and
2.691 for [H$_2$] > 18 811.25 ppm (i.e., once a solid μm of ZrHx has formed).

The terms are defined as:

Q_H = activation energy for the diffusion of hydrogen in zirconium, 9135.6 cal/mol,
$Q_{Hillner}$ = Hillner ex-reactor activation energy, 24 825 cal/mol,
F_{H2} = hydrogen enhancement factor,
[H$_2$] = metal-oxide interface hydrogen content, ppm, and
F_{rim} = hydrogen enhancement factor at solid ZrHx rim (i.e. [H$_2$] = 18 811.25 ppm).

The effect of the hydrogen term in PFCC on the predicted oxide profiles is non-existent for the plena positions (no heat flux) and small for low to moderate burnups (limited corrosion hydrogen). At rod oxide thicknesses of about 60 μm, the model predicts a solid rim of ZrHx. This calculated result is consistent with the hot cell results discussed above, thus validating the accuracy of the model calculation. After a solid hydride rim is formed, the hydride effect on corrosion should remain constant with increasing oxide thickness. The thickness of the rim increases but the hydrogen content at the interface stays constant at the value for ZrHx. At intermediate oxide thickness, the hydrogen effect increases with exposure and is magnified by the thermal feedback of the growing oxide insulator.

If the incremental hydrogen effect term in the PFCC code is set to zero (no hydrogen contribution), the predicted profile is that presented in Fig. 7. The measured profile data for this rod are compared to provide perspective on the importance of considering the effect of redistributed hydrogen on the corrosion of high burnup Zircaloy cladding. The cumulative effect of hydride is very large and is the dominant effect on corrosion at high burnups.

Metallurgical Variables

Within the last 20 years, several metallurgical factors have been shown to affect the corrosion behavior of Zircaloy-2 and -4. Some of the research results have been documented in earlier proceedings of this symposia series. Important factors found by various workers are: (1) tin content, (2) precipitate size and distribution, (3) thermomechanical processing variables, (4) carbon content, and (5) silicon content. The effects of tin content, precipitate size and distribution, and thermomechanical processing variables have been broadly characterized and are incorporated in the present PFCC code for quantification. Carbon and silicon increase the uniformity of alloy chemistry (mostly Fe and Cr in Zircaloy-4 and Fe, Cr, and Ni in Zircaloy-2) in the alloy matrix by reducing the Widmanstatten platelet sizes during quenching of Zircaloy billets from the β-phase [27]. Smaller platelet sizes serve to refine the precipitate sizes and increase the uniform distribution of precipitates when combined with proper thermomechanical process control. Since the effects of carbon and silicon are most likely due to their effects on the distribution of Fe and Cr and the Zr(Cr,Fe)$_2$ precipitate, the PFCC code does not treat these elements separately.

Tin Content—The results in Refs *1–4* suggest that improved low-tin cladding can reduce the oxide thickness at high burnups by about 20 to 40%. The published data include tin content from ~1.2 to 1.4 wt%. The role of tin in the oxide growth mechanism is not clear, and there has been little discussion in published research results.

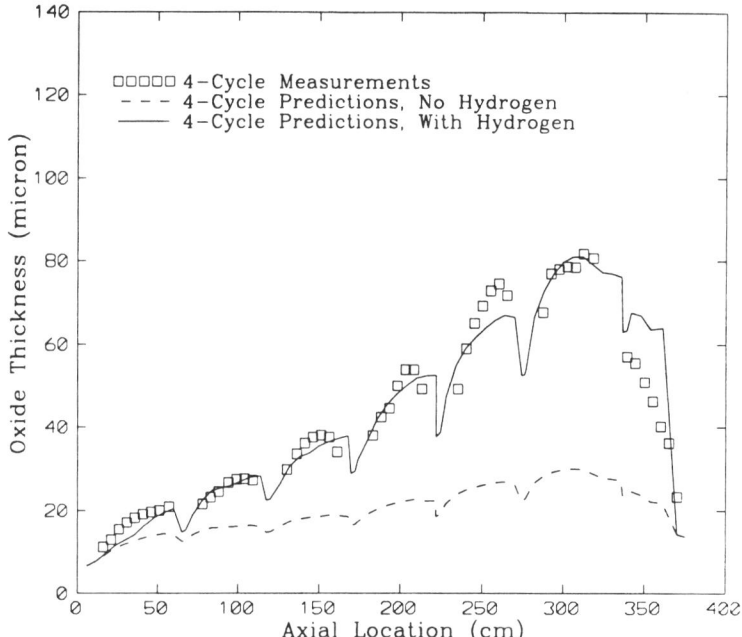

FIG. 7—*Illustration of PFCC prediction of the separate effect of radial thermal hydrogen redistribution factor.*

The high burnup oxide measurements studied in PFCC included those for cladding with low-tin content (<1.4 wt%) and standard tin content (1.45 to 1.6 wt%). A direct comparison was possible. A pre-exponential factor dependent on tin content for the PFCC code was determined by iterative calculations with trial formulations until the best fit for all 17 benchmark rods was obtained. This also involved repeating iterations as other factors in the predictor were refined.

The empirical and predicted oxide profiles for low-tin and high-tin cladding on two comparable rods irradiated together in Plant A are presented in Fig. 8. PFCC accounts for an important tin content effect that influences the entire thin and thick film profile. The difference between the peak oxide thicknesses as measured and predicted for the low-tin and standard rods is 20 to 30%. This range is consistent with the fuel cladding oxide results reported earlier for low versus standard tin content [1–4].

Intermetallic Particles—The size and distribution or the volume density of intermetallic precipitates has been shown to have strong effects on the nodular corrosion resistance of Zircaloy by Cheng et al. [28] and on the uniform corrosion behavior by Garzarolli et al. [29] and Schemel et al. [30]. In the earlier mechanistic studies by Cheng and Adamson [31], Zircaloy corrosion was demonstrated to be controlled by the concentrations of solute elements, mainly Fe, Ni, and Cr, in the alloy matrix. Based on the corrosion mechanism proposed by Cheng and Adamson, a matrix supersaturated with Fe and Ni will have a higher uniform corrosion rate. Garzarolli et al. later demonstrated such a correlation using the cumulative annealing parameter ΣA_i as an indicator [29]. Zircaloy cladding processed using a ΣA_i of less than 10^{-19} h usually possesses excellent nodular corrosion resistance but has a relatively higher uniform corrosion

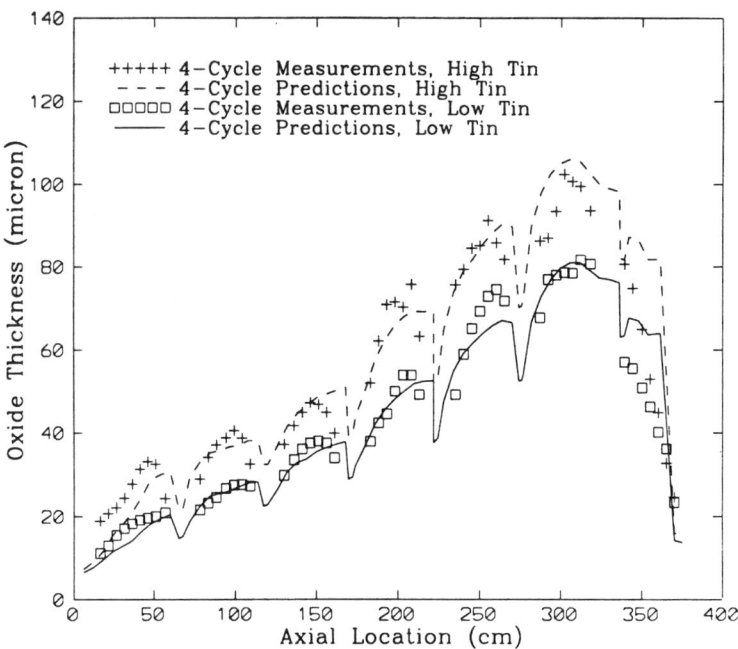

FIG. 8—*Illustration of PFCC prediction and experimental observation of the separate effect of tin variation factor.*

rate. Fabrication to produce $\Sigma A_i < 10^{-19}$ h is currently used by boiling water reactor (BWR) cladding manufacturers. With a ΣA_i greater than 10^{-18} h, Zircaloy cladding normally possesses poor resistance to nodular corrosion, but has better resistance to uniform corrosion. Thus, a ΣA_i of 10^{-18} to 10^{-17} h has been adopted by PWR cladding manufacturers.

A consequence of using low ΣA_i for BWRs is that the intermetallic particles become smaller, mostly less than 0.1 μm in size, and the number density is high. The larger ΣA_i for PWRs results in larger particles, mostly in the range of 0.1 to 0.5 μm. In power reactor environments, the size of the particles can strongly influence the corrosion performance because the particles, particularly the small ones, can dissolve due to fast neutron irradiation, as reported by Cheng et al. [32], Yang et al. [33], and Griffith et al. [34] and others in subsequent studies. Post-irradiation electron microscopy studies by the researchers cited demonstrated that individual intermetallic particles are changed starting with the formation of outer rings of amorphous material and working inward until the particle is dissolved. Motta [35] quantified the ring formation progression as about 1 μm per 10^{23} *nvt* fast neutron fluence and presented evidence that the dissolution phenomenon was irradiation exposure-temperature dependent. Also, amorphization and subsequent dissolution of the particles result from diffusion of Fe from the intermetallic lattices into the alloy matrix, where irradiation defects are produced to allow low-temperature diffusion and supersaturation of Fe in Zircaloy. However, most results on amorphization and dissolution of intermetallic particles have been obtained on BWR cladding maintained at a temperature of ~290 to 310°C (563 to 583 K) during in-reactor operation. Motta's results suggest that the rate of amorphization will decrease with increasing temperature, probably because of a lesser degree of excess radiation-produced lattice defects with increasing temperature.

The formulation for the intermetallic particle resolution effect used in PFCC is based on a temperature-dependent dissolution rate. Motta's findings can be expressed as:

$$R = k \cdot \phi \cdot t$$

where R is the thickness of the amorphous ring (μm), $k = 1 \times 10^{-23}$ μm per nvt, ϕ is the fast neutron flux (n/cm^2-s), and t is the exposure time (s). This formula implies that all particles less than $2R$ in diameter will be completely dissolved at a fluence of $\phi \cdot t$. Given the particle size distribution and the fast flux, the fraction of the particles dissolved at any time interval during the Zircaloy cladding exposure can be calculated, the finer particles dissolving first. Since the dissolution process is a temperature-dependent phenomenon, the Motta expression was modified to make k a function of absolute temperature.

The temperature-dependent formulation of k in PFCC, shown as Fig. 9, was estimated by extrapolation of Hood and Schultz's diffusion data [36] for iron in zirconium at temperatures above the athermal-thermal transition temperature (303°C, 576 K) for Zr(Cr,Fe)$_2$ suggested by Motta [35]. In the lower temperature regime, the formulation was set to correctly predict the amorphous rim thickness values measured from the electron micrographs published by Griffith [34] for particles irradiated to various reported fluences at fixed irradiation temperatures. In the higher temperature regime, amorphization would be retarded and its effect on corrosion reduced. The resulting formulation calculates both the fraction of particles dissolved and the increase in inter-particle midpoint iron content. Iterative calculations determined that the benchmark rod oxide profiles could best be satisfied by setting the incremental corrosion rate increase term proportional to the fraction of particles dissolved.

Note that the temperature of the metal-oxide interface varies along the length of a PWR

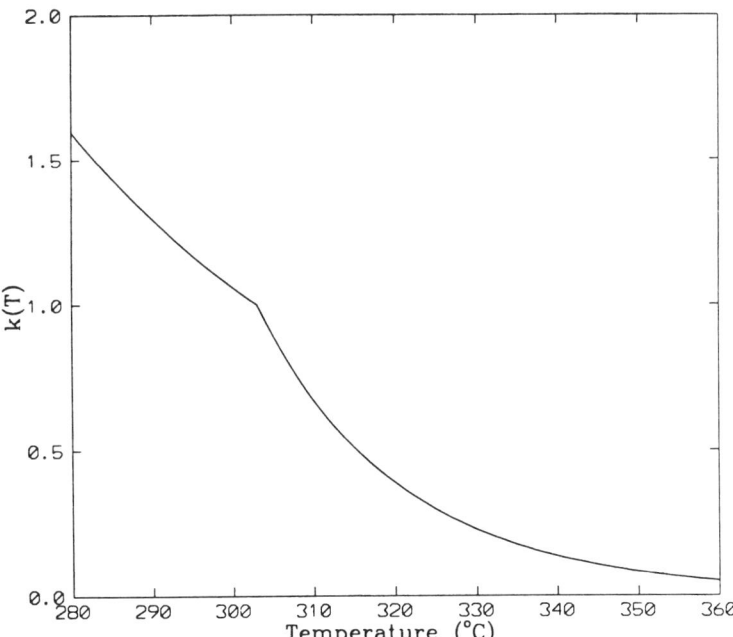

FIG. 9—Temperature-dependent k(T) factor in in-reactor dissolution of intermetallic particles.

cladding tube. The range of interface temperature along a single cladding tube over its exposure to 60 GWD/t was found in the PFCC studies to be 288 to 360°C (561 to 633 K). Thus, the particle dissolution must be computed for each time interval for the appropriate temperature at that time and integrated over time.

The temperature dependence in the formulation leads to predicting that little incremental corrosion effect of intermetallic particle dissolution occurs at the hottest locations along a PWR fuel rod. At intermediate locations, especially early in exposure, the effect on corrosion is large, and, due to thermal feedback, persists throughout life. The incremental effect is predicted to be highest at the coldest inlet peaks. Increasing the importance rating of the intermetallic particle dissolution effect in PFCC increased predicted inlet span oxide thicknesses with little effect on predicted highest oxide thicknesses, i.e., increasing the relative weighting flattened the predicted profile.

The effect of setting the intermetallic particle term to zero (assumption of no effect on oxidation) is illustrated in Fig. 10a for the same oxide profile presented throughout this discussion. The cladding from Plant A has an estimated ΣA_i of $\sim 8 \times 10^{-18}$ h, and the particle size is relatively large. Thus, the particle dissolution effect on corrosion is small, particularly near the peak location, where the temperature is higher ($>340°C$, 613 K). The corrosion effect due to particle dissolution is important near the inlet, where metal-oxide temperature is much lower ($<300°C$, 573 K).

Figure 10b shows the results of a simulation study to illustrate the effect of particle size distribution or ΣA_i on the axial corrosion profile. Increasing the ΣA_i value from 8×10^{-18} to 6×10^{-17} h only slightly decreases the oxide thickness at the peak power location. On the other hand, a low ΣA_i of 3×10^{-19} h is predicted to significantly increase the corrosion rate along the entire rod. The results in Fig. 10b indicate that PFCC predictions for different particle size distributions do not vary greatly for values in the range 10^{-17} to 10^{-18} h, but are significantly higher for fine particles ($\sim 10^{-19}$). This difference is consistent with the observations reported in Refs 29 and 37 and with the observations of unacceptably high corrosion of experimental beta-quenched cladding ($\Sigma a_i \ll 10^{-19}$ h) after only one cycle of operation [38].

Lithium Hydroxide Exposure

LiOH at a lithium concentration of less than ~ 2.5 ppm has been generally used in PWRs at startups to maintain the coolant pH ($\sim 300°C$, 573 K) within 6.9 to 7.4 in order to control the corrosion of primary system materials (iron- and nickel-based alloys) and minimize corrosion product transport within the primary system. The lithium concentration normally decreases in coordination with decreasing boric acid concentration (1200 to 2000 ppm B at startups) as time progresses and is normally reduced to ~ 0.6 ppm at the end of a fuel cycle when the boron concentration is depleted. For longer fuel cycle operations with higher target discharge burnups, the startup concentrations of boric acid, and thus lithium hydroxide, need to be increased. The effect of higher startup lithium concentration (particularly its effect on high burnup fuel and fuel with transition boiling and crud deposition) has been of interest to fuel users. As mentioned earlier, corrosion models by Sabol [4] and Billot [9] have attributed a significant portion of in-reactor Zircaloy cladding corrosion to an enhancement effect of lithium. However, recent assessment of corrosion data taken from several plants operated with various startup concentrations of lithium, ranging from 2.2 to 3.5 ppm, have indicated no easily discernible effect of lithium on Zircaloy cladding corrosion when operated to an oxide thickness of less than 70 μm in PWRs [11,39].

Laboratory and ex-reactor heat transfer loop studies of the effects of lithium hydroxide/boric acid additions on Zircaloy-4 corrosion have been reported [8–11,40]. Evidence from these

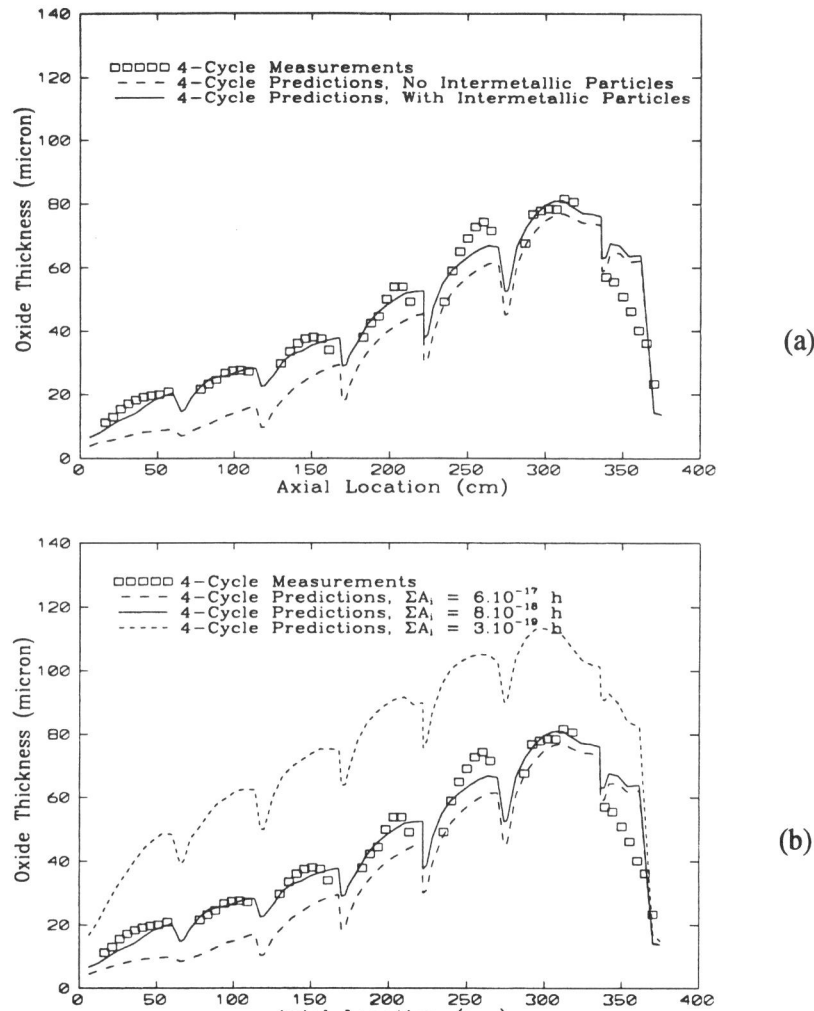

FIG. 10—(a) *Illustration of PFCC prediction of the separate effect of the intermetallic particle dissolution factor;* (b) *illustration of PFCC prediction of the separate effect of varying ΣA_i (particle-size distribution).*

studies for a lithium-related acceleration under PWR water chemistry conditions is somewhat conflicting.

In loop tests in the core of the Halden test reactor (4.0 to 4.5 ppm lithium, 700 to 1000 ppm B, and ≤1% voids), no remarkable effect of lithium on Zircaloy cladding corrosion was observed in 425 days for most portions of the rod as 40-μm oxide films grew to 95 μm [*12*].

The lack of clear evidence of PWR cladding corrosion enhancement due to lithium, unlike the cases involving metallurgical and hydride effects discussed earlier, has limited the ability to quantify the effect of lithium in PWRs. The approach taken in the PFCC studies is to evaluate available laboratory and in-core simulation test results to formulate the lithium effect and

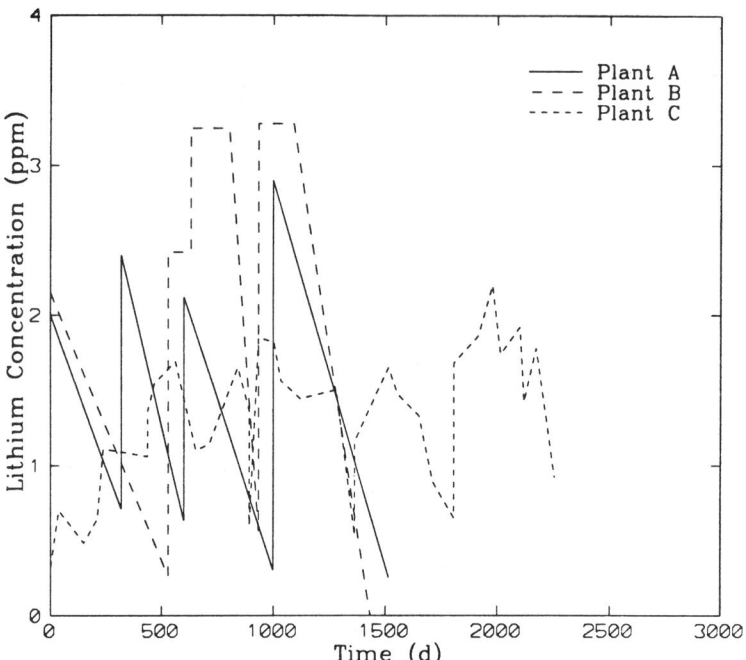

FIG. 11—*Lithium histories for Plants A, B, and C.*

calibrate the weight of the lithium term using the 17 benchmark rods. The 17 rods had the various lithium operation histories shown in Fig. 11. Plant B was operated with lithium concentrations up to 3.5 ppm during part of two cycles to study the effect of high lithium on crud transport.

The formulation of a Li history-dependent pre-exponential factor in the PFCC studies has involved numerous trial expressions. Having accounted for thermal feedback, tin content, hydrogen thermal redistribution, and intermetallic particle resolution effects, a lithium effect was developed. The mathematical expression is as follows:

$$F_{Li2} = \exp[10.17[Li] - 20.4[Li]/T*f]$$

where

f = fitting constant.

This expression is similar to, but much less intense than, the in-reactor single-effect expression proposed by Billot [8,9]. This formulation is consistent with the low heat flux, heat transfer loop data, and with Halden test reactor data.

The 17 benchmark high burnup oxide profiles include those for six fuel rods in Plant B, which experienced lithium up to 3.5 ppm for parts of two cycles of exposure. The other eleven rods were operated under two different low lithium histories, as shown in Fig. 11. The PFCC predictions agree with both the high and low lithium histories, as shown in Fig. 4.

The integrated effect of lithium on Zircaloy cladding corrosion is predicted using the Plant A rod shown in Fig. 12 by assuming no lithium effect (setting the lithium factor equal to unity). PFCC predicts little effect of lithium near the inlet locations and a modest effect at the peak power location, in part because the lithium effect is formulated to be temperature dependent.

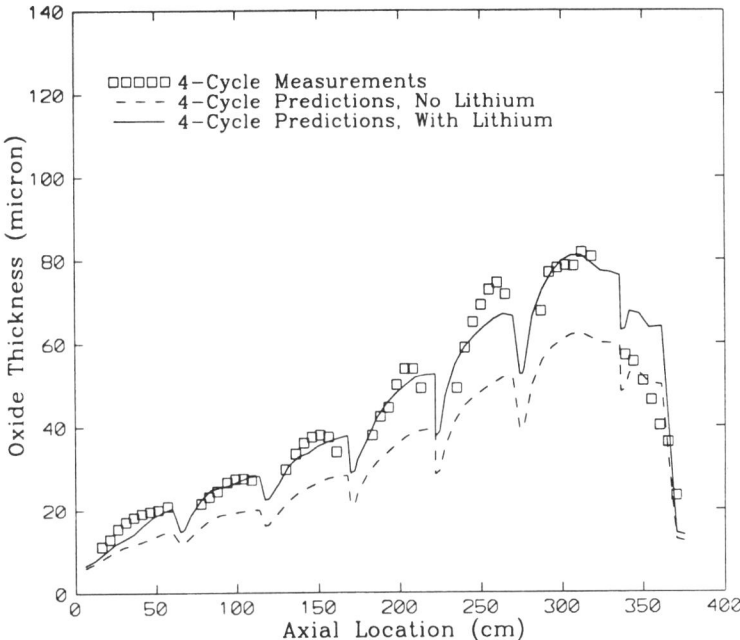

FIG. 12—*Illustration of PFCC prediction of the separate effect of the lithium factor.*

The integrated lithium effect is estimated to be about 20% for this rod with a burnup of 60 GWD/t.

Special computations were made to determine the effect of operating fuel under various lithium operation strategies, including the 2.2 ppm Li/pH = 6.9 coordinated and the 2.2 to 2.4 ppm Li/pH = 7.4 modified chemistry [41]. Using the reference low-tin rod from Plant A, PFCC predicts that the peak oxide will increase by 12% from 81 to 91 μm when the chemistry is changed from a coordinated to a modified strategy.

Conclusions

Computer simulation of both the magnitude and shape of oxide axial profiles measured for Zircaloy-4 cladding for 17 high burnup fuel rods in three different PWRs is improved by adding factors to account for radial redistribution of corrosion hydrogen, metallurgical variables, and lithium hydroxide exposure history to the previously published [6] EPRI corrosion model. The improved model, EPRI PFCC, predicts that the effect on cladding corrosion of hydrogen thermal redistribution to the oxide-metal interface, coupled with thermal feedback, is the predominant factor in determining the magnitude and shape of cladding oxide profiles at high burnup for PWR fuel rods. Tin content, coupled with thermal feedback, is also found to have a very important effect on the predicted magnitude of oxide profiles at all burnups. Radiation-induced dissolution of fine $Zr(Cr,Fe)_2$ intermetallic particles in modern, larger particle size Zircaloy-4 cladding are predicted to primarily affect the shape of the oxide profiles at the colder inlet region of the fuel rods. Lithium hydroxide exposure history and radiation effects in the oxide film are predicted to play non-negligible, secondary roles in determining oxide profiles at all burnups.

Acknowledgment

Dr. J. C. Gillis of S. Levy Incorporated provided invaluable assistance with the mathematics of the hydrogen thermal redistribution simulation. Mr. J. M. Sorensen of S. Levy Incorporated provided expert assistance with the single channel thermal hydraulic model employed. The authors would like to thank Dr. Rosa Yang of EPRI for her encouragement throughout this study.

References

[1] Van Swam, L. F., Garzarolli, F., and Steinberg, E., "Advanced PWR Cladding," *Proceedings*, 1994 International Topical Meeting on Light Water Reactor Fuel Performance, West Palm Beach, FL, American Nuclear Society, La Grange Park, IL, 17–21 April 1994.
[2] Garde, A. M., Pati, S. R., Krammen, M. A., Smith, G. P., and Endter, R. K., "Corrosion Behavior of Zircaloy-4 Cladding with Varying Tin Content in High-Temperature Pressurized Water Reactors," *Zirconium in the Nuclear Industry: Tenth International Symposium, ASTM STP 1245*, A. M. Garde and E. R. Bradley, Eds., American Society for Testing and Materials, West Conshohocken, PA, 1994, pp. 760–778.
[3] Romary, H., "Corrosion and Hydriding of Fuel Rods in EdF Reactors," *Proceedings*, EPRI Utility Workshop on PWR Fuel Corrosion, Washington, DC, B. Cheng, Ed., Electric Power Research Institute, Palo Alto, CA, 28–30 July 1993.
[4] Sabol, G. P., Comstock, R. J., Weiner, R. A., McAtee, K. R., Leech, W. J., Correal-Pulver, O. A., Miller, R. S., and Stanutz, R. N., "In-Reactor Corrosion Performance of ZIRLO™ and Zircaloy-4 and Related Corrosion Modelling," *Proceedings*, 1994 International Topical Meeting on LWR Fuel Performance, 17–21 April 1994.
[5] Sabol, G. P., Comstock, R. J., Weiner, R. A., Larouere, P., and Stanutz, R. N., "In-Reactor Corrosion Performance of ZIRLO and Zircaloy-4," *Zirconium in the Nuclear Industry: Tenth International Symposium, ASTM STP 1245*, A. M. Garde and E. R. Bradley, Eds., American Society for Testing and Materials, West Conshohocken, PA, 1994, pp. 724–744.
[6] Garzarolli, F., Jung, W., Shoenfeld, H., Garde, A. M., Parry, G. W., and Smerd, P. G., "Waterside Corrosion of Zircaloy Fuel Rods," EPRI NP-2789, Research Project 1250-1, Kraftwerk Union, AG and Combustion Engineering, Inc., Electric Power Research Institute, Palo Alto, CA, December 1982.
[7] Gilmore, P. M., Klepfer, H. H., and Sorensen, J. M., *EPRI PWR Fuel Rod Corrosion Model, Theory and User's Manual*, to be issued by the Electric Power Research Institute, Palo Alto, CA.
[8] Billot, P. and Giordano, A., "Comparison of Zircaloy Corrosion Models from the Evaluation of In-Reactor and Out-of-Pile Loop Performance," *Zirconium in the Nuclear Industry: Ninth International Symposium, ASTM STP 1132*, American Society for Testing and Materials, West Conshohocken, PA, 1991, pp. 539–565.
[9] Billot, P., Robin, J. C., Giordano, A., Peybernes, J., Thomazet, J., and Amanrich, H., "Experimental and Theoretical Studies of Parameters that Influence Corrosion of Zircaloy-4," *Zirconium in the Nuclear Industry: Tenth International Symposium, ASTM STP 1245*, A. M. Garde and E. R. Bradley, Eds., American Society for Testing and Materials, West Conshohocken, PA, 1994, pp. 351–377.
[10] Bramwell, I. L., Parsons, P. D., and Tice, D. R., "Corrosion of Zircaloy-4 PWR Cladding in Lithiated and Borated Water Environments," *Zirconium in the Nuclear Industry: Ninth International Symposium, ASTM STP 1132*, C. M. Eucken and A. M. Garde, Eds., American Society for Testing and Materials, West Conshohocken, PA, 1991, pp. 628–642.
[11] Polley, M. V. and Evans, H. E., "A Comparison of Zircaloy Oxide Thicknesses on Millstone-3 and North Anna-1 PWR Fuel Cladding," EPRI TR-102826, Electric Power Research Institute, Palo Alto, CA, August 1993.
[12] Karlsen, T. and Vitanza, C., "Effects of Pressurized Water Reactor (PWR) Coolant Chemistry on Zircaloy Corrosion Behavior," *Zirconium in the Nuclear Industry: Tenth International Symposium, ASTM STP 1245*, A. M. Garde and E. R. Bradley, Eds., American Society for Testing and Materials, West Conshohocken, PA, 1994, pp. 779–789.
[13] Garde, A. M., "Enhancement of Aqueous Corrosion of Zircaloy-4 Due to Hydride Precipitation at the Metal-Oxide Interface," *Zirconium in the Nuclear Industry: Ninth International Symposium, ASTM STP 1132*, C. M. Eucken and A. M. Garde, Eds., American Society for Testing and Materials, West Conshohocken, PA, 1991, pp. 566–594.
[14] Kim, Y. S., Rheem, K. S., and Min, D. K., "Phenomenological Study of In-Reactor Corrosion of Zircaloy-4 in Pressurized Water Reactors," *Zirconium in the Nuclear Industry: Tenth International*

Symposium, ASTM STP 1245, A. M. Garde and E. R. Bradley, Eds, American Society for Testing and Materials, West Conshohocken, PA, 1994, pp. 745–759.

[15] Cheng, B., Kruger, R. M., and Adamson, R. B., "Corrosion Behavior of Irradiated Zircaloy," *Zirconium in the Nuclear Industry: Tenth International Symposium, ASTM STP 1245*, A. M. Garde and E. R. Bradley, Eds., American Society for Testing and Materials, West Conshohocken, PA, 1994, pp. 400–418.

[16] Kido, T., "A Study on Enhanced Uniform Corrosion of Zircaloy-4 Cladding During High Burnup Operation in PWRs," *Proceedings*, Sixth Symposium on Environmental Degradation of Materials in Nuclear Power Systems-Water Reactors, R. E. Gold and E. P. Simonen, Eds., National Association of Corrosion Engineers, Houston, TX, 1993, pp. 449–454.

[17] Shepard, K. D., Spayer, D. M., Chan, Y. Y., Frankl, I., and Strasser, A. A., "Analysis of Zircaloy Oxide Thickness Data from PWR's," EPRI NP-6698, Research Project 1250-18, S. M. Stoller Corporation, Pleasantville, NY, February 1990.

[18] Kolstad, E., "In-Pile Zircaloy Corrosion Tests With High Coolant Lithium Content," EPRI TR-104516, Electric Power Research Institute, Palo Alto, CA, May 1995.

[19] Hillner, E., "Corrosion of Zirconium-Base Alloys—An Overview," *Zirconium in the Nuclear Industry: Third Conference, ASTM STP 633*, A. L. Lowe, Jr. and G. W. Perry, Eds., American Society for Testing and Materials, West Conshohocken, PA, 1977, pp. 211-235.

[20] Cox, B., "Assessment of PWR Waterside Corrosion Models and Data," EPRI NP-4287, Electric Power Research Institute, Palo Alto, CA, October 1985.

[21] Evans, H. E. and Polley, M. V., "A Review of the NFIR-I Zircaloy Corrosion Projects, Volume 3: Code Predictions of In-Reactors Corrosion," EPRI NP-7320-D, Vol. 3, Nuclear Electric PLC, Berkeley, UK, September 1992.

[22] Smith, G. P., Pirek, R. C., Freeburn, H. R., and Schrire, D., "Hot Cell Examination of Extended Burnup Fuel from Calvert Cliffs-1," Vol. 1, EPRI TR-103302, Electric Power Research Institute, Palo Alto, CA, July 1994.

[23] Kunishi, H., Wilson, H., and Stanuts, R. N., "Evaluation of Fuel Rod Leakage Mechanisms," EPRI TR-104721/ESEERCO EP89-31, Electric Power Research Institute, Palo Alto, CA, December 1994.

[24] Smith, G. P., Pirek, R. C., and Griffiths, "The Evaluation and Demonstration of Methods for Improved Nuclear Fuel Utilization," DOE/ET/34013-15 UC-523, CEND-432, August 1994.

[25] Blat, M. and Noel, D., "Detrimental Role of Hydrogen on the Corrosion Rate of Zirconium Alloys," this publication.

[26] White, A. J., Sawatzky, A., and Woo, C. H., "A Computer Model for Hydride-Blister Growth in Zirconium Alloys," AECL-8386, Whiteshell Nuclear Research Establishment, Pinawa, Manitoba, ROE 1LO, June 1985.

[27] Walker, J., Private Communication, November 1980.

[28] Cheng, B, Levin, H. A., Adamson, R. B., Marlowe, M. O., and Monroe, V. L. in *Zirconium in the Nuclear Industry: Seventh International Symposium, ASTM STP 939*, R. B. Adamson and L. F. P. Van Swam, Eds., American Society for Testing and Materials, West Conshohocken, PA, 1994, pp. 257–283.

[29] Garzarolli, F., Steinberg, E., and Weidinger, H. G., "Microstructure and Corrosion Studies for Optimized PWR and BWR Zircaloy Cladding," *Zirconium in the Nuclear Industry: Eighth International Symposium, ASTM STP 1023*, L. F. P. Van Swam and C. M. Eucken, Eds., American Society for Testing and Materials, West Conshohocken, PA, 1989, pp. 202–212.

[30] Schemel, J. H., Charquet, J. H., and Adier, J. F., "Influence of the Manufacturing Process on the Corrosion Resistance of Zircaloy-4 Fuel Cladding," *Zirconium in the Nuclear Industry: Eighth International Symposium, ASTM STP 1023*, American Society for Testing and Materials, West Conshohocken, PA, 1989, pp. 141–142.

[31] Cheng, B. and Adamson, R. B. in *Zirconium in the Nuclear Industry: Seventh International Symposium, ASTM STP 939*, R. B. Adamson and L. F. P. Van Swam, Eds., American Society for Testing and Materials, West Conshohocken, PA, 1994, pp. 387–416.

[32] Cheng, B., Adamson, R. B., Bell, W. L., and Proebstle, R. A., "Corrosion Performance of Some Zirconium Alloys Irradiated in the Steam Generating Heavy Water Reactor-Winfrith," *Proceedings*, International Symposium on Environmental Degradation of Materials in Nuclear Power Systems-Water Reactors, Myrtle Beach, SC, 22–25 Aug. 1983, National Association of Corrosion Engineers, Houston, TX, p. 274.

[33] Yang, W. J. S., Tucker, R. P., Cheng, B., and Adamson, R. B., "Precipitates in Zircaloy: Identification and the Effects of Irradiation and Thermal Treatment," *Journal of Nuclear Materials*, Vol. 138, 1986, p. 185.

[34] Griffiths, M., Gilbert, R. W., and Carpenters, G. J. C., "Phase Stability, Decomposition, and Redistribution of Intermetallic Precipitates in Zircaloy-2 and Zircaloy-4 During Neutron Irradiation," *Journal of Nuclear Materials*, Vol. 150, 1987, pp. 53–66.

[35] Motta, A. T., LeFebvre, F., and Lemaignan, C., "Amorphitization of Precipitates in Zircaloy under Neutron and Charged-Particle Irradiation," *Zirconium in the Nuclear Industry: Ninth International Symposium, ASTM STP 1132*, C. M. Eucken and A. M. Garde, Eds., American Society for Testing and Materials, West Conshohocken, PA, 1991, pp. 718–739.

[36] Hood, G. M. and Schultz, R. J., "Diffusion of 3D Transition Elements in μ Zr and Zirconium Alloys," *Zirconium in the Nuclear Industry: Eighth International Symposium, ASTM STP 1023*, L. F. P. Van Swam and C. M. Eucken, Eds., American Society for Testing and Materials, West Conshohocken, PA, 1989, pp. 435–450.

[37] Van Swam, L. F. P. and Shann, S. H., "The Corrosion of Zircaloy-4 Fuel Cladding in Pressurized Water Reactors," *Zirconium in the Nuclear Industry: Ninth International Symposium, ASTM STP 1132*, American Society for Testing and Materials, West Conshohocken, PA, 1991, pp. 758–781.

[38] Evans, H. E., Bale, M. G., and Polley, M. V., "A Review of the NFIR-I Zircaloy Corrosion Projects, Vol. 2: Evaluation of the Corrosion Data," EPRI NP-7320-D, Vol. 2, Electric Power Research Institute, Palo Alto, CA, August 1991.

[39] Bergman, C. A. et al., "PWR Primary System Chemistry Coolant Experienced with Elevated pH at Millstone Unit 3," EPRI TR-105245, Electric Power Research Institute, Palo Alto, CA, to be published.

[40] McDonald, S. G., Sabol, G. P., and Sheppard, K. D. in *Zirconium in the Nuclear Industry: Seventh International Symposium, ASTM STP 824*, D. G. Franklin and R. B. Adamson, Eds., American Society for Testing and Materials, West Conshohocken, PA, 1984, pp. 519–530.

[41] Cheng, B., "Effect of Primary Coolant Li/pH on Fuel Cladding Corrosion in PWRs," issued to EPRI-PWR Water Chemistry Guidelines Committee, 10 May 1995.

DISCUSSION

P. Billot[1] (written discussion)—(1) Concerning the effect of hydriding on the enhancement of corrosion rates, the level of hydride at the metal/oxide interface is an important parameter. Is the enhancement factor due to hydriding the same at the plenum level with no heat flux compared to that obtained in the fissile column (subjected to thermal gradient)? (2) How was it determined that the forms of the kinetics equation, taking the influence of varied parameters into account, in particular some coupled effects, can be explained (lithium effect and Sn effect)?

B. Cheng et al. (authors' closure)—Due to the absence of heat flux at the plenum region, hydrides will never concentrate at the metal-oxide interface. Similarly, guide tubes in fuel assemblies are also not affected by the "hydride enhancement" factor. Based on data in Fig. 7, the hydride rim effect alone contributes about 51 μm of the total 80 μm of a high-burnup, low-Sn rod. If one further removes the thermal feedback effect, an adjacent guide tube exposed to the otherwise similar operating condition will have a thickness of ~10 μm. This is about the range of oxide thickness we have seen in some high-exposure guide tubes and grids. (2) The effect of Li on low-Sn and high-Sn Zircaloy is about the same, percentage wise, within the range of interest. However, since the oxide thickness is a result of multiple-step calculations, the coupling effect of all variables is a pure empirical mathematical "miracle." Our kinetic equation is a result of thousands of reiterations using some well-characterized fuel rod oxide/material data.

B. Cox[2] (written discussion)—If you are going to include a LiOH effect in your model, you can't ignore the way the boric acid concentration changes. With high boric acid concentrations at the beginning of cycle, you should not see any significant acceleration by LiOH. However, at the end of a cycle, when H_3BO_3 is low or absent, there could be a rapid increase in corrosion rate.

B. Cheng et. al. (authors' closure)—Yes, I fully agree with you. I have checked the Li/B operation histories of many PWRs. It appears that by the time the boron concentration is reduced to <100 ppm, the Li concentration is reduced to well below 1 ppm (mostly at ~0.5 to 0.6 ppm). Since all laboratory simulation tests on Li effect have shown that the effect of Li on Zircaloy corrosion becomes negligible with boron >100 ppm, it is thus evident that the effect of Li on Zircaloy cladding under normal operation conditions will not be significant. Our model reflects an effect smaller than several other models with an Li term. The magnitude of the Li effect is within measurement accuracy in oxide <10 μm.

A. T. Motta[3] (written discussion)—(1) Concerning your equation for the concentration of Fe in the matrix, there is a change in the Fe depletion as the critical temperature for amorphization is exceeded; how does your model take into account the difference between dissolution with and without amorphization of precipitation? (2) There have been results in the literature showing an improvement in corrosion resistance of neutron-irradiated material when the oxide is taken off. How do you account for this in your model?

B. Cheng et al. (authors' closure)—(1) At present, we have only taken into account the fraction of precipitates that are dissolved by irradiation without including amorphization. This

---

[1] Commissariat a l'Energie Atomique, St. Paul lez Durance, France.
[2] University of Toronto, Toronto, Ontario, Canada.
[3] Pennsylvania State University, University Park, PA.

certainly will result in underestimating the total Fe and Cr that are dissolved into the matrix. Fortunately, this can be taken care of by adjusting the sensitivity factors. (2) I guess you might be referring to the paper published by Cheng, Kruger, and Adamson in the last symposium in Baltimore. In that paper, the "nodular" corrosion resistance was shown to be improved by irradiation. The uniform corrosion resistance of irradiation Zircaloy in 400°C steam was shown to increase with increasing fluence. The results on uniform corrosion are consistent with the results of this PFCC model.

M. Griffiths[4] (written discussion)—Regarding the effect of temperature on dissolution of intermetallic precipitates, you say that the Fe dissolution is diminished at temperatures >360°C. Do you mean to say that the state of the redistributed Fe is different, i.e., more finely disposed at lower temperatures with the possibility of a higher dynamic Fe concentration in the matrix during irradiation? Significant Fe, Cr, and Ni dispersion is apparent at high temperature (≥400°C), but secondary precipitation is also significant. I would agree that the concentration of these elements in solution is probably less at higher temperatures.

B. Cheng et al. (authors' closure)—Yes. We fully agree with you on the temperature effect on precipitate dissolution and reprecipitation. It is a dynamic situation. We believe the corrosion and possible hydriding rates of irradiated Zircaloy are controlled by the concentration of Fe, Cr, and Ni in the alloy matrix. Thus, we use the net effect of irradiation on precipitate dissolution as an indicator for the concentration of the alloying elements in Zircaloy.

---

[4] Atomic Energy of Canada, Ltd., Chalk River, Ontario, Canada.

Corrosion Mechanisms—II

Yoichi Ito[1] *and Takemi Furuya*[1]

Correlation Between Electrochemical Properties and Corrosion Resistance of Zirconium Alloys

REFERENCE: Ito, Y. and Furuya, T., "**Correlation Between Electrochemical Properties and Corrosion Resistance of Zirconium Alloys,**" *Zirconium in the Nuclear Industry: Eleventh International Symposium, ASTM STP 1295,* E. R. Bradley and G. P. Sabol, Eds., American Society for Testing and Materials, 1996, pp. 163–180.

ABSTRACT: The electrochemical behavior of some zirconium alloys including Zry-2 with various ΣAi from 2.5×10^{-20} to 1.2×10^{-17} (h), modified Zry-2 with iron contents of 0.15, 0.25, and 0.5%, and standard Zry-4 was studied by measuring anodic polarization curves in sulfuric acid solution. The results of these electrochemical tests were compared with those of steam autoclave tests.
In Zry-2, the current peak was observed at 1250 mV (versus SCE) on the anodic polarization curve, and this peak area increased with ΣAi and with the size of secondary precipitates. Also, this peak was closely correlated with nodular corrosion resistance as expected from the above results. As iron contents in modified Zry-2 increased, the current peak at 1250 mV decreased and a new peak at 1900 mV appeared. The former peak disappeared and the latter peak increased further at 0.5% iron. In Zry-4, the current peak was observed at 1900 mV, but not at 1250 mV, and this behavior was the same as that of modified Zry-2 of 0.5% iron.
AEM observation of secondary precipitates in these zirconium alloys indicated that the peak at 1250 mV appeared when the Fe/Cr ratio of $Zr(Fe,Cr)_2$ was below 1.0 and the peak at 1900 mV appeared when the ratio was above 1.5. In order to clarify the mechanism of these peak behaviors, the anodic polarization properties of intermetallic compounds that precipitated in zirconium alloys were studied. The results suggest that the peak at 1250 mV is related to the reaction $Cr^{3+} \rightarrow Cr^{6+}$ (soluble in solution) and the peak at 1900 mV is related to the reaction $Fe^{3+} \rightarrow Fe^{6+}$ (soluble in solution) and oxygen generation on iron-rich precipitates.

KEYWORDS: electrochemical, zirconium alloys, nodular corrosion, precipitates, analytical electron microscopy, anodic polarization curve

In light water reactors, Zry-2 and Zry-4 are used as cladding tubes in BWR and PWR, respectively. It is known that the corrosion resistance of these Zircaloys are influenced by alloying elements such as Sn, Fe, Cr, Ni, and Nb and annealing conditions in the fabrication process [1–3]. It was also reported that the size, number density, and composition of secondary phase particles such as $Zr(Fe,Cr)_2$ and $Zr_2(Fe,Ni)$ in the zirconium alloys, which depend on the kinds and quantities of alloying elements and annealing conditions, affected corrosion resistance [4–5]. Recently, studies on the interaction between precipitates and oxide film when these precipitates were taken into the oxide film, such as the dissolution of Fe, Cr, and Ni into the oxide film, have been conducted [6–8]. However, it is very important to investigate the electrochemical behavior of zirconium alloys and the precipitates in them because the corrosion reaction is inherently an electrochemical reaction between metals and the environment, and,

---

[1] Research engineer and general manager, respectively, Nuclear Fuel Industries Ltd., Muramatsu, Tokai-mura, Ibaraki-ken 319-11.

therefore, many studies on the electrochemical behaviors of zirconium alloys have been conducted [9–10].

In austenitic stainless steels, it is known that the susceptibility to intergranular corrosion depends on the concentration of Cr along grain boundaries, that is, the degree of sensitization. This degree of sensitization can be measured by the anodic polarization curves in sulfuric acid solution, and it is used practically to evaluate the susceptibility to intergranular corrosion as an electrochemical potentiokinetic reactive method (EPR method).

It is known that the nodular corrosion resistance of Zircaloy is influenced by the size of secondary phase precipitates, which depend on the accumulated annealing parameter (ΣAi) defined by Eq 1. ΣAi expresses an index of the total amount of heat received during annealing in the α-region after β or ($\alpha + \beta$) quenching, and this parameter (ΣAi) is used widely as a standard for evaluation of the corrosion resistance. However, this parameter neglects the difference in the size of precipitates just after quenching as an initial condition. Moreover, the activation energy used in Eq 1 was originally evaluated from the annealing kinetics of cold-worked materials [11]. From these reasons, ΣAi defined by Eq 1 may not be suitable as an index of corrosion resistance and its physical meaning is also not clear. The authors have shown that ΣAi defined by Eq 1 is not always a suitable index for corrosion resistance [12]:

$$\Sigma Ai = \Sigma ti \cdot \exp(-Q/RTi) \quad (1)$$

where

Ti = annealing temperature in Process i, K,
ti = annealing time in Process i, h, and
Q/R = 40 000 K.

Therefore, if the corrosion resistance of zirconium alloys can be evaluated from their electrochemical properties, it will be extremely useful because the corrosion resistance can be evaluated by measuring electrochemical properties, which is simple and easy compared with autoclave tests, which require a long time and the measurement of precipitate diameter by using SEM and TEM.

In this study, the correlation between the anodic polarization properties and corrosion resistance was investigated for various zirconium alloys with various alloying elements and annealing parameters.

Experimental Procedure

Materials

The materials used in this study were Zry-2 cladding tube for BWR with six kinds of ΣAi from 2.5×10^{-20} to 1.2×10^{-17} (h), conventional Zry-4 cladding tube for PWR ($\Sigma Ai \approx 1.2 \times 10^{-17}$ (h)), and modified Zry-2 sheet with iron content of 0.15, 0.25, and 0.5 wt%, which were fabricated by simulating the cladding tube manufacturing process for BWR ($\Sigma Ai \approx 1.2 \times 10^{-17}$ (h)). The chemical compositions of these materials are shown in Table 1, and the manufacturing processes of Zry-2 cladding tube are shown in Fig. 1.

Three kinds of intermetallic compounds $Zr(Fe,Cr)_2$ with various Fe/Cr ratios as shown in Table 1 were prepared by vacuum arc melting in order to examine the electrochemical properties of $Zr(Fe,Cr)_2$ because the anodic polarization properties of zirconium alloys depend mainly on the state of $Zr(Fe,Cr)_2$.

Electrochemical Testing

In order to measure the anodic polarization curves, the specimens were ground with No. 1000 emery paper and then pickled in a solution of 5% HF : 35% HNO_3 : 60% H_2O. The

TABLE 1—*Chemical compositions of specimens (wt%).*

Specimens	Material Type	Character	Zr	Sn	Fe	Cr	Ni
Zry-2	Cladding tube	$\Sigma Ai(h)$: 2.5×10^{-20} $\sim 1.2 \times 10^{-17}$	Bal.	1.3	0.18	0.09	0.07
Modified Zry-2	Sheet	...	Bal.	1.5	0.15	0.1	0.05
		...	Bal.	1.5	0.25	0.1	0.05
		...	Bal.	1.5	0.50	0.1	0.05
Zry-4	Cladding tube	...	Bal.	1.3	0.2	0.1	...
		Fe/Cr ratio:					
$Zr(Fe_1,Cr_1)_2$	Button	1.0	46	...	28	26	...
$Zr(Fe_{1.2},Cr_{0.8})_2$	Button	1.5	46	...	33	21	...
$Zr(Fe_{1.33},Cr_{0.67})_2$	Button	2.0	46	...	37	17	...

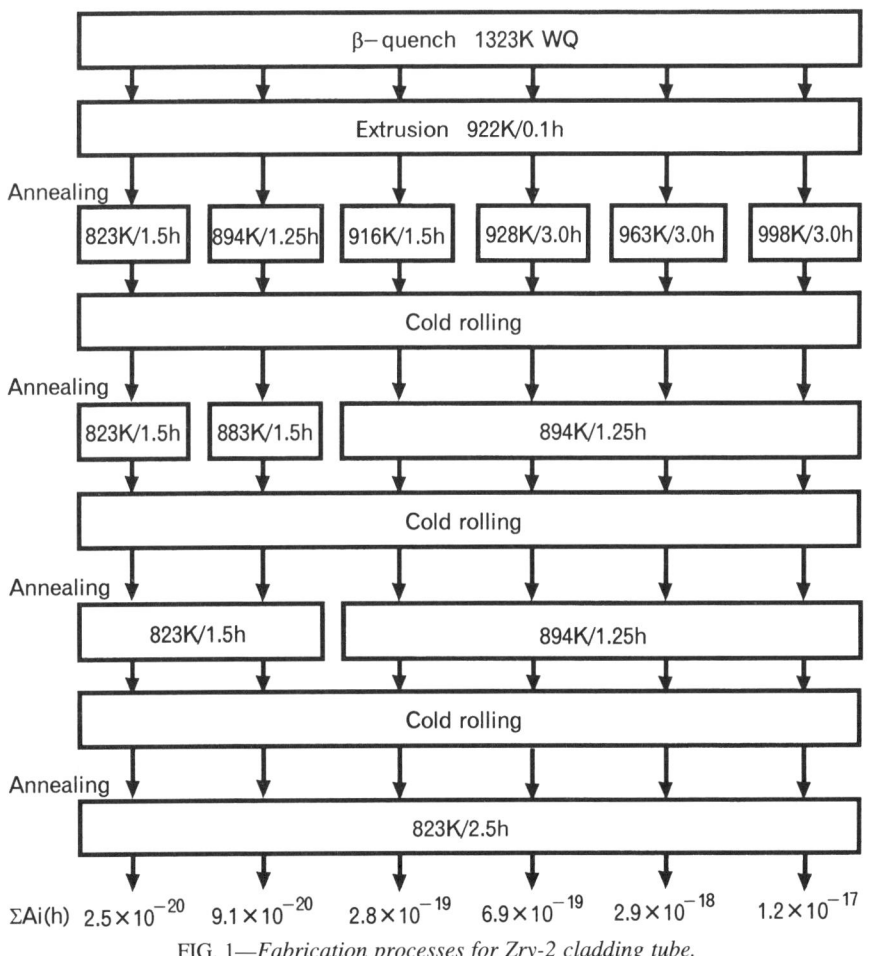

FIG. 1—*Fabrication processes for Zry-2 cladding tube.*

peripheral region of the specimen for electrochemical testing was covered with insulating resin, leaving an area of 1 cm² after ultrasonic cleaning in acetone. The equipment consisted of a potentiostat, function generator, computer system, and an electrolysis cell, and the anodic polarization program was run and the data were acquired by the computer system. A platinum (Pt) electrode and saturated calomel electrode (SCE) were used for the counter electrode and reference electrode, respectively. The anodic polarization curves were measured at 298 K in the electrolyte of 5% sulfuric acid. The sweeping rate of potential was 50 mV/min. High-purity argon gas was passed as a cover gas while measuring to prevent oxygen entering the electrolyte. Contamination of the electrolyte by chloride ion carryover from the calomel electrode was avoided by use of a 5% H_2SO_4 salt bridge. Successful control of chloride carryover into the electrolyte was gaged by the visual absence of pitting on the samples and by stable electrochemical characteristics.

In order to elucidate the anodic polarization behaviors of Zry-2 and Zry-4, the surface of specimens interrupted at the potential just before the current peak, just at the current peak, and just after the current peak in the anodic polarization curves were observed by SEM. The specimens for observation were etched in 10% HF : 10% HNO_3 : 80% glycerine solution. Regarding the intermetallic compound of $Zr(Fe,Cr)_2$ with various ratios of Fe/Cr, anodic polarizations were interrupted at the potential immediately before the current increased remarkably at about 800 mV (SCE), and then passive films on the surface were analyzed by X-ray photoelectron spectroscopy (XPS).

Autoclave Testing

To evaluate the nodular corrosion resistance of Zry-2 cladding tubes with various ΣAi and modified Zry-2 sheet, the corrosion tests were carried out in high-pressure steam of 10.3 MPa at 798 K for 24 h in a refreshed autoclave. The concentration of oxygen and hydrogen in water was controlled to 200 and 5 ppb, respectively, and the pH and conductivity of the water were monitored during testing. The corrosion resistance was evaluated from the weight gain during the corrosion test.

Precipitate Characterization

For the Zry-2 cladding tube, the precipitate was observed by a scanning electron microscope (SEM; JSM-5400), and the mean particle diameter of precipitate was measured by an image analyzer. The chemical composition of $Zr(Fe,Cr)_2$ and $Zr_2(Fe,Ni)$ precipitates was analyzed by energy dispersive X-ray spectroscopy (EDS; EMAX-3000) combined with transmission electron microscope (TEM; H-800) for the Zry-2 cladding tube with ΣAi of 2.4×10^{-19} as a reference, Zry-4 cladding tube, and the modified Zry-2 sheet.

Specimens for SEM analysis were etched in a 10% HF : 10% HNO_3 : 80% glycerine solution, and then the surface was rinsed in 60% HNO_3. The thin foil for TEM was prepared by the twin-jet polishing technique with a solution of 5% $HClO_3$ and 95% C_2H_5OH after thinning to about 0.1 mm by chemical etching in a solution of 35% HNO_3, 5% HF, and 60% H_2O. The polishing was conducted at 233 K, and the voltage was 20 V.

Results

Anodic Polarization Properties

Figure 2 shows the anodic polarization curves of Zry-2 and Zry-4. These similarly reached a passive state at about 0 mV (SCE), and the passive current density is about 5 ~ 10 $\mu A/cm^2$.

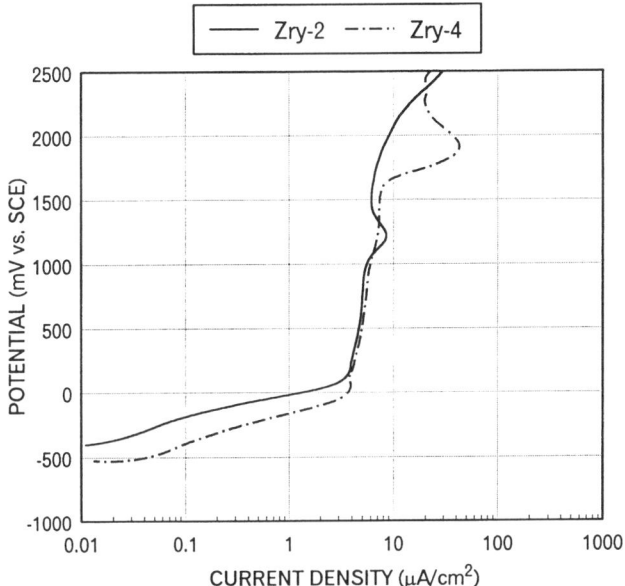

FIG. 2—*Anodic polarization curves of Zry-2 and Zry-4 in 5% sulfuric acid.*

In Zry-2, the current density rapidly increases from about 1000 mV (SCE) and a current peak exists at about 1250 mV (SCE). In Zry-4, the current density increases rapidly from about 1600 mV (SCE), and a current peak exists at about 1900 mV (SCE). The corrosion potential of Zry-2 is slightly higher than that of Zry-4.

The anodic polarization curves of Zry-2 cladding tube with various ΣAi are shown in Figs. 3a and 3b. It is clear that the current peak of 1250 mV (SCE) becomes higher with increasing ΣAi from the magnified current peak of Fig. 3b. However, other properties such as corrosion potential and passive current density are very similar. The current density is linearized to obtain the area of the peak, and the areas enclosed by the current peak and base line are measured by the image analyzer. The relation between ΣAi and peak area is shown in Fig. 4, where the material with ΣAi of 2.5×10^{-20} (h) is used as a standard value. The peak area increases linearly with increasing ΣAi, demonstrating the correlation between the anodic polarization properties and ΣAi.

Figure 5 shows the anodic polarization curves of modified Zry-2 sheet with various iron contents. This indicates that the anodic polarization properties of Zry-2 are greatly influenced by the iron content, that is, the current peak of 1250 mV (SCE) becomes lower and that of 1900 mV (SCE) becomes higher as the iron content increases. The corrosion potential is slightly noble at high iron content.

Corrosion Properties

The results of the corrosion tests at 798 K on Zry-2 cladding tube with various ΣAi are shown in Table 2. Nodular corrosion resistance deteriorates with increasing ΣAi, and this was confirmed by observing the surface appearance of specimens after the corrosion test, revealing many nodules on the edge side at large ΣAi. As for the modified Zry-2, the nodular corrosion resistance deteriorates as iron content decreases.

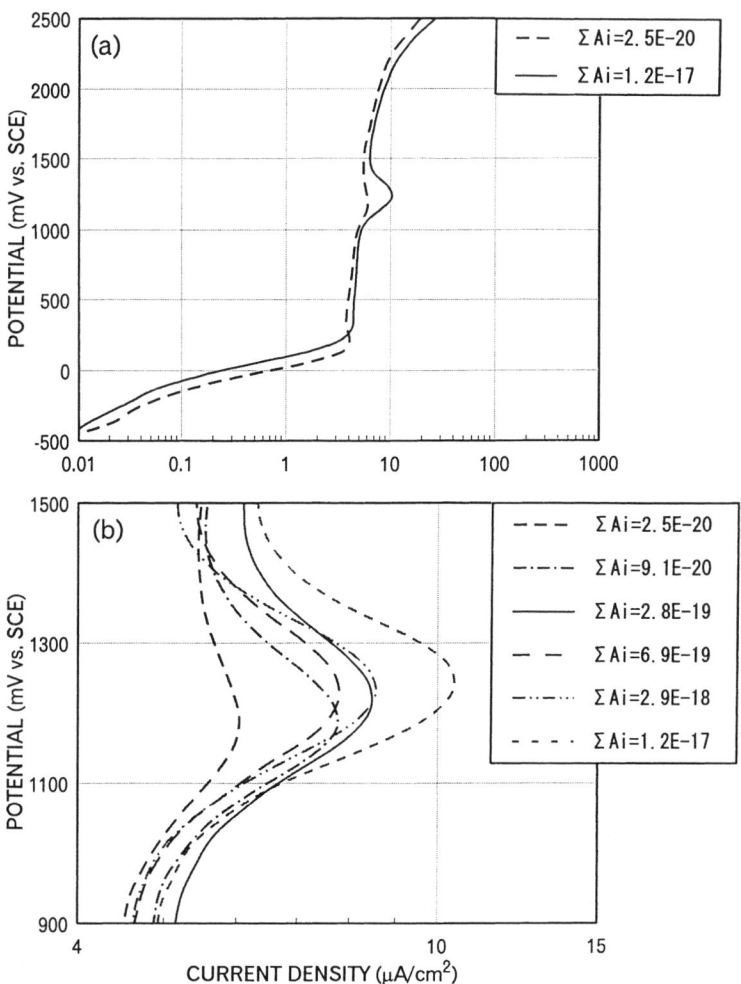

FIG. 3—*Anodic polarization curves of Zry-2 as a function of ΣAi in 5% sulfuric acid: (a) the full curves with two typical ΣAi; (b) the magnified current peak with various ΣAi.*

Precipitate Characterization

The results of precipitate analysis in Zry-2 cladding tube, Zry-4 cladding tube, and modified Zry-2 sheet are shown in Table 2. Typical TEM photographs and EDS spectrums of modified Zry-2 sheet with iron content of 0.15 and 0.5% are shown in Fig. 6 and Fig. 7, respectively. Hcp-$Zr(Fe,Cr)_2$ and bct-$Zr_2(Fe,Ni)$ were observed in Zry-2, and hcp-$Zr(Fe,Cr)_2$ only were observed in Zry-4, as reported previously [*13*]. It was found from the analyses of precipitates that the Fe/Cr ratio of $Zr(Fe,Cr)_2$ and the Fe/Ni ratio of $Zr_2(Fe,Ni)$ in Zry-2 and modified Zry-2 with low iron content were below 1.0 and below 1.3, respectively, and that those in modified Zry-2 with high iron content and Zry-4 were above 1.5, and the Fe/Ni ratio of $Zr_2(Fe,Ni)$ was above 2.6.

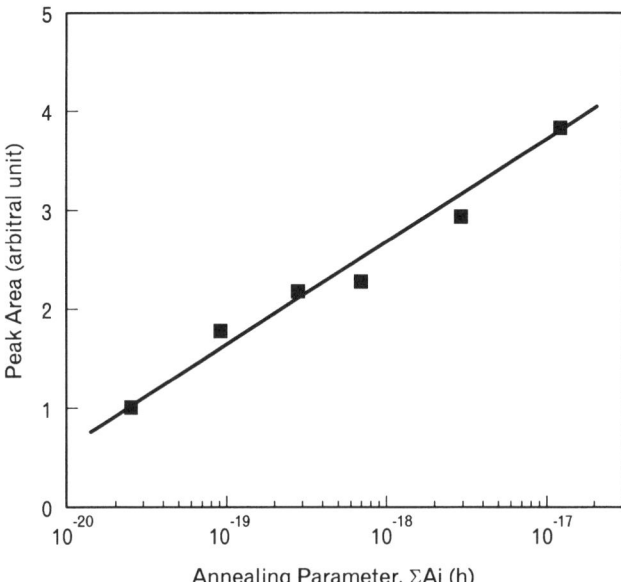

FIG. 4—*Current peak area of Zry-2 at 1250 mV in 5% sulfuric acid as a function of annealing parameter (ΣAi).*

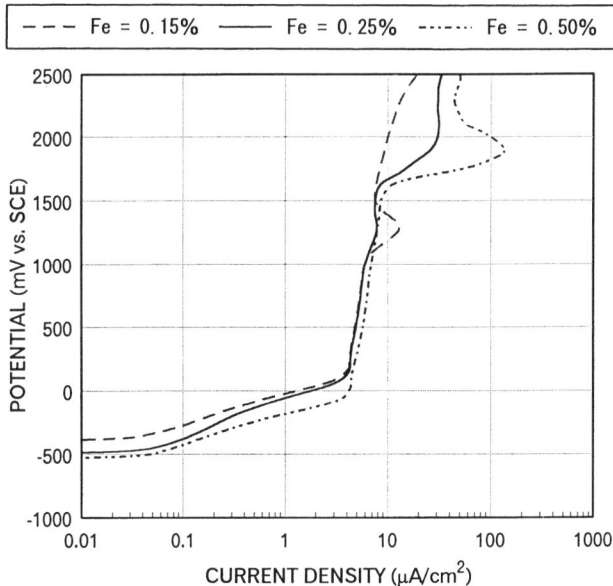

FIG. 5—*Anodic polarization curves of modified Zry-2 sheet in 5% sulfuric acid.*

TABLE 2—*Corrosion data (798 K), precipitate diameter, composition of precipitate, and anodic polarization properties.*

Specimens	Character	Weight gain, (mg/dm^2)	Precipitate Diameter, μm	Composition of Precipitate Fe/Cr Ratio of Zr(Fe,Cr)^2	Fe/Ni Ratio of Zr^2(Fe,Ni)	Anodic Polarization Properties
Zry-2 cladding tube	ΣAi (h):					
	2.5×10^{-20}	56.9	0.092	...	...	
	9.1×10^{-20}	57.9	0.105	...	...	
	2.8×10^{-19}	58.9	0.111	0.97	1.24	1250-mV peak
	6.9×10^{-19}	60.9	0.122	...	...	
	2.9×10^{-18}	73.8	0.154	...	...	
	1.2×10^{-17}	86.6	0.167	...	...	
Modified Zry-2 sheet	Fe (wt%):					
	0.15	1742	...	0.83	1.28	1250-mV peak
	0.25	156.0	..	1.50	2.62	1250/190-mV peak
	0.50	51.7	...	2.47	6.01	1900-mV peak
Zry-4 cladding tube	...	...	...	1.92	...	1900-mV peak

The particle size of precipitate in Zry-2 cladding tubes with various ΣAi is shown in Table 2. This indicates that the particle size of precipitate grows almost linearly as ΣAi increases.

Surface Appearance

Figures 8 and 9 show the state of the precipitate after polarizing up to various potentials and before polarization. In Zry-2, the precipitates are normally observed in the specimens without polarization and interrupted at the potential just before the current peak (1000 mV). On the other hand, some traces of precipitates (indicated by arrows in Figs. 8c and 8d) as well as precipitates are observed in the specimens interrupted at the potential just at and just after the current peak. The appearance of precipitates in Zry-4 is basically the same as those of Zry-2, although the potential at the current peak is higher than that of Zry-2. It is not clear whether traces of precipitates exist or not in the specimen interrupted at the potential at the current peak (1900 mV) because the whole surface of the specimens is covered with a thick oxide film.

Discussion

Anodic Polarization Properties

As shown in Fig. 2 and Fig. 5, two kinds of current peaks at 1250 and 1900 mV were observed in the anodic polarization curves of Zry-2, modified Zry-2, and Zry-4 in 5% sulfuric acid solution (pH $\simeq$ 0). The current peak at 1250 mV was observed clearly in Zry-2 and modified Zry-2, and an extremely small peak seems to exist in Zry-4. This current peak area at 1250 mV increased with increasing ΣAi in Zry-2 and decreased with increasing iron content in modified Zry-2. On the other hand, the Fe/Cr ratio of Zr(Fe,Cr)$_2$ in modified Zry-2 increased with increasing iron content, as shown in Table 2. These results suggest that the current peak at 1250 mV appears only when the Fe/Cr ratio is below about 1.0. The current peak at 1900 mV was observed clearly in Zry-4 and modified Zry-2 with high iron content, but was not clear

ITO AND FURUYA ON ELECTROCHEMICAL PROPERTIES 171

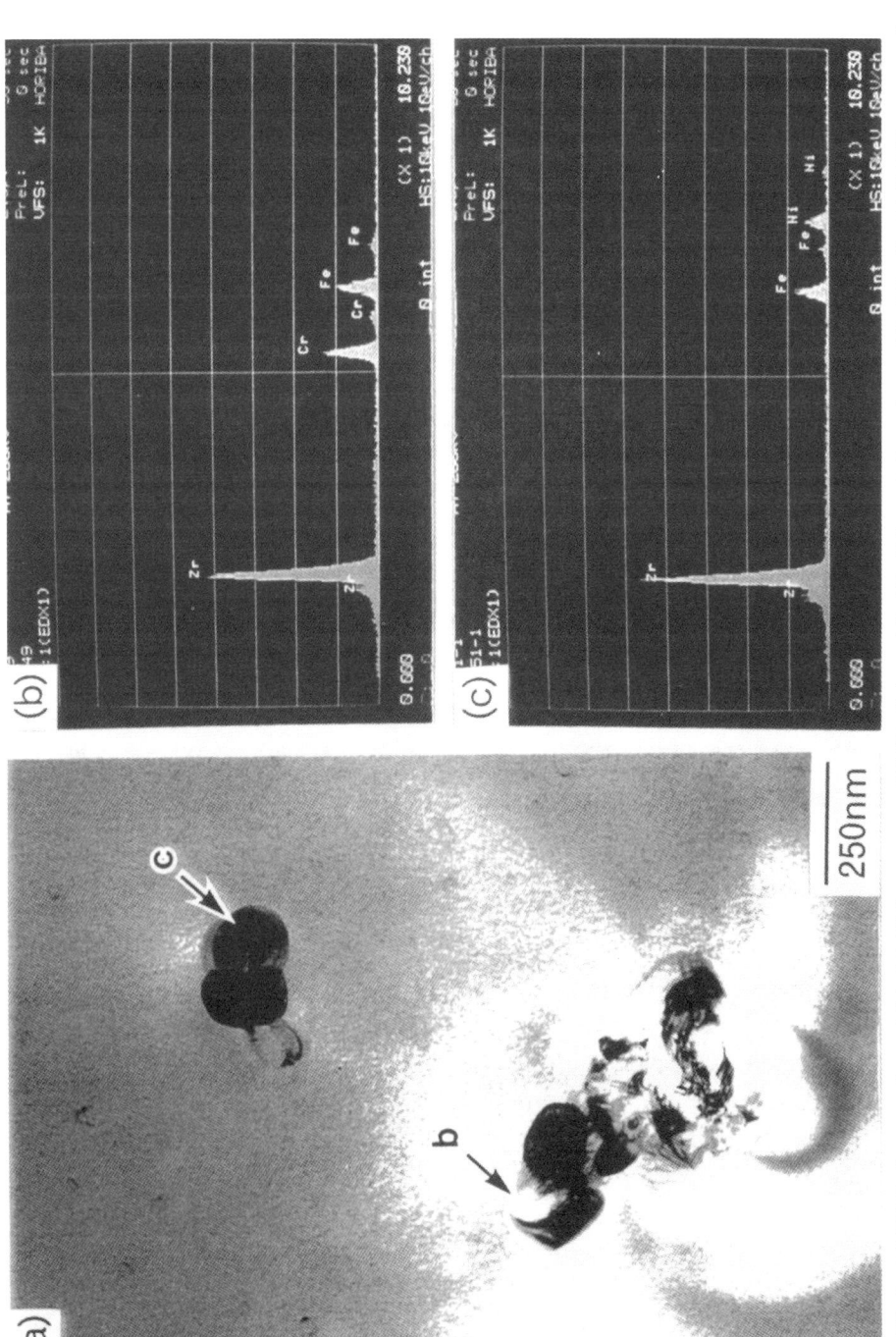

FIG. 6—(a) TEM photographs and (b), (c) EDS spectrum of $Zr(Fe,Cr)_2$ and $Zr_2(Fe,Ni)$ precipitates in modified Zry-2 with 0.15% content of iron.

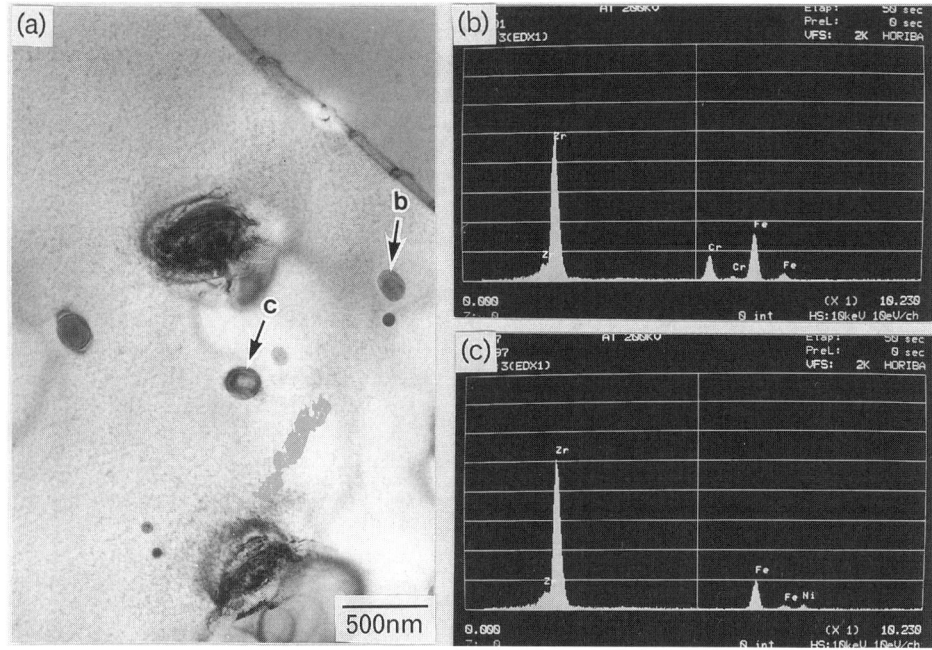

FIG. 7—(a) TEM photographs and (b), (c) EDS spectrum of $Zr(Fe,Cr)_2$ and $Zr_2(Fe,Ni)$ precipitates in modified Zry-2 with 0.5% content of iron.

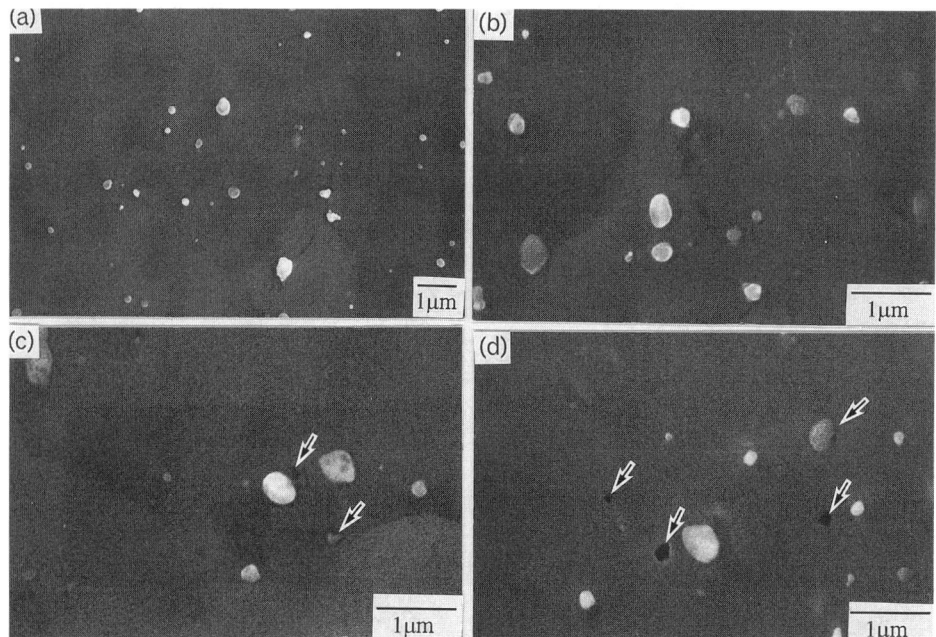

FIG. 8—SEM micrographs of the state of precipitates after polarization up to various potentials (Zry-2 with $\Sigma Ai = 1.2 \times 10^{-17}$ (h)): (a) as etched, (b) before the peak (1000 mV), (c) at the peak (1250 mV), and (d) after the peak (1500 mV).

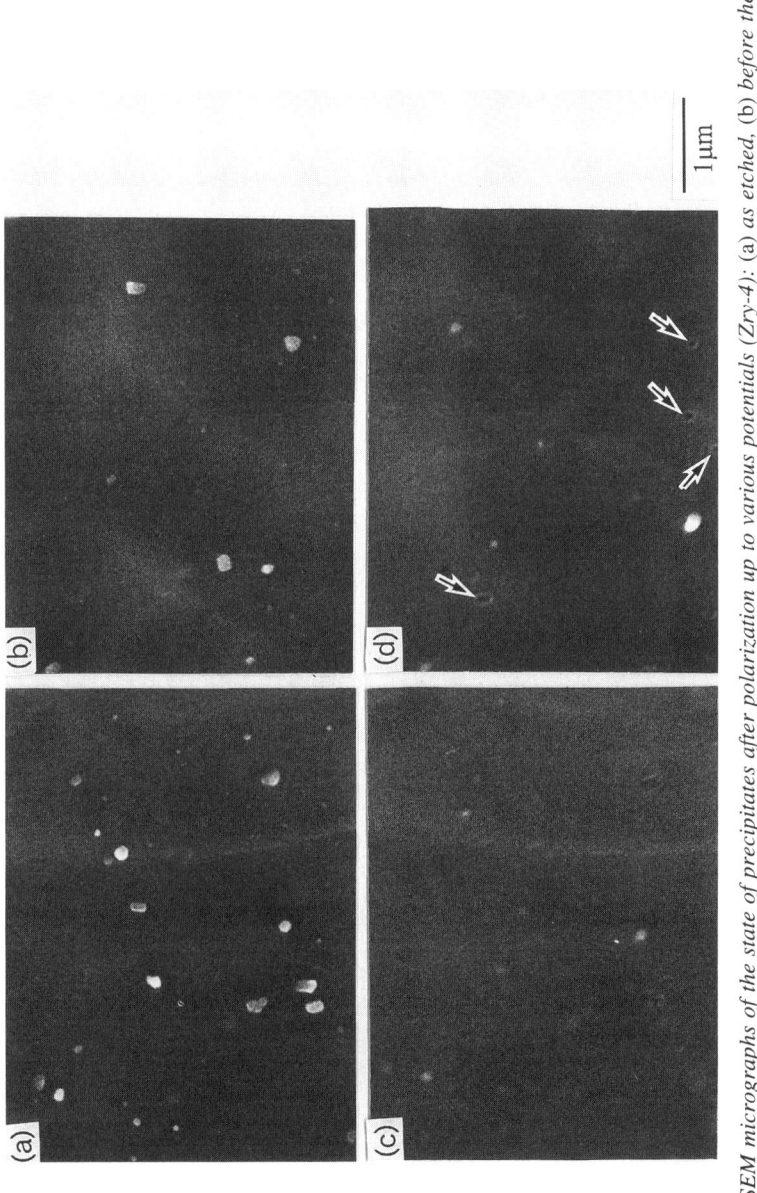

FIG. 9—SEM micrographs of the state of precipitates after polarization up to various potentials (Zry-4): (a) as etched, (b) before the peak (1600 mV), (c) at the peak (1900 mV), and (d) after the peak (2500 mV)

in Zry-2 and modified Zry-2 with low iron content. These results indicate that the current peak at 1900 mV appears when the Fe/Cr ratio is greater than about 1.5. Therefore, it is concluded that the anodic polarization behavior of Zry-2 and Zry-4 depends mainly on the Fe/Cr ratio of $Zr(Fe,Cr)_2$, that is, the current peak at 1250 mV appears when the Fe/Cr ratio is less than 1.0, and the current peak at 1900 mV appears when the Fe/Cr ratio is more than 1.5. In order to confirm the dependence of anodic polarization behavior on the Fe/Cr ratio of precipitate of $Zr(Fe,Cr)_2$, the anodic polarization behavior of intermetallic compounds $Zr(Fe,Cr)_2$ with various Fe/Cr ratios was examined. We found two current peaks, one at 1000 ~ 1500 mV and the other at over 2000 mV, and the height of the current peak at 1000 ~ 1500 mV increased while that at over 2000 mV decreased with decreasing Fe/Cr ratio as shown in Fig. 10. These behaviors seem to agree qualitatively with those of Zry-2 and Zry-4. Moreover, it is considered from Fig. 11 that the fraction of ZrO_2 and Cr_2O_3 in the oxide film on the $Zr(Fe,Cr)_2$ decreases and that of Fe_2O_3 increases with increasing Fe/Cr ratio of the intermetallic compound.

It is necessary to elucidate the mechanism of the appearance of the current peak in the anodic polarization curve in order to explain the dependence of each peak area on ΣAi and iron content. According to Pourbaix potential-pH diagram [14], the potential of 1250 mV at the current peak in this study is approximately in accordance with the potential for the reaction $Cr^{3+} \rightarrow Cr^{6+}$ in a solution of pH $\simeq$ 0. It is known that the solubility of Cr^{6+} in a strong acid solution increases remarkably compared with Cr^{3+}. Furthermore, Fig. 8, which is the scanning electron micrograph of the specimen surface polarized up to the potential just before and just after the current peak, indicated that some precipitates of $Zr(Fe,Cr)_2$ came out into the solution at the current peak potential as a result of the reaction $Cr^{3+} \rightarrow Cr^{6+}$. Similarly, the potential of 1900 mV at the current peak is closed to the potential of the reaction $Fe^{3+} \rightarrow Fe^{6+}$ in a strong acid solution of pH $\simeq$ 0, according to Pourvaix potential-pH diagram and also the potential of oxygen gas generation on iron. From Fig. 9, which is the scanning electron micrograph of the specimen surface polarized up to the potential just before and just after the current peak, it is

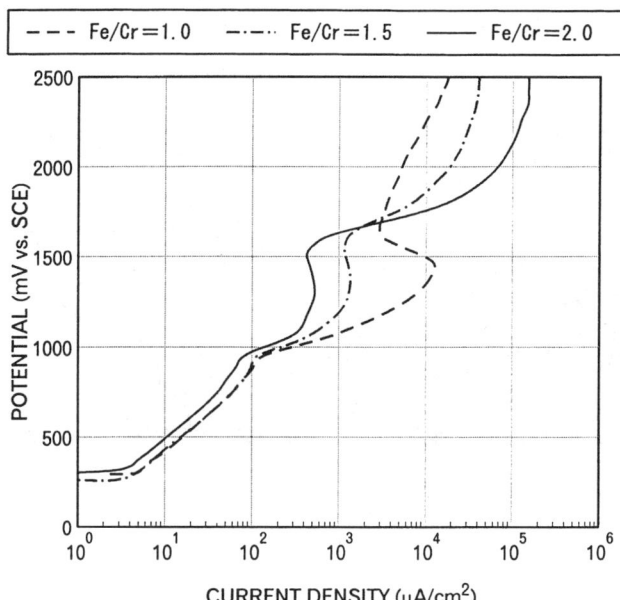

FIG. 10—*Anodic polarization curves of intermetallic compounds ($Zr(Fe,Cr)_2$ alloys) in 5% sulfuric acid.*

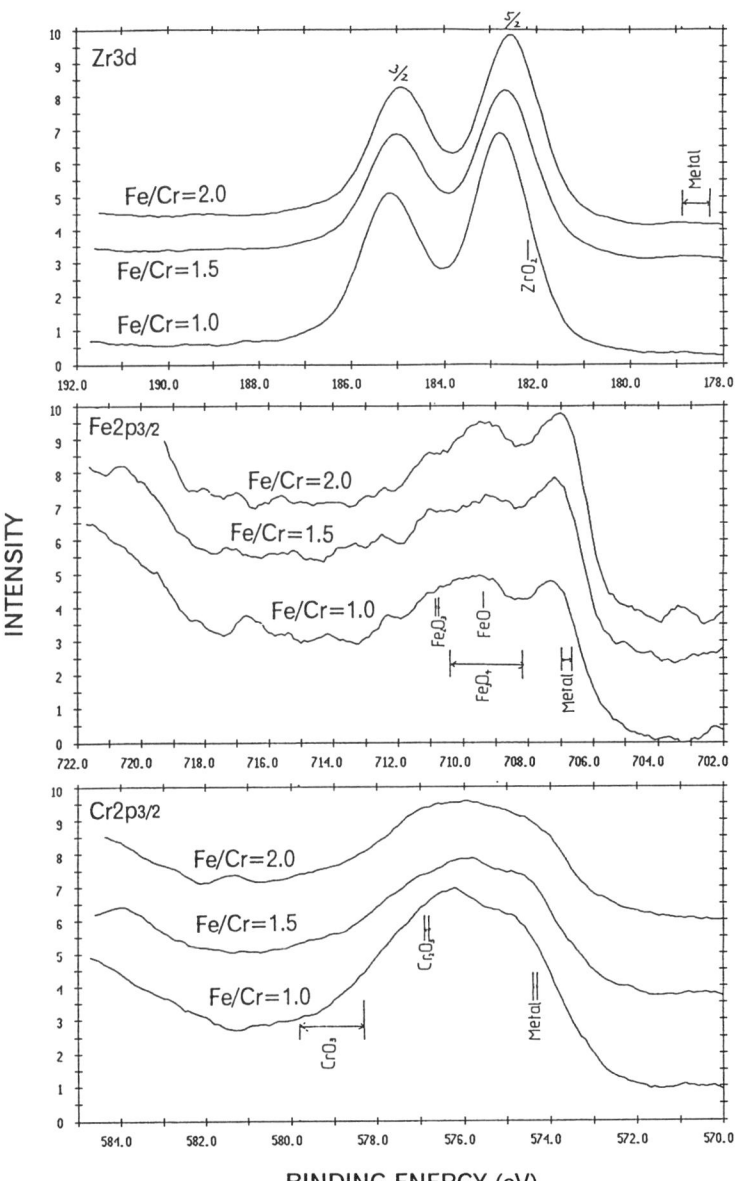

FIG. 11—*XPS spectrum on passive film of Zr(Fe,Cr)$_2$ alloys after anodic polarization up to 800 mV.*

considered that some precipitates of Zr(FeCr)$_2$ dissolved into the strong acid solution (pH ≃ 0) as the result of the reaction Fe^{3+} → Fe^{6+}, and oxygen gas simultaneously may have been generated because the surface of the specimens in which the anodic polarization was interrupted at the current peak potential was covered with a thick oxide film.

The mechanism for the occurrence of the current peak in the anodic polarization curves of

Zry-2, modified Zry-2, and Zry-4 is speculated as shown in Fig. 12 for Zry-2 and Fig. 13 for Zry-4. Regarding Zry-2, a mixed oxide film of ZrO_2, Cr_2O_3, and Fe_2O_3 is formed on the precipitates of $Zr(Fe,Cr)_2$ at a potential of less than 1000 mV, and very small amounts of these mixed oxides may dissolve in an electrolyte of 5% sulfuric acid because of the low pH (pH ≃ 0). The Cr^{3+} is then oxidized to Cr^{6+} when the potential goes over 1000 mV, and the Cr^{6+} easily dissolves in the electrolyte as follows:

$$Cr_2O_3 + 4H_2O \rightarrow Cr_2O_7^{2-} + 8H^+ + 6e^- \quad (2)$$

$$Cr_2O_3 + 5H_2O \rightarrow 2HCrO_4^- + 8H^+ + 6e^- \quad (3)$$

and the current also increases. When most of the Cr_2O_3 in contact with the electrolyte has dissolved, the current decreases to the level before the peak. Some of the precipitates of $Zr(Fe,Cr)_2$ then come out into the electrolyte as a result of the dissolution of Cr^{6+} around the precipitates as shown in Fig. 12c. At the potential of 1900 mV, the increase of current is not remarkable because the Fe/Cr ratio in the precipitates is less than 1.0, and the iron content in the precipitates is low in Zry-2. In the case of Zry-4, the mixed oxide of ZrO_2, Cr_2O_3, and Fe_2O_3 is similarly formed on the precipitates of $Zr(Fe,Cr)_2$ at a potential of less than 1000 mV, but the increase of current is not remarkable at the potential of 1250 mV because the Fe/Cr ratio is more than 1.5, that is, the Cr content in precipitates is low in Zry-4. When the potential approaches 1900 mV, Fe^{3+} is oxidized to Fe^{6+} as follows:

$$Fe_2O_3 + 5H_2O \rightarrow 2HFeO_4^{2-} + 8H^+ + 6e^- \quad (4)$$

and Fe^{6+} is easily dissolved in the electrolyte, and therefore the current increases remarkably. At the same time, oxygen gas is generated on the precipitate, and the whole surface of the specimen is covered with an oxide film as shown in Fig. 13. When most of the Fe_2O_3 in contact with the electrolyte has dissolved, the current decreases near to the level before the peak. The dependence of ΣAi on the peak area at 1250 mV in Zry-2 can be explained by this mechanism. It is known that the diameter of precipitates increases with ΣAi. It is considered that the amount of reaction $Cr^{3+} \rightarrow Cr^{6+}$ increases, and the current also increases as the contact area of the precipitates with the electrolyte increase with the increasing diameter of precipitates. The anodic polarization behavior of modified Zry-2 with various iron content can be understood by con-

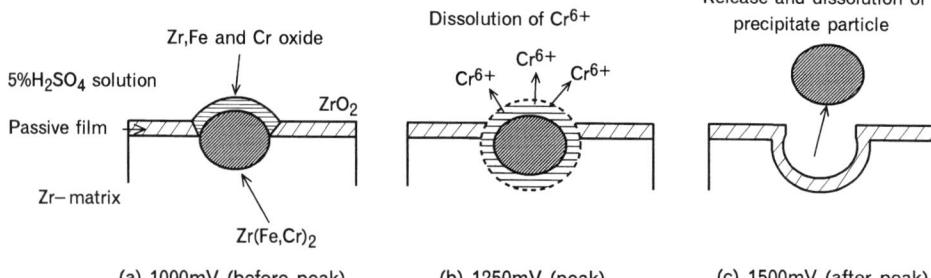

FIG. 12—Anodic polarization mechanism of Zry-2: (a) formation of Zr, Fe, and Cr oxide on precipitate particles; (b) oxidization of $Cr^{3+} \rightarrow C_r^{6+}$ around the precipitate particles ($Cr_2O_3 + 4H_2O \rightarrow Cr_2O_7^{2-} + 8H^+ + 6e^-$ and/or $Cr_2O_3 + 5H_2O \rightarrow 2HCrO_4^- + 8H^+ + 6e^-$) (c) release and dissolution of some precipitate particles.

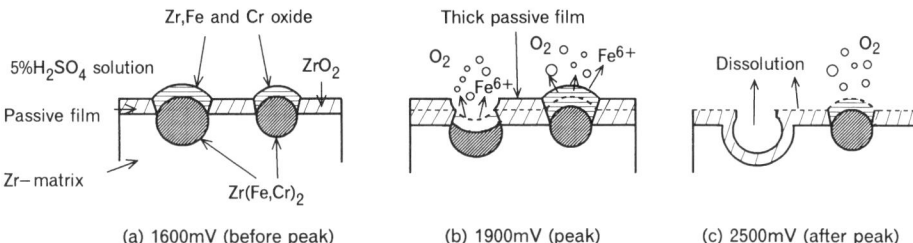

FIG. 13—*Anodic polarization mechanism of Zry-4:* (a) *formation of Zr, Fe, and Cr oxide on precipitate particles;* (b) *oxidization of* $Fe^{3+} \rightarrow Fe^{6+}$ *and oxygen gas generation on Fe-oxide* ($Fe_2O_3 + 5H_2O \rightarrow 2HFeO_4^{2-} + 8H^+ + 6e^-$ *and* $2H_2O \rightarrow O_2 + 4H^+ + 4e^-$); (c) *dissolution of the passive film and some precipitate particles.*

sidering the Fe/Cr ratio of $Zr(Fe,Cr)_2$ in modified Zry-2. When the iron content in Zry-2 increases, the Fe/Cr ratio of the precipitates of $Zr(Fe,Cr)_2$ also increases. Therefore, the amount of reaction $Cr^{3+} \rightarrow Cr^{6+}$ at 1250 mV decreases and the amount of reaction $Fe^{3+} \rightarrow Fe^{6+}$ at 1900 mV increases. As a result, the current peak at 1250 mV decreases and that at 1900 mV increases with increasing iron content in Zry-2.

Correlation Between Electrochemical Properties and Corrosion Resistance

Figure 4 shows the extremely good correlation between the current peak area at 1250 mV and ΣAi. Since ΣAi is used widely as an index for the corrosion resistance of Zry-2, the current peak area may also be used as an index for the corrosion resistance of Zry-2 instead of ΣAi. The weight gains in autoclave tests at 798 K for 24 h and the precipitate diameter as a function of the current peak area are shown in Fig. 14, indicating that the current peak area is correlated with the corrosion resistance and precipitate diameter as expected. Therefore, it is concluded that the current peak area at 1250 mV can be used as an index for the precipitate diameter and corrosion resistance of Zry-2.

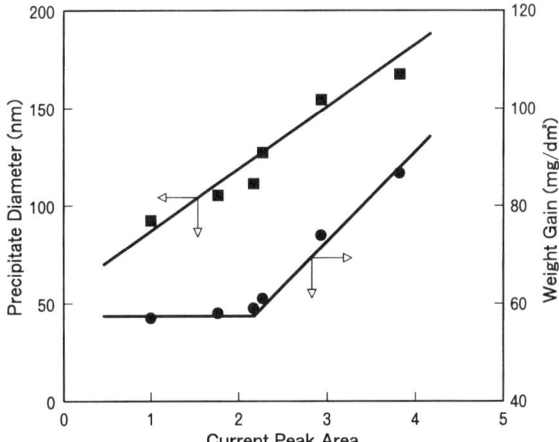

FIG. 14—*Precipitate diameter and nodular corrosion properties as a function of current peak area at 1250 mV.*

It seems that nodular corrosion resistance is improved by decreasing the current peak area at 1250 mV by reducing precipitate diameter and/or by increasing the iron content in order to reduce the Cr content in $Zr(Fe,Cr)_2$, but it is not clear why the nodular corrosion resistance of Zry-4 is generally worse in spite of the extremely small current peak at 1250 mV. The Ni in Zry-2 may play some role in the nodular corrosion process.

Conclusions

1. The current peak appears at about 1250 mV in Zry-2 and low-Fe modified Zry-2 on the anodic polarization curves in 5% sulfuric acid solution. On the other hand, the current peak appears at 1900 mV in Zry-4 and high-Fe modified Zry-2. It was found that the peak of 1250 mV appears when the Fe/Cr ratio of the $Zr(Fe,Cr)_2$ in Zircaloy is less than 1.0 and the peak of 1900 mV appears when the Fe/Cr ratio is more than 1.5.

2. The current peak at 1250 mV is thought to be caused by the oxidization reaction $Cr^{3+} \rightarrow Cr^{6+}$ in the mixed oxide film of ZrO_2, Cr_2O_3, and Fe_2O_3 formed on the precipitates of $Zr(Fe,Cr)_2$ with the Fe/Cr ratio of less than 1.0. Similarly, the current peak at 1900 mV seems to be caused by the oxidization reaction $Fe^{3+} \rightarrow Fe^{6+}$ in the mixed oxide film formed on the precipitates of $Zr(Fe,Cr)_2$ with the Fe/Cr ratio of more than 1.5. Oxygen gas may be also simultaneously generated at the iron oxide at near 1900 mV.

3. The current peak area at 1250 mV in the anodic polarization curve in 5% sulfuric acid can be used as an index for the nodular corrosion resistance of Zry-2 instead of ΣAi. Since the current peak can be measured very quick and simply, this index is very useful.

Acknowledgment

The authors gratefully acknowledge the help of Dr. K. Fujiwara (KOBELCO Research Institute, Inc.) for useful discussion in the electrochemical field.

References

[1] Garzarolli, F., Schumann, R., and Steinberg, E. in *Zirconium in the Nuclear Industry: Tenth International Symposium, ASTM STP 1245*, A. M. Garde and E. R. Bradley, Eds., American Society for Testing and Materials, West Conshohocken, PA, 1994, pp. 709–723.

[2] Rudling, P., Mikes-Lindbäck, M., Lethinen, B., Andrén, H.-O., and Stiller, K. in *Zirconium in the Nuclear Industry: Tenth International Symposium, ASTM STP 1245*, A. M. Garde and E. R. Bradley, Eds., American Society for Testing and Materials, West Conshohocken, PA, 1994, pp. 599–614.

[3] Anada, H., Nomoto, K., and Shida, Y. in *Zirconium in the Nuclear Industry: Tenth International Symposium, ASTM STP 1245*, A. M. Garde and E. R. Bradley, Eds., American Society for Testing and Materials, West Conshohocken, PA, 1994, pp. 307–327.

[4] Andersson, T. and Thorvaldsson, T. in *Zirconium in the Nuclear Industry: Seventh International Symposium, ASTM STP 939*, R. B. Adamson and L. F. P. Van Swam, Eds., American Society for Testing and Materials, West Conshohocken, PA, 1987, pp. 321–327.

[5] Bangaru, N. V., Busch, R. A., and Schemel, J. H. in *Zirconium in the Nuclear Industry: Seventh International Symposium, ASTM STP 939*, R. B. Adamson and L. F. P. Van Swam, Eds., American Society for Testing and Materials, West Conshohocken, PA, 1987, pp. 341–363.

[6] Kubo, T. and Uno, M. in *Zirconium in the Nuclear Industry: Ninth International Symposium, ASTM STP 1132*, C. M. Eucken and A. M. Garde, Eds., American Society for Testing and Materials, West Conshohocken, PA, 1991, pp. 476–498.

[7] Pêcheur, D., Lefebvre, F., Motta, A. T., Lemaignan, C., and Wadier, J. F. in *Journal of Nuclear Materials*, Vol. 189, 1992, p. 318.

[8] Pêcheur, D., Lefebvre, F., Motta, A. T., Lemaignan, C., and Charquet, D. in *Zirconium in the Nuclear Industry: Tenth International Symposium, ASTM STP 1245*, A. M. Garde and E. R. Bradley, Eds., American Society for Testing and Materials, West Conshohocken, PA, 1994, pp. 687–708.

[9] Hettiarachchi, S., Lndig, M. E., and Cubicciotti, D., *Proceedings,* Sixth International Symposium on Environmental Degradation of Materials in Nuclear Power Systems-Water Reactors, R. E. Gold and E. P. Simonen, Eds., The Minerals, Metals and Materials Society, Warrendale, PA, 1993, p. 443.
[10] Shirvington, P. J. in *Journal of Nuclear Materials,* Vol. 50, 1974, pp. 183–199.
[11] Steinberg, E., Weidinger, H. G., and Schaa, A. in *Zirconium in the Nuclear Industry: Sixth International Symposium, ASTM STP 824,* D. G. Franklin and R. B. Adamson, Eds., American Society for Testing and Materials, West Conshohocken, PA, 1984, pp. 106–122.
[12] Ito, Y. and Furuya, T., *Journal of Nuclear Science and Technology,* Vol. 32, No. 11, 1995, pp. 1118–1126.
[13] KTG Specialist Workshop, "Intermetallic Particles and Matrix Properties of the Zircaloys," Kerntechnische Gesellschaft e. V., Bonn, Germany, 1985.
[14] Franklin, J. A., *Atlas of Electrochemical Equilibria in Aqueous Solutions,* Pergamon Press Ltd., New York, 1966.

DISCUSSION

J. Piippo[1] (written discussion)—Your measurements were performed at room temperature and in sulfuric acid. Zircaloys are used normally at high temperatures in neutral water solutions. Is your environment relevant?

Y. Ito et al. (authors' closure)—In this study, we evaluated the electrochemical properties of Zircaloy in sulfuric acid solution at room temperature. However, we could indirectly evaluate the corrosion resistance of Zry-2 from the anodic polarization properties because there is a correlation between the anodic polarization properties and the precipitate diameter.

[1] VTT Manufacturing Technology, Finland.

H. Göhr,[1] J. Schaller,[1] H. Ruhmann,[2] and F. Garzarolli[2]

Long-Term In Situ Corrosion Investigation of Zr Alloys in Simulated PWR Environment by Electrochemical Measurements

REFERENCE: Göhr, H., Schaller, J., Ruhmann, H., and Garzarolli, F., "**Long-Term In Situ Corrosion Investigation of Zr Alloys in Simulated PWR Environment by Electrochemical Measurements,**" *Zirconium in the Nuclear Industry: Eleventh International Symposium, ASTM STP 1295*, E. R. Bradley and G. P. Sabol, Eds., American Society for Testing and Materials, 1996, pp. 181–202.

ABSTRACT: The corrosion behavior of Zircaloy-type alloys with different tin contents of 1.55, 0.70, and 0.55 wt% was studied at 350°C and 17 MPa in an environment similar to PWR primary water. For this non-interrupted test, a special autoclave system was used that was equipped with electrical feed that allowed followup on the growth of oxide layers by impedance spectroscopy and corrosion potential measurement at high temperature and pressure. As a reference electrode, a platinum wire was used that works as a hydrogen electrode according to the hydrogenated environment established during the start-up procedure. The test ran without interruption for 471 days.

Impedance spectra were taken at time intervals and evaluated for thickness and morphology of the oxide layer as well as for its electrical resistance. The tests without any temperature and pressure cycling showed similar oxidation behavior with repeated transitions as in discontinuously performed standard autoclave tests. Early in the pre-transition range, a dense oxide layer is formed, and fast changes of corrosion potential and electrical resistance are observed. The dense layer increases in thickness and homogeneity up to the transition, where a sudden breakdown occurs. Abrupt changes of the corrosion potential and electrical resistance were observed also at those points. After transition, a new dense layer is built up.

The corrosion potential changes are caused by a decrease of the electrical corrosion current with increasing oxide layer thickness, by the formation of a potential drop over the high-resistance dense oxide layer, and by structural changes at the transition points.

In general, alloys with different tin contents show similar behavior. However, they show differences in the times to transition, the kinetic constants deduced from their impedance spectra, and in the ionic and electronic resistance of the dense inner layer controlling corrosion.

KEYWORDS: zirconium alloys, corrosion, oxide layer thickness, impedance spectroscopy, in situ measurement

In light water reactors (LWR), Zr alloys are used for fuel element claddings and structural components in the high flux range. The corrosion behavior of the Zr alloys is an important parameter for optimizing fuel element economy. Modern PWR cladding tube variants such as low-tin Zry-4, optimized Zry-4, and Duplex ELS all have a low tin content [1]. To support empirical correlations used for material optimization, some insight into the mechanism of the corrosion process is also necessary. Therefore, several mechanism-related examinations have been performed over the last few years [2–4]. Special attention was given to the question of

---

[1] University of Erlangen, Germany.
[2] Siemens AG, Power Generation Group (KWU), 91050 Erlangen, Germany.

how Sn influences the corrosion mechanism. In the past, in literature and from studies in the Siemens laboratory, many different alloy compositions were investigated under different environments. The most impressive point observing the growth of corrosion layers as a function of time, for many alloys, is the change of the corrosion kinetics from a non-linear sub-parabolic to a more or less linear behavior. The point, or more precisely time interval, at which this kinetic transition occurs is called *transition*. One of the outstanding deviations from ideality is the observation of a totally missing transition (as very often found for zirconium-niobium alloys) or the observation of a sequence of repeated cycles of cube root non-linear kinetics (cyclic corrosion behavior). Examples for what may be found are shown in Fig. 1.

The characteristic changes of corrosion kinetics have been subject to numerous investigations. The identified reasons for this behavior were changes in the microstructure of the corrosion layer and the buildup of compressive stresses in the oxide originating from the difference between the lattice constants of metal and oxide (Pilling Bedworth ratio of 1.52). In releasing these stresses, cracks and pores in the adherent oxide layer are generated. Microstructure investigations of post-transition oxides by electron microscopy (TEM) gave some insights into the pore systems [4], as did earlier work that correlated electrical conductivity changes under high-pressure mercury to intrinsic porous oxides on zirconium metal [5]. The discussion never stopped about the origin of the observed cyclic behavior, and the suspicion has never been

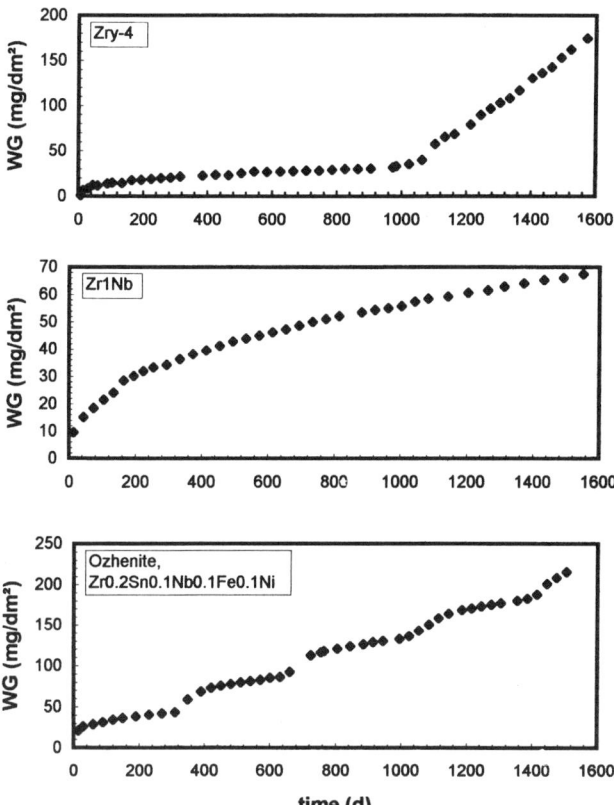

FIG. 1—*Typical corrosion kinetics of zirconium-based alloys in 350°C, 17-MPa water (Siemens laboratory data).*

finally rejected that cyclic kinetics are artificially induced by the periodically thermal cycles that the specimens are suffering from performing weight gain measurements. This could be accomplished only by measuring the increase of the corrosion layer in situ in pressurized water in a non-interrupted corrosion test at constant temperature and pressure.

One experimental method appropriate to allow in situ insights to the growing structure of the oxide layer is impedance spectroscopy (IS). Impedance spectroscopy gives the information on the complex a-c resistance (impedance) of a metal/metal oxide interface as a function of frequency of the applied alternating voltage. Earlier work very often has been constrained to the a-c resistance measured at only one arbitrarily chosen frequency.

Since the classical work on a-c impedance of zirconium/zirconium oxide interfaces by Wanklyn et al. going back to the early 1950s and 1960s [6] and references quoted therein, only little has been reported in the literature. Rosecrans [7] discussed results measured on Zry-4 and Zry-2 specimens with different heat treatments on the background of a review of earlier data. Only pre-transition oxides on Zry-2 were examined by Kubo and Uno by impedance spectroscopy [8]. In Ref 2, IS results from different materials oxidized under different conditions were presented and correlated to the microstructure of the investigated materials. A similar approach is made in a recent work [9] where the importance of manufacturing parameters of the investigated materials to the pore structure of the corrosion layer is emphasized. Up to now, only IS results from irradiated materials are reported by Gebhardt [10].

The general procedure of all authors mentioned above was to compare corrosion layer thickness determined by impedance spectroscopy to a total oxide thickness calculated from weight gains (WG) obtained by exposing the specimens to different aggressive environments for different times. The common finding of all authors was a general agreement of the IS-determined thickness to the total oxide thickness only up to values of some micrometers. Afterwards, increasing differences between the IS and WG scale thickness are observed. The point where the deviation becomes significant was correlated to the transition and the final IS scale thickness to the corrosion rate. These findings led to the imagination of a dense layer that slows down the corrosion process until the development of pores reduces its effective thickness. If linear corrosion kinetics are observed, the dense layer stayed at a more or less constant thickness value. A barrier layer thickness below 1 μm is typical for the inner dense sublayer in adherent uniform corrosion layers, whereas much thinner values were found in extremely porous nodular oxide layers.

All findings mentioned before are characterized by the experimental fact that the samples are not investigated under the environment where they have been corroded. The specimens were taken out of autoclaves after cooling down, exposed to air, and resoaked again into an electrolyte, where the impedance measurements were performed mostly at RT and atmospheric pressure. Systematic experiments reported in the literature showed that significant times are necessary to resoak porous corrosion layers [2, 9–11]. It was often questioned whether the fine pores are really filled by the electrolyte under these test conditions and whether this technique gives really relevant values for the thickness of the dense sublayer and its resistance. The only possibility to eliminate this uncertainty is to perform IS testing in the corrosive environment "in situ." Such a technique was developed and successfully applied in this study.

Experimental

Experimental Arrangement for In Situ IS at High Temperature and Pressure

The in situ impedance spectroscopy measurements have been performed alternating on specimens arranged according to Fig. 2 around a Pt reference electrode. This symmetrical geometry has been utilized to minimize inhomogenieties of the electrical field that may influence the

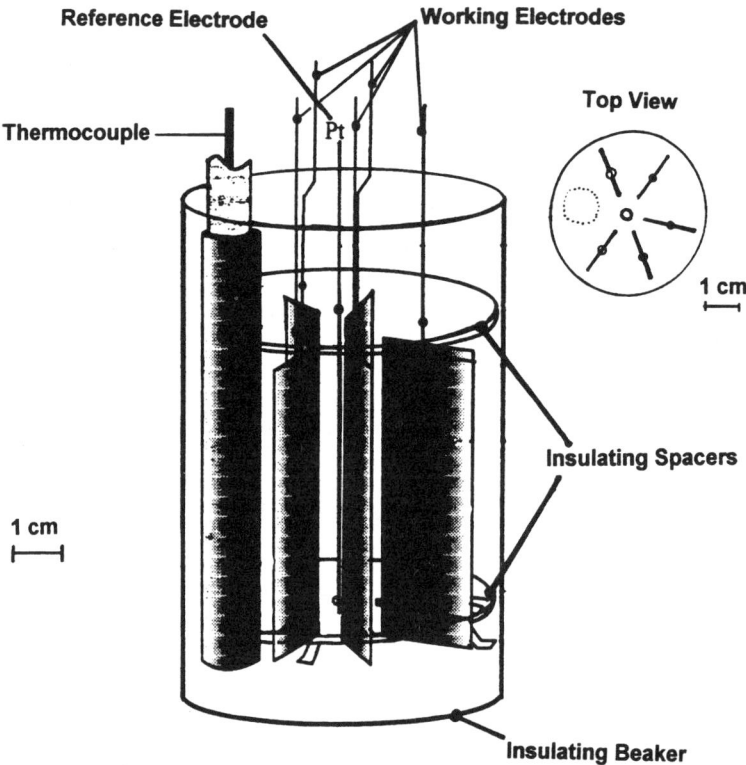

FIG. 2—*Experimental setup for the in situ impedance spectroscopy experiment.*

impedance spectra at high frequencies. During the measurement, the two specimens adjacent to the one switched as working electrode were operating as counter electrodes. The specimens were located inside of a 500-mL volume stainless steel autoclave in an insulating beaker. As isolating materials in the autoclave pre-oxidized, Zry-4 was used. Prior to startup, the autoclave was purged with argon and pressurized at room temperature with a 3-bar argon/hydrogen mixture. The lid of the autoclave was equipped with commercial electrical feedthrough (Conax) with soapstone sealing for the Zry-4 wires welded to the specimen. The autoclave was filled with water as a corrosive medium simulating coolant of a pressurized water reactor by adding 500 ppm boron as boric acid and 1.5 ppm Li as lithium hydroxide. Unfortunately, the electrical conductivity of this electrolyte was too low, even at high temperature, to allow a reliable registration of impedance spectra. Therefore, to increase the electrical conductivity 300 ppm, potassium as potassium sulfate had to be added. After this, the electrolyte resistance measured in the cell was 12 Ohm at 350°C. The autoclave ran under static conditions without any interruption continuously for 471 days.

Materials

The specimens were cut from cladding tube segments consisting of three different zirconium-based alloys. Alloy composition was kept constant with the exception of the tin content, which was varied from 0.55 to 1.55 wt% as shown in Table 1.

TABLE 1—*Alloy composition of the specimens.*

Sample ID	Sn, %	Fe, %	Cr, %
333-01	0.55	0.2	0.1
333-06	0.92	0.2	0.1
333-15	1.55	0.2	0.1

The cladding tubes were fabricated according to standard manufacturing routines. For the impedance experiment, tube segments were mechanically flattened without significant reduction of the original wall thickness of 0.725 mm. Rectangular 15 by 58-mm specimens were degreased, pickled in hydrofluoric acid, and welded to support wires from Zry-4 that were used also for the electrical connection. The exposed total specimen area was 18.4 cm^2.

One material was duplicated in the test to check disturbing influences of different specimen positions. But, in fact, no relevant differences between the impedance spectra of the two otherwise identical specimens were found experimentally. Sample duplication allowed for also checking the reproducibility of the experimental results, which proved to be excellent.

Impedance Spectroscopy Measurements

Impedance spectra were recorded with commercially available equipment (IM5d, Zahner-Electric, Kronach, Germany). The computer system integrated into this device was used for the data aquisition and the data evaluation. All impedance spectra were plotted in the frequency range from 0.2 Hz to 100 kHz.

The impedance spectra were measured at the specimens' zero current potential, superimposing a low-amplitude a-c voltage signal of 10 mV, which was swept in frequency from 0.2 Hz to 100 kHz. In this way, the disturbance of the material electrode system was low enough to ensure a linear response in the form of an alternating current signal with the same frequency but shifted by phase angle ω with respect to the applied a-c voltage. The total complex impedance of the material electrode system is recorded as a function of the applied frequency, f. The recorded impedance spectra are treated as reported in Refs *12* and *13* by

$$\log \{Z(\omega)\} = \log\{|Z|(\omega)\} + j\omega(\omega) \log e \qquad (1)$$

The main advantage of this logarithmic treatment is that, independent of the value of Z, the relative deviation is represented by the same difference in the entire range of plotted values. This is important, in particular, at large ranges of several orders of magnitude. In such a case in the complex plane of impedance, small values disappear against large values and different small values can scarcely be discerned. These facts generally should be considered [*14*]. Since the real part of log Z according to Eq 1 is the logarithm of modulus Z and the imaginary part is the phase angle ω, it is obvious to utilize Bode diagrams (log $|Z|$ versus log f, log ω versus log f) if frequency dependencies shall explicitly be represented. The advantages of Bode plots and the loss of information resulting from the Nyquist plot was also convincingly demonstrated by Mansfeld [*15*].

Influence of the Counter Electrode on the Measured Impedance Spectra

In the realized experimental arrangement, the overall cell impedance (the impedance of working electrode, the resistance of the electrolyte, and the impedance of the counter electrode connected in series) is not measured. Instead of the cell impedance exclusively, the impedance of the working electrode, including only a small fraction of the electrolyte resistance, is registrated against a platinum reference electrode. This type of measurement is necessary and indispensable to obtain high-precision results about the behavior of a single electrode (our specimen). The size and nature of the counter electrode therefore has no observable influence on the measured impedance spectra.

General Evaluation Procedure for Impedance Spectra

The objective of impedance spectroscopy is to elucidate the electrochemical properties of the investigated interface and to evaluate kinetic parameters. By the measurement of the complex impedance as a function of frequency, the dynamic behavior of the specimen electrode can be analyzed as shown in the flow diagram of Fig. 3. The experimental spectra have to be modeled by a set of interconnected impedance elements (resistance, capacitance), each being a characteristical representation of a physical feature or process at the interface. A derivation and compilation of different impedance elements and their physical meaning have been given in Ref *16*. For reasonable results, the following rules for the creating of equivalent circuits have to be followed:

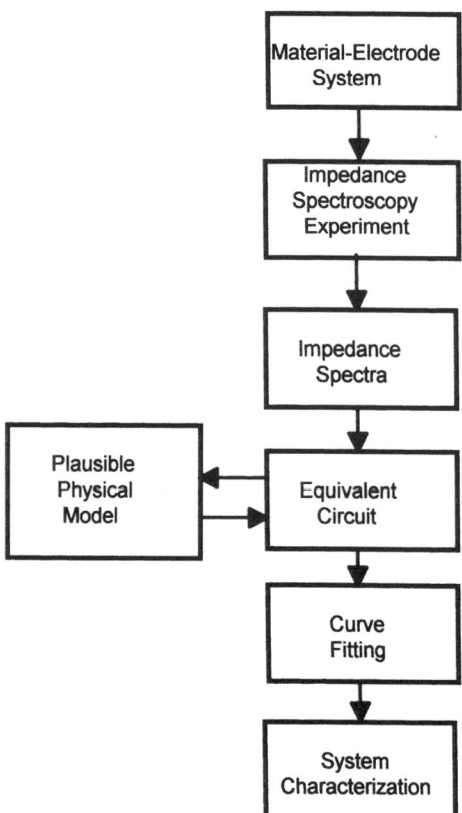

FIG. 3—*Flow chart for IS characterization of an interface being part of a material electrode system.*

1. Physical model on, at least, an intuitive basis.
2. Minimal number of impedance elements (no redundant elements are allowed).
3. Proof of the validity (variation of one physical parameter must give consistent values).

In the next step, the measured set of spectra has to be fitted by the frequency response of the defined equivalent circuit. This is performed by a computer simulation and fitting program [*17*]. The reliability of the final system characterization is highly dependent on the accuracy (minimal deviation between measured and fitted curve) of this procedure.

Results and Discussion

IS Results of the Non-Oxidized Specimen Surface and Effects of Heat-Up Procedure

After closing the autoclave, impedance spectra were recorded at room temperature (25°C). Then, after an equilibrating time of 23 h, the temperature was increased within 1 h up to 200°C. At this temperature, IS measurements were again performed, and, finally, after an interruption of 1 h, the temperature was set to get the target value of 350°C, which was reached within 3 h. Somewhat before this temperature was reached, impedance spectra were recorded at 334°C. The results of this sequence is plotted in Fig. 4 for Specimen 333-06, representative for all samples that showed no differences in their behavior. The logarithm of the modulus of impedance $|Z|$ and the phase angle ω, respectively, are plotted as a function of the logarithm of the frequency f measured in cycles per second (Hertz). At room temperature, the impedance shows high values at low frequencies with a linear decrease with increasing frequency. The phase angle accordingly falls non-linearly from values around 85° at low frequency to 15° at 100 kHz. As demonstrated in the diagram, the spectra can be fitted perfectly by an appropriate set of interconnected impedance elements. The continuous lines in the plot correspond to values obtained by the computer simulation, while plotted points are measured values. No significant deviation is observed between the experimental points and the simulated curves.

In the case of room-temperature measurements, a nearly ideal impedance behavior of the capacitance of a dense oxide layer is observed. Therefore, a quite simple equivalent circuit was sufficient, as shown in Fig. 5, consisting of a parallel circuit with a Young impedance and a Faradayic impedance in series to a resistance representing the electrolyte. The advantage of a Young impedance is the fact (in contrast to later-discussed constant phase elements like loss capacitances) that physical information can be obtained directly from the parameters of the element without any further physical calibration. Therefore, the measured capacitance, C_Y, can be directly converted to a scale thickness. The validity of this concept was proven by an earlier investigation [*11*] where scale thickness from IS was compared to values determined from weight gain and infrared spectroscopy results on a set of ex situ autoclave-corroded samples. From the measured C_Y values from spectra of different specimen, we obtain as a mean value:

$$C_Y = 37 \pm 3 \ \mu F$$

which can be converted to a scale thickness

$$d = \epsilon_0 \ \epsilon \ A \ r/C_Y = 10.6 \pm 0.8 \ nm$$

(dielectric constant $\epsilon = 22$, specimen area $A = 18.5 \ cm^2$, and a surface roughness $r = 1.2$). This thickness is in agreement with the thickness of a native oxide on zirconium alloys determined by photo electron spectroscopy [*3*].

As shown in Fig. 4, significant changes are observed in IS spectra measured at elevated

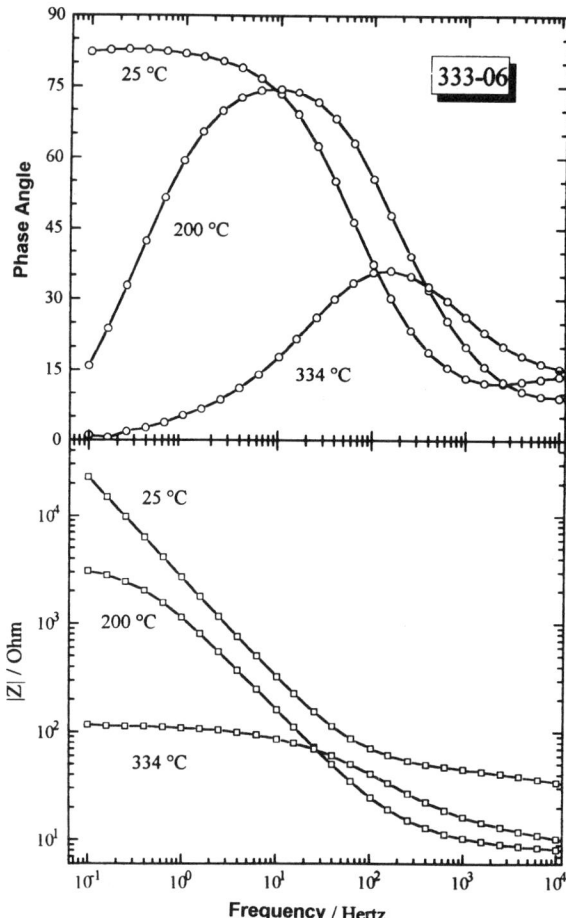

FIG. 4—*Impedance spectra of Specimen 333-06 at 25°C and during heatup.*

temperatures. The impedance at low frequencies decreases by two orders of magnitude, and the phase angle is developing a maximum that shifts to higher frequencies with increasing temperature. Even visually, one would conclude an obviously more complex behavior of the zirconium alloy interface at high temperature than at room temperature. As a consequence, the high-temperature spectra can no longer be simulated by the simple equivalent circuit used for the simulation of room temperature spectra. In fact, the successful simulation shown in Fig. 4 for the high-temperature spectra (agreement between plotted lines of fitted spectra and measured points) is obtained only with the equivalent circuit presented in Fig. 6. Unfortunately, the simulation was not possible using Young impedance elements. As shown in Fig. 6, the circuit includes loss capacitance, C_v, that cannot directly be converted to a physical property, like scale thickness, without calibration by an independently measured value. Nevertheless, as will be discussed later in detail, a calibration and a reasonable physical interpretation of the interconnected impedance elements is successfully performed.

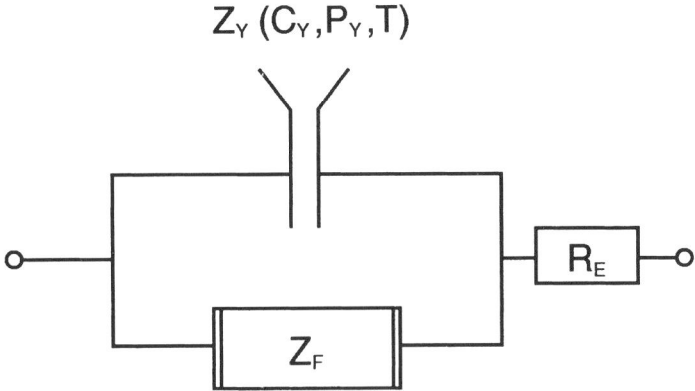

FIG. 5—*Equivalent circuit for the simulation of impedance spectra measured on a non-oxidized specimen.*

IS Results at 350°C and Changes as a Function of Time

After reaching 350°C, impedance spectra were measured at certain time intervals over a total exposure time of 471 days. The effects are discussed first phenomenologically and later on a more quantitative basis. In Fig. 7, impedance spectra of Sample 333-15 are plotted for the first 300 days of exposure to 350°C, 17-MPa water. Both the impedance and the phase angle do not show constant behavior during progressive exposure. For both values, alternating decreases and increases are visible. The time-dependent changes of the impedance spectra are demonstrated in Fig. 8. With increasing time, the impedance $|Z|$ increases and, after passing a maximum value, decreases again. Accordingly, the phase angle maximum is further shifted and a shoulder is developed.

The time at which the first impedance minimum is reached is different for the different alloys

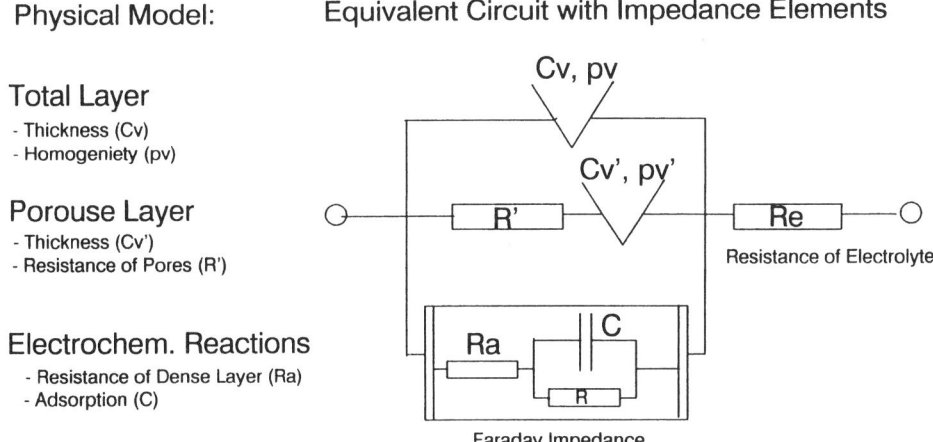

FIG. 6—*Correlation between equivalent circuit and physical model for impedance spectra measured at high temperatures on zirconium alloy interfaces.*

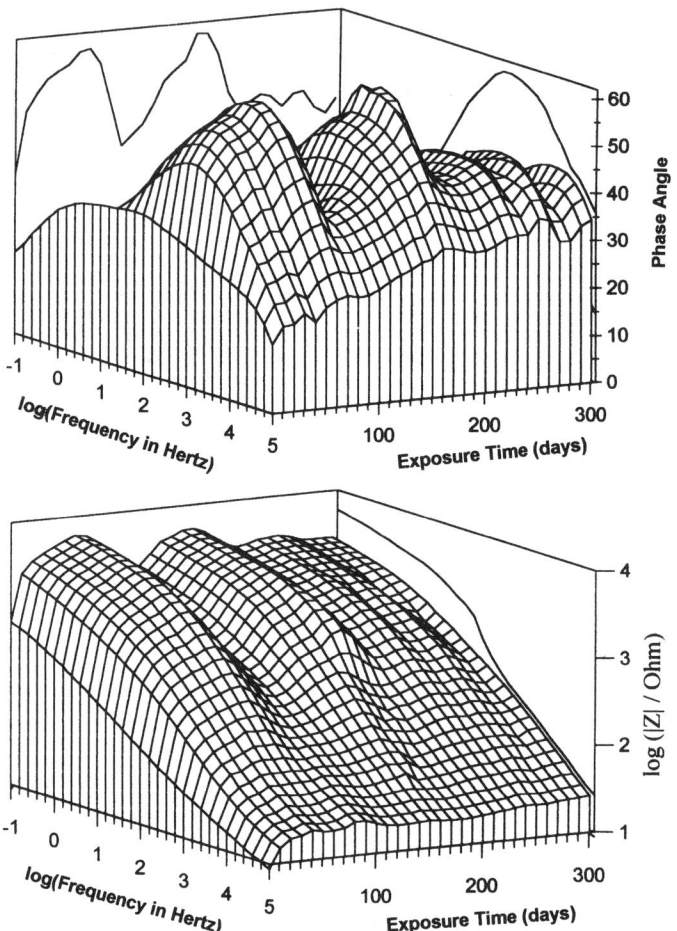

FIG. 7—*Change of the measured impedance spectra with time of exposure for Specimen 333-15 at 305°C, 17-MPa water.*

and can be correlated to the tin content of the specimen as shown in Fig. 9. Into the same diagram, time-to-transition values measured for the same alloys in a standard 350°C, 17-MPa water autoclave test are plotted. The good agreement between the periodicity in the IS and the classical gravimetric autoclave experiment is obvious. In the following paragraphs, further congruities between the ex situ corrosion behavior and the in situ results are presented.

Analysis of the High-Temperature Impedance Spectra and Correlation of Equivalent Circuit Impedance Elements to a Physical Model

The elements of the equivalent circuit necessary for fitting of the measured impedance spectra at 350°C are shown in Fig. 6. It consists of three parallel branches in series to a resistor, R_e, representing the resistance of the electrolyte.

The topmost impedance element in the equivalent circuit consists of a loss capacitance, C_v,

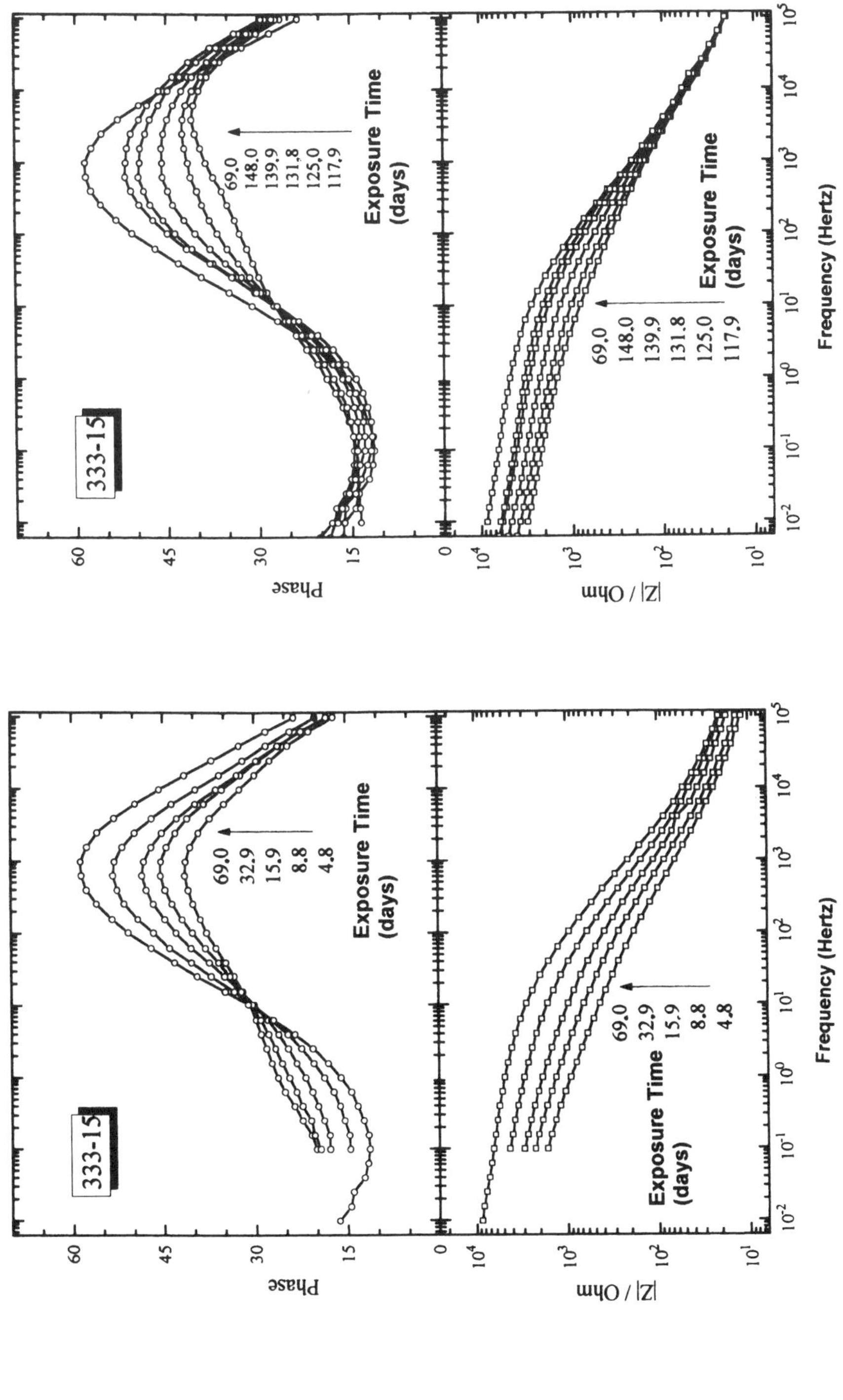

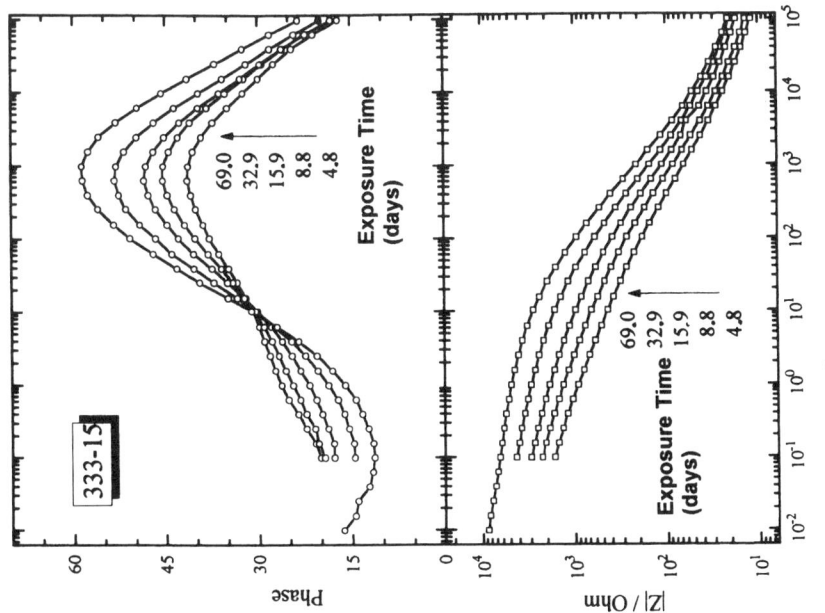

FIG. 8—*Impedance and phase angle variation as a function of exposure time for Sample 333-15.*

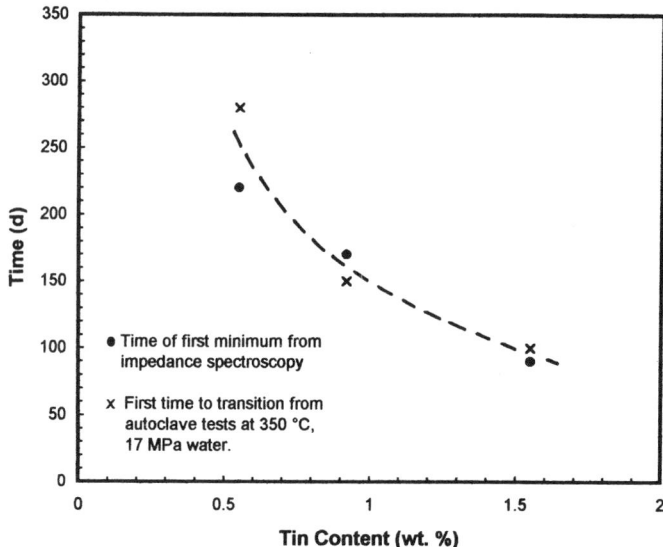

FIG. 9—*Correlation of the time to reach the first minimum of impedance and phase angle to the tin contents of the specimens in comparison to the first time to the transition from weight gain measurements.*

with the loss parameter, p_v (Fig. 6). This impedance element is allocated to the capacitance of the total oxide layer growing at the interface. Constant phase elements like loss capacitances cannot be directly converted to a physical property. The reason is that they include a normalizing frequency, $f°$, that can be arbitrarily chosen. The measured reciprocal capacitance, $1/C_v$, normalized to different values of $f°$ was compared to weight gain curves of the same materials, and it was found that only with $f° = 30$ kHz can physically consistent curves be obtained. Therefore, this frequency was used for further data evaluation.

A plot of $1/C_v$ versus exposure time is shown in Fig. 10 and compared to weight gain curves of the same materials from a 350°C, 17-MPa water autoclave test. The agreement between the in situ and the ex situ results is obvious. Even the cycling behavior of Specimen 333-15 with the highest tin content is reproduced in the data from the non-interrupted in situ corrosion experiment. Therefore, the question about the origin of the often-observed cyclic behavior in weight gain curves appears to be an intrinsic behavior of the material and not induced by temperature-cycle-induced changes of the corrosion layer.

The loss parameter, p_v, expresses the deviation in the phase angle from an ideal capacitor. Therefore, its physical meaning is correlated to the homogeneity of the total layer with a thickness represented by $1/C_v$. A decrease of p_v is correlated to an increase of the layer homogeneity and vice versa. The time dependence of p_v is plotted in Fig. 11 for the three specimens with different tin contents (see Table 1). At the times correlating to transition points, maximal values of p_v are observed that have to be understood as a loss of homogeneity of the layer at these kinetic transformation states.

The second branch in the parallel circuit consists of the resistance, R', and a loss capacitance, C'_v. These elements play no role in the circuit at the beginning of the corrosion experiment at low-scale thickness, but they are of significance and indispensable for exposure times at and beyond the transition. These impedance elements are attributed to a physical scale property

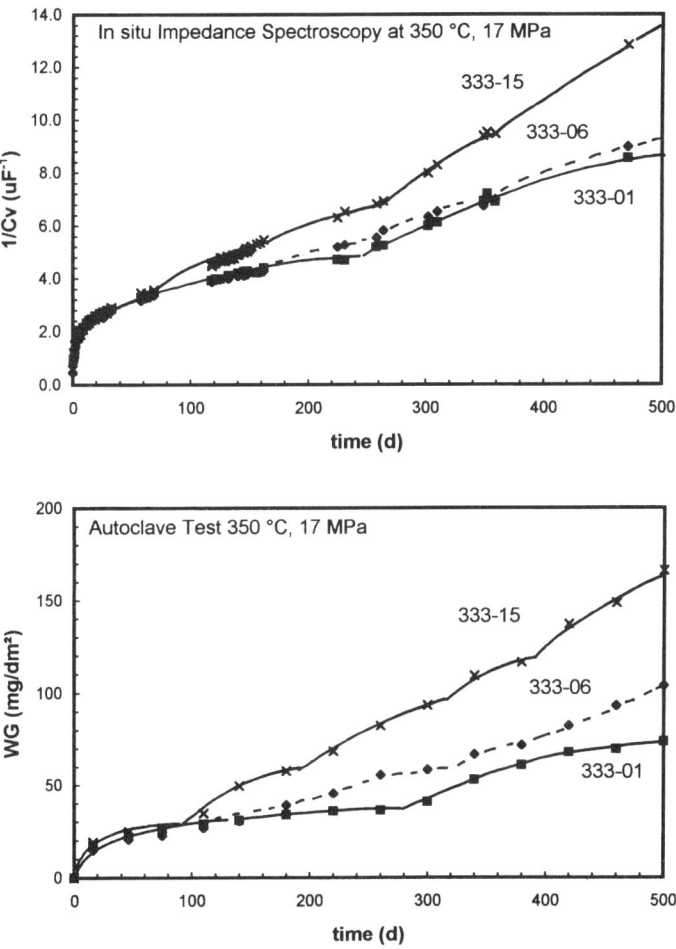

FIG. 10—*Comparison of the reciprocal capacity, C_v, to weight gain as a function of time for alloys with different tin contents.*

that develops at transition. What develops at transition is a porosity of the corrosion layer. The capacitor, C'_v, consequently represents the dense part of the total oxide layer. Due to the fact that C_v and C'_v are parallel capacitors, the dense part of the corrosion layer is represented by the sum of C_v and C'_v. In other words, the reciprocal value $1/(C_v + C'_v)$ is proportional to the physical thickness of the dense layer. The resistance of the electrolyte incorporated into the pore system is represented by resistance R'. In Fig. 12, $1/(C_v + C'_v)$ is plotted as a function of exposure time for the investigated specimens with different tin contents. Starting from zero, the thickness of the dense layer increases with exposure time up to a certain thickness, where an abrupt decrease occurs followed by a new increase up to the next transition. The time at which the first breakdown occurs increases with decreasing tin content of the alloy and corresponds well to the first transition known from weight-gain curves. Different values of the reciprocal capacitance (thickness) are reached depending on the tin content of the materials.

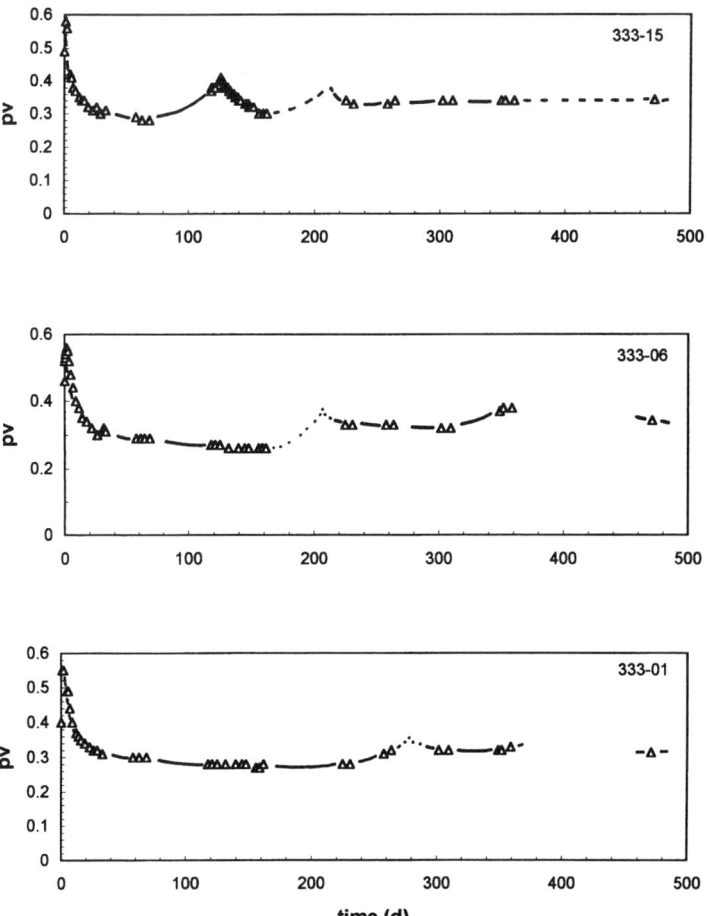

FIG. 11—*Loss parameter, p_v, as a function of exposure time for alloys with different tin contents.*

Lower values are observed for high-tin rather than low-tin material. After the first transition, the thickness of the dense layer shows additional abrupt changes (333-15, 1.5% tin) and levels out to an almost constant value during the final part of exposure. These values are not very different for alloys with different compositions.

If the reciprocal capacitance ($1/C_v$) correlated to total scale thickness (Table 2) obtained at the end of the in situ experiment is calibrated to scale thicknesses measured with the identical samples in an ex situ autoclave corrosion experiment, the physical thicknesses of the total and dense oxide layers as a function of exposure time can be calculated. This was performed, and the result is plotted in Fig. 13 for Specimen 333-15.

As can be seen in Fig. 13, the thickness calculated from impedance spectra correlates well to the weight gain curve. At low exposure times, impedance spectroscopy shows somewhat higher values than are derived from weight gains. Responsible for this may be an additional effect of temperature. Changes in the impedance spectra were observed during cooling down at the end of the experiment, which cannot be attributed to pore filling or soaking effects. They have to be attributed to temperature-activated, non-reversible morphological changes in the

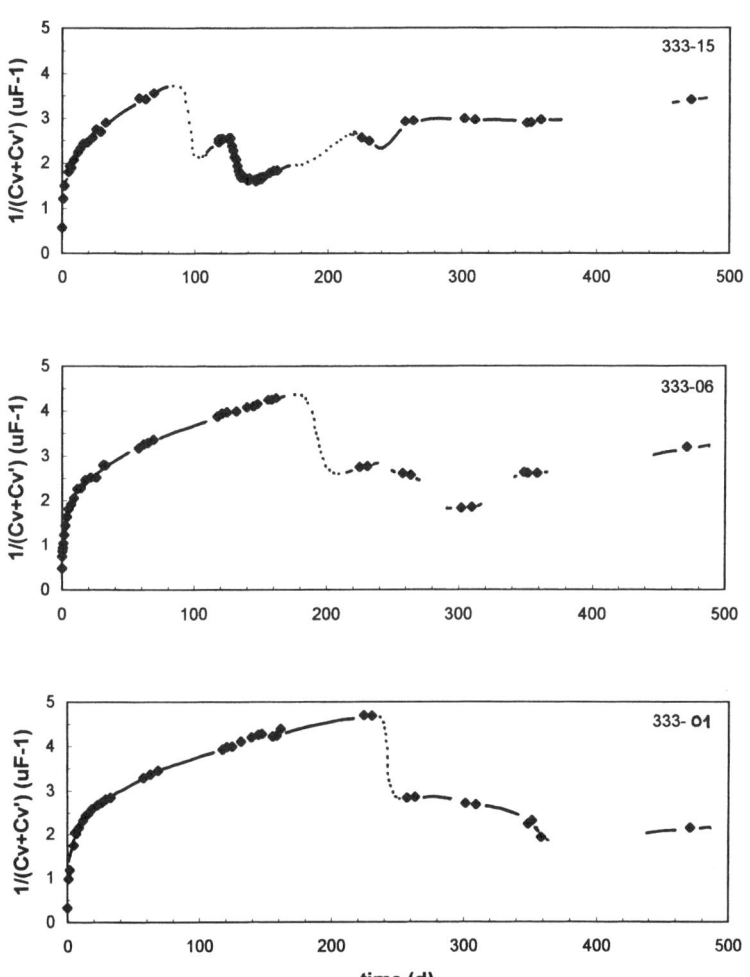

FIG. 12—*The reciprocal sum of* $C_v + C'_v$ *corresponding to the dense layer thickness as a function of time for alloys with different tin contents.*

TABLE 2—*Comparison of oxide scale thickness determined by different methods after 471-day exposure time for samples with different tin contents.*

Sample ID	Tin, wt%	Weight Gain, mg/dm^2	d (WG), μm	d (IR), μm	$1/C_v$, (μF)$^{-1}$
333-15	1.55	152.9	10.4	10.09	12.8
333-06	0.92	95.6	6.5	5.88	9.0
333-01	0.55	71.6	4.8	4.68	8.6

NOTE: d (WG): Scale thickness calculated with the density of bulk Zr oxide.

d (IR): Oxide thickness measured with fourier transformed infrared spectroscopy.

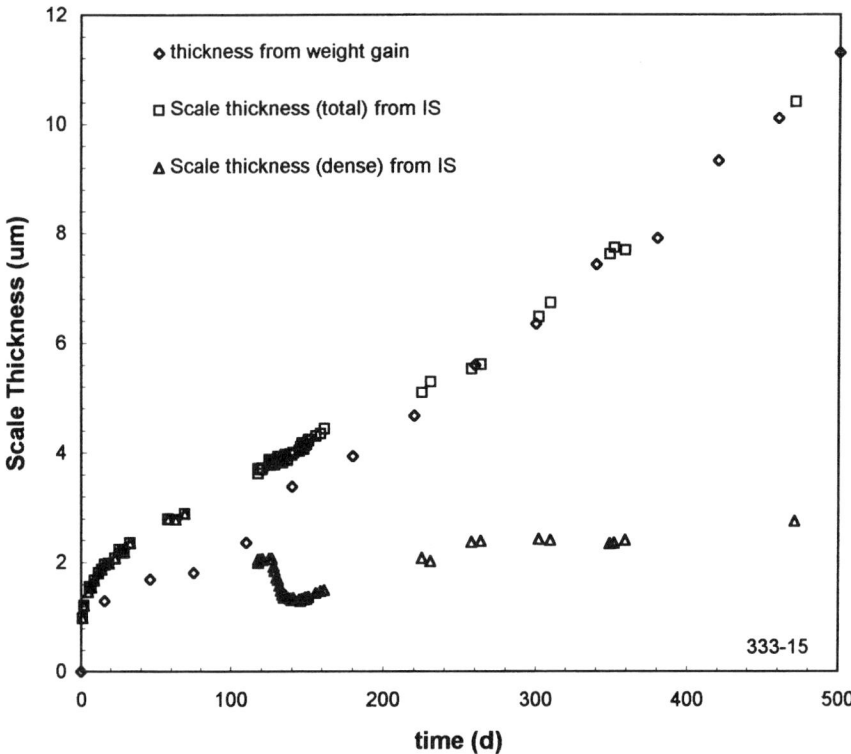

FIG. 13—*Comparison of scale thickness from weight gain and impedance spectroscopy as a function of exposure time.*

oxides as reported in Ref *11*. The thickness of the dense layer is comparable to values reported in an earlier investigation [2] for uniform oxides on zirconium-based alloys.

The third branch in the parallel circuit represents a Faradayic impedance element. This impedance element covers all those processes connected to changes of concentrations of chemical species being transported onto the metal/metal-oxide interface or within the interface. Examples of those processes are the transportation of charge carriers within the layer or the formation of adsorptive layers at the surface immersed to the electrolyte. The resistance, R_a, represents the electrical resistance of the dense layer. The values of Capacitor C and the Parallel Resistor R, if at all significant, are typical for layer-induced adsorptive processes at the metal/electrolyte interface. The time dependence of the dense layer resistance, R_a, is plotted in Fig. 14. A cyclic behavior with exposure time is observed showing lower values for high-tin than for low-tin alloys. With increasing time and scale thickness, an increase of the resistance is also observed. If the scale resistance is, in fact, connected to the transportation of charge carriers necessary for the chemical corrosion reaction (a decrease of the resistance is equivalent to an increase of corrosion current and herewith corrosion rate), sudden changes should also be seen in the corrosion potential. That there is actually a connection can be concluded from Fig. 15. The corrosion potentials for alloys with different tin contents are plotted as a function of the exposure time. Points of changing corrosion rate (equivalent to transitions) are connected to abrupt changes in the corrosion potential.

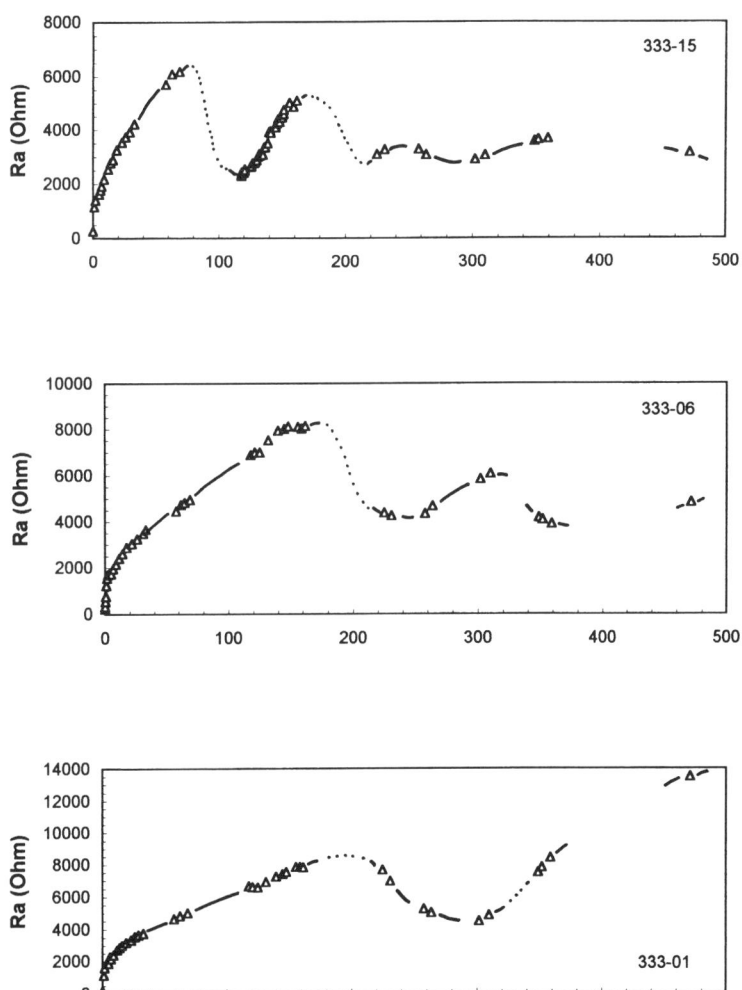

FIG. 14—*Resistance, R_a, of the dense layer as a function of exposure time for alloys with different tin contents.*

Discussion and Conclusions Related to the Corrosion Mechanism

The in situ IS examinations confirm the following model: oxide layers grown on zirconium alloys consist of a dense and porous part. Early in the corrosion process, a dense layer grows to a thickness of about 2 μm, where a rate transition occurs. Afterwards, the oxide layer consists of two different sublayers, an outer porous one and an inner dense one. Normally, the corrosion rate is governed by a solid state transport of charge carriers through the dense inner sublayer. Corrosion rate should depend, therefore, on the thickness of this dense layer. The thickness values of the different layers and their properties was deduced from in situ IS.

If the corrosion rate is controlled by the transport of charge carriers, the resistance (or its

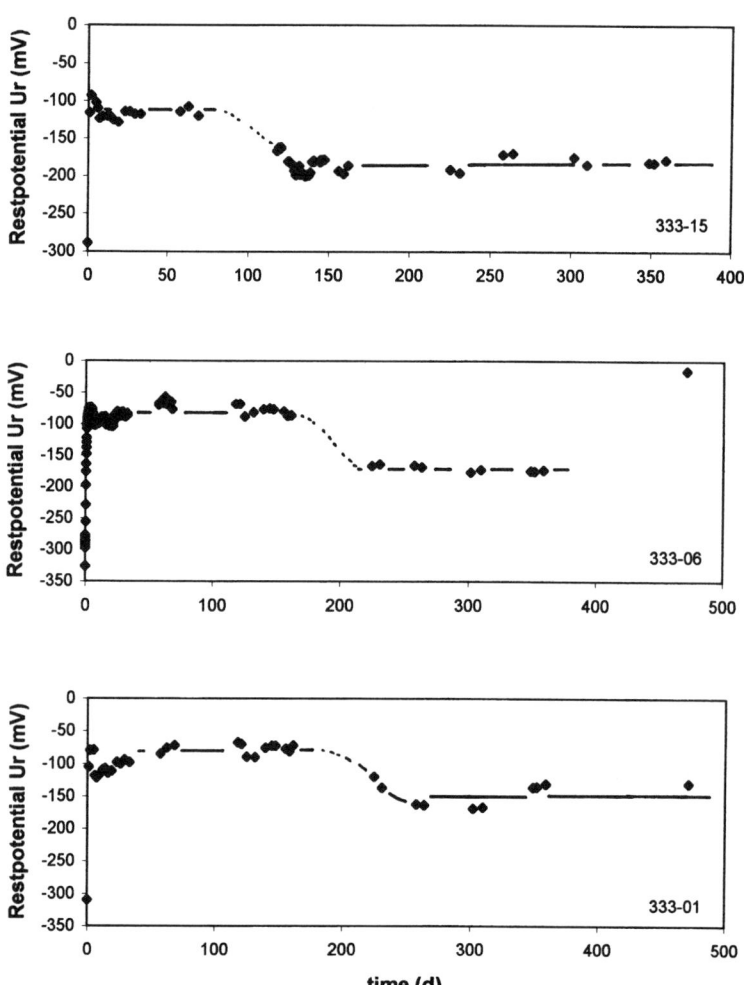

FIG. 15—*Rest potential, U_R, as a function of exposure time for alloys with different tin contents.*

inverse value, conductivity) of the dense layer should be correlated to the corrosion behavior. At constant area of the samples, the measured value of R_a is proportional to the specific resistance of the rate-controlling dense oxide layer. In Fig. 16, the quotient of R_a and the thickness of the dense layer $1/(Cv + Cv')$ is plotted as a function of the exposure time for the alloys with different tin contents. As can be seen in the diagram, the specific resistance of the dense layer increases with time. At transition points, sudden decreases are observed, and only for rather long exposure times are nearly constant values obtained for the alloys with 0.92 and with 1.55% tin contents. These findings show that not only the thickness but also the "quality" of the dense layer depend on the exposure time. Therefore, a comparison to post-transition corrosion layers is reasonable only for steady-state values of the resistance R_a.

Table 3 summarizes characteristic data for the rate transition and time-averaged values for the post-transition period for the three investigated alloys with different tin contents.

The resistance, R_a, deduced by IS governs the electronic current as well as the ionic current.

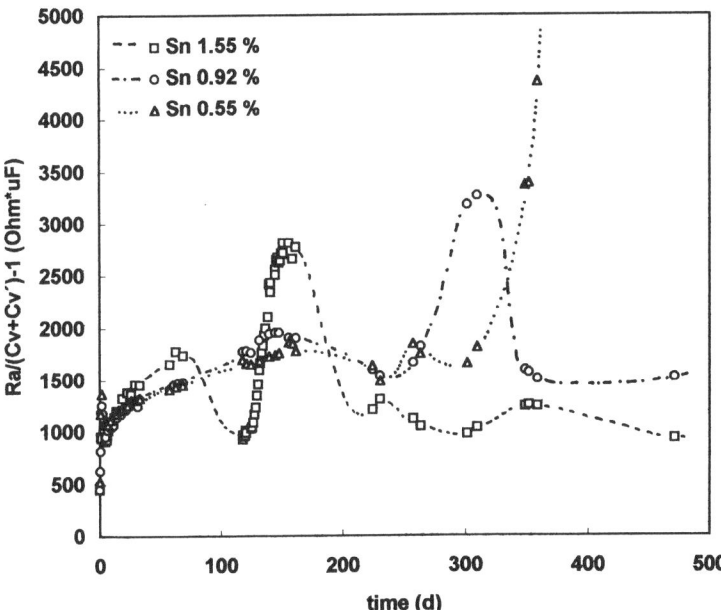

FIG. 16—*Quotient of the resistance, R_a, and the thickness $d \propto [1/(C_v + C'_v)]$ as a function of the exposure time.*

Conclusions on the electronic resistance can be deduced from the zero current potential. This potential is determined by two electrochemical processes, the anodic oxidation of the zirconium metal and the cathodic reduction of H^+. At low-oxide layer thickness, the potential is controlled by the electrochemical processes. The zero current potential becomes more positive with increasing oxide thickness (decreasing corrosion rate). With an increase of the dense oxide layer (decreasing corrosion rate), the zero current potential approaches more and more the electrochemical potential of the hydrogen electrode. At larger dense layer thickness, the potential drop over the high-resistant oxide layer plays the major role. This has been shown by Beie et al. [4] using the zero current potential results at the beginning of the experiment reported here.

According to the post transition data of Table 3, at increasing tin content the corrosion rate increases and the zero current potential is shifted to more negative values. Simultaneously, the resistance R_a of the dense layer decreases. The thickness of this layer, however, remains nearly

TABLE 3—*Characteristic data for the rate transition and time-averaged values for the post-transition period of the three alloys with different tin contents.*

Tin Content, %	Time to Transition, d	Corrosion Rate, μm	Dense Layer Thickness, relative	Dense Layer Resistance, R_a (Ohm)	Zero-Current Potential, mV, HE
0.55	250	0.009	2.6	9000[a]	−150
0.92	200	0.013	2.7	4500	−175
1.55	100	0.021	2.5	3500	−190

NOTE: d = days.
[a] Constant value not yet reached.

constant. Therefore, a change of this resistance can only be due to a reverse change of the conductivity in this region. Assuming that corrosion in the present case is controlled mainly by transport of charge carriers through the dense part of the layer, the overall conductivity of this region necessarily depends on the tin content in such a manner that at least one of the two conductivities, the electronic or the ionic conductivity, increases with increasing tin content.

Zero current potentials of corroding specimens are controlled by kinetic restrains of the occurring anodic and cathodic processes. Neglecting concentration changes of species determining the potential, a shift of the zero current potential is, on the one hand, equivalent to an equal shift of the overpotential of the anodic process, which is connected with transport of ions through the layer. On the other hand, it is equivalent to a reverse shift of the overpotential of the cathodic process, which is connected with the transport of electronic charge carriers. Therefore, the increase of the cathodic electronic current, corresponding to the corrosion rate at increasing tin content, is at least partially caused by the observed increase of the cathodic overpotential. No increase of the electronic conductivity is necessary for this increase. Then, the higher layer conductivity observed at higher tin content will be caused mainly by an increase of the ionic conductivity. This will be the main reason for the increase of the anodic ion current in spite of a lower anodic overpotential.

It has often been speculated that the hydrogen concentration within the oxide affects the protective property of the oxide layer [2] and the hydrogen pickup may be driven by this potential drop over the oxide. It has been reported that low-tin alloys behave differently in different environments [18] and show relatively high hydrogen pickup fractions [1]. The potentially negative effect of a high electronic resistance together with the beneficial effect of the high ionic resistance of dense oxide layers may be the mechanistic explanation for this behavior. The reason why the ionic conductivity increases with increasing tin contents is still an open question. Beie et al. [4] have not seen any obvious change of the microstructure of the oxide with varying tin contents. The only obvious difference that he found was a lower fraction of tetragonal oxide in low-tin material. Tetragonal oxide is often assumed to have a lower electrical resistance than monoclinic zirconium-oxide [19]. Probably the fraction of tetragonal oxide that increases with tin is therefore the real explanation of the tin effect on corrosion.

Another important influence on the electronic resistance of the dense layer probably comes from the size and distribution of intermetallic precipitates. No conclusion can be drawn from this experiment on the effect of intermetallics because the Fe and Cr contents and the size of the intermetallics were identical for all tested alloys. This question has to be answered by similar experiments.

Conclusions

1. The in situ impedance spectroscopy investigation on zirconium alloy cladding showed a cyclic behavior of impedance elements that is correlated to physical and kinetic properties of the corrosion layer. The cyclic behavior is proved to be an intrinsic property of the growth process and not induced by temperature or pressure changes.

2. Scale thickness of a dense and porous part of the corrosion layer is evaluated. A time-dependent cyclic growth and deterioration of the dense corrosion layer is observed.

3. Abrupt changes of the corrosion potential are correlated to changes of the electrical resistance of the dense corrosion layer.

4. The influence of the alloy composition (tin contents) is in agreement to observations from standard autoclave tests. Increasing tin contents decreases the ionic and electronic resistance of the dense layer, but has nearly no effect on its thickness.

References

[1] Seibold, A., Garzarolli, F., and Steinberg, E., "Optimized Zry-4 with Enhanced Fe and Cr Contents and DUPLEX Cladding: The Answer to Corrosion on PWR," *Proceedings*, International KTG/ENS Topical Meeting on Nuclear Fuel, TOPFUEL 95, Würzburg, Germany, 12–15 March 1995, Vol. II, p. 117.
[2] Garzarolli, F. et al., "Oxide Growth Mechanism on Zirconium Alloys," *Zirconium in the Nuclear Industry (Ninth Volume), ASTM STP 1132*, American Society for Testing and Materials, West Conshohocken, PA, 1991, pp. 395–415.
[3] Döbler, U. et al., "On the Initial Corrosion Mechanism of Zirconium Alloy," *Zirconium in the Nuclear Industry (Tenth Volume), ASTM STP 1245*, American Society for Testing and Materials, West Conshohocken, PA, 1994, pp. 644–662.
[4] Beie, H.-J. et al., "Examinations of the Corrosion Mechanism of Zirconium Alloys," *Zirconium in the Nuclear Industry (Tenth Volume), ASTM STP 1245*, American Society for Testing and Materials, West Conshohocken, PA, 1994, pp. 615–643.
[5] "Corrosion of Zirconium Alloys in Nuclear Power Plants," IAEA TECDOC-684.
[6] Wanklyn, N. J. et al., "Influence of Environment on the Corrosion of Zirconium and Its Alloys," *Journal of the Electrochemical Society*, Vol. 110, IAEA, Vienna, 1963, p. 856.
[7] Rosecrans, P. M. in *Zirconium in the Nuclear Industry (Sixth International Symposium), ASTM STP 824*, American Society for Testing and Materials, West Conshohocken, PA, 1984, pp. 531–553.
[8] Kubo, T. and Uno, M., "Electrical Resistivity of Oxide Films and its Relevance to Corrosion Resistance," *Proceedings,* Fourth International Symposium on Environmental Degradation of Materials in Nuclear Power Systems-Water Reactors, Palo Alto, CA, 1989, pp. 10-1 to 10-14.
[9] Cox, B. and Yamaguchi, Y., "The Development of Porosity in Thick Zirconia Films," *Journal of Nuclear Materials*, Vol. 210, 1994, pp. 303–317.
[10] Gebhardt, O., "Investigation of In-Pile Formed Corrosion Films on Zircaloy Fuel Rod Cladding by Impedance Spectroscopy and Galvanostatic Anodization," *Journal of Nuclear Materials*, Vol. 203, 1993, pp. 17–26.
[11] Göhr, H. et al., "Impedance Studies of the Oxide Layer on Zircaloy after Previous Oxidation in Water Vapor at 400°C," *Electrochimica Acta*, Vol. 38, 1993, pp. 1961–1964.
[12] Meissner, W., "Eine neue Methode der Messung und Auswertung von Impedanzspektren an Elektroden . . .," thesis, University of Erlangen, FRG, 1974.
[13] Göhr, H., "Dechema Monographien," *Verlag Chemie Weinheim*, Vol. 90, 1981, pp. 1–13.
[14] Macdonald, D. D. in *Electrochimica Acta*, Vol. 35, 1990, p. 1509.
[15] Mansfeld, F., "Technical Note: Concerning the Display of Impedance Data," *Corrosion Science*, Vol. 44, No. 8, 1988, pp. 558–559.
[16] Göhr, H., "Über Beiträge einzelner Elektrodenprozesse zur Impedanz," *Berichte der Bunsengesellschaft für Physikalische Chemie*, Vol. 85, 1981, pp. 274–280.
[17] Göhr, H., "Faraday-Impedanz als Verknüpfung von Impedanzelementen," *Zeitschrift für Physikalische Chemie N. F.*, Vol. 148, 1986, pp. 105–124.
[18] Garzarolli, F. et al., "Comparison of the Long Time Corrosion Behavior of Certain Zr Alloys in PWR, BWR and Laboratory Tests," this publication.
[19] Saario, T., Tähtinen, S., Piippo, J., and Kukkonen, J. J. V., "In Situ Measurements of the Effect of LiOH on the Stability of Fuel Cladding Oxide Film in Simulated PWR Primary Water Environment," *Proceedings*, Corrosion 95, NACE, Orlando 1995.

DISCUSSION

J. Piippo[1] (written discussion)—470 days in a static stainless steel autoclave is a long time. Were you able to maintain any water chemistry parameters during the measurement period?

H. Göhr et al. (authors' closure)—A water quality monitoring system was not installed in the autoclave, but changes in the water chemistry would have been detected by the impedance spectra measured in situ during the course of the experiment. No significant changes of the conductivity of the electrolyte was observed from our impedance measurement; therefore, we conclude that the "quality" of the electrolyte did not change significantly.

J. Piippo[1] (written discussion)—Various measurements performed for Alloy 600 in PWR environment show that the major electrochemical reaction taking place is the hydrogen redox reaction, which disturbs the electrochemical measurements. Can this also happen when measuring Zircaloys?

H. Göhr et al. (authors' closure)—This reaction obviously did not play an important role. The overall amount of the hydrogen in the system did not change during the experiment. At the end of the test, the remaining pressure was the same as established at the beginning. The amount of hydrogen released by the corrosion reaction can be neglected in that respect. All impedance measurements were performed at the zero current potential (rest potential), which is given in Fig. 15 as a function of time with respect to the platinum reference electrode. The potential of the platinum electrode is determined by the hydrogen in the system.

---

[1] VTT Manufacturing Technology, Finland.

Takeshi Isobe,[1] *Takuya Murai*,[1] *and Yoshiharu Mae*[1]

Anodic Protection Provided by Precipitates in Aqueous Corrosion of Zircaloy

REFERENCE: Isobe, T., Murai, T., and Mae. Y., **"Anodic Protection Provided by Precipitates in Aqueous Corrosion of Zircaloy,"** *Zirconium in the Nuclear Industry: Eleventh International Symposium, ASTM STP 1295,* E. R. Bradley and G. P. Sabol Eds., American Society for Testing and Materials, 1996, pp. 203–217.

ABSTRACT: Alloying elements such as Fe and Cr are generally considered to be effective even in small quantities for corrosion resistance of Zircaloy-4. The maximum total solubility of Fe + Cr in a Zr-Sn matrix has been reported to be very low [1]. Therefore, most of these elements are observed in the form of ternary Zr-Fe-Cr–type precipitates.

To clarify the effects of precipitates on corrosion property, Zr-1.3 Sn-(Fe,Cr) alloys containing Fe + Cr from 45 up to 180 ppm (the Fe to Cr ratio is about 2) were melted from pure zirconium (X-bar Zr and EB-Zr) and pure alloying elements. They were subjected to corrosion testing in 633 K water and microstructural analysis. It was found that precipitate-free materials showed much larger weight gains than precipitate-containing materials even at the same alloy compositions.

Subsequently, a corrosion test on the precipitate-free material galvanically coupled with a noble intermetallic compound of $Zr(Fe_{0.66}Cr_{0.33})_2$ was performed. It was found that the precipitate-free material, having very poor corrosion resistance in itself, was covered with thin and adherent black film under galvanically coupled conditions. In addition, its oxide grain structure was almost the same as that of the precipitate-containing material Zircaloy. From these results, it was concluded that the good corrosion resistance of Zircaloy-4 is attributed to the anodic protection provided by precipitates in the alloy.

KEYWORDS: anodic protection, precipitate, intermetallic compound, uniform corrosion, corrosion resistance, X-bar zirconium, EB zirconium, zirconium alloys, Zircaloy-4

Zircaloy-4 has been used widely as a cladding material in pressurized water reactors (PWRs). However, the recent trend toward extended burnup in PWR fuel has led to the necessity of improving the waterside corrosion resistance of Zircaloy-4 fuel claddings.

To improve the mechanistic understanding of the uniform corrosion behavior of zirconium alloys, for example Zircaloy-4, microstructural studies on precipitates in metal substrates and in growing oxides and on the oxide layer morphology have been performed in detail [2–18]. The size and distribution of precipitates were associated with corrosion resistance. In recent years, the crystallinity and the grain structure of the oxide layer, especially in the vicinity of the metal-oxide interface, have been regarded as controlling factors of the corrosion rate.

Only a few studies [19], however, have been done on the substantial function of intermetallic compounds on the corrosion resistance of Zircaloys in spite of its importance. Therefore, in this study, zirconium-tin alloys containing and not containing precipitates were prepared using high-purity zirconium and alloying elements and were subjected to a corrosion test. Subsequently, the effect of galvanically coupling precipitate-free zirconium-tin alloy with noble intermetallic compounds of $Zr(Fe,Cr)_2$ on the corrosion property was investigated.

---

[1] Research engineer, research engineer, and manager, respectively, Central Research Institute, Mitsubishi Materials Corporation, 1-297 Kitabukuro-cho, Omiya, Saitama 330, Japan.

TABLE 1—*Impurity levels of high-purity zirconium used (unit: ppm).*

Element	Zr-X-Bar[2]	EB-Zr[1,2]	Zr Sponge
Fe	6	2	~950
Cr	3	<0.01	65
Al	2	<0.01	25
Ti	<0.1	0.05	25
Si	0.2	30	40
Cu	0.1	0.02	5
W	<0.1	0.2	25
Hf	26	80	75
Mg	<0.1	<0.01	—
O	70	600	1100
C	25	90	110
N	<10	30	45

[1] EB-Zr: refined from sponge zirconium by EB-melting.
[2] Analyzed by GDMS.

Experimental Procedures

Materials

Charquet et al. [*1*] reported that the highest total solubility of Fe + Cr was 150 ppm at 1083 K in an alpha-phase of a Zr-1.4 Sn alloy with a Fe to Cr ratio similar to that in Zircaloy-4 and that solubility decreased with decreasing temperature. Sponge zirconium usually contains some hundreds ppm of iron, and materials made from it contain many precipitates. Therefore, both crystal bar (X-bar) zirconium and EB-zirconium were prepared as the raw materials. The former is a high-purity zirconium produced by a chemical vapor deposition, and the latter is one refined from sponge zirconium by electron beam melting in a high-vacuum (under 10^{-3} Pa) chamber. Table 1 shows the results of their chemical analysis by glow discharge mass spectroscopy (GD-MS). Total contents of Fe + Cr of X-bar zirconium and EB-zirconium were as low as about 10 and 2 ppm, respectively. EB-zirconium contained larger amounts of Si, Hf, and gaseous elements (O, C, and N) as impurities than X-bar zirconium.

Zr-1.3 Sn-(Fe,Cr) alloys with three levels of Fe + Cr contents ranging from 45 up to 180 ppm (the Fe to Cr ratio is about 2) were melted into about 200-g buttons by Ar arc melting from the above-mentioned high-purity zirconium, 99.99% Sn 99.9% Fe, and 99.9% Cr. Almost equal amounts of X-bar and EB-zirconium were used for each button. Table 2 summarizes the results of chemical analysis of the button ingots by GD-MS. They are designated 30/15, 70/40,

TABLE 2—*Chemical compositions of samples with low Fe and Cr contents.*

Sample ID	Alloy Composition			
	Sn, % by weight	Fe, ppm	Cr, ppm	O[1], % by weight
30/15	1.3	30	15	0.07
70/40	1.3	70	40	0.04
120/60	1.3	120	60	0.04

[1]Analyzed after final solution heat treatment.

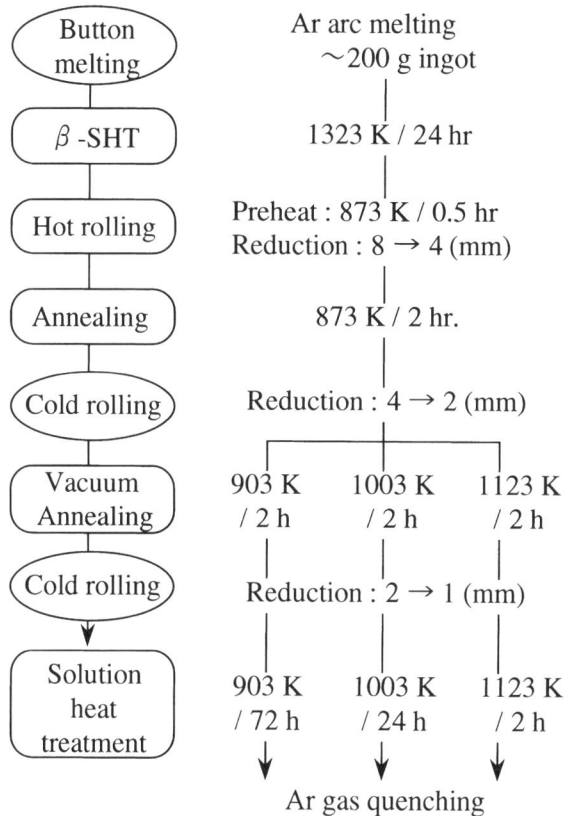

FIG. 1—*Processing sequence of sheet materials with low Fe and Cr contents.*

and 120/60, respectively, based on their Fe + Cr contents in ppm unit. For example, 30/15 means the zirconium-1.3 Sn alloy that contains 30 ppm Fe and 15 ppm Cr.

Button ingots were then rolled to about 0.8-mm-thick sheets according to a processing sequence, shown in Fig. 1, consisting of beta solution heat treatment, alpha hot rolling, cold rolling with intermediate alpha annealing, and final solution heat treatments at 903, 1003, and 1123 K followed by cooling in argon gas. Holding times of the final heat treatments were varied from 72 h at 903 K to 2 h at 1123 K, considering the thermodynamic equilibrium and the absorption of oxygen gas.

Oxygen analyses were performed on materials heat treated at 1123 K for 2 h, and results are shown in Table 2. The oxygen content of the 30/15 alloy (0.07 wt%) was slightly higher than those of the others (0.04 wt%). However, no severe oxygen uptake during the material processing was confirmed.

Experiments

Analytical Electron Microscopy

Transmission electron microscopy (TEM) and energy dispersive X-ray analysis (EDX) were conducted on the starting materials at 200 kV of accelerated voltage in a H-800/H-8010 analytical electron microscope (AEM). Thin foils were prepared by chemical milling in an aqueous solution of 5% hydrogen fluoride/45% nitric acid and electropolishing in a solution of 10% perchloric acid/90% methanol at about 15 V and 233 K.

Corrosion Test

Specimens were cut from 1-mm-thick sheets into 20 by 25-mm coupons for a corrosion test and 10 by 30-mm coupons for a galvanic coupling test. Corrosion tests were performed in water at 633 K at a pressure of 19 MPa in the static autoclave in accordance with the ASTM Practice for Aqueous Corrosion Testing of Samples of Zirconium and Zirconium Alloys (G2).

Oxide Microstructure

Thin oxide films were obtained by mechanically grinding one side of each oxidized sheet specimen after corrosion tests and then chemically dissolving the rest of the metal substrate in an aqueous solution of 5% hydrogen fluoride/45% nitric acid. The cross sections of the self-fractured oxide film were observed by high-resolution scanning electron microscopy (HR-SEM) (Hitachi S-900).

Results and Discussion

TEM Analysis

As shown in Fig. 2 the 30/15 and 120/60 alloys, solution heat treated both at the highest (1123 K) and at the lowest (903 K) temperatures, indicated as circles, were subjected to AEM observations. TEM micrographs of the two samples of the 30/15 alloy heat treated at 903 K and the 120/60 alloy heat treated at 1123 K are shown in this figure. Although not so many, ternary Zr-Fe-Cr—type precipitates were observed in the former material. Naturally, more precipitates with the same composition were observed in the 120/60 alloy treated at 903 K. However, no precipitates could be observed in the specimens treated at 1123 K in both the 30/15 and 120/60 alloys over a 20-μm^2 area for each sample.

Optical microscopy observations were also conducted on the materials heat treated at 1123 K besides the TEM analysis; however, acicular structures and the Fe and/or Cr rich phase in the grain boundary could not be observed. It seems that alpha + beta phase was not produced even by the highest temperature treatment.

Corrosion Testing

Figure 3a shows the relationship between corrosion weight gains in 633 K water for 15 days, final heat treatment temperatures, and contents of Fe + Cr. Materials heat treated at 1123 K, in which no precipitates could be observed by the TEM analysis, showed a poor corrosion resistance, although corrosion weight gains tended to decrease with increasing Fe + Cr contents. Weight gains of the lowest Fe + Cr containing alloy, namely the 30/15 (1123 K), were underestimated because of the oxide spalling off.

An additional corrosion test for 15 days was conducted on samples heat treated at lower temperatures (903 and 1023 K). Results are shown in Fig. 3b. Corrosion weight gains tended to decrease with increasing Fe + Cr contents and decreasing final-solution heat-treatment temperatures.

It was found that the precipitate-free samples had a poor corrosion resistance compared with the precipitate-containing samples, even at the same alloy composition. In addition, the cor-

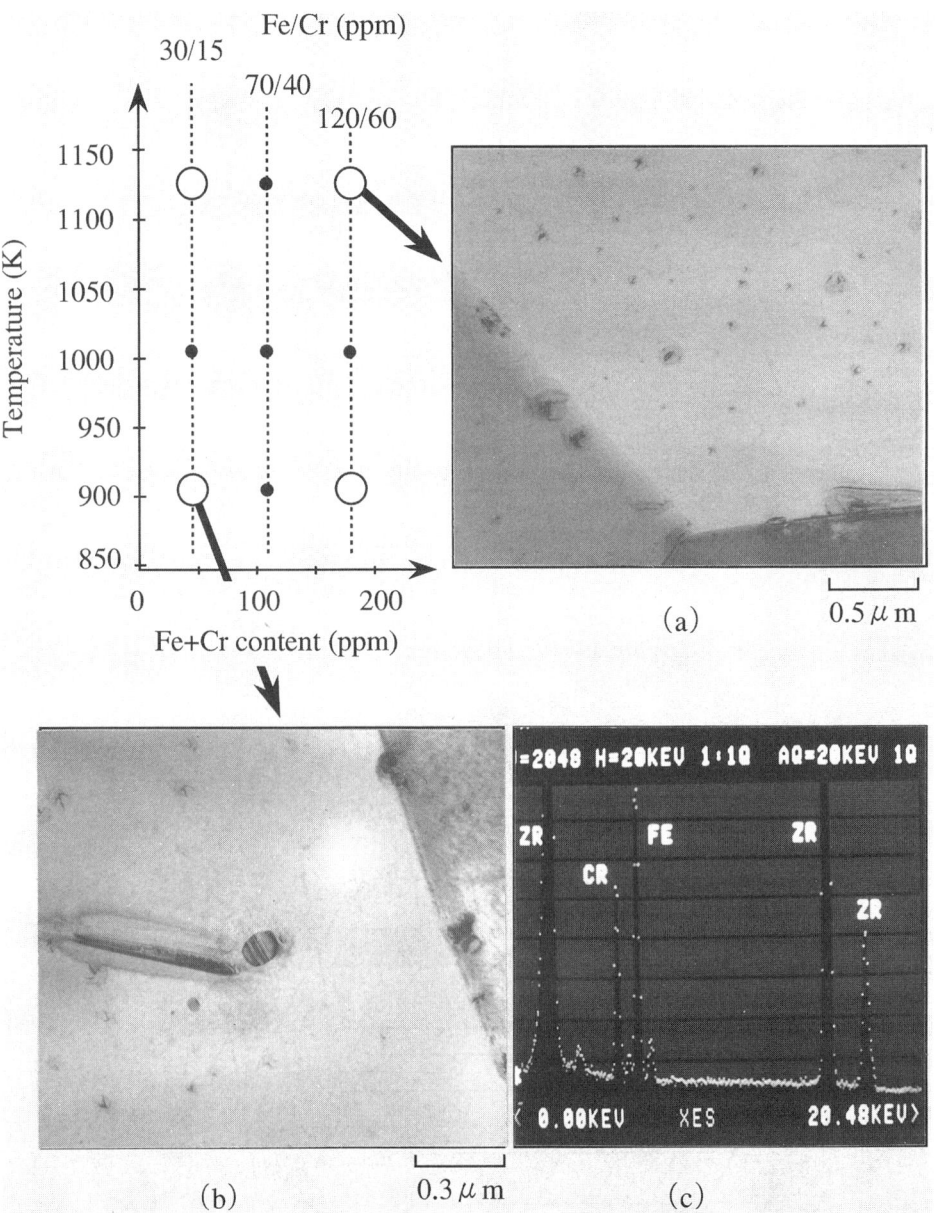

FIG. 2—(a) *TEM micrograph of 120/60 heat treated at 1123 K;* (b) *TEM micrograph; and* (c) *EDX spectrum of the precipitate of 30/15 heat treated at 903 K.*

rosion resistance tended to increase with an increasing quantity of precipitates, that is, the presence of precipitates improved the corrosion resistance of the alloy drastically.

A short-term corrosion test in 633 K water for three days was conducted on the specimens heat treated at 1123 K to observe the oxide appearances before the accelerated corrosion.

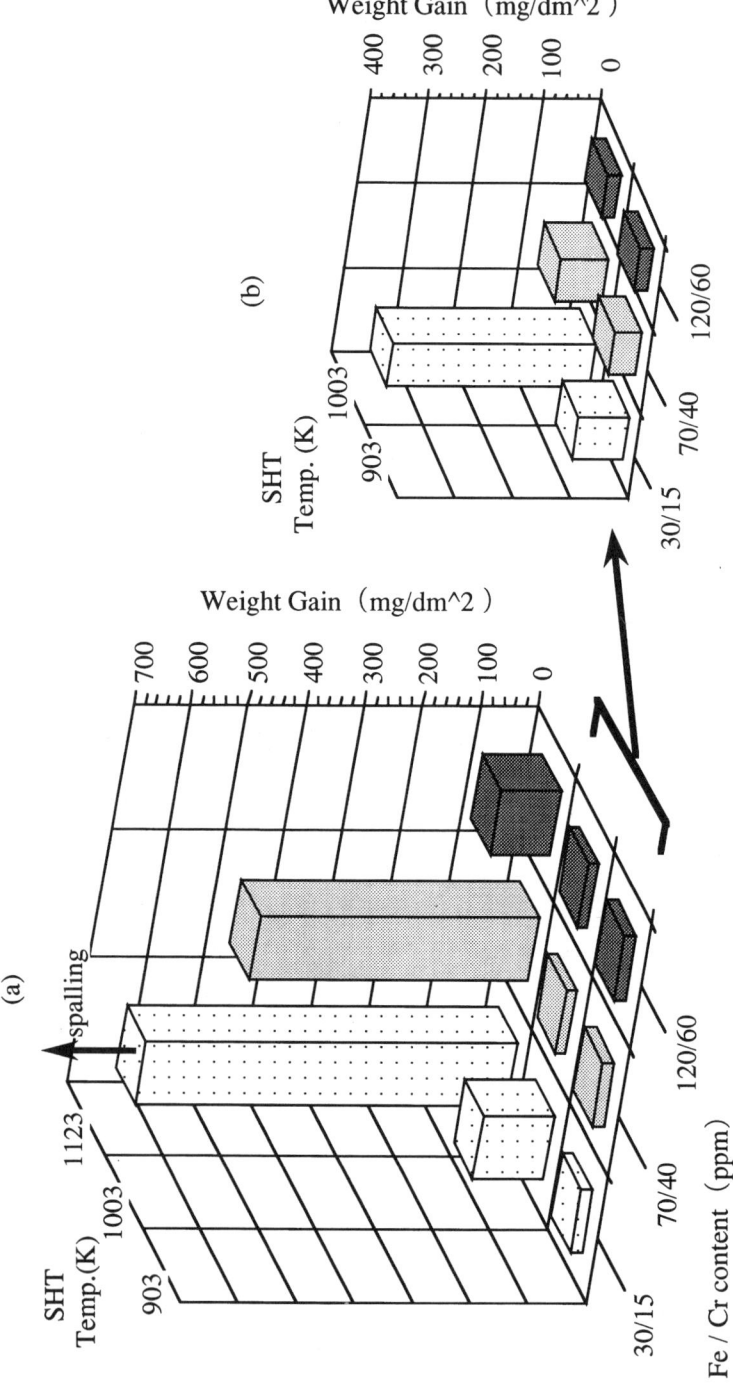

FIG. 3—*Relationship between Fe + Cr content, solution heat-treatment temperature, and corrosion weight gains in 633 K water for (a) 15 days and (b) 30 days.*

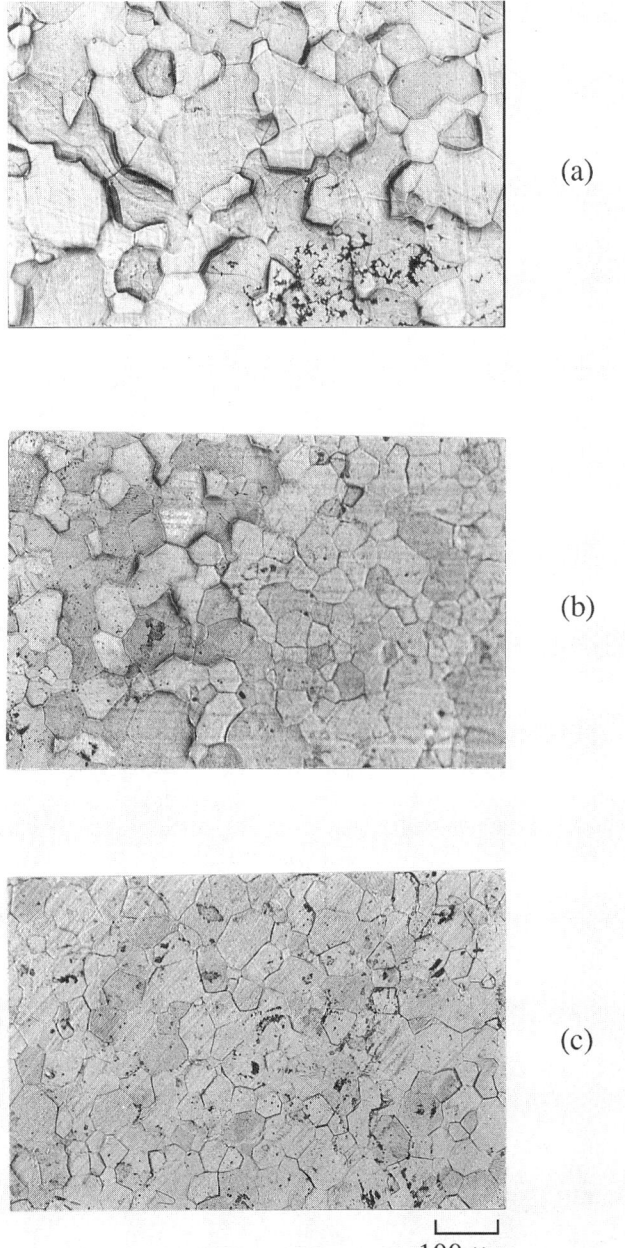

FIG. 4—*Optical micrographs (polarized light) of oxide films after three-day exposure of* (a) *30/15,* (b) *70/40, and* (c) *120/60 alloy heat treated at 1123 K.*

Average weight gains (two for each specimen) were about 260, 30, and 10 mg/dm^2 for the 30/15, 70/40, and 120/60 alloys, respectively. Figures 4a to 4c show the optical micrographs (polarized light) of the oxide surface. Well-developed heterogeneous oxide growth related to macroscopic structures of the substrate was observed, especially in the 30/15 alloy (1123 K) of the poorest corrosion resistance. Moreover, small transgranular cracks in the oxide were also observed in the 30/15 alloy. Non-uniform oxide growth could be related to the anisotropy of the underlying grains, as described by Charquet et al. [20]. These observations indicate that an accelerated corrosion already occurs even in the early stage.

Galvanic Coupling Test

Rest potentials and corrosion weight gains of X-bar zirconium, Zircaloys (Zircaloy-2, Zircaloy-4), and intermetallic compounds were measured by Weidinger et al. [19]. It has been reported that intermetallic compounds were more noble than X-bar zirconium and Zircaloys at 298 and 523 K and that they had larger weight gains than Zircaloys in 673 and 773 K steams.

Similar results on rest potentials were obtained in our experiment on pure zirconium and $Zr(Fe_{0.66}Cr_{0.33})_2$ intermetallic compound in 0.5 M potassium sulfate solution. In addition, as mentioned above, the precipitate-free zirconium-tin alloy, which corresponds to the alpha-zirconium matrix in Zircaloy, showed much inferior corrosion resistance to Zircaloy-4, that is, both intermetallic compounds and alpha-zirconium matrix alone reveal poor corrosion resistance, but, once coupled, good corrosion resistance might be expected just like that observed for Zircaloy.

To clarify the effect of galvanically coupling alpha-zirconium matrices with noble intermetallic compounds on the corrosion property, the precipitate-free sheet was welded with a flake of $Zr(Fe,Cr)_2$ intermetallic compound according to a processing sequence shown in Fig. 5. The 30/15 alloy heat treated at 1123 K, which showed very poor corrosion resistance, and the $Zr(Fe_{0.66}Cr_{0.33})_2$ intermetallic compound with the same composition as the representative precipitate in Zircaloy-4, were used, respectively. The two materials were coupled by arc welding under a partial pressure of argon gas. In addition, the precipitate-free sheets were also welded to each other to confirm the effect of the welding heat on the corrosion property.

A corrosion test in 633 K water was conducted for 15 days including the non-coupled precipitate-free sheet material as a corrosion reference. Figures 6a and 6b show a schematic representation of the samples and their appearances after corrosion testing, respectively. The precipitate-free sheet materials, both coupled with each other and non-coupled, were corroded severely just like the sample 30/15 (1123 K) in Fig. 3a, while the one coupled with the intermetallic compound was covered with a protective black oxide film even at a location far from the welding zone. Evidently, electrical contact with the noble intermetallic compound made the oxide film that formed on the precipitate-free matrix more protective.

Zirconium is one of the transition elements and can be passivated easily even in the air at room temperature. But, in high-temperature water, this spontaneous passivation becomes insufficient, and other passivation mechanisms are needed. These results indicate that a large intermetallic compound with the same composition as the representative precipitate in Zircaloy-4 can serve as an effective cathode site during anodic oxidation of the alpha-zirconium matrix and thus enhance its passivation, as also discussed by Weidinger et al. [19]. It is thought that the superior corrosion resistance of precipitate-containing alloys was achieved by anodic protection due to the presence of more noble precipitates.

On the other hand, as shown in Fig. 6b, the intermetallic compound on the galvanic coupling sample had been partially lost. At least two mechanisms of the brittle fracture by hydrogen absorption and the dissolution by anodic oxidation can be taken into consideration to explain this phenomenon.

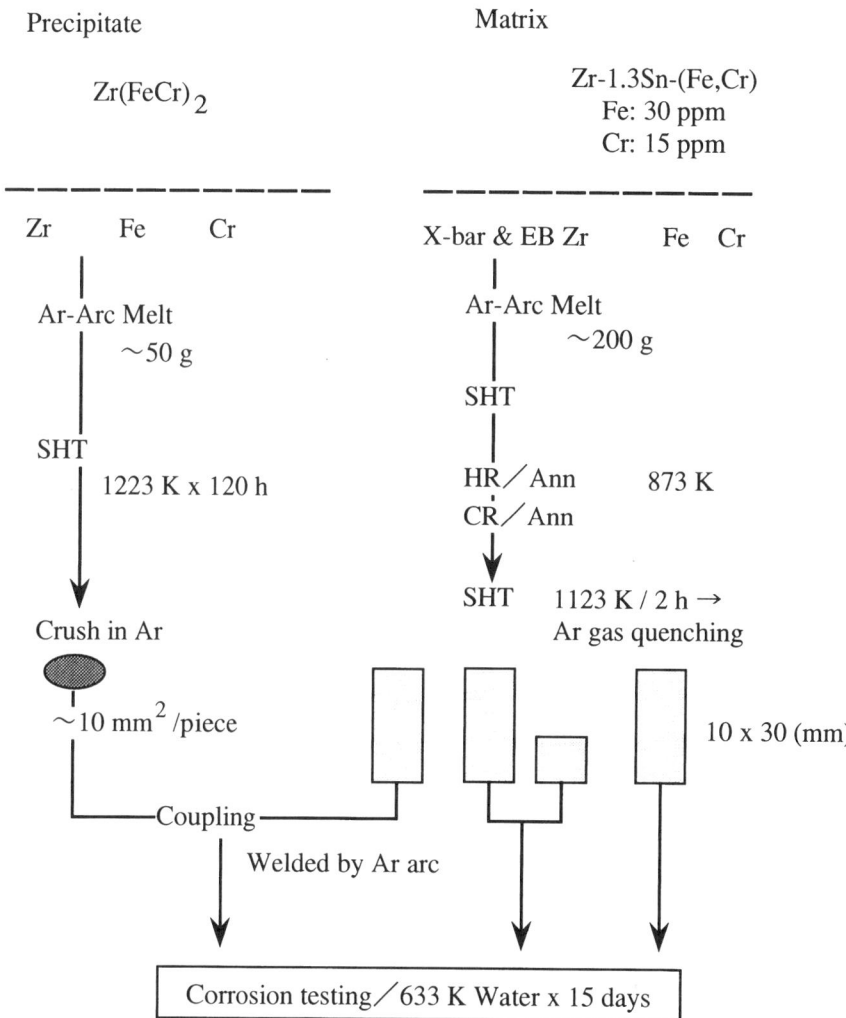

FIG. 5—*Processing sequence of the galvanically coupled material and reference materials.*

As mentioned above, hydrogen is reduced on a surface of the intermetallic compound because it can serve as a cathode site. Hydrogen analyses were conducted on the galvanically coupled sample after the corrosion test. Both the intermetallic compound and the precipitate-free matrix contained a relatively large amount of hydrogen, about one thousand and some hundreds ppm, respectively. Very high hydrogen uptake of the matrix covered with the protective film suggests that the intermetallic compound worked as a window for the hydrogen absorption, in other words, it served as a cathode site. However, no hydride peaks were actually observed from the X-ray diffraction pattern of the intermetallic compound. The hydrides formed in the intermetallic compound in the early stage might be fractured and lost as the corrosion reaction proceeds.

Moreover, in the later stage of the corrosion, the electrochemical polarization of the intermetallic compound is thought to change from cathodic to anodic because the passivated matrix

(a) Schematic representation of specimens

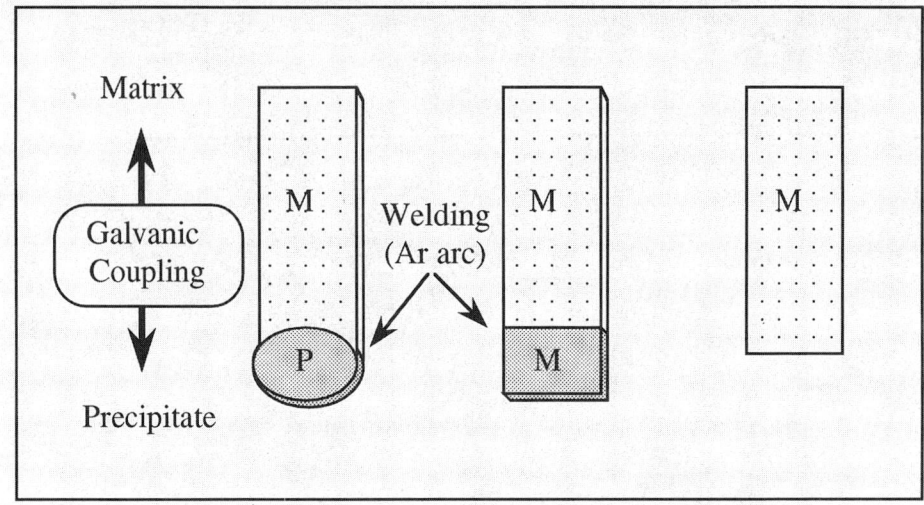

M : 30 / 15 (Zr-1.3 %Sn - 30 ppmFe - 15 ppmCr) solution heat treated at 1123 K.

P : Intermetallic compound of Zr(FeCr)2

(b) After corrosion testing (in 633 K water for 15 days)

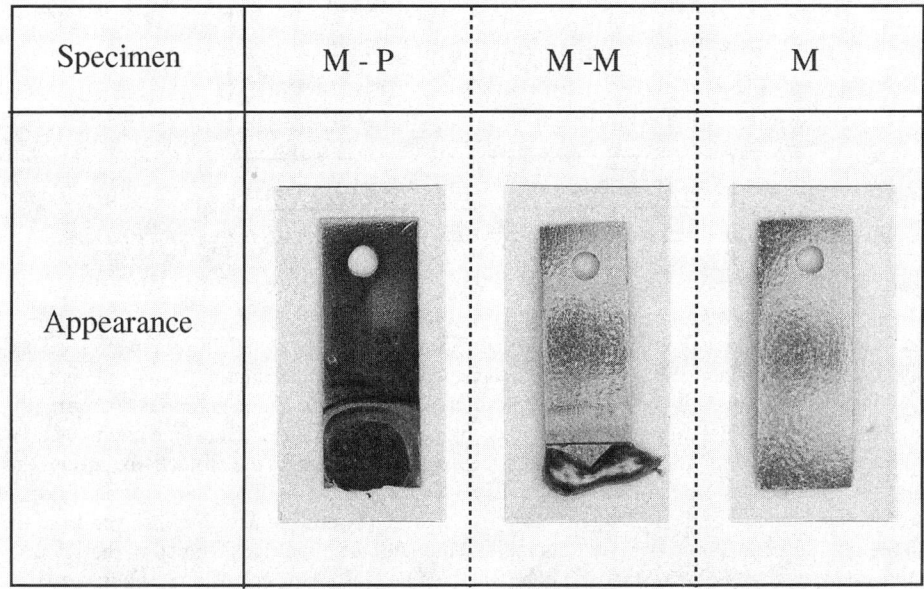

FIG. 6—(a) *Schematic representation of specimens and* (b) *appearance after corrosion testing in 633 K water for 15 days.*

is expected to be more noble than the intermetallic compound. Consequently, the loss of the intermetallic compound can be caused by the oxidation as anodic reaction. Similar oxidation of precipitates has been observed in the oxide film of Zircaloy-4 by cross-sectional TEM analysis [14,18].

Both mechanisms mentioned above should be considered for the depletion of the intermetallic compound; however, more detailed studies on their electrochemical properties are needed to identify each contribution.

Oxide Morphology

Comparison of oxide microstructures of the 30/15 alloy heat treated at 1123 K (non-coupled), the same alloy coupled with the $Zr(Fe_{0.66}Cr_{0.33})_2$ intermetallic compound, and the 120/60 alloy heat treated at 903 K (non-coupled) were made using HR-SEM. Results are shown in Figs. 7a to 7c, respectively. Observations were conducted using the oxide films of the middle part of each sheet as shown schematically in this figure.

The 30/15 alloy (precipitate-free and non-coupled), having poor corrosion resistance, revealed very fine and porous equiaxed grain structure with an average diameter of about 30 nm, while the others, having good corrosion resistance, revealed mainly well-developed columnar grain structure. Thus, both the precipitate-containing material, in which precipitates were finely dispersed, and the precipitate-free material, coupled with the macroscopic intermetallic compound of the same composition as the precipitates, exhibited similar oxidation behavior.

Garzarolli et al. [10] performed careful studies on microstructures of oxide films formed on Zircaloys and reported that an equiaxed grain structure was found in nodular oxide while a well-developed columnar structure was observed for uniform oxide. In this study, the precipitate-free material (the 30/15 alloy heat treated at 1123 K) was covered with only equiaxed grains in spite of the lower-temperature test condition. It is thought that the absence of the precipitate serving as an effective cathode site, which leads to random distribution of both anode and cathode sites on the surface (metal-oxide interface), resulted in the interruption of the continuous growth of oxide crystallites.

Future Development

To improve the understanding of the corrosion mechanism of zirconium alloys from the electrochemical point of view, additional studies on electrochemical properties, such as polarization characteristics of intermetallic precipitates and alpha zirconium matrices, are considered very important. Interesting results of the polarization characteristics of several intermetallics have been already obtained [21]. The effect of the composition, size, and distribution of intermetallic precipitates on the anodic protection mechanism will be examined in the future.

Conclusions

1. Corrosion tests and microstructural analysis were conducted for Zr-1.3 Sn-(Fe,Cr) alloys containing Fe + Cr from 45 up to 180 ppm. Precipitate-free materials showed much larger weight gains than precipitate-containing materials even at the same alloy composition.

2. Galvanically coupling a precipitate-free sheet to a piece of intermetallic compound improved corrosion resistance. A thin black protective oxide film covered the precipitate-free sheet even at a location far from the intermetallic compound.

3. The oxide structure on precipitate-free material consisted of very fine equiaxed grains, while the structure of precipitate-containing material and precipitate-free material coupled with the intermetallic compound consisted mainly of well-developed columnar grains.

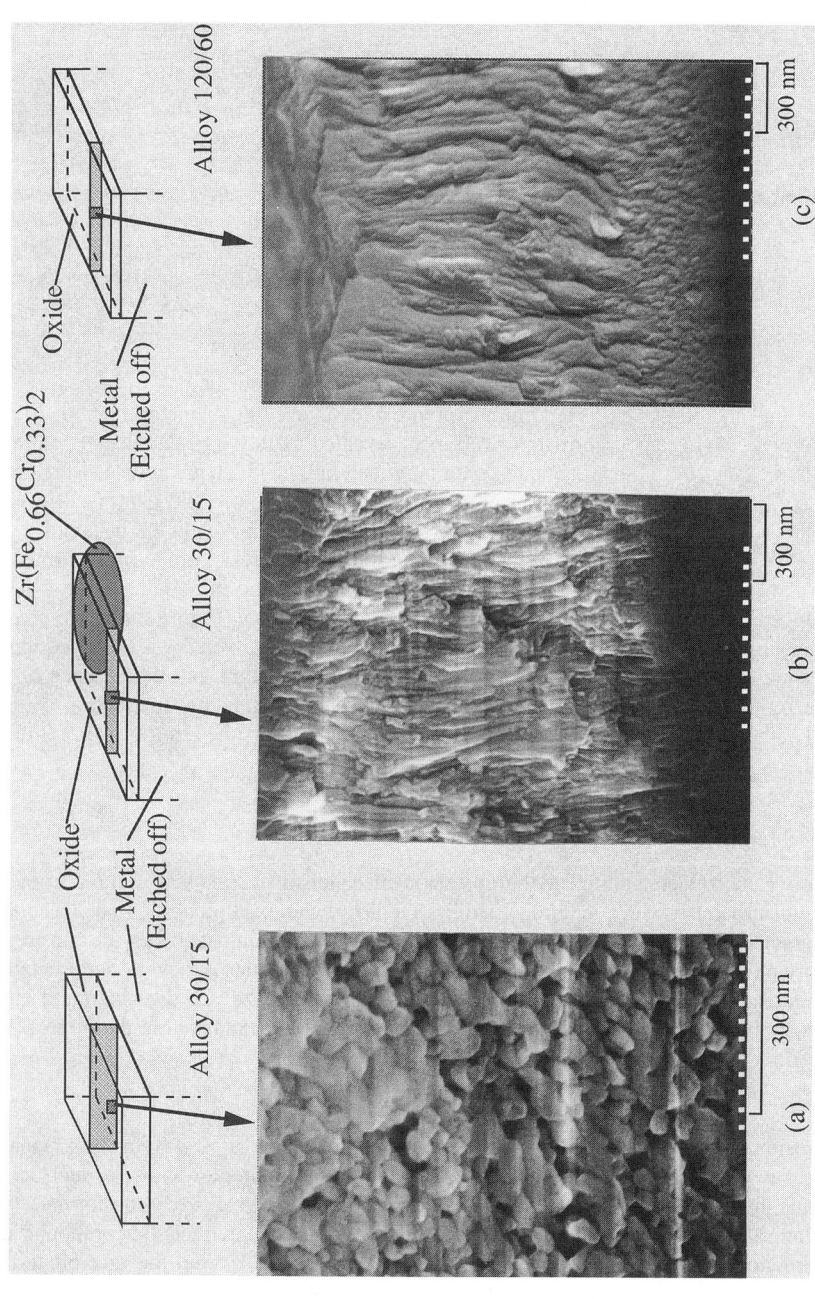

FIG. 7—*Oxide microstructures of (a) 30/15 alloy heat treated at 1123 K (non-coupled), (b) one coupled with $Zr(Fe_{0.66}Cr_{0.33})_2$ intermetallic compound, and (c) 120/60 alloy heat treated at 903 K (non-coupled).*

4. From these results, it is concluded that the superior corrosion resistance of precipitate-containing alloys can be achieved by an anodic protection provided by more noble precipitates.

References

[1] Charquet, D. et al., "Solubility Limit and Formation of Intermetallic Precipitates in ZrSnFeCr Alloys," *Zirconium in the Nuclear Industry: Eighth International Symposium, ASTM STP 1023*, American Society for Testing and Materials, West Conchohocken, PA, 1989, pp. 405–422.

[2] Charquet, D. et al., "Influence of Variations in Early Fabrication Steps on Corrosion, Mechanical Properties, and Structure of Zircaloy-4 Products," *Zirconium in the Nuclear Industry: Seventh International Symposium, ASTM STP 939*, American Society for Testing and Materials, West Conchohocken, PA, 1987, pp. 431–447.

[3] Thorvaldsson, T. et al., "Correlation between 400°C Steam Corrosion Behavior, Heat Treatment, and Microstructure of Zircaloy-4 Tubing," *Zirconium in the Nuclear Industry: Eighth International Symposium, ASTM STP 1023*, American Society for Testing and Materials, West Conchohocken, PA, 1989, pp. 128–140.

[4] Garzarolli, F. et al., "Microstructure and Corrosion Studies for Optimized PWR and BWR Zircaloy Cladding," *Zirconium in the Nuclear Industry: Eighth International Symposium, ASTM STP 1023*, American Society for Testing and Materials, West Conchohocken, PA, 1989, pp. 202–212.

[5] Rudling, P. et al., "Correlation Performance of Zircaloy-2 and Zircaloy-4 PWR Fuel Cladding," *Zirconium in the Nuclear Industry: Eighth International Symposium, ASTM STP 1023*, American Society for Testing and Materials, West Conchohocken, PA, 1989, pp. 213–226.

[6] Zhou, B.-X. "Electron Microscopy Study of Oxide Films Formed on Zircaloy-2 in Superheated Steam," *Zirconium in the Nuclear Industry: Eighth International Symposium, ASTM STP 1023*, American Society for Testing and Materials, West Conchohocken, PA, 1989, pp. 374–391.

[7] Gros, J. P. and Wadier, J. F., "Precipitate Growth Kinetics in Zircaloy-4," *Journal of Nuclear Materials*, Vol. 172, 1990, pp. 85–96.

[8] Foster, J. P. and Worcester, S., "Influence of Final Recrystallization Heat Treatment on Zircaloy-4 Strip Corrosion," *Journal of Nuclear Materials*, Vol. 173, 1990, pp. 164–178.

[9] Harada, M. et al., "Effect of Alloying Elements on Uniform Corrosion Resistance of Zirconium-Based Alloys in 360°C Water and 400°C Steam," *Zirconium in the Nuclear Industry: Ninth International Symposium, ASTM STP 1132*, American Society for Testing and Materials, West Conchohocken, PA, pp. 368–391.

[10] Garzarolli, F. et al., "Oxide Growth Mechanism on Zirconium Alloys," *Zirconium in the Nuclear Industry: Ninth International Symposium, ASTM STP 1132*, American Society for Testing and Materials, West Conchohocken, PA, pp. 395–415.

[11] Godlewski J. et al., "Raman Spectroscopy Study of the Tetragonal-to Monoclinic Transition in Zirconium Oxide Scales and Determination of Overall Oxygen Diffusion by Nuclear Microanalysis of O^{18}," *Zirconium in the Nuclear Industry: Ninth International Symposium, ASTM STP 1132*, American Society for Testing and Materials, West Conchohocken, PA, pp. 416–434.

[12] Kubo, T. and Uno, M., "Precipitate Behavior in Zircaloy-2 Oxide Films and Its Relevance to Corrosion Resistance," *Zirconium in the Nuclear Industry: Ninth International Symposium, ASTM STP 1132*, American Society for Testing and Materials, West Conchohocken, PA, pp. 476–498.

[13] Warr, B. D. et al., "Oxide Characteristics and Their Relationship to Hydrogen Uptake in Zirconium Alloys," *Zirconium in the Nuclear Industry: Ninth International Symposium, ASTM STP 1132*, American Society for Testing and Materials, West Conchohocken, PA, pp. 740–757.

[14] Pecheur, D. et al., "Precipitate Evolution in the Zircaloy-4 Oxide Layer," *Journal of Nuclear Materials*, Vol. 189, 1992, pp. 318–332.

[15] Anada, H. et al., "Corrosion Behavior of Zircaloy-4 Sheets Produced under Various Hot-Rolling and Annealing Conditions," *Zirconium in the Nuclear Industry: Tenth International Symposium, ASTM STP 1245*, American Society for Testing and Materials, West Conchohocken, PA, pp. 307–327.

[16] Wadman, B. et al., "Microstructure of Oxide Layers Formed During Autoclave Testing of Zirconium Alloys," *Zirconium in the Nuclear Industry: Tenth International Symposium, ASTM STP 1245*, American Society for Testing and Materials, West Conchohocken, PA, pp. 579–598.

[17] Beie, H.-J. et al., "Examinations of the Corrosion Mechanism of Zirconium Alloys," *Zirconium in the Nuclear Industry: Tenth International Symposium, ASTM STP 1245*, American Society for Testing and Materials, West Conchohocken, PA, pp. 615–643.

[18] Pecheur, D. et al., "Oxidation of Intermetallic Precipitates in Zircaloy-4: Impact of Radiation," *Zirconium in the Nuclear Industry: Tenth International Symposium, ASTM STP 1245*, American Society for Testing and Materials, West Conchohocken, PA, pp. 687–708.

[19] Weidinger, H. G. et al., "Corrosion-Electrochemical Properties of Zirconium Intermetallics," *Zirconium in the Nuclear Industry: Ninth International Symposium, ASTM STP 1132,* American Society for Testing and Materials, West Conchohocken, PA, pp. 499–534.

[20] Charquet, D. et al., "Heterogeneous Scale Growth During Steam Corrosion of Zircaloy-4 at 500 C," *Zirconium in the Nuclear Industry: Eighth International Symposium, ASTM STP 1023,* American Society for Testing and Materials, West Conchohocken, PA, pp. 374–391.

[21] Murai, T. et al., "Polarization Curves of Precipitates in Zirconium Alloys," *Journal of Nuclear Materials,* to be published.

DISCUSSION

H. G. Weidinger[1] (written discussion)—I appreciate very much this very interesting investigation. Do you have direct experimental evidence for the curve of passivation for pure Zr?

T. Isobe et al. (authors' closure)—We could not have the curve of passivation for precipitate-free zirconium in water containing 2.2 ppm Li and 500 ppm B at 250°C, which is the maximum temperature in our measuring system, but we are sure that an active peak can be observed for it at higher temperatures, for example, 360°C, judging from its much inferior corrosion resistance to precipitate-containing alloys like Zircaloy.

---

[1] The S. M. Stoller Corporation, Erlangen, Germany.

Olaf Gebhardt,[1] *Armin Hermann,*[1] *Gerhard Bart,*[2] *Hubert Blank,*[1] *Friedrich Garzarolli,*[3] *and Ian L. F. Ray*[4]

Investigation of In-Pile Grown Corrosion Films on Zirconium-Based Alloys

REFERENCE: Gebhardt, O., Hermann, A., Bart, G., Blank, H., Garzarolli, F., and Ray, I. L. F., **"Investigation of In-Pile Grown Corrosion Films on Zirconium-Based Alloys,"** *Zirconium in the Nuclear Industry: Eleventh International Symposium, ASTM STP 1295,* E. R. Bradley and G. P. Sabol, Eds., American Society for Testing and Materials, 1996, pp. 218–241.

ABSTRACT: In-pile grown corrosion films on different fuel rod claddings (standard Zircaloy-4, extra low tin Zircaloy (ELS), and Zr2.5Nb) have been studied using a variety of experimental techniques. The aim of the investigations was to find out common features and differences between the corrosion layers grown on zirconium alloys having different composition. Methods applied were scanning and transmission electron microscopy (SEM, TEM), electrochemical impedance spectroscopy (EIS), and electrochemical anodization. The morphology and topography of these oxide layers are, in general, of common nature. However, morphological differences have been observed between the specimens that could explain the irradiation enhancement of corrosion of Zircaloy-4. The features of the compact oxide close to the oxide/metal interface have been characterized by electrochemical methods. The relationship between the thickness of this protective oxide and the overall oxide thickness has been investigated by EIS. It was found that this relation is dependent on the location of the oxide along the fuel rod and on the corrosion rate.

KEYWORDS: zirconium-based alloys, corrosion films, pressurized water reactors, electron microscopy, impedance spectroscopy

Evaluations of oxide layer measurements from PWR fuel rods and corrosion coupons have revealed that corrosion of Zircaloy-4 in PWRs is the same as out-of-pile corrosion up to a thickness of 4 to 5 μm and increases afterwards by a factor of about 4 (= irradiation enhancement factor) [1,2]. In-reactor and out-of-reactor tests with different Zr alloys have shown that corrosion can vary quite significantly among different alloys, particularly in PWRs [2,3]. Sn was found to increase as well as out-of-pile corrosion as the irradiation enhancement. The binary alloy Zr2.5Nb was observed to corrode quite similarly in PWR as out of pile. To verify these empirical observations, a knowledge of mechanistic aspects of the corrosion process is important.

Many studies have been published on the mechanism of uniform out-of-pile corrosion [4–9], and the mechanistic aspects of the effect of the alloying elements were discussed [7–9]. It has been shown that the oxide layer is dense up to a thickness of about 2 μm. At this thickness, a transition from a cubic rate law to an almost liner rate law occurs. After this transition, the outer part of the oxide layer becomes porous and the corrosion is governed mostly by a thin dense interlayer at the metal/oxide interface. This has been shown by experiments on the

---

[1,2] Senior scientists and head, respectively, Laboratory for Materials Behavior, Paul Scherrer Institute, CH-5232 Villigen, Switzerland.

[3] Deputy director, Nuclear Fuel Cycle Joint Technology and Materials Issues, Siemens AG, Power Generation Group (KWU), D-91050 Erlangen, Germany.

[4] Senior scientist, EU Institute for Transuranium Elements, D-76149 Karlsruhe, P.O. Box 2340, Germany.

penetration behavior of a liquid electrolyte into the oxide layer by electrochemical impedance spectroscopy (EIS) [10–16]. Certainly the individual impedance measurement suffered from the preparation effects and also from the effect of partially filling up the pores, but the principal conclusions from these measurements were certainly correct. Electron microscopic examinations of the oxide layer have shown that the uniform oxide consists of thin columnar grains growing from the metal/oxide interface. Periodically, the growth of these columnar crystals becomes interrupted and equiaxed grains with intergranular pores are formed in thin coherent zones perpendicular to the growth direction.

The analyses of oxide topography and morphology at the oxide/metal interface by metallography, SEM, and TEM [15–19] have shown that this interface is quite rough and exhibits a complex topography. At a relatively low magnification (~X5000), one recognizes more or less steep "hills" or "ridges" and, in between, "valleys," "canyons," and "hollows." The diameters or lengths of topographical elements hills and ridges (L_1) are usually found to be in the range between 5 and 25 μm. The differences in height between the top of the hills and the bottom of the canyons (H_1) vary roughly in the range 0.5 to 5 μm. At higher magnifications, it can be observed that the surface of this "landscape" is densely covered by mini-hills (or flat cups) whose diameters (L_2) usually vary in the range 0.2 to 0.8 μm. The mini-hills of scale L_2 are resolved into finer details. These are often heaps of "sticks," probably columnar grains, with diameters L_3 in the range 20 to 100 nm and lengths of about 0.4 to 1 μm.

Not much information has been gained on the question how the alloying elements and irradiation influence this process. Reference 8 reported that the fraction of tetragonal oxide is lower in an extra-low-Sn Zircaloy-4 type alloy (ELS), but the crystallite size is not significantly affected by varying the Sn content between 0.6 and 1.5%. Reference 9 concluded recently from in situ impedance measurements during the corrosion of alloys with varying Sn content in an autoclave that Sn increases corrosion rate because it decreases the ionic and electronic resistance.

The study reported here had the goal to find out the mechanistic reasons for the effect of irradiation on corrosion and the differences observed for three zirconium alloys: standard Zircaloy-4, an extra low tin alloy (ELS 0.6, a Zircaloy-type alloy with 0.6% Sn), and the binary alloy Zr2.5Nb.

Experimental

Materials and Corrosion Parameters

For this study, cladding samples were taken from different experimental fuel rods with different cladding materials irradiated in the PWR Gösgen for two to four cycles. The specimen sources and the conditions for in-reactor and autoclave corrosion are listed in Table 1. The thickness of the corrosion layer along the rod was obtained by eddy current measurements. Later, the exact oxide thickness was determined on metallographic cross sections. The fuel rods were cut at defined positions, and the fuel of the rod segments was removed by drilling and scraping. The defueled cylindrical cladding samples were used to prepare specimens for microscopic and electrochemical investigations. In addition, several unirradiated samples exposed in autoclaves to water at 350°C and to 0.01 M LiOH solution at 350°C were selected for comparative examinations to find out the effect of irradiation.

Scanning Electron Microscopy (SEM)

SEM images of the oxide were taken from the outer surface of the oxide, from the metal/oxide interface after dissolution of the metal in 10% Br_2/ethylacetate, and from fracture surfaces through the oxide in the axial direction. A thin conductive Pt film of 10 nm in thickness was

TABLE 1—Specimen sources, materials, and corrosion conditions.

No.	Specimen Source	Cladding Material	Alloying Elements, Sn, Fe, Cr, %	Annealing Parameter, ΣA_i, s	Abbreviation	Corrosion Environment	Period of Corrosion, days	Oxide Thickness, μm
1	Fuel Rod A	Zircaloy-4	1.45; 0.21; 0.12	$4 \cdot 10^{-17}$	SZ4	in-pile KKG 3 cycles	969 fpd	102*
2	Fuel Rod A[1]	Zircaloy-4	1.52; 0.21; 0.11	$3 \cdot 10^{-17}$	SZ4[1]	in-pile KKG 4 cycles	1226 fpd	105*
3	Fuel Rod B	Extra low tin Zircaloy	0.56; 0.22; 0.10	$3 \cdot 10^{-17}$	ELS	in-pile KKG 2 cycles	609 fpd	14*
4	Fuel Rod C	Zr2.5Nb	...	...	ZN	in-pile KKG 3 cycles	927 fpd	16*
5	Cladding tubing	Zircaloy-4	1.40; 0.20; 0.10	$9 \cdot 10^{-17}$	SZ4AW	autoclave pure water 350°C	1340	25.6
6	Cladding tubing	Zircaloy-4	1.50; 0.22; 0.10	$1 \cdot 10^{-17}$	SZ4AL	autoclave 0.01 M LiOH 350°C	234	68

NOTE: KKG = PWR Gösgen; fpd = full power days; * = maximum thickness.
[1] Equals A'.

sputtered on the surface of the specimens to avoid charging effects during electron bombardment in the DSM 962 scanning electron microscope.

Transmission Electron Microscopy (TEM)

Cross sections of the specimens were thinned mechanically by polishing and dimple grinding until a certain thickness of about 10 μm was reached. Then the specimens were etched by an Ar^+ ion beam at a flat angle of about 10°. This ion milling was stopped when electron transparency was reached at selected parts near the metal/oxide interface of the specimens. A transmission electron microscope (TEM) with a 200-kV acceleration voltage was used to obtain the images.

Electrochemical Impedance Spectroscopy (EIS)

With this technique, the behavior of the oxide films is studied in response to an applied a-c or d-c voltage. The aim of this examination was to get information on the consistency of the oxide layer [20–22].

For EIS, cylindrical tubing specimens were prepared as shown in Fig. 1. The specimens were positioned between two watertight overlapping plastic end pieces. This design prevents the penetration of the electrolyte to the inner surface of the specimen. All specimens possessed an effective outer surface of about 6.1 cm^2. The impedance measuring unit, IM5d, was used. Details about the measuring procedure and the network synthesis program of this unit as well as of the used impedance elements can be found in Refs *23* and *24*. For the potentiostatic measurements, a conventional three-electrode measuring cell was used, and a drive voltage of 10 mV was applied at the zero d-c current potential. The specimens were immersed at room temperature in 1 M H_2SO_4 electrolyte solution, which penetrates the porous fraction of the oxide. It has been proven that no differences can be found in the EIS data for different electrolytes.

The EIS characteristics were measured under quasi-stationary conditions (negligible changes of impedance parameters due to soaking effects during the time of measurement for all considered frequency points). Generally, significant time-dependent changes of the impedance parameters are finished after about 24 h. In addition, quasi-stationary impedance characteristics can be obtained to study long-time soaking effects for a broad band of frequencies. If frequencies from 10^{-4} to 10^6 Hz are used, the measurement time exceeds 12 h and quasi-stationary conditions appear to be not easily achieved, especially after soaking.

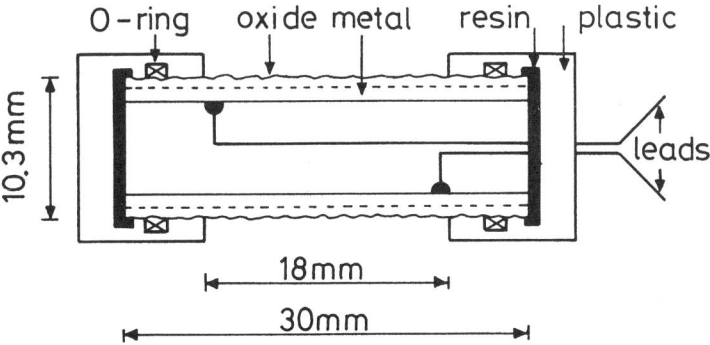

FIG. 1—*Schematic of a specimen prepared for electrochemical examination in a hotcell.*

Diagrams that present the total impedance and the phase angle versus logarithms of frequency (Bode diagrams) have been preferred. The experimental points of impedance and phase have been fitted with the parameters of an equivalent circuit. The fitting procedure developed for the IM5d unit is based on the Kramers-Kronig relationship [25] and has been briefly described in Ref 23.

Galvanostatic Anodization (GA)

In this technique, the behavior of the oxide films is studied in response to an applied anodic current [10,26–29]. The galvanostatic anodization properties of the specimens were obtained after impedance measurements in situ using the same experimental equipment as for the impedance measurements. The samples were charged with an anodic current of 0.23 mA, and the time dependence of the anodic voltage across the specimens was recorded.

Results

Scanning Electron Microscopy

SEM images were taken from the outer (waterside) surface, the metal/oxide interface after dissolution of the metal, and from fractured cross sections. Table 2 gives characteristic data for the examined PWR and autoclave specimens. The outer surface looks dense even at a magnification of X20 000. Large pores were seen only occasionally. From fractured cross sections, no new information has been deduced: the whole oxide scale consists of several intermediate layers each about 2 μm in thickness (Fig. 2). Analysis of oxide topography and morphology at the oxide/metal interface has shown that this interface is quite rough.

Views from irradiated standard Zircaloy-4 fuel rod cladding with an oxide layer thickness of 5 and 60 μm, irradiated extra low Sn cladding, irradiated Zr2.5Nb cladding, a Zircaloy-4 oxide sample grown out of pile in 350°C water, and a Zircaloy-4 oxide sample grown out of pile accelerated in 0.01 M LiOH at 350°C are shown in Figs. 3a–f at a magnification of ~X2000, respectively. The diameters or lengths of topographical elements hills and ridges (L_1) were measured on samples from different positions along two fuel rods with standard Zircaloy-4 cladding, and the maximum values are listed in Table 3.

The evaluated L_1 data are between 5 and 27 μm (see Table 3) and seem closely related to the temperature that determines the corrosion rate. The irradiated extra low Sn and Zr2.5Nb samples exhibit the lowest roughness. The irradiated Zircaloy-4 samples with an oxide layer thickness of 5 and 60 μm and the out-of-pile water sample with 26 μm oxide have a similar appearance at ~X2000 magnification. The oxide layer grown in LiOH shows pores at grain boundaries of the coagulates, indicating that the active zone of this oxide layer may not be dense.

Even more pronounced differences at the metal/oxide interface were seen among the different samples at higher magnification (~X20 000, Figs. 4a–d). Figures 4a, 4c, and 4d (samples from the out-of-pile corrosion test in water, the extra low Sn cladding, and the Zr2.5Nb cladding) show all thin platelets growing up from the metal/oxide interface (probably arrays of columnar grains) and a relatively smooth interface at this magnification, similar to that reported before [17]. Figure 4b from the standard Zircaloy-4 PWR fuel rod exhibits a different structure. Here the interface is composed of spherical crystal agglomerates. This particular structure was seen at both examined positions in spite of the large difference in oxide layer thickness. A similar structure was reported before [30] for thick oxide layers from fuel rods from two PWRs and a BWR. Obviously, this type of structure is typical for thick oxide layers of Zry-2/4 if grown in the reactor to a certain thickness.

TABLE 2—*Specimen characteristics*.

No.	Specimen Designation	Specimen Source No. on Table 1	Position on Fuel Rod,[a] mm	Temperature of Corrosion, °C	Oxide Thickness, μm	Corrosion Rate, μm/day
1	A-M	1	1818	364[b]	60	0.062
2	A-V	1	268	303[b]	5	0.005
3	B-K	3	2912	353[b]	14	0.023
4	B-S	3	842	329[b]	5	0.008
5	C-K	4	2912	352[b]	13	0.014
6	C-S	4	842	329[b]	5	0.005
7	SZ4AW	5	...	350	25.6	0.019
8	SZ4AL	6	...	350	65	0.28

[a] From the lower end.
[b] Temperature at the oxide/metal interface at the end of irradiation, calculated.

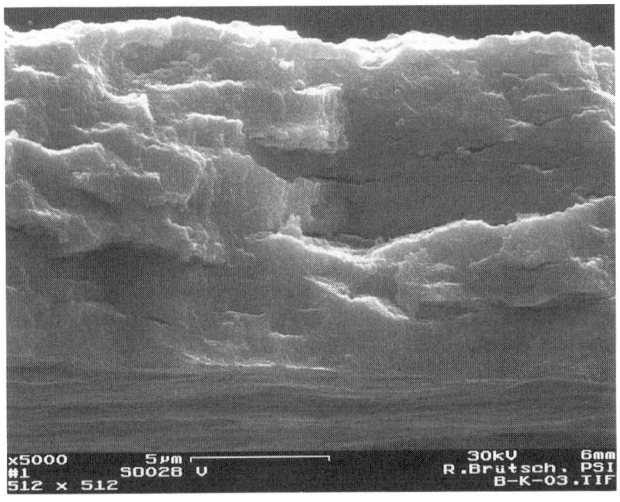

FIG. 2—*Fractured cross section of Sample A-V (SEM image, magnification ~X500) showing the metal and several intermediate oxide layers.*

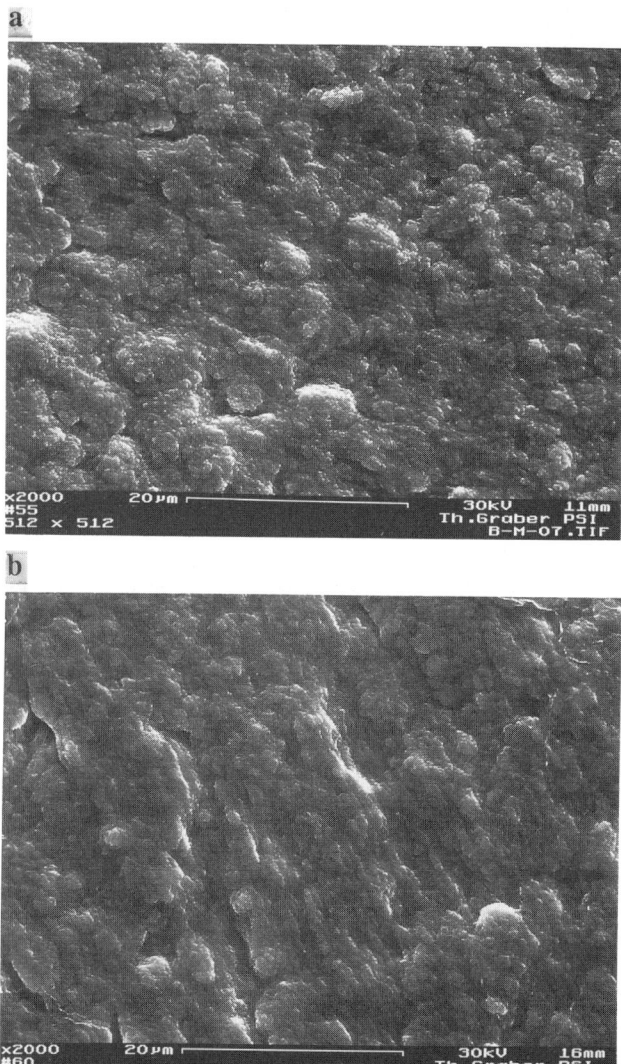

FIG. 3—*SEM images (magnification ~X2000) of the oxide at the oxide/metal interface:* (a) *Sample A-V (SZ4), 5 μm oxide;* (b) *Sample A-M (SZ4), 60 μm oxide;* (c) *Sample B-K (ELS), 14 μm oxide;* (d) *Sample C-K (ZN), 13 μm oxide;* (e) *Sample SZ4-AW (SZ4), 26 μm oxide;* (f) *Sample SZ4-AL (SZ4), 65 μm oxide.*

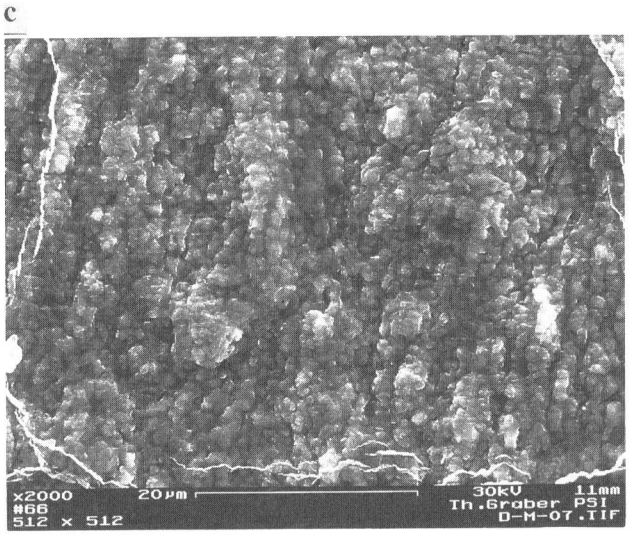

FIG. 3—*Continued*

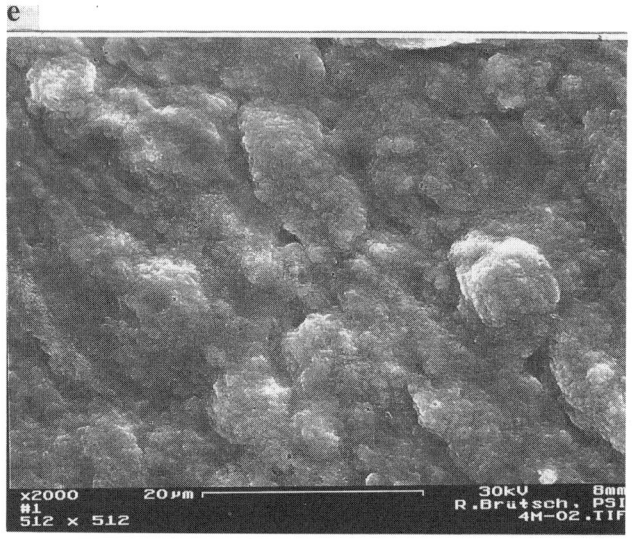

FIG. 3—*Continued*

TABLE 3—*Changes of characteristic length L_{lm} and H_2 of oxide morphology at the oxide/metal interface along the axis of standard Zircaloy-4 rods.*

Rod	Position from Lower Rod End, mm	Temperature[a] at Oxide/Metal Interface, °C	Oxide Thickness, μm	Corrosion Rate, μm/day	L_{lm},[b] μm	H_2, general impression
A'	2593	357	68	0.055	~23	large
	1546	342	42	0.034	~20	large
	1010	333	27	0.022	~10	smaller
	270	303	8	0.0064	~5.5	small
A	1818	364	60	0.062	~27	large
	268	303	5	0.005	~4.5	small

NOTE: A' = 4-cycle rod; A = 3-cycle rod.
[a] Temperature after Cycle 3 for both rods, calculated.
[b] Maximum dimension.

Transmission Electron Microscopy

The cross section of the oxide layer of experimental fuel rods with an extra low Sn and normal Sn cladding were examined by TEM (Rods A and B, respectively, see Table 1). The characteristic data for the sample positions are given in Table 2. The examination revealed a preferentially columnar structure of the oxide close to the metal/oxide interface for both materials (Figs. 5 and 6). Directly at the metal/oxide interface, the low tin sample showed an equiaxed or amorphous band of oxide with a depth of about 30 nm. The observed structure is very similar to the structure found [5,7–8] in unirradiated autoclaved samples from similar materials. At this high magnification, looking to only very small areas, there is no obvious difference between the faster-corroding Zircaloy-4 and the extra low Sn cladding and also no difference between autoclave samples and samples from the investigated fuel rods. Only a fraction of equiaxed grains seems to be somewhat higher in the Zircaloy-4 PWR sample than in the extra low Sn PWR sample.

Electrochemical Analyses

Impedance Spectroscopy—Two segments each from Rods A, B, and C were investigated by EIS. In addition, a reference specimen (standard Zircaloy-4, designated as Zry-4a) with an air-formed oxide film on the outer surface was investigated. Procedures interpreting EIS data obtained from oxide layers on zirconium-based alloys are described in previous papers [10–11,15–16].

The EIS data of all rod segments and the Specimen Zry4a are plotted in Figs. 7a–c. The EIS data of the in-pile oxidized specimens are quite different from Specimen Zry4a. The slope of the total impedance graph of Specimen Zry4a is close to -1 in the Bode diagram (characteristics of a pure capacitance). The total thickness, δ_t, of the air-formed oxide can be calculated from these characteristics. A layer thickness of about $\delta_t \approx 5$ nm was found.

The total impedance of the oxides on the irradiated specimens is about one or two orders of magnitude higher than that of the air-formed oxide at high frequencies $f > 10^5$ Hz. It should be noted that the in-pile formed oxides show clear differences in their spectra.

Specimen A-V—This specimen shows the typical form of a spectrum (see Fig. 7a) that can be classified as Type A [15,16]. The impedance graph of this spectrum possesses a slope about -1 at high frequencies ($f > 10^5$ Hz) and a slope of about -0.5 at medium frequencies (10^0 Hz $\leq f \leq 10^5$ Hz). The part of the impedance spectrum appearing at the highest frequencies represents the capacitance of the total oxide carrying the total transmitted displacement current.

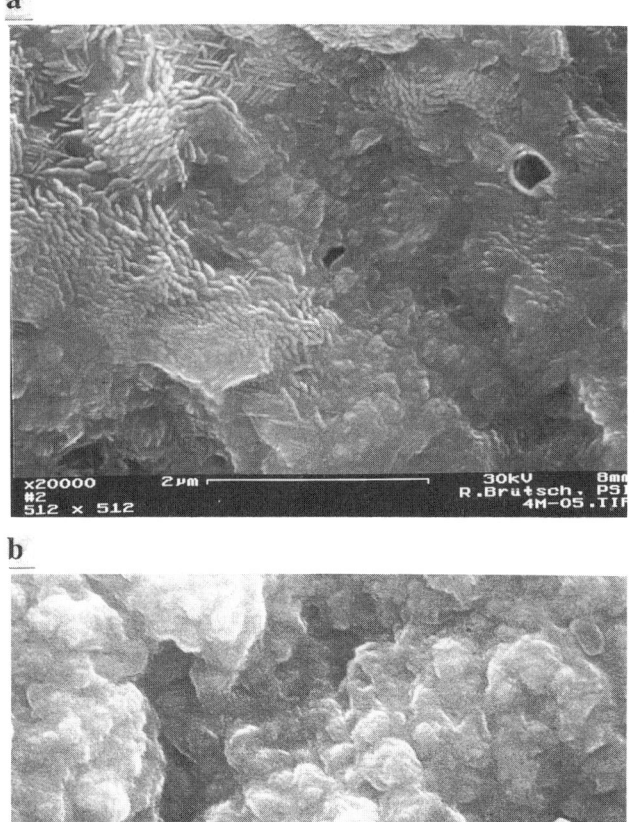

FIG. 4—*SEM images (magnification ~X20 000) of the oxide at the oxide/metal interface:* (a) *Sample SZ4-AW;* (b) *Sample A-V;* (c) *Sample B-K;* (d) *Sample C-K.*

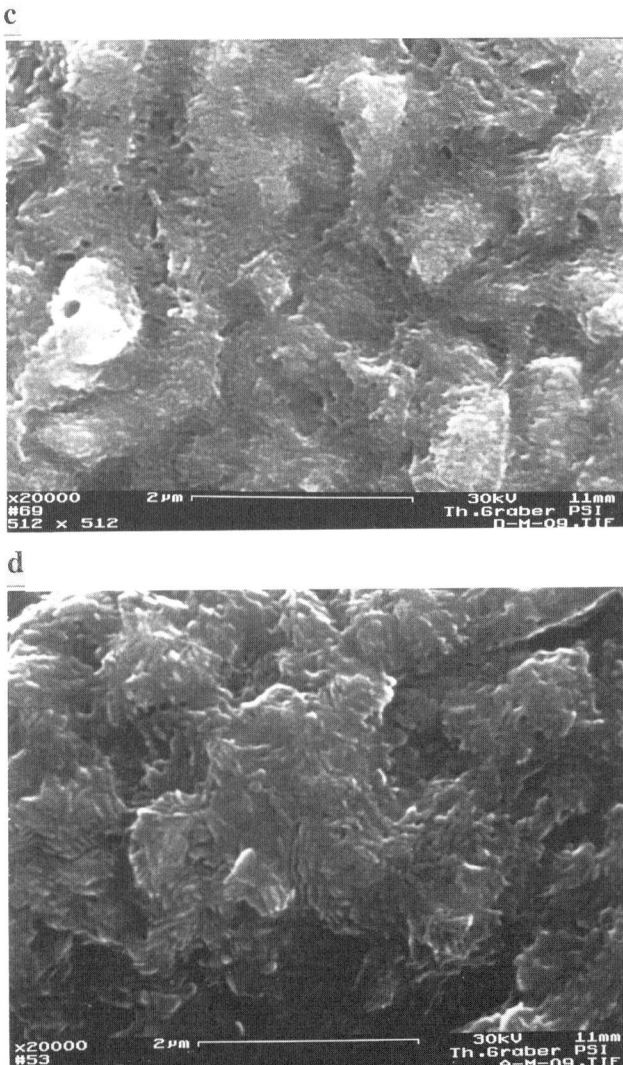

FIG. 4—*Continued*

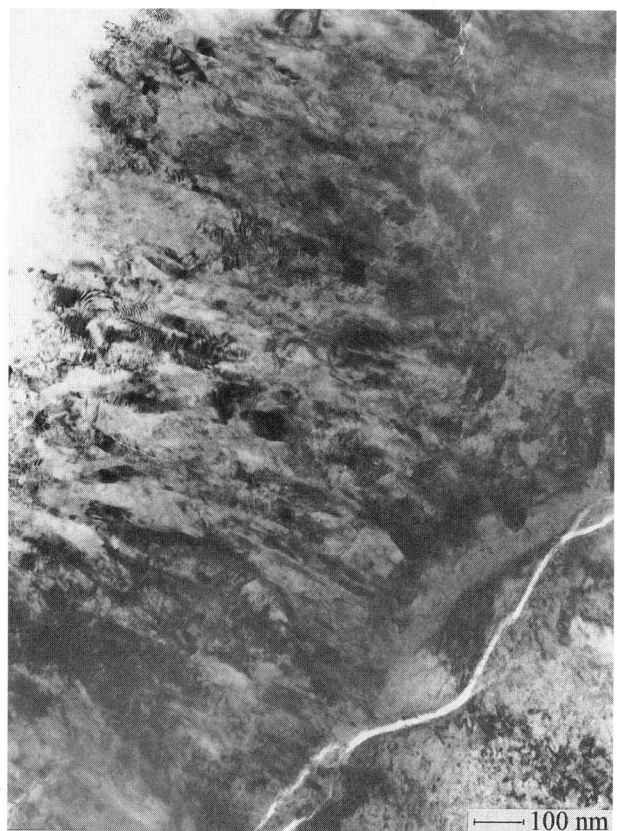

FIG. 5—*TEM image of oxide at the oxide/metal interface, Sample B-K (ELS), magnification ~X130 000.*

If the relative or effective dielectric constant ϵ_{eff} of the oxide film that is partially filled by 1 M sulfuric acid as electrolyte is known, the thickness δ_t of the total oxide can be calculated [16]. On the other hand, ϵ_{eff} can be evaluated if the correct oxide thickness has been determined by other methods. For Specimen A-V, a value of $\epsilon_{eff} \approx 38$ has been determined, which indicates that the porous part of the oxide is partially filled by electrolyte.

The slope of about -0.5 of the impedance graph appearing at medium frequencies is caused by electrically conductive domains spatially distributed in the oxide layer. These electrically conductive domains can be caused by electrolytic paths and/or by components of intermetallics incorporated partially unoxidized in the oxide layer as observed by TEM [30–32] and SIMS techniques [33–34].

At frequencies $f < 1$ Hz, a resistive part appears possessing a slope of about 0 in this diagram. This part arises from electronic conduction in the porous oxide layer, which is due mainly to electrolytic paths.

Specimen A-M—This specimen shows impedance characteristics (see Fig. 7a) that can be classified as Type B [15–16]. Unfortunately, the upper end of the frequency spectrum does not show the impedance of the total oxide. This is caused by limitations of the measuring unit ($f \leq 1$ MHz). In the frequency interval $10^3 \leq f \leq 10^5$ Hz, a highly dissipative part of the spectrum

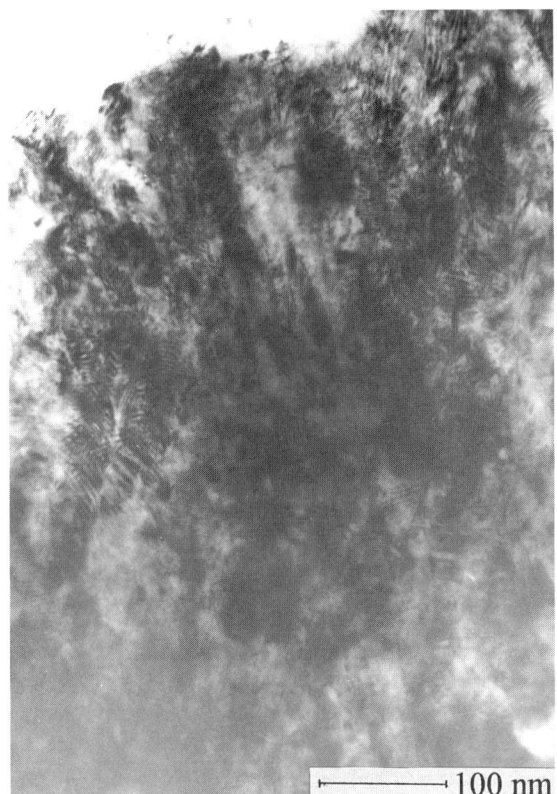

FIG. 6—*TEM image of oxide near the oxide/metal interface, Sample A-V, magnification ~X270 000.*

is obtained (slope about −0.2 to −0.3). Because the total impedance and the slope of the impedance graph in this frequency interval is strongly dependent on the time of soaking, this section can be related to a porous part of the oxide that is well filled by electrolyte. Thus, the impedance characteristics in this frequency interval probably represent an outer part of the oxide layer that is porous. This part of the oxide layer contains interconnected pores and cracks oriented laterally to the metal/oxide interface. Such lateral pores and cracks in the outer oxide layer have been found by SEM studies of polished cross sections of similar specimens [15]. In addition, the time dependence of impedance characteristics during the soaking process support the assumption that the electrolyte fills lateral pores and cracks. It has been maintained that the impedance graph is displaced nearly parallel with an increase of soaking time, which indicates that the electrolyte penetrates successive planes in the oxide layer that form electrically equipotential areas [11–12].

At medium frequencies ($10^{-3} \leq f \leq 10^3$ Hz), a section of the impedance graph appears that is close to the graph of Specimen Zry4a and that shows rather capacitive characteristics (slope about −0.95). Obviously, this section of the impedance graph represents thin and rather compact oxide. This compact oxide, which is thought to control the corrosion rate, possesses a columnar grain structure and has been obtained close to the metal/oxide interface by different authors using TEM studies [15–16,30–32]. This compact oxide layer does not contain inter-

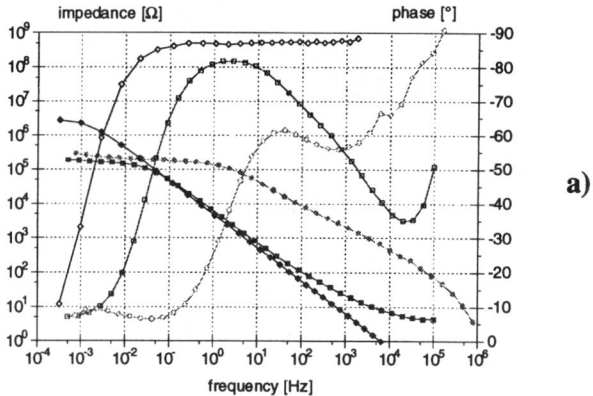

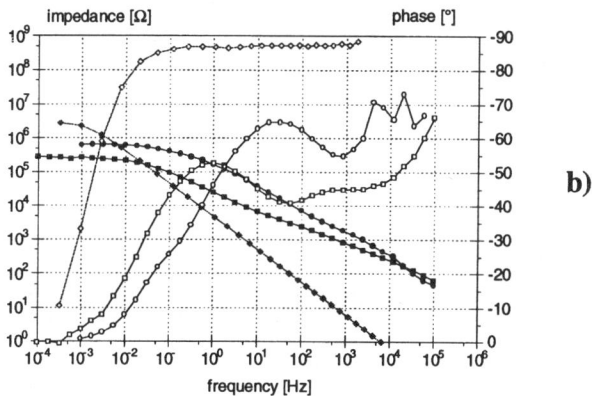

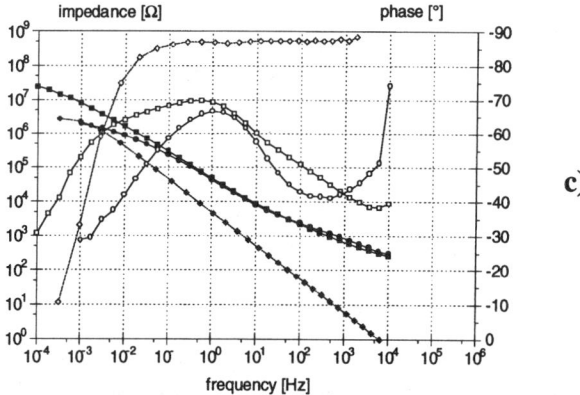

FIG. 7—*Experimentally obtained EIS characteristics of the specimens in the well-soaked state:*
(a) *Specimen A-M:* ■ *impedance,* □ *phase; Specimen A-V: impedance* ●, *phase* ○; (b) *Specimen B-K:* ■ *impedance,* □ *phase; Specimen B-S: impedance* ●, *phase* ○; (c) *Specimen C-K:* ■ *impedance,* □ *phase; Specimen C-S: impedance* ●, *phase* ○. *In* (a), (b), *and* (c), *Specimen Zry4a:* ♦ *impedance,* ◊ *phase.*

connected lateral pores or cracks, but these lateral cracks appear at a defined distance from the metal/oxide interface [*31*]. Because the layer is chemically active, it can be called "active layer" (A-layer).

The thickness δ_a of the A-layer can be calculated from the impedance of an electrical circuit element carrying the capacitively displaced current through this layer. This impedance element should be of rather a complex nature, reflecting physical properties of the oxide layer [*12,16*]. Because the A-layer is rather compact and obviously contains only a few electrolyte in fine pores, an effective dielectric constant of $\epsilon_{eff} = 22$ can be used for these calculations. Nevertheless, the impedance characteristics of the oxide layers have to be modeled in the full interval of frequencies that have been measured by an equivalent circuit. In this case, the data of the individual impedance elements can be fitted to the experimental data with high accuracy by the computer program [*11–12,15–16*].

At the lowest frequencies ($f < 10^{-2}$ Hz), the impedance graph of Specimen A-M possesses rather resistive characteristics in the well-soaked state. At these frequencies, the soaking process is characterized by a decrease of the slope of the impedance graph until nearly resistive characteristics are obtained. This behavior indicates that the electrolyte probably may penetrate fine vertically directed pores in the A-layer even at room temperature [*11–12,15–16*].

Specimen B-S—The characteristics of this specimen arising from Rod B (see Fig. 7*b*) are similar to those of Specimen A-V (see Fig. 7*a*). Thus, an intrinsic compact layer (A-layer) cannot be distinguished from these characteristics. For the effective dielectric constant, a value of $\epsilon_{eff} \approx 32$ has been calculated.

Specimens B-K, C-K, and C-S—Specimens B-K from Rod B (Fig. 7*b*) and Specimens C-K and C-S from Rod C (Fig. 7*c*) show similar behavior as Specimen A-M from Rod A (Fig. 7*b*, impedance characteristics of Type B) and can be discussed in the same manner. Thus, the thickness δ_a of a compact oxide layer close to the metal/oxide interface (A-layer) can also be calculated for these specimens.

Galvanostatic Anodization (GA)—If the voltage in the regions of the thinnest barrier oxide exceeds the anodization voltage U_a (a field strength greater than 10^8 V/m causes oxygen transport), the oxide grows in these regions and the capacitance of the oxide layer decreases [*10,26*]. If a homogeneous increase of the oxide (constant pore area) occurs, a nearly constant slope of the second part of the anodization curve exists. By extrapolating the second part of the anodization curves to $Q = 0$, one obtains the voltage U_a, which is used to calculate the thickness δ_b of a barrier oxide that impedes transportation of oxygen. Because anodic oxide grows at a rate of 2.3 nm/V [*15*], the thickness of the barrier oxide is given by $\delta_b = 2.3\ U_a$.

The anodization graphs of the irradiated samples are shown in Fig. 8. It should be noted that the Slopes S of the nearly linear parts of the anodization curves differ considerably. This suggests that they contain additional information about certain properties of the oxide. Because of leakage currents, the deviation from linearity increases with increase of voltage.

Results of Electrochemical Analyses—In Table 4, the results of studies by EIS and GA are presented. In Figs. 9*a–b*, the thickness δ_a of the A-layer as well as the thickness δ_b of the barrier oxide are plotted versus the oxide layer thickness and versus the position of segments along the fuel rods. It is obvious that the standard Zircaloy-4 rod possesses the thinnest A-layer as well as the thinnest barrier oxide at the position of highest corrosion rate.

Discussion

General Morphology of Oxide Scales

Analyses of samples from the oxide layers on PWR fuel claddings by metallography, SEM, and TEM suggest the morphology described in Refs *15* to *18* and shown schematically in Fig. 10. The surface oxide at the water side is the result of the pre-transition oxidation period. Then

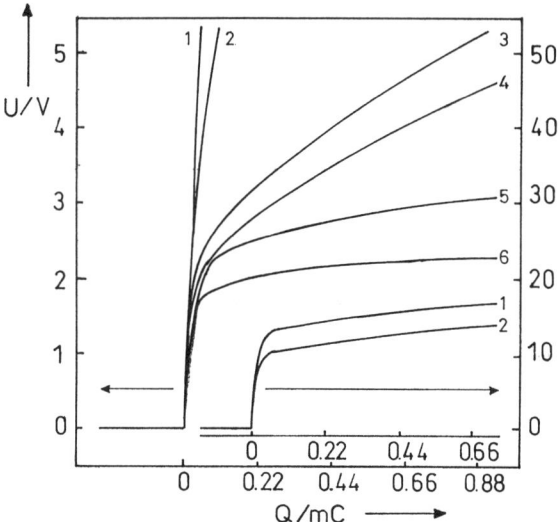

FIG. 8—*Experimentally obtained anodization graphs. Specimens: 1 = A-V (standard Zircaloy-4), 2 = B-S (ELS), 3 = C-S (Zr2.5Nb), 4 = C-K (Zr2.5Nb), 5 = B-K (ELS), and 6 = A-M (standard Zircaloy-4) (see Tables 1 and 2).*

follows a series of intermediate oxide layers with 1 to 2 µm thickness, and, finally, adjacent to the alloy phase, one has the last oxide layer at which the currently active oxidation is occurring. This last layer is thought to control the corrosion rate and can be called "active layer" or A-layer. Previously, this layer was called "barrier layer" in the literature.

As mentioned above, the total oxide scale is composed of several layers. This can be observed clearly after fracturing the oxide scale of an irradiated or autoclaved cladding tube. The existence of a layer structure for oxide scales on Zircaloy has been known for a long time. It is revealed by dark/bright fringe systems that can be observed under polarized light at X200 or X500 on mechanically polished cross sections of autoclave-oxidized and irradiated PWR cladding materials. It is further accentuated by the presence of apparent circumferential cracks and porosity. These layers are probably produced by the change in growth mechanism between pre- and post-transition growth and subsequently between the various post-transition periods. The layers are separated by interfaces that consist of lattice defects like sheets of pores on lateral

TABLE 4—*Calculated thickness of characteristic layers of the oxide scale at different rod positions. Thickness of the total oxide, δ_t, of the compact oxide layer, δ_a (from EIS studies) and thickness of the barrier oxide, δ_b (from GA studies, probably the oxide at the bottoms of vertically directed pores in the compact oxide layer). S = segment, P = position of segments along the rod (bottom = zero).*

	Rod A					Rod B					Rod C			
S	P, mm	δ_t, µm	δ_a, nm	δ_b, nm	S	P, mm	δ_t, µm	δ_a, nm	δ_b, nm	S	P, mm	δ_t, µm	δ_a, nm	δ_b, nm
V	268	≈5	...[a]	28	S	842	≈5	...[a]	23	S	842	≈5	183	6.5
M	1818	60	7.2	3	K	2912	14	59.9	5.3	K	2912	13	224	5.5

[a] EIS characteristics of Type A [15–16].

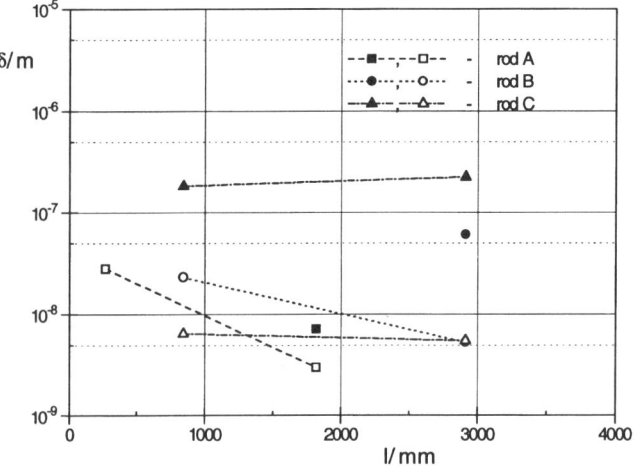

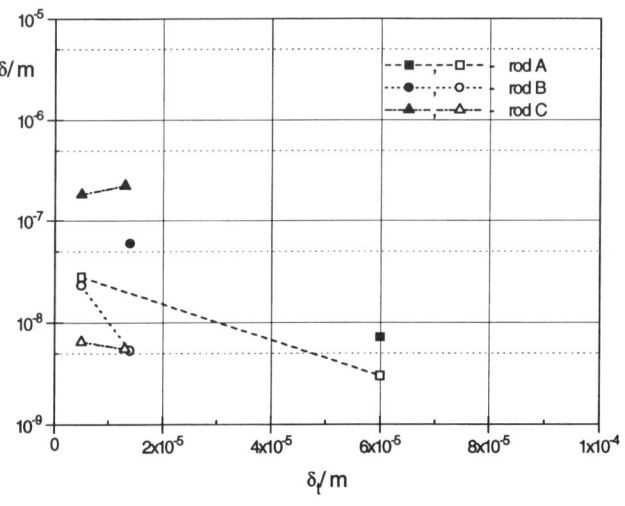

FIG. 9—*Thickness δ_a of the compact oxide (full symbols) and thickness δ_b of the barrier oxide (open symbols) for Rod A (●, ○), Rod B (■, □), and Rod C (▲, △): (a) versus position of the segments along the fuel rod (bottom = zero); (b) versus the thickness δ_t of the total oxide.*

grain boundaries. TEM investigations indicate that new crystals are formed at these positions that often show a high fraction of equiaxed grains in these porous lateral zones.

In the oxide scales on cladding tubes, in addition, a system of radial cracks develops in the pre-transition and intermediate sublayers. These are likely produced by the mechanical tension that develops with increasing oxide thickness in the outer layers of the scale. The existence of these radial and lateral cracks is readily demonstrated by impedance spectroscopy.

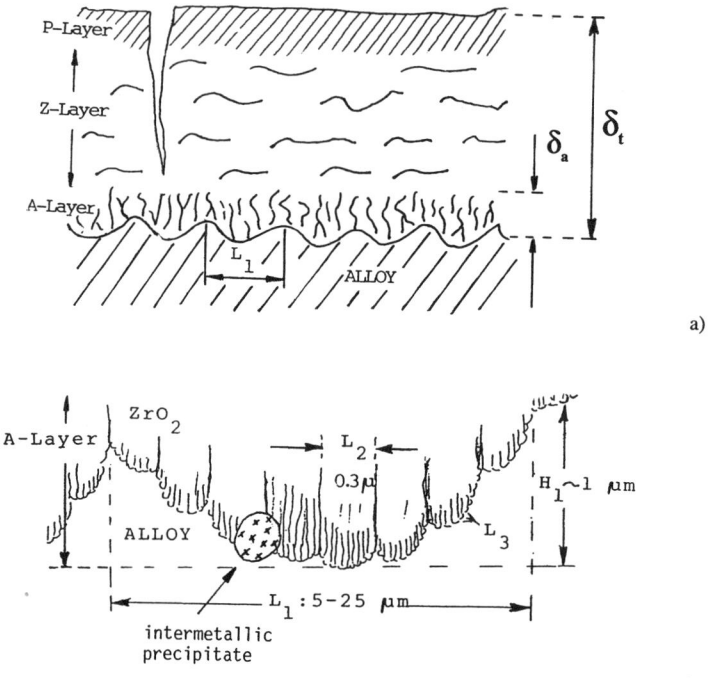

FIG. 10—*Semi-schematic representation of the oxide morphology on zirconium-based alloys after uniform corrosion: (a) layer structure: P = surface oxide layer, Z = intermediate oxides layers, A = active layer; (b) columnar grain structure of the active layer at Levels L_1, L_2, and L_3.*

Morphology of the Active Layer

Analysis of oxide topography and morphology at the oxide/metal interface by SEM has shown that this interface is quite rough and exhibits more or less steep "hills" and "ridges" and, in between, "valleys," "canyons," and "hollows." The diameters or lengths of the topographical elements, hills and ridges (L_1), are in the range between 5 and 25 µm as found before [17–18]. The differences in height between the top of the hills and the bottom of the canyons (H_1) vary roughly in the range of 0.5 to 5 µm. At high magnifications, it can be observed that the surface of this "landscape" is densely covered by mini-hills (or flat cusps) whose diameters (L_2) usually vary in the range 0.2 to 0.8 µm. The mini-hills of Scale L_2 are resolved into finer details as shown schematically in Fig. 10. These are columnar grains with diameters, L_3, in the range 20 to 100 nm and lengths of about 0.4 to 1 µm.

The general topography and morphology described above are, in principle, the same for zirconia scales grown on different zirconium alloys in and out of pile. Nevertheless, some differences can be observed concerning the structural elements L_i. The topographical element L_1 increases with increasing oxide thickness from the lower to the upper end of standard Zircaloy fuel rods (Table 3). For Zr2.5Nb, this has not been found. On the contrary, L_2 remains approximately constant with oxide thickness as for standard Zircaloy-4 and increases slightly for Zr2.5Nb. Also, there are differences in changes of L_1 and L_2 in the pre- and post-transition periods [18].

It has to be considered that the topography of the oxide/metal interface and the morphology of the oxide at the interface might be closely related to the mechanism by which the alloy is transformed into oxide. Therefore, much attention was given to the question of whether or not the oxide/metal interface differs between out-of-pile and in-PWR grown oxide and between fast- and slowly corroding cladding materials. The structure differences observed at the metal/oxide interface by SEM for slow-corroding PWR fuel with alternate Zr alloy claddings and out-of-pile samples on the one side and the Zircaloy-4 PWR cladding samples on the other side can probably explain the mechanism leading to the irradiation enhancement of Zircaloy-4 corrosion and the reason why the extra-low-Sn and the Zr2.5Nb claddings corrode less in PWR than Zircaloy-4.

Figure 11 summarizes all available SEM information with respect to structure at the metal/oxide interface. It reveals that Zircaloy-4 always shows spherical crystal agglomerates if the oxide layer is formed in PWR. In an oxide layer having a thickness <10 μm, the structure is often still mixed, but, at higher thickness, all available data indicate this partial structure. This structure has also been seen with the fast-corroding nodular oxide out of pile and with oxide grown in BWRs. A different smoother surface has been seen for all examined out-of-pile uniform oxide layer samples (up to the maximum thickness of 25 μm examined) and both low-

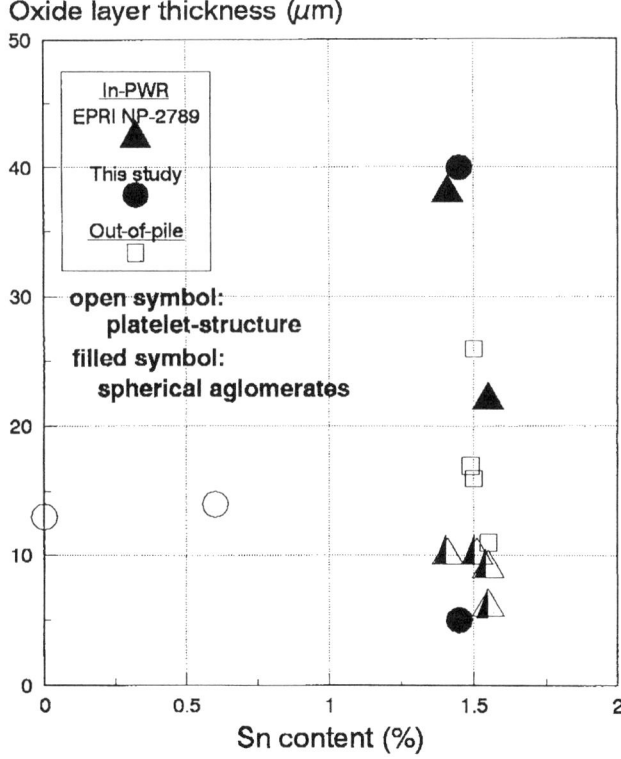

FIG. 11—*Type of structure observed at metal/oxide interface of uniform oxide formed in PWR and out of pile on different Zr alloys.*

corrosion PWR claddings examined here (with an oxide layer thickness of 13 to 14 µm). The spherical crystal agglomerates are probably formed at the end of "fast diffusion" paths (or very tiny pores) through the active layer.

This conclusion conforms to the results from the impedance measurements, which indicate only a very small thickness of the compact active oxide layer for the in-pile corroded Zircaloy-4 sample. A similar small barrier thickness was reported also for out-of-pile grown nodular oxide, which exhibits this particular structure at the metal oxide interface, too. Thus, it is concluded that irradiation of Zircaloy-4 in PWR forms at an oxide layer thickness of ≥ 5 µm fast diffusion paths in the barrier layer (inhomogeneous zones). The average size of the spherical crystal agglomerates is about 300 nm. TEM frequently shows close to the metal/oxide interface slight changes of the growth direction of the columnar crystals for the irradiated Zircaloy-4 sample (Fig. 6) and to a lesser degree in the irradiated extra low Sn sample (Fig. 5). TEM examinations in Ref 8 on the same types of material with out-of-pile grown oxides, however, have shown a much more uniform growth direction at the metal oxide interface. It is not certain yet whether these differences are statistically significant. Frequent changes of growth direction could be the result of the spherical agglomerate oxide growth mode and probably also the cause of it. Higher oxygen ion mobility can be expected at positions with changing growth direction of the columnar crystals. Thus, the spherical agglomerate oxide growth mode seen in oxide layers grown in PWR on Zircaloy-4 fuel cladding may well be a self-inducing process.

Recently, the differences in oxide growth of the different alloys have been explained by H. Blank [18] assessing the structure and properties of the metal phase layer adjacent to the oxide/metal interface. In this layer, an (6Zr+O)-complex (α-Zr(O) phase) is of special chemical stability, and the difference in corrosion behavior can be linked to the different destabilizing effects of the alloying elements in the oxidation process.

Thickness of the Active Layer

It has been speculated that the cladding corrosion behavior depends mostly on the average thickness of the compact oxide layer (A-layer) close to the metal/oxide interface. If one assumes a cubic relation between average rate and thickness of the compact oxide layer as observed in the pre-transition regime, a corrosion enhancement by a factor of 4 would correlate to a 60% thinner compact oxide layer. The thickness of the A-layer deduced from impedance measurement of the irradiated Zircaloy-4 sample is only 7 nm. This is almost a factor of 10 lower than the values of δ_a measured for unirradiated oxide by EIS by other authors [5]. However, in the EIS method, the geometrical parameters have to be deduced from the electrolytic and electrical properties of the oxide via a theory based on electrical circuit analysis. Hence, apart from all theory, all contributions to the average electrical properties of a layer would have to be known with high precision in order to interpret the assessed geometrical parameters. This means that the calculated parameters of different authors cannot be easily compared.

Thus, it has to be postulated that the thickness, δ_a, of the active layer deduced from impedance measurements is not necessarily the only factor that determines the corrosion rate. However, the ranking between thickness of the compact layer deduced from impedance of oxide layers of irradiated Zircaloy-4, irradiated extra low Sn, and irradiated Zr2.5Nb claddings correlates to the ranking of in-PWR corrosion of these claddings.

Conclusions

1. SEM investigations of the oxide layers grown on standard Zircaloy-4 on extra low tin (ELS) Zircaloy and on Zr2.5Nb in PWR as well as oxide layers grown on Zircaloy-4 in an autoclave revealed the same topographical and morphological elements of the oxide, especially

at the oxide/metal interface. With higher magnification (~X20 000), it was recognized that the fast-corroding standard Zircaloy-4 in PWR is formed by spherical crystal agglomerates, whereas all other slowly grown oxides show crystal platelets.

2. TEM investigations show that the same kind of columnar grains appear in the oxide near the oxide/metal interface grown in PWR on the specimens of standard Zircaloy-4 and ELS fuel rods. The crystals of ELS oxide on the ELS rod tend to show a more uniform growth direction as compared to the oxide on the standard Zircaloy-4 rod. This may be linked to the difference in crystal agglomeration mentioned above. The findings indicate fast diffusion paths that are formed through the active layer of in-PWR-grown oxide on standard Zircaloy-4, causing a higher corrosion rate.

3. Electrochemical EIS and anodization results confirm Conclusions 1 and 2. The thinnest compact oxide layer close to the metal/oxide interface and the thinnest barrier oxide was measured with an oxide grown in PWR on a standard Zircaloy-4 fuel rod.

Acknowledgments

The interest of Dr. R. Bodmer in this investigation and the support of NPP Gösgen is gratefully acknowledged. We would like to thank R. Brütsch, M. Gehringer, Th. Graber, H. P. Linder, and H. Schweikert, Paul Scherrer Institute, for assistance with sample preparation and SEM investigations and K. Huber and H. Thiele, Institute for Transuranium Elements, for help in performing TEM investigations.

References

[1] Garzarolli, F., Bodmer, R. P., Stehle, H., and Trapp-Pritsching, S., "Progress in Understanding PWR Fuel Rod Waterside Corrosion," *Proceedings,* ANS Topical Meeting on LWR Fuel Performance, Orlando, FL, Vol. I, 1985, pp. 3–55.
[2] Garzarolli, F., Broy, Y., and Busch, R. A., "Comparison of the Long Time Behavior of Certain Zr-alloys in PWR, BWR and Laboratory Tests," *Zirconium in the Nuclear Industry: Eleventh International Symposium,* Garmisch-Partenkirchen, 11–14 Sept. 1995.
[3] Fuchs, H. P., Garzarolli, F., Weidinger, H. G., Bodmer, R. P., Meier, G., Besch, O., and Lisdat, R., "Cladding and Structural Material Development for the Advanced Siemens PWR Fuel FOCUS," *Proceedings,* ANS-ENS International Topical Meeting on LWR Fuel Performance, Avignon, France, 1992, pp. 682–690.
[4] Cox, B. in *Advances in Corrosion Science and Techniques,* Fontana and Staehle, Eds., Vol. 5, Plenum Press, New York, 1976, p. 137.
[5] Garzarolli, F., Seidel, H., Tricot, R., and Gros, J. P., "Oxide Growth Mechanism on Zirconium Alloys," *Zirconium in the Nuclear Industry: Ninth International Symposium, ASTM STP 1132,* 1991, pp. 395–415.
[6] Godlewski, J., Gros, J. P., Lambertin, M., Wadier, J. F., and Weidinger, H., "Raman Spectroscopy Study of the Tetragonal-to-Monoclinic Transition in Zirconium Oxide Scales and Determination of Overall Oxygen Diffusion by Nuclear Microanalysis of O^{18}," *Zirconium in the Nuclear Industry: Ninth International Symposium, ASTM STP 1132,* 1991, pp. 416–434.
[7] Wadman, B., Lai, Z., Andrén, H.-O., Nyström, A.-L., Rudling, P., and Petersson, H., "Microstructure of Oxide Layers Formed During Autoclave Testing of Zirconium Alloys," *Zirconium in the Nuclear Industry: Tenth International Symposium, ASTM STP 1245,* 1994, pp. 579–598.
[8] Beie, H. J., Mitwalsky, A., Garzarolli, F., Ruhmann, H., and Sell, H.-J., "Examinations of the Corrosion Mechanism of Zirconium Alloys," *Zirconium in the Nuclear Industry: Tenth International Symposium, ASTM STP 1245,* 1994, pp. 615–643.
[9] Göhr, H., Schaller, J., Ruhmann, H., and Garzarolli, F., "Long Term In Situ Corrosion Investigation of Zr Alloys in Simulated PWR Environment by Electrochemical Measurements," *Zirconium in the Nuclear Industry: Eleventh International Symposium,* Garmisch-Partenkirchen, 11–14 Sept. 1995.
[10] Gebhardt, O., "Investigation of In Pile Formed Corrosion Films on Zircaloy Fuel Rod Claddings by Impedance Spectroscopy and Galvanostatic Anodization," *Journal of Nuclear Materials,* Vol. 203, 1993, pp. 17–26.

[11] Gebhardt, O., "A Phase Reference Procedure for Interpretation of Impedance Spectroscopy Experiments," *Electrochimica Acta*, Vol. 38, No. 5, 1993, pp. 633–641.
[12] Göhr, H., Schaller, J., and Schiller, C. A. "Impedance Studies of the Oxide Layer on Zircaloy after Previous Oxidation in Water Vapour at 400°C," *Electrochimica Acta*, Vol. 38, 1993, pp. 1961–1964.
[13] Cox, B. and Yamaguchi, Y., "The Development of Porosity in Thick Zirconia Films," *Journal of Nuclear Materials*, Vol. 210, 1994, pp. 303–317.
[14] Cox, B., Wong, Y. M., and Hoang, Th., "Electrically Conducting Paths in Thick (>10 μm) Zirconia Films," *Journal of Nuclear Materials*, Vol. 223, 1995, pp. 201–209.
[15] Gebhardt, O., Gehringer, M., Graber, Th., and Hermann, A., "Investigation of Corrosion Films on Zirconium Based Alloys by Electrochemical and Microscopic Methods," *Materials Science Forum*, Vol. 192–194, 1995, pp. 587–598.
[16] Gebhardt, O. and Hermann, A., "Microscopic and Electrochemical Impedance Spectroscopy Analyses of Zircaloy Oxide Films Formed in Highly Concentrated LiOH Solution," *Proceedings*, Third International Conference on Electrochemical Impedance Spectroscopy, EIS'95, Nieuwpoort, Belgium, 7–12 May 1995; *Electrochimica Acta*, Vol. 41, 1996, pp. 1181–1190.
[17] Blank, H., Bart, G., and Thiele, H., "Structural Analysis of Oxide Scales Grown on Zirconium Alloys in Autoclaves and in a PWR," *Journal of Nuclear Materials*, Vol. 188, 1992, pp. 273–279.
[18] Blank, H., "On the Origin of Different Rates of Inpile Post-Transition Oxidation for Zr-based Cladding Alloys," *Zeitschrift fuer Metallkunde*, Vol. 85, 1994, pp. 645–657.
[19] Hutchinson, B. and Lehtinen, B., "A Theory of the Resistance of Zircaloy to Uniform Corrosion," *Journal of Nuclear Materials*, Vol. 217, 1992, pp. 243–249.
[20] Macdonald, J. R., *Impedance Spectroscopy*, John Wiley and Sons Inc., New York, 1987.
[21] Macdonald, D. D. and Kubre, M. C. H., "Impedance Measurements in Electrochemical Systems," *Modern Aspects of Electrochemistry*, No. 14, Plenum Press, New York, 1982.
[22] Park, J. R. and Macdonald, D. D., "Impedance Studies of the Growth of Porous Magnetite Films on Carbon Steel in High Temperature Aqueous Systems," *Corrosion Science*, Vol. 23, 1983, pp. 295–315.
[23] Göhr, H., "Aufnahme und Auswertung von Impedanzspektren zur Analyse von Elektrodenprozessen," *DECHEMA Monographien*, Vol. 90, 1981, pp. 1–13.
[24] Göhr, H., "Ueber Beiträge einzelner Elektrodenprozesse zur Impedanz," *Zeitschrift fuer Physikalische Chemie*, Vol. 85, 1981, pp. 274–280.
[25] Kronig, R. de L., "On the Theory of Dispersion of X-Rays," *Journal of the Optical Society of America and Review of Scientific Instruments*, Vol. 12, 1926, p. 547; Kronig, R. de L., "Dispersionstheorie im Röntgengebiet," *Physikalische Zeitschrift*, Vol. 30, 1929, p. 521; Kramers, H. A., "Die Dispersion und Absorption von Röntgenstrahlen," *Physikalische Zeitschrift*, Vol. 30, 1929, p. 522.
[26] Ramasubramanian, N. and Ling, V. C., "Localized Impedance-Anodization Measurements to Characterize Corrosion Films on Irradiated Zirconium Alloys," *Journal of Nuclear Materials*, Vol. 183, 1992, pp. 226–228.
[27] Cox, B., "Factors Affecting the Growth of Porous Anodic Oxide Films on Zirconium," *Journal of Electrochemical Society*, Vol. 117, 1970, pp. 654–663.
[28] Urquhart, A. W. and Vermilyea, D. A., "Characterization of Zircaloy Oxidation Films," *Zirconium in Nuclear Applications, ASTM STP 551*, 1974, pp. 463–478.
[29] Howest, V. R., "The Effect of the Growth Rate on Anodically Formed Zirconium Oxide Films," *Corrosion Science*, Vol. 14, 1974, pp. 491–502.
[30] Pecheur, D., Lefebvre, F., Motta, A. T., Lemaignan, C., and Charquet, D., "Oxidation of Intermetallic Precipitates in Zircaloy-4: Impact of Irradiation," *Zirconium in the Nuclear Industry: Tenth International Symposium, ASTM STP 1245*, 1994, pp. 687–708.
[31] Anada, H., Nomoto, K., and Shida, Y., "Corrosion Behavior of Zircaloy-4 Sheets Produced Under Various Hot-Rolling and Annealing Conditions," *Zirconium in the Nuclear Industry: Tenth International Symposium, ASTM STP 1245*, 1994, pp. 307–327.
[32] Wadman, B., Lai, Z., Andrén, H.-O., Nyström A. L., and Petterson, H., "Microstructure of Oxide Layers Formed During Autoclave Testing of Zirconium Alloys," *Zirconium in the Nuclear Industry: Tenth International Symposium, ASTM STP 1245*, 1994, pp. 579–598.
[33] Gebhardt, O., Aerne, E. T., Martin, M., and Wittmaack, K., "SIMS Depth Profile and Line Scan Analyses at the Metal/Oxide Interface of Corrosion Films on Zirconium Based Alloys," *Secondary Ion Mass Spectrometry, SIMS X*, A. Benninghoven et al., Eds., Wiley, Chichester, England, and New York, in press.
[34] Gebhardt, O., Aerne, E. T., Martin, M., Tao, S., Loibl, N., and Wittmaack, K., "High Lateral Resolution Imaging at the Metal/Oxide Interface of Zirconium Based Alloys by Secondary Ion Mass Spectrometry," *Surface and Interface Analysis*, in press.

DISCUSSION

B. Lehtinen[1] (written discussion)—As we also have observed platelets of the oxide-metal interface, I am interested in your interpretation. Are the platelets a true effect or are they as mentioned before, by Brian Cox, just artifacts.

O. Gebhardt (authors' closure)—We do not think these platelets or needles are artifacts for several reasons: (1) Such an oxide morphology was found not only with bromine/ethylacetate dissolution of the metal phase but also using other mixtures; (2) It would be difficult to explain why artifacts arise with one kind of oxide only and not in all cases where the same method of dissolution of the metal was applied; and (3) It is even more difficult to claim that the logical dependencies of oxide morphology from alloy composition and irradiation and corrosion conditions in our and your papers were randomly achieved.

---

[1] Institute of Metals Research, Stockholm, Sweden.

Xavière Iltis,[1] Florence Lefebvre,[1] and Clément Lemaignan[2]

Microstructure Evolutions and Iron Redistribution in Zircaloy Oxide Layers: Comparative Effects of Neutron Irradiation Flux and Irradiation Damages

REFERENCE: Iltis, X., Lefebvre, F., and Lemaignan, C., **"Microstructure Evolutions and Iron Redistribution in Zircaloy Oxide Layers: Comparative Effects of Neutron Irradiation Flux and Irradiation Damages,"** *Zirconium in the Nuclear Industry: Eleventh International Symposium, ASTM STP 1295,* American Society for Testing and Materials, 1996, pp. 242–264.

ABSTRACT: To understand the acceleration of the Zircaloy corrosion kinetics in PWR conditions, TEM microstructural characterizations of oxide layers grown in an autoclave or directly in-reactor have been performed. To separate the influence on the oxidation process of the irradiation damage in the alloy from the dynamic effect of neutron flux, oxide layers have also been grown in an autoclave on previously neutron-irradiated cladding. The comparative characterization of these oxide layers leads to the following results: the nucleation and growth process are observed to be similar on oxides formed in-autoclave and significantly different on oxides grown directly in-reactor, indicating that this process is essentially affected by neutron irradiation or, more generally, parameters specific to the reactor environment. Concerning grain growth phenomena, it appears that the high microstructural instability noticed in oxides formed in-reactor is also the consequence of parameters specific to the reactor environment such as neutron irradiation or the lithium concentration gradient. Finally, the iron distribution in the oxide is almost the direct image of the iron distribution in the metal.

The impact of each experimental statement on the oxidation kinetics is discussed. The role of neutron irradiation on the structure of the oxide layer appears to be the most determining.

KEYWORDS: Zircaloy-4, irradiation effects, oxide microstructure, oxidation kinetics, alloying element redistribution, analytical transmission electron microscopy

Zircaloy-4 is used mainly as fuel cladding material in pressurized water reactors (PWR). Its corrosion rate in these conditions, increasing with operating time, constitutes a limitation to the extension of fuel rod burnup. When the thickness of the oxide layer reaches several micrometres, the corrosion rates appear significantly higher in-reactor compared to out-of-pile tests [1]. Although the thermo-hydraulic and chemical conditions of reactor operation obviously play a great part in such an acceleration, as shown by a comparison between autoclave and out-of-pile loop test results, they appear not to be sufficient to account, for first, the in-reactor enhancement factor and, second, the end-of-life breakaway [2]. The effects of irradiation either on the coolant or on the material could also play an important role.

The influence of the microstructural state, chemical composition, nature, size, and distribution of intermetallic precipitates on the corrosion behavior of non-irradiated Zircaloys has

---

[1] Research engineer and [2]research director, CEA-Grenoble, DRN/DTP/SECC, 17 rue des Martyrs, 38054 Grenoble Cédex 9, France.

already been well established [3–4]. Since the microstructural evolutions induced by irradiation in Zircaloy contribute to a continuous change of the microstructural state of the oxidizing material [5], an impact on the corrosion kinetics is not unexpected.

In a previous work, the microstructures of oxide layers formed in autoclaves and in the BR3 reactor during three cycles have been compared and clear differences have been shown between these two types of layers [6–7]. When the oxide was grown in an autoclave on non-irradiated cladding, the nucleation and growth process preferentially led to strongly oriented oxide crystallites. When the oxide was grown under irradiation, many more areas formed of nanocrystallites of random orientation were found. A considerable grain growth of zirconia grains was observed through virtually the entire thickness of the layer. The redistribution of the iron contained in the intermetallic precipitates in the oxide layer appeared to be considerably enhanced, and iron-rich areas (up to 1 at%), systematically associated with the presence of a significant amount of tetragonal zirconia, were even found.

These particularities of oxide layers grown in reactor conditions raise two questions:

1. What is the origin of these experimental features and, more precisely, what part of each phenomenon can be attributed to the presence of irradiation defects in the oxidizing alloy specifically and what part is the direct consequence of the reactor environment (water chemistry, irradiation flux, and thermal flux) on the oxidation process?
2. Can a cause and effect relationship be established between these experimental observations and the oxidation kinetics acceleration?

These questions remained difficult to discuss as long as it was not possible in this first approach to separate the mixed effects of the reactor environment affecting directly the oxidation process from those linked specifically to the irradiation-induced microstructural evolutions of the oxidizing alloy.

This work is devoted to distinguishing this second metallurgical effect: on the one hand, we completed the previous observations [6–7] by the characterization of an oxide grown during four cycles in the BR3 reactor; on the other hand, we studied cladding specimens that had already been irradiated in this reactor, that had their oxide layers mechanically removed, and that were re-oxidized in a shielded autoclave. Owing to the aim of this work, all the samples have been chosen from similar ingots in the same metallurgical state and with the same irradiation history in the BR3 reactor. They are thus fully satisfactory even though this reactor is not completely representative of PWRs.

In this paper, we first present the microstructural observations related to oxide layers formed in-reactor. A detailed description of the four-cycle layer is presented, and the main evolutions between three and four cycles are highlighted. In the second part of the paper, we present the results concerning the re-oxidation of irradiated cladding. We give a detailed description of two specimens, one irradiated during three cycles and re-oxidized for 60 days and the other irradiated during two cycles and re-oxidized for 210 days. These results are discussed according to the two preceding questions.

Experimental Procedures

All the cladding tubes used for this study are high-oxygen and high-tin recrystallized Zircaloy-4 irradiated in the BR3 reactor during one, two, three, or four cycles. The composition and final heat treatment of all the tubes are similar and are given in Table 1. The irradiation temperature and the fluence evaluated for each portion of the tubes are given in Table 2.

For the re-oxidation experiments, cladding tube sections 25 mm long were cut, and the

TABLE 1—*Chemical composition and final heat treatment of the studied cladding.*

Material	Composition					Final Heat Treatment
	Zr	Sn	Fe	Cr	O	
Zircaloy-4 cladding, wt%	bal.	1.65	0.21	0.12	0.17	3 h at 848 K
Zircaloy-4 cladding, at%	bal.	1.27	0.34	0.21	0.97	
Zircaloy-4 cladding, Equivalent at% in ZrO_2	...	0.42	0.11	0.07	...	

external oxide layer was mechanically removed in a hot cell. Oxidation was performed in a static autoclave containing superheated steam at 673 K under a pressure of 15 MPa. The oxidation times were 60, 210, and 320 days.

Thin foils taken parallel to the oxide growth front were prepared in various locations (near the oxide-water interface, in the middle part of the layer, or close to the metal-oxide interface) according to the procedure described in Ref 8. All the observations were performed using a transmission electron microscope (TEM) JEOL 1200EX equipped with a LINK AN10000 system for energy dispersive X-ray spectroscopy (EDX). The microanalysis procedure, described in Ref 7, leads to a detection threshold for iron in zirconia of 0.1 at% (corresponding to a concentration in α-Zr of 0.3 at%).

Results on Oxide Layers Formed In-Reactor

Four-Cycle Cladding

The thickness of the oxide layer formed during four irradiation cycles in the BR3 reactor reaches 30 μm. This thickness compared to that of the three-cycle sample (6 μm) indicates that most of the layer was formed during the fourth cycle even though the power and temperature operation remained low (see Table 2). The microstructure of the alloy was characterized by a large amount of $\langle c \rangle$ component dislocation loops, an advanced dissolution of intermetallic precipitates (mean diameter $\leq$ 50 nm), and a homogeneous iron enrichment in the Zr matrix of about 0.35 at%.[3] During the preparation of thin foils, this oxide layer showed very brittle

TABLE 2—*Irradiation conditions and tests performed on the different cladding tubes.*

Samples and Number of EFPD	Local Fast Fluence,[a] 10^{25} n · m^{-2}	Number of Cycles and Irradiation Conditions: Power, W · cm^{-1} (Mean Cladding Temperature, K)				Tests Performed and Examinations
		1	2	3	4	
(EX001) 360	3	220 (588)				Re-oxidation
(EX002-AA177) 700	4	220 (588)	180 (578)			Re-oxidation + TEM
(AA190) 1061	6.5	320 (613)	195 (583)	210 (588)		TEM, Re-oxidation + TEM
(AF159) 1315	8	305 (613)	215 (588)	230 (593)	60 (553)	SEM + TEM

[a] The local fast fluence is calculated for neutrons with an energy $\geq$1 MeV.

[3] This value, corresponding to the complete dissolution of the iron contained in the intermetallic precipitates, seems to be slightly overestimated. Such an effect is not unexpected with EDX analysis.

behavior (high spalling tendency); thus, a specific approach was adopted to characterize it. SEM observations were first performed, and then thin foils were prepared to characterize structures representative of different locations through the thickness of the layer.

SEM Observations—Most of the oxide layer thickness corresponds to thin strata piled up (Fig. 1*a*). This structure explains the brittle behavior mentioned above and makes it difficult to precisely identify the location of thin foils: one can be completely located within a stratum, while another can include different strata (Fig. 1*b*).

In the following observations, we will distinguish only the external quarter of the oxide layer, corresponding more or less to the oxide formed during the first three cycles, and the inner part of the oxide layer, corresponding to the oxide formed during the fourth cycle, which shows the same characteristics through all its thickness, i.e., ~25 μm.

Close to the Water-Oxide Interface—Keeping apart observations made in numerous oxide areas highly polluted by exogenous elements (Fe, Ni, and Cr) coming from the primary water contamination, the oxide microstructure can be described as the result of a progressive structural evolution of the oxide obtained after three cycles. Oxide grains have undergone clear grain growth. The resulting microstructure is very heterogeneous, especially in terms of grain sizes that range from 100 nm to several μm. These grains, sometimes unstable under the electron beam, are exclusively monoclinic and frequently show a poor cohesion. Their iron enrichment reaches systematically one to several tenths of an at%.

The Inner Part of the Oxide Layer—This part of the layer has a highly stratified structure, as previously shown by SEM. At the scale of TEM examinations, it can be described up to the metal-oxide interface with two major types of microstructures. Thus, the immediate vicinity of the metal-oxide interface can hardly be called "dense" anymore.

The first type of microstructure (Fig. 2) is characterized by electron diffraction patterns showing marked preferred orientations of zirconia grains. These grains, mainly monoclinic, generally have good cohesion (confirmed by the quite systematic observation of Moiré fringes), even in areas where they reach several hundred nanometers in size (Fig. 2*a*). Tin is homogeneously distributed and has a concentration around 0.4 at%, in good agreement with the initial composition of the alloy and with the analyses made in the three-cycle layer. The zirconia iron enrichment is generally low, around a tenth of an at%. All precipitates encountered are amorphous, and their first neighbor distance corresponds to amorphous zirconia [7,9]. They have been oxidized while already highly iron depleted. No iron enrichment is thus to be measured in the surrounding zirconia, in good agreement with our analyses. In some peculiar areas, with lateral dimensions on the order of 1 μm, a correlation between a significantly higher iron enrichment ([Fe] ~ 0.5 to 1 at%) and an oxide microstructure characterized by a poor cohesion can be noticed. If indications of preferential orientations are still visible on diffraction patterns, the oxide has a granular aspect (Fig. 2*b*). Diffraction patterns also frequently reveal the presence of a small fraction of tetragonal zirconia in these areas.

The second type of microstructure (Figs. 3 and 4) is characterized by electron diffraction patterns showing no preferred orientations of zirconia grains and by a tin concentration always lower than 0.4 at% and even sometimes under the detection threshold, i.e., 0.1 at%. This microstructure seems highly unstable since it shows significant evolutions close to the metal-oxide interface and moreover exhibits a clear tendency towards extra grain growth under the electron beam. In strata located in the immediate vicinity of the metal-oxide interface, it is composed of partly amorphous zirconia and of zirconia with nanocrystallites (mean diameter ≤ 5 nm) mainly tetragonal (see Fig. 3). In strata located further away from the interface, these randomly oriented grains reach several hundred nanometers, have a very poor cohesion, and are always monoclinic (see Fig. 4). They also often contain dislocations or twins. The iron distribution is globally the same as in the first type of microstructure: its concentration, variable from place to place, ranges from 0.1 to 1 at%.

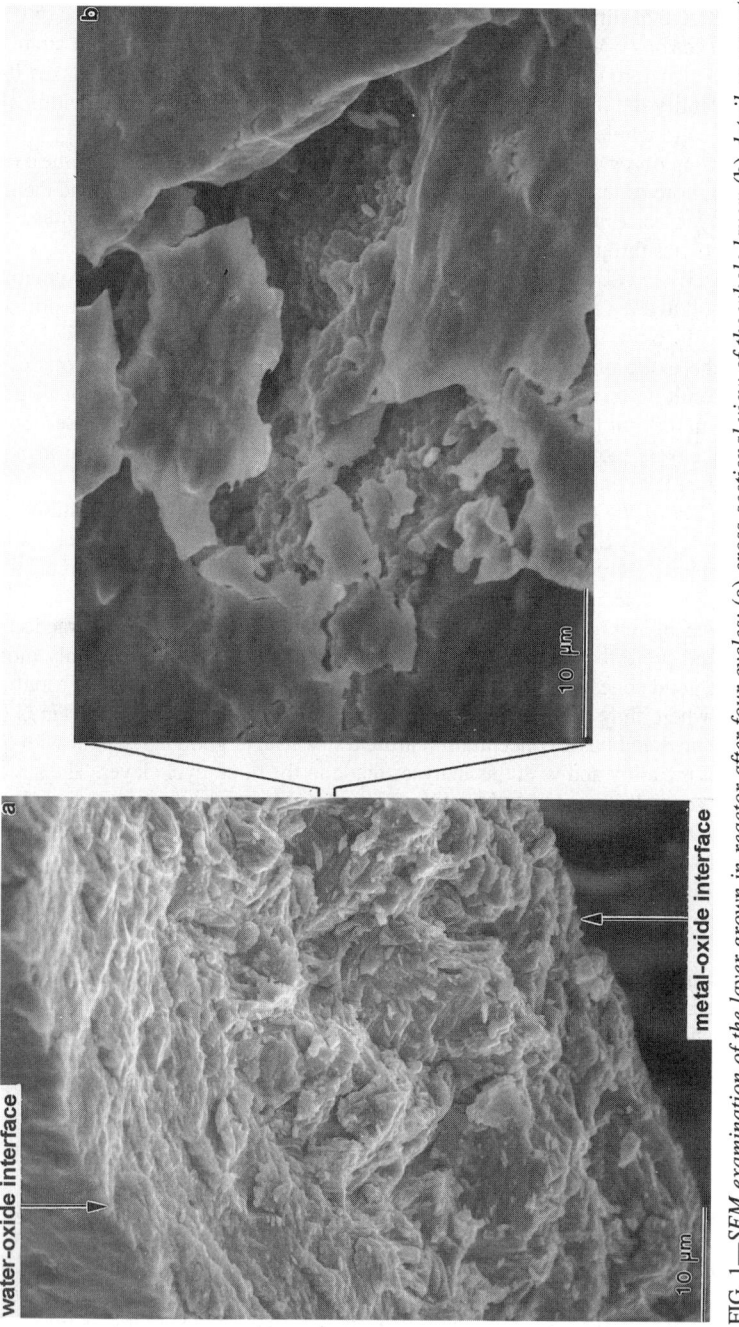

FIG. 1—*SEM examination of the layer grown in-reactor after four cycles: (a) cross-sectional view of the whole layer; (b) detail on a partly ion-thinned foil taken parallel to the oxide growth front in the middle part of the layer.*

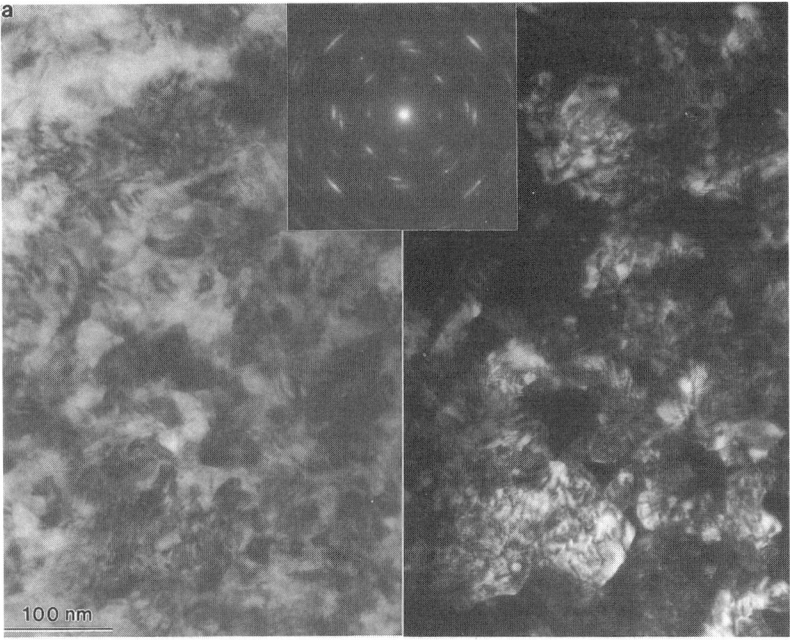

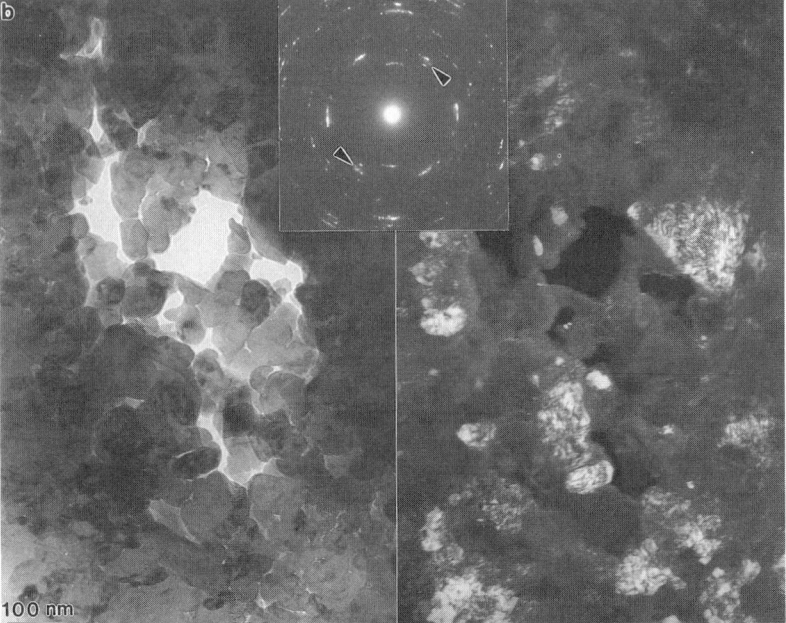

FIG. 2—*First type of microstructure found in the inner part of the layer grown in-reactor after four cycles:* (a) **typical area:** *mainly characterized by zirconia grains with good cohesion, as indicated by the Moiré fringes and marked preferred orientations;* (b) **peculiar area:** *mainly characterized by a significant iron enrichment (up to 1 at%), a poor microstructural cohesion, and the presence, evidenced on diffraction patterns, of a small fraction of tetragonal zirconia (arrowed diffraction spots).*

FIG. 3—*Second type of microstructure found in the inner part of the layer grown in-reactor after four cycles: in the immediate vicinity of the metal-oxide interface, it is mainly characterized by nanocrystallites, essentially tetragonal, having a random orientation.*

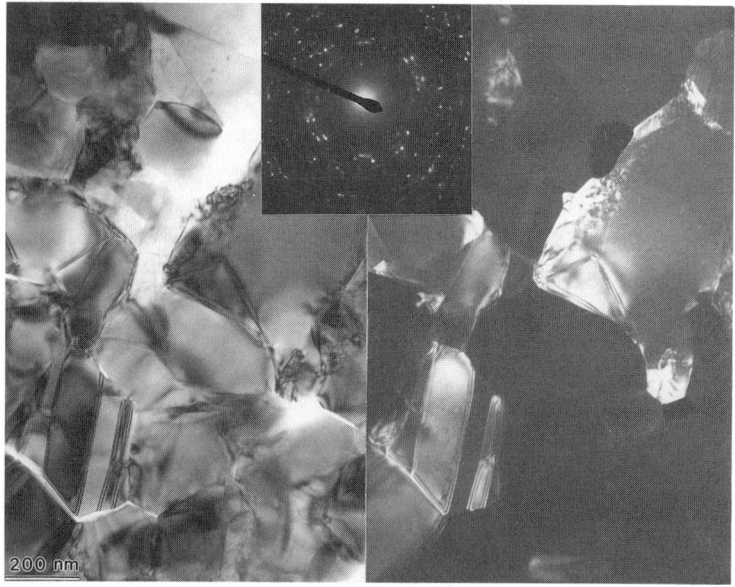

FIG. 4—*Second type of microstructure found in the inner part of the layer grown in-reactor after four cycles: a few micrometres away from the metal-oxide interface, it is mainly characterized by zirconia grains with a poor cohesion and no preferred orientation.*

Summary and Main Evolutions Between Three and Four Cycles

Certain tendencies that appeared through the observations of the three-cycle oxide layer [6–7] are clearly confirmed on the four-cycle one. They concern:

1. The trend toward an extensive grain growth of the zirconia grains through virtually the entire thickness of the layer.
2. The redistribution process of iron through the oxide layer and its impact on the zirconia microstructure. In the three-cycle layer, when sufficiently concentrated, iron clearly led to a local stabilization of tetragonal zirconia. In the four-cycle one, these areas seem to have evolved and are characterized by a very poor cohesion of the grains.

The modification, with irradiation, of the nucleation and growth process of zirconia grains from the metal-oxide interface is also confirmed. The processes involved in the three-cycle layer and in the four-cycle one cannot, however, be described in the same way. In the three-cycle layer, a trend towards the preferential formation of nanocrystallized tetragonal zirconia, with no preferential orientation, instead of strongly textured crystallites, was found [6]. In the case of the four-cycle layer, these two modes of nucleation seem to alternate, leading to a very brittle stratified structure. Tin redistribution may be implicated in the mechanism of stratification.

Results on Oxide Layers Formed in Autoclave on Previously Irradiated Claddings

Five irradiated samples and three reference samples (i.e., non-irradiated) have been re-oxidized in an autoclave. Their respective oxide thicknesses have been measured by metallography. The resulting oxidation kinetics are presented in Fig. 5. A slight but significant difference appears between irradiated samples and non-irradiated ones after 210 days of oxidation.

Two irradiated samples have been chosen to be examined and compared to non-irradiated ones. The microstructural state of the metal after re-oxidation has systematically been characterized in order to determine the degree of recovery of the irradiated microstructure, especially in terms of iron distribution.

Three-Cycle, Autoclaved 60 Days

The thickness of the oxide layer grown on the three-cycle sample for 60 days is similar to the thickness of the non-irradiated sample, i.e., 3 μm. Thin foils have been prepared on both samples close to the metal-oxide interface.

Metal Microstructure—After 60 days at 673 K, no significant recovery of dislocation loops is visible. Iron has come back to the amorphous $Zr(Fe,Cr)_2$-type precipitates, the Fe/Cr ratio of which increased from 0.2 before re-oxidation to 0.4 or 0.5, while the mean matrix enrichment has decreased to about 0.2 at%. No evidence for secondary precipitation is found.

Oxide Close to the Metal Interface—The microstructure of the oxide is very similar to the one observed on the non-irradiated sample: most areas are characterized by strong preferred orientations of zirconia crystallites in relation with the underlying metal. Precipitates are found amorphous with a Fe/Cr ratio of about 0.4 (Fig. 6), in good agreement with the observations made in the metal. Almost no iron is detected in the oxidized matrix except in some areas at the metal interface, which can contain up to 0.6 at% of this element. Tin is homogeneously distributed. Its concentration is around 0.4 at%.

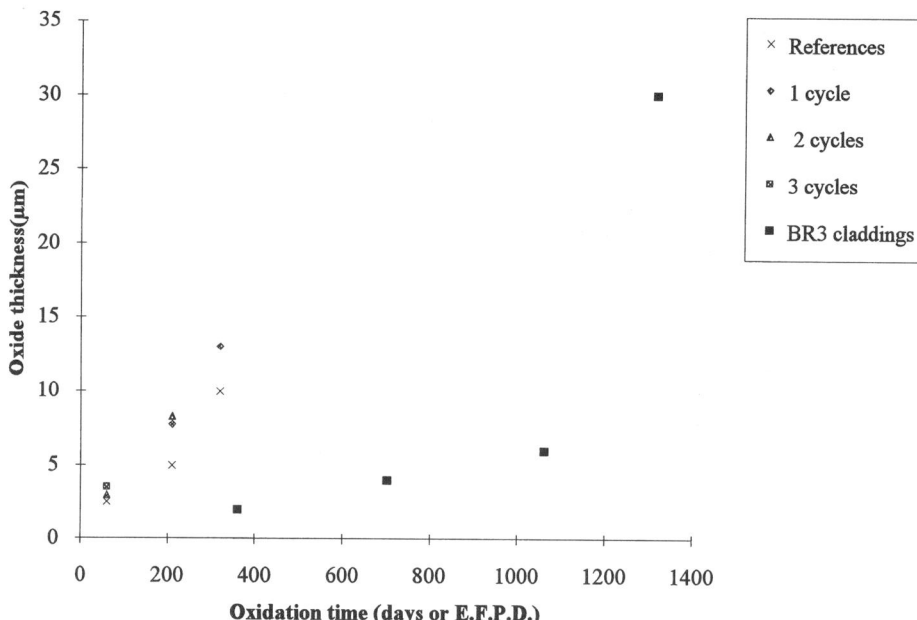

FIG. 5—*Comparative oxidation kinetics in autoclave (673 K) on references and on previously irradiated claddings and in-reactor (~590 K).*

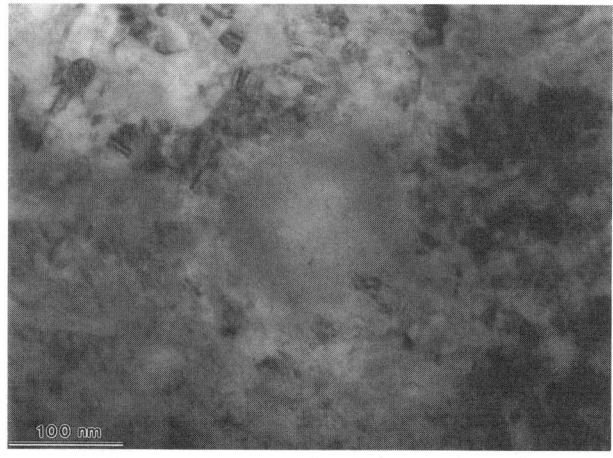

FIG. 6—*Microstructure found close to the metal-oxide interface of the layer grown on the three-cycle cladding, autoclaved 60 days. The observed amorphous precipitate is characterized by a Fe/Cr ratio of 0.4.*

Two-Cycle, Autoclaved 210 Days

The thickness of the oxide layer grown on the two-cycle sample for 210 days is 8 ± 1 μm. It appears significantly different from the thickness measured on the non-irradiated sample, i.e., 5 ± 1 μm. The structure of the metal has been studied, and thin foils have been prepared in the oxide layer in different locations.

Metal Microstructure—The dislocation loop population does not seem significantly affected by 210 days of annealing at 673 K. If the amorphous parts of intermetallic precipitates remain fully amorphous, their Fe/Cr ratios now range between 0.6 and 1, indicating a large iron return from the matrix to the precipitates. Moreover, a thin and heterogeneous precipitation of Fe- and Cr-rich particles ($\phi \sim 10$ nm) can be observed, the density of which varies from grain to grain and is clearly more pronounced in the grain boundaries (Fig. 7a). The corresponding matrix iron enrichment varies from 0 (non-detectable) to 1 to 2 tenths of an at%.

Oxide Close to the Metal Interface—The structure of the oxide is very similar to that formed on the non-irradiated sample. Most areas, indeed, show oxide crystallites with preferential orientations, and only a few are nanocrystallized and randomly oriented. So, the proportion of both nucleation modes is rather standard [6]. No sign of extra grain growth and no peculiarity concerning tin concentration are observed. One major difference between non-irradiated and previously irradiated samples is observed in this part of the oxide layer. It lies, in fact, in the distribution of iron. The $Zr(Fe,Cr)_2$-type precipitates are found in the same state as in the metal: their amorphous ring is characterized by a Fe/Cr ration ranging between 0.6 and 1, and their crystalline core has a Fe/Cr ratio of about 2 (Fig. 7). In the zirconia matrix, the iron distribution is very similar to that observed in the underlying metal. In some areas, no iron is detected, while in others, the iron content can reach 1 at%. High iron enrichment is frequently associated with chromium detection (Fig. 7b) and can be related to the areas of the metal containing secondary precipitates (Fig. 7a). These precipitates can't, however, be imaged in the oxidized matrix either because they are already dissolved or more probably because they can't be distinguished from the zirconia crystallites.

The heterogeneous iron distribution in the oxidized matrix has no influence on the zirconia microstructure in terms of tetragonal zirconia stabilization, this variety being frequently observed in this part of the layer [7].

Oxide in the Middle Part of the Layer and at the Steam Interface—The greatest proportion of oxide consists of monoclinic grains showing a marked texture. This type of microstructure is again very similar to the ones observed in the non-irradiated sample. Precipitates remain amorphous with a first neighbor spacing corresponding to amorphous zirconia. They generally do not lead to significant enrichment in their neighborhood. Nevertheless, numerous areas showing a poor cohesion are also found. They are always characterized by a high content of iron (up to 0.6 at%), a quasi absence of any preferred orientation, and signs of tetragonal zirconia present. Their microstructure appears similar to that shown on Fig. 2b.

Summary of the Re-Oxidation Experiments

After 60 days of oxidation, no differences exist between the oxide thicknesses of non-irradiated and previously irradiated samples (3 μm). After 210 days, a significant difference is measured between the thicknesses of the previously irradiated sample (8 μm) and the reference one (5 μm).

The TEM characterization of the oxide layers formed, either after 60 or 210 days, either on non-irradiated or on previously irradiated material, reveals no significant differences in the structure of the oxides (taking apart local observations concerning iron-rich oxide areas with poor microstructural cohesion). Grains with marked preferential orientations are commonly

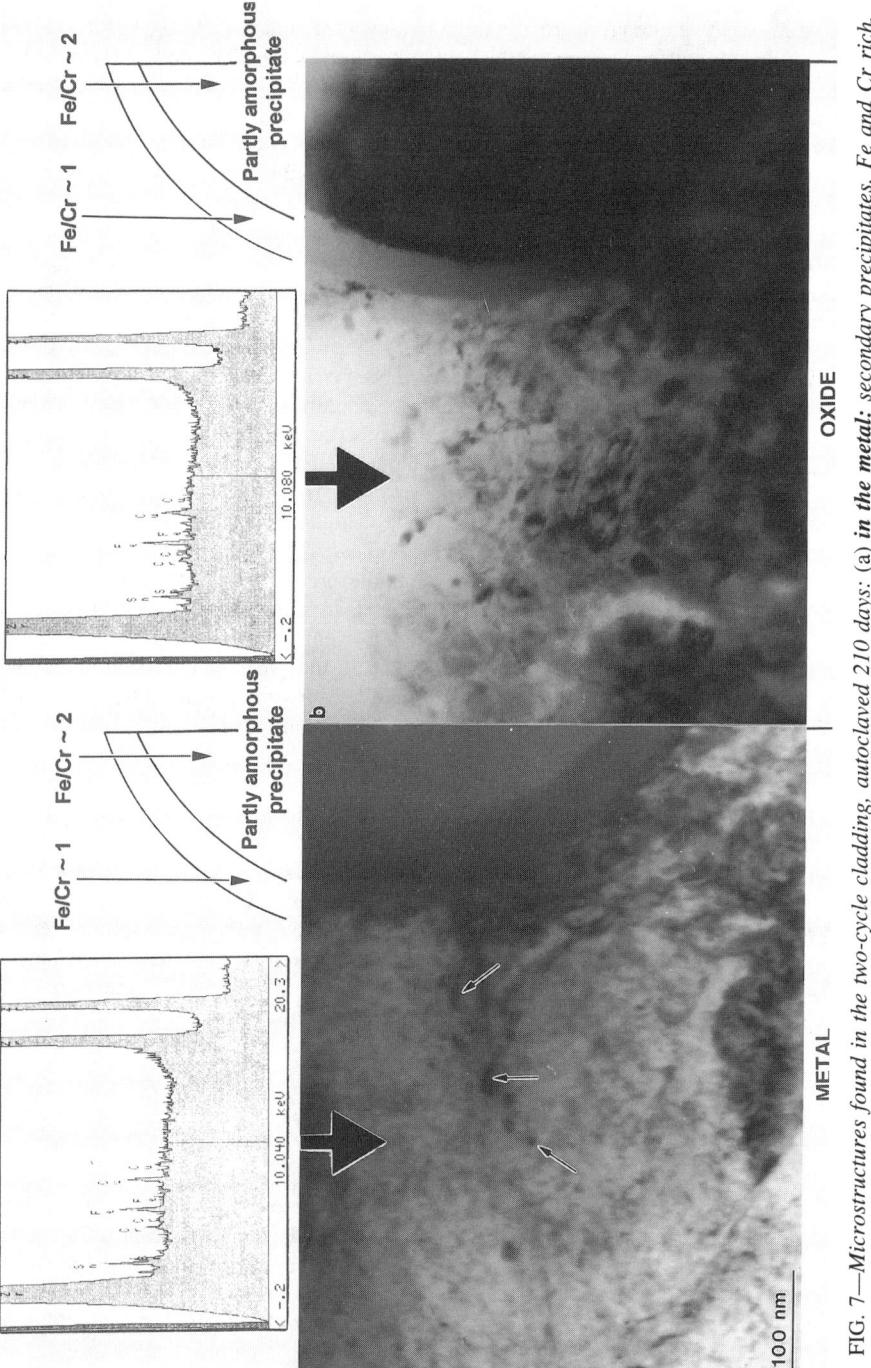

FIG. 7—*Microstructures found in the two-cycle cladding, autoclaved 210 days:* (a) **in the metal:** *secondary precipitates, Fe and Cr rich, are arrowed;* (b) **in the oxide close to the metal-oxide interface:** *secondary precipitates are not visible, but the oxide is significantly enriched in iron and chromium.*

observed. They are tetragonal and monoclinic close to the metal-oxide interface and mostly monoclinic away from this interface. No signs of extra grain growth within the oxide layer are visible. The main difference between previously irradiated and non-irradiated samples begins to appear after 60 days of oxidation and becomes obvious after 210 days: it lies in the iron distribution. Close to the metal-oxide interface of the previously irradiated samples, in the dense part of the oxide layer, a high content of iron, heterogeneously distributed, is detected. On the other hand, the dense layer of reference samples consists of zirconia without iron detected and intermetallic precipitates containing the iron.

Discussion

The oxide microstructures formed on irradiated materials either in an autoclave or directly under irradiation show significant differences, summarized in Table 3, related to the zirconia crystallites nucleation and growth process from the metal-oxide interface, to the grain growth process within the thickness of the layer, and to the redistribution of tin and iron.

Role of Irradiation Defects

One first question is to determine which part of these processes is to be specifically attributed to the presence of irradiation defects in the oxidizing alloy.

Nucleation and Growth Process—Two types of oxide microstructures can be observed close to the metal-oxide interface. One is characterized by tetragonal and monoclinic grains showing strong preferred orientations in relation with the underlying metal. The other consists of smaller nanocrystallites, preferentially tetragonal, randomly oriented. The first nucleation mode seems to prepare for a regular columnar growth to develop [10], while the second would be the first step, according to our observations, of the development of a highly porous structure resulting from the drastic structural destabilization of the tetragonal zirconia as observed in some ceramics [11].

Close to the metal-oxide interface, on the oxides formed in an autoclave on previously irradiated materials, the relative proportion of areas corresponding to these two nucleation and growth modes is rather classical. It globally corresponds to the proportion encountered on reference oxide layers (i.e., formed in an autoclave on non-irradiated Zircaloy). So, most areas are textured. Thus, the establishment of preferred orientation relations between the metal and its oxide appears to be as easy on non-irradiated Zircaloy as on a previously irradiated one, as long as the oxidation takes place in an autoclave in steam. Irradiation defects such as dislocation loops would then have little impact on the nucleation and growth process.

On the oxides formed in-reactor, areas with nanocrystallites randomly oriented or even quasi-amorphous zirconia are far more abundant than textured ones. The oxidation temperature (about 590 K at the metal-oxide interface, in-reactor, and 673 K in-autoclave) is to be considered first to discuss such an experimental observation. Indeed, it is well established that when decreasing the oxidation temperature from 673 to 523 K the nucleation and growth process is slightly modified: the lower the temperature, the smaller the oxide crystallites and the harder the epitaxial relationships [12]. However, in autoclaves such a tendency remains limited, while in-reactor it undergoes major changes: in the three-cycle layer, a large number of areas with much smaller nanocrystallites randomly oriented coexist with fewer textured areas, and in the four-cycle layer they even seem to constitute periodically a complete layer that evolves structurally very quickly. This mode of nucleation and growth appears, then, to be slightly favored by decreasing the oxidation temperature, but it seems to be essentially enhanced by the reactor conditions. On the other hand, as long as the oxidation kinetics remain moderate, the existence of a thermal gradient through the oxide layer or the presence of a lithiated environment are

TABLE 3—*Main structural and chemical features pointed out in the studied oxide layers.*

	Nucleation and Growth Process from the Metal-Oxide Interface	Grain Growth Process within the Thickness of the Layer	Iron Redistribution	Tin Redistribution
In autoclave, on non-irradiated samples	Predominantly oriented. Announcing a columnar growth.	No extra grain growth. Good cohesion and good stability of the microstructure.	Detected a few micrometres away from the metal-oxide interface, i.e., away from the dense inner layer.	Not detected
In autoclave, on previously irradiated samples	Predominantly oriented. Announcing a columnar growth.	No extra grain growth. Good cohesion and good stability of the microstructure.	Detected in the direct vicinity of the metal-oxide interface, i.e., in the dense inner layer.	Not detected
In-reactor, after three cycles	Predominantly disoriented. Announcing an equiaxed microstructure.	Extra grain growth. Poor cohesion and high instability of the microstructure.	Detected in the direct vicinity of the metal-oxide interface, i.e., in the dense inner layer.	Not detected
In-reactor, after four cycles	Periodically, completely disoriented or mainly oriented. Announcing a stratified structure.	Extra grain growth. Poor cohesion and high instability of the microstructure.	Detected in the direct vicinity of the metal-oxide interface, i.e., in the dense inner layer.	Detected (alternating strata, with different Sn content)

reported to have no influence on the oxide microstructure close to the metal-oxide interface [*13*]. Thus, the neutron irradiation flux, by creating atomic displacements and disordering the metal-oxide interface, seems to be at the origin of such an observation by inhibiting the establishment of preferred orientation relations as shown in Ref *14*. However, the imbalance between the two types of nucleation and growth processes tends to increase dramatically between the third and the fourth cycles. Since the fast neutron flux and the concentration of mobile defects are considered similar during these two cycles, another parameter becoming more and more active with time in-reactor must be involved. One can think of an accumulation effect close to the metal-oxide interface of elements coming either from the metallic alloy (iron or tin) [*15–16*] or from the coolant (hydrogen [*17–18*] or lithium [*19–20*]). Metallographic examinations performed on the three-cycle and the four-cycle samples show, in both cases, a relatively low concentration of hydrides with a similar distribution (Fig. 8). Thus, it seems that a drastic acceleration of the corrosion rate due to a hydrogen accumulation at the metal-oxide interface can be excluded in this peculiar case. On the other hand, certain similarities between the microstructures observed after an acceleration of the oxidation process, in-reactor during the fourth cycle or in-autoclave after 200 days in an highly lithiated environment [*13*], are to be underlined. Thus, an accumulation of lithium hydroxide assisted by the existence of a progressively more severe thermal gradient through the oxide layer is not unexpected [*13,21*]. An interfacial instability could then result from the periodical degradation of the inner diffusional barrier [*22*]. Indeed, in addition to the irradiation flux effect, a purely kinetic effect would be involved. The dense inner layer being periodically degraded, the oxidation rate would be periodically drastically increased, leading to a disoriented nucleation and growth mode.

Grain Growth Within the Oxide Layers—On non-irradiated Zircaloy, when sampling thin foils parallel to the growth front, oxide grain sizes range between about 10 nm near the metal-oxide interface to a few tens of nm at a distance of several micrometres from this interface [*6*]. These results are in good agreement with other studies using X-ray diffraction experiments [*23*]. At the external surface of oxide layers, grains with large lateral dimensions can sometimes be encountered [*12*]. This situation, induced by surface energy anisotropies, is easily identified. Occasionally, important grain growth cases are also reported within the thickness of the layer up to the metal-oxide interface [*24*]. These observations could, however, be due to a possible mechanical destabilization during the thin foil preparation. When characterizing the oxide layers grown in autoclaves either on non-irradiated samples or on previously irradiated ones, this situation has been encountered extremely rarely. On the contrary, in oxide layers formed in-reactor, during three or four cycles, numerous areas were observed with large-size crystallites (100 nm or more), occasionally with a tendency to further crystallization under the electron beam. This particular feature was found, irrespective of the crystallographic or chemical properties of the oxide, through virtually the entire thickness of these two layers and was especially strong in the four-cycle one in strata coming from nanocrystallized tetragonal zirconia. Whether this grain growth occurs during thin foil preparation or during irradiation, it reveals a high degree of instability of the microstructures formed under irradiation. In connection with this last point, it's worth noting that no extensive grain growth is observed in oxide layers formed in loops [*13*], which tends to show that irradiation is directly implicated in this phenomenon.

Iron Redistribution—In oxide layers formed on non-irradiated Zircaloy, iron redistribution profiles can be detected around oxidized precipitates at several micrometres from the metal-oxide interface [*7*]. The corresponding local iron enrichments, however, remain low and have no effect on the zirconia crystallography. Close to the metal-oxide interface, due to the delayed oxidation process of intermetallic precipitates [*8,25*], iron remains entirely located in the precipitates. No iron redistribution is thus detected in the so-called dense inner layer. In the oxide layers formed on irradiated claddings, in autoclaves as well as directly in-reactor, the iron redistribution appears to be modified, especially in that dense inner layer.

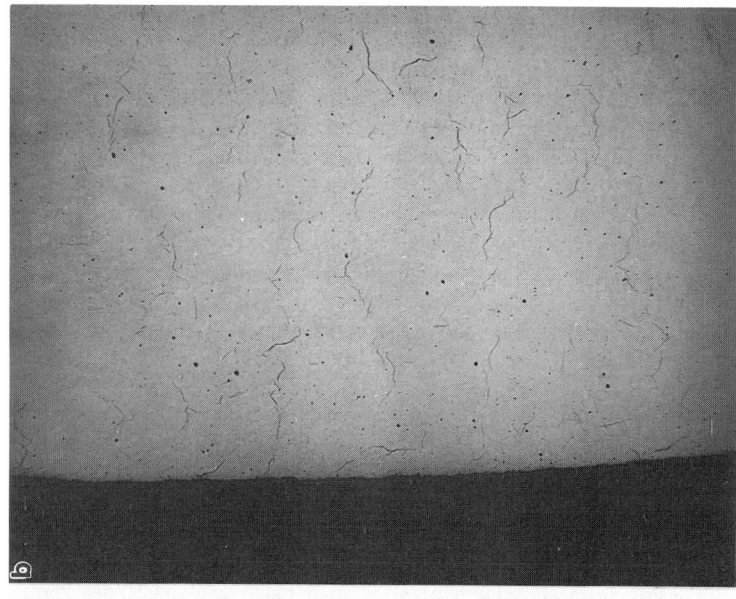

FIG. 8—*Metallographic examination of the hydrides distribution in the studied cladding specimens corroded in the BR3 reactor: (a) three-cycle sample; (b) four-cycle sample.*

Oxide Formed in Autoclaves on Previously Irradiated Claddings—One difference with the above reference case is the observation at several micrometres from the metal-oxide interface of highly iron-enriched areas with poor microstructural cohesion, which tends to indicate that the amorphization process of the $Zr(Fe,Cr)_2$ precipitates under irradiation permits an easier iron redistribution in zirconia [7], even if the oxidation is carried out without irradiation.

The detection of significant amounts of iron in the dense inner sub-layer constitutes, in our point of view, the major difference with the reference cases. Since in this part of the layer the oxide was just formed, no depletion from the precipitates was likely to have occurred. The iron detected in the dense layer must then come either from the α-Zr solid solution or from secondary precipitates.

In order to characterize the behavior of the iron-zirconium solid solution, a complementary approach has been studied on zirconium samples implanted with iron [26]. The redistribution of iron induced by very brief oxidation treatments of the implanted areas was followed using surface spectroscopy techniques. Simulations of the corresponding thermal cycles were also performed under vacuum. It was found that, in oxidizing conditions, the implanted iron did not precipitate and was not rejected ahead of the oxide growth front but was easily incorporated into the oxide layer. The distribution profile of iron at the metal-oxide interface was shown to depend on the respective iron solubilities and mobilities in the metal and in the oxide and on the interfacial velocity [15,27]. Although in that experiment the amount of implanted iron was of the order of one to several at%, its conclusions can be useful for the discussion of our present observations.

On the sample irradiated for three cycles and oxidized for 60 days, iron starts to return to the precipitates, and its remaining concentration in α-Zr is about 0.2 at%. Iron is thus generally not detected in the oxide (its expected content is under our detection threshold [7]) except very close to the metal-oxide interface, where it tends to accumulate as observed in the complementary experiment described above. On the sample irradiated for two cycles and oxidized for 210 days, the distribution of iron in the dense layer appears as the image of the distribution observed in the metal: Large amounts of iron can be found locally in the inner part of the oxide layer, in complete agreement with the heterogeneous distribution of iron in the metal. This behavior can, in fact, be linked directly to the observed secondary precipitation. It appears also to be in excellent agreement with the results published by Cheng et al. concerning similar re-oxidation experiments [28].

Oxide Formed In-Reactor—As in the preceding case, in the outer part of the oxide layer, iron-rich areas have been found. An obvious relationship can be established between iron-rich areas with a significant proportion of tetragonal zirconia reported in the middle part of the three-cycle layer [7] and those found in the four-cycle one, which are also iron rich and highly porous. One can, indeed, assume that the tetragonal zirconia destabilization results in this porous structure.

After three cycles in-reactor, virtually no iron is detected in the dense inner layer except in some accumulation sites occasionally found close to the metal-oxide interface [7]. This result is fully consistent since a large amount of iron is in solid solution in the metal and thus easily incorporated in the oxide layer [26]. In most areas this amount becomes undetectable because of our EDX detection threshold [7] except in some accumulation places. After four cycles in-reactor, the inner layer can hardly be called a dense layer. Nevertheless, relatively large amounts of iron can be found locally near the metal-oxide interface. Dynamic accumulation conditions of this element in zirconia are indeed likely to occur since the growth mode of the layer seems to be periodically markedly modified.

To Summarize—In autoclaves on non-irradiated materials, the redistribution of iron in oxide layers is linked to the progressive iron depletion of oxidized precipitates; thus, it occurs several micrometres away from the metal-oxide interface. In our two experiments, such a depletion mechanism appears to be rather similar even if the presence either of neutron irradiation flux or of previous irradiation damage seems to facilitate the diffusion of iron in zirconia and consequently the formation of highly porous areas in the outer sub-layer.

On the contrary, the redistribution of iron in the dense oxide layer is specific to irradiated samples. In both oxides formed under irradiation and formed in-autoclave on previously irradiated materials, this redistribution appears to be the result of the redistribution of iron in the metal before oxidation. Thus, it appears to be clearly the result of irradiation damage.

Tin Redistribution—In oxides grown in-autoclave, tin is always found homogeneously distributed through the entire layer. In the oxide layers formed in-reactor, two cases have been encountered. After three cycles, tin is also homogeneously distributed, while, after four cycles and a drastic acceleration of the corrosion rate, its concentration appears to depend clearly on the local oxide microstructure, i.e., the nature of the stratum. Tin concentration is either around 0.4 at%, in the first type of stratum coming from a textured zirconia, or significantly lower in the second type of stratum coming from randomly oriented tetragonal nanocrystallites and leading to extra grain growth. One should, however, notice that this distribution gives a global tin mass unbalance and has then to be taken into account very carefully. Most published results concerning tin distribution in Zircaloy oxide layers point out an homogeneous distribution of this element in the layer, either for oxides grown in the presence of irradiation [29] or in out-of-pile tests [30]. To our knowledge, Ref *31* is the only one to mention a tin segregation effect. It is, however, important to notice that the analysis techniques used in these studies have relatively low resolution in depth (i.e., ≥ 1 μm) compared to the thickness of the strata observed in the four-cycle layer (≤ 1 μm). According to our results, tin redistribution in the oxide layer happens only in reactor conditions after a severe acceleration of the oxidation kinetics. Tin distribution in oxide layers formed in loops should, however, be studied to complete these results. Finally, a more general study of the relationships between the observed tin alternate localization and the main causes of the oxidation breakaway in-reactor should be of great interest.

Consequences on the Oxidation Kinetics

The previous discussion allowed us to separate the origins of the various microstructural differences observed in the oxide layers grown in-pile or out-of-pile. One question, however, is still pending concerning the impact of these microstructural differences on the global oxidation kinetics.

As illustrated in Fig. 5, our two experiments (autoclave and reactor) can hardly be compared directly since the oxidation conditions (temperature, water chemistry, thermal gradient) are different in both cases. Each one separately can, however, bring information. In the re-oxidation experiments, in accordance with results reported in Ref *28*, previously irradiated samples corrode faster than reference samples. In-reactor, a drastic oxidation acceleration takes place in the fourth cycle even if the mean coolant temperature and the generated power are lower than during the third cycle (Table 2).

Nucleation and Growth Process—The main difference between oxides grown without irradiation (in-autoclave or in-loop) and oxides grown under irradiation is the preferential structure of the oxide at the metal interface, strongly textured in the first case or nanocrystallized and randomly oriented in the latter case. Since this second type of microstructure leads to the formation of a highly porous structure, one can imagine that the observation of numerous areas corresponding to this type of nucleation and growth process would be linked directly with a clear oxidation acceleration to come. After three cycles in-reactor, a large proportion of disoriented areas is observed at the metal-oxide interface, while the oxidation rate remains moderate. It is then reasonable to assume that this microstructural specificity of oxide layers grown in-reactor accounts for a part of the global in-reactor enhancement factor. After the fourth cycle, the oxidation rate has drastically increased and the nucleation and growth process appears to be completely unstable. In this case, instead of the coexistence of the two types of nucleation and growth processes obtained in the three-cycle sample, we observe the alternate occurrence of each process over the whole sample. Thus, it seems that as long as sufficient areas still oxidize according to the oriented nucleation mode followed by a regular columnar growth, the oxidation kinetics remain low [10]. As soon as randomly distributed nanocrystallites are observed to cover the whole interface, the oxidation kinetics break away. This correlation between the oxide growth mode and its formation kinetics is similar to that reported in Ref 32.

A tentative explanation for the alternating nature of the phenomenon can be proposed. When disoriented oxide crystallites nucleate at the metal interface, it corresponds to a high growth rate progressively decreasing. When the oxidation rate has sufficiently decreased (at about 1 μm from the interface), the "stable" nucleation and growth process can occur. After a while, the interface would again be destabilized due, for example, to lithium hydroxide accumulation, inducing a degradation process [21–22]; then disoriented oxide crystallites would nucleate again.

Even if this simple explanation is only a scientifically sound hypothesis, the observed inner layer is no longer to be considered as a dense protective layer. This effect is thus of major importance regarding the global oxidation kinetics.

Grain Growth Within the Oxide Layers—The tendency for extra grain growth, indicating a high degree of instability of the oxide microstructure and enhancing the formation of micropores, appears to be specific to oxide layers formed under irradiation. Moreover, it appears to be easier on the nanocrystallized grains that preferentially nucleated under irradiation. As long as this crystallization process takes place in the outer layer, commonly called "porous" layer, this phenomenon has little impact on the global kinetics, essentially controlled by the thickness of the dense inner layer acting as a diffusion barrier. This is the case for the three-cycle sample where significant grain growth is observed essentially in the middle part and at the water side of the oxide layer. In the four-cycle sample, this tendency is observed even at the immediate vicinity of the metal-oxide interface. Thus, a very porous structure seems to form very near the metal interface, making the diffusion barrier layer ineffective. This effect is then detrimental for the global kinetics.

Iron Redistribution—The iron redistribution concerns two different effects in the oxide layer. The first one takes place a few micrometres away from the metal-oxide interface: It is the redistribution process around oxidized precipitates. On non-irradiated materials, it remains limited and has no or little effect on the oxide structure. On previously irradiated material, and more clearly on materials oxidized under irradiation, it becomes sufficient to locally stabilize the tetragonal zirconia, giving after destabilization highly porous areas. Anyhow, since this effect concerns the outer part of the oxide layer, it should have only limited impact on the corrosion kinetics.

The second effect is more specific to irradiated materials and concerns the inner dense layer. As long as most of the iron is trapped in the $Zr(Fe,Cr)_2$ precipitates, it does not spread through this dense layer; however, when it is almost completely dissolved in the Zr solid solution (up to 0.3 at% after three cycles) or distributed in a fine precipitation, it is found dispersed in the dense inner layer. This effect is observed in the three-cycle and four-cycle samples oxidized in-reactor as well as in the two and three-cycle samples re-oxidized in-autoclave. Since this iron redistribution is the major peculiarity of the re-oxidized samples, it clearly appears to have some impact on the oxidation kinetics (which can be understood in terms of a modification of the electrical and diffusional properties of zirconia [33–34]). However, at our test temperature, some dissolved iron reprecipitates, and, due to our TEM resolution, it is not possible to tell whether these fine precipitates are incorporated in the oxidized matrix dissolved or not. If not, iron would have no influence on the electrical nor diffusional properties of zirconia. Thus, we cannot exclude in our experiment a limitation of the iron-enhancing effect due to its progressive precipitation in the metallic matrix.

Summary and Conclusion

In this paper, we wanted to separate the effects of irradiation damages on the oxidation process of Zircaloy-4 from those of all the parameters specific to a reactor environment.

We have first characterized the oxide layers formed in-reactor:

1. The modification, essentially with irradiation, of the nucleation and growth process of zirconia grains from the metal-oxide interface constitutes the major observed effect.
2. An extensive grain growth of the zirconia grains through virtually the entire thickness of the layers is observed.
3. The redistribution process of iron through the oxide layer induces essentially local stabilization of the tetragonal zirconia, resulting after destabilization in highly porous areas.
4. A tin redistribution is observed associated with a stratification mechanism of the oxide layer.

We have then characterized irradiated claddings mechanically polished and autoclaved:

1. Previously irradiated samples are shown to corrode more than reference ones.
2. The main microstructural difference between the two types of samples lies in the iron distribution. Indeed, a high content of iron, heterogeneously distributed, is detected in the dense inner part of the oxide layer formed on irradiated claddings.

These experimental observations highlight the following conclusions:

1. Irradiation damages in the oxidizing alloy are observed, after an autoclave re-oxidation, to affect essentially the iron redistribution. A significant but limited impact on the corrosion kinetics is shown.
2. The direct consequences of the reactor environment (water chemistry, thermal flux, and irradiation flux) on the oxidation process concern, on the one hand, the nucleation and growth mode, and, on the other hand, the grain growth within the layer.

As long as the oxidation kinetics remain moderate (i.e., before the breakaway), irradiation is shown to favor the formation of zirconia nanocrystallites randomly oriented, mostly tetragonal with a clear tendency to destabilization, leading thus to a highly unstable growth mode instead of oriented nuclei preparing a regular columnar growth. Moreover, a high degree of instability always characterizes the microstructures formed under irradiation, allowing extra grain growth and pore formation throughout the oxide layer.

These two effects, specific to irradiation, are thus of great importance regarding the global oxidation kinetics and can account for the global in-reactor enhancement factor.

However, they cannot explain the oxidation breakaway occurring after several irradiation cycles, when there is no more efficient inner diffusion barrier layer. Indeed, all the destabilization tendencies observed on the three-cycle layer are drastically more pronounced in the four-cycle layer, although neutron irradiation flux remained similar. Thus, in addition to the neutron irradiation flux, an oxidation rate instability is involved, which could be linked to in-depth lithium hydroxide penetration through the layer.

Acknowledgments

The authors would like to give warmest thanks to F. Gomez, C. Regnard, and L. Rouillon for their support in the SEM and TEM work. They also want to thank D. Pêcheur and R. Salot for their contribution in the re-oxidation experiments.

In addition, the authors are grateful to B. Cox from the University of Toronto for stimulating discussions. Finally, they would like to thank FRAMATOME and EDF for financial support.

References

[1] Garzarolli, F. and Holzer, R., "Waterside Corrosion Performance of Light Water Power Reactor Fuel," *Nuclear Energy,* Vol. 31, 1992, pp. 65–86.
[2] "Corrosion of Zirconium Alloys in Nuclear Power Plants," IAEA-TECDOC-684, IAEA, Vienna, 1993, pp. 73–127.
[3] Charquet, D., "Influence of Precipitates on the Corrosion of Zircaloy-4," *Journal of Nuclear Materials,* Vol. 211, 1994, pp. 259–261.
[4] Anada, H., Nomoto, K., and Shida, Y., "Corrosion Behavior of Zircaloy-4 Sheets Produced Under Various Hot-Rolling and Annealing Conditions," *Zirconium in the Nuclear Industry: Tenth International Symposium, ASTM STP 1245,* A. M. Garde and E. R. Bradley, Eds., American Society for Testing and Materials, West Conshohocken, PA, 1994, pp. 307–327.
[5] Gilbon, D. and Simonot, C., "Effect of Irradiation on the Microstructure of Zircaloy-4," *Zirconium in the Nuclear Industry: Tenth International Symposium, ASTM STP 1245,* A. M. Garde and E. R. Bradley, Eds., American Society for Testing and Materials, West Conshohocken, 1994, pp. 521–548.
[6] Iltis, X., Lefebvre, F., and Lemaignan, C., "Microstructural Study of Oxide Layers Formed on Zircaloy-4 in Autoclave and In Reactor, Part I," *Journal of Nuclear Materials,* Vol. 224, 1995, pp. 109–120.
[7] Iltis, X., Lefebvre, F., and Lemaignan, C., "Microstructural Study of Oxide Layers Formed on Zircaloy-4 In Autoclave and In Reactor, Part II," *Journal of Nuclear Materials,* Vol. 224, 1995, pp. 121–130.
[8] Pêcheur, D., Lefebvre, F., Motta, A. T., Lemaignan, C., and Wadier, J. F., "Precipitate Evolution in the Zircaloy-4 Oxide Layer," *Journal of Nuclear Materials,* Vol. 189, 1992, pp. 318–332.
[9] Livage, J., Doi, K., and Mazières, C., "Nature and Thermal Evolution of Amorphous Hydrated Zirconia," *Journal of the American Ceramic Society,* Vol. 51, 1968, pp. 349–353.
[10] David, G., Geschier, R., and Roy, C., "Etude de la croissance de l'oxyde sur le zirconium et le Zircaloy-2," *Journal of Nuclear Materials,* Vol. 38, 1971, pp. 329–339.
[11] Fu, Y., Evans, A. G., and Kriven, W. M., "Microcrack Nucleation in Ceramics Subject to a Phase Transformation," *Journal of the American Ceramic Society,* Vol. 67, 1984, pp. 626–630.
[12] Iltis, X., Viennot, M., David, D., Hertz, D., and Michel, H., "Special Kinetic and Microstructural Features Associated with Zircaloy-4 Oxidation Between 520 and 620 K in the Post-Discharge of an Argon-Oxygen Post-Discharge," *Journal of Nuclear Materials,* Vol. 209, 1994, pp. 180–190.
[13] Pêcheur, D., Picard, E., Billot, P., Thomazet, J., and Amanrich, H., "Microstructure of Oxide Films Formed During the Waterside Corrosion of Zircaloy-4 Cladding in a Lithiated Environment," this volume.
[14] Ojima, K. and Taneda, Y., "Electron Microscopic Study of the Oxidation of β-tin Irradiated by Electron Beam," *Journal of Materials Science,* Vol. 25, 1990, pp. 563–566.
[15] Smeltzer, W. W., "Diffusional Growth of Multiphase Scales and Subscales on Binary Alloys: A Review," *Materials Science and Engineering,* Vol. 87, 1987, pp. 35–43.
[16] Zhang, M. X. and Chang, Y. A., "Stability of an Alloy/Oxide Interface with Oxygen Being the Dominant Diffusing Species in the Oxide Scale," *Acta Metallurgica et Materialia,* Vol. 41, 1993, pp. 739–746.
[17] Kim, Y. S., Rheem, K. S., and Min, D. K., "Phenomenological Study of In-Reactor Corrosion of Zircaloy-4 in Pressurised Water Reactors," *Zirconium in the Nuclear Industry: Tenth International Symposium, ASTM STP 1245,* A. M. Garde and E. R. Bradley, Eds., American Society for Testing and Materials, West Conshohocken, PA, 1994, pp. 745–759.
[18] Blat, M., Kerrec, O., and Noël, D., "Detrimental Role of Hydrogen on the Corrosion Rate of Zirconium Alloys," this volume.
[19] Billot, P., Robin, J. C., Giordano, A., Peybernès, J., Thomazet, J., and Amanrich, H., "Experimental and Theoretical Studies of Parameters that Influence Corrosion of Zircaloy-4," *Zirconium in the Nu-*

clear Industry: Tenth International Symposium, ASTM STP 1245, A. M. Garde and E. R. Bradley, Eds., American Society for Testing and Materials, West Conshohocken, PA, 1994, pp. 351–377.

[20] Ramasubramanian, N. and Balakrishnan, P. V., "Aqueous Chemistry of Lithium Hydroxide and Boric Acid and Corrosion of Zircaloy-4 and Zr-2.5Nb Alloys," *Zirconium in the Nuclear Industry: Tenth International Symposium, ASTM STP 1245,* A. M. Garde and E. R. Bradley, Eds., American Society for Testing and Materials, West Conshohocken, PA, 1994, pp. 378–399.

[21] Cox, B., "What Is Wrong with Current Models for In-Reactor Corrosion," *IAEA Technical Committee Meeting on Fundamental Aspects of Corrosion on Zirconium Based Alloys in Water Reactor Environment, IAEA IWGFPT/34,* International Atomic Energy Agency, Vienna, 1990, pp. 167–173.

[22] Cox, B. and Wu, C., "Dissolution of Zirconium Oxide Films in 300°C LiOH," *Journal of Nuclear Materials,* Vol. 199, 1993, pp. 272–284.

[23] Barberis, P., "Zirconia Powders and Zircaloy Oxide Films: Tetragonal Phase Evolution During 400°C Autoclave Tests," *Journal of Nuclear Materials,* Vol. 226, 1995, pp. 34–43.

[24] Sabol, G. P., McDonald, S. G., and Airey, G. P., "Microstructure of the Oxide Films Formed on Zirconium-Based Alloys," *Zirconium in the Nuclear Industry: Second International Symposium, ASTM STP 551,* American Society for Testing and Materials, West Conshohocken, PA, 1974, pp. 435–448.

[25] Pêcheur, D., Lefebvre, F., Motta, A. T., Lemaignan, C., and Charquet, D., "Oxidation of Intermetallic Precipitates in Zircaloy-4: Impact of Irradiation," *Zirconium in the Nuclear Industry: Tenth International Symposium, ASTM STP 1245,* A. M. Garde and E. R. Bradley, Eds., American Society for Testing and Materials, West Conshohocken, PA, 1994, pp. 687–708.

[26] Iltis, X. and Lefebvre, F., to be published.

[27] Kofstad, P., "High Temperature Corrosion," Elsevier Applied Science Publishers LTD, England, 1988, pp. 342–388.

[28] Cheng, B. C., Kruger, R. M., and Adamson, R. B., "Corrosion Behavior of Irradiated Zircaloy," *Zirconium in the Nuclear Industry: Tenth International Symposium, ASTM STP 1245,* A. M. Garde and E. R. Bradley, Eds., American Society for Testing and Materials, West Conshohocken, PA, 1994, pp. 400–418.

[29] Van Der Linde, A., "Some Observations on In-Flux Corrosion of Zirconium Alloys Exposed in a Pressurised Water Loop," *Canadian Metallurgical Quarterly,* Vol. 11, 1972, pp. 7–19.

[30] Douglass, D. L., "Corrosion Mechanism of Zirconium and Its Alloys—III. Solute Distribution Between Corrosion Films and Zirconium Alloy Substrates," *Corrosion Science,* Vol. 5, 1965, pp. 347–360.

[31] Greenbank, J. C. and Harper, S., "Solute Distribution in Oxidized Zirconium Alloys," *Electrochemical Technology,* Vol. 4, 1966, pp. 142–148.

[32] Wadman, B., Lai, Z., Andrén, H. O., Nyström, A. L., Rudling, P., and Petterson, H., "Microstructure of Oxide Layers Formed During Autoclave Testing of Zirconium Alloys," *Zirconium in the Nuclear Industry: Tenth International Symposium, ASTM STP 1245,* A. M. Garde and E. R. Bradley, Eds., American Society for Testing and Materials, West Conshohocken, PA, 1994, pp. 579–598.

[33] Eloff, G. A., Greyling, C. J., and Viljoen, P. E., "The Role of Space Charge in the Oxidation of Zircaloy-4 Between 350 and 450°C in Air," *Journal of Nuclear Materials,* Vol. 199, 1993, pp. 285–288.

[34] Harding, J. H., "The Effect of Alloying Elements on Zircaloy Corrosion," *Journal of Nuclear Materials,* Vol. 202, 1993, pp. 216–221.

DISCUSSION

R. B. Adamson[1] *(written discussion)*—(1) Was there any indication that the Cr (from dissolved precipitates) was incorporated into the oxide inner layer? (2) Do you mean to imply that if Li had not been present, the corrosion acceleration during the fourth cycle would not have occurred? (3) *Comment:* We have observed accelerated corrosion at high fluence in conditions where neither hydrides nor Li could have been involved.

X. Iltis et al. (authors' closure)—(1) In the inner part of the oxide layers formed in-reactor, only iron is detected out of the amorphous precipitates. In the inner part of the oxide layer formed on the two-cycle sample re-oxidized in-autoclave for 210 days, chromium is occasionally detected very locally and always associated with iron. We think that this is correlated with the secondary precipitation observed in the metallic matrix in this sample. (2) and (3) No. In our discussion, we want to discuss the origin of the corrosion acceleration during the fourth cycle in relation to our microstructural observations. As part of this approach, we mean only to imply that the hypothesis of a lithium effect appears in good agreement with our TEM observations, while the hypothesis of an hydrogen effect does not seem reasonable. We cannot exclude, of course, any other hypothesis related to the reactor operating conditions, for example.

G. P. Sabol[2] *(written discussion)*—Your mechanism implies that Li is the cause of the in-reactor enhanced corrosion in PWRs. Earlier this morning, Bo Cheng indicated that high hydrogen (and hydrides) at the metal-oxide interface was the primary cause of accelerated in-reactor corrosion. Will you please comment on which factor, Li or hydrides, is most responsible for the accelerated corrosion?

X. Iltis et al. (authors' closure)—In our discussion, we try to examine the potential factors responsible for the corrosion acceleration during the fourth cycle. Two main factors have been highlighted in this congress: Li and hydrogen. Surprisingly, the hydride concentrations observed in the three-cycle and in the four-cycle samples are not drastically different (see Fig. 8 in our paper). The observed microstructures are not very different from the microstructures reported by D. Pêcheur et al. in the same session. Thus, we consider that the hypothesis of a lithium accelerating effect is more relevant for our samples.

S. K. Yagnik[3] *(written discussion)*—(1) Your in-pile samples, corroded at 320°C, and your out-of-pile previously irradiated samples, corroded at 400°C, exhibited larger (~100 nm) equiaxed grains in the former case than in the latter case (~10 nm). Yet it is believed that lower temperature will promote small crystallites to nucleate. This appears to be a contradiction that needs explanation. (2) Have you found conclusive proof of compact columnar growth and disoriented equiaxed alternatively through a thick oxide layer as proposed in your schematic mechanism?

X. Iltis et al. (authors' closure)—(1) In-pile, at 320°C, zirconia grains nucleate at the metal-oxide interface mainly in the form of randomly oriented nanocrystallites ($\phi \leq 5$ nm). A few micrometres away from this interface, the grains have undergone significant growth ($\phi \approx 100$ nm) as shown on Figs. 3 and 4 of our paper. Out-of-pile, previously irradiated samples oxidized at 400°C behave similarly to non-irradiated samples: zirconia grains nucleate at the metal-oxide interface mainly in the form of strongly oriented grains ($\phi \approx 20$ nm). A few micrometres away

---

[1] GE Nuclear Energy, Fuel Materials Technology, Vallecitos Nuclear Center, Pleasanton, CA 94566.
[2] Westinghouse Electric Corporation, Nuclear Manufacturing Division, Pittsburgh, PA 15230.
[3] EPRI, 3412 Hillview Avenue, Palo Alto, CA 94303.

from this interface, the grains have undergone only limited evolutions: they remain highly textured and their average diameter is below 50 nm. Thus, at the metal-oxide interface, a lower temperature effectively promotes small crystallites to nucleate. However, when getting away from this interface, the grain-size evolution observed in-pile appears not to be the result of a classical thermally activated grain growth process but, more likely, the result of an irradiation induced one. (2) The comparison between various microstructural studies carried out by TEM (published or performed in our laboratory) either on thin foils taken parallel to the oxide growth front or on cross-sectional ones shows that a highly textured oxide, when observed parallel to the growth front, is associated, in cross sections, with a columnar growth mode, while a randomly oriented oxide, when observed parallel to the growth front, is associated, in cross sections, with an equiaxed growth mode. In the four-cycle sample, several "parallel" thin foils have been taken in various locations through the oxide layer. Both types of microstructure (highly textured or randomly oriented) were found irrespective of the location. So, we propose a schematic mechanism leading to a stratified structure.

B. D. Warr,[1] *P. A. W. Van Der Heide,*[2] *and M. A. Maguire*[3]

Oxide Characteristics and Corrosion and Hydrogen Uptake in Zr-2.5 Nb CANDU Pressure Tubes

REFERENCE: Warr, B. D., Van Der Heide, P. A. W., and Maguire, M. A., "**Oxide Characteristics and Corrosion and Hydrogen Uptake in Zr-2.5 Nb CANDU Pressure Tubes,**" *Zirconium in the Nuclear Industry: Eleventh International Symposium, ASTM STP 1295,* E. R. Bradley and G. P. Sabol, Eds., American Society for Testing and Materials, 1996, pp. 265–291.

ABSTRACT: Oxides on removed pressure tubes from Pickering Unit 3 after 13.4 effective full power years (EFPY) have been examined to investigate the cause of variability in bulk alloy deuterium contents in outlet regions in order to improve predictions and minimize deuterium uptake in operating CANDU reactors. Secondary ion mass spectroscopy (SIMS) and electrochemical impedance spectrometry (EIS) were used for characterization with minimal sample preparation and modification. Two SIMS techniques were used for quantification: (1) the relative sensitivity factor (RSF) method, which requires a reference material and is subject to matrix effects as a result of variation in the secondary ion intensities of a species when different materials are sputtered; and (2) the SIMS infinite velocity (IV) method, which circumvents matrix effects by extrapolating all secondary ion intensity data to infinite velocity. A novel ^{13}C oxide dating technique was used to determine oxide growth kinetics and ensure that oxide spalling had not occurred in the regions examined.

Pressure tubes with high bulk alloy deuterium contents showed characteristics near the metal-oxide interface in inside surface oxides that were not present in oxides on tubes with low deuterium contents. In samples with high bulk alloy deuterium content, the inside surface corrosion rate, determined by the ^{13}C dating method, may have increased from ~ 0.3 to 1 μm/EFPY about five years before tube removal. A constant rate of corrosion was inferred in samples with low deuterium contents. The inner regions of inside surface oxides in tubes with high deuterium contents, corresponding to the faster growing oxide (up to ~ 5 μm from the interface) showed relatively higher porosity (inferred from the ^{2}H profile) and almost constant levels of lithium. These oxides also showed a low value of the electrical resistance term in one of the EIS responses, which has been interpreted as being due to the presence of a larger number of water penetration routes.

In order to investigate possible matrix effects in relatively thick inside surface oxides, through-oxide thickness concentration profiles for ^{2}H and ^{12}C, obtained by SIMS RSF and IV methods, were compared. Reasonable agreement was obtained between these methods for ^{2}H concentration profiles. However, evidence of a significant matrix effect for ^{12}C quantification, up to 5 μm from the inside surface metal-oxide interface, was found in tubes with high bulk alloy deuterium uptake. Further work is required to understand the reason for this matrix effect and its implications with respect to SIMS RSF quantification for other elements and analysis of excess ^{13}C profiles.

Outside surface oxides generally showed similar characteristics for all tubes. Very low constant rates of corrosion of 0.1 μm/EFPY were inferred from excess ^{13}C profiles. Apparent substoichiometry (O:Zr $\sim$ 1) was found by SIMS IV analysis in outside surface oxides on a tube with

---

[1] Principal scientist, Materials Unit, Ontario Hydro Technologies, Toronto, Ontario, Canada M8Z 5S4.
[2] Research scientist, Surface Science Western, University of Western Ontario, London, Ontario, Canada N6A 5B7.
[3] Research scientist, Atomic Energy of Canada, Chalk River Nuclear Laboratories, Chalk River, Ontario, Canada.

high deuterium content that may be related to breakdown of the efficacy of the oxide as a deuterium permeation barrier. Thus, although present results correlate deuterium uptake with inside surface corrosion effects, a contribution from the gas annulus cannot be ruled out.

KEYWORDS: hydrogen uptake, corrosion, characterization, Zr-2.5 Nb oxide, secondary ion mass spectrometry (SIMS), infinite velocity (IV), carbon dating, electrochemical impedance spectroscopy (EIS)

Zr-2.5 Nb pressure tubes in CANDU reactors operate at similar temperatures (~250 to 310°C), although in the absence of significant heat flux, compared to Zircaloy fuel cladding in boiling or pressurized water reactors, where relatively high rates of corrosion and hydrogen uptake are found. Extensive monitoring of pressure tubes in operating CANDU reactors has generally shown relatively low hydrogen (deuterium) concentrations in the bulk alloy and thin uniform oxides on inside and outside surfaces [1]. The most extensive data base on corrosion and deuterium uptake in pressure tubes from operating reactors has been developed from analysis of rings cut from ~400 Pickering Unit 3 pressure tubes during the large-scale fuel channel replacement (LSFCR) program after 13.4 effective full power years (EFPY)[4] of operation. Bulk alloy deuterium contents were generally highest in outlet regions where heat transport system (HTS) coolant temperatures were highest, although considerable variability in deuterium uptake was also found in these regions as shown in Fig. 1.

Despite the relatively low deuterium concentrations observed to date in CANDU pressure tubes, minimizing the uptake rate will help maximize tube longevity and is a high priority. Consequently, a program of material characterization has been ongoing to determine the causes of variability in deuterium uptake rates to improve the control and prediction of deuterium

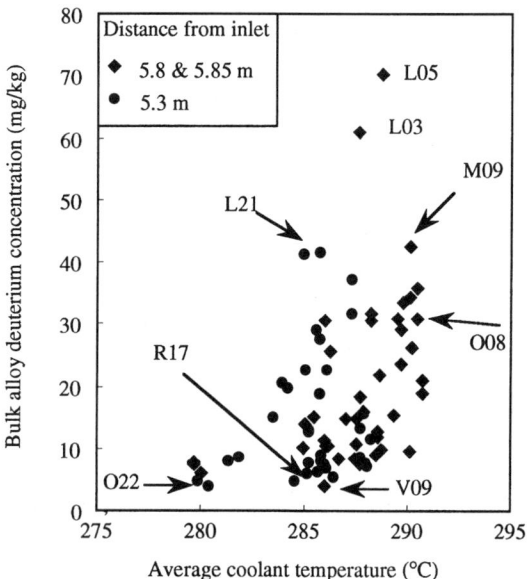

FIG. 1—*Relationship between bulk alloy deuterium concentration and average PHTS coolant temperature in outlet region of Pickering 3 pressure tubes after 13.4 EFPY.*

---

[4] EFPY is equivalent to 8760 effective full power hours (EFPY).

uptake in operating reactors. Because test-reactor and out-reactor experiments are limited in their ability to simulate real pressure tube operating environments, the best source of material for examining the origins of variability is the pressure tubes themselves. Samples from Pickering 3 pressure tubes have been used in the present characterization program. Since deuterium ingress into the bulk alloy during operation may arise from permeation of deuterium via both inside and outside surfaces [1], characterization of both of these near-surface regions is required. Considerable effort has been directed towards the development of optimum characterization techniques as part of the research program in this area supported by the CANDU Owners Group (COG). These techniques include secondary ion mass spectroscopy (SIMS) and electrochemical impedance spectroscopy (EIS), both of which can be applied to the surfaces of removed tubes with minimal sample preparation or modification, thereby allowing for comparisons of microcompositional and electrochemical characteristics of near-surface regions of a number of tubes with a range of deuterium uptake.

Experimental Procedure

Material Selection and Exposure Conditions

Small specimens (typically ~10 by 10 by 1 mm) were cut from rings taken from different axial locations of Pickering Unit 3 pressure tubes with high (L05 and L03) and low (O22 and V09) bulk alloy deuterium contents and from samples exposed out-reactor in high-temperature aqueous/gaseous environments. In addition to the rings from which samples were taken, other rings were selected for characterization by EIS and oxide thickness measurement by Fourier transform infrared spectroscopy (FTIR) [2]. Salient data on the pressure tubes selected (Fig. 1) are summarized in Table 1. Bulk alloy deuterium contents vary from <6 mg/kg in O22, V09, and R17 to between 31 to 70 mg/kg for Tubes O08, L21, M09, L05, and L03. Since diffusion of deuterium axially along the pressure tubes is relatively slow during reactor operation, these observed differences in bulk alloy deuterium content are due to differences in local rates of deuterium uptake. Carbon contents in these alloys, based on ingot analyses, varied from 143 to 220 mg/kg as shown in Table 1. Note that pairs of tubes—R17 and L21, and O22 and L03— were from the same ingots and hence show similar bulk alloy carbon contents. These rings were also taken from pressure tubes with different power levels, which is reflected in the fast neutron flux at a given axial location in Table 1. HTS coolant temperatures shown in Table 1

TABLE 1—*Summary of all samples examined.*

Tube Designation[a]	Distance from Inlet, m	Techniques Applied	AGS Flow, %[b]	Fast Flux, n/cm^2/s $\times 10^{12}$	Coolant Temperature, °C	D[c] mg/kg	Ingot C, mg/kg
022[e]	5.3	SIMS/EIS/FTIR	105	7.1	280	4	143
V09	5.85	SIMS/EIS/FTIR	115	3.7	286	4	170
R17[d]	5.3	EIS/FTIR	99	6.9	286	6	170
O08	5.85	EIS/FTIR	0	7.2	290	31	160
L21[d]	5.3	EIS/FTIR	97	6.8	285	41	170
M09	5.8	EIS/FTIR	0	8.6	290	42	183
L03[e]	5.8	SIMS/EIS/FTIR	0	6.1	288	61	143
L05	5.85	SIMS/EIS/FTIR	81	6.7	289	70	220

[a] Lattice position of channel in reactor.
[b] AGS flow rates vary from 0% (fully blocked) to 126% (least blocked).
[c] Bulk Alloy D content measured by hot-vacuum-extraction mass spectrometry (HVEMS).
[d,e] Tubes from same ingots.

represent average values based on location in the reactor and fueling history of the channel. Annulus gas system (AGS) flow, measured during LSFCR, is shown for each channel as a percentage of the estimated design flow. Oxygen additions were made to the Pickering 3 AGS about six months prior to tube removal. Blockage of gas annuli may have led to the development of local environments that were insufficiently oxidizing and may have prevented oxygen from entering some gas annuli, possibly leading to degradation of the normally protective outside surface oxide. All rings had spent about four years in the storage bay in light water at room temperature prior to sampling.

Secondary Ion Mass Spectroscopy (SIMS) Analysis

SIMS is a well-known technique for the characterization of the microchemistry of near-surface regions of removed pressure tubes [3–8] and fuel cladding material [9,10]. Dynamic SIMS allows for the identification and mapping in three dimensions of all elemental constituents present within any solid sample. Isotope identification is carried out by mass spectrometrically separating the ions emanating from the sample (secondary ions) produced by ion bombardment (primary ions). For these studies, the SIMS instrument, a Cameca IMS-3f, and technique have been described in a previous publication [11]. Under ideal conditions, the depth and lateral spatial resolutions of this instrument are ~ 1 nm and ~ 1 μm, respectively.

In this study, the primary (Cs^+) ion beam (accelerating voltage, 10 kV, and beam current, 250 to 1000 nA) is rastered over an area of 250 by 250 μm, and the ions analyzed are collected from within an area 60 μm in diameter centered in the middle of the rastered area for the depth profiles. Sputtering rates in the oxide and metal phases are ~ 0.01 and ~ 0.016 Å $\cdot$ nA^{-1} $\cdot$ s^{-1}, respectively. Images were collected from the central 150 μm in diameter area. Two SIMS techniques were used for quantification: the relative sensitivity factor (RSF) method, which uses a reference material, and the infinite velocity (IV) method, which does not require a reference standard.

The RSF method of quantification is a calibration procedure in which the secondary ion intensity emanating from a sample containing a known content of the isotope of interest in a reference material is multiplied by a calibration factor to reproduce the known (predefined) concentration profile [11]. This method was used to quantify most of the SIMS data on inside surface oxides. However, it is known that the secondary ion intensity of a species may vary when different materials are sputtered (excluding those arising from concentration variations), a phenomenon commonly referred to as the matrix effect. Matrix effects and the ionization mechanism still elude complete theoretical descriptions.

The IV method is a newly developed procedure for SIMS quantification that circumvents the matrix effect by extrapolating all secondary ion intensity data (after correction for instrument transmission and sputter yield effects) to infinite velocity. Since matrix effects have been shown to decrease with increasing ion emission velocity [12,13], concentration data can be calculated for specific elements regardless of the matrix being sputtered through. To perform quantification via this method, the electron multiplier sensitivity to mass must also be defined (this is not necessary for the RSF method). This was accomplished by noting the correction (calibration) factor required to reproduce the known concentrations of the implant reference materials as shown in Fig. 2. A more detailed discussion of the IV procedure can be found elsewhere [12,13].

The IV method of SIMS quantification was applied largely to outside surfaces with relatively thin oxides since they were found to be immune to electrical charging (a prerequisite of the IV method) once gold coated. By comparison, thicker oxides on inside surfaces were found to charge by as much as 50 V, but such charging was not found to significantly influence the RSF results because of the high (>200 V) offset voltages used. The SIMS IV technique was applied

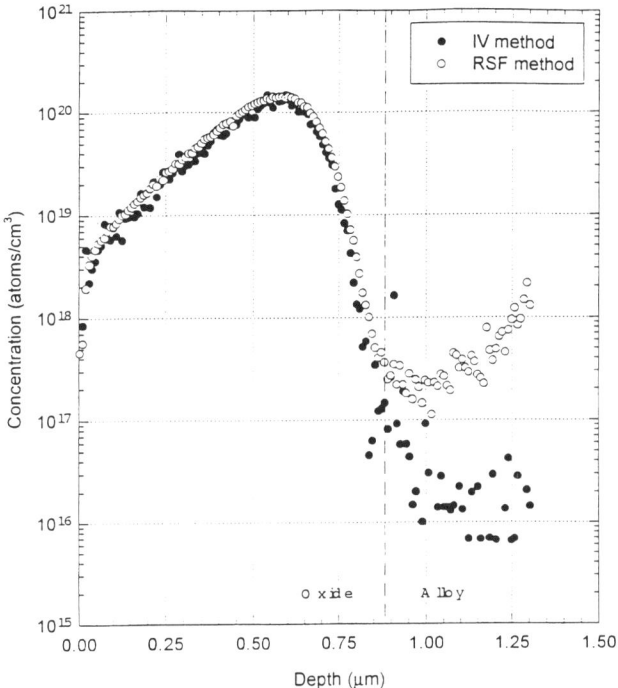

FIG. 2—*Comparison of ^2H quantitative depth profiles for ion-implanted standard material by SIMS RSF and IV methods.*

to thicker inside surface oxides close to the metal-oxide interface, where charging (measured during the analysis) was found to be acceptably low. The conditions (offsets) used in the depth profile analysis by IV and RSF methods are shown in Table 2. The offsets used in the image acquisition mode were fixed at 50 V since significant offsets both reduce the lateral resolution and distort the image in microscope-type instruments.

Electrochemical Impedance Spectroscopy (EIS)

This technique is used to observe and model individual dielectric responses or relaxations associated with the surface oxide film. A section of the pressure tube is masked to define a 3-

TABLE 2—*Conditions used in the SIMS depth profile analyses to quantify the secondary ion intensities.*

Element	Mass	Offset (IV)	Offset (RSF)
Deuterium	2	40, 55, 70, 85 V	200 V
Lithium	7	40, 55, 70, 85 V	200 V
Carbon	12	40, 55, 70, 85 V	400 V
Carbon	13	40, 55, 70, 85 V	400 V
Oxygen	18	70, 85, 100, 115 V	400 V
Zirconium	90	50, 75, 100, 125 V	200 V
Zirconium Oxide	106	NA	400 V

cm² area and covered with electrolyte using vacuum impregnation. A three-electrode electrochemical cell is used so that a constant d-c potential can be maintained while a 50-mV a-c stimulus is imposed across the oxide film. The EIS spectra are generated by step scanning the a-c frequency, seven points per decade, from 100 MHz to 10 mHz. The resulting cell current and phase shift at each frequency is measured using a frequency response analyzer. The response is interpreted using an equivalent circuit analysis of a Maxwell-Wagner "effective medium" model [14]. The oxide film is considered to be a continuous phase (ZrO_2) with a number of features, such as porosity and grain boundaries, having quite different dielectric responses. Each response is represented in its simplest form as an RQ element or generalized Debye response [15]. Multiple responses are modeled using a parallel combination of individual RQ elements [14].

The constant phase element, Q, is a mathematical way to express the observed broadening of the dielectric response [16]. It is described using two parameters: the capacitive admittance, B, and a parameter, n, related to the amount of broadening of the response as compared to that of an ideal capacitor. Broadening is interpreted as a measure of size or surface uniformity of the described feature. A value for n of 1 represents an ideal uniform feature, e.g., a parallel plate capacitor. As the value of n becomes less than 1, the characteristic variation in the observed capacitance increases. This model adequately describes the physical attributes of the oxide film, yields realistic values for the electrical parameters, and can be easily adjusted to fit the number of observed responses. A schematic of the features of the oxide film and the corresponding equivalent circuit model is shown in Fig. 3. The oxide film is comprised of two

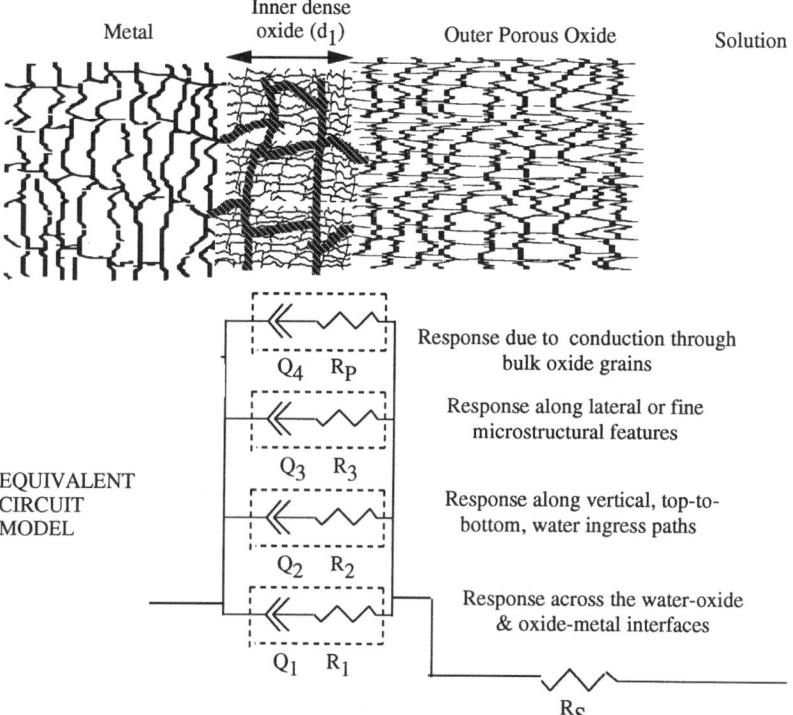

FIG. 3—*Structure of the oxide film as related to the equivalent circuit model: a parallel combination of Debye responses.*

regions: an outer oxide with interconnected porosity, and an inner, less porous oxide adjacent to the metal-oxide interface. This representation of the observed structure has been widely reported for oxide films on Zircaloys [17–19]. The structure of the oxide next to the interface is based on work on thin oxides (~1 μm) on Zr-2.5 Nb pressure tube material [5,6,20–22].

EIS is relatively insensitive to the outer porous oxide because it contains a sufficient amount of water that it cannot be distinguished from the solution conductivity. EIS analysis yields two types of information on the inner oxide: the thickness of the oxide and information on contained porosity, and/or grain boundary features. These can be ranked in order of the series resistance. In the model shown in Fig. 3, the first response arises from the dielectric response across the water-oxide and oxide-metal interfaces. The thickness of the inner oxide layer, d_1, is calculated from Q_1 and the dielectric constant ($\epsilon = 35$) of the dense oxide. The dielectric constant has been adjusted to account for water residing in the oxide and hence is higher than the value previously quoted [23]. The second response refers to the coarse interconnected porosity that forms a path for water from the outer porous oxide to the metal-oxide interface. The rapid change in this response after contact with water provides evidence of the water permeation. The third response refers to finer porosity that, for example, may be associated with oxide grain boundaries. R_P is associated with conductance via oxide grain interiors [24].

Results

The surface condition of the samples was quite variable, and there was evidence for damage incurred during tube and sample removal. Samples from outside surfaces showed grinding marks from fabrication, and those from inside surfaces showed topographical features typical of sandblasted surfaces and axial scratches at the bottom of the tube from fueling that were avoided during analyses (Figs. 4a and 4b). A representative metallographic cross section of oxides on inside surfaces showing an apparently unspalled layer is shown in Fig. 4c.

Oxide Thickness

Oxide thickness measured by FTIR on inside and outside surfaces at the top (12 o'clock) and bottom (6 o'clock) of all rings is shown in Table 3. With the exception of Tubes M09 and O08, measured inside surface oxides are <13 μm. As noted, there is a greater uncertainty in measurements of thicker oxides, and in some cases FTIR spectra were difficult to obtain. In general, oxides are thicker on the bottom of the tube, which is slightly hotter than the top during operation—a similar trend has been found for other Pickering 3 removed tubes [25]. Although M09 and O08 also show the highest calculated coolant temperature, in general there is no simple relationship between oxide thickness and temperature. Oxide thickness found by metallographic examination of cross sections of pressure tubes is generally in good agreement with the results of FTIR measurements. In the present study, with the exception of the ring from Tube M09, outside surface oxides are consistently <3 μm.

Through-Thickness Concentration Profiles

Concentration profiles for ^{2}H, ^{7}Li, and ^{12}C, obtained from RSF quantification of inside surface oxides from different spots on samples from the top of Tubes L05 and L03 (high bulk alloy deuterium content), are shown in Fig. 5a. The location of the metal-oxide interface, given by the start of the dropoff of the ^{18}O or ^{106}ZrO intensity, is shown in this and subsequent figures. Concentrations and through-thickness profiles for each isotope showed the same behavior in samples taken from the top and bottom of each ring provided SIMS profiling was carried out at locations well away from the fueling scratches observed at the bottom of some rings. Con-

FIG. 4—*Typical oxide morphology of inside surfaces of removed tubes showing:* (a) *fueling scratches,* (b) *surface condition typical of sandblast material, and* (c) *cross section.*

TABLE 3—*Oxide thickness measured by FTIR.*

Tube	D, mg/kg	Coolant Temperature, °C	Oxide Thickness, μm			
			Inside Bottom	Inside Top	Outside Bottom	Outside Top
O22	4	280	8.4	9	2.4	2
V09	4	286	6	6.2	1.6	1.3
R17	6	286	12.5	8.1	2.4	1.7
O08	31	290	21.5^a	14.8	2.5	1.9
L21	41	285	11.7	8.5	2.4	1.5
M09	42	290	8.6^b	17.2^a	4.2	2.7
L03	61	288	11.5^b	10.6^b	2.3	1.8
L05	70	289	12.1^b	10.2	2.4	1.7

^{a} For thicker oxides (>15μm), there is greater uncertainty in thickness measurements.
^{b} These spectra were difficult to obtain.

centration profiles obtained from samples from the top of Tubes V09 and O22, with low bulk alloy deuterium content, are shown in Fig. 5b. As shown in these figures, up to ~9 μm from the surface in both sets of samples, through-oxide-thickness profiles for each element are alike. Under optimal conditions, differences in concentration of ±~10% are found between successive SIMS analyses performed on thin conducting films.

Samples from the tubes with high bulk alloy deuterium contents showed thicker inside surface oxides based on the start of the dropoff of the ^{18}O or ^{106}ZrO intensity as shown in these figures. Inside surface oxides on these samples also contained an inner region, commencing ~4 to 5 μm from the start of the metal-oxide interface, with higher ^{12}C concentrations and relatively constant concentrations of ^{2}H and ^{7}Li. This region was not present in oxides from samples from tubes with low deuterium contents. In relatively thin outside surface oxides, through-oxide-thickness concentration profiles were obtained by IV for ^{2}H, ^{12}C, ^{18}O, and ^{90}Zr. As for inside surface oxides, these profiles were largely reproducible for different spots in each sample.

Oxide Stoichiometry

Through-thickness concentration profiles for ^{18}O and ^{90}Zr found by the SIMS IV method were used to determine the O:Zr ratio; for a fully stoichiometric oxide, an O:Zr ratio of 2 should be found. The ^{18}O and ^{90}Zr concentrations are divided by their natural isotope abundance values such that they represent the true concentrations of these elements. As expected for inside surface oxides, where sample charging was not a factor, O:Zr ratios were consistent with fully stoichiometric oxides.

Outside surfaces of all tubes showed stoichiometric oxides with the exception of a sample taken from L05. In Fig. 6, representative ^{90}Zr and ^{18}O concentration profiles are shown for outside surface oxides in Tubes L03 and L05. The start of the metal-oxide interface given by the dropoff in the ^{106}ZrO intensity is also shown in this figure. The L03 sample shows the expected ^{90}Zr and ^{18}O concentration profiles, which are consistent with a fully stoichiometric oxide also found in other samples. The sample from L05, on the other hand, shows a gradually decreasing O:Zr ratio as the metal-oxide interface is approached, with pronounced substoichiometry at the interface—this result was reproducible in all five spots analyzed. These results could not be attributed to effects of sample charging since it is observed that charging dimin-

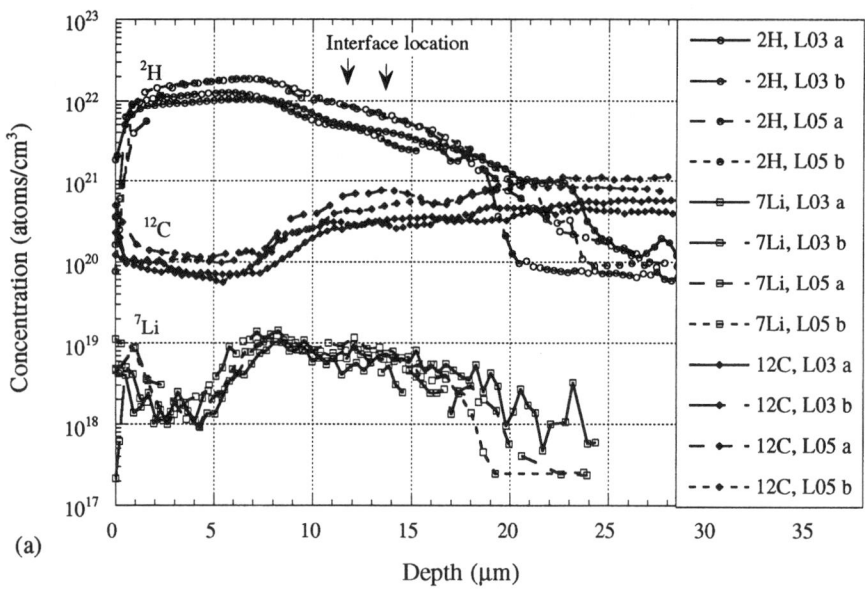

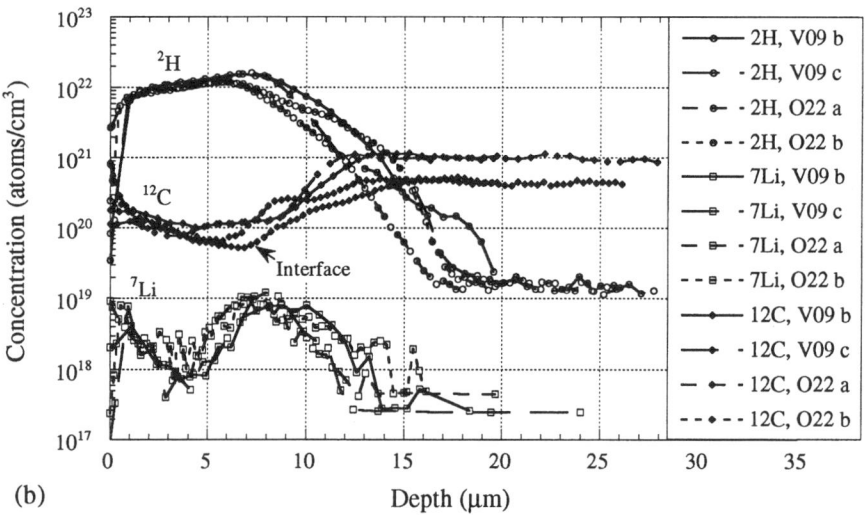

FIG. 5—*Quantitative depth profiles by SIMS RSF of ^{2}H, ^{7}Li, and ^{12}C in inside surface oxides from the top of tubes with:* (a) *high (L03 and L05) and* (b) *low (O22 and V09) bulk alloy deuterium uptake.*

ishes as the metal-oxide interface is approached and the O:Zr ratio increases; this is the reverse of the trend observed.

Zr-2.5 Nb samples in the prefilmed state (400°C steam, 24 h) were exposed to high-temperature vacuum environments (10^{-7} Torr, 500°C, ~70 h) in order to determine whether similar results to those above could be obtained. Samples exposed to the same conditions have pre-

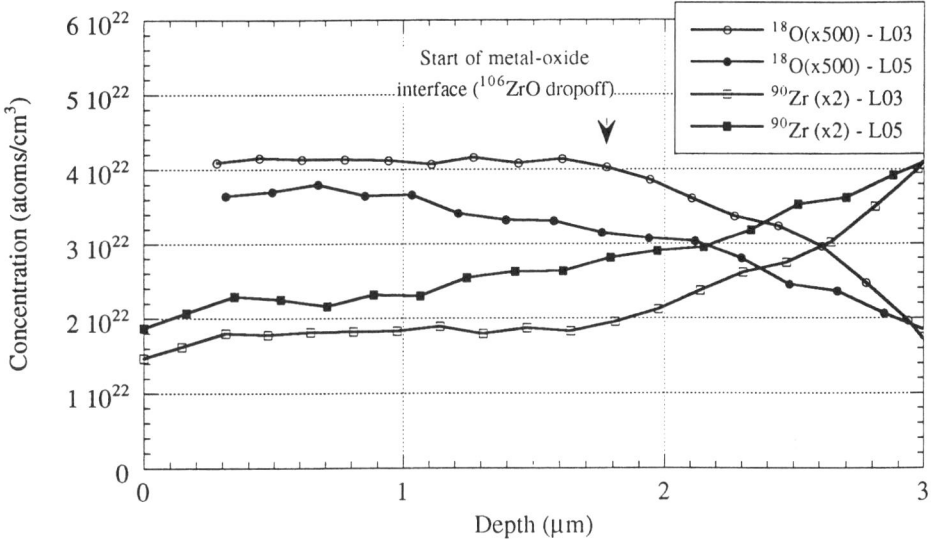

FIG. 6—*Representative quantitative SIMS IV depth profiles for ^{18}O (X500) and ^{90}Zr (X2) in outside surface oxides from the bottom of tubes with high bulk alloy deuterium contents (L03 and L05).*

viously shown substoichiometry by X-ray photoelectron spectroscopy (XPS) and increased bulk alloy uptake [26]. SIMS analyses of these samples, however, showed no evidence of substoichiometry, with profiles similar to those found for the sample from Tube L03 in Fig. 6.

^2H Concentration and Distribution

Typical ^2H concentrations obtained by SIMS RSF analyses of inside surface oxides from high and low bulk alloy deuterium content tubes are shown in Fig. 7a in which both scales are linear. ^2H concentration in all samples examined is approximately constant (or increases slightly) in the outer regions of inside surface oxides, with maximum concentrations of 1 to 2 × 10^{22} atoms/cm^3 (600 to 1200 mg/kg by weight) at ~7 μm from the surface. In samples from tubes with low uptake, the ^2H concentration drops off sharply to the metal-oxide interface at ~8 to 10 μm from the surface. In samples from high uptake tubes, the ^2H concentration also decreases ~7 to 8 μm from the surface to ~50% of its maximum value, then decreases gradually between ~10 μm from the surface and the metal-oxide interface at 11 to 14 μm from the surface. ^2H concentration gradients at the metal-oxide interface (~0.5 to 5 × 10^{21} atoms/cm^3/μm) are up to X10 lower than in tubes with low bulk alloy deuterium uptake. A representative ^2H concentration profile from Zr-2.5 Nb samples following ~1100 days exposure at 310°C (pH ~ 10.5) is also shown in Fig. 7a. ^2H contents in these oxides are ~X10 lower with a maximum concentration of ~2 × 10^{21} atoms/cm^3.

^2H concentrations found by IV in these inside surface oxides (where sample charging was minimal) show similar maximum levels in the oxide, but lower concentrations at the interface and in the metal compared to the RSF method. The RSF method used in generating the concentration data, however, is no longer applicable over the metal region since the ionization probability of ^2H is enhanced over metallic regions (a matrix effect). Although ^2H concentration gradients inferred near the interface using data obtained by the SIMS RSF method may be

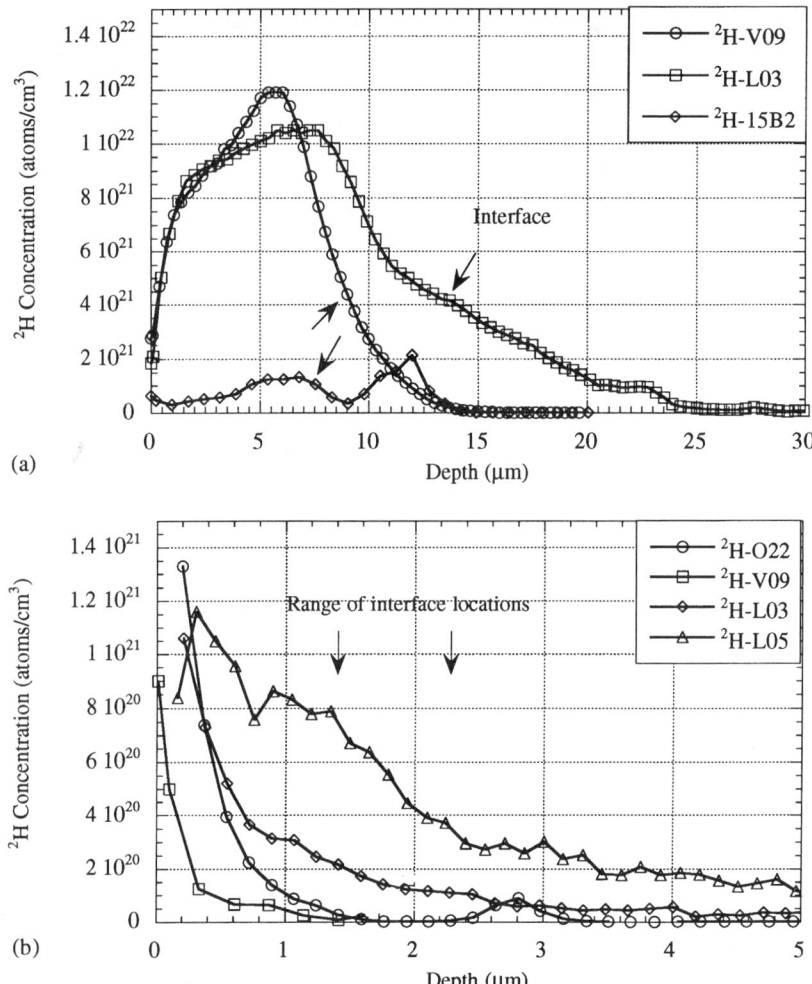

FIG. 7—*Quantitative ^2H depth profiles for:* (a) *inside of L05 and V09 and autoclave sample (310°C, ~1100 days) by SIMS RSF and* (b) *outside of L03, L05, V09, and O22 by SIMS IV (arrows mark start of metal-oxide interface).*

affected (i.e., lower than actual) by these matrix effects, relative differences found between high and low uptake samples should still be valid.

Concentration profiles for ^2H through outside surface oxides are significantly different from inside surface profiles and have much lower ^2H concentrations at the metal-oxide interface. ^2H concentration profiles in outside surface oxides on samples from tubes with high and low bulk alloy deuterium contents (L03, L05, O22, and V09) are shown in Fig. 7b. ^2H concentrations of $\sim 10^{21}$ atoms/cm^3 found near the surface of these (and other samples examined) are similar to those found previously in other pressure tubes samples [3–8]. ^2H contents within the oxide and at the interface are higher in the case of the samples from tubes with high bulk alloy deuterium contents. ^2H concentrations measured in the bulk alloy by IV range from ~ 0.7 to 2 $\times 10^{20}$ atoms/cm^3 in samples from tubes with high bulk alloy concentrations (63 to 70 mg/kg

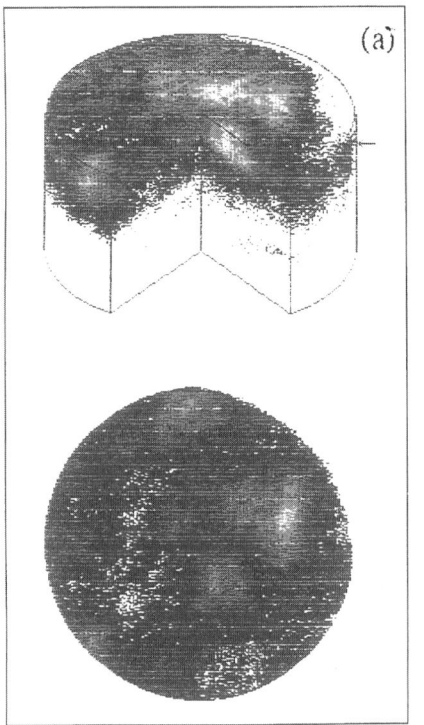 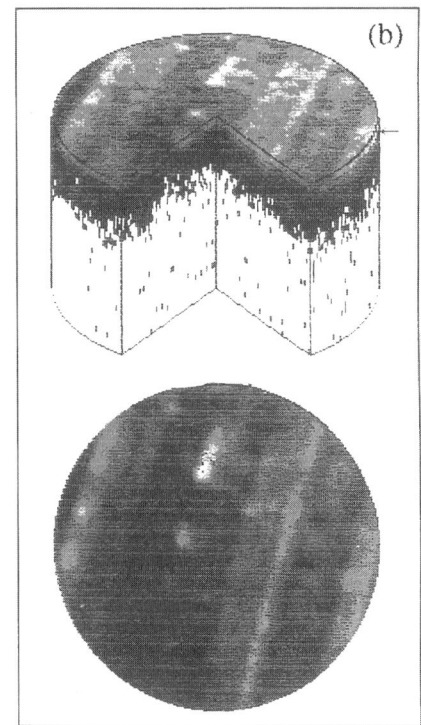

FIG. 8—*Three-dimensional representation of distribution of ^2H in* (a) *inside, and* (b) *outside surface oxides in Tube L05 with high bulk alloy deuterium uptake. Notes: surface topography not accounted for; diameter of cylinder = 150 µm; plan view shown is at arrow; highest intensity = white; lowest intensity = black.*

^{2}H) to ~2 to 4 × 10^{18} atoms/cm^3 in samples from low uptake tubes (4 mg/kg ^{2}H). The SIMS bulk alloy ^{2}H data are in reasonable agreement with those measured by analyses of bulk alloy samples by hot-vacuum-extraction mass spectrometry (HVEMS) shown in Table 1.

Using the SIMS scanning mode, significant heterogeneities in ^{2}H distribution within the oxide were observed. Typical three-dimensional representations of the distribution of ^{2}H in inside and outside surface oxides in Tube L05 with high bulk alloy deuterium uptake are shown in Figs. 8a and 8b, respectively. Surface topography is not accounted for in these figures. Also, the diameter of the cylinder shown is 150 µm, whereas the total depth shown is ~50 and ~20 µm for Figs. 8a and 8b, respectively. Inside surface oxides show local regions of relatively high ^{2}H intensity (concentration) close to the metal-oxide interface. In outside surface oxides, striations are created from alternately high and low ^{2}H intensity that may be related to the alignment of grinding marks from the fabrication process. No correlation was found between these features and bulk alloy deuterium content of the sample.

Li in Inside Surface Oxides

Representative through-thickness ^{7}Li profiles obtained by SIMS RSF are shown in Fig. 9 for samples from tubes with high (L05) and low (V09) deuterium contents and from the sample exposed out-reactor, for which ^{2}H profiles are shown in Fig. 7a. Since the ^{7}Li secondary ion

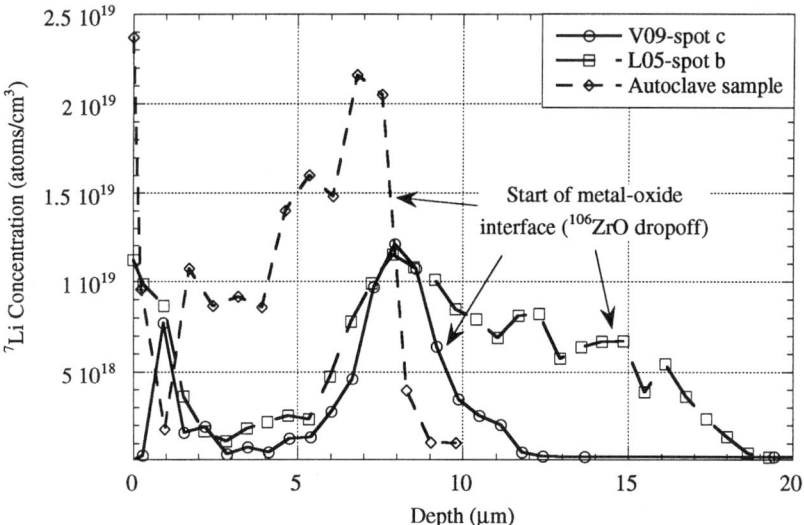

FIG. 9—*Representative quantitative ^7Li depth profiles by SIMS RSF for inside surface oxides from removed Tubes L05 and V09 and Zr-2.5 Nb samples exposed in autoclaves at 310°C for 1100 days.*

intensity was not sufficiently high to allow use of the SIMS IV method, matrix effects are not accounted for in these results. ^7Li concentrations of up to 1 to 2 x 10^{19} atoms/cm^3 (~10 mg/kg) are observed at ~7 μm from the surface in all samples examined in the present study. Samples from tubes with low deuterium uptake show a single peak in the ^7Li profile. In the case of samples from tubes with high deuterium uptake, the Li concentration is relatively constant in the inner oxide (8 to 14 μm from the surface) before dropping off after the metal-oxide interface. In samples exposed out-reactor, maximum ^7Li concentrations are ~X2 higher within the oxide and decrease sharply just before the interface. The increase in ^7Li content near the surface of some samples is probably due to the presence of Li-containing deposits. In outer regions of oxides (~1 to 4 μm from the surface) in samples from removed tubes, ^7Li concentrations (~2 x 10^{18} atoms/cm^3) are ~X10 lower than the peak value closer to the interface. A similar although less pronounced trend is observed in the unirradiated sample.

Distribution of Carbon Isotopes

The distribution of ^{12}C and ^{13}C isotopes can be used to study corrosion kinetics and to investigate possible relationships between carbon distribution and bulk alloy deuterium uptake. Typical ^{12}C concentration profiles obtained by IV in outside surface oxides from tubes with high and low bulk alloy deuterium contents are shown in Fig. 10a. All samples showed similar profiles regardless of bulk alloy deuterium uptake. Concentrations of ^{12}C are higher at ~3 to 4 μm from the oxide surface in all samples, probably as a result of surface contamination. Therefore, the minimum observed ^{12}C concentration in the oxide (~3 to 8 x 10^{19} atoms/cm^3) in all spots examined most closely reflects the original ^{12}C present in the bulk alloy. This is in agreement with the expected ^{12}C concentration in the oxide given the level in the bulk alloy and allowing for volume expansion during oxide formation (Pilling-Bedworth ratio = 1.56). As expected, ^{12}C content increases at the start of the metal-oxide interface to 1 to 3 x 10^{20} atoms/cm^3 (~350 to 1000 mg/kg) in the metal; this is somewhat higher than that found in

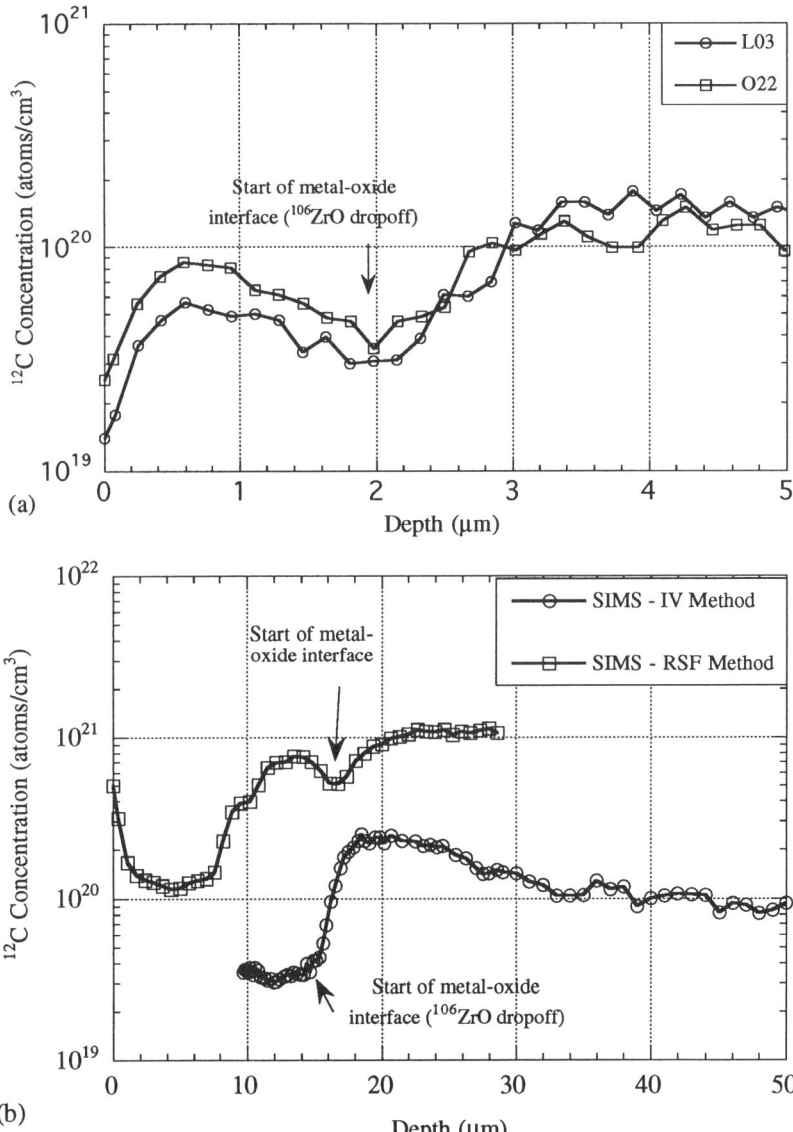

FIG. 10—*Representative quantitative ^{12}C depth profiles for:* (a) *outside surface oxides by SIMS IV and* (b) *inside surface oxides in tubes with high bulk alloy deuterium content by SIMS RSF and IV methods.*

offcuts (rings cut from the ends of pressure tubes during installation) by conventional analyses (143 to 220 mg/kg), shown in Table 1.

Typical ^{12}C concentration profiles obtained by SIMS RSF and IV techniques in inside surface oxides on tubes with high bulk alloy deuterium uptake are shown in Fig. 10b. As discussed previously in relation to Fig. 5a, in all samples from tubes with high deuterium content, the

^{12}C concentration found by RSF increases unexpectedly before the metal-oxide interface is reached (at ~8 μm from the surface) as shown in Fig. 10b. SIMS IV results, however, also shown in this figure do not indicate an increase in the ^{12}C content within the oxide, but at the interface. The increase in concentration is greater, however, than that expected based on the Pilling-Bedworth ratio. These IV results indicate that the apparent ^{12}C increase within the oxide found by RSF is due to a matrix effect in the oxide near the interface. Since the RSF method is based on a reference material using a typical steam-formed oxide, these results appear to indicate the presence of an oxide next to the interface that is significantly different from that used for the reference material.

The IV analyses also show a ~X10 increase in ^{12}C between the oxide and metal, which is greater than that found for outside surface oxides. In the typical profile shown in Fig. 10b, the ^{12}C concentration is highest in the metal just below the interface and continues to decrease in the metal to ~30% of the interface concentration ~30 μm from the interface into the metal. This concentration in the metal of 8 x 10^{19} is equivalent to ~250 mg/kg, which is close to the bulk alloy (ingot) analysis for this tube (L05) of 220 mg/kg shown in Table 1. Note that the SIMS IV technique is not affected by differences in ionization efficiencies between the metal and oxide phases.

Inferring Corrosion Kinetics by Carbon Dating

Excess ^{13}C concentrations at a given depth, d, were calculated from the ^{12}C and ^{13}C concentrations found by RSF in inside and outside surface oxides as follows:

$$[\text{excess }^{13}\text{C}]_d = [^{13}\text{C}]_d - 0.011*[^{12}\text{C}]_d$$

Representative excess ^{13}C concentration depth profiles for outside surface oxides of two tubes with high (L05) and low (O22) bulk alloy deuterium uptake are shown in Fig. 11a. Similar representative profiles for inside surface oxides of the two tubes with low bulk alloy deuterium uptake (V09 and O22) that were exposed to different fast neutron fluxes are shown in Fig. 11b. A typical excess ^{13}C concentration depth profile for inside surface oxides from a tube with high bulk alloy deuterium uptake (L05) is shown in Fig. 11c. The profiles show a number of general features that are typical of all excess ^{13}C profiles determined to date for irradiated pressure tubes:

1. A near surface region exhibiting either a rapidly rising (outside oxides) or slowly falling (inside oxides) concentration of excess ^{13}C.
2. A relatively flat portion corresponding to an approximately constant excess ^{13}C concentration.
3. A portion of steadily declining excess ^{13}C concentration starting at a depth of about 1 to 3 μm and extending to the oxide-metal interface.

These features may be explained as follows. Excess ^{13}C (i.e., ^{13}C above the level expected from the indigenous carbon impurity) is produced by the fast neutron ^{16}O(n,alpha)^{13}C reaction. For a pressure tube oxide, this production occurs for that part of the exposure *after* the oxide has formed. Hence, only the prefilmed oxide "sees" the full operating exposure (EFPY) of the tube and exhibits a maximum ^{13}C production. The ^{13}C is formed initially with a recoil energy of about 0.5 MeV. Upon thermalization, the "hot" ^{13}C atoms are trapped at oxide lattice sites within their recoil range (~0.5 μm) and are assumed to undergo very little thermal diffusion thereafter; hence, the accumulated ^{13}C traces the local oxide growth as a function of time and depth [27].

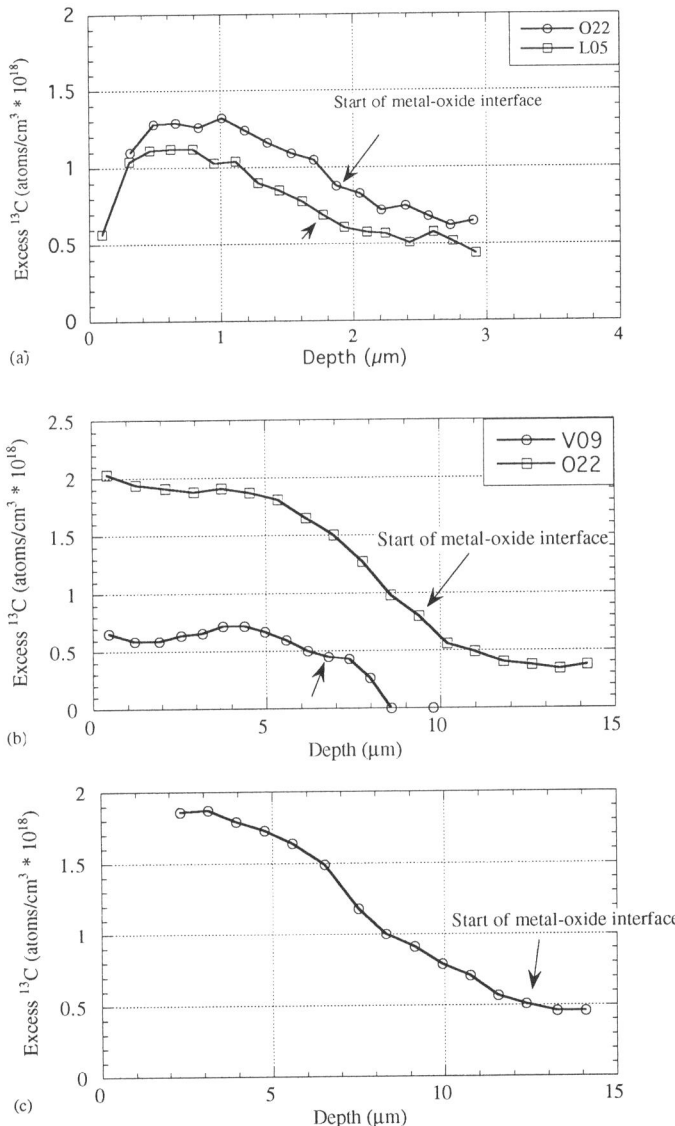

FIG. 11—*Excess ^{13}C through-thickness concentration profiles in:* (a) *outside surface oxides from tubes with high (L05) and low (O22) deuterium uptake, and inside surface oxides from tubes with* (b) *low and* (c) *high uptake.*

At the oxide surface, recoil into or from the external environment (N_2 annulus gas or HTS coolant D_2O) adjacent to the pressure tube occurs. This results in loss of ^{13}C from the near-surface region of the outside surface oxide and implantation of ^{13}C into the near-surface region of the inside surface oxide. Beyond this region, the flat portion of the profile corresponds to ^{13}C production in the pre-filmed oxide with no loss or gain to or from the external environment. The absolute ^{13}C concentration differs from tube to tube and along the length of a given tube because of the local variations in the fast neutron flux throughout the reactor core. The fast fluxes (>1 MeV) to the O22 and V09 samples were 0.71 x 10^{13} and 0.37 x 10^{13} n/cm²/s, respectively, i.e., the fluxes were approximately in the ratio 2:1. The observed excess ^{13}C concentrations shown in the flat regions of the profiles for these samples (see Fig. 11b) are 1.9 x 10^{18} (O22) and 0.75 x 10^{18} (V09) atoms/cm³, which is slightly higher than the expected 2:1 ratio.

The ^{13}C concentration depth profiles beyond about 1 μm for the outside surface oxides (Fig. 11a) and about 4 μm for the inside surface oxides (Figs. 11b and 11c) correspond to regions of in-reactor growth of these oxides. The slopes of the profiles are an inverse function of oxidation rate. The profiles may therefore be converted into kinetic plots shown in Fig. 12 by utilizing the fact that the excess ^{13}C concentration at a given depth, d, is proportional to the length of time oxide was present at that depth, t_d, i.e.,

$$[^{13}C]_d = k \cdot t_d \text{, and } [^{13}C]_{max} = k \cdot t_{EFPY}$$

where k is a constant, and t_{EFPY} is the total time of exposure of the tube. Hence the time, t, for the oxide to grow to depth, d, is equal to $t_{EFPY} - t_d$ with

$$t = 1/k \, ([^{13}C]_{max} - [^{13}C]_d)$$

Clearly $[^{13}C]_{max}$ defines the "zero" of a time axis from which a kinetic plot may be constructed. This procedure was used to obtain the kinetic plots shown in Figs. 12a-c from the excess ^{13}C concentration depth profiles in Figs. 11a-c.

As shown in Fig. 12a, outside surfaces of pressure tubes appear to show a constant low oxidation rate of ~0.1 μm/EFPY regardless of bulk alloy deuterium content. Inside surface oxides on tubes with low bulk alloy uptake (O22) show remarkably low and constant corrosion rates of ~0.3 μm/EFPY for the majority of the operating life (Fig. 12b). In samples from tubes with high bulk alloy uptake, however, the inferred rate of inside surface corrosion increases after ~9 EFPY (Fig. 12c). Corrosion rates of ~0.3 μm/EFPY and ~1 μm/EFPY are found before and after the transition to a higher rate, respectively.

Impedance Spectroscopy of Inside Surface Oxides

Spectra of inside surface oxides from the top (12 o'clock) location yielded the most consistent results. Values for the capacitive admittance, B, varied from 10 to 500 nF. Values for the broadening factor, n, ranged from ~0.8 to 0.9 for the first two responses (n_1 and n_2), and from 0.35 to 0.55 for n_3. The thickness of the inner dense oxide region, d_1, ranged from 0.2 to 1 μm.

The two resistance terms, R_2 and R_3, were strongly correlated (>90%) to bulk alloy deuterium uptake as shown in Fig. 13. For all tubes with high bulk alloy deuterium content, R_2 is consistently low; however, a significant unexplained variation in R_2 is found in the two tubes with low bulk alloy deuterium content (O22 and V09). With reference to Fig. 3, a low value of R_2 associated with high bulk alloy deuterium uptake is interpreted as indicating the presence of a larger size or number of porous channels extending from the inner water-oxide interface toward the oxide-metal interface. As shown in Fig. 13, R_3 is directly proportional to bulk alloy deuterium uptake, being >X5 higher in the case of the tubes with high deuterium contents. Since the R_3 term generally shows much higher resistance than R_2, this is associated with either finer porosity or grain boundaries within the dense oxide.

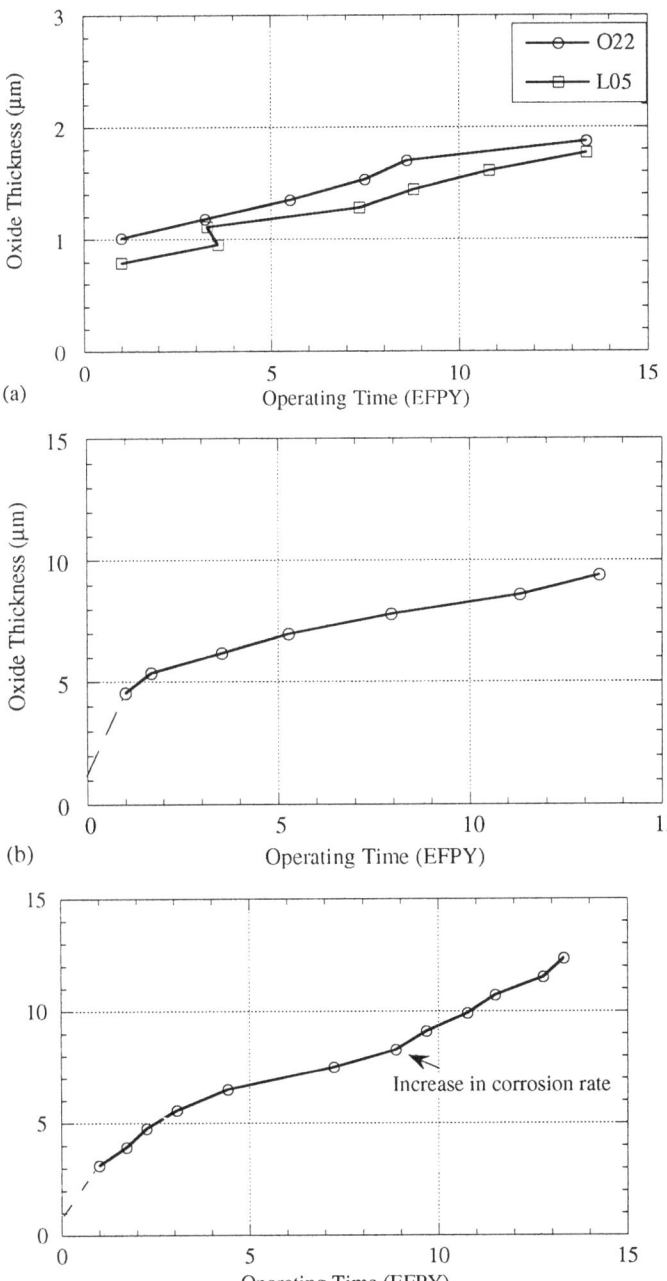

FIG. 12—*Inferred corrosion kinetics for:* (a) *outside surfaces from tubes with high (L05) and low (O22) bulk alloy deuterium uptake, and inside surfaces from tubes with* (b) *low and* (c) *high deuterium uptake.*

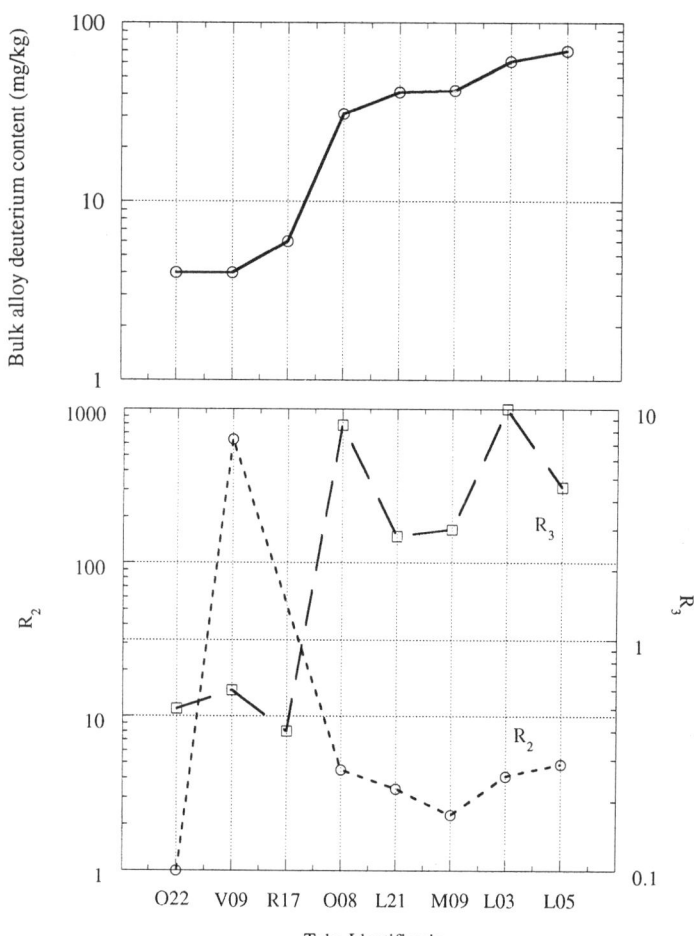

FIG. 13—*Resistance of dense inner layer (next to metal-oxide interface) of inside surface oxides from tubes with low (O22, V09, R17) and high (O08, L21, M09, L03, L05) bulk alloy deuterium content.*

Discussion

The techniques applied here have been able to identify differences in near-surface composition and electrochemical response between pressure tubes with high and low bulk alloy deuterium uptake with minimal sample preparation/alteration. The SIMS technique has been successfully applied using both the RSF method, employing ion-implanted standard material, and the IV method, which is unaffected by the difference in secondary ion yields between the oxide and metal. The SIMS IV technique has proven useful in providing more accurate ^{2}H and ^{12}C concentrations free from matrix effects in thin oxides and in the bulk alloy on outside surfaces, and near the metal-oxide interface and in the bulk alloy on inside surfaces. Analysis of excess ^{13}C through-thickness profiles provides a novel technique for inferring corrosion kinetics in these oxides. EIS results may be related to those obtained by SIMS.

As stated previously, the objective of this work is to determine the routes (via inside or outside surfaces) and mechanism of deuterium uptake, and thereby account for the observed variability in bulk alloy deuterium content from tube to tube. With respect to identification of the predominant uptake route, based on these results for the limited number of samples examined in this study, deuterium uptake arising from the inside surface corrosion reaction is likely to be significantly higher in tubes with high bulk alloy deuterium content.

The mechanism of high deuterium uptake appears to be related to increased porosity near the metal-oxide interface. Samples from these tubes showed different characteristics near the metal-oxide interface in inside surface oxides, which are consistent with the expectation of higher percentage theoretical deuterium uptake present. In samples with high bulk alloy deuterium content, the inside surface corrosion rate, determined by ^{13}C dating techniques, may have increased from ~0.3 to 1 μm/EFPY about five years before tube removal. Also, the inner regions of these oxides with higher corrosion rates (up to ~5 μm from the interface) showed relatively higher porosity (inferred from the ^{2}H profile) and almost constant levels of lithium. A low value of one of the electrical resistance terms derived from EIS responses was also found for all samples with bulk alloy deuterium contents > 30 mg/kg; this has been interpreted as being due to the presence of a larger number of water penetration routes that is likely related to increased deuterium uptake. These features were not generally found in inside surface oxides on tubes with low deuterium contents.

Excess ^{13}C through-thickness profiles have provided an effective means of determining corrosion kinetics based on RSF data for ^{12}C and ^{13}C in the present work. In all cases, these profiles showed minimal spalling to have occurred. Very low rates of corrosion, of 0.1 μm and up to 1 μm/EFPY on outside and inside surfaces, respectively, were inferred from these profiles. In samples with high bulk alloy deuterium content, the inside surface corrosion rate may have increased from ~0.3 to 1 μm/EFPY about five years before tube removal. Further work is required to determine the influence of the significant matrix effect for ^{12}C quantification, in the same region of oxide where this increased rate is inferred, with respect to analysis of excess ^{13}C profiles. A constant rate of corrosion was inferred on inside surfaces of samples with low deuterium contents. Corrosion kinetics were generally typical of those expected in zirconium alloys, with cubic to parabolic behavior in the pre-transition regime [28]. Compared to typical rates found in Zircaloy fuel cladding material for example [29], inferred rates of inside surface corrosion are extremely low for all samples examined.

As a result of these low corrosion rates, high percentage theoretical deuterium uptakes would be required if inside surface corrosion alone were to account for all deuterium that entered the bulk alloy in tubes with high deuterium concentrations. For tubes with low deuterium uptake (V09 and O22), average percentage theoretical deuterium uptake from inside surface corrosion is ~3%, assuming negligible uptake from the outside surfaces. This is of the same order as percentage uptakes found in out-reactor aqueous exposures of Zr-2.5 Nb samples at 310°C [5] for which ^{2}H and ^{7}Li through-thickness profiles are shown in Figs. 7 and 9, respectively. Assuming a conservative value of theoretical uptake of ~10% for Tube L05 prior to the inferred increase in corrosion rate, bulk alloy deuterium contents would not exceed ~10 mg/kg in these tubes in up to ~9 EFPY operation. In order for all the deuterium that enters the bulk alloy to arise from the inside surface corrosion reaction, therefore, high percentage theoretical uptakes (~100%) are required over the last 4.5 EFPY operation (~4 μm oxide growth) for the bulk alloy deuterium content to reach the observed value (70 mg/kg).

As a result of the high values required for percentage uptake and the relatively low rates of corrosion inferred, it is not clear whether uptake from this source is the sole cause of high deuterium uptake. In Zircaloys (fuel cladding and pressure tubes), increasing rates of corrosion and associated increases in percentage of theoretical hydrogen uptake (in some cases up to 100%) are observed both in-reactor and out-reactor [29,30]. Although there is some evidence that rates of corrosion of Zr-2.5 Nb may increase during in- and out-reactor exposures [31,32], there is presently no evidence that percentage hydrogen uptake also increases. Also, in the present work, high percentage uptake cannot be explained on the basis of spalling in Zr-2.5 Nb and its effects on measured oxide thickness because ^{13}C dating shows oxide loss to be negligible.

The increased porosity (based on ^{2}H concentration gradients in Fig. 7a), inferred near the interface in inside surface oxides on tubes with high deuterium contents, is consistent with the observed increase in corrosion rates. This may also be the cause of high percentage deuterium uptake in these tubes. In Zircaloys, increasing rates of corrosion (and hydrogen uptake) arise from (structural) breakdown of the oxide film. This breakdown leads to increased porosity and reduction in the effective thickness of the non-porous oxide barrier near the interface [17,19].

There are several experimental data that suggest the ^{2}H distribution in these oxides is related to oxide porosity. ^{2}H present in these oxides may readily be exchanged with ^{1}H at elevated temperature [5,7]. Also, since through-thickness ^{2}H gradients remain unchanged after long operating times with minimal oxide growth, ^{2}H probably does not enter bulk oxide grains but remains as physisorbed OD groups on the surface of pores and microcracks in these oxides [5]. Although other techniques are not available to provide independent measurements of porosity distribution, qualitative agreement has been found via electron microscopy [6]. Thus, for a given environment, i.e., equivalent ^{2}H partial pressure and exposure temperature, ^{2}H concentration profiles are probably related to the density of active surface sites in the oxide for adsorption of deuteroxyl groups. A steeper ^{2}H concentration gradient at the interface, found in tubes with low bulk alloy deuterium contents, is consistent with a less porous and more effective barrier layer oxide being present close to the interface.

The relatively constant lithium concentration found in the oxide within ~5 μm of the metal-oxide interface in tubes with high deuterium contents may also be related to increased porosity, increased rates of corrosion, and relatively high deuterium uptake rates. In PWR Zircaloy fuel cladding, where thicker oxides are found (~100 μm), deleterious effects of Li on corrosion and hydrogen uptake are associated with higher levels of Li in the oxide (~100 mg/kg) [9,10]. The detrimental effect of Li has been associated with increasing oxide porosity [29]. In tubes with low deuterium uptake (O22 and V09), the ^{7}Li content peaks at ~8 μm from the oxide surface and decreases more sharply before the metal-oxide interface, implying an oxide next to the interface that contains relatively low Li contents. Observed maximum levels of ~10^{19} atoms/cm^3 in all inside surface oxides in the present work correspond to ~10 mg/kg in ZrO_2, which is in agreement with previous results [7]. It should be noted, however, that all Li analyses to date were obtained using the SIMS RSF technique that may be affected by a changing oxide matrix near the interface.

Isotope exchange experiments suggest that lithium is largely present within pores and on pore surfaces in these oxides [7]. Although the ^{7}Li concentration at the metal-oxide interface is similar for both high and low uptake samples, the difference in concentration gradient is indicative of the presence of higher Li contents in the oxide adjacent to the metal in tubes with high deuterium contents. Although not shown here, the ^{7}Li:^{6}Li ratio found throughout these inside surface oxides reflects the expected isotopic ratio. Similar results have been found in analyses of other removed tubes and explained on the basis of continual exchange of Li from the coolant throughout the oxide during operation [7]. Relatively low Li contents near the surface of all inside surface oxides on removed tubes may represent the "equilibrium" concentration in pores assuming access and exchange of coolant within these relatively porous outer regions of the oxides; confirmation of this would require measurements of pore density. The fact that Li concentrations are higher near the interface shows that, in the absence of a significant matrix effect on Li quantification, Li is concentrating in this region; however, the concentration mechanism is not understood.

The results of EIS analysis are also consistent with increased porosity next to the metal-oxide interface in tubes with high deuterium content as discussed above. In analyzing the EIS spectra, a clear distinction can be made between the outer oxide with interconnected pores and a relatively non-porous inner oxide near the metal-oxide interface. The estimated thickness (d_1) of this region of the oxide varies from 0.2 to 1 μm. No correlation was found between the value of d_1 and bulk alloy deuterium content. EIS results suggest that this region of the oxide present on all tubes with high uptake contains a large number of water penetration routes, which is likely to result in higher percentage uptake. The reason for the relatively high resistance of the R_3 term found on samples with high uptake is less clear. It may be explained on the basis of a mechanism by which deuterium (0.1 nm atomic radius) generated by the reduction reaction remains trapped in these fine pores near the oxide-metal interface for longer periods, possibly increasing the chance of its entering the bulk alloy.

The presence of a different oxide next to the interface in inside surface oxides on high uptake tubes (L03 and L05) is also suggested by the differences in ^{12}C profiles obtained by different SIMS methods. In the absence of structural information at present on this inner oxide, it can only be inferred that these differences are associated with the other features discussed above. The decrease in carbon concentration into the metal found by the SIMS IV method on inside surfaces suggests carbon segregation ahead of the metal-oxide interface during corrosion. No correlation has been found between the extent of segregation and bulk alloy deuterium content, although further analyses are required to determine if this is one of the factors influencing bulk alloy deuterium uptake.

The evidence obtained in the present work, for a significant (higher than that expected from corrosion) contribution to bulk alloy deuterium uptake from the annulus gas, is limited to the observed apparent substoichiometry in one sample from Tube L05. In this sample, the O:Zr ratio is close to 1 next to the metal-oxide interface. It has been proposed that at high temperatures and insufficiently oxidizing environments during operation, a pressure tube oxide may lose its efficacy as a deuterium permeation barrier [1]. Oxide dissolution into the metal and oxide substoichiometry, where metallic regions exist within the oxide, has been observed in Zr-2.5 Nb samples following exposures at high temperatures and high vacuum conditions [20]. In the present work, however, substoichiometry has not been observed in samples exposed to vacuum at high temperature. Also, since a constant rate of oxidation was inferred for this sample (L05), substoichiometry is unlikely to result from insufficient oxidant being present (in the annulus gas) during operation. These observations imply that if a degradation process is present in gas annuli of operating reactors, it may be quite different from that previously hypothesized, and further work is required to confirm this result and elucidate the associated microstructure and degradation mechanism. Oxygen additions are now made to the gas annuli of operating reactors to maintain the protective nature of outside oxides and ensure that deuterium uptake from this source is minimized.

Conclusions

1. SIMS and EIS techniques have been able to detect differences in composition and electrochemical response in oxides on removed pressure tubes from Pickering Unit 3 after 13.4 effective full power years (EFPY). These results have implications regarding the cause and mechanism of variability in bulk alloy deuterium contents.

2. Tubes with high bulk alloy deuterium contents showed different characteristics near the metal-oxide interface in inside surface oxides, which were not present in oxides on tubes with low deuterium contents. Based on these limited results, deuterium uptake arising from the inside surface corrosion reaction is likely to be significantly higher in tubes with high bulk alloy deuterium content. The mechanism of high deuterium uptake appears to be related to the generation of increased porosity near the metal-oxide interface.

3. Excess ^{13}C through-thickness profiles have provided an effective means of determining corrosion kinetics and the extent of spalling. By this technique, it has been shown that these pressure tubes studied experienced very low rates of (pretransition) corrosion of ~0.1 μm and up to 0.3 μm/EFPY on outside and inside surfaces, respectively. In samples with high bulk alloy deuterium content, however, the inside surface corrosion rate may have increased to ~1 μm/EFPY about five EFPY years before tube removal.

4. The inner regions of inside surface oxides in tubes with high deuterium contents, corresponding to the faster growing oxide (up to ~5 μm from the interface), showed relatively higher porosity (inferred from the ^{2}H profile) and almost constant levels of lithium. These oxides also consistently showed a low value of one of the electrical resistance terms derived from EIS responses, which has been interpreted as being due to the presence of a larger number of water penetration routes.

5. The SIMS infinite velocity (IV) method, a new standardless technique, has provided evidence of a matrix effect for ^{12}C quantification (by the RSF technique) up to 5 μm from the inside surface metal-oxide interface in tubes with high bulk alloy deuterium uptake. Further work is required to understand the reason for this matrix effect and its implications with respect to SIMS RSF quantification of other elements, ^{13}C oxide dating, and corrosion and deuterium uptake.

6. Outside surface oxides generally showed similar characteristics for all tubes with very low constant rates of corrosion of 0.1 μm/EFPY inferred from excess ^{13}C profiles. Because some evidence for substoichiometry was found in these oxides on a tube with high deuterium content, a contribution from the gas annulus cannot be ruled out.

Acknowledgments

Funding for the above work was provided by Ontario Hydro and the CANDU Owners Group (COG). The authors acknowledge extensive contributions of Dr. F. Greening to both the technique development for measurements of the excess ^{13}C profiles and to their interpretation in this paper. We also thank him for providing significant constructive criticism of the remainder of the paper. Other important contributions to this work are gratefully acknowledged from staff at Ontario Hydro Technologies (OHT) and Atomic Energy of Canada Limited, Chalk River Nuclear Laboratories (AECL-CRL), and the University of Western Ontario, Surface Science Western (SSW), including Drs. M. Leger, V. Urbanic, N. Ramasubramanian, M. Elmoselhi, Y.-P. Lin, S. McIntyre, and Mr. G. Mount. The assistance of the Ontario Hydro Nuclear staff was required for the operating information. Dr. P. Chan (AECL-CRL) provided the sample exposed to high-temperature vacuum.

References

[1] Warr, B. D., Ramasubramanian, N., Elmoselhi, M. B., Greening, F. R., Lin, Y-P., and Lichtenberger, P. C., "Hydrogen Ingress in Pressure Tubes," *Ontario Hydro Research Review*, No. 8, August 1993.
[2] Ramasubramanian, N., Ling, V. C., Shankula, M. H., and Chenier, R. J., "Hot-Cell Examination of Corrosion Films on Zirconium Alloys Using Infrared Spectroscopy, Impedance, and Anodization Techniques," *Proceedings*, International Atomic Energy Agency, Technical Committee Meeting on Post-Irradiation Examination Techniques for Water Reactor Fuel, Portland, OR, 11-14 Sept. 1990, IAEA, Vienna, Austria, pp. 124-135.
[3] McIntyre, N. S., Weisener, C. G., Davidson, R. D., Brennenstuhl, A., and Warr, B. D., "Analysis of Zr-Nb Fuel Channel Surfaces for Hydrogen and Other Elements Using Secondary Ion Mass Spectroscopy (SIMS)," *Journal of Nuclear Materials*, Vol. 178, 1991, pp. 80-92.
[4] Ramasubramanian, N. and Balakrishnan, P. V., "Aqueous Chemistry of Lithium Hydroxide and Boric Acid and Corrosion of Zircaloy-4 and Zr-2.5 wt% Nb Alloys," *Zirconium in the Nuclear*

Industry: Tenth International Symposium, ASTM STP 1245, A. M. Garde and E. R. Bradley, Eds., American Society for Testing and Materials, West Conshohocken, PA, 1994, pp. 378–397.

[5] Elmoselhi, M. B., Warr, B. D., and McIntyre, N. S., "A Study of the Hydrogen Uptake Mechanism in Zirconium Alloys," *Zirconium in the Nuclear Industry: Tenth International Symposium, ASTM STP 1245*, A. M. Garde and E. R. Bradley, Eds., American Society for Testing and Materials, West Conshohocken, PA, 1994, pp. 62–77.

[6] Warr, B. D., Elmoselhi, M. B., Brennenstuhl, A. B., Lichtenberger, P. C., McIntyre, N. S., Newcomb, S. B., "Oxide Characteristics and Their Relationship to Hydrogen Uptake in Zirconium Alloys," *Zirconium in the Nuclear Industry: Ninth International Symposium, ASTM STP 1132*, C. M. Euken and A. M. Garde, Eds., American Society for Testing and Materials, West Conshohocken, PA, 1991, pp. 740–755.

[7] Ramasubramanian, N., "Lithium Uptake and the Corrosion of Zirconium Alloys in Aqueous Lithium Hydroxide Solutions," *Zirconium in the Nuclear Industry: Ninth International Symposium, ASTM STP 1132*, C. M. Euken and A. M. Garde, Eds., American Society for Testing and Materials, West Conshohocken, PA, 1991, pp. 613–625.

[8] McIntyre, N. S., Weisener, C. G., Davidson, R. D., Lennard, W. N., Massoumi, G. R., Mitchell, I., Brennenstuhl, A. B., and Warr, B. D., "Analysis of Zr-Nb Pressure Tube Surfaces for Hydrogen Using Secondary Ion Mass Spectrometry (SIMS)," *Surface and Interface Analysis*, Vol. 15, 1990, pp. 591–597.

[9] Pecheur, D., Giordano, A., Picard, E., Billot, P., and Tomazet, J., "Effect of Elevated Lithium on the Waterside Corrosion of Zircaloy-4: Experimental and Predictive Studies," *Proceedings*, International Atomic Energy Agency, Technical Committee Meeting on Influence of Water Chemistry on Fuel Cladding Behaviour, Rez, Czech Republic, 4–8 Oct. 1993, IAEA, Vienna, Austria.

[10] Billot, P., Robin, J.-C., Giordano, A., Peybernes, J., Thomazet, J., and Amanrich, H., "Experimental and Theoretical Studies of Parameters that Influence Corrosion of Zircaloy-4," *Zirconium in the Nuclear Industry: Tenth International Symposium, ASTM STP 1245*, A. M. Garde and E. R. Bradley, Eds., American Society for Testing and Materials, West Conshohocken, PA, 1994, pp. 351–375.

[11] Benninghoven, A., Rudenauer, F. G., and Werner, H. W., *Secondary Ion Mass Spectrometry—Basic Concepts, Instrumental Aspects, Applications and Trends*, John Wiley & Sons, New York, 1987.

[12] Van der Heide, P. A. W., Zhang, M., Mount, G. R., and McIntyre, N. S., "The Infinite Velocity Method, a New Method for SIMS Quantification," *Surface and Interface Analysis*, Vol. 21, No. 11, 1994, pp. 747–757.

[13] Van der Heide, P. A. W., "Variations in the Value of the Characteristic Velocity of the Zr^- Secondary Ion Emissions Across the Oxide/Metal Interface," *Nuclear Instrumentation and Methods in Physics Research B*, Vol. 93, 1994, pp. 421–426.

[14] MacDonald, J. R., *Impedance Spectroscopy: Emphasizing Solid Materials*, John Wiley & Sons, New York, 1987, p. 346.

[15] Jonsher, A., *Dielectric Relaxation in Solids*, Chelsea Dielectrics Press, London, 1983, p. 380.

[16] Liu, C. and Kaplan, T., "The Theory of AC Response of Rough Interfaces," *Fractals in Physics*, E. T. L. Pietronero, Ed., Elsevier Science Publishers, New York, 1986, pp. 383–389.

[17] Sabol, G. P., McDonald, S. G., and Airey, G. P., "Microstructure of the Oxide Films Formed on Zirconium-Based Alloys," *Zirconium in Nuclear Applications*, J. H. Schemel and H. S. Rosenbaum, Eds., American Society for Testing and Materials, West Conshohocken, PA, 1974, pp. 435–448.

[18] Urquhart, A. W. and Vermilyea, D. A., "Characterization of Zircaloy Oxidation Films," *Zirconium in Nuclear Applications*, J. H. Schemel and H. S. Rosenbaum, Eds., American Society for Testing and Materials, West Conshohocken, PA, 1972, pp. 463–478.

[19] Wadman, B., Lai, Z., Andrén, H., Nystrom, A., Rudling, P., and Pettersson, H., "Microstructure of Oxide Layers Formed during Autoclave Testing of Zirconium Alloys," *Zirconium in the Nuclear Industry: Tenth International Symposium, ASTM STP 1245*, A. M. Garde and E. R. Bradley, Eds., American Society for Testing and Materials, West Conshohocken, PA, 1994, pp. 579–596.

[20] Newcomb, S. B., Warr, B. D., and Stobbs, W. M, "The TEM Characterization of Oxidation-Reduction Processes in Zr-Nb Alloys," *EMAG, Institute of Physics Conference Series*, F. J. Humphreys, Ed., No. 119, pp. 221–224.

[21] Warr, B. D., Rasile, E. M., Brennenstuhl, A. M., Elmoselhi, M. B., McIntyre, N. S., Newcomb, S. B., and Stobbs, W. M., "The Microscopical and Compositional Analyses of Oxides on Zr-2.5 Nb in Relation to Their Role as Permeation Barriers to Deuterium Uptake in Reactor Applications," *Proceedings*, Microscopy of Oxidation: Proceedings of the International Conference held at the University of Cambridge, U.K., M. J. Bennett and G. W. Lorimer, Eds., The Institute of Metals, London, UK, 1991, pp. 292–298.

[22] Lin, Y. P., unpublished results.

[23] Rosecrans, P. M., "Applications of Alternating-Current Impedance Measurements to Characterize

Zirconium Alloy Oxidation Film," *Zirconium in the Nuclear Industry: Sixth International Symposium, ASTM STP 824*, D. G. Franklin and R. B. Adamson, Eds., American Society for Testing and Materials, West Conshohocken, PA, 1984, pp. 531–553.

[24] Bonanos, N., Steele, B. C. H., and Butler, E. P., "Applications of Impedance Spectroscopy: Interpretation of the Impedance Spectra of Materials," *Impedance Spectroscopy: Emphasizing Solid Materials and Systems*, J. R. Macdonald, Ed., John Wiley & Sons, New York, 1987, pp. 215–238.

[25] Ramasubramanian, N., unpublished results.

[26] Chan, P. K., Irving, K. G., and Urbanic, V. F., "Studies of Microstructural Effect on Zr-2.5 Nb Oxide Degradation and Deuterium Ingress by XPS," *Proceedings*, 34th Annual Conference of Metallurgists, Metallurgical Society of Canadian Institute of Mining and Metallurgy, Vancouver, B.C., Canada, 22–24 Aug. 1995.

[27] Greening, F. and Ramasubramanian, N., "Post-Irradiation Investigations of Corrosion and Deuterium Pickup by Zr-2.5 wt% Nb Alloy Pressure Tubes: ^{14}C, ^{13}C and ^{11}B Tracers in Outside Surface Oxides," Letter to the Editor, *Journal of Nuclear Materials*, Vol. 226, 1995, pp. 263–271.

[28] Cox, B., "Oxidation of Zirconium and Its Alloys," *Advances in Corrosion Science and Technology*, Plenum Press, NY, Vol. 5, 1976, p. 173.

[29] International Atomic Energy Agency, "Corrosion of Zirconium Alloys in Nuclear Power Plants," IAEA-TECDOC-684, January 1993, IAEA, Vienna, Austria.

[30] Urbanic, V. F., Cox, B., and Field, G. J., "Long-Term Corrosion and Deuterium Uptake in CANDU-PHW Pressure Tubes," *Zirconium in the Nuclear Industry: Seventh International Symposium, ASTM STP 939*, R. B. Adamson and L. G. P. Van Swam, Eds., American Society for Testing and Materials, West Conshohocken, PA, 1987, pp. 189–205.

[31] Sabol, G. P., Schoenberger, G., and Balfour, M. G., "Improved Fuel Cladding," *Proceedings*, International Atomic Energy Agency, Technical Committee Meeting on Materials for Advanced Water Cooled Reactors, Plzen, Czechoslovakia, 14-17 May 1991, IAEA TECDOC-665, September 1992, IAEA, Vienna, Austria.

[32] Warr, B. D., unpublished results.

DISCUSSION

S. K. Yagnik[1] (written discussion)—The in-reactor D-pickup data show a tremendous scatter when plotted against exposure time under apparently identical conditions. Could you comment on what parameters control the D-pickup significantly. How do your SIMS and carbon dating analyses help in understanding the basic mechanism of D-pickup in pressure tubes?

B. D. Warr et al. (authors' closure)—(1) The reasons for the observed scatter in deuterium uptake between pressure tubes in the same unit is not fully understood and is the subject of a more extensive research program of which the work reported here forms a part. Microstructural and microchemical examination of the pressure tube alloy in the as-received condition and following in-reactor exposures are being performed to develop a fuller understanding of the relationship between tube material and corrosion and deuterium uptake; (2) Inferences about the basic mechanisms of deuterium uptake can be drawn from SIMS analyses and carbon dating and the EIS analyses performed in the present work. For example, increases in corrosion rates determined from carbon dating together with the evidence for increased oxide porosity in the same regions of the oxide may imply that the two observations are related and may be the cause of higher percentage deuterium uptake.

[1] Electric Power Research Institute, Palo Alto, CA.

Corrosion and Hydrogen Effects

H. Richard Peters[1] *and John L. Harlow*[1]

The Effect of the Trace Impurity Uranium on PWR Aqueous Corrosion of Zircaloy-4

REFERENCE: Peters, H. R. and Harlow, J. L., "**The Effect of the Trace Impurity Uranium on PWR Aqueous Corrosion of Zircaloy-4,**" *Zirconium in the Nuclear Industry: Eleventh International Symposium, ASTM STP 1295*, E. R. Bradley and G. P. Sabol, Eds., American Society for Testing and Materials, 1996, pp. 295–318.

ABSTRACT: Commercial Zircaloy-4 specifications control natural uranium to <3.5 ppm by weight. Long-term in-pile corrosion tests were conducted on Zircaloy-4 coupons containing 0.1, 2.1, 10.6, and 24.1 ppm by weight of natural uranium. The objective of this testing was to determine the effect of the trace impurity uranium on the in-pile corrosion of Zircaloy-4. The tests were conducted in the advanced test reactor in Idaho at test temperatures of 355, 332, 310, and 274°C and at fast-neutron (>1 MeV) fluxes of $>1.0 \times 10^{14}$ n/cm²-s. Limited testing was also conducted at lower fluxes.

The pre-transition corrosion rates were observed to increase with increasing uranium content and flux level. The transition times were doubled at 2.1 ppm and quadrupled at 10.6 and 24.1 ppm uranium compared to the 0.1 ppm control ingot. Transition time differences were observed to diminish with decreasing fast-neutron flux. Post-transition corrosion rates also increased with uranium levels. The net effect on long-term corrosion is a reduction (X2 to X5) in corrosion film thickness for Zircaloy-4 containing uranium levels at the specification limit (3.5 ppm) compared to the 0.1 ppm control ingot. Based on the data presented, significant differences in PWR corrosion will develop as a result of varying natural uranium levels in ingots.

KEYWORDS: Zircaloy-4, in-pile, PWR, corrosion, impurities, uranium, thorium, pretransition, transitory, steady-state, post-transition

Commercial fuel utilization has increased to the point where Zircaloy-4 corrosion and hydrogen uptake have reached limiting values. Corrosion thicknesses are seen to vary by a factor of up to X4 at high utilizations [1]. Attempts to model the corrosion behavior accounting for metal oxide interface temperatures and the in-pile environment reduce the observed variation to ≈X2 [2]. A significant amount of research is being conducted to understand this variability with the objective of optimizing Zircaloy-4 in-pile corrosion. The reduction of tin within specification levels (1.2 to 1.7 w/o) has been found to affect the corrosion of Zircaloy-4 beneficially [3–5]. Thermomechanical processing has also been found to affect corrosion [5,6].

In-pile corrosion testing of Zircaloy-4 ingots in the advanced test reactor (ATR) has shown that in a pressurized water reactor (PWR) environment at least part of this variability is a result of the uranium and thorium levels within the ingot. The test results indicated that post-transition corrosion thicknesses at 310°C varied by ±30% at low fast-neutron (>1 MeV) fluxes and varied by ±80% at high fast-neutron fluxes ($>1.0 \times 10^{14}$ n/cm²-s). The observation of increased variability as a function of fast-neutron flux suggested that the variability was related to unaccounted-for nuclear reactions. It was found that the trace impurities uranium and thorium levels (0 to 3 ppm) could be correlated with corrosion thickness, where higher uranium and thorium levels were associated with the lowest corrosion.

---

[1] Senior engineers, Lockheed Martin, P.O. Box 1072, Schenectady, New York 12301-1072.

This paper presents the results from an in-pile corrosion test of Zircaloy-4 that was conducted in the ATR corrosion loops to isolate the effect of uranium from other impurities.

Experimental Procedure

Materials

The Zircaloy-4 ingots containing 0.1, 2.1, 10.6, and 24.1 ppm by weight of natural uranium used in this test were melted using the same starting crystal bar stock that was selected for its low level of impurities. Natural uranium was added to obtain the three upper levels of uranium. The corrosion coupons used in this investigation were manufactured from recrystallized and stress-relieved sheet and had final dimensions of 24 by 49 by 1 mm. The coupons were pre-filmed for three days at 360°C in a static water autoclave. The final alloying levels for each of the ingots are given in Table 1.

Irradiation Conditions

The irradiations were carried out in the ATR pressurized water-forced circulation corrosion loops, which are operated at a pH of $\approx$10.2 at 25°C without boron additions. The loops are operated with a hydrogen overpressure giving a 40 cm^3/kg concentration and a dissolved oxygen level of <20 ppb. Fast-neutron (>1 MeV) fluxes from 0.0 to 2×10^{14} n/cm^2-s are available. The irradiations were conducted primarily at a high fast-neutron flux (>1×10^{14} n/cm^2-s) to determine the maximum uranium effect. Two tests were also exposed to low (2–3 $\times 10^{13}$ n/cm^2-s) and very low (7–9 $\times 10^{11}$ n/cm^2-s) fast-neutron fluxes to confirm the expected flux effect. The thermal-to-fast-neutron flux ratio in the ATR is approximately 1.0 at all positions. The test temperatures were 355, 332, 310, and 274°C.

Four corrosion coupons having uranium levels of 0.1, 2.1, 10.6, and 24.1 ppm were irradiated side by side in a corrosion coupon holder at each temperature and flux level. Each set of coupons, therefore, saw identical environmental conditions for the entire test. The corrosion weight gain of each coupon was determined after decrudding at intervals of 100 to 300 days. The repeatability of these measurements is ±4 mg/dm^2 over the duration of the test based on control samples.

Uranium Measurement Technique

Trace impurities of natural uranium and thorium in zirconium base alloys were measured by an isotope dilution technique. In this procedure, known amounts of Th^{230} and U^{233} tracers are added to a known weight of sample. The thorium and uranium are extracted from the combined sample by precipitation techniques and then characterized for isotopic abundance using a mass spectrometer. The known abundance of the tracers in the extract and the known natural isotopic abundance allow quantification of the impurities' natural uranium and thorium.

TABLE 1—*Alloying constituents.*

Ingot	U, ppm	Th, ppm	Fe, w/o	Cr, w/o	Sn, w/o
K412	0.1	<0.05	0.22	0.08	1.38
K413	2.1	<0.05	0.22	0.09	1.32
K414	10.6	<0.05	0.22	0.10	1.35
K415	24.1	<0.05	0.23	0.09	1.38

TABLE 2—*Zircaloy-4 corrosion data at different natural uranium levels.*

U = 0.1 ppm		U = 2.1 ppm		U = 10.6 ppm		U = 24.1 ppm		Time, Days	Flux[c]	Temperature, °C/°F	Fluence[d]
ΔW[a]	Rate[b]	ΔW[a]	Rate[b]	ΔW[a]	Rate[b]	ΔW[a]	Rate[b]				
13.3		11.9		12.9		13.8	← PREFILM				
28.4	0.17	28.2	0.18	33.2	0.23	37.6	0.27	88.8	1.18	331/627	9.1
36.7	0.07	35.9	0.06	45.6	0.10	58.5	0.17	214.9	1.24	332/630	22.6
77.6	0.24	42.6	0.04	57.1	0.07	79.3	0.12	383.6	1.23	336/637	40.6
478.9	1.07	248.3	0.55	81.0	0.06	105.3	0.07	759.2	1.29	339/642	82.7
687.3	1.91	470.4	2.04	199.1	1.08	112.1	0.06	868.1	1.52	337/639	97.0
1229.9	1.93	1101.8	2.25	890.5	2.46	786.7	2.40	1149.2	1.34	336/636	129.7
1473.9	1.75	1415.6	2.25	1235.4	2.47	1146.7	2.58	1288.7	1.10	337/639	143.0
0.0		0.0		0.0		0.0	← PREFILM				
21.1	0.24	18.6	0.21	22.6	0.25	19.9	0.22	88.8	.007	332/630	0.1
31.5	0.08	29.1	0.08	33.2	0.08	31.0	0.09	214.9	.009	333/632	0.2
75.4	0.14	71.7	0.13	75.9	0.13	72.4	0.13	536.5	.008	337/639	0.4
12.6		12.8		12.5		11.6	← PREFILM				
25.3	0.06	27.2	0.07	32.5	0.10	43.3	0.16	202.0	1.38	312/594	24.2
27.0	0.02	29.0	0.02	39.7	0.08	53.3	0.11	292.8	1.58	309/589	36.6
39.3	0.07	36.9	0.04	49.7	0.06	68.9	0.09	470.2	1.63	312/593	61.6
56.5	0.07	37.6	0.00	51.4	0.01	77.9	0.04	700.4	1.54	310/590	92.3
136.0	0.38	55.4	0.09	56.8	0.03	81.0	0.01	907.5	1.45	311/591	118.3
309.3	1.92	63.9	0.09	60.9	0.05	83.1	0.02	997.6	1.72	312/594	131.7
740.9	2.03	174.7	0.52	71.7	0.05	89.9	0.03	1210.3	1.60	312/593	161.2
1296.7	2.20	733.3	2.21	111.3	0.16	110.9	0.08	1463.5	1.67	313/595	197.9
12.2		12.6		11.2		11.8	← PREFILM				
29.2	0.08	29.9	0.09	31.6	0.10	34.5	0.11	202.0	0.22	316/600	4.0
33.4	0.05	34.9	0.06	35.7	0.05	41.2	0.07	292.8	0.26	315/599	6.0
51.3	0.10	43.5	0.05	46.6	0.06	57.0	0.09	470.2	0.25	316/600	10.0
77.3	0.11	51.5	0.03	48.1	0.01	69.0	0.05	700.4	0.23	316/600	14.6
101.3	0.12	71.8	0.10	54.1	0.03	76.1	0.03	907.5	0.19	316/600	18.2
114.8	0.15	79.7	0.09	56.1	0.02	80.1	0.04	997.6	0.33	316/600	20.8
11.4		12.1		11.2		11.5	← PREFILM				
15.2	0.03	16.7	0.03	19.9	0.06	26.8	0.11	141.3	1.35	272/522	16.6
21.8	0.03	20.9	0.02	31.3	0.04	50.6	0.09	395.1	1.95	274/526	59.4
26.9	0.02	22.2	0.01	36.3	0.02	61.5	0.04	642.1	1.32	271/520	87.6
40.0	0.04	27.2	0.01	41.7	0.01	75.0	0.04	1006.2	1.37	273/523	130.8
47.5	0.04	29.6	0.01	43.8	0.01	79.4	0.03	1174.0	1.35	276/529	150.4
10.0				11.9	← PREFILM						
55.6	0.55			61.5	0.60			83.1	1.14	359/679	8.2
137.1	0.79			73.6	0.12			186.8	1.20	356/672	19.0
203.5	1.09			88.2	0.24			247.7	1.33	358/676	26.0
292.6	1.16			105.9	0.23			324.7	1.36	353/667	35.1
422.9	1.19			118.3	0.11			434.1	1.46	356/673	49.0
502.1	1.14			128.1	0.14			503.7	1.17	353/667	56.0
591.4	1.07			180.7	0.63			587.1	1.14	352/666	64.3
692.3	1.23			285.0	1.27			669.4	1.31	356/672	73.6
822.1	1.23			432.2	1.39			775.0	1.15	353/668	84.2
886.5	1.14			520.1	1.55			831.6	1.41	356/672	91.1
		11.6				12.3	← PREFILM				
		51.6	0.48			71.6	0.71	83.1	1.14	360/680	8.0
		140.8	0.86			87.8	0.16	186.8	1.20	356/673	19.0
		216.3	1.24			98.0	0.17	247.7	1.34	358/677	26.1
		319.2	1.34			134.3	0.47	324.7	1.36	353/668	35.2
		467.8	1.80			178.0	0.53	407.4	1.50	357/674	46.0

(Continued)

TABLE 2—Continued

U = 0.1 ppm		U = 2.1 ppm		U = 10.6 ppm		U = 24.1 ppm		Time, Days	Flux[c]	Temperature, °C/°F	Fluence[d]
ΔW[a]	Rate[b]	ΔW[a]	Rate[b]	ΔW[a]	Rate[b]	ΔW[a]	Rate[b]				
		545.3	1.11			236.3	0.84	477.0	1.17	353/667	53.0
		648.1	1.23			360.3	1.49	560.3	1.14	352/666	61.3
		759.5	1.35			506.2	1.77	642.6	1.31	356/672	70.7
		901.9	1.35			686.8	1.71	748.2	1.15	353/668	81.2
		977.5	1.33			786.2	1.75	804.9	1.41	356/672	88.1

[a] Weight gain in mg/dm².
[b] Corrosion rate in mg/dm²/day.
[c] Fast neutron flux (>1 MeV) in units of 10^{14} n°/cm²-s.
[d] Fast neutron fluence (>1 MeV) in units of 10^{20} n°/cm².

Experimental Results

The specific weight gain and corrosion rate data from the test program are compiled in Table 2 and are discussed below.

Irradiations at 274°C

The weight gains from the four corrosion coupons irradiated at 274°C and high fast-neutron (>1 MeV) flux are shown in Fig. 1. The fast neutron fluxes were 1.3 to 1.4 × 10^{14} n/cm²-s for all interims except the second, where it was 2.0 × 10^{14} n/cm²-s. A fast-neutron fluence of 150 × 10^{20} n/cm² was achieved. The temperature range during the test was 271 to 276°C. The pre-transition corrosion rates are seen to increase with uranium content. The control ingot transitioned at 300 to 400 days. The 2.1 ppm ingot may have transitioned after 600 days. The 10.6 and 24.1-ppm ingots are still exhibiting pre-transition kinetics at 1174 days, indicating an increase in time to transition of at least 400% compared to the 0.1-ppm control ingot.

Irradiations at 310°C

The weight gains from the four corrosion coupons irradiated at 310°C and high fast-neutron (>1 MeV) flux are shown in Fig. 2. The fast-neutron fluxes were 1.4 to 1.7 × 10^{14} n/cm²-s during the test. A fast-neutron fluence of 198 × 10^{20} n/cm² was achieved. The temperature range during the test was 309 to 313°C. The effect of increasing uranium levels on the time to transition is the most obvious. The transition times were increased 50% at 2.1 ppm and 300% at 10.6 ppm uranium compared to that observed for the 0.1 ppm control ingot. The increases in the pre-transition corrosion rates are not as obvious as seen in the 274°C test data.

The weight gains from the four corrosion coupons irradiated at 310°C and low flux are shown in Fig. 3. The fast-neutron fluxes were 0.2 to 0.3 × 10^{14} n/cm²-s during the test. A fast-neutron fluence of 21 × 10^{20} n/cm² was achieved. The temperature range during the test was 314 to 316°C. The transition time was doubled at 2.1 ppm compared to that observed for the 0.1 ppm control ingot. Transition is not yet obvious at the two higher uranium levels at 1000 days when the test was stopped. Little to no pre-transition acceleration is noted at 2.1 and 10.6 ppm relative to the control ingot, but the high-uranium ingot is showing increased corrosion rates relative to the control ingot.

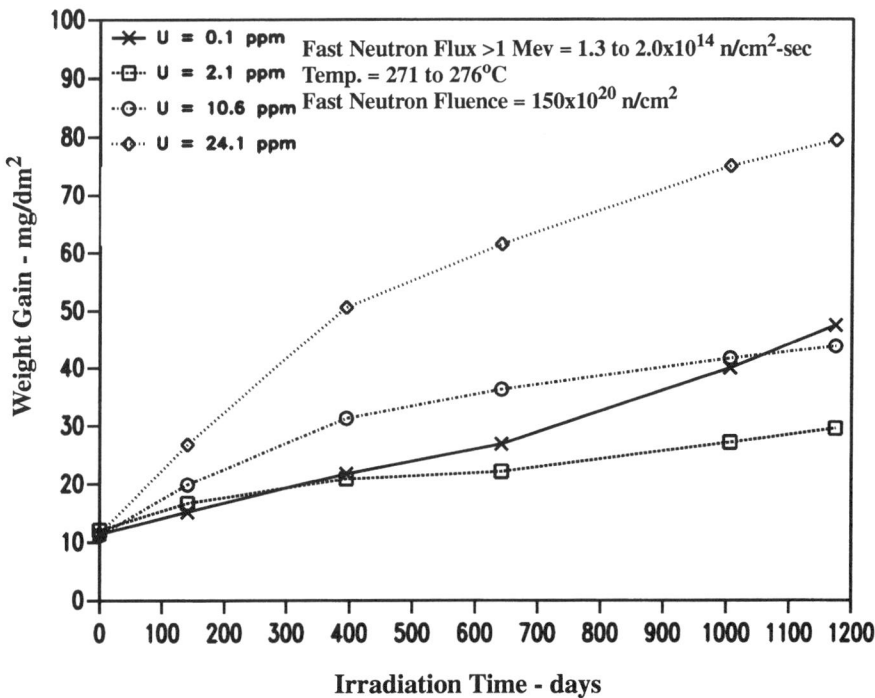

FIG. 1—*Effect of natural uranium impurities on in-pile corrosion of Zircaloy-4 at 274°C and high flux.*

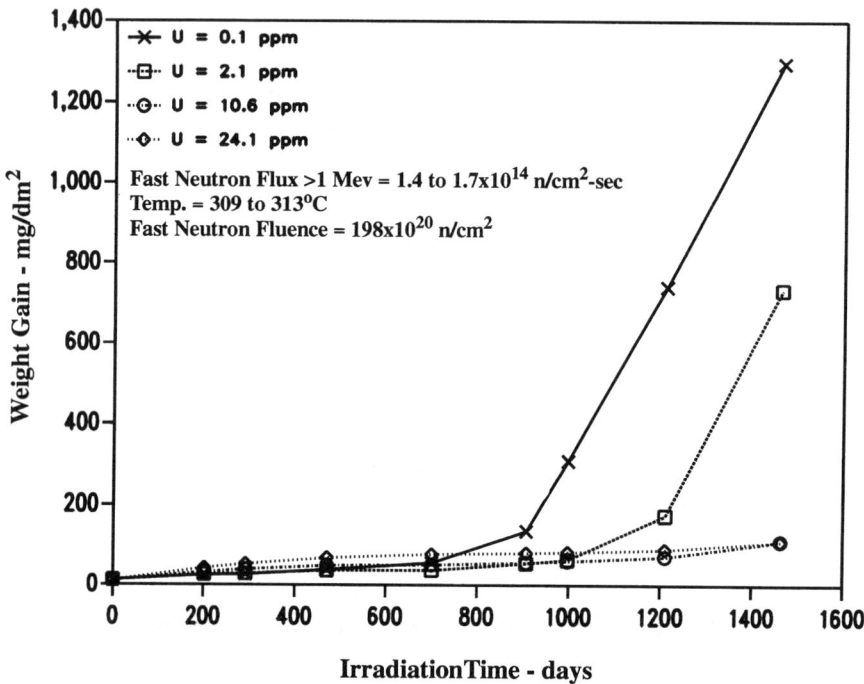

FIG. 2—*Effect of natural uranium impurities on in-pile corrosion of Zircaloy-4 at 310°C and high flux.*

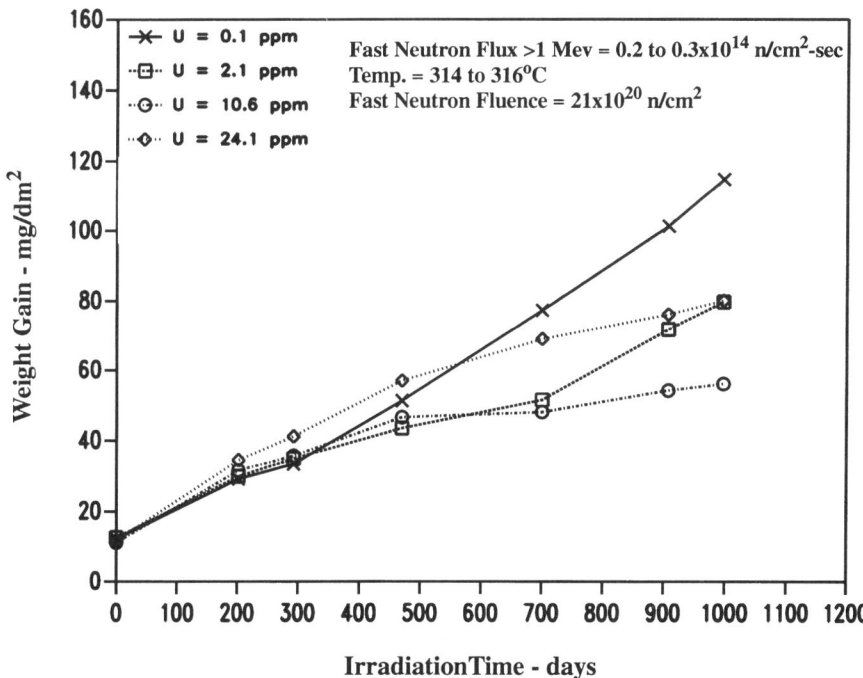

FIG. 3—*Effect of natural uranium impurities on in-pile corrosion of Zircaloy-4 at 310°C and low flux.*

Irradiations at 332°C

The weight gains from the four corrosion coupons irradiated at 332°C and high fast-neutron (>1 MeV) flux are shown in Fig. 4. The fast-neutron fluxes were 1.1 to 1.3 × 10^{14} n/cm²-s for all interims except the fifth, where it was 1.5 × 10^{14} n/cm²-s. A fast-neutron fluence of 143 × 10^{20} n/cm² was achieved. The temperature range during the test was 331 to 339°C. The effect of increasing uranium levels on the time-to-transition is the most obvious. The transition time was doubled at 2.1 ppm and quadrupled at 10.6 and 24.1 ppm uranium compared to that of the 0.1 ppm control ingot. Less obvious are the increases in the pre-transition and post-transition corrosion rates.

The weight gains from the four corrosion coupons irradiated at 332°C and very low flux are shown in Fig. 5. The fast-neutron fluxes were 7 to 9 × 10^{11} n/cm²-s during the test. A fast-neutron fluence of 0.4 × 10^{20} n/cm² was achieved. The temperature range during the test was 332 to 337°C. No effect of uranium is seen in these data affirming the expected flux dependency. Additionally, all four coupons have essentially the same weight gains, indicating that there are no other major differences between the four coupons such as processing or impurities.

Irradiations at 355°C

The weight gains from the four corrosion coupons irradiated at 355°C are shown in Fig. 6. The fast-neutron (>1 MeV) fluxes were 1.1 to 1.5 × 10^{14} n/cm²-s. A fast-neutron fluence of 91 × 10^{20} n/cm² was achieved. The temperature range during the test was 352 to 359°C.

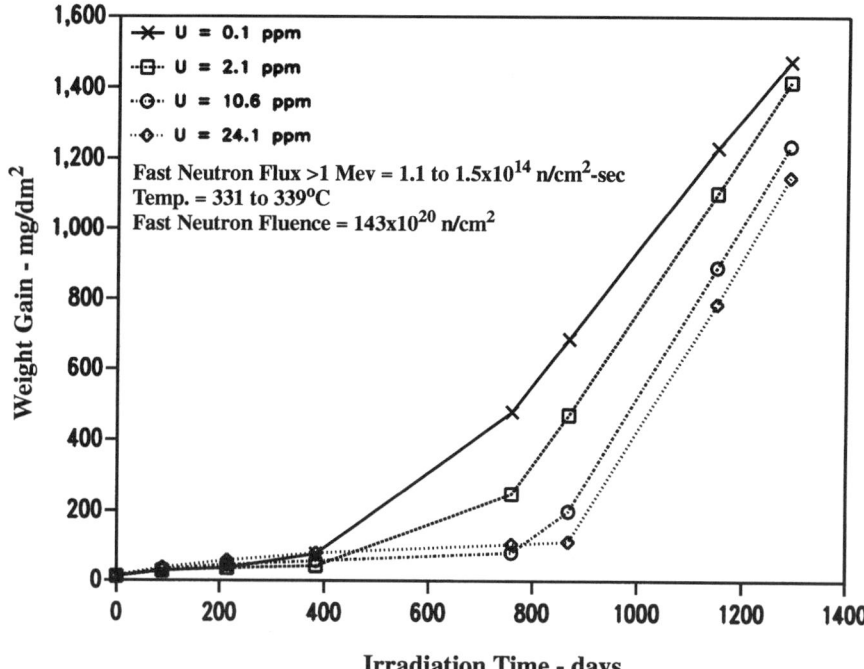

FIG. 4—*Effect of natural uranium impurities on in-pile corrosion of Zircaloy-4 at 332°C and high flux.*

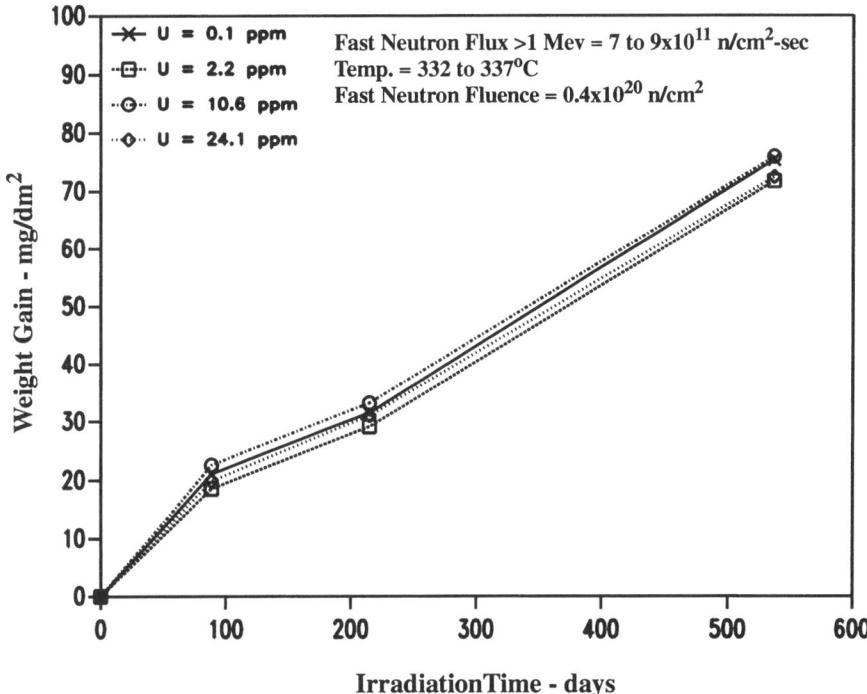

FIG. 5—*Effect of natural uranium impurities on in-pile corrosion of Zircaloy-4 at 332°C and very low flux.*

Transition occurred during the first interim (83 days) for the control ingot and the 2.1 ppm ingot. Based on the identical weight gains at levels <200 mg/dm^2, these two ingots transitioned at the same time. At lower temperatures, the 2.1 ppm ingot exhibited increased transition times and accelerated rates compared to the control ingot.

The increased pre-transition corrosion rates at 355°C, coupled with lower fission rates expected at short times (lower fission density in the oxide), has apparently reduced the effect of uranium sufficiently so as to not change the transition time of the 2.1-ppm ingot. At longer times, the 2.1 ppm uranium has clearly accelerated the post-transition rates relative to the control ingot. This would be expected since an equilibrium fission rate is now established as a result of breeding plutonium[2,39]. The two highest uranium ingots are showing increased transition times relative to the control ingot. This would be expected since they have 5 to 20 times the fission rate of the 2.1 ppm ingot. It is interesting to note that the 24.1 ppm ingot transitioned before the 10.6 ppm ingot at 355°C. This behavior has not been explained. The post-transition rates at 355°C are seen to increase with uranium levels consistent with 332 and 310°C observations.

Scanning Electron Microscope (SEM) and Metallographic Examinations

Metallographic and SEM examinations were performed on the corrosion coupons, which were irradiated at 310°C and low fast-neutron (>1 MeV) flux (see Fig. 4). SEM examination of the oxide surfaces revealed that the oxide films on the 0.1 and 2.1 ppm uranium ingots were

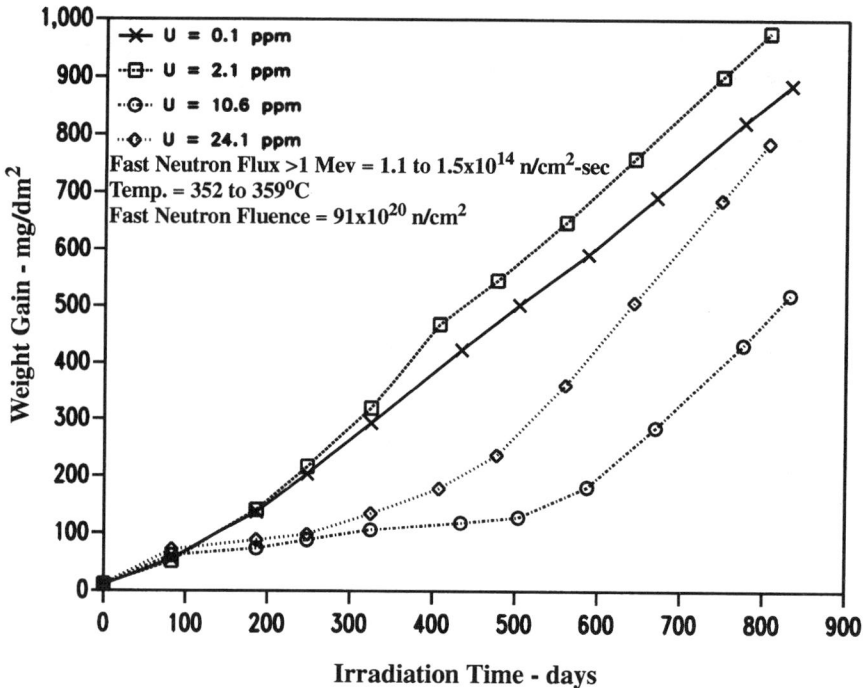

FIG. 6—*Effect of natural uranium impurities on in-pile corrosion of Zircaloy-4 at 355°C and high flux.*

intact with no cracking, whereas the oxides on the higher uranium level (10.6 and 24.1 ppm) samples exhibited cracking. Metallographic examination of the oxide on the highest uranium ingot showed that the cracks penetrate up to 50% of the oxide layer and do not reach to the oxide metal interface. This observation indicates that the oxide formed early in the test (outer oxide) shrunk or that the base metal grew relative to its original dimensions. The base metal can grow by irradiation growth or hydrogen swelling. Since these coupons only experienced a fast-neutron fluence of 20×10^{20} n/cm²-s and had weight gains <120 mg/dm² (low hydrogen pickup), neither of these growth mechanisms would be expected to cause the observed cracking. In addition, the cracking observed on the two highest uranium corrosion coupons was not observed to be associated with prior metal boundaries and is randomly oriented, whereas cracking caused by irradiation growth would be expected to be aligned perpendicular to the direction of growth. Therefore, the observed cracking is thought to be due to the outer oxide layer shrinking.

Analysis of Data

The corrosion behavior of Zircaloy-4 was previously characterized as having three regimes: pre-transition, transitory, and steady-state post-transition [7]. As shown in Fig. 7, the pre-transition regime is controlled by diffusion. The transitory regime occurs subsequent to transition and is characterized by linearly increasing corrosion rates as a function of increasing weight gain. The steady-state post-transition regime is characterized by a constant corrosion rate. The effect of uranium on corrosion rates and the weight gain at transition become more

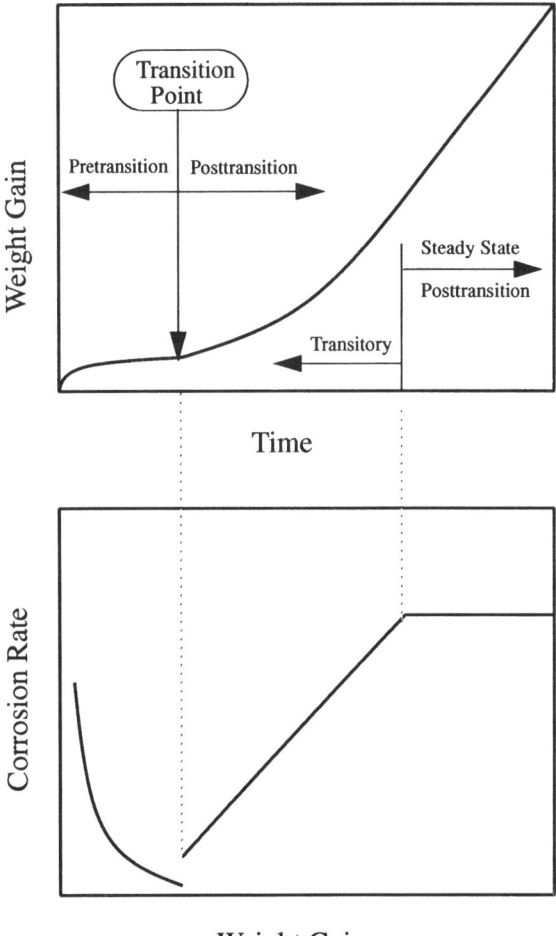

FIG. 7—*Zircaloy-4 corrosion regimes as defined in Ref 7.*

obvious when interim corrosion rates are plotted against the interim average weight gains [7]. The corrosion data from 355, 332, and 310°C have achieved steady-state post-transition corrosion. Figures 8, 9, and 10 compare transitory and steady-state post-transition corrosion rates at these temperatures with the out-of-pile transitory and steady-state post-transition rates from Ref 7. These data clearly show that in-pile steady-state post-transition corrosion rates are accelerated compared to out-of-pile rates and that the rates increase with increasing uranium contents relative to the control ingot. The amount of rate increase due to fissioning is of the same order of magnitude as that caused by fast-neutron flux alone, which is determined by the difference between the control ingot rates and the out-of-pile rates. In-pile corrosion of Zircaloy-4 is clearly seen to have accelerated the transitory corrosion regime relative to that seen in out-of-pile corrosion [7].

The very low fast-neutron (>1 MeV) flux test at 332°C, the low flux test at 310°C, and high fast-neutron (>1 MeV) flux test at 274°C did not achieve steady-state post-transition after

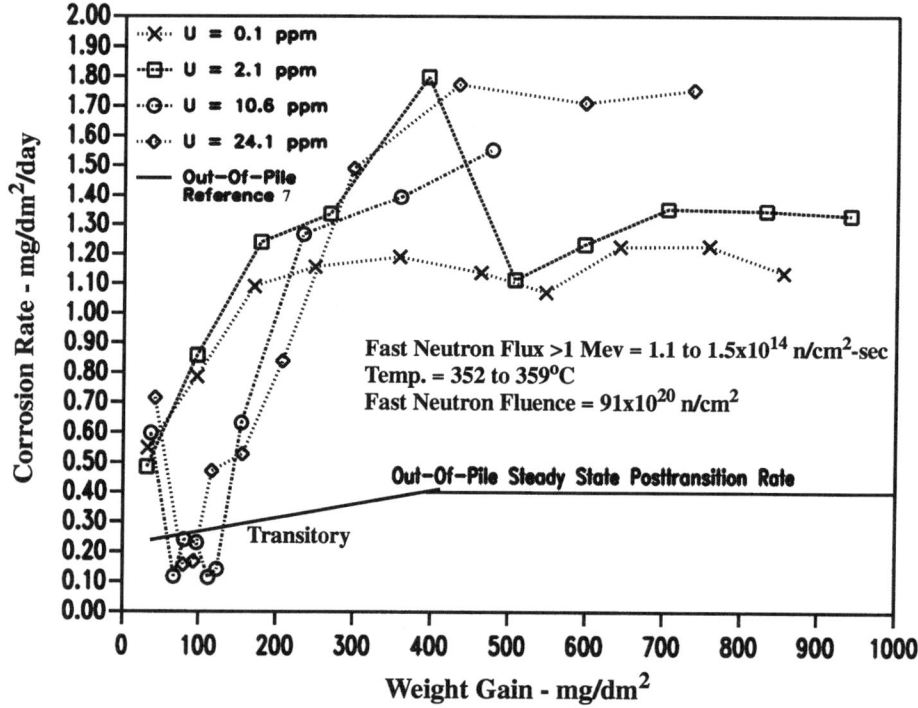

FIG. 8—*Effect of natural uranium impurities on in-pile corrosion rate of Zircaloy-4 at 355°C.*

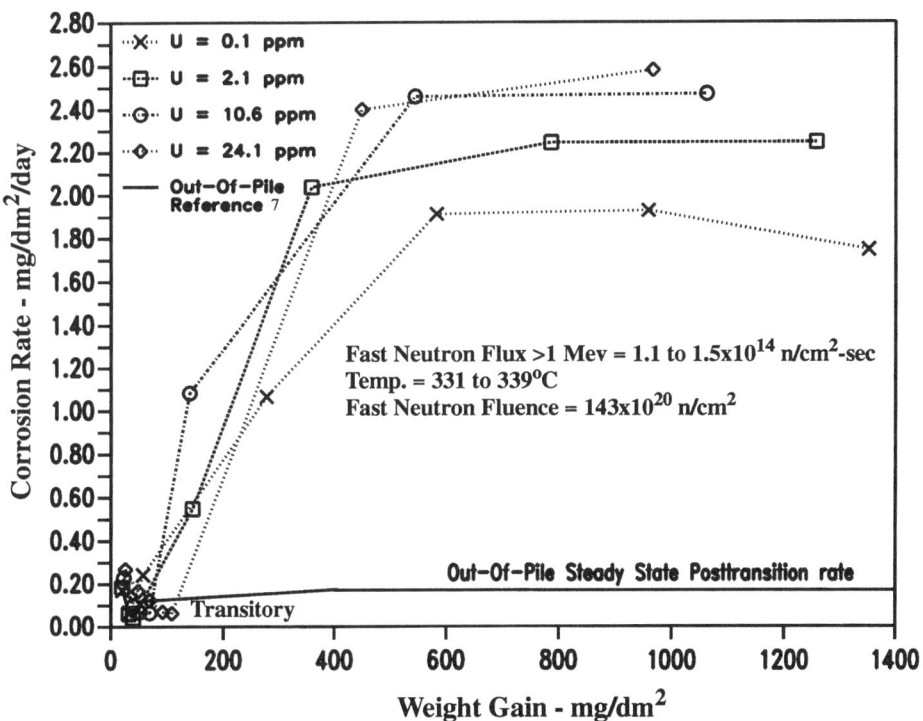

FIG. 9—*Effect of natural uranium impurities on in-pile corrosion rate of Zircaloy-4 at 332°C.*

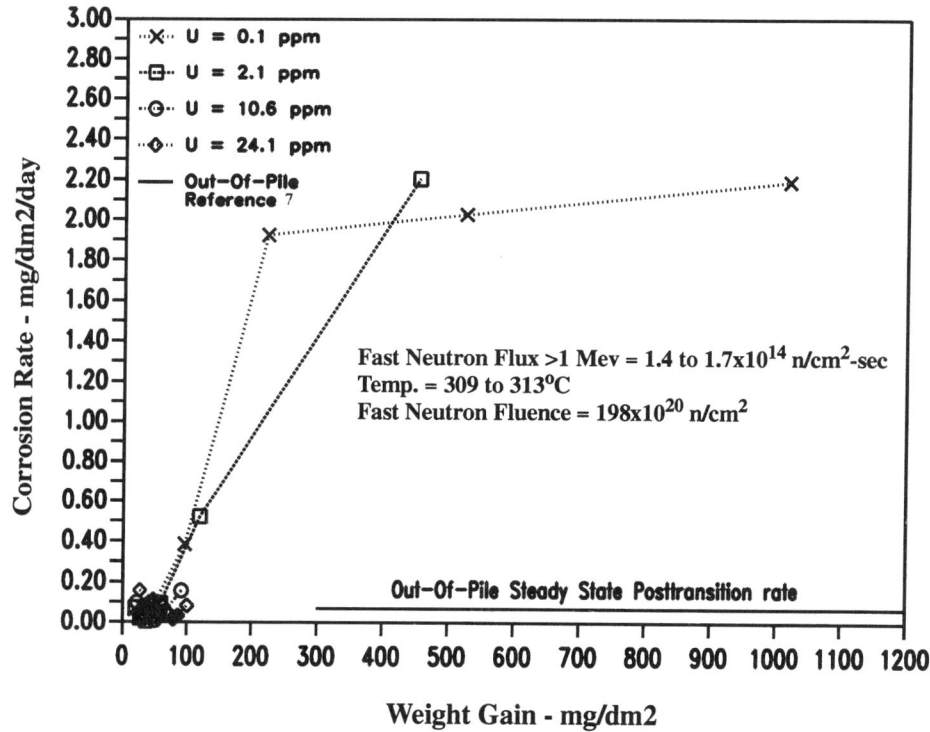

FIG. 10—*Effect of natural uranium impurities on in-pile corrosion rate of Zircaloy-4 at 310°C.*

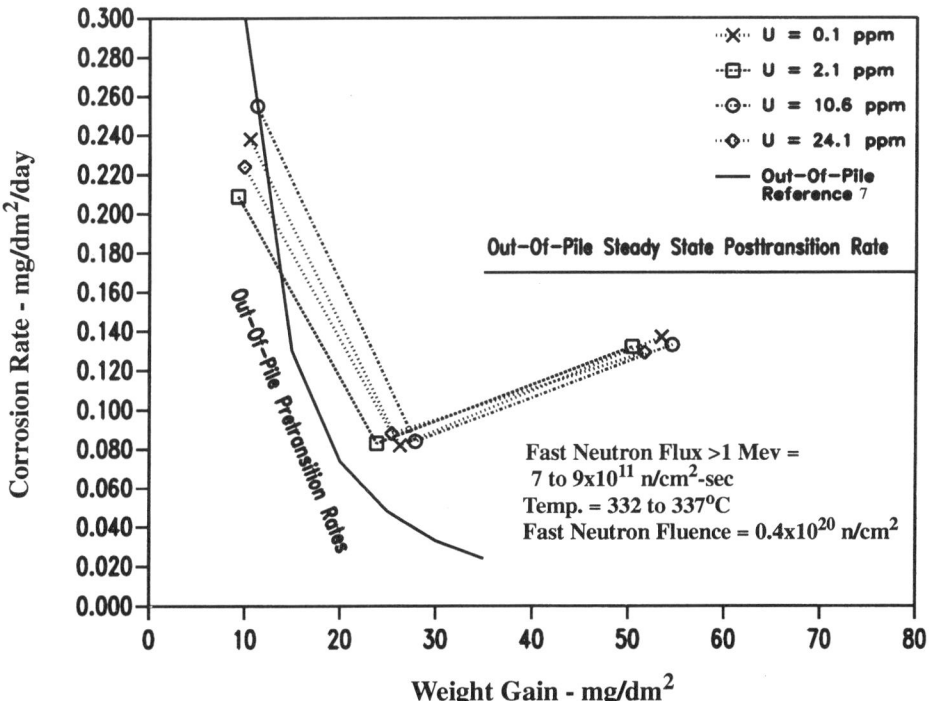

FIG. 11—*Effect of natural uranium impurities on in-pile corrosion rate of Zircaloy-4 at 332°C and very low flux.*

≈1200 critical days in test. The corrosion rate data from these tests are shown in Figs. 11, 12, and 13. The very low flux test at 332°C (Fig. 11) is exhibiting thermal pre-transition corrosion rates and has transitioned at greater than 30 mg/dm². No effect of uranium content on transition weight gain is seen. The corrosion rate data from the low flux 310°C test is shown in Fig. 12. These data clearly show accelerated pre-transition corrosion rates compared to the thermal rates seen in out-of-pile testing [7]. The pre-transition rates are seen to increase with uranium content. The control ingot transitioned between 33 to 51 mg/dm² and the 2.1 ppm ingot has transitioned between 52 to 72 mg/dm². The two higher uranium content ingots have not transitioned yet with weight gains of 56 and 80 mg/dm². The corrosion rate data from the high-flux 274°C test (Fig. 13) also shows that pre-transition rates increase with uranium content and that the rates are accelerated compared to the thermal rates [7].

Transition weight gains also increase with uranium content. Figures 14, 15, and 16 are enlargements of Figs. 8, 9, and 10 showing pre-transition corrosion rate and weight gain at transition detail for high flux at 355, 332, and 310°C. The corrosion rate data in these figures also clearly show that pre-transition rates increase with uranium content and that the rates are accelerated compared to the out-of-pile rates [7]. Transition weight gains also increase with uranium content.

Estimates of the weight gain at transition made from Figs. 13 to 16 and Table 2 are plotted on Fig. 17 for all four irradiation temperatures. Weight gain at transition trends, as a function of uranium content for each temperature, is indicated by dashed lines. The effect of uranium content on transition becomes increasingly stronger as the temperature increases.

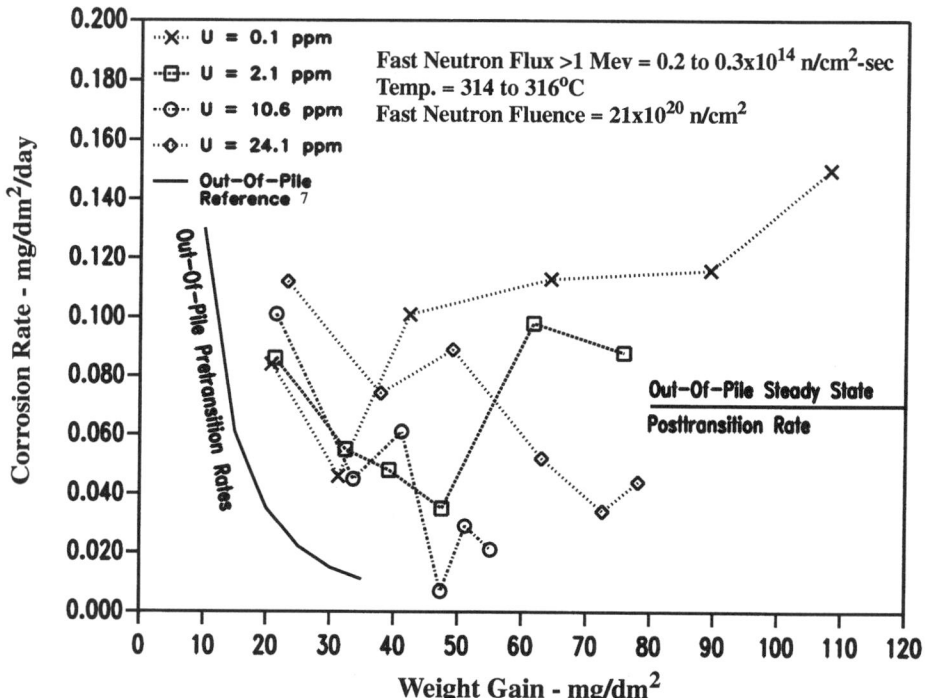

FIG. 12—*Effect of natural uranium impurities on in-pile corrosion rate of Zircaloy-4 at 310°C and low flux.*

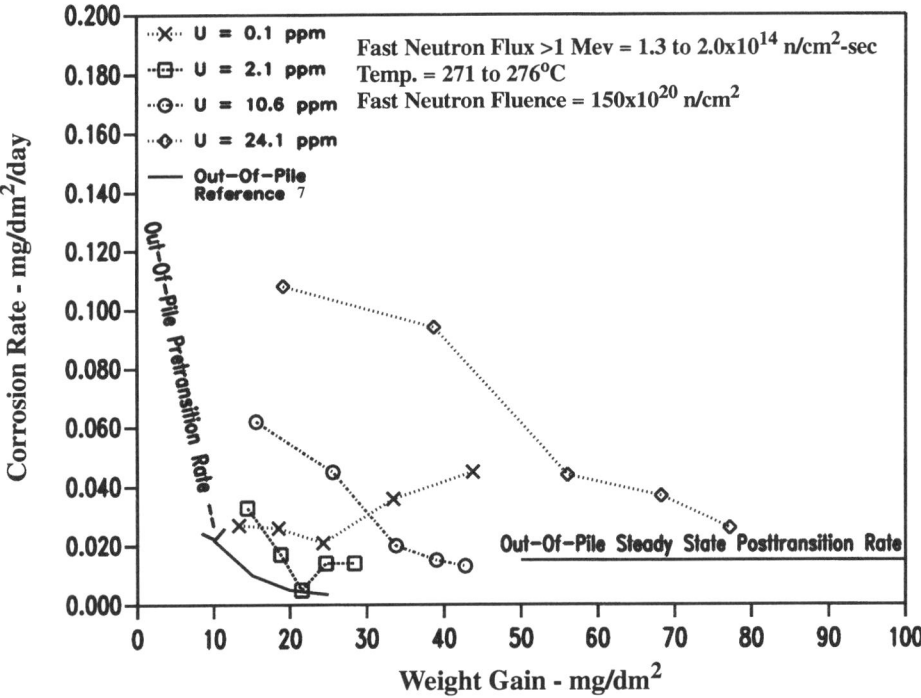

FIG. 13—*Effect of natural uranium impurities on in-pile corrosion rate of Zircaloy-4 at 274°C.*

Discussion

In-reactor corrosion of Zircaloy-4 has been found to be significantly affected by trace levels of fissile impurities. This effect becomes more beneficial as the natural uranium content increases through the range of 0.1 to 24.1 ppm. The fissile material occurs partially as the uranium^{235} isotope of natural uranium trace impurity and partially as a result of breeding of the fertile thorium^{232} and uranium^{238} to uranium^{233} and plutonium^{239}, respectively.

Fissioning of the fissile impurities yields energetic fission fragments that tear through the growing oxide and affect the corrosion process by (1) accelerating the pre-transition and post-transition corrosion rates, and (2) delaying the transition from pre-transition to post-transition kinetics. The net effect on long-term corrosion is a reduction (X2 to X5) in corrosion film thickness for Zircaloy-4 uranium levels at the specification limit (3.5 ppm) compared to the uranium free alloy. The pre-transition and post-transition corrosion rates increase as a function of the uranium content. This is thought to be due to irradiation-enhanced diffusion of the oxidizing species as a result of the fissioning within the oxide along with easier diffusion paths created by the fission tracks. A secondary effect may be the relaxation of the compressive stresses in the oxide as the result of the formation of a cubic phase in the oxide. Relaxed compressive stresses may also enhance the diffusion of the oxidizing species [2] and/or modify the tetragonal to monoclinic phase change [8], which may change the transition weight gain.

The beneficial effect of fissioning of fissile impurities on in-pile corrosion has been found to be a direct result of the delay in transition. This beneficial effect is postulated to arise either from or in combination with the formation of a different crystalline phase in the growing zirconium dioxide (monoclinic to cubic) [9–11]. The alteration of the growing oxide from the

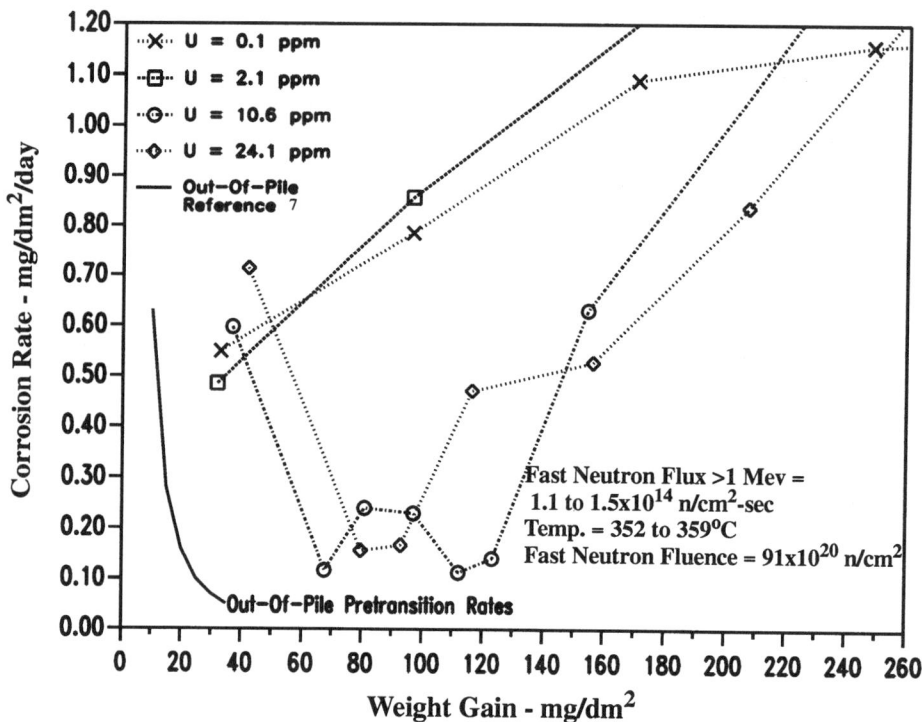

FIG. 14—*Effect of natural uranium impurities on in-pile transition weight gain of Zircaloy-4 at 355°C.*

energy deposition of the fission fragment is thought to occur either by electronic excitation or by displacement or a combination of both. Without regard to what specific mechanism is active, experimental evidence has demonstrated that a fission event displaces at least a million atoms [9,10]. The range of a fission fragment in a typical oxide solvent material is of the order of 4 to 6 μm [12]. Thus, each fissioned atom can send fragments far from the point of origin, offering the potential to effect a relatively large volume of material. Localized transfer of the fission fragment energy to the lattice has been demonstrated to cause a major phase conversion.

Figure 18 shows data supporting nearly 100% phase conversion of monoclinic zirconia to cubic zirconia at fission densities of about 10^{16} fissions per cm^3 of oxide volume [9,10]. As a result of this phase transformation, after any given exposure time the local density of the oxide is increased significantly from its value for the monoclinic phase to a value weighted by the presence of higher density cubic phase. This change in density in turn is hypothesized to bring about a significant relief of the compressive stress and/or strain known to be present in the zircaloy oxide, which, in turn, affects the kinetics of oxide growth and transition time.

Reference 13 provides a detailed analytic examination of the deposition of energy and increase in volume of cubic phase in the growing oxide. Reference 13 also discusses and accounts for the annealing of the cubic phase back to monoclinic. The process of formation and annealing of the cubic phase will result in an equilibrium conversion rate that will depend on the temperature and fission rate. The fission rate will be dependent on the fast and thermal neutron fluxes (neutron spectrum). A fission rate of 3 to 5 $\times$ 10^7 fissions per second per cm^3 of oxide at 310°C has been calculated for 1 ppm of natural uranium at a thermal and fast-neutron flux (>1 MeV) of 1.0×10^{14} n/cm^2-s in the ATR. Based on the conversion data shown in Fig. 18,

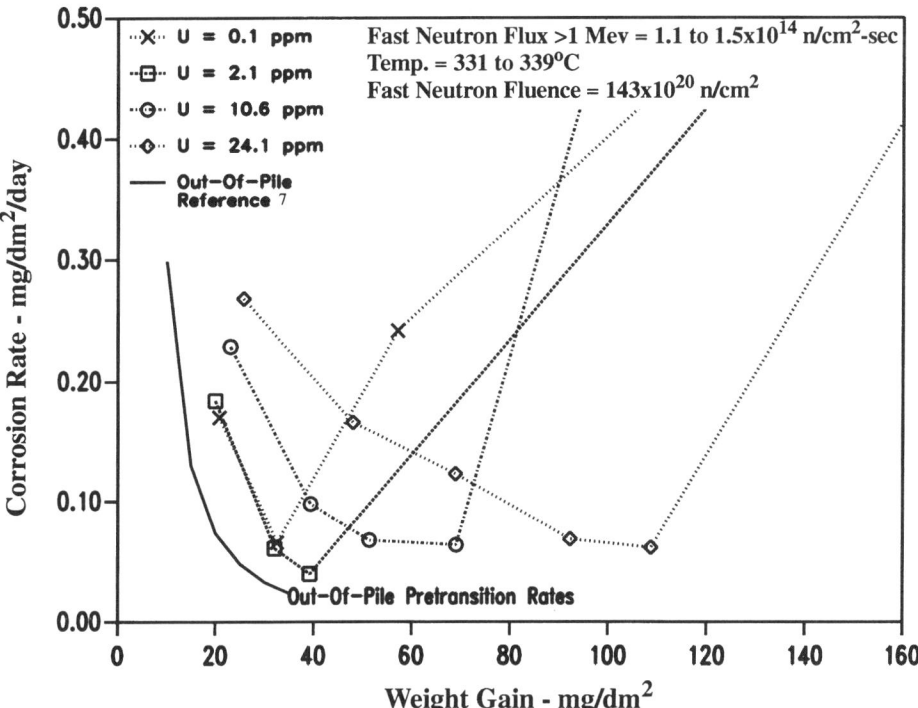

FIG. 15—*Effect of natural uranium impurities on in-pile transition weight gain of Zircaloy-4 at 332°C.*

a 10% percent conversion of monoclinic to cubic would occur in 300 days in the ATR at a flux of 1.0×10^{14} n/cm²-s, not accounting for annealing.

Metallographic and scanning electron microscope (SEM) examinations support conversion to cubic since the outer oxide layer appears to have shrunk. This would be expected since the outer oxide would have the highest percentage conversion to cubic, whereas the oxide at the oxide metal interface would have the least. If the conversion to cubic was 100%, the expected shrinkage would be $\approx 11\%$. The amount of shrinkage based on the surface cracking in the 24.1 ppm sample is estimated to be 1 to 3%. The expected conversion to cubic for this sample would be $\approx 100\%$ without annealing; therefore, annealing is apparently occurring at 310°C.

Uranium levels in ingots can vary within specification from 0 to 3.5 ppm uranium. Thorium levels are not controlled, but are typically <0.5 ppm. Using the data in Figs. 1 and 3, an estimate of the maximum variability expected in a PWR at high-power positions has been made (Table 3).

The weight gains given in Table 3 for 0.1 and 2.1 ppm are data, whereas the values for 4.0 ppm are interpolated from the 2.1 and 10.6 ppm data. As can be seen from Table 3, corrosion levels in a PWR are expected to vary by up to X5 depending on temperature and the amount of corrosion.

The steady-state post-transition corrosion rates at 332 and 310°C exhibit no temperature sensitivity. Furthermore, the observed steady-state post-transition rates at 355°C are lower than those observed at 332 and 310°C. This would indicate that steady-state post-transition corrosion rates are controlled by the fast-neutron flux and associated radiolysis within the film [14]. It also indicates that the corrosion rates are source limited rather than diffusion limited. The

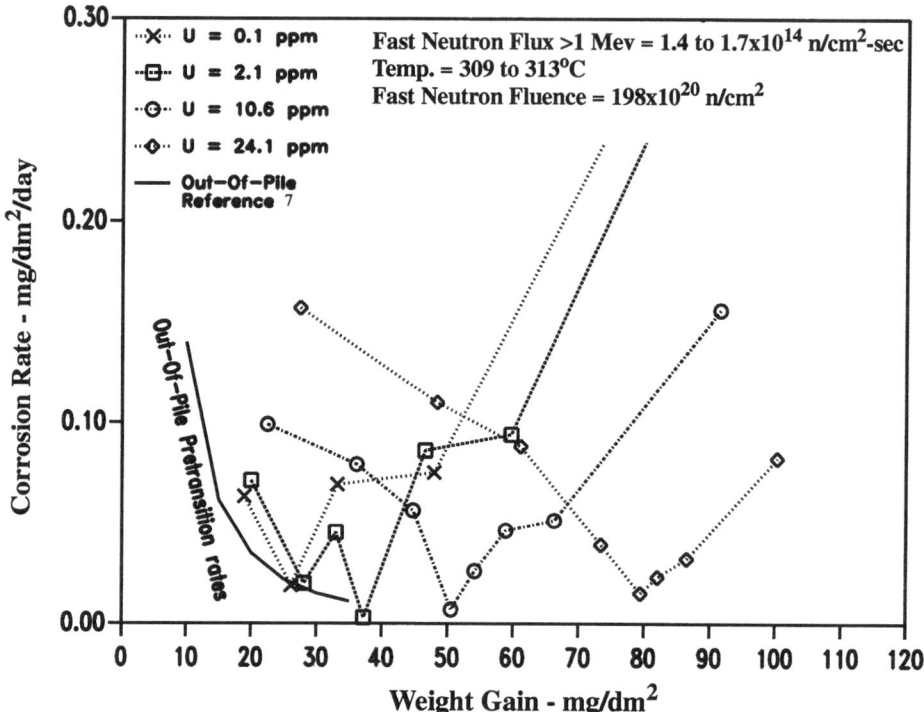

FIG. 16—*Effect of natural uranium impurities on in-pile transition weight gain of Zircaloy-4 at 310°C.*

reduced corrosion rates at 355°C are thought to be caused by the lower fluence at 355°C (compared to the 332 and 310°C data), which results in lower steady-state post-transition corrosion rates [15].

Conclusions

Natural uranium impurities in Zircaloy-4 ingots will result in larger corrosion variability in PWRs than observed in out-of-pile corrosion testing. Uranium impurities at the limit of the specification (3.5 ppm) are expected to reduce the corrosion by a factor of 2 to 5 relative to no uranium impurities at PWR temperatures and maximum PWR fast-neutron flux levels.

The beneficial effect of uranium impurities on in-pile corrosion is a direct result of a delay in transition. This beneficial effect of the fissioning of uranium in the oxide is postulated to arise either from, or in combination with, the formation of a different crystalline phase in the growing zirconium dioxide (monoclinic to cubic).

In-pile pre-transition corrosion rates are slightly accelerated relative to out-of-pile rates at 0.1 ppm uranium. The in-pile pre-transition corrosion rate increases significantly as a function of uranium content.

In-pile transitory and steady-state post-transition corrosion rates are significantly increased relative to out-of-pile rates even with no uranium additions. The in-pile post-transition corrosion rates also increase with uranium content.

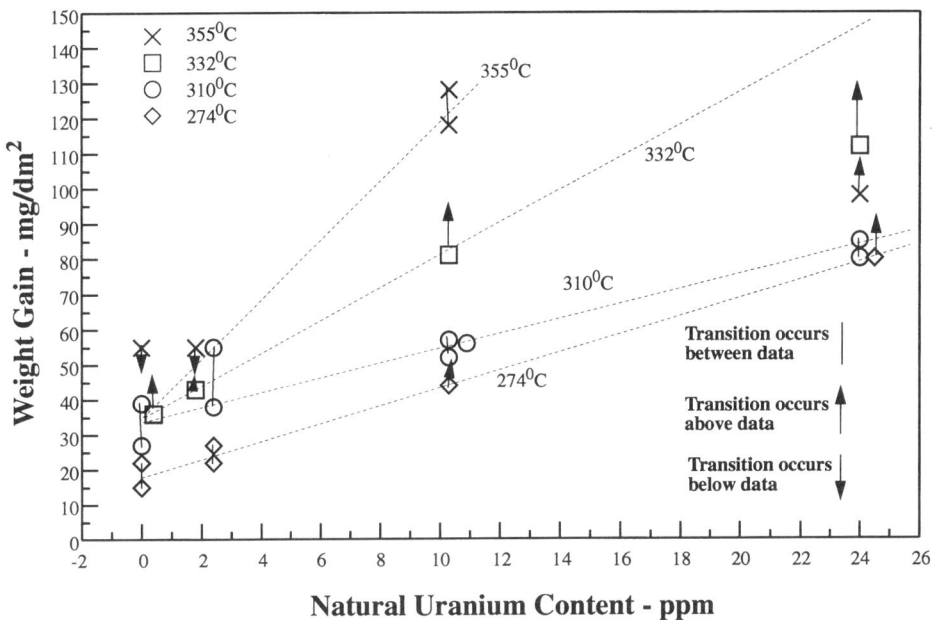

FIG. 17—*Weight gain at transition trends as a function of natural uranium content and temperature at fast-neutron fluxes $>1.0 \times 10^{14}$ n/cm^2-s.*

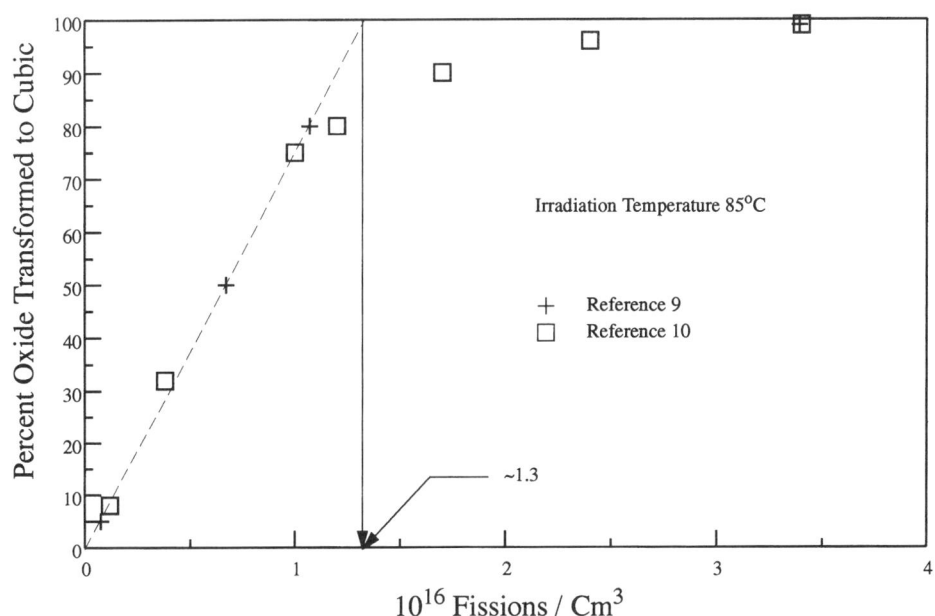

FIG. 18—*Conversion of monoclinic to cubic zirconia by fissioning of internal fissile impurities.*

TABLE 3—*Weight gain reduction relative to 0.1 ppm.*

	310°C				332°C			
	998 days		1464 days		759 days		1149 days	
U, ppm	ΔW^a	Factor	ΔW^a	Factor	ΔW^a	Factor	ΔW^a	Factor
0.1	309	X1.0	1297	X1.0	479	X1.0	1230	X1.0
2.1	64	X4.8	733	X1.8	248	X1.9	1101	X1.1
4.0	63	X4.9	580	X2.2	200	X2.4	1050	X1.2

^{a} Weight gain in mg/dm^2.

Acknowledgment

The authors would like to thank Gary W. Dansfield, who designed and initiated the in-reactor test, for his helpful discussions.

References

[1] Van Swam, L. F. P. and Shann, S. H., *Zirconium in the Nuclear Industry: Ninth International Symposium, ASTM STP 1132,* American Society for Testing and Materials, West Conshohocken, PA, 1991, pp. 758–781.
[2] Billot, P. and Giordano, A., *Zirconium in the Nuclear Industry: Ninth International Symposium, ASTM STP 1132,* American Society for Testing and Materials, West Conshohocken, PA, 1991, pp. 539–565.
[3] Eucken, C. M., Finden, P. T., Trapp-Pritschins, S., and Weidinger, H. G., *Zirconium in the Nuclear Industry: Eighth International Symposium, ASTM STP 1023,* American Society for Testing and Materials, West Conshohocken, PA, 1989, pp. 113–127.
[4] Garde, A. M. et. al., *Zirconium in the Nuclear Industry: Tenth International Symposium, ASTM STP 1245,* American Society for Testing and Materials, West Conshohocken, PA, 1993, pp. 760–778.
[5] Garzarolli, F., Schumann, R., and Steinberg, E., *Zirconium in the Nuclear Industry: Tenth International Symposium, ASTM STP 1245,* American Society for Testing and Materials, West Conshohocken, PA, 1993, pp. 709–723.
[6] Garzarolli, F., Steinberg, E., and Weidinger, H. G., *Zirconium in the Nuclear Industry: Eighth International Symposium, ASTM STP 1023,* American Society for Testing and Materials, West Conshohocken, PA, 1989, pp. 202–212.
[7] Peters, H. R., *Zirconium in the Nuclear Industry: Sixth International Symposium, ASTM STP 824,* American Society for Testing and Materials, West Conshohocken, PA, 1982, pp. 507–518.
[8] Godlewski, J., *Zirconium in the Nuclear Industry: Tenth International Symposium, ASTM STP 1245,* American Society for Testing and Materials, West Conshohocken, PA, 1993, pp. 663–683.
[9] Wittels, M. C. and Sherrill, F. A., "Fission Fragment Damage in Zirconia," *Physical Review Letters,* Vol. 3, 1959, pp. 176–177.
[10] Wittels, M. C., Steiger, J. O., and Sherrill, F. A., "Radiation Effects in Uranium-Doped Zirconia," *Reactor Science and Technology (Journal of Nuclear Energy Parts A/B),* Vol. 16, 1962, pp. 237–244.
[11] Vance, E. R. and Bolard, J. N., "Fission Fragment Irradiation of Single Crystal Monoclinic ZrO_2," *Radiation Effects,* Vol. 37, 1978, p. 237.
[12] Denschlag, H. and Weber, M., "Experimental Evaluation of a Simple Formula for the Ranges of Heavy Ions in Compounds," *Radiochemica Acta,* Vol. 15, 1971, pp. 9–12.
[13] Bobone, R., "Space Time Distribution of Fission-Induced Monoclinic to Cubic Transformation in Zirconia During In-pile Corrosion," *Proceedings,* Corrosion 84/185, National Association of Corrosion Engineers, Houston, TX.
[14] Johnson, A. B., Jr., "Thick Film Effects on Zircaloy-2 Under Irradiation," *Proceedings,* IAEA Technical Committee on Fundamental Aspects of Corrosion on Zirconium Base Alloys in Water Reactor Environments, Portland, OR, 11–15 Sept. 1989.
[15] Cheng, B., Kruger, R. M., and Adamson, R. B., *Zirconium in the Nuclear Industry: Tenth International Symposium, ASTM STP 1245,* American Society for Testing and Materials, West Conshohocken, PA, 1993, pp. 400–418.

DISCUSSION

S. Yagnik[1] (written discussion)—(1) Could you elaborate on your explanation of the observed effects of uranium in the metal on the monoclinic to cubic transformation in the oxide? (2) Have you estimated the U levels tested in your program with those expected in LWR rods (from recoil uranium from pellets)?

H. Richard Peters et al. (authors' closure)—(1) The uranium in the metal affects the monoclinic to cubic transformation in the oxide in the following two ways: (a) the uranium incorporates into the oxide during the oxidation process and subsequently fissions or (b) ~50% of the fission recoils originating in the metal 8 µm below the oxide will end in the oxide or pass through it. The resulting fission track density in the oxide will result in the transformation. (2) Fuel pellets do not contribute to the fission track density in the oxide since the recoil range of a fission fragment in Zircaloy-4 is ≤10 µm, which is significantly less than the 20 mil clad on PWR fuel rods.

B. Warr[2] (written discussion)—Kinetic behavior of samples from the same ingot can show significant variability. Did you look at several samples from the same ingot in order to investigate the variability of kinetics for a given U content?

H. Richard Peters et al. (authors' closure)—Accounting for uranium in in-pile corrosion correlations will result in an overall standard deviation of ~13%, which includes ingot variability, measurement error of flux, temperature, and weight gain at all fluxes and temperatures.

Y. S. Kim[3] (written discussion)—How do you confirm the existence of cubic zirconia in the oxide? Up to now, there seems to be only tetragonal phase identified in the oxide even on the irradiated Zircaloy-4 cladding.

H. Richard Peters et al. (authors' closure)—We have not yet confirmed the existence of cubic zirconia in the oxide but are planning X-ray diffraction and Raman examinations.

R. Holt[4] (written discussion)—Are you attributing the effect of U to the displacement damage caused by the fission products or to impurity effects associated with the presence of the fission products? If the former, how does the displacement damage from fission compare with that from the past neutrons?

H. Richard Peters et al. (authors' closure)—The increasing pre-transition rates as a function of increasing uranium are attributed to radiation-enhanced diffusion as a result of the fissioning. The delay in transition time as the uranium increases is attributed to a new higher density phase being formed as a result of the fissioning. Based on the high and low flux data taken at 310°C for the 0.1-ppm ingot, there is no difference in transition time between the two flux levels; therefore, I conclude that fast neutron and/or fluence at the time transition is not affecting transition.

---

[1] Electric Power Research Institute, Palo Alto, CA.
[2] Ontario Hydro, Toronto, Ontario, Canada.
[3] Korea Atomic Energy Research Institute, Daejom, Korea.
[4] Atomic Energy of Canada Ltd., Chalk River, Ontario, Canada.

B. Cox[5] (written discussion)—There are a couple of questions that arise because your results have shown major differences between your tests where the fission fragments come from within the material and our early work where the fission fragments come from the solution. We both see increased pre-transition oxide growth, but we saw reduced times to transition whereas you saw the reverse. From TEM replicas, we also saw that these films were full of pores. Can you suggest a reason for these differences and did you examine your pre-transition oxides at high magnification?

H. Richard Peters et al. (authors' closure)—The early work you referenced was performed in uranyl sulfate solution where transition occurred after 10 to 20 days of irradiation at ~20 mg/dm^2. At these high corrosion rates, the fission density in the oxide would be too low for any significant conversion of monoclinic to cubic as seen in the 355°C data presented in this paper. Additionally, in the work referenced, spalling oxide limited the length of the tests. This fact would indicate that there is a fundamental difference in oxide character between the referenced tests and those reported in this paper, which are in water and have not spalled up to >1400 mg/dm^2 versus the referenced work, which spalled at ≤60 mg/dm^2. Pre-transition oxides have not been examined at high magnification for pores since our corrosion samples typically are not removed until much higher weight gains.

---

[5] University of Toronto, Toronto, Ontario, Canada.

Martine Blat[1] *and Didier Noel*[1]

Detrimental Role of Hydrogen on the Corrosion Rate of Zirconium Alloys

REFERENCE: Blat, M. and Noel, D., "**Detrimental Role of Hydrogen on the Corrosion Rate of Zirconium Alloys,**" *Zirconium in the Nuclear Industry: Eleventh International Symposium, ASTM STP 1295*, E. R. Bradley and G. P. Sabol, Eds., American Society for Testing and Materials, 1996, pp. 319–337.

ABSTRACT: Recent studies have suggested that hydride precipitation at the metal/oxide interface could play a detrimental role on the waterside corrosion rate. Nevertheless, the mechanism of that detrimental role is not completely understood, and two hypotheses were investigated to understand the mechanism that controls the role of the hydrides. The first hypothesis is based on a mechanical effect: the hydrides precipitate at the metal/oxide interface and destroy the physical integrity of the "barrier" oxide layer. The second hypothesis is a modification of the transport properties of the oxide grown on the hydrided metal.

The detrimental role of hydrides on the corrosion rate was studied by charging unirradiated Zircaloy-4 cladding material with hydrogen to a level higher than the limit of solubility at 400°C. Both gaseous and cathodic charging techniques were used. Static corrosion tests were carried out in autoclave with steam at 400°C on an as-received and hydrided sample. The detrimental role of hydrides is confirmed for the post-transition corrosion rate, and that effect is more significant for high cathodic charging.

The results of the metallurgical examinations are discussed to provide an understanding of the mechanism. No relationship between hydrides, physical defects in the oxide, and local corrosion rate enhancement was found. Therefore, our results do not support the hypothesis of a mechanical effect at the scale of the performed examinations, but more detailed work is required to confirm this.

KEYWORDS: Zircaloy-4, autoclave corrosion, steam, cathodic charging, gaseous charging, hydride precipitation, metal/oxide examination

The kinetics of corrosion on Zircaloy-4 in reactor show an acceleration when burnup is over 30 to 40 GWd/tU (GigaWatt day per ton of Uranium). This acceleration is independent of the effect of the temperature rise at the metal/oxide interface as the oxide thickness increases. The main hypotheses proposed to explain this behavior are:

1. An irradiation effect changes the structure of the material or the redox conditions in the pores of a thick oxide [*1–2*].
2. Lithium is incorporated into the oxide [*3–5*].
3. Hydrides precipitate at the metal/oxide interface [*6–7*].

This work investigated the last hypothesis. The objective was to find out whether hydrides have an adverse effect on the corrosion kinetics of zirconium alloys and to obtain a better understanding of the effect.

---

[1] Research engineer and section group manager, respectively, EDF, R & D Division, Département Etudes des Matériaux, Les Renardières, BP 1, 77250 Moret sur Loing, France.

Recent works [6–7] have shown that a high hydrogen concentration in the vicinity of the metal/oxide interface can be correlated with the acceleration of corrosion kinetics. According to A. M. Garde [6], the hydride precipitation, which occurs at the metal/oxide interface when the hydrogen solubility limit is reached, causes oxide degradation by a mechanical process. Promoting the access of the water near the metal would accelerate corrosion kinetics. Furthermore, T. Kido [7] reported that charging with hydrogen prior to corrosion induced an increase of corrosion in water at 360°C when the initial hydrogen content was greater than 600 to 1000 ppm (Fig. 1). This effect increased by increasing the hydrogen concentration.

This acceleration may be due to the effect of hydrides on the integrity of the "barrier" oxide layer (Garde's hypothesis), but it could also possibly be due to a change in the electronic or ionic properties of the oxide as a result of the presence of hydrogen concentration. This hypothesis suggests that Zircaloy corrosion would not be dependent on a liquid phase diffusion in the oxide pores but on solid phase transport via a compact barrier layer. These mechanisms are discussed in a previous paper [8].

The work reported herein provides a better understanding of the hydride contribution to waterside corrosion. Cold-worked stress-relieved (CWSR) Zircaloy-4 fuel cladding was charged to different hydrogen levels using gaseous and cathodic techniques. Static corrosion tests were carried out in steam at 400°C on both as-received and hydrided specimens. The oxide and metal/oxide interfaces were examined by optical microscopy and SEM to determine the effect of the hydrides on the oxide. The results are presented and discussed in relation to the theories on the effect of hydrides on corrosion.

Experimental Procedures

Materials

The study was carried out using mainly one stress-relieved Zircaloy-4 cladding tube (1.5% Sn). This material was taken from a lot of Zircaloy-4 tubing purchased according to the current specifications of EDF fuel cladding and introduced in a french PWR. Additionally, one plate in the recrystallized state was used only for some tests, such as glow discharge optical spectroscopy (GDOS), to characterize the hydrided state before corrosion tests. The samples were

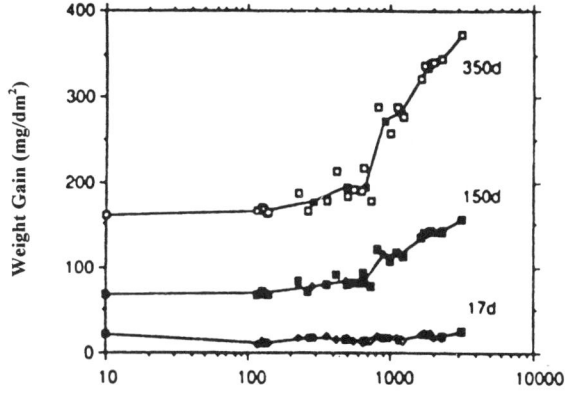

FIG. 1—*360°C corrosion weight gain at 17, 150, and 350 days as a function of initial hydrogen concentration (from Kido [5]).*

TABLE 1—*Chemical composition (wt%) of Zicaloy 4 samples.*

	Final Heat Treatment	Alloying element								
		Sn	Ni	Fe	Cr	Fe + Cr	C, ppm	H, ppm	N, ppm	O, ppm
Tube	stress-relieved 5 h at 470°C $10^{-17} < \Sigma A < 10^{-16}$	1.48	<0.005	0.21	0.11	0.32	135	3–5	29	1350
Sheet	recrystallized 3 h at 650°C	1.38	<0.005	0.21	0.13	0.33	122	14	41	1528

in the form of specimens measuring 30 or 50 mm in length and 9.5 mm in diameter for the tube. Their chemical compositions and final heat treatment are listed in Table 1. Their microstructures are shown in Fig. 2.

Hydrogen Charging

Approach—The aim was to obtain unoxidized samples with hydrogen contents above and below the hydrogen solubility limit, which is estimated to be about 200 ppm at 400°C [9], the temperature of the corrosion tests. The distribution and the morphology of the hydrides formed were to be representative at room temperature of those obtained in static autoclave tests, which generally showed a banded distribution as previously observed on this cladding material [8]. The hydride distribution was also to be homogeneous over the length of the sample.

Two charging methods were used: cathodic charging and gaseous furnace charging. Those two techniques each have their own specific features. Cathodic charging generates locally, at the interface, a high pressure of hydrogen. Gaseous charging is used at moderated hydrogen pressure and generates a different interaction between gas and metal at the interface.

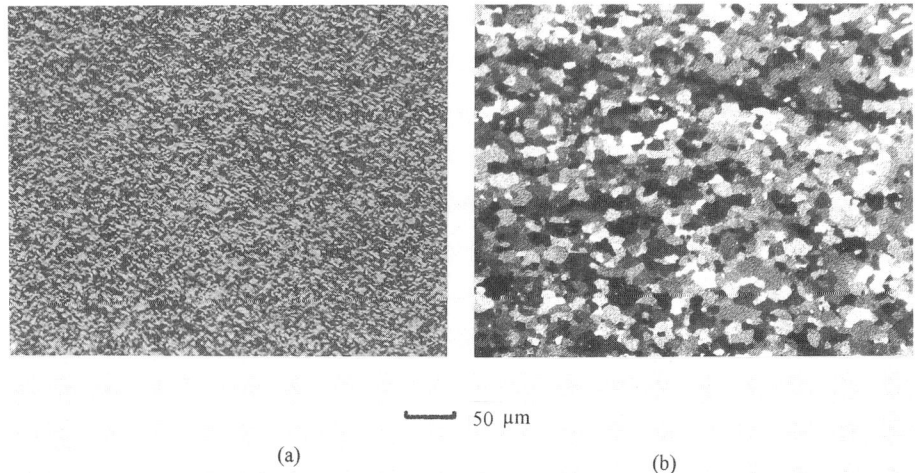

FIG. 2—*Optical micrographs of the material's microstructure on:* (a) *a tube specimen, and* (b) *a sheet specimen. The observation was performed on a transversal section with polarization of light.*

Cathodic Charging—After optimizing the operating conditions, cathodic charging was carried out on pickled samples at room temperature in sulfuric acid (1 N, aerated medium). An intensity of -5mA/cm^2 was applied to previously pickled samples (HF/HNO$_3$) for 72 h. Average hydrogen contents ranged from about 100 to 500 ppm but with a high hydrogen concentration at the outer surface. This technique did not allow the charging of hydrogen by the inner surface of the tubing sample. After hydrogen charging, specimens were kept in liquid nitrogen to avoid desorption. Then, as the charging method was performed at room temperature and to homogenize the hydrogen initially located at the outer surface before corrosion, an argon heat treatment was carried out for 26 h at 430°C; afterwards, the samples were cooled slowly in the furnace. After heat treatment, specimens were held at room temperature until the corrosion test since desorption measurements performed on some of these specimens demonstrated that no further evolution occurred. Hydride distribution and morphology before and after the heat treatment will be discussed in the results. Nevertheless, the hydrogen content introduced by cathodic charging was poorly reproducible, and hydrogen analyses were performed systematically, before corrosion tests, on a part of each sample. The main parameters used for cathodic charging are summarized in Table 2.

Gaseous Charging—Gaseous charging was performed to produce samples with higher hydrogen contents (400 to 1000 ppm). The unpickled samples were put in a quartz enclosure heated to 470°C with a vacuum of 10^{-6} mbar for degassing. Hydrogen was introduced at an overpressure of 3 to 4 mbar. After 24 h of hydrogen absorption, the samples were cooled slowly (1°C/min). The total time for degassing samples and hydrogen charging was about 40 h. The main parameters used for gaseous charging are summarized in Table 2. Before corrosion testing, the hydrogen content was systematically analyzed on a specimen cut on each sample for both cathodic- and gaseous-charged samples.

Corrosion Tests

The corrosion tests were performed in a static autoclave on fuel cladding samples already gaseously and cathodically charged with hydrogen and on reference samples pickled but not precharged (six samples). The homogenizing heat treatment (26 h at 430°C) was performed on each cathodically charged sample. Corrosion tests were carried out for periods of 3 days, 30 days, and 60 days in steam [400°C, 10.3 MPa, as recommended by ASTM Test Method for Corrosion Testing of Products of Zirconium, Hafnium, and Their Alloys in Water at 680°F or in Steam at 750°F (G2)]. The autoclave used was made of stainless steel and had a capacity of 7 L.

Hydrogen Content Measurement

The hydrogen content of all the samples was measured using a HMAT 2205 (Ströhlein) type analyzer by melting the samples at 1500°C in the presence of a carrier gas (argon). After

TABLE 2—*Hydrogen-charging parameters.*

Method of Charging	Environment	T, °C	Charging Duration, h	[H], ppm
Cathodic	1 N H$_2$SO$_4$ aerated $I = -5$ mA/cm^2	20	72	<400
Gaseous	< 100 mbar H$_2$	470	40 or 70[a]	400–1000

[a] For [H] < 800 ppm, 70 h for [H] > 800 ppm.

calibration, the hydrogen content was measured by comparing the conductivity of the argon/hydrogen mixture with the one of pure argon. This method measured a global hydrogen value (metal and oxide) and did not take into account the local concentration gradients.

Results

Characterization of the Hydrogen Distribution Before Corrosion Test

Cathodic Charging—To validate this charging method, some preliminary tests were carried out in the as-charged state before and after the homogenizing heat treatment (26 h at 430°C). Before homogenizing treatment, a massive surface hydride layer was observed on all samples and only some small hydrides in the bulk (Fig. 3a). This is obviously due to high hydrogen pressure produced at the outer specimen surface by cathodic potential and low hydrogen diffusivity at room temperature. The surface layer has been identified as a δ hydride layer from

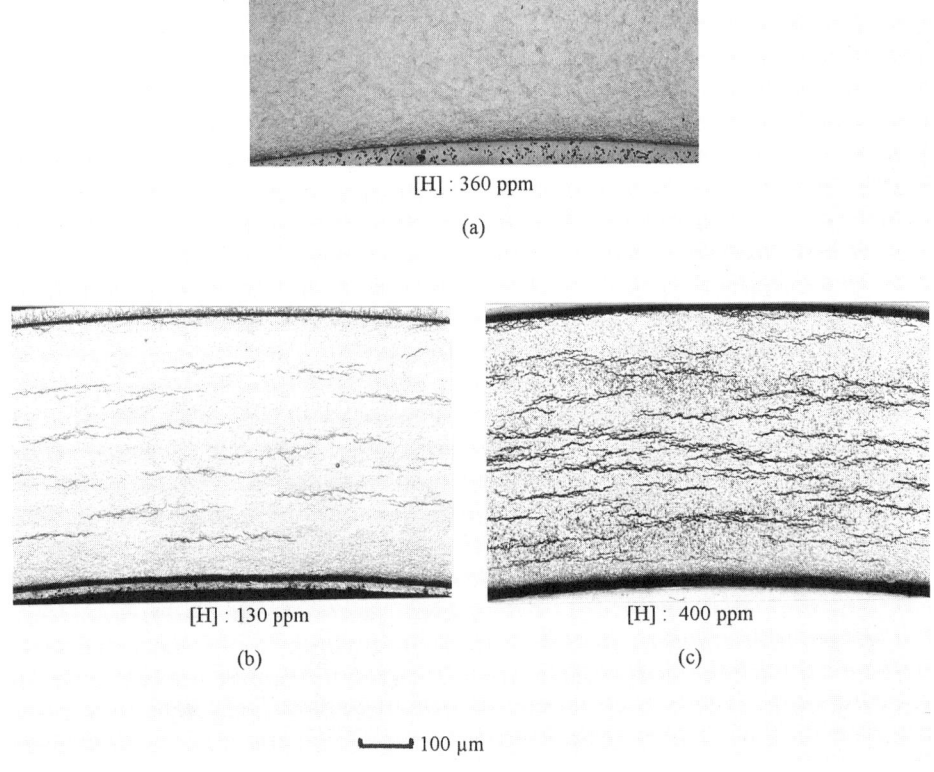

FIG. 3—*Optical micrographs on etched samples: hydride distribution after* (a) *cathodic charging and* (b) *and* (c) *cathodic charging and heat treatment.*

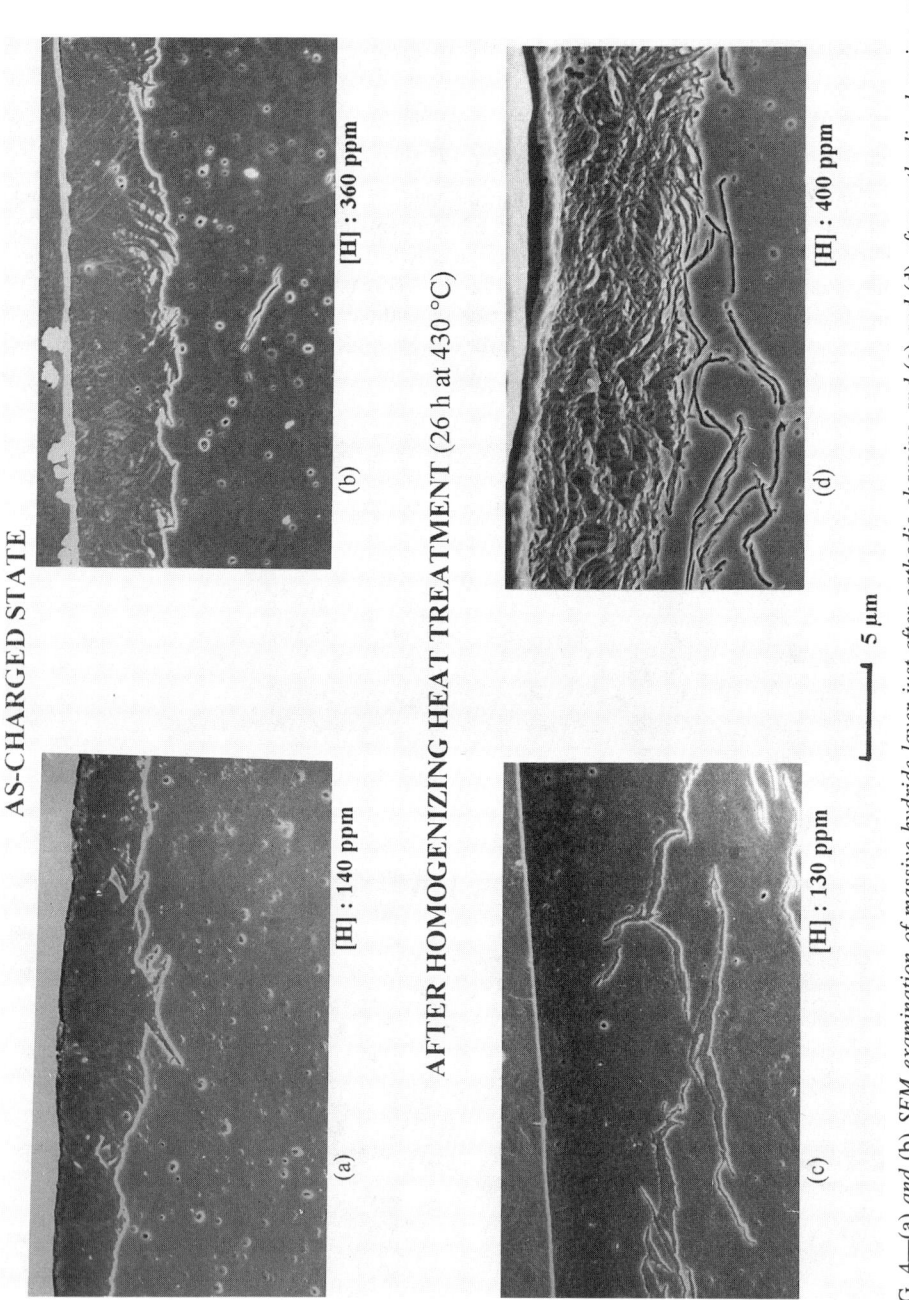

FIG. 4—(a) and (b) SEM examination of massive hydride layer just after cathodic charging and (c) and (d) after cathodic charging and homogenizing heat treatment (26 h at 430°C).

X-ray diffraction. The SEM examinations (Figs. 4a and 4b) showed its morphology as a thickness of about 5 μm. This layer is more irregular for a sample with lower average hydrogen content (Fig. 4a).

The heat treatment (26 h at 430°C) performed in furnace modified this surface hydride layer in two ways depending of the average hydrogen content in the sample:

1. When this average content is below the solubility limit in the material at the temperature of the heat treatment (about 270 ppm H at 430°C), the massive surface hydride layer disappeared after heat treatment (Fig. 4c). The hydrides are then precipitated in bands preferentially oriented in the hoop direction and in the entire thickness of the sample (Fig. 3b).
2. When the average hydrogen content is higher than the solubility limit at 430°C, only a part of the hydrogen contained in the hydride layer diffused in the bulk to form hydride platelets. So, after heat treatment at 430°C, a residual surface hydride layer was observed together with a δ hydride platelet distribution in the bulk (Figs. 3c and 4d).

These observations are schematically summarized in Fig. 5. To check the hydrogen distribution in the material, some glow discharge optical spectroscopy (GDOS) analyses were performed on flat samples (RXA Zircaloy-4 material) with a hydrogen content of 175 ppm. Before the heat treatment (26 h at 430°C), the elementary profiles showed that a hydrogen overconcentration existed near the sample surface (Fig. 6a). That is in agreement with the massive hydride layer observed by SEM. After heat treatment at 430°C, the hydrogen overconcentration was eliminated. This is consistent with the fact that the hydrogen average content is lower than 300 ppm for this sample. Optical microscope observations showed only some hydride platelets precipitated in the bulk. Some increased concentration in hydrogen, carbon, and oxygen elements remained only on the first 0.5-μm outer layer (Fig. 6b). In fact, the GDOS lateral resolution was not accurate enough to reveal the hydride platelet precipitation.

Taking into account the massive hydride layer thickness and the average value of hydrogen in the bulk (after machining this hydride layer), the calculation indicates that the real hydrogen

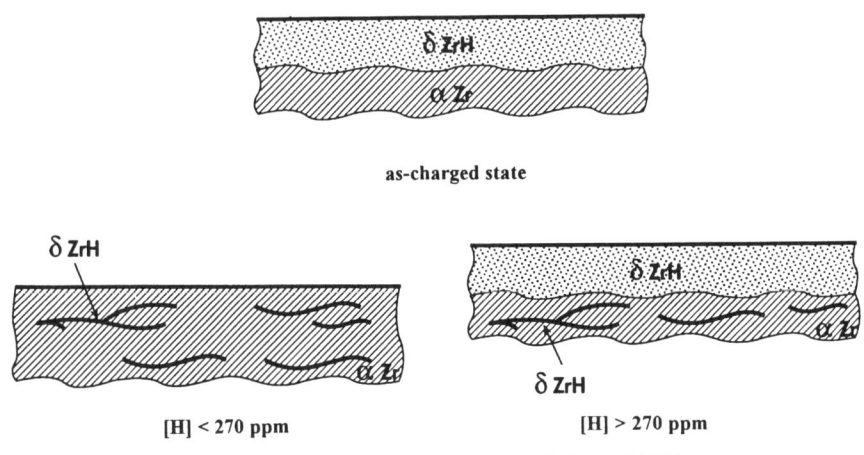

FIG. 5—*Schematic view of the outer massive hydride layer evolution before and after homogenizing heat treatment with respect to the average hydrogen content.*

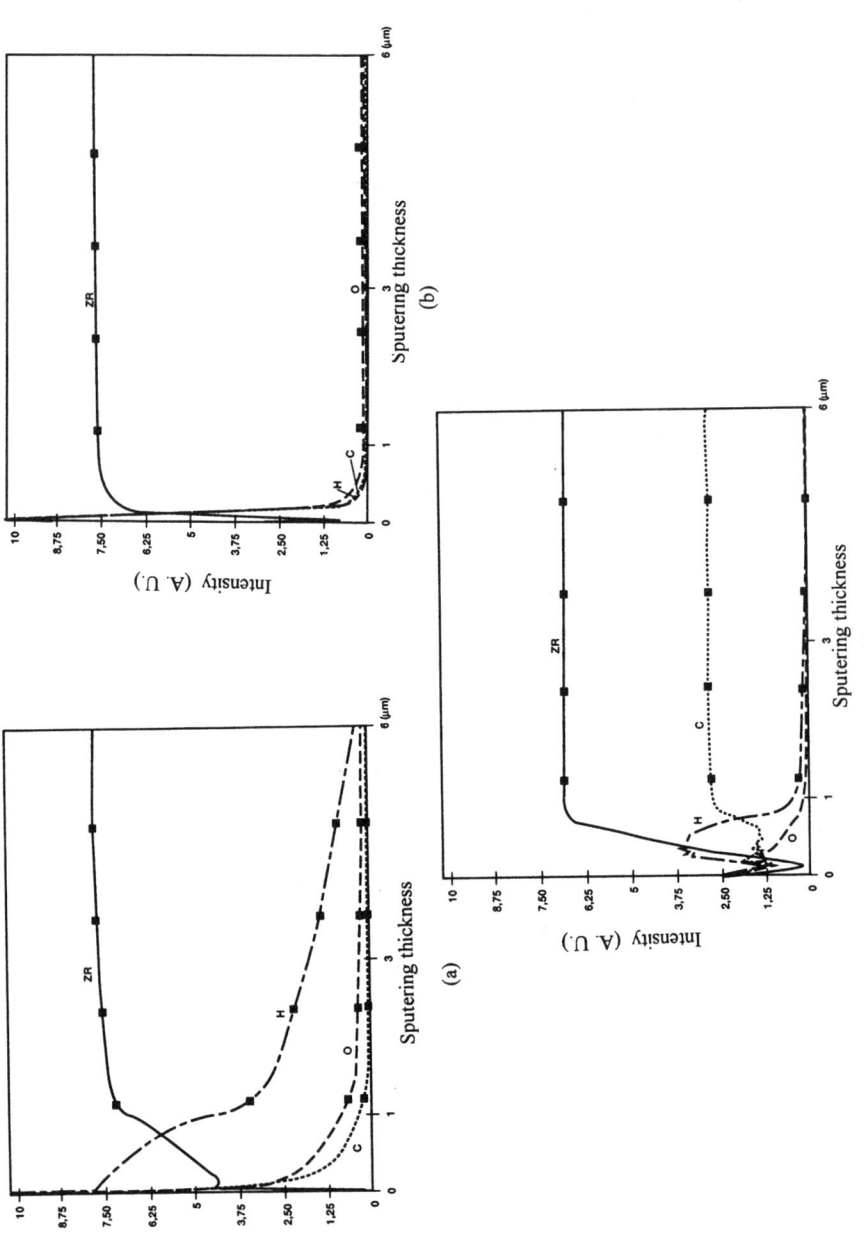

FIG. 6—*Glow discharge optical spectroscopy profiles of the Zr, H, O, and C distribution on the sample surface: (a) cathodic charging, plate specimen, 175 ppm H; (b) cathodic charging and heat treatment at 430°C, plate specimen, 175 ppm H; (c) gaseous charging, tube specimen, 400 ppm H.*

content in this hydride layer should be about 12 500 ppm. According to the equilibrium diagram Zr-H [*10*], the expected hydrogen content in a δ hydride phase would be 17 000 ppm. So, some residual αZr is supposed to be present in this layer.

Gaseous Charging—The samples were charged with hydrogen in a quartz enclosure for 40 h at 470°C under a hydrogen overpressure of 3 to 4 mbar. The gaseous charging led to a homogeneous hydride distribution in the bulk (Figs. 7*a* and 7*b*). In this case, the hydrides were shorter and finer than for cathodic charging. Even for an average hydrogen content higher than the limit of solubility at 470°C (about 365 ppm), no massive hydride layer was observed underneath the surface by SEM examination (Fig. 7*c*). The GDOS analyses performed on a tube charged to a level of 200 ppm showed an increase of the hydrogen level to a depth of less than 1 μm (Fig. 6*c*). On an as-received specimen and before oxidation, a local hydrogen peak may be found but on a thickness of less than 0.5 μm (same appearance as the profile on Fig. 6*b*). So, it is not impossible—but not proved—that a very local hydrogen overconcentration has occurred during gaseous charging underneath the surface. Even in that case, this hypo-

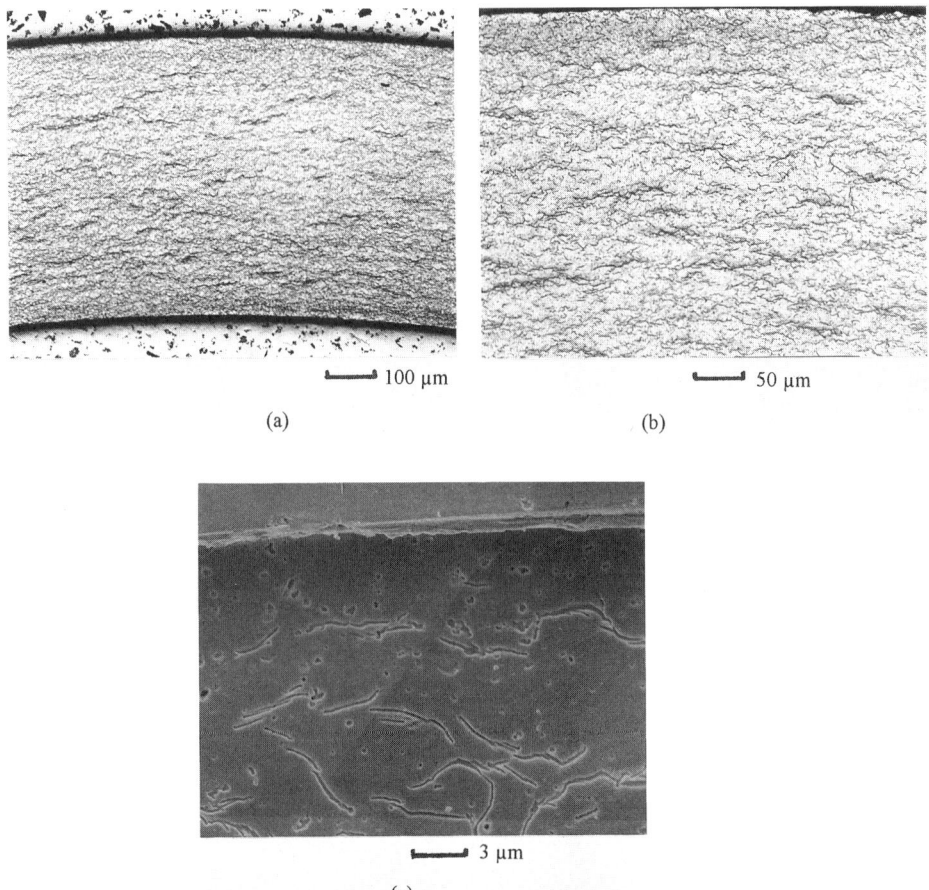

FIG. 7—*Optical and SEM examination of the hydride distribution after gaseous charging (etched sample: HF, HNO_3, H_2SO_4).*

thetical hydride layer should have disappeared quickly by corrosion (less than 30 days, as a weight gain of 42mg/dm² corresponds to a loss of metal thickness of about 1.8 μm).

Effect of the Initial Hydrogen Content on the Weight Gain after Corrosion Test

Steam corrosion tests were performed at 400°C on prehydrided and reference samples. The results are presented in Fig. 8 and Table 3. The curves on this figure represent the total weight gain on the sample as a function of the average value of the initial hydrogen content.

After three-day tests, the corrosion was still in the pre-transition phase (<15 mg/dm²) (Table 3). The results did not show any clear effect of the initial hydrogen content on the corrosion rate (Fig. 8). After 30 days of testing, the corrosion on the reference samples was slightly beyond the kinetic transition value (40 mg/dm²). For this test duration, the effect of a cathodic

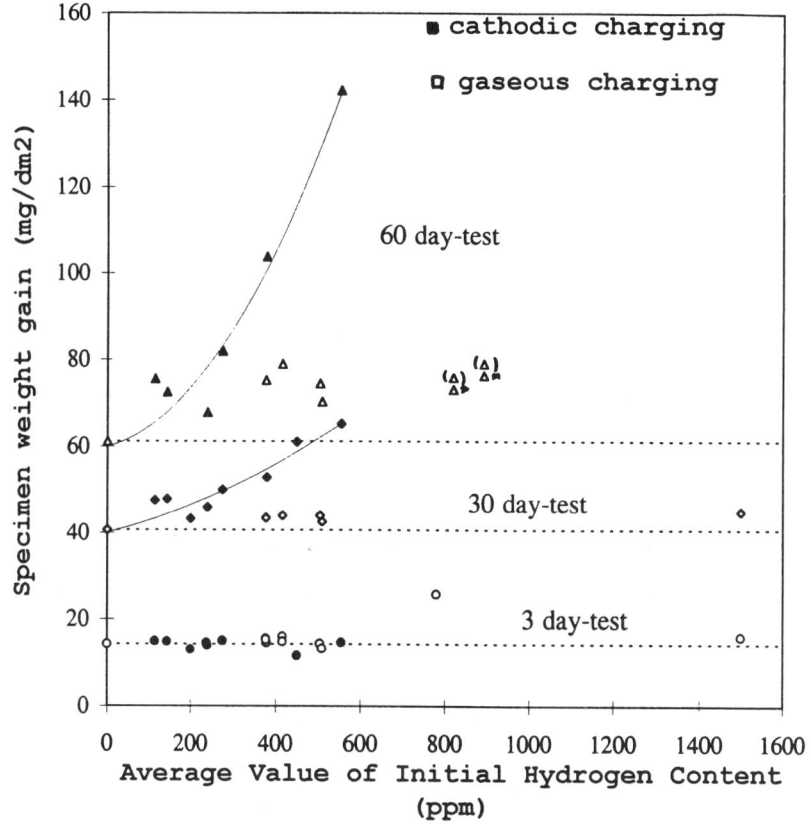

*: points for 57 day-test ; ( ) : points between brackets are corrected for 60 day-test.

FIG. 8—*Influence of the initial hydrogen content on the weight gain of Zircaloy-4 after 400°C steam corrosion tests.*

TABLE 3—*Corrosion tests: weight gain and hydrogen content measured.*

Charging Method	Initial Hydrogen Content, ppm	3-day Test		30-day Test*		60-day Test*	
		ΔW, mg/dm^2	[H], ppm	ΔW, mg/dm^2	[H], ppm	ΔW, mg/dm^2	[H], ppm
Reference	5	14.3 $s = 0.4$	17	40.6 $s = 0.7$	65	61.1 $s = 2.2$	122
Cathodic	115	15.0	...	47.6	...	75.4	229
	140	14.9	...	47.8	...	72.4	...
	200	13.1	...	43.3	250	...	...
	240	14.6	195	...	...	...	...
	240	14.0	...	45.9	...	67.8	...
	275	15.0	...	50.1	...	81.8	...
	380	14.6	...	52.9	...	103.8	...
	450	11.6	...	61.0	420	...	...
	555	14.6	...	65.1	...	142.2	555
Gaseous	375	15.5	...	43.6	...	75.2	...
	375	15.3	...	43.4	420	...	...
	415	16.0	(460)	...	...	...	...
	415	14.9	...	44.1	...	78.9	...
	505	14.4	...	44.1	...	74.4	...
	510	14.2	...	42.6	...	70.2	622
	780	25.7	732	...	...	...	...
	820	...	...	40.9*	...	73.0*	...
	890	...	...	43.0*	...	76.2*	...
	1500	16.0	...	45.0	1350	...	...

NOTE: s = standard deviation; w = weight gain.
* These weight gains were measured for 27 and 57 days of testing instead of 30 and 60 days. The data for 30- and 60-day tests have been estimated by adding the average kinetic of the reference sample in post-transition (0.68 mg · dm^{-2} · d^{-1}) (see Fig. 8).

charging is quite clear. It occurred for hydrogen values higher than about 250 ppm and increased with initial hydrogen content (Fig. 8). The hydrogen effect was less marked when precharging had been performed by the gaseous method, but it did still occur. An increase in corrosion of over 10% was observed for contents over 400 ppm. This was greater than the scatter observed in the weight gain data on the as-received materials, which was less than 2% (Table 3). Nevertheless, the effect was not very linear at this time of testing, probably because the corrosion level reached was at the limit of the one required to show the effect of hydrogen introduced by the gaseous method.

The detrimental effect of hydrogen on corrosion kinetics increased with the test duration. After 60 days of testing, this effect was still very strong for cathodic charging and occurred at a hydrogen level as low as 115 ppm. For cathodic-charged samples with an average hydrogen content greater than 300 ppm, a massive hydride layer was present underneath the outer surface. So the corrosion reaction did not occur on an α matrix with some hydride platelets but on a massive δ hydride layer with only some percent of αZr. In Fig. 9, the mean hydrogen content at the specimen surface (12 000 ppm <[H]< 17 000 ppm) is correlated to the outer oxide thickness measured on the sample. After the 60-day test, the corrosion rate was about four times greater than for the reference samples. So, δ hydride induces a great corrosion enhancement compared to αZr. But, the effect of δ hydride obviously did not explain the enhancement observed at levels lower than 200 ppm.

For gaseous charging, this effect was less pronounced than for cathodic charging, but the

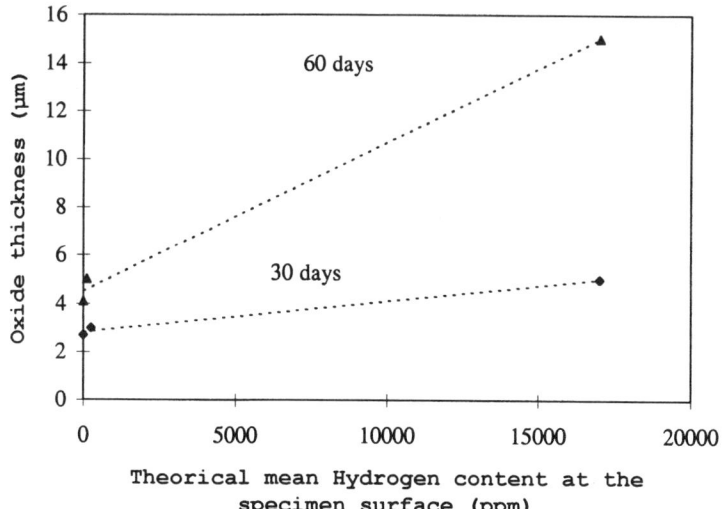

FIG. 9—*Influence of the massive hydride layer on the oxide thickness after 30- and 60-day tests in 400°C steam corrosion tests.*

corrosion rate increases with the hydrogen content. The enhancement factor is about 20% after the 60-day test, far beyond scatter of data on the reference sample or any difference between pickled and unpickled materials. Moreover, the gaseous charging at low pressure is not supposed to induce a large local perturbation near the surface. These results agree with those of T. Kido [7], who found no effect of hydrogen for times of tests inducing weight gains of less than 30 mg/dm^2, but a clear effect for longer times. The possible reasons for the difference in behavior between the two charging methods will be discussed later.

Hydride Distribution and Morphology after Corrosion Tests

The hydrogen contents measured in the metal and oxide after 3- and 30-day corrosion tests were, in the majority of cases, slightly lower than those measured before testing (Table 3). This result showed that some of the hydrogen may have desorbed during the corrosion tests. In fact, there was no correlation between the test duration and the desorbed hydrogen fraction, only with the initial hydrogen content (Fig. 10). So, this phenomenon probably occurred in the initial phase of the test, namely when the autoclave was coming up to temperature and as the oxide was thin.

From 3-day to 60-day tests, the optical and SEM examinations of the hydrides in the bulk of the samples showed the same distribution and morphology as before the corrosion test (Figs. 3, 7, and 11).

For high cathodically charged samples, the outer massive hydride layer was still present but with a decreased thickness and hydride density (Fig. 12c). For the only one sample cathodically charged and examined after the three-day test (with a low hydrogen level of 195 ppm), no hydride concentration was observed in the vicinity of the metal/oxide interface. However, SEM examinations exhibited, underneath the oxide, some defects that could be identified as small hydrides or cavities (Fig. 12a). At this date, an exact identification has not been done. But, these defects could have been induced by the massive hydride layer present at this location before the homogenizing heat treatment (26 h at 430°C) or induced by high local pressure

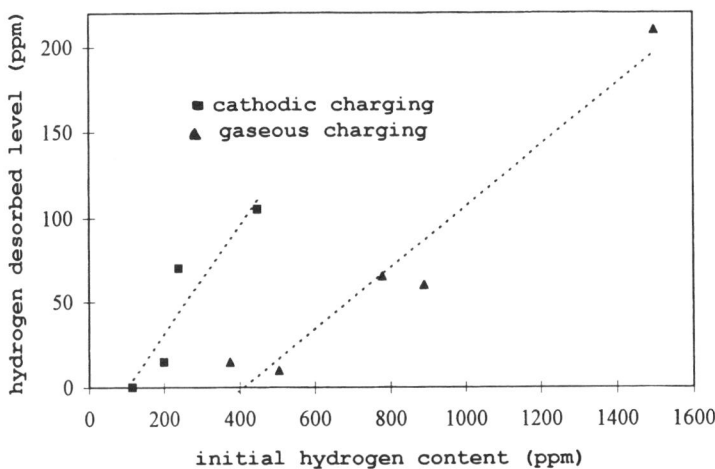

FIG. 10—*Hydrogen desorption during the first step of corrosion. The hydrogen-desorbed level is estimated by taking account the hydrogen absorbed during corrosion tests of reference samples.*

during cathodic charging. After 30 days of testing, these defects had disappeared (Fig. 12b). They were probably included by corrosion in the oxide.

For gaseous-charged samples, no massive hydride layer was observed underneath the oxide (Fig. 12d) in spite of local local hydrogen overconcentration observed on GDOS profile (less than 1 μm).

SEM Oxide Observations—After corrosion testing, the same oxide morphology was observed for all samples, reference or hydrogen-charged samples (Fig. 13). As already described in several previous works [8,11], circumferential bands were found with a periodicity distance of about 1 μm for all samples. No particular defects or decohesion at the oxide-metal interface were found, and no correlation with the hydrogen level was observed.

Discussion

The static corrosion tests performed in steam confirmed the detrimental role of hydrides on the corrosion kinetics. That effect appears after the kinetic transition. It is quite moderate for gaseous charging for the test duration performed and more important for cathodic charging, especially when an outer massive hydride layer was observed. A key question is why and how the hydrogen can modify the corrosion behavior of Zircaloy.

The first hypothesis involves a mechanical effect of hydrides. According to Ref 6, the hydrides that precipitate at the metal/oxide interface destroy the integrity of the oxide "barrier" layer. This hypothesis will be examined through three situations:

1. First, for cathodic charging with a high hydrogen level, a massive δ hydride layer was observed at the surface and a strong effect on corrosion rate was induced. At the scale of the SEM examination, the metal/oxide interface was not modified. The oxide did not have a higher density of defects such as cracks (Fig. 13). So, the acceleration seems due to a higher corrosion rate of the massive hydride and not to defects in the oxide.
2. For cathodic charging with a low hydrogen level, no surface hydride layer was observed and only hydrides platelets were distributed in the bulk. The acceleration effect remained, and the oxide did not have a higher density of defects than the reference sample.

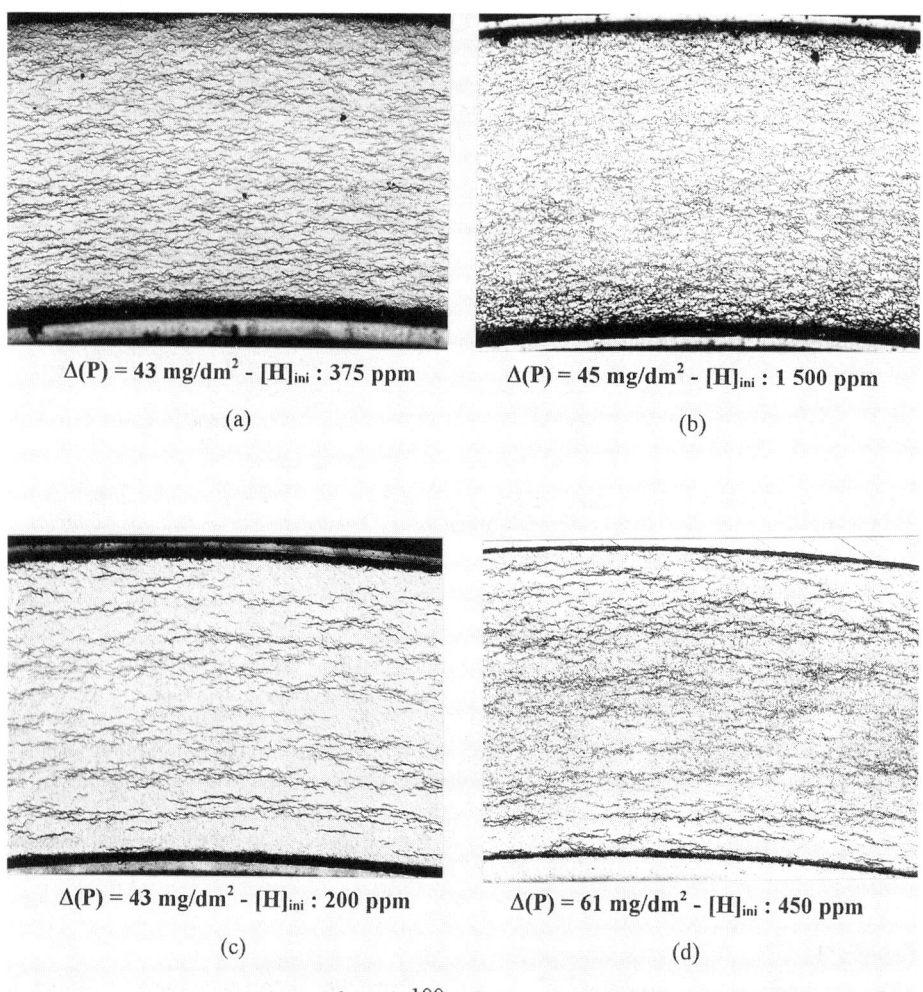

FIG. 11—*Optical examination of the hydride distribution after a 30-day test:* (a) *and* (b) *gaseous charging;* (c) *and* (d) *cathodic charging.*

3. For gaseous charging, no hydride surface layer was observed (only hydrides platelets in the bulk) and the acceleration effect remains. As previously, no higher density of defects was present in the oxide (Fig. 13c) and no thicker oxide on superficial hydrides was observed (Fig. 12d).

At the scale of the present observations, in all cases no change of appearance in the oxide was observed. So, the observations do not support the hypothesis of a hydride mechanical effect. Others hypotheses that may be taken into consideration are:

1. An accelerated corrosion rate of a massive surface hydride. The hypothesis could be supported by observations on high cathodically charged samples. However, this does not

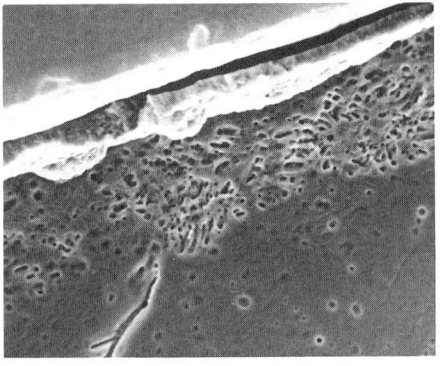
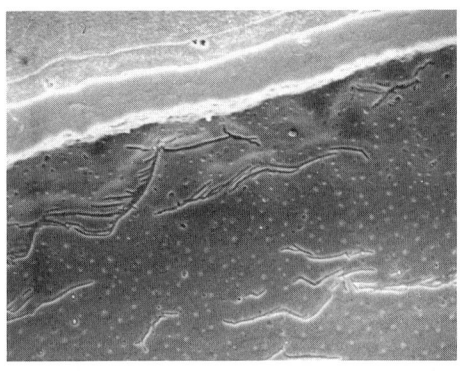
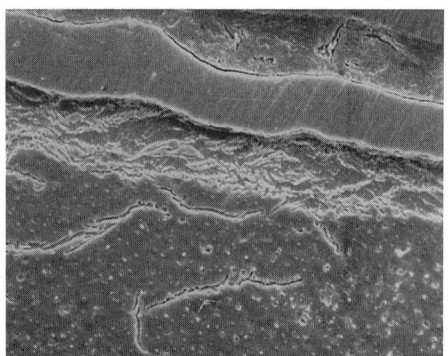
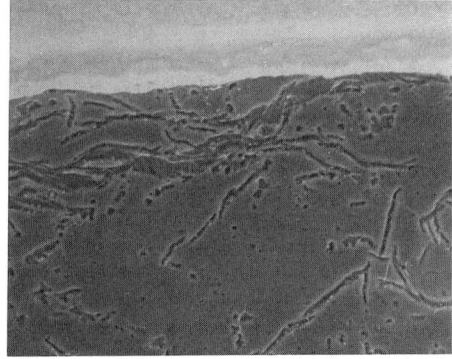

3 day-test : $\Delta(P) = 15$ mg/dm^2 - [H]$_{ini.}$: 240 ppm   30 day-test : $\Delta(P) = 43$ mg/dm^2 - [H]$_{ini.}$: 200 ppm

(a)   5 µm   (b)   5 µm

30 day-test : $\Delta(P) = 61$ mg/dm^2 - [H]$_{ini.}$: 450 ppm   30 day-test : $\Delta(P) = 45$ mg/dm^2 - [H]$_{ini.}$: 1500 ppm

(c)   5 µm   (d)   5 µm

FIG. 12—*SEM examination of the metal/oxide area for hydrogen-charged samples after a corrosion test: (a), (b), and (c) after cathodic charging, and (d) after gaseous charging.*

explain the acceleration observed at hydrogen levels lower than 270 ppm, nor the one obtained with gaseous charging.

2. The acceleration could also be due to damage in the material produced by the previous massive hydride present for cathodic charging before thermal treatment. This could be consistent with SEM examination, which exhibited defects in the metal after the three-day corrosion test for a low-hydrogen cathodic-charged sample (Fig. 12a). These defects could be created by the hydride layer formed before the homogenizing treatment or by the high local hydrogen pressure induced by the cathodic charging. So, this surface material damaging could be responsible for the corrosion acceleration but does not clearly explain why that effect increases with test duration. It also does not explain the acceler-

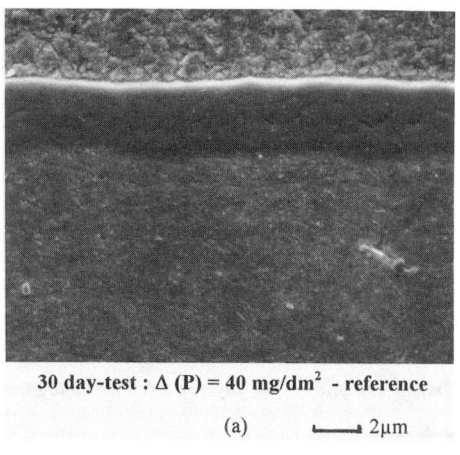

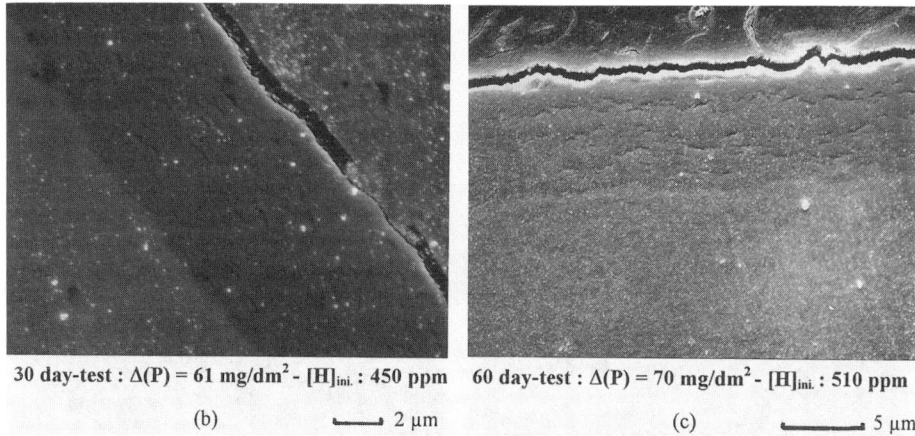

FIG. 13—*SEM oxide examination of reference samples and a hydrided sample:* (a) *reference sample after a 30-day test,* (b) *cathodic-charged sample after a 30-day test, and* (c) *a gaseous-charged sample after a 60-day test.*

ation obtained on gaseous-charged specimens since this charging method is not supposed to induce high local hydrogen pressure.
3. Finally, another strong hypothesis is that hydrides could change the protective properties of the barrier layer in the metal/oxide area without a mechanical effect. Further work should be performed to better understand how the hydrides changed the properties of the oxide "barrier" layer. This effect of hydrogen could be a modification of the transport properties in the oxide and/or a modification of the oxide morphology. High-temperature electrochemistry investigations could check this hypothesis.

Conclusions

This work confirmed that hydrogen precharging has an adverse effect on the corrosion kinetics of zirconium alloys. Large hydrogen contents (from 200 to 1000 ppm) have been introduced in Zircaloy-4 fuel cladding by gaseous charging and cathodic charging before static corrosion tests in steam at 400°C. These first results show a detrimental role of hydrogen when the corrosion is beyond the transition. This effect is far greater for high cathodic charging, as a massive hydride layer is underneath the surface. In that case, this hydride layer obviously modified the material behavior. For cathodic charging to a level less than 300 ppm, no massive hydride layer remained, but the accelerating effect is less but still present. For gaseous charging at a hydrogen level > 300 ppm, the hydrogen detrimental effect remains but is less important for this test duration. No hydrides have been observed in that case by SEM or optical microscope examination in spite of a very local hydrogen overconcentration underneath the surface (less than 1 μm thick).

Up to now, the reasons of that detrimental effect with or without hydride layer are not clearly understood. At the scale of SEM observation, the hypothesis of a mechanical degradation of the integrity of the oxide is not supported. No difference of appearance in the oxide was observed between hydrided and reference samples. Another working hypothesis is that the oxide properties (transport properties and/or morphology) are modified as the oxide grows on the massive hydride or on the metal with a high hydrogen level. Further work should be performed to obtain a better understanding.

Acknowledgment

The authors acknowledge Geraldine Mazeau for her intensive work on specimens during her training period at EDF.

References

[1] Johnson, A. B., "Thick Film Effect in the Oxidation and Hydriding of Zirconium Alloys," *Proceedings*, IAEA Technical Committee Meeting on Fundamental Aspects of Corrosion of Zirconium-Base Alloys in Water Reactor Environments, Portland, 11–15 Sept. 1989, IAEA, Vienna, 1990.
[2] Lemaignan, C., "Impact of β^- Radiolysis and Transient Product on Irradiation Enhanced Corrosion of Zirconium Alloys," *Journal of Nuclear Materials*, Vol. 187, 1992, pp. 122–130.
[3] Perkins, R. A. and Busch, R. A., "Corrosion of Zircaloy in the Presence of LiOH," *Zirconium in the Nuclear Industry: Ninth Volume, ASTM STP 1132*, C. M. Eucken and A. M. Garde, Eds., American Society for Testing and Materials, West Conshohocken, PA, 1991, pp. 595–612.
[4] Ramasubramanian, N., "Lithium Uptake and the Corrosion of Zirconium Alloys in Aqueous Lithium Hydroxide Solutions," *Zirconium in the Nuclear Industry: Ninth Volume, ASTM STP 1132*, C. M. Eucken and A. M. Garde, Eds., American Society for Testing and Materials, West Conshohocken, PA, 1991, pp. 613–627.
[5] Billot, Ph., Robin, J. C., Giordano, A., Peybernès, J., Thomazet, J., and Amanrich, H., "Experimental and Theorical Studies of Parameters that Influence Corrosion of Zircaloy-4," *Zirconium in the Nuclear Industry: Tenth Volume, ASTP STP 1245*, A. M. Garde and E. R. Bradley, Eds., American Society for Testing and Materials, West Conshohocken, PA, 1994, pp. 351–377.
[6] Garde, A. M., "Enhancement of Aqueous Corrosion of Zircaloy-4 Due to Hydride Precipitation at the Metal-Oxide Interface," *Zirconium in the Nuclear Industry: Ninth Volume, ASTM STP 1132*, C. M. Eucken and A. M. Garde, Eds., West Conshohocken, PA, 1991, pp. 566–594.
[7] Kido, T., "A Study on Enhanced Uniform Corrosion of Zircaloy-4 Cladding During High Burnup Operation in PWRs," *Proceedings*, Sixth International Symposium on Environmental Degradation of Materials in Nuclear Power Systems, San Diego, 1–5 Aug. 1993.
[8] Blat, M., Kerrec, O., Bourgoin, J., Vrignaud, E., and Amanrich, H., "Comportement à la corrosion de materiaux de gainage en Zircaloy-4 pour de forts taux de combustion dans les REP d'EDF," *Proceedings*, SFEN, Paris, International Symposium of Fontevraud III, 12–16 Sept. 1994.
[9] Kearns, J. J., "Terminal Solubility and Partitioning of Hydrogen in the Alpha Zirconium, Zircaloy 2 and Zircaloy-4," *Journal of Nuclear Materials*, Vol. 22, 1967, p. 292.
[10] Douglass, D. L., "The Metallurgy of Zirconium," *IAEA Review* (supplement), 1971.
[11] Bryner, J. S., "The Cyclic Nature of Corrosion of Zircaloy-4 in 633 K Water," *Journal of Nuclear Materials*, Vol. 82, 1979, pp. 84–101.

DISCUSSION

Bo Cheng[1] (written discussion)—You show a clear enhancement effect of hydrogen/hydride on Zircaloy-4 corrosion at 400°C. Can you compare the difference in the reaction rates of zirconium hydrides and Zircaloy-4 at 400°C and compare the results of Kido at 360°C?

M. Blat and D. Noel (authors' closure)—From our results, we conclude that we have observed a strong detrimental effect of massive δ hydrides on the corrosion rate (an enhancement factor of about 4 compared to the reference sample). That detrimental effect is less significant when the corrosion occurs on αZr and δ hydride platelet microstructure. The results of T. Kido at 360°C show that the acceleration of the weight gain becomes significant on samples with an initial hydrogen content greater than 600 ppm and for a test duration longer than 150 days ($\Delta W > 70$ mg/dm^2 for the reference sample). Except for cathodic charging with high hydrogen content, it is quite difficult to compare our results with Kido's after a 60-day test. In fact, if we put our results (initial hydrogen content and weight gain) on Kido's curves, they are at the beginning of the corrosion rate increase observed.

B. Cox[2] (written discussion)—Did you reprepare your surfaces after hydriding so that they were all comparable? If not, you have an unknown contributing effect from different surface states and impurities. For instance, you seem to have been reducing organic impurities in your cathodic charging and leaving a surface contaminated with carbon.

M. Blat and D. Noel (authors' closure)—We do not reprepare our surface after hydriding. The depth of the metal that could be affected by a different hydriding method is less than 1 μm (GDOS analyses), and we do not think that surface pollution could increase the corrosion rate with respect to test duration.

P. Rudling[3] (written discussion)—Do you believe that the hydrides at the metal/oxide interface may accelerate the corrosion rate by destroying the epitaxy between metal and oxide, thereby forming equiaxed less protective grains (instead of the protective columnar grains)?

M. Blat and D. Noel (authors' closure)—Some TEM examinations will be performed to investigate the oxide morphology. Therefore, we cannot answer this question. Nevertheless, the proposed assumption could explain the detrimental effect of the hydrides on the corrosion rate.

V. Urbanic[4] (written discussion)—Did you perform any hydrogen analyses of the samples after the corrosion tests to determine what effect the hydrogen content of the base metal has on subsequent hydrogen pickup during corrosion?

M. Blat and D. Noel (authors' closure)—Hydrogen analyses were carried out on all specimens before the corrosion test and on the specimens removed from the autoclave. However, it is difficult to determine the actual hydrogen pickup for the high corrosion rate (cathodic charging with high H content). Indeed, for these specimens, the hydride layer is localized on the external surface, and, after the corrosion test, we have external and internal oxydation. Furthermore, for all prehydrided specimens, it seems that hydrogen desorption occurs in the first steps of the corrosion. Indeed, our evaluation of the hydrogen pickup is quite similar for reference and prehydrided specimens (about 30% hydrogen pickup after a 60-day test in steam).

---

[1] Electric Power Research Institute, Palo Alto, Canada.
[2] University of Toronto, Toronto, Ontario, Canada.
[3] ABB Atom, Sweden.
[4] Atomic Energy of Canada, Ltd., Chalk River, Ontario, Canada.

J. B. Bai[5] (written discussion)—Do you obtain the same hydrides by cathodic and gaseous hydriding? How about the residual stresses generated by the above two hydriding methods and their effect on the detrimental effect mentioned?

M. Blat and D. Noel (authors' closure)—The cooling rate after gaseous charging (470°C) and homogenizing heat treatment for cathodic hydriding (430°C) are quite similar and induce δ hydride precipitation. The nature of the massive hydride layer has been identified by X-ray diffraction analysis: δ hydride phase is confirmed. The residual stresses generated by the two hydriding methods have not yet been determined.

---

[5] Lab. MSS/MAT, Ecole Centrale Paris, France.

Bruce F. Kammenzind,[1] David G. Franklin,[1] H. Richard Peters,[2] and Walter J. Duffin[1]

Hydrogen Pickup and Redistribution in Alpha-Annealed Zircaloy-4*

REFERENCE: Kammenzind, B. F., Franklin, D. G., Peters, H. R., and Duffin, W. J., "**Hydrogen Pickup and Redistribution in Alpha-Annealed Zircaloy-4,**" *Zirconium in the Nuclear Industry: Eleventh International Symposium, ASTM STP 1295,* E. R. Bradley and G. P. Sabol, Eds., American Society for Testing and Materials, 1996, pp. 338–370.

ABSTRACT: Zircaloy-4, which is used widely as a core structural material in pressurized water reactors (PWR), picks up hydrogen during service. Hydrogen solubility in Zircaloy-4 is low, and hydrides precipitate after the Zircaloy-4 matrix becomes supersaturated with hydrogen. These hydrides embrittle the Zircaloy-4. To study hydrogen pickup and concentration, a post-irradiation nondestructive radiographic technique for measuring hydrogen concentration was developed and qualified. Experiments on hydrogen pickup were conducted in the advanced test reactor (ATR). Ex-reactor tests were conducted to determine the conditions in which hydrogen would dissolve, migrate, and precipitate. Finally, a phenomenological model for hydrogen diffusion was indexed to the data. This presentation describes the equipment and the model, presents the results of experiments, and compares the model predictions to experimental results.

KEYWORDS: Zircaloys, hydrogen, corrosion, diffusion, hydrogen redistribution, hydrogen solubility

During service, Zircaloys pick up hydrogen, primarily from hydrogen liberated as part of the oxidation of Zircaloy by water [1]. Hydrogen in solution will diffuse down a thermal gradient and concentrate in cold regions of Zircaloy components [2]. Due to the low solubility of hydrogen in Zircaloys, zirconium hydrides form at relatively low concentrations of hydrogen, especially on cooling to near room temperature. These hydrides are brittle and at low concentration begin to degrade the mechanical properties of Zircaloys [3]. These changes in properties need to be considered in component evaluations. Fundamental data and models that describe hydrogen pickup and redistribution can be used in this evaluation process. To provide the bases for such evaluations, in-reactor and out-of-reactor experiments were conducted to develop a database for a fundamental model that provides insight into the phenomena involved.

The low solubility of hydrogen in zirconium alloys and the detrimental effects of hydrides were recognized in the 1950s [4] during the early development of zirconium alloys for nuclear

---

[1] Bettis Atomic Power Laboratory, Westinghouse Electric Corporation, West Mifflin, PA 15122.
[2] Knolls Atomic Power Laboratory, Martin Marietta Corp., Schenectady, NY 12301-1072.
* This report was prepared as an account of work sponsored by the United States Government. Neither the United States, nor the United States Department of Energy, nor any of their employees, nor any of their contractors, subcontractors, or their employees, makes any warranty, express or implied, or assumes any legal liability or responsibility for the accuracy, completeness or usefulness of any information, yapparatus, product or process disclosed, or represents that its use would not infringe privately owned rights.

application. Under normal conditions, the influx of hydrogen in-reactor is quite low. Further, it diffuses rapidly in the Zircaloy matrix, and the hydride does not form at the corroding surface and propagate into the metal as does the oxide. Rather, hydrides precipitate heterogeneously as delta hydride, Zr_2H_3, throughout the metal at sites that can best accommodate the 17% volume increase on formation of the hydride. Due to the protective nature of zirconium oxide, corrosion and hydrogen pickup are limited. As long as the environment is not too reducing, the only hydrogen pickup in the metal is from the corrosion process [5]. Therefore, hydrogen is not a concern to clad integrity under normal operating conditions, at least until much corrosion has occurred or hydrogen has concentrated in cold regions.

One of the first modes of fuel rod failure in commercial power reactors was from hydride blisters, which consisted of a high concentration of hydrides in a local cladding region [6]. Hydride blisters formed either from internal moisture in the fuel rod, e.g., moisture in the fuel-pellet open porosity, or from coolant entering through fuel-cladding defects. Hydride failures were almost eliminated in the 1970s by special drying methods for fuel pellets and by reductions in defects of other types.

Performance of Zircaloy-clad fuel rods has been excellent since 1980 [7]. However, isolated instances of hydride failures have occurred. Steep local temperature gradients may have caused concentration of hydrogen in cladding on PWR fuel rods [8]. These concentrations of hydrides may have been sufficient to embrittle the Zircaloy-4 metal and promote clad cracking. More recently, use of zirconium liners on fuel cladding may have promoted both hydrogen pickup in defected fuel rods due to the relatively poor corrosion resistance of zirconium liners and high stresses from the liner volume expansion as it oxidizes [9–11]. An intermediate step may be the preferential hydride formation in the zirconium liner over that of the adjacent Zircaloy-2 despite the higher liner temperature due to the lower strength of the liner (Fig. 14 of Ref 10). Hydrides form preferentially in regions of low strength due to the volume expansion during hydride formation [12]. This expansion may increase stresses in the Zircaloy-2 prior to full liner oxidation. Subsequent oxidation of the liner removes evidence of this step. This failure mode illustrates the complicated nature and importance of hydrogen diffusion and hydride formation. Understanding hydrogen behavior becomes more important as fuel burnups are increased, which results in increased corrosion and hydrogen pickup [13].

Fundamental experiments and modeling of diffusion kinetics were reported during early development of zirconium alloys for nuclear applications. Appropriate references are provided below in the discussion of the results presented herein. Recent work primarily addressed hydrogen in Zr-2.5Nb pressure tubes, as reported by Canadian researchers [5,14–17]. These experiments are characterized by studies by hydrogen concentration profiles rather than bulk hydrogen pickup measurements. Little recent work on Zircaloys has been reported. These data and associated models are required to predict the pickup, movement, and precipitation of hydrogen. As shown by the discussion above on the impact of hydrogen on fuel performance, such a capability is important if it can be used to avoid designing and operating fuel elements in a region where large concentrations of hydrides and stresses develop.

Hydrogen Pickup

Experimental

Hydrogen pickup was measured on alpha-annealed Zircaloy-4 coupons that were irradiated in the advanced test reactor (ATR) at 310°C (590°F) and 360°C (680°F). A hydrogen overpressure is maintained in the cooling water to control the oxygen level in the coolant. Under normal conditions, this molecular hydrogen does not act as a hydrogen source, as the oxide film is essentially impervious to the molecule. The source of the hydrogen entering the Zircaloy is that produced in the corrosion reaction. A nondestructive method of measuring hydrogen content was developed and qualified. This method measured the neutron attenuation by absorbed hydrogen as a neutron beam passed through a specimen (Fig. 1). A neutron beam with a white noise spectrum was filtered by passing it through indium-cadmium material. This removed most neutrons with energies that activated indium atoms, resulting in the spectrum shown in Fig. 1 as the incident notched spectrum. Passing this beam through a thin foil of indium (in Detector D_1 in Fig. 1) activated this foil in proportion to the remaining number of neutrons at the indium activation energies. This foil was counted to produce the signal value D_1. The neutron beam then passed through the Zircaloy-4 sample. The neutrons interacted with hydrogen atoms in the Zircaloy-4, which reduced the energy of many of the neutrons. This neutron scattering resulted in many neutrons having their energies decreased to the energy bands for indium activation. This new neutron beam energy spectrum is shown in Fig. 1 as the moderated notched spectrum. The moderated beam passed through another indium foil (In detector D_2 in Fig. 1) and activated the indium in proportion to the numbers of neutrons at the indium energies. This foil was counted to determine D_2. The relative difference between D_2 and D_1 was proportional to the amount of hydrogen in the specimen, $(D_2 - D_1)/D_1$. The attenuation of the beam was calibrated to the hydrogen content of Zircaloy-4 by the use of

I. INERT GAS FUSION OF CORROSION COUPON SECTIONS

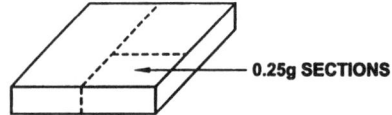

II. NEUTRON BEAM ATTENUATION

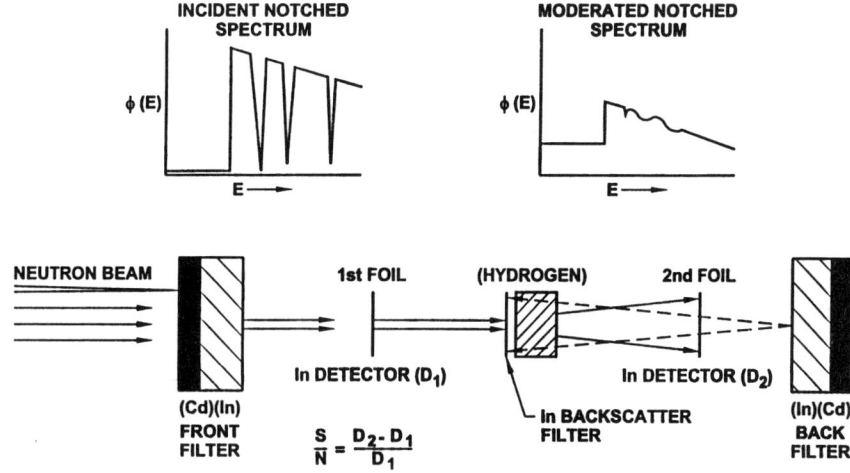

FIG. 1—*Methods used to measure the hydrogen pickup.*

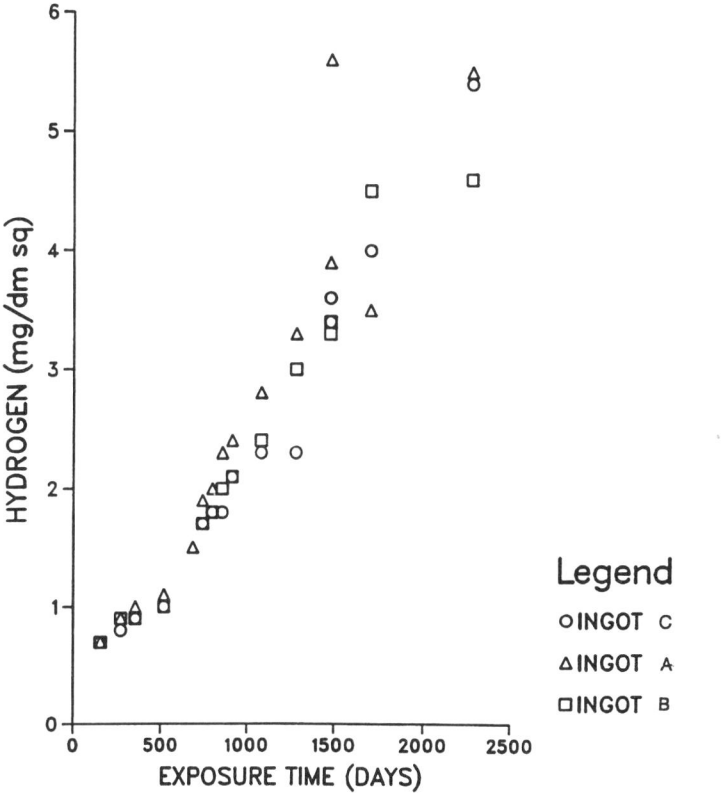

FIG. 2—*Hydrogen pickup of alpha-annealed Zircaloy-4 exposed in autoclave at 316°C.*

standards. In addition, each set of specimens included a standard with a known concentration of hydrogen. Since the method was nondestructive, interim measurements were made on coupons using an on-site neutron beam. This allowed the hydrogen pickup of individual samples to be followed as a function of radiation exposure. Out-of-reactor hydrogen pickup was measured after autoclave exposure by destructive hot vacuum extraction to obtain reference data.

Results

Out-of-reactor reference pickup data on Zircaloy-4 coupons from three ingots are shown in Fig. 2 for 316°C exposure. Autoclave water was degassed and deionized at pH 207 with a resistivity above one million ohm centimeters. Pressure was held above that for saturation at 316°C, about 10.65 MPa. Similar in-reactor pickup data are shown in Figs. 3 and 4 for exposure at 360 and 310°C, respectively, for three neutron flux levels: high flux or $\phi > 10^{18}$ n/m²/s**, medium flux or 10^{18} n/m²/s $> \phi > 10^{17}$ n/m²/s, and low flux or $\phi < 10^{17}$ n/m²/s. The data are

** All fluxes are for $E > 1$ MeV.

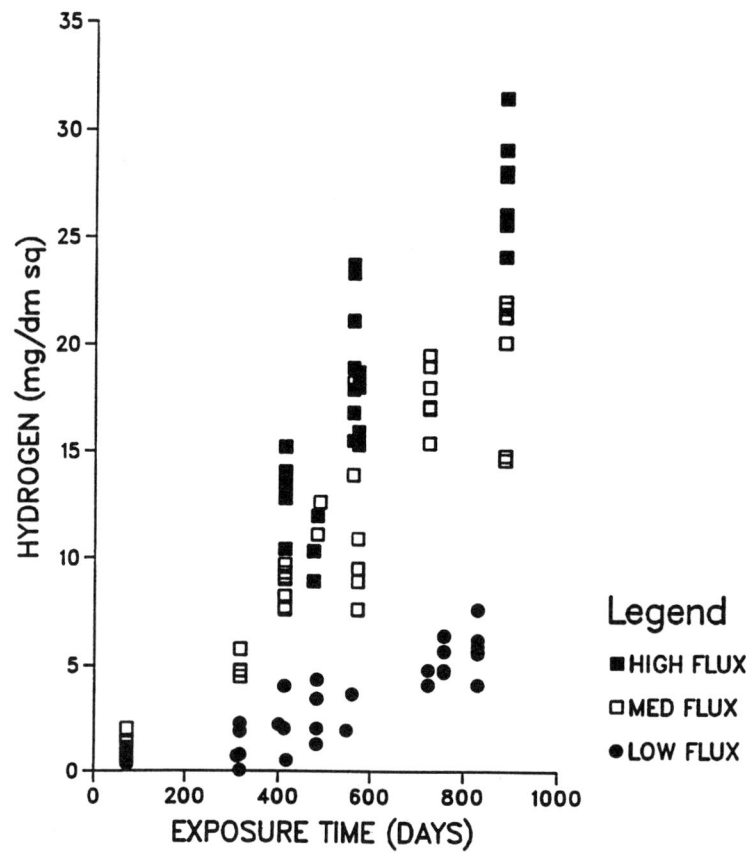

FIG. 3—*Hydrogen pickup of alpha-annealed Zircaloy-4 exposed in the ATR at 360°C.*

listed in Table 1 for autoclave exposures and Table 2 for in-reactor exposures. Table 3 lists the chemistry of the Zircaloy-4 used in the testing. To determine if the hydrogen pickup was correlated with sample oxidation, the measured hydrogen content was plotted as a function of measured oxygen concentration as determined by weight-gain measurements and as shown in Fig. 5. This suggested that most of the hydrogen absorbed during service was from the oxidation. Further, it supported use of a factor to estimate hydrogen pickup from measured or predicted oxidation. In general, the hydrogen pickup is correlated to the oxidation. Above 1000 (mg/dm^2), the pickup ratio ($\Delta H/\Delta O$) tends to deviate in both directions from the average value. This is due to a sensitivity of the hydrogen pickup ratio ($\Delta H/\Delta O$) to temperature and fluence as suggested in Fig. 6.

Hydrogen Content of Oxide Films

Experimental

The hydrogen concentration in the oxide film may be different from the hydrogen concentration in the Zircaloy-4 metal. Several studies have shown this to be the case [*14,18*]. High or

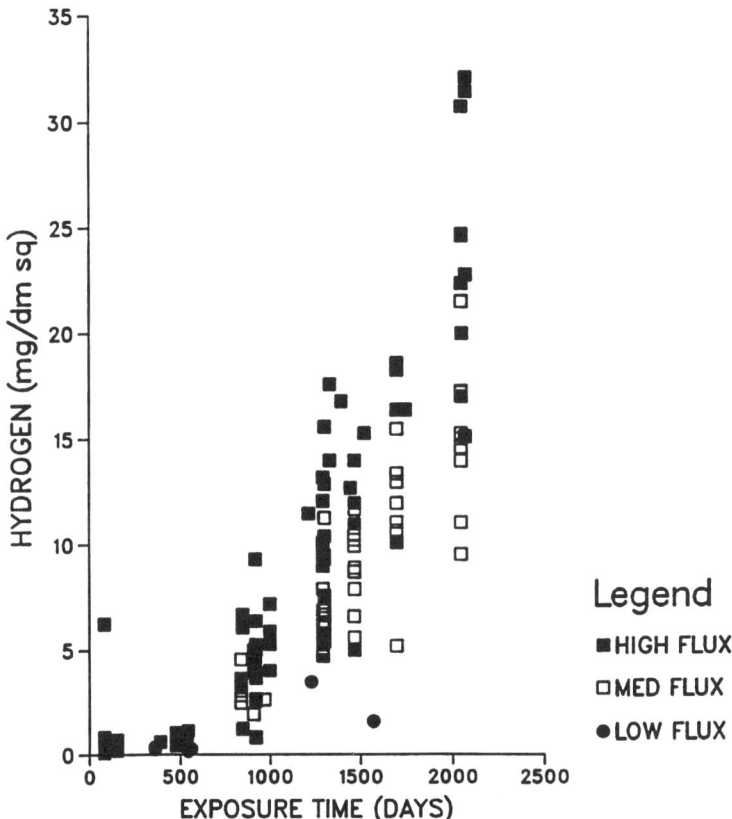

FIG. 4—*Hydrogen pickup of alpha-annealed Zircaloy-4 exposed in the ATR at 310°C.*

low hydrogen concentrations in the oxide can have an impact on interpretation of hydrogen measurements to determine $\Delta H/\Delta O$ ratios and hydrogen migration. To determine the impact of this effect, secondary ion mass spectroscopy (SIMS) was used to determine the concentration distribution through the oxide and metal of an autoclave Zircaloy-4 specimen. The hydrogen distribution is shown in Fig. 7 in which hydrogen is uniformly dispersed in the oxide and concentrated in hydrides in the metal.

To quantify the hydrogen concentrations in the oxide and metal, autoclaved and in-reactor corrosion coupons were tested for hydrogen concentration, with and without the oxide film, as shown in Fig. 8. The oxide was removed with silicon carbide paper or with a Dremel tool. Both were used under flowing water to avoid heating the material. The results from autoclave specimens tested at 360°C, 19.2 MPa in deionized degassed water are given in Table 4. The results from in-reactor testing are shown in Figs. 9 and 10. Out-of-reactor samples exhibited similar hydrogen concentrations in the oxide and metal. However, in-reactor samples had hydrogen concentrations in the oxide much greater than that in the metal. Figures 9 and 10 show how the in-reactor hydrogen concentrations in the oxide varied with fast fluence (>1 MeV) and oxide thickness. There is considerable scatter in both and no sign of a simple dependence of thickness, fluence, or fast flux. The hydrogen content of the oxide saturated quickly and

TABLE 1—*Weight gain of alpha-annealed Zircaloy-4 exposed at 316°C (589 K) in autoclave (mg/dm sq).*[a]

Spec	Ingot	Days	O-WTGN	H-WTGN	H/O
1	A	1470.0	95.30	5.60	0.0588
2	A	854.0	49.70	2.30	0.0463
3	A	910.0	52.60	2.40	0.0456
4	A	1078.0	63.20	2.80	0.0443
5	A	266.0	21.10	0.90	0.0427
6	A	1276.0	76.70	3.30	0.0430
7	A	798.0	48.00	2.00	0.0417
8	A	1470.0	95.00	3.90	0.0411
9	A	742.0	47.10	1.90	0.0403
10	A	154.0	18.30	0.70	0.0383
11	A	350.0	26.00	1.00	0.0385
12	A	686.0	40.50	1.50	0.0370
13	A	2267.0	155.40	5.50	0.0354
14	A	518.0	27.90	1.10	0.0394
15	A	1694.0	111.40	3.50	0.0314
16	B	742.0	38.30	1.70	0.0444
17	B	910.0	52.90	2.10	0.0397
18	B	266.0	23.10	0.90	0.0390
19	B	1694.0	116.40	4.50	0.0387
20	B	854.0	52.00	2.00	0.0385
21	B	1078.0	62.60	2.40	0.0383
22	B	798.0	48.20	1.80	0.0373
23	B	1276.0	81.90	3.00	0.0366
24	B	1470.0	97.50	3.40	0.0349
25	B	518.0	29.00	1.00	0.0345
26	B	1470.0	97.60	3.30	0.0338
27	B	154.0	21.30	0.70	0.0329
28	B	350.0	28.10	0.90	0.0320
29	B	2267.0	150.70	4.60	0.0305
30	C	910.0	53.90	2.10	0.0390
31	C	1470.0	92.30	3.60	0.0390
32	C	1078.0	60.70	2.30	0.0379
33	C	798.0	49.20	1.80	0.0366
34	C	1470.0	91.50	3.40	0.0372
35	C	1694.0	106.90	4.00	0.0377
36	C	2267.0	147.70	5.40	0.0366
37	C	350.0	24.10	0.90	0.0359
38	C	742.0	47.30	1.70	0.0359
39	C	854.0	51.20	1.80	0.0352
40	C	154.0	20.30	0.70	0.0345
41	C	266.0	24.20	0.80	0.0331
42	C	518.0	30.00	1.00	0.0333
43	C	1276.0	76.70	2.30	0.0300

[a] Surface initially etched.

TABLE 2—*Weight gains (mg/dm^2) of alpha-annealed Zircaloy-4 exposed in the advanced test reactor.*

				360°C (633.1 K)					
Spec	Ingot	Film	Days	AVFLUX	Fluence	O-WTGN	H-WTGN	H/O	TEMP (K)
1	B	11.18	71.7	0.539	3.3	50.4	1.1	0.0218	630.4
2	B	11.17	886.3	0.760	58.2	1010.6	21.4	0.0212	626.5
2	B	11.17	723.8	0.782	48.9	814.3	17.0	0.0209	626.9
2	B	11.17	412.1	0.797	28.4	445.1	9.0	0.0202	628.3
3	B	10.35	71.7	0.873	5.4	54.8	1.3	0.0237	630.9
4	B	10.76	886.2	1.261	96.6	1326.4	26.1	0.0197	627.6
5	B	11.18	412.1	1.199	42.7	547.2	12.8	0.0234	629.6
5	B	11.18	558.6	1.261	60.9	773.5	23.7	0.0306	628.9
6	B	0.00	71.7	0.539	3.3	59.8	1.0	0.0167	630.4
7	B	0.00	71.7	0.873	5.4	63.7	2.0	0.0314	630.9
8	C	11.59	412.1	1.199	42.7	492.0	15.2	0.0309	629.6
8	C	11.59	558.6	1.261	60.9	701.4	23.4	0.0334	628.9
9	C	11.59	886.2	1.261	96.6	1209.6	28.1	0.0232	627.6
10	L	1.65	481.5	0.024	1.0	262.5	1.9	0.0074	630.9
11	C	11.59	886.2	0.760	58.2	933.9	21.3	0.0228	626.5
11	C	11.59	723.8	0.782	48.9	749.0	19.5	0.0260	626.9
11	C	11.59	412.1	0.797	28.4	409.9	7.6	0.0185	628.3
12	L	1.24	481.5	1.224	50.9	605.2	11.1	0.0183	628.7
13	D	10.35	886.2	0.760	58.2	1008.6	20.1	0.0199	626.5
13	D	10.35	723.8	0.782	48.9	814.7	17.0	0.0209	626.9
13	D	10.35	412.1	0.797	28.4	446.0	9.3	0.0209	628.3
14	D	9.93	886.2	1.261	96.6	1318.3	29.1	0.0221	627.6
15	D	9.93	412.1	1.199	42.7	546.8	14.0	0.0256	629.6
15	D	9.93	558.6	1.261	60.9	776.9	21.1	0.0272	628.9
16	L	14.08	611.8	1.388	73.4	792.6	15.8	0.0200	628.4
17	E	12.01	412.1	1.199	42.7	539.8	12.8	0.0237	629.6
17	E	12.01	558.6	1.261	60.9	754.7	18.9	0.0250	628.9
18	E	10.76	886.2	0.760	58.2	1003.8	21.7	0.0216	626.5
18	E	10.76	723.8	0.782	48.9	811.2	18.0	0.0222	626.9
18	E	10.76	412.1	0.797	28.4	441.7	9.7	0.0220	628.3
19	E	10.75	886.2	1.261	96.6	1306.9	29.1	0.0223	627.6
20	F	11.59	71.7	0.539	3.3	46.0	1.4	0.0305	630.4
21	F	12.42	412.1	1.199	42.7	543.9	13.2	0.0243	629.6
21	F	12.42	558.6	1.261	60.9	777.2	16.8	0.0216	628.9
22	F	11.59	71.7	0.873	5.4	49.7	1.4	0.0282	630.9
23	F	11.59	886.2	0.760	58.2	1022.5	22.0	0.0215	626.5
23	F	11.59	723.8	0.782	48.9	820.0	19.0	0.0232	626.9
23	F	11.59	412.1	0.797	28.4	447.6	9.1	0.0203	628.3
24	F	10.76	886.2	1.261	96.6	1319.8	27.9	0.0211	627.6
25	F	0.41	71.7	0.873	5.4	54.2	1.1	0.0203	630.9
26	F	0.83	71.7	0.539	3.3	55.5	1.0	0.0180	630.4
27	G	12.83	886.2	0.760	58.2	953.4	20.1	0.0211	626.5
27	G	12.83	723.8	0.782	48.9	770.0	17.1	0.0222	626.9
27	G	12.83	412.1	0.797	28.4	418.5	8.2	0.0196	628.3
28	G	12.84	886.2	1.261	96.6	1241.3	31.5	0.0254	627.6
29	G	12.83	412.1	1.199	42.7	513.9	13.8	0.0269	629.6
29	G	12.83	558.6	1.261	60.9	735.0	23.3	0.0317	628.9
30	D	8.82	570.2	1.282	63.2	811.9	15.3	0.0188	628.2
31	D	11.41	570.2	0.218	10.7	432.6	9.5	0.0220	626.2
32	D	12.29	757.9	0.000	0.0	232.1	6.4	0.0276	627.2
33	E	11.41	570.2	1.282	63.2	793.7	15.9	0.0200	628.2
34	E	9.65	570.2	0.218	10.7	437.3	10.9	0.0249	626.2
35	E	11.41	757.9	0.000	0.0	245.3	5.7	0.0232	627.2
36	C	11.41	570.2	1.282	63.2	746.3	18.7	0.0251	628.2
37	N	1.66	723.8	0.000	0.0	283.1	4.8	0.0170	631.0

(Continued)

TABLE 2—Continued

360°C (633.1 K)

Spec	Ingot	Film	Days	AVFLUX	Fluence	O-WTGN	H-WTGN	H/O	TEMP (K)
38	C	12.18	570.2	0.218	10.7	410.4	8.9	0.0217	626.2
39	N	1.66	829.6	0.000	0.0	261.6	4.1	0.0157	628.1
40	C	12.35	757.9	0.000	0.0	218.5	5.7	0.0261	627.2
41	N	2.07	1581.2	0.001	0.1	391.0	8.8	0.0226	618.1
42	N	1.66	481.5	0.055	2.3	345.5	4.3	0.0123	630.9
43	G	10.58	570.2	1.282	63.2	756.9	18.0	0.0238	628.2
44	G	11.46	570.2	0.218	10.7	425.7	10.9	0.0256	626.2
45	G	11.46	757.9	0.000	0.0	228.9	4.8	0.0210	627.2
46	B	12.35	570.2	1.282	63.2	804.3	18.4	0.0229	628.2
47	B	10.53	570.2	0.218	10.7	437.6	7.6	0.0174	626.2
48	B	11.47	757.9	0.000	0.0	240.4	4.7	0.0196	627.2
49	G	2.48	558.6	0.916	44.2	640.1	13.9	0.0217	628.3
50	G	2.08	481.5	1.002	41.7	581.4	12.6	0.0217	629.2
51	G	2.48	481.5	0.000	0.0	193.7	2.9	0.0150	629.6
51	G	2.48	829.6	0.000	0.0	271.0	5.6	0.0207	628.1
52	O	0.83	829.6	0.000	0.0	288.1	5.9	0.0205	627.2
53	O	14.08	886.2	0.315	24.1	779.7	14.8	0.0190	627.1
54	O	17.38	412.1	0.104	3.7	260.0	4.0	0.0155	627.9
55	O	13.66	829.6	0.000	0.0	269.8	7.6	0.0282	627.2
56	O	13.66	71.7	0.000	0.0	15.2	0.3	0.0198	632.6
57	O	14.49	71.7	0.175	1.1	38.6	0.5	0.0130	629.8
58	O	14.90	412.1	1.147	40.8	550.4	13.4	0.0243	629.7
58	O	14.90	886.2	1.210	92.6	1284.2	25.6	0.0199	628.0
58	O	14.90	558.6	1.215	58.6	798.0	17.9	0.0224	629.1
59	O	14.08	412.1	0.000	0.0	103.1	2.0	0.0194	630.0
60	O	14.91	71.7	0.851	5.3	54.4	1.4	0.0257	630.9
61	P	1.66	829.6	0.000	0.0	267.6	6.2	0.0232	628.1
62	Q	0.00	829.6	0.000	0.0	268.1	5.7	0.0213	627.9
63	Q	−2.48	558.6	0.000	0.0	237.2	3.7	0.0156	631.4
64	Q	0.41	723.8	0.000	0.0	328.8	4.1	0.0125	630.6
65	Q	1.24	558.6	1.000	48.3	763.5	18.3	0.0240	629.4
66	Q	−0.83	481.5	0.937	39.0	652.3	12.0	0.0185	630.6
67	Q	0.83	558.6	0.000	0.0	177.9	3.6	0.0202	629.6
68	Q	2.07	481.5	0.000	0.0	153.6	3.4	0.0223	629.5
69	B	2.49	417.5	0.000	0.0	89.9	0.5	0.0056	625.8
70	B	2.06	474.1	1.110	45.5	480.6	8.9	0.0185	625.5
71	B	2.48	311.7	0.000	0.0	121.3	0.7	0.0058	628.9
71	B	2.48	547.2	0.000	0.0	159.2	1.9	0.0121	628.9
72	B	2.48	474.1	1.180	48.3	486.8	10.3	0.0212	625.5
73	I	12.34	317.4	0.163	4.5	190.9	4.8	0.0249	624.0
74	I	11.84	317.4	0.020	0.5	159.1	0.0	0.0003	628.4
75	H	12.60	317.4	0.163	4.5	187.6	5.8	0.0307	624.0
76	H	13.73	317.4	0.020	0.5	154.8	0.8	0.0050	628.4
77	B	12.16	317.4	0.163	4.5	183.5	4.5	0.0244	624.0
78	B	13.20	317.4	0.020	0.5	165.4	2.2	0.0135	628.4
79	V	15.36	317.4	0.020	0.5	134.1	1.9	0.0139	628.4
80	V	14.71	317.4	0.163	4.5	181.6	4.7	0.0261	624.0
81	B	0.82	400.7	0.001	0.0	145.1	2.2	0.0151	628.5
82	A	10.65	71.7	0.539	3.3	51.8	0.6	0.0116	630.4
83	A	11.44	886.2	0.760	58.2	1039.4	14.6	0.0140	626.5
83	A	11.44	723.8	0.782	48.9	837.2	15.4	0.0184	626.9
83	A	11.44	412.1	0.797	28.4	455.2	7.7	0.0169	628.3
84	A	11.46	71.7	0.873	5.4	57.1	1.0	0.0175	630.9
85	A	11.47	886.2	1.261	96.6	1276.3	24.1	0.0189	627.6
86	A	9.55	412.1	1.199	42.7	553.3	10.4	0.0188	629.6
86	A	9.55	558.6	1.261	60.9	776.2	15.5	0.0200	628.9

TABLE 2—Continued

360°C (633.1 K)

Spec	Ingot	Film	Days	AVFLUX	Fluence	O-WTGN	H-WTGN	H/O	TEMP (K)
87	A	0.41	71.7	0.539	3.3	57.2	0.9	0.0157	630.4
88	A	−1.06	71.7	0.873	5.4	52.9	0.7	0.0132	630.9

310°C (583 K)

Spec	Ingot	Film	Days	AVFLUX	Fluence	O-WTGN	H-WTGN	H/O	TEMP (K)
89	B	12.00	1569.9	0.000	0.0	41.3	1.6	0.0378	580.1
90	B	10.25	997.5	1.471	126.7	208.2	7.2	0.0346	584.4
91	F	11.10	1210.3	1.471	153.8	943.8	11.5	0.0122	584.1
92	U	16.17	804.2	1.489	103.5	163.3	1.9	0.0116	582.1
93	D	8.77	997.5	1.661	143.1	337.7	4.0	0.0118	586.1
94	D	11.41	907.7	0.594	46.6	213.5	4.7	0.0220	588.4
94	D	11.41	1464.3	0.609	77.0	513.8	10.2	0.0199	588.2
95	E	9.66	997.5	1.661	143.1	326.5	5.5	0.0168	586.1
96	E	10.44	907.7	0.594	46.6	192.3	4.6	0.0239	588.4
96	E	10.44	1464.3	0.609	77.0	480.8	10.5	0.0218	588.2
97	C	11.41	997.5	1.661	143.1	229.3	5.3	0.0231	586.1
98	C	11.41	907.7	0.594	46.6	153.8	1.9	0.0124	588.4
99	C	11.41	1464.3	0.609	77.0	411.0	8.7	0.0212	588.2
100	G	9.70	997.5	1.661	143.1	509.9	5.9	0.0116	586.1
101	G	10.59	907.7	0.594	46.6	175.0	3.9	0.0223	588.4
101	G	10.59	1464.3	0.609	77.0	568.9	6.6	0.0116	588.2
102	B	12.35	997.5	1.661	143.1	319.5	5.8	0.0182	586.1
103	B	11.46	907.7	0.594	46.6	217.2	5.0	0.0230	588.4
103	B	11.46	1464.3	0.609	77.0	502.8	9.9	0.0198	588.2
103	B	11.46	2107.6	0.623	113.4	961.1	19.5	0.0202	588.0
104	I	11.70	838.2	1.697	122.9	98.3	3.6	0.0365	584.6
104	I	11.70	999.8	1.713	148.0	283.0	4.4	0.0155	584.2
105	I	12.37	999.8	0.300	25.9	103.5	3.8	0.0368	580.1
105	I	12.37	838.2	0.304	22.0	76.1	2.8	0.0368	580.7
106	H	13.18	999.8	1.713	148.0	396.6	8.6	0.0218	584.2
107	H	12.91	999.8	0.300	25.9	106.0	3.7	0.0347	580.1
107	H	12.91	838.2	0.304	22.0	75.4	2.4	0.0325	580.7
108	J	12.81	999.8	1.713	148.0	470.2	5.7	0.0122	584.2
109	J	14.49	999.8	0.300	25.9	95.3	4.2	0.0445	580.1
109	J	14.49	838.2	0.304	22.0	68.4	2.1	0.0304	580.7
110	B	11.73	838.2	1.697	122.9	108.6	3.2	0.0292	584.6
110	B	11.73	999.8	1.713	148.0	237.4	5.9	0.0249	584.2
111	V	13.96	838.2	1.697	122.9	162.8	3.5	0.0216	584.7
111	V	13.96	999.8	1.713	148.0	480.2	8.3	0.0173	584.2
112	V	15.01	999.8	0.300	25.9	116.3	3.3	0.0281	580.1
112	V	15.01	838.2	0.304	22.0	84.5	4.6	0.0539	580.7
113	B	2.05	1406.5	1.388	168.7	744.7	14.7	0.0198	586.2
114	S	13.67	804.2	1.489	103.5	113.0	1.5	0.0130	582.1
115	W	10.00	543.3	0.000	0.0	12.4	0.1	0.0121	588.4
116	W	10.80	543.3	1.355	63.6	57.4	1.1	0.0193	588.7
117	W	10.40	155.3	1.340	18.0	13.0	0.7	0.0522	586.7
118	W	10.80	155.3	1.340	18.0	13.9	0.2	0.0151	586.7
119	W	10.00	387.9	1.340	44.9	28.1	0.6	0.0206	588.2
120	B	8.90	538.0	0.528	24.5	37.2	0.8	0.0212	584.4
121	B	9.92	1745.4	1.196	180.4	1051.5	16.4	0.0156	587.3
121	B	9.92	2048.0	1.198	212.0	1421.6	20.0	0.0141	587.2
122	B	9.80	538.0	1.271	59.1	32.7	0.5	0.0153	587.2
123	A	1.46	2045.7	0.465	82.2	722.2	9.6	0.0132	588.2
124	A	9.75	1297.3	1.537	172.3	394.2	5.4	0.0137	584.9
124	A	9.75	2045.7	1.585	280.1	1215.5	17.0	0.0140	584.6
125	A	9.75	1297.3	0.437	49.0	336.2	5.6	0.0167	588.3

(Continued)

TABLE 2—Continued

360°C (633.1 K)

Spec	Ingot	Film	Days	AVFLUX	Fluence	O-WTGN	H-WTGN	H/O	TEMP (K)
125	A	9.75	2045.7	0.465	82.2	727.1	11.1	0.0152	588.2
126	F	10.57	1297.3	0.437	49.0	351.1	11.3	0.0322	588.3
126	F	10.57	2045.7	0.465	82.2	848.5	24.6	0.0290	588.2
127	F	9.76	1297.3	1.537	172.3	923.0	15.6	0.0169	584.9
127	F	9.76	2045.7	1.585	280.1	2279.8	30.8	0.0135	584.6
128	F	0.98	2045.7	0.465	82.2	811.3	21.5	0.0265	588.2
129	B	0.65	1297.3	1.537	172.3	665.4	9.3	0.0140	584.9
130	B	0.74	1297.3	0.437	49.0	352.8	5.4	0.0153	588.3
130	B	0.74	2045.7	0.465	82.2	776.7	14.0	0.0180	588.2
131	D	8.94	1297.3	1.537	172.3	631.7	12.9	0.0204	584.9
131	D	8.94	2045.7	1.585	280.1	1927.3	24.7	0.0128	584.6
132	D	8.94	1297.3	0.437	49.0	337.1	7.1	0.0211	588.3
132	D	8.94	2045.7	0.465	82.2	793.9	15.0	0.0189	588.2
133	D	0.81	2045.7	0.465	82.2	785.2	15.3	0.0195	588.2
134	E	8.95	1297.3	1.537	172.3	568.3	7.5	0.0132	584.9
135	E	8.94	1297.3	0.437	49.0	338.0	6.7	0.0198	588.3
135	E	8.94	2045.7	0.465	82.2	823.4	17.2	0.0209	588.2
136	C	8.94	2045.7	1.585	280.1	1479.2	22.4	0.0151	584.6
137	C	0.98	1297.3	1.537	172.3	461.1	9.5	0.0206	584.9
138	C	0.98	1297.3	0.437	49.0	287.0	6.1	0.0213	588.3
138	C	0.98	2045.7	0.465	82.2	669.0	14.6	0.0218	588.2
139	G	8.95	1297.3	1.537	172.3	958.9	10.4	0.0108	584.9
140	G	8.94	1297.3	0.437	49.0	329.8	7.6	0.0230	588.3
140	G	8.94	2045.7	0.465	82.2	910.4	17.3	0.0190	588.2

NOTE: FILM = oxide prefilm (mg/dm^2). Negative values caused by measurement error. FLUX = (n/cm^2/sec) × 10. FLUENCE = (n/cm^2) × 10

TABLE 3—*Chemistry of the Zircaloy-4[a] exposed in the advanced test reactor (ppm by weight).*

Ingot	Melts	Sn	Fe	O	CR	Al	C	Cu	Hf	Mn	Ni	Nb	Si	Ta	Ti
A	2	14400	1500	760	90	35	101	<40	<125	<25	15	<100	55	<200	<29
B	2	14300	2100	1400	1100	<35	<135	<25	93	<25	<35	<50	<60	<100	<25
C	2	13200	2300	1400	1100	<35	<135	<25	<80	<25	<35	<50	<60	<100	<25
D	2	14500	1900	1400	1200	<35	<135	28	92	<25	<35	<50	<60	<100	<25
E	2	14900	2000	1300	1100	<35	<135	<25	89	<25	<35	<50	<60	<100	<25
F	2	13700	2100	1255	1000	30	150	<10	73	<10	...	<100	<40	<200	15
G	2	14600	2030	960	1000	<35	80	14	56	<25	<35	...	50	...	<25
H	2	16600	2100	1200	1100	48	149	26	81	<25	<35	<50	74	<100	<25
I	3	15100	2100	1400	1000	43	<100	32	84	<25	<35	<50	75	<100	<25
J	3	14400	2100	1300	1000	40	<100	27	<80	<25	<35	<50	71	<100	<25
L	2	15500	2100	1400	1100	44	106	<25	<80	<25	<35	<50	90	<100	<25
M	2	15400	2100	1525	1050	42	103	<25	<80	<25	<35	<50	82	<100	<25
N	2	15000	2100	1400	1000	42	110	<25	<80	<25	<35	<50	95	<100	<25
O	2	15200	2100	1400	1100	46	130	<25	<80	<25	<35	<50	78	<100	<25
P	2	15500	2100	1400	1100	43	133	<25	<80	<25	<35	<50	92	<100	<25
Q	2	15300	2100	1400	1100	45	133	<25	<80	<25	<35	<50	97	<100	<25
R	3	15900	1810	1400	1100	58	50	27	65	7	26	20	15	8	13
S	3	15900	1810	1400	1100	58	50	27	65	7	26	20	15	8	13
T	3	15800	2320	1400	1100	46	160	10	41	6	29	19	120	12	12
U	3	15800	2320	1400	1100	46	160	10	41	6	29	19	120	12	12
V	3	14400	2200	1300	1000	50	<100	<25	<80	<25	<35	<50	69	<100	<25
W	3	15100	2100	1200	900	28	123	18	42	<10	17	...	42	<100	<20

[a] The ''A'' parameter $A = \Sigma t_i$ loop $(-79\,480/RT_i)$ where t_i and T_i are post beta quench anneal times, and temperatures for the alpha-annealed coupons produced from these ingots are $A = 2.7 \times 10^{-16}$.

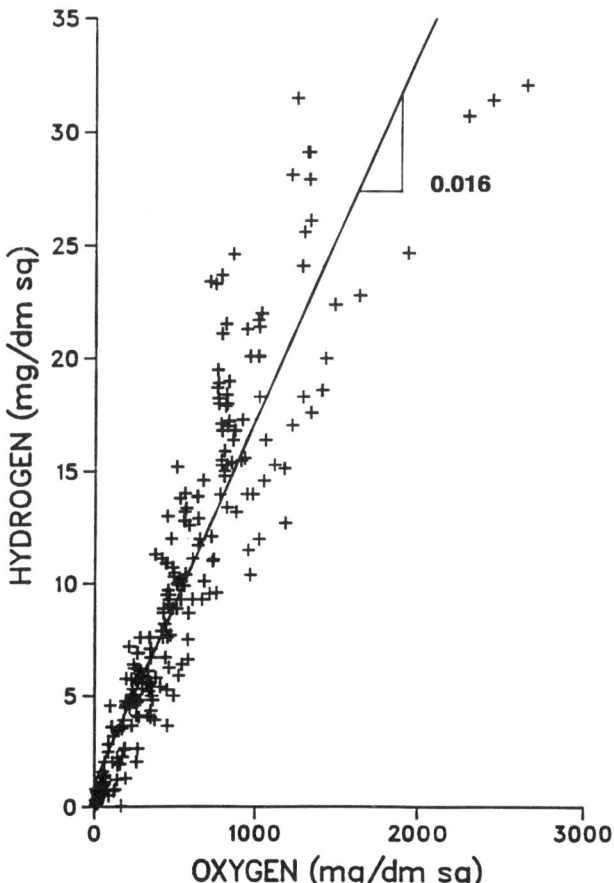

FIG. 5—*Hydrogen pickup of alpha-annealed Zircaloy-4 exposed in the ATR at 310 and 360°C versus oxygen weight gain.*

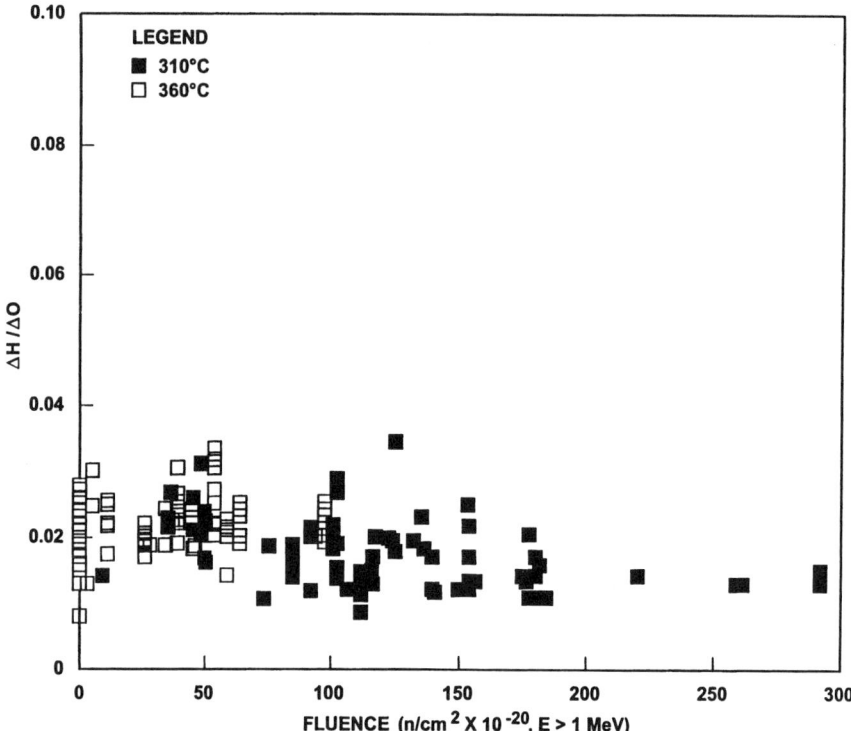

FIG. 6—*Hydrogen-to-oxygen weight gain ratio of alpha-annealed Zircaloy-4 exposed in the ATR versus fluence.*

thereafter was independent of oxide thickness, fluence, or flux. The original coupon thickness was about 1000 μm. Since the hydrogen concentration in the oxide in-reactor after about 20 μm of oxide was independent of exposure and oxide thickness (Figs. 9 and 10), the relative amounts of hydrogen in the oxide and metal depended on the relative volumes of oxide and hydride as well as the amount in the metal, which does depend on exposure. In thin coupons, the hydrogen in the oxide can be a significant fraction of the total hydrogen content of the oxidized coupon.

Hydrogen Migration in Zircaloy-4

Hydrogen migration down temperature gradients results in a concentration of hydrogen in hydrides when the metal hydrogen concentration exceeds solubility in the cold regions of a zirconium alloy. However, hydrogen migration models cannot be based on equilibrium, principally because volume mismatch between metal and hydride densities results in lattice distortion on precipitation. The energy required to induce this distortion delays precipitation until the metal hydrogen concentration exceeds the equilibrium concentration. A model for the migration of hydrogen in Zircaloy [27], modified to accommodate a difference in the solubility limits for dissolution and precipitation of the hydride, was used to describe the process. The

360°C WATER FILM

FILM THICKNESS 25μm

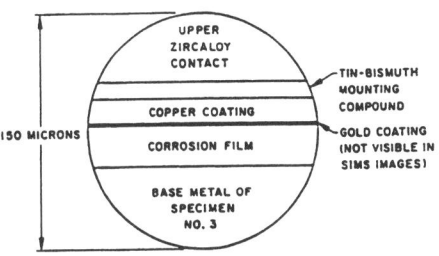

TRANSVERSE SECTION OF SPECIMEN AS SEEN
IN A 150μm FIELD SIMS IMAGE

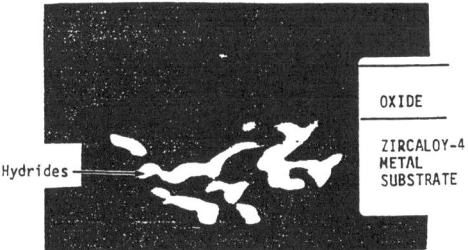

1H SIMS IMAGE OF A SECOND PORTION OF SPECIMEN #3.
Cs PRIMARY BEAM, 150 MICRON FIELD, 60 SECOND DATA COLLECTION.

FIG. 7—*Hydrogen distribution in the oxide film and metal substrate.*

metal concentration at equilibrium between metal and hydrides was used for dissolution of the hydrides, C_{eq}. Precipitation of hydrides requires a supersaturation of hydrogen relative to a solubility limit for precipitation C_{PT}, which is somewhat higher than C_{eq}. When the hydrogen concentration in solution in the metal is between C_{eq} and C_{PT}, hydride will not precipitate and existing hydrides will not grow. Hydrides dissolve in equilibrium with the metal matrix at the equilibrium concentration C_{eq}. The model is summarized in Table 5. Qualification experiments were designed to confirm the model and to provide data for indexing constants.

Experimental

To determine the solid-solution diffusion coefficient, diffusion couples were made by precharging half of the sample (Fig. 11) and then annealing the sample at temperatures between 260 and 482°C, e.g., for 3.9 days at 427°C (Fig. 12). The material used was that of Ingot B in Table 3. Specimens were precharged with hydrogen by gaseous equilibration when the desired

INERT GAS FUSION + OF SAMPLES WITH AND WITHOUT OXIDE FILMS

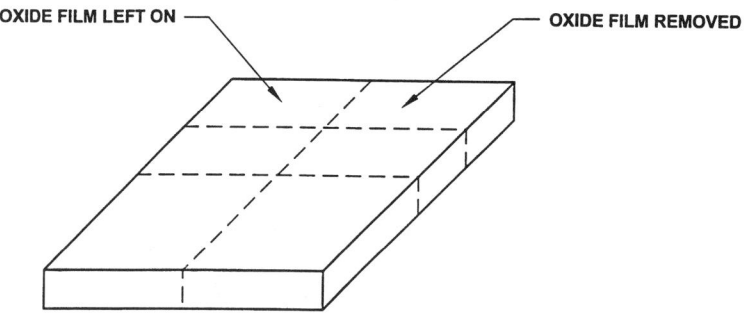

CONCENTRATION OF OXIDE FILM FOUND AS FOLLOWS

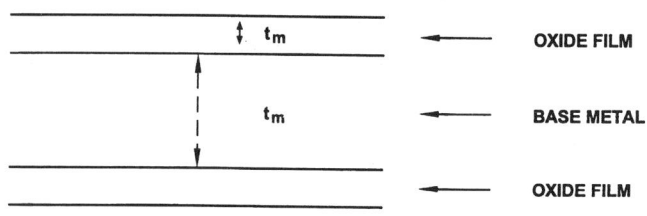

$$C_T(t_m \rho_m + 2t_o \rho_o) = C_m t_m \rho_m + 2C_o \rho_o t_o \tag{1}$$

WHERE C_T -IS THE TOTAL HYDROGEN CONCENTRATION OF THE SAMPLE MEASURED WITH THE OXIDE FILM LEFT ON (PPM BY WEIGHT)
C_m -IS THE HYDROGEN CONCENTRATION OF THE METAL (PPM BY WEIGHT)
C_o -IS THE HYDROGEN CONCENTRATION OF THE OXIDE (PPM BY WEIGHT)
t_m -IS THE BASE METAL THICKNESS (CM)
t_o -IS THE OXIDE FILM THICKNESS (CM)
ρ_m -IS THE DENSITY OF ZIRCALOY (6.545 G/CM^3)
ρ_o -IS THE DENSITY OF THE OXIDE FILM (5.65 G/CM^3)

+ SAMPLE MELTED AND VOLUME OF HYDROGEN RELASED DETERMINED.

FIG. 8—*Method used to determine hydrogen in the oxide film.*

TABLE 4—*Hydrogen retained in oxide autoclaves samples.*

Sample	Sample Thickness	Weight Gain	Hydrogen Content, ppm	
			Oxide On[a]	Oxide Off[b]
1	890 μm	642 mg/dm²	669 ± 11	660 ± 7
2	890 μm	616 mg/dm²	751 ± 9	750 ± 4
3	1550 μm	1091 mg/dm²	574 ± 8	589 ± 6
4	1450 μm	993 mg/dm²	558 ± 10	552 ± 6
5	1650 μm	204 mg/dm²	177 ± 3	185 ± 6
6	2890 μm	1052 mg/dm²	337 ± 3	338 ± 3

[a] Measurements of hydrogen concentration in both metal and oxide.
[b] Measurements of hydrogen concentration in metal only.

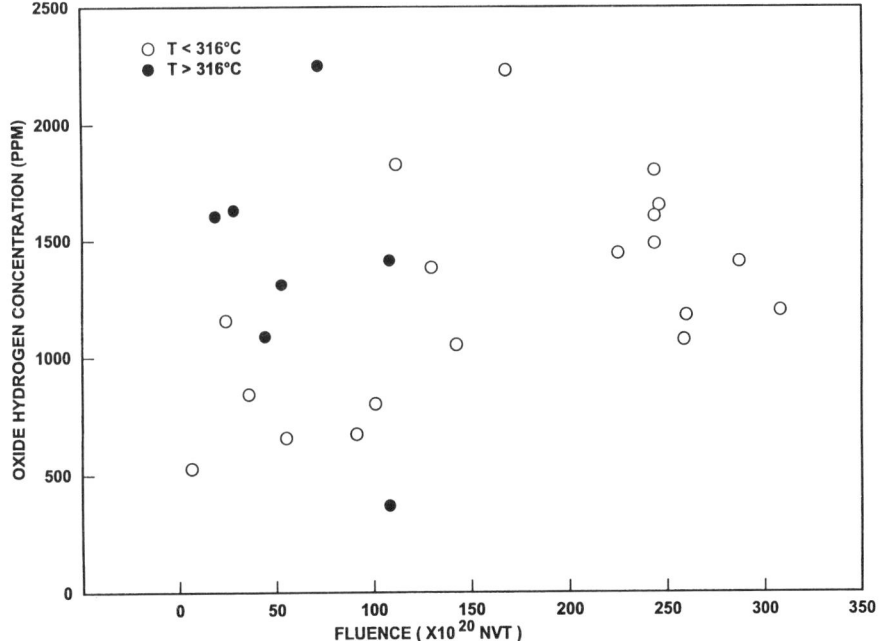

FIG. 9—*Hydrogen concentration in the oxide film as a function of fluence.*

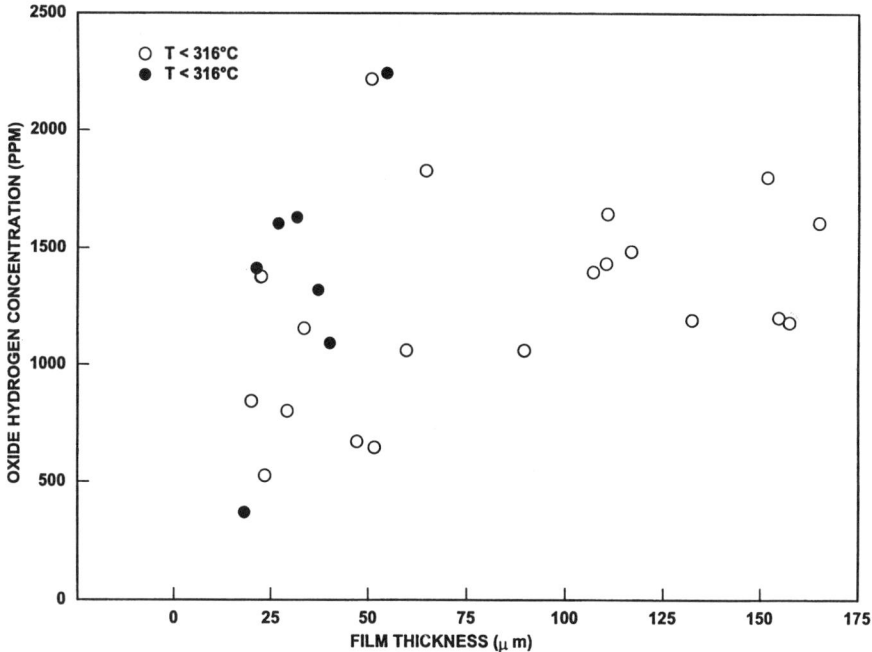

FIG. 10—*Hydrogen concentration in the oxide film as a function of oxide film thickness.*

hydrogen distribution was uniform. Specimens where only a portion was to be precharged were cathodically precharged by partial immersion in 1 N sulfuric acid poisoned against hydrogen recombination with 0.25 g/L of sodium arsenate. The current was 100 mA/cm^2, and the temperature was 80°C.

Results

The diffusion coefficients determined from these studies are shown in Fig. 13 and are in reasonable agreement with the results of Kearns [19] but below results reported by Someno [20], Sawatzky [21,26], and Mallett and Albrecht [22]. Data in the literature have demonstrated that a hysteresis exists in the precipitation and dissolution solvi (Fig. 14).

Hydrogen Dissolution

Experimental

To determine the hydride dissolution solvus, similar diffusion couples were made, but the two sides were allowed to come to equilibrium at each temperature (Fig. 15). The hydride dissolution solvus as a function of temperature was determined by measuring the postanneal hydrogen concentration of the initially uncharged half of the samples.

TABLE 5—*Model used for the migration of hydrogen.*

I. Hydrogen diffuses through the alpha (hcp) phase in response to temperature and solid solution concentration gradients.

$$\vec{J} = -D\vec{\nabla}C - \frac{DQ^*C}{RT^2}\vec{\nabla}T$$

where

D = diffusion coefficient of hydrogen through α Zircaloy,
C = solid solution concentration of hydrogen in α Zircaloy,
Q^* = heat of transport, and
T = temperature.

II. The rate of change in hydrogen concentration at a point is

$$\frac{\delta C_T}{\delta t} = -\vec{\nabla} \cdot \vec{J}$$

The rate of change in hydrogen concentration solid solution in the α Zircaloy is

$$\frac{\delta C}{\delta t} = -\vec{\nabla} \cdot \vec{J} - \alpha^2(C - C_{PT}), \; C > C_{PT}$$

$$\frac{\delta C}{\delta t} = \frac{\delta C}{\delta t} = -\vec{\nabla} \cdot \vec{J}, \; C < C_{PT}, \; \text{or}$$

where

$\frac{\delta C}{\delta t} = 0$, $C = C_{eq}$ if hydrides are dissolving,
C_T = total hydrogen concentration,
$C_{PT} = C_{PTO} \exp(-Q_P/RT)$ concentration for precipitation,
$C_{eq} = C_{eqo} \exp(-Q_{eq}/RT)$ concentration for dissolution, and
$\alpha^2 = \alpha_0^2 \exp(-2 Q_\alpha/RT)$ fitted rate parameter for precipitation.

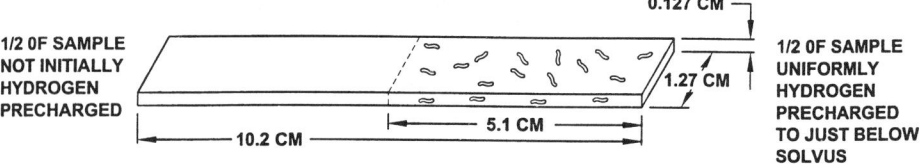
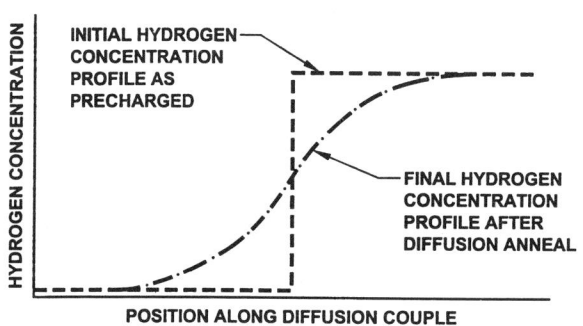
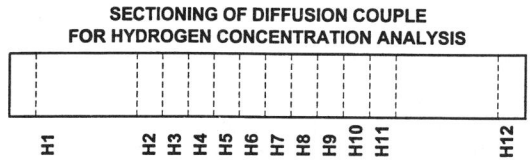

FIG. 11—*Preparation and testing sequence of samples to determine the hydrogen-diffusion kinetics and hydrogen solvus.*

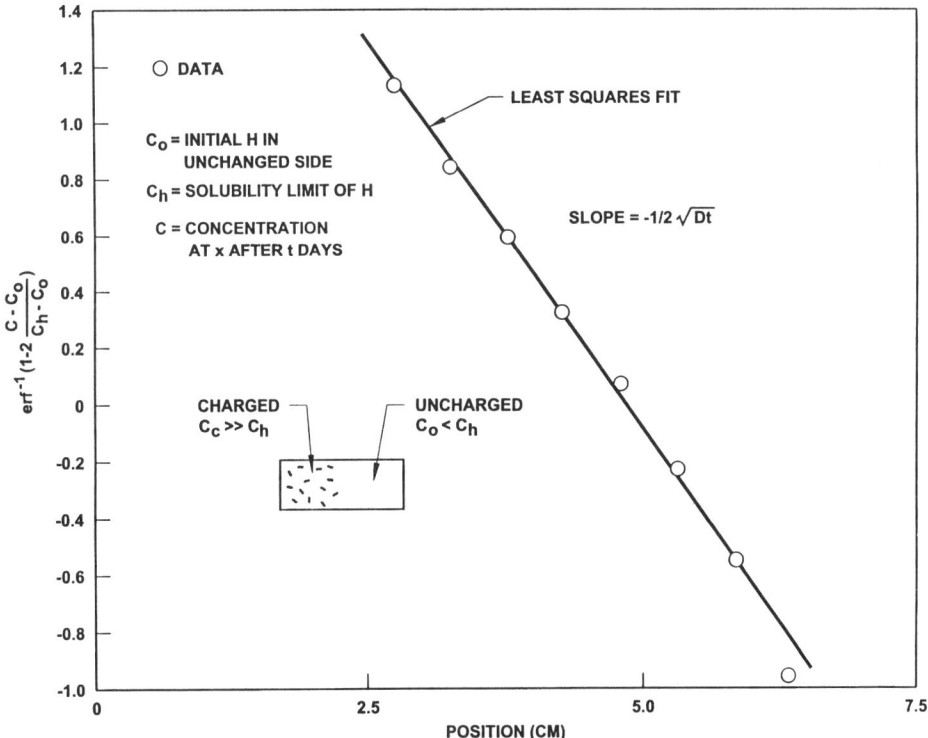

FIG. 12—*Alpha-annealed Zircaloy-4 diffusion couple results after exposure of 3.9 days at 427°C.*

Results

The results are shown in Fig. 16 and compared to the results of other researches. The results agree well with those of Sawatzky [21], Slattery [23], and Kearns [24], who used two different techniques. They do not agree with the results of Erickson and Hardie [25]. Based on the agreement of three investigators, it is concluded that the data presented herein accurately reflect dissolution of zirconium hydrides in Zircaloy. In addition, there appears to be no difference between the dissolution solvi for Zircaloy-2 and Zircaloy-4.

Hydrogen Precipitation

Experimental

To determine the hydride precipitation solvus, uniformly hydrogen-charged specimens were annealed in a temperature gradient that varied linearly and monotonically from one end of each sample to the opposite end (Fig. 17). Hydrogen precharge levels were chosen such that the entire sample was in the single-phase state at the beginning of the test. The cold ends of each sample were held at either 260, 316, 371, or 427°C, and the temperature gradients were either 6.6 or 8.7°C/mm. Hydrogen diffuses down the temperature gradient until the concentration

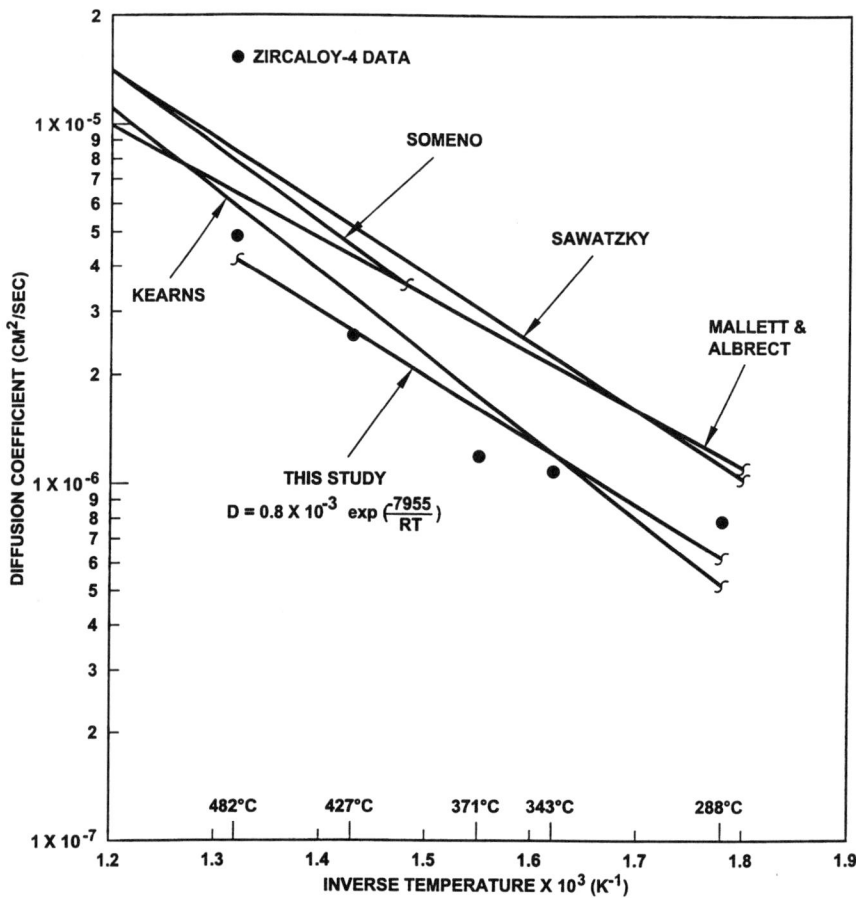

FIG. 13—*Diffusion coefficient of hydrogen in alpha-zirconium as a function of temperature.*

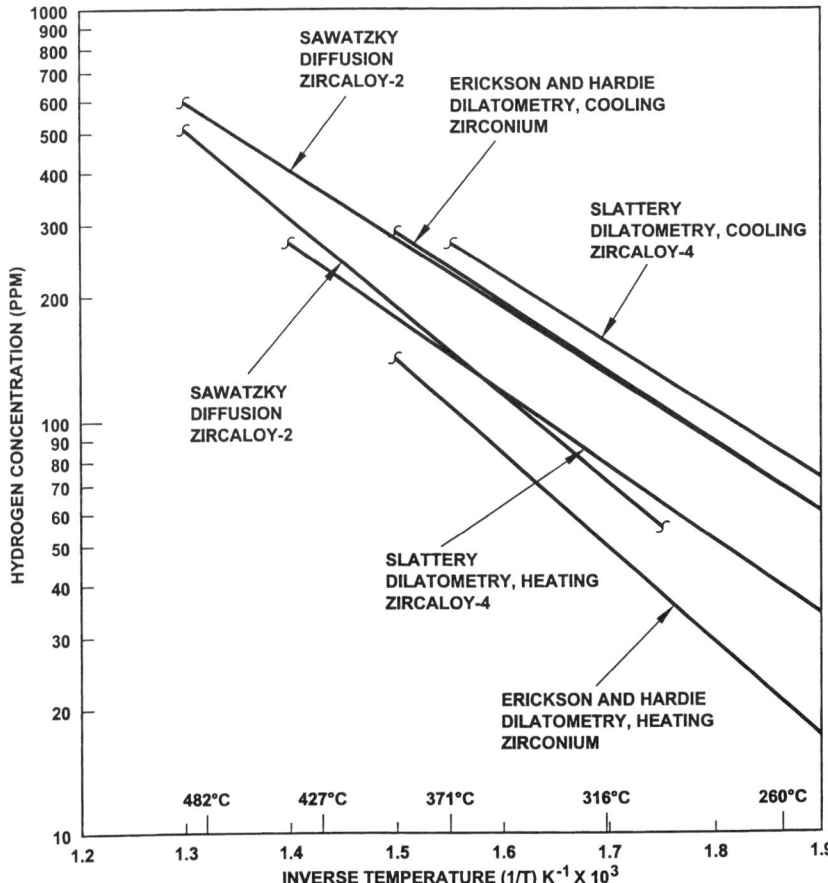

FIG. 14—*Hydrogen solvus in zirconium alloys as a function of temperature.*

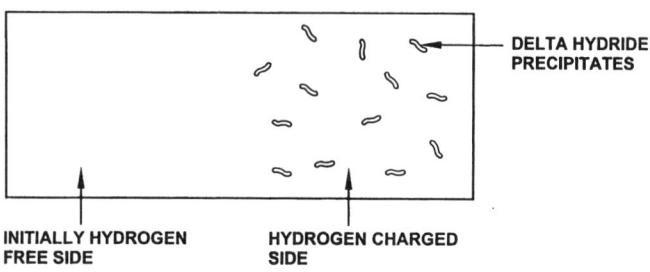

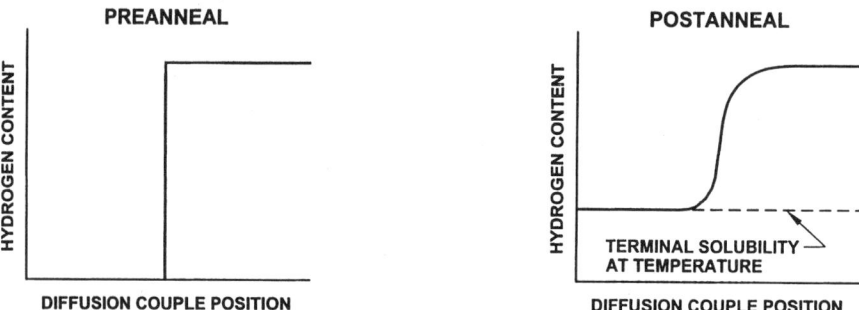

FIG. 15—*Determination of the solvus on dissolution by the diffusion method.*

gradient provides sufficient counter driving force to balance the thermal migration. At equilibrium, the hydrogen concentration in solution in the Zircaloy matrix will have the distribution

$$C(T) = C_0 \exp(Q^*/RT(x))$$

This will be the temperature and spatial distribution, as $T = T(x)$, of the total or measured concentration when the total concentration is below that required for precipitation of the hydride. When the total concentration is above $C(T)$, there has been precipitation. For example, the hydrogen concentration profile when a profile close to $C(T)$ has developed except at the cold end is shown in Fig. 18. In this case, the hydrogen concentration at the cold end of the specimen is well above $C(T)$, indicating precipitation. The concentration C_{eq} for dissolution of the hydride is also shown by the dashed line, which at 371°C is well below that for precipitation, C_{PT}.

Results

The results from each test are shown in Fig. 19, where they are compared with literature data. At temperatures above 316°C, there is reasonable agreement among the results of the three literature references and the data reported herein. Our data may have a slightly lower slope. At 260°C, the precipitation concentration reported herein is greater than that reported by Erickson and Hardie or by Slattery.

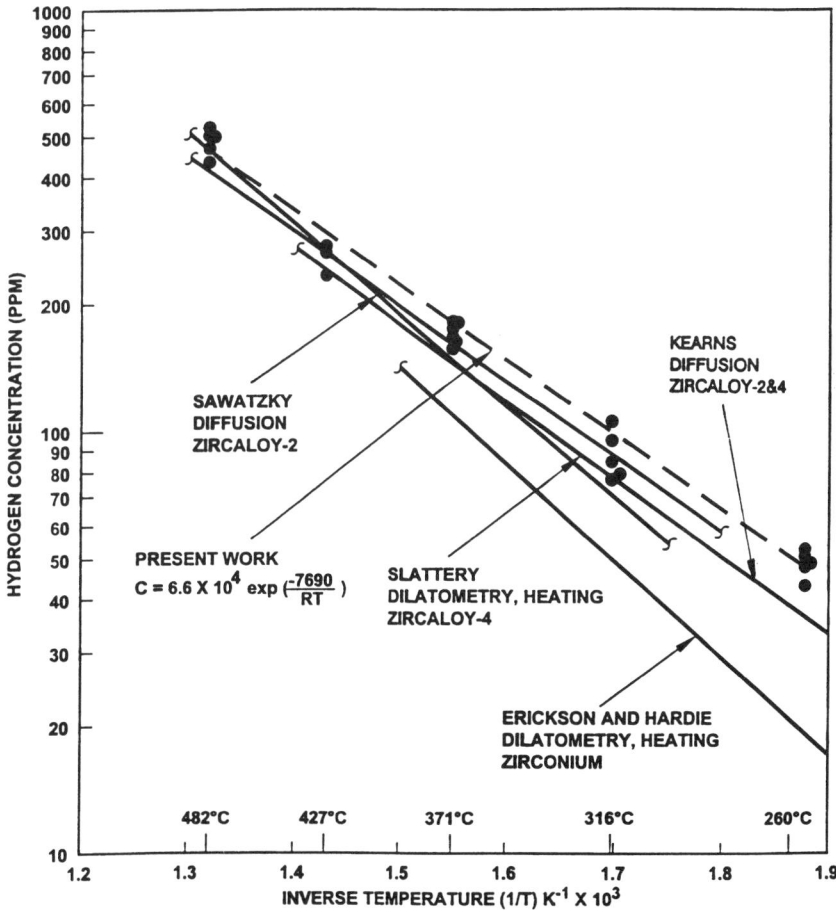

FIG. 16—*Hydride dissolution solvus of zirconium alloys as a function of temperature.*

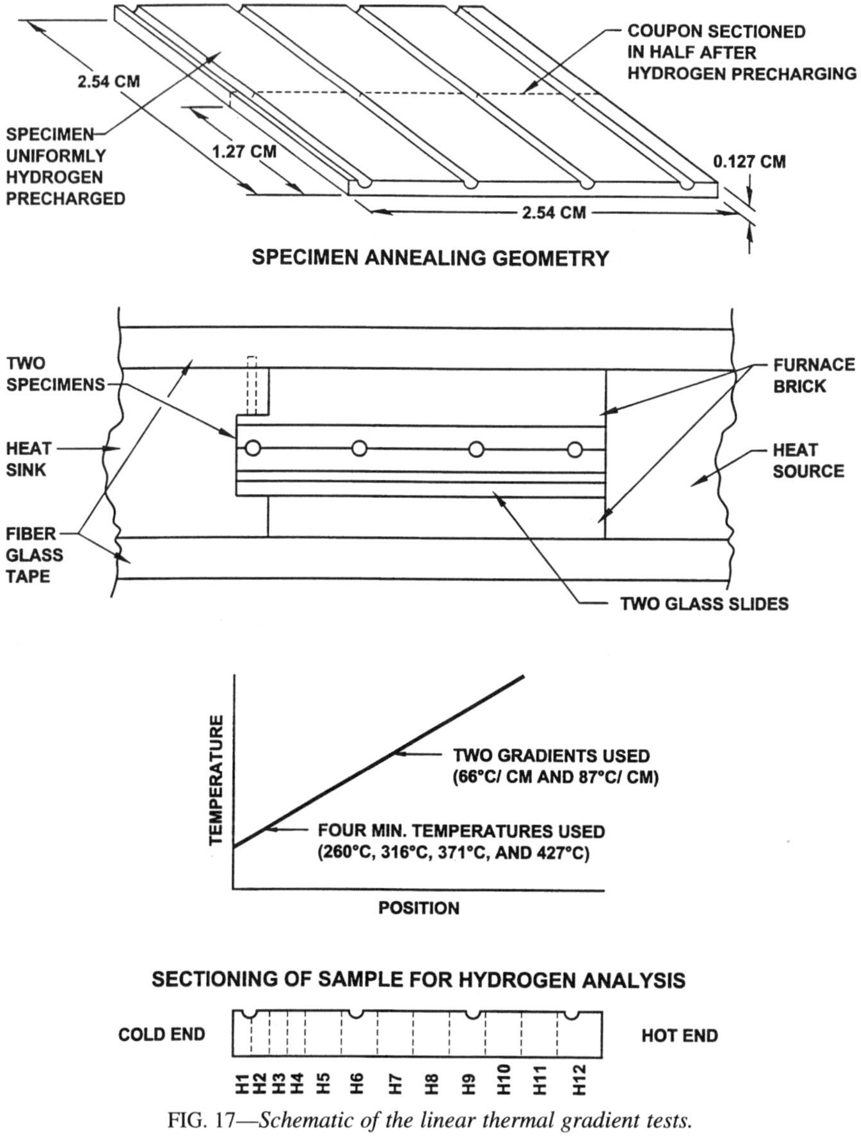

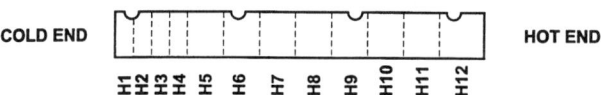

FIG. 17—*Schematic of the linear thermal gradient tests.*

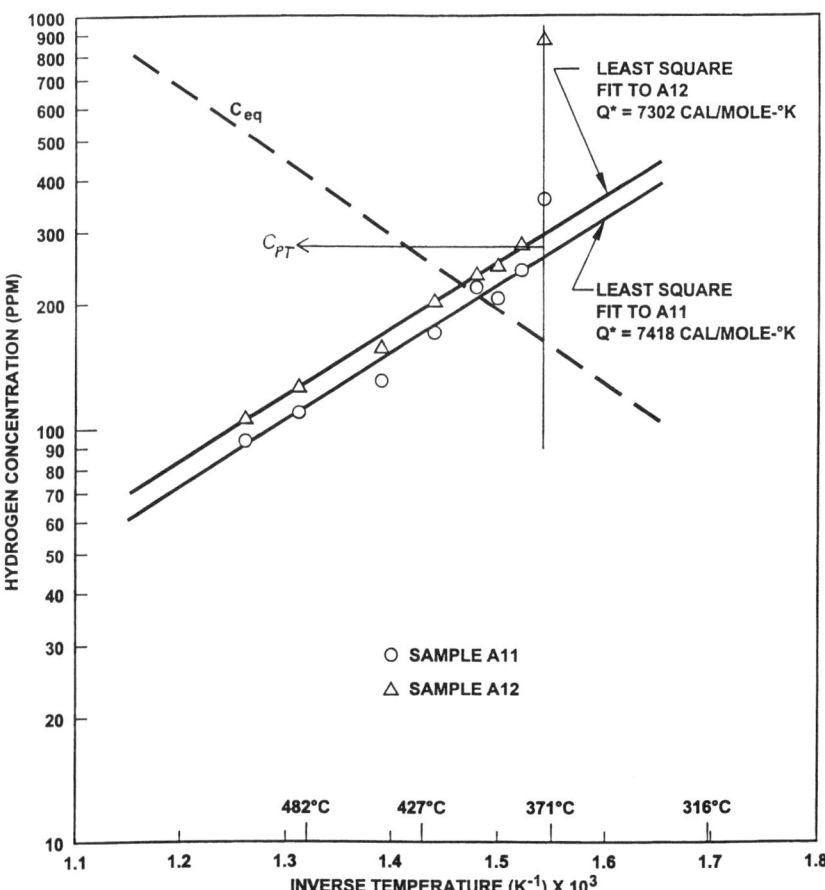

FIG. 18—*Hydrogen concentration as a function of temperature in a 66°C/cm linear thermal gradient test, cold end at 371°C.*

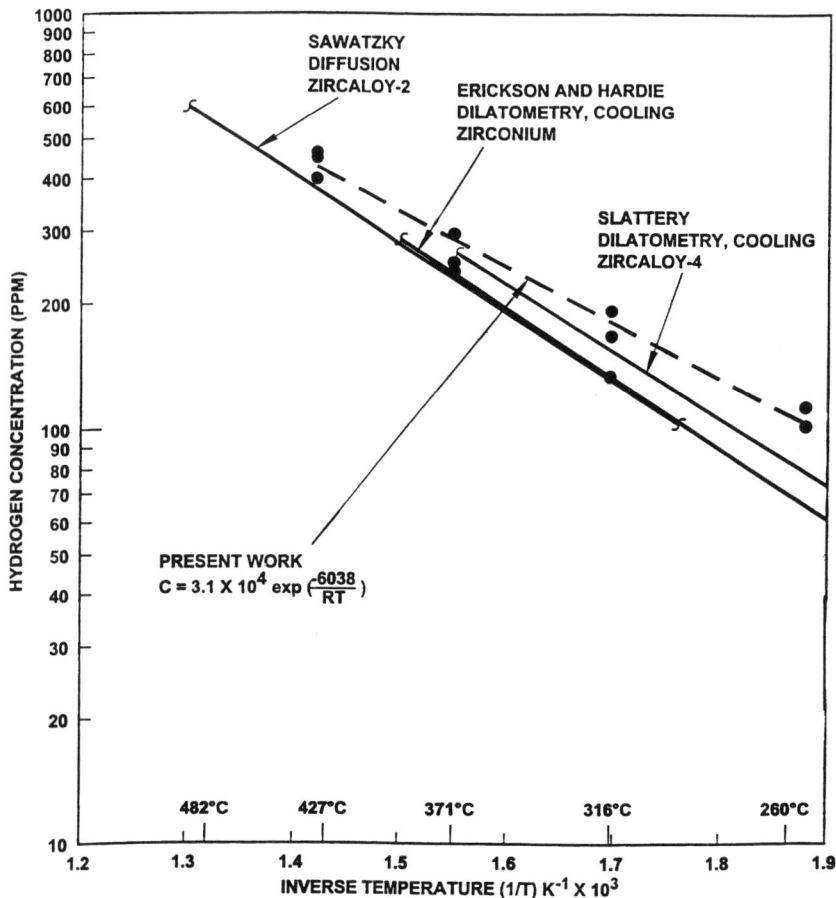

FIG. 19—*Hydrogen precipitation solvus of zirconium alloys as a function of temperature.*

Discussion

The heat of transport, Q^* in Eq 1 in Table 5, was determined for each sample from the slopes of log ([H]) versus inverse temperature, as shown in Fig. 18, for a sample with a cold end at 371°C. The results are shown in Fig. 20. The reprecipitation constant, α in Eq 2 in Table 5, determines the reprecipitation kinetics in this model and can be determined by measuring the surface and midplane precipitated hydrogen concentration, as shown in Fig. 21. Samples were corroded in concentrated lithiated water at temperatures between 288 and 360°C in autoclaves at pressures high enough to maintain temperature below the saturation temperature, 2.79 MPa and 18.65 MPa, respectively. The results are shown in Fig. 22.

This model for hydrogen diffusion was benchmarked to several out-of-reactor experiments. One sample (Fig. 23) was placed in a steep temperature gradient, and the other sample (Fig. 24) was placed in a shallow temperature gradient. After annealing, the samples were sectioned, and hydrogen concentrations were measured by hot vacuum extraction. The resulting hydrogen-concentration profiles are shown in Figs. 23 and 24, along with the temperature profiles and the model predictions of the hydrogen concentrations. The model predicts the observed con-

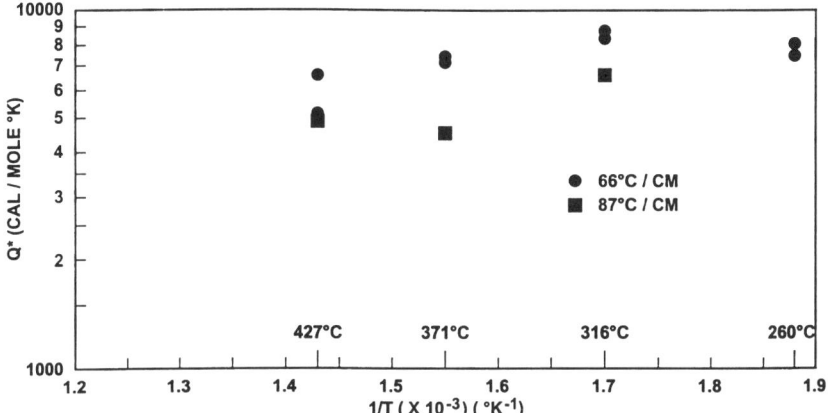

NOTED TEMPERATURE IS COLD TEMPERATURE IN SAMPLES CONTAINING LINEAR THERMAL GRADIENTS OF 300°F / INCH AND 400°F / INCH

FIG. 20—*Heat of transport of hydrogen in alpha-annealed Zircaloy-4 as a function of temperature.*

SUPERCHARGING OF ISOTHERMAL CORROSION COUPONS REF. (28)

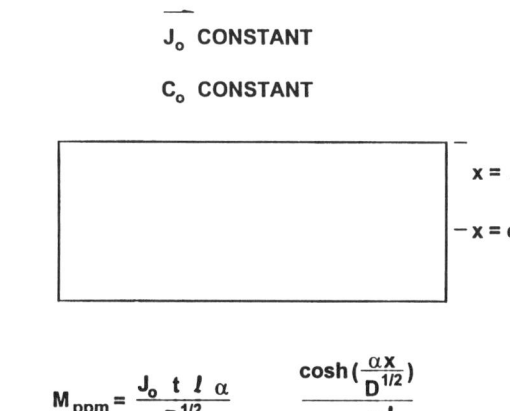

$$M_{ppm} = \frac{J_o \, t \, l \, \alpha}{D^{1/2}} \quad \frac{\cosh(\frac{\alpha x}{D^{1/2}})}{\sinh(\frac{\alpha l}{D^{1/2}})}$$

J_o = SURFACE HYDROGEN FLUX
D = DIFFUSION COEFFICIENT
l = SPECIMEN HALF THICKNESS
x = DISTANCE FROM MIDPLANE
M = QUANTITY OF HYDROGEN PRECIPITATED

$MX_1 / MX_o = \cosh(\alpha x / D^{1/2})$

FIG. 21—*Method used to determine the precipitation rate parameter α.*

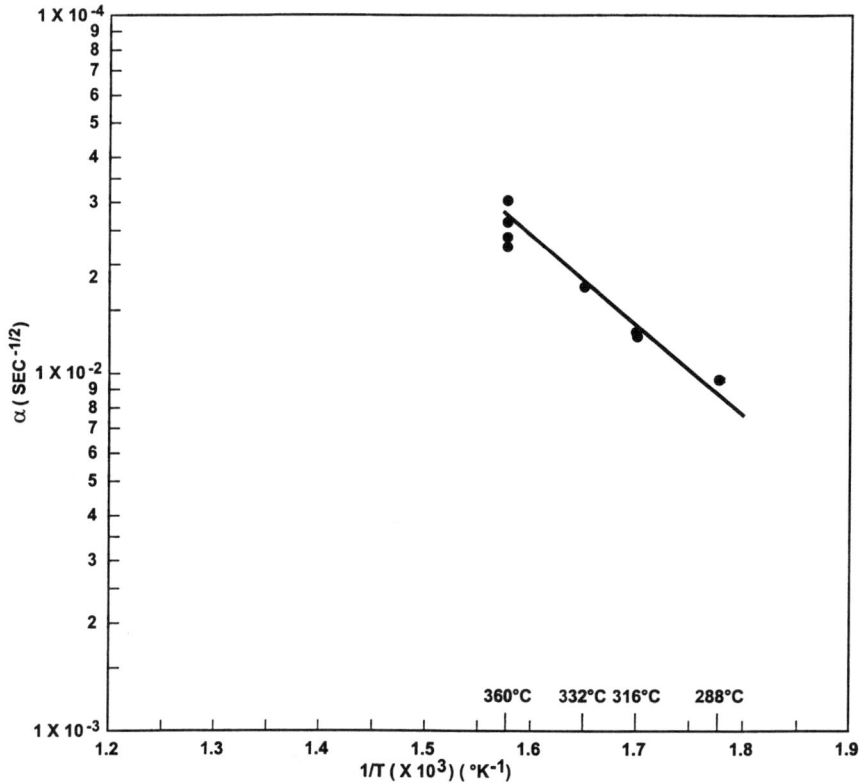

FIG. 22—*Precipitation rate parameter α as a function of temperature.*

centrations adequately for the sample with the steep temperature gradient (Fig. 23) but not for the sample with the shallow temperature gradient, for which the model overpredicts hydrogen concentration in the colder regions. The most likely cause of the model misprediction is hydrogen trapping at lattice defects or a second phase. Trapping would result in higher required driving forces, i.e., thermal gradients, to redistribute the hydrogen.

For application to fuel designs, the above migration model must be combined with a model for hydrogen pickup. Based on the discussion in the previous section, the pickup model must properly reflect the relative amounts of hydrogen in the metal and oxide.

Conclusions

To describe the embrittlement of zirconium alloys due to formation of hydrides, the following aspects must be understood: hydrogen pickup rate, holdup of hydrogen in the oxide, diffusion rates in a matrix with hydrogen traps, hydrogen concentration in the metal for hydride dissolution, and metal hydrogen concentration for hydride precipitation and dissolution. The experiments described provide qualification data for each aspect, although additional data are needed in some areas, e.g., no holdup of hydrogen in the oxide. The model described is adequate for steep gradients but overpredicts hydrogen redistribution for shallow temperature gradients. This deficiency is thought to be due to the lack of a description of hydrogen trapping at lattice

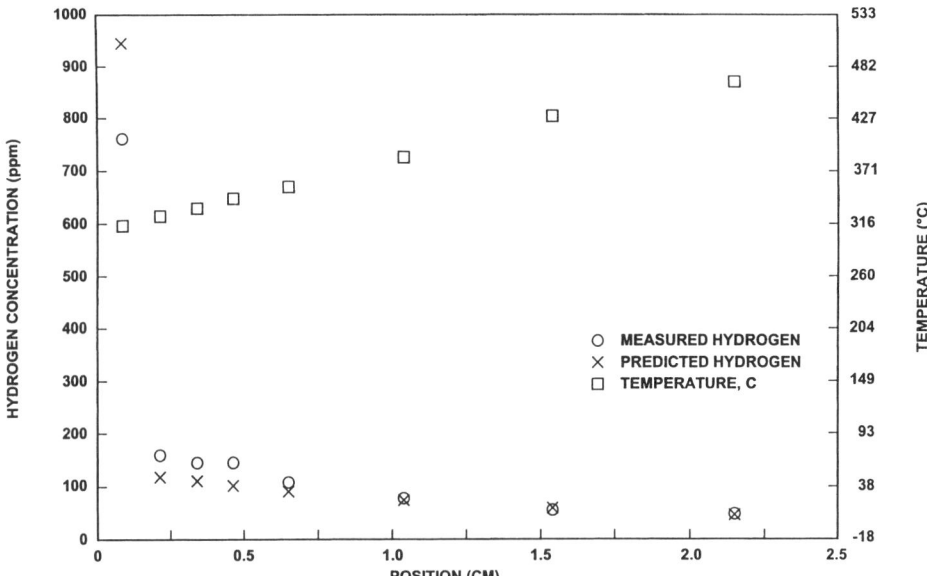

FIG. 23—*Predicted and observed hydrogen concentration profile after 30 days in a thermal gradient.*

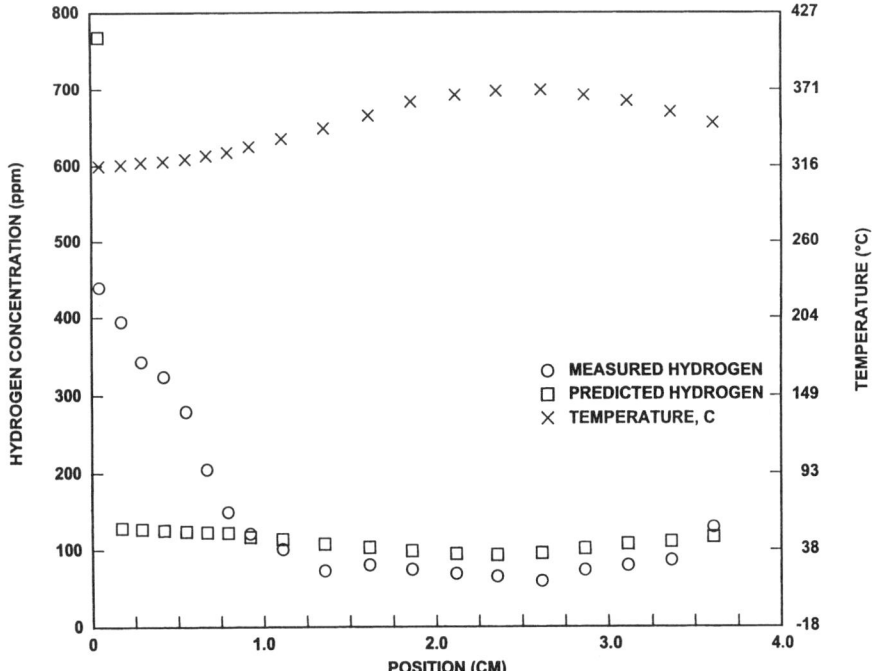

FIG. 24—*Predicted and observed hydrogen concentration profile after 200 days in a thermal gradient—the effect on an internal temperature maximum.*

defects, i.e., the model should reflect the binding energies between the hydrogen strain field and lattice defect strain fields.

Acknowledgement

This work was supported by DOE contract DE-AC11-93PN38195.

References

[1] Cox, B, "Oxidation of Zirconium and Its Alloys," *Advances in Corrosion Science and Technology,* Vol. 5, 1976, pp. 173–391.
[2] Markowitz, J. M., "The Thermal Diffusion of Hydrogen in Alpha-Delta Zircaloy-2," *Transactions of the Metallurgical Society of AIME,* Vol. 221, 1961, pp. 819–824.
[3] Kreyns, P. H., Bourgeois, W. F., Charpentier, P. L., White, C. J., Kammenzind, B. F., and Franklin, D. G., "Lifetime Embrittlement of Reactor Core Materials," WAPD-T-3046, this publication.
[4] Mudge, W. L. Jr., "Effect of Hydrogen on the Embrittlement of Zirconium and Zirconium Tin Alloys," *Zirconium and Zirconium Alloys,* American Society for Metals, Cleveland, OH, 1953, pp. 146–167.
[5] Elmoselhi, M. B., Warr, B. D., and McIntyre, S., "A Study of the Hydrogen Uptake Mechanism in Zirconium Alloys," *Zirconium in the Nuclear Industry: Tenth International Symposium, ASTM STP 1245,* A. M. Garde and E. R. Bradley, Eds., American Society for Testing and Materials, West Conshohocken, PA, 1994, pp. 62–79.
[6] Proebstle, R. A. et al., "The Mechanism of Defection of Zircaloy-Clad Fuel Rods by Internal Hydriding," *Proceedings,* Joint Topical Meeting on Commercial Nuclear Fuel Technology Today, Toronto, Canada, 28–30 April 1975, 75-CNA/ANS-100, American Nuclear Society, La Grange Park, IL, pp. 2-15 to 2-34.
[7] Preble, E. A. et. al., "Fuel Performance Annual Report for 1990," NUREG/CR-3950, PNL-5210, Vol. 8, Pacific Northwest Laboratory, Richland, WA 99352, November 1993.
[8] Yang, R., Ozer, O., and Klepfer, H. H., "Fuel Performance Evaluation for EPRI Program Planning," *Proceedings,* International Topical Meeting on LWR Fuel Performance, Avignon, France, 21–24 April 1991, American Nuclear Society, La Grange Park, IL, pp. 2-58 to 2-71.
[9] Schrire, David et al, "Secondary Defect Behavior in ABB BWR Fuel," *Proceedings,* 1994 International Meeting on Light Water Reactor Fuel Performance, West Palm Beach, FL, 17–21 April 1994, American Nuclear Society, La Grange Park, IL, pp. 398–409.
[10] Armijo, J. S., "Performance of Failed BWR Fuel," *Proceedings,* 1994 International Meeting on Light Water Reactor Fuel Performance, West Palm Beach, FL, 17–21 April 1994, American Nuclear Society, La Grange Park, IL, pp. 410–422.
[11] Yang, R. L. et al., "EPRI Failed Fuel Degradation R&D Program," *Proceedings,* 1994 International Meeting on Light Water Reactor Fuel Performance, West Palm Beach, FL, 17–21 April 1994, American Nuclear Society, La Grange Park, IL, pp. 435–446.
[12] Puls, M. P., "Effects of Crack Tip Stress States and Hydride-Matrix Interaction Stresses on Delayed Hydride Cracking," *Metallurgical Transactions A,* Vol. 21A, 1990, pp. 2905–2917.
[13] Baur, Karl et al., "Fuel Behavior in High Performance PWRs," *Proceedings,* 1994 International Meeting on Light Water Reactor Fuel Performance, West Palm Beach, FL, 17–21 April 1994, American Nuclear Society, La Grange Park, IL, pp. 22–30.
[14] Warr, B. D. et al., "Oxide Characteristics and Their Relationship to Hydrogen Uptake in Zirconium Alloys," *Zirconium in the Nuclear Industry: Ninth International Symposium, ASTM STP 1132,* C. M. Eucker and A. M. Garde, Eds., American Society for Testing and Materials, West Conshohocken, PA, 1991, pp. 740–756.
[15] Laursen, T. et. al., "Hydrogen Ingress into Oxidized Zr-2.5Nb," *Journal of Nuclear Materials,* Vol. 209, 1994, pp. 52–61.
[16] Laursen, T., Leslie, J. R., and Tapping, R. L., "Corrosion of Zr-2.5 wt% Nb Pressure Tube Material in D_2O Steam: Deuterim Depth Distributions Measured by Nuclear Reaction Analysis," *Journal of Nuclear Materials,* Vol. 182, 1991, pp. 151–157.
[17] Laursen, T., Leslie, J. R., and Tapping, R. L., "Deuterium Depth Distributions in Oxidized Zr-2.5 wt% Nb Measured by Neutron Reaction Analysis," *Journal of the Less-Common Metals,* Vol. 172–174, 1991, pp. 1306–1312.
[18] Woolsey, I. S. and Morris, J. R., "A Study of Zircaloy-2 Corrosion in High Temperature Water

Using Ion Beam Methods," Paper 178, *Proceedings*, presented at the 3–7 March 1980 Conference of NACE, Chicago; also in *Corrosion 37*, 1981, pp. 575–585.

[19] Kearns, J. J., "Diffusion Coefficient of Hydrogen in Alpha Zirconium, Zircaloy-2, and Zircaloy-4," *Journal of Nuclear Materials*, Vol. 43, 1972, pp. 330–338.

[20] Someno, M., "Determination of the Solubility and Diffusion Coefficient of Hydrogen in Zirconium," *Nihon Kinzoku Gakkaishi*, Vol. 24, 1960, pp. 249–253.

[21] Sawatzky, A., "The Diffusion and Solubility of Hydrogen in the Alpha Phase of Zircaloy-2," *Journal of Nuclear Materials*, Vol. 2, 1960, pp. 62–68.

[22] Mallet, M. W. and Albrecht, W. M., "Low Pressure Solubility and Diffusion of Hydrogen in Zirconium," *Journal of the Electrochemical Society*, Vol. 104, 1957, p. 142.

[23] Slattery, G. F., "The Terminal Solubility of Hydrogen in Zirconium Alloys between 30 and 400°C," *Journal of the Institute of Metals*, Vol. 95, 1967, pp. 43–47.

[24] Kearns, J. J., "Terminal Solubility and Partitioning of Hydrogen in the Alpha Phase of Zirconium, Zircaloy-2 and Zircaloy-4," *Journal of Nuclear Materials*, Vol. 22, 1967, pp. 292–303.

[25] Erickson, W. H. and Hardie, D., "The Influence of Alloying Elements on the Terminal Solubility of Hydrogen in α-Zirconium," *Journal of Nuclear Materials*, Vol. 13, 1964, pp. 254–262.

[26] Sawatzky, A. and Wilkins, B. J .S., "Hydrogen Solubility in Zirconium Alloys Determined by Thermal Diffusion," *Journal of Nuclear Materials*, Vol. 22, 1967, pp. 304–310.

[27] Marino, G. P., "A 2-Dimensional Computer Program for Migration of Interstitial Solutes of Finite Solubility in a Thermal Gradient," WAPD-TM-1157, National Technical Information Service, Springfield, VA, June 1974.

[28] Marino, G. P., "Hydrogen Supercharging in Zircaloy," *Materials Science Engineering*, Vol. 7, 1971, pp. 335–341.

DISCUSSION

A. T. Motta[1] (written discussion)—One would expect hydrogen to be associated with lattice defects such as point defects and/or dislocations. (1) Do you plan to consider that in your modeling? (2) Under irradiation, the concentration of defects will be very different and could potentially affect H concentration. Do you intend to model the concentration of H under radiation?

B. F. Kammenzind et al. (authors' closure)—Future models will include these effects if sufficient data become available to quantify their impact.

H. Ruhmann[2] (written discussion)—Is the method for H determination sensitive to the chemical nature of the hydrogen (charge on the atom)? This would be allowed to distinguish between H and H+, OH$^-$ in the oxide layer.

B. F. Kammenzind et al. (authors' closure)—The nondestructive measurement technique did not distinguish among the various chemical forms of hydrogen.

B. Cox[3] (written discussion)—What was the water chemistry in the ATR loop used for preparing your specimens. Was it the same in the autoclave tests and did it involve the use of LiOH?

B. F. Kammenzind et al. (authors' closure)—ECP and pH were controlled in the normal PWR range for these types of tests with the exception that some autoclave tests were conducted with neutral water chemistry. The specific methods of chemistry control were sufficiently standard to not affect the results reported herein.

G. K. Shek[4] (written discussion)—Have the effects of irradiation on fundamental properties such as TSS been looked at? Based on the work in Zr-2.5 Nb pressure tube material, is TSS for precipitation affected by yield strength?

B. F. Kammenzind et al. (authors' closure)—Such effects are beyond the scope of this presentation and do not significantly affect the results presented.

---

[1] Pennsylvania State University, University Park, PA.
[2] Siemens AG, Erlangen, Germany.
[3] University of Toronto, Toronto, Ontario, Canada.
[4] Ontario Hydro Technologies, Toronto, Ontario, Canada.

Ex-Reactor Mechanical Behavior

Patrick Delobelle,[1] Pascal Robinet,[1] Pol Bouffioux,[2] Philippe Geyer,[2] and Isabelle Le Pichon[2]

A Unified Model to Describe the Anisotropic Viscoplastic Behavior of Zircaloy-4 Cladding Tubes

REFERENCE: Delobelle, P., Robinet, P., Bouffioux, P., Geyer, P., and Le Pichon, I., "**A Unified Model to Describe the Anisotropic Viscoplastic Behavior of Zircaloy-4 Cladding Tubes,**" *Zirconium in the Nuclear Industry: Eleventh International Symposium, ASTM STP 1295,* E. R. Bradley and G. P. Sabol, Eds., American Society for Testing and Materials, 1996, pp. 373–393.

ABSTRACT: This paper presents the constitutive equations of a unified viscoplastic model and its validation with experimental data. The mechanical tests were carried out in a temperature range of 20 to 400°C on both cold-worked stress-relieved and fully annealed Zircaloy-4 tubes. Although their geometry (14.3 by 1.2 mm) is different, the crystallographic texture was close to that expected in the cladding tubes. To characterize the anisotropy, mechanical tests were performed under both monotonic and cyclic uni- and bi-directional loadings, i.e., tension-compression, tension-torsion, and tension-internal pressure tests. The results obtained at ambient temperatures and the independence of the ratio $R^p = \varepsilon^p_{\theta\theta} / \varepsilon^p_{zz}$ with respect to temperature would seem to indicate that the set of anisotropy coefficients does not depend on temperature. Zircaloy-4 material also has a slight supplementary hardening during out-of-phase cyclic loading.

We propose to extend the formulation of a unified viscoplastic model, developed and identified elsewhere for other initially isotropic materials, to the case of Zircaloy-4. Generally speaking, anisotropy is introduced through fourth order tensors affecting the flow directions, the linear kinematical hardening components, as well as the dynamic and static recoveries of the forementioned hardening variables. The ability of the model to describe all the mechanical properties of the material is shown. The application of the model to simulate mechanical tests (tension, creep, and relaxation) performed on true CWSR Zircaloy-4 cladding tubes with low tin content is also presented.

KEYWORDS: Zircaloy-4 alloys, mechanical properties, anisotropy, multiaxial loadings, modeling, internal variables model

In France, more than 75% of the electricity is generated by nuclear energy. Therefore, the French PWR participate in the grid regulation, and load following and remote control are now normal operating modes. In addition, EDF (the French utility) has decided to increase the discharge burnup of the fuel sub-assemblies to about 60 GWd/tU. Those operating conditions lead to numerous power transients [1]. In order to improve the accuracy of the stress-strain levels induced by the pellet clad mechanical interaction (PCMI) within the cladding, a unified model to describe the anisotropic viscoplastic behavior of the Zircaloy cladding tubes has been developed.

---

[1] Research engineer and graduate student, respectively, Laboratoire de Mécanique Appliquée R. Chaléat, U.A. CNRS, UFR Sciences, Route de Gray, La Bouloie, BESANCON, 25030, Cedex, France.

[2] Research engineers and graduate student, respectively, Electricité de France, Direction des Etudes et Recherches, Les Renardières, Route de Sens, BP 1, 77250 MORET/LOING, France.

The present paper describes the experimental approach (materials and mechanical tests) used to optimize the formalism of the model. The constitutive equations are fully described in this paper. The ability of the model to simulate the viscoplastic behavior of CWSR Zircaloy-4 cladding tubes at 350°C under uniaxial and biaxial loading conditions is also discussed.

Experimental Techniques

The Materials

The material used in this study has been produced by two vendors and covered metallurgical conditions ranging from cold-worked stress-relieved (CWSR) to fully recrystallized (R). The mechanical tests have been carried out on three types of Zircaloy-4 tubes:

1. CWSR (final heat treatment: 460°C) and R (final treatment: ≈700°C) tubes with the nominal dimensions: 14.3 mm outside diameter and 1.2-mm wall.
2. CWSR cladding tubes (called CT-CWSR; final heat treatment: ≈450°C) with the nominal dimensions: 9.5 mm outside diameter and 0.57-mm wall.

The chemical compositions measured on both ingots are given in Table 1. The content of alloying elements is consistent with ASTM Specification for Wrought Zirconium and Zirconium Alloy Seamless and Welded Tubes for Nuclear Service (B 353), although the tin content is close to the lower limit. The microstructure after recrystallization was composed of small equiaxial grains (the grain diameter is about 2 to 5 μm) surrounded by layer grains of up to 15 to 18 μm [2]. For the CWSR material, the microstructure was composed of fine elongated grains in the axial direction.

The texture analysis reveals a fairly pronounced anisotropy [3]. The characteristics of the {0002} pole figures are mentioned in Table 2. They show a basal pole intensity with two maximum at ±30° around the radial direction in the r,θ plane for the R material (Fig. 1a). Four peak tilt angles are observed for the CWSR materials, two at ±34° (CWSR) and at ±40° (CT-CWSR) in the r,θ plane and two others at about ±15° in the r,z plane (Fig. 1b). The Kearns factors exhibit the radial trend for the crystallographic texture.

Experimental Method

The determination of the anisotropic viscoplastic behavior of the tubes has required uniaxial and biaxial tests, i.e., tension-compression, tension-torsion, and tension-internal pressure tests. Monotonic and cyclic tests at imposed strain rates have been performed on commercial machines having hydraulic and electric actuators. The machines were controlled on the basis of signals delivered by high-temperature extensometers (longitudinal, axial, and tension-torsion) directly attached to the specimen. Uni- and multi-axial creep tests were performed on dead

TABLE 1—*Weight composition of Zircaloy-4.*

	Alloying Elements (wt%)					Impurities, ppm					
	Cr	Fe	Sn	O	Zr	C	N	H	Hf	Al	Si
R and CWSR	0.10	0.21	1.30	0.135	Bal.	151	24	9	56	98	99
CT-CWSR	0.12	0.22	1.25	0.161	Bal.	120	31	<2	...	...	...

TABLE 2—*Characteristics of the pole figures {0002}.*

	I_{max} in the plane (r, θ)	I_{max} in the plane (r,z)	Kearns factors, f_R, f_T, f_A		
Cold-worked tubes (CW)	±40°	+17°/−15°	0.56	0.31	0.12
Cold-worked stress-relieved tubes (CWSR)	±34°	±14°	0.58	0.32	0.10
Recrystallized tubes (R)	±30°	0	0.56	0.33	0.11
Cladding tubes (CT-CWSR)	±40°	±15°	0.57	0.35	0.08

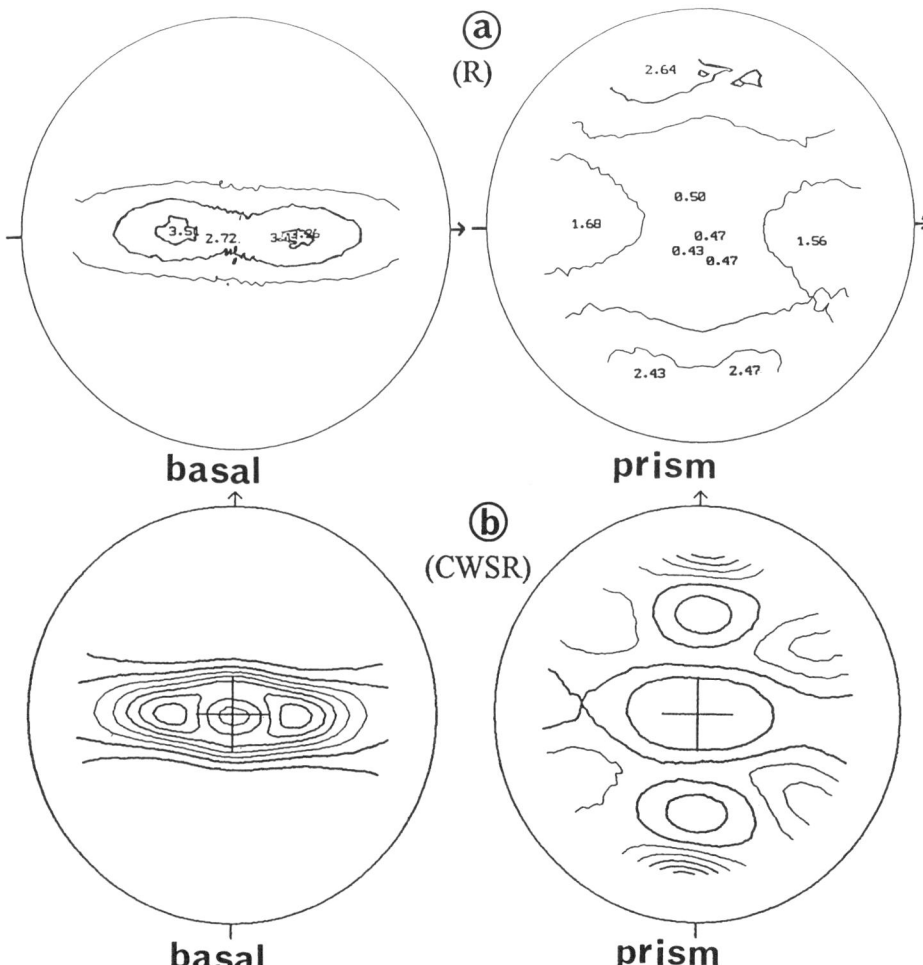

FIG. 1—*The basal and prism pole figures for the two materials: (a) recrystallized state (R); (b) cold-work stress-relieved state (CWSR).*

weight machines with the internal pressure obtained using inert gases. As before, the strains were measured using extensometers directly attached to the specimens.

Experimental Results and Analysis

In the text, the subscripts T, e, and p mean, respectively, total, elastic, and plastic components, and zz, $\theta\theta$, and $z\theta$ denote the axial, hoop, and shear components.

Uniaxial Tests

Monotonic Tension Tests—Different monotonic tension tests have been performed between ambient temperature and 450°C for two very different loading rates to quantify the viscosity effects as a function of the temperature. An example of a test at 350°C for $\dot{\varepsilon}_{zz}^T = 6.6 \cdot 10^{-7}\ s^{-1}$ and for the two metallurgical states is shown in Fig. 2a. If σ_{ec} is the flow stress of the material for the two states, the variation of σ_{ec} with total strain at different temperatures at the two applied loading rates showed (Fig. 3) a decrease between 20 and 270°C followed either by a flattening off or a slight maximum at approximately 300°C. Above 400°C, this decrease becomes more rapid with decreasing strain rates (Fig. 3). The presence of a maximum, frequently observed in solid solutions, is caused by the dynamic strain aging phenomena [4,5]. Above 350°C, thermally activated deformation mechanisms appear and lead to a further decrease in σ_{ec}.

The variation of the strain rate sensitivity with temperature can be characterized by the parameter Δ, defined by:

$$\Delta = [\sigma_{ec}(\dot{\varepsilon}_{zz}^T \max)/\sigma_{ec}(\dot{\varepsilon}_{zz}^T \min)]_{e,T} \qquad (1)$$

Indeed, since the ratio between the maximum and minimum strain rates is relatively large (10^3), this parameter Δ seems to be fairly representative. Figure 4 shows that Δ has a distinct minimum around 300°C ($\Delta \cong 1.0$) and tends toward higher values at higher temperatures (at 450°C $\Delta \cong 1.5$ for the R state and $\Delta \cong 1.7$ for the CWSR state with $\tilde{\varepsilon}_{zz} = 1\%$). The value of $\Delta \cong 1$ means that the material is strain-rate insensitive at 300°C, which is consistent with maximum in σ_{ec} corresponding to dynamic strain aging. It can be noted that Zircaloy-4 is more viscous at ambient temperatures [$\Delta = 1.15$ (R) and $\Delta = 1.09$ (CWSR)] than at 300°C.

During the imposed strain rate tests, the flow direction in the $\{\theta\theta\text{-}zz\}$ plane is determined by drawing the curve $\varepsilon^T_{\theta\theta} = f(\varepsilon^T_{zz})$ (Fig. 2b). Indeed, when $\varepsilon^p_{ij} \gg \varepsilon^e_{ij}$ (plastic and elastic strains), the slope of the experimental curves allows the anisotropy ratio (plastic Poisson's ratio) to be determined:

$$R^p = \varepsilon^p_{\theta\theta}/\varepsilon^p_{zz} \qquad (2)$$

In a more conventional way, one can derive the traditional CSR, R from the parameter R^p:

$$R = -R^p/1 + R^p$$

The numerical analysis of the derivative of these representations allows the variation of R^p with temperature and strain to be determined. The numerical values obtained by this method show that R^p seems to have a slight maximum around 350°C and is quasi-independent of the strain over the strain range 0.3 to 5%. Thus, in a first approximation, we will adopt the mean values: $R^p = -0.63 \pm 0.04$ ($R = 1.70$), $R^p = -0.56 \pm 0.03$ ($R = 1.27$) for $20 < T < 400°C$, and $R^p = -0.60 \pm 0.02$ ($R = 1.5$) for $T = 350°C$, respectively, for the R, CWSR, and CT-CWSR states.

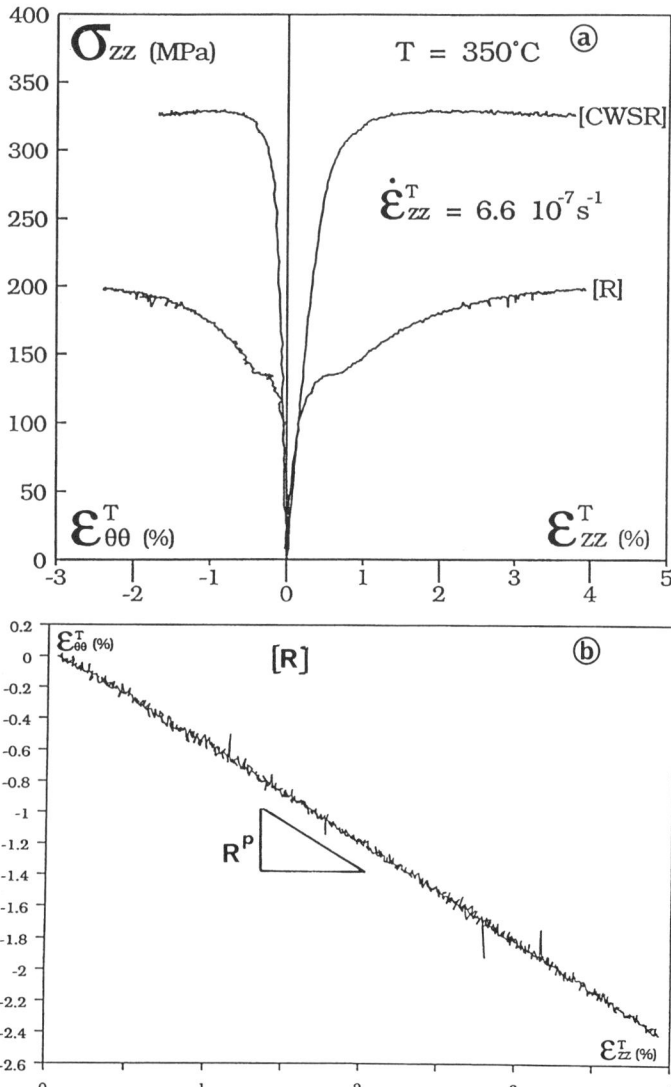

FIG. 2—(a) *Examples of monotonic stress-strain curves for the R and CWSR materials:* $T = 350°C$, $\dot{\varepsilon}_{zz}^T = 6.6 \cdot 10^{-7} s^{-1}$; *(b) example of representation* $\varepsilon^T{}_{\theta\theta} = f(\varepsilon_{zz}^T)$. *Determination of the plastic Poisson's ratio:* $T = 350°C$, $\dot{\varepsilon}_{zz}^T = 6.6 \cdot 10^{-7} s^{-1}$, *R material.*

The analysis of the crystallographic texture (in particular, the Kearns factors) also allows the anisotropy to be quantified [6]. We obtain $R^p = -0.63$ (R), $R^p = -0.64$ (CWSR), and $R^p = -0.62$ (CT-CWSR). For the R and CT-CWSR materials, the values are similar to those given by the mechanical tests. For the CWSR material, the two results are slightly different. This difference remains to be explained.

Relaxation Tests—To quantify the viscoplastic properties of Zircaloy-4 and to confirm the results obtained for the parameter Δ, different multiple relaxation tests have been performed at

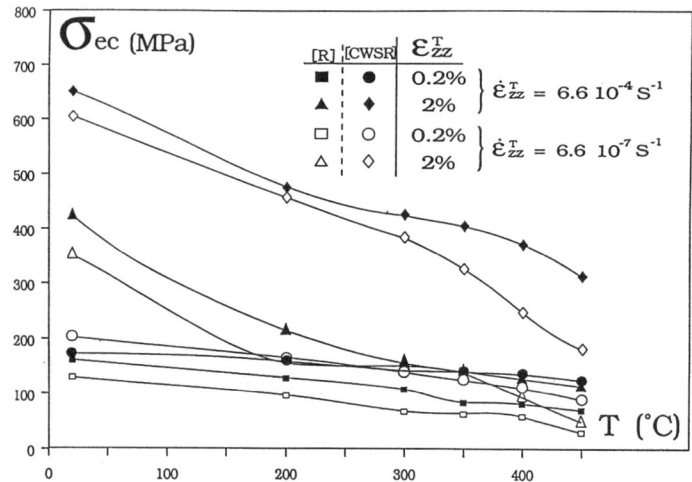

FIG. 3—*Variation of the flow stress σ_{ec} with the temperature for $\dot{\varepsilon}_{zz}^T = 6.6 \cdot 10^{-4}$ s^{-1} and $6.6 \cdot 10^{-7}$ s^{-1} and for the two materials R and CWSR.*

different temperatures on the R state and for 350°C on the CWSR and CT-CWSR states. Figure 5 shows the relaxed amplitude $\Delta\sigma_{rel}$ ($\Delta\sigma_{rel} = \sigma_o - \sigma_t$; σ_o is the flow stress before the start of relaxation and σ_t the stress at the end of the relaxation) normalized with respect to the flow stress σ_o as a function of the temperature. For the R material at 0.4% strain, there is a maximum in the value of $\Delta\sigma_{rel}/\sigma_o$ between 200 and 300°C. For the other strain amplitudes, the observations reported above for the parameter Δ on the R state are confirmed. As a consequence, during the phenomenological modeling, only a viscoplastic approach is used in the temperature range 20 to 450°C.

Uniaxial Creep Tests—The variation of the creep strain with the temperature in the R material has a distinct minimum around 300°C, a very rapid increase in strain beyond 350°C (creep recovery), and a fairly significant strain at 200°C (logarithmic creep). This behavior is similar to that observed by Fidleris [7] for Zircaloy-2 and corroborates the results of Yi et al. [8] for Zircaloy-4. According to these authors and in agreement with our interpretation, dynamic aging is responsible for these behavioral anomalies. The values calculated for the anisotropic coefficient R^p during the creep tests are in agreement with those reported for the imposed strain rate tests. So, the R^p parameter is relatively independent of the test temperature and strain rate, which corroborates the results of Beauregard et al. [9] and Murty [10].

The sensitivity parameter, n, of the rate to the stress is defined by: $n = (\partial \text{Ln}\dot{\varepsilon}^{st}/\partial \text{Ln}\sigma)_T$. Values of n have been determined and, at the lowest stresses, had values of 7.5 (R), 4.0 (CWSR), and 3.5 (CT-CWSR)). The values then increase with increasing stress [10].

During the creep tests, the size of the effective stress component σ_v ($\sigma_v = \sigma - \alpha$, α being the internal or back stress) is estimated for each stress level by using the "stress dip-test technique" [11]. Several "strain dip tests" were also performed as a complement to the previous technique. Although these two techniques are experimentally imprecise, they allow the relationship between strain rate and the effective stress component, $\dot{\varepsilon} = f(\sigma_v)_T$, to be determined. From the experimental data, the following form of the state equation was finally proposed:

$$|\dot{\varepsilon}_{zz}| = \dot{\varepsilon}_o(\sinh(\sigma_v/N))^{n*} \qquad (3)$$

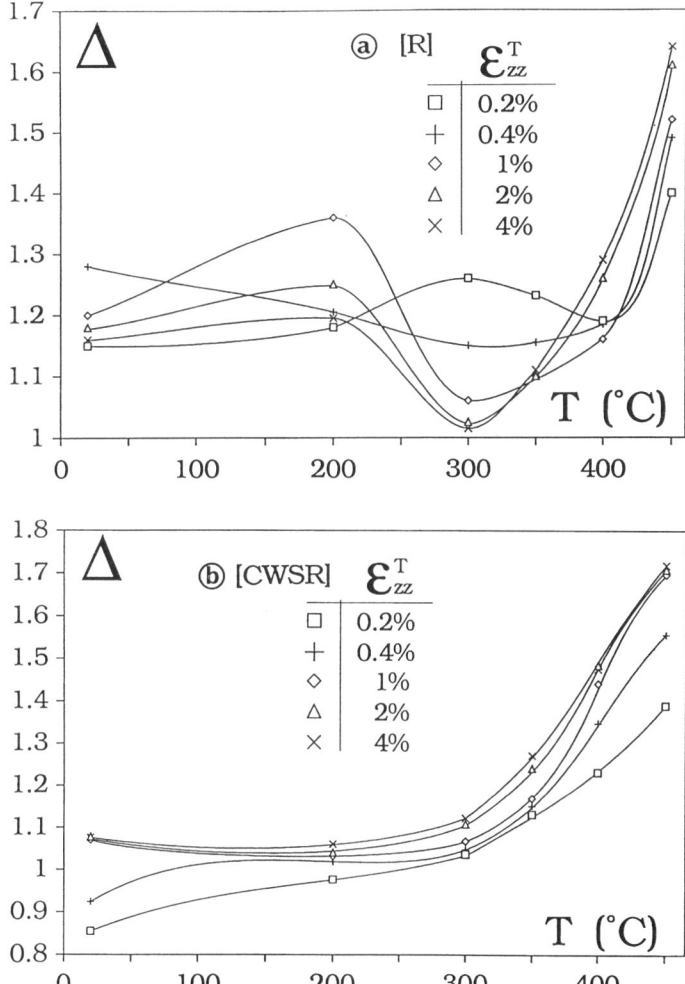

FIG. 4—*Variation of the Δ parameter with the temperature, for different strain levels:* (a) *recrystallized state (R);* (b) *cold-worked stress-relieved state (CWSR).*

where $\dot{\varepsilon}_o$, N, and n^* are constants. The values of these coefficients are given in Table 3. Note that the values of n and n^* are very different because they respectively correspond to the sensitivity parameter of the rate to the stress and of the rate to the effective stress [*12*]. The physical signification of these two parameters is also different [*12*].

Cyclic Uniaxial Test—Different cyclic tests between imposed strains have been performed at the same temperatures as the monotonic tests in order to obtain the cyclic curves $\Delta\sigma_{zz}/2 = f(\Delta\varepsilon_{zz}^T/2)$ (Fig. 6). The variation of the maximum and minimum stresses as a function of the number of cycles N [*13*] shows:

1. The cyclic stability of this alloy since only a very slight softening is observed on the CWSR material.

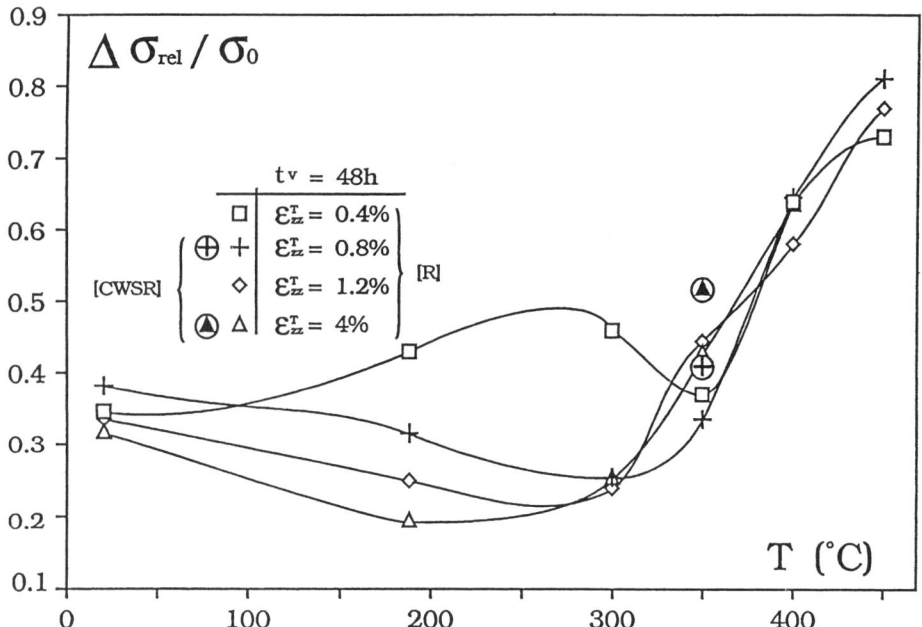

FIG. 5—*Evolution of the normalized relaxed amplitude $\Delta\sigma_{rel}/\sigma_o$ in 48 h as a function of the temperature for different strain levels (R material and CWSR material at 350°C).*

2. The quasi-symmetry of the cycles ($\sigma_{max} \cong |\sigma_{min}|$) for all temperatures, except at 20°C.
3. The absence of strain history memory; the cyclic amplitude depends on the amplitude of the imposed strain in a univocal fashion.

Multiaxial Tests

In order to complete the characterization of the anisotropy in the tubes, different multiaxial tests in tension-torsion and tension-internal pressure were performed at imposed strain rates or stresses. Tests at only two temperatures were studied, 20 and 350°C for the imposed strain rate tests and 350°C for the multiaxial creep tests.

From a phenomenological point of view, the results were analyzed by considering, in a first step, an orthotropic anisotropy such as Hill's [*14*] that can be written in the matrix form:

$$\bar{\sigma} = ((3/2)(M_{ij}\,\sigma'_i\,\sigma'_j))^{1/2} \qquad (4)$$

$\bar{\sigma}$ is the equivalent stress, M_{ij} the matrix of anisotropy, and σ'_i the deviatoric components of the stress tensor.

Tension-Torsion Tests and Monotonic Tests at Imposed Strain Rates—In these different tests, the equivalent strain rate in the von Mises sense is imposed, and the ratio between the shear and axial strain rates was varied. For the R material, as for the tension tests, an instability in the two stress components was found; hence, there was some certain imprecision in the definition of the elastic limit. Nevertheless, the elastic surfaces defined at $\bar{\varepsilon}^T = 2 \cdot 10^{-4}$ can be plotted in the ($\sigma_{z\theta}, \sigma_{zz}$) plane for the two temperatures and for the two states (Fig. 7). Taking the tension tests as a reference and plotting the equivalent plasticity limit in the von Mises

TABLE 3—Values of model coefficients identified at T = 350°C.

	R	CWSR
Elasticity	$E = 78\,000$ MPa, $v = 0.4$	$E = 73\,000$ MPa, $v = 0.42$
State equation	$N = 6.5$ MPa, $n^* = 2.4$, $\dot{\varepsilon}_o = 6.58 \cdot 10^{-8}$ s^{-1}	$N = 15$ MPa, $n^* = 2.4$, $\dot{\varepsilon}_o = 1.5 \cdot 10^{-6}$ s^{-1}
Kinematical variables	$p = 8000, p_1 = 600, p_2 = 65$ $r_m = 2.4 \cdot 10^{-12}$ MPa s^{-1}, $\alpha_o = 10$ MPa, $m_o = 8$	$p = 9000, p_1 = 400, p_2 = 10$, $r_m = 1.53 \cdot 10^{-8}$ MPa s^{-1}, $r_{m1} = 1 \cdot 10^{-43}$ MPa s^{-1} $\alpha_o = 10$ MPa, $m_o = 3.7, m_1 = 28$
Scalar variables	$Y_o = 69$ MPa, $Y^{\text{sat}}_\theta = 13$ MPa, $b_\theta = 35$	$Y_o = 177$ MPa, $Y^{\text{sat}} = -40$ MPa, $Y^{\text{sat}}_\theta = 30$ MPa, $b = 2.5$ $b_\theta = 5$
Coefficients of anisotropy	$M_{11} = 0.398, M_{22} = 0.620,$ $M_{33} = 0.666,$ $M_{13} = -0.222, M_{23} = -0.444,$ $M_{12} = -0.176, M_{44} = 3.8$ $N_{11} = 1.038, N_{22} = 0.720, N_{33} = 0.666$ $N_{13} = -0.492, N_{23} = -0.174, N_{12} = -0.546$ $N_{44} = 0.295$ $Q_{11} = Q_{22} = Q_{33} = \frac{2}{3}, Q_{44} = 1$ $Q_{13} = Q_{23} = Q_{12} = -\frac{1}{3}$ $R_{11} = 0.303, R_{22} = 0.637, R_{33} = 0.666$ $R_{13} = -0.166, R_{23} = -0.5, R_{12} = -0.137$ $R_{44} = 2.07$	$M_{11} = 1.1, M_{22} = 1.2, M_{33} = 0.666$ $M_{13} = -0.28, M_{23} = -0.38,$ $M_{12} = -0.82, M_{44} = 3.4$ $N_{11} = 0.699, N_{22} = 0.562, N_{33} = 0.666$ $N_{13} = -0.401, N_{23} = -0.265, N_{12} = -0.297$ $N_{44} = 0.33$ $Q_{11} = Q_{22} = Q_{33} = \frac{2}{3}, Q_{44} = 1$ $Q_{13} = Q_{23} = Q_{12} = -\frac{1}{3}$ $R_{11} = 0.626, R_{22} = 0.880, R_{33} = 0.666$ $R_{13} = -0.206, R_{23} = -0.460, R_{12} = -0.420$ $R_{44} = 2.75$

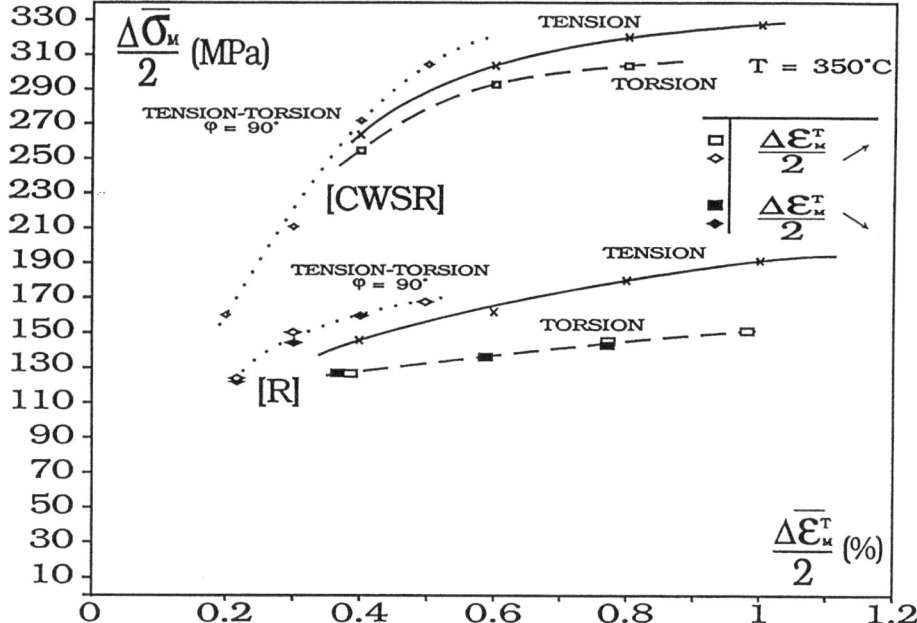

FIG. 6—*Cyclic curves $\Delta\bar{\sigma}_M/2 = f(\Delta\varepsilon^T_M/2)$ (von Mises equivalent stress and strain) for tension, torsion, and out-of-phase tension-torsion tests (R and CWSR materials).*

sense, it can be shown that the R material possesses a much smaller resistance to tension-torsion than that predicted by the von Mises equation.

In the tension-torsion case, Eq 4 was written as:

$$\bar{\sigma} = \left(\frac{3}{2} M_{33}\left(\sigma_{zz}^2 + \frac{M_{44}}{M_{33}} \sigma_{z\theta}^2\right)\right)^{1/2} \quad (5)$$

with $\sigma_{zz} = \sigma_3$ and $\sigma_{z\theta} = \sigma_4$. Given the equivalence imposed in tension, we postulate $M_{33} = 2/3$. The coefficient M_{44} thus allowed the position of the experimental points plotted in Fig. 7 to be adjusted. For the two isotherms under study, we obtained $M_{44} \cong 3.8$ (R) and $M_{44} = 3.2$ (CWSR), values that are greater than that for isotropic material ($M_{44} = 2.0$).

Tension-Torsion Creep Tests—With the aid of three extensometers mounted on the specimen, the strains ε^T_{zz}, $\varepsilon^T_{\theta\theta}$, and $\varepsilon^T_{z\theta}$ were recorded simultaneously. The experimental sequence consisted in applying the same equivalent von Mises stress for all the tests and to varying, for each test, the ratio $\sigma_{z\theta}/\sigma_{zz}$.

Given the slope of the flow direction and in accordance with the normality rule:

$$\dot{\varepsilon}^p_{z\theta}/\dot{\varepsilon}^p_{zz} = M_{44}\sigma_{z\theta}/2M_{33}\sigma_{zz} = 3M_{44}\sigma_{z\theta}/4\sigma_{zz} \quad (6)$$

it was possible to determine M_{44} for each of the tests. Even with some scatter, average values of $M_{44} = 3.7 \pm 0.2$ (R) and $M_{44} = 3.5 \pm 0.2$ (CWSR) can be reported. Note that in tension, according to Eq 4:

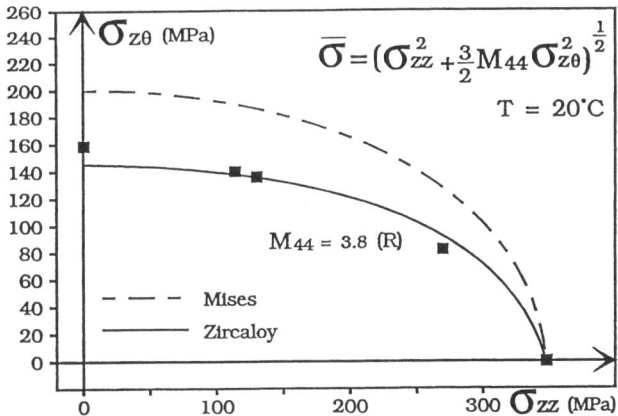

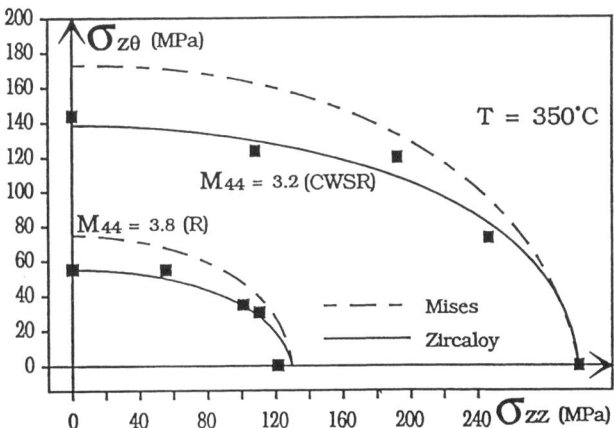

FIG. 7—*Loci of the yield stress in the plane ($\sigma_{z\theta}$, σ_{zz}) for the two studied materials. Determination of M_{44} for T = 20°C and T = 350°C (——— = predicted curves).*

$$R^p = \varepsilon^p_{\theta\theta}/\varepsilon^p_{zz} = M_{23}/M_{33} = 3M_{23}/2 \tag{7}$$

This leads to $M_{23} = -0.42$ (R), $M_{23} = -0.38$ (CWSR), and $M_{23} = -0.40$ (CT-CWSR) values, which are different than that for an isotropic material ($M_{23} = -\frac{1}{3}$). If the equivalent von Mises rates, $\dot{\bar{\varepsilon}}$, are plotted as a function of the equivalent stress, it is clearly seen that the rates are always (R and CWSR) greater than those of creep in tension, confirming the low torsional resistance already mentioned during the study of the elastic threshold surfaces.

Cyclic Torsion Tests—Cyclic torsion tests allow the torsional cyclic hardening curves to be determined by performing tests with successively increasing, followed by decreasing, strain levels. The maximum and minimum stress on each cycle for all the tests are seen to be perfectly symmetric. The cyclic torsion curves are reported in Fig. 6, $\Delta\bar{\sigma}_M/2 = f(\Delta\varepsilon^T_M/2)$, where these two quantities are, respectively, the maximum of the equivalent von Mises stresses and strains. It is shown, for the two states, that the torsional stress levels are smaller than those in tension.

Out-of-Phase Cyclic Tension-Torsion Tests—In order to account for an eventual effect due to the non-radiality of the loading [15,16], out-of-phase cyclic tension-torsion tests with a phase shift equal to 90° are performed. Knowing the strain and stress for each level, the cyclic curve $\Delta\bar{\sigma}_M/2 = f(\Delta\bar{\varepsilon}_M^T/2)$ can be determined. Figure 6 shows that the R and CWSR material possess a slight supplementary hardening due to the non-radiality of the loading, the curves for $\varphi = 90°$ being greater than the cyclic tension and torsion curves.

Tension-Internal Pressure Tests and Monotonic Tests with Imposed Strain Rates—A procedure similar to that employed in the tension-torsion case was adopted here for all the tension-internal pressure tests. Thus, an equivalent von Mises strain rate was imposed, and the

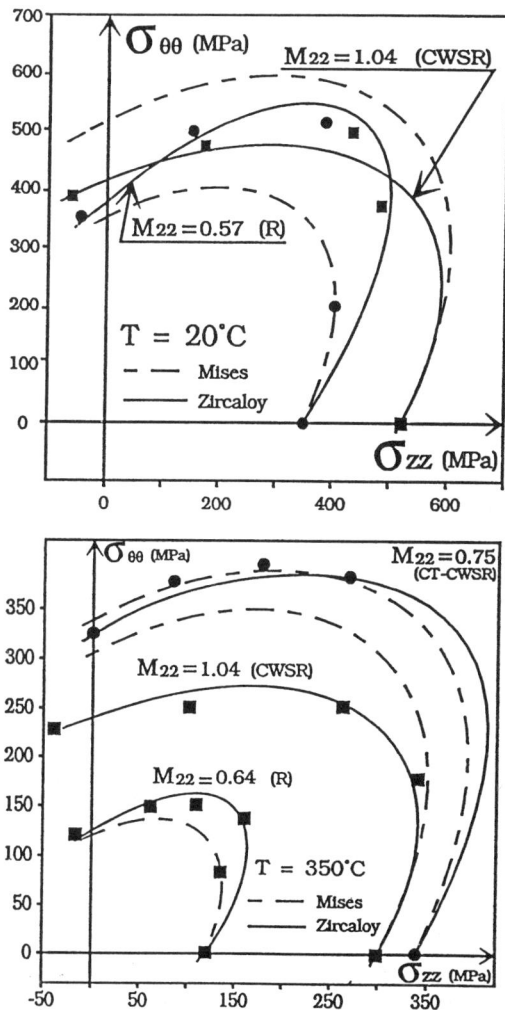

FIG. 8—*Loci of the yield stress in the plane* ($\sigma_{\theta\theta}$, σ_{zz}) *for the two studied materials. Determination of M_{22} for T = 20°C and T = 350°C (——— = predicted curves).*

ratio between the diametral and axial rates was varied. Figure 8 shows the elasticity surfaces in the $\sigma_{\theta\theta},\sigma_{zz}$ plane determined at $\varepsilon^T = 2.10^{-4}$ for the two metallurgical states. By taking the tension test as a reference and plotting the equivalent von Mises plasticity threshold, it is shown that the R material has a much higher strength than that predicted by the von Mises theory. We observe an opposite situation for the CWSR material, that is, the strength is smaller than that predicted by the von Mises criterion. Given the crystallographic texture, the form of these surfaces agrees with those reported by Stehle et al. [17], Baty et al. [18], and Daugherty and Murty [19]. Phenomenologically, by adopting Hill's formulation and neglecting σ_{rr}, the equation of the threshold surfaces can be written:

$$\bar{\sigma} = \left(\frac{3}{2} M_{22} \sigma^2_{\theta\theta} + \frac{3}{2} M_{33} \sigma^2_{zz} + 3 M_{23}\, \sigma_{\theta\theta} \sigma_{zz} \right)^{1/2} \quad (8)$$

($\sigma_{rr} = \sigma_1$, $\sigma_{\theta\theta} = \sigma_2$, $\sigma_{zz} = \sigma_3$). At present, the value of M_{33} is imposed and values for M_{23} have been determined in tension, thus allowing M_{22} to be evaluated by curve fitting the experimental points. In accord with Fig. 8, we obtain slightly different values between ambient temperature and 350°C, that is, $M_{22} \cong 0.57$ at 20°C, $M_{22} \cong 0.64$ at 350°C (R), and $M_{22} \cong 1.04$ (CWSR), $M_{22} \cong 0.75$ (CT-CWSR). Globally, M_{22} is smaller than the isotropic value ⅔ for the R state and greater for the CWSR state.

Creep Tests under Tension-Internal Pressure—As with the tension-torsion creep tests, the same equivalent stress was applied for all tests, and the ratio $\sigma_{\theta\theta}/\sigma_{zz}$ was varied from one test to another. Knowing the slope of the flow directions for each test, M_{22} can be evaluated using the normality rule:

$$\dot{\varepsilon}^p_{\theta\theta}/\dot{\varepsilon}^p_{zz} \cong (M_{22}\sigma_{\theta\theta} + M_{23}\sigma_{zz})/(M_{23}\sigma_{\theta\theta} + M_{33}\sigma_{zz}) \quad (9)$$

Average values of approximately 0.62 (R) and 1.1 (CWSR) were obtained.

The plot of equivalent strain rates, $\bar{\dot{\varepsilon}}$, versus equivalent stresses, $\bar{\sigma}$, shows that they are always lower than those obtained in tension for the R state and always greater than those in tension for the CWSR and CT-CWSR materials. The alloy thus has a better resistance to creep in tension-internal pressure than in tension for the R state and a lower resistance for the two CWSR states.

Conclusions Concerning the Anisotropy of the Tubes

In conclusion, by adopting the formulation of Eq 4 applied to an orthotropic material and taking into account the different determinations presented above along the symmetry relations, we propose M_{ij} as follows for the R, CWSR, and CT-CWSR conditions:

$M_{11} = 0.40, 1.00, 0.61$, $M_{12} = -0.18, -0.72, -0.35$, $M_{13} = -0.22, -0.28, -0.26$,
$M_{22} = 0.62, 1.10, 0.75$, $M_{23} = -0.44, -0.38, -0.40$, $M_{33} = 0.66, 0.66, 0.66$,
$M_{44} = 3.80, 3.40, ?$, $M_{55} = ?$, $M_{66} = ?$

The kinematic nature of the hardening is clearly observable in the cyclic testing as well as in the monotonic tests followed by an unloading sequence, indicating the insufficiency of Hill's formulation. The following equation is proposed to replace Eq 4:

$$\bar{\sigma} = \{(3/2)M_{ij}(\sigma'_i - \alpha'_i)(\sigma'_j - \alpha'_j)\}^{1/2} \qquad (10)$$

The α'_i represent the deviatoric components of the internal stress tensor whose anisotropic laws remain to be specified. This aspect will be examined in the following paragraph concerning the modeling.

Modeling of Zircaloy-4

This section discusses the application of a unified viscoplastic model with internal variables, developed and identified elsewhere on other materials [20], to Zircaloy-4 in the conditions used in the tests. A detailed description of the global structure of this model is beyond the scope of this paper, and only the introduction of anisotropy to describe the multiaxial loadings will be presented.

Presentation of the Anisotropic Model Particularized to the Case of Zircaloy 4

In a general manner, two second-rank tensors can be related either by a scalar function, $X_{ij} = MY_{ij}$, which is the case of the isotropic model, or by a fourth rank tensor, $X_{ij} = M_{ijkl}Y_{kl}$. It is this second possibility that is used here, and as a consequence only the tensor quantities will be affected by the anisotropy. Note that a similar approach has already been proposed by D. Nouailhas, in the case of the ONERA model, to simulate the behavior of monocrystals of superalloy having cubic symmetry [21]. The model of Lee et al. [22] also presents some similarities.

Adopting the matrix notation of Voigt, we have:

$$[\sigma]^t = [\sigma_1 = \sigma_{11}, \sigma_2 = \sigma_{22}, \sigma_3 = \sigma_{33}, \sigma_4 = \sigma_{23}, \sigma_5 = \sigma_{13}, \sigma_6 = \sigma_{12}]$$
$$[\varepsilon]^t = [\varepsilon_1 = \varepsilon_{11}, \varepsilon_2 = \varepsilon_{22}, \varepsilon_3 = \varepsilon_{33}, \varepsilon_4 = 2\varepsilon_{23}, \varepsilon_5 = 2\varepsilon_{13}, \varepsilon_6 = 2\varepsilon_{12}]$$
$$[\Delta]^t = [1, 1, 1, 0, 0, 0]$$

The general formulation of the model requires the introduction of four fourth rank tensors, denoted [M], [N], [Q], and [R] affecting the flow rules (Eqs 14, 15, and 17). Given the crystallographic texture of the tubes, the material axes are assumed to be identical with the anisotropic axes. The three incompressibility relations allow the number of independent unknowns to be reduced from nine to six. Thus, each of the four tensors has the same symmetry as [M], that is:

$$[M] = \begin{bmatrix} M_{11} & M_{12} & M_{13} & & & \\ M_{12} & M_{22} & M_{23} & & 0 & \\ M_{13} & M_{23} & M_{33} & & & \\ & & & M_{44} & & \\ & & & & M_{55} & \\ & 0 & & & & M_{66} \end{bmatrix} \quad \text{with:} \begin{array}{l} M_{11} + M_{12} + M_{13} = 0 \\ M_{12} + M_{22} + M_{23} = 0 \\ M_{13} + M_{23} + M_{33} = 0 \end{array} \qquad (11)$$

The model equations are written according to:

$$[\dot{\varepsilon}^T] = [\dot{\varepsilon}^e] + [\dot{\varepsilon}] \tag{12}$$

$$[\dot{\varepsilon}^e] = \frac{1+v}{E}[\dot{\sigma}] - \frac{v}{E}[\Delta]'[\dot{\sigma}][\Delta] \tag{13}$$

$$[\dot{\varepsilon}] = \frac{3}{2}\dot{\bar{\varepsilon}}\frac{[M][\sigma' - \alpha']}{\bar{\sigma} - \bar{\alpha}}, \text{ with } \dot{\bar{\varepsilon}} = \dot{\varepsilon}_o\left\{\sinh\left(\frac{\overline{\sigma - \alpha}}{N}\right)\right\}^{n^*} \tag{14}$$

$[\dot{\varepsilon}^T]$, $[\dot{\varepsilon}^e]$, and $[\dot{\varepsilon}]$ are, respectively, total, elastic, and viscoplastic strain rate vectors. $[\sigma]$ and $[\sigma']$ are the applied stress and deviatoric stress vectors, and $[\alpha]$ and $[\alpha']$ are the kinematic hardening variable and deviatoric kinematic hardening variable vectors.

In these equations, we have:

$$[\sigma'] = [\sigma] - \frac{1}{3}[\Delta]'[\sigma][\Delta], [\alpha'] = [\alpha] - \frac{1}{3}[\Delta]'[\alpha][\Delta], \text{ and}$$

$$\overline{\sigma - \alpha} = \left\{\frac{3}{2}[\sigma' - \alpha']'[M][\sigma' - \alpha']\right\}^{1/2}$$

The three kinematical hardening variables $[\alpha']$, $[\alpha'^{(1)}]$, and $[\alpha'^{(2)}]$ are defined by:

$$\left.\begin{aligned}[\dot{\alpha}'] &= p\left(\frac{2}{3}Y^*[\underline{N}][\dot{\varepsilon}] - [Q][\alpha' - \alpha'^{(1)}]\dot{\bar{\varepsilon}}\right) - f(\bar{\alpha})[\underline{N}][\underline{R}]\frac{[\alpha']}{\bar{\alpha}}\\
\text{where } f(\bar{\alpha}) &= r_m\left(\frac{\bar{\alpha}}{\alpha_o}\right)^{m_o} \text{ for the R material and } f(\bar{\alpha}) = \left(r_m\left(\frac{\bar{\alpha}}{\alpha_o}\right)^{m_o} + r_{m_1}\left(\frac{\bar{\alpha}}{\alpha_o}\right)^{m_1}\right)\\
\text{for the CWSR material,}&\\
[\dot{\alpha}'^{(1)}] &= p_1\left(\frac{2}{3}Y^*[\underline{N}][\dot{\varepsilon}] - [Q][\alpha'^{(1)} - \alpha'^{(2)}]\dot{\bar{\varepsilon}}\right)\\
[\dot{\alpha}'^{(2)}] &= p_2\left(\frac{2}{3}Y^*[\underline{N}][\dot{\varepsilon}] - [Q][\alpha'^{(2)}]\dot{\bar{\varepsilon}}\right) \text{ with: } [\alpha^{(k)}]_0 = 0\end{aligned}\right\} \tag{15}$$

In these equations, $\bar{\alpha}$ is defined by: $\bar{\alpha} = \left\{\frac{3}{2}[\alpha']'[\underline{R}][\alpha]\right\}^{1/2}$, and Y^* represents the sum of the contributions of the different scalar variables to the global hardening of the material. In Eq 16, Y_o is a constant for a given temperature, Y_{iso} and Y_θ are scalar variables, respectively, associated with the increase of the total dislocation density and of the dislocation density when a phase shift exists between the loadings. Thus:

$$\begin{aligned}Y^* &= Y_o + Y_{iso} + Y_\theta \text{ with:}\\
\dot{Y}_{iso} &= b(Y^{sat} - Y_{iso})\dot{\bar{\varepsilon}} \text{ for CWSR material and } \dot{Y}_{iso} = 0 \text{ for R material}\end{aligned} \tag{16}$$

$$\dot{Y}_\theta = b_\theta(Y^{sat}_\theta\, f(\theta) - Y_\theta)\bar{\dot{\varepsilon}}, \text{ with: } Y_\theta(0) = 0 \text{ and}$$

$$\cos\theta = \frac{3}{2}\frac{[\alpha']^t[R][\dot{\alpha}']}{\bar{\alpha}\,\bar{\dot{\alpha}}}, f(\theta) = (1 - |\cos\theta|) \tag{17}$$

The supplementary hardening observed during out-of-phase loadings is described by the function Y_θ.

Finally, Eqs 12 to 17 constitute the anisotropic version of the general model in the case of Zircaloy-4.

Identification and Application of the Model in the Case of Zircaloy-4

The model identification has only been performed at 350°C. However, the experimental study presented in the first section shows that the measured anisotropic parameter R^p is relatively insensitive to the temperature. This cannot be applied to all components of the anisotropy tensors since the P parameter defined as the CSR in the uniaxial hoop test is sensitive to the test temperature and applied strain rate. As a consequence, a few model parameters are affected by temperature, in particular those related to the anisotropy, to the viscosity, and to the static recovery. At a fixed temperature, the determination of all coefficients remains the most difficult part and requires a specific methodology that has been described by Delobelle and Robinet [13].

All of the model coefficients are regrouped in Table 3.

Numerical Simulations

Numerical simulations of all tests have been performed, but only a few figures are presented to illustrate the results.

Figure 9a represents the simulations of the tension test for different strain rates for the two metallurgical states (R and CWSR). Figure 9b shows results for the cladding tubes in a biaxial tension-internal pressure test with $\alpha = \sigma_{zz}/\sigma_{\theta\theta} = 0$. The agreement is acceptable except in the neighborhood of the plastic instability for the R state. Figure 10 gives four examples of creep curves, respectively, for the creep tests in pure tension (CT-CWSR), in pure torsion (R), and in tension-internal pressure (R and CWSR). Note that there is good agreement between the test results and the calculated values.

In conclusion, all of the multiaxial tests performed at 350°C for the three metallurgical states are fairly well modeled; hence, there is confirmation of the potential of this model to describe the anisotropic viscoplastic behavior of Zircaloy-4.

Conclusions

The anisotropic viscoplastic properties of the recrystallized and cold-worked stress-relieved Zircaloy-4 tubes have been partially studied between 20 and 400°C with the aid of monotonic and cyclic multiaxial tests. At 350°C, the anisotropy has been quantified in a detailed fashion. The results obtained at room temperature, as well as the independence of the ratio R^p with respect to the temperature, indicate that the anisotropic coefficients do not depend on temperature. However, the fluidity of Zircaloy-4 has a minimum in the neighborhood of 300°C, being caused by the dynamic strain aging frequently observed in interstitial solid solutions. During tension-torsion out-of-phase cyclic loading, Zircaloy-4 shows a slight supplementary hardening.

An extension of the unified viscoplastic model, developed and identified on other initially isotropic materials, is proposed for the case of Zircaloy-4. In a general way, the introduction

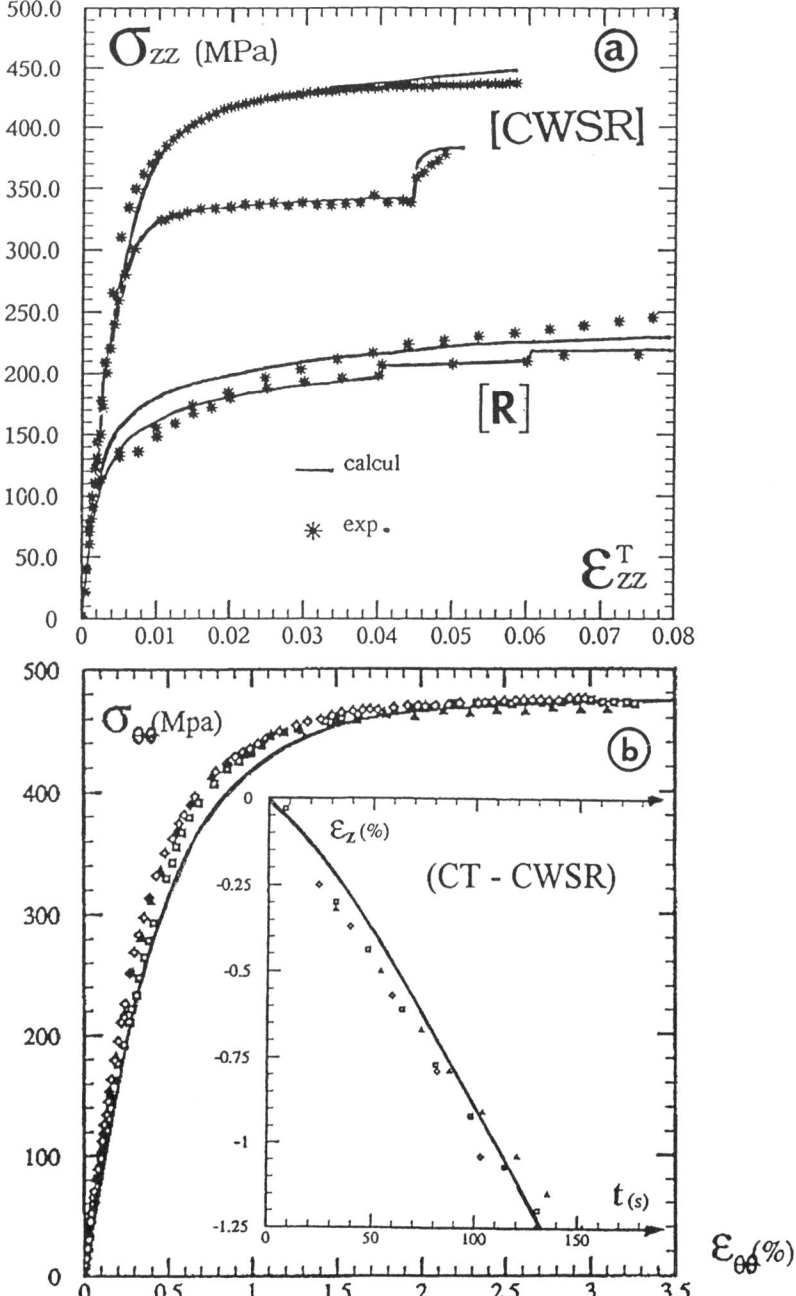

FIG. 9—(a) *Modeling of the tension tests for different incremental strain rates,* $\dot{\varepsilon}_{zz}^T = 6.6 \cdot 10^{-7}$ s^{-1} *and* $6.6 \cdot 10^{-4}\ s^{-1}$ *for the R and CWSR materials;* (b) *Modeling of biaxial tension-internal pressure test for the cladding tubes (CT-CWSR), with* $\sigma_{zz}/\sigma_{\theta\theta} = 0$.

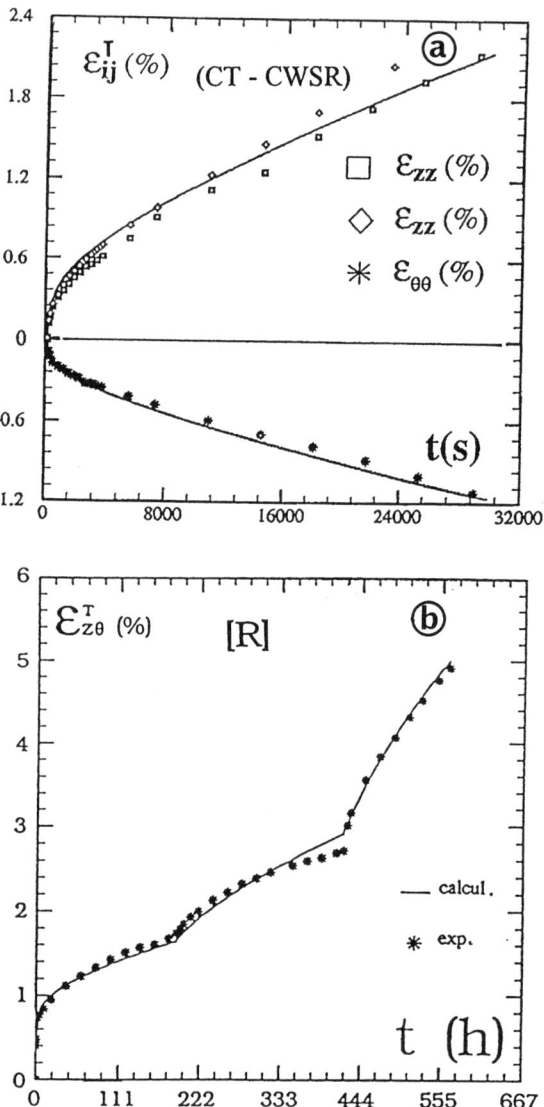

FIG. 10—(a) *Modeling of a uniaxial creep test performed on CT-CWSR tube with* $\sigma_{zz} = 400$ *MPa;* (b) *Modeling of a pure incremental torsion test for R material with increasing stresses:* $\sigma_{z\theta} = 72.8$ *MPa and 86.6 MPa;* (c) *Modeling of tension-internal pressure test performed on R material with increasing stresses:* $\sigma_{\theta\theta}/\sigma_{zz} = 0.8$, $\sigma = 164.7$, 183.0, *and* 207.5 *MPa;* (d) *Modeling of tension-internal pressure test performed on CWSR material with increasing stresses:* $\sigma_{\theta\theta}/\sigma_{zz} = \infty$, $\sigma = 150$ *and 244 MPa.*

of the anisotropy in this model is made via the four fourth rank tensors affecting the flow directions [M], the linear parts of kinematical hardening [N], as well the dynamic and static recoveries, [Q] and [R], of these same hardening variables.

The identification of this model has been performed at 350°C, and it is shown that it is adequate for representing the mechanical characteristics of Zircaloy-4 at the test temperatures.

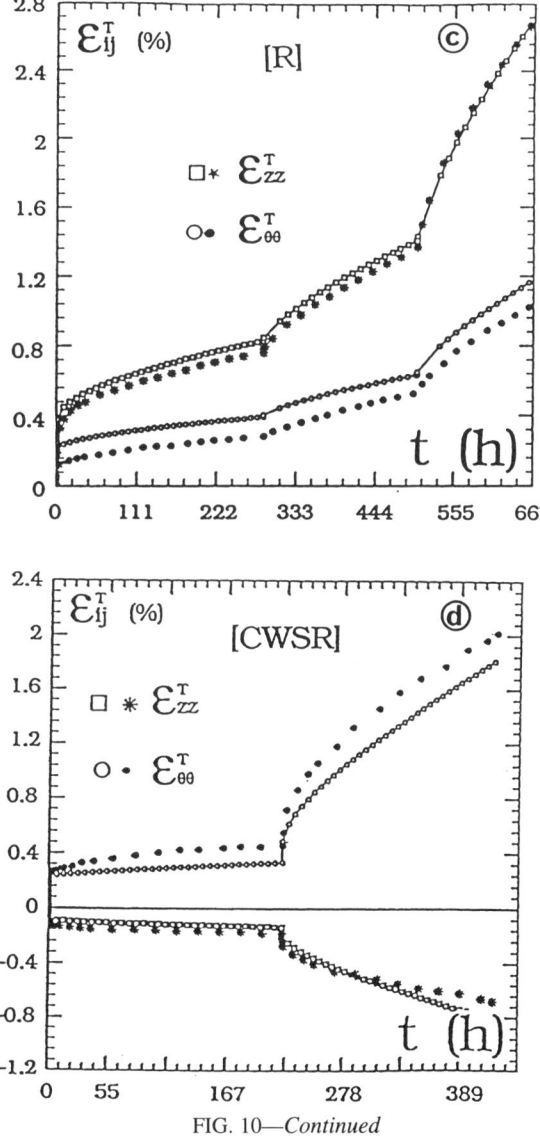

FIG. 10—Continued

Acknowledgments

This study has been entirely financed by E.D.F. under the Contract No. T43/7A754/RNE 319.

References

[1] Baron, D. and Bouffioux, P., "Le Crayon Combustible des Réacteurs à Eau Pressurisée de Grande Puissance," Rapport EDF HT M2/88-27, 1989.

[2] Bouffioux, P., "Caractérisation des Tubes en Zircaloy 4," Doc. Techn. EDF, No. DTI-CB-03, 1991.
[3] Bouffioux, P., "Analyse de Texture Cristallographique des Produits en Zircaloy 4," Doc. Techn. EDF, No. DTI-CB-04, 1991.
[4] Miller, A. K. and Sherby, O. D., "A Simplified Phenomenological Model for Non-Elastic Deformation: Predictions of Pure Aluminium Behavior and Incorporation of the Solute Strengthening Effects," *Acta Metallurgica,* Vol. 26, 1978, p. 289.
[5] Cribb, W. R. and Reed-Hill, R. E., "Static Strain Aging in Nickel 200 Between 373 K and 473 K," *Metallurgical Transactions,* Vol. 8A, 1977, p. 71.
[6] Van Swam, L. F. P., Knorr, D. B., Pelloux, R. M., and Shewbridge, J. F., "Relationship Between Contractile Strain Ratio R and Texture in Zirconium Alloy Tubing," *Metallurgical Transactions,* Vol. 10A, 1979, p. 483.
[7] Fidleris, V., "The Effect of Texture and Strain Aging on Creep of Zircaloy-2," Applications Related Phenomenon in Zirconium and Its Alloys, *ASTM STP 458,* American Society of Testing and Materials, West Conshohocken, PA, 1969, p. 17.
[8] Yi, J. K., Park, H. B., Park, G. S., and Lee, B. W., "Yielding and Dynamic Strain Aging Behavior of Zircaloy-4 Tube," *Journal of Nuclear Materials,* Vol. 189, 1992, p. 353.
[9] Beauregard, R. J., Clevinger, G. S., and Murty, K. L., "Effect of Annealing Temperature on the Mechanical Properties of Zircaloy-4 Cladding," *Proceedings,* SMIRT IV, 1977, Paper C3/5.
[10] Murty, K. L., "Applications of Crystallographic Textures of Zirconium Alloys in Nuclear Industry," Zirconium in the Nuclear Industry, 8th Symposium, *ASTM STP 1023,* American Society for Testing and Materials, West Conshohocken, PA, 1989, p. 570.
[11] Ahlquist, C. N. and Nix, W. D., "A Technique for Measuring Mean Internal Stress During High Temperature Creep," *Scripta Metallurgica,* Vol. 3, 1969, p. 679.
[12] Ahlquist, C. N. and Nix, W. D., "The Measurement of Internal Stresses During Creep of Al and Al-Mg-alloys," *Acta Metallurgica,* Vol. 19, 1971, p. 373.
[13] Delobelle, P. and Robinet, P., "Etude du Comportement et de la Modélisation Viscoplastique du Zircaloy 4 Recristallisé sous Chargements Monotones et Cycliques Uni-et Multiaxes," *Journal of Physics III,* Vol. 4, 1994, p. 1347.
[14] Hill, R., *The Mathematical Theory of Plasticity,* Clarendon Press, Oxford, 1950.
[15] McDowell, D. L., "A Two Surface Model for Transient Non Proportional Cyclic Plasticity," *Journal of Applied Mechanics,* Vol. 52, 1985, p. 298.
[16] Benallal, A. and Marquis D., "Constitutive Equations for Non-Proportional Cyclic Elastoviscoplasticity," *Journal of Enginering and Material Technology,* Vol. 109, 1987, p. 326.
[17] Stehle, H., Steinberg, E., and Tenckhoff, E., "Mechanical Properties, Anisotropy, and Microstructure of Zircaloy Canning Tubes," Zirconium in the Nuclear Industry (3rd Conference), *ASTM STP 633,* American Society for Testing and Materials, West Conshohocken, PA, 1977, p. 486.
[18] Baty, D. L., Pavinich, W. A., Dietrich, M. R., Clevinger, G. S., and Papazoglou, T. P., "Deformation Characteristics of Cold-Worked and Recrystallized Zircaloy-4 Cladding," Zirconium in the Nuclear Industry (6th Symposium), *ASTM STP 824,* American Society Testing and Materials, West Conshohocken, PA, 1984, p. 306.
[19] Daugherty, W. L. and Murty, K. L., "Application of Texture in Predicting Nuclear Fuel Cladding Creep," *Nuclear Technology,* Vol. 80, 1988, p. 443.
[20] Delobelle, P., "Synthesis of the Elastoviscoplastic Behavior and Modelization of an Austenitic Stainless Steel Over a Large Temperature Range, Under Uniaxial and Biaxial Loadings," *International Journal of Plastics,* Vol. 9, 1993, p. 87.
[21] Nouailhas, D. and Freed, A. D., "A Viscoplastic Theory for Anisotropic Materials," *Journal of Engineering and Materials Technology,* Vol. 114, 1992, p. 97.
[22] Lee, D., Zaverl, J. R., and Plaza-Meyer, E., "Development of Constitutive Equations for Nuclear Reactor Core Materials," *Journal of Nuclear Materials,* Vol. 88, 1980, p. 104.

DISCUSSION

M. Nakatsuka[1] (written discussion)—Mechanical tests carried out on irradiated material by NFD do not emphasize the effect of the strain rate as shown in the presentation. How do you explain this?

P. Delobelle et al. (authors' closure)—The results presented deal with an unirradiated material. The effect of strain data is significant because hardening recovery is activated due to the appearance of secondary creep outperforming plasticity. The propensity to creep of the irradiated material being less significant, recovery mechanisms are certainly less activated, which could explain a smaller sensitivity to strain rate of the irradiated material.

M. Nakatsuka[1] (written discussion)—Based on the proposed model, significant strain rate dependence of strength was expected. I could not find such large strain rate dependence in stress-strain curves for irradiated Zry cladding tubes. This is due to less strain hardening with irradiation damage. I recommend that this kind of phenomena be included in your model.

P. Delobelle et al. (authors' closure)—The model described in the paper has been validated against the experimental database established on unirradiated material. The strain rate dependence of the flow stress has been experimentally observed and therefore taken into account in the modeling. The irradiation damage will certainly decrease the strain rate effect due to less dynamic recovery of the irradiated material. The phenomenon will be included in the second phase of the model development, which will deal with the description of the irradiation effect.

Y. S. Kim[2] (written discussion)—Can you explain how the high-temperature biaxial tests were done? It seems to me that during these high-temperature tests, the gauging part of the specimen is heated and the diameter change is measured in situ. If so, I wonder what the gripping effect caused by the end grip would be on the measured value of the strain?

P. Delobelle et al. (authors' closure)—The biaxial mechanical tests at room and elevated temperatures are carried out on a machine involving internal pressure and axial load. The tubular specimen is gripped with Swagelock connectors. Its length is such that no grip end effect is expected, and the temperature gradient does not exceed $\pm 1°C$ over a 50-mm gauge length. The diametral elongation is measured by four extensometers having a point contact at the mid-length of the specimen. An axial extensometer is also attached to the specimen. For the tests not conducted to rupture, the diameter of the removed specimen is measured with a LASER micrometer at different axial locations within the gauge length. If the differences between the measurements show uniformity of the hoop strain, the average value of the diametral elongation is then compared to the one measured by the extensometers. The good agreement certifies the validity of the test with respect to an eventual strain recovery.

J. Harbottle[3] (written discussion)—Have you carried out any experiments on stress reversal (stress cycling) and have you extended your model to include the charges in behavior that this would introduce?

P. Delobelle et al. (authors' closure)—Cyclic mechanical tests under uniaxial tension and tension-torsion have been carried out for different strain rates. The hardening or softening of the material induced by the cyclic loading as well as the Bauschinger effect are fully described in the kinematic hardening variables of the model. The effect of a pre-cyclic loading and stress reversal on the monotonic behavior of the material is simulated by the model.

[1] Motoyoshida-cho 1360-3, Mito, Ibaraki-Ken 310, Japan.
[2] Korea Atomic Energy Research Institute, Daejom, Korea.
[3] The S. M. Stoller Corp. Erlangen, Germany.

P. Efsing[1] and K. Pettersson[1]

The Influence of Temperature and Yield Strength on Delayed Hydride Cracking in Hydrided Zircaloy-2

REFERENCE: Efsing, P. and Pettersson, K., "**The Influence of Temperature and Yield Strength on Delayed Hydride Cracking in Hydrided Zircaloy-2,**" *Zirconium in the Nuclear Industry: Eleventh International Symposium, ASTM STP 1295,* E. R. Bradley and G. P. Sabol, Eds., American Society for Testing and Materials, 1996, pp. 394–404.

ABSTRACT: To determine if delayed hydride cracking (DHC) can be the cause of the long axial cracks occasionally found in BWR fuel cladding, a systematic study of DHC in Zircaloy cladding has begun. In the initial stage of the project, a test technique was developed and applied to unirradiated samples of Zircaloy. The present study includes an investigation of the influence of the yield strength and temperature on the crack growth rate and the threshold stress intensity that must be exceeded before cracking begins.

Recrystallized (RXA) Zircaloy-2 has been compared to stress relief annealed (SRA) Zircaloy-2 with similar texture and composition. The results show that the crack propagation rate increases with increasing yield strength at similar stress intensity levels by as much as a decade when the yield strength is tripled. The maximum crack propagation rate measured in this study is $\sim 6 \times 10^{-7}$ m/s. The threshold stress intensity, K_{IH}, was found to decrease with increasing yield stress. The measured threshold values are in the range of 13.5 to 7.5 MPa. These figures are close to theoretically derived values using a critical fracture stress criterion of the hydrides as the limiting factor. The incubation period before cracking begins is found to be longer at 200°C than it is at 300°C.

KEYWORDS: Zircaloy, delayed hydride cracking, crack growth, yield stress, hydrides, stress intensity factor, threshold stress intensity

Delayed hydride cracking (DHC) of zirconium alloys has long been recognized as a potentially important failure mechanism of high-strength pressure tubes. It has not been considered seriously as a failure mechanism of light water reactor (LWR) fuel cladding since significant tensile stresses when present, for instance, in connection with power ramps are of short duration. There is, of course, in that case also an alternative failure mechanism: stress corrosion cracking caused by fission products. However, recent observations in connection with long axial cracks on failed Zircaloy fuel cladding have suggested that DHC might have been the mechanism behind the growth of the long cracks.

The post-irradiation examination revealed that the long cracks were secondary defects formed as a result of a small-debris-induced primary defect that had leaked water into the fuel rod. The crack tip region of the long cracks was heavily hydrided in a way that suggested that hydrogen had diffused to the crack tip area and precipitated as hydrides oriented perpendicular to the tensile stress. In addition to this, fractography of the failed fuel rods revealed striation markings on the crack surfaces and that the cracks had propagated in both directions from the crack initiation point [1,2].

---

[1] Graduate student and professor in mechanical metallurgy, respectively, Department of Materials Science and Engineering, Division of Mechanical Metallurgy, Royal Institute of Technology, S-100 44 Stockholm, Sweden.

In view of the possibility that DHC might be a mechanism of secondary failure of Zircaloy clad fuel rods, the present investigation was begun with the objective of determining DHC crack growth rates of irradiated Zircaloy as a function of the stress intensity factor, hydrogen concentration, and temperature. Such data may subsequently be used in modelling the degradation of failed fuel rods. The present paper covers the initial phase of the project with the development of a test method that can be used on fuel cladding and application of the method to unirradiated cladding with different strengths.

Delayed hydride cracking is caused by nucleation and growth of hydrides at a crack tip in a stressed component. When the hydride reaches a certain size, it cracks and the macroscopic flaw gets an increment of growth and becomes sharp. The process repeats itself at the new crack tip. This step-wise crack growth is frequently revealed by striations on the fracture surface. The driving force for the nucleation and growth of hydrides at a stressed crack tip is that the molar volume of the hydride is larger than the molar volume of the material from which it is formed. There will thus be a difference in the chemical potential of hydrogen in a hydride under a tensile stress in comparison with the potential of hydrogen in an unstressed hydride.

If each hydride is in local equilibrium with the matrix, there will thus be a driving force for diffusion of hydrogen in the matrix so that hydrogen will be transported from the unstressed hydride to the hydride at the crack tip under tensile stress. It has also been pointed out by Eadie and Coleman [3] that in cases where the molar volumes are nearly equal, as is the case in zirconium, the stress-driven diffusion of hydrogen in the matrix will lead to a supersaturation at locations of high-tensile stresses and subsequent precipitation of hydrides. In such a case, stress-induced nucleation of hydrides may take place even if all hydrides are dissolved in the unstressed state.

A complication to the problem is that the matrix and possibly also the hydride have anistropic elastic properties. However, it seems that at least the diffusion phenomena can be modelled fairly accurately. The aspect least understood is what makes the crack tip hydride crack, or, in other words, what fracture criterion should be used for the crack tip hydride?

For the process to occur, it appears that the applied stress must be large enough so that locally in the plastic zone in front of the crack there will be enough displacement to crack the hydride. Hydrides are generally reported to need some plastic deformation of the matrix before cracking [4–6] unless they are large, 50 to 100 μm, when they can crack with negligible plastic deformation [5]. In the core of a nuclear reactor, the required hoop stresses for DHC in the fuel cladding may be caused by, for example, fuel oxidation due to a primary defect and the adjacent expansion of the pellet onto the cladding.

The discontinous crack growth behavior has been observed by several researchers investigating crack growth rate in commercial zirconium-based alloys [7]. In tubing of pressure tube dimensions, crack growth rates of up to 5×10^{-7} m/s have been found for Zircaloy-2 [8]. These crack growth rates have been found at stress intensities of 16 MPa$\sqrt{\text{m}}$ and above. Below this point, the crack growth rate is strongly dependent on the stress intensity. Below approximately 10 MPa$\sqrt{\text{m}}$, there is a threshold level of the stress intensity where no crack growth at all has been observed, K_{IH}, regardless of the chosen zirconium alloy. This level has also been established theoretically [9,10]. The crack growth rate can be divided into two separate regions, near the threshold stress intensity and above. In Stage I crack growth, near the threshold, the crack growth rate is strongly dependent on the stress intensity. In Stage II, above this level of stress intensity, the crack growth rate is almost independent of the applied stress intensity.

Experimental Procedures

The material chosen for the testing was Zircaloy-2 tubing, 12.25-mm outer diameter and 0.82-mm wall thickness, from Sandvik-Sweden, in recrystallized (RXA) and stress relief annealed (SRA) conditions. The recrystallization was performed at 565°C for 30 min. The basal pole textures of the RXA and SRA materials are fairly similar for these Sandvik tubes with basal pole maxima at about 30° from the radial direction. The grain size in the recrystallized condition is typically 4 μm. The grain size before recrystallization was not determined. The recrystallized tubing was gaseously hydrided at 400°C to 500 and 1000 wt-ppm hydrogen nominally in two different test series. These relatively high-hydrogen concentrations were chosen since the fuel cladding where the long cracks had been observed had similar concentrations.

After the hydriding, the specimens containing 500 ppm H showed an uneven layer of hydrides at the surface of the specimens and hydride packages in the matrix. The surface layer was markedly denser for the specimens containing 1000 ppm H. In addition to the surface hydrides and the hydrides in the matrix, there was also a densely hydrided layer 1/10 of the wall thickness from the outer surface. These specimens were homogenized during ten days while cycling the temperature between room temperature and 400°C in an attempt to make the hydride distribution more uniform. Metallographic examination confirmed that the treatment had been successful.

The stress relief annealed material with the highest yield strength was electrolytically hydrided in 0.5% sulphuric acid at about 85°C to approximately 1000 ppm hydrogen as measured by weight gain. The material was homogenized at 400°C in ten days to even out a dense layer of surface hydrides, after which they were furnace cooled. This made the hydride distribution more uniform, but a thin surface layer still remained. Two test series of SRA material with lower yield strength were gaseously hydrided to approximately 1000 ppm hydrogen nominally. These showed structures similar to what was found in the 500 ppm RXA material and thus were not homogenized.

In order to produce conditions similar to those found in nuclear fuel cladding, a test method was developed where a center-cracked tension type of specimen of a 10-mm-wide segment of a half tube was used. A fatigue-sharpened, spark-eroded crack was used as a precrack. The axial (with reference to the original tube) length of the spark-eroded crack was approximately 2 mm (Fig. 1). The specimen geometry also allows reasonable estimates of the stress intensity factor using the formula for a central crack in a thin plate with finite size and a defined stress state at the boundary, loaded in Mode I crack-opening mode under plane stress.

The specimen was clamped at its two ends between two tools shaped to follow the circumference of the specimen. Pull rods were connected in threaded holes in the tools so that the

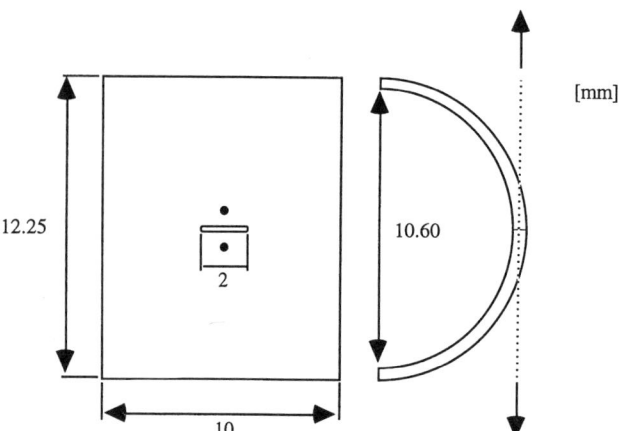

FIG. 1—*Specimen design: the spark-eroded crack has the dimension 0.2 by 2 mm². The filled dots represent the position of the welded measuring electrodes for potential drop. The arrows indicate the positioning of the applied load.*

TABLE 1—*Measured maximum crack propagation rates, measured and calculated threshold stress intensities before crack growth starts in hydrided Zircaloy-2.*

Material	σ_y, MPa	Temperature, °C	V_c, m/s	Measured K_{IH}, MPa$\sqrt{m}$	Calculated K_{IH}, MPa$\sqrt{m}$	ppm, H	Incubation Period, h
RXA	270	200	1×10^{-8}	13.8 ± 0.8	7.7	500	36–48
RXA	230	300	5×10^{-8}	14.2 ± 0.6	9.5	500	26–28
RXA	230	300	6.2×10^{-8}	14.2 ± 0.5	9.5	1000	22–28
SRA	490[a]	300	3×10^{-7}	8.2 ± 0.4	5.4	1000	22–26
SRA	550[a]	300	7×10^{-7}	8.0 ± 0.2	5.2	1000	19–23
SRA	620[a]	300	5×10^{-6}	7.5 ± 0.3	4.9	1000	18–20

[a] The yield strength of the SRA material was determined by using Vickers indentation.

specimen could be loaded in tension with the tensile force perpendicular to the crack plane. The holes in the tools are positioned so that there is no bending moment in the specimen when the load is applied. The fatigue crack growth was performed at low stress intensities where the final K_{max} values were lower than the applied stress intensity at the start of the test and the R-value was kept positive. To determine the crack growth rate of delayed hydride cracking, the specimens were tested under constant load condition.

The extension of the crack was monitored by electrical d-c potential drop measurements over the crack. The current was fed through the clamping arrangement of the specimen holder, and the potential drop was measured over electrodes spot-welded to the points shown in Fig. 1. The observed potential drop was compared to a master curve that had been obtained during fatigue testing of similar specimens. The range of potential drop between the beginning and end of the crack growth normally deviated from the range observed for the corresponding crack lengths in the calibration run. The potential drop of each experiment was therefore multiplied by a factor to make the range equal to that observed in the calibration run.

It was also assumed that the crack growth was double ended and symmetric. Post-test examination of the final crack confirmed that the crack growth was reasonably symmetric. Apart from the scatter in the absolute level of potential drop, the method is very sensitive to small changes in crack length since a relatively large potential drop, typically 2 mV, is obtained with a current of 4 A. The crack velocity was evaluated from the crack length as a function of time.

Huang and Mills [8] have reported that there is a transition temperature, a direction of approach temperature (T_{DAT}) above which the test temperature has to be reached from above. This temperature was given as 180°C for Zircaloy-2 tubing. Thus, before each crack growth test, a thermal cycle was performed to dissolve some hydrides and to promote crack growth. For testing at 300°C, the solution temperature was 340°C initially but was increased to 360°C after the preliminary tests since the latter temperature was observed to work better in promoting the start of crack growth. At 200°C, the overshoot was 50°C. The increased temperature was kept for 4 h, and loading of the specimen was performed at the end of the cycle. Specimens that did not exhibit any crack growth at 300°C within 72 h were unloaded and subjected to a new thermal cycle before loading with a higher load. The specimens tested at 200°C were allowed an incubation period of 96 h. This way, an approximate figure of the threshold stress intensity, K_{IH}, was found as well as the incubation period before a critical hydride length had been formed (Table 1).

Results and Discussion

The RXA specimens containing 500 ppm H were tested at both 200 and 300°C. The maximum crack growth rate at 300°C was found to be of the order of 5×10^{-8} m/s, while the specimens tested at 200°C showed a maximum velocity of 10^{-8} m/s (Table 1). The crack growth rate at 300°C is close to previously published data on unirradiated Zircaloy-2 tubing and lower than what has been found in Zr-2.5Nb, while the crack growth rate at 200°C is slightly lower than previously published results of Zircaloy-2. The maximum crack growth rate in these tests may be slightly underestimated at both 200 and 300°C since the transition point Stage I/Stage II was not passed with 100% certainty. It was impossible to test at higher stress intensities with the RXA material with the present specimens since these will then fail by plastic collapse due to the low strength of the material.

The threshold stress intensity level, K_{IH}, was found to be approximately 14 MPa at 300°C and slightly lower, about 13 MPa, at 200°C (see Table 1). This is a little higher compared to what has been found in other investigations. It cannot be ruled out that this can be an effect of the specimen design in the sense that the K-correlation used is not totally appropriate for a half tube specimen. Good agreement was, however, achieved between the different specimens in the present tests. The specimens containing 1000 ppm H showed crack growth rates close to those found for 500 ppm. The maximum crack velocity was found to be slightly less than 6.2×10^{-8} m/s. This is almost within the error range of the crack growth rates in 500 ppm H. The threshold stress intensity level was found to be similar to that found at 500 ppm H.

The SRA specimens were tested only at 300°C. As can be seen in Table 1 and Fig. 2, the crack propagation rates are significantly higher than in the recrystallized state. Four different yield stress levels, including RXA, were tested, and the results show that the crack growth rate increases with yield stress. The crack propagation was increased more than half a decade when

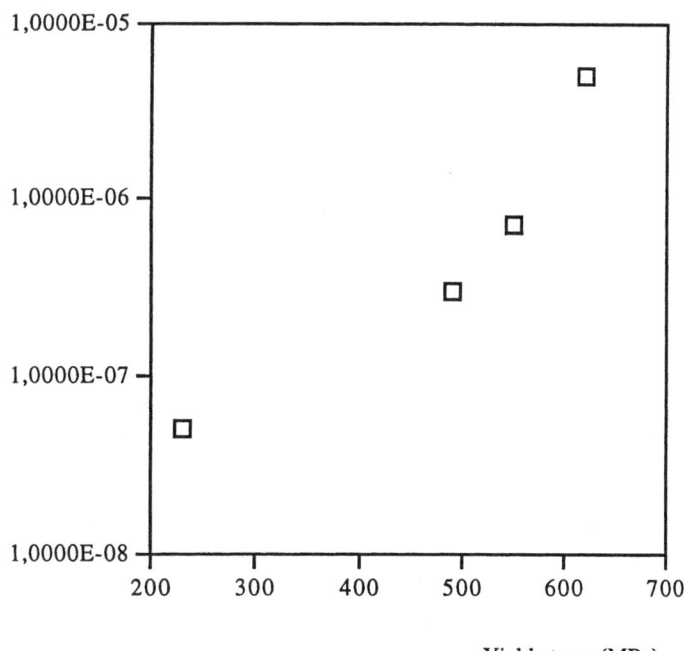

FIG. 2—*Crack growth rates versus yield stress of the matrix at 300°C. The maximum crack growth rate is increased as the yield strength of the matrix is increased.*

the yield strength was increased 50% from the recrystallized state to 3×10^{-7} m/s and a full decade more when the yield strength was doubled to 5×10^{-6} m/s. The small difference in basal pole texture between the recrystallized and the stress relief annealed states might influence the crack growth rate and thus be an additional factor. We believe this effect to be small since the texture difference is not greater than the difference between the present materials and those of other investigations that have resulted in similar crack growth rates.

The test series that included a high enough stress intensity range exhibits a marked Stage I/ Stage II behavior in the crack growth rate as a function of the applied K (Fig. 3). Note that the points in Fig. 3 include results from several tests. The K-range of Stage I crack growth is very small, which is similar to what has been found in other investigations [8]. Once the stress intensity level passes the transition point, the crack growth rate is fairly constant with increasing K level. Due to the specimen design and the low yield stress of the recrystallized specimens, it was not possible to test at stress intensities large enough to evaluate if the crack propagation rate is constant with increasing stress intensity in Stage II.

The incubation period before cracking started was determined as the time that elapsed between loading the specimen and the start of cracking. At 200°C, the incubation period varied between 36 and 48 h. At 300°C, the incubation period was shorter; it varied between 18 and 30 h. The ranges of incubation periods are given in Table 1. It can be seen that the average incubation period was decreased as the yield strength of the material was increased. It can be noted that, even though there is a large scatter in the test results, there was little overlap in incubation time between the different test series.

Shi and Puls [11] have derived a theoretical expression of the threshold stress intensity before crack initiation by delayed hydride cracking initiates. Their basic assumption is that there exists a critical threshold stress for hydride fracture that is supposed to be a material property of the hydride. They also assume that the hydride is affected only by the applied effective stress and the stress that arises inside the hydride due to its formation. The resulting stress field is assumed to be sufficient over most of the hydride to enable it to crack entirely. This leads to the conclusion that a hydride would crack at the instant the fracture stress is locally exceeded at any point in the hydride. They derive an expression for K_{IH} that is dependent on this hydride fracture stress, σ_f^h, the thickness of the hydride, t, the yield strength of the matrix, σ_y, and the stress free strain of the hydride in the normal direction $\varepsilon_\perp$

$$(K_{IH})^2 = \frac{Et}{1-v^2} \cdot \left(\frac{E\varepsilon_\perp}{8\pi(1-v^2)\left[\frac{1}{1-2v} - \frac{\sigma_f^h}{\sigma_y}\right]} - \frac{\Delta}{2} \sigma_y \right) \tag{1}$$

We have used $t = 4$ μm, which was determined by back-scattered electron imaging in a scanning electron microscope on a polished surface, to avoid etch effects on observed thickness. The constants $\varepsilon_\perp = 0.054$, Poisson's ratio $v = 0.35$, and $\sigma_f^h = 590$ MPa were taken from Shi and Puls [10]. Calculations were performed with $\sigma_y = 230/270/490/550$ and 620 MPa, respectively, and Young's modulus $E = 80$ GPa. The latter value was chosen as being typical of Zircaloy tubing of the present texture based on closed end burst testing performed in a number of previous investigations [12]. Δ is a constant dependent on the thickness of the hydride and the stress state in front of the crack, as

$$\Delta = \frac{1}{2\pi} \exp\left(-6.518 \cdot \frac{x}{t}\right) \tag{2}$$

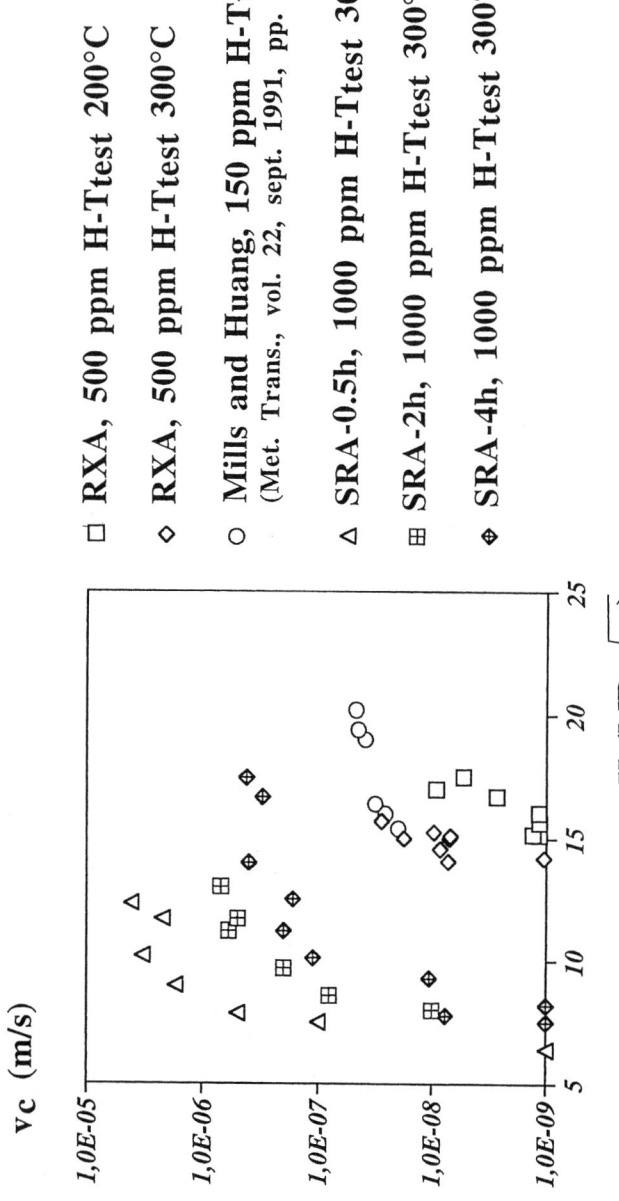

FIG. 3—*Crack growth rates versus applied stress intensity. The points at 1×10^{-9} m/s are from specimens that have showed some indication of crack growth, but the uncertainty in the measurements is large.*

where x is the distance from the crack tip to the location of maximum stress inside the plastic zone according to Rice and Johnson. Shi and Puls estimate this distance, which is a function of the applied stress intensity, the yield strength of the matrix, and Young's modulus, to be $0.20 \times t$. This gives Δ a value of approximately 0.05, which should account for its maximum value. The calculated values of the threshold stress intensities K_{IH} are found in Table 1. The calculated values are lower than the measured values, but, apart from a constant factor, their dependence of strength is almost the same as the observed values. Also, in other investigations, the observed values of K_{IH} have been about twice as large as calculated values [11]. It can be noted in the case of Zr-2.5Nb that the results from Shi and Puls are half the result from Zheng et al. [9] derived under the assumption of an energy balance in the elasto-plastic crack growth. Zheng's model was not applied to the present case since it contains parameters that are specific to Zr-2.5Nb.

It has been suggested that the cause of the long axial cracks in fuel cladding is unstable crack growth due to hydride embrittlement or even low fracture toughness of the irradiated Zircaloy [13]. We feel confident, however, that the mechanism for crack growth in the present tests is delayed hydride cracking. We base this conclusion on the following observations:

1. The crack path is covered by hydrides along most of the crack (Fig. 4). Note that these hydrides are oriented differently from the majority of hydrides in unstressed material. A few of the original hydrides can be seen in Fig. 5 as fairly wide plates with an irregular contour.
2. Along the crack there is a depleted zone where very few, if any, hydrides can be found.
3. In front of the crack tip, there is also evidence of uncracked hydrides (Fig. 5).
4. The crack surface is that of a brittle fracture with markings perpendicular and parallel to the main propagation direction that show signs of plastic deformation (Fig. 6). It is also fairly convincing that our observations of threshold stress intensity factors and crack growth rates are in fair agreement with previously published results.

The present work was performed to develop a way of testing irradiated fuel cladding so that its sensitivity to delayed hydride cracking can be determined. The observed effects of strength indicate that even though delayed hydride cracking may not be a problem to take into account

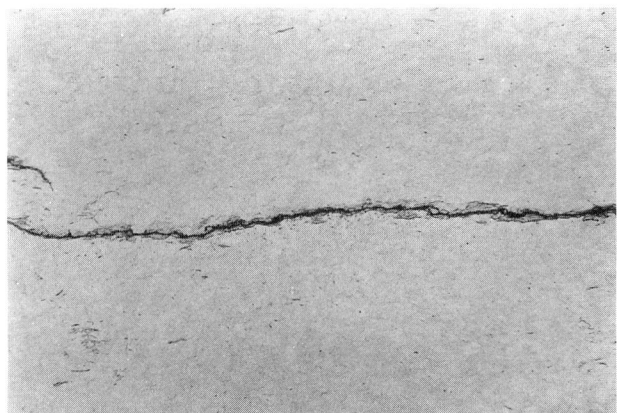

FIG. 4—*Metallographic cross section of the crack path of a specimen subjected to testing at 200°C. Most of the crack is covered by hydrides with a typical thickness of 4 μm.*

FIG. 5—*Metallographic cross section of crack tip, SRA annealed for 4 h ($\sigma_y \approx 490$ MPa). Note the uncracked hydride a short distance ahead of the crack and the absence of hydrides in the rest of the vicinity of the crack tip (×500).*

in the unirradiated state, it may be assumed that irradiated cladding with its high yield strength will indeed be sensitive to DHC.

Conclusions

1. Cracks in hydrided Zircaloy-2 cladding can grow by the delayed hydride cracking mechanism.
2. The tests in this study indicate that increasing the yield strength of the material will result in higher crack growth rates at a given stress intensity.
3. In agreement with earlier investigations, the threshold stress intensity level of DHC was decreased when higher-strength materials were used.

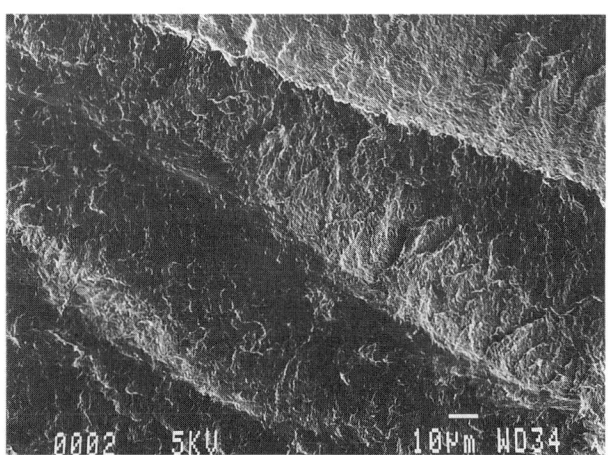

FIG. 6—*Fractography of the crack path. The specimen is an RXA specimen that has been subjected to testing at 200°C at an initial $K = 16$ MPa$\sqrt{m}$.*

4. Within the range of hydrogen concentrations used, 500 to 1000 wt-ppm, there is no effect of hydrogen level on the crack growth rate due to DHC.

Acknowledgments

This project is sponsored by ABB Atom AB, OKG AB, The Swedish Nuclear Power Inspectorate, Sydkraft AB, and Vattenfall AB, whose support is gratefully acknowledged. The Zircaloy tube is provided by Sandvik Steel AB, Sandviken, Sweden.

References

[1] Lundholm, L., Efsing, P., Lysell, G. and Schrire, D., "Secondary Fuel Failure Crack Propagation Mechanisms," *Proceedings,* Enlarged Halden Group Meeting at Storefjell, Norway, HPR-343, Vol. II, Institutt for energitettuik, OECD Holden reactor project, Holden, Norway, 1993.
[2] Lin, K.-F., Chung, C.-S., Yeh, J.-J., Chen, J.-H., Chu, S.-S., and Lin, L.-F., "Investigation on the Post-Defect Detorition of Nonbarrier BWR Failed Rods," *Proceedings,* 1994 International Topical meeting on Light Water Reactor Fuel Performance in West Palm Beach, FL, American Nuclear Society, La Grange Park, IL, 1994.
[3] Eadie, R. L. and Coleman, C. E., "Effect of Stress on Hydride Precipitation in Zirconium-2.5% Niobium and on Delayed Hydride Cracking," *Scripta Metallurgica,* Vol. 23, 1989, pp. 1865–1870.
[4] Simpson, L. A., "Criteria for Fracture Initiation at Hydrides in Zirconium-2.5 pct Niobium Alloy," *Metallurgical Transactions,* Vol. 12A, December 1981, pp. 2113–2124.
[5] Puls, M. P., "The Influence of Hydride Size and Matrix Strength on Fracture Initiation at Hydrides in Zirconium Alloys," *Metallurgical Transactions,* Vol. 19A, June 1988, pp. 1507–1522.
[6] Yunchang, F. and Koss, D. A., "The Influence of Multiaxial States of Stress on the Hydrogen Embrittlement of Zirconium Alloy Sheet," *Metallurgical Transactions,* Vol. 16A, April 1985, pp. 675–681.
[7] Northwood, D. O. and Kosasih, U., "Hydrides and Delayed Hydride Cracking in Zirconium and Its Alloys," *International Metallurgical Review,* Vol. 28, No. 2, 1983, pp. 92–121.
[8] Huang, F. H. and Mills, W. J., "Delayed Hydride Cracking Behaviour for Zircaloy-2 Tubing," *Metallurgical Transactions,* Vol. 22A, September 1991, pp. 2049–2060.
[9] Zheng, X. J., Lou, L., Metzger, D. R., and Sauve, R. G., "A Unified Model of Hydride Cracking Based on Elasto-Plastic Energy Release Rate Over a Finite Extension," *Journal of Nuclear Materials,* Vol. 218, 1995, pp. 174–188.
[10] Sagat, S., Coleman, C. E., Griffiths, M., and Wilkins, B. J. S., "The Effect of Fluence and Irradiation Temperature on Delayed Hydride Cracking in Zr-2.5 Nb," *Zirconium in the Nuclear Industry: Tenth International Symposium, ASTM STP 1245,* American Society of Testing and Materials, West Conshohocken, PA, 1993, pp. 35–61.
[11] Shi, S.-Q. and Puls, M. P., "Criteria for Fracture Initiation at Hydrides in Zirconium Alloys: I. Sharp Crack Tips," *Journal of Nuclear Materials,* Vol. 208, 1994, pp. 232–242.
[12] Pettersson, K., unpublished results.
[13] Armijo, J. S., "Performance of Failed BWR Fuel," *Proceedings,* 1994 International Topical Meeting on Light Water Reactor Fuel Performance in West Palm Beach, FL, American Nuclear Society, La Grange Park, IL, 1994.

DISCUSSION

B. Cheng (written discussion)—(1) The photomicrographs you showed do not appear to have as much hydride as the 500 ppm you indicate; (2) Did you look into the effect of hydride orientation on your test results? (3) If you already have charged 500 to 1000 ppm hydrides, then you already have a lot of hydrides for crack propagation. Why do you need a DHC process to propagate a crack?

P. Efsing et al. (authors' closure)—(1) Since we section in the plane of the original hydrides, they are not easily seen, but a few relatively large hydride plates are actually present in the micrographs. (2) No. (3) Perhaps not, but crack growth as a time-dependent phenomenon is in a view DHC.

H. M. Chung (written discussion)—Were the fractographic characteristics of the brittle-type axial cracks (produced in reactor) and those of the hydrided unirradiated specimens (simulated in laboratory) similar? On the fracture surface of the former, did you observe features indicating cleavage of hydrides? In our previous study, we observed striations that are characteristic of low-energy fracture associated with impurity segregation rather than with hydrides.

P. Efsing et al. (authors' closure)—We have not examined the irradiated material in the detail required to answer the first question. We did not see obvious signs of cleavage of hydrides in our unirradiated specimens.

P. Bouffioux (written discussion)—In the presented tensile test on the slot half-ring specimen, it exists in the plane-strain condition. Does the author take into account that condition in the stress analysis? The author confirms that the plane-strain condition has been taken into account, nevertheless he does not describe the method used for the combination of the stresses and strain.

P. Efsing et al. (authors' closure)—We intend to perform FEM analysis on the specimen in order to improve the K-a correlation. (We interpret the question as referring to the restraint on axial deformation imposed by the clamping).

E. Kohn (written discussion)—Comment: Recent experience at Ontario Hydro has shown evidence of delayed hydride cracking in fuel bundle end plates. This fuel was part of a test where the fueling direction was changed, resulting in irradiated end plate material (Zircaloy-4) being placed in a stressed condition. The test included a cool-down transient. Delayed hydride cracking does occur in Zircaloy-4 when the material's yield strength is increased and other conditions for DHC are met.

J. Harbottle (written discussion)—Would you comment on the possibility that you may, in fuel rod cladding, be seeing "hydrogen-assisted cracking" and not classical DHC as you have described.

P. Efsing et al. (authors' closure)—If you refer to other hydrogen embrittlement phenomena, like hydrogen-enhanced localized plasticity as a cracking mechanism, we regard that as an interesting possibility, especially since we do not see cleavage of hydrides on the fracture surfaces.

G. Shek (written discussion)—(1) Was there a correlation between striation spacing or hydride thickness and crack velocity or yield strength? (2) Comment on Cheng's question: Hydride concentration from metallography depends on hydride orientation, especially with layering. Fewer hydrides will show if the platelet is in the same plane as the metallographic section.

P. Efsing et al. (authors' closure)—We have not looked into these specific correlations.

In-Reactor Mechanical Behavior

John Schemel Award Paper

A. M. Garde,[1] G. P. Smith,[1] and R. C. Pirek[1]

Effects of Hydride Precipitate Localization and Neutron Fluence on the Ductility of Irradiated Zircaloy-4*

REFERENCE: Garde, A. M., Smith, G. P., and Pirek, R. C., "**Effects of Hydride Precipitate Localization and Neutron Fluence on the Ductility of Irradiated Zircaloy-4,**" *Zirconium in the Nuclear Industry: Eleventh International Symposium, ASTM STP 1295*, E. R. Bradley and G. P. Sabol, Eds., American Society for Testing and Materials, 1996, pp. 407–430.

ABSTRACT: The ductility of highly irradiated Zircaloy-4 material was evaluated by conducting tube burst, tube tensile, and ring tensile tests on fuel cladding and guide tubes irradiated in two PWRs. The specimen fluence ranged between 9 and 12.3×10^{21} n/cm^2 ($E > 1$ MeV), and test temperatures ranged from 313 to 673 K. The average thickness of the waterside oxide layer on the specimens ranged from 12 to 114 μm. Specimens with an oxide thickness greater than about 100 μm contained regions of spalling oxide and local areas of oxide significantly thicker than the specimen average. The corresponding average hydrogen contents ranged from 40 to 674 ppm for specimens without spalling oxide and estimated to be greater than 950 ppm with spalling. Non-uniform hydride distributions were observed in the specimens due to temperature gradients during operation.

The residual ductility for these high-fluence specimens is on the order of 1% uniform plastic strain for all the specimens except for two specimens with average concentration of hydrogen greater than 700 ppm and spalled oxide. The reduction in the material ductility due to radiation damage appears to be synergistically affected by localized hydride distributions and the orientation of hydride precipitates relative to the loading direction. The extent of ductility reduction due to hydride precipitates appears to be the most for the burst tests among the three tests investigated. The tensile specimens showed different fracture modes depending on deformation temperature, hydrogen concentration, local hydride volume fraction, and hydride orientation.

Neckdown and spiral fractures were observed. Examination of fracture surfaces indicated ductile failure in the metallic ligaments separating the hydride precipitates that appeared to have failed in a brittle fashion. The ductility data are analyzed by treating the material as a composite of relatively ductile metal phase separated by more brittle hydride platelets. Localization of hydride phase with a reduced presence of metallic ligaments in the composite results in reduction of ductility. A local hydride volume fraction greater than a critical value is needed to initiate and propagate fracture across the specimen cross section to thereby reduce the ductility below a set value. A model is proposed to suggest a possible ductility minima at intermediate fluences and the effect of hydride precipitates on ductility.

KEYWORDS: Zircaloy-4, ductility, radiation damage, hydriding, hydride localization, mechanical properties, composite, hydride volume fraction.

One of the key material properties of Zircaloy is its ductility after long exposure to neutron bombardment. The ductility of irradiated Zircaloy is lower than that of the unirradiated, as-fabricated material due to two factors: radiation-induced damage to the material microstructure and hydrogen embrittlement due to the hydrogen uptake from the waterside corrosion reaction.

---

[1] ABB Combustion Engineering Nuclear Operations, a business unit of Combustion Engineering, Inc., Windsor, CT 06095.
* Copyright 1995 by Combustion Engineering, Inc.

The literature data indicate that the former is a primary factor, while the latter is a secondary factor. However, the results described in this paper show that with significant hydride localization, the effect of hydrogen concentration could also be significant. The mechanical property data for Fort Calhoun fuel cladding published six years ago [1] indicated that ductility could decrease to uniform elongations of the order of 1%. Since then Combustion Engineering Nuclear Operations (ABB CENO) has completed two hot-cell examinations dealing with the ductility evaluations of Zircaloy-4 components irradiated to fluences greater than those of the Fort Calhoun cladding.

The first hot-cell examination was jointly sponsored by EPRI, the Baltimore Gas and Electric Co., and ABB CENO. This program examined a five-cycle assembly cage and fuel rods irradiated in Calvert Cliffs 1 [2,3]. The second hot-cell examination was sponsored by the U.S. Department of Energy (DOE) dealing with five-cycle fuel rods irradiated in Arkansas Nuclear One Unit 2 (ANO-2) [4]. The material ductility results of these two hot-cell examinations are presented, analyzed, and compared to the results from Fort Calhoun cladding.

In this paper, the ductility of Zircaloy-4 material with fluences and hydrogen levels greater than previously investigated are evaluated to enhance the understanding of ductility reduction in irradiated Zircaloy-4. For this reason, the recent hot-cell examinations included hydride distribution characterization, fractography, and transmission electron microscopy.

Experimental Procedure

Fuel assemblies were irradiated in the Calvert Cliffs Unit 1 (CC-1) and Arkansas Nuclear One Unit 2 (ANO-2) pressurized water reactors (PWRs). Selected fuel rods were examined non-destructively in the spent fuel pools of the respective reactors and destructively in the hot cells of the Chalk River Laboratories (CRL). Guide tubes from a CC-1 fuel assembly were also examined. The chemistry of the cladding and guide tube material were consistent with ASTM Alloy Designation UNS R60804 (Zircaloy-4) and contained 1.40 to 1.58 w/o Sn, 0.19 to 0.21 w/o Fe, and 57 to 100 ppm Si. The annealing parameter was about 1.5×10^{-17} h [with an activation energy (Q/R) value of 40 000 K]. The initial hydrogen concentrations were typically less than 15 ppm. The fuel cladding was in a cold-worked and stress-relief-annealed condition. The basal pole density distribution of the fuel cladding was concentrated in the plane containing the tube radii and tangent, with the density peaks at $\pm 30°$ from the radial direction. The guide tube material was in the recrystallized annealed state.

Mechanical tests were conducted on irradiated Zircaloy-4 fuel rod cladding and guide tube specimens taken from different axial elevations. Metallographic examinations were conducted to determine oxide thickness and hydride orientation in the irradiated Zircaloy-4 cladding. Hydrogen concentrations were measured by a hot-vacuum-extraction, mass spectrometric (HVEMS) analysis to derive a correlation between oxide thickness and average hydrogen concentration in the cladding. Cladding hydride volume fractions were determined by a point count method applied to photomicrographs. Fracture surfaces of selected irradiated mechanical specimens were examined by scanning electron microscope (SEM) to determine fracture mode. Samples from CC-1 fuel cladding were examined by transmission electron microscopy (TEM) to evaluate microstructural changes and irradiation damage.

Cladding burst tests were performed at 588 K at a strain rate of 6.7×10^{-5} s^{-1}. A three-zone furnace was used to heat the specimen. A 0.32-cm hole was drilled through the centerline of the fuel in the burst specimens (20 cm in length) to assure oil accessibility to the entire gage length. A system expansion curve (pressure versus volume) was generated for the specific volume expansion rate and temperature parameters using a heavy-walled non-deforming specimen of approximately the same internal void volume as the specimen. An expansion curve for the irradiated specimen was derived from the experimentally measured curve by subtracting the system expansion. The burst fracture region was photographed, and the region of maximum strain was sectioned and mounted as a transverse metallographic section to determine the failure strain. The uniform plastic diametral strain was calculated using the volume change in the specimen at the point of maximum load determined from the specimen expansion curve [5]. The yield and tensile stresses were calculated using the formula for a thin-walled cylinder and a tube cross section excluding the oxide layer.

Cladding tube tensile specimens with a total length of 12.7 cm were defueled. The specimen gage length was 7.6 cm. The specimens were tested in axial strain at temperatures of 313, 573, and 673 K. The specimen elongation was measured by using an extensometer at a strain rate of 4.2×10^{-5} s^{-1} until past the 0.2% yield point. At this point, the extensometer was disconnected, and the strain rate to failure was increased to 2.8×10^{-4} s^{-1}. For stress calculations, the cross-sectional area without the oxide layer was used. The uniform elongation is the plastic strain associated with the maximum load point. The total plastic elongation was measured from the combined length within the gage marks of the two fractured specimen pieces.

Ring tensile specimens 0.635 cm long were prepared from CC-1 cladding and guide tube materials. The cladding specimens were defueled. The specimens were expanded circumferentially at a strain rate of 4.2×10^{-5} s^{-1} at a test temperature of 573 K for the cladding specimen and strain rate of 1.0×10^{-4} s^{-1} and temperature of 323 K for the guide tube tests. Yield and tensile strengths were calculated from the specimen load-displacement curve, correcting the specimen thickness for the oxide layer. The uniform and total plastic strains were determined from the measured displacement and dividing by the gage length. The gage length was taken to be 20% of the ring circumference [6] based on strain measurements on specimens with photo-sensitive grid coating.

Results

Burst Ductility Results

The burst ductility data for cladding specimens from fuel rods irradiated in CC-1 and ANO-2 are presented in Tables 1 and 2, respectively. The fast fluence for these specimens was approximately 1.1 to 1.2×10^{22} n/cm^2 ($E > 1$ MeV). The burst strength values varied between 860 and 1010 MPa except for the single specimen with spalled oxide, where a lower value of 480 MPa was measured. The average hydrogen content in each specimen was estimated from the local average oxide layer thickness and the correlation between cladding oxide thickness and measured hydrogen concentration. Except for the single specimen from Rod 5 irradiated

TABLE 1—*Burst ductility of CC-1 fuel cladding tube specimens at 588 K.*[b]

Fuel Rod Designation	Uniform Plastic Elongation, %	Total Plastic Circumferential Elongation, %	Average Hydrogen Content, ppm
1	1.47	2.69	207
2	1.89	6.47	229
3	2.60	3.30	322
4	1.49	5.04	387
1	2.11	2.41	423
1	2.11	2.22	573
1	0.77	3.16	595
5	0.05[a]	0.58[a]	731

[a] Specimen included regions with spalled oxide.
[b] Data from EPRI Project RP2905-02.

TABLE 2—*Burst ductility of ANO2 fuel cladding tube specimens at 588°K.*

Fuel Rod Designation	Uniform Plastic Elongation, %	Total Circumferential Plastic Elongation, %	Bulk Hydrogen Content, ppm
7	0.67	2.28	162
8	1.25	2.08	178
9	1.16	2.28	217
10	0.84	1.64	233
11	1.50	2.58	233
7	0.97	2.06	273
10	0.95	1.73	289
8	1.23	1.47	320
9	1.36	2.45	336

in CC-1 that contained spalled oxide, the total plastic strain for all specimens was greater than 2.0%. Localized deformation bands were not observed on the surface of any of the burst-tested specimens. The fracture region of each specimen was a split along the tube axial direction.

Tube Tensile Ductility Results

Tube tensile ductility data for cladding specimens from fuel rods irradiated in CC-1 and ANO-2 are presented in Tables 3 and 4, respectively. Except for the two CC-1 specimens, one with 430 ppm of hydrogen tested at 313 K and the other with spalling oxide and 645 ppm of hydrogen tested at 573 K, the total plastic strains were greater than 2%. The yield and tensile strengths varied with deformation temperature but did not show significant dependence on hydrogen content at any deformation temperature. For all the tube tensile specimens tested at 313 K, the fracture surface was generally perpendicular to the tube axis (tensile loading direction), and localized deformation bands were not observed on these specimens. Localized de-

TABLE 3—*Tensile ductility of CC-1 fuel cladding tube specimens.*[c]

Fuel Rod	Test Temperature, K	Uniform Plastic Elongation, %	Total Plastic Elongation, %	Average Hydrogen Content, ppm
2	313	2.95	8.39	150
3	313	0.92	8.61	279
3	313	0.44	1.08	430
6[b]	573	2.49	14.24	110
3	573	1.25	4.74	171
3	573	1.01	7.89	279
1	573	0.75	7.89	487
3	573	1.02	6.82	159
5	573	0.37[a]	1.74[a]	645
5	573	0.37[a]	3.62[a]	674
6[b]	673	2.21	14.84	128
2	673	1.82	16.03	150
3	673	1.65	9.11	293
3	673	1.63	4.55	430

[a] Specimen included regions with spalled oxide.
[b] Non-fueled rod.
[c] Data from EPRI Project RP2905-02.

TABLE 4—*Tensile ductility of ANO2 fuel cladding tube specimens.*

Fuel Rod Designation	Test Temperature, K	Uniform Plastic Elongation, %	Total Plastic Elongation, %	Bulk Hydrogen Content, ppm
7	313	1.51	10.1	130
7	313	1.33	6.2	217
7	313	1.12	7.0	281
11	313	1.14	6.8	289
10	313	0.94	8.1	320
8	573	1.28	4.3	305
11	573	1.20	3.4	313
8	573	1.18	3.5	352
9	673	2.15	14.0	170
9	673	1.99	7.2	233
9	673	0.91	3.0	305
11	673	1.62	2.8	344
10	673	1.52	2.5	360

formation bands were observed on the surface of all ANO-2 specimens tested at 573 and 673 K, and the fracture surface orientation was approximately 45° to the tube tensile axis.

For CC-1 specimens tested at 573 and 673 K, localized deformation bands were observed at intermediate hydrogen concentrations of 200 to 450 ppm. Outside this range of hydrogen concentrations, bands did not form. The orientation of the fracture surface for the 573 and 673 K specimens varied according to the hydrogen content and whether the localized deformation bands formed or not. At hydrogen concentrations less than 200 ppm, the fracture surface was about 45° to the tube tensile axis but was not associated with a deformation band since localized deformation bands did not form at this low hydrogen level. For hydrogen concentrations between 200 and 450 ppm, the fracture surface was oriented about 45° to the tube tensile axis, and fracture occurred within one of the deformation bands. A typical example is presented in Fig. 1. For an average hydrogen concentration greater than 600 ppm, the orientation of the

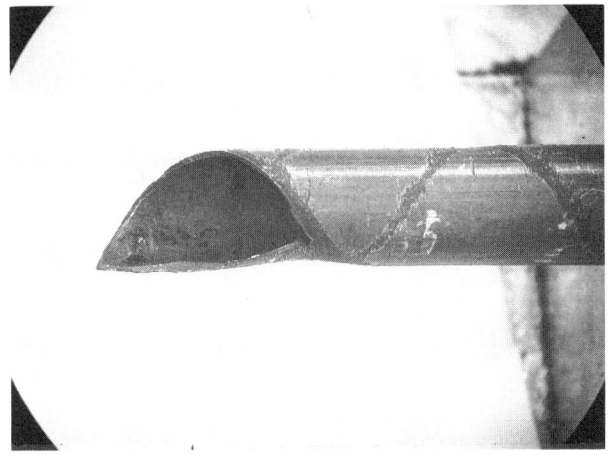

FIG. 1—*Specimen appearance after tensile testing at 673 K showing spiral fracture.*

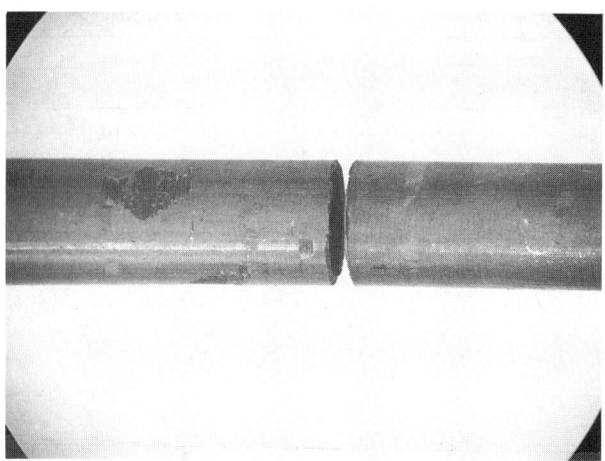

FIG. 2—*Specimen appearance after tensile testing at 573 K.*

fracture surface was perpendicular to the tube tensile axis. The specimen surface did not show deformation bands. A typical example of this is shown in Fig. 2.

Ring Tensile Ductility Results

Ring tensile ductility data for specimens from CC-1 fuel cladding deformed at 573 K are presented in Table 5. Except for the specimen with an average hydrogen concentration of 760 ppm, the total plastic strains for these specimens were relatively high and greater than 8%. Localized deformation bands were observed on specimens deformed at 573 K except for the specimens with the lowest and highest hydrogen concentrations of 110 and 760 ppm. The orientation of the fracture surface was approximately 45° to the loading direction (i.e., 45° to the tube longitudinal axis) for all specimens except the one with 760 ppm hydrogen. The orientation of the fracture surface for that specimen was perpendicular to the loading direction, i.e., parallel to the tube longitudinal axis.

Ring tensile results from the guide tube specimens were similar. Guide tube specimens tested at 323 K with hydrogen concentration in the range 40 to 50 ppm showed total plastic strain

TABLE 5—*Tensile ductility of CC-1 fuel cladding ring specimens at 573 K.*[c]

Fuel Rod Designation	Uniform Plastic Elongation, %	Total Plastic Elongation, %	Average Hydrogen Content, ppm
6[b]	3.03	13.12	110
3	1.56	8.50	186
3	3.27	15.28	308
3	2.63	10.64	451
5	2.63	13.37	674
5	0.44[a]	0.44[a]	760

[a] Specimen included regions with spalled oxide.
[b] Non-fueled rod.
[c] Data from EPRI Project RP2905-02.

values greater than 2.8%. Localized deformation bands were not observed on the guide tube ring specimens, and fracture surface orientation was 45° to the loading direction.

Hydrogen Analysis Results

The results of quantitative hydrogen analyses of cladding specimens irradiated in each reactor were used to derive correlations between the hydrogen concentration (averaged over the clad wall thickness) and the local thickness of the waterside corrosion layer for each reactor data set. The local oxide thickness was determined from metallographic examination of cladding specimens taken adjacent to the specimen analyzed for hydrogen content. The correlations were then used to estimate the hydrogen concentration at a measured oxide thickness from non-destructive measurements using an eddy current technique when local hydrogen measurements were not available. The measured or estimated average hydrogen level in each mechanical specimen is listed in Tables 1 to 5. The precision on the reported average concentration values is estimated to be ±8%.

Metallographic Examination Results

In transverse fuel cladding cross sections with a uniform oxide layer (without local spalling), the zirconium hydride platelets were circumferentially oriented with a greater concentration near the tube outer surface due to its lower temperature. In a longitudinal section (Fig. 3), the hydride platelet cross sections appear as long lines parallel to the tube axis with a greater concentration near the tube outer surface due to the radial thermal gradient. At pellet/pellet interfaces, and to a greater extent at pellet/pellet gap locations, the extent of hydride localization was significant due to axial thermal gradients. In addition, the hydride distribution is significantly modified by local spalling of the oxide. When a fraction of the oxide layer spalls, a local cold spot is created due to a smaller insulation effect of the thin oxide in the spalled region. The local cold spot attracts hydrogen from the surrounding, higher-temperature regions, resulting in an extremely high concentration of hydrides in the spalled oxide region (Fig. 4).

Hydride Volume Fraction Measurement Results

The cladding wall thickness, as viewed in photomicrographs, was divided into three regions of equal thickness, and the volume fraction of the hydrides was measured by a point count method [7]. The hydride volume fraction data are listed in Table 6. Area 1 is the inner third of the cladding wall thickness, which experiences the highest temperatures in operation, while Area 3 is the outer third of the cladding wall thickness, which experiences the lowest temperatures in operation. As expected, the volume fraction of hydride increases significantly underneath a spalled oxide region, as seen by the data from fuel Rod 5 specimen in Table 6.

Fractography Results

The fracture surface of a cladding tube tensile specimen with 279 ppm H and fractured at 573 K showed a typical dimpled appearance associated with ductile failure (Fig. 5). This fracture surface was a combination of several different segments, each oriented about 45° to the tube axis (not continuation of one segment, which leads to a spiral fracture). Note that the hydride platelets, expected to be present at a 45° orientation to the fracture surface, do not show a classic brittle fracture appearance in this figure.

The fracture surface of another cladding tube tensile specimen with a higher hydrogen concentration of 674 ppm was also examined. The fracture surfaces of this specimen are shown

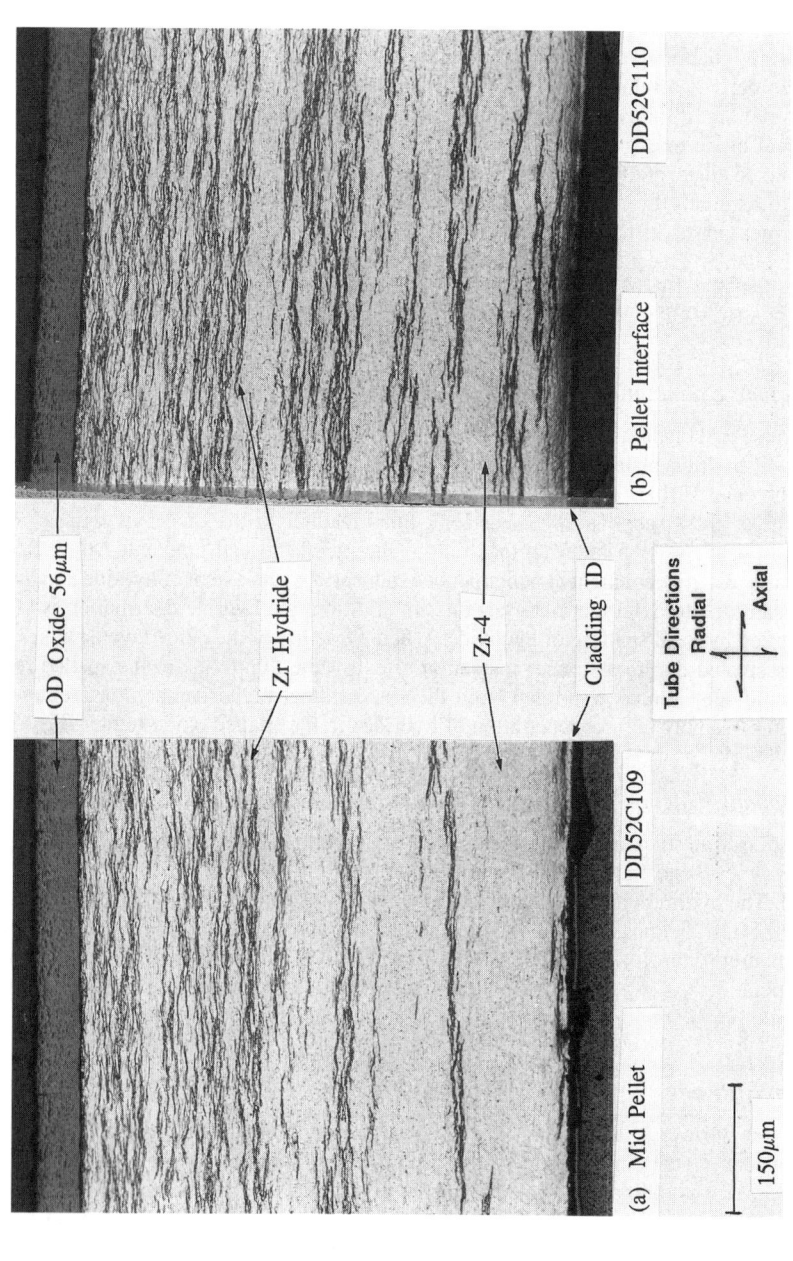

FIG. 3—*Etched appearance of zirconium hydride concentrations and orientation at mid-pellet and pellet-to-pellet interface location in fuel Rod 11.*

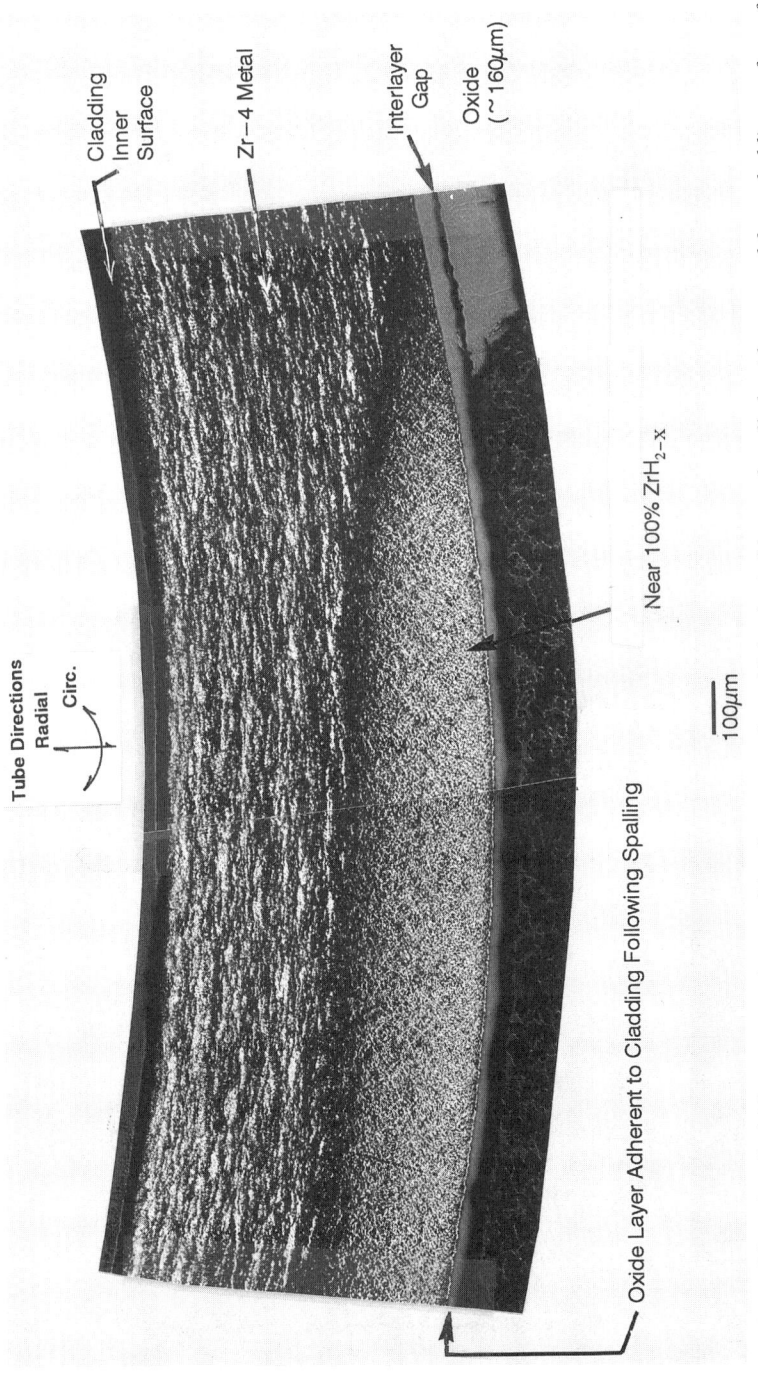

FIG. 4—*Zirconium hydride concentration resulting from operation with laminated layer of spalled oxide removed from cladding surface of fuel Rod 5.*

TABLE 6—Determination of volume fraction of ZrH_{2-x} in Zircaloy-4 CC-1 cladding specimens.

Rod Designation	Sample Surface and Location	Avg Oxide Thickness, μm	Avg[a] H, ppm	Area 1 (Innermost) Volume Fraction of ZrH_{2-x}	Area 2 (Central) Volume Fraction of ZrH_{2-x}	Area 3 (Outermost) Volume of Fraction of ZrH_{2-x}
6[b]	Transverse	20	85	12.9	12.1	14.4
1	Transverse at Pellet Int.	92	600	27.3	32.2	43.4
1	Transverse at Midpellet	91	595	3.5	14.7	47.6
3	Transverse at Midpellet	65	410	1.4	7.0	42.7
3	Transverse at Pellet Int.	65	410	10.5	35.7	50.3
5	Transverse Near Pellet Int. Full Ox. Thick.	140	950	52.3	83.0	92.0
5	Trans. at Pellet Int. at Spalled Oxide	Spalled	>950	81.8	88.6	100.0

NOTE: Int. = interface; Ox. = oxide; Trans. = transverse.
[a] From correlation between average oxide thickness and H content.
[b] Non-fueled Rod.

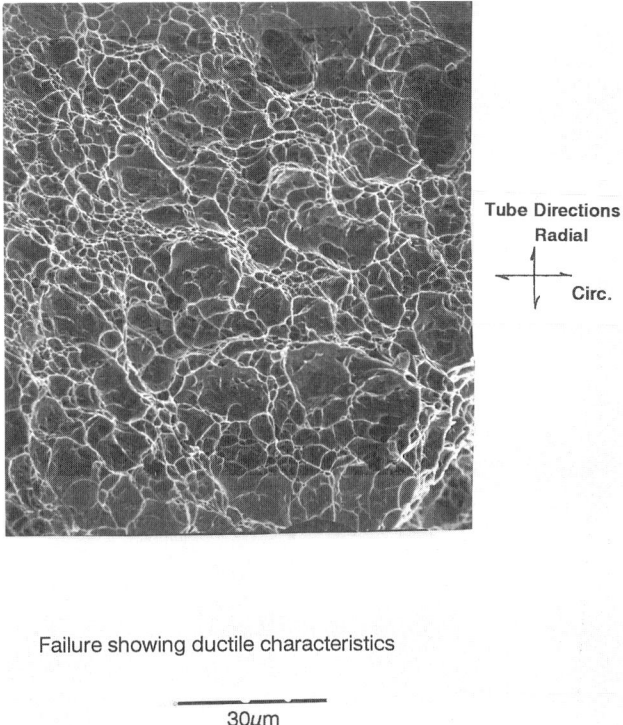

Failure showing ductile characteristics

30μm

FIG. 5—*Fracture surface of tube tensile specimen of fuel Rod 3. Estimated average hydrogen content of 279 ppm.*

in Fig. 6. The inner half of the cladding wall thickness of this specimen had fracture surfaces inclined at 45° to the tube axis and had a mixed appearance, i.e., dimples associated with the metallic ductile ligaments separated by non-ductile hydrides (Fig. 6a). The outer half of the cladding wall thickness of this specimen had a fracture surface perpendicular to the tube axis and had a typical brittle appearance (Fig. 6b) with no dimples.

The fracture surface appearance of a cladding ring tensile specimen with 760 ppm H is shown in Fig. 7. The surface close to the inner tube surface shows shallow dimples (Fig. 7a), and the fracture surface close to the outer tube surface exhibits brittle features. Note the axially oriented cracks in Fig. 7b corresponding to the hydride precipitates.

Discussion

Zircaloy-4 Metal/Hydride Composite

The microstructure of highly irradiated Zircaloy components includes a metallic phase of irradiated Zircaloy and a relatively brittle zirconium hydride phase, probably of the composition $ZrH_{1.66}$. Although the metallic phase is embrittled due to radiation damage and has a ductility much less than that of unirradiated Zircaloy, the irradiated Zircaloy metallic phase is ductile compared to the irradiated zirconium hydride phase. Therefore, the irradiated Zircaloy exposed to high-neutron fluences in light water reactors can be treated as a composite of a relatively ductile metallic phase containing more brittle zirconium hydride platelets. The mechanical

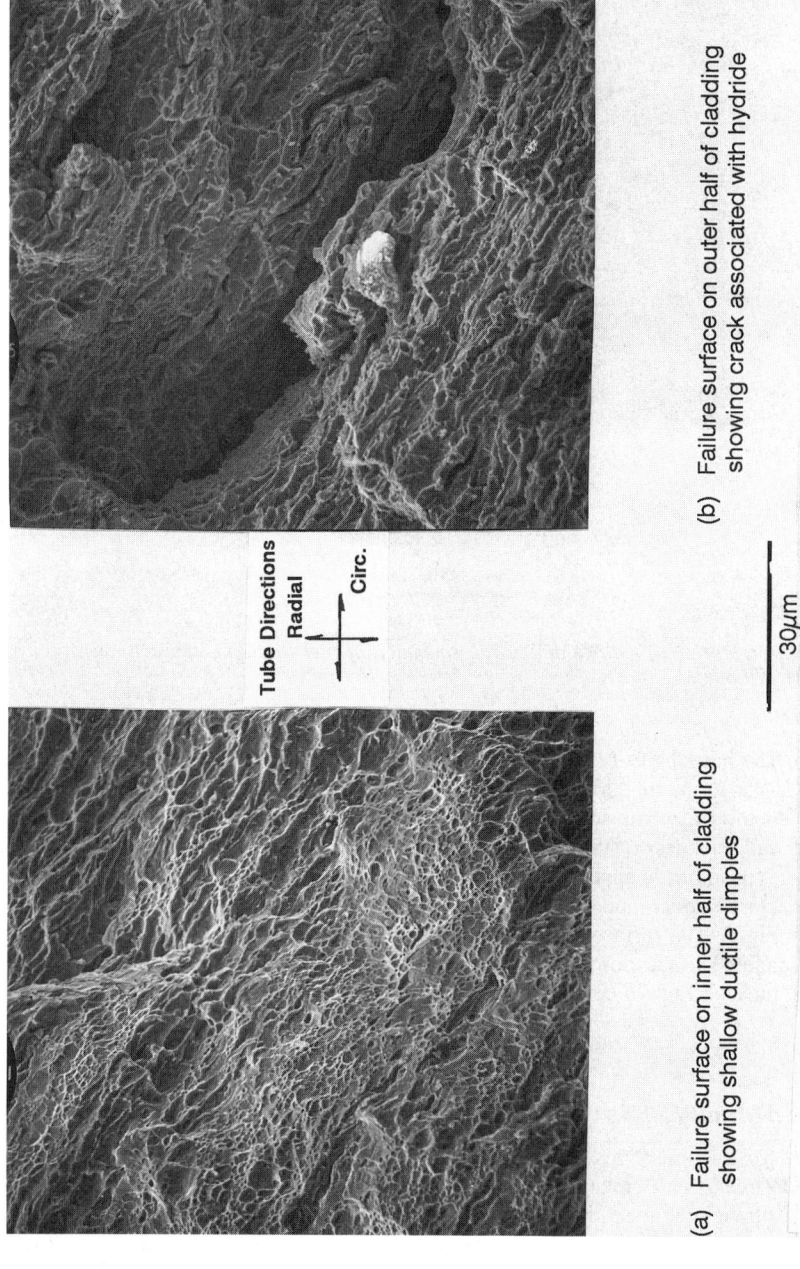

FIG. 6—*Fracture surface of tube tensile specimen of fuel Rod 5. Estimated average hydrogen content of 674 ppm.*

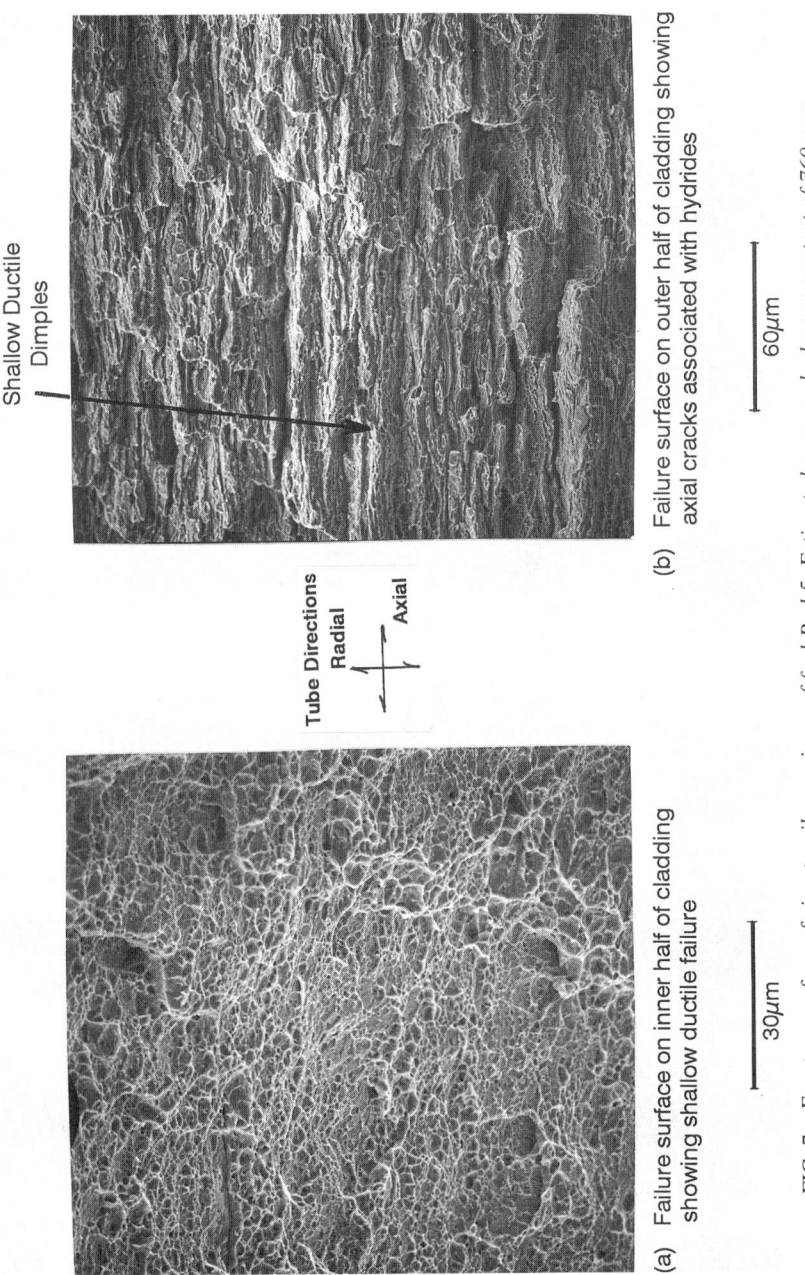

FIG. 7—*Fracture surface of ring tensile specimen of fuel Rod 5. Estimated average hydrogen content of 760 ppm.*

properties of this composite will depend on the properties of the irradiated metallic phase and the orientation of the hydride platelets with respect to the applied loading direction.

Application of composite theory [8] to irradiated Zircaloy is difficult for two reasons. The properties of irradiated zirconium hydride are not available, and the precipitate/metal interface is not "inert" as in the case of a typical reinforced composite. Rather, the interface has coherency and chemical bonding properties. The orientation of hydrides influences the ductility of the composite according to the orientation of the hydride precipitate geometric shape with respect to the fracture surface orientation. As a result of the starting texture of Zircaloy, the hydrides precipitate in a tube material as disk-shaped platelets with the flat surface of the platelet aligned closely to the tube circumferential direction. The length of the platelet in the tube axial direction is significantly larger compared to the tube circumferential direction. The effect of hydride platelets on the tube tensile ductility is expected to be less severe as the platelets are oriented perpendicular to the fracture surface and the platelet cross section within the fracture surface is small. On the other hand, for the tube-burst or the ring tensile tests, the long axis of the hydride platelet has an orientation close to the fracture surface and the hydride platelets might be expected to have a greater impact on the material ductility.

The ductility of the metallic phase affects the ductility of the composite. The ductility of the metallic phase depends on the extent of the radiation damage and the relationship of the applied stress with respect to the texture of Zircaloy. Although irradiation damage reduces the ductility of Zircaloy, at high fluences dislocation channeling (mainly due to a-type dislocations) initiates, resulting in localized deformation. Although the material has low ductility on a macro scale, the fracture surface appearance is typically ductile with dimples [1,3,9]. The ductility associated with the dislocation channels can be reduced due to hydride precipitates with a habit plane orientation (basal plane) similar to the channel plane orientation [1,10]. On the other hand, the strain localization within a single channel can be reduced by introduction of c-type defects that can provide "forest" dislocation-type work hardening to the a-type dislocations in the channel. This may explain the ductility minimum observed at an intermediate fluence in the absence of hydride precipitates [11].

Specimen ductility depends on the relationship of texture and the stress state imposed on the specimen. In a burst test (with biaxial loading), the normal prism slip (a-dislocations moving on the prism slip plane) is unable to operate to achieve the wall thickness reduction necessary to achieve circumferential expansion. The material ductility is therefore low. In the uniaxial tube tensile and ring tests, specimen elongation in the loading direction is accompanied by a reduction in the width direction and ductility is higher. In order to show the synergistic effects of hydrides and radiation damage on the ductility of Zircaloy, the effect of the two individual parameters are first discussed separately.

Effect of Hydrides on the Ductility of Unirradiated Zircaloy

The solubility of hydrogen in Zircaloy is limited at low temperatures [12,13] to approximately 80 ppm at 588 K, approximately 65 ppm at 573 K, and less than 0.01 ppm at 293 K. Hydride precipitation occurs when the hydrogen level exceeds the solubility limit. Precipitated hydrides reduce the ductility of Zircaloy [14–16] deformed in the temperature range of 293 to 616 K. Uniaxial tensile tests at room temperature show that the addition of 615 ppm H decreases the fracture strain of Zircaloy-2 from more than 80% to about 20% [14]. The fracture surface appearance with this hydrogen addition exhibits dimples, indicating ductile failure of the metallic phase initiated by void formation resulting from strain-induced fracture of zirconium hydride.

Hydrogen embrittlement is more severe under biaxial loading conditions than under uniaxial loading conditions [14]. In another investigation dealing with the room temperature ductility of Zircaloy-4 [15–16], the addition of approximately 1200 ppm H decreased the elongation to almost 0%. The total elongation at a concentration of 650 ppm H was about 5%. At a deformation temperature of 623°K, even the addition of 1500 ppm hydrogen did not embrittle Zircaloy-4 [15–16]. The tube tensile test data for Zircaloy-4 with an addition of up to 700 ppm H [17] show significant total elongations of about 14% at 616 K. These results show that although hydrogen addition reduces the ductility of unirradiated Zircaloy, at a deformation temperature of about 600 K, Zircaloy retains significant ductility (elongation > 12%) even with 700 ppm H. Moreover, the fracture surface appears mostly ductile in the metallic phase even with such hydrogen additions with a relatively uniform distribution of hydride precipitates on the fracture surface that show some evidence of brittle failure.

Effect of Radiation Damage on the Ductility of Non-Hydrided Zircaloy

Tensile tests were conducted on irradiated Zircaloy-2 material with neutron fluence levels up to 4×10^{21} n/cm^2 ($E > 1$ MeV) [9]. The specimens were encapsulated with helium gas during irradiation to avoid hydrogen charging of specimens due to waterside corrosion. Tensile testing was conducted over the temperature range 298 to 673 K. Increasing neutron fluence reduces the material ductility. However, even for the highest fluence levels, for the deformation temperature range 298 to 616 K, the failure elongations are 3 to 5% and the appearance of the fracture surface is a typical ductile fracture with dimples. In another similar investigation [11], encapsulated specimens were irradiated to a fluence range of 5×10^{20} to 1.1×10^{22} n/cm^2 ($E > 1$ MeV) and subjected to uniaxial tensile testing at 293 and 588 K. Total elongation at room temperature ranged between 3 and 5%, with a minimum at the intermediate fluence value of 1.5×10^{21} n/cm^2 ($E > 1$ MeV). At a deformation temperature of 588 K, the total elongation values range between 6 and 8%, with a minimum elongation at a similar intermediate value of fluence. These data indicate that without the presence of hydride precipitates, at a fluence level of 1×10^{22} n/cm^2 ($E > 1$ MeV), irradiated Zircaloy has residual ductility of about 4% and the material shows ductile fracture features. Moreover, there appears to be a ductility minimum at the intermediate fluence level of about 2×10^{21} n/cm^2 ($E > 1$ MeV).

Comparing the relative impact of individual factors, hydride precipitates, and neutron fluence on the ductility of Zircaloy, it is clear that the major ductility decrease is due to the neutron fluence, and the presence of hydrides makes a secondary contribution.

Effect of Hydrides on the Ductility of Irradiated Zircaloy

Zircaloy components used in LWRs experience radiation damage to the microstructure as well as waterside corrosion at the same time. A fraction of hydrogen generated from the corrosion reaction is absorbed in the metal. The charged hydrogen precipitates as hydride when the hydrogen solubility is exceeded. The mechanical tests on Zircaloy components exposed in LWRs, therefore, represent an influence of both factors (radiation damage and hydrides) together on the ductility of Zircaloy. Before presenting an analysis of the data collected in the current investigation, similar data from the published literature are reviewed for comparison.

Burst test data [1,18] on fuel cladding irradiated in Fort Calhoun to fluence values of roughly 1×10^{22} n/cm^2 ($E > 1$ MeV) and tested at 588 K showed total plastic strains on the order of 1 to 4%. Tube tensile tests showed higher elongation values. Dislocation channeling was observed in the failure of the tube tensile specimens. Local "average" hydrogen levels were estimated to be 330 to 400 ppm. However, hydride morphology was not characterized on the mechanical specimens to characterize the extent of hydride localization. The waterside oxide layer thickness was up to 52 μm, and the oxide did not show evidence of spalling.

Tube tensile data on Zircaloy-4 cladding irradiated in Oconee 1 [17] to a fluence of roughly

9×10^{21} n/cm^2 ($E > 1$ MeV), and hydrogen concentrations of up to 400 ppm and deformed at 616 K showed total elongations of 5 to 15%. Ring tensile tests at the same temperature showed total elongations of 4 to 8%. Oxide spalling, unusual oxide thickening due to localized crud deposition, and dislocation channeling were observed. The oxide layer thickness was up to 66 μm in non-spalled regions.

Burst test data on fuel cladding irradiated in Zion [19] to a fluence of 1.3×10^{22} n/cm^2 ($E > 1$ MeV) with a waterside oxide thickness of about 37 μm with no evidence of spalling, a hydrogen concentration of about 170 ppm, and subjected to a burst test at 588 K showed total elongation values of 0.8 to 2.7%. The 0.8% value was associated with a specimen taken from a fuel rod section with heavy localized hydriding due to a missing pellet chip at that axial elevation in the pellet stack.

Cladding tensile data for fuel cladding irradiated to a fluence of 8.6×10^{21} n/cm^2 ($E > 1$ MeV) [20] in Ohi 1, with an oxide thickness of about 60 μm (hydrogen level not listed but estimated to be about 400 ppm, assuming a 15% pickup fraction) and tested at 658 K showed a total elongation of about 9%.

Summarizing these prior data, it is seen that the burst test data show lower ductility values, and the lowest value observed in the Zion data is associated with a region of heavy, localized hydriding due to a cold axial spot. The effect of hydride volume fraction on the ductility of irradiated Zircaloy is described in the next section dealing with the analysis of hydride volume fraction data from the current investigation.

Effect of Hydride Localization on the Ductility of Irradiated Zircaloy

Since the ductility of the zircaloy/hydride composite is likely to depend on the local hydride volume fraction, local volume fractions of hydride phase were measured on each third of the clad wall thickness. As expected, the maximum value of hydride volume fraction occurred in the outer third of the wall, which has lower operating temperature. Since the hydride phase is more brittle than the metal phase, the hydrides are expected to fail first and can be assumed to fail at the uniform strain. Therefore, a correlation was attempted between the volume fraction of hydride in the outer third wall and the uniform strain (Fig. 8) to evaluate the effect of hydrides on the fracture initiation process. The data indicate a trend of decreasing uniform strain with increasing hydride volume fraction, and a minimum set value of uniform strain can be assured if the local hydride volume fraction is below a critical value. The necking strain, defined as the difference between the total and uniform strains, can be associated with the crack propagation process through the inner two thirds of the clad wall.

A correlation between the necking strain and average hydride volume fraction in the two inner layers would be suitable to evaluate the effect of hydrides on the fracture propagation process. Such a correlation is presented in Fig. 9. Similar critical average hydride volume fractions can be defined to assure a specific minimum value of the necking strain. The effect of average hydride volume fraction (averaged over the metallographically examined cross section, i.e., an average of the three values measured for the three sections of the wall in the examined section—it is not a specimen-averaged hydride volume fraction) on the total plastic strain (a measure of fracture initiation and propagation together) measured in different tests conducted in the temperature range of 573 to 588 K is presented in Fig. 10.

Replots of Figs. 8 and 9 with a change in the X-axis of the plots to hydride volume fraction averaged over the entire fracture surface (averaged value over the three wall segments listed in Table 6) gave results similar to Figs. 8 and 9, respectively, with a change in slopes of different lines for data set for each type of mechanical test. The data show that the burst test is a more severe test than the other two tests in that a given hydride volume fraction results in the lowest total plastic strain. The current data together with the earlier published data show that the

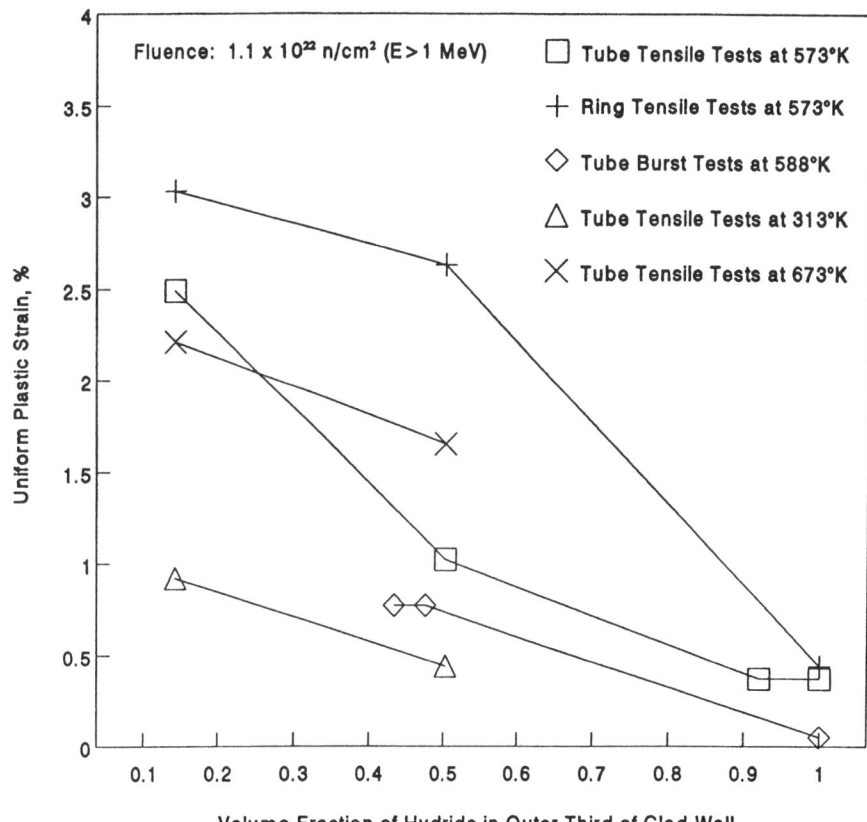

FIG. 8—*Uniform plastic strain of irradiated Zircaloy-4 as a function of hydride volume fraction in the examined metallographic section in the outer third of the clad wall.*

important variable affecting the ductility of irradiated Zircaloy is the *local* hydride volume fraction at the fracture location. Correlation of ductility with the local hydride volume fraction is a more refined approach compared to the correlation with the specimen average hydrogen concentration.

A comparison of uniform and total strain values given in Tables 1 and 2 show that strains are lower for ANO-2 specimens than for CC-1 specimens in spite of the lower hydrogen levels in ANO2 specimens. It is possible that the difference in the specimen geometry between ANO2 and CC-1 data sets (i.e., different tube diameter to wall thickness ratio) may have contributed to this strain difference. Such an effect can be anticipated from the observations of Adamson et al. [21].

Model for the Ductility of Irradiated and Hydrided Zircaloy

The ductility of irradiated Zircaloy at 600 K is limited by the localized deformation associated with dislocation channeling. Initiation of channeling requires a minimum fluence of roughly 3×10^{20} n/cm^2 ($E > 1$ MeV) [22]. Although channeling has been reported on both the basal [23,24] and prism planes [21], the localized deformation bands observed in the present inves-

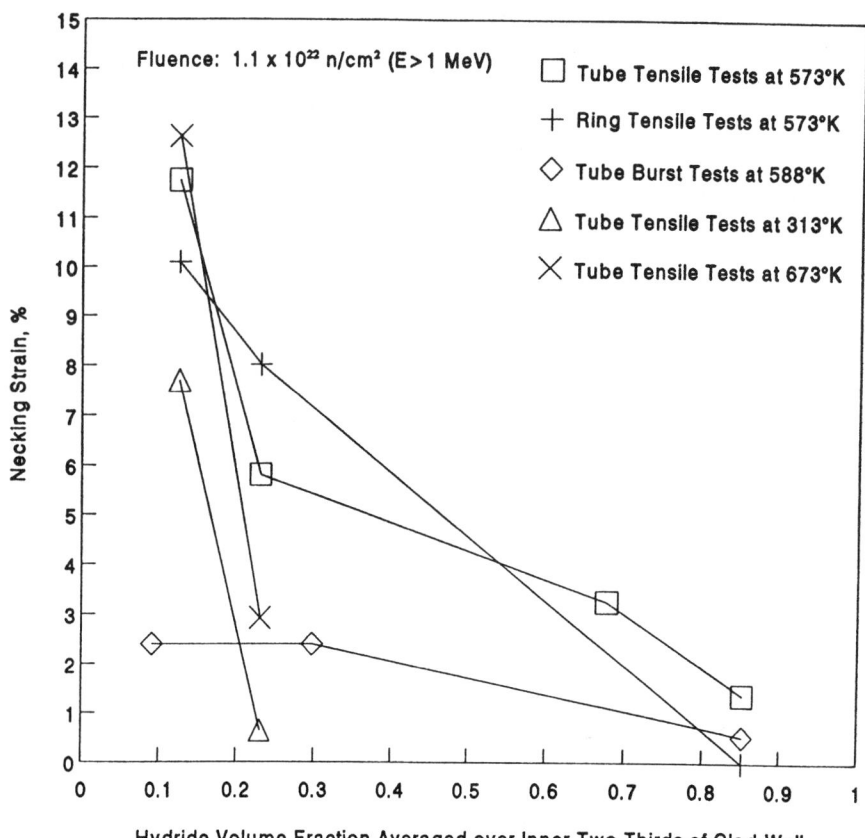

FIG. 9—*Necking plastic strain of irradiated Zircaloy-4 as a function of average hydride volume fraction in the examined metallographic section in the inner two thirds of the clad wall.*

tigations appear to be consistent with the basal plane association with the dislocation channels. Moreover, the channel formation is believed to be due to the movement of a-type dislocations on the basal plane. The movement of a-type dislocations in the basal plane is more likely to produce the spiral dislocation channels observed in the textured Zircaloy used as fuel cladding than the motion of a-type dislocations on the prism plane.

Another microstructural change observed in highly irradiated Zircaloy is the generation of c-type dislocation defects and loops [25,26]. These c (or $c+a$) defects appear in recrystallized Zircaloy-4 at fluence values greater than 3×10^{20} n/cm^2 ($E > 1$ MeV) [26] and continuously increase with fluence in cold-worked material. A stress-relief annealed cladding material is expected to behave in between these two behaviors. A c-type or $c+a$-type dislocation cannot move in a basal plane. Therefore, when c-type dislocations appear in the microstructure, they may provide a "forest" dislocation type obstacle to the a-type dislocations moving in a channel. This interaction can provide some work hardening in the channel, reducing the strain localization within the channel. This will nucleate a new channel, and overall ductility of irradiated Zircaloy will increase by a small amount with increasing fluence. The model is consistent with the fact that channels are initiated at a lower fluence than the c-type dislocations. The CC-1 and ANO-2 materials had sufficient fluence to initiate both channeling and c-type defects. The

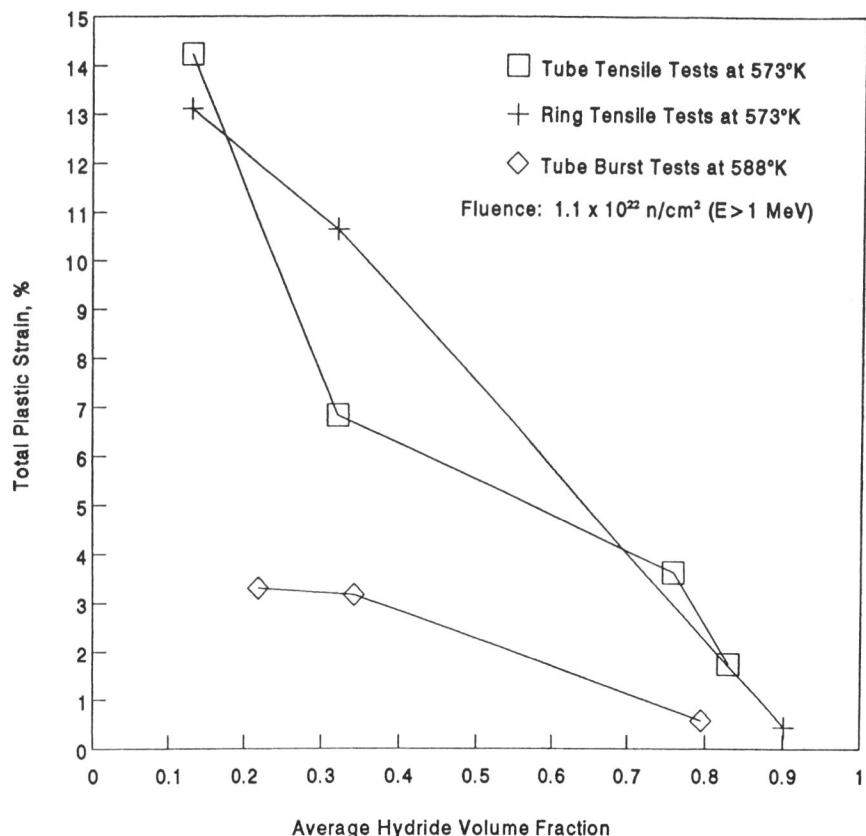

FIG. 10—*Total plastic strain of irradiated Zircaloy-4 as a function of volume fraction of hydride averaged over the entire wall thickness metallographic section close to the fracture surface.*

observed localized deformation band configurations on these different specimens also appear to agree with the above model. Single orientation bands were observed on Fort Calhoun specimens, while localized multiple orientation bands were observed on the specimens from the CC-1 and ANO-2 fuel rod cladding.

Prior to the generation of the c-loop dislocations, the hydride precipitates (with the same habit plane as the basal channel plane) can initiate fracture within the channel [1] and thereby reduce ductility. In fact, hydride precipitates within a dislocation channel of irradiated Zircaloy (at a fluence level insufficient to initiate c or $c+a$ defect generation) had been observed [10]. The initiation of dislocation channeling in the current investigation requiring a minimum hydrogen level may be another indication of the association of channeling and hydride precipitation.

All the above discussion applies when the volume fraction of hydrides in the fracture plane is less than the critical value to avoid the localized brittle failure due to hydride localization. If the local hydride volume fraction is greater than the critical value, channeling will not take place. The proposed model is summarized in Table 7. Although the absolute values of strain for different rows in Table 7 do not vary significantly, the following points are worth noting: (a) the fracture surface appearance is ductile due to dislocation channeling when the hydrogen

TABLE 7—*Model for the ductility of irradiated Zircaloy charged with hydrogen.*

Fluence	Volume Fraction of Hydride	Expected Total Plastic Elongation	Appearance of Fracture Surface
$<f_d$	$<V_H$	~4%	Ductile
	V_H	~2%	Mixed
	$>>V_H$	<1%	Brittle
$>f_d, <f_c$	$<V_H$	~3%	Ductile
	V_H	~2%	Mixed
	$>>V_H$	<1%	Brittle
$>f_c$	$<V_H$	~3–4%	Ductile
	V_H	~2%	Mixed
	$>>V_H$	<1%	Brittle

NOTE: f_d = Critical fluence necessary to initiate dislocation channeling.
f_c = Critical fluence necessary to produce significant c-defects.
V_H = Local critical volume fraction of hydrides to achieve 2% total strain.

level is less than the critical hydride volume fraction V_H; (b) a ductility minimum is anticipated at the intermediate fluence level due to interaction between basal plane dislocation channels and c-type defect generation at high fluences, and (c) the fracture is brittle when significant hydride localization occurs as indicated by a local hydride volume fraction above V_H.

The applicability of the ductility data presented in this paper to modern low-tin Zircaloy-4 material needs to be evaluated by conducting a hot-cell program on high-burnup, low-tin Zircaloy-4 cladding. Specifically, the following points need to be addressed:

1. What are the threshold fluence values to initiate channeling in low-tin Zircaloy-4 and in high-tin Zircaloy-4?
2. What are the threshold fluence values to initiate c-type defect generation in low-tin and high-tin Zircaloy-4?
3. Do the mechanical properties of hydride depend on the tin content of Zircaloy-4?
4. Mechanical properties of irradiated zirconium hydride phase.

Conclusions

The following conclusions can be drawn from this investigation:

1. The ductility of Zircaloy-4 irradiated to fluence levels of 1.2×10^{22} n/cm^2 ($E > 1$ MeV) at LWR operating temperatures of roughly 600 K is about 3 to 4% and depends on the hydride precipitate local volume fraction.
2. As long as the local hydride volume fraction is below the critical volume fraction, the ductility can be maintained above a certain minimum value. The important parameter affecting the ductility is the *local* hydride volume fraction, which provides a better correlation than the average hydrogen concentration level.
3. Considering the operating deformation modes and the hydride precipitate orientation, a burst test is a more severe test to evaluate the ductility of irradiated Zircaloy than the tube tensile or ring tensile test.
4. A model is proposed that suggests a ductility minimum at intermediate fluence based on the *a*-type dislocations moving in the channels and the *c*-type defect generated at high fluences providing obstacles to these *a*-type dislocations.

Acknowledgments

The authors would like to thank EPRI (Dr. O. Ozer) and the Baltimore Gas and Electric Co. for co-sponsoring examination of the CC-1 materials. The sponsorship of the ANO-2 hot cell program by the U.S. Department of Energy (Dr. P. Lang) is acknowledged. In addition, the efforts of the staff of the following organizations in the execution of different phases of the investigation are gratefully acknowledged: EPRI, BG&E, AECL Chalk River and Whiteshell Laboratories, USDOE and Entergy Operations, Inc. Technical discussions with Dr. M. Griffiths, AECL are acknowledged.

References

[1] Garde, A. M., "Effects of Irradiation and Hydriding on the Mechanical Properties of Zircaloy-4 at High Fluence," *Zirconium in the Nuclear Industry: Eighth International Symposium, ASTM STP 1023*, L. F. P. Van Swam and C. M. Eucken, Eds., American Society for Testing and Materials, West Conshohocken, PA, 1989, pp. 548–569.
[2] Smith, G. P., Jr., Ruzauskas, E. J. Pirek, R. C., and Griffiths, M., "Hot Cell Examination of Extended Burnup Fuel from Calvert Cliffs-1," RP2905-02, EPRI TR-103302-V1, CE NPSD-780-P, Electrical Power Research Institute, Palo Alto, CA, November 1993.
[3] Smith, G. P., Jr., Pirek, R. C., and Griffiths, M., "Hot Cell Examination of Extended Burnup Fuel from Calvert Cliffs-1," RP2905-02, EPRI TR-103302-V2, CE NPSD-780-P, Electrical Power Research Institute, Palo Alto, CA, May 1994.
[4] Smith, G. P., Jr., Pirek, R. C., Freeburn, H. R., and Schrire, D., "The Evaluation and Demonstration of Methods for Improved Nuclear Fuel Utilization, Final Report," DOE/ET/34013-15 UC-523, CEND-432, Department of Energy, Washington, DC, August 1994.
[5] Lowry, L. M.. et al, "Evaluating Strength and Ductility of Irradiated Zircaloy Task 5," Experimental Data, Final Report, NUREG/CR-1729, BMI-2066, Vol. 1, May 1981.
[6] Steward, K. P., "The Properties of Cold-Worked Zirconium-2.5% Niobium Fuel Sheathing," Report AECL-2250, Atomic Energy of Canada Ltd., March 1965.
[7] Hilliard, J. E., Chapter 3 in *Quantitative Microscopy*, R. T. DeHoff and F. N. Rhines, Eds., McGraw-Hill, New York, 1968, pp. 45–76.
[8] Ashby, M. F. and Jones, D. R. H., Chapter 25 in *Engineering Materials 2*, Pergamon Press, Elmsford, NY, 1992, pp. 241–254.
[9] Yasuda, T., Nakatsuka, M., and Yamashita, K., "Deformation and Fracture Properties of Neutron-Irradiated Recrystallized Zircaloy-2 Cladding Under Uniaxial Tension," *Zirconium in Nuclear Industry: Seventh International Symposium, ASTM STP 939*, R. B. Adamson and L. F. P. Van Swam, Eds., American Society for Testing and Materials, West Conshohocken, PA, 1987, pp. 734–747.
[10] Chung, H. M., Materials Science Division Light-Water-Reactor Safety Research Program: Quarterly Progress Report, January-March 1982, NUREG/CR-2970, Vol. 1, ANL-82-41, Vol. 1, Argonne National Laboratory, Argonne, IL, p. 86.
[11] Morize, P., Baicry, J., and Mardon, J. P., "Effect of Irradiation at 588 K on Mechanical Properties and Deformation Behavior of Zirconium Alloy Strip," *Zirconium in the Nuclear Industry: Seventh International Symposium, ASTM STP 939*, R. B. Adamson and L. F. P. Van Swam, Eds., American Society for Testing and Materials, West Conshohocken, PA, 1987, pp. 101–119.
[12] Douglass, D. L., *The Metallurgy of Zirconium*, Atomic Energy Review, IAEA Supplement 1971, International Atomic Energy Agency, Vienna, Austria, 1971, p. 160.
[13] Ells, C. E., "Hydride Precipitates in Zirconium Alloys," *Journal of Nuclear Materials*, Vol. 28, 1968, pp. 129–151.
[14] Yunchang, F. and Koss, D. A., "The Influence of Multiaxial States of Stress on Hydrogen Embrittlement of Zirconium Alloy Sheet," *Metallurgical Transactions A*, Vol. 16A, April 1985, pp. 675–681.
[15] Bai, J. B., "Influence of Oxide Layer on the Hydride Embrittlement in Zircaloy-4," *Scripta Metallurgica*, 1993, pp. 617–622.
[16] Bai, J. B., Prioul, C., and Francois, D., "Hydride Embrittlement in Zircaloy-4 Plate: Part I. Influence of Microstructure on the Hydride Embrittlement in Zircaloy-4 at 20°C and 350°C," *Metallurgical and Materials Transactions A*, Vol. 25A, June 1994, pp. 1185–1197.
[17] Newman, L. W., "The Hot-Cell Examination of Oconee 1 Fuel Rods after Five Cycles of Irradiation," Babcock and Wilcox Report BAW-1874, DOE/ET/34212-50, Department of Energy, Washington, DC, October 1986.

[18] Garde, A. M., "Hot-Cell Examination of Extended Burnup Fuel from Fort Calhoun," Combustion Engineering Report CEND-427, DOE/ET/34030-11, Department of Energy, Washington, DC, September 1986.

[19] Nayak, U. P., Kunishi, H., and Smalley, W. R., "Hot-Cell Examination of Zion Fuel Cycle 5," Research Report EP80-16, Empire State Electric Energy Research Corporation, New York, Final Report, June 1985.

[20] Doi, S., Abeta, S., Irisa, Y., and Inoue, S., "High Burnup Experience of PWR Fuel in Japan," *Proceedings,* International ANS-ENS Topical Meeting on LWR Fuel Performance, Avignon, France, April 1991, Vol. 2, pp. 588–597.

[21] Adamson, R. B., Wisner, S. B., Tucker, R. P., and Rand, R. A., "Failure Strain for Irradiated Zircaloy Based on Subsized Specimen Testing and Analysis," *Use of Small-Scale Specimens for Testing Irradiated Material, ASTM STP 888,* W. R. Corwin and G. E. Lucas, Eds., American Society for Testing and Materials, West Conshohocken, PA, 1986, pp. 171–185.

[22] Onchi, T., Kayano, H., and Higaschiguchi, Y., "Inhomogeneous Plastic Deformation and Its Relevance to Iodine Stress Corrosion Cracking Susceptibility in Irradiated Zircaloy-2 Tubing," *Journal of Nuclear Materials,* Vol. 116, 1983, pp. 211–218.

[23] Pettersson, K., "Evidence for Basal or Near-Basal Slip in Irradiated Zircaloy," *Journal of Nuclear Materials,* Vol. 105, 1982, pp. 341–344.

[24] Adamson, R. B., and Bell, W. L. "Effects of Neutron Irradiation and Oxygen Content on the Microstructure and Mechanical Properties of Zircaloy," *Microstructure and Mechanical Behavior of Materials,* Vol. 1, Proceedings of the International Symposium at X'ian, Republic of China, October 1985, Engineering Materials Advisory Services Ltd., Warley, United Kingdom, pp. 237–246.

[25] Gilbon, D. and Simonot, C., "Effects of Irradiation on the Microstructure of Zircaloy-4," *Zirconium in the Nuclear Industry: Tenth International Symposium, ASTM STP 1245,* A. M. Garde and E. R. Bradley, Eds., American Society for Testing and Materials, West Conshohocken, PA, 1994, pp. 521–548.

[26] Griffiths, M., Holt, R. A., and Rogerson, A., "Microstructural Aspects of Accelerated Deformation of Zircloy Nuclear Reactor Components During Service," presented at the 17th ASTM Symposium on Irradiation Effects on Materials, Sun Valley, Idaho, June 1994 *Journal of Nuclear Materials,* Vol. 225, 1995, pp. 245–258.

DISCUSSION

G. P. Sabol[1] (written discussion)—(1) When solid hydride formed at the OD surface, there was usually oxide spallation. At what oxide film thickness does spalling begin? (2) Also, to move or accumulate the hydrogen at the OD, a heat flux to support a temperature gradient is necessary. What were the temperature gradients in your sample?

A. M. Garde et al. (authors' closure)—Spallation of oxide layer on high-tin Zircaloy-4 fuel cladding was observed for oxide thickness generally greater than 100 µm. The temperature gradient between the inside and outside surface of the cladding wall during normal operation ranged between 40 to 80°F (22 to 45 K).

A. Hermann[2] (written discussion)—It has been known at least since the last symposium on Zr in the Nuclear Industry that the oxide layer influences the mechanical behavior of the cladding. French authors have shown a decrease in measured ductility with increasing thickness of an oxide layer on samples. Indeed, you have a composite of three components. Would you comment on this?

A. M. Garde et al. (authors' closure)—Zircaloy oxidized in an aqueous environment consists of a composite of three phases: substrate base metal, hydride platelets distributed in the metallic substrate, and a surface oxide layer. If oxidation is conducted in the carbon dioxide environment, the carbide phase is formed instead of the hydride phase. The influence of each of the three phases on the ductility of the composite is different since the ductilities of the individual phases are different and the geometrical configuration of each phase is different. The oxide phase is the most brittle of the three phases and is present on the surface of the specimen by itself (not interdispersed with a ductile component). As a result, the surface oxide layer is normally cracked even prior to the mechanical deformation of the composite. Therefore, the oxide layer, per se, is not expected to affect the ductility of the deformed composite. The hydride (or carbide) phase has intermediate ductility and is distributed within the metallic phase. The effect of hydride (or carbide) phase on the ductility of the composite depends on the local volume fraction of hydride (or carbide) in the metallic phase as shown in our work. The influence of the hydride (or carbide) phase is different from that of oxide partly because of the distribution difference. The oxide phase is continuous, while the hydride (carbide) is interdispersed in ductile metallic phase. The metallic phase is the most ductile of the three component phases, and its ductility generally controls the resultant ductility of the composite. Non-uniform distribution of the hydride (carbide) phase in the metallic layer due to the imposed heat flux during corrosion or a corrosion rate faster than the diffusion rate of hydrogen (or carbon) away from reaction interface can result in ductility reduction of the composite. It is believed that such non-uniform distributions of hydrides (or carbides) in the substrate metallic phase were most likely misinterpreted in earlier investigations as "the effect of oxide layer" in reducing the ductility of the composite.

S. Yagnik[3] (written discussion)—The measured hydrogen levels do not appear to agree with your proposed model of the outer one-third radius being a solid hydride, say, at 50% volume fraction hydride. The hydrogen mass balance should be satisfied, of course, assuming solid δ-hydride at 17 000 ppm hydrogen—the measured hydrogen ought to be much higher than reported. Please comment.

A. M. Garde et al. (authors' closure)—The possible reasons for the difference listed above are: (1) the measured hydrogen concentration values refer to a larger specimen volume and

[1] Westinghouse NMD, Energy Conter, Pittsburgh, PA.
[2] Paul Scherrer Institute, Switzerland.
[3] Electric Power Research Institute, Palo Alto, CA.

therefore represent a concentration value averaged over a larger volume; (2) the hydride volume fraction values refer to a metallographic cross section and therefore represent a smaller region over which averaging is conducted; (3) the hydride volume fraction is not measured on the fracture surface but on a metallographic section close to but some distance away from the fracture surface; (4) the fracture process, particularly the fracture initiation process, is a "localized" process where the mechanical properties of the "local" region are important while the hydrogen concentration measurements refer to a relatively larger volume; (5) subdividing the tube wall into many smaller regions for the hydride volume fraction measurement rather than the three sections employed here and measuring "local" hydrogen concentration (if at all possible) rather than the bulk average may reduce the disagreement in the two values.

J. B. Bai[4] (written discussion)—Do you have any data concerning the mechanical properties of the highly localized hydrides (massive hydrides) and the oxide layer, or the mixture of the two?

A. M. Garde et al. (authors' closure)—We did not measure the mechanical properties of oxide or hydride phases. Because of the inherent brittleness of both of these phases, especially at low temperatures associated with the LWR fuel cladding normal operation, it is difficult to conduct mechanical tests on these two phases. Some of the literature references on mechanical properties of hydride phase are listed as follows: (1) Simpson, L. A. and Cann, C. D. in *Journal of Nuclear Materials,* Vol. 87, 1979, pp. 303–316; (2) Beck, R. L. and Mueller, W. M. in *Nuclear Metallurgy, VII,* AIME, 1960, p. 63; (3) Beevers, C. J. and Barraclough, K. G. in *Journal of Materials Science,* Vol. 4, 1969, pp. 802–808; (4) Barraclough, K. G. and Beevers, C. J. in *Journal of Materials Science,* Vol. 4, 1969, pp. 518–525; (5) Ambler, J. E. R., AECL 2538, 1966.

R. B. Adamson[5] (written discussion)—(1) When you compare the failure strains of the various specimens tested, did you take into account the specific gage lengths of each specimen design? (2) Did you experimentally determine the primary plane of dislocation channeling? Normally we find that the channels occur on prism planes with occasional channeling on basal planes. (3) We have found by post-irradiation annealing experiments that the effect of c-component dislocations on mechanical properties is small. Do you have experimental evidence that c-components actually cause work hardening in the channels?

A. M. Garde et al. (authors' closure)—(1) The specific gauge length of each specimen design was accounted for in the failure strain calculations. Burst test failure strain was calculated from the circumferential measurement at the maximum strain location by metallography using the original undeformed circumference associated with the outer diameter. Tube tensile test failure strain was calculated by measuring distance between gauge marks of pieced-together fractured specimens and dividing by the original gauge length. For ring tensile tests, the failure strain was calculated from the crosshead separation distance at failure divided by 20% of the ring outer circumference. (2) We did not experimentally determine the orientation of the primary dislocation channeling plane. Although channeling with "a" burger's vector has been reported on both the basal and prism planes, the postulated work hardening in the "a" dislocation channels at high fluences due to "c" defect loops will be more likely if the "a" dislocations are moving on the basal planes rather than on the prism planes. (3) The response of c-defects to the post-irradiation annealing would depend on the stability of c-defects with respect to the additional thermal energy. The hypothesis of work hardening in the "a" dislocation channels moving on basal planes and c-defects is proposed on the basis of the inability of c-defects to move on the basal planes. We do not have direct evidence of the proposed work-hardening mechanism.

[4] Lab. MSS/MAT, Ecole Central, Paris, France.
[5] GE Nuclear Energy, Pleasanton, CA.

V. Grigoriev,[1] B. Josefsson,[2] and B. Rosborg[3]

Fracture Toughness of Zircaloy Cladding Tubes

REFERENCE: Grigoriev, V., Josefsson, B., and Rosborg, B., "**Fracture Toughness of Zircaloy Cladding Tubes,**" *Zirconium in the Nuclear Industry: Eleventh International Symposium, ASTM STP 1295,* E. R. Bradley and G. P. Sabol, Eds., American Society for Testing and Materials, 1996, pp. 431–447.

ABSTRACT: The fracture toughness of Zircaloy-2 cladding has been estimated by means of the recently developed pin-loading (PL) tension test. Axially notched ring specimens, cut directly from different cladding (annealed, cold-worked, hydrided, and irradiated), have been tested in a way similar to that used for compact tension specimens.
The results of the PL tension tests, performed at temperatures of 293 and 573 K, revealed for actual cladding all main phenomena observed earlier for Zircaloy materials. A threshold hydrogen content of 600 to 700 wtppm, above which only brittle fracture occurred at ambient temperature, was observed for unirradiated cladding. Existence of a continuous hydride network in the cladding facilitated brittle fracture. There was no obvious influence of the hydrogen content of about 900 wtppm on the fracture toughness at 573 K. The irradiation changed the fracture toughness in accordance with its known influence on the ductility and strength of the cladding. Intensive load serrations occurred for both irradiated and unirradiated cladding during plastic deformation of the notched ring specimens at 573 K.
A maximum-load fracture toughness of about $J_{max} \approx 100$ kN/m, obtained for irradiated cladding at 573 K, is comparable with fracture toughness of the cold-worked unirradiated cladding and reasonably agrees with published results for pressure tube materials.
The PL tension test was shown to be an effective method for evaluating the fracture toughness of actual cladding after hydriding or irradiation.

KEYWORDS: Zircaloy cladding, test method, fracture toughness, hydrogen embrittlement, irradiation

During recent years, much attention has been devoted to the problems connected with the degradation of mechanical properties of zirconium alloys due to irradiation hardening combined with hydrogen embrittlement. For the investigation of pressure tube materials, the most successful approach appeared to be an evaluation method where small compact tension (CT) specimens cut from the pressure tube are used to evaluate unstable failure of the actual pressure tube [1–3].
For nuclear fuel cladding, the application of fracture mechanics concepts is restricted by at least two main circumstances. First, the geometry of the cladding does not offer any possibility to construct tension specimens that comply with standard requirements for plane-strain conditions. Moreover, it has been shown that values of crack propagation and threshold stress intensities obtained using CT specimens cannot be extrapolated to the behavior of thin-walled cladding [4]. An effect of the thickness was also noticed for the fracture behavior of hydrided Zircaloy plates when the ductile-brittle transition occurred at different hydrogen contents for the thicknesses 0.5 and 3.1 mm [5].

---

[1] Senior scientist, [2]section head, and [3]technical director, Studsvik Material AB, S-611 82 Nyköping, Sweden.

Specimens for tension testing may be produced from previously flattened cladding tube pieces [4], but the flattening brings about some uncertainties in the test results due to changes in the microstructure of the material and a new distribution of internal stresses. In some cases, as for irradiated cladding tubes, flattening is not possible due to the brittleness of the material.

Second, testing of specimens with the same geometry as the original tubing presents a number of difficulties. Uniaxial tension testing of notched ring specimens, which were cut from Zircaloy cladding tubes, revealed difficulties in creating the correct stress-state conditions at elevated temperatures [6]. Also, uniaxial tension testing of the notched specimens is extremely difficult, especially for irradiated or other low-ductility material, because of the very small displacements that are involved.

The presence of the notch in the specimen is needed since it changes the fracture behavior of the material. For instance, the room temperature ductile-brittle transition in notched 2-mm-thick Zircaloy specimens was found at a lower hydrogen content compared to smooth specimens [7].

The above considerations clearly provide a necessity for fracture toughness evaluation of real cladding, but until now any *J*-resistance measurements have been reported neither on hydrided nor irradiated cladding. Thus, a new loading mode and test method, called the Pin-Loading Tension Test, has been developed for evaluation of the fracture toughness of thin-walled tubing [8]. This paper presents experimental results obtained for fuel cladding by means of the new method and illustrates the main possibilities of the method for fracture toughness evaluation of actual hydrided and irradiated cladding.

Description of the Pin-Loading Tension Test

A sharply notched thin-walled tubular specimen (Fig. 1a) is subjected to a mode of loading that is similar to the pin loading (PL) of a compact tension specimen for fracture mechanics testing. Compared to uniaxial tension of the specimen (Fig. 1b), the PL tension localizes the deformation and fracture processes to only a part of the specimen cross section (Fig. 1c), thus reducing both the load necessary for deformation and the elastic strains stored in the loading system and eliminating spontaneous failure of the thin-walled specimen. Displacements measured at the load-line are increased in comparison with those measured under uniaxial tension.

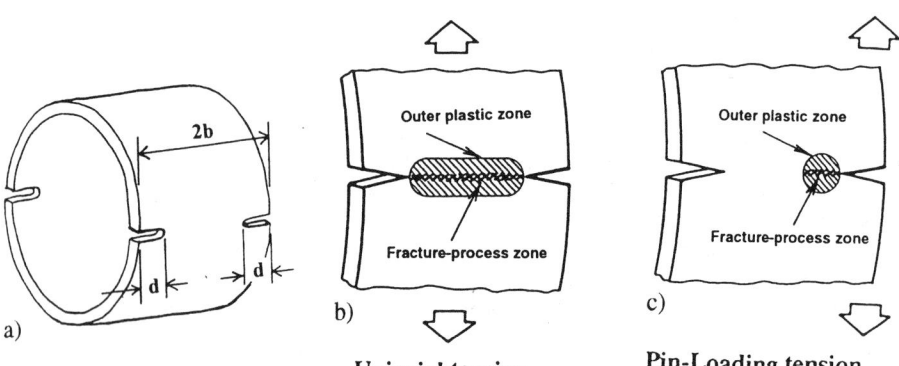

FIG. 1—*Configuration of the axially notched ring specimen* (a) *and schematically shown notch-tip deformation zones in such a specimen under uniaxial,* (b) *pin-loading,* (c) *tension.*

Specimen Configuration

The specimen for a PL tension test is prepared directly from the thin-walled tubing, thus fully representing its processing, thermal, and irradiation history. The proportions for the specimen dimensions d and $2b$ (see Fig. 1a) were selected to prevent axial contraction in the deformed area and to have the dimension $2b$ unchanged during testing. For the same reason, the notches at the unloaded side of the specimen are needed.

Fixture Design

A special fixture has been designed for a specimen with axial notches to provide loading according to Fig. 1c. The fixture consists of two halves, which when placed together form the cylindrical holder (Fig. 2a). This cylindrical holder has a diameter that allows it to be inserted into the ring specimen while maintaining a minimal interfacial gap.

The fixture halves, being loaded in tension through the pins, have the capability of mutual rotation around the axis determined by a small pin placed between the fixture halves at the outside end of the cylindrical holder (Fig. 2b). The rotation of the fixture halves is similar to the rotation of compact specimen halves under tension with only one difference: in the designed fixture, the rotation axis does not change its position when the crack in the specimen is growing.

Specimen and fixture have closely interrelated configurations and, when combined, create an assembly (Fig. 2c) characterized by the ratio W/a, where W is the distance between the load-line and the rotation axis of the fixture halves, and a is the distance between the load-line and the notch tip in the specimen. In the assembly, all four notches of the specimen should be situated at the contact plane of the fixture halves. The rotation axis passes through the notch tips of the back, unloaded, specimen side. Under PL tension testing, deformation takes place mainly at the notch tips at the front side of the specimen next to the load line.

Test Records

During a PL tension test, the load applied to the specimen is recorded versus the load-line displacement. Specific values of load and load-line displacement obtained from the test record give the necessary information for evaluation of the fracture properties of the material.

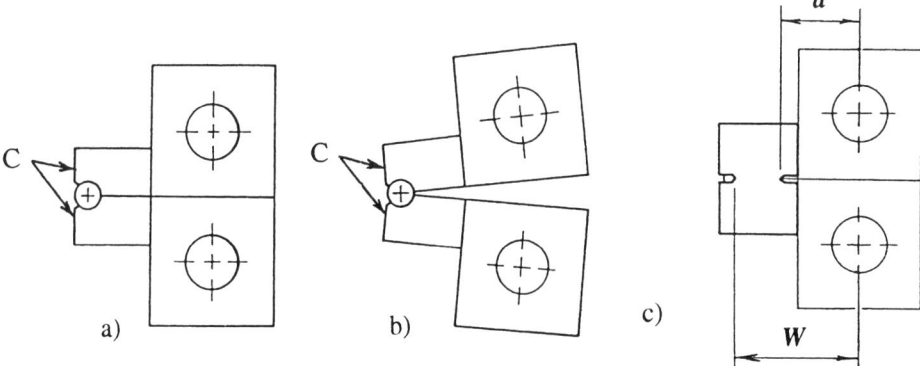

FIG. 2—*Pin-loading fixture design for an axially notched specimen,* (a) *mutual rotation of fixture halves,* (b) *specimen-fixture assembly before a pin-loading tension test,* (c) *where* C = *the cylindrical part of the holder to be inserted into tubular specimen,* a = *the distance between the load-line and notch-tip in the specimen, and* W = *the distance between the load-line and rotation axis of the fixture halves.*

Experimental Details

Materials

Different Zircaloy-2 cladding material variants were selected for evaluation under PL tension testing: (1) cold-worked cladding (CW) with a final reduction of about 80%; (2) annealed cladding (ANN); (3) corrosion specimens from autoclave tests; and (4) three different metallurgical states of irradiated fuel cladding from the Oskarshamn-2 BWR (irradiation temperature 566 to 613 K, depending on heat fluxes during cladding lifetime: fluence $\approx 6 \cdot 10^{25}$ n/m^2, $E >$ 1 MeV).

Before autoclave tests, the corrosion specimens were either in the cold-worked condition or had different heat treatments (according to Table 1), which together with autoclaving at 723 or 773 K provided weight gains of 150 to 950 mg/dm^2 and, consequently, hydriding in the range 240 to 900 wtppm.

All three types of irradiated cladding, here designated as LK0, LK1, and LK2, were produced from the same ingot. The production route up to and including the first cold rolling was the same for all three types of fuel cladding. The subsequent process for Types LK0 and LK1 cladding included a number of cold-rolling and recrystallization annealing steps. The only major difference between LK0 and LK1 was the annealing temperature being 110°C lower for the LK1 type of cladding. Both annealing temperatures were in the alpha-phase range for the Zircaloy-2 alloy. In the production of the LK2 type of cladding, the first recrystallization anneal was replaced by a beta quench. The production route provided to the LK0, LK1, and LK2 cladding resulted in the Vickers hardness values of 164, 168, and 180, respectively [9]. The initial metallurgical states of the irradiated cladding used in this investigation were identical to materials characterized earlier [10].

Material Examination

The hydrogen content of the cladding was evaluated by the hot vacuum extraction technique, both for autoclaved and for irradiated specimens, without removing the surface oxide. The extraction was made from material taken from the specimens after PL tension testing. For irradiated materials, the hydrogen content was evaluated in only test per specimen.

The hydride morphology was examined by optical metallography.

PL Tension Test Procedure

Irradiated specimens were tested at 573 K and all other materials at 293 and 573 K. When the supply of material was sufficient, PL tension testing was done repeatedly.

All specimens had an outside diameter of 12.25 mm, a wall thickness of 0.80 to 0.82 mm, and nominal dimensions of $d = 2$ mm and $2b = 9$ mm (see Fig. 1a) based on the results and experience from previous investigations [8]. Every specimen had notches made by a diamond wheel saw. The notches at the back or unloaded side of the specimen in all cases had a width of 0.5 mm. At the front or loaded side, which is next to the load line, the specimens had a notch width of either 0.5 or 0.16 mm.

Unirradiated specimens had a notch width of 0.5 mm (notch tip radius ≈ 0.25 mm) and were tested with fixtures having $W = 15$ or 25 mm (Tables 2 and 3). Irrespective of the W-value, the initial displacement rate (V_t) at the notch tip in every test was the same due to different crosshead displacement speeds (V_D) calculated from the ratio $V_t/V_D = (W - a)/W$. Some of the cold-worked and annealed specimens (Table 2) had a notch width of 0.16 mm (notch tip radius ≈ 0.08 mm) to estimate the effect of the notch width.

All irradiated specimens, including the cold-worked unirradiated Specimen 2C6 used as a

TABLE 1—*Zircaloy-2 corrosion specimens used for PL tension test.*

Corrosion Specimen	Treatment Before Corrosion Test	Corrosion Test, K/MPa/h	Weight Gain, mg/dm^2	Hydrogen content,[a] (wtppm)
E23	CW	723/10/24	353	471 (408 to 577)
E24	CW	723/10/24	314	403 (396 to 411)
C29	948 K/16 h	723/10/24	151	241 (241 to 242)
B13	848 K/0.5 h	773/10/24	384	660 (633 to 687)
C8	798 K/0.5 h	733/10/24	385	332 (330 to 333)
C14	1048 K/0.5 h	733/10/24	348	386 (382 to 391)
C22	798 K/0.5 h	733/10/24	954	911 (789 to 1068)
B24	948 K/16 h	733/10/24	366	558 (525 to 597)
B51	848 K/16 h	773/10/14	660	722 (676 to 770)

[a] The average and the range of values are shown.

reference, had a notch width of 0.16 mm. Testing of irradiated specimens was performed for $W = 15$ mm.

The load-displacement curve was recorded for every specimen by means of Instron machine facilities without using a clip gage. Typical load-displacement curves for different materials and test temperatures are shown in Figs. 3a and 3b, together with the external view of the tested specimens.

Fracture Toughness Characteristics Obtained from PL Tension Test

For pressure tube materials, it has been shown that zirconium alloys are capable of accommodating a large amount of slow stable crack growth under increasing loads [*11*]. Therefore,

TABLE 2—*The PL tension test results for annealed and cold-worked Zircaloy-2 cladding.*

Material/Specimen	Test Temperature	W, mm	Δ_t, mm	P_{max}, N	J_{max}, kN/m	J_{pl}, kN/m	J_{el}, kN/m
ANN/1	293 K	15	0.46	887	...	...	...
8[a]		15	0.57	898	218	181	37
9		15	0.56	918	203	175.5	27.5
ANN/2	573 K	25	0.72	255	132	115.5	16.5
3		15	0.77	536	...	...	...
5		15	0.65	439	146	129	17
7[a]		15	0.73	530	...	...	...
CW/1	293 K	15	0.11	1020	...	...	...
2		25	0.14	617	...	...	...
3		25	0.12	561	...	...	...
4		25	0.13	581	...	...	...
5		25	0.12	622	91	45	46
6		25	0.15	643	101.5	53	48.5
8		25	0.12	617	88.5	42.5	46
13[a]		15	0.13	1214	99	44.5	54.5
CW/7	573 K	25	0.27	449	...	...	...
9		25	0.31	413	105	73	32
11		15	0.27	839	128	83	45
12[a]		15	0.25	856	103	70	33

[a] Specimen had a notch width of 0.16 mm.

TABLE 3—*The PL tension test results for hydrided corrosion specimens.*

Corrosion Specimen	Test Temperature, K	W, mm	Δ_t, mm	P_{max}, N	J_{max}, kN/m	J_{pl}, kN/m	J_{el}, kN/m
E23	293	25	0.25	471	97.5	61.5	36
	573	15	0.58	398	105	81.5	23.5
E24	293	15	0.30	625	111	71.5	39.5
	573	15	0.70	503	140	115.5	24.5
C29	293	25	0.29	476	108.5	77.5	31
		15	0.26	765	...	...	...
B13	293	25	0.20	365	63	32.5	30.5
		25	0.08	403	...	...	...
	573	25	0.82	131	103.5	91	12.5
C8	293	25	0.15	408	68	40	28
	573	15	0.68	564	139.5	120.5	19
C14	293	15	0.08	386	30.5	14.5	16
	573	15	0.77	648	150.5	126.5	24
C22	293	25[a]	...	...	...	≈0	...
	573	15	0.69	508	167.5	125	42.5
B24	293	25	0.09	403	...	...	...
		15	0.22	826	72	31	41
	573	25	0.81	241	108	98.5	9.5
B51	293	15	0.01	617	...	...	...
		25	0.005	230	10.5	≈0	10.5

[a] The specimen was broken when placed into testing machine.

the carrying capacity of the cladding tube could be evaluated at the point of maximum load, as has been done for pressure tube materials [*3*]. Such an approach circumvented the need to detect crack initiation in the PL specimen.

The ductility of the notched cladding specimens tested under PL tension was characterized by the parameter Δ_t. This parameter is similar to the crack-opening displacement, represents the plastic part of the notch-tip opening displacement at the maximum load, P_{max}, of the load-displacement curve, and can be calculated from the ratio $\Delta_t/\Delta_D = (W - a)/W$, where Δ_D is the load-line displacement at the maximum load P_{max} (Fig. 3c).

Also, the *J*-integral evaluation for the cladding tested under PL tension has been done on the basis of the following equation developed for compact tension specimens [*12*]:

$$J_I = \beta A_T/B b_o \tag{1}$$

where
A_T = the total area under the load versus load-line displacement curve (see Fig. 3c),
B = specimen thickness, and
b_o = the uncracked ligament length.

The coefficient β depends on the ratio a_o/b_o, where a_o is the initial crack length.
The *J*-integral values were considered to be represented by:

$$J_I = J_{el} + J_{pl} \tag{2}$$

where J_{el} and J_{pl} are the elastic and plastic components of J_I, respectively. If the load displacement is purely elastic, A_1 and A_2, shown in Fig. 3c, are zero, and Eq 1 reduces to $J_I = J_{el}$.

To apply Eq 1 to the PL test configuration, one should know the β-coefficient value. For a

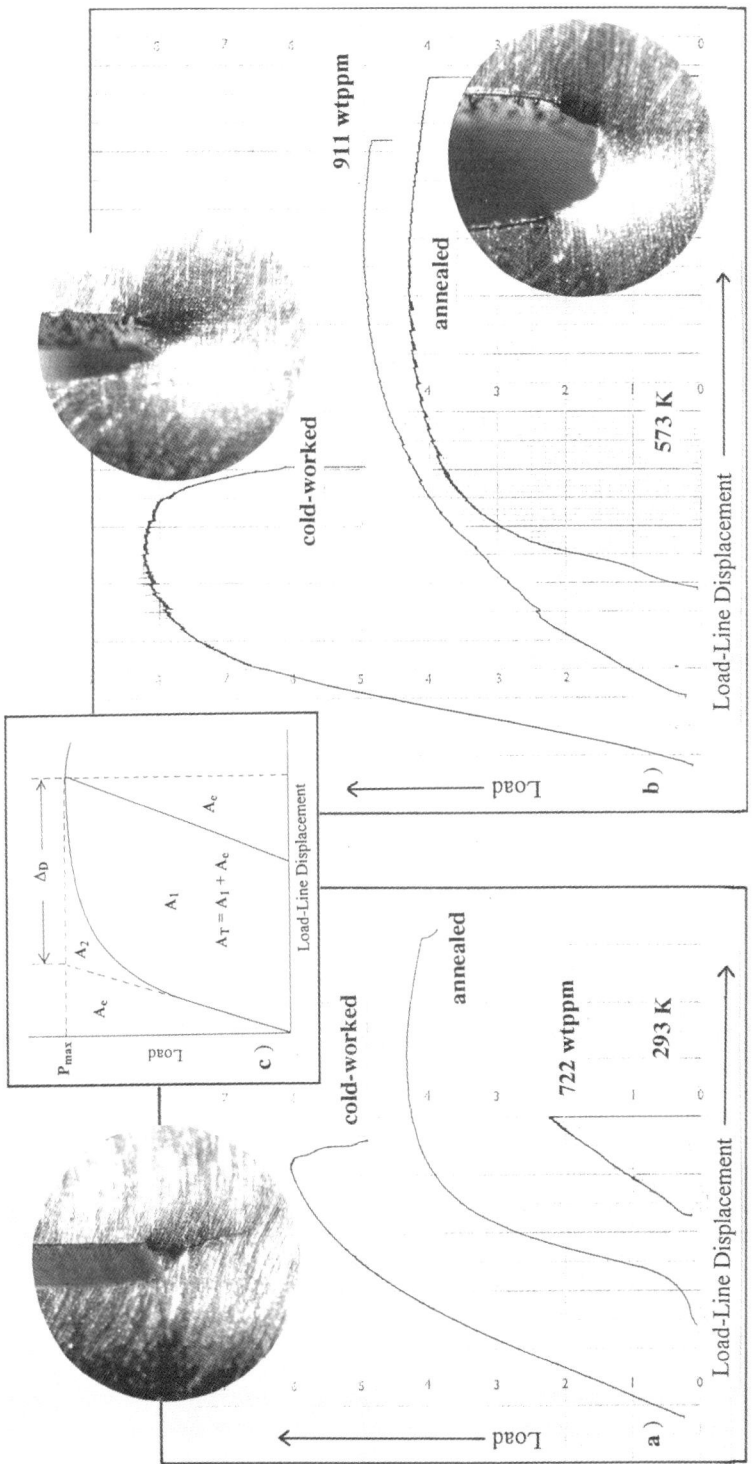

FIG. 3—(a) Typical load-displacement curves for annealed, cold-worked, and hydrided (722 and 911 wtppm) Zircaloy-2 tubing under PL tension at room temperature; (b) at 573 K. The load scale is $2 \cdot 10^3$ N for cold-worked material tested at 293 K and $1 \cdot 10^3$ N for all other specimens; (c) load versus load-line displacement schematic illustrating the areas used for J estimation calculations and for evaluation of the notch-tip opening displacement at maximum load P_{max}.

compact tension specimen, the coefficient β is connected to the position of the rotation center and, thus, determines the uncracked ligament area subjected to the tensile stress. It is accepted here that $\beta = 1$ since all uncracked ligament in the PL specimen are subjected to the tensile stress.

The J-integral values (J_{max}, J_{pl}, J_{el}), as calculated from Eq 1 and Eq 2, and notch-tip opening displacement values Δ_t, obtained at P_{max} of the load-displacement curves in the PL tension test, were then compared with hydrogen content and hydride morphology of tested materials. The fracture toughness of irradiated cladding were compared with cold-worked unirradiated material.

Results and Discussion

PL Testing of Unirradiated Specimens

The results on notched specimen ductility for annealed and cold-worked materials are summarized in Table 2. The reproducibility of the method could be evaluated from the results for cold-worked cladding: Δ_t-values for eight specimens tested at room temperature are in the interval 0.11 to 0.15 mm with an average value 0.13 mm, i.e., within $\pm 15\%$. For annealed and cold-worked cladding, the specimens with notch widths of 0.5 and 0.16 mm showed similar values of the measured properties. The difference would most probably be higher for less ductile materials, such as irradiated cladding.

The results for corrosion specimens are collected in Table 3 and in Fig. 4. For annealed material, the cladding ductility at room temperature decreases with increasing hydrogen content (Fig. 4a). Completely brittle fracture occurred at hydrogen concentrations above 700 wtppm. Such behavior observed in the PL test of actual cladding is very similar to that obtained for flat specimens under tension testing. A ductile-brittle transition at room temperature was earlier found in annealed Zircaloy-4 for hydrogen contents between 400 and 800 wtppm [13] and about 750 wtppm [5,14].

There is no influence of the hydrogen content on the ductility of the cladding at 573 K: the specimen with a hydrogen content of about 900 wtppm behaves similar to non-hydrided material. Such behavior also agrees with earlier observation of a brittle-ductile transition in hydrided Zircaloy-4 between 293 and 623 K [5,15].

Partial annealing of cold-worked cladding during corrosion testing leads to an increase in the ductility characterized by notch-tip opening displacement values Δ_t (Fig. 4b).

In some cases, specimens with a similar hydrogen content showed different ductilities at room temperature, for example Specimens B13 and B51 or E23, E24, C8, and C14. Apart from the heat treatment and the corrosion test temperature, which were different (see Table 1), the hydride morphology could also influence the results of PL tests. In general, the more pronounced the continuity of the hydride network, the lower the ductility (Fig. 5). Such a relation has also been noticed in earlier investigations of hydrided zirconium [16] and Zircaloy-4 [5].

The PL Testing of Irradiated Cladding

The external view of some of the irradiated specimens before testing are shown in Fig. 6, and the corresponding load-displacement curves, recorded during testing, are presented in Fig. 7.

The PL testing of every specimen was interrupted when the load, after having passed a maximum value P_{max}, had decreased to 70 to 80% of P_{max}. After that, the specimen was unloaded and the crack surface tinted. After cooling to room temperature, the specimen was broken to reveal the fracture surface created during PL testing at 573 K. In all cases, the average

GRIGORIEV ET AL. ON ZIRCALOY CLADDING TUBES 439

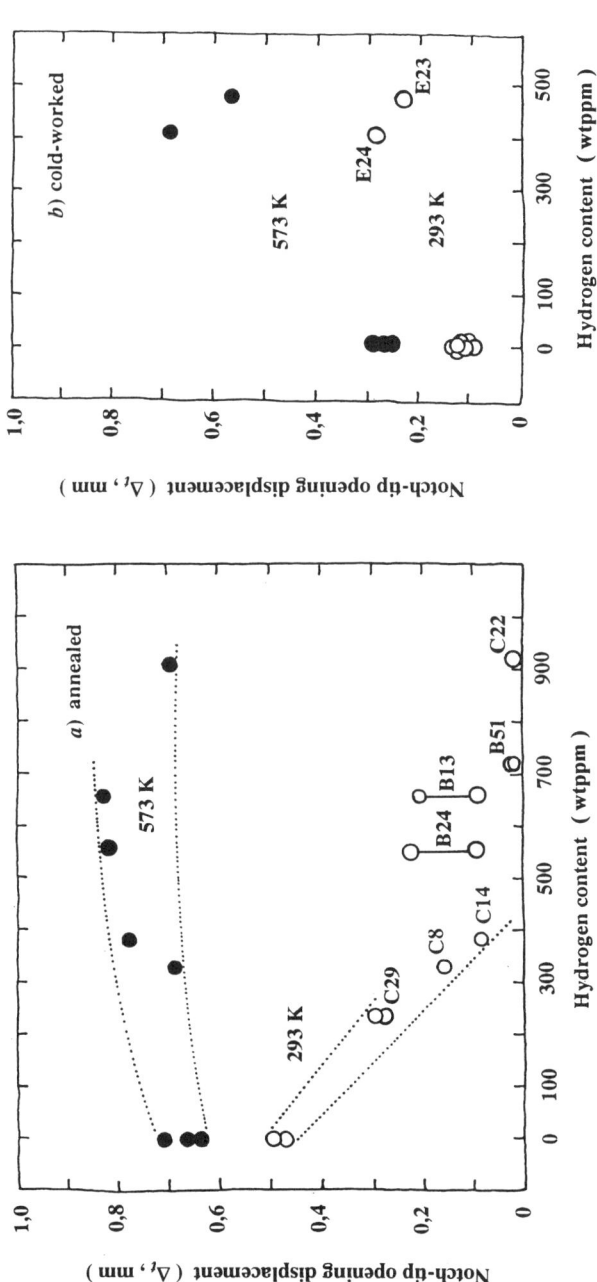

FIG. 4—*Notched specimen ductility characterized by NTOD (Δ_t) versus hydrogen content for (a) annealed and (b) cold-worked Zircaloy-2 cladding tested under PL tension at 293 and 573 K.*

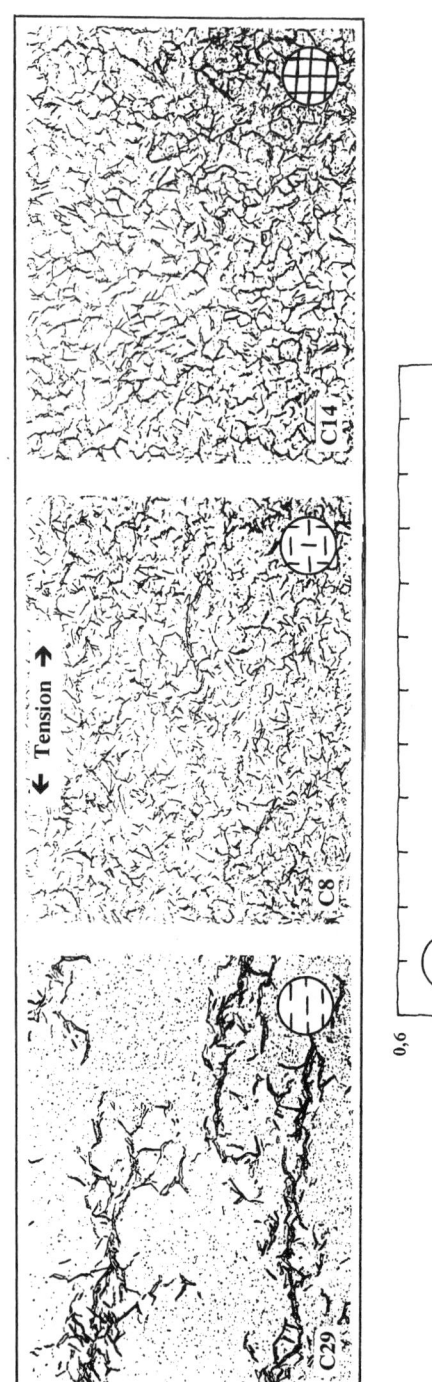
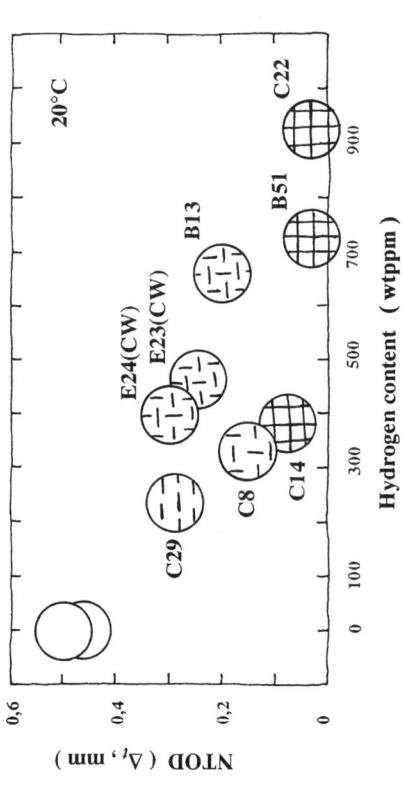

FIG. 5—*Influence of the hydride morphology on the ductility of Zircaloy-2 cladding characterized by the notch-tip opening displacement (NTOD) in the PL tension test at 293 K.*

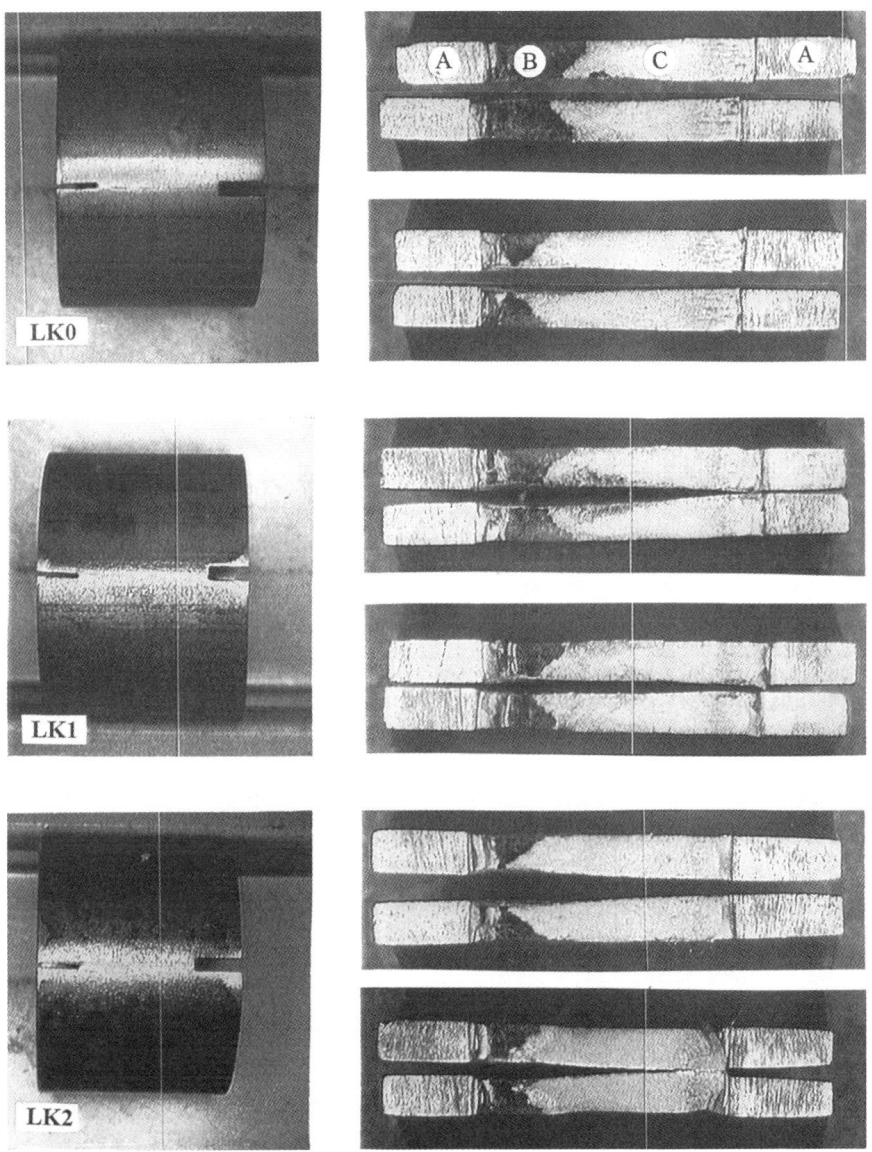

FIG. 6—*Irradiated specimens before PL tension test and fracture surfaces after testing: A = notch surfaces, B = high-temperature fracture surface, C = room temperature fracture surface (specimen's length = 9 mm; the back or unloaded side of the specimen is on the right).*

length of the propagated crack was 1.2 to 1.5 mm and was always longer at the inner surface of the specimen (see Fig. 6).

Examples of the load-displacement curves, shown in Fig. 7, include the curve for the reference cold-worked unirradiated material (Specimen 2C6) tested in the same set together with irradiated cladding (LK0, LK1, LK2).

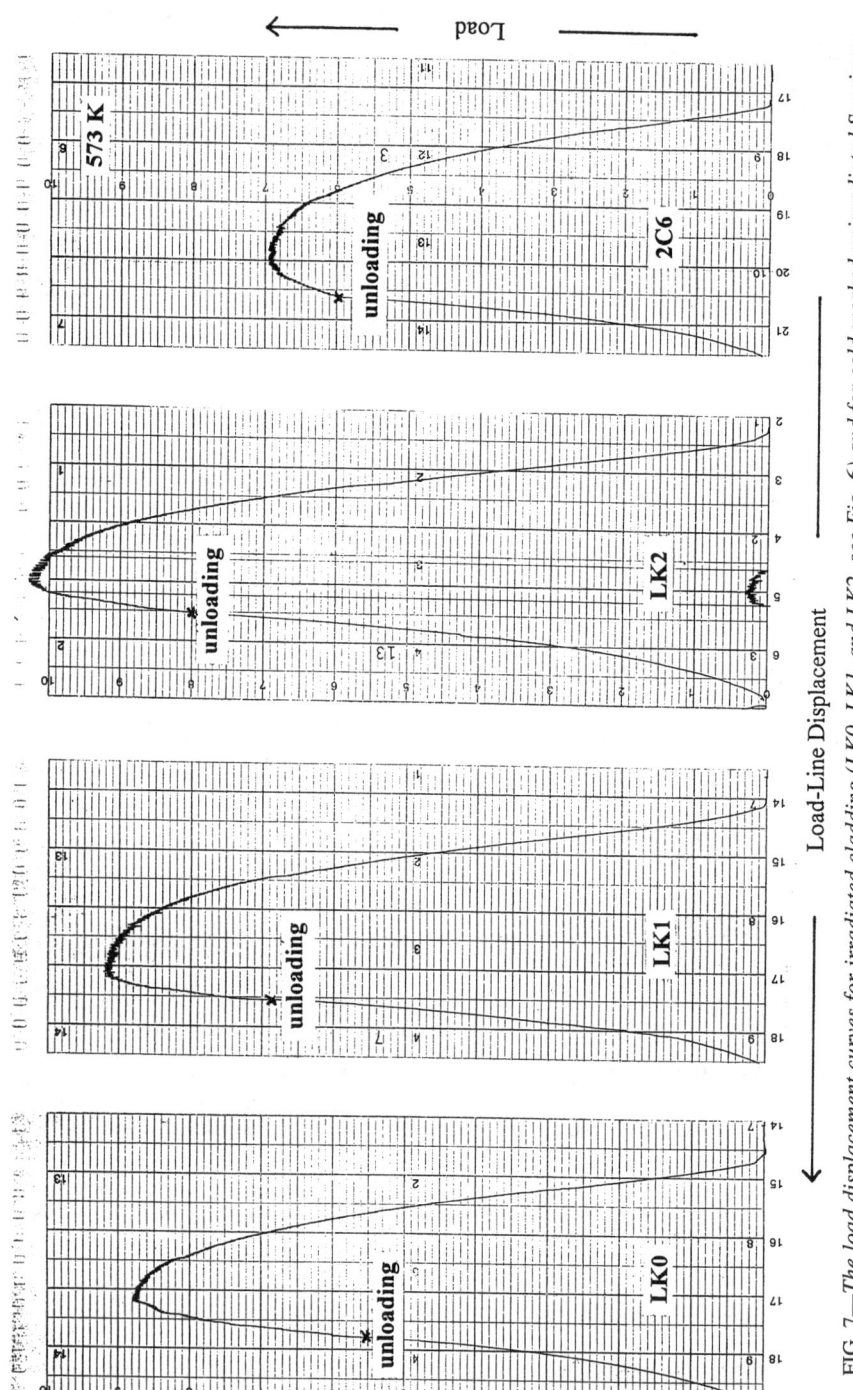

FIG. 7—*The load displacement curves for irradiated cladding (LK0, LK1, and LK2, see Fig. 6) and for cold-worked unirradiated Specimen 2C6 used as a reference in PL tension testing at 573 K.*

TABLE 4—*The PL tension test results for irradiated Zircaloy-2 specimens at 573 K.*

Specimen	Hardness[a]	Hydrogen, wtppm	Δ_t, mm	P_{max}, N	J_{max}, kN/m	J_{pl}, kN/m	J_{el}, kN/m
LK0-1	164	203	0.13	1 113	87	47	40
LK0-2			0.13	1 094	87	48	39
LK1-1	168	96	0.19	1 150	116	74	42
LK1-2			0.18	1 200	113	69	44
LK2-1	180	94	0.13	1 287	104	56	46
LK2-2			0.14	1 181	98	54	44
2C6	...	as-received	0.20	869	85	59	26

[a] Vickers hardness of irradiated cladding before irradiation.

The values of the notch-tip opening displacement (Δ_t) and the maximum load (P_{max}), obtained from the curves and presented in Table 4, are in a good agreement with the well-known effects of neutron flux: irradiation hardening and decrease of ductility compared to the unirradiated state of annealed cladding (see Table 2). Irradiation has not changed the initial ranking of the materials according to their strength: the higher the hardness of the cladding before irradiation, the higher is P_{max} in post-irradiation PL tests (see Table 4).

Load Serrations

Intensive load serrations were observed for all specimens (annealed, cold-worked, hydrided, and irradiated) tested at 573 K, but serrations were absent in the room temperature tests (see Fig. 3 and Fig. 7). For the PL testing conditions, there is some probability of specimen slippage on the grips as the opening angle of the split grip increases. However, the behavior observed in the PL tension tests of actual cladding agrees well with earlier results for flat, tension specimens of Zircaloy-4 [7], where no slippage was possible. In the latter investigation, load serrations were not observed in smooth (unnotched) specimens. In the notched specimens, serrations did not occur at room temperature but only at 473 and 573 K and were enhanced by the existence of hydrides.

Fracture Toughness of Cladding Tubes

The J-integral values obtained from the PL tension tests were evaluated at the maximum load by Eq 1 and Eq 2. The calculated values of J_{max}, J_{el}, and J_{pl} are included in Tables 2 through 4.

In general, the J-integral values J_{max} and J_{pl} for the corrosion specimens at room temperature (Fig. 8) follow the trend, which is observed for the notch-tip opening displacement (Δ_t) versus hydrogen content dependence (see Fig. 4a), i.e., $J_{max} \sim J_{pl} \approx \Delta_t$. At the same time, the dependence of J_{el} on the hydrogen content, as well as dependence of P_{max} on the hydrogen content, is much weaker (see Table 3). This means that $J_{el} \approx P_{max}$ and the J_I-integral values mainly depend on the cladding ductility.

A further confirmation of the correlations $J_{pl} \approx \Delta_t$ and $J_{el} \approx P_{max}$ can be obtained from the results for irradiated cladding (see Table 4), where the minimum and maximum values are situated as shown in Table 5.

A more detailed quantitative investigation of the fracture toughness of the irradiated cladding LK0, LK1, and LK2 and its relation to the initial metallurgical state of the cladding will be

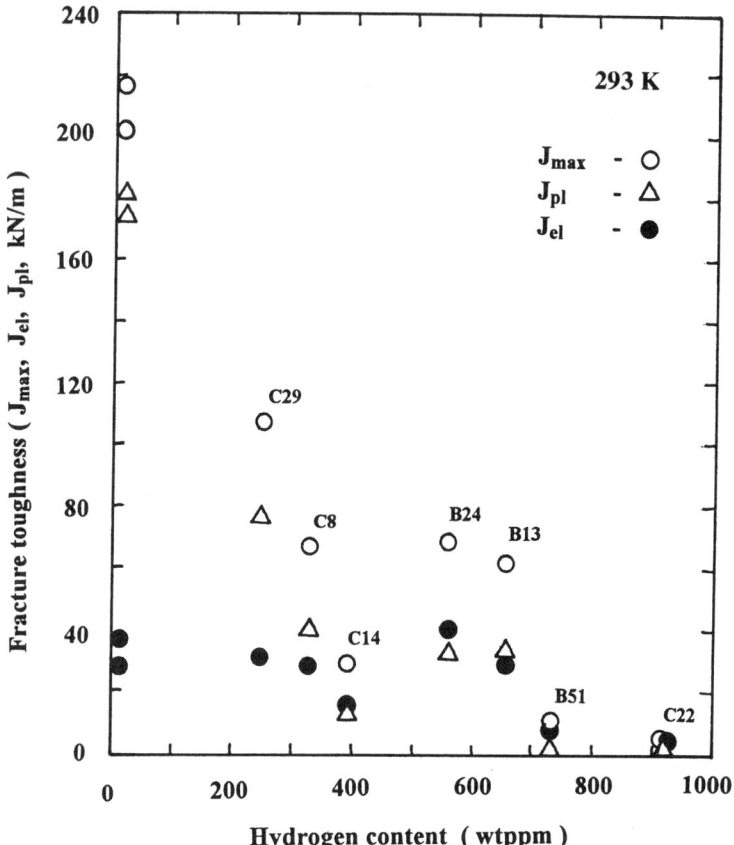

FIG. 8—*The J_I-integral values for the maximum load point (J_{max}) and its elastic (J_{el}) and plastic (J_{pl}) components evaluated from the PL tension test at room temperature versus the hydrogen content in unirradiated Zircaloy-2 cladding.*

done later. Currently, it can be concluded that none of the materials LK0, LK1, and LK2 showed significant embrittlement in the PL tension testing at 573 K after in-reactor use as fuel cladding and that their fracture toughness is comparable with that for unirradiated cold-worked cladding.

The actual J-integral values obtained for the thin-walled cladding under PL tension test can be compared with those obtained for pressure tube materials. Thus, the values of $J_{max} \approx 100$

TABLE 5—*Correlations $J_{pl} \approx \Delta_t$ and $J_{el} \approx P_{max}$ observed for irradiated cladding data (see Table 4).*

	NTOD,[a] Δ_t	P_{max}	J_{el}	J_{pl}
LK0	min	min	min	min
LK1	max	...	...	max
LK2	...	max	max	...

[a] Notch-tip opening displacement.

kN/m obtained for irradiated and cold-worked Zircaloy-2 cladding agree reasonably well with the results for pressure tube materials [3,17,18], which are somewhat lower, probably due to plane-strain conditions, which are typical for the fracture processes in a pressure tube due to its larger wall thickness.

Conclusions

A new pin-loading tension test, which allows the use of fracture mechanics concepts for thin-walled tube materials, has been shown to be an effective method for evaluating the fracture toughness of actual Zircaloy-2 cladding after hydriding and irradiation.

Axially notched ring specimens have been tested in a way similar to that used for compact tension specimens. The J-integral values and the notch-tip opening displacements were obtained experimentally at the maximum load of the load-displacement curves for different cladding: annealed, cold-worked, hydrided, and irradiated. A fracture toughness of $J_{max} \approx 100$ kN/m obtained for irradiated and cold-worked Zircaloy-2 cladding agrees reasonably well with results obtained for pressure tube materials.

The experimental results obtained by means of the pin-loading tension test have revealed for actual cladding all main phenomena earlier observed for Zircaloy materials:

1. A ductile-brittle transition occurs in the fracture behavior of unirradiated hydrided cladding at room temperature and at hydrogen concentrations of 600 to 700 wtppm.
2. There is no influence of the hydrogen content on the fracture toughness of unirradiated cladding at 573 K and at hydrogen concentrations up to 900 wtppm.
3. Plastic deformation of every notched cladding specimen tested at 573 K is accompanied by intensive load serrations.
4. Irradiation up to fluences of about $6 \cdot 10^{25}$ n/m^2 ($E > 1$ MeV) has not changed the initial ranking of the cladding according to its strength; at a temperature of 573 K, the fracture toughness of irradiated cladding containing 100 to 200 wtppm of hydrogen is comparable with that for cold-worked unirradiated cladding.

Acknowledgments

This work was partly sponsored by Vattenfall AB, OKG Aktiebolag, Southern Sweden Power Supply, and Studsvik AB. The experimental assistance of Y. Haag is gratefully acknowledged.

References

[1] Simpson, L. A. and Clark, C. in *Elastic-Plastic Fracture, ASTM STP 668.* American Society for Testing and Materials, West Conshohocken, PA, 1979, pp. 643–662.
[2] Chow, C. K. and Simpson, L. A. in *Fracture Mechanics: Eighteenth Symposium, ASTM STP 945,* American Society for Testing and Materials, West Conshohocken, PA, 1988, pp. 419–439.
[3] Asada, T., Kimoto, H., Chiba, N., and Kasai, Y. in *Zirconium in the Nuclear Industry: Ninth International Symposium, ASTM STP 1132,* American Society for Testing and Materials, West Conshohocken, PA, 1991, pp. 99–118.
[4] Rebak, R. B. and Szklarska-Smialowska, Z. in *Corrosion.* Vol. 50, No. 5, 1994, pp. 378–393.
[5] Bai, J. B., Prioul, C., and François, D. in *Metallurgical and Materials Transactions A,* Vol. 25A, June 1994, pp. 1185–1197.
[6] Andersson, T. in *Journal of Nuclear Materials,* Vol. 62, 1976, pp. 95–104.
[7] Huang, J.-H. and Huang, S.-P. in *Journal of Nuclear Materials,* Vol. 208, No. 1, 1994, pp. 166–179.
[8] Grigoriev, V., Josefsson, B., Lind, A., and Rosborg, B. in *Scripta Metallurgica et Materialia,* Vol. 33, No. 1, 1995, pp. 109–114.

[9] Rudling, P. and Lethinen, B., "Mechanistic Understanding of Nodular Corrosion. Material Characterization and In-Reactor Program," EPRI TR-103396 Final Report, November 1993.
[10] Rudling, P., Pettersson, H., Andersson, T., and Thorvaldsson, T. in *Zirconium in the Nuclear Industry: Eighth International Symposium, ASTM STP 1023,* American Society for Testing and Materials, West Conshohocken, PA, 1989, pp. 213–226.
[11] Simpson, L. A., "Initiation COD as a Fracture Criterion for Zr-2.5%Nb Pressure Tube Alloy," *Fracture 1977,* Vol. 3, pp. 705–711.
[12] Clark, G. A. and Landes, J. D. in *Journal of Testing and Evaluation,* Vol. 7, No. 5, 1979, pp. 264–269.
[13] Lin, S.-C., Hamasaki, M., and Chuang, Y.-D. in *Nuclear Science and Engineering,* Vol. 71, 1979, pp. 251–266.
[14] Bai, J. B., Prioul, C., Lansiart, S., and François, D. in *Scripta Metallurgica et Materialia,* Vol. 25, 1991, pp. 2559–2563.
[15] Bai, J. B. and François, D. in *Journal of Nuclear Materials,* Vol. 187, 1992, pp. 186–189.
[16] Gill, B. J., Bailey, J. E., and Cotterill, P. in *Journal of Less-Common Metals,* Vol. 40, 1975, pp. 129–138.
[17] Chow, C. K., Coleman, C. E., Hosbons, R. R., Davis, P. H., Griffiths, M., and Choubey, R. in *Zirconium in the Nuclear Industry: Ninth International Symposium, ASTM STP 1132,* American Society for Testing and Materials, West Conshohocken, PA, 1991, pp. 246–275.
[18] Wallace, A. C., Shek, G. K., and Lepik, O. E. in *Zirconium in the Nuclear Industry: Eighth International Symposium, ASTM STP 1023,* American Society for Testing and Materials, West Conshohocken, PA, 1989, pp. 66–88.

DISCUSSION

P. Davies[1] (written discussion)—Please clarify the stress field experienced by the remaining ligament of the PL specimen. For example, the loading provided by the grips appears to be combined tension and bending, but the *J*-analysis equation appears to assume tensile loading.

V. Grigoriev et al. (authors' closure)—The main components of the tensile stress in the notched ring specimen under PL tension are similar to those that exist in a compact tension specimen. Additional stress components can arise at the notch tip as a result of bending due to the initial gap between the specimen and cylindrical holder. Due to this, careful attention should be given to minimizing the initial gap of the specimen holder. A numerical analysis is needed to quantify this additional component versus gap values. As we observed in the experiments, the radial gaps of 0.02 to 0.07 mm did not change fracture parameters significantly.

---

[1] Atomic Energy of Canada, Ltd., Chalk River, Ontario, Canada.

Magnus Limbäck[1] and Thomas Andersson[2]

A Model for Analysis of the Effect of Final Annealing on the In- and Out-of-Reactor Creep Behavior of Zircaloy Cladding

REFERENCE: Limbäck, M. and Andersson, T., "**A Model for Analysis of the Effect of Final Annealing on the In- and Out-of-Reactor Creep Behavior of Zircaloy Cladding,**" *Zirconium in the Nuclear Industry: Eleventh International Symposium, ASTM STP 1295,* E. R. Bradley and G. P. Sabol, Eds., American Society for Testing and Materials, 1996, pp. 448–468.

ABSTRACT: The creep behavior of Zircaloy cladding materials depends on materials texture, degree of recrystallization, and chemical composition. This study is devoted mainly to the analysis of the effect of the final annealing (i.e., the degree of recrystallization) on the creep characteristics. For this purpose, data from a series of thermal creep tests are presented and evaluated. In addition, the in-reactor creep data presented by Franklin et al. are used to evaluate the effect of irradiation on cladding creep performance. The out-of-reactor tests are performed under internal pressurization, and the test matrix covers seven conditions with temperatures from 330 to 400°C and hoop stresses between 80 and 160 MPa. Three lots of Zircaloy-2 claddings and one lot of Zircaloy-4 are considered. The difference between the three Zircaloy-2 lots is in their final annealing conditions. The claddings are either stress relief annealed (SRA), recrystallization annealed (RXA), or partially recrystallization annealed (PRXA). The materials used when fabricating the Zircaloy-2 claddings are from the same ingot, and the chemical compositions of the three types of claddings are almost identical. The Zircaloy-4 cladding included in the test is SRA, and the tin content in this material is similar to that in the Zircaloy-2 materials. The creep data are analyzed by separating the primary (transient) and the secondary (steady-state) creep. In this analysis, the Matsuo creep model, which accounts for both primary and secondary creep, is modified, calibrated, and verified using the new thermal creep data. Based on in-reactor data, the thermal creep model is extended to cover also the creep behavior under irradiation. The claddings considered in the in-reactor test were of both SRA and RXA types, and the experiments were made under external pressure. It is observed that for moderate hoop stresses (<120 MPa) the measured hoop creep rates, both out-of-reactor and in-reactor, decrease with increasing final annealing temperature (increasing degree of recrystallization). However, for higher stresses, the steady state creep rate has a minimum for partially recrystallized claddings. Moreover, the thermal creep tests show that the creep performance of Zircaloy-2 and Zircaloy-4 are similar when the chemical compositions and the fabrication procedures are similar.

KEYWORDS: zirconium, Zircaloy, tubes, creep, heat treatment, final annealing, in-reactor

Zircaloy-2 and Zircaloy-4 are zirconium-based alloys used as structural materials in nuclear fuel, e.g., as material for fuel rod claddings and spacer grids. Accordingly, the mechanical properties of these materials are important for the integrity of fuel rods and fuel assemblies.

At high temperatures, the amount of time-dependent plastic strain can be much larger than the strain produced instantaneously during the time interval a load is placed on a specimen.

---

[1] Staff engineer, ABB Atom AB, Nuclear Fuel Division, S-721 63 Västerås, Sweden.

[2] Manager, product and process development, AB Sandvik Steel, R & D Centre, Special Metals Division, S-811 81 Sandviken, Sweden.

The time-dependent plastic strain observed in stressed materials is called creep. The operating temperatures of a nuclear reactor enables creep of the Zircaloy materials.

Creep can be demonstrated directly by a creep curve that represents graphically the function between creep strain and time. This type of creep curve generally exhibits three characteristic regions. In the first region, the region of primary (or transient) creep, the specimen begins to deform relatively quickly. However, the creep rate decreases with time as resistance to continued deformation builds up through strain hardening, and the deformation rate eventually decreases to some near constant value. The constant creep rate defines the second region, the region of secondary (or steady-state) creep. At this stage, the thermal recovery offsets strain hardening. In the third region, the region of tertiary creep, the specimen deformation rate accelerates to relatively high levels until the specimen fails.

The creep behavior of zirconium-based alloys has been thoroughly described by Franklin et al. [1]. Their review includes descriptions of diffusional creep mechanisms, such as Nabarro-Herring and Coble creep, dislocation climb-controlled creep or dislocation climb-glide creep, and glide-controlled creep. According to Franklin et al., it appears that dislocation climb-glide creep predominates at high stresses and dislocation glide creep at intermediate stresses. At low stresses, Coble creep predominates. However, in the stress-temperature region of interest for Zircaloy components in water-cooled reactors, the dislocation climb-glide creep mechanism dominates [1].

Out-of-reactor and in-reactor creep behavior of Zircaloy cladding materials is studied in this work. The study of out-of-reactor, or thermal, creep behavior is based on experimental data in combination with a model for thermal creep presented by Matsuo [2]. The evaluation of in-reactor creep is based on in-reactor data presented by Franklin, Lucas, and Bement [1]. The in-reactor data are used to extend the thermal creep model to also cover the in-reactor conditions. When applying the thermal creep model to in-reactor conditions, the irradiation-hardening effect on creep strength is accounted for, and a term relating to the irradiation-induced creep is included in the model.

Cladding Materials

Materials in Out-of-Reactor Creep Tests

The materials used in the thermal creep tests comprise Zircaloy-2 claddings from three different lots and cladding tubes from one lot of Zircaloy-4. The materials used when fabricating the Zircaloy-2 claddings are from the same ingot, and the chemical compositions of the three types of claddings are almost identical. The difference between the three Zircaloy-2 lots is in their final annealing conditions. The claddings are either stress relief annealed (SRA), recrystallization annealed (RXA), or partially recrystallization annealed (PRXA). The Zircaloy-4 cladding included in the test is SRA, and the tin content in this material is similar to that in the Zircaloy-2 materials. Tables 1a, 1b, and 1c present material characterization data, i.e., cladding chemical compositions, annealing parameters, final reduction, final annealing conditions, degrees of recrystallization, and tension test data for the materials tested out-of-reactor.

The cladding tubes considered in the thermal creep tests were fabricated by AB Sandvik Steel (ABSS). The nominal inner and outer diameters of the tubes are 8.357 and 9.5 mm, respectively. The final pilgering step was made using a cylindrical mandrel and a degree of reduction of 80%. The cladding basal pole texture is characterized by Kearns texture factors, f_r, f_t, and f_a, being about 0.65, 0.30, and 0.05, respectively. Moreover, the angle between the direction of maximum intensity of basal poles and the radial direction in the radial-circumferential plane of Zr polycrystal (ϕ_{max}) is 30 ± 2°.

TABLE 1a—*Materials chemical composition.*

ABSS Lot No.	Clad Type	Sn, wt%	Fe, wt%	Cr, wt%	Ni, wt%	Si, ppm	O, ppm	C,[a] ppm
81408	SRA[b] Zr-4	1.34	0.20	0.10	...	80	1340	130–140
81416	SRA Zr-2	1.36	0.16	0.10	0.05	75	1295	120–140
81428	PRXA[c] Zr-2	1.37	0.16	0.09	0.05	70	1245	120–140
81417	RXA[d] Zr-2	1.37	0.16	0.09	0.05	70	1215	120–140

[a] The carbon concentration is from chemical analysis of the ingot. The other concentrations are from analysis of the final product, i.e., the cladding tube.
[b] SRA = stress relief annealed.
[c] PRXA = partially recrystallization annealed.
[d] RXA = recrystallization annealed.

TABLE 1b—*Materials fabrication.*

ABSS Lot No.	Clad Type	Annealing Parameters[a]			Final Reduction[b]	Final Heat Treatment	Degree of Recrystallization
		Log A_S	Log A_A	Log $SOCAP$			
81408	SRA Zr-4	−16.77	−13.07	−20.64	80%	486°C/3.5 h	4%
81416	SRA Zr-2	−16.77	−13.07	−20.64	80%	486°C/3.5 h	12%
81428	PRXA Zr-2	−16.77	−13.07	−20.64	80%	502°C/3.5 h	49%
81417	RXA Zr-2	−16.77	−13.07	−20.64	80%	565°C/1.5 h	100%

[a] The influence of heat treatments on Zircaloy mechanical and corrosion properties have been discussed and modeled [3–5]. The dependence of different heat treatments has been interpreted by the normalized annealing time (i.e., annealing parameter). Different annealing parameter correlations have been used to relate the heat treatment with the material's mechanical properties [3], oxide weight gain [3,4], and precipitate growth [5]:

$$A_S = \sum_{i=1}^{n} t_i \cdot \exp[(-Q_1/(R \cdot T_i)] \quad \text{(according to Charquet et al. [3])}$$

$$A_A = \sum_{i=1}^{n} t_i \cdot \exp(-Q_2/(R \cdot T_i)) \quad \text{(according to Thorvaldsson et al. [4])}$$

$$SOCAP = \sum_{i=1}^{n} t_i \cdot \frac{K_3}{T_i^2} \cdot \exp[-Q_3/(R \cdot T_i)] \quad \text{(according to Gross and Wadier [5])}$$

where i = different heat treatments,
t_i = annealing time (in hours for A_S and A_A and in seconds for $SOCAP$),
T_i = temperature, K,
Q_1/R = activation energy divided by the molar gas constant = 40 000 K,
Q_2 = activation energy = 63 000 cal/mol,
Q_3/R = activation energy divided by the molar gas constant = 18 700 K,
$K_3 = 1.11 \times 10^{-11}$ m³ K²/s, and
R = the molar gas constant, cal/mol K.

[b] The final pilgering step was made using a cylindrical mandrel and a degree of reduction of 80%, resulting in a cladding basal pole texture characterized by Kearns texture factors of $f_r = 0.65$, $f_t = 0.30$, and $f_a = 0.05$. The angle between the direction of maximum intensity of basal poles and the radial direction in the radial-circumferential plane of Zr polycrystal (ϕ_{max}) is 30 ± 2°.

TABLE 1c—Data from tension tests at room temperature and at 385°C.

ABSS Lot No.	Clad Type	Final Heat Treatment	Degree of Recrystallization	Room Temperature			385°C		
				$R_{p0.2}$,[a] MPa	R_m,[b] MPa	A_{50},[c] %	$R_{p0.2}$,[a] MPa	R_m,[b] MPa	A_{50},[c] %
81408	SRA Zr-4	486°C/3.5 h	4%	576	775	18	358	452	20
81416	SRA Zr-2	486°C/3.5 h	12%	567	769	16	354	458	20
81428	PRXA Zr-2	502°C/3.5 h	49%	458	648	23	239	345	29
81417	RXA Zr-2	565°C/1.5 h	100%	376	547	30	148	254	43

[a] $R_{p0.2}$ = yield strength.
[b] R_m = tensile strength.
[c] A_{50} = elongation (measured on 50-mm gage length).

Materials in In-Reactor Creep Tests

The basis for the in-reactor clad creep model is cladding creep data from tests performed by Babcock and Wilcox (B & W) in the Duke Power's Oconee reactor [1]. The claddings considered are of both SRA (S-1 tubings in Ref 1) and RXA (S-2 claddings in Ref 1) types, both manufactured by Sandvik. These types of SRA and RXA claddings are very similar to the cladding tubes included in the thermal creep tests. The characteristics of the cladding materials used in the in-reactor experiments are provided in Table 2.

The Thermal Creep Tests

The test matrix consists of a series of tests performed at seven different combinations of temperature and stress (Table 3). The temperature is varied from 330 to 400°C for a hoop stress of 120 MPa, and the hoop stress is varied from 80 to 160 MPa at 385°C.

The duration of six of the seven tests was 480 h. The test that was performed at the lowest-considered temperature (330°C) was prolonged to 960 h. The creep deformation was measured after 120, 240, 360, and 480 h. In addition, the creep deformation at 330°C was measured after 600, 720, and 960 h.

The tests are performed under internal pressurization. Accordingly, the ratio between the tensile hoop stress, σ_θ, and the axial stress, σ_z, is 2.0 for all the considered experiments.

Description of Creep Measurements

The procedure used when testing the creep properties of the cladding materials is:

1. The outer diameter of the cladding is measured at the axial midheight of the tube section and at ±25 mm from this position. The diameter at each one of these axial positions is the average of four measurements. These four measurements are performed between the following points on the cladding periphery: 0 to 180°, 45 to 225°, 90 to 270°, and 135 to 315°. The diameter is measured with a micrometer having an accuracy of ±1 μm.
2. The cladding wall thickness is measured at four different points (0, 90, 180, and 270°) in each end of the tube specimen. The average wall thickness is calculated from the eight measured values. The wall thickness is measured with a micrometer having an accuracy of ±1 μm.
3. The length of the specimen is measured. The nominal length is 160 mm.
4. The specimen is weighed.

TABLE 2—*Selected properties of cladding materials considered in the Oconee experiment [1].*

Designation in Ref 1	Clad Type	Tube Supplier	Final Annealing, Temperature/Time	ϕ_{max},[a]	C, ppm	$R_{p0.2}$ (RT),[b] MPa
S-1	SRA Zr-4	Sandvik	485°C/2.0 h	28°	163	547
S-2	RXA Zr-4	Sandvik	579°C/2.0 h	28°	97	369

[a] ϕ_{max} = angle between the direction of maximum intensity of basal poles and the radial direction in the radial-circumferential plane of Zr polycrystal. Moreover, the cladding basal pole textures are characterized by Kearns texture factors of about $f_r = 0.65$, $f_t = 0.30$, and $f_a = 0.05$.

[b] $R_{p0.2}$ (RT) = yield strength at room temperature.

5. The cladding wall thickness is calculated from the diameter, length, and weight of the specimen.
6. The internal pressure, P_i, is calculated from the measured average outer diameter, the measured average wall thickness, and the specified hoop stress, σ_θ, according to:

$$P_i = \frac{\sigma_\theta \cdot 2 \cdot W}{D_o - 2 \cdot W} \tag{1}$$

where P_i is the internal pressure (MPa), σ_θ the hoop stress (MPa), W the average wall thickness (mm), and D_o the average outer diameter (mm).

In cases where more than one sample was tested, the calculation is based on the sample having the thickest wall. If more than one specimen has the maximum observed wall thickness, the one with the smallest outer diameter is chosen. Accordingly, all tested specimens will as a minimum have the specified hoop stress.

7. Each sample is welded to a Zircaloy support tube, and a Zircaloy end plug is welded to the bottom end of the specimen.
8. For the testing, a high-temperature burst and creep testing equipment is used. This equipment includes an oven with three zones where each zone has its individual temperature regulator (EUROTERM). The maximum temperature equals 1100°C. The pressurization equipment includes a membrane compressor (AMINCO) working in one step with a final pressure of 90 MPa. The high-pressure section is connected to a second section where the maximum pressure is 40 MPa. Finally, the cladding tubes are internally pressurized with argon, and the maximum pressure is 35 MPa. The thermocouples, pressure transducers, and temperature and pressure regulators are controlled through a computerized system. When the specimens have been loaded into the oven, the temperature is

TABLE 3—*The creep test matrix.*

Temperature, °C	Minimum Hoop Stress, MPa			
	80	120	140	160
330	...	960	...	...
360	...	480	...	...
385	480	480	480	480
400	...	480	...	...

NOTE: Values presented in the matrix represent testing time in hours.

increased to the testing temperature and the cladding tubes are internally pressurized. The temperature is kept within ±3°C of the specified temperature, and the internal pressure is within ±0.15 MPa of the specified pressure.

9. At the end of the testing period, the internal pressure is decreased to the level of the external pressure, whereupon the full set of specimens is removed from the oven to be cooled in room-temperature air. This procedure allows the specimen temperature to decrease relatively quickly, and thereby recovery of creep strain is avoided at this stage.
10. After each testing period, the outer diameters of the samples are measured using the method described under Point 1, above. The creep strain is calculated from:

$$\varepsilon = 100 \cdot \frac{D_o(t) - D_o(t_0)}{D_o(t_0)} \quad (2)$$

where ε is the hoop (or diametral) creep strain (%), t the time (h), D_o the outer diameter (mm) after t hours, and $D_o(t_0)$ is the initial outer diameter (mm).

It should be noted that the same internal pressure is applied to all samples in the test. Accordingly, due to variations in cladding dimensions, the true hoop stress deviates from the specified hoop stress. In addition, as the creep deformation increases, the cladding diameter increases and the cladding wall thickness decreases. Hence, the applied hoop stress increases. This observation indicates that we have to account for the increasing hoop stress when we evaluate measured creep data from this type of test.

Analysis of the Thermal Creep Tests

Figures 1a and 1b show the measured creep strains as functions of time for one of the considered cladding materials (SRA Zr-2). Figure 1a depicts the data from the four tests made with different minimum hoop stresses at a temperature of 385°C, while Fig. 1b shows the data from the four tests made at different temperatures and with a minimum hoop stress of 120 MPa. The experimental data show the typical behavior observed in creep tests, including both primary and secondary creep. The creep data are analyzed by separating the primary (transient)

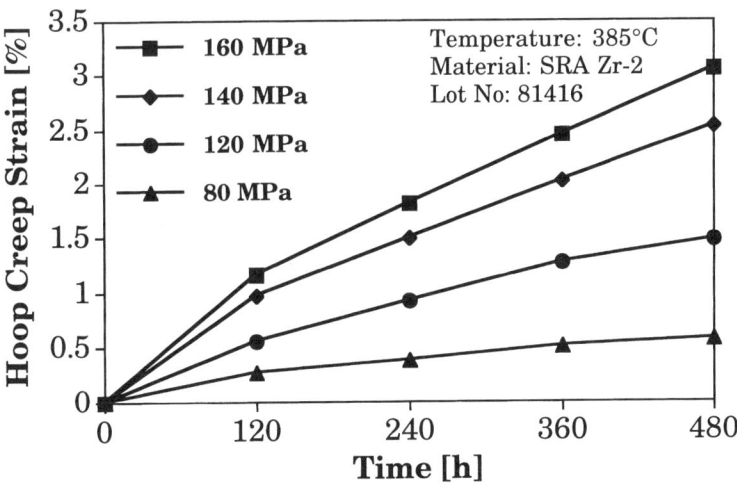

FIG. 1a—*Creep strains versus time for the SRA Zircaloy-2 cladding in the tests made at 385°C.*

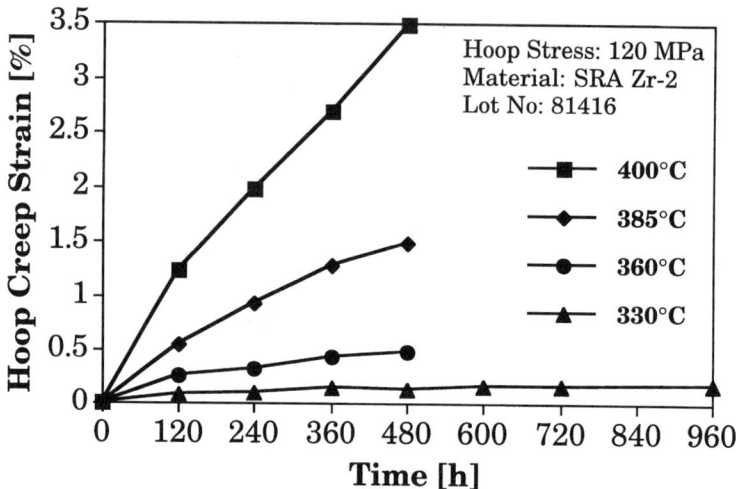

FIG. 1b—*Creep strains versus time for the SRA Zircaloy-2 cladding in the tests made with a minimum hoop stress of 120 MPa.*

and the secondary (steady-state) creep. Subsequently the Matsuo creep model [2], which accounts for both primary and secondary creep, is modified and recalibrated to correctly predict the data obtained from the thermal creep tests.

Regarding a plot of creep strain versus time, the secondary creep rate is determined by fitting a straight line to the linear part of the creep curve. Knowing the secondary creep rate, the saturated primary creep strain is easily determined as it is the strain at which the linear secondary creep rate fit intersects the strain axis.

The main results from the different creep tests are presented in Tables 4a and 4b, that is, the saturated primary creep strains and the steady-state creep rates for the considered cladding materials under the different test conditions. Some corrections have been accounted for when determining the measured data. These corrections are described parallel to the description of the Matsuo creep model, below.

The Matsuo Thermal Creep Model

The starting point for the analysis of the thermal creep data is Matsuo's creep description with undetermined constants [2]:

Steady-state hoop creep rate (h^{-1}):

$$\dot{\varepsilon}_s = A \cdot \frac{E}{T} \cdot \left(\sinh \frac{a \cdot \sigma_\theta}{E} \right)^n \cdot e^{-Q/R \cdot T} \tag{3}$$

Saturated primary hoop creep strain:

$$\varepsilon_p^s = B \cdot \dot{\varepsilon}_s^b \tag{4}$$

Total hoop creep strain:

$$\varepsilon = \varepsilon_p^s [1 - \exp(-C \cdot \sqrt{\dot{\varepsilon}_s \cdot t})] + \dot{\varepsilon}_s \cdot t \tag{5}$$

TABLE 4a—*Saturated primary creep strains [-] for the materials in the thermal creep tests.*

Temperature, °C	Hoop Stress, MPa	SRA Zr-4 81408	SRA Zr-2 81416	PRXA Zr-2 81428	RXA Zr-2 81417
330	120	1.32E-03	1.24E-03	8.49E-04	8.65E-04
360	120	1.90E-03	1.93E-03	1.47E-03	1.51E-03
385	120	4.73E-03	4.15E-03	3.33E-03	3.22E-03
400	120	6.38E-03	5.91E-03	4.80E-03	5.33E-03
385	80	2.31E-03	1.98E-03	1.32E-03	8.40E-04
385	140	6.43E-03	5.95E-03	5.27E-03	7.00E-03
385	160	7.00E-03	6.71E-03	6.26E-03	...

Young's modulus (MPa):

$$E = 1.148 \times 10^5 - 59.9 \cdot T \tag{6}$$

where A, a, B, b, and C are constants, Q is the activation energy of creep (J/mol), R is the molar gas constant (8.314 J/(mol K)), σ_θ is the hoop stress (MPa), t is the time (h), and T is the temperature (K).

Stress Dependency

Matsuo has chosen to express the stress dependency by a power-law hyperbolic-sine function (Eq 3). Garofalo [6] showed that the stress dependency of the steady-state creep rate was different in high- and low-stress regions. It was not self evident which data in the test matrix could eventually be considered as low-stress data. Thus, Eq 3 was used for evaluating the stress dependency of our data.

As mentioned above, the tests were performed under constant internal pressure, which results in a hoop stress that (i) is dependent of the dimensions of the clad specimen, and (ii) increases with increasing deformation. The data were analyzed by assuming that the hoop stress is constant. Consequently, a correction factor has to be applied. This factor has been determined and used as follows:

1. The average cladding outer diameter during a testing period is determined from the measured values before and after the period. The testing period is defined as the period between two consecutive measurements.

TABLE 4b—*Secondary creep rates [1/h] for the materials in the thermal creep tests.*

Temperature, °C	Hoop Stress, MPa	SRA Zr-4 81408	SRA Zr-2 81416	PRXA Zr-2 81428	RXA Zr-2 81417
330	120	1.11E-06	7.86E-07	6.38E-07	6.36E-07
360	120	6.85E-06	6.68E-06	4.48E-06	4.41E-06
385	120	2.21E-05	2.29E-05	1.64E-05	1.74E-05
400	120	5.44E-05	5.89E-05	4.27E-05	4.48E-05
385	80	7.39E-06	7.90E-06	4.61E-06	2.90E-06
385	140	3.90E-05	4.16E-05	3.20E-05	4.52E-05
385	160	5.32E-05	5.17E-05	4.53E-05	...

2. The average hoop stress applied to the sample during each separate testing period is calculated according to:

$$\sigma_{\theta av} = \frac{D_i \cdot P_i - D_o \cdot P_o}{D_o - D_i} \quad (7)$$

where P_i is the internal pressure (MPa), P_o is the outer pressure (1 atm = 0.101 325 MPa), D_o is the cladding average outer diameter (mm) determined according to Point 1, and D_i is the cladding average inner diameter (mm) calculated according to:

$$D_i = \sqrt{D_o^2 - \frac{4 \cdot A_{cs}}{\pi}} \quad (8)$$

and

$$A_{cs} = \frac{W}{L \cdot \rho} \quad (9)$$

where A_{cs} is the cross section area (mm^2), W is the specimen weight (g), L the specimen length (mm), and ρ is the Zircaloy density (6.57 × 10^{-3} g/mm^3).

The weight and length of the specimens were measured prior to the tests. Observations presented by Matsuo show that the axial strain is small in comparison with the diametric one (1/100 to 1/20), and that the axial strain can be regarded as zero [2]. Consequently, specimen length changes during the creep tests were neglected.

3. The creep deformation during the test period is normalized to a predefined hoop stress according to:

$$\Delta\varepsilon_{\theta c} = \Delta\varepsilon_m \cdot \left(\frac{\sinh(a \cdot \sigma_{\theta c}/E)}{\sinh(a \cdot \sigma_{\theta av}/E)}\right)^n \quad (10)$$

where $\Delta\varepsilon_{\theta c}$ is the corrected incremental creep deformation (-, non-dimensional), $\Delta\varepsilon_m$ is the measured incremental creep deformation (-), $\sigma_{\theta c}$ is the constant hoop stress (MPa) used for normalizing the creep strain data, $\sigma_{\theta a,v}$ is the average hoop stress (MPa) during the test period, E is Young's modulus (MPa), and a and n are constants to be determined. The hoop stresses used when normalizing the creep strain data are presented in Table 5.

The creep deformation is given by adding the incremental terms calculated for each separate testing period.

4. The creep data are analyzed by separating the primary (transient) and secondary (steady-state) creep. First, the steady-state creep rate is determined from the data that seem to have attained the steady state. Then the steady-state creep rate is extrapolated to $t = 0$ to calculate the primary creep strain.

5. The predicted secondary creep rates are compared with the measured values by determining the quotients between predicted and measured values. For a specific material, the standard deviation of these quotients is divided by the average value of the same quotients, and this value is designated by StdDev/Av. The values of the constants a and n are chosen to minimize the value of StdDev/Av. If the new values of a and n are different from the ones used in point 3, we return to point 3. On the other hand, if the new values of a and

TABLE 5—*Stresses used when normalizing the measured creep strains.*

Temperature, °C	Minimum Hoop Stress, MPa			
	80	120	140	160
330	...	127	...	...
360	...	127	...	...
385	83	127	151	173
400	...	127	...	...

NOTE: Values presented in the matrix represent hoop stresses, MPa, used when normalizing the measured creep strains.

n equal the values used in point 3; we have determined the stress dependency for the considered cladding material.

The stress dependency constants presented in Table 6 were determined according to the procedure described above.

Temperature Dependency

Like most creep models, the Matsuo model expresses the temperature dependency by an Arrhenius type expression (Eq 3). In addition, Matsuo includes the temperature in the linear term (the E/T term in Eq 3) and in the hyperbolic-sine function by including Young's modulus.

The activation energy, Q, for each separate material was determined by plotting $\ln \dot{\varepsilon}_s + \ln \frac{T}{E} - n \cdot \ln [\sinh(a \cdot \sigma_\theta/E)]$ versus $1/T$.

Figure 2 is an example of least square curve fitting used to determine the activation energy. Table 6 presents the activation energies determined for each separate material.

The data in Table 6 result in an average activation energy of 197 kJ/mol with a standard deviation of 8 kJ/mol. Apart from the materials presented in this paper, the thermal creep test included additional Zircaloy materials and zirconium-based alloys with chemical compositions outside the ranges specified by ASTM for Zircaloy-2 and Zircaloy-4. Including those materials, an average activation energy of 201 kJ/mol and a standard deviation of 7 kJ/mol is calculated. Accordingly, the variations in the measured activation energy are small, and an activation energy of 201 kJ/mol was chosen.

TABLE 6—*Values of the constants in the expression for creep stress dependency* (a, n) *and creep activation energies* (Q) *for the different cladding materials.*

ABSS Lot No.	Clad Type	Final Heat Treatment	Degree of Recrystalization %	Sn, wt%	O, ppm	a (-)	n (-)	Q, kJ/mol
81408	SRA Zr-4	486°C/3.5 h	4	1.34	1340	650	2.0	185
81416	SRA Zr-2	486°C/3.5 h	12	1.36	1295	650	2.0	205
81428	PRXA Zr-2	502°C/3.5 h	49	1.37	1245	650	2.3	199
81417	RXA Zr-2	565°C/1.5 h	100	1.37	1215	650	3.5	198

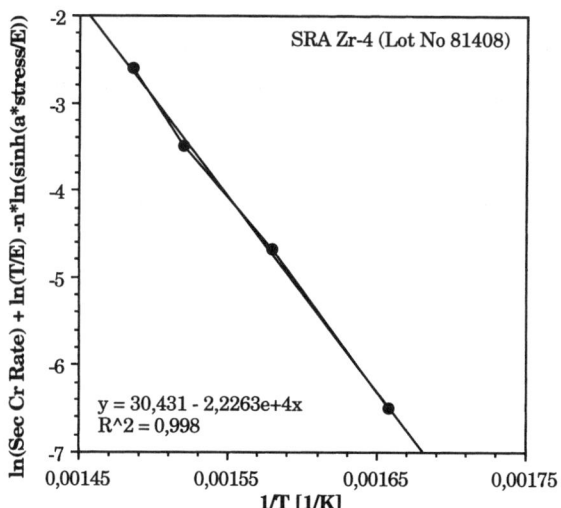

FIG. 2—*Determination of creep activation energy for material from Lot 81408. An activation energy of 185 kJ/mol is calculated from 22 263 × 8.314 = 185 kJ/mol.*

The Value of the Parameter A

After determining the values of the constants in the stress-dependency factor (a and n) and the activation energy, the only constant value in the expression for steady-state creep rate that remains to be determined is the value of the constant A in Eq 3.

The value of the Parameter A has been determined by adjusting A until a best fit is achieved, that is, for each material the value of A has been tuned until the average value of the quotients between predicted and measured secondary creep strain rates from all the tests equals 1.0. Table 7 presents the value of A for the different materials. In addition, Table 7 gives the standard deviation (s) for the considered quotients between predicted and measured creep rates.

The Expression for Primary Creep

The saturated primary creep and the primary creep as a function of time were given in Eq 4 and as the first term of Eq 5, respectively. The unknown parameters in the expressions for primary creep are b, B, and C. In a first step, the values of the constants b and B were determined

TABLE 7—*The value of the constant A for the different cladding materials and the standard deviation of the quotients between predicted and measured steady-state creep rates. Note that an activation energy of 201 kJ/mol has been used for all cladding materials.*

ABSS Lot No.	Clad Type	Final Heat Treatment	Degree of Recrystalization, %	Sn, wt%	O, ppm	A, K/MPa/h	s (-)
81408	SRA Zr-4	486°C/3.5 h	4	1.34	1340	1.08e9	0.145
81416	SRA Zr-2	486°C/3.5 h	12	1.36	1295	1.06e9	0.110
81428	PRXA Zr-2	502°C/3.5 h	49	1.37	1245	7.06e8	0.086
81417	RXA Zr-2	565°C/1.5 h	100	1.37	1215	5.47e8	0.044

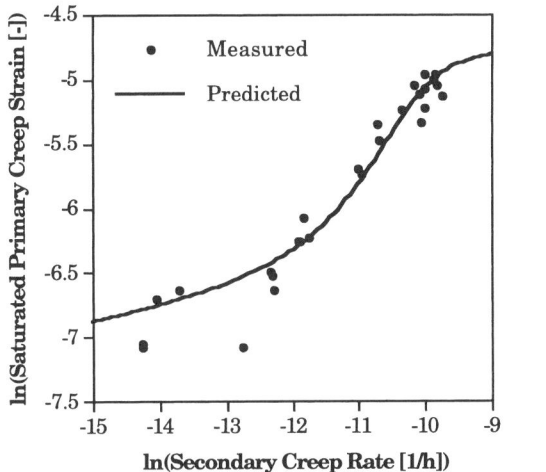

FIG. 3—*Relationship between saturated primary creep strain and steady-state (secondary) creep rate for the revised Matsuo correlation.*

by plotting $\ln \varepsilon_p^s$ versus $\ln \dot{\varepsilon}_s$ and making least square fits to the data. However, it was concluded that Eq 4 had to be modified to describe the behavior of our data. Accordingly, an additional factor was included in the expression for saturated primary creep strain:

$$\varepsilon_p^s = B \cdot \dot{\varepsilon}_s^b \cdot [2 - \tanh(D \cdot \dot{\varepsilon}_s)]^d \quad (11)$$

where the values of the factors B (0.0216 h^b) and b (0.109) were chosen in accordance with the Matsuo model [2], and the values of the Constants D (35 500 h) and d (-2.05) were determined through curve fitting.

Constant C in Eq 5 determines the time dependency of the primary creep. As mentioned by Matsuo [2], C is a complex function of the test conditions (temperature and stress), but it will not be a cause of any large error if a constant value is used for C. The value used here is the same as the one used by Matsuo [2], i.e., $C = 52$.

Figure 3 shows measured and predicted relations between saturated primary creep strain and secondary creep rate. As mentioned above, in the region of secondary creep, there is a balance between the rate of strain hardening and the rate of thermal recovery. The point of balance is dependent on the applied stress and the temperature. The relation between saturated primary creep strain and secondary creep rate, presented in Fig. 3, indicates that the point of balance is reached when a certain amount of strain hardening has been obtained from primary creep, that is, the saturated primary creep strain can, as in Eq 11, be expressed as a function of the secondary creep rate.

Measured Versus Predicted Creep Strains

Figures 4a to 4d show plots of measured versus predicted creep strains for the four different cladding lots. The figures present all data points, i.e., from all seven combinations of temperature and stress (see Table 3). The model used is described by Eqs 3, 5, 6, and 11. In the expression for secondary creep rate (Eq 3), the activation energy is set to 201 kJ/mol, while the values of the constants A, a, and n are presented in Tables 6 and 7. The constants in the equations that describe the primary creep (Eqs 5 and 11) were presented above.

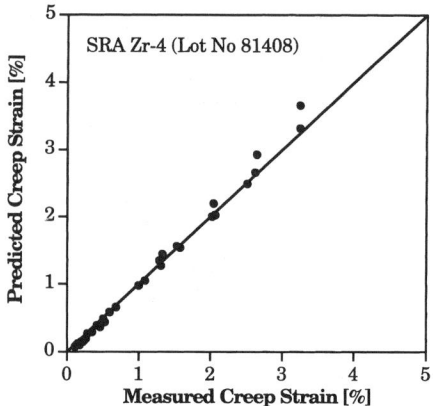

FIG. 4a—*Predicted versus measured thermal hoop creep strains for the SRA Zircaloy-4 cladding (Lot 81408).*

From the results presented in Figs. 4a to 4d, it is concluded that the predicted creep strains are in good agreement with the measured values.

Analysis of the In-Reactor Creep Tests

Modifications of the Thermal Creep Model

Irradiation hardening affects the thermal creep of the Zircaloy cladding. The hardening can be described as irradiation-induced defect structures, such as vacancy and interstitial clusters, vacancy and interstitial loops, and damage or depleted zones, which can act as obstacles to dislocation motion [1]. When applying the thermal creep model, presented above, to in-reactor conditions, the irradiation-hardening effect on creep strength is modeled by an evolution of the parameter a, with the fast fluence:

$$a_i = a \cdot \{1 - A_1 \cdot [1 - \exp(-A_2 \cdot \Phi^{A_3})]\} \tag{12}$$

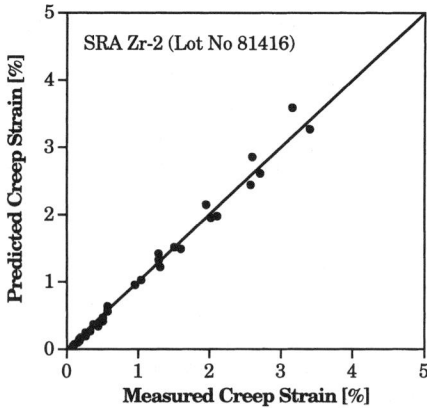

FIG. 4b—*Predicted versus measured thermal hoop creep strains for the SRA Zircaloy-2 cladding (Lot 81416).*

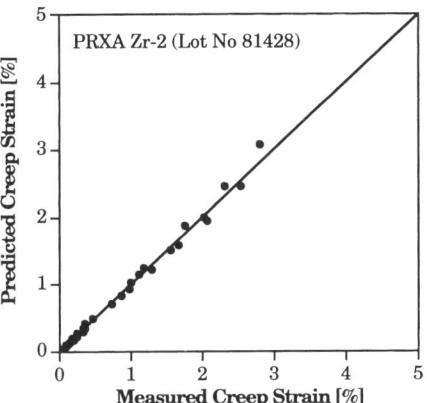

FIG. 4c—*Predicted versus measured thermal hoop creep strains for the PRXA Zircaloy-2 cladding (Lot 81428).*

where, a_i is the in-reactor value of the parameter a in Eq 3, Φ is the fast neutron (>1 MeV) fluence (n/cm^2), and A_1, A_2, and A_3 are constants.

This type of irradiation-hardening model has been used by Hoppe [7]. However, the constant A_3 has been introduced in this study.

Irradiation-Induced Creep

The model for irradiation-induced creep is based on the model presented by Hoppe [7], that is, the irradiation creep strain rate is simply related to the current fast flux and stress:

$$\dot{\varepsilon}_i = C_0 \cdot \phi^{C_1} \cdot \sigma_\theta^{C_2} \tag{13}$$

where $\dot{\varepsilon}_i$ is the irradiation-induced creep rate (1/h), ϕ is the fast neutron (>1 MeV) flux (n/m^{2}s), σ_θ is the hoop stress (MPa), and C_0, C_1, and C_2 are constants.

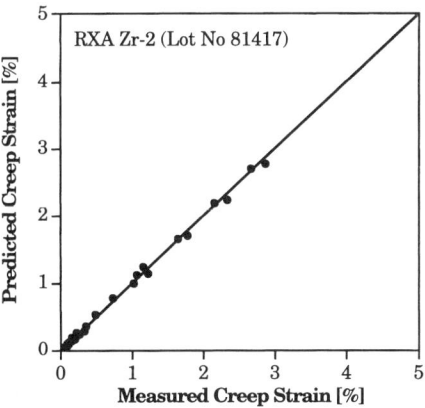

FIG. 4d—*Predicted versus measured thermal hoop creep strains for the RXA Zircaloy-2 cladding (Lot 81417).*

TABLE 8—*The values of the constants used when applying the creep model to in-reactor conditions.*

Clad Type	A_1 (-)	A_2, (n/cm^2)$^{-A_3}$	A_3 (-)	C_0 (n/m^{2}s)$^{-C_1}$ (MPa)$^{-C_2}$/h	C_1 (-)	C_2 (-)
SRA	0.56	1.4e–27	1.3	3.557e–24	0.85	1.0
RXA	0.56	1.4e–27	1.3	1.654e–24	0.85	1.0

Equations 4, 11, and 5 show that the primary creep strain can be expressed as a function of secondary creep rate. When modeling the creep under in-reactor conditions, the secondary creep rate used in Eqs 5 and 11 is the sum of the thermal secondary creep rate, from Eq 3, and the irradiation-induced creep rate, from Eq 13.

Calibration of the In-Reactor Creep Model

The in-reactor creep model has been calibrated against data for SRA and RXA claddings presented by Franklin et al. [*1*]. For this calibration Eqs 3, 5, 6, 11, 12, and 13 were implemented in the Microsoft program Excel, and the required input data, fast flux, cladding stress, cladding temperature, and irradiation time were all taken from Ref *1*. The outcome was the calculated hoop strain versus time, and the data have been evaluated by analyzing the predicted (*P*) over measured (*M*) value of hoop strain. The values of the constants in Eqs 12 and 13, as they are calibrated to fit the data presented by Franklin et al. [*1*], are presented in Table 8.

The predicted creep strains for SRA and RXA claddings are plotted versus measured creep strains in Figs. 5a and 5b, respectively. The data set depicted in these figures covers hoop stresses between 70 and 105 MPa, fast neutron fluxes from 3.7 to 23 × 10^{20} n/m^2 h, temperatures in the region of 297 to 310°C, and irradiation times between 10 000 and 23 000 h.

Summary of the Analysis

The creep model presented in this paper consists of two main parts, primary and secondary creep. The saturated primary creep strain is a function of the secondary creep rate according

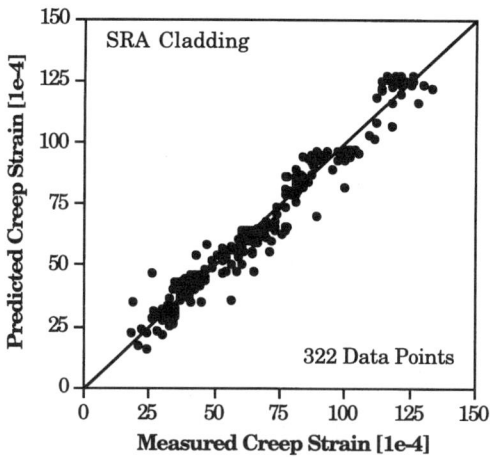

FIG. 5a—*Predicted versus measured in-reactor hoop creep strains for the SRA claddings.*

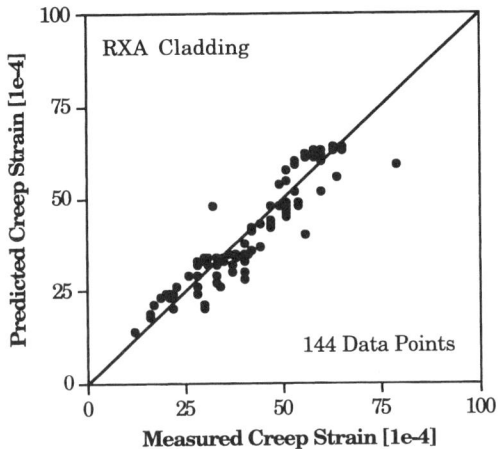

FIG. 5b—*Predicted versus measured in-reactor hoop creep strains for the RXA claddings.*

to Eq 11, and the time dependence of the saturation of the primary creep is given by the first term of Eq 5. The thermal steady-state creep is given by Eq 3. When applied to in-reactor conditions, the secondary creep rate is the sum of the thermal steady-state creep and an athermal, irradiation-induced term. More precisely, under in-reactor conditions the thermal secondary creep is given by Eq 3 combined with Eq 12, and the total steady-state creep is determined by adding the irradiation-induced secondary creep, given by Eq 13, to the thermal creep.

Discussion

The materials used in the thermal creep tests covered in this study are fabricated with final annealing temperatures from 486 to 565°C, while the in-reactor data cover materials manufactured with final annealing temperatures of either 485 or 579°C. The thermal creep tests presented here cover test temperatures from 330 to 400°C and hoop stresses between 80 and 160 MPa, while temperatures between 297 and 310°C and hoop stresses from 70 to 105 MPa are covered in the in-reactor tests. The duration of the thermal creep tests were 480 or 960 h, while the in-reactor tests covered irradiation periods of 10 000 to 23 000 h.

Figure 6 shows thermal and irradiation-induced creep rates as functions of temperature for a hoop stress of 100 MPa. The fast neutron flux is 2×10^{21} n/m² h, and three different fast neutron fluences are considered 0, 5×10^{20}, and 4×10^{21} n/cm². The figure shows how the thermal creep rate decreases with increasing fast neutron fluence and thereby how the threshold temperature between irradiation-induced creep and thermal creep increases with irradiation as the irradiation-induced hardening decreases the thermal creep rate.

Figures 7a and 7b show calculated thermal secondary creep rates and saturated primary creep strains as functions of hoop stress at 385°C. From these figures it is concluded that for moderate hoop stresses (<120 MPa) the secondary creep rate and the saturated primary creep strain decrease with increasing final annealing temperature (increasing degree of recrystallization), that is, in this stress region the SRA and the RXA claddings obtain the largest and smallest creep deformations, respectively. For higher stresses, the steady-state creep rate as well as the saturated primary creep strain have minima for PRXA claddings, and, for stress levels in excess of 140 MPa, the RXA material has the highest creep rate and the largest saturated primary creep strain. Since the temperature dependence of creep is similar for the claddings included

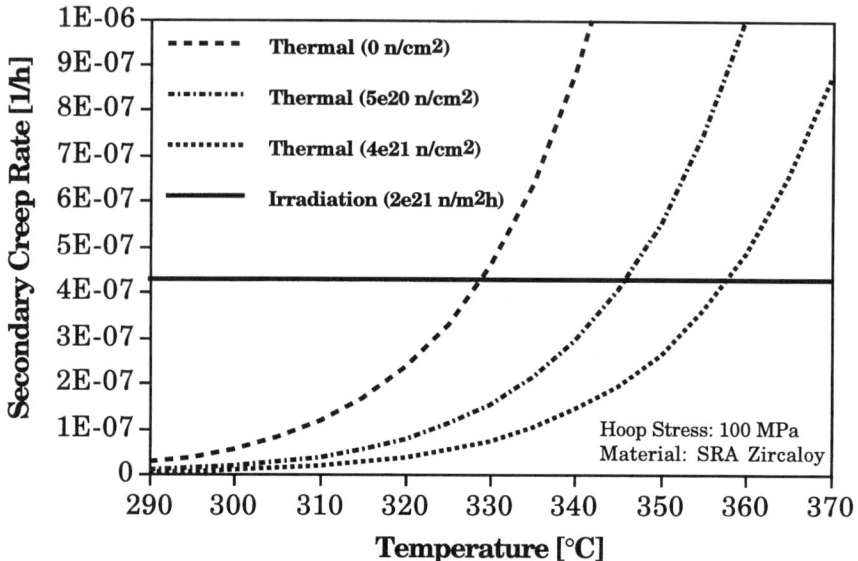

FIG. 6—*Comparison between thermal and irradiation-induced secondary creep rates as functions of temperature.*

in this study, the stress dependencies observed at 385°C are, at least, valid in the temperature region covered by the thermal creep tests (330 to 400°C).

The effects of the final annealing temperature on cladding creep properties have been studied by others [8–11].

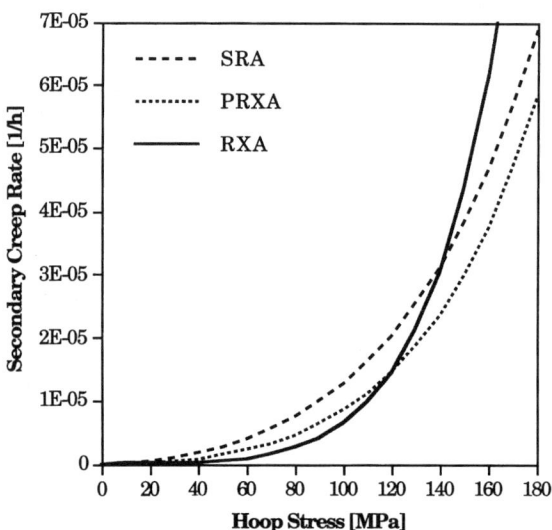

FIG. 7a—*Calculated thermal secondary creep rates as functions of hoop stresses at 385°C.*

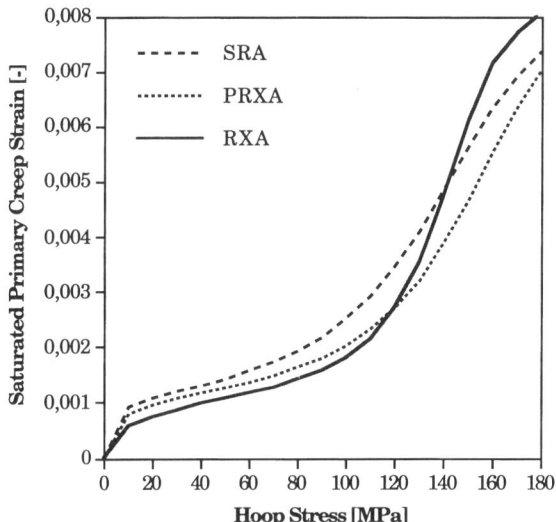

FIG. 7b—*Calculated thermal saturated primary creep strains as functions of hoop stresses at 385°C.*

Frenkel and Weisz evaluated the transverse creep strength of Zircaloy-4 at 400°C using cladding materials fabricated with final annealing temperatures in the range of 440 to 650°C [8]. They observed that, for low and intermediate hoop stresses (lower than 133 MPa), the transverse creep strength increases as the final annealing temperature increases from 440 to 650°C. They did, however, observe that this effect can be obscured when using too high creep stresses that can be, above a certain final annealing temperature, of the order of the yield stress. They also stated that in-reactor tests confirmed the results from their thermal creep tests [8].

Källström et al. investigated the effects of cold work and final annealing on the creep properties of Zircaloy-4 claddings [9]. The claddings were cold worked 50 or 80% to a series of textures and final annealed at between 475 and 575°C. The tube specimens were creep tested at 400°C under internal pressure, resulting in hoop stresses of about 152 MPa. Källström et al. concluded that annealing at 510 to 520°C, in the range of partial recrystallization, gives maximum transverse creep strength [9].

Stehle et al. compared the creep performance of Zircaloy-4 claddings manufactured with four different final annealing sequences, either 450, 500, 550, or 600°C, for 2.5 h [10]. The experimental conditions covered temperatures between 320 and 400°C, uniaxial stresses from 75 to 127 MPa, and biaxial stress conditions (internal pressurization) with hoop stresses in the region of 109 to 173 MPa. They observed that, whereas the primary creep decreases as the final annealing temperature increases, the steady-state creep rate increases at stress levels approaching the yield strength. They concluded that the steady-state creep rate has a minimum for moderately stress-relieved material. As a consequence, they stated that short- and long-time creep experiments can lead to an arbitrary assessment of creep resistance, and in long-time creep the stress-relieved condition is the condition with the lowest strain [10].

Murty et al. studied the biaxial creep of Zircaloy in the cold-worked, stress-relieved, and recrystallized conditions in terms of loci at constant creep dissipation [11]. The difference in the deformation behavior of the two types of materials were concluded to arise from changes in slip mode, basal slip and prism slip being dominant in the stress-relieved and recrystallized materials, respectively. The dominance of basal slip in the stress-relieved material was believed to be due to the hardening of the prismatic planes during fabrication [11].

It can be concluded that the present study elucidates the effect of the final heat treatment on the stress dependency of the cladding hoop creep strength, and with one exception the results presented here are in accordance with the studies presented by Frenkel and Weisz [8], Källström et al. [9], and Stehle et al. [10], the exception being the saturation of the primary creep strain. Stehle et al. conclude that the primary creep decreases as the final annealing temperature increases independent of the actual stress level [10]. However, according to our data and the model presented above, the saturated primary creep strain follows the stress dependency of the secondary creep rate. The difference between the two observations may be explained by differences in cladding texture. The final pilgering step was made using different degrees of reduction, 80% in this study, and 65% in the study of Stehle et al. [10]. When the degree of reduction decreases, the temperature required for recrystallization increases, and consequently the RXA cladding used by Stehle et al. has been final annealed at a higher temperature (600°C) than the RXA cladding used in this study (565°C). During recrystallization some changes of the texture can occur, for example, rotation of the crystals around the basal poles and concentration of the basal poles towards the radial direction. These textural changes are enhanced by higher final annealing temperature, and for the RXA cladding used by Stehle et al., the basal poles are concentrated towards the radial direction of the cladding tube [10]. On the other hand, the basal pole orientations of the RXA, PRXA, and SRA claddings used in this study are similar, $\phi_{max} = 30 \pm 2°$ (see Table 1b). Hence no concentration of the basal poles towards the radial direction has occurred during the final heat treatment of the RXA cladding covered here. Accordingly, due to the difference in final reduction and the different final annealing temperatures, there are substantial differences between the texture of the RXA claddings used in this study and in the work presented by Stehle et al. [10], respectively, and these differences affect the creep behavior.

Conclusions

1. The Matsuo thermal creep model, which accounts for both primary and secondary creep, has been modified, calibrated, and verified for three different types of Zircaloy, stress relief annealed (SRA), partially recrystallization annealed (PRXA), and recrystallization annealed (RXA). Based on in-reactor data, the thermal creep model has been extended to cover also the creep behavior under irradiation. The predicted creep strain rates, out-of-reactor as well as in-reactor, are in good agreement with the measured creep data.

2. In the considered stress-temperature space (80 to 160 MPa, 330 to 400°C), the temperature dependence of creep is similar for the claddings included in this study, and the dependence is characterized by a constant activation energy of 201 kJ/mol.

3. For moderate hoop stresses (<120 MPa) and temperatures between 330 and 400°C, the thermal creep tests in combination with the model analyses show that the secondary creep rate and the saturated primary creep strain decrease with increasing final annealing temperature (increasing degree of recrystallization), that is, in this region of stress-temperature space, the SRA and the RXA claddings obtain the largest and smallest creep deformations, respectively. For higher stresses, the steady-state creep rate as well as the saturated primary creep strain have minima for PRXA claddings, and for stress levels in excess of 140 MPa, the RXA material has the highest creep rate and the largest saturated primary creep strain.

4. The thermal creep tests show that the creep performance of Zircaloy-2 and Zircaloy-4 are similar when the chemical compositions and the fabrication procedures are similar.

5. The in-reactor data, which cover hoop stresses between 70 and 105 MPa, confirm the

thermal creep data; i.e., for these stress levels, the creep rate is lower for RXA materials than for SRA claddings.

Acknowledgments

We especially want to mention Ali Massih, of ABB Atom, for initializing the study and for continuous inspiration, and Jan Gunnarsson, of AB Sandvik Steel, who carried out the thermal creep tests.

References

[1] Franklin, D. G., Lucas, G. E., and Bement, A. L. in *Creep of Zirconium Alloys in Nuclear Reactors, ASTM STP 815,* American Society for Testing and Materials, West Conshohocken, PA, 1983.
[2] Matsuo, Y., "Thermal Creep of Zircaloy-4 Cladding under Internal Pressure," *Journal of Nuclear Science and Technology,* Vol. 24, No. 2, February 1987, pp. 111–119.
[3] Charquet, D., Steinberg, E., and Millet, Y., "Influence of Variations in Early Fabrication Steps on Corrosion, Mechanical Properties, and Structure of Zircaloy-4 Products," *Zirconium in the Nuclear Industry (Seventh International Symposiums), ASTM STP 939,* American Society for Testing and Materials, West Conshohocken, PA, 1987, pp. 431–447.
[4] Thorvaldsson, T., Andersson, T., Wilson, A., and Wardle, A., "Correlation Between 400°C Steam Corrosion Behavior, Heat Treatment, and Microstructure of Zircaloy-4 Tubing," *Zirconium in the Nuclear Industry (Eighth Symposium), ASTM STP 1023,* American Society for Testing and Materials, West Conshohocken, PA, 1989, pp. 128–140.
[5] Gross, J. P. and Wadier, J. F., "Precipitate Growth Kinetics in Zircaloy-4," *Journal of Nuclear Materials,* Vol. 172, 1990, pp. 85–96.
[6] Garofalo, F., "Fundamentals of Creep and Creep-Rupture in Metals," Macmillan Co., New York, 1965.
[7] Hoppe, N. E., "Engineering Model for Zircaloy Creep and Growth," *Proceedings,* ANS-ENS International Topical Meeting on LWR Fuel Performance, Avignon, France, 21–24 April 1991, pp. 201–209.
[8] Frenkel, J. M. and Weisz, M., "Effect of the Annealing Temperature on the Creep Strength of Cold-Worked Zircaloy-4 Cladding," *Zirconium in Nuclear Applications, ASTM STP 551,* American Society for Testing and Materials, West Conshohocken, PA, 1974, pp. 140–144.
[9] Källström, K., Andersson, T., and Hofvenstam, H., "Creep Strength of Zircaloy Tubing at 400°C as Dependent on Metallurgical Structure and Texture," *Zirconium in Nuclear Applications, ASTM STP 551,* American Society for Testing and Materials, West Conshohocken, PA, 1974, pp. 160–168.
[10] Stehle, H., Steinberg, E., and Tenckhoff, E., "Mechanical Properties, Anisotropy, and Microstructure of Zircaloy Canning Tubes," *Zirconium in the Nuclear Industry (Third International Conference), ASTM STP 633,* American Society for Testing and Materials, West Conshohocken, PA, 1977, pp. 486–507.
[11] Murty, K. L., Mahmood, S. T., and Adams, B. L., "Upper and Lower Bound Plastic Modelling of Creep Loci of Zircaloy," *Zirconium in the Nuclear Industry: Seventh International Symposium, ASTM STP 939,* American Society for Testing and Materials, West Conshohocken, PA, 1987, pp. 120–135.

DISCUSSION

N. Christodoulou[1] (written discussion)—Was a subset of data used for the derivation of the equation constants and then the equation used to predict all the data? Or were all the data used for the derivation of constants? If the latter, then the exercise is a fit, not a prediction.

M. Limbäck and T. Andersson (authors' closure)—This work can be divided into two sections, out-of-reactor creep and in-reactor creep. The data used to model the out-of-reactor, or thermal, creep covers a matrix with seven different combinations of temperature and stress. The idea behind using such a database is to cover fairly large regions of temperature and stress and to derive the temperature and stress dependencies that are valid within the considered stress-temperature space. That is, all available data from the out-of-reactor tests were used to derive the constants in the thermal creep model. In addition, if the model is a good mechanistic model, one could expect that it should be valid not only in the stress-temperature space covered by the data, but also in a somewhat larger stress-temperature space.

The in-reactor data used in this study are from one single, but quite extensive, series of creep tests. The data resulting from these creep tests are consistent, and dividing the data into one data set that is used for calibration and another set that is used for validation of the model would not be any more correct than to use all available data when calibrating the model. Consequently, all available data from the considered in-reactor creep test were used when calibrating the model. Finally, it should be mentioned that after the completion of this work the model has been further validated against additional in-reactor creep data, and the results from this work have been encouraging.

N. Christodoulou[1] (written discussion)—What is the physical basis for using a flux exponent of 0.85 and not 1?

M. Limbäck and T. Andersson (authors' closure)—The values of the flux exponent presented in the open literature typically vary from 0.25 to 1.0, and, for the creep data considered in our study, a flux exponent of 0.85 results in good agreement between calculated and measured creep strains. In addition, Dollins has presented a theory where the flux dependency varies with the actual flux such that the exponent is 0.25 at low fluxes (10^{12} n/cm^2 s), increases with increasing flux, and reaches an asymptotic value of 1.0 at high fluxes (10^{14} n/cm^2 s) [C. C. Dollins, "Irradiation Creep Associated with Dislocation Climb," *Radiation Effects*, Vol. 11, 1971, pp. 123–131]. This theory is based on the assumption that edge dislocations are pinned by Seeger zones. A Seeger zone emanates from damages to the lattice structure caused by fast neutron irradiation and consists of a volume depleted of matrix atoms. For the fast neutron flux levels covered by the data considered in our study (1×10^{13} to 6.4×10^{13} n/cm^2 s, with 90% of the data in the region 3×10^{13} to 6.4×10^{13} n/cm^2 s), a flux exponent of 0.85 is in fairly good agreement with the flux exponent resulting from the work by Dollins.

---

[1] Atomic Energy of Canada, Chalk River, Ontario, Canada.

C. K. Chow,[1] C. E. Coleman,[2] M. H. Koike,[3] A. R. Causey,[2] C. E. Ells,[2] R. R. Hosbons,[2] S. Sagat,[2] V. F. Urbanic,[2] and D. K. Rodgers[2]

Properties of an Irradiated Heat-Treated Zr-2.5Nb Pressure Tube Removed From the NPD Reactor

REFERENCE: Chow, C. K., Coleman, C. E., Koike, M. H., Causey, A. R., Ells, C. E., Hosbons, R. R., Sagat S., Urbanic, V. F., and Rodgers, D. K. **"Properties of an Irradiated Heat-Treated Zr-2.5Nb Pressure Tube Removed From the NPD Reactor"** *Zirconium in the Nuclear Industry: Eleventh International Symposium, ASTM STP 1295,* E. R. Bradley and G. P. Sabol, Eds., American Society for Testing and Materials, 1996, pp. 469–491.

ABSTRACT: Some pressure tubes in reactors moderated by heavy water have been made from heat-treated (HT) Zr-2.5Nb. One such tube was removed from the NPD nuclear reactor after 20 years of operation. An extensive program was carried out jointly by AECL and PNC to evaluate the condition and properties of this pressure tube. The investigations include irradiation creep, tensile, corrosion, delayed hydride cracking (DHC), fatigue, and fracture properties.
  Results show that: (1) the in-reactor elongation rate is much lower and the transverse strain rates are slightly larger than in cold-worked (CW) Zr-2.5Nb tubes; (2) the tensile properties, hydrogen pickup, threshold stress intensity factor for DHC initiation, DHC velocity, and fatigue crack growth rates were similar to those of the CW Zr-2.5Nb material; (3) the fracture toughness of this tube, as measured by curved compact toughness specimens and burst tests, is slightly higher than the CW tubes. The results were also compared with other heat-treated Zr-2.5Nb materials irradiated in the Fugen reactor.
  The tube was in excellent condition when removed from the reactor and would have been satisfactory for further service.

KEYWORDS: heat-treated Zr-2.5Nb, irradiation deformation, creep, oxidation, deuterium, delayed hydride cracking, DHC, tensile properties, fatigue, fracture toughness, pressure tube

The use of Zr-Nb alloys in water-cooled power reactors was first suggested in Russian papers [1,2] at the 1958 Geneva conference. From these papers, and probably from private conversations with the Russian delegates, W. B. Lewis decided that a development program at Atomic Energy of Canada Limited (AECL) would concentrate on the composition Zr-2.5Nb. From the Russian work, it was clear that the material in a heat-treated (HT) condition could have a significantly higher tensile strength than the cold-worked (CW) Zircaloy-2 then in use for pressure tubes in CANDU®[4] reactors. Thus, with thinner tubes an improvement in neutron economy could be attained. In this paper, we use the term "HT Zr-2.5Nb" to describe the metallurgical condition resulting from the sequence of: extrusion, solution treatment in the α

---
[1] AECL, Whiteshell Laboratories, Pinawa, Manitoba, R0E 1L0, Canada.
[2] AECL, Chalk River Laboratories, Chalk River, Ontario, K0J 1J0, Canada.
[3] Power Reactor and Nuclear Fuel Development Corporation, O-Arai Engineering Centre, 4002, O-Arai, Higashi-Ibaraki, Ibaraki-Ken, 311-13, Japan.
[4] CANDU is a registered trademark of Atomic Energy of Canada, Ltd.

+ β phase, water quenching, cold working about 15%, aging 24 h at 500°C. The normal final autoclaving at 400°C has little effect on properties of the bulk tube. All of the pressure tubes in the KANUPP, Fugen, and Gentilly-1 reactors are in this metallurgical condition. A few full-diameter tubes were irradiated in NRU at AECL Chalk River Laboratories, WR-1 reactor at AECL Whiteshell Laboratories, SGHWR in the U.K., and one tube, the subject of this paper, in the NPD reactor. The tubes in the KANUPP and Fugen reactor have seen extensive service, those in the Gentilly-1 reactor less so but for reasons not associated with the tubes.

Early in the laboratory tests, the quench and age heat treatment described above was established [3]. When production of test batches of tubes was started, it was found that cold-working between the quench and age was required to achieve the required dimensional tolerances, and with this cold-work some additional benefit in corrosion resistance was derived. With production of HT tubes and associated irradiation programs underway, AECL diverted some attention to the production of tubes of CW Zr-2.5Nb. It was felt that tubes in the latter metallurgical condition could be fabricated with greater confidence in the uniformity of their mechanical properties and that they would be produced at lower cost.

Leading to the specification of the Zr-2.5Nb tubes for Pickering-3 (and subsequent CANDU reactors), there was considerable debate within AECL and Ontario Hydro on the relative merits of the CW and HT conditions. The choice of HT Zr-2.5Nb tubes for the KANUPP reactor was made at a time when it was felt that the HT tubes would be superior to Zircaloy-2, and the development of the HT tubes was further advanced than the CW Zr-2.5Nb tubes. Alternatively, the choice for the CANDU-BLW Gentilly-1 reactor was heavily influenced by the reactor physics requirement to have the mass of neutron parasitic material in the reactor core as small as possible. The Fugen reactor core was being designed by Power Reactor and Nuclear Fuel Development Corporation (PNC) [4] at the same time as Gentilly-1, and there was close contact between the two groups of designers. Largely because of this association, it was logical for PNC to follow the same fabrication route for their pressure tubes. When the final choice for Pickering-3 was made, it was possible to specify the CW condition with some confidence.

The work needed to install both a CW and a HT tube in the NPD reactor was started in 1965, and the decision to do this led to the irradiation which is the subject of this paper. The HT tube was installed in 1967 and remained in the NPD reactor until its final shutdown in 1987 [5]. The tube, identified as Tube 589, was located in Position H-06 in the NPD reactor, installed in May 1967 and removed in January 1990. The horizontal fuel channels were about 4 m long, with an inside diameter of 82.5 mm and a wall thickness of 2.13 mm. The ingot analysis of the Zr-2.5Nb used indicated a Nb concentration of 2.48 wt% and 0.099 wt% O. The fabrication process was:

1. Extruded at 840°C.
2. Solution treated at about 870°C, water quenched.
3. Cold drawn about 15%.
4. Tempered at 500°C for 24 h.
5. Machined and pickled.
6. Autoclaved 72 h at 400°C.

This procedure produced a mixture of primary-α grains in a matrix of martensite and a more random texture than the CW materials [3,6], which has a concentration of basal normals in the transverse direction. During operation in the NPD reactor, the temperature of the inlet and outlet of the primary heat transport system was 252 and 280°C, respectively, while the hoop stress was about 132 MPa. The calculated neutron fluence was roughly cosine-shaped along the tube with a maximum value of about 4.7×10^{25} n/m² close to the center of the channel. All fluence values reported are for $E > 1$ MeV. Throughout the irradiation period, the pH (D) control of the primary coolant varied between use of LiOD and ND_3.

The removal of the HT Zr-2.5Nb pressure tube allowed a full evaluation of this material after a long irradiation under power reactor operating conditions. The main interests were the factors that control pressure tube lifetime deformation and fracture. During operation, a pressure tube elongates and its diameter increases. It sags when used in the horizontal mode. These deformations must be low enough to be accommodated by the reactor structure and do not lead to operational problems with the fuel. Fracture is avoided by eliminating flaws and other stress concentrators and minimizing the chance of crack initiation. The most likely cracking mechanisms are fatigue and delayed hydride cracking (DHC). To avoid the former requires knowledge of stress cycles, while the latter requires information on the concentration of hydrogen isotopes to ensure that hydrides cannot form at flaws. If a flaw does propagate, the reactor needs to be protected by establishing that leak before break is valid; thus, one also needs information on crack growth rate and the critical crack length.

In this paper, we describe the measurements and tests done on the HT Zr-2.5Nb Pressure Tube 589 removed from the NPD reactor and place measurements from Fugen and its surveillance specimens in the context of the NPD tube. An unirradiated pressure tube identified as Tube 585 with similar production history was tested to compare with the irradiated tube. Whenever possible, the properties of the HT tube were compared with a CW Zr-2.5Nb tube, identified as Tube 582, installed in the NPD reactor at the same time. The fabrication process for CW Zr-2.5Nb tubes was: extruded at 840°C, cold-drawn 25%, and autoclaved 24 h at 400°C.

Experimental Details

Irradiation Deformation

The length and inside diameter of the pressure tube were measured periodically to determine the axial and transverse strain rates. Diameter measurements were taken when the tube was installed in the reactor (0 h) and eight times during the irradiation period. Length measurements were started after the tube had been installed for 5500 h and seven more times subsequently.

The length was measured by inserting a gaging tube inside the pressure tube and using a boroscope to photograph the ends of the pressure tube with respect to the scales scribed on the gauging tube. The accuracy of the length measurements is about ±0.25 mm. Inside diameters are measured by passing an LVDT-based gaging tool through the empty tube. Six readings taken at 30° intervals around the tube are averaged to give one diameter every 12.7 mm of axial travel [7]. The reproducibility of the inside-diameter measurements is consistent from gaging to gauging, and the measurement errors due to misalignment, temperature, and transient conditions are estimated to be less than ±0.01 mm.

Tensile Properties

Two types of tension specimens were used in the present study on the irradiated tube. The longitudinal specimens had a gage length of 12.7 mm, width of 3.7 mm, and the full wall thickness (2.13 mm). They were pulled at a nominal strain rate of 3×10^{-4}/s. Transverse properties were evaluated from 6.7-mm-wide rings cut from the tube and pulled in a yoke grip with a crosshead speed of 0.21 mm/s. With this yoke technique, the complication either of flattening the specimens or using miniature specimens is avoided, but the only reliable tensile properties obtained are ultimate tensile stress (UTS) and reduction in area.

Corrosion and Deuterium Ingress

The corrosion and deuterium concentrations were measured on ring specimens taken from six axial locations along the tube. Oxidation of both the inside and outside surfaces was measured metallographically at four locations around the circumference. Deuterium concentrations were measured on 500-mg pellets hydraulically punched from areas adjacent to those selected for metallographic examination.

DHC

Notched cantilever beam specimens [8] with a crack on the radial-longitudinal plane and the direction of crack propagation in the radial direction were prepared at 2.66 to 2.71 m from the inlet (west) end of the tube. This section of the tube was irradiated to a fluence of about 4.5×10^{25} n/m² at a temperature of 270°C. The total hydrogen isotope concentration in the tube was between 0.175 and 0.23 at% (19 and 25 µg/g hydrogen equivalent). Tests on unirradiated material have been performed on Pressure Tube 585. The total hydrogen isotope concentration in this section was between 0.19 and 0.23 at% (21 and 25 µg/g hydrogen equivalent).

The DHC velocity is sensitive to the temperature history; the maximum value is attained by cooling from a temperature higher than the solvus temperature of hydrogen in the specimen. Thus, a standard procedure for measuring the crack velocity was developed [8]. Prior to all tests, the specimens were heated to 250°C, held 1 h, and then cooled to the test temperature. The specimens were loaded in bending, and cracking was detected by acoustic emission. Crack velocity was derived from the average crack depth divided by the time over which steady cracking occurred.

Tests to measure the threshold stress intensity factor, K_{IH}, were performed in the same manner except that the load was gradually decreased until cracking stopped. K_{IH} was obtained from the final average crack depth and the final load.

Fatigue

We have evaluated both the irradiated and unirradiated HT Zr-2.5Nb pressure tubes using the tensile fatigue test and the crack growth rate tests. Tension fatigue tests were performed in air using an hour-glass specimen [9] punched out of the pressure tube with the tensile direction parallel to the axial direction of the tube. The length of the specimen was 82.5 mm. The specimen thickness was that of the tube, and the radius of curvature of the hour-glass section was 34 mm going from a 19-mm-wide specimen head to a 2.8-mm-wide waist. Crack growth tests were conducted on modified, curved 34-mm compact toughness specimens. The crack plane was the radial-axial plane of the tube. The in-plane dimensions of the specimen were the same as for ASTM Test Method for J_{IC}, a Measure of Fracture Toughness (E 813) except for two modifications. The first was the thickness, which was the tube thickness, and the second was the spacing of the holes. These were 4.75 mm in diameter and 5 mm from the crack plane to limit the bending moment on the curved specimens.

Tension fatigue tests were performed using an MTS electrohydraulic testing machine with a load ratio, R, of 0.1 at a frequency of 5 Hz until complete failure of the specimen occurred. Most of the test time was taken up in initiation of a crack, and crack propagation took relatively few cycles. Fatigue crack growth specimens were fatigue pre-cracked at a stress intensity range, ΔK, of 12.5 MPa$\sqrt{m}$ for a crack growth distance of 2 mm. A commercially available "Krak-

Gage"[5] was bonded to the specimen to monitor the crack length at the surface of the specimen. The 1-Hz cyclic load used gave an initial ΔK of 15 MPa$\sqrt{m}$. As the crack extended, the value of ΔK increased until at the end of the test it was approximately 30 MPa$\sqrt{m}$. Data were ignored for values of a/W greater than 0.65, where a is the crack length and W is the width of the specimen. The crack length was checked optically after the completion of each test. The test results were fitted to the Paris' equation relating crack growth per cycle, da/dN, to the amplitude of the stress intensity factor, ΔK:

$$da/dN = C(\Delta K)^n$$

where C and n are constants.

Fracture Toughness

One of the advantages of a pressure tube reactor is the ability to test the pressure vessel at full size, out-reactor, after a period of service. In practice, a few sections of tubing are internally pressurized to failure to confirm the value of the critical crack length (CCL) based on measurements to fracture toughness from many small specimens. Curved compact toughness specimens are the preferred configuration for the small specimens [10].

Curved Compact Toughness Specimens—The J-resistance curve was used to measure the fracture toughness of the specimens. The curvature of the tube was retained because flattening the material to produce flat specimens would be difficult due to irradiation embrittlement and the possibility that any dislocation movement induced by the flattening process would destroy the defect structure produced by irradiation. A procedure was developed for testing 17-mm curved compact tension specimens [11], and it was shown that because of the small size of the specimens, the curvature of the specimen did not introduce significant errors [10].

Detailed experimental procedures are given in Ref 11. The crack plane was the radial-axial plane with the direction of crack propagation parallel to the axial direction of the tube. Before testing, the specimens were fatigue-precracked for about 2 mm at room temperature, such that a/W was about 0.5. A constant displacement rate of 0.25 mm/min was used. This is approximately equal to an initial stress intensity factor rate of 0.2 MPa$\sqrt{m}$/s. A d-c potential drop (PD) method was used to measure the crack growth. The tests continued until the PD indicated that the crack had propagated for about 3 mm or when the specimen fractured. The fracture surface was photographed after the test, and the calibration constant for the PD was derived for each specimen. The load, test temperature, load-point displacement, and PD values were recorded and used to calculate the J-resistance curve.

Slit Burst Tests—Slit burst tests were done on 460-mm-long sections of the pressure tube. The experimental details are given in Ref 12. A machined and fatigue-sharpened axial flaw was used because the hoop stress is the largest component of stress imposed by internal pressure. Crack extension was estimated from surface crack-propagation gages. For the tests at room temperature, the crack was extended to rupture by fatigue while, for the test at 300°C, the hoop stress was raised monotonically at about 4 MPa$\sqrt{s}$ by pressurizing with nitrogen. After the tests, the fracture surface was photographed to confirm the crack length and observe any unusual features.

---

[5] Krak-Gage is a registered trademark of Hartrun Corporation.

Results and Discussion

Irradiation Deformation

The length data are shown as length change versus fluence in Fig. 1. The length change between 1968 June (5500 h) and 1988 January (124 300 h) is 1.5 mm. Axial rates have been calculated for the complete irradiation period (0 to 124 300 h). The measured axial strain rate is 2.9×10^{-9} /h (1.16×10^{-29} m^2/n) at average conditions over the length of 265°C, 132-MPa hoop stress and flux of 0.62×10^{17} n/(m^{2}s). This rate is considerably lower than the rate of 2.1×10^{-8} /h (7.5×10^{-29} m^2/n) at an average condition of 265°C, 101-MPa hoop stress, and flux of 0.82×10^{17} n/(m^{2}s) for the CW Zr-2.5Nb tube [13], also irradiated in NPD. The data are shown in Fig. 1.

The transverse strains averaged over the middle 25 mm of each fuel bundle position are plotted against fast-neutron fluence in Fig. 2. The strains are almost linear with fluences. The transverse strain rate is about 1.27×10^{-28} m^2/n at mid-channel for fluences $>1 \times 10^{25}$ n/m^2. The maximum measured strain rate (with respect to time) for the HT tube is 5.9×10^{-8}/h at mid-channel. The transverse strain rate for the Fugen tube (mid-channel) is 7.6×10^{-8}/h [285°C, 100 MPa, 2.2×10^{17} n/(m^{2}s)] [14]. Both NPD and Fugen data agree with the extrapolations by the Ross-Ross equation [15].

The transverse strain rates for the HT tube are compared with the rates of the NPD irradiated CW Zr-2.5Nb Tube 582, extrapolated to the HT operating conditions in Fig. 3. The rates for the HT tube are slightly higher than the CW tube at the coolant inlet end and about equal at the coolant outlet end of the tube. The transverse-strain rates at the coolant-outlet end of the pressure tube are higher than the coolant-inlet end even though the fast-neutron flux at the inlet end of the tube is higher than at the outlet end. The pressure tube was installed in the reactor with its back end, i.e., the end exiting from the extrusion press last, at the coolant-outlet end.

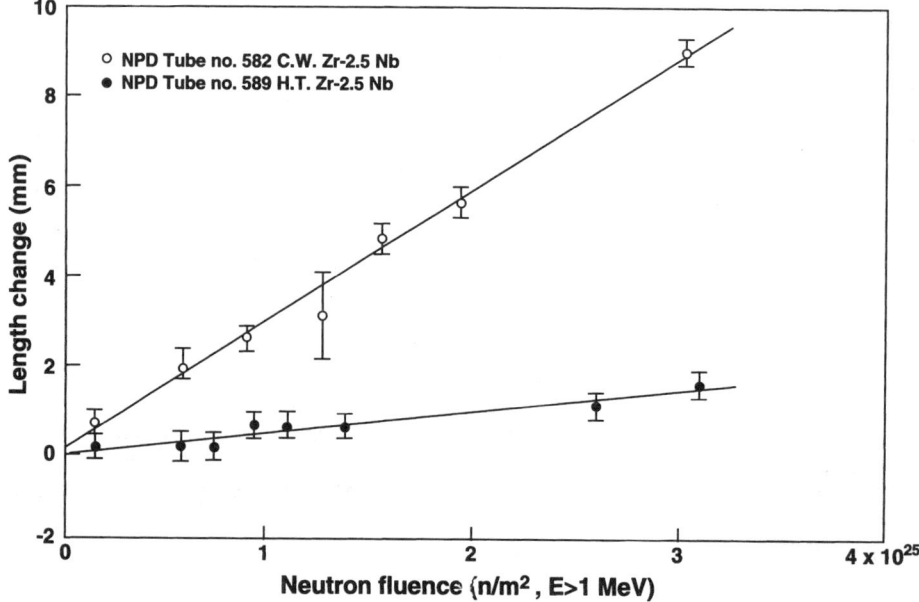

FIG. 1—*Length change as a function of fluence for HT Tube 589. Data for the CW Tube 582 are presented for comparison.*

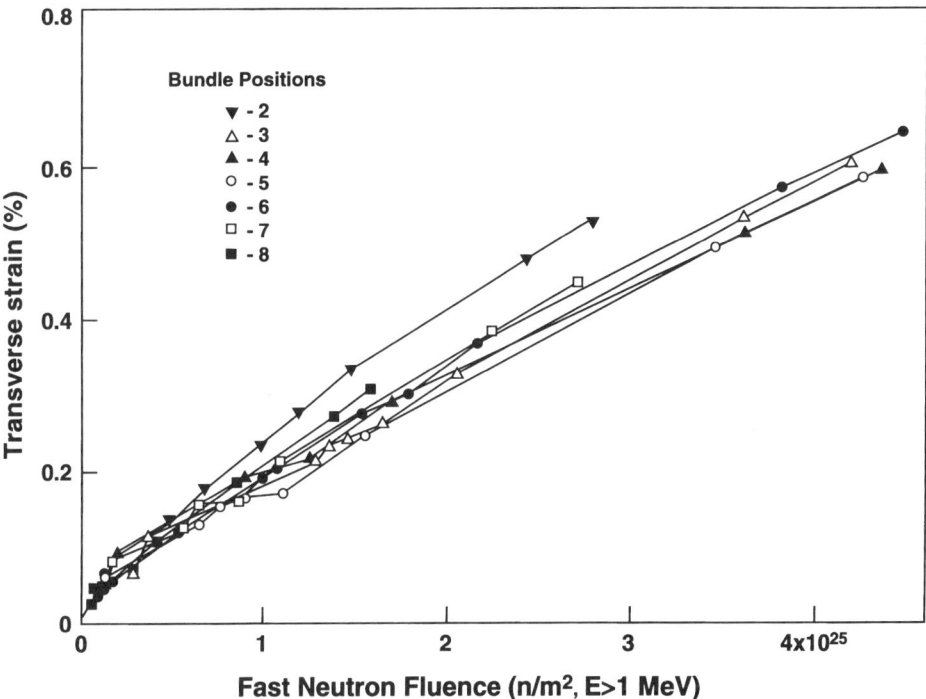

FIG. 2—*Transverse strain at various bundle positions versus fast-neutron fluence for HT Zr-2.5Nb Pressure Tube 589. Bundle 1 is the coolant outlet end.*

CW Zr-2.5Nb tubes in CANDU power reactors also display higher strains at the back-end outlet [7]. For these tubes, this behavior is attributed to end-to-end variations in the material's creep and growth properties of the tube.

The higher transverse and lower axial strain in the HT Zr-2.5Nb pressure tube than the CW tube are consistent with the known effects of crystallographic texture on anisotropy of the in-reactor deformation of zirconium alloy tubes [6,16].

Tensile Properties

The published data on the tensile properties of HT Zr-2.5Nb after irradiation at about 300°C are presented in Fig. 4. This figure, based on a previous compilation, Fig. 7 in Ref *17* with the addition of the Tube 589 data and some data reported in Ref *18*, shows a significant increase in strength and a corresponding decrease in ductility with increase in irradiation. The strength of these HT Zr-2.5Nb materials is similar to the CW Zr-2.5Nb material after a fluence of 10^{25} n/m² [*12*].

As noted in Refs *19* and *20*, the strength of irradiated HT Zr-2.5Nb is little changed after a fluence of about 1×10^{25} n/m² is reached. Estimates of the average increment at a fluence of about 1×10^{25} n/m² are listed in Table 1, showing a reasonable agreement of about 200 MPa. Probably because of differences in specimen size and shape, there is much wider scatter in ductility values. In tests at 300°C, Pickles [*20*] reported, for irradiated material, reduction in area (RA) values of 79% for transverse specimens, whereas on some ring specimens Langford

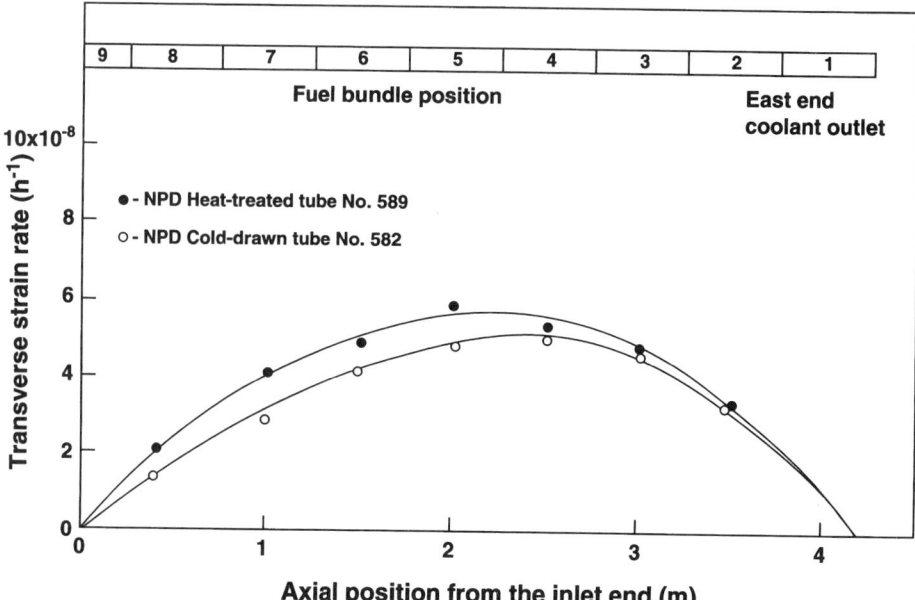

FIG. 3—*Transverse strain rate versus axial position for the HT and CW Zr-2.5Nb pressure tubes.*

et al. [21] reported values of 12%. We are unable to reconcile this wide divergence, but the tests on both the Fugen surveillance specimens and Tube 589 indicate that the material retains acceptable tensile ductility.

Corrosion and Deuterium Pickup

The pressure tube surfaces were in contact with high-temperature, high-pressure heavy water on the inside and reactor vault air on the outside. Corrosion reactions on both surfaces produced a stable oxide film. The reaction with heavy water on the inside surface generated deuterium, a fraction of which was picked up by the tube during its service in the reactor.

Oxidation was greater on the inside surface in contact with water than on the outside surface in contact with air. Oxidation on the inside surface increased slightly along the length of the pressure tube, varying from 6 to 8 μm at the inlet end to 8 to 10 μm at the outlet end. No significant variation in oxide thickness was seen around the circumference of the tube at any of the axial locations. Measurements on the outside surfaces showed that in all cases the oxides were thin and uniform, ranging from 1 to 2 μm thick along the entire length of the tube. At all axial locations, the variation in deuterium concentration around the circumference of the tube was small and within ± 0.007 at% (1.5 μg/g) of the calculated mean concentrations. The deuterium concentration was found to gradually increase from the inlet to the outlet end, with the maximum deuterium concentration varying from 0.04 at% (8 μg/g) at the inlet end to 0.07 at% (15 μg/g) at the outlet end of the pressure tube.

The gradual increase in deuterium concentration along the pressure tube was consistent with the small change in oxide thickness measured from the inlet to the outlet end of the tube. The absence of any measurable peak in both the oxide thickness and deuterium concentration near the center of the tube suggested that the increase in temperature along the length of the tube,

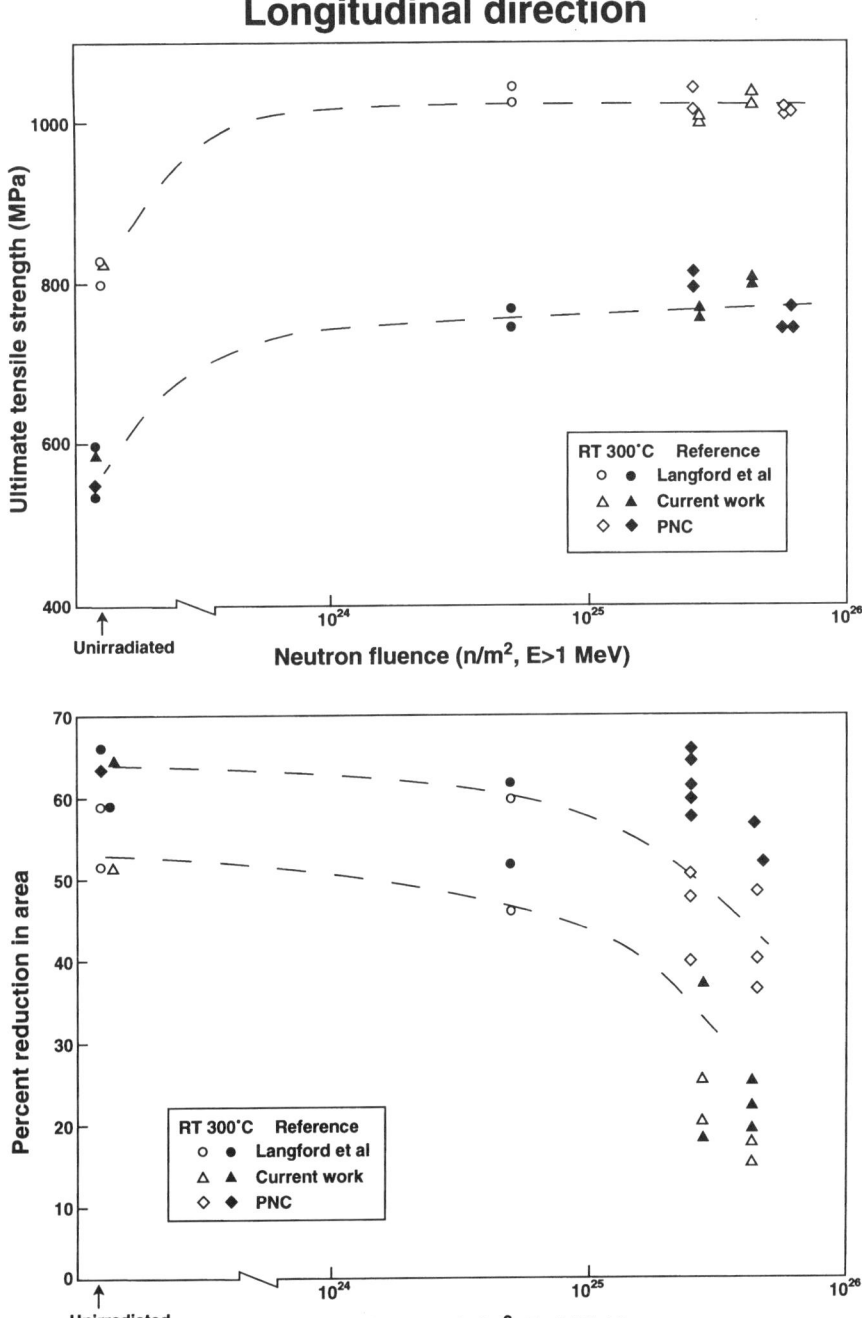

FIG. 4—*Tensile properties at about 300°C for irradiated HT Zr-2.5Nb materials. Data from PNC* [17,22], *Langford et al.* [18,21], *and Pickles* [20] *are included for comparison.*

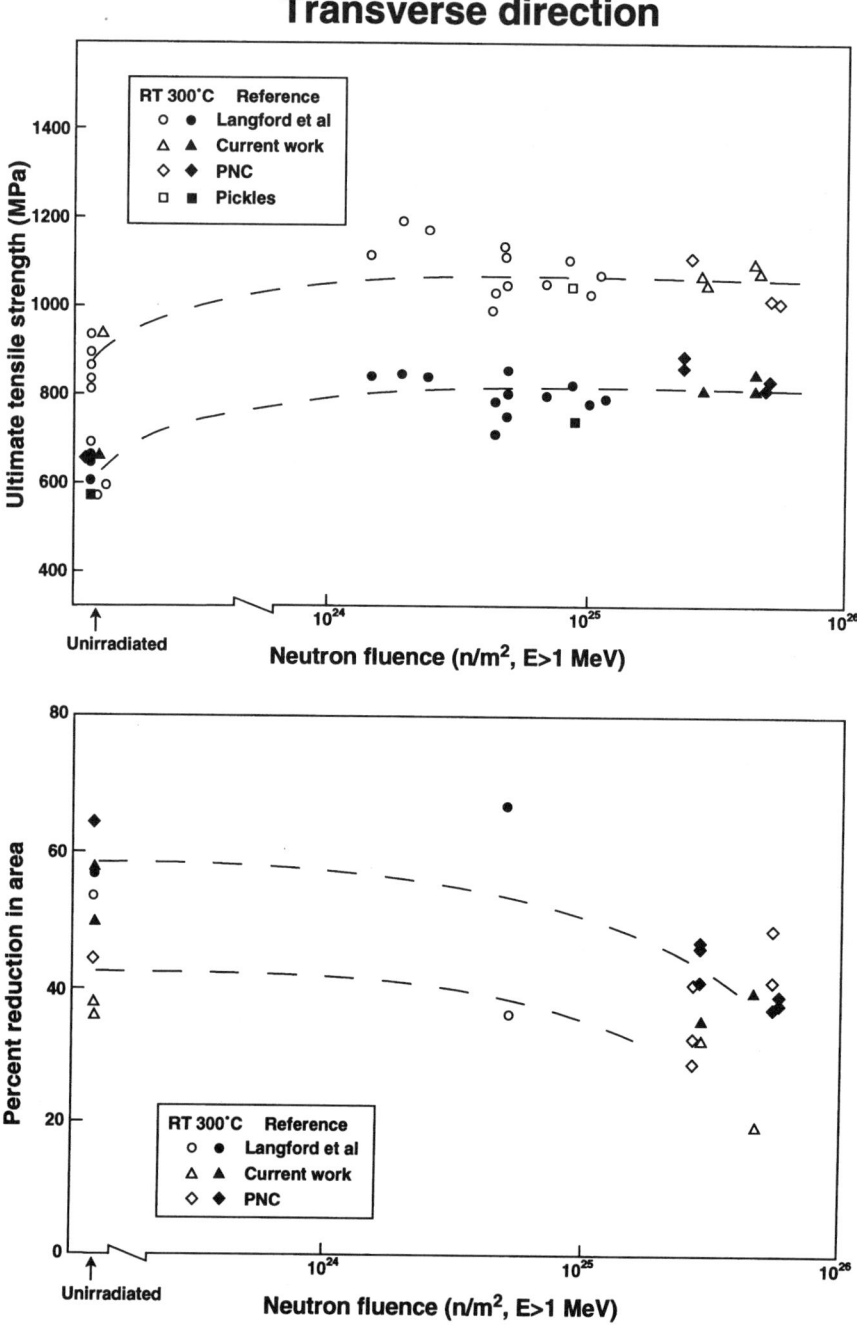

FIG. 4—continued

TABLE 1—Increments in tensile properties due to irradiation (MPa).

Data Source	Property	Circumferential		Axial	
		RT	300°C	RT	300°C
NPD	UTS	172	171	189	192
PNC [17,22]	UTS		218		191
Pickles [20]	UTS	242	182		
Langford [21]	UTS	172	130		
NPD	0.2% YS			263	227
PNC [17,22]	0.2% YS		220		239

rather than the fast neutron flux over the same region, is largely responsible for the end-to-end variation in deuterium pickup. Deuterium pickup from the annulus gas surrounding the tube is very unlikely since the oxidizing nature of the gas (air) would maintain the outside surface oxide as a barrier to any hydrogen isotopes in the annulus gas.

The maximum oxidation measured on the inside surface of Tube 589 is compared with other Zr-2.5Nb tubes and specimens as a function of time in Fig. 5. Most of the data correspond to irradiation at 270 to 280°C with the exception of CANDU-PHW pressure tube data, which correspond to irradiation temperatures of 290 to 300°C. The oxidation of heat-treated tubes in NRX/NRU reactors is consistent with pressure tube specimens irradiated in the JMTR and Fugen reactors. Specimens irradiated in the SGHWR oxidized at a much higher rate, most

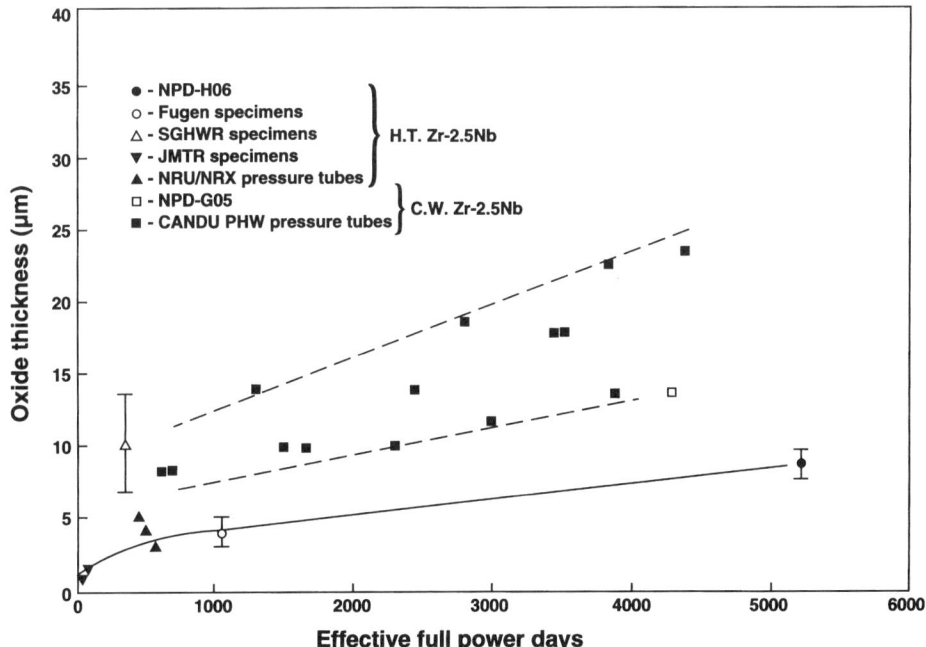

FIG. 5—The maximum oxide thickness measured on the inside surface of Tube 589 is compared with other Zr-2.5Nb tubes and specimens as a function of time.

likely because of the more oxidizing water chemistry in this reactor. Oxidation of Tube 589 is lower than that for cold-worked Zr-2.5Nb tubes after a similar exposure time. When the Fugen and the Tube 589 data are compared, a linear oxidation rate of about 1.3×10^{-3} μm of oxide per day is calculated, about two to three times lower than that for CW Zr-2.5Nb tubes in CANDU-PHW reactors. The precipitation of β-Nb precipitates and the concomitant reduction of Nb in the matrix during both fabrication and irradiation are believed to be largely responsible for the lower corrosion compared with cold-worked pressure tubes. The extremely low oxidation (<2 μm) on the outside surface of the tube demonstrates the exceptional resistance of heat-treated pressure tube material to dry air, the gas annulus environment, during irradiation in the NPD reactor.

The maximum deuterium pickup measured in Tube 589 is compared, in terms of equivalent hydrogen, with other Zr-2.5Nb tubes and specimens as a function of time in Fig. 6. Hydrogen pickup is expressed in terms of mg/dm^2 to account for differences in thickness of the various tubes and specimens. Only the cold-worked tubes in commercial CANDUs that have operated with a CO_2 gas annulus are included for comparison in Fig. 6 since it is believed that essentially all the hydrogen picked up during service in those tubes came from the corrosion reaction with the coolant. Tubes that have operated with a N_2 gas annulus have been excluded from the comparison because of suspected contributions to the hydrogen ingress from the gas annulus [23]. The equivalent hydrogen pickup in heat-treated Zr-2.5Nb is low (<1.5 mg/dm^2 over 5000 full power days) and similar to that measured in the CW Zr-2.5Nb tube irradiated in the NPD reactor at the same time and with HT material irradiated in Fugen. The lower pickup in these tubes compared with CW Zr-2.5Nb tubes in commercial CANDU reactors may be a consequence of both lower temperatures and lower fluxes in the NPD reactor.

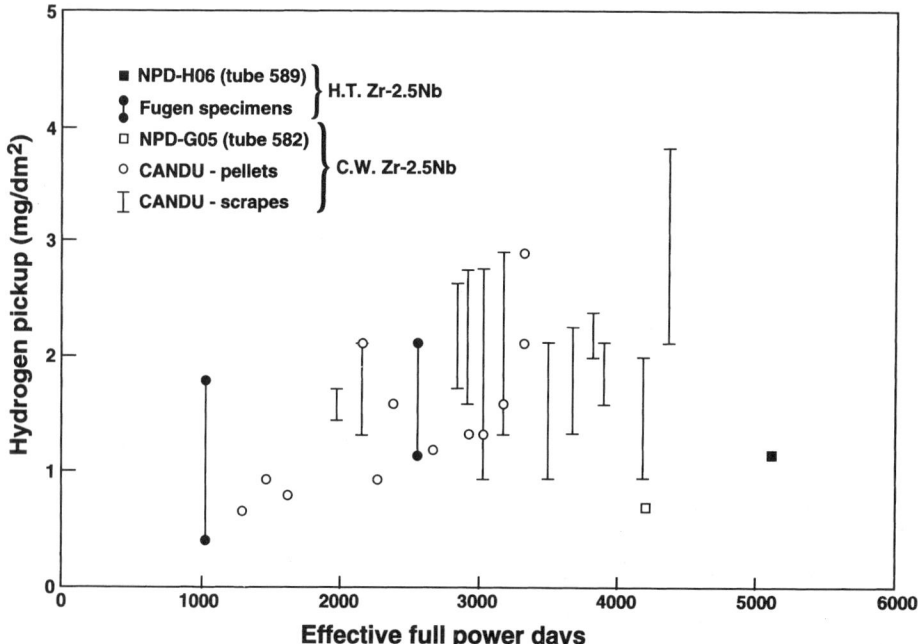

FIG. 6—*The maximum deuterium pickup measured in Tube 589 is compared, in terms of equivalent hydrogen, with other Zr-2.5Nb tubes and specimens as a function of time.*

DHC Behavior

To initiate and grow a DHC crack, the terminal solid solubility of hydrogen in the bulk must be exceeded and there must be a sufficiently high, normal tensile stress acting on the crack-tip hydride to cause it to fracture [*24*]. Fracture in zirconium alloys by DHC occurs in three stages: (1) crack initiation, (2) stable crack growth, and (3) unstable crack growth. The stress intensity factor, K_I, derived from linear elastic fracture mechanics, is used to describe the crack driving force. The DHC velocity rises quickly when the stress intensity factor is higher than the threshold value, K_{IH}. K_{IH} has important practical implications because it allows one to estimate how big a flaw can be tolerated in a component without jeopardizing its performance. Once the crack is initiated, it will grow at a steady rate that is insensitive to K_I for $K_I > K_{IH}$. This steady state crack velocity is an important parameter in evaluating the leak-before-break time between through-wall crack penetration (leak) and attainment of the critical crack length (break).

Threshold Stress Intensity Factor for DHC, K_{IH}—The threshold stress intensity factors in irradiated and unirradiated heat-treated tubes are similar and do not depend on the temperature between 130 and 180°C. The values range from 6.0 to 8.6 MPa$\sqrt{m}$ with an average value of 6.9 MPa$\sqrt{m}$. A lack of temperature dependence of K_{IH} was also observed in CW Zr-2.5Nb pressure tube material between 80 and 150°C [*24*]. K_{IH} values in the HT NPD tubes are similar to those in irradiated and unirradiated CW Zr-2.5Nb pressure tubes. For irradiated CW Zr-2.5Nb pressure tubes, K_{IH} is about 7 MPa$\sqrt{m}$ with a 95% confidence interval of ±2.2 MPa$\sqrt{m}$ [*8*]. Thus, the flaw tolerance in HT Zr-2.5Nb pressure tubes is about the same as in CW Zr-2.5Nb pressure tubes.

DHC Velocity—Crack velocity results from both irradiated Tube 589 and unirradiated Tube 585 are shown in Fig. 7. Crack velocity tests have been performed at an initial K_I of 17.5 MPa$\sqrt{m}$; the value of K_I may have slightly increased or decreased during the test. The least squares linear regression analysis of the data gave the following temperature dependencies:

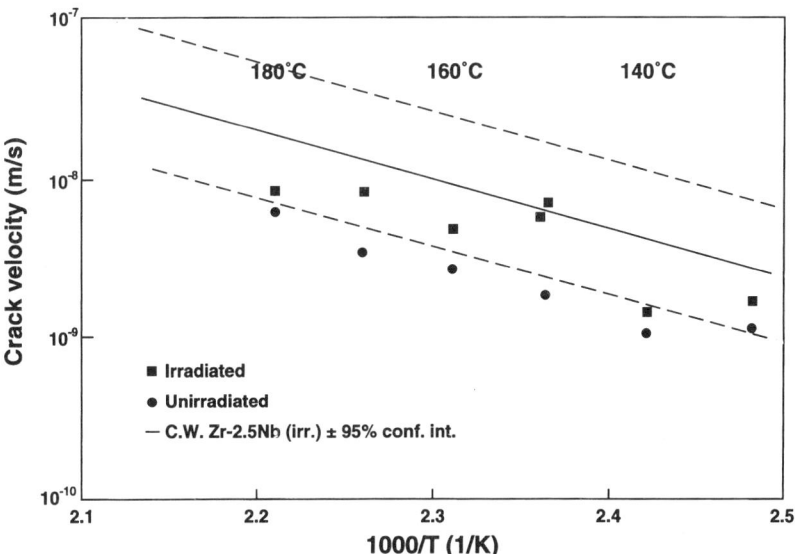

FIG. 7—*DHC velocity of irradiated and unirradiated HT Zr-2.5Nb pressure tube. Data from Ref 8 for CW Zr-2.5Nb material are included for comparison. The dotted lines represent the ±95% confidence interval for the CW material.*

$V = 2.699 \times 10^{-2} \exp(-6655/T)$ for irradiated material, and
$V = 8.285 \times 10^{-3} \exp(-6452/T)$ for unirradiated material

where V = the crack velocity in m/s, and T = the temperature in K.

Crack velocities in the irradiated, HT Zr-2.5Nb pressure tube are about two times higher than those in the unirradiated pressure tube (Fig. 7). The increase of the crack velocity in the irradiated material has been attributed to the irradiation hardening [8]. The HT Zr-2.5Nb material has marginally lower crack velocities than the CW Zr-2.5Nb material (Fig. 7). This is mainly because the HT material starts with already decomposed β-phase, whereas the CW material starts with continuous β-phase; materials with a continuous β-phase have higher crack velocities than those with a decomposed β-phase [25].

Fatigue Behavior

Tensile Fatigue Tests—Specimens were tested at either room temperature or 300°C, and the results of tests on both irradiated and unirradiated material are shown in Fig. 8. There seems to be only a slight irradiation effect on the fatigue stress. The fatigue stress amplitude at 10^6 cycles at 300°C is approximately 100 MPa, which is lower than the 140 MPa measured on CW Zr-2.5Nb at 300°C. However, the tests on CW Zr-2.5 Nb were tensile-compression tests with an R ratio of -1, whereas the tests on HT material had an R ratio of 0.1. The different R ratios may account for some of the difference in the results. The fatigue stress amplitude at 10^6 cycles of unirradiated material at room temperature appears to be higher, approximately 175 MPa, but the data near that fatigue limit are sparse and this figure may not be accurate. Irradiation slightly lowers the fatigue life, and the fatigue stress amplitude at 10^6 cycles is approximately 150 MPa. There are no data on room temperature fatigue of CW Zr-2.5Nb with which to compare these results.

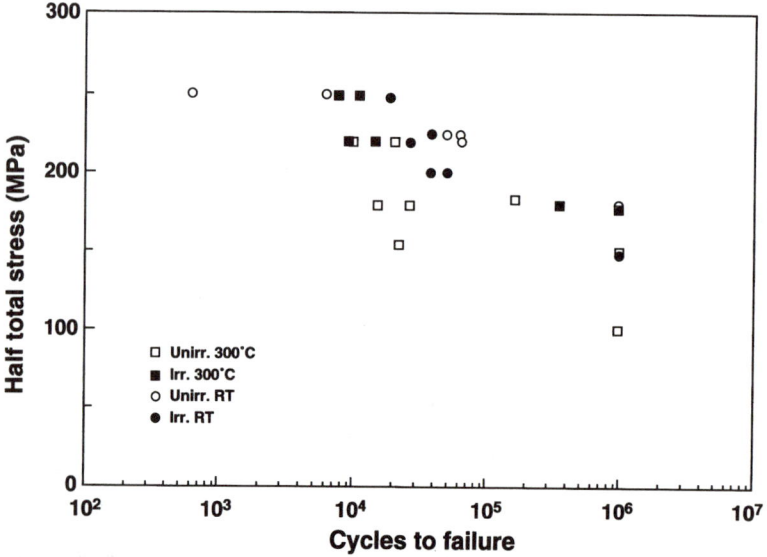

FIG. 8—*Fatigue properties of HT Zr-2.5Nb materials at room temperature and 300°C.*

Fatigue Crack Growth Rate Tests

Specimens were tested in three conditions: (1) unirradiated (Tube 585), (2) low fluence (2 × 10^{25} n/m^2), and (3) high fluence (4 × 10^{25} n/m^2). The cyclic load used in the crack-growth-rate tests was chosen to give an initial ΔK of 15 MPa$\sqrt{\text{m}}$. As the crack extended, the value of ΔK increased until, at the end of the test, it was approximately 30 MPa/m. The values of da/dN and ΔK for the duplicate tests were then combined and fitted to the Paris' equation.

At room temperature, there is no effect of irradiation for the HT Zr-2.5Nb pressure tube (Fig. 9a). These data are almost identical to the unirradiated CW Zr-2.5Nb [26]. The coefficients of

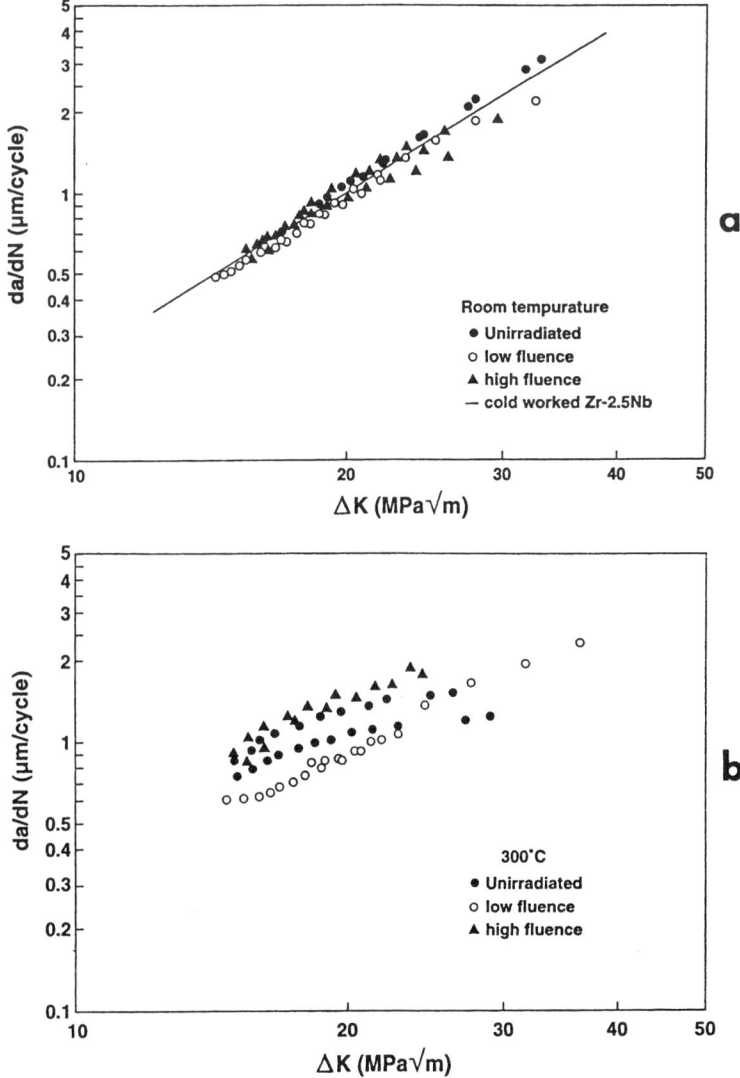

FIG. 9—*Fatigue crack growth rate at* (a) *room temperature and* (b) *at 300°C.*

TABLE 2—*Coefficients of Paris' equation for HT Zr-2.5Nb.*

Condition	Temperature	C	n
Unirradiated	Room	1.47×10^{-9}	2.16
Low fluence	Room	1.56×10^{-9}	2.18
High fluence	Room	1.87×10^{-9}	2.09
Unirradiated	300°C	5.16×10^{-9}	1.73

the Paris' equation fit are given in Table 2. The fracture surface of a typical specimen tested at room temperature is flat. Unlike the room temperature tests, there seems to be an irradiation effect for tests done at 300°C (Fig. 9b). The fatigue crack growth rate of the high-fluence specimens were about a factor of 2 higher than the unirradiated specimens. The data from irradiated specimens show a marked curvature, and the Paris' equation does not represent the data. Similar observations were made with the materials from the Fugen reactor [17]. The fatigue crack surface changes into shear for tests at 300°C. Broek [27] cites examples where the fatigue crack growth rate slows down as the fracture mode changes to shear. Thus, Paris' equation has only been fitted to the data from the unirradiated specimens, and the coefficients are given in Table 2.

Fracture Toughness

Tests on Curved Compact Toughness Specimens—The J-integral was calculated according to ASTM E 813 (1981) or ASTM Test Method for Determining *J-R* Curves (E 1152) using the single-specimen method. Details of the J-integral calculations and test methods have been reported [10,11,28]. The J-integral for crack initiation, $J_{0.2}$, was taken as the J value at 0.2-mm crack extension. The average slope of the J-resistance curve between the 0.15 and 1.5-mm exclusion lines, as defined by ASTM E-813 (1981), was termed average *dJ/da,* and it has been shown that it is a sensitive parameter for the toughness of the zirconium pressure tube materials [10]. The $J_{0.2}$ and the average *dJ/da* will be used to compare the toughness of various materials.

Unirradiated Material—Table 3 summarizes the results on unirradiated material. In the unirradiated material, crack jumps were observed during room temperature tests only; no crack jumps were observed at 300°C. Crack jump is a sudden extension of the crack before it arrests.

Irradiated Material—During the tests on the irradiated material using small specimens, stable crack propagation was usually mixed with a series of small crack jumps, similar to CW materials [29]. The lower the test temperature, the longer the jumps. In four tests in this test series, the crack jumping resulted in fracture of the specimen: three at room temperature and one at 70°C. In all other cases, the crack arrested and propagated stably until the test stopped.

The J-integral at crack initiation, $J_{0.2}$, at room temperature ranged from 10 to 22 kN/m with

TABLE 3—*Fracture toughness of unirradiated pressure tube No. 585.*

Specimen I.D.	Test Temperature, °C	$J_{0.2}$, kN/m	Average dJ/da MPa
PT585.C1	28	40.1	113.0
PT585.C2	31	51.0	108.1
PT585.C5	29	47.0	44.9
PT585.C3	272	65.0	218.0

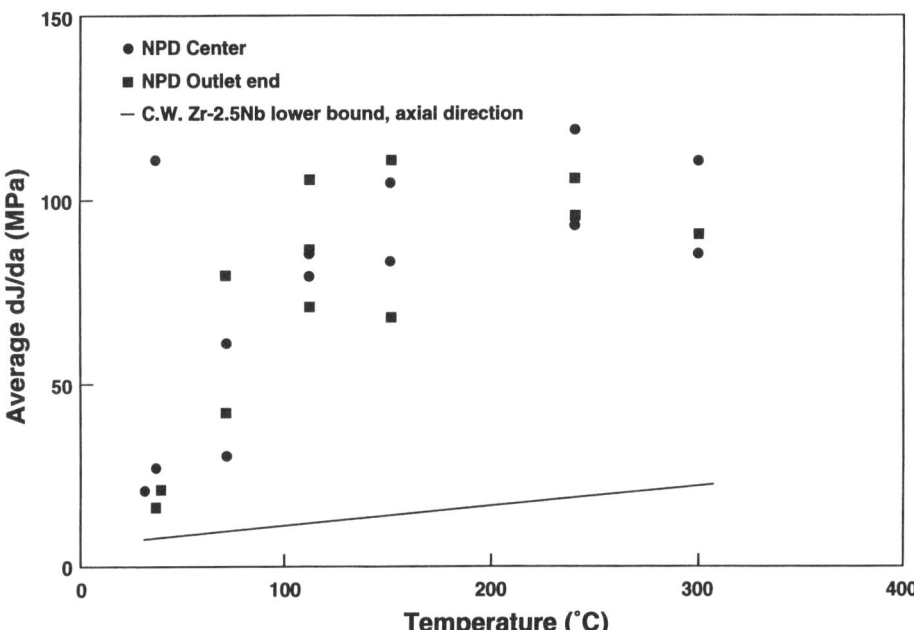

FIG. 10—*Fracture toughness as a function of temperature for irradiated HT Zr-2.5Nb material. The lower bound curve for CW Zr-2.5Nb pressure tube materials is included for comparison.*

a mean value of 15 kN/m. At 300°C, $J_{0.2}$ ranged from 44 to 52 kN/m with a mean value of 48 kN/m. Figure 10 shows the average dJ/da as a function of temperature. It shows that there is a transition from low to high fracture toughness at about 100°C, with a possible maximum at about 200°C. The data show that there is no effect of the fluence in the range tested: 2.5×10^{25} n/m² for the outlet end and 4.5×10^{25} n/m² for the center of the tube. The lower bound curve for CW Zr-2.5Nb is given in Fig. 10 for comparison. Note that most of the irradiated CW Zr-2.5Nb pressure tubes gave higher toughness values than the lower bound shown in Fig. 10 [*12*]. NPD Tube 589 has slightly higher toughness than the average irradiated CW Zr-2.5Nb materials.

Slit Burst Tests

The traditional way to present results of burst tests is to plot the hoop stress required to propagate the crack. These values can be converted to fracture toughness parameters, K or J, using the Dugdale crack model [*30*] with a Folias correction [*31*] for bending at the crack tip. For the test at 300°C, a J-resistance curve was derived from the indications of the crack propagation gages, and this curve was analyzed as described for the compact specimens.

The results of the slit burst tests are summarized in Table 4. The cracks in specimens H06-4 and H06-3 propagated unstably at room temperature by fatigue earlier than expected. This knowledge was exploited with Specimen H06-2 to develop a fatigue-sharpened crack for the test at 300°C. In this test, $J_{0.2}$ was 40 kN/m, and the average dJ/da was about 103 MPa compared with 88 to 109 MPa for the same range of crack length in small specimens. The J-resistance curve of the curved compact toughness specimens and the burst tests are different (see Fig. 11). Although the initial parts of the J-resistance curves were similar, the slope of the

TABLE 4—*Slit burst tests on Pressure Tube No. 589.*

Specimen I.D.	Length of Spark-Machined Flaw, mm	Length of Crack after Fatigue Cycling, mm	Test Temperature, °C	Maximum Pressure, MPa	Hoop Stress at Failure MPa	Apparent K_{IC} at Failure MPa$\sqrt{m}$
H06-4	21	29	30	6.5	116	53
H06-3	40	49	30	3.0	54	50
H06-2	40	43	300	15.0	275	323

burst data became higher than the compact specimen. The higher slope means higher fracture resistance. The differences in values are because of differences in specimen size and loading mode (bending in the compact specimens and tensile in the tubes), but elucidation of the relative contributions of these factors requires further work. Some recent work in this area is presented in this symposium by Davies and Shewfelt [*32*].

The *J*-resistance curve from the burst test at 300°C should be used with caution because it

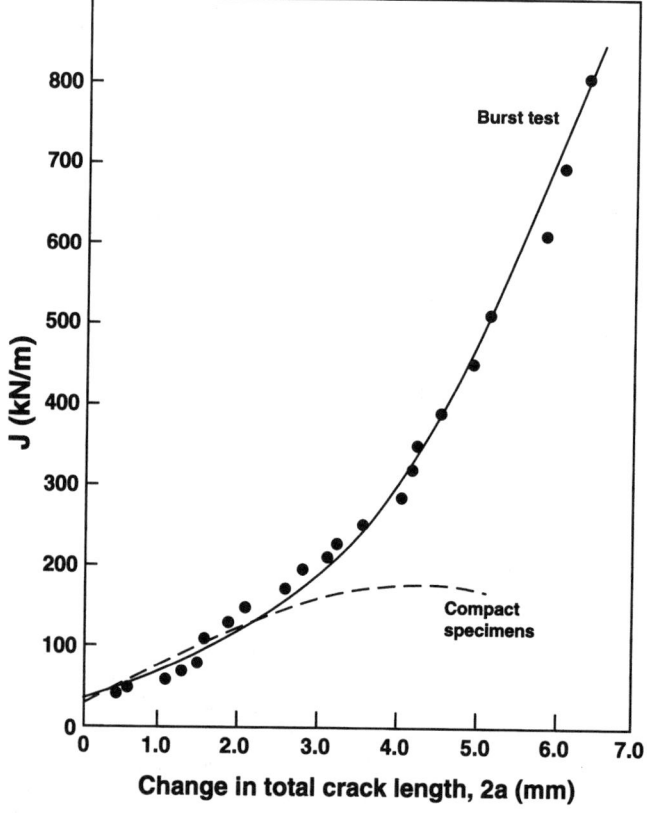

FIG. 11—*J-resistance curve at 300°C measured using curved compact toughness specimen and slit burst.*

is based on surface measurements, which, in tough materials, probably represent the position of the plastic zone rather than the true position of the crack tip. Thus, the values of J may be overestimated. However, the extent of stable crack growth, apparent on the fracture surface, at each end of the crack was about 2.5 mm, while that indicated by the gages was about 3 mm, suggesting that the error is not large.

The fracture surface morphology of the irradiated compact specimens was studied with a scanning electron microscope. The fracture surface consisted of a flat, triangularly shaped crack at the center, with shear lips (slant fracture) on both sides. Dimples were found on the surfaces, indicating that void growth and coalescence had taken place at both room temperature and at 300°C. For tests at room temperature, there were regions where the fracture surface looks like fractured hydrides. The fracture surface of the shear lips formed at the sides of the specimen were planar with very small dimples, indicating that the failure mechanism was highly localized. Similar observations were made on CW Zr-2.5Nb material [29].

The fractography on the tube specimens was limited to optical macrographs. At room temperature, the unstable fracture was mostly flat, in the same plane as the starter crack, with small shear lips at each surface. No indications of stable cracking were detected. At 300°C, a triangle of flat fracture, surrounded by shear fracture, emanates from the fatigue crack, but all the subsequent fracture was slant, shear fracture.

There are few data available on the fracture toughness of HT Zr-2.5Nb material. The most extensive data were generated by PNC [17,22,33]. In these reports, the fracture toughness K_c was measured using small curved bend specimens for unirradiated and irradiated HT material irradiated up to a fluence of 5.7×10^{25} n/m^2 ($E > 1$ MeV). If we assume that K_c was the stress intensity factor at crack initiation, we can compare these values with the results from this test series. From our J-resistance curves, we assumed that the J-integral at crack initiation is $J_{0.2}$, then converted the J-integral to K using: $K = \sqrt{(JE)}$, where E is the Young's modulus of zirconium (E in GPa $= 95.9 - 9.57\ T$, and T is the temperature in °C). Honda et al. [33] reported that K_c at room temperature for unirradiated material was between 40 and 90 MPa$\sqrt{\text{m}}$ (Fig. 7 of Ref 33), which is in good agreement of our data: 62 to 75 MPa$\sqrt{\text{m}}$. The results for irradiated material are plotted in Fig. 12 with data from Koike et al. [17] also showing good agreement, except at 70°C, which may indicate a ~50°C shift in transition temperature to a higher value with increased irradiation. The current results confirm the previous implication that once an initial toughness loss due to irradiation is sustained, in the first 0.4×10^{24} n/m^2, the fracture properties deteriorate no further [12,22]. Our results are in good agreement with those obtained at much lower neutron fluences (Fig. 12).

Conclusions

A HT Zr-2.5Nb pressure tube removed from NPD has been examined after 20 years of service. The results were compared with CW Zr-2.5Nb pressure tube material, for which a much larger database exists. The results show that:

1. The in-reactor elongation rate is much lower and the transverse strain rates are slightly larger than in CW Zr-2.5Nb tubes.
2. Tensile properties of this irradiated tube are similar to the irradiated CW tubes.
3. The hydrogen pickup in HT Zr-2.5Nb is similar to that measured in the CW Zr-2.5Nb tube irradiated in the NPD reactor at the same time.
4. The radial DHC velocity of the irradiated HT Zr-2.5 Nb material is about two times higher than that in the unirradiated material. DHC velocity in the HT material is similar to that in the CW material.

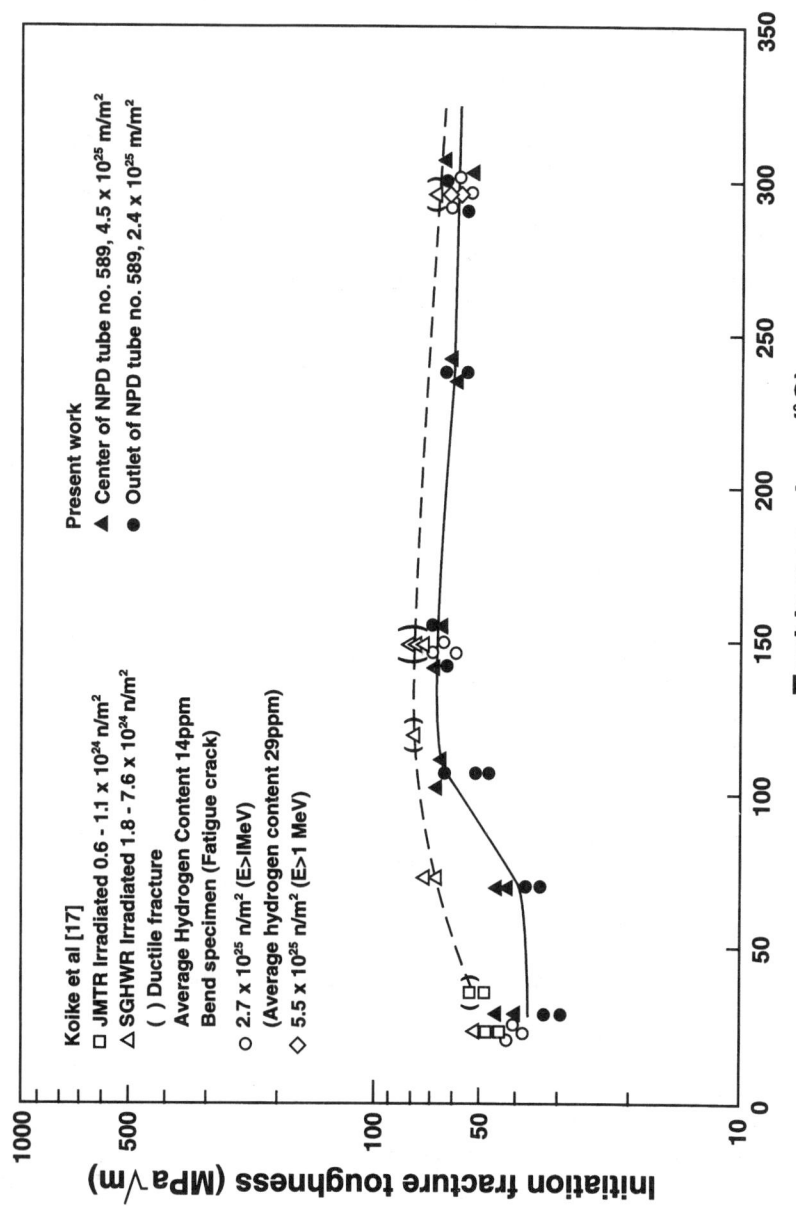

FIG. 12—*Crack initiation fracture toughness of HT Zr-2.5Nb as a function of test temperature. Data generated by PNC are included for comparison.*

5. The threshold stress intensity factor, K_{IH}, is little affected by temperature and irradiation. The average K_{IH} in the HT Zr-2.5Nb is 6.9 MPa$\sqrt{\text{m}}$, similar to those in the CW Zr-2.5Nb material.

6. The fatigue crack growth rates were almost identical to those in the CW Zr-2.5Nb material.

7. The fracture toughness of this tube, as measured by curved compact specimens and burst tests, is slightly higher than in the CW tubes.

The tube was in excellent condition when removed from the reactor, and would have been satisfactory for further service.

Acknowledgments

The authors would like to thank their many colleagues from CRL and WL for technical assistance. Y. Hayamizu, Y. Kasai, and T. Asada of PNC are also acknowledged for their efforts in managing and co-ordinating this joint project. Funding was supplied by PNC and the CANDU Owners Group (COG).

References

[1] Ambartsumyan, R. S., Kiselev, A. A., Grebennikov, R. V., Myshkin, V. A., Tsuprun, L. J., and Nikulina, A. F., "Mechanical Properties and Corrosion Resistance of Zirconium and Its Alloys in Water, Steam and Gases at High Temperature," *Proceedings,* Second International Conference on the Peaceful Uses of Atomic Energy, United Nations, Geneva, Vol. 5, 1958, pp. 12–33.

[2] Ivanov, O. S. and Grigorovich, V. K., "Structure and Properties of Zirconium Alloys," *Proceedings,* Second International Conference on the Peaceful Uses of Atomic Energy, United Nations, Geneva, Vol. 5, 1958, pp. 34–51.

[3] Ells, C. E., Dalgaard, S. B., Evans, W., and Thomas, W. R., "Development of Zirconium-Niobium Alloys," *Proceedings,* Third International Conference on the Peaceful Uses of Atomic Energy, United Nations, Geneva, 1964, Vol. 9, 1965, pp. 91–99.

[4] Fugen HWR, *Nuclear Engineering International,* Vol. 24, Vol. 289, 1979, pp. 33–39.

[5] Coleman, C. E., Cheadle, B. A., Causey, A. R., Chow, C. K., Davies, P. H., McManus, M. D., Rodgers, D. K., Sagat, S., and van Drunen, G., "Evaluation of Zircaloy-2 Pressure Tubes from NPD," *Zirconium in the Nuclear Industry: Eighth International Symposium, ASTM STP 1023,* L. F. P. Van Swam and C. M. Eucken, Eds., American Society for Testing and Materials, West Conshohocken, PA, 1989, pp. 35–49.

[6] Ibrahim, E. F. and Holt, R. A., "Anisotropy of Irradiation Creep and Growth of Zirconium Alloy Pressure Tubes," *Journal of Nuclear Materials,* Vol. 91, 1980, pp. 311–321.

[7] Causey, A. R., Fidleris, V., MacEwen, S. R., and Schulte, C. W., "In-Reactor Deformation of Zr-2.5 wt% Nb Pressure Tubes," *Influence of Radiation on Material Properties: Thirteenth International Symposium, ASTM STP 956,* F. A. Gardner, C. H. Henager, Jr., and N. Igata, Eds., American Society for Testing and Materials, West Conshohocken, PA, 1987, pp. 54–68.

[8] Sagat S., Coleman C. E., Griffiths M., and Wilkins B. J. S., "The Effect of Fluence and Irradiation Temperature on Delayed Hydride Cracking in Zr-2.5Nb," *Zirconium in the Nuclear Industry, Tenth International Symposium, ASTM STP 1245,* A. M. Garde and E. R. Bradley, Eds., American Society for Testing and Materials, West Conshohocken, PA, 1994, pp. 35–61.

[9] ASTM Practice for Conducting Constant Amplitude Axial Fatigue Tests of Metallic Materials (E 466–82), Fig. 4.

[10] Chow, C. K. and Simpson, L. A., "Determination of the Fracture Toughness of Irradiated Reactor Pressure Tubes Using Curved Compact Specimens," *Fracture Mechanics: Eighteenth Symposium, ASTM STP 945,* D. T. Read and R. P. Reed, Eds., American Society for Testing and Materials, West Conshohocken, PA, 1988, pp. 419–439.

[11] Simpson, L. A., Chow, C. K., and Davies P. H., "Standard Test Method for Fracture Toughness of CANDU Pressure Tubes," AECL Report COG-89-110-1, 1989 September.

[12] Chow, C. K. Coleman, C. E., Hosbons, R. R., Davies, P. H., Griffiths, M., and Choubey, R., "Fracture Toughness of Irradiated Zr-2.5Nb Pressure Tubes from CANDU Reactors," *Zirconium in the Nuclear Industry, Ninth International Symposium, ASTM STP 1132,* C. M. Eucken and A. M. Garde, Eds., ASTM, West Conshohocken, PA, 1991, pp. 246–275.

[13] McManus, M. D., unpublished results, AECL, Chalk River Laboratories.
[14] Koike, M. and Asada, T., "Irradiation Creep and Growth of Pressure Tubes in HWR Fugen," *Journal of Nuclear Materials*, Vol. 159, 1988, pp. 62–74.
[15] Ross-Ross, P. A. and Fidleris, V., "Design Basis for Creep of Zirconium Alloy Components in a Fast Neutron Flux," Paper C216, *Proceedings*, International Conference on Creep and Fatigue in Elevated Temperature Applications, 1973, pp. 216.1–216.7; also published as AECL-04626.
[16] Causey, A. R., Elder, J. E., Holt, R. A., and Fleck, R. G., "On the Anisotropy of In-Reactor Creep of Zr-2.5Nb Tubes," *Zirconium in the Nuclear Industry, Tenth International Symposium, ASTM STP 1245*, A. M. Garde and E. R. Bradley, Eds., American Society for Testing and Materials, West Conshohocken, PA, pp. 202–220.
[17] Koike, M. H., Akiyama, T., Nagamatsu, K., and Shibahara, I., "Changes of Mechanical Properties by Irradiation and Evaluation of the Heat-treated Zr-2.5Nb Pressure Tube," *Zirconium in the Nuclear Industry, Tenth International Symposium, ASTM STP 1245*, American Society for Testing and Materials, West Conshohocken, PA, 1994, p. 183.
[18] Langford, W. J., Mooder, L. E., and Bryson, J. G., "Metallurgical Properties of Heat-treated Zr-2.5Nb Pressure Tubes Irradiated Under Power Reactor Conditions," *Canadian Metallurgical Quarterly*, Vol. 11, 1972, p. 147.
[19] Ells, C. E. and Cheadle, B. A., "The Anisotropy of Fracture Ductility in Flat Tension-Test Bars of Alpha Zirconium Alloy," *Applications Related Phenomena in Zirconium and its Alloys, ASTM STP 458*, American Society for Testing and Materials, West Conshohocken, PA, 1970, p. 68.
[20] Pickles, B. W., "Embrittlement of Heat-Treated Zr-2.5Nb Pressure Tubes," *Canadian Metallurgical Quarterly*, Vol. 11, 1972, p. 139.
[21] Langford, W. J., Mooder, L. E., and Bryson, J. G., unpublished results at AECL CRL related to Fig. 5 in Ref *18*.
[22] Koike, M., Asada, T., Miyamoto, F., Taniyama, S., and Komine, Y. in *PNC Technical Review*, No. 66, 1988 (in Japanese), pp. 20–35.
[23] Urbanic, V. F., Warr, B. D., Manolescu, A., Chow, C. K., and Shanahan, M. W., "Oxidation and Deuterium Uptake of Zr-2.5Nb Pressure Tubes in CANDU-PHW Reactors," *Zirconium in the Nuclear Industry: Eighth International Symposium, ASTM STP 1023*, American Society for Testing and Materials, West Conshohocken, PA, pp. 20–34, 1989.
[24] Coleman, C. E. and Ambler, J. F. R., "Susceptibility of Zirconium Alloys to Delayed Hydrogen Cracking," *Zirconium in the Nuclear Industry, ASTM STP 633*, A. L. Lowe, Jr. and G. W. Parry, Eds., American Society for Testing and Materials, West Conshohocken, PA, 1977, pp. 589–607.
[25] Simpson, L. A. and Cann, C. D., "The Effect of Microstructure on Rates of Delayed Hydride Cracking in Zr-2.5 Nb Alloy," *Journal of Nuclear Materials*, Vol. 126, 1984, pp. 70–73.
[26] Wilkins, B. J. S. and Reich, A. R., "Probabilistic Aspect of Fatigue Crack Propagation Data for Zirconium-2.5%Niobium," AECL Report AECL-5529, November 1976.
[27] Broek, D., *Elementary Engineering Fracture Mechanics*, 3rd ed., Martinus Nijhof, Boston, 1984, p. 259.
[28] Simpson, L. A. and Chow, C. K., "Effect of Metallurgical Variable and Temperature on the Fracture Toughness of Zirconium Alloy Pressure Tubes," *Zirconium in the Nuclear Industry, Seventh International Symposium, ASTM STP 939*, R. B. Adamson and L. F. P. Van Swan, Eds., American Society for Testing and Materials, West Conshohocken, PA, 1987, pp. 579–596.
[29] Chow, C. K. and Simpson, L. A, "The Effect of Fracture Micromechanism on Crack Growth Resistance Curves of Irradiated Zr-2.5 wt% Nb Alloy," *Nonlinear Fracture Mechanics: Volume II—Elastic-Plastic Fracture, ASTM STP 995*, J. D. Landes, A. Saxena, and J. G. Merkle, Eds., American Society for Testing and Materials, West Conshohocken, PA, 1989, pp. 537–562.
[30] Dugdale, D. S., "Yielding of Steel Sheets Containing Slits," *Journal of the Mechanics and Physics of Solids*, Vol. 8, 1970, pp. 100–108.
[31] Folias, E. S., "On the Theory of Fracture of Curved Sheets," *Engineering Fracture Mechanics*, Vol. 2, 1970, pp. 151–164.
[32] Davies, P. H. and Shewfelt, R. S. W., "Link Between Results of Small- and Large-Scale Toughness Tests on Irradiated Zr-2.5Nb Pressure Tube Material," this publication.
[33] Honda, S., Asada, T., and Takeuchi, M., "Fracture toughness of Zr-2.5 wt% Nb Pressure Tubes," PNC N341, 1983, pp. 83–11.

DISCUSSION

A. T. Motta[1] (written discussion)—Given that the dose rate in this experiment is lower than that found in conventional pressure tubes, have you applied the irradiation deformation models developed at Chalk River and, if so, do the irradiation creep and growth strains correspond?

C. K. Chow et al. (authors' closure)—We have applied the models (see Ref 6 in the main text) relating to the crystallographic texture with the anisotropy of irradiation creep and growth to the measurements of elongation and transverse strain rate of the NPD cold-worked Zr-2.5Nb and heat-treated Zr-2.5Nb. We find that the model and measurements agree very well even though the neutron flux was much lower in NPD than in a modern CANDU reactor. Based on the crystallographic texture, the calculated contributions to strain from creep and growth are summarized in Table D-1 (below).

Thus, the ratio of the calculated elongations (cold-worked Zr-2.5Nb/heat-treated Zr-2.5Nb) is 5.8 and the ratio of the calculated transverse deformations is 0.8, in close agreement with the observations presented in Figs. 1 and 3 in the main text.

B. Cox[2] (written discussion)—My recollection is that the CW Zr-2.5% Nb tube in NPD showed white oxide patches. Were there any signs of similar patches on the heat-treated tube?

C. K. Chow et al. (authors' closure)—The white patches on the cold-worked Zr-2.5Nb tube referred to in the question were initiated at scratches caused by inspection equipment. These oxide patches were up to 55 μm thick. In the heat-treated Zr-2.5Nb tube, the maximum oxide thickness was 10 μm. Some oxide had spalled off in these areas, but these features were not associated with inspection marks. No thick oxide patches were observed.

TABLE D-1—*Normalized contributions of creep and growth in the axial and transverse directions in cold-worked Zr-2.5Nb and heat-treated Zr-2.5Nb pressure tubes.*

Material	Creep Contribution		Growth Contribution		Transverse Deformation	Elongation
	Transverse Direction	Axial Direction	Transverse Direction	Axial Direction		
Cold-worked Zr-2.5Nb	12.7	3.4	0.8	9.3	13.5	12.7
Heat-treated Zr-2.5Nb	17.5	1.3	−0.5	0.9	17.0	2.2

---

[1] Pennsylvania State University, University Park, PA.
[2] University of Toronto, Toronto, Ontario, Canada.

Pauline H. Davies[1] and Robert S. W. Shewfelt[2]

Link Between Results of Small- and Large-Scale Toughness Tests on Irradiated Zr-2.5Nb Pressure Tube Material

REFERENCE: Davies, P. H. and Shewfelt, R. S. W., "**Link Between Results of Small- and Large-Scale Toughness Tests on Irradiated Zr-2.5Nb Pressure Tube Material,**" *Zirconium in the Nuclear Industry: Eleventh International Symposium, ASTM STP 1295,* E. R. Bradley and G. P. Sabol, Eds., American Society for Testing and Materials, 1996, pp. 492–517.

ABSTRACT: The link between the results of small- (curved compact) and large-scale (burst) toughness tests on irradiated Zr-2.5Nb pressure tube material was investigated using material from tubes of different toughness values. Comparison between the crack growth resistance (deformation *J-R*) curves from the small- and large-scale specimens reveals the material dependence of geometry effects and shows that the crack-tip constraint in the compact specimen is generally higher than that in the burst specimen. For higher toughness material, the crack-extension region over which there is good correspondence between the *J-R* curves from the small- and large-scale specimens is in agreement with current knowledge of validity requirements for *J*-controlled crack growth of bend-type specimens. Fractographic studies were conducted and the results shown to be consistent with the observed geometry effects, a larger proportion of highly constrained, flat fracture being produced with the small bend-type specimens than with the burst tests. The results are discussed using a volume-controlled fracture model for bend-type specimens in which it is assumed that the toughness is governed by the development of the plastic zone associated with an intermediate-constraint, transition region between the central, flat-fracture zone and surface-shear or slant-fracture zone. Applying scaling factors from the volume-controlled fracture model, good agreement is obtained between scaled values of the maximum pressure/load toughness from the small- and large-scale tests over a range of normalized plastic zone size of 0.4 to 1.

KEY WORDS: fracture toughness, Zr-2.5Nb, irradiation, pressure tubes, crack growth resistance *(J-R)* curve

In CANDU®[3] nuclear reactors, the primary containment for the uranium dioxide fuel is provided by thin-walled pressure tubes (nominally 6.3 m long, 103 mm in diameter, and 4.2 mm thick) of cold-worked Zr-2.5Nb. Heavy water flows through the tubes to cool the fuel, operated at an internal pressure of about 10 MPa and at a temperature ranging from about 250 to 265°C (inlet ends) to about 290 to 315°C (outlet ends). As part of a defence-in-depth approach to fitness for service, the tubes are operated to satisfy a leak-before-break (LBB) criterion [1]. This requires that any undetected flaw in a pressure tube, if it grows across and along the tube, must produce sufficient leakage of primary coolant for detection and safe shutdown of the reactor before growing to the critical crack length (CCL).

The structural integrity of the pressure tubes is periodically assessed by means of mechanical testing of surveillance tubes removed from service [2]. Such testing provides information on any degradation in the mechanical properties due to the effects of irradiation damage and

---

[1] Research scientist, AECL, Chalk River Laboratories, Chalk River, Ontario, Canada K0J 1J0.
[2] Research scientist, AECL, Whiteshell Laboratories, Pinawa, Manitoba, Canada R0E 1L0.
[3] CANDU = CANada Deuterium Uranium: registered trademark.

deuterium pickup, as well as providing safety margins for continued tube operation. For example, the CCL may be determined from burst tests on 500-mm tube sections with axial, through-wall defects [3–5]. Estimates of CCL may also be obtained from 17-mm-wide curved compact (toughness) specimens machined directly from the tube material [6,7]; such specimens generally producing conservative results [4].

The objective of the present work was to study the relationship between the toughness (CCL) results from both small- (curved compact) and large-scale (burst test) specimens with the aim of obtaining more realistic estimates of CCL from the small specimens. Previous work has demonstrated the material dependence of geometry (constraint) effects in results from different curved specimens of pressure tube material [5]. Therefore, tests were conducted on material from irradiated Zr-2.5Nb pressure tubes removed from Pickering Nuclear Generating Station (PNGS) A, Units 3 and 4 having different toughness values. Crack growth resistance or deformation J-R (J-integral versus crack-extension) curves were obtained in the operating temperature regime (250°C). The results are discussed with reference to fractographic studies on the two different test configurations and the use of scaling factors based on a volume-controlled fracture model.

Experimental

Material

The test program used material from full-length pressure tubes removed from PNGS A Units 3 and 4 CANDU reactors after about 18 years of operation and irradiation to a fast-neutron fluence of up to 11×10^{25} n.m^{-2} ($E > 1$ MeV). The tubes were all fabricated as standard cold-worked (about 26%) Zr-2.5Nb pressure tubes in the late 1960s according to the specification in Table 1 [8]. Such tubes can exhibit significant variability in toughness due to the presence of microsegregated species (Zr-Cl-C complex) and particles (carbides and phosphides) [9–11]. The current test matrix of 15 tubes also showed a wide variation in toughness, principally as a result of variations in the chlorine and phosphorus concentrations (Table 2). Three of the tubes (491, 566, and 508 from Fuel Channels P4L17, P3R13, and P4Q17, respectively) were

TABLE 1—*Chemical specification for CANDU Zr-2.5Nb pressure tubes used in present study (tubes fabricated before 1987).*

Element	Specification (up to 1987)
Niobium	2.4–2.8 wt%
Oxygen	900–1300 ppm[a]
Carbon	<270 ppm
Chlorine	...
Chromium	≤200 ppm
Hydrogen	≤25 ppm[b]
Iron	≤1500 ppm
Nickel	≤70 ppm
Nitrogen	≤65 ppm
Phosphorus	...
Silicon	≤120 ppm
Tantalum	≤200 ppm
Zirconium and other impurities	Balance

[a] ppm by weight.
[b] 20 ppm hydrogen for ingot, 25 ppm hydrogen for final tube.

TABLE 2—*Summary of chemical analysis results of PNGS A Units 3 and 4 tubes used in present study.*

Fuel Channel	Tube ID	Ingot	GDMS[a] Tube Sample No.	Cl, ppm	P, ppm
P3M11	98	375195	95/96/97[b]	10/8/10	15/12/13
P3J17	544	377088	542[b]	8	15
P3K05	7	374951	7	9	20
P3M12	256	375790	258[b]	8	34
P3L03	95	375195	95	10	15
P3O07	258	375790	258	8	34
P3M03	65	375044	65	12/8[c]	20/19[c]
P3H07	1963	390597	1963	6	13
P3M09	119	375392	119	10/9[c]	12/11[c]
P3N07	192	375583	192	10/9[c]	30/26[c]
P3S13	682	377607	682	5/4[c]	19/18[c]
P3N08	323	376210	323	0.9	57
P4L17	491	376914[d]	491	0.2/0.16[c]	57/39[c]
P3R13	566	377325[d]	566	0.4	42
P4Q17	508	376914[d]	508	0.1/0.2[c]	57/13.5[c]

[a] Glow discharge mass spectroscopy.
[b] Sister tube.
[c] Repeat analysis.
[d] Ingot fabricated from 100% recycled material.

fabricated from ingots made up from 100% recycled material. The chlorine concentration of such tubes is low, and they were anticipated to have high toughness values.

The mechanical tests generally used material sampled from the mid-length of each tube close to the position of highest neutron flux in-reactor, i.e., 2000 to 4000 mm from the inlet end. The fast-neutron fluence of the tube sections varied from 9.0 to 11.1×10^{25} n.m^{-2}, and the irradiation temperature varied from 257 to 279°C. The tests were all conducted at the same test temperature of 250°C. This temperature, used previously for all small specimen testing of early Pickering material [9,10], was selected to avoid annealing out any irradiation damage and corresponded to the lowest (operating) irradiation temperature of the reactor tubes. The maximum total equivalent hydrogen concentration (hydrogen + 0.5 deuterium concentration in weight fraction) of the tube sections was 24 ppm (0.22 at%) so that the hydrogen isotopes would have been in solution during mechanical testing at 250°C. [The terminal solid solubility for hydride dissolution at 250°C is about 29 ppm (0.26 at%).]

Burst Tests

Each burst test used a 500-mm-long section of tube with an axial, through-wall notch (about 1 mm wide) spark-machined along the mid-length. The sections were sealed using a patch of 0.81-mm-thick Teflon sheet, 0.25-mm-thick stainless steel shim, and 0.41-mm-thick aluminum sheet, the patch being secured inside in the region of the notch with silicone rubber. Zr-2.5Nb voltage leads were spot-welded 16 mm above and below the notch centerline and copper potential drop current leads (17 A constant current) attached to the specimen 90° around the circumference from the centerline of the notch (180° from each other). This configuration produces a linear relationship between the change in voltage and change in notch or crack length [5]. Finally, each section was sealed with mechanical end caps, attached to the pressurizing system, and placed inside a protective bell jar for testing (Fig. 1).

Before testing, the axial notches were extended approximately 5 mm at each crack tip by

FIG. 1—*Burst specimen with mechanical caps and potential drop leads.*

fatigue by pressure cycling at the hot cell temperature (about 30°C). Water was used as the pressurizing medium, and a maximum stress intensity factor of 15 MPa$\sqrt{m}$ was applied during the last 2.5 mm of crack growth to minimize the residual compressive stresses produced at each crack tip. Each test section was then heated to 250°C, using external heating coils inside the bell jar, and pressurized monotonically to failure with argon gas using a rate corresponding to an initial rate of increase of the stress intensity factor of about 0.75 MPa$\sqrt{m}.s^{-1}$.

Deformation J-R (J-integral versus crack-extension) curves were determined for each burst test using the following strip yield equation for an axial, through-wall defect in a pressurized tube [12]:

$$J = \frac{K_I^2}{E} = \frac{8}{\pi} \frac{\sigma_f^2}{E} a \ln\left[\sec\left(\frac{\pi}{2} \frac{M\sigma_h}{\sigma_f}\right)\right] \tag{1}$$

where

σ_f = flow stress (mean of the yield stress and ultimate tensile strength),
E = Young's modulus,
$2a$ = total crack length,
σ_h = hoop stress, pr_i/t,
p = internal pressure,
r_i = internal radius,
t = wall thickness, and
M = Folias bulging correction factor [13], given approximately by Ref 12:
$M = \sqrt{\{1 + 1.255[a^2/(r_m t)] - 0.0135[a^4/(r_m t)^2]\}}$, where r_m is the mean radius.

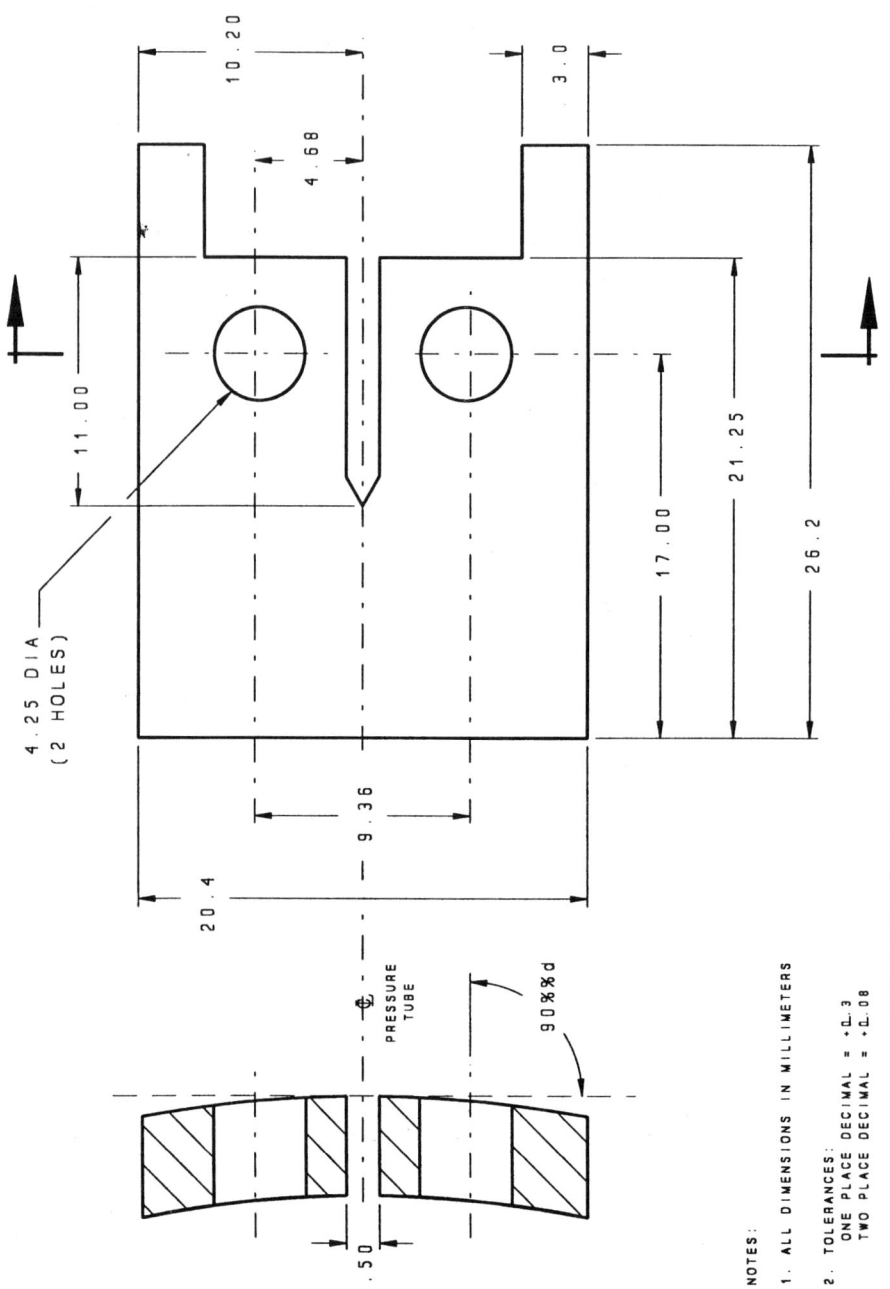

FIG. 2—*Standard (17-mm-wide) curved compact specimen configuration.*

The J-integral value at the intersection of the 3.0-mm offset line and J-R curve, $J_{3.0}$, and the maximum-pressure toughness based on the instantaneous crack size, J_{mpi}, were also determined.

Standard (17-mm-Wide) Curved-Compact Specimen Tests

These tests were conducted following the standard method for fracture toughness testing of CANDU reactor pressure tubes [7]. Standard (17-mm-wide) curved compact specimens (Fig. 2) were spark-machined directly from the tube sections using a cookie cutter electrode, with the specimens oriented for crack growth in the axial direction on the radial-axial (RA) plane [6,7]. Each specimen was fatigue precracked approximately 1.8 mm at cell temperature using an R ratio of 0.1 and a maximum stress intensity factor of 15 MPa$\sqrt{m}$ to produce an initial relative crack length, a_0/W, of 0.5. The specimens were then heated to 250°C in an air furnace and loaded in stroke control at a constant displacement rate of 0.5 mm.min^{-1}, corresponding to an initial rate of increase of stress intensity factor of about 1 MPa$\sqrt{m}$.s^{-1}. During testing, the d-c potential drop method (4 A of constant current) was used to monitor any stable crack growth, and each test was terminated after about 3 to 4 mm of crack growth to achieve a final relative crack length, $a_f/W \leq 0.75$. Finally, the specimens were partially unloaded and heat tinted to mark the final crack front position before breaking open. The initial and final crack length were then used to calibrate the potential drop method on an individual basis by matching the change in voltage to the crack extension since previous work has shown a linear relationship between the voltage and crack length over the crack-size range of interest, a/W of 0.5 to 0.75 [7]. Finally, the J-R curves were calculated following the procedures in Ref 7. The J-integral value at the intersection of the 3.0-mm offset line and J-R curve, $J_{3.0}$, and the maximum-load toughness, J_{ml}, were also noted.

Test Program

Initially, burst tests were conducted on four sections from two tubes of different toughness values, a lower toughness tube removed from Fuel Channel P3L03 and an intermediate toughness tube from Fuel Channel P3N08. Notches of 25, 40, 55, and 70 mm length were spark machined into the four different test sections from each tube with the aim of producing initial crack lengths (after fatigue precracking) of about 35, 50, 65, and 80 mm, respectively. The results from these tests (reported in Ref *14*) indicated little effect of initial crack length on the J-R curves or on the toughness parameters, $J_{3.0}$ or J_{mpi}. For all subsequent testing, the crack size for the burst tests was standardized, i.e., each burst specimen was machined with a notch 45 mm long and fatigue precracked to produce an initial crack length of about 55 mm. Such standard burst tests were conducted on sections from the mid-length of all the remaining 13 tubes.

After burst testing, two standard curved compact specimens were subsequently machined from each test section and tested. The transverse flow stress at 250°C was also determined using two enlarged end specimens machined from pressure tube blanks. Each crack in the burst test section was positioned at the 3 o'clock position, with the compact and tensile specimens being machined diametrically opposite at 9 o'clock (Fig. 3). The aim was to minimize any material variability around the tube circumference as a result of variations in fuel channel temperature between the top (12 o'clock) and bottom (6 o'clock) positions of the tube.

Experimental Results

Table 3 summarizes the results from the 13 matched pairs of standard burst- and curved-compact-specimen tests, together with those from Tubes P3L03 and P3N08 using the 50-mm-

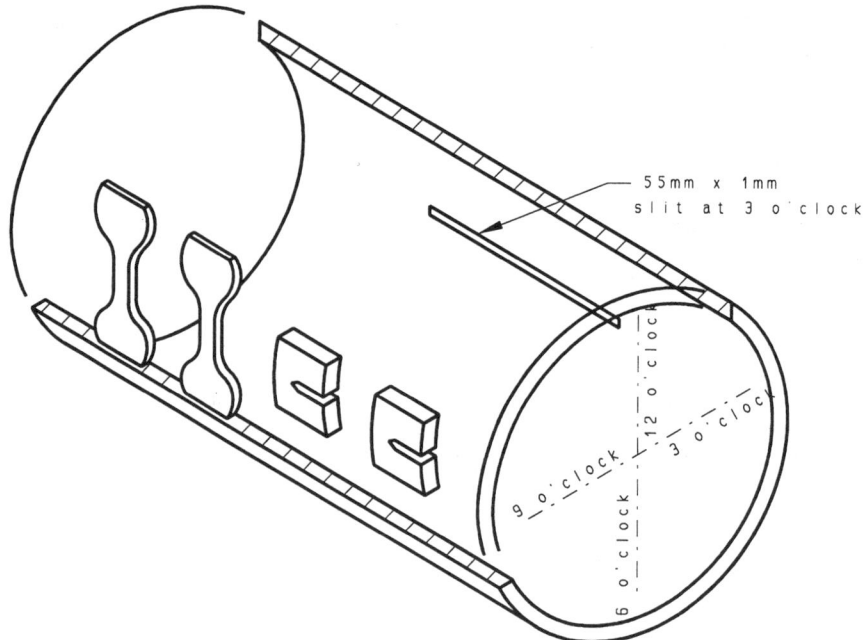

FIG. 3—*Schematic diagram indicating small specimen sampling from burst test section.*

long cracks. The transverse flow stress of the different sections is also included. For irradiated, cold-worked Zr-2.5Nb, the ultimate tensile strength (UTS) generally occurs at very low plastic strains, indicative of the low work-hardening behavior of this material [4,9,10]. This results in there being little distinction between the 0.2% offset yield stress and the UTS and therefore the flow stress of the irradiated material. As anticipated, tubes from fuel channels P4L17, P3R13, and P4Q17, fabricated from 100% recycled material (low chlorine concentration), exhibited the highest toughness values.

The normalized burst stress based on the instantaneous crack size, σ_{mp}/σ_{plc} (see Table 3 for definition), for the tubes of different toughness values varied from 0.30 (lowest toughness tube, P3M11) to 0.67 (highest toughness tube, P4Q17), i.e., the maximum burst stress was less than 70% of the plastic collapse stress in all cases. This indicates that the use of the small-scale yielding equation based on the Dugdale strip yield model (Eq 1) is a reasonable approximation. In comparison, the normalized maximum load, P_{ml}/P_{ll}, for the small specimens varied from 0.60 (P3M11) to 1.11 (P4Q17) based on the plane-stress limit load, with these values being reduced by 14% using the plane-strain limit load. Thus, the small specimens of highest toughness failed at loads at or close to the limit load.

Typical pressure versus crack-extension and *J-R* curves showing the range in results obtained for the different tubes are given in Figs. 4 and 5, respectively, where results from the tubes of lowest toughness values (P3M11), intermediate toughness values (P3N07 and P3N08), and highest toughness values (P4Q17) are included. A comparison of the *J-R* curves obtained from matched sets of standard burst- and compact-specimen tests is given in Figs. 6, 7, and 8 for Tubes P3M11, P3N08, and P4Q17, respectively.

After about 3 mm of stable extension at each crack tip, there was little further increase in pressure, with final crack instability occurring after varying amounts of further stable crack

TABLE 3—Summary of results from burst test and curved compact specimens tested at 250°C.

Tube ID	Fluence, 10^{25} n·m^{-2}	Irradiation Temp., °C	σ_f, MPa	Burst Test Results					Curved Compact Specimen Results				
				Δa_{mp},[a] mm	σ_{mp}[b]$/\sigma_{plc}$[c]	$J_{3.0}$, kJ·m^{-2}	J_{mpi}, kJ·m^{-2}		Δa_{ml},[d] mm	P_{ml}/P_{ll},[e] pl. stress	P_{ml}/P_{ll},[e] pl. strain	$J_{3.0}$, kJ·m^{-2}	J_{ml}, kJ·m^{-2}
P3M11	9.61	257	896	3.65	0.301	68.4	80.5		1.03	0.595	0.511	45.4	24.5
P3J17	10.50	267	886	3.82	0.331	86.6	97.2		1.38	0.747	0.639	60.5	40.1
P3K05	10.70	272	843	9.00	0.446	105	186		1.49	0.843	0.722	72.4	46.0
P3M12	10.36	268	876	11.7	0.460	95.9	239		1.91	0.944	0.805	68.8	48.6
P3L03	9.74	265	870	5.96	0.424	108	159		1.63	0.786	0.671	65.0	39.2
P3O07	10.20	266	881	6.51	0.433	110	181		1.27	0.654	0.562	54.5	36.2
P3M03	9.84	266	827	9.39	0.469	108	220		1.51	0.916	0.785	83.9	51.0
P3H07	9.26	266	794	9.88	0.512	113	244		1.24	0.895	0.767	89.7	55.8
P3M09	10.10	275	735	7.44	0.546	127	231		1.65	1.031	0.882	86.6	56.8
P3N07	10.24	268	847	10.2	0.583	135	363		1.64	0.887	0.758	72.0	46.8
P3S13	10.30	272	771	6.75	0.647	227	361		1.25	1.008	0.865	102	66.2
P3N08	10.20	273	862	5.40	0.599	246	336		1.07	0.950	0.817	160	84.8
P4L17	10.92	267	867	5.38	0.626	320	398		0.716	1.020	0.878	263	103
P3R13	10.27	266	840	3.94	0.617	327	353		0.830	1.007	0.867	192	88.5
P4Q17	11.12	267	852	3.99	0.671	423	458		0.619	1.113	0.961	406	156

[a] Δa_{mp} = average stable crack growth per crack tip at maximum pressure.
[b] σ_{mp} = hoop stress at maximum pressure.
[c] σ_{plc} = hoop stress at plastic collapse based on instantaneous crack size at maximum pressure, i.e., $\sigma_{plc} = \sigma_f/M$, where σ_f = flow stress (mean of the yield stress and ultimate tensile strength).
[d] Δa_{ml} = crack extension at maximum load.
[e] P_{ml}/P_{ll} = maximum load/limit load based on instantaneous crack size.

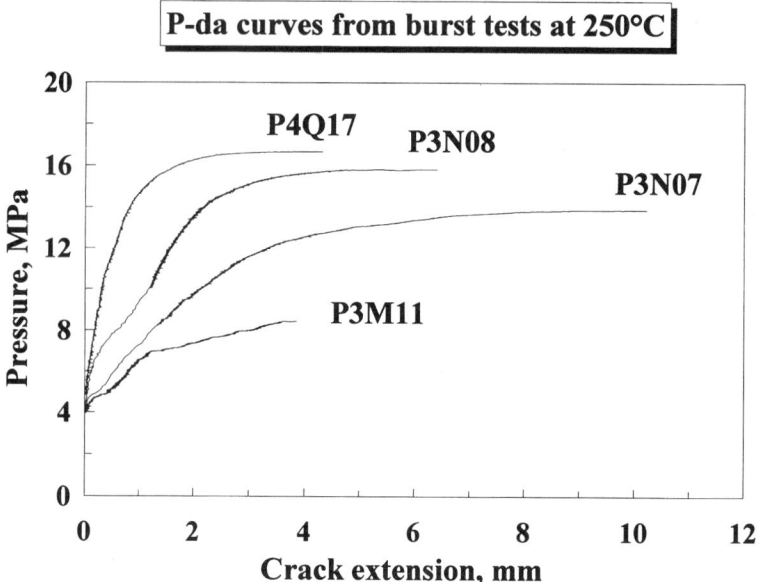

FIG. 4—*Pressure versus crack-extension curves from burst specimens of different toughness (250°C).*

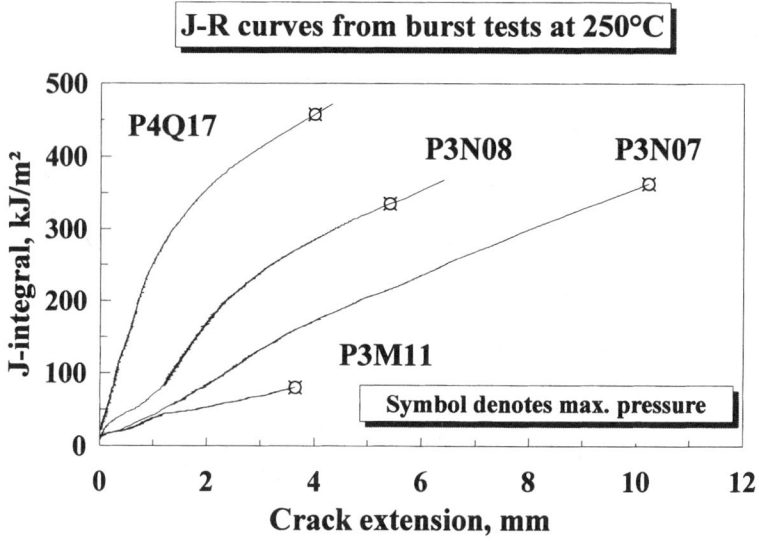

FIG. 5—*Crack growth resistance (J-R) curves from burst specimens of different toughness (250°C).*

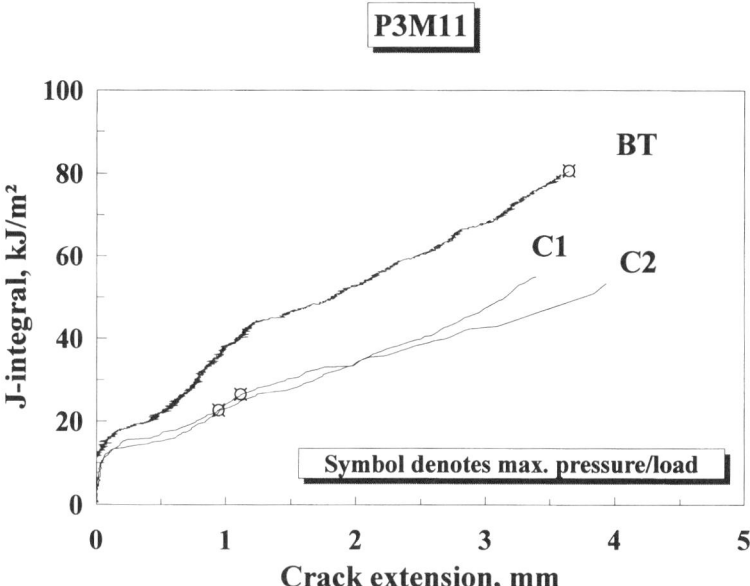

FIG. 6—*Comparison of* J-R *curves from burst test and curved compact specimens (lowest toughness Tube P3M11, 250°C).*

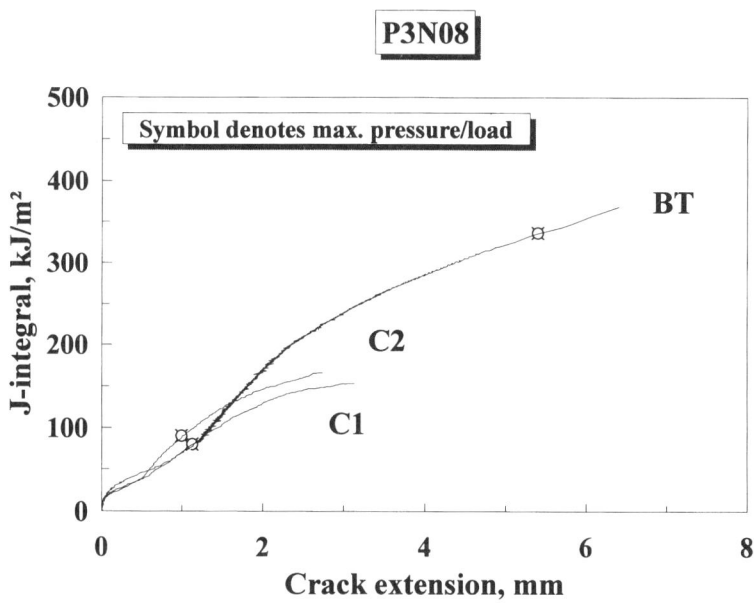

FIG. 7—*Comparison of* J-R *curves from burst test and curved compact specimens (intermediate toughness Tube P3N08, 250°C).*

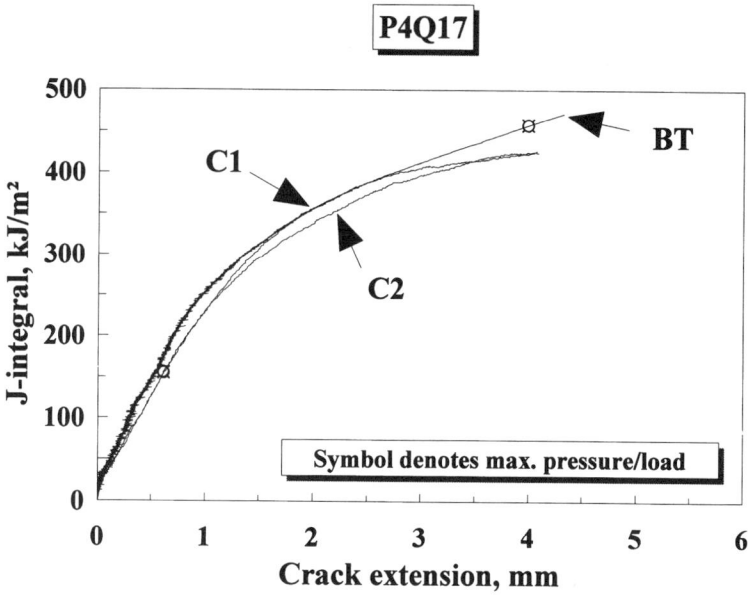

FIG. 8—*Comparison of J-R curves from burst test and curved compact specimens (highest toughness Tube P4Q17, 250°C).*

growth according to the toughness level. For example, the two tubes of lowest toughness (P3M11, P3J17) and the two tubes of highest toughness (P3R13, P4Q17) failed after 3.6 to 4 mm of crack growth, with tubes of intermediate toughness exhibiting more variation as well as larger amounts of stable crack growth before instability, Δa_{mp}, of up to 12 mm. The variability in Δa_{mp} for the tubes of intermediate toughness values is believed to be due, in part, to the difficulty in defining a true crack instability point when the pressure versus crack-extension curve is relatively flat over a significant crack-size range (see Fig. 4). Since none of the burst tests exhibited crack instability at 250°C until after a minimum of 3 mm of stable crack growth, the *J*-integral value at the 3-mm offset, $J_{3.0}$, was selected as an alternative parameter to the maximum pressure toughness, J_{mpi}, to characterize the *J-R* curves at larger crack extensions.

The majority of the *J-R* curves showed evidence of three-stage crack growth behavior with a significant reduction in slope at short crack extensions (Stage 1) followed by an increase in slope (Stage 2) and subsequent decrease in slope close to instability (Stage 3) (Type A crack growth behavior). However, with increasing toughness, the extent of Stage 1 decreased until the *J-R* curves for the tube of highest toughness, P4Q17, exhibited only a steadily decreasing slope at all crack extensions (Type B crack growth behavior). Although the shape of the *J-R* curves from the burst tests and small specimens showed similar trends, the maximum-load toughness, J_{ml}, was achieved at much shorter crack extensions of 0.6 to 1.9 mm for the small specimens (see Table 3) than the range of 3.6 to 11.7 mm achieved at maximum pressure in the burst tests (see Table 3). In fact, the majority of *J-R* curves from the small specimens were well below those from the corresponding burst tests (Figs. 6 and 7). However, the region of correspondence improved with increasing toughness such that there was excellent agreement between the curves from the two different specimens for the highest toughness tube, P4Q17 (Fig. 8).

This trend of improved agreement between the *J-R* curves from small- and large-scale spec-

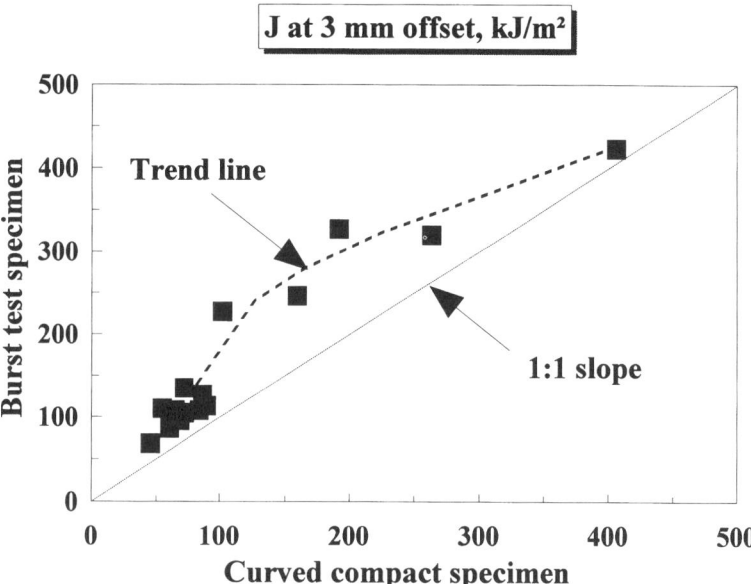

FIG. 9—*Comparison of $J_{3.0}$ toughness parameter determined from burst tests and curved compact specimens (250°C).*

imens at shorter crack extensions and higher toughness levels is reflected in the $J_{3.0}$ parameter (Fig. 9). In particular, there appears to be a maximum difference in the values for the two different types of specimens at intermediate toughness levels. This difference in *J-R* curve behavior for tubes of intermediate toughness values is also reflected in the maximum pressure toughness, J_{mpi}, plotted versus the maximum load toughness from the small specimens, J_{ml}, in Fig. 10. (Note that in this case the results do not revert to a 1:1 correspondence at higher toughness values as a result of the different crack extensions up to the maximum pressure/load toughness.) The shape of the trend line suggests a transition in crack growth behavior for tubes of intermediate toughness for which the stable growth at instability is a maximum value. Again, the scatter in the values of J_{mpi} for these tubes is partly due to the uncertainty in defining the crack instability point.

Fractography

A selection of fracture surfaces from burst and curved compact specimens of different toughness values were examined in the scanning electron microscope.

Burst Specimens

Scanning electron fractographs showing tilt angle and normal views from tubes of the lowest (P3M11), intermediate (P3S13), and the highest (P4Q17) toughness values are shown in Figs. 11 to 13. In all cases, a tapered, triangularly shaped, flat-fracture zone developed at the specimen mid-section with 45° shear lips at the surfaces. This flat-fracture zone extended about 5 to 9 mm ahead of the fatigue crack tip, the zone length being longer for the intermediate toughness

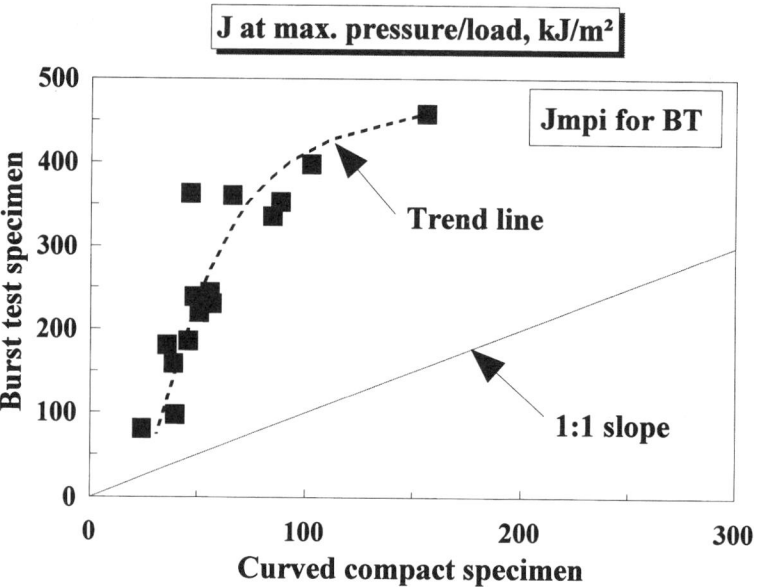

FIG. 10—*Comparison of maximum pressure/load toughness (J_{mpi}, J_{ml}) determined from burst tests and curved compact specimens (250°C).*

specimens. Microscopic examination revealed that failure was by ductile fibrous fracture in all cases.

For the material of lowest toughness (P3M11), there was a sharp transition between the flat-fracture zone and the large shear lips that developed at the specimen surfaces immediately from the fatigue crack (Fig. 11a). There was little evidence of through-thickness yielding (Fig. 11b), and at about 4 mm of crack growth (onset of instability) the fracture surface consisted of about 30% flat and 70% shear fracture. After 6 mm of crack extension, the fracture was predominantly shear. Both the flat- and shear-fracture areas showed evidence of dimple alignment parallel to the crack growth (axial) direction characteristic of "fissures," which were separated either by equiaxed (flat zone) or fine shear (shear region) dimples with a few regions of micron-sized particles and particle clusters. Such fissures are due to preferential void nucleation at micro-segregated species (Zr-Cl-C complex) [11]. They are especially damaging to the toughness of cold-worked Zr-2.5Nb pressure tube material since they effectively divide the specimen into a series of thinner specimens or ligaments, each of which may then fail at a lower fracture strain [9,11].

With increasing toughness, the central, flat-fracture zone became cup-shaped and lengthened, and there was more evidence of yielding through the thickness. In addition, only narrow shear lips developed initially at the fatigue crack tip, as shown in Fig. 12a for Tube P3S13. At larger crack extensions (about 4 to 6 mm of crack growth, corresponding to the position of maximum through-thickness yield (Fig. 12b) and the start of the plateau of the pressure versus crack-extension curve), the shear lip thickness increased rapidly until at instability the fracture consisted of between 50 to 80% shear fracture. Full-shear fracture generally developed at much larger crack extensions of 10 to 12 mm. Microscopic examination also revealed a less "woody" appearance to the fracture surfaces than observed for the tubes of lowest toughness, with fewer fissures and particles, especially in the transition zone between the central, flat fracture at the mid-section and the surface shear lips, i.e., along the sides of the "cup."

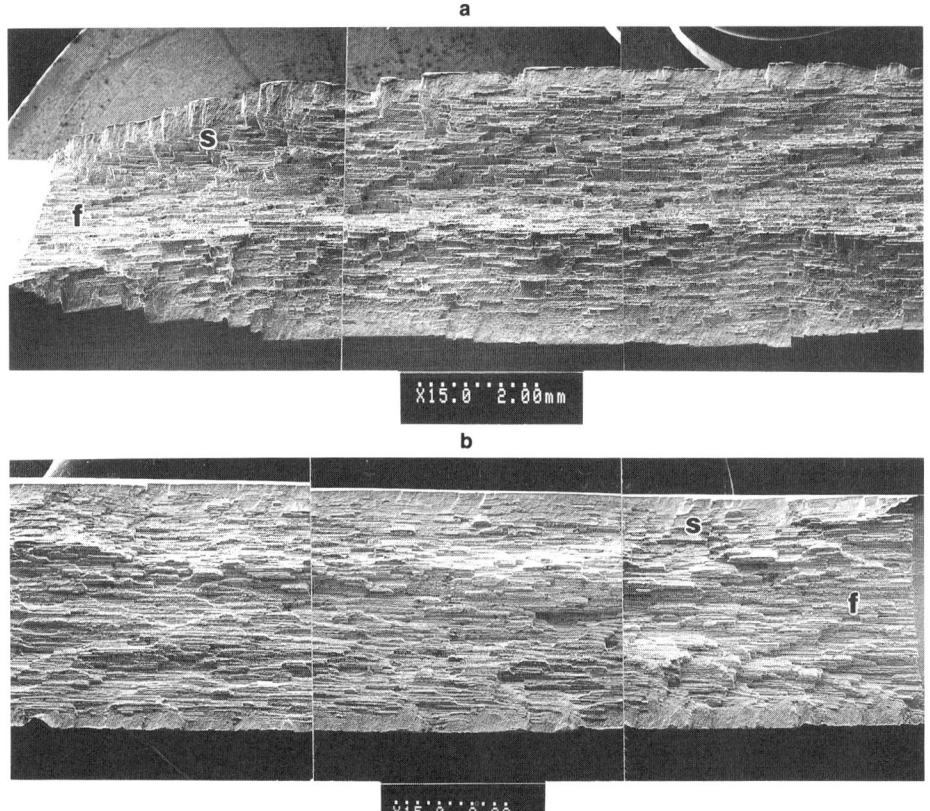

FIG. 11—*Scanning electron fractograph of burst specimen from lowest toughness Tube P3M11 showing triangular flat fracture zone at mid-section (F) and shear or slant fracture developing at surfaces (S):* (a) *tilt angle view at 45°, and* (b) *normal view.*

The fractures from the three tubes of highest toughness values (P4L17, P3R13, and P4Q17) exhibited the deepest cup-shaped fractures, with wide transition zones between narrow zones of flat fracture at the mid-section and the surface shear lips, as shown in Fig. 13a for Tube P4Q17. This is consistent with the significant through-thickness yielding exhibited by these specimens (see Fig. 13b), which increased to a maximum of 15, 18, and 22% for Tubes P4L17, P3R13, and P4Q17, respectively, at about 2.5 mm from the fatigue crack tip. This position also corresponded to the crack length at which there was a sharp increase in the thickness of the shear lips developing from the surface, as well as the start of the plateau of the pressure versus crack-extension curve. At the onset of instability (at about 4 mm of crack growth for P3R13 and P4Q17 and 5 mm of crack growth for P4L17), the fracture consisted of about 50 and 75% shear fracture, the higher value being obtained for Tube P4Q17, which exhibited a longer plateau in the pressure versus crack-extension curve before instability. Full shear developed after about 6 to 7 mm of crack growth.

Closer examination of these three fractures revealed no evidence of fissures but a high density of particles and particle alignments at the bottom of arrays of dimples aligned in the crack growth (axial) direction of the tube. At the macroscopic level, the dimple alignments bore a

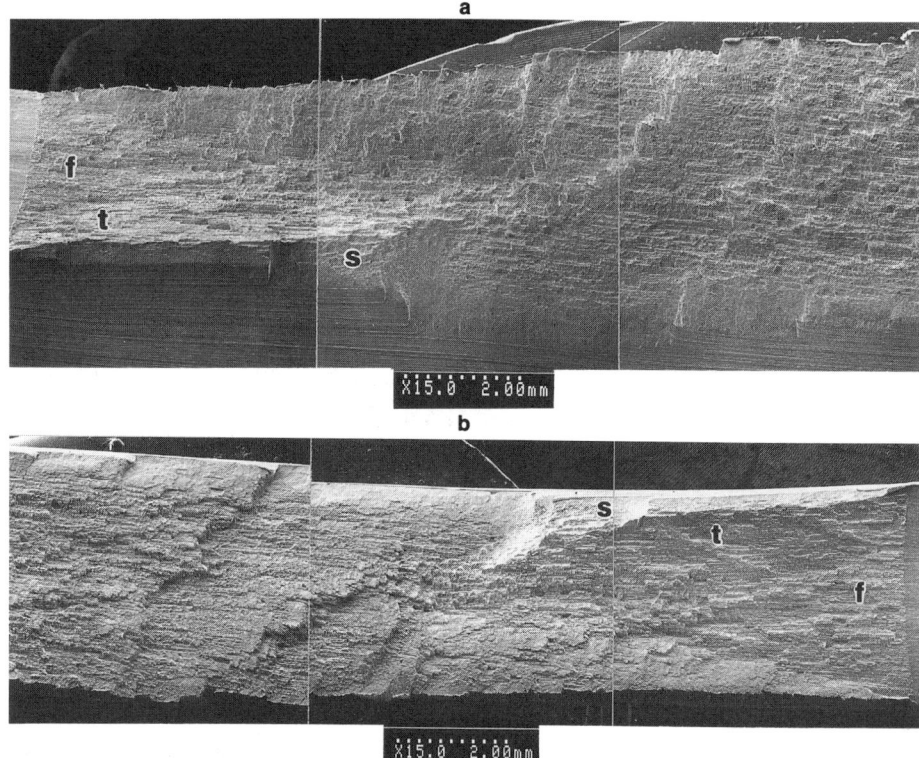

FIG. 12—*Scanning electron fractograph of burst specimen from intermediate toughness Tube P3S13 showing triangular flat fracture zone at mid-section (F), cup-shaped transition zone (T), and shear or slant fracture developing at surfaces (S): (a) tilt angle view at 45° and (b) normal view.*

remarkable resemblance to fissures, but unlike fissures there was evidence of a particle (void nucleation site) at the center of each dimple. Such particles have been identified as zirconium phosphides, which is consistent with the low chlorine concentration but high phosphorus concentration of these early reactor tubes fabricated from 100% recycled material (see Table 2) [10]. The particles were most evident in the narrow zone of flat fracture at the mid-section (region of highest crack-tip constraint) and less evident in the transition zone closer to the specimen surfaces (region of intermediate crack-tip constraint). There was also more evidence of elongated and tearing dimples in this transition zone, indicative of the high-energy absorbing capacity of the material in this region. However, the fine shallow dimples observed in the shear-fracture zones were similar to those observed along the shear lips of the lower toughness tubes.

In summary, the fractographic examination of the burst test sections suggests that the onset of instability is controlled by the geometric requirement of a change in fracture plane from one in the radial-axial plane (normal to the hoop-stress direction) to one close to the planes of maximum shear stress for plane-stress fracture, i.e., planes at 45° to the radial-axial and transverse-axial planes. This occurs as the crack front tunnels forward at the specimen mid-section and shear fracture develops, with increasing crack length, in the region of lowest constraint at the specimen surfaces. However, the formation of a full shear or slant fracture does not appear to be a necessary requirement for such instability to be initiated, merely the development of an

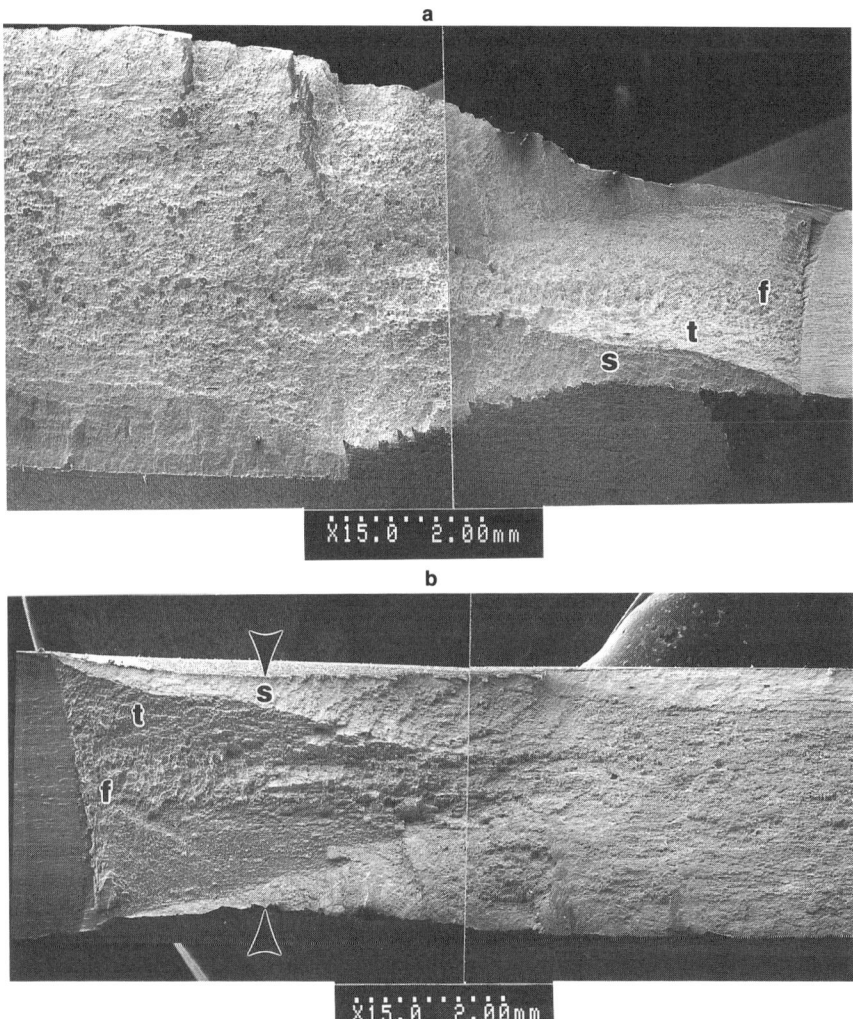

FIG. 13—*Scanning electron fractograph of burst specimen from highest toughness Tube P4Q17 showing triangular flat fracture zone at mid-section (F), wide cup-shaped transition zone (T), and shear or slant fracture developing at surfaces (S): (a) tilt angle view at 45° and (b) normal view. Note significant through-thickness yielding with this specimen indicated by arrows in (b).*

angled fracture with a sufficiently high proportion of slant fracture (between 50 and 80%) to precipitate the sliding-off mechanism. Tube sections of lower or higher toughness values showed a tendency to develop this shear fracture mode at shorter crack extensions, consistent with the less-extended plateaus observed with the pressure versus crack-extension records. The fine, shallow dimples observed in the shear region were also consistent with the shear fracture mode in irradiated Zr-2.5Nb, requiring little further expenditure of energy once initiated.

Variations in toughness values of the different tubes appear to be governed mainly by the ability of the crack front to tunnel forward in the initial stages of crack growth (3 to 4 mm of crack growth) and produce this partial slant fracture. Specimens exhibiting a more cup-shaped fracture with a wider transition zone between the central, flat-fracture region (mid-section) and the surface-shear lips exhibited the highest toughness values, i.e., the initiation of the slant fracture required higher pressure levels. The reduction in fissure and particle density and the increase in incidence of elongated and tearing dimples in this transition zone, compared with the mid-thickness region, was consistent with the development of a higher energy absorbing (toughness) fracture mode in this zone.

Curved Compact Specimens

The fracture surfaces of the curved compact specimens showed similar characteristics to those of the burst specimens from which they were machined. After 3 to 4 mm of stable crack extension, the crack front had tunneled forward at the mid-section, producing a central, flat-fracture zone with narrow shear lips starting to develop at the surfaces. However, at the same equivalent crack extension, the width of the flat-fracture zone was generally wider and the shear lips narrower than observed for the corresponding burst specimen. For example, at 3 mm of crack extension, the widths of the flat- and shear-fracture zone for small specimen P3M11-C1 were about 66 and 33% of the original wall thickness (Fig. 14), respectively, compared with about 32 and 65% for the burst specimen (see Fig. 11b). This suggests a significantly higher, local crack-tip constraint being developed over a larger portion of the wall thickness for the smaller curved compact specimen than for the burst specimen.

A difference in the proportions of flat and shear fracture was also noted for higher toughness materials with slightly more flat fracture observed for the small specimens. For example, the total width of the flat- and shear-fracture zones for small specimen P4Q17-C1 was 62 and 19% of the original wall thickness, respectively (Fig. 15), compared with 53 and 27% for the burst specimen (see Fig. 13b). However, in the case of material of higher toughness, the flat-fracture zone was made up of two distinct regions as mentioned previously: a central, flat-fracture zone

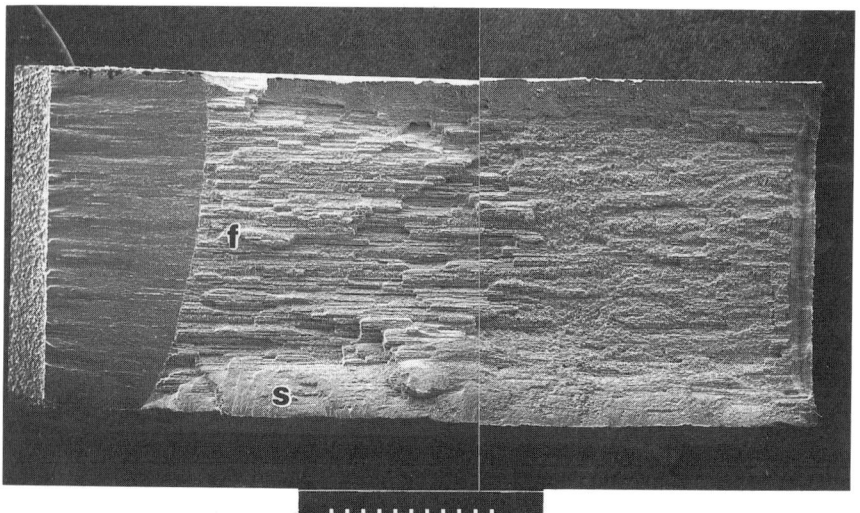

FIG. 14—*Scanning electron fractograph of curved compact specimen from lowest toughness Tube P3M11 showing wide flat fracture zone at mid-section (F) and narrow shear lips developing at surfaces (S). (Normal view.)*

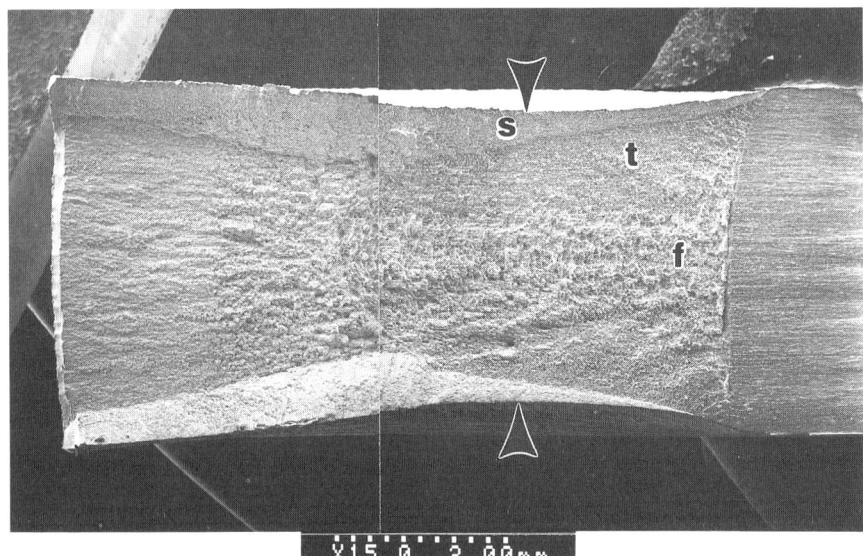

FIG. 15—*Scanning electron fractograph of curved compact specimen from highest toughness Tube P4Q17 showing flat fracture zone at mid-section (F), wide cup-shaped transition zone (T), and narrow shear lips developing at surfaces (S). Note significant through-thickness yielding with this specimen indicated by arrows. (Normal view.)*

at the mid-section and a cup-shaped transition zone closer to the surface. More careful examination revealed that the widths of the flat and transition zones were similar at about 31% of the original wall thickness for the small specimen compared with 22 and 31%, respectively, for the burst specimen. Thus, although there was a difference in the relative proportions of flat and shear fracture for the small- and large-scale specimens, the extent of through-thickness yielding (about 20% of the original wall thickness) and the width of the cup-shaped transition zone (about 31%) were similar.

Such observations suggest that discrepancies observed between the *J-R* curves from the two different specimens can be related to differences in the energy-absorbing capacities and relative proportions of the three different fracture modes observed in the irradiated Zr-2.5Nb pressure tube material, i.e.,

  a. Flat-fracture zone at the mid-section in the region of highest constraint (Stage 1).
  b. Transition zone consisting of a flat/cup-shaped fracture between the mid-section and surface in the region of intermediate constraint (Stage 2).
  c. Shear or slant-fracture at the surface in the region of lowest constraint (Stage 3).

For the material of highest toughness (Tube P4Q17), the *J-R* curves from the small- and large-scale specimens were in good agreement. Thus, the high-energy absorbing capacity of the transition zone (b. above) probably dominated with this fracture such that the discrepancy between the relative proportions of the other two fracture modes (a. and c. above) had a negligible effect on the overall fracture behavior. However, further work is required to quantify the energy-absorbing capacities of these different fracture modes, as well as to conduct more detailed fractographic studies on the development of the fractures as the crack tunnels forward at the mid-section.

Discussion

Size Effects

The earlier test results on the tube of lower toughness, P3L03, and intermediate toughness, P3N08, revealed little effect of initial crack size on the resultant J-R curves [14]. This suggests that in each case a similar level of crack-tip constraint was achieved for the tube sections of different material. In a burst test, the tensile axial stress should raise the local crack-tip stress, and bulging should have a similar effect by producing a sharp stress gradient (bending) ahead of the crack tip. Since bulging increases with increasing crack length, the results suggest that the axial stress has a stronger influence on the transverse hoop stress than bulging in raising the crack-tip constraint, i.e., that the crack-tip constraint is mainly governed by the biaxial stress state in the thin-walled tubes. However, further tests are required on tubes of higher toughness values, for which the bulging is expected to have the greatest influence due to the larger crack openings, to confirm the absence of a crack-size effect on the J-R curve behavior over the entire toughness range.

Geometry Effects

Comparison between the J-R curves obtained from the small- and large-scale specimens has clearly demonstrated an effect of material toughness on the observed geometry effects. For example, for the majority of tubes, there was little or no region of agreement (<0.1 mm of crack extension) between the J-R curves from the two different geometries. In contrast, for the higher toughness tubes (P3N08, P4L17, and P3R13), the J-R curves were in reasonable agreement up to 0.7 to 1 mm (about 10 to 15% of the remaining ligament, $b = W - a$), this region increasing to 2 to 2.5 mm (about 25 to 30% of b) for the highest toughness tube, P4Q17. In fact, the J-R curves for irradiated Tube P4Q17 are very similar to those obtained previously for unirradiated Tube C70 tested at room temperature, for which there was agreement between the J-R curves from the burst test and curved compact specimens over a similar crack extension range of about 2 mm [5]. Thus, in most cases the crack-tip stress state ahead of the growing crack in the small specimen appears to be very different from that in the burst test, with the crack-tip constraint in the uniaxial, bend-type specimen being generally very much higher than that in the biaxial, tension-type specimen. The fractographic evidence is consistent with this, with a larger proportion of highly constrained, flat fracture being produced with the small, bend-type specimen.

The material dependence of the extent of geometry independence of the J-R curves is consistent with the known validity requirements for J-controlled crack growth (JCCG), i.e., for a Hutchinson, Rice, and Rosengren (HRR) deformation-type stress and strain field to be maintained at the crack tip [15,16]. For example, the requirement that the increase in external load be sufficiently large for the plastic strain increments to be predominantly proportional (ω parameter = $bdJ/daJ > 10$ for bend-type specimens) could only be met over a reasonable J- or crack-extension range for the compact specimens of highest toughness. In addition, although there is a requirement on the maximum crack extension range for JCCG to ensure limited unloading at the crack tip (α parameter = $\Delta a/b < 0.06$ to 0.10), more recent work by others has demonstrated that this limit on JCCG is not unique, but dependent on material type [17]. This is in agreement with the large α-values of 25 to 30% obtained here for irradiated Tube P4Q17, as well as for unirradiated Tube C70 [5].

In fact, a high-strength, low work-hardening material, such as cold-worked Zr-2.5Nb pressure tube material, is expected to develop more significant geometry effects compared with a lower-strength, higher work-hardening material due to premature breakdown of the HRR deformation field [18]. The low-energy fractures observed in the flat- and shear-fracture zones of irradiated material are also likely to be more sensitive to stress state than the ductile fibrous fracture observed in the transition zone, as evidenced by the sensitivity of the onset of shear or strain localization to the presence of free boundaries [19]. These factors account for the significant geometry effects observed in the majority of cases with the present irradiated material, e.g., the development of microscopic shear fractures between closely spaced fissures and the development of macroscopic shear or slant fractures between the transition zone and the specimen surfaces. The large difference in the toughness parameters, $J_{3.0}$ and J_{mpi}, noted previously for the tubes of intermediate toughness, for which the stable crack growth at instability is largest, is likely to be due to these tubes exhibiting the maximum deviation from both small-scale and fully plastic-yielding conditions for JCCG.

Due to the stringent size limitations for JCCG, there is now considerable work underway on developing a two-parameter characterization of the crack-tip stress field for lower-constraint geometries [20–23]. For example, the approach of Shih [20] uses J for scaling the crack-tip process zone (of large stress and strain) and a second parameter, the Q-stress (or the elastically equivalent T-stress), for scaling the near-tip stress distribution relative to a reference high-triaxiality stress state. The Q-stress refers to the stress acting in the crack growth direction, which may be tensile (positive) and raise the local crack-tip stress (increase constraint) or compressive (negative) and lower the crack-tip constraint. Within this framework, bend-type specimens exhibit a high crack-tip constraint (positive or tensile Q-stress) compared with the low constraint of a uniaxial or biaxial center-cracked tensile specimen (negative or compressive Q-stress) [20]. Wide-spread plasticity reduces the overall level of crack-tip constraint in both cases. The current experimental results for the curved compact and biaxial burst specimens are generally consistent with these numerical studies.

Scaling Fracture-Toughness Parameters

An alternative approach to size and geometry effects has been adopted by Turner [24–26]. In this approach, the plastic work dissipation is governed by parameters related to the work per unit area (plane-strain fracture) and the work per unit volume (plane-stress fracture). For a given small specimen geometry, this plastic work dissipation may then be limited by some limiting toughness or characteristic dimension (e.g., the specimen thickness, B, or remaining ligament, b), resulting in the observed size and geometry effects. Turner has argued that by scaling one or both of the axes of the J-R curve by this characteristic dimension, it should be possible to correlate small- and large-scale specimen results. For volume-controlled fracture, Turner derived the following relationship for deeply-cracked bend specimens [24]:

$$\frac{dJ}{da} = \frac{2\gamma_{vol} s^2}{Bb} \quad (2)$$

where

γ_{vol} = work per unit volume, and
s = width of shear lip.

Turner's model is based on the development of high-energy-absorbing shear lips, which is governed by the plane-stress plastic zone size, r_y [24]. Therefore, substituting r_y for the shear lip width, s, the above volume-controlled model becomes:

$$\frac{dJ}{da} = \frac{2\gamma_{vol}r_y^2}{Bb} \qquad (3)$$

In contrast, the toughness of irradiated Zr-2.5Nb pressure tube material appears to be dominated by the development of the transition zone in the region of intermediate constraint (Stage 2). However, it is assumed a similar volume-controlled fracture model may be applied, albeit that the development of the plane-stress plastic zone, r_y, is now associated with this transition zone, which is a precursor to the formation of the low-energy-absorbing shear lips.

In fact, compact specimens of different widths of unirradiated Zr-2.5Nb pressure tube material were tested previously [27], and good agreement was obtained between the experimental results and Eq 3. The unirradiated material exhibited distinct flat/cup-shaped fractures with significant through-thickness yielding and negligible shear lips. In particular, it was demonstrated that, for a given toughness level, the plane-stress plastic zone radius, taken as $EJ/2\pi\sigma_f^2$, is limited by the remaining ligament [27], being directly proportional to $\sqrt{(bB)}$, or more specifically $\sqrt{b}$, since the specimen thickness was not varied. Experimental results from a recently developed curved compact specimen having a width of 34 mm are also generally consistent with such a $\sqrt{b}$ dependence [5,28] in spite of some additional, out-of-plane bending associated with the difficulty of testing wide, curved specimens. Therefore, for the small specimen results, $\sqrt{(bB)}$ appears to be the most promising scaling factor.

For the burst specimens, there was no significant evidence of a crack-size effect for toughness levels up to that of the tube of intermediate toughness, P3N08 [14]. Therefore, any characteristic dimension (or scaling parameter) is likely to be related to dimensions such as the radius, r, and wall thickness, t, rather than the crack length, a. In fact, the best agreement between the scaled, maximum-pressure/load parameters, J_{mpi} and J_{ml}, is obtained using the parameter $\sqrt{(2r_m t)}$, where r_m is the mean radius. This is shown in Fig. 16, which is a normalized (dimensionless)

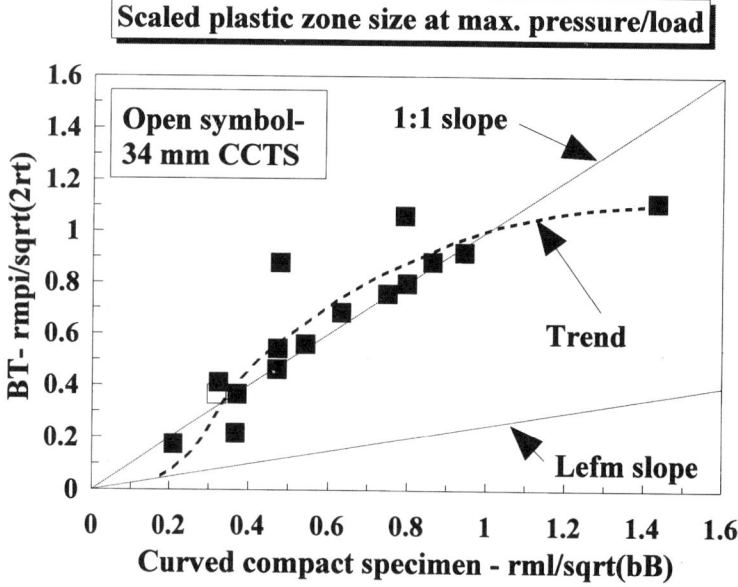

FIG. 16—*Comparison of scaled plastic zone size at maximum pressure/load (r_{mpi}, r_{ml}) from burst tests and curved compact specimens (250°C).*

plot based on the corresponding plastic zone sizes, r_{mpi} and r_{ml}, where for consistency with the strip-yield Eq 1, the Dugdale small-scale yielding expression, $\pi EJ/8\sigma_f^2$, is used in both cases. Figure 16 includes the mean value of four results obtained from irradiated Tube P3L03 using the wider (34-mm) curved compact specimen [28]. The scaling parameter $\sqrt{(2r_m t)}$ might be considered a measure of the degree of constraint resulting from biaxial stress and/or bulging effects, the latter being proportional to $\sqrt{(rt)}$ [13].

The results in Fig. 16 suggest that, for the majority of tubes, the scaling parameters of $\sqrt{(bB)}$ for the small specimens and $\sqrt{(2r_m t)}$ for the burst tests produce reasonable correspondence between the small- and large-scale specimens over a range of normalized plastic zone size of about 0.4 to 1. However, deviations from such a relationship might be expected, especially at lower toughness levels. For example, in the linear elastic fracture regime, the maximum load toughness from the small specimens, J_{mi}, should equal the maximum pressure toughness from the burst tests, J_{mpi}. The trend line in Fig. 16 indicates that the experimental results deviate from the linear elastic fracture mechanics (LEFM) slope at J values close to the limiting level for a valid linear elastic fracture toughness as provided in ASTM Test Method for Plane-Strain Fracture Toughness of Metallic Materials (E 399), i.e., $K_I \leq \sigma_y \sqrt{(B/2.5)} \leq 35$ MPa$\sqrt{m}$ or equivalently, $J = K^2(1 - \nu^2)/E \leq 11$ kJ.m^{-2} or $(\pi EJ/8\sigma_f^2)/\sqrt{(bB)} \leq 0.1$.

At higher toughness levels, $(r_{mpi}/\sqrt{(2r_m t)}$ or $r_{ml}/\sqrt{(bB)} > 1)$, the results suggest a further deviation of results for the tube of highest toughness, P4Q17. This could be due to the onset of a crack-size dependency for the maximum-pressure toughness, J_{mpi}, with the burst test as a result of increased bulging. However, a change in the ligament dependency of the maximum load toughness, J_{ml}, for the small specimen is also possible. For example, at lower toughness levels, the plastic zone size is limited by the remaining ligament due to the plastic hinge action ahead of the crack tip in bend-type specimens, the formation and reformation of the hinge with crack growth being determined by the toughness and ratio of the ligament size to the wall thickness, b/B [24]. However, for higher toughness material, the width of the plastic zone in which hinging occurs should be governed more by the wall thickness, as fully plane-stress conditions predominate. In fact, previous work on compact specimens of unirradiated material of different width demonstrated a reduction in the ligament dependency at higher toughness levels [27]. Such potential size and geometry effects for higher toughness material require further investigation.

In theory, once the scaling factors have been determined for the two different configurations, it should be possible to scale the J-R curve obtained from the curved compact specimen to produce that of the corresponding burst test. This approach was investigated using one of the small specimen J-R curves from an intermediate toughness tube, P3M03. The results are shown in Fig. 17, where both the J-integral and crack-extension results for small specimen C1 are scaled by the parameter $\sqrt{(2r_m t/bB)}$, where b is taken as the instantaneous remaining ligament. Scaling the J-R curve determined from the small specimen produces excellent agreement with the burst test results up to a crack extension of about 4.5 mm. However, at larger crack extensions the scaled, small specimen results show an increase in slope compared with a reduction in slope for the results from the burst test before crack instability at 9.4 mm. These slope changes suggest the onset of additional size and geometry effects before the maximum load-bearing capacity of the different specimens is achieved. For example, a reduction in slope before crack instability was observed with the majority of burst test and corresponded to the start of the plateau in the pressure versus crack-extension curve. This is probably due to an increase in the out-of-plane bending (bulging) just before failure. However, such crack growth behavior indicates the difficulty in scaling J-R curves from small specimens, where scaling both axes may be required. Under such circumstances, scaling the maximum-pressure/load toughness alone may prove a more viable approach for failure prediction.

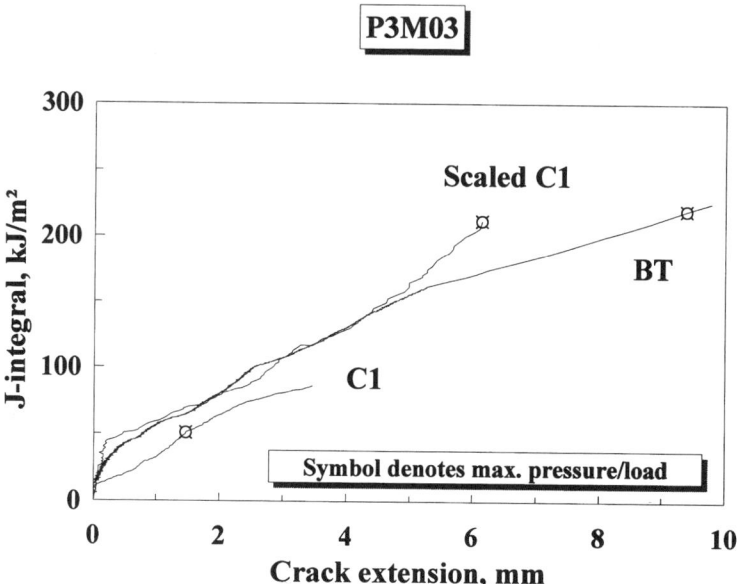

FIG. 17—*Comparison of the J-R curve from a burst test with those from a curved compact specimen before and after scaling (intermediate toughness Tube P3M03, 250°C).*

Summary

The link between the results of small- (curved compact) and large-scale (burst test) toughness tests on irradiated Zr-2.5Nb pressure tube material has been investigated. The results are summarized below.

1. Burst tests on tubes of low and intermediate toughness values revealed no significant effect of initial crack size on the *J-R* curves, suggesting that the axial stress had a stronger influence on the transverse hoop stress than bulging in raising the crack-tip constraint.
2. *J-R* curves from small- and large-scale specimens of material of different toughness values confirmed the material dependence of geometry effects and showed that the crack-tip constraint in the uniaxial, bend-type (compact) specimen is generally much higher than that in the biaxial, tension-type (burst test) specimen. For higher toughness material, the crack extension region over which there is good correspondence between the *J-R* curves from the small- and large-scale specimens is in agreement with current knowledge of validity requirements for *J*-controlled crack growth of bend-type specimens.
3. The fractographic evidence was consistent with the geometry effects being due to variations in the relative proportions of three different fracture modes:
   a. Flat-fracture zone at the mid-section in the region of highest constraint (Stage 1).
   b. Transition zone consisting of a flat/cup-shaped fracture between the mid-section and surface in the region of intermediate constraint (Stage 2).
   c. Shear or slant-fracture zone at the surface in the region of lowest constraint (Stage 3).

   In particular, a larger proportion of highly constrained, flat fracture was produced with the small, bend-type specimen than with the burst specimen.
4. Assuming the toughness to be dominated by the development of the plastic zone asso-

ciated with the transition zone (b., above), a volume-controlled fracture model for bend-type specimens was applied to the results from the small specimen tests. The approach produced good agreement between scaled values of the maximum pressure/load toughness determined from small- and large-scale tests over a range of normalized plastic zone size of 0.4 to 1, suggesting that scaling small specimen toughness values may prove a viable method for predicting pressure tube behavior.

Further work is required on tubes of lower and higher toughness (normalized plastic zone size <0.4 and >1.0) to investigate deviations from the observed material behavior. In particular, tests are required on material of high toughness to determine any crack-size dependence of J-R curves at higher toughness values.

Acknowledgments

Thanks are due to R. Behnke, G. R. Brady, and A. R. Reich for their technical assistance. Special thanks are due to A. K. Järvine, R. R. Bawden, and J. C. Owens for arranging shipments of material and the hot cell staff at WL for conducting the tests. Funding of this work by the CANDU Owner's Group (COG) under Work Packages 3192, 6506, and 6511 is gratefully acknowledged.

References

[1] Moan, G. D., Coleman, C. E., Price, E. G., Rodgers, D. K., and Sagat, S. in *International Journal of Pressure Vessels and Piping,* Vol. 43, 1990, pp. 1–21.
[2] Cheadle, B. A., Coleman, C. E., Rodgers, D. K., Davies, P. H., Chow, C. K., and Griffiths, M., AECL-9710, AECL Research, November 1988.
[3] Langford, W. J. and Mooder, L. E. J. in *International Journal of Pressure Vessels and Piping,* Vol. 6, 1978, pp. 275–309.
[4] Chow, C. K., Coleman, C. E., Hosbons, R. R., Davies, P. H., Griffiths, M., and Choubey, R. in *Zirconium in the Nuclear Industry: Ninth International Symposium, ASTM STP 1132,* American Society for Testing and Materials, West Conshohocken, PA, 1992, pp. 246–275.
[5] Davies, P. H. Shewfelt, R. S. W., and Järvine, A. K. in *Constraint Effects in Fracture, Theory and Applications: Second Volume, ASTM STP 1244,* American Society for Testing and Materials, West Conshohocken, PA, 1995, pp. 392–424.
[6] Chow, C. K. and Simpson, L. A. in *Fracture Mechanics: Eighteenth Symposium, ASTM STP 945,* American Society for Testing and Materials, West Conshohocken, PA, 1988, pp. 419–439.
[7] Simpson, L. A., Chow, C. K., and Davies, P. H., CANDU Owner's Group Report No. COG-89-110-1, AECL Research, September 1989.
[8] Cheadle, B. A., Coleman, C. E., and Licht, H. in *Nuclear Technology,* Vol. 57, 1982, pp. 413–425.
[9] Davies, P. H., Hosbons, R. R., Griffiths, M., and Chow, C. K. in *Zirconium in the Nuclear Industry: Tenth International Symposium, ASTM STP 1245,* American Society for Testing and Materials, West Conshohocken, PA, 1994, pp. 135–167.
[10] Davies, P. H., Aitchison, I., Himbeault, D. D., Järvine, A. K., and Watters, J. F. in *Fatigue and Fracture of Engineering Materials and Structures,* Vol. 18, No. 7/8, 1995, pp. 789–800.
[11] Aitchison, I. and Davies, P. H. in *Journal of Nuclear Materials,* Vol. 203, 1993, pp. 206–220.
[12] Kiefner, J. F., Maxey, W. A., Eiber, R. J., and Duffy, A. R. in *Progress in Flaw Growth and Fracture Toughness Testing, ASTM STP 536,* American Society for Testing and Materials, West Conshohocken, PA, 1973, pp. 461–481.
[13] Folias, E. S. in *International Journal of Fracture Mechanics,* Vol. 1, No. 2, 1965, pp. 104–113.
[14] Davies, P. H. and Shewfelt, R. S. W., CANDU Owner's Group Report No. COG-95-3, AECL RC-1352, March 1995.
[15] Hutchinson, J. W. in *Journal of Mechanics and Physics of Solids,* Vol. 16, 1968, pp. 3–31 and pp. 337–347.
[16] Rice, J. R. and Rosengren, G. F. in *Journal of Mechanics and Physics of Solids,* Vol. 16, 1968, pp. 1–12.

[17] Jones, R. L, Gordon, J. R., and Challenger, N. V. in *Fatigue and Fracture of Engineering Materials and Structures,* Vol. 14, 1991, pp. 777–788.
[18] Shih, C. F. in *Journal of Mechanics and Physics of Solids,* Vol. 29, 1981, pp. 305–326.
[19] Mudry, F., *The Assessment of Cracked Components by Fracture Mechanics, EGF4,* Mechanical Engineering Publications, London, 1989, pp. 133–160.
[20] Shih, C. F., O'Dowd, N. P., and Kirk, M. T. in *Constraint Effects in Fracture, ASTM STP 1171,* American Society for Testing and Materials, West Conshohocken, PA, 1993, pp. 2–20.
[21] Hancock, J. W., Reuter, W. G., and Parks, D. M. in *Constraint Effects in Fracture, ASTM STP 1171,* American Society for Testing and Materials, West Conshohocken, PA, 1993, pp. 21–40.
[22] Roos, E., Eisele, U., and Silcher, H. in *Constraint Effects in Fracture, ASTM STP 1171,* American Society for Testing and Materials, West Conshohocken, PA, 1993, pp. 41–63.
[23] Brocks, W. and Schmitt, W. in *Constraint Effects in Fracture, ASTM STP 1171,* American Society for Testing and Materials, West Conshohocken, PA, 1993, pp. 64–78.
[24] Turner, C. E., *Size Effects,* Mechanical Engineering Publications, London, 1986, pp. 25–31.
[25] Gibson, G. P., Druce, S. G, Turner, C. E. in *International Journal of Fracture,* Vol. 32, 1987, pp. 219–240.
[26] Turner, C. E. and Braga, L. in *Constraint Effects in Fracture, ASTM STP 1171,* American Society for Testing and Materials, West Conshohocken, PA, 1993, pp. 158–175.
[27] Davies, P. H. and Smeltzer, J. M. in *Fracture Mechanics: Twenty-Second Symposium, ASTM STP 1131,* American Society for Testing and Materials, West Conshohocken, PA, 1992, pp. 93–104.
[28] Järvine, A. K., Longhurst, G. C., and Smeltzer, J. M., CANDU Owner's Group Report No. COG-94-179, AECL Research, April 1994.

DISCUSSION

C. Lemaignan[1] (written discussion)—For leak-before-break safety analysis, the amount of stored energy is a critical parameter. Could you explain why you used argon pressurization instead of oil or water, a condition that is closer to the reactor environment? Have you any idea of the effect of gas-pressure-stored energy on axial split enhancement?

P. H. Davies et al. (authors' closure)—Argon gas, rather than oil or water, was used for pressurization in the hot cells for practical reasons, i.e., due to the reduction in clean-up operations and the disposal of contaminated fluids required after testing. Although the use of argon gas will clearly promote a fast running (dynamic) crack in a tube once the crack instability (maximum pressure) condition is achieved (due to the high-energy storage), it is believed that the pressurization fluid should have little effect on the achievement of that crack instability (maximum pressure) condition, which is the factor of primary interest here. To simulate reactor operating conditions more closely, tests are also being conducted using the CRACLE (Chalk River Active Channel Leak Evaluation) test facility. This facility uses tube sections in which axial, through-wall cracks are grown by delayed hydride cracking to failure with leakage. An objective of the large-scale testing program on pressure tube sections at AECL is to compare the crack instability condition (critical crack length) obtained by the two different techniques.

I. J. O'Donnell[2] (written discussion)—The fractography clearly demonstrates differences between the "centre-cracked" tube and the bend specimen. Can you rationalize your results on the basis of the developing J-T or J-Q technology?

P. H. Davies et al. (authors' closure)—As stated in the paper, bend-type specimens are known to exhibit a high crack-tip constraint (positive or tensile T- or Q-stress) compared with the low constraint of a uniaxial or biaxial center-cracked tensile specimen (negative or compressive T- or Q-stress) [Ref 20 in paper], the current experimental results for the curved geometries being generally consistent with these numerical results. However, attempts to determine the T-stress for an internally pressurized tube with an axial, through-wall crack have proven extremely difficult. This was indicated by B. W. Leitch (AECL, Whiteshell Laboratories, Pinawa) during a presentation on this topic at the ASTM 27th National Symposium on Fatigue and Fracture Mechanics, Williamsburg, VA, 27–29 June 1995. The difficulty of analysis appears to be related to the significant out-of-plane bending associated with an axial, through-wall crack in a tube.

---

[1] CEA; Grenoble, France.
[2] AEA Technology, England.

N. Christodoulou,[1] A. R. Causey,[1] R. A. Holt,[1] C. N. Tomé,[2]
N. Badie,[3] R. J. Klassen,[1] R. Sauvé,[3] and C. H. Woo[2]

Modeling In-Reactor Deformation of Zr-2.5Nb Pressure Tubes in CANDU Power Reactors

REFERENCE: Christodoulou, N., Causey, A. R., Holt, R. A., Tomé, C. N., Badie, N., Klassen, R. J., Sauvé, R., and Woo, C. H., "**Modeling In-Reactor Deformation of Zr-2.5Nb Pressure Tubes in CANDU Power Reactors**," *Zirconium in the Nuclear Industry: Eleventh International Symposium, ASTM STP 1295*, E. R. Bradley and G. P. Sabol, Eds., American Society for Testing and Materials, 1996, pp. 518–537.

ABSTRACT: Changes in shape of internally pressurized tubes caused by operating temperatures and pressures are enhanced by fast neutron irradiation. Lengths and diameters of Zr-2.5Nb pressure tubes in CANada Deuterium Uranium-Pressurized Heavy Water (CANDU-PHW) power reactors and test reactors have been monitored periodically over the past 20 years. Axial and transverse strain rates have been evaluated in terms of operating variables and the crystallographic texture and anisotropic microstructure of the extruded and cold-drawn tubes. The anisotropic deformation occurring during steady-state irradiation creep and growth is described by a self-consistent model that takes into account the presence of intergranular stresses without building up any discontinuities of strain and stress at the grain boundaries. In this model, it is assumed that climb-assisted glide of dislocations on prismatic, basal, and pyramidal planes is the dominant creep mode and that growth occurs by net fluxes of interstitials and vacancies to a non-random distribution of dislocations and grain boundaries. The predictions from a deformation equation based on data from the Pickering and Point Lepreau Nuclear Generating Stations and the WR1, Osiris, DIDO, and NRU test reactors are in good agreement with measurements of pressure tubes in Bruce units. The equation has been employed as a material subroutine in the 3-D finite element code H3DMAP for predicting the detailed shape change of pressure tubes. The prediction from H3DMAP is a more complete description of shape change than that obtained from the closed-form expression.

KEYWORDS: zirconium alloys, nuclear industry, pressure tubes, in-reactor deformation, irradiation creep, modeling, Zr-2.5Nb, self-consistent, grain interaction stresses, texture, dislocation structure, deformation equation, finite element code

The cold-worked Zr-2.5Nb pressure tubes in service in CANDU reactors undergo irradiation-enhanced changes in shape. Several equations have been proposed to account for the changes in length, diameter, and sag of CANDU pressure tubes in terms of the operating environment and the microstructures produced during fabrication [1–4]. These equations have generally described the anisotropic deformation using the concepts of additive separable components [4,5] of in-reactor thermal creep, irradiation-induced creep (shape change due to irradiation and

---

[1] Scientist, senior scientist, branch manager, and scientist, respectively, Reactor Materials Research Branch, Chalk River Laboratories, Chalk River, Ontario, K0J 1J0, Canada.

[2] Scientist and senior scientist, respectively, Reactor Materials Research Branch, Whiteshell Laboratories, Pinawa, Manitoba, R0E 1L0, Canada.

[3] Senior research engineer and principal research engineer, respectively, Ontario Hydro Technologies, 800 Kipling Ave., Toronto, Ontario, M8Z 5S4.

applied stress at constant volume), and irradiation growth (shape change due to irradiation in the absence of applied stress at constant volume).

The six main differences between the equation to be presented here and earlier ones [6–8] are: (1) In the past, the anisotropic polycrystalline behavior was predicted using "lower-bound" [1,7] and "upper-bound" [4,8–10] models, which allowed discontinuities of strain or stress at the grain boundaries due to the lack of self-consistency between the deformation of the individual grains and the deformation of the polycrystal. In the development of the present deformation equation, the anisotropy was derived using a self-consistent deformation model [11–15] in which there are no strain or stress discontinuities at the grain boundaries; (2) Due to the lack of irradiation growth data from tests on specimens with the same crystallographic texture and dislocation structure as the power reactor pressure tubes, the relative amounts of creep and growth were not accounted for accurately in the earlier deformation equations; (3) The present equation is based on a data set that includes measurements from both power reactor pressure tubes and small specimens of Zr-2.5Nb pressure tube material in test reactors to much higher fluences than earlier equations; (4) The temperature dependence of the growth term in the present equation was derived from specimens in test reactors and was shown to be negative in contrast to that used in previous equations; (5) The amount of thermal creep is less in this analysis than proposed earlier; (6) The end-to-end variation was determined independently and was not based on an empirical fit to power reactor data.

In this work, strain-producing mechanisms are assumed to operate in each individual crystal, and the sum of strains from all crystals is equal to the observed total deformation measured in the polycrystal. Thus, to predict the total deformation of the polycrystal, a deformation law describing the behavior of single crystals is needed. Three single crystal deformation laws are used: a creep law with a stress exponent larger than 1, a creep law that is linear in stress, and a deformation term due to growth. The self-consistent deformation model was employed in two steps: (1) the thermal creep component of the polycrystal was determined by using a power law describing the single crystal deformation, and (2) the irradiation creep and growth terms were calculated by using a linear creep law and growth law for the single crystal as described in Ref 15.

Pressure tubes in CANDU reactors are typically 6 m in length with an inside diameter of 0.104 m and a wall thickness of 0.0042 m. The internal pressure is about 10 MPa; therefore, under normal operating conditions, the tubes are under an applied hoop stress of about 120 MPa and an axial stress of ~60 MPa. Also, the inlet and outlet temperature in high power channels is about 520 and 562 K, respectively. The data used to derive the values of the constants involved in the equation are measurements of dimensional changes in pressure tubes from the Pickering, Bruce, and Point Lepreau Nuclear Generating Stations (NGS). Creep and growth results from test reactors (i.e., NRU, WR1, OSIRIS in France and DIDO in England) were also used. The equation gives the functional relationships between temperature, T, stress, σ, fast neutron flux, ϕ, and microstructural parameters such as texture, grain size, and dislocation density and can be used to predict the deformation to fluences beyond the envelope of data for current tubes and the behavior of tubes fabricated by different procedures and operated at different conditions.

Pressure Tube Materials

Fabrication procedures and the resulting microstructures of the extruded and cold-worked Zr-2.5Nb tubes have been described in detail elsewhere [4,16]. The major parameters that affect the deformation behavior of the tubes in-reactor are the crystallographic texture and the density and Burgers vectors of the dislocations. The standard pressure tube has the majority of basal poles in the transverse (circumferential) direction in the radial-transverse plane (Fig. 1). The

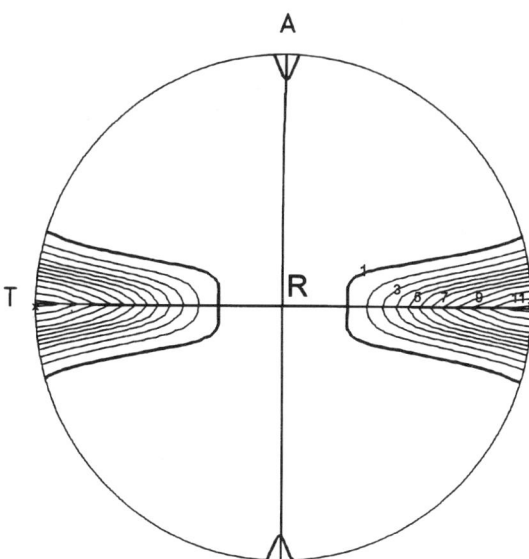

FIG. 1—*A full (0002) pole figure of an "average" Pickering pressure tube. R, T, and A stand for radial, transverse, and axial directions, respectively.*

resolved fractions of basal plane normals in the radial, transverse, and axial directions and the dislocation density of pressure tubes as measured by X-ray line broadening are given in Ref 17.

The computer code SELFPOLY, based on the self-consistent model needed for deriving the thermal and irradiation creep and growth anisotropy factors, requires as input the crystallographic texture and the average grain shape of the materials considered in the analysis [14,15]. Details of the crystallographic texture are presented later. The average grain shape is described in SELFPOLY by the ratio of lengths of unit vectors parallel to the thickness, width, and length of grains in pressure tube materials. The average grain shape used in the derivation of the deformation equation was 0.2/1.0/5.0, respectively, and was established by counting the dimensions of a large number of grains from TEM micrographs [17].

Deformation Measurements

The elongation of pressure tubes in CANDU power reactors is measured using equipment and procedures described in Ref 4. A typical plot of elongation against fluence of a large number of tubes from the Bruce Unit 2 NGS is shown in Fig. 2. The average elongation rate of Pickering reactors used in deriving the constants of the deformation equation is 11.2×10^{-29} m²/n. The average elongation rates of Bruce and Point Lepreau units, i.e., 11.8 and 13×10^{-29} m²/n, respectively, were used for comparison with those calculated from the deformation equation to be presented next. The axially averaged fast flux in each type of reactor is 1.96, 2.4, and 2.35×10^{17} n/m²/s,* respectively. The fast neutron flux peaks at about 2.5, 4, and 3.5×10^{17} n/m²/s in Pickering, Bruce, and Point Lepreau stations, respectively.

Diameteral profiles along the lengths of the tubes are measured during reactor shutdowns by

---

* The fast neutron fluxes and fluences mentioned throughout the text are for neutrons with $E > 1$ MeV.

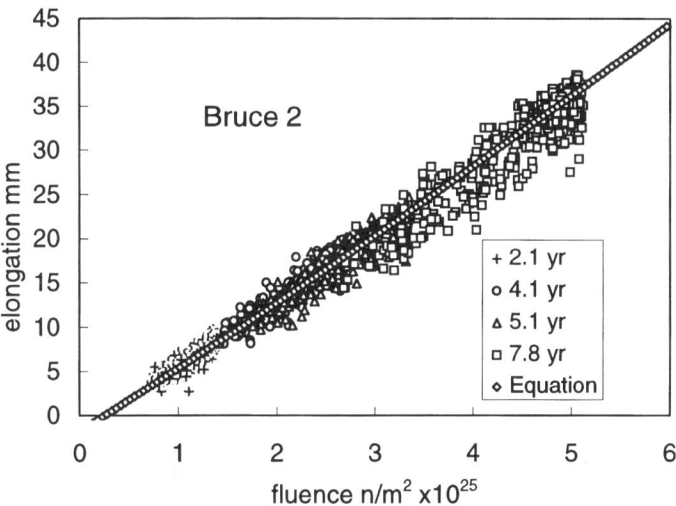

FIG. 2—*Elongation versus fluence of pressure tubes in Bruce Unit 2 power reactor. The line represents the prediction of the deformation equation presented in the Analysis section, as discussed in the Predictions section. The data were not used in deriving the equation.*

passing a gauging tool through the tubes as described in Ref 4. The strain at the position of the middle of the twelve fuel bundles was plotted against time, and a least squares fit was used to obtain the steady-state strain rates. From the fitting, the strain intercept due to primary creep was also obtained and was related to the steady-state rate by means of a linear expression. A typical example for two Pickering pressure tubes that operated for about 16 years is shown in Fig. 3. Note that tubes with their back ends (i.e., the ends extruded last) at the coolant outlet of the channel exhibit pronounced peaks near the outlet, whereas tubes with their back end at

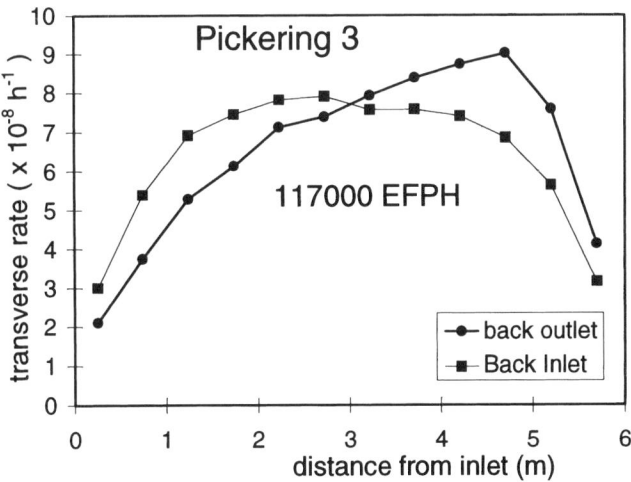

FIG. 3—*Transverse strain rate profile of two Pickering pressure tubes showing two typical types of behavior.*

the coolant inlet have a more uniform strain rate profile [4]. As is evident in Fig. 3, the average strain rate increases along the tube from inlet (~523 K) to outlet (~567 K) and is higher in the high flux region of the tube.

Reference irradiation creep rates and creep compliances have been determined experimentally from shear (i.e., springs loaded in tension), bent-beam stress relaxation experiments, and internally pressurized creep capsules [15]. In addition, Causey et al. [18] measured experimentally irradiation creep rates and creep compliances as a function of stress and temperature on internally pressurized capsules from Zr-2.5Nb micropressure tube material that had the same crystallographic texture and dislocation density as typical pressure tubes. These capsules were irradiated to higher fluences than those described in Ref 15 in the OSIRIS reactor at a fast flux of about 1.8×10^{18} n/m²/s. The results from these tests were used to verify the predictions from the deformation equation described next.

Irradiation growth data were obtained from Ref 19, and a typical example (Fig. 4) shows the axial growth behavior of two specimens from a pressure tube irradiated in the DIDO test reactor. Although a linear expression appears to reproduce the measurements fairly accurately, a statistically better fit to the data is obtained if it is assumed that the growth rate changes somewhat with fluence. This apparent increase of growth rate with fluence has also been observed in Zircaloy-2 pressure tubes and has been attributed to an increase with irradiation of the $\langle c \rangle$ component dislocation density as discussed by Griffiths et al. [20]. Thus, the present equation allows for the possibility that the growth rate depends on fluence, unlike the previous one [4] where the growth rate was assumed to remain constant with fluence.

Thermal creep is a small, but not a negligible component of strain during irradiation. Two sets of data were used to estimate the magnitude of the in-reactor thermal creep term. For the temperature range between 520 to 570 K, the set of data used here was obtained from the periodic gauging of a pressure tube installed in the NRU reactor and operated for about 54 000 h. The strain profile of the tube included sections where the flux was nearly zero and the creep rate was a minimum [3] due to radiation hardening from which the thermal component could be established. Figure 5 shows the evolution of strain with time at the minimum creep locations

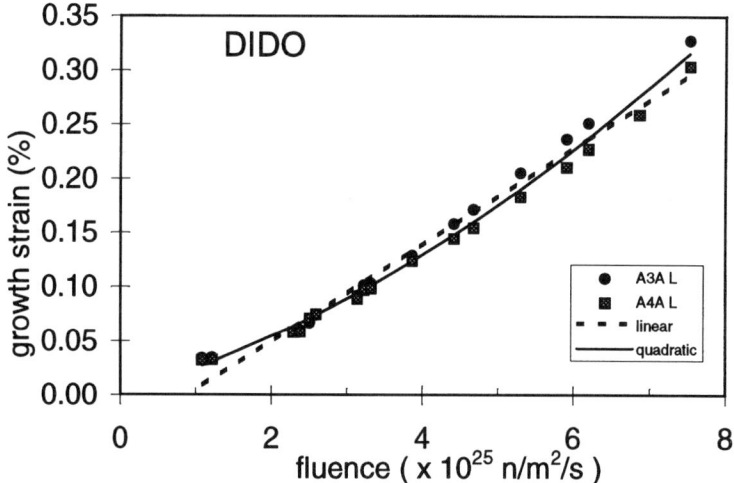

FIG. 4—*Dependence of growth strain on fluence of two pressure tube samples from the longitudinal direction irradiated at DIDO [19]. The best linear and quadratic fits to the data are also shown.*

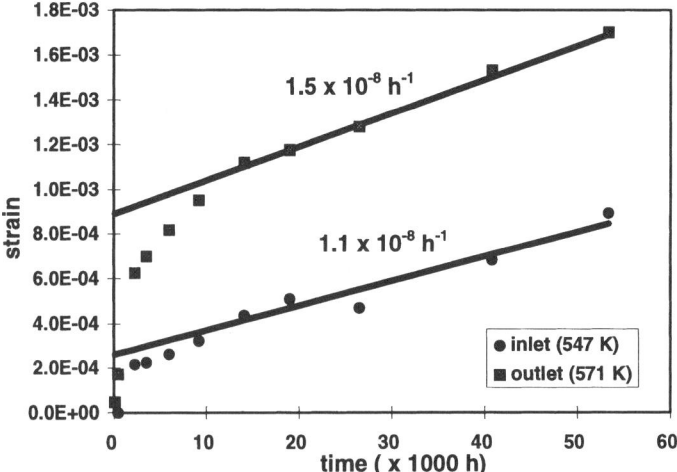

FIG. 5—*Thermal creep strain as a function of time at the inlet and outlet of a NRU pressure tube. The creep rates indicated on the figure were used in Eq 1.*

at the inlet and outlet ends. The steady-state thermal creep rates quoted in the figure were the values used to derive the parameters for the thermal component for the temperature range mentioned above. The contribution due to primary creep is less than 5% of the total strain, and it was neglected. For temperatures in the range of 570 to 680 K, creep rates from pressure tubes operated in the WR1 reactor at Whiteshell and from uniaxial creep tests in NRU were used [4].

Analysis

The complex interaction between the effects of temperature and fast neutron flux on the deformation of zirconium alloys has led to the development of analyses that assume that long-term steady-state deformation consists of separable, additive components from thermal creep, irradiation creep, and irradiation growth [3–5]. All components are anisotropic and contribute to length as well as diameter changes. The equation describing the deformation of pressure tubes has the form:

$$\dot{\varepsilon}_d = \dot{\varepsilon}_d^{\text{thermal}} + \dot{\varepsilon}_d^{\text{creep}} + \dot{\varepsilon}_d^{\text{growth}} \tag{1}$$

where

$$\dot{\varepsilon}_d^{\text{thermal}} = [K_1 C_1^d \sigma_1 + K_2 C_2^d \sigma_2^2] \exp(-Q_1/T) + K_3 C_1^d \sigma_1 \exp(-Q_3/T) \tag{1a}$$

$$\dot{\varepsilon}_d^{\text{creep}} = K_c K_4(x) C_4^d(x) \sigma(x) \phi [\exp(-Q_4/T) + K_5] \tag{1b}$$

$$\dot{\varepsilon}_d^{\text{growth}} = K_g K_6(x,\phi t) C_6^d(x) \phi \exp(-Q_6/T) \tag{1c}$$

The in-reactor thermal creep component has two terms [3,4] that dominate at temperatures above and below 570 K, respectively. The last two terms describe flux-dependent creep and irradiation growth, respectively. The stress exponent for thermal creep varies with stress: for stresses below 120 MPa, the stress exponent is 1, while for stresses between 120 and 200 MPa, the stress exponent increases to 2 [3]. The parameters in Eq 1 are defined as follows:

$\dot{\varepsilon}_d$ = strain rate in a direction d (i.e., radial, transverse, axial), h^{-1},
$\dot{\varepsilon}K_1, K_2$ = constants for high-temperature in-reactor thermal creep,
K_3 = constant for low-temperature in-reactor thermal creep,
$K_4(x)$ = a function describing the variation of irradiation creep due to variations of microstructure along the length of the tube,
$K_6(x,\phi t)$ = a function describing the variation of irradiation growth due to variations of microstructure along the length of the tube as a function of fluence,
C_1^d, C_2^d = anisotropy factors due to texture for in-reactor thermal creep in a given direction d and for Stress Exponents 1 and 2, respectively,
$C_4^d(x), C_6^d(x)$ = anisotropy factors due to texture for irradiation creep and growth, respectively, in a given direction d along the length of the tube,
K_c, K_g = constants for irradiation creep and growth, respectively,
Q_1, Q_3, Q_4, Q_6, K_5 = activation temperatures and constant, respectively,
σ_1, σ_2 = effective stresses for thermal creep and Stress Exponents of 1 and 2, MPa,
$\sigma(x)$ = effective stress for irradiation creep, MPa,
T = temperature in K,
ϕ = fast flux, n/m^2/s ($E > 1$ MeV), and
t = irradiation time in s.

The equivalent stresses σ_1, σ_2, and $\sigma(x)$ are related to the radial, axial, and transverse stress σ_r, σ_a, and σ_t, respectively, by means of the Hill's anisotropy constants [21], namely:

$$\sigma_i = [F_i (\sigma_a - \sigma_t)^2 + G_i (\sigma_t - \sigma_r)^2 + H_i (\sigma_r - \sigma_a)^2]^{1/2} \quad (2)$$

The subscript i stands for 1 (i.e., $n = 1$), 2 ($n = 2$), or in the case of irradiation creep $\sigma_i = \sigma(x)$. The Hill's anisotropy constants for irradiation creep depend on the distance, x, from the back end of the tubes, and for a 6-m tube this dependence is given by:

$$F(x) = F^b + (F^f - F^b) x/6$$
$$G(x) = G^b + (G^f - G^b) x/6 \quad (3)$$
$$H(x) = 1.5 - F(x) - G(x)$$

where F^b, F^f, G^b, and G^f are the values of Hill's anisotropy constants F and G at the back and front end of the tube. The dependence of Hill's anisotropy constants F_i, G_i, and H_i ($i = 1, 2$) on x was neglected because of the relatively small magnitude of the thermal component. Using the terminology employed in Eq 2, the anisotropy factors due to texture for in-reactor thermal or irradiation creep are given by:

$$C_i^r = [H_i (\sigma_r - \sigma_a) - G_i (\sigma_t - \sigma_r)]$$
$$C_i^t = [G_i (\sigma_t - \sigma_r) - F_i (\sigma_a - \sigma_t)] \quad (4)$$
$$C_i^a = [F_i (\sigma_a - \sigma_t) - H_i (\sigma_r - \sigma_a)]$$

Here $i = 1, 2,$ or (x). The coefficient describing the end-to-end effect of irradiation creep along the length of the tube is given by:

$$K_4(x) = K_{4-1} + K_{4-2} x \tag{5}$$

The growth coefficient describing the end-to-end effect and the dependence of growth on fluence is given by:

$$K_6(x,\phi\ t) = (K_{6-1} + K_{6-2}\ x)(1 + C/B\ [\phi\ t]) \tag{6}$$

where K_{4-1}, K_{4-2}, K_{6-1}, and K_{6-2} are defined later. The growth anisotropy factors are given by:

$$C_6^a(x) = G_a^b + (G_a^f - G_a^b)x/6$$

$$C_6^t(x) = G_t^b + (G_t^f - G_t^b)x/6 \tag{7}$$

$$C_6^r(x) = -C_6^a(x) - C_6^t(x)$$

Here (G_a^b, G_t^b) and (G_a^f, G_t^f) are the growth anisotropy constants in the back and front end, and in the axial and transverse direction of the tube, respectively. It should be noted that there is a systematic variation of crystallographic texture along the length of pressure tubes. The dependence of the creep and growth anisotropy factors $C_4^d(x)$ and $C_6^d(x)$ on x is due only to measured texture variations. The axial dependence of $K_4(x)$ and $K_6(x)$ represents the effects of microstructural variations, e.g., dislocation density and grain size. $K_4(x)$ was determined from experimental data and $K_6(x)$ from a growth model.

Derivation of Constants

Hill's Anisotropy Constants

The computer code SELFPOLY, based on the self-consistent model described in Refs *11* to *14*, was used to derive the creep constants (F^b, F^f, G^b, G^f) in Ref *15*. These constants are used in Eqs 3 and 4 to calculate the creep anisotropy constants and coefficients. SELFPOLY requires, as input, the crystallographic texture of pressure tubes in the form of a crystallite orientation distribution function (CODF). The (0002), (10$\bar{1}$0), (11$\bar{2}$0), (11$\bar{2}$2), and (10$\bar{1}$1) pole figures were determined by X-ray diffraction from 23 Pickering tubes and then were used as input to a computer code that calculates three Euler angles as a function of the volume fraction [22]. The Euler angles were calculated according to Bunge's notation [23], and they relate the coordinate system associated with each grain to that of the pressure tube [14]. The average value of the resolved basal pole fraction in the radial, transverse, and axial directions, $f_R, f_T,$ and f_L, in the back end and front end of pressure tubes is 0.36, 0.60, 0.05 and 0.30, 0.64, 0.06, respectively. The eigenvalues of the single crystal creep compliance tensor describing pyramidal, prismatic, and basal climb-assisted glide of dislocations in pressure tube materials during in-reactor deformation were derived in Ref *15*, and they are 0.284, 7.086, and 2.84 × 10^{-30} m^2/n/MPa, respectively. Experimental data from internally pressurized capsules, stress relaxation specimens, and data from the Pickering NGS were used to derive these values in Ref *15*.

A non-linear self-consistent code based on a model described in Ref *24* was used to calculate Hill's thermal creep anisotropy constants $F_i, G_i,$ and $H_i,$ where $i = 1, 2$. As was mentioned above, the single crystal creep law assumed to describe the behavior of the material during

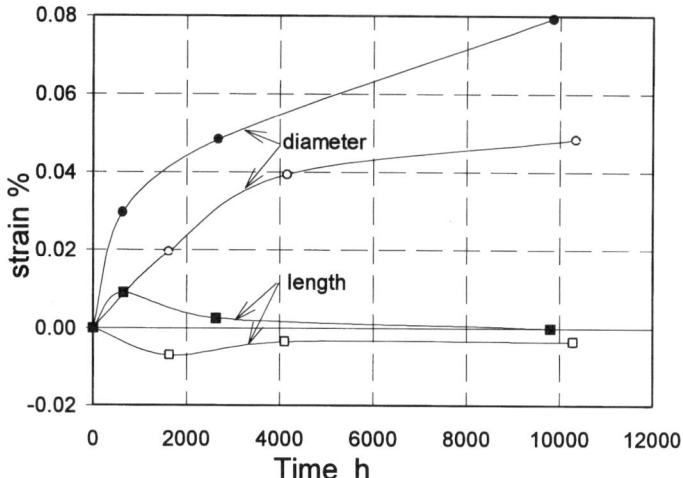

FIG. 6—*Out-reactor creep behavior of internally pressurized capsules from pressure tube material tested at 573 K. Solid symbols correspond to a hoop stress of 175 MPa, while open symbols correspond to a hoop stress of 100 MPa. Note that the strain in the axial direction is much lower than the strain in the transverse direction.*

thermal creep is a power law [24]. When the stress is below 120 MPa, the term with $n = 1$ dominates, while when the stress is larger than 200 MPa, the term with $n = 2$ is dominant. The single crystal creep parameters needed for the derivation of Hill's anisotropy constants are the values of critical resolved shear stresses (CRSS) for dislocation glide on prismatic, basal, and pyramidal planes. The values of these three parameters used here are 100, 120, and 240 MPa, respectively. These constants were derived so that the predicted behavior of the polycrystalline material is consistent with out-reactor thermal creep experiments on internally pressurized capsules from small tubes with textures and microstructures like pressure tube materials. Results from out-reactor creep experiments are shown in Fig. 6.

Growth Anisotropy Constants

The values of growth constants (G_t^b, G_t^f, G_a^b, G_a^f) that define the anisotropy factors in Eq 7 were calculated by using SELFPOLY as well. The growth anisotropy of pressure tube materials irradiated in test reactors [19] was predicted by using the CODF defined earlier, the single crystal creep compliances mentioned above, and a single crystal growth tensor equal to (-1.42, 2.13, -0.71) $\times$ 10^{-29} m^2/n.

End-to-End Effect

The function $K_4(x)$ in Eq 5 was determined from in-reactor bent-beam stress relaxation tests of pressure tube materials obtained from different locations along the tube. In these tests, the measured deformation rate is due only to irradiation creep and not growth [25]. The axial variation in $K_4(x)$ with respect to the middle of the tube was determined from the best fit to the data shown in Fig. 7 to be about $\pm17\%$. The end-to-end variation in $K_6(x)$ (i.e., the first term in Eq 6) was calculated from the variation in grain thickness along the length of the tube and the dependence of growth rate on grain thickness [26]. The average grain thickness in the front

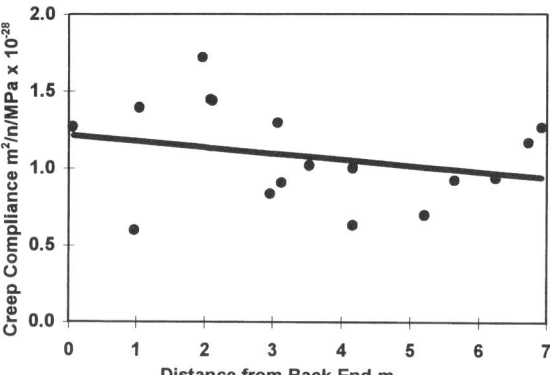

FIG. 7—*Dependence of the axial creep compliance with distance from the back end of the tube.*

end of pressure tubes used here is about 0.39 μm, while that in the back end is about 0.27 μm. The dependence of growth rate on grain thickness is derived from Fig. 18 in Ref 26 and is shown here in Fig. 8. Using the average grain thickness for the front and the back of a pressure tube, the variation of growth rate along the tube with respect to the middle of the tube is about ±8%. The end-to-end effect for creep and growth is shown in Fig. 9. Finally, the dependence of K_6 on ϕt is determined from a number of samples from pressure tube materials mentioned in Ref 19. The constants (K_{4-1}, K_{4-2}) and (K_{6-1}, K_{6-2}) in Eqs 5 and 6 are the intercept and the slope, respectively, of the two lines in Fig. 9.

Creep and Growth Constants

After determining the above-mentioned constants, the creep and growth constants K_c and K_g were calculated by fitting the diametral and elongation data available from the Pickering NGS. These values were used in Eq 1 to predict the creep compliance in the axial and transverse

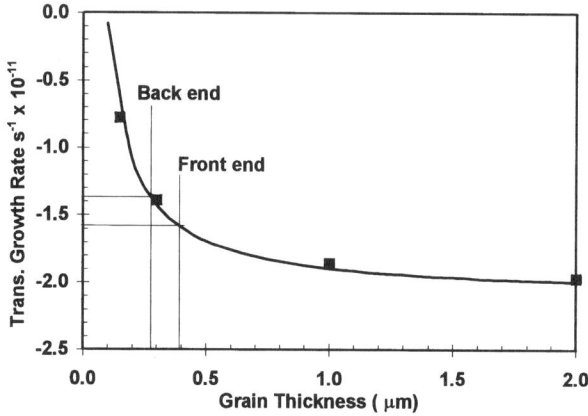

FIG. 8—*Dependence of the transverse growth rate on grain thickness (derived from Ref 26).*

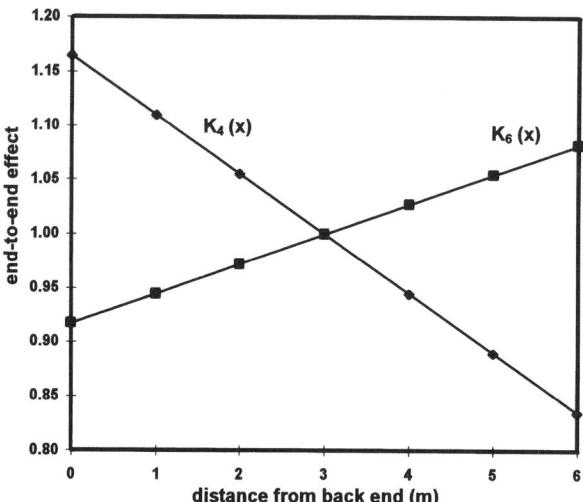

FIG. 9—*End-to-end effect along the length of a pressure tube for the creep and the growth terms in Eq 1.*

directions of micropressure tube material tested in OSIRIS [*18*] and the growth rate of pressure tube materials tested in OSIRIS and DIDO [*19*]. The agreement of the predictions with the measurements was good.

Temperature Dependence

The high-temperature thermal creep term in Eq 1 reflects the high-temperature dependence of creep rate exhibited by the pressure tubes in the WR1 reactor and by the uniaxial creep specimens tested in the NRU reactor (Fig. 1 in Ref *3*). The value of Q_1 that fits the data best is equal to 17 000 K. The value of Q_3 in the low-temperature thermal creep term is derived from Fig. 5 and is equal to 1000 K.

The temperature dependence of creep was derived from power reactor data and that for growth was derived from specimens irradiated in the OSIRIS reactor [*19*] after first removing the thermal creep contribution. Q_4 and K_5 were found equal to 9900 K and 1.1×10^{-7}, respectively, whereas Q_6 is about -3000 K. The temperature dependence of creep does not differ significantly from that in Ref *4*; however, that of the growth component is strongly negative compared to a positive value that was proposed in Ref *4* on the basis of growth data for cold-worked Zircaloy-2 pressure tube material.

Predictions

Using the parameters mentioned above, the predicted deformation rate from Eq 1 is compared to measured values as shown in the following figures. Figure 10 shows the predicted diametral profile of a Pickering tube with its back end in the outlet that was gauged at various time intervals. Figure 11 shows the prediction of the diametral profile of a Pickering tube with its back end in the inlet. In both cases, the calculated rate profile compares well with the measurements. Figure 12 depicts the predicted versus the measured diametral rates for a number

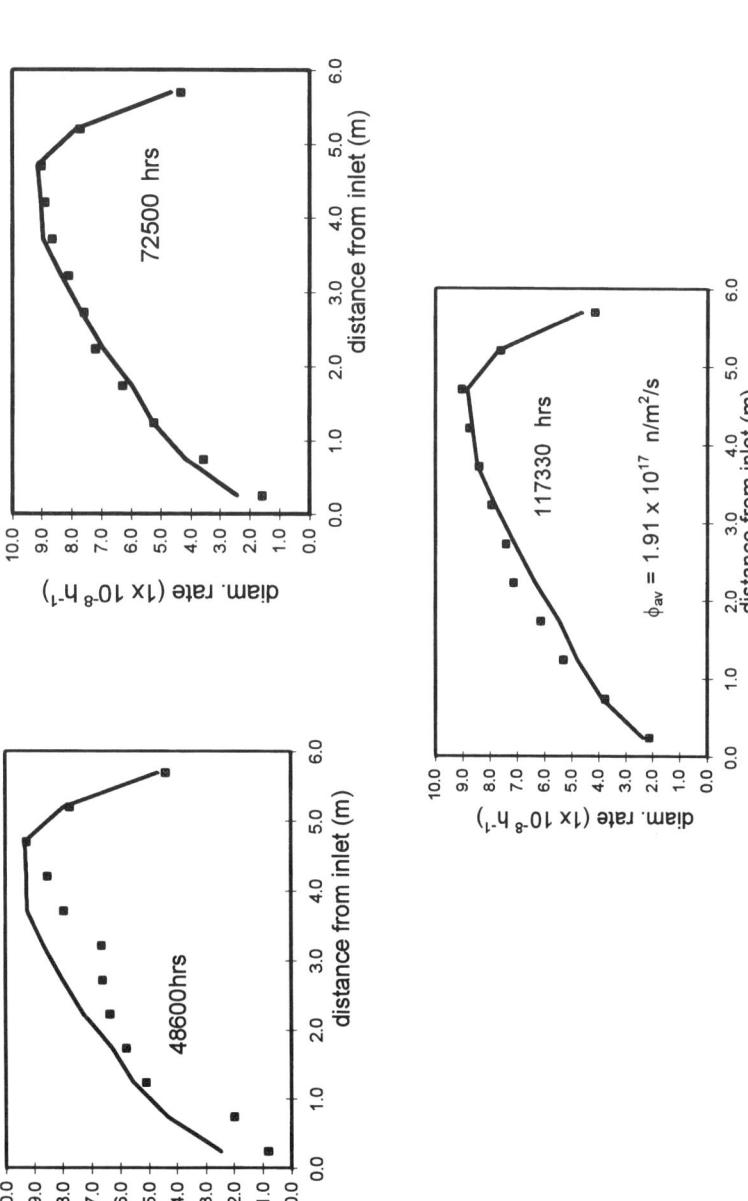

FIG. 10—*Comparison of the predicted diametral rate profile at various time intervals with measurements from a Pickering pressure tube with its back end in the outlet.*

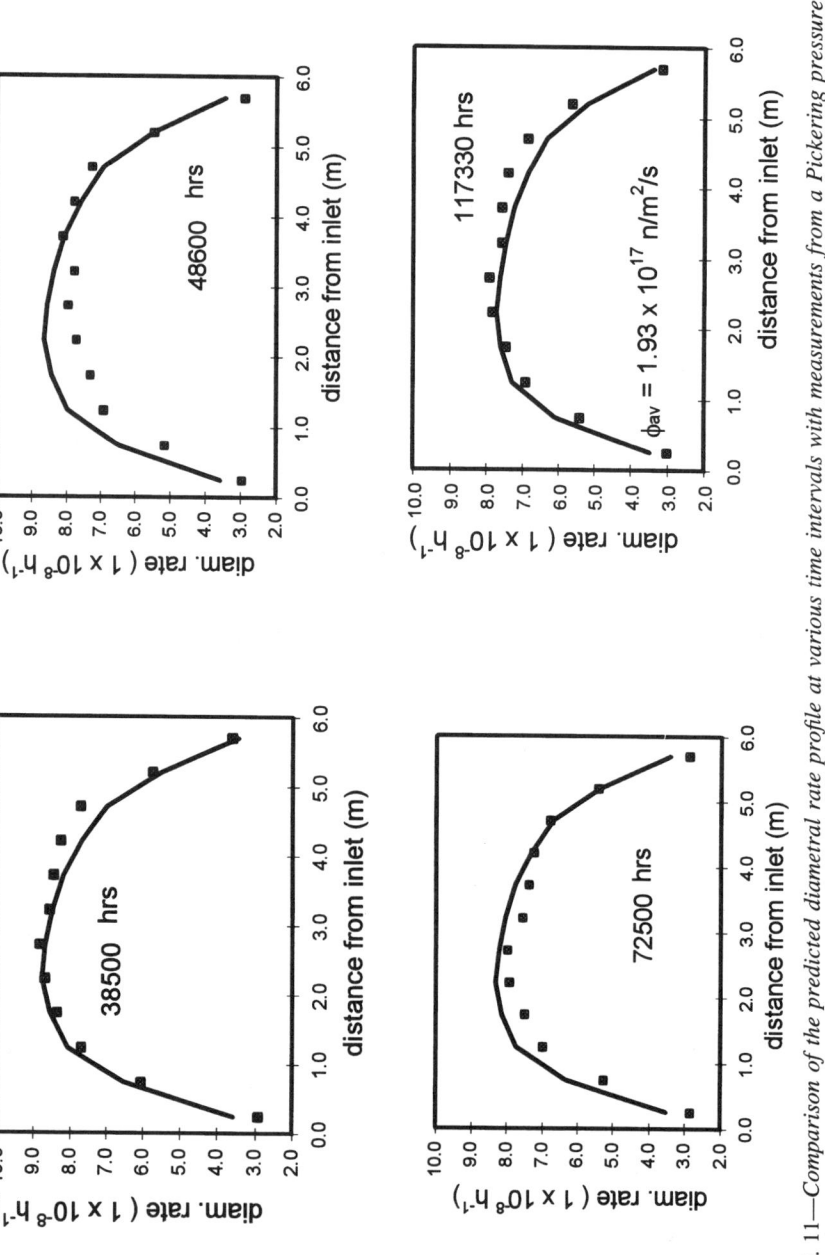

FIG. 11—*Comparison of the predicted diametral rate profile at various time intervals with measurements from a Pickering pressure tube with its back end in the inlet.*

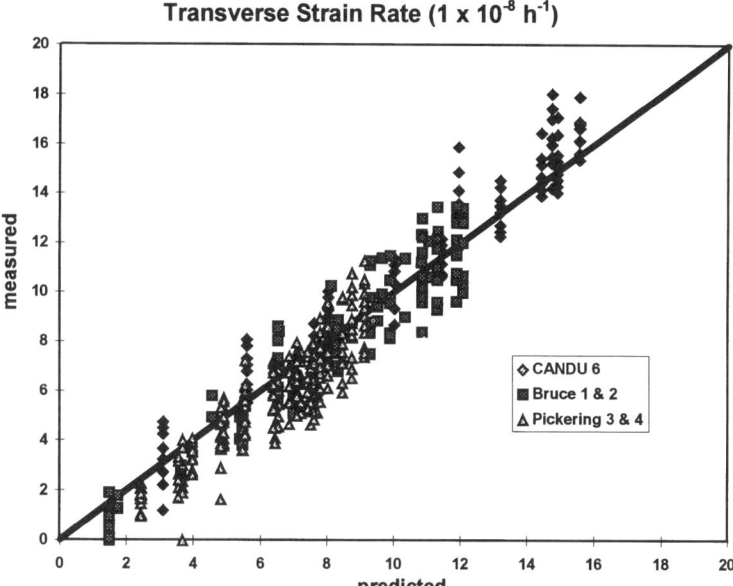

FIG. 12—*Predicted against measured diametral rate in Pickering, Bruce, and CANDU 6 units.*

of tubes from CANDU 6, Bruce, and Pickering units. It should be noted, however, that, for a number of tubes in channels with a low average flux, the prediction is consistently higher than the measurements. This is shown in Fig. 13. Figure 14 shows the measured versus the calculated diametral rate of the pressure tubes irradiated in the WR1 reactor. Finally, Fig. 2 summarizes the elongation of a number of Bruce unit 2 tubes as compared with the elongation predicted from Eq 1. It appears that, in most cases, the predicted deformation rate agrees well with the measurements.

The predictive capability of Eq 1 was also tested by calculating the creep and growth anisotropy factors from SELFPOLY by using the individual crystallographic texture of each tube that was available. Generally, the calculated diametral profile of the deformation rate along the tube was improved as compared to that calculated by employing the coefficients of the "average" tube. In the worst cases, a quantitative overall fit to the deformation rate could be obtained by modifying the amounts of creep and growth, i.e., the values of K_c and K_g in Eqs 1b and 1c by the order of ±10% of the values used for the "average" tube. Therefore, the present deformation equation can be used to predict the behavior of individual tubes to within about ±10% if the creep and growth anisotropy factors of the tube are determined by SELFPOLY and subsequently used in the equation.

3-D Finite Element Code H3DMAP

A finite element code (FEC) H3DMAP developed earlier [27] was used to simulate a complete fuel channel, namely, the pressure tube, calandria tube, part of the end-fitting of the channel, and the spacers that are placed along the tube. The deformation Eq 1 was incorporated into H3DMAP for predicting the detailed shape change after irradiation of a typical pressure tube. In addition to the diametral and length changes, the predictions from the FEC also yield

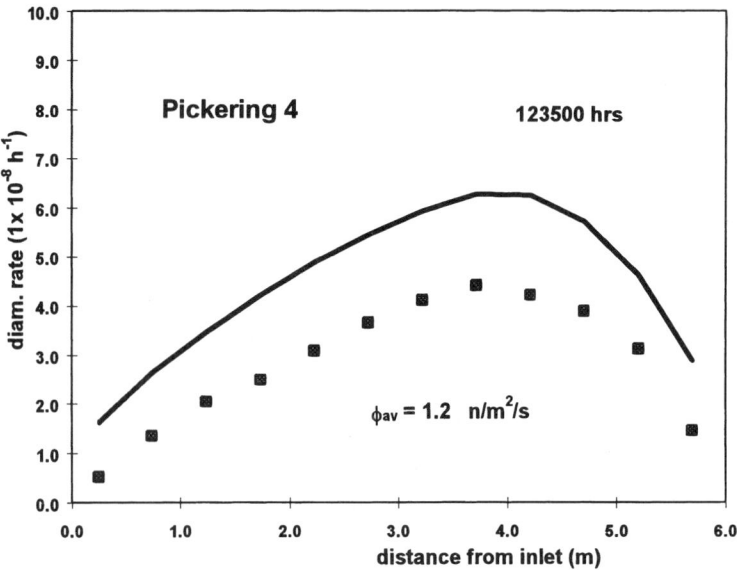

FIG. 13—*Prediction of diametral rate of a low flux channel in Pickering Unit 4. This is selected to show how a low-flux channel does not agree well with the prediction. Other tubes in high-flux channels have a good agreement.*

predictions about the sag of calandria tubes (i.e., deflection from the horizontal plane), information not directly available from the closed form expression (Eq 1).

A comparison between the calculated and the diametral profile measured in situ in a Pickering fuel channel after ten years of operation is shown in Fig. 15a. The calculation shows agreement

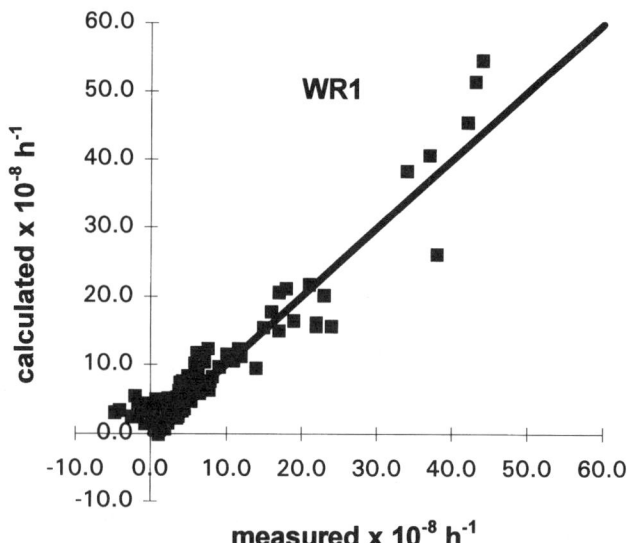

FIG. 14—*Measured versus calculated diametral rate of the WR1 pressure tubes.*

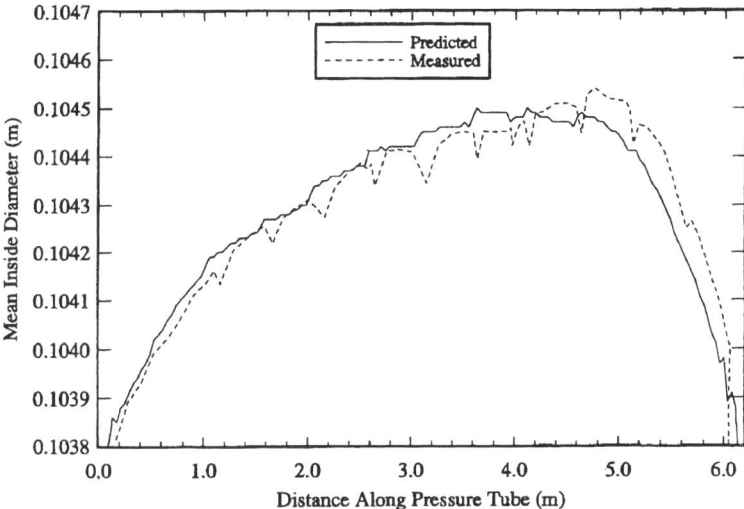

FIG. 15a—*Diametral profile of a Pickering pressure tube after ten years of operation. The calculated curve was determined from H3DMAP with Eq 1 as a materials routine.*

with both the average profile (Fig. 15a) and the ovality (difference between the diameter at the horizontal and vertical planes) at ~2.5 and 4.0-m positions, developed where the pressure tube is supported by the spacers separating it from the calandria tube (Fig. 15b). Note that the measurement shows residual ovality as installed along the whole length of the tube, whereas the calculation shows only the ovality developed during service. Also, from Fig. 15b, the section of the tube not supported by the spacers appears to develop a small negative ovality, probably as a response to the large positive values observed at the location of the spacers.

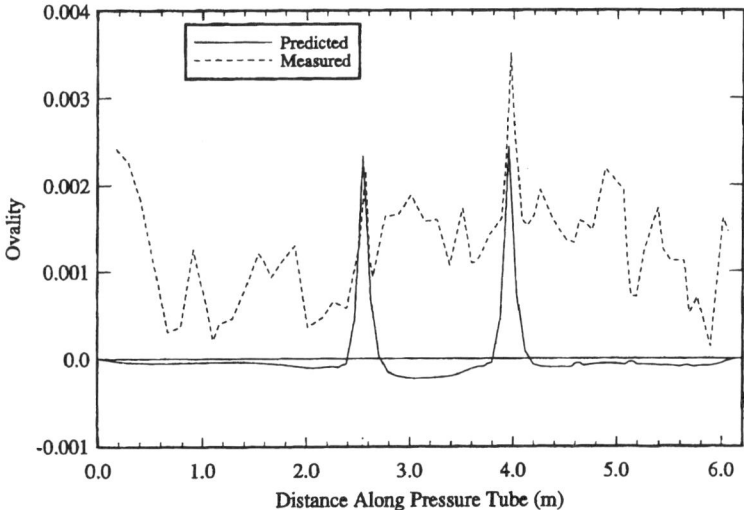

FIG. 15b—*Ovality profile for the pressure tube depicted in Fig. 15a.*

Discussion

Equation 1 is similar in form and was derived in a comparable manner to that described in Ref 4. Deformation due to irradiation creep and growth and due to thermal creep is assumed to be the result of strain-producing steps due to climb and glide of dislocations primarily on prismatic systems [15] and to some degree on pyramidal systems. As in Ref 4, in the present equation the creep and growth anisotropy factors were derived from a model that takes into account the crystallographic texture, the interactions between grains, and the stresses that develop as a result of these interactions. When grain interactions are considered, specimens with randomly oriented grains have the minimum average strain rates. In contrast, materials with anisotropic single crystal properties and with non-random texture are strengthened in one or two directions and are weakened in the other directions. The major difference between the anisotropy model used for the derivation of Ref 4 and Eq 1 is that in the former analysis all grains were subjected to the same amount of deformation, namely, that of the polycrystal. As a result of this assumption, there are stress discontinuities that develop from grain to grain. In the self-consistent model used here [11-15] for the derivation of the present equation, the weighted average of the sum of strains (or stresses) from all the crystals is equal to the overall strain (or imposed stress) of the polycrystal. In this fashion, there are no stress or strain discontinuities from grain to grain.

The approach developed here results in a different set of creep and growth anisotropy factors from that derived in Ref 4. In contrast to the equation in Ref 4, the temperature dependence of growth is strongly negative in the present model, whereas it was positive in Ref 4. Furthermore, the relative amounts of creep and growth are different in the present equation compared to that proposed earlier. For instance, in the previous version of the deformation equation, axial growth accounted for about 89% of the total axial strain (Table 2 in Ref 28), whereas growth contributes only 24% to the total axial rate in Eq 1. Also, both the negative contribution of growth and the positive contribution of creep to the total diametral strain rate are smaller with Eq 1 than with the earlier model, i.e., in Eq 1 transverse growth and creep are about 24 and 67%, respectively, of the values predicted by the equation in Ref 4. These differences change the predicted effect of stress on the deformation of the pressure tubes. For example, Eq 1 predicts that a 1.9-kN (20%) higher compressive stress than the earlier model must be applied in order to eliminate the axial elongation completely in a Pickering tube after 117 000 h of operation than the value estimated from the previous equation, and the effect of a small increment of internal pressure on the diametral strain rate is 10 to 15% higher with Eq 1 than with the earlier model.

The present equation is derived on the basis of a single set of parameters based on experimentally determined textures and deformation data from a large number of Pickering 3 and 4 pressure tubes. In the earlier version, different sets of parameters were employed for different reactors. The present model is reasonably successful in predicting deformation rates in other reactors and in predicting the deformation behavior of individual tubes when the creep and growth anisotropy constants for the tube are employed instead of those for the ‘'average'' tube. Hence, the predictions for individual reactors could probably be improved by using a statistically significant sampling of textures of tubes from each reactor.

As mentioned above, it appears that the prediction of the diametral strain rates is consistently higher than the measurements in some cases, especially for channels with a low average flux or when the measurement was obtained after a short irradiation period. This may be attributed to an overestimate of the contribution of thermal creep, which was derived by assuming that the thermal creep rate varies linearly from the inlet to outlet between the values indicated in Fig. 5. However, the degree of suppression of the thermal creep rate in the presence of a high fast flux has not been established, and it is possible that thermal creep is reduced further with a high fast neutron flux than at the edges of the core where the fast flux is as low as 1% of that at the center of the tube.

Unlike the earlier model [4], the effects of the variation in dislocation density and grain size along the length of the tubes on the creep and growth rates was based on independent experimental measurements and models and not on an empirical fit to the power reactor data. These effects are of the order of ±17 and ±8% for creep and growth, respectively, and their magnitude was sufficient to reproduce the difference between the diametral profiles of tubes with their back ends in the inlet or outlet flow of the channels.

The finite element formulation provides detailed information about various aspects of both the pressure tube and calandria tube deformation, such as the ovalization of the pressure tube at the spacers. Some of the results can be compared with in-reactor or post irradiation measurements, while others can be used to gain insight into aspects of deformation that are not routinely examined (e.g., horizontal deflection) or difficult to measure (e.g., stress distributions).

Conclusions

1. A database of changes in length and diameter of Zr-2.5Nb pressure tubes during service in power and test reactors was correlated to the operating conditions and microstructure of the tubes. This database extends to much higher fast neutron fluences than were previously available [4].

2. A self-consistent polycrystalline model proposed earlier has been employed to calculate the anisotropy of creep and growth from the preferred orientation of grains and the resulting grain interactions under a fast neutron flux.

3. A deformation equation describing anisotropic thermal creep and irradiation creep and growth under normal operating conditions in CANDU reactors has been derived using the data mentioned in Conclusion 1 and the polycrystalline model mentioned in Conclusion 2.

   a. The thermal creep strain is assumed to be the result of slip of $\langle a \rangle$ type dislocations gliding on prismatic planes and $\langle c+a \rangle$ type dislocations gliding on pyramidal planes. The dependence of the thermal creep rate on stress has two contributions, i.e., at low values of stress, a stress exponent of 1 dominates, and at higher values a stress exponent of 2 dominates.
   b. The irradiation creep strain is assumed to be the result of climb-assisted glide of $\langle a \rangle$ and $\langle c+a \rangle$ type dislocations operating on prismatic, basal, and pyramidal planes, and the irradiation creep rate is linearly dependent on stress.
   c. The growth rate was based on new growth data obtained from CANDU pressure tube material irradiated to high fluences in DIDO and OSIRIS. The polycrystalline growth anisotropy factors were calculated by means of the self-consistent model.

4. The effect of variations of dislocation density and grain size along the length of the tube were determined independently from the power reactor data.

5. The equation has been incorporated in a finite element code and was used for predicting the shape change of the tubes in more detail.

Acknowledgments

The authors would like to thank the CANDU Owners Group, Working Party 32, for financial support.

References

[1] Holt, R. A. and Ibrahim, E. F. in *Acta Metallurgica*, Vol. 27, 1979, p. 1319.
[2] Ibrahim, E. F. and Holt, R. A. in *Journal of Nuclear Materials*, Vol. 91, 1980, p. 311.
[3] Holt, R. A., Causey, A. R., and Fidleris, V. in *Proceedings of the British Nuclear Society*, London, 1983, p. 175.
[4] Causey, A. R., Fidleris, V., MacEwen, S. R., and Schulte, C. W. in *Influence of Radiation on Material Properties: Thirteenth International Symposium, ASTM STP 956*, American Society for Testing and Materials, West Conshohocken, PA, 1988, p. 54.
[5] Nichols, F. A. in *Journal of Nuclear Materials*, Vol. 30, 1969, p. 249.
[6] MacEwen, S. R. and Carpenter, G. J. C. in *Journal of Nuclear Materials*, Vol. 90, 1980, p. 108.
[7] Causey, A. R., Holt, R. A., and MacEwen, S. R. in *Zirconium in the Nuclear Industry (Sixth International Symposium) ASTM STP 824*, American Society for Testing and Materials, West Conshohocken, PA, 1985, p. 269.
[8] Holt, R. A. and Causey, A. R. in *Journal of Nuclear Materials*, Vol. 150, 1987, p. 306.
[9] Savino, E. J. and Laciana, C. E. in *Journal of Nuclear Materials*, Vol. 90, 1980, p. 89.
[10] Savino, E. J. and Harriague, S. in *Effect of Radiation on Materials: Twelfth International Symposium, ASTM STP 870*, American Society for Testing and Materials, West Conshohocken, PA, p. 667.
[11] Woo, C. H. in *Journal of Nuclear Materials*, Vol. 131, 1985, p. 105.
[12] Woo, C. H. in *Proceedings of the British Nuclear Society*, London, 1987, p. 65.
[13] Causey, A. R., Woo, C. H., and Holt, R. A. in *Journal of Nuclear Materials*, Vol. 159, 1988, p. 225.
[14] Tomé, C. N., So, C. B., and Woo, C. H. in *Philosophical Magazine*, Vol. 67, 1993, p. 917.
[15] Christodoulou, N., Causey, A. R., Woo, C. H., Tomé, C. N., Klassen, R. J., and Holt, R. A., *Effects of Radiation on Materials (Sixteenth Volume), ASTM STP 1175*, American Society for Testing and Materials, West Conshohocken, PA, 1994, p. 1111.
[16] Holt, R. A. in *Journal of Nuclear Materials*, Vol. 59, 1976, p. 234.
[17] Chow, P. C. K., Coleman, C. E., Holt, R. A., Sagat, S., and Urbanic, V., *Effects of Radiation on Materials (Sixteenth Volume), ASTM STP 1175*, American Society for Testing and Materials, West Conshohocken, PA, 1994, p. 1077.
[18] Causey, A. R., Elder, J. E., Holt, R. A., and Fleck, R. G., *Zirconium in the Nuclear Industry (Twelfth Volume), ASTM STP 1245*, American Society for Testing and Materials, West Conshohocken, PA, 1994, p. 202.
[19] Fleck, R. G., Elder, J. E., Causey, A. R., and Holt, R. A., *Zirconium in the Nuclear Industry (Twelfth Volume), ASTM STP 1245*, American Society for Testing and Materials, West Conshohocken, PA, 1994, p. 168.
[20] Griffiths, M., Holt, R. A., and Rogerson, A. in *Journal of Nuclear Materials*, in press.
[21] Hill, R., *The Mathematical Theory of Plasticity*, Oxford University Press, London, 1954.
[22] Root, J. H. and Holden, T. M., "Programs to Determine the CODF from Diffraction Intensity Pole Figures," March 1990, unpublished report ANDI-29, available from the Scientific Document Distribution Office, AECL, Chalk River, Ontario, K0J 1J0.
[23] Bunge, H.-J., *Texture Analysis in Materials Science, Mathematical Methods*, Butterworths & Co., London, 1982.
[24] Hutchinson, J. W. in *Proceedings of the Royal Society of London*, Vol. A 348, 1976, p. 101.
[25] Kreyns, P. H. and Burkart, M. W. in *Journal of Nuclear Materials*, Vol. 26, 1968, p. 87.
[26] Holt, R. A. in *Journal of Nuclear Materials*, Vol. 159, 1988, p. 310.
[27] Sauvé, R. G. and Badie, N., *Pressure Vessel and Piping*, Vol. 265, 1993, p. 267.
[28] Holt, R. A. and Fleck, R. G. in *Zirconium in the Nuclear Industry (Ninth Volume), ASTM STP 1132*, American Society for Testing and Materials, West Conshohocken, PA, 1991, p. 218.

DISCUSSION

A. T. Motta[1] (written discussion)—(1) Assuming there is a departure from linearity in the growth coefficient showing a dependence on fluence, what is the reason for assuming that such dependence is linear? Or non-linear? (2) What is the reason for a negative temperature dependence of growth?

N. Christodoulou et al. (authors' closure)—(1) A possibility for a non-linear fit to the growth data is the generation of $<c>$-component dislocation loops that can result in an increase in the growth rate during irradiation. However, from a physical model, the growth rate may saturate at large fluences. (2) This observation stems from experimental measurements (see Holt and Fleck, ASTM STP 1132, page 218, Fig. 8). A physical explanation for the fluence dependence of the growth rate was given by Holt and Fleck in ASTM STP 1023, page 705, Table 4.

S. Yagnik[2] (written discussion)—Would your model (DAD) also predict non-linear growth in CWSRA Zr-4 at higher temperatures of irradiation (~600 to 625 K)?

N. Christodoulou et al. (authors' closure)—Very likely, acceleration would occur in material with relatively low levels of cold work and fewer c than a dislocations initially. However, in a heavily cold-worked material, it is conceivable that the initial c dislocation density could be high enough that little acceleration or even a deceleration could occur.

A. T. Motta[1] (written discussion)—Does the fact that you have used DAD to explain your results mean that one does not need the production bias model to rationalize the irradiation growth data?

R. A. Holt et al. (authors' closure)—In the temperature range of interest, we think that DAD and not "production bias" is the dominant mechanism; however, the principle that the growth rate will be a maximum when the sink strength of two competing types of sinks are equal should also apply to production bias.

---

[1] Pennsylvania State University, University Park, PA.
[2] Electric Power Research Institute, Palo Alto, CA.

Effect of Irradiation on Microstructure

F. Garzarolli,[1] W. Goll,[1] A. Seibold,[1] and I. Ray[2]

Effect of In-PWR Irradiation on Size, Structure, and Composition of Intermetallic Precipitates of Zr Alloys

REFERENCE: Garzarolli, F., Goll, W., Seibold, A., and Ray, I., "**Effect of In-PWR Irradiation on Size, Structure, and Composition of Intermetallic Precipitates of Zr Alloys,**" *Zirconium in the Nuclear Industry: Eleventh International Symposium, ASTM STP 1295*, E. R. Bradley and G. P. Sabol, Eds., American Society for Testing and Materials, 1996, pp. 541–556.

ABSTRACT: The corrosion behavior of Zr alloys depends on the kind, size, and distribution of the intermetallic second-phase particles. TEM examinations of Zr-Sn-Fe-Cr alloys irradiated in PWRs at temperatures between 300 and 370°C and fast fluences in the range of 5E21 to 1.3E22 cm^{-2} have been performed to study the irradiation-induced effects on the precipitates. The alloys contained different types of second-phase particles such as $Zr(Fe,Cr)_2$, $Zr_2(Fe,Si)$, and Zr_3Fe before irradiation. The influence of irradiation was found to depend on temperature and type of second-phase particles.

At temperatures below 310°C, the Laves phase $Zr(Fe,Cr)_2$, which normally is the most frequent precipitate in Zry-4, depletes in Fe, becomes amorphous, and dissolves completely at higher fluences. With increasing temperature, the rate of Fe depletion and dissolution decreases and new Zr-Fe phases are formed. At temperatures above 370°C, the Laves phase remains stable or even grows under irradiation.

In Fe-containing Zr alloys with little or no Cr, rather large Zr_3Fe precipitates are the most frequent particles. These particles are not dissolved by irradiation even at low temperatures. This was confirmed by annealing after irradiation.

As a hypothesis, it was assumed that the different behavior of the various precipitates can be related to their melting or decomposition temperatures by using the homologous temperature (i.e., the temperature under consideration in K normalized to the melting or decomposition temperature in K). This interrelationship has been found to apply for irradiation-induced amorphization. The empirical approach to describe the thermal ripening behavior of second-phase particles before irradiation and to describe the transition from irradiation-induced dissolution to irradiation-induced growth by a normalized (homologous) temperature led to reasonable results.

KEYWORDS: zirconium alloys, precipitates, precipitate size, precipitate distribution, precipitate type, neutron irradiation, precipitate amorphization, precipitate dissolution, radiation effects, nuclear application

In light water reactors (LWR), Zr alloys are used for fuel rod cladding and structural parts of the fuel assembly in the high flux range. The corrosion behavior of these alloys is one of the limiting factors of burnup extension. In ZrSn(TM)-type alloys (TM = transition metal), the corrosion behavior depends on the kind, size, and distribution of the second-phase particles (SPP) precipitated in the matrix material. The type of the SPP depends on the chemical composition of the alloy. In ZrSn(TM) alloys, the most frequent types of SPP are the Laves phase $Zr(Fe,Cr)_2$, and $Zr_2(Fe,Ni)$. In the β-range (>950°C) of ZrSn(TM)-type alloys, the SPP-forming

---

[1] Siemens AG, Power Generation Group (KWU), 91050 Erlangen, Germany.
[2] Institute for Transuranium Elements, 76125 Karlsruhe, Germany.

transition elements Fe, Cr, and Ni are completely dissolved in the ZrSn matrix at concentration levels of ≤1%, which are typically used for alloying. In the α-range at temperatures below 840 to 800°C, these elements are almost insoluble [1] and precipitate as second-phase particles. The size and distribution of SPP depends on their type, the quenching rate from the β-temperature (which is usually applied in the fabrication of Zr alloys at an intermediate product dimension), and on the subsequent process and annealing temperatures in the α-range. For Zircaloy production with an intermediate fast β-quenching step, the effect of subsequent annealing can be described and controlled by the A-parameter [2]. The A(accumulated annealing)-parameter is given by $\Sigma t_i \cdot \exp(-Q/RT_i)$ with t_i and T_i being the time (h) and temperature (K) of the annealing steps i after β-quenching. Q/R is 40 000 K. In-reactor, the SPP are affected by neutron irradiation. The Laves phase and $Zr_2(Fe,Ni)$ SPP can dissolve at temperatures of about 300°C [3,4], whereas, at peak temperatures of PWR fuel cladding of about 350°C, an irradiation-induced growth of the Laves phase [5] or at least much less dissolution [6] was observed. The purpose of this investigation was to confirm the data on $Zr(Fe,Cr)_2$ SPP, to extend the knowledge of irradiation effects to other types of SPP, and to find an approach to normalize and predict the behavior of the different types of SPP to be considered in alloy development for high burnup application.

Characterization of SPP Before Irradiation

ZrSn(TM) alloys with up to 1.7% Sn, up to 1% Fe and Cr, up to 120 ppm Si, and up to 50 ppm Ni were examined. The main characteristics of these SPP are summarized in Table 1. In complex alloys, the SPP are formed by more than two constituents in most cases. The binary intermetallics, considered to be the basis of the ternary precipitates, are also given in Table 1.

In Zr-Sn-Fe-Cr alloys with a Fe/Cr ratio <4, almost all SPP are of the $ZrCr_2$ or $Zr(Fe,Cr)_2$ type. If the ratio Fe/Cr is >4, Zr_3Fe is formed besides the Laves phase $Zr(Fe,Cr)_2$. But this ratio of 4 is not an absolute limit. Sometimes Zr_3Fe can even be formed at a Fe/Cr ratio of <4 [9], whereas $Zr(Fe,Cr)_2$ was observed also at a Fe/Cr ratio of 10 besides Zr_3Fe. Sometimes

TABLE 1—*Characteristics of SPP according to literature [1–8] and own results.*

Type	Remark	Melting or Decomposition Temperature	Crystal Structure (a and c in nm)	Growth Rate (relative)
$Zr(Fe,Cr)_2$	most frequent SPP in Zry-4, Fe/Cr = 0 to 4		in $\alpha+\beta$-range: fcc, $a = 0.71$ in α:hex, $a = 0.5$, $c = 0.83$	medium
$ZrCr_2$		1620°C	>950°C: fcc, $a = 0.72$ else: hex, $a = 0.51$, $c = 0.83$	slow
$ZrFe_2$		1675°C	fcc, $a = 0.71$	
$Zr_2(Fe,Ni)$	appears sometimes		bct, $c = 0.65$, $a = 0.55$	medium-fast
Zr_2Ni		1150°C	bct, $c = 0.66$, $a = 0.53$	
Zr_2Fe		974°C	bct, $c = 0.64$, $a = 0.54$	
$Zr_2(Fe,Si)$	appears if Si > 50–80 ppm		bct, $c = 0.66$, $a = 0.53$	very slow
Zr_2Si		2110°C	bct, $c = 0.66$, $a = 0.53$	
Zr_3Fe	appears if Fe/Cr > 4	885°C	orthorhombic, $a = 0.33$, $b = 1.1$, $c = 0.88$	very fast
Zr_3Si	formed in the β-range		tetr, $a = 1.1$, $c = 0.55$	

other types of Fe containing precipitates with little or no Cr can be seen in addition to the $Zr(Fe,Cr)_2$ SPP at Fe/Cr < 4. These precipitates often also contain some Si or Ni to form $Zr_2(Fe,Ni)$ or $Zr_2(Fe,Si)$. The size of $Zr_2(Fe,Si)$, which is observed at Si contents above 50 to 80 ppm, is generally very fine. The crystal structure indicates that this is based on Zr_2Si.

All SPP are very fine after β-quenching but grow at the expense of their number density during the subsequent hot deformation and annealing steps. The determination of the average size of SPP is sensitive to the sample preparation and to the resolution of the electron microscope used. The arithmetic mean of all SPP may not be the same if measured under different conditions. The SPP size taken for the analysis in this paper is the most frequent size and is called modal size. In Fig. 1, the SPP size distribution of two different Zircaloy-4 material types is given by the number density in μm^{-2} versus the size in μm. The number density per unit area was determined from the measured values of SPP in areas within a foil thickness of 100 ± 20 nm. Type A represents several ZrSn (TM) heats with the same Fe/Cr ratio and a very low Si content. All type A-heats were fabricated according to the same process routine and nominally the same A-parameter. Type B has the same Fe/Cr ratio as Type A, but has a larger Si content of 60 to 110 ppm and a slightly lower A-parameter. The graphs are the result from

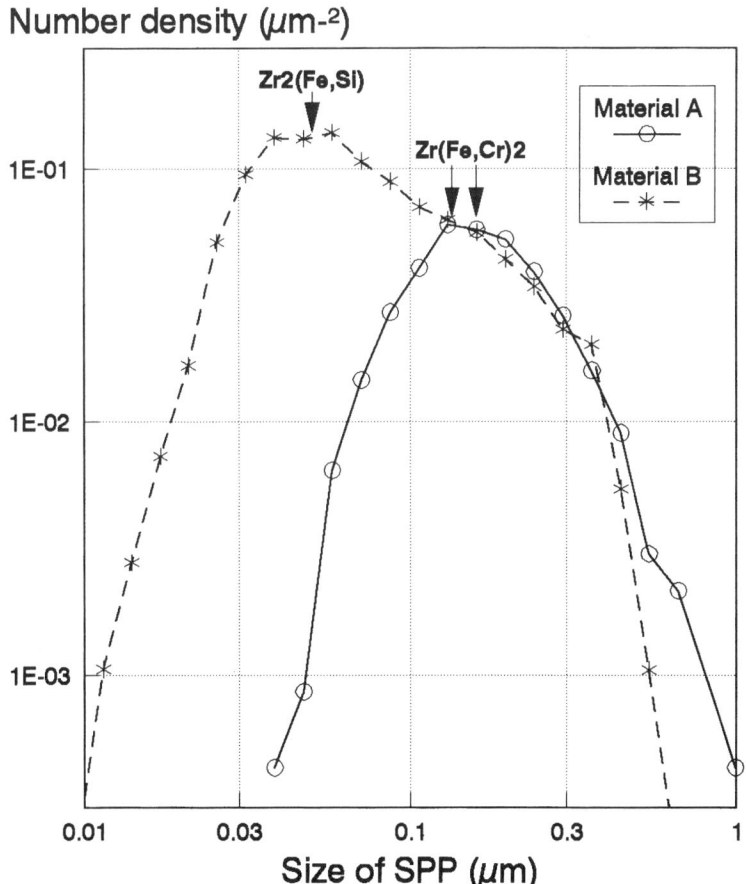

FIG. 1—*Size distribution of SPP in Materials A and B.*

several examinations of several lots of the same type of material. Material A has a mono-modal distribution and only $Zr(Fe,Cr)_2$ SPP. Material B has a bi-modal SPP size distribution with a high number density of very fine $Zr_2(Fe,Si)$ SPP and about the same number density and size distribution of $Zr(Fe,Cr)_2$ SPP as Material A. Both materials were also examined after irradiation, as will be shown later.

To get more insight into the behavior of the different types of SPP, their growth characteristics as a function of heat treatment were determined. In Fig. 2, the sizes of different types of precipitates are plotted versus the A-parameter. At comparable temperature/time histories, Zr_3Fe SPP exhibit the largest size, $Zr(Fe,Cr)_2$ SPP exhibit a medium size, and $ZrCr_2$ and $Zr_2(Fe,Si)$ exhibit the smallest size. Laves-phase SPP with a Fe/Cr ratio of ≤ 0.1 grow slower compared with those having a Fe/Cr ratio of ≥ 1. These observations led to the hypothesis that a relationship between growth behavior and stability of SPP, expressed by the melting or decomposition temperature, exists. Therefore, as an attempt to normalize the different growth behavior of different types of SPP, the particle sizes obtained after similar annealing times but different annealing or fabrication temperatures were plotted versus the homologous process temperatures, i.e., the process temperatures were related to the respective melting or decomposition temperatures. As can be seen from Fig. 3, a reasonable correlation exists.

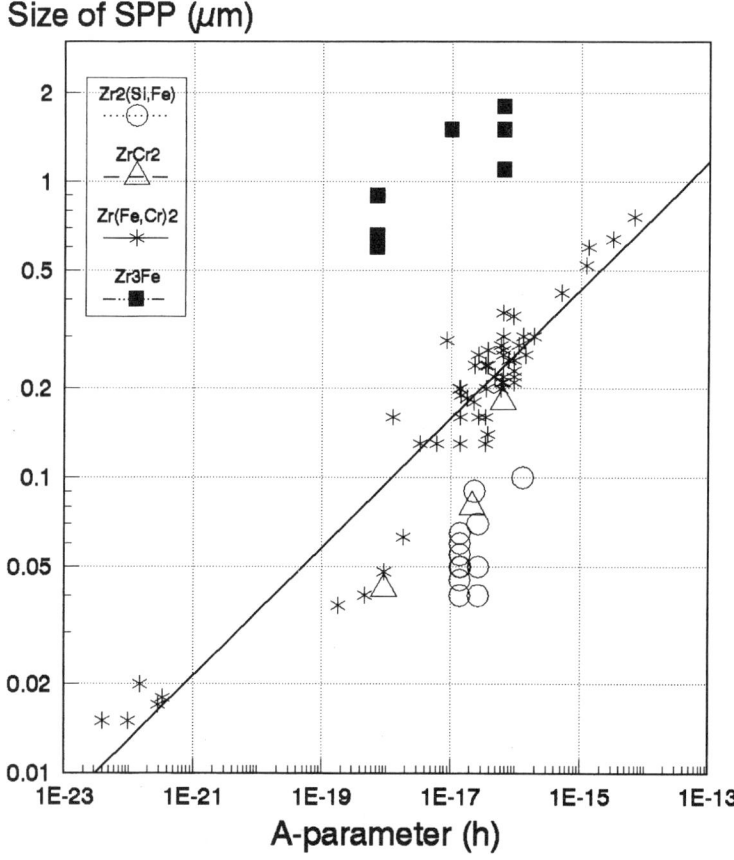

FIG. 2—*Correlation between A-parameter and SPP size.*

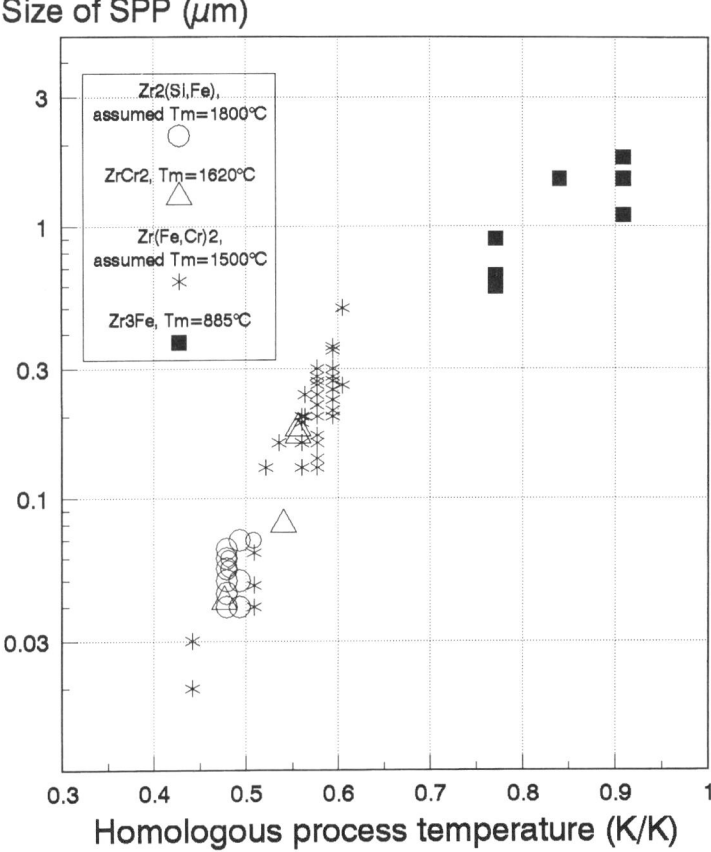

FIG. 3—*Correlation of SPP size versus ratio of process temperature and melting/decomposition temperature.*

The general applicability of the homologous temperature for description and normalization of the different growth behavior, however, has to be verified by additional investigations. A prerequisite for the verification is the thorough determination of the melting/decomposition temperatures of the different SPP.

Characterization of SPP after Irradiation

Materials and Techniques

The characteristic data of the irradiated samples are shown in Table 2. Samples A1 to A3, made of Material A with <30 ppm Si, are taken from a single five-cycle PWR fuel rod irradiated for 1543 effective full power days (EFPD) up to a maximum fluence of 1.3×10^{22} cm^{-2} ($E > 1$ MeV). The three irradiated samples from three different axial positions of the PWR fuel rod represent different operating temperatures between 300 and 350°C. Compared to Zircaloy-4, this material has a lower Sn content. The investigation of unirradiated Type A material samples from different heats with different Sn contents demonstrated that the difference in Sn content does not affect the size of the Laves phase significantly as long as the fabrication process

TABLE 2—*Irradiated samples.*

			Chemical Composition				Irradiation Condition		
Sample	Material	Sample Type	Sn, wt%	Fe, wt%	Cr, wt%	Si, ppm	Time, EFPD	Fluence, cm^{-2}, >1 MeV	Temp., °C
A1	Type A	Fuel rod	0.61	0.23	0.10	<30	1543	4.9E21	300
A2	Type A	Fuel rod	0.61	0.23	0.10	<30	1543	1.2E22	325
A3	Type A	Fuel rod	0.61	0.23	0.10	<30	1543	1.3E22	350
B1	Type B	Fuel rod	1.32	0.22	0.11	73	625	4.9E21	325
B2	Type B	Fuel rod	1.32	0.20	0.11	77	625	4.9E21	325
B3	Type C	Fuel rod	1.3	0.22	0.11	88	625	4.9E21	325
C1	Type D	Strip sample	1.58	0.21	0.12	n.m.	1244	8E21	315
C2	Type E	Strip sample	0.24	0.53	0.05	n.m.	1244	8E21	315
C3	Type F	Strip sample	0.24	0.21	...	n.m.	1246	8E21	315

n.m. = not measured.

(ΣA-parameter) is the same. Thus, the Material A graph was used for comparison. The irradiated samples B1 to B3 represent different heats of Zircaloy-4 Materials B and C containing 73 to 88 ppm Si. The samples were taken from three 2-cycle fuel rods each at the same axial position. The operation temperature at the sample position was 325°C for all samples. Samples C1 to C3 are from strip coupons fabricated out of three different materials: D, E, and F. All three coupons were inserted in one water rod in a PWR fuel assembly and irradiated at a temperature of 315°C for four cycles up to a fluence of 8×10^{21} cm^{-2} ($E > 1$ MeV). C1 is a Zircaloy-4 sample. C2 is Zr-0.2Sn-0.5Fe-0.05Cr with a Fe/Cr ratio = 10, and C3 is Zr-0.2Sn-0.2Fe with no Cr addition. Samples for transmission electron microscopy were taken about 0.1 mm from the outer surface of the cladding tubes and in the middle of the wall of the corrosion coupons.

Determination of type and size of precipitates was performed at the European Institute for Transuranium Elements in Karlsruhe in a 200-kV Hitachi H700 HST transmission electron microscope (TEM) that has been modified for handling radioactive samples. Standard techniques were applied for sample preparation and thinning. SPP composition was analyzed using a Tracor Northern TN 5500 energy dispersive X-ray analyzer (EDX) attached to the TEM. For determination of the SPP structure, the electron diffraction pattern was taken and checked for consistency with published structures.

Particle measurements were made taking care to cover a representative area of the sample. In many cases, the particle contrast was low and different operation modes of the TEM as well as EDX had to be used to confirm the existence of the phases. For SPP sizes of >1 µm, the scanning electron microscope (SEM) mode was used to examine larger areas not transparent to the electron beam. The number of particles per unit area was in this case determined from the particles seen on the surface of the thinned TEM samples. The number densities derived from both examination techniques were combined.

Examination Results

The frequency of the SPP in Material A before and after irradiation to a very high fluence at different temperatures (Samples A1 to A3) is shown in Fig. 4. In Table 3, the total number density of SPP and their composition is listed. Fine SPP were formed after irradiation at 325 and 350°C in addition to the larger ones observed before irradiation. The size and number

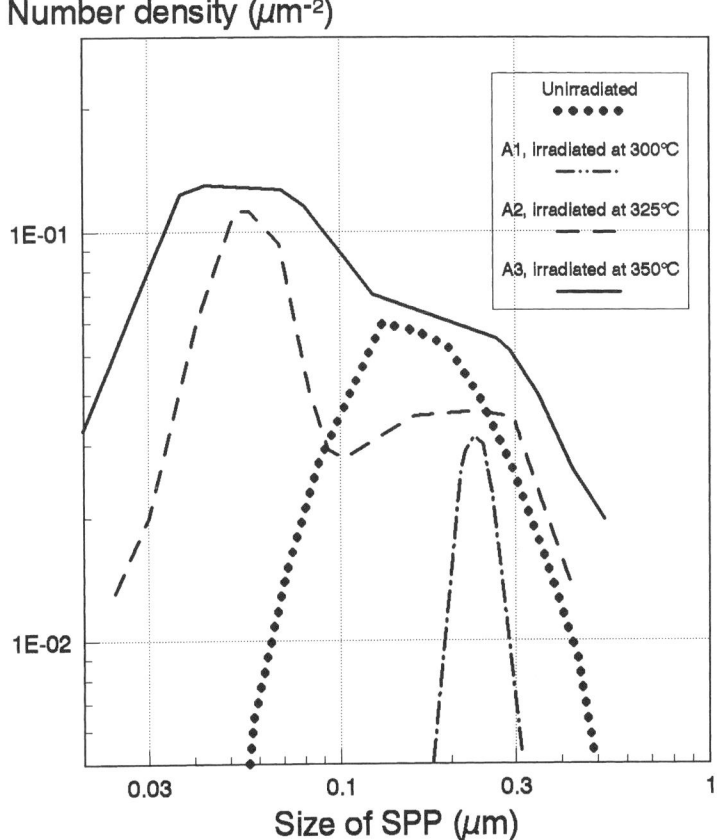

FIG. 4—*Size distribution in Material A after irradiation at different temperatures.*

density of the large SPP population seems to be higher after irradiation at 350°C than before irradiation. This may be due to internal dissolution, as will be discussed later, or due to data scatter (for Sample A3 only an area of 160 μm^2 with 210 SPP was examined). Irradiation at 325°C (Sample A2) definitely decreases the number density of the larger SPP. After irradiation at 300°C (Sample A1), only a few SPP with sizes of 0.2 to 0.3 μm and none in the range of 0.03 to 0.2 μm were found. The majority of the SPP was obviously dissolved at this temperature. Not only the size and distribution, but also the type comprising stoichiometry, chemical composition, and lattice structure of the precipitates was changed by irradiation. In Sample A1, the major constituent of the SPP is Fe besides Zr (Fe/Cr ratio = 9). It is assumed that almost all of the Cr and most of the Fe is more or less uniformly distributed in the matrix after irradiation at 300°C. The Cr-depleted SPP remaining at 300°C have a weak contrast and do not show any diffraction spots but also no diffuse ring patterns. It is assumed that they are different from the Laves phase existing before irradiation. Some Zr-Fe SPP were also observed in Sample A3. These Fe-Zr SPP, however, have a very distinct contrast and a clear diffraction pattern. According to electron diffraction, the structure of these Zr-Fe SPP is the orthorhombic Zr_3Fe. Before irradiation, Zr_3Fe SPP had not been observed in Material A. All precipitates in A2 and most of those in A3 contain Fe and Cr and are of the type $Zr(Fe,Cr)_2$. They generally have a

TABLE 3—*Summary of results from TEM/SEM examinations.*

Sample	Fluence, ($E > 1$ MeV) cm^{-2}	Temp., °C	SPP Density, μm^{-2}	Most Frequent Size Zr$_2$(Fe,Si), μm	Zr(Fe,Cr)$_2$, μm	Zr$_n$Fe, μm	Zr$_3$Fe, μm	Mean Fe/Cr Ratio of Zr(Fe,Cr)$_2$
Material A	...	...	0.3	...	0.16^a	...	...	2^a
A1	4.9E21	300	0.05	...	...	0.25	...	...
A2	1.2E22	325	0.65	...	0.06 + 0.2	...	...	0.19
A3	1.3E22	350	1.3	...	0.06 + 0.2	...	0.1^a	1.1
Material B	...	...	1.3	0.05^a	0.14^a	...	...	1.8^a
B1	4.9E21	325	1.3	...	0.06	...	0.06^a	0.62
B2	4.9E21	325	1.35	...	0.07	...	0.08^a	0.32
Material C	...	...	1.2	0.05^a	0.14^a	...	...	1.9^a
B3	4.9E21	325	1.1	...	0.09	...	0.06	0.57
Material D	...	...	0.6	...	0.03 + 0.3^a	...	...	2.5^a
C1	8E21	315	0.2	...	0.1^b	...	n.e.^{a}	0.26
Material E	...	...	0.075	...	0.26^a	...	1.5^a	4^a
C2	8E21	315	0.065	...	...	...	1.6^a	...
Material F	...	...	0.036	...	...	...	0.2/1.5^a	...
C3	8E21	315	0.03	...	...	...	0.25/1.6^a	...
C3	8E21 + annealed	315 + 600/th	no change	...	...	...	no change	...

^{a} crystalline.
^{b} amorphous.
^{c} n.e. = not estimated.

weak contrast and do not show a clear diffraction pattern. The Fe/Cr ratio of these SPP has decreased due to irradiation, especially in Sample A2. Whereas in Sample A2 all SPP exhibited a similar Fe/Cr ratio, Sample A3 revealed quite large variations in the Fe/Cr ratio (between 1.9 and 0.8). Figure 5 shows a size distribution of the two different SPP observed in Sample A3. The size distribution is nearly the same for both types of SPP. Samples A2 and A3 revealed a total number density of SPP that is much higher than before irradiation (Table 3). This increase is due to the fine SPP that have been formed in-reactor at 325 and 350°C. A significant reduction of the total number density of SPP was only observed for Sample A1 irradiated at 300°C. Obviously the Laves-phase-type SPP lose Fe and become dissolved in the temperature range between 300 and 350°C. New fine SPP (most probably Laves-phase type) and Zr$_3$Fe SPP are formed. The degree of dissolution and Fe depletion (reduction of the Fe/Cr ratio) decreases with increasing temperature, as can be seen from Table 3.

Materials B and C with higher Si content (Samples B1, B2, B3) have a large number density of fine Zr$_2$(Fe,Si) SPP before irradiation. The effect of irradiation to a medium fluence on SPP size distribution (average of Samples B1 and B2) at 325°C is shown in Fig. 6. Si was not detected any more in the fine SPP. The Zr$_2$(Fe,Si) SPP either dissolve or release Si in-reactor at 325°C. Furthermore, new Zr-Fe SPP with clear contrast and diffraction patterns were formed in Samples B1, B2, and B3, probably Zr$_3$Fe, as found before in Sample A3. The Zr(Fe,Cr)$_2$ SPP lose a large fraction of the Fe at this temperature and show significantly reduced Fe/Cr ratios (1.3 to 0.2). They have a weak contrast and do not show clear diffraction spots. The size of the larger Zr(Fe,Cr)$_2$ SPP does not seem to change much in spite of the Fe depletion. This seems to be a contradiction, but the fact that the particles show quite weak contrasts (Fig. 7) and a relatively low TM to Zr ratio (EDX) even for the large SPP indicates that the Zr(Fe,Cr)$_2$

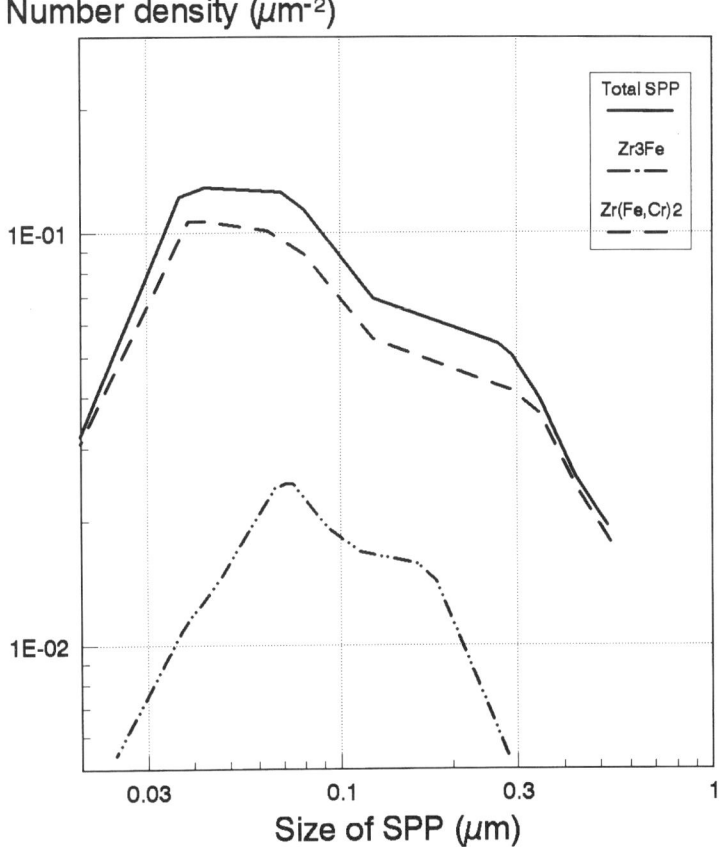

FIG. 5—*Size distribution of different SPP types of Material A after irradiation at 350°C.*

SPP dissolve mainly from the interior without becoming smaller. This is different from the observations at 300°C, where the SPP were found to shrink in size and finally dissolve under irradiation [5,10].

Material D (Sample C1) is also a Zircaloy-4 with a rather large SPP size before irradiation. Irradiation to a rather high fluence at 315°C has caused a significant reduction in number density and size of SPP within this material. The Fe/Cr ratio was reduced from 2.5 before irradiation to 0.26 (0.15 to 0.32) after irradiation. Again, some new Zr-Fe SPP have been formed. Clear diffraction patterns of the Zr-Fe SPP were consistent with the structure of Zr_3Fe. The Zr-Fe-Cr SPP again exhibited no clear diffraction pattern after irradiation.

Both Materials E and F (Samples C2 and C3) containing high Fe and low Cr additions showed primarily large Zr_3Fe SPP besides a few $Zr(Fe,Cr)_2$ SPP before irradiation. Because of the large spread of the SPP size in these materials, the TEM and SEM mode was used for the examination. After irradiation, only Zr_3Fe SPP were observed. Figure 8 shows SEM micrographs of Zr_3Fe SPP before and after irradiation. No irradiation-induced effect on the Zr_3Fe, no change of SPP size, and almost no change of SPP number density was observed, as can be seen from Table 3. Annealing of irradiated Sample C3 at 600°C for 1 h did not change the SPP

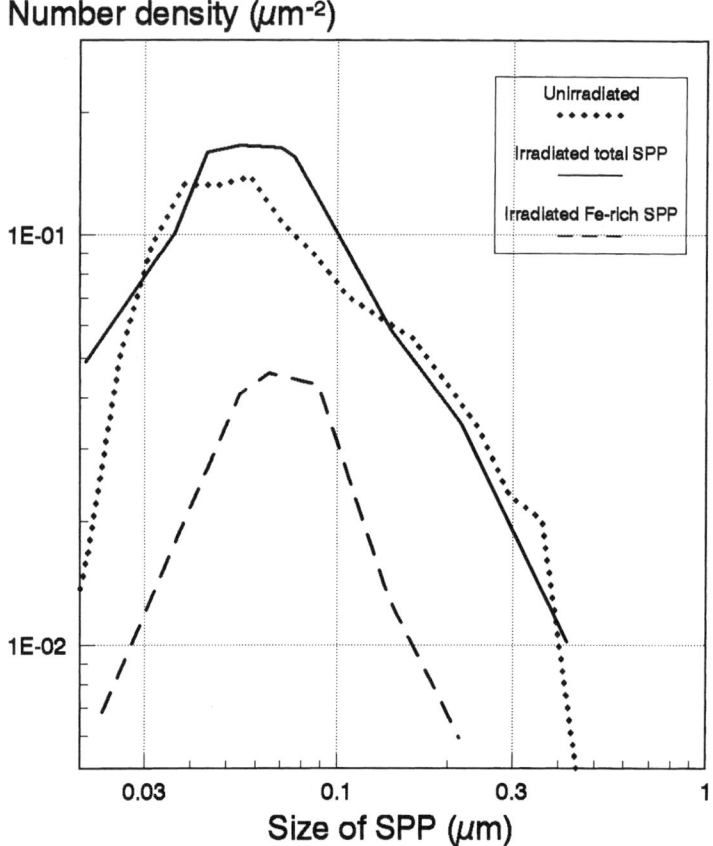

FIG. 6—*Size distribution of SPP in Material B after irradiation at 325°C.*

and did not create any new fine SPP. This indicates that no Fe was in super-saturated solution in the α-matrix after irradiation at 315°C in this material.

Discussion

In ZrSn (TM) alloys, different types of SPP such as $Zr(Fe,Cr)_2$, $Zr_2(Fe,Ni)$, $Zr_2(Fe,Si)$, and Zr_3Fe are formed depending on the alloying content. The different types of SPP have a different thermal growth behavior. The growth behavior should depend on the energy balance between the surface increase of the growing particle and the dissolution (stability) of the dissolving particle, as well as on diffusion of TM atoms in the Zr matrix. Stability of a particle might be expressed by its melting temperature or decomposition temperature. As an attempt to assess the relationship between in-reactor stability and melting temperature or decomposition temperature of the SPP, the homologous temperature (ratio of process and melting or decomposition temperature) was used to normalize the different growth behaviors of the different types of SPP. For the investigation range, a reasonable correlation was found, but as a prerequisite for the verification of this concept the melting/decomposition temperatures of the different SPP have to be thoroughly determined.

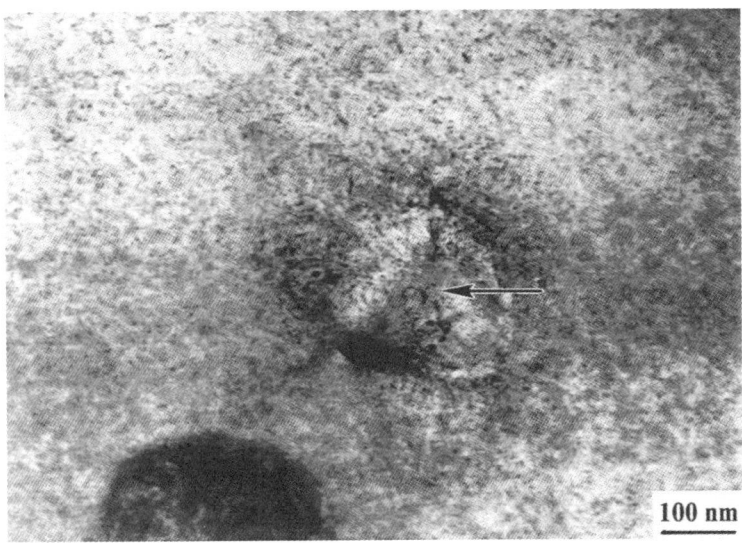

FIG. 7—$Zr(Fe,Cr)_2$-type SPP in Sample B1 (Material B, 325°C, 4.9E21 cm^{-2}).

The influence of irradiation in light water reactors (LWR) on SPP in Zr-Sn-Fe-Cr-(Si) alloys depends on the type of SPP and irradiation temperature. Zr_3Fe SPP were found to be quite stable under irradiation up to 8E21 cm^{-2} at 315°C and to be even formed after dissolution of other types of SPP at temperatures of ≥315°C (Table 3). This finding is in agreement with former results [10], which indicated a small size increase of Zr_3Fe SPP (from 0.31 to 0.35 μm) after irradiation up to 1.5E21 cm^{-2} at 288°C and no additional SPP formation after a post-irradiation annealing at 575°C.

As regards $Zr(Fe,Cr)_2$ SPP, it is reported [2,3,5,11] that irradiation at ≤300°C causes a depletion of Fe, a subsequent amorphization, and shrinking of the precipitates. The number density of the $Zr(Fe,Cr)_2$ SPP decreases with increasing fluence, and they finally disappear completely. The amorphous SPP appear bright in the irradiation-damaged Zr matrix and show diffuse ring electron diffraction patterns. The Fe depletion and amorphization process starts from the periphery of the SPP at this temperature. The thickness of the affected rim and the diameter of the SPP change almost linearly with fluence. The boundary between the amorphous SPP and the Zr matrix is sharp at this temperature. The irradiation-induced change of the $Zr(Fe,Cr)_2$ SPP at 315°C was found in this study to be very similar as reported previously for 300°C. Irradiation at 325 and 350°C, however, affects the $Zr(Fe,Cr)_2$ SPP differently. The contrast of the SPP becomes weak, the boundaries between the SPP and the matrix become diffuse, and the ratio between Zr and TM probably increases. Furthermore, the number density of these types of SPP increases after irradiation in this temperature range. In addition, neither the diffuse ring electron diffraction pattern, usually seen after irradiation at ≤315°C, nor a clear diffraction pattern were observed. Irradiation-induced Fe depletion occurs also in this temperature range. However, the degree of Fe depletion decreases with increasing temperature. Dissolution of the SPP in this temperature range is not accompanied by shrinkage but rather by growth, indicating a dissolution from the interior. Besides the reprecipitation of fine Laves-phase type Zr-Fe-Cr SPP, Zr_3Fe SPP also form in parallel. At temperatures above 350°C, there seems to be a transition from Fe depletion to an irradiation-induced ripening of the $Zr(Fe,Cr)_2$ SPP. Reference 5 found no Fe depletion but a ripening at 370°C and [6] some Fe

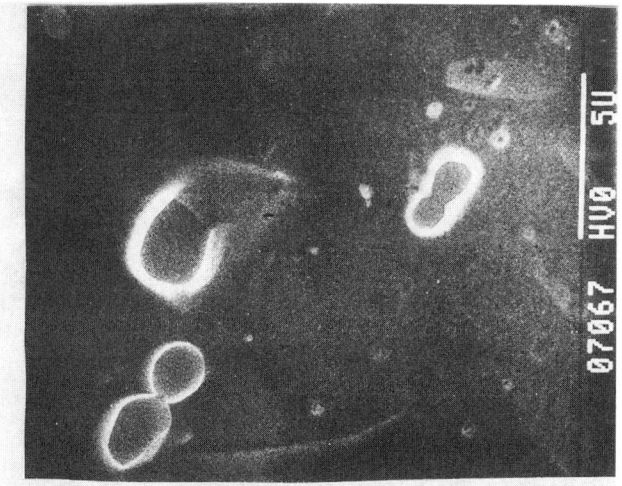

a) Unirradiated

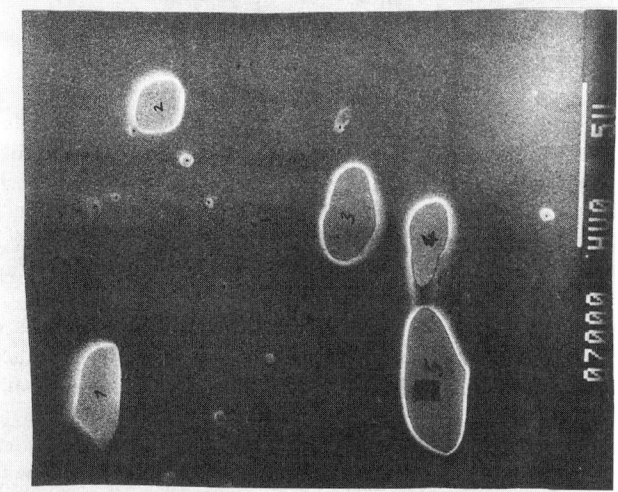

b) Irradiated at 315°C

FIG. 8—*SEM micrographs of Zr_3Fe SPP in Material E.*

depletion, but crystalline structure at 380°C. Figure 9 summarizes the findings of the present investigation and former results on irradiation behavior of the Laves phase. Even if the results deviate slightly from each other, neutron-irradiation-induced effects in the temperature range of the PWR operation can be described quite satisfactorily.

The fine $Zr_2(Fe,Si)$ SPP existing in Zircaloy-4 lots containing 60 to 120 ppm Si (Materials B and C) had disappeared after irradiation at 325°C either due to dissolution or to Si release. To normalize the different amorphization behavior of different types of SPP as a consequence of irradiation, the homologous irradiation temperature (the irradiation temperature in K divided by the melting or decomposition temperature in K) has been used by several authors [12]. As shown before, the approach to normalize the thermal growth behavior of different SPP types

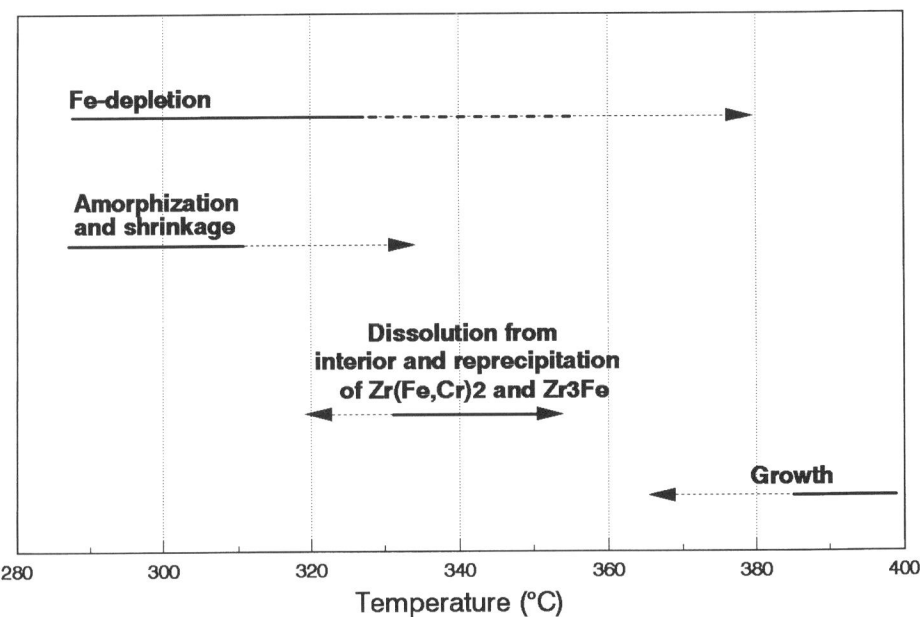

FIG. 9—*Effect of LWR irradiation on Zr(Fe,Cr)₂ (Laves phase) SPP.*

by using the homologous process temperature led to a reasonable correlation. Therefore, the hypothesis was made that a behavior connected with the displacement of SPP constituents— irradiation-induced internal dissolution and irradiation-induced growth—can be described with the homologous temperature concept. As a first attempt, the homologous temperature ranges for different irradiation effects were deduced from the behavior of the Laves-phase $Zr(Fe,Cr)_2$ (see Table 4) and applied to the other types of SPP observed in the Zr-Sn (TM) alloys under investigation. The results from this irradiation and the PIE program on the different types of SPP support the prediction of dissolution or Si release of $Zr_2(Fe,Si)$ and growth for Zr_3Fe at PWR operating temperatures. However, to verify the hypothesis, more data on Zr_3Fe and $Zr_2(Fe,Si)$ irradiated in different homologous temperature regions are needed. If the homologous temperature concept is applicable, it would provide a tool to predict the in-pile behavior of different types of SPP. This could be useful for alloy development, including chemical composition as well as fabrication.

TABLE 4—*Temperature ranges (°C) for the different irradiation-induced changes of different types of SPP.*

SPP Type	Amorphization and Shrinkage	Internal Dissolution	Growth
$Zr(Fe,Cr)_2$	<320	330–360	>370
Homolog. temp.[a]	<0.33	0.34–0.36	>0.365
$Zr_2(Fe,Si)$[b]	<420	430–470[c]	480[c]
Zr_3Fe[b]	110[c]	120–145[c]	>150

[a] Assuming $Tm = 1500°C$.
[b] Predicted using the homologoue temperature.
[c] Not relevant for reactor operation.

Conclusions

1. The effect of neutron irradiation in a light water reactor depends on type of SPP and temperature.
2. Depending on irradiation temperature, the irradiation-induced effects on $Zr(Fe,Cr)_2$ can either be (a) Fe depletion, amorphization, and shrinkage, or (b) Fe depletion and internal dissolution, or (c) growth.
3. The temperature ranges where phase changes occur differ significantly between the different types of SPP.
4. The homologous temperature concept might be a tool to predict the temperature ranges for the different irradiation-induced effects on different types of SPP. For the assumption of the melting or decomposition temperatures made, the behavior of the SPP under investigation supports the applicability of the homologous temperature concept. However, to verify the hypothesis, additional data on the irradation behavior of SPP in the different homologous temperature regions are needed.

References

[1] Charquet, D., Hahn, R., Ortlieb, E., Gros, J. P., and Wadier, J. F., "Solubility Limits and Formation of Intermetallic Precipitates in ZrSnFeCr Alloys," *Zirconium in the Nuclear Industry: Eighth International Symposium, ASTM STP 1032*, American Society for Testing and Materials, West Conshohocken, PA, 1989, p. 405.

[2] Charquet, D., Steinberg, E., and Millet, Y., "Influence of Variations in Early Fabrication Steps on Corrosion, Mechanical Properties, and Structure of Zircaloy-4 Products," *Zirconium in the Nuclear Industry: Seventh International Symposium, ASTM STP 939*, American Society for Testing and Materials, West Conshohocken, PA, 1987, pp. 431–447.

[3] Yang, W. J. S., Tucker, R. P., Cheng, B., and Adamson, R. B., "Precipitates in Zircaloy: Identification and the Effects of Irradiation and Thermal Treatment," *Journal of Nuclear Materials*, Vol. 138, 1986, pp. 185–195.

[4] Griffiths, M., Gilbert, R.W., and Carpenter, G. J. C., "Phase Instability, Decomposition and Redistribution of Intermetallic Precipitates in Zircaloy-2 and -4 During Neutron Irradiation," *Journal of Nuclear Materials*, Vol. 150, 1987, pp. 53–66.

[5] Garzarolli, F., Dewes, P., Maussner, G., and Basso, H. H., "Effects of High Neutron Fluences on Microstructure and Growth of Zircaloy-4," *Zirconium in the Nuclear Industry: Eighth International Symposium, ASTM STP 1032*, American Society for Testing and Materials, West Conshohocken, PA, 1989, p. 641.

[6] Gilbon, D. and Simonot, C., "Effect of Irradiation on the Microstructure of Zircaloy-4," *Zirconium in the Nuclear Industry: Tenth International Symposium, ASTM STP 1245*, American Society for Testing and Materials, West Conshohocken, PA, 1994, p. 521.

[7] Charquet, D. and Alheritiere, E., "Precipitates in Zircaloys," *Proceedings*, KTG Workshop on Second-phase Particles and Matrix Properties of the Zircaloys, Erlangen, FRG, 1–2 July 1985.

[8] Eucken, C. M., Finden, P. T., Trapp-Pritsching, S., and Weidinger, H. G., "Influence of Chemical Composition on Uniform Corrosion of Zirconium Base Alloys in Autoclave Tests," *Zirconium in the Nuclear Industry: Eighth International Symposium, ASTM STP 1032*, American Society for Testing and Materials, West Conshohocken, PA, 1989, pp. 113–127.

[9] Ruhmann, H., Manzel, R., Sell, H.-J., and Charquet, D., "In BWR and Out-of-Pile Nodular Corrosion Behavior of Zry 2/4 Type Melts with Varying Fe, Cr, and Ni Content and Varying Process History," this publication.

[10] Kruger, R. M. and Adamson, R. B., "Precipitate Behavior in Zirconium-Based Alloys in BWRs," *Journal of Nuclear Materials*, Vol. 205, 1993, pp. 242–250.

[11] Etoh, Y. and Shimada, S., "Neutron Irradiation Effects on Intermetallic Precipitates in Zircaloy as a Function of Fluence," *Journal of Nuclear Materials*, Vol. 200, 1993, pp. 59–69.

[12] Motta, A. T., Lefebvre, F., and Lemaignan, C., "Amorphization of Precipitates in Zircaloy Under Neutron and Charged Particle Irradiation," *Zirconium in the Nuclear Industry: Ninth International Symposium, ASTM STP 1132*, American Society for Testing and Materials, West Conshohocken, PA, 1991, p. 718.

DISCUSSION

C. Lemaignan[1] *(written discussion)*—(1) When you have Zr_3Fe reprecipitation under irradiation, have you found any epitaxy of these precipitates on the Zr matrix? (2) Have you performed a mass balance for Fe or any other measurements to evaluate any increase in Fe solubility in Zr under irradiation.

A. Seibold et al. (authors' closure)—(1) No epitaxy was observed between the Zr matrix and the reprecipitated Zr_3Fe SPP. (2) No, we have not. Such calculations give—even for unirradiated samples where an estimation of SPP size distribution and number density is much more accurate—in most cases noticeable deviations from the value expected theoretically.

G. Bart[2] *(written discussion)*—(1) How significant are the conclusions drawn from the experiments and what were the statistics for finding the different particle species? (2) The irradiation-induced changes of SPPs are also dependent on the neutron flux apart from n-fluence. Could you comment on this fact?

A. Seibold et al. (authors' closure)—(1) The areas examined (32 to 160 μm^2 by TEM and 400 to 2700 μm^2 by SEM) were not as large and the numbers of SPP detected were not as high as one would like, however, as long as only the most frequent size is considered, accuracy is still good. (2) The neutron flux (damage rate) is also considered to have a strong influence. The samples we examined were all irradiated at about the same flux in high power positions of modern, efficient PWRs. Therefore, the reported results are conservative regarding PWR operating conditions.

B. Cheng[3] *(written discussion)*—Dissolution of Fe into the Zircaloy matrix by irradiation will affect the corrosion resistance and $<c>$ dislocation and stacking faults. Your results show that, with 80 ppm Si in Type B material, the $Zr_2(Fe,Si)$ precipitate will dissolve by irradiation. Did you examine whether the $Zr_2(Fe,Si)$ ppts points will dissolve faster than the Laves phase, hence causing an effect on the material properties. What is your view of the current industry trend of controlling Si in the range of ~80 ppm? Any potential harm? How about the effect of carbon?

A. Seibold et al. (authors' closure)—We cannot answer the questions with respect to the Si containing SPP and carbon from our examinations because all our samples had rather low carbon contents and only high doses. However, we would like to emphasize that the $Zr_2(Fe,Si)$ SPP are generally very fine. Thus, they probably dissolve early. On the other hand, they could play a role in the renucleation of the Zr_3Fe SPP. The current trend of controlling Si at 80 ppm is based on fuel rod irradiations and analysis of the corrosion and other measurements and not on theoretical mechanistic considerations.

S. Yagnik[4] *(written discussion)*—(1) The conclusive evidence of particle growth above 370°C is rather high for PWR conditions. Your own experience of limited growth was below 350°C. For the "other work" you cited for 370°C, was this a special irradiation? Has that work been published? (2) How great are the statistics of particle dissolution from the interior that you presented?

A. Seibold et al. (authors' closure)—(1) To our knowledge, there is only one publication

---

[1] CEA—Grenoble, France.
[2] Paul Scherrer Institute, Switzerland.
[3] Electric Power Research Institute, Palo Alto, CA.
[4] EPRI.

that definitely reported growth of the Laves-type SPP. This is Ref 5. The samples studied in this reference came from experimental fuel rods with thick oxide layers irradiated with thick oxide layers at high heat fluxes. (2) Different modes of irradiation-induced dissolutions for different SPP types were reported previously [5]. The contemporary examinations showed that Zr-Fe-type SPP neither become amorphous nor shrink due to irradiation at 300°C like the ZrFeCr SPP but form lamellas of zirconium within the precipitate. The conclusion "dissolution from interior" is based (first) on the observed diffuse boundary between the SPP and the matrix, where sometimes even Zr islands can be seen, (second) on the relatively low TM to Zr ratio (EDX) of all large SPP, whereas before irradiation at least some of the large SPP exhibit a ratio close to the theoretical value, and (third) on the finding that no shrinkage but rather growth was observed in spite of the Fe depletion. This type of dissolution could be caused by the formation of submicroscopic Zr islands or by a change of the stoichiometry between Zr and TM.

M. Inagaki[5] *(written discussion)*—Do you think that such phenomena as you presented can be reproduced by a high voltage electron irradiation experiment?

A. Seibold et al. (authors' closure)—High voltage electron irradiations do not allow studies at damage rates and irradiation times effective in PWR fuel cladding. Thus, only amorphization and dissolution at high damage rates can be studied and probably not reprecipitation and growth.

---

[5] Hitachi Research Laboratory, Ibavaki-keu, Japan.

Arthur T. Motta,[1] Joseph A. Faldowski,[1] Lawrence M. Howe,[2] and Paul R. Okamoto[3]

In Situ Studies of Phase Transformations in Zirconium Alloys and Compounds Under Irradiation

REFERENCE: Motta, A. T., Faldowski, J. A., Howe, L. M., and Okamoto, P. R., "**In Situ Studies of Phase Transformations in Zirconium Alloys and Compounds Under Irradiation,**" *Zirconium in the Nuclear Industry: Eleventh International Symposium, ASTM STP 1295*, E. R. Bradley and G. P. Sabol, Eds., American Society for Testing and Materials, 1996, pp. 557–579.

ABSTRACT: The High Voltage Electron Microscope (HVEM)/Tandem facility at Argonne National Laboratory has been used to conduct detailed studies of the phase stability and microstructural evolution in zirconium alloys and compounds under ion and electron irradiation. Detailed kinetic studies of the crystalline-to-amorphous transformation of the intermetallic compounds $Zr_3(Fe_{1-x},Ni_x)$, $Zr(Fe_{1-x},Cr_x)_2$, Zr_3Fe, and $Zr_{1.5}Nb_{1.5}Fe$, both as second phase precipitates and in bulk form, have been performed using the in situ capabilities of the Argonne facility under a variety of irradiation conditions (temperature, dose rate). Results include a verification of a dose rate effect on amorphization and the influence of material variables (stoichiometry x, presence of stacking faults, crystal structure) on the critical temperature and on the critical dose for amorphization.

Studies were also conducted of the microstructural evolution under irradiation of specially tailored binary and ternary model alloys. The stability of the ω-phase in Zr-20%Nb under electron and Ar ion irradiation was investigated as well as the β-phase precipitation in Zr-2.5%Nb under Ar ion irradiation. The ensemble of these results is discussed in terms of theoretical models of amorphization and of irradiation-altered solubility.

KEYWORDS: Zircaloy, intermetallic compounds, amorphization, charged-particle irradiation, phase transformations, irradiation precipitation, Laves phases, omega phase, beta phase

Notation

$\dot{A}$ Annealing rate (annealed defects per atom/s)
a_o Interplanar distance, cm
C_1 Proportionality constant
C_j Concentration of Defect j (atom fraction)
C_s, C_d Sink densities expressed in atom fraction
D_j Diffusion coefficient of Defect j, cm$^2 \cdot$s^{-1}
D_{sd} Diffusion coefficient for slowest defect, cm$^2 \cdot$s^{-1}

---

[1] Assistant professor and graduate student, respectively, Department of Nuclear Engineering, 231 Sackett Building, The Pennsylvania State University, University Park, PA, 16802.
[2] Senior research scientist, AECL Research, Reactor Materials Research Branch, Chalk River Laboratories, Chalk River, Ontario, Canada, K0J 1J0.
[3] Senior research scientist, Materials Science Division, Argonne National Laboratory, 9700 South Cass Avenue, Argonne, IL 60439.

D_{Nb} Nb diffusion coefficient, cm$^2\cdot$s^{-1}
E_j Migration energy of Defect j, eV
E_d displacement energy, eV
f Fraction of freely migrating defects
$\dot{G}_{eff}$ Effective defect generation rate, dpa$\cdot$s^{-1}
$\dot{G}$ Defect production rate, dpa$\cdot$s^{-1}
h foil thickness, cm
k_B Boltzman's constant, eV$\cdot$K^{-1}
L_{crit} Critical level of damage for amorphization in dpa
S_{kj} Sink strength of Sink k for Defect j, cm^{-2}
S_s Surface sink strength, cm^{-2}
T_c Critical temperature for amorphization, K
t_{irr} Irradiation time, s
t_{irr}^{am} Irradiation time to amorphization, s
t_{irr}^i Irradiation time under irradiation Type i, s
$<x>$ Diffusion length, cm
z_{sd} Dislocation bias factor for slowest defect
ε Ratio of total sink strength to the dislocation sink strength
Φ Flux of damaging particles, particles$\cdot$s$^{-1}\cdot$cm^{-2}
χ_i Ratio of Nb diffusion length under irradiation Type i to that under neutron irradiation at 770 K to 0.62 dpa
$\nu_{NRT}(T)$ Number of displacements caused by ion energy T, according to NRT model
ν_j Vibrational frequency of Defect j, s^{-1}
ρ_d Dislocation density, cm^{-2}
σ_d Displacement cross section, barn
τ Time to steady state, s

Introduction

Various zirconium alloys have been employed during the past decades for cladding, tubing, and structural materials in nuclear power reactor fuel elements. Among such alloys are Zircaloy-2 used in BWR, Zircaloy-4 used in PWR, the Canadian Zr-2.5%Nb used in CANDU pressure tubes and calandria tubes, the Zr-1%Nb alloy used in VVER and RBMK reactors, and other newer alloys such as ZIRLO. The alloying additions and optimized fabrication microstructures given by specified thermomechanical treatments give those alloys excellent resistance to high-temperature corrosion, very good resistance to in-reactor deformation, and good mechanical strength. These properties, combined with zirconium's low thermal neutron absorption cross section, has allowed for superior performance of fuel cladding under the harsh conditions found in the cores of nuclear power reactors.

A great deal of knowledge has been accumulated during the last decades on in-reactor behavior of zirconium alloys [1–3]. Fabrication and irradiation procedures have been tightened and made more reproducible so that during normal operation there is a reasonable expectation of near zero cladding failures [4]. Most of this experience, however, is based on the operation up to 30 GWd/ton (three years in-reactor) with a fuel cycle of around one year and well-defined reactor temperature and water chemistry. Any major deviation in this combination of operational parameters puts the cladding in uncharted territory and makes its behavior less predictable. This is especially true for high-fluence components, such as pressure tubes in CANDU reactors, and high-burnup cladding in LWR. At 30 GWd/ton, the microstructure of the zirconium alloy components is still evolving [3] so that a breakaway regime could have its onset at 45 or 60 GWd/ton. In general, it simply is not possible to have an experimental database that can comprise the possible combinations of temperature, flux, flux spectrum, fluence, temperature, material composition, microstructure, and water chemistry. The only hope of extending the existing database beyond its current limits is by understanding the mechanisms of radiation damage and microstructural evolution and developing mechanistic models that can be applied in a more general sense.

In that regard, the use of charged-particle irradiation under controlled conditions for the study of mechanisms of irradiation damage has several benefits [5]. The higher dose rates afforded by charged particle (electron and ion) irradiation allow us to reach damage levels in displacements per atom (dpa) comparable to those obtained in neutron irradiation in much less time. We also have greater control of experimental parameters such as temperature and dose rate under charged particle irradiation than under neutron irradiation. It should be emphasized that charged particle irradiation should not be seen as a "simulation" of neutron irradiation per se, but as a different irradiation altogether. This is not a drawback, but a positive aspect of these irradiations since they allow us to explore different areas of phase space than is possible with neutron irradiation. For example, it is possible to study the influence that displacement cascades have on a given process by irradiating the material with electrons since electrons do not produce damage in displacement cascade but in isolated Frenkel pairs. It is necessary, however, to couple the experiments with a theoretical understanding of the processes in order to draw any significant conclusions on operating mechanisms.

Among the possible means of irradiation with charged particles, a particularly useful one is the use of in situ irradiation with high-energy electrons and ions in an electron microscope. This means of irradiation has the additional advantage of allowing the detailed and systematic study of irradiation kinetics. This paper reports on such a study conducted in collaboration at the three institutions involved with the goal of understanding the mechanisms and kinetics of phase transformations under electron and ion irradiation. We focus on the crystalline-to-amorphous transformation (amorphization) in Zr-based intermetallic compounds, the stability of ω-phase precipitates in Zr-Nb alloys, and on the destabilization of Zr-Nb solid solutions with respect to β-phase precipitation.

Intermetallic precipitates in zirconium alloys have been extensively studied. The precipitates normally found in Zircaloys are of the type $Zr(Cr,Fe)_2$ (with a fcc C14 or hcp C15 structure) and $Zr_2(Ni,Fe)$ (bct C16 structure) [6]. Zr_3Fe (orthorombic)-based precipitates have been observed in alloys containing excess Fe [7] or after annealing of neutron-irradiated material [8]. The Fe/Cr and Fe/Ni ratios in Zircaloys can affect the alloy microstructure and behavior. We study here the influence of internal stoichiometry in the pseudo-binary compound on amorphization and irradiation-induced β-precipitation in Zr-2.5 Nb. These issues can have impact on cladding behavior. The amorphization of precipitates in Zircaloy has been linked to faster precipitate dissolution with consequent changes to the alloy microchemistry that impacts on atomic transport properties [44] and corrosion resistance [45]. In the same way, irradiation-induced β-phase precipitation in Zr-Nb alloys has been linked to improved corrosion resistance [46]. After reporting our results, we place them in the context of other experimental results and establish some guidelines for theoretical modeling.

Experimental Methods

For the amorphization studies, model alloys were prepared at AECL, Chalk River Laboratory and The Pennsylvania State University by arc melting from pure components (Zr 99.8%, Cr 99.99%, Fe 99.98%, Ni 99.98%, Nb 99.9%) followed by appropriate heat treatment as described in Ref 9. Samples suitable for examination by transmission electron microscopy (TEM) were prepared by mechanical grinding, punching, or spark cutting followed by electropolishing with a 10% perchloric acid solution in methanol. The alloys prepared were Zr_3Fe, $Zr_3(Fe_{0.9},Ni_{0.1})$, $Zr_3(Fe_{0.5},Ni_{0.5})$, $ZrCr_2$, $ZrFe_2$, and $Zr_{1.5}Nb_{1.5}Fe$.

The other alloys were prepared as follows: Zr-20%Nb plate material was annealed at 1123 K for 3 h, cooled to room temperature, and then annealed at 673 K for 24 h to form an even dispersion of the omega phase. For the β-precipitation study, samples of Zr-2.5%Nb pressure tube material were annealed for 1 h at 970 K. Finally, the Zr-1%Nb alloy was made by arc melting. Slices of the alloy were annealed for 2 h at 1223 K, then 17.7 h at 1023 K, followed by vacuum quenching from 848 K. Following the heat treatment, TEM samples were prepared by a similar electropolishing method, as above.

The heat treatment and fabrication process resulted in three types of alloys:

1. $Zr_3(Fe_{1-x},Ni_x)$: This alloy was formed with the structure of the orthorombic Zr_3Fe phase [10] for $x = 0, 0.1$, and 0.5, with similar lattice parameters ($a = 3.32$ Å, $b = 10.99$ Å, and $c = 8.81$ Å) as verified by electron diffraction and energy dispersive X-ray (EDX).
2. $ZrCr_2$ and $ZrFe_2$: Both of these alloys exhibited a mixture of bcc Fe or hcp Zr and the corresponding intermetallic phase. Diffraction patterns from both intermetallic phases were consistent with a C15 $MgCu_2$ Laves phase face-centered-cubic structure [11,12]. The compositional analysis performed by EDX was consistent with the reported stoichiometry within the margin of error ($\pm 2\%$). Those alloys exhibited two different microstructures for the intermetallic compound. Figure 1 shows the general aspect of a $ZrCr_2$ alloy. In this alloy and in the corresponding $ZrFe_2$ alloy, the intermetallic was found both as a "bulk" phase (Fig. 1b) with large grains and a few stacking faults (B) (Fig. 1c) or as part of an intimate mixture of the intermetallic compound (A) (Fig. 1a) with a solid solution of zirconium in iron. The intermetallic phase designated "SF" in Fig. 2 had a high density of stacking faults. This is also seen in $ZrFe_2$ (Fig. 2). Both bulk and "SF" phases had the C15 crystal structure.
3. h-($Zr_{1.5}Nb_{1.5}Fe$): By introducing Nb in the place of Zr, we formed a compound that had the approximate stoichiometry $Zr_{1.5}Nb_{1.5}Fe$ as verified by EDX. The sample also contained another compound of the type $(Zr,Nb)_2Fe$. The diffraction patterns from the $Zr_{1.5}Nb_{1.5}Fe$ phase could all be indexed assuming a hexagonal crystal structure with $a = 5.4$ Å and $c = 8.8$ Å. There are two possibilities reported in the literature for the identification of this phase. Woo et al. [13] have reported a $(Zr,Nb)_3Fe$ hcp phase with these lattice parameters, and Shishov et al. [43] have recently reported a $Zr(Nb,Fe)_2$ with a C14 structure. Although our stoichiometry is close to the first structure, the second structure C14 is a known Zr-Fe phase. At present we cannot definitely distinguish between these two possibilities for sample identification. Another study of Zr-Nb-Fe precipitates in Zircaloy also found a hexagonal structure with similar lattice parameters [14], although in that case the stoichiometry was different.

These compounds were examined before and after irradiation in a Philips CM-30 TEM at Chalk River Laboratory, a Philips 420 TEM at the Materials Characterization Laboratory at Penn State, and a JEOL-100CX and Philips 420 TEM at the Center for Electron Microscopy at Argonne National Laboratory. Specific areas in the thin foils were identified for later irradiation and studied with diffraction and EDX. Electron irradiations at 0.9 MeV were conducted at the High Voltage Electron Microscope (HVEM) facility at the Center for Electron Microscopy at Argonne National Laboratory. The electron current can be measured with a Faraday cup, and calibrations have been performed to determine the exact gaussian shape of the beam, as shown in Fig. 3. The HVEM has an ion beam attachment that allows for in situ ion irradiation with a wide range of different ions and energies. These were used in the present study of irradiation-induced precipitation and dissolution in zirconium alloys.

Alloys 1 through 3 were irradiated with electrons until amorphous at temperatures ranging from 25 to 250 K. Bright field and dark field micrographs were taken at regular intervals during

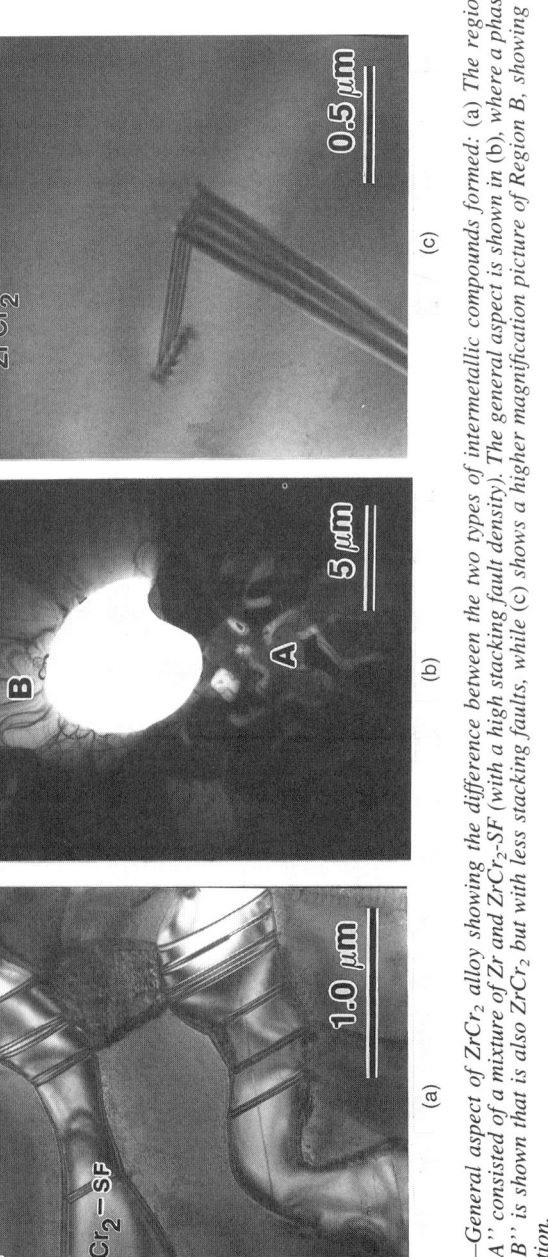

FIG. 1—*General aspect of ZrCr$_2$ alloy showing the difference between the two types of intermetallic compounds formed: (a) The region marked "A" consisted of a mixture of Zr and ZrCr$_2$-SF (with a high stacking fault density). The general aspect is shown in (b), where a phase marked "B" is shown that is also ZrCr$_2$ but with less stacking faults, while (c) shows a higher magnification picture of Region B, showing a 2-b condition.*

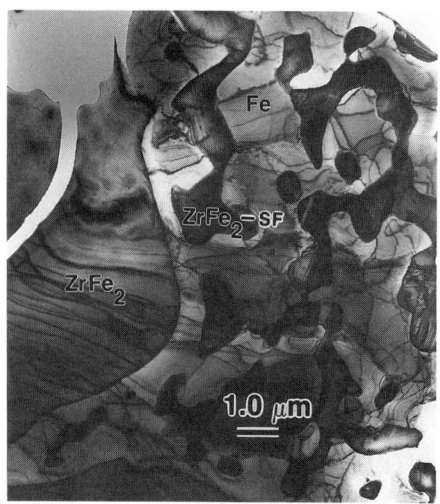

FIG. 2—*Bright field (BF) of $ZrFe_2$ alloy showing the two types of C15 $ZrFe_2$: a bulk phase designated $ZrFe_2$ and a high stacking fault density phase designated $ZrFe_2$-SF.*

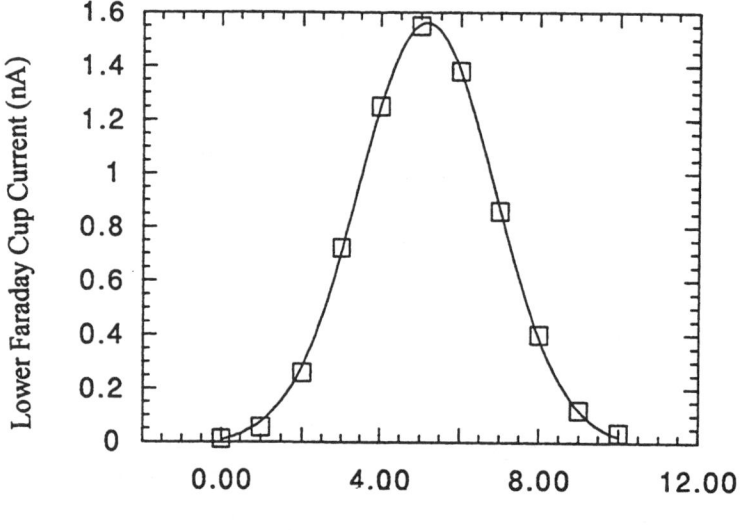

Lower Faraday Cup Position (Arbitrary Units)

FIG. 3—*Gaussian shape of the electron beam as measured by the Faraday cup. The precise determination of dose allowed the study of dose rate effects.*

the irradiation to record the progress of the transformation. The amorphization process was also followed by recording the change of the spot pattern in the diffraction pattern into a ring pattern.

The two zirconium alloys, Zr-20%Nb containing the ω-phase and the Zr-2.5%Nb, were irradiated with 350-kV Ar ions and electrons at various temperatures ranging from 300 to 773

K to determine whether the ω-phase precipitates were destabilized or dissolved in the first case and whether the β-phase precipitated out of solution in the second case. The Ar ion energy was chosen so that the peak in the damage distribution as calculated by TRIM 92 occurred within the thin foil. Vacuum during these experiments was on the order of 10^{-7} torr. For the ω-phase samples, the progress of the irradiation was followed by recording dark field pictures using a reflection from the ω-phase.

Results

Amorphization of Intermetallic Compounds

The amorphization process is shown in the bright field sequence in Fig. 4. This particular example refers to amorphization of Zr_3Fe at 180 K. As the dose is increased, first the higher-order bend contours are distorted (Fig. 4b), then weaken and eventually disappear (Fig. 4d). With continued irradiation, the lower-order bend contours disappear as well, while an amorphous ring is formed in diffraction. Finally, a dose is reached where using the smallest dif-

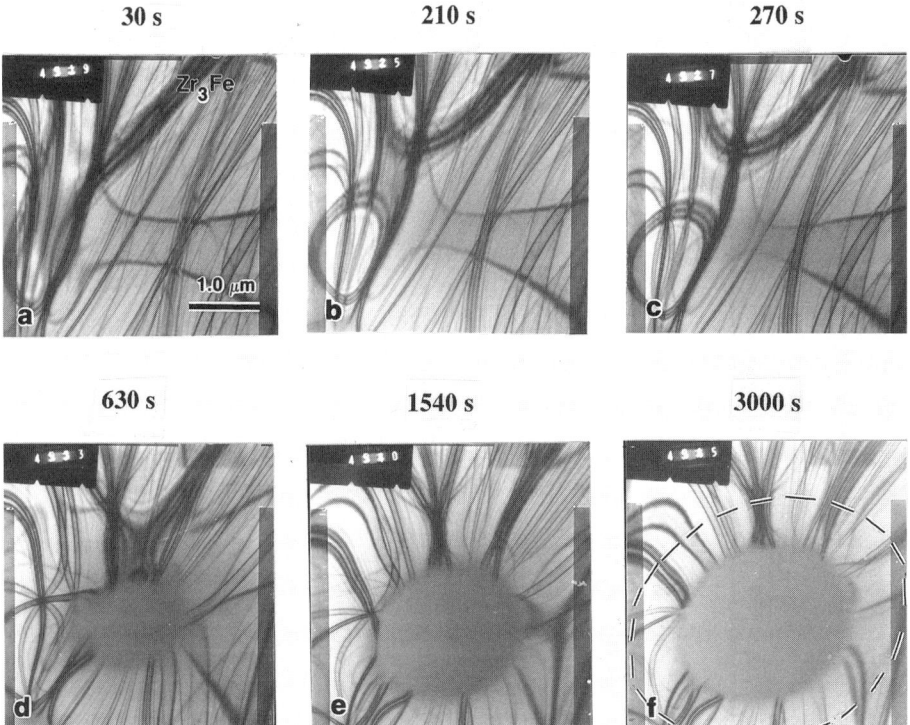

FIG. 4—*Amorphization of Zr_3Fe under electron irradiation at 180 K. Only a slight discoloration is present after 30-s irradiation (a). After 210 s (b), some higher order contours disappear and others become thinner and distorted. At 630 s (d), there is the onset of amorphization. The amorphous radius increases until it saturates at 3000 s at a value smaller than the beam size, shown approximately by the dotted line. The experiment was taken to 3600 s with no change in the size of the amorphous region.*

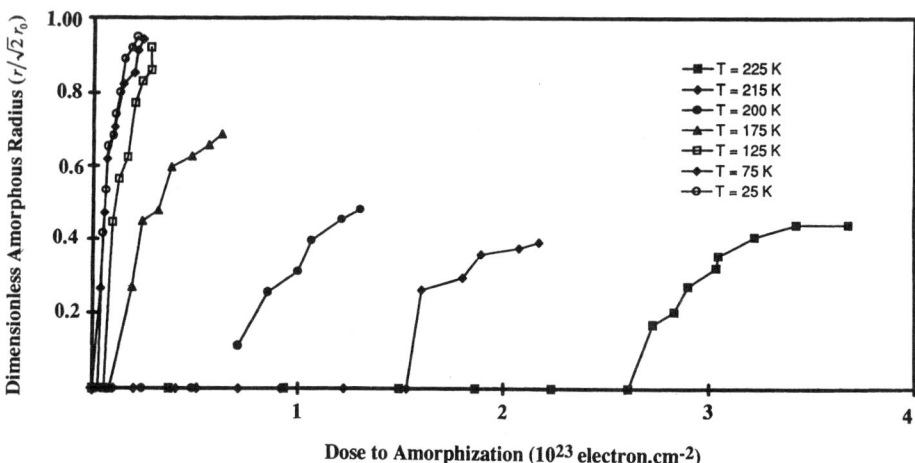

FIG. 5—*Dimensionless amorphous radius versus dose (electron/cm²) for electron irradiation of $Zr_3(Fe_{0.9},Ni_{0.1})$.*

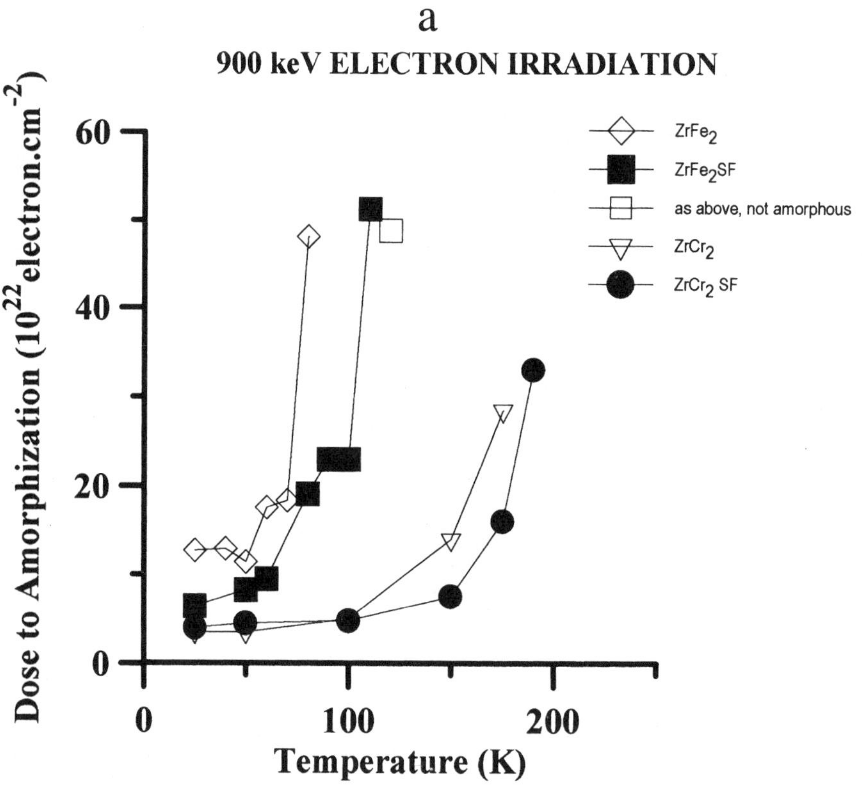

FIG. 6—*Dose to onset of amorphization under electron irradiation for (a) the $ZrFe_2$-$ZrCr_2$ system and (b) o-$Zr_3(Fe_{1-x},Ni_{1-x})$, h-$Zr_{1.5}Nb_{1.5}Fe$, Zr_2Fe.*

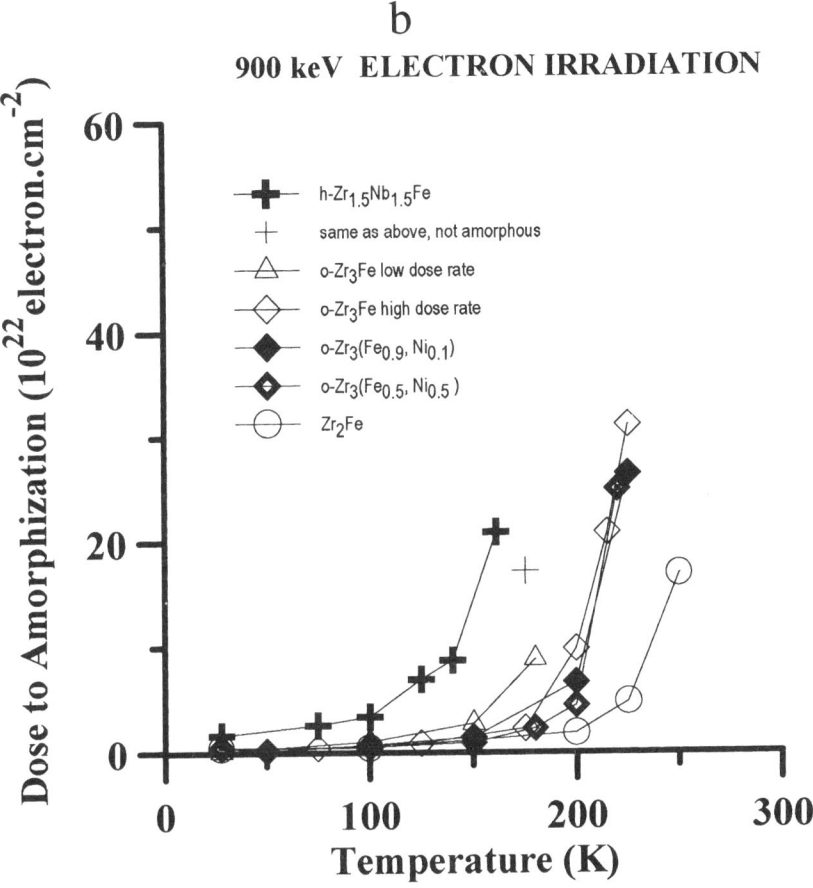

FIG. 6—Continued

fraction aperture it is possible to obtain an amorphous ring without any diffraction spots. This is defined as the dose for the onset of amorphization. In Fig. 4, it occurs around 630 s. At the end of the amorphization process (which was taken to 3600 s without further changes in radius), the radius saturates at a value smaller than the beam radius as shown in Fig. 4f.

The growth of this amorphous zone is then tracked as a function of dose. Figure 5 shows the amorphous radius measured from the negatives as a function of irradiation time in $Zr_3(Fe_{0.9},Ni_{0.1})$ for several temperatures. It can be seen that the radius remains at zero until the transformation is achieved. At the onset of amorphization, the amorphous radius increases abruptly and continues to increase with dose. The increase is abrupt at low temperature, showing that at low temperature there is no dose rate effect: for all dose rates the dose to amorphization is the same. This dose is defined as the critical dose. At higher temperatures, the situation is different; as the dose rates decrease with increasing radii, the dose to amorphization increases and the amorphous radius saturates at a radius smaller than the beam size. Thus, at high temperature, the dose to amorphization increases with decreasing dose rate. In fact, there is a critical dose rate (corresponding to the saturation radius) at which the dose to amorphization goes to infinity. From the full kinetic information displayed in Fig. 5, taking successive iso-dose-rate cuts, it is possible to obtain the variation of the dose to amorphization with temperature as shown in Ref. 9. Plotting the onset of amorphization against temperature, we obtain the critical temperature for amorphization, shown in Fig. 6.

Critical Temperature for Amorphization

These data were obtained for all the compounds of interest. The results are shown in Figs. 6a and 6b. The curves are very reproducible, as we verified in repeating some of these experiments. As shown in the previous section, the exact temperature at which amorphization ceases is dependent on the dose rate. The critical temperatures reported here are for the peak dose rate, but they do not vary much with the location of the cut in the kinetic curves (Fig. 5) as long as the cut is made within the first 30% of the radius.

There are several interesting features of the critical temperature that are described in the following and analyzed more thoroughly in the discussion section.

1. The lowest critical temperature is that of $ZrFe_2$ (around 80 K), followed by h-$Zr_{1.5}Nb_{1.5}Fe$ (~150 K), $ZrCr_2$ (180 K), o-$Zr_3(Fe_{0.9},Ni_{0.1})$, o-$Zr_3(Fe_{0.5},Ni_{0.5})$, and o-Zr_3Fe (all at 220 K), and Zr_2Fe (260 K). It is interesting to note that $ZrCr_2$ and $ZrFe_2$ have the same crystalline structure but a difference of 100 K in critical temperature. The critical dose for $ZrFe_2$ is also double that of $ZrCr_2$. Both of these results indicate that $ZrFe_2$ is more difficult to amorphize than $ZrCr_2$. In the $Zr(Fe_{1-x},Cr_x)_2$ system, the cubic phase C15 is stable for $x > 0.9$ and $x < 0.1$, while for $0.1 < x < 0.9$, the hexagonal C14 structure is stable [15]. Therefore, while it would be interesting to measure the dose to amorphization for intermediate x, the results would not be directly comparable to those for $x = 0$ and $x = 1$.
2. Another interesting feature is that the critical temperatures for $ZrCr_2$ and $ZrFe_2$ are different for the stacking faulted phase and the bulk phase (A and B in Fig. 2). It can be seen in Fig. 7 that for both $ZrCr_2$ and $ZrFe_2$ a higher stacking fault density increases the critical temperature by approximately 10 K. A higher density of stacking faults in $ZrFe_2$ reduces the critical dose by half. It should also be noted that the dose to amorphization versus temperature for the stacking-faulted $ZrFe_2$ phase exhibits a "step" (two-fold increase) to a higher plateau at a temperature corresponding to the critical temperature for the low-stacking fault density $ZrFe_2$ phase.
3. For the $Zr_3(Fe_{1-x},Ni_x)$ system, the critical temperature is 220 K for $x = 0, 0.1$, and 0.5. The curves overlap within experimental error for the full temperature range studied. There is, thus, no effect of internal stoichiometry on the susceptibility to amorphization in this system. This result is somewhat unexpected since, while Zr_3Fe is the stable phase at low temperature, this is not true in the Zr-Ni system where a mixture of Zr and Zr_2Ni would be stable at that stoichiometry. There is, therefore, some value of x at which $Zr_3(Fe_x,Ni_{1-x})$ becomes unstable with respect to $Zr_2(Fe_y,Ni_{1-y})$ + Zr, so one would expect that additions of Ni would affect phase stability. We saw no evidence of this change in stability under low temperature in this work.
4. By contrast, the introduction of Nb in a $Zr_{3-x}Nb_xFe$ alloy had a large effect on the critical temperature of amorphization. As x varies from 0 to 1.5, the critical temperature decreases from 220 to about 150 K. Clearly a major difference in this case is that the crystalline structure has changed from orthorombic at $x = 0$ to hexagonal at $x = 1.5$. This means that the substitution of Nb for either Zr or Fe (depending on whether we take the crystal structure in Ref *13* or in Ref *43*) has a major effect on crystal stability. It is interesting to note that the Nb_3Fe phase is not stable with respect to a mixture of Nb and the NbFe compound [16]. The highest critical temperature obtained was that of Zr_2Fe, which was found to be about 260 K. This phase, formed by phase separation during cooling from the melt, is metastable at low temperature.

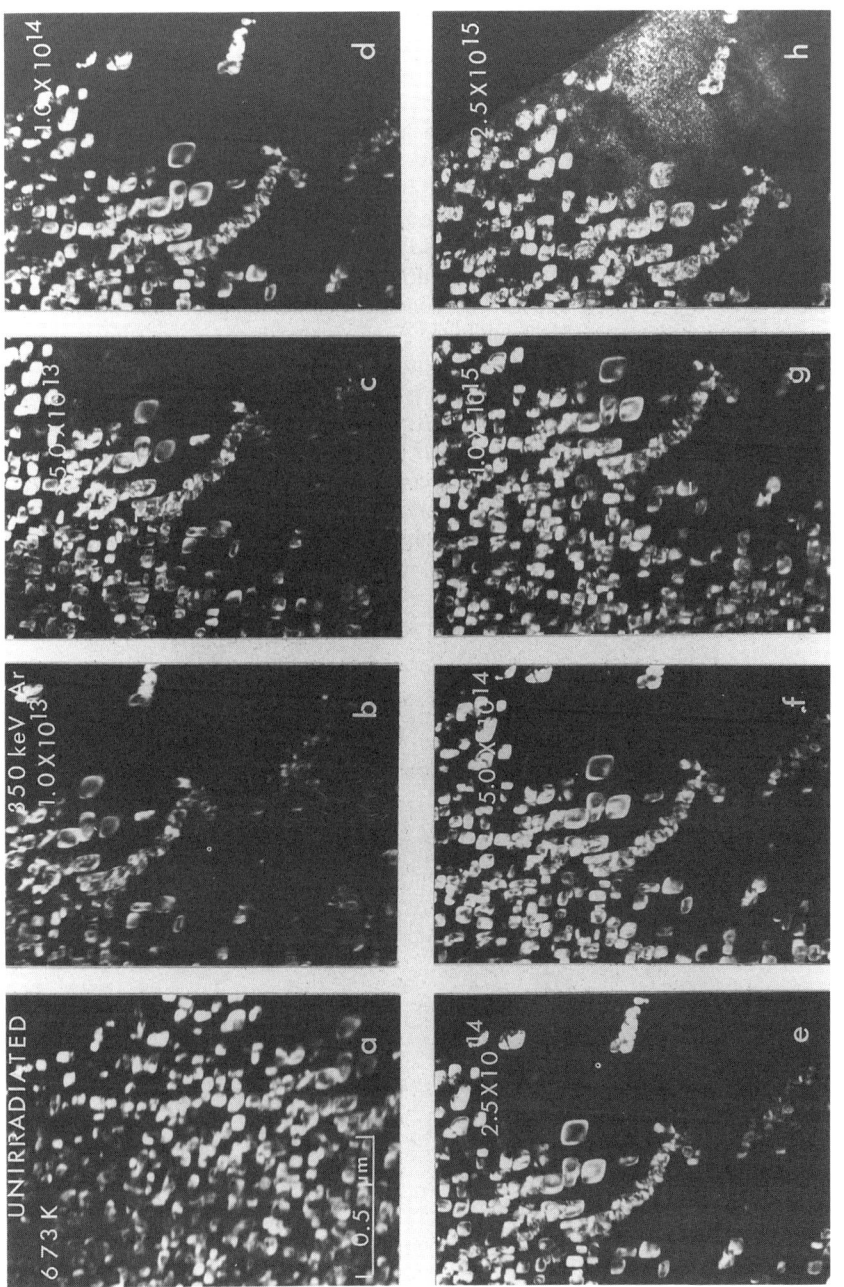

FIG. 7—*In situ sequence for Ar ion irradiation of ω-phase in Zr-20%Nb at 673 K. We use a ω-phase reflection for the dark field to check for ω-phase precipitation in the β-matrix.*

Stability of ω-Phase During Irradiation

An in situ irradiation sequence of ω-phase with Ar ions is shown in Fig. 7. The microstructure obtained after the heat treatment utilized, but before irradiation, is shown in the first frame. Cuboidal ω-particles are seen within the β-phase matrix. The cuboidal phase has a lower Nb content than the matrix and has the crystal structure of ω-CrTi [17]. The dark-field sequence is obtained using a ω-phase reflection. As the irradiation progresses, no ω-phase precipitates out in the matrix except for the last frame, which we attribute to contamination. The post-irradiation examination of this same sample shows the ω-phase particles still intact and little evidence of precipitation in the β-matrix (Fig. 8). It is not possible to rule out that ω-phase precipitation has occurred in the β-matrix since irradiation causes the appearance of many defects such as dislocation loops, which confuse the contrast. It is possible, therefore, that precipitation on the order of <100 nm would not have been detected. With the preceding caveats, the results from extensive experiments conducted on the stability of the ω-phase under different irradiation conditions can be summarized simply as that there were no effects observed of the 350-KeV Ar ion irradiation in the temperature range 573 to 673 K to 5.8 dpa and 400-KeV electrons at 623 K to 5 dpa on the ω-phase.

The experimental results are shown in Table 1. The results obtained in this work directly contradict those obtained by Nuttall and Faulkner [18], especially the electron irradiation experiment, which was conducted under the same conditions. We did observe a loss of contrast akin to the mottled contrast reported in their paper (see Figs. 7 and 8), but we ascribe it to surface contamination. Detailed post-irradiation analysis confirmed this last hypothesis: it was

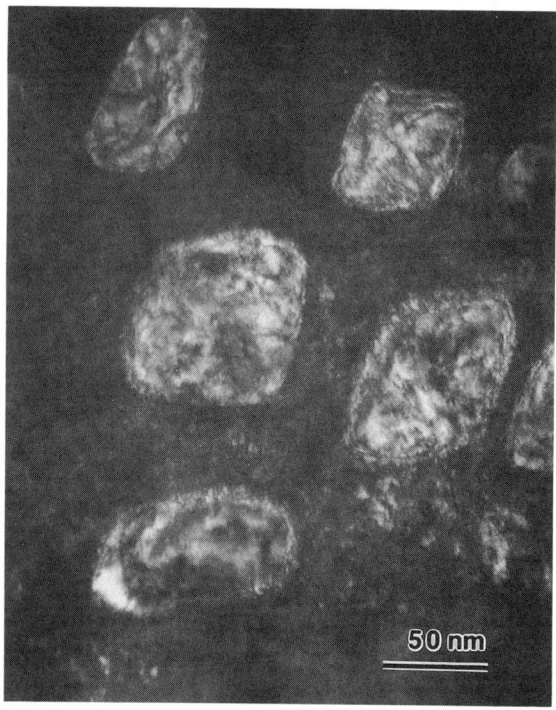

FIG. 8—*Dark field micrograph of ω-particles in Zr-20%Nb after irradiation to a fluence of 2.5 × 10^{15} Ar ion/cm^2 showing no breakup.*

TABLE 1—*Irradiations of ω-phase in Zr-Nb alloys.*

Alloy	Particle	T, K	Dose, dpa	Result	Reference
Zr 12%Nb	1-MeV electrons	623	1–8	Disintegration and reprecipitation	[18]
Zr 12%Nb	3-MeV Ni	698	10.8	No major change	[19]
Zr 20%Nb	350-KeV Ar	573	3	No major change	This work
Zr 20%Nb	350-KeV Ar	673	5.8	No major change	This work
Zr 20%Nb	900-KeV electrons	623	~5	No major change	This work

not possible to light up any of the "particles" that made up this contrast using the dark field reflections from the cuboidal ω-particles as seen in Fig. 8c. These results are in agreement with those of Hernandez and Potter [19], who did not observe any effect on the ω-phase after irradiation to 10.8 dpa with 3-MeV Ni ions at 425°C. We also observed the same oxide superlattice reflections in the $(100)_\beta$ diffraction pattern as observed in Ref 19, indicating that even at 10^{-7} torr there are sample contamination problems.

Precipitation of β-Phase During Irradiation

For Nb contents above 0.9, the bcc high-temperature β-phase is stabilized at room temperature. Precipitation of the β-phase from solid solution in the α-matrix in Zr-2.5%Nb has been observed under neutron [3,20], proton [21], and electron [22] irradiation. In this work, we attempted to reproduce these results using in situ Ar ion irradiation and monitoring the possible appearance of the β-phase in the α-phase by setting up the correct dark field conditions from the bulk β-phase. Figure 9 shows a dark field for a β-phase stringer in Zr-1%Nb after irradiation to 2.5 × 10^{14} Ar ion/cm². No precipitation is visible in the matrix. The Zr-2.5%Nb samples were irradiated to a fluence of 10^{15} ion/cm² (2.33 dpa) at temperatures of 573, 673, and 733 K. The Zr-1%Nb sample was irradiated to 2.5 × 10^{15} ion/cm² (5.8 dpa) at 723 K. In both cases, the matrix exhibited a high-defect concentration at the end, but no β-phase precipitation was observed. The contrast after 2.5 × 10^{15} ion/cm² is made more confused by the presence of oxide stringers and small dislocation loops (Fig. 10), which do not allow us to completely rule out that some fine precipitation may have taken place.

Discussion

Amorphization

Amorphization under irradiation occurs when the accumulation of damage caused by the incident particles makes it favorable for the material to exchange the defected long-range order of the irradiated crystal for the short-range order of the amorphization structure. Pure metals and metallic solid solutions are not susceptible to amorphization because, when irradiated, they can only store topological defects (point defects, dislocations), whereas intermetallic compounds can store anti-site defects (chemical disorder) in addition.

The ordered nature of the crystalline structure of intermetallic compounds originates from the imperative of maximizing the number of unlike atom pairs [23]. This is especially true for compounds that have a large negative heat of mixing such as those studied in this work. The root cause of amorphization is the need to maintain a high concentration of unlike pairs in the material even under irradiation. As point defects and anti-site defects are created by irradiation, the number of unlike pairs in the irradiated solid decreases until it becomes favorable for its

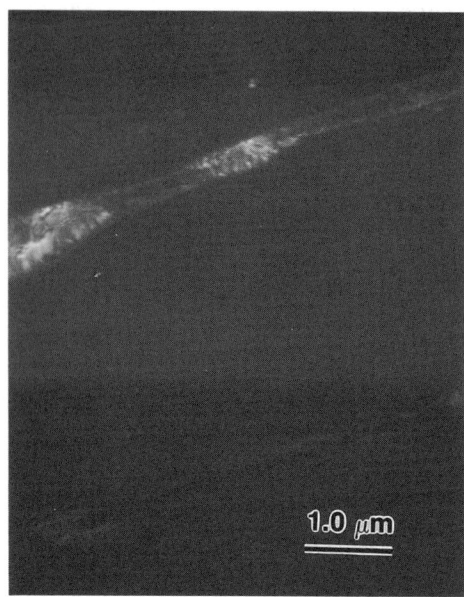

FIG. 9—*Dark field micrograph of β-phase filament (already present in the unirradiated material) in α-matrix in Zr-1%Nb, after irradiation to 2.5×10^{14} ion cm^{-2}. No additional β-precipitation is seen in the α-matrix.*

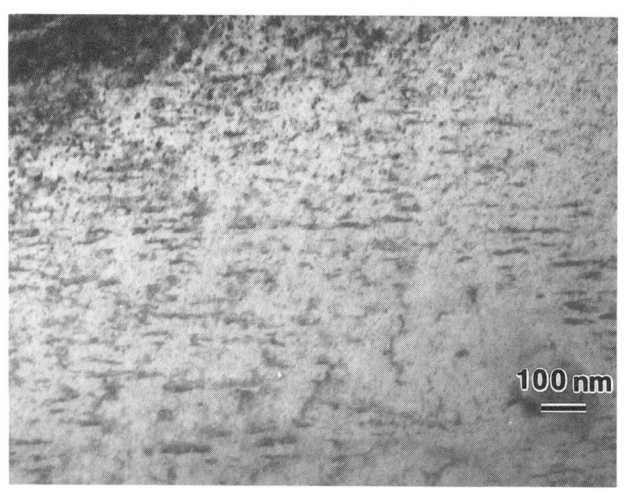

FIG. 10—*Bright-field micrograph of the fine-scale damage (oxide stringers, oxide particles, small dislocation loops) in Zr-1%Nb after in situ irradiation to 2.5×10^{15} Ar ion cm^{-2}. It is difficult to rule out precipitation on a scale finer than 100 Å.*

atoms to rearrange themselves in an amorphous structure where the requirements of chemical bonding can be more closely met, even at the expense of destroying crystallinity.

In the amorphous material, the local environment or short-range order is very similar to that in the undefected crystalline material [24], indicating the material recovers the short-range order to compensate for the long-range order it loses as it amorphizes. Amorphization can be thus seen as a compromise between the need to minimize disruption to chemical and topological order and the need to follow the kinetic demands imposed on the material by irradiation.

There are two aspects to the amorphization process. One is the accumulation of damage creating the necessary conditions for amorphization. The other is the actual rearrangement of atoms attendant upon the transformation. Taking the second point first, there is evidence that the transformation occurs fast compared to the total irradiation time [25], possibly by a catastrophic collapse induced by an elastic instability of the damaged structure [26]. That being the case, the rate-controlling step for irradiation-induced amorphization is the accumulation of enough damage in the structure.

The amorphization process depends, then, on the relative rates of damage accumulation and annealing. The two processes occur in parallel under irradiation, their relative importance changing with temperature. At very low temperature, the point defects responsible for annealing are immobile, and damage accumulates as fast as it is produced. At higher temperature, different defects become mobile. The annealing from the motion of these defects is proportional to the defects' concentration and to their mobility, $v \exp(-E_j/k_B T)$, where E_j is the migration energy of Defect j and v the vibration frequency. The level of damage necessary for amorphization has been modeled by an increase in the free energy of the irradiated solid equal to the difference in free energy between the crystal and the amorphous [25] or by an increase in the mean-square displacement of the atoms in the defected crystal relative to the pristine one, as specified by the generalized Lindemann criterion [27]. If the critical level of damage is estimated by one of the methods above L_{crit}, then we can write the amorphization condition as:

$$L_{crit} = (\dot{G} - \dot{A}) \, t_{irr}^{am} \qquad (1)$$

where $\dot{G}$ is the damage rate, $\dot{A}$ the annealing rate, and t_{irr}^{am} the irradiation time to amorphization.

In this formulation, the critical temperature T_c is the temperature at which $\dot{G} = \dot{A}$, so that at T_c, t_{irr}^{am} is infinite.

$\dot{A}$ is given by

$$\dot{A} = C_1 \sum_j C_j \, \nu_j \, e^{-E_j/k_B T} \qquad (2)$$

where C_1 is a constant, v_j is the vibration frequency of defect, and C_j is the concentration of Defect j. The damage rate, $\dot{G}$, is given by

$$\dot{G} = \Phi \, \sigma_d, \qquad (3)$$

where Φ is the particle flux, and σ_d is the displacement cross section.

Within this framework, the increase in the dose to amorphization occur at temperatures at which a certain type of defect becomes mobile, thereby increasing the annealing rate. If the increase is not enough to match damage production, it will still be possible to amorphize, but it will take longer; hence, a "step" is observed. Equation 1 implies that the higher the rate of damage, the higher the temperature at which $\dot{A} = \dot{G}$. This means that the higher the rate of damage the higher is T_c, in agreement with experiments [9,25]. However, the difference is not large; increasing the dose rate by a factor of six increases the critical temperature by approximately 20 K [9].

For a given dose rate, changing the damage mechanism changes the critical temperature for amorphization [5,28]. The biggest difference is between cascade-producing irradiation (ion and neutron) and electron irradiation. In the case of Zr_3Fe, the difference between the T_c for electron and Ar ion irradiation is approximately 350 K [29]. This difference is similar to that observed in the critical temperature for amorphization of $Zr(Cr,Fe)_2$ precipitates in Zircaloy when induced by electrons (300 K) and neutrons or ions (650 K) [30].

This work also shows that the presence of stacking faults can change the critical temperature. It is possible that the presence of stacking faults changes point defect mobility or the defect diffusion modes. Another possible explanation is that the stacking faults increase the energy stored in the lattice, thereby decreasing the amount of damage necessary to amorphization. We attribute the difference between the dose to amorphization of $ZrFe_2$ and $ZrFe_2$-SF at low temperature to a decrease in L_{crit} (Eq 1) caused by the presence of stacking faults rather than to a change in $(\dot{G} - \dot{A})$.

The difference in the critical temperatures of $ZrCr_2$ and $ZrFe_2$ is, by contrast, likely to be caused by different migration energies of defects in the two structures. The higher dose to amorphization at low temperature for $ZrFe_2$ as compared to $ZrCr_2$ indicates that annealing mechanisms are much more efficient in $ZrFe_2$ than $ZrCr_2$.

Previously published research on $Zr(Cr,Fe)_2$ precipitates in Zircaloy [5,25] showed the critical temperature to be around 300 K for 1.5-MeV electron irradiation. The discrepancy with T_c for $ZrCr_2$ measured in this work (180 to 200 K) is not great since in the previous study: (1) a beam-heating correction of 20 to 40 K was included so the effective T_c was 260 K, (2) the dose rates were higher than in the present study by a factor of three, and (3) the irradiation was taken to much higher values of dpa (up to three times as much). There is, however, a large difference between the T_c in $Zr(Cr,Fe)_2$ and $ZrFe_2$, suggesting that the migration energy of Fe is affected by the presence of Cr in $Zr(Cr,Fe)_2$. It is interesting to note that the T_c's for Zr_2Fe and Zr_2Ni [32] are very similar.

The steps found in the dose to amorphization versus temperature curve for $ZrFe_2$ are of great interest. The presence of the steps in $ZrFe_2$ indicates that two types of defects become mobile, one at 60 K, one at 80 K. Similar steps have been previously seen in CuTi [33] and Zr_3Fe [34]. The interest lies in using the amorphization process to study the properties of defects in intermetallic compounds and comparing them to the properties of defects determined by molecular dynamics. This should enable us to discern which defects are responsible for annealing and what their dependence is on stoichiometry.

Irradiation-Altered Solubility

In a binary alloy of a given overall composition, there is a preferred combination of phases of set compositions that minimizes the overall free energy of the system at each temperature. These are the equilibrium phases. The terminal solid solubility (TSS) in a given phase is the maximum amount of solute that can be held in solid solution within a primary phase. This solubility limit is a thermodynamic quantity and is dependent only on temperature. By establishing a limit for the amount of solute in solution, the TSS effectively controls the relative amounts of matrix and second phase formed.

When we try to apply these thermodynamic principles to commercial alloys in nuclear power reactors, we run into two types of difficulties that can alter phase equilibria: those related with the state of the material and those related with irradiation.

The state of the material can alter solubility in several ways. The TSS is measured for a well-annealed, binary alloy in equilibrium. All of these conditions are violated for commercial alloys. The addition of other alloying elements can change the apparent solubility of a given solute, for example, by solute-impurity trapping. The presence of cold work can also change the overall amount of solute contained in the matrix, for example, by decreasing the amount of solute in solution because of enhanced precipitation at dislocations. Finally, the fabrication processes used in commercial alloys often do not produce equilibrium microstructures. For example, the β-quench process results in a finer distribution of second-phase precipitates and a higher alloying content in the matrix than in the α-recrystallized material.

Irradiation can also alter phase stability. Indeed, in a strict sense, it is not possible to speak of thermodynamically stable phases under irradiation, as several of the conditions necessary for thermodynamic equilibrium are not satisfied [35,36]. However, by describing the kinetics of the irradiation processes, it is possible to discern the direction of variation of the material structure under irradiation. Irradiation can alter phase equilibria in two distinct ways: there can be irradiation *enhancement* of phase transformations or irradiation *inducement* of phase transformations. Irradiation can thus either accelerate the appearance of the thermodynamically stable phase or induce the appearance of new phases not observed outside of irradiation.

β-phase Precipitation

It has been argued that the precipitation of β-phase within the α-phase of Zr-2.5%Nb should be classified as a radiation-enhanced transformation [22]. This is because the precipitate morphology is similar to that observed in β-quenched and aged Zr-2.5%Nb [43] and because post-irradiation annealing of neutron-irradiated Zr-2.5%Nb either coarsened or left unchanged the precipitate distribution. If we accept the framework above, the precipitation of the β-phase in Zr-2.5%Nb is classified as a radiation-enhanced phase transformation. The fact that β-phase precipitation is observed under neutron, proton, and electron irradiation indicates that no irradiation-specific process, such as cascade production, or specific secondary defect structures are essential to β-phase precipitation enhancement.

β-phase precipitation is, therefore, likely to be controlled by diffusion of Nb atoms in α-Zr and should be favored for high values of the typical diffusion length $\langle x \rangle$, given by

$$\langle x \rangle = \sqrt{D_{Nb} t_{irr}} = \sqrt{C_j D_j t_{irr}} \qquad (4)$$

where D_{Nb} is the Nb diffusion coefficient, C_j is the defect responsible for Nb diffusion, and D_j is the defect diffusion coefficient. The calculation of defect concentrations under irradiation has been reviewed by Sizman [38]. Following his work, by determining the time to steady state in each of the above irradiations and the regime of operation (sink-dominated or recombination-dominated), we can estimate C_j. The time to steady-state τ is given by:

$$\tau = \frac{1}{\sum_k S_{kj} D_{sd}} = \frac{1}{(z_{sd} \rho_d + S_s) D_{sd}} \qquad (5)$$

where S_{kj} is the strength of Sink k for the slowest defect, D_{sd} is the diffusion coefficient, z_{sd} is the bias factor, ρ_d is the dislocation density, and S_s is the surface sink strength when spread over the bulk. For the parameters in Table 2, τ is smaller than 1 s for all the irradiations considered, which means that steady state is established as soon as the dislocation structure is fully developed.

TABLE 2—*Parameters for calculations.*

ρ_d = dislocation density, 10^{10} cm^{-2}
h = foil thickness, (cm):
  Bulk electron irradiation and proton irradiation = 5×10^{-2}
  Neutron irradiation = 0.1
  Ar ion irradiation = 1×10^{-5}
a_o = interplanar distance, 3 Å
z_j = bias factor for Defect j, 1

We define the parameter ε as:

$$\varepsilon = 1 + \left(\frac{S_s}{S_d}\right) = 1 + \frac{C_s}{C_d} = 1 + \frac{2a_o/h}{\rho_d a_o^2} = 1 + \frac{2}{\rho_d a_o h} \simeq 660 \qquad (6)$$

where C_s and C_d are the surface and dislocation sink concentrations when spread over the material, h is the foil thickness, and a_o is the interplanar distance. The parameter ε is the ratio of the total sink strength to the dislocation sink strength. For the values in Table 2, applicable to a foil thinned for TEM with a dislocation density of 10^{10} cm^{-2}, $\varepsilon \simeq 660$, which means that even when fully developed, the dislocation sink is negligible compared to the surface sink. For a 1-mm-thick disk as used in Ref 20 or the 0.5-mm-thick disks used in Refs 21 and 22, ε is, respectively, 1.07 and 1.93. We can estimate the regime of point defect behavior (sink-dominated or recombination-dominated) and hence calculate the defect concentration using rate theory [37,38]. All the irradiations listed in Table 3 were performed in conditions corresponding to a sink-dominated regime and where a steady state is quickly obtained. In that case, for a solid containing dislocation sinks and a surface sink,

$$\langle x \rangle = \sqrt{\frac{\dot{G}_{\text{eff}} D_j t_{\text{irr}}}{\sum_k S_{kj} D_j}} = (S_{dj})^{1/2} \left(\frac{\dot{G} f t_{\text{irr}}}{\varepsilon}\right)^{1/2} \qquad (7)$$

where S_{kj} is the strength of Sink k for Defect j, and Subscript d stands for dislocation. We calculated the value of $\langle x \rangle$ from Eq 7 using the parameters in Table 2. The fraction of freely migrating defects, f, produced by each type of irradiation is a matter of current research interest [42]. Because of intra-cascade recombination, the actual amount of defects that survive the cascade and are free for long-range migration is much smaller than the number calculated from the Norgett-Robinson-Torrens formula $\nu_{\text{NRT}}(T) = 0.8\ T/2\ E_d$ [ASTM Practice for Neutron Radiation Damage Simulation by Charged-Particle Irradiation (E 521-83)]. Here $\nu_{\text{NRT}}(T)$ is the number of displacements caused by an atom energy, T, and E_d is the displacement energy. We use here the relative efficiency values proposed in Ref 42, assuming 100% efficiency for electrons ($f = 1$), 50% efficiency for protons ($f = 0.5$), and 5% efficiency ($f = 0.05$) for neutrons and heavy ions. The reason for the difference is the sharp decrease of f with increasing mean recoil energy. The results obtained are summarized in Table 3, presented in the form of the ratio χ_i:

$$\chi_i = \frac{\langle x \rangle_i}{\langle x \rangle_n} = \frac{(\sqrt{Dt_{\text{irr}}})_i}{(\sqrt{Dt_{\text{irr}}})_n} = \left[\frac{\dot{G}_i f_i t_{\text{irr}}^i \varepsilon_n}{\dot{G}_n f_n t_{\text{irr}}^n \varepsilon_i}\right]^{1/2} \qquad (8)$$

TABLE 3—β-precipitation in Zr-Nb α-phase.

Alloy	Particle	Dose, dpa	$\dot{G}$, dpa/s	T_{irr}, K	t, s	f, %	ε	χ	Precipitate Size, nm	Reference
Zr 2.5%Nb	Neutron	0.62	1.28×10^{-7}	770	4.86×10^6	5	1	1	400	[20]
Zr 2.5%Nb	Neutron	0.74	1.52×10^{-7}	670	4.86×10^6	5	1	1.09	30	[20]
Zr 2.5%Nb	Neutron	0.8	1.65×10^{-7}	570	4.86×10^6	5	1	1.13	Not visible	[20]
Zr 2.5%Nb PT	Neutron	5.4	1.88×10^{-7}	570	2.88×10^6	5	1	3.05	< 10	[1]
Zr 2.5%Nb	3.6-MeV Protons	0.94	9.9×10^{-7}	720	9.52×10^5	50	1	3.78	5–30	[21]
Zr 2.5%Nb	10-MeV Electrons	1.2	8.17×10^{-7}	710	1.22×10^6	100	1	6.04	70	[22]
Zr 2.5%Nb	10-MeV Electrons	0.6	4.5×10^{-7}	713	1.33×10^6	100	2.3	4.27	30	[22]
Zr 2.5%Nb	10-MeV Electrons	0.6	4.5×10^{-7}	733	1.33×10^6	100	2.3	4.27	15	[22]
Zr 2.5%Nb	350-KeV Ar	2.3	1.27×10^{-3}	573	1.83×10^3	5	660	7.7×10^{-2}	No precipitation bigger than 10 nm	this work
Zr 2.5%Nb	350-KeV Ar	2.3	3.35×10^{-3}	673	6.95×10^2	5	660	7.7×10^{-2}	"	this work
Zr 2.5%Nb	350-KeV Ar	2.3	3.67×10^{-3}	773	6.33×10^2	5	660	7.7×10^{-2}	"	this work
Zr 1%Nb	350-KeV Ar	5.8	1.53×10^{-3}	723	3.8×10^3	5	660	1.22×10^{-1}	"	this work
Zr 2.5%Nb	3.6-MeV H_1, H_2, H_3	N/A	N/A	720	N/A	50	1.3	N/A	15–40	[31]

where $\langle x \rangle_i$ refers to Irradiation i and $\langle x \rangle_n$ refers to neutron irradiation to 0.62 dpa at 770 K.[4] χ_i is close to 1 for neutron irradiation, slightly larger than 1 for electron and proton irradiation, while for Ar ion irradiation it is about 10^{-2}. In this analysis, the higher χ_i is, the higher the concentration of freely migrating defects, and the more β phase should precipitate. In heavy ion irradiation of thin foils, the presence of the free surface combined with the low f depresses the defect concentration below the level necessary to induce enough Nb transport to cause β-precipitation. This is shown graphically in Fig. 11. That is the likely reason for the absence of β-precipitation under Ar irradiation in this work: the typical diffusion length of Nb is much lower than in neutron, electron, or photon irradiation due to the lower concentration of freely migrating defects which mediate the diffusion process.

It is questionable whether β-phase precipitation in the α-phase requires Nb transport from the β- to α-phase. If that were the case, the thin foil geometry would further reduce the possibility of precipitation. However, the fact that preferential β-precipitation near β-α grain boundaries was not observed [22] argues for precipitation to occur using the Nb already in the α-phase.

The absence of β-precipitation near grain boundaries during bulk electron irradiation [22] can be qualitatively explained by the depression in the defect concentration caused by the proximity to the grain boundary defect sink. It would be interesting to investigate whether bulk heavy ion irradiation (at a lower dose rate) or thin foil electron irradiation could also produce β-precipitation. We should note that for bulk Ar ion irradiation ε is much smaller than in the thin foil case due to the absence of the surface sink, and precipitation may occur. However,

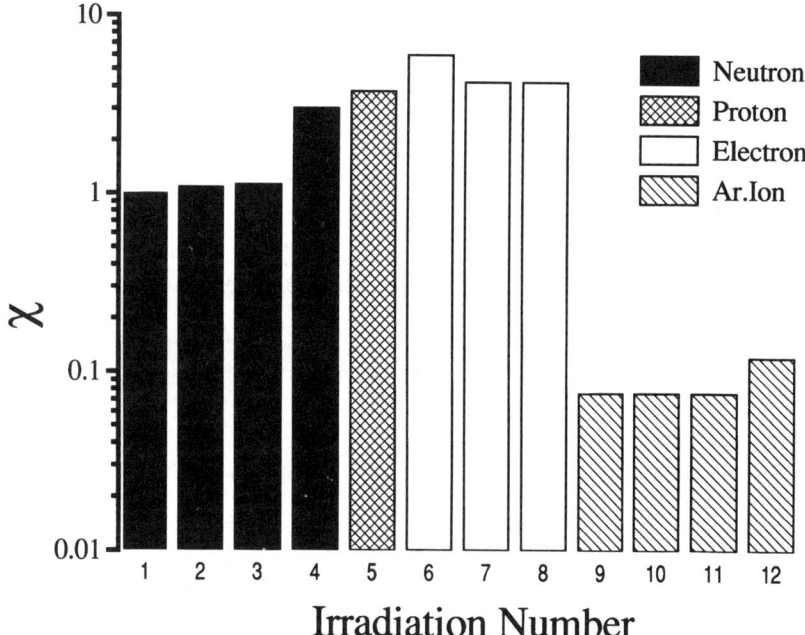

FIG. 11—*The parameter χ for the various irradiations listed in Table 3. The irradiation number corresponds to the order the experiments were listed in Table 3. A high χ correlates with β-phase precipitation, so precipitation occurs in the neutron, electron, and proton irradiations, but no precipitation is observed in the Ar ion irradiations.*

---

[4] Notice that, according to Eq 8, as long as the experiment occurs in a sink-dominated regime, the defect concentration does not depend on temperature [35].

one effect not considered here, namely the spatial superposition of cascades along the ion track during Ar ion irradiation, could further reduce f and hamper precipitation.

ω-Phase Dissolution

In this work, neither a direct attempt to reproduce ω-phase dissolution with electron irradiation nor other attempts to cause it to occur with ion irradiation in this work and in Ref *19* were successful. These results call into question the results obtained by Nuttall and Faulkner showing ω-phase dissolution and refinement under electron irradiation.

The dissolution of phases should, in general, be favored under ion irradiation relative to electron irradiation because the presence of cascades enables such processes as recoil resolution [*39*], interfacial mixing [*40*], and disordering and amorphization [*41*], which favor precipitate dissolution. In the present case, since the ω-phase has a lower Nb content than the β-phase in order for the precipitates to dissolve, there would need to be some mixing of Nb and Zr atoms, which would be driven by the processes above. It is, therefore, unlikely that electron irradiation would destabilize precipitates while ion irradiation would not.

Conclusions

The amorphization of several Zr-based intermetallic compounds and the stability of specific alloys to precipitation and dissolution were studied using in situ charged-particle irradiation. The use of in situ irradiation is shown to be a useful means of obtaining kinetic data that can be used to extend our knowledge of material behavior in irradiation environments. The following points are emphasized:

1. The critical temperature for amorphization T_c in the compounds studied increases with the density of pre-existing stacking faults and with increasing dose rate. For $ZrFe_2$, the athermal dose is reduced by a factor of two when more stacking faults are present.

2. A marked compositional effect on T_c was noted in the $ZrCr_2$-$ZrFe_2$ system with the same C15 crystal structure, where changing Fe for Cr increases T_c by 100 K and increases the athermal dose by a factor of 2.

3. The increase in the Nb concentration in $Zr_{1.5}Nb_{1.5}Fe$ compared to either $ZrFe_2$ or Zr_3Fe decreased T_c, while the addition of Ni to $Zr_3(Fe_{1-x}Ni_x)$ up to $x = 0.5$ had no effect on its amorphization behavior.

4. The above results can be rationalized with a kinetic model that predicts that amorphization occurs when the accumulation of radiation damage opposed by thermal annealing reaches a critical limit.

5. No β-precipitation was observed during irradiation of Zr-2.5%Nb with Ar ions at several temperatures. The discrepancy with other experiments is rationalized based on a simple model, based on irradiation-enhanced diffusion of Nb, which shows that the concentration of freely migrating defects is lower in our irradiations than in the other irradiations considered.

6. The ω-phase present in Zr-20%Nb was found to be stable under Ar ion and electron irradiation. The results agree with those in Ref *19* and contradict those in Ref *18*. An explanation for the discrepancy based on sample contamination is proposed.

Acknowledgments

The authors would like to thank D. Phillips and H. Plattner of Chalk River Laboratories and Stan Ockers, Ed Ryan, and Loren Funk of Argonne National Laboratory for their expert technical assistance. Joseph Faldowski is grateful for the support for his thesis research from a Lab-Grad Fellowship from the Division of Educational Programs at Argonne. The interest and support of the CANDU Owners Group (COG) Working Party 32 is gratefully acknowledged.

References

[1] Jostons, A., Kelly, P. M., Blake, R. G., and Farrell, K. in *Zirconium in the Nuclear Industry, Third Symposium, ASTM STP 633*, American Society for Testing and Materials, West Conshohocken, PA, 1979, pp. 46–61.
[2] Griffiths, M. in *Journal of Nuclear Materials*, Vol. 159, 1988, p. 190.
[3] C. Lemaignan and Motta, A. T., "Zirconium Alloys in Nuclear Applications," *Nuclear Materials*, B. R. T. Frost, Ed., Vol. 10B, Material Science and Technology Series, VCH, New York, R. W. Cahn, P. Haasen and E. J. Kramer, Eds., 1994, pp. 1–51.
[4] Strasser, A., O'Boyle, D., and Yang, R. in *Proceedings*, International Topical Meeting on LWR Fuel Performance, West Palm Beach, American Nuclear Society, La Grange, IL, 1994, pp. 3–14.
[5] Motta, A. T., Lefebvre, F., and Lemaignan, C. in *Zirconium in the Nuclear Industry, Ninth Volume, ASTM STP 1132*, American Society for Testing and Materials, West Conshohocken, PA, 1991, pp. 718–739.
[6] Charquet, D. and Alheritière, E. in *Proceedings*, Workshop on Second-Phase Particles in Zircaloys, Kerntechnische Gesellschaft, Erlanger, Germany, 1985, pp. 5–11.
[7] Seibold, A. and Woods, K. N. in *Proceedings*, International Topical Meeting on LWR Fuel Performance, West Palm Beach, American Nuclear Society, La Grange, IL, 1994, pp. 633–642.
[8] Gilbon, D. and Simonot, C. in *Zirconium in the Nuclear Industry (Tenth International Symposium), ASTM STP 1245*, A. M. Garde and E. R. Bradley Eds., American Society for Testing and Materials, West Conshohocken, PA, 1994, pp. 521–548.
[9] Motta, A. T., Howe, L. M., and Okamoto, P. R. in *Journal of Nuclear Materials*, Vol. 205, 1993, pp. 258–266.
[10] Aubertin, F., Gonser, V., Campbell, S. J., and Wagner, H. G. in Zeitschrift fur Metallkunde, Vol. 76, 1985, p. 237.
[11] Arias, D. and Abriata, J. P. in *Bulletin of Alloy Phase Diagrams*, Vol. 9, No. 5, 1988, p. 597.
[12] Arias, D. and Abriata, J. P. in *Bulletin of Alloy Phase Diagrams*, Vol. 7, No. 3, 1986, pp. 237–243.
[13] Woo, O. T. and Carpenter, G. J. C. in *Proceedings*, Twelfth International Congress of Electron Microscopy, 1990, San Francisco Press, p. 132.
[14] Sabol, G. P., Comstock, R. J., Weiner, R. A., Larouere, P., and Stanutz, R. N. in *Zirconium in the Nuclear Industrty (Tenth International Symposium), ASTM STP 1245*, A. M. Garde and E. R. Bradley, Eds., American Society for Testing and Materials, West Conshohocken, PA, 1994, pp. 724–746.
[15] Shaltiel, D., Jacob, I., and Davidov, D. in *Journal of Less-Common Metals*, Vol. 53, 1976, pp. 117–131.
[16] Paul, E. and Swartzendruber, L. J. in *Bulletin of Alloy Phase Diagrams*, Vol. 7, No. 3, 1986, pp. 248–254.
[17] Abriata, J. P. and Boleieh, J. C. in *Bulletin of Alloy Phase Diagrams*, Vol. 3, No. 1, 1982, pp. 1711–1712.
[18] Nuttall, K. and Faulkner, D. in *Journal of Nuclear Materials*, Vol. 67, 1977, pp. 131–139.
[19] Hernandez, O. G. and Potter, D. I. in "Phase Stability Under Irradiation," *Proceedings*, AIME Symposium, J. R. Holland et al., Eds., Metals Park, OH, 1980, pp. 601–612.
[20] Coleman, C. E., Gilbert, R. W., Carpenter, G. J. C. Carpenter, and Wetherly, G. C. in *Phase Stability Under Irradiation, AIME Symposium Proceedings*, J. R. Holland et al., Eds., Metals Park, OH, 1980, pp. 581–599.
[21] Cann, C. D., So, C. B., Styles, R. C., and Coleman, C. E. in *Journal of Nuclear Materials*, Vol. 205, 1993, pp. 267–272.
[22] Woo, O. T., Hutcheon, R. M., and Coleman, C. E. in *Materials Research Society Symposium Proceedings*, Vol. 373, I. M. Robertson, L. E. Rehn, S. J. Zinkle, and W. J. Phythian, Eds., 1995, pp. 189–194.
[23] Westbrook, J. H., Ed., *Intermetallic Compounds*, Wiley, New York, 1967, Chapter 1.
[24] Hausleitner, C. and Hafner, J. in *Journal of Non-Cryst. Solids*, Vol. 144, 1992, pp. 175–186.
[25] Motta, A. T., Olander, D. R., and Michaels, A. J. in *Effects of Radiation on Materials: Fourteenth International Symposium, ASTM STP 1046*, N. H. Packan, R. E. Stoller, and A. S. Kumar, Eds., American Society for Testing and Materials, West Conshohocken, PA, 1989, pp. 457–469.
[26] Koike, J. in *Physics Review B*, Vol. 47, No. 13, 1993, pp. 7700–7704.
[27] Lam, N. Q. and Okamoto, P. R. in *Materials Research Society Bulletin*, Vol. 7, 1994, pp. 41–46.
[28] Koike, J., Okamoto, P. R., Rehn, L. E., and Meshii, M. in *Metallurgical Transactions*, Vol.. 21A, 1990, p. 1799.
[29] Howe, L. M., Phillips, D., Motta, A. T., and Okamoto, P. R. in *Surface and Coatings Technology*, Vol. 66, 1994, pp. 411–418.
[30] Motta, A. T., Howe, L. M., and Okamoto, P. R. Okamoto in *MRS Symposium Proceedings*, Vol. 279, 1993, M. Nastasi, L. R. Harriott, N. Herbots, and R. S. Averback, Eds., pp. 517–522.

[31] C. D. Cann, unpublished results.
[32] Xu, G. B., Meshii, M., Okamoto, P. R., and Rehn, L. E. in *Journal of Alloys and Compounds*, Vol. 194, No. 2, 1993, pp. 401–405.
[33] Xu, G. B. in *Proceedings of the 47th Electron Microscopy Society of America*, 1989, p. 658.
[34] Howe, L. M. Howe, McCooeye, D. P., Rainville, M. H., Bonnett, J. D. Bonnett, and Phillips, D. in *Nuclear Instruments and Methods in Physics Research*, Vol. B59/60, 1991, p. 884.
[35] Frost, H. J. and Russell, K. C. in *Phase Transformations During Irradiation*, F. V. Nolfi, Ed., Applied Science Publications, New York, 1983, pp. 75–114.
[36] Martin, G. in *Physics Review B*, Vol. 30, No. 3, 1984, pp. 1424–1436.
[37] Wiedersich, H. in *Radiation Effects*, Vol. 12, 1972, pp. 111–125.
[38] Sizmann, R. in *Journal of Nuclear Materials*, Vols. 69 and 70, 1978, pp. 386–412.
[39] Nelson, R. A., Hudson, J. A., and Mazey, D. J. Mazey in *Journal of Nuclear Materials*, Vol. 44, 1972, pp. 318–330.
[40] Johnson, W. L., Cheng, Y. T., Van Rossum, M., and Nicolet, M.-A. in *Nuclear Instruments and Methods in Physics Research*, Vol. 67, No. 8, 1985, p. 657.
[41] Motta, A. T. and Lemaignan, C. in *Journal of Nuclear Materials*, Vol. 195, 1992, pp. 277–285.
[42] Rehn, L. E. and Birtcher, R. C. in *Journal of Nuclear Materials*, Vol. 205, 1993, pp. 31–39.
[43] Shishov, V. N., Nikulina, A. V., Markelov, V. A., Peregud, M. M., Kozlov, A. V., Averin, S. A., Kolbenkow, S. A., and Novoselov, A. E., this symposium.
[44] King, A. D., Hood, G. M., and Holt, R. A., *Journal of Nuclear Materials*, Vol. 185, 1991, pp. 174–181.
[45] Etoh, Y., Kikuchi, K., Yasuda, T., Koizumi, S., and Oishi, M. in *Proceedings*, International Topical Meeting on LWR Fuel Performance, Avignon, France, American Nuclear Society, La Grange, IL, 1991, pp. 691–700.
[46] Urbanic, V. F. and Gilbert, R. W. in *High Temperature Oxidation and Sulphidation Processes*, J. D. Embury Ed., Pergamon Press, Elmsford, NY, 1990, p. 182.

DISCUSSION

M. Griffiths[1] (written discussion)—The assignment of (Zr,Nb)$_3$Fe as hexagonal close-packed is different from the work of Dr. Shishov et al. They state that (Zr,Nb)$_3$Fe is orthorhombic and Zr(Nb,Fe$_2$) is hexagonal close-packed. Are you sure you had a single phase in your arc-melted material? And are you confident that you are not analyzing Zr(Nb,Fe)$_2$?

A. T. Motta, et al. (authors' closure)—We always have multiple phases in our arc-melted samples and identify specific grains of the desired phases by diffraction analysis and EDX during preirradiation characterization. The same grains are then later irradiated at the HVEM. In our arc-melted samples, we find the following phases: α-Zr, (Zr,Nb)$_2$Fe, and an hcp Zr-Nb-Fe phase that was studied in the present work. The phase irradiated in this work was a single grain of hcp Zr$_{1.5}$Nb$_{1.5}$Fe. What we know about the crystalline structure of hcp Zr$_{1.5}$Nb$_{1.5}$Fe phase is that it is hexagonal with lattice parameters a = 5.4 Å and c = 8.8 Å. Not being aware of Dr. Shishov's work [43] when preparing the paper, we assigned the phase to the published work of Woo et al. [13], which reported hexagonal (Zr,Nb)$_3$Fe. Both published structures in Ref 13 and Ref 43 agree with our results. While the structure proposed by Woo et al. matches our stoichiometry more closely, the assignment proposed by Shishov et al. is to an existing phase in the Zr-Cr-Fe system, namely C14 hcp Laves phase Zr(Cr,Fe)$_2$. On the other hand, atomic size and chemical compatibility considerations would indicate a substitution of Nb for Zr rather than Fe. Our experimental data cannot distinguish between those two.

The main conclusion of the paper, that the different crystalline structure of the hcp Zr$_{1.5}$Nb$_{1.5}$Fe phase is responsible for the change in amorphization behavior is valid in either case. If the correct assignment is Zr(Nb, Fe)$_2$, then by inserting Nb atoms into the Fe sublattice we raise the critical temperature of ZrFe$_2$ from 100 K to 150 K. If the correct assignment is (Zr,Nb)$_3$Fe, then the insertion of Nb into the crystal structure decreased T_c by 70 K as compared to orthorhombic Zr$_3$Fe.

[1] Atomic Energy of Canada, Ltd., Chalk River, Ontario, Canada.

M. Griffiths,[1] *J. F. Mecke,*[2] *and J. E. Winegar*[2]

Evolution of Microstructure in Zirconium Alloys During Irradiation

REFERENCE: Griffiths, M., Mecke, J. F., and Winegar, J. E., "**Evolution of Microstructure in Zirconium Alloys During Irradiation,**" *Zirconium in the Nuclear Industry: Eleventh International Symposium, ASTM STP 1295*, E. R. Bradley and G. P. Sabol, Eds., American Society for Testing and Materials, 1996, pp. 580–602.

ABSTRACT: X-ray diffraction (XRD) and transmission electron microscopy (TEM) have been used to characterize microstructural and microchemical changes produced by neutron irradiation in zirconium and zirconium alloys. Zircaloy-2, Zircaloy-4, and Zr-2.5Nb alloys with differing metallurgical states have been analyzed after irradiation for neutron fluences up to 25×10^{25} n.m^{-2} ($E > 1$ MeV) for a range of temperatures between 330 and 580 K.

Irradiation modifies the dislocation structure through nucleation and growth of dislocation loops and, for cold-worked materials in particular, climb of existing network dislocations. In general, the **a**-type dislocation structure tends to saturate at low fluences ($<1 \times 10^{25}$ n.m^{-2}). The c-component dislocation structure, however, may evolve over long periods of irradiation (for fluences $>10 \times 10^{25}$ n.m^{-2} in some cases).

The phase structure is also modified by irradiation. The common alloying/impurity elements, Fe, Cr, and Ni, are relatively insoluble in the α-phase but are dispersed into the α-phase during irradiation irrespective of the state of the phase initially containing these elements, i.e., metastable β-phase or stable intermetallic precipitate. The stable intermetallic particles may undergo structural changes dependent on their composition and the temperature. For the metastable dual-phase α/β-alloys (Zr-2.5Nb alloy), the β-phase structure is modified during irradiation, but the change is complex, being a combination of thermal decomposition and radiation-induced mixing.

KEY WORDS: zirconium, niobium, Zircaloy, zirconium alloys, microstructure, microchemistry, neutron irradiation, radiation damage, dislocation density, lattice parameters

The physical and chemical properties of Zr alloys are modified by the microstructural changes that occur in the reactor environment. It is therefore necessary to assess how the microstructure evolves during service in the nuclear reactor to predict future trends in properties. It is particularly important for structural components expected to remain in the reactor core for the life of the reactor (or until refurbishment occurs). In this paper, the microstructure of structural components from power reactors will be characterized as a function of fluence. All fluences quoted will be for neutrons with $E > 1$ MeV.

The two main alloy types used for nuclear reactor structural components are either based on Zr-Sn (Zircaloy-2 and -4) or Zr-Nb (Zr-2.5Nb). Zircaloy-2 was used for pressure tubes in the Winfrith Steam Generating Heavy Water Reactor (SGHWR) and in early CANDU[3] reactors; however, the current standard material for pressure tubes in CANDU reactors is Zr-2.5Nb. Zircaloy-2 is still used in CANDU reactors as guide tubes for control rods and flux detectors and for the calandria tubes that separate the hot Zr-2.5Nb pressure tubes from the cool mod-

---

[1] Staff scientist and [2] technologists, Atomic Energy of Canada Limited, Chalk River Laboratories, Chalk River, Ontario, Canada, K0J 1J0.

[3] *CANada Deuterium Uranium.*

erator in the calandria vessel. Most of the structural core components in pressurized water reactors (PWRs) and boiling water reactors (BWRs) are fabricated from Zircaloy-2 or Zircaloy-4.

The Zircaloys are predominantly Sn rich (about 1.5 wt% Sn) α-phase materials with Fe, Cr, and Ni present in intermetallic precipitates. Zr-2.5Nb alloys are normally fabricated with a dual phase α/β structure, the α-phase being supersaturated in Nb (0.5 to 1 wt% Nb); the remaining Nb (and Fe or Cr impurities) is present in a metastable β-phase (about 20 wt% Nb, 1 wt% Fe, and 0.2 wt% Cr). The Zr-2.5Nb-alloy is also sometimes used in a heat-treated state and then reduced to a primarily α-phase structure containing precipitates of β-Nb and a few ZrNbFe-type intermetallic precipitates such as $(Zr,Nb)_3Fe$, $(Zr,Nb)_2Fe$, or $Zr(Nb,Fe)_2$ [1].

Experimental

X-Ray diffraction (XRD) and transmission electron microscopy (TEM) have been used to characterize microstructural and microchemical changes produced by neutron irradiation of various Zr alloy components. Both unirradiated and irradiated materials were prepared and examined. For the irradiated materials, care had to be taken to minimize the volume of material used because of the residual radioactivity in the samples.

XRD specimens that were about 1 cm^2 area by 0.5 mm thick were prepared by cutting slices out of the component perpendicular to each of three principal component axes. These axes are illustrated for typical tubular components made from Zircaloy and Zr-2.5Nb in Fig. 1, together with basal plane pole figures and schematic representations showing the preferred orientation of the α-grains. The specimens were labeled according to the direction of the normal to each slice, i.e., longitudinal normal (LN), transverse normal (TN), and radial normal (RN). Each specimen was then chemically polished to remove any damaged layer introduced by the preparation. In this case at least 0.025 mm was removed, that being the depth of damage for the diamond wheel used to cut the specimen. TEM specimens were prepared by punching 3-mm-diameter disks from 0.1-mm-thick slices. Thin foils were then made by electropolishing using a Materials Science Northwest twin-jet apparatus with a solution of 10% perchloric acid in methanol at about 230 K and a current density of about 1 A.cm^{-2}.

The network dislocations in the hexagonal close-packed (hcp) α-phase of Zr alloys primarily have Burgers vectors of $\frac{1}{3}\langle 11\bar{2}0\rangle$ (**a**-type) or $\frac{1}{3}\langle 11\bar{2}3\rangle$ (**c**+**a**-type) and [0001] (**c**-type). The latter **c**-component network dislocations have large Burgers vectors and consist of double half planes. Irradiation results in the formation of dislocation loops that have a single half-plane and Burgers vectors of: $\frac{1}{3}\langle 11\bar{2}0\rangle$ (**a**-type), especially at low fluences and temperatures; $\frac{1}{6}\langle 20\bar{2}3\rangle$ (**c**/2+**p**-type) or $\frac{1}{2}[0001]$ (**c**/2-type), especially at high fluences and temperatures. The latter radiation-induced **c**-component dislocations can also be produced by climb of existing **c**-component network dislocations, primarily on basal planes. Dislocation densities were determined from line-broadening analysis using those X-ray diffraction lines having the maximum intensity for each specimen orientation. Typical errors for dislocation densities, determined from repeat measurements of the same specimen, were <4%. For the hcp α-phase, measurements were made from the Type I $\{10\bar{1}0\}$ and Type II $\{11\bar{2}0\}$ prism planes (**a**-component dislocations) and from basal (0002) planes (**c**-component dislocations) [2]. Because intergranular residual stresses also affect line broadening [2,3] and therefore the accuracy in terms of an absolute value of dislocation density, integral breadths will be used to show trends in line-broadening behavior that represent changes in dislocation densities. Typical errors for integral breadths, determined from repeat measurements of the same specimen, were <1%. Values of dislocation densities, where quoted, then represent an estimate based on prior calibrations [2].

Crystallographic planar spacings were determined from the peak positions using the average of peak maximum and center of gravity measurements. The α-phase **a**-type lattice parameters

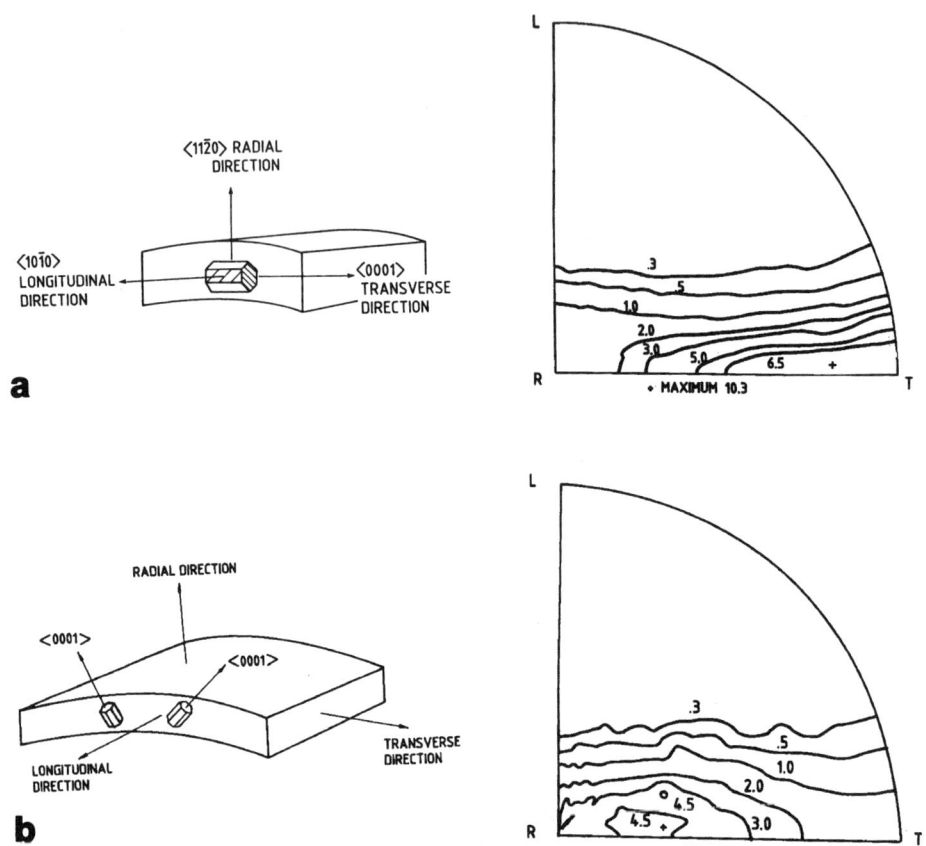

FIG. 1—*Schematic diagrams and basal pole figures illustrating the orientation of most grains in typical Zr alloy reactor core components: (a) Zr-2.5Nb pressure tube; (b) Zircaloy-4 guide tube.*

were calculated from the {11$\bar{2}$0} or {10$\bar{1}$0} diffraction peaks, and the c-type lattice parameters were calculated from the {0002} diffraction peaks. The β-phase lattice spacings were calculated from the (200) and (110) diffraction peaks. For Zr-2.5Nb pressure tubes, the Nb concentration in the body-centered-cubic (bcc) β-phase was then estimated from the β-phase diffraction peaks [3]. Typical errors for lattice parameters, determined from repeat measurements of the same specimen, were <1%. The various diffraction peaks were measured using a Rigaku or a Siemens diffractometer. CuK$_\alpha$ radiation (wavelength = 0.154 nm) was used in each case. The specimens were rotated in the focusing plane to increase the scanning area and sample as many grains as possible.

The state of the dislocation structure, the phase morphology, and the microchemical distribution were studied by TEM. Analysis was performed using a Philips CM 30 (300-kV) electron microscope. Microchemical analyses were obtained by energy dispersive X-ray analysis (EDXA) using a Link ISIS analyzer system and super atmospheric thin window (SATW) detector and by parallel electron energy loss spectrometry (PEELS) using a Gatan Peels analyzer. The spatial resolution for microchemical analysis was between 10 to 40 nm depending on specimen thickness. After correcting for background and self-generated X-ray signals (from neutron-irradiated materials), relative errors for chemical composition based on repeat measurements of the same sample were generally <5%.

Results

Dislocation Structure

For all zirconium alloys, irradiation results in the formation of **a**-type loops. They are intrinsic to Zr, forming in the pure material as well as in the alloys. Their size and density are primarily dependent on the irradiation temperature. Both interstitial and vacancy loops are formed, but the relative proportions of each type are determined by the irradiation temperature and the presence of other microstructural features (grain boundaries, network dislocations, **c**-component loops, etc.) [4]. Recent analysis of loops in single-crystal Zr irradiated at 573 K shows that equal numbers of vacancy and interstitial loops are formed. For unalloyed polycrystalline Zr, the relative numbers of vacancy **a**-type loops decreases as the temperature is increased above 573 K [5]. The same observation applies to annealed Zircaloy-2 and -4 irradiated at temperatures at and above 573 K [4].

For annealed Zircaloy-2 components operating at about 330 to 350 K (calandria tubes and guide tubes in CANDU reactors), the radiation damage is in the form of **a**-type loops that are <5 nm in diameter. There is some evidence from post-irradiation annealed specimens to suggest that interstitial loops are more prevalent at these low temperatures (<350 K) [4], although there are no direct analyses to confirm this. From X-ray diffraction line broadening, calculations indicate that the **a**-type dislocation density increases from about 0.1 (as-received) to about 1 x 10^{14} m^{-2} after a fluence of about 3×10^{25} n.m^{-2}. There is no significant increase in the measured **c**-component dislocation density (typically remaining within the range of 0.01 to 0.1 $\times 10^{14}$ m^{-2}), and there is no evidence of **c**-component loop formation from TEM analysis for fluences up to about 2.5×10^{26} n.m^{-2}. For the same type of alloy operating at about 570 K (Zircaloy-4 guide tubes in PWRs and channel boxes in BWRs), the **a**-type loops are larger, about 10 to 20 nm in diameter, and are the primary form of radiation damage at low fluences. X-ray diffraction line-broadening calculations indicate that the **a**-type dislocation density in PWR guide tubes increases from about 0.1 (as-received) to about 8×10^{14} m^{-2} (irradiated) after a fluence of about 1×10^{25} n.m^{-2} at about 570 K. The higher apparent **a**-type dislocation density at the higher temperature, compared with material irradiated at about 350 K, may reflect differences in the strain fields from the bounding dislocations when the loops are small, there being a lower net strain per unit length of dislocation line at the lower temperatures. In contrast to the lower temperature case, **c**-component defects in the form of basal plane dislocation loops having vacancy character are observed in annealed Zircaloy-4 irradiated at about 570 K after an incubation period of about 3×10^{25} n.m^{-2} [6,7]. In this latter case, the **c**/2 dislocation density calculated from X-ray line broadening increases from an initial value of about 0.01 to 0.1×10^{14} m^{-2} (pre-breakaway) at a rate of about 5×10^{-13} n^{-1} (post-breakaway). The **c**-type dislocation density evolves over much longer periods compared with the saturating **a**-type dislocation density as illustrated by plotting prism and basal plane line broadening as a function of fluence (Fig. 2). Saturation in the prism plane broadening (**a**-type dislocation loop density) occurs after a low fluence ($\leq 1 \times 10^{25}$ n.m^{-2}); however, there are insufficient high fluence data to show when the saturation in basal plane broadening (**c**-type dislocation loop density) will occur. There is an acceleration in the rate of irradiation growth corresponding with the increase in **c**-type dislocation density [8], and theoretical calculations show that this acceleration will continue until the **a**- and **c**-type dislocation densities are approximately equal [9].

For cold-worked Zircaloy-2 and -4, the **a**-type loops appear from TEM observations to be formed in similar numbers compared with the annealed materials at temperatures between about

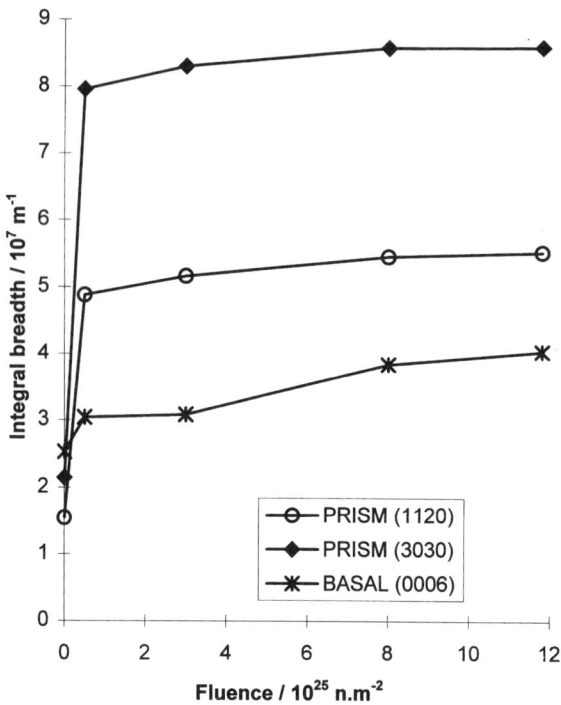

FIG. 2—*Variation in integral breadths as a function of neutron fluence for Zircaloy-4 guide tubes irradiated at 560 to 580 K using RN(1120), LN(2020), and TN(0004) specimens cut perpendicular to the radial, longitudinal, and transverse axes of the tube, respectively (see text).*

330 and 580 K. There are no data concerning XRD analysis of **a**-type dislocations and dislocation loops in 25% cold-worked Zircaloy-2 irradiated at 330 K; however, basal plane line-broadening measurements indicate that the **c**-component dislocation density increases from about 0.4×10^{14} m^{-2} to about 0.8×10^{14} m^{-2} after a fluence of 7.4×10^{25} n.m^{-2}, assuming that the additional dislocations are the same as the **c**-component network dislocations and have double half-planes. TEM analysis shows that there is a corresponding increase in the number of basal plane segments of **c**-component dislocations in the neutron-irradiated material (Fig. 3). Electron irradiation experiments [10,11] have shown that these basal plane segments can be attributed to helical climb on screw dislocations that had been generated initially by cold working. On this basis, they are likely to have single half-planes, i.e., having a **c**/2-component (**c**/2 or **c**/2+**p**), and the increment in dislocation density will be four times the value calculated as if the dislocations were **c** or **c** + **a** [2,9]. The screw-type **c**-component network dislocations provide nucleation sites for the climb on the basal plane, producing helices. This is especially clear when one examines grains with and without pre-existing **c**-type screw dislocations. Grains without **c**-component network dislocations do not contain basal plane **c**-component loops. This is consistent with the results for annealed material irradiated at the same low temperature, i.e., there are no new **c**-component dislocation loops nucleated at low temperatures (<350 K).

For 25 to 30% cold-worked Zircaloy-2 pressure tube material irradiated at about 550 to 580 K, the **a**-type dislocation density increases from about 2×10^{14} m^{-2} (as-received) to about 8×10^{14} m^{-2} after a fluence of about 1×10^{25} n.m^{-2}. The steady-state density of **a**-type defects

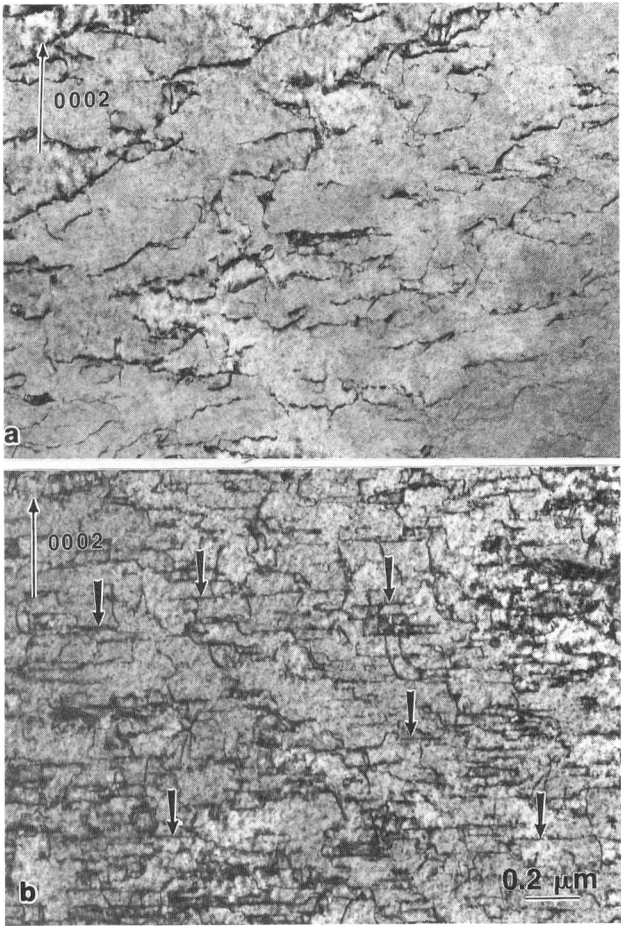

FIG. 3—*Comparison of* **c**-*type dislocation structures in Zircaloy-2* (a) *before and* (b) *after irradiation to a fluence of 7.2 × 10^{25} n.m^{-2} at about 330 K. Edge segments (arrowed) are produced during irradiation.*

in the irradiated cold-worked material is similar to that observed for annealed material at the same temperature. The development of basal plane c-component dislocation segments (by helical climb) that is observed at about 330 to 350 K in cold-worked Zircaloy-2 is also observed for cold-worked Zircaloy-2 irradiated at about 550 to 580 K [8,9]. As with the low temperature case, this is explained by helical climb on pre-existing c-component screw dislocations. Contrary to the indications of previous work [4]; c-type dislocation loop formation, i.e., separate from the screw network dislocations, can also occur at the higher temperature (550 to 580 K) and is especially apparent close to intermetallic precipitates in grains where the original c-component network dislocation density is low. This preferential formation of c-component loops close to Zr(Cr,Fe)$_2$ intermetallic precipitates is similar to that observed for the annealed material [8]. Therefore, there are two mechanisms for increasing c-component dislocation density: (1) by helical climb on existing network dislocations; (2) by dislocation loop formation.

The former occurs at the onset of irradiation, whereas the latter may be delayed if there is an incubation period for loop formation. The two contributions are apparent when comparing the behavior of cold-worked and annealed material. The relative change in basal line broadening (c-type dislocation density) with increasing fluence for annealed and cold-worked Zircaloy-2 irradiated at about 550 to 580 K is shown in Fig. 4. The overall rate of increase in calculated c/2-component dislocation density for the cold-worked material in Fig. 4 is about 2.5×10^{-12} n^{-1} and is about an order of magnitude greater than the post-breakaway rate calculated for the annealed material (5×10^{-13} n^{-1}).

There are few published data concerning the effect of irradiation on dislocation structures in annealed Zr-2.5Nb because, in the majority of cases, Zr-2.5Nb is used in a cold-worked condition. The main use of cold-worked Zr-2.5Nb is in pressure tubes in CANDU reactors. They contain the hot primary coolant operating at 520 to 570 K and are separated from the cold moderator (350 K) by a gas gap and a Zircaloy-2 calandria tube. Irradiation results in the production of a high density of small (about 15 nm in diameter) a-type dislocation loops. There is no clear evidence for c-component loop formation during irradiation at about 520 to 570 K. X-ray diffraction analysis shows that there is an increase in a-type (prism plane) line broadening with neutron fluence, i.e., in the core of the reactor, whereas very little appears to be happening to the c-type (basal) line broadening (Fig. 5). There is a variation in prism plane broadening along the tubes [*12*] that can be attributed to a-type dislocation loop formation with a higher density of loops at the cooler inlet end (about 520 K) compared with the hotter outlet end (about 570 K). Calculations show that the corresponding dislocation densities are about 8 ×

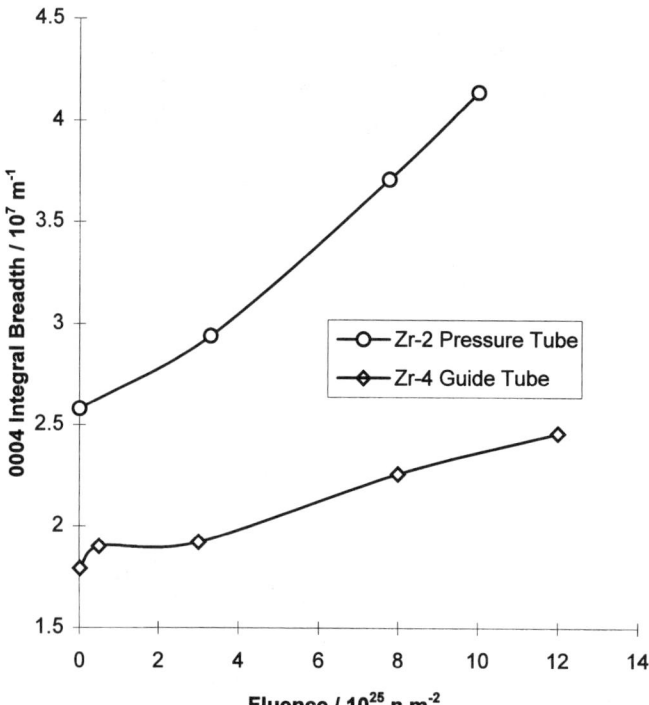

FIG. 4—*Variation in basal line broadening as a function of neutron fluence at about 570 K for annealed Zircaloy-4 guide tubes and cold-worked Zircaloy-2 pressure tubes using RN(1120).*

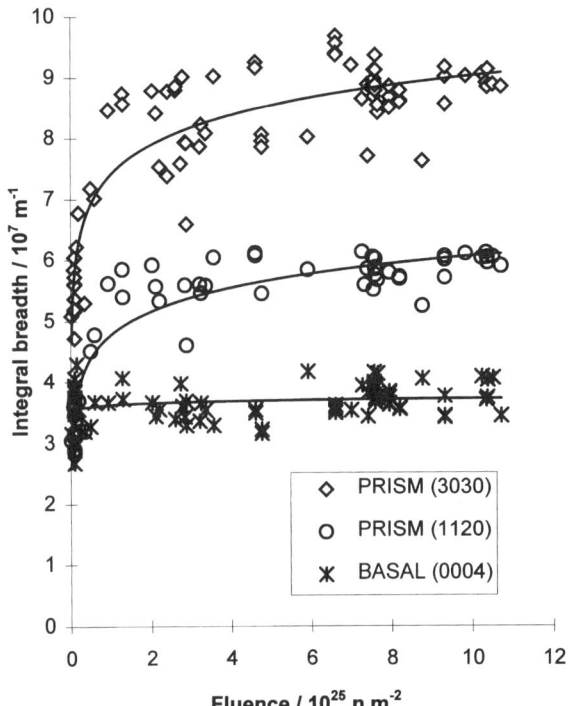

FIG. 5—*Variation in integral breadths as a function of neutron fluence in Zr-2.5Nb pressure tubes at about 520 to 570 K using RN(1120), LN(2020), and TN(0004), where RN, LN, and TN refer to specimens cut perpendicular to the radial, longitudinal, and transverse axes of the tube (see text).*

10^{14} and 6×10^{14} m^{-2}, respectively, being similar to the values obtained from cold-worked Zircaloy-2 pressure tubing. There is an apparent slight sharpening in the basal lines in the irradiated part of the tubes that can be explained by splitting of **c**-component dislocations. This type of splitting is readily observed during in situ irradiations in a high voltage electron microscope (HVEM) [10,11]. In general, the **a**-type dislocation density changes tend to saturate by fluences of about 1×10^{25} n.m^{-2} and are somewhat similar to that observed in 25 to 30% cold-worked Zircaloy-2 [9] for the majority of the grains. However, the basal plane line broadening shows an increase for the Zircaloy-2 pressure tube material, indicating an increase in **c**-type dislocation density not apparent in the Zr-2.5Nb pressure tube material. This may be related to differences in susceptibility to basal plane climb or to differences in behavior of grains as a function of their orientation [9].

Second Phase Structure

At about 330 to 350 K, the $Zr(Cr,Fe)_2$ and $Zr_2(Ni,Fe)$ intermetallic precipitates, and also silicides, in Zircaloy-2 or -4 become amorphous. The amorphous transformation occurs at low fluences ($<1 \times 10^{25}$ n.m^{-1}) [13].

At irradiation temperatures of about 570 K, the $Zr_2(Ni,Fe)$ precipitates remain crystalline, whereas the $Zr(Cr,Fe)_2$ precipitates become amorphous [13–17]. In the latter case, the amorphous transformation begins at the edge of the precipitate and advances inwards with increasing fluence. There is also a depletion of Fe coincident with the amorphous zones and a corresponding increase in concentration of Cr and Zr (Fig. 6). The concentration profile within the precipitates has two distinct zones corresponding with two distinct phases (crystalline core and amorphous periphery). There is a difference in phase transformation characteristics of $Zr(Cr,Fe)_2$ precipitates in the Zircaloys in some cases. For precipitates in Zircaloy-2 irradiated in the DIDO reactor, the particle centers can also become amorphous, in addition to the amorphous peripheral layer, even at low doses (Fig. 7). In this case, because the particles become

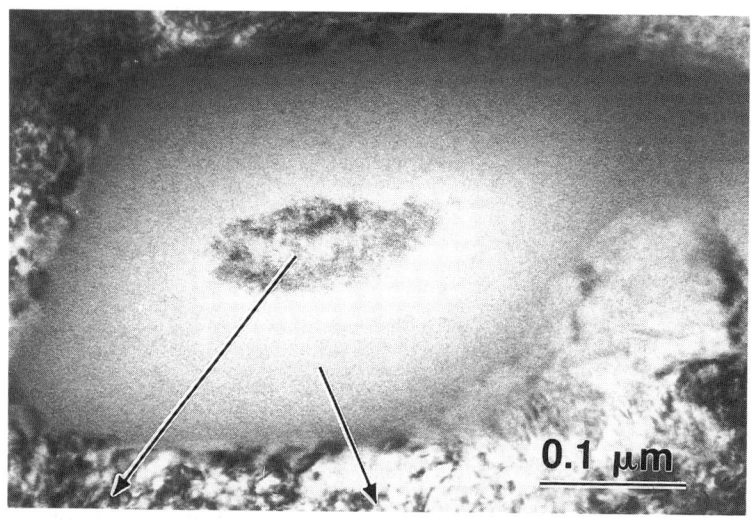

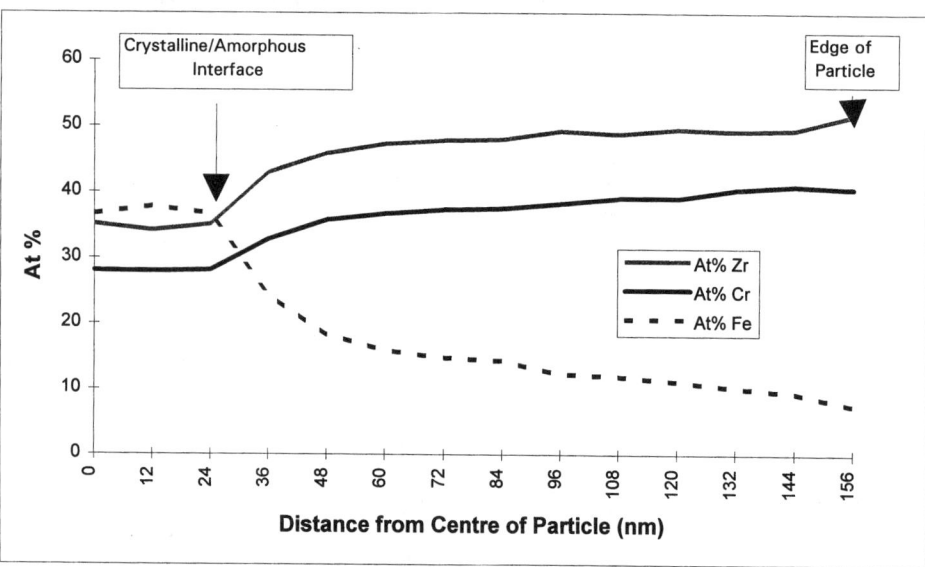

FIG. 6—*Micrograph and composition profile for a $Zr(Cr,Fe)_2$ precipitate in Zircaloy-4 irradiated at about 580 K to a fluence of 12×10^{25} $n.m^{-2}$. The crystalline core retains the original composition, and the amorphous peripheral layer is depleted in Fe.*

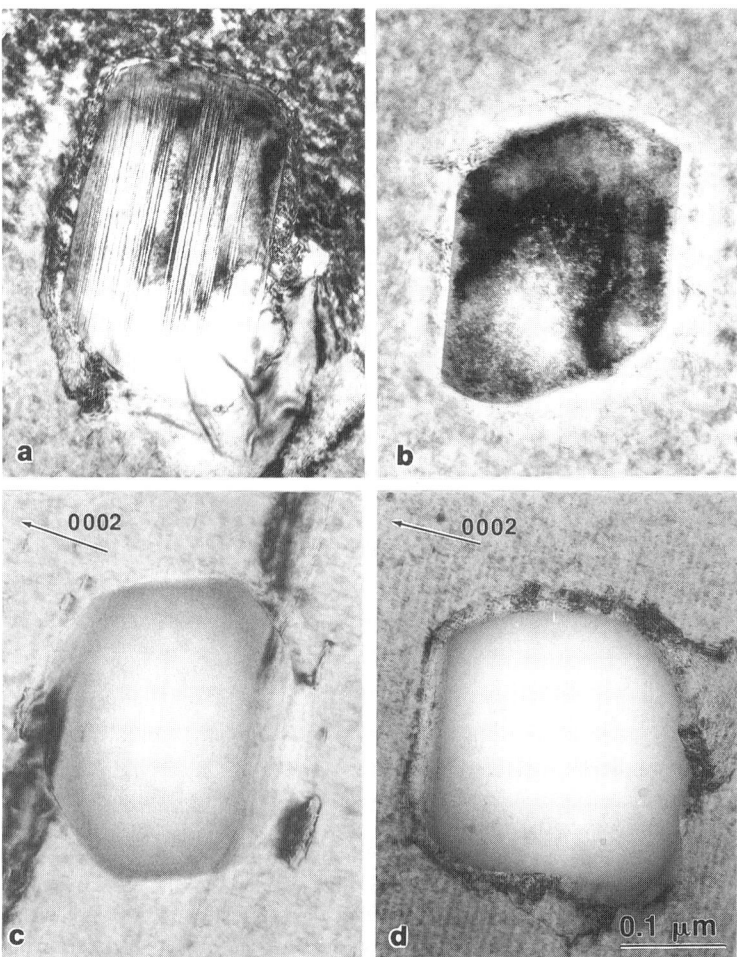

FIG. 7—*Crystalline-amorphous transformation of $Zr(Cr,Fe)_2$ intermetallic precipitates in Zircaloy-2 after irradiation at 553 K to a fluence of:* (a) 0×10^{25} n.m^{-2}; (b) 0.4×10^{25} n.m^{-2}; (c) 0.9×10^{25} n.m^{-2}; and (d) 2.0×10^{25} n.m^{-2}.

amorphous at a low dose, the Fe-depletion profile within the precipitate (Fig. 8) does not have the same step shape as that exhibited by the duplex crystalline/amorphous precipitates (Fig. 6). In each case, the dispersed Fe appears to be in some finely divided metastable state because post-irradiation thermal treatment below the amorphous phase recrystallization temperature results in a replenishment of Fe within the amorphous phase as the Fe diffuses back from the matrix [18,19]. The difference in dose to amorphous transformation may be related to the composition of the $Zr(Cr,Fe)_2$ precipitates, precipitates in Zircaloy-2 having a higher Cr,Fe ratio compared with Zircaloy-4 [20]. The flux may also be a factor, the precipitates shown in Fig. 7 having been irradiated in a materials test reactor (MTR) with a higher nominal flux (about 8×10^{17} n.m^2.s^{-1}) relative to a normal PWR or BWR reactor (about 6×10^{17} n.m^2.s^{-1}). This is supported by the fact that similar (Zircaloy-2) material irradiated in a lower flux CANDU reactor (about 3×10^{17} n.m^2.s^{-1}) did not become fully amorphous at the same low fluences.

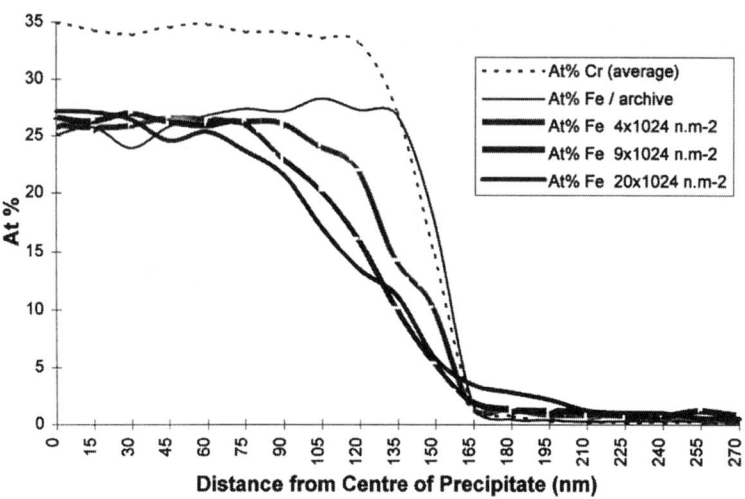

FIG. 8—*Composition profiles as a function of fluence for $Zr(Cr,Fe)_2$ precipitates in Zircaloy-2 irradiated at about 553 K. The precipitates are completely amorphous by a fluence of 0.9×10^{25} $n.m^{-2}$.*

In addition to the depletion of Fe from within the $Zr(Cr,Fe)_2$ precipitates, there is erosion at the edges of the $Zr_2(Ni,Fe)$ and $Zr(Cr,Fe)_2$ precipitates [13–17]. The erosion has been observed primarily on surfaces intersecting the basal plane for $Zr(Cr,Fe)_2$ precipitates [21]. The same directionality may exist for $Zr_2(Ni,Fe)$ precipitates also [15–17]. For the $Zr(Cr,Fe)_2$ precipitates, the directional erosion can be related to anisotropic dispersion of the dissolved elements (Cr and Fe), the concentration of Cr-rich precipitates forming after post-irradiation annealing being highest in regions adjacent to the $Zr(Cr,Fe)_2$ precipitates in directions parallel with the basal plane [20]. These same regions tend to contain a higher number of the basal plane c-component loops, and this may therefore be related to the local change in chemistry around the precipitates. The association between the dispersed elements and c-component loops is also apparent from the coincidence of the Cr and Fe, concentrated in discrete layers parallel with the basal plane, and c-component loops [8]. This localized association with the $Zr(Cr,Fe)_2$ precipitates may be linked to the fact that Cr is relatively immobile compared with Fe and Ni [22] and therefore maintains a higher concentration closer to the $Zr(Cr,Fe)_2$ intermetallic precipitates. There does not appear to be any association between the amorphous transformation and c-component loop formation, there being no evidence for c-component loops in the vicinity of the amorphous $Zr(Cr,Fe)_2$ precipitates shown in Fig. 7d.

The β-phase in Zr-2.5Nb pressure tubes is originally a single bcc phase non-equilibrium structure containing about 20 wt% Nb. It transforms to Nb- and Fe-depleted, hcp ω-phase precipitates embedded in an enriched bcc β-phase (20 to 50 wt% Nb, about 1 wt% Fe) during the final stress-relief treatment of 24, 48, or 72 h at 673 K prior to installation in the reactor (Fig. 9). This partially decomposed metastable phase continues to change during service. The concentration of Nb in the β-phase (based on the average of $\beta(200)$ and $\beta(110)$ lattice parameter measurements) can be used as a measure of the degree of decomposition and is plotted as a function of position for pressure tubes removed from service after periods ranging from 2 to 14 effective full power years in Fig. 10. A schematic flux profile is also shown in Fig. 10. The temperature increases gradually from the inlet (about 520 K) to the outlet (about 570 K), but

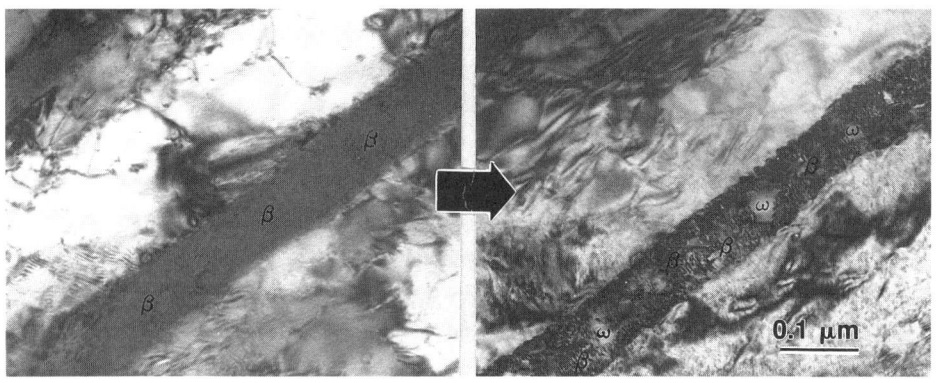

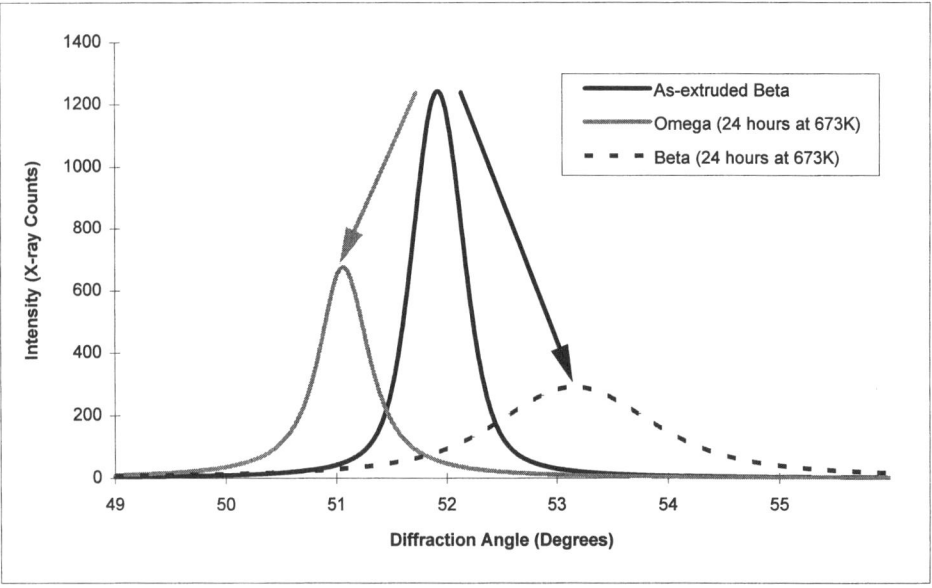

FIG. 9—*XRD diffraction lines and electron micrographs showing change in β-phase composition as a function of autoclaving heat-treatment for 24 h at 673 K. The initial single β-phase decomposes to give Nb-depleted ω-phase precipitates embedded in a Nb-enriched matrix.*

is not plotted for clarity. The scatter in the data represents variations in service time and initial tube-to-tube microstructure variations. There is an apparent Nb enrichment of the β-phase in the out-of-flux sections at each end of the tubes, and this enrichment is more pronounced at the outlet. The effect of irradiation appears to be suppression of any further thermal decomposition of the β-phase. Although there are insufficient data to make a rigorous analysis of the effects of irradiation and temperature, inspection of the data at any one position shows that there is a flux and temperature-dependent steady-state value for the composition of the β-phase achieved after about two to three years of operation. There appears to be a balance between the effects of irradiation and temperature that, of course, does not apply in the out-of-flux sections (temperature dependence only). This view is supported by measurements on specimens

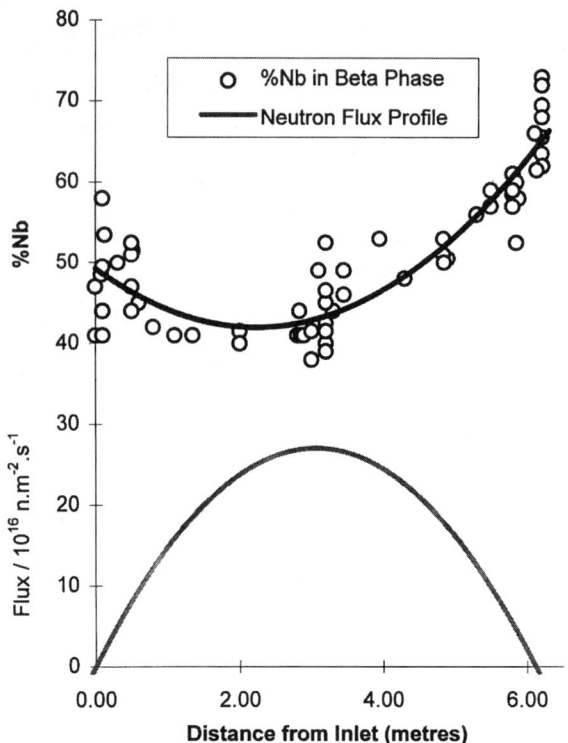

FIG. 10—*Variation in Nb composition in the β-phase as a function of position relative to the inlet for Zr-2.5Nb pressure tubes after 2 to 14 years of service. There is additional decomposition from the as-fabricated state that is apparent in the out-of-flux sections at the inlet (about 520 K) and outlet (about 570 K) but appears to be suppressed in the irradiated central section. A combined average of the RN(200) and LN(110) diffraction lines is plotted to compensate for intergranular stresses.*

that had been heat-treated for 72 h at 673 K prior to irradiation at 523 K. In this case, there was a reversal of the effects of thermal decomposition due to the low temperature irradiation (Fig. 11). The structure of the transformed β-phase is too complex to analyze using TEM; however, analysis of the X-ray peaks shows that there is an apparent decrease in volume fraction of ω-phase concurrent with the decrease in concentration of Nb in the remaining β-phase (Fig. 12). This observation is consistent with a ballistic process involving sputtering of the ω-phase precipitates. The sputtering effect appears to be more pronounced at lower temperatures, there being more evidence for increased thermal decomposition at the outlet ends (about 570 K) compared with the inlet ends (about 520 K) of the pressure tubes (Fig. 10). The net effect of temperature and neutron flux is consistent with the mechanism for precipitate redistribution proposed by Nelson et al. [23] involving a balance between the rate of radiation-induced dissolution and the rate of growth due to radiation-enhanced and thermal diffusion.

The Fe in the β-phase is dispersed during irradiation at temperatures between 520 and 570 K to the point where it is indistinguishable from the Fe content in the α-phase after a fluence of about 3×10^{25} n.m^{-2}. Fe is retained in the β-phase in the out-of-flux sections; Fe-rich phases are even observed after long periods of service, particularly in the region of the outlet rolled joints (Fig. 13a) [12]. Post-irradiation annealing for 1 h at 773 K reverses the flow of Fe in the

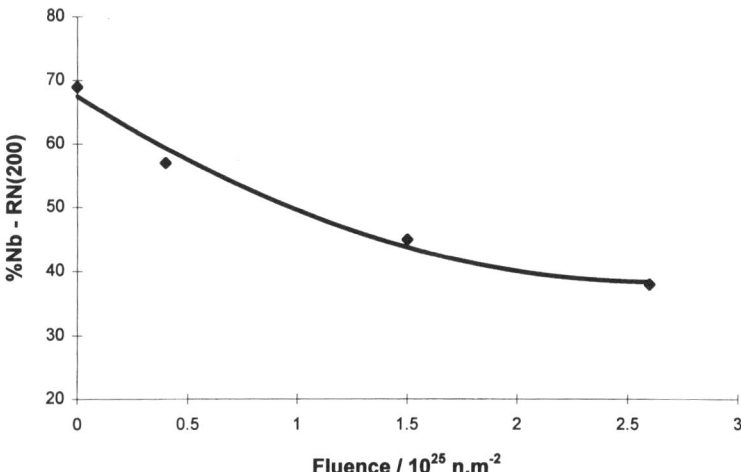

FIG. 11—*Variation in Nb composition in the β-phase (from the RN(200) diffraction lines) as a function of neutron fluence for Zr-2.5Nb pressure tube material irradiated at about 520 K in an MTR.*

irradiated material, the Fe returning to the β-phase (Fig 13b). The Fe dispersion is not determined by the metastable state of the β-phase because the same effect is observed for stable (Zr,Nb,Fe,Cr)-containing precipitates in heat-treated Zr-2.5Nb pressure tube material during irradiation at about 573 K. In this case, the precipitates behave somewhat similarly to the $Zr(Cr,Fe)_2$ precipitates in the Zircaloys in that there is a preferential depletion of Fe during irradiation (Fig. 14). Preliminary analysis has shown that the precipitates initially appear to be $(Zr,Nb)_3(Fe,Cr)$, consistent with the observations of Shishov et al. [1]. Analysis of 0.2-μm-diameter precipitates irradiated to a fluence of 1.5×10^{25} n.m^{-2} at 570 K shows that the Fe concentration is uniformly depleted, decreasing from about 20 at% in the unirradiated material to about 4 at% after irradiation, which is somewhat faster (as a function of fluence) than the depletion observed for amorphous $Zr(Cr,Fe)_2$ precipitates in Zircaloy-2 irradiated at a similar temperature (Fig. 8). There are corresponding structural changes, the precipitates decomposing to give a polycrystalline structure (Fig. 15).

Summary of Results

The significant new results presented in this paper are summarized below:

1. The dislocation loop structure in Zircaloy-2 and -4 irradiated with neutrons at low temperatures (330 to 350 K) is **a**-type only for fluences up to 2.5×10^{26} n.m^{-2}. Although **c**-component loops are not apparent in annealed material up to high fluences, irradiation of cold-worked Zircaloys containing **c**-component network dislocations (about 0.4×10^{14} m^{-2} initially) results in an additional **c**-component, i.e., **c**/2, dislocation density increase of about 1.6×10^{14} m^{-2} after a fluence of about 7.5×10^{25} n.m^{-2} due to helical climb on the existing network.

2. The **a**-type dislocation loop structure generated in annealed Zircaloy-4 at about 560 to 580 K increases rapidly in the early stages of irradiation ($<0.5 \times 10^{25}$ n.m^{-2}) from about 0.1×10^{14} m^{-2} to achieve approximate steady-state values of about 8×10^{14} m^{-2}. The **c**-type dislocation loop structure develops over a longer period. After an initial incubation

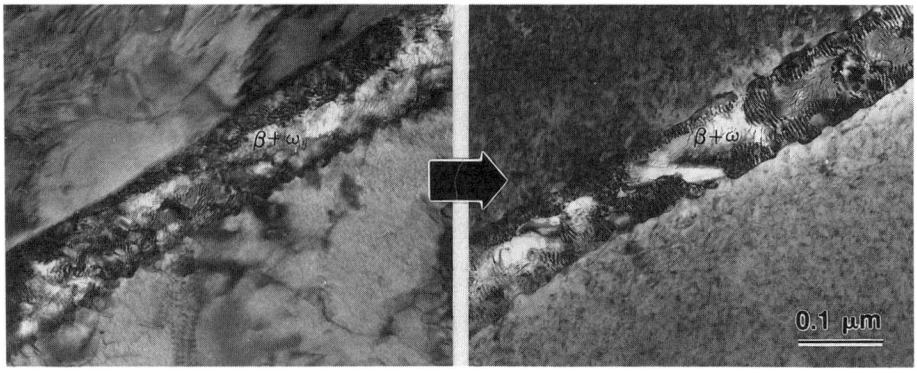

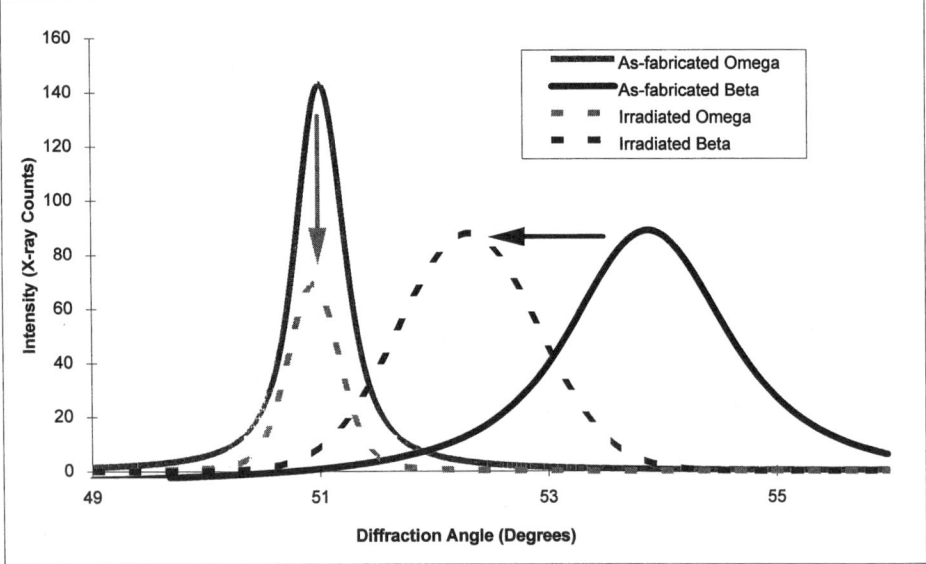

FIG. 12—*XRD diffraction lines and electron micrographs showing change in β-phase composition (for material that had been heat-treated for 72 h at 673 K) as a function of irradiation at about 520 K to a fluence of 2.6×10^{25} n.m^{-2}. The ω-phase volume fraction decreases coincidentally with a decrease in Nb concentration in the remainder of the filament.*

period lasting about 3×10^{25} n.m^{-2}, the **c**-component loop density increases from an initial value of about 0.01 to 0.1×10^{14} m^{-2} (pre-breakaway) at a rate of about 5×10^{-13} n^{-1} (post-breakaway).

3. The **a**-type dislocation loop structure generated in 25% cold-worked Zircaloy-2 pressure tubes at about 560 K is similar to that generated in 25% cold-worked Zr-2.5Nb pressure tubes at the same temperature and achieves a steady state value of about 8×10^{14} m^{-2}, similar to that of annealed Zircaloys irradiated at similar temperatures. The **c**-component dislocation structure measured in grains corresponding to the peak basal pole texture (radial orientation for Zircaloy-2 pressure tubes and transverse orientation for Zr-2.5Nb pressure tubes) is different in each case. Whereas the **c**-component, i.e., **c**/2, dislocation

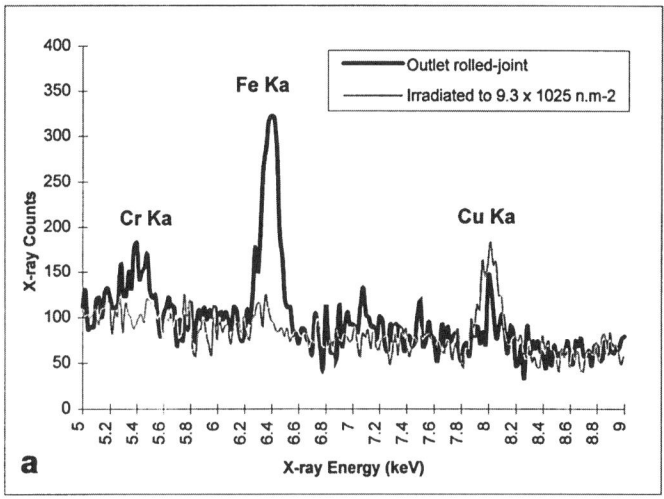

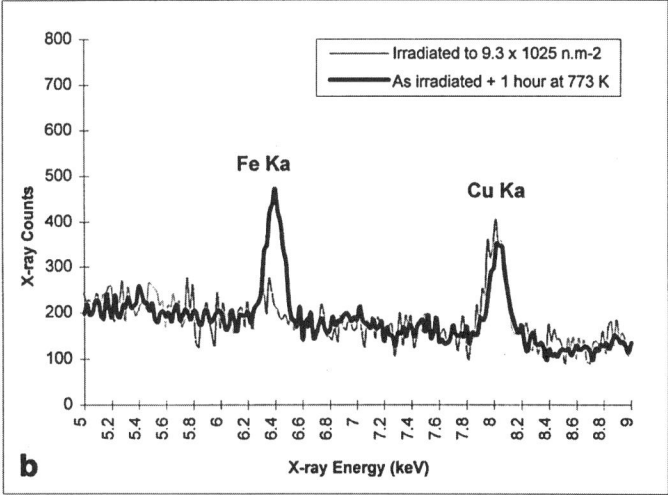

FIG. 13—*EDX spectra showing:* (a) *Fe concentration in the β-phase in an out-of-flux (at a temperature of about 570 K) and in-flux section (irradiated to a fluence of 9.3×10^{25} at 544 K) of a cold-worked Zr-2.5Nb pressure tube;* (b) *Fe concentration in the β-phase for an in-flux section of the same tube before and after post-irradiation annealing for 1 h at 773 K.*

density increases at a rate of about $2.5 \times 10^{-12}\,\mathrm{n}^{-1}$ in the Zircaloy-2 case, the Zr-2.5Nb material exhibits little, if any, change.

4. The transformed β-phase in as-fabricated Zr-2.5Nb pressure tubes consists of hcp ω-precipitates embedded in an Nb-enriched bcc matrix. Neutron irradiation at about 520 K with a flux of about $2 \times 10^{18}\,\mathrm{n.m^{-2}.s^{-1}}$ results in mixing of the two components.

5. Irradiation of Zr alloys results in the redistribution of small-sized impurity or alloying elements such as Fe, Ni, and Cr from second phases into the primary α-phase. This

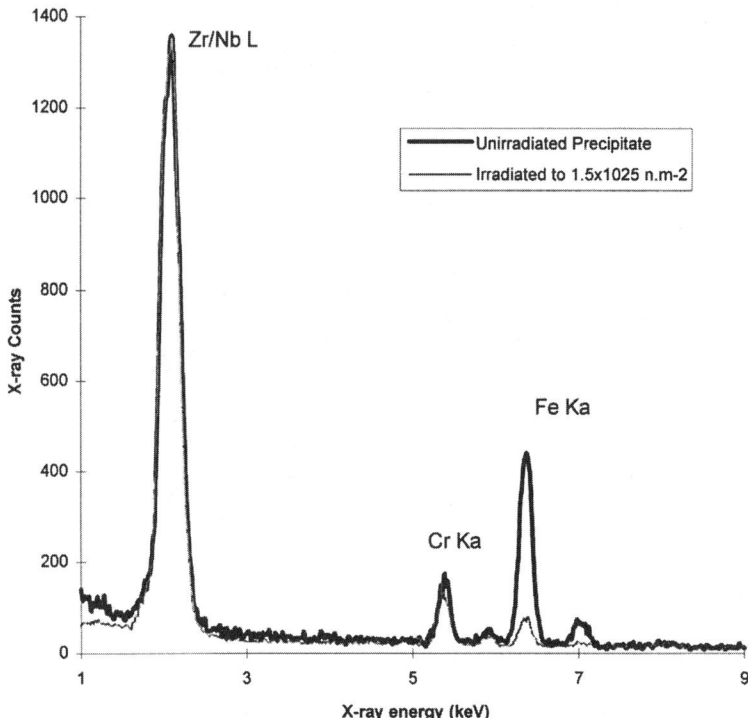

FIG. 14—*EDX spectra from the center of (Zr, Nb, Fe, Cr)-containing intermetallic precipitates in heat-treated Zr-2.5Nb before and after irradiation to a fluence of 1.6×10^{25} at 570 K.*

redistribution occurs irrespective of the form of the phase containing the impurity elements, and the redistribution can be reversed by post-irradiation annealing. For Fe, the rate of depletion from (Zr,Nb,Fe,Cr)-containing intermetallics in Zr-2.5Nb alloys is higher than that observed for Zr,Cr,Fe-type intermetallics in Zircaloy-2.

Discussion

The microstructural changes that occur in nuclear reactor components can have a profound effect on the service life of the individual component or even the reactor itself. Many of the important physical properties that characterize the performance of any one component, for example deformation (creep and growth), corrosion, fracture toughness, and hydrogen uptake, can change as a result of microstructure evolution during service.

For Zr-2.5Nb pressure tube materials, there is a change in fracture properties with irradiation corresponding with the evolution of the **a**-type dislocation structure [24]. The formation of a high density of **a**-type dislocation loops is important for modifying fracture and tensile properties. In general, the increased loop density results in a harder material with a corresponding lower fracture toughness compared with the unirradiated state. The dislocation density, strength, and fracture toughness each approach a saturation value for fluences less than 1×10^{25} n.m^{-2} [24]. The fact that annealed Zircaloy-4 also exhibits the same rapid increase in **a**-type dislocation at low fluences to achieve a saturated steady-state value, as shown by the line-broadening data (Fig. 2), suggests that this material will also exhibit the same trend in tensile and fracture

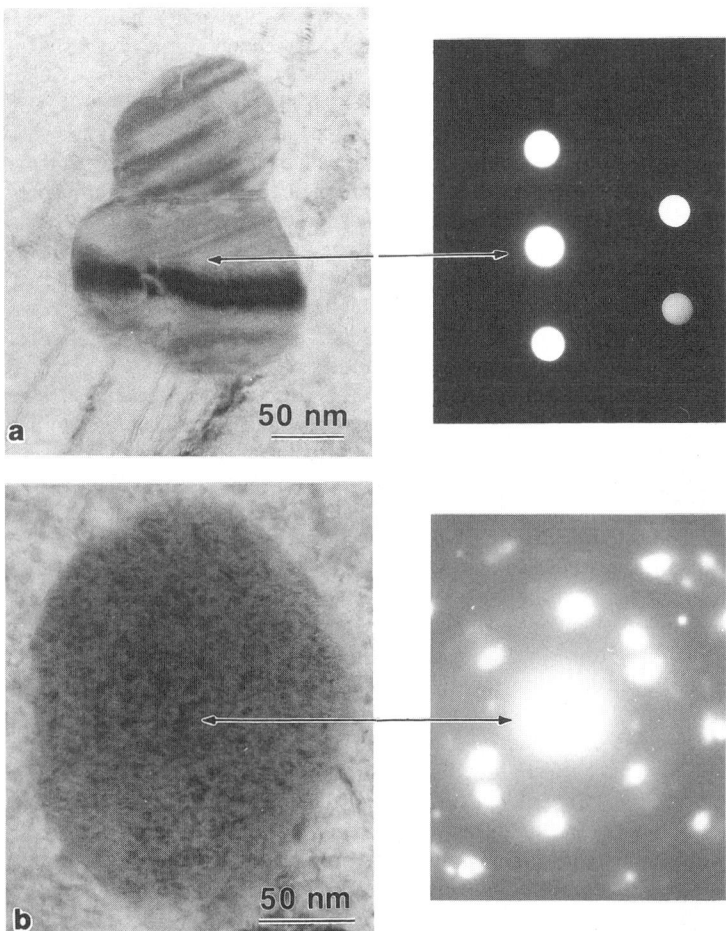

FIG. 15—*Structure of (Zr, Nb, Fe, Cr)-containing intermetallic precipitates in:* (a) *unirradiated heat-treated Zr-2.5Nb;* (b) *heat-treated Zr-2.5Nb irradiated to a fluence of* 1.6×10^{25} $n.m^{-2}$ *at 570 K.*

properties, achieving saturation at the same time as the **a**-type dislocation density. The role of radiation-induced **c**-component defects on tensile and fracture properties of Zr-2.5Nb and the Zircaloys has not been clearly established, there being insufficient data as a function of fluence to show whether there is a change in mechanical properties corresponding with the evolution of the **c**-component loop structure.

The dislocation structure is also important for irradiation creep and growth. Although creep rates are enhanced in a neutron flux [25], post-irradiation testing of irradiated material shows that thermal creep rates are reduced in pre-irradiated material, especially for material that was originally in an annealed state, and can be related to the effect of hardening. This hardening effect saturates at fluences of about 0.4×10^{25} $n.m^{-2}$ [25] and is consistent with the saturation in the **a**-type dislocation loop density. There is some evidence to indicate that in-reactor creep rates for cold-worked Zr-2.5Nb at about 570 K may increase with increasing fluence at high stresses (246 MPa) [25]; however, this has not been observed at lower stresses in later experiments [26], indicating that non-linear creep (strain localization) may be occurring at the higher stresses. There is also some low-temperature data to suggest that in-reactor creep rates at 330 K are increased for Zircaloy-4 pre-irradiated to high fluences at about 580 K, i.e., in a post-breakaway state [27]. The latter observation suggests that the evolving c-component loop structure in irradiated Zircaloys may contribute to a creep enhancement; however, there are insufficient data to be conclusive. Although there are few data concerning creep variation with an evolving dislocation microstructure, there is unequivocal evidence to show that irradiation growth is very susceptible to changes in the c-component dislocation structure [8].

There are few data concerning accelerated growth in Zr-2.5Nb alloys except at high temperatures, >640 K [28], and in this case the higher growth rates can be directly related to appearance of basal plane faulted dislocation loops in the material. For annealed Zircaloys, there is often an initial transient strain that saturates at low fluences that may be related to a saturation in the a-type dislocation loop density [4] or relaxation of residual stresses [29]. This is followed by a secondary stage of low growth rates and can then develop into higher (accelerated) growth rates at increasing fluences [8]. For cold-worked Zircaloys, the initial growth rates are high compared with annealed material but also appear to increase steadily with increasing fluence [8,9]. The phenomenon of accelerated growth is associated with an increased c-component dislocation density that develops either as a result of loop nucleation or due to climb of existing network dislocations [8,9]. The c-component loop formation often varies from one material to another [28]. In any one material, there are also grain-to-grain variations that may be related to variations in local chemistry, stress, or grain orientations [8,9,28]. For accelerated irradiation growth in annealed Zircaloys, the existence of an incubation period is linked with the delayed development of the c-component loop structure. The c-component loop formation is initially observed close to the $Zr(Cr,Fe)_2$ intermetallic precipitates and can be related to the increase in Cr and Fe in the matrix surrounding intermetallic precipitates [8]. Another factor that could affect growth is the Sn concentration within the α-phase [30]. At the normal operating temperature of Zircaloy components (about 560 to 580 K), however, there is no evidence for changes in Sn concentration within the α-phase that could affect irradiation growth, although Sn precipitation is observed at higher temperatures (>600 K) [4,17,20].

Apart from the potential effect on the dislocation structure and therefore the mechanical and deformation properties of irradiated Zr alloys, the microchemical changes induced by irradiation can also affect corrosion. For Zircaloys, the main effect of irradiation at about 570 K appears to be an increase in corrosion rate; this has been related to an increase in the solute (Fe,Cr, and Ni) concentration in the matrix during irradiation [18]. The dissolution and dispersion of alloying elements or impurities such as Fe, Cr, and Ni is always a potential factor in modifying alloy properties at ever increasing fluences. It has already been established that for Zircaloys irradiated at about 570 K there is a preferential radiation-induced dissolution of Fe from within $Zr(Cr,Fe)_2$ intermetallic precipitates [13–21] and also from within $Zr_2(Ni,Fe)$ precipitates in some cases [17,31]. The Fe depletion is coincident with an amorphous phase transformation in the $Zr(Cr,Fe)_2$ precipitates, and the flow is reversed during post-irradiation annealing—some Fe returns to the recrystallized precipitate at high temperatures (873 K) [17] or to the amorphous phase at lower temperatures (<773 K) [19,31]. In addition to the internal depletion of Fe, there is erosion and dispersion of Cr and Fe from $Zr(Cr,Fe)_2$ precipitates [13–21] and Ni and Fe from $Zr_2(Ni,Fe)$ precipitates [15–17,20]. These elements diffuse into the α-phase and appear to be uniformly dispersed in some finely divided form.

The effect of irradiation on corrosion in Zr-2.5Nb pressure tubes is opposite to the Zircaloy case, there being a marked decrease in corrosion rate for material that has been irradiated at 520 to 570 K compared with unirradiated material [32]. The improvement has been related to the decrease in Nb content in the α-phase due to Nb precipitation during irradiation [32].

Precipitates are observed in the α-phase of cold-worked Zr-2.5Nb pressure tubes after fluences of about 1×10^{25} n.m^{-2} [12,33,34]. They are generally very small (5 to 10 nm) and of insufficient volume to give coherent X-ray diffraction. EDX analysis has shown that they are Nb-rich [33] as one might expect for precipitates forming from a solid solution supersaturated with Nb. There are no equivalent observations of Nb precipitation in the α-phase in out-of-flux sections of pressure tubes [12], indicating that the precipitates are the result of irradiation.

The β-phase in Zr-2.5Nb pressure tubes has a complex structure consisting of hcp, Nb- and Fe-depleted, ω-phase precipitates in a bcc Nb- and Fe-enriched matrix. The analysis of XRD data indicates that the concentration of Nb in the bcc phase decreases and the volume fraction of ω-phase decreases during irradiation at about 520 K. The net effect on the ω-phase can be described as a balance between the rate of radiation-induced dissolution and the rate of growth due to radiation-enhanced and thermal diffusion. A similar mechanism has been postulated to explain the decomposition of large ω-phase precipitates in a Zr-12Nb alloy during electron irradiation [35]. In addition to the internal structural changes, Fe depletion and dispersion from the metastable β-phase has been observed in irradiated Zr-2.5Nb pressure tubes at temperatures between 520 to 570 K; however, unlike the Zircaloys, the effects of dispersed Fe on physical properties has not been established. The Fe dissolution is not necessarily dependent on the stability of the β-phase because the same phenomenon is observed for stable (Zr,Nb,Fe,Cr)-containing intermetallic (Zr,Nb)$_3$(Fe,Cr) precipitates at 573 K (Fig. 14). The Fe depletion occurs more rapidly in the latter case compared with amorphous phases in the Zircaloys irradiated at a similar temperature (Fig. 8).

The dissolution and dispersion of Fe can occur for many different Zr-alloy phases containing Fe. A number of hypotheses have been proposed to explain the phenomenon [13–17,36], and all depend to a large extent on diffusion of Fe into the matrix, i.e., the precipitate-matrix interface is not a barrier to diffusion. The depletion of Fe during irradiation and replenishment of Fe during post-irradiation annealing for the Zr(Cr,Fe)$_2$ intermetallic precipitates in the Zircaloys [19] and the β-phase in Zr-2.5Nb alloys (Fig. 13) indicates that the Fe can migrate across the interface by diffusion and that it is in a metastable state (supersaturated solid solution or in small solute clusters) in the α-phase. The recent data of Fig. 6 show that, for the Zr(Cr,Fe)$_2$ precipitates in the Zircaloys, the decrease in Fe content is coincident with an increase in Zr and Cr in approximately equal proportions. This indicates that the Fe diffuses into the matrix without being replaced by another element, e.g., by a vacancy exchange mechanism.

Conclusions

X-ray diffraction (XRD) and transmission electron microscopy (TEM) have been used to characterize microstructural and microchemical changes produced by neutron irradiation in Zircaloy-2 and -4 and Zr-2.5Nb nuclear reactor components.

1. Irradiation modifies the dislocation structure initially by nucleation and growth of **a**-type dislocation loops and, for cold-worked materials, climb of existing network dislocations. In general, the **a**-type dislocation structure tends to saturate at low fluences ($<1 \times 10^{25}$ n.m^{-2}). The **c**-component dislocation structure, however, can evolve over long periods of irradiation (for fluences $> 10 \times 10^{25}$ n.m^{-2} in many cases).

2. In Zircaloy-2 and -4, all precipitates become completely amorphous at low fluences ($<1 \times 10^{24}$ n.m^{-2}) during low-temperature neutron irradiation (about 330 K) with no associated chemical composition change. At higher temperatures (about 573 K), a duplex amorphous-crystalline structure is produced in the Zr(Cr,Fe)$_2$ precipitates, all other precipitate types remaining unchanged. The Zr(Cr,Fe)$_2$ precipitates often retain a crystalline core surrounded by a peripheral amorphous layer that advances inwards with increasing fluence. The amorphous

outer layer is coincident with a depletion of Fe that is dispersed into the surrounding hcp α-phase matrix. In some cases, depending on composition or flux, the precipitate centers can also become amorphous even at low doses. Subsequent post-irradiation heat-treatment below the amorphous phase recrystallization temperature results in the back-diffusion of Fe into the amorphous phase.

3. For Zr-2.5Nb materials, irradiation results in the formation of Nb-rich precipitates within the α-phase with a corresponding decrease of Nb in solid solution. The metastable transformed β-phase structure is modified during irradiation, but the change is complex, being a combination of thermal decomposition and radiation-induced mixing. There is dispersion of Fe from the β-phase during irradiation at temperatures between 520 and 570 K and, after a fluence of about 3×10^{25} n.m^{-2}, the β-phase Fe content cannot be distinguished from that in the α-phase matrix. Subsequent post-irradiation heat treatment, for times and temperatures that do not induce significant recrystallization of the α- or β-phase, results in the back-diffusion of Fe into the β-phase.

Acknowledgments

The authors would like to thank R. A. Holt, B. A. Cheadle, C. Lemaignan, D. Gilbon, and Y. de Carlan for many useful discussions. The authors would also like to thank R. R. Hosbons and P. H. Davies for the supply of some specimens, G. P. Smith of ABB-Combustion Engineering and O. Ozer of EPRI for permission to use unpublished data (Project No. RP-2905-02), and the CANDU Owners Group for funding under Work Package 32-3325.

References

[1] Shishov, V. N., Nikulina, A. V., Markelov, V. A., Peregud, M. M., Kozlov, A. V., Averin, S. A., Kolbenkov, S. A., and Novoselov, A. E., "Influence of Neutron Irradiation on Dislocation Structure and Phase Composition in Zr-Base Alloys," this publication.

[2] Griffiths, M., Winegar, J. E., Mecke, J. F., and Holt, R. A., "Determination of Dislocation Densities in HCP Metals using XRD and TEM," *Advances in X-ray Analysis*, C. S. Barrett et al., Eds., Plenum Press, New York, 1991, p. 593.

[3] Griffiths, M., Winegar, J. E., Mecke, J. F., Holden, T. M., and Holt, R. A., "A Comparison of X-ray and Neutron Diffraction for the Determination of Residual Stress and Chemical Composition in Zr-Alloy Tubes," *Advances in X-ray Analysis*, C. S. Barrett et al., Eds., Plenum Press, New York, 1991, p. 475.

[4] Griffiths, M., "A Review of Microstructure Evolution in Zr-Alloys during Irradiation," *Journal of Nuclear Materials*, Vol. 159, 1988, p. 190.

[5] Jostsons, A., Kelly, P. M., Blake, R. G., and Farrell, K., "Neutron-Induced Defect Structures in Zirconium," *Effects of Irradiation on Structural Materials (Ninth Conference), ASTM STP 683*, 1979, p. 46.

[6] Holt, R. A. and Gilbert, R. W., "<c> Component Dislocations in Annealed Zircaloy Irradiated at about 570 K," *Journal of Nuclear Materials*, Vol. 137, 1986, p. 185.

[7] Griffiths, M. and Gilbert, R. W., "The Formation of c-Component Defects in Zirconium Alloys during Neutron Irradiation," *Journal of Nuclear Materials*, Vol. 150, 1987, p. 169.

[8] Griffiths, M., Holt, R. A., and Rogerson, A., "Microstructural Aspects of Accelerated Deformation of Zircaloy Nuclear Reactor Components during Service," *Journal of Nuclear Materials*, Vol. 225, 1995, p. 245.

[9] Holt, R. A., Causey, A. R., Christodoulou, N., Griffiths, M., Ho, E. T. C., and Woo, C. H., "Non-Linear Deformation of Cold-Worked Zircaloy-2 During Irradiation," this publication.

[10] Griffiths, M., Styles, R. C., Woo, C. H, Philipp, F., and Frank, W., "Study of Point Defect Mobilities in Zirconium during Electron Irradiation in a HVEM," *Journal of Nuclear Materials*, Vol. 208, 1994, p. 324.

[11] Griffiths, M., Gilbon, D., Regnard, C., and Lemaignan, C., "HVEM Study of the Effects of Alloying Elements and Impurities on Radiation damage in Zr-Alloys," *Journal of Nuclear Materials*, Vol. 205, 1993, p. 273.

[12] Griffiths, M., Chow, C. K., Coleman, C. E., Holt, R. A., Sagat, S., and Urbanic, V. F., "Evolution of Microstructure in Zr-Alloy Core Components during Service," *Effects of Irradiation on Materials (Sixteenth Volume), ASTM STP 1175*, 1993, p. 1077.
[13] Gilbert, R. W., Griffiths, M., and Carpenter, G. J. C., "Amorphous Intermetallics in Neutron Irradiated Zircaloys After High Fluences," *Journal of Nuclear Materials*, Vol. 135, 1985, p. 265.
[14] Yang, W. J. S., Tucker, R. P., Cheng, B., and Adamson, R. B., "Precipitates in Zircaloy: Identification and the Effects of Irradiation and Thermal Treatment," *Journal of Nuclear Materials*, Vol. 138, 1986, p. 185.
[15] Garzarolli, F., Dewes, P., Maussner, G., and Basso, H. H., "Effects of High Neutron Fluences on Microstructure and Growth of Zircaloy-4," *Zirconium in the Nuclear Industry (Eighth International Symposium), ASTM STP 1023*, 1989, p. 641.
[16] Etoh, Y. and Shimada, S., "Neutron Irradiation Effects on Intermetallic Precipitates in Zircaloy as a Function of Fluence," *Journal of Nuclear Materials*, Vol. 200, 1993, p. 59.
[17] Gilbon, D. and Simonot, C., "Effect of Irradiation on Microstructure of Zircaloy-4," *Zirconium in the Nuclear Industry (Tenth Volume), ASTM STP 1245*, 1994, p. 521.
[18] Cheng, B., Kruger, R. M., and Adamson, R. B., "Corrosion Behaviour of Irradiated Zircaloy," *Zirconium in the Nuclear Industry (Tenth Volume), ASTM STP 1245*, 1994, p. 400.
[19] Griffiths, M., de Carlan, Y., Lefebvre, F., and Lemaignan, C., "A TEM Study of the Stability of Intermetallic Precipitates in Zircaloy Nuclear Reactor Components," to be published in *Micron*.
[20] Griffiths, M., Gilbert, R. W., and Carpenter, G. J. C., "Phase Stability, Decomposition and Redistribution of Intermetallic Precipitates in Zircaloy-2 and -4 During Neutron Irradiation," *Journal of Nuclear Materials*, Vol. 150, 1987, p. 53.
[21] Yang, W. J. S., "Precipitate Stability in Neutron Irradiated Zircaloy-4," *Journal of Nuclear Materials*, Vol. 158, 1988, p. 71.
[22] Hood, G. M., "Point Defect Diffusion in α-Zr," *Journal of Nuclear Materials*, Vol. 159, 1988, p. 149.
[23] Nelson, R. S., Hudson, J. A., and Mazey, D. J., "The Stability of Precipitates in an Irradiation Environment," *Journal of Nuclear Materials*, Vol. 44, 1972, p. 318.
[24] Davies, P. H., Hosbons, R. R., Griffiths, M., and Chow, C. K., "Correlation Between Irradiated and Unirradiated Fracture Toughness of Zr-2.5Nb Pressure Tubes," *Zirconium in the Nuclear Industry (Tenth Volume), ASTM STP 1245*, 1994, p. 135.
[25] Fidleris, V., "The Irradiation Creep and Growth Phenomenon," *Journal of Nuclear Materials*, Vol. 159, 1988, p. 22.
[26] Causey, A. R., Elder, J. E., Holt, R. A., and Fleck, R. G., "On the Anisotropy of In-Reactor Creep of Zr-2.5Nb Tubes," *Zirconium in the Nuclear Industry (Tenth Volume), ASTM STP 1245*, 1994, p. 202.
[27] Causey, A. R., Fidleris, V., and Holt, R. A., "Acceleration of Creep and Growth of Annealed Zircaloy-4 by Pre-Irradiation to High Fluences," *Journal of Nuclear Materials*, Vol. 139, 1986, p. 277.
[28] Griffiths, M., Gilbert, R. W., and Fidleris, V., "Accelerated Irradiation Growth of Zirconium Alloys," *Zirconium in the Nuclear Industry (Eighth International Symposium), ASTM STP 1023*, 1989, p. 658.
[29] Causey, A. R., Woo, C. H., and Holt, R. A., "The Effect of Intergranular Stresses on the Texture Dependence of Irradiation Growth of Zr-Alloys," *Journal of Nuclear Materials*, Vol. 159, 1988, p. 225.
[30] Rogerson, A., "Irradiation Growth in Zirconium and Its Alloys," *Journal of Nuclear Materials*, Vol. 159, 1988, p. 43.
[31] Kruger, R. M. and Adamson, R. B., "Precipitate Behaviour in Zirconium-Based Alloys in BWR's," *Journal of Nuclear Materials*, Vol. 205, 1993, p. 242.
[32] Urbanic, V. F. and Griffiths, M., "Corrosion Response of Pre-Irradiated Zr-2.5Nb Pressure Tube Material," *Proceedings, Seventeenth International Symposium on the Effects of Irradiation on Materials*, to be published.
[33] Griffiths, M. and Mullejans, H., "A TEM Study of α-Phase Stability in Zr-2.5Nb Pressure Tubes Following Neutron Irradiation," to be published in *Micron*.
[34] Coleman, C. E., Gilbert, R. W., Carpenter, G. J. C., and Weatherly, G. C., "Precipitation in Zr-2.5 Nb during Neutron Irradiation," *Stability during Irradiation*, J. R. Holland, L. K. Mansur, and D. I. Potter, Eds., AIME, 1980, p. 587.
[35] Faulkner, D. and Nuttall, K., "The Effect of Irradiation on the Stability of Precipitates in Zr-2.5Nb Alloys," *Journal of Nuclear Materials*, Vol. 67, 1977, p. 131.
[36] Motta, A. T. and Lemaignan, C., "A Ballistic Mixing Model for the Amorphization of Precipitates in Zircaloy under Neutron Irradiation," *Journal of Nuclear Materials*, Vol. 195, 1992, p. 277.

DISCUSSION

A. T. Motta[1] (written discussion)—In what state is the Fe in the matrix after leaving the precipitates in Zr (Cr, Fe)$_2$? If you believe that the Fe is in solid solution in the matrix, is it your belief that the solubility of Fe under irradiation in the matrix is enhanced relative to the solubility outside of irradiation?

M. Griffiths et al. (authors' closure)—All that can be said is that the Fe is in a finely dispersed form. The fact that some of the dispersed Fe returns to the parent phase during post-irradiation annealing suggests that it is in a metastable state within the α-phase. One could argue that this state is a supersaturated solid solution, although some secondary precipitation would also be expected.

S. Yagnik[2] (written discussion)—Since Fe reappears back in the particles upon modest annealing (500°C, ~ 1 h), it appears reasonable that Fe is in a metastable state not too far from the particle itself. Yet we have seen evidence of entirely new Fe-bearing particles forming, presumably quite a distance from the original particles, in another paper in this session. This appears to be an apparent contradiction. Please comment.

M. Griffiths et al. (authors' closure)—Not necessarily. The dispersed Fe (and Cr or Ni) will be available to form secondary precipitates, and such precipitates are readily observed at higher temperatures of irradiation (>673 K). It is likely that small secondary precipitates do exist close to the parent intermetallic particle. We do not know, however, whether the Fe that returns to the parent phase comes from re-dissolved (metastable) secondary precipitates or from a supersaturated solid solution.

B. Lehtinen[3] (written discussion)—Are there any "extra peaks" in the EDS spectrum due to the radiated specimen? If so, do you subtract them from the spectrum or are you using an energy filtering method?

M. Griffiths et al. (authors' closure)—Yes, self-generated X-ray peaks in neutron-irradiated Zr alloys are a problem and have to be subtracted prior to analysis of the spectrum by a hole-count method. The main extra peaks are Mn K, Zr K, and Nb K generated by election capture.

Y. Etoh[4] (written discussion)—Is there any correlation between Cr distribution and c-component dislocation loops near amorphous Zr-Fe-Cr particles? Is the c-dislocation a fast diffusion path for Cr?

M. Griffiths et al. (authors' closure)—Previous work has shown a correlation between impurities, such as Fe and Cr, and c-component dislocation loops. I believe that the Cr is important for enhancing c-component loop stability near intermetallic particles because it is less mobile than Fe (or Ni) and a higher local concentration is achieved. I do not know of any data to indicate that c-dislocations enhance the diffusion of Cr specifically, although one might anticipate enhanced diffusion (of all elements) along any dislocation.

---

[1] Pennsylvania State University, University Park, PA.
[2] Electric Power Research Institute, Palo Alto, CA.
[3] Institute of Metals Research, Stockholm, Sweden.
[4] Nippon Nuclear Fuel Development Co., Ltd., Ibataki-ken, Japan.

Vyatcheslav N. Shishov,[1] Antonina V. Nikulina,[1] Vladimir A. Markelov,[1] Mikhail M. Peregud,[1] Alexander V. Kozlov,[2] Sergey A. Averin,[2] Stanislav A. Kolbenkov,[2] and Andrey E. Novoselov[3]

Influence of Neutron Irradiation on Dislocation Structure and Phase Composition of Zr-Base Alloys

REFERENCE: Shishov, V. N., Nikulina, A. V., Markelov, V. A., Peregud, M. M., Kozlov, A. V., Averin, S. A., Kolbenkov, S. A., and Novoselov, A. E., "**Influence of Neutron Irradiation on Dislocation Structure and Phase Composition of Zr-Base Alloys,**" *Zirconium in the Nuclear Industry: Eleventh International Symposium, ASTM STP 1295*, E. R. Bradley and G. P. Sabol, Eds., American Society for Testing and Materials, 1996, pp. 603–622.

ABSTRACT: Studied were evolution of dislocation structure, phase, and element composition of binary alloys Zr-1Nb and Zr-2.5Nb and multicomponent alloys Zr-1Nb-1.2Sn-0.4Fe and Zr-1.2Sn-0.4Fe under neutron irradiation. The investigations were carried out using cladding and pressure tubes before and after irradiation to a fluence of $\sim 10^{26}$ n/m^2 ($E \geq 0.1$ MeV) in experimental and commercial reactors at 300 to 350°C using TEM, EDX, and XRD. In most cases, irradiation-induced defects are in the form of dislocation loops with Burgers vector $\frac{1}{3}\langle 11\bar{2}0 \rangle$. The density of dislocations with a $\langle c \rangle$ component is less than 2×10^{14} m^{-2}. A higher fluence or the presence of strain results in the ordering of the dislocation structure of $\langle c \rangle$ component and $\langle a \rangle$-type dislocation loops. Before irradiation, the multicomponent alloys contain fine precipitates of Zr-Nb-Fe composition, and the matrix is depleted in Fe. Under irradiation, recrystallization proceeds intensively (as distinct from Zr-Nb alloys), changes take place in size, distribution, and composition of precipitates (with a relative decrease of Fe content compared to Nb), and the Fe content of α-Zr matrix is increased. None of the materials studied showed any significant evidence of secondary phase particle amorphization. The density of dislocations with $\langle a \rangle$ and $\langle c \rangle$ components and irradiation-induced defects, their mean size, the extent of ordering, and the planes of their occurrence were determined. A comparison was made between irradiation-induced evolutions of microstructures of the different alloys.

KEYWORDS: zirconium alloys, neutron irradiation, irradiation growth, dislocations, precipitates, composition, grains, loops

For several years we have been engaged in studying zirconium alloy microstructure changes induced by irradiation. Our first results were reported in 1992 [1]. We assume the stability of zirconium alloys under neutron irradiation to be related primarily to their microstructures. In other papers [2–4] attempts are also made to relate the macroscopic in-pile behavior of zirconium components to microscopic changes detected by analytical electron microscopy and,

---

[1] Senior scientific officer, leader of laboratory, leading scientific officer, senior scientific officer, respectively, A. A. Bochvar All-Russia Scientific Research Institute of Inorganic Materials, Rogov st. 5, 123060, Moscow, Russia.

[2] Leader of laboratory, senior scientific officer, scientific officer, respectively, Sverdlovsk Branch, Research and Designe Institute of Power Engineering, Zarechny, 624051, Ekaterinbourg region, Russia.

[3] Senior scientific officer, V. I. Lenin Scientific Research Institute of Atomic Reactors, 433510, Dimitrovgrad 10, Ulyanov region, Russia.

based on this, to predict the behavior of components during their long-term operation. A small number of elements (Sn,Nb,Fe,Cr,Ni) are used to alloy commercial zirconium alloys. However, the combinations of these elements, their concentration, and different mechanical working and heat treatment can substantially alter the microstructure and properties. Changes in the geometrical shapes of components, their corrosion, and tensile properties under neutron irradiation to a fluence of $\geq 10^{26}$ n/m^2 at moderate temperatures ($<\sim$350°C) are related to irradiation-induced defects such as dislocation loops of the $\bar{a}$- and $\bar{c}$-types, as well as alteration of the phase and microchemical compositions [4].

In our previous work [1] we emphasized specific features of microstructural changes in Zr-1Sn-1Nb-0.4Fe alloy under neutron irradiation. The in-pile performance of this alloy is described by a high resistance to creep, growth, and nodular corrosion. As distinct from binary Zr-Nb and Zircaloys, the $\bar{c}$-component dislocations are rarely seen under neutron irradiation and the alloy reaches an equilibrium state rather easily and quickly (with respect to recrystallization, secondary phase composition). The features we observed were new and needed corroboration, particularly the lack of evolution of a $\bar{c}$-component dislocation structure. This paper discusses the results of the continued studies of irradiation-induced changes in the microstructure of zirconium alloys using in Russia.

Materials Studied and Experimental Procedure

Materials

Investigations were carried out using Zr-alloy samples cut out of fuel rod claddings, specifically Zr-1.2Sn-1Nb-0.4Fe (E635) in comparison to Zr-1Nb (E110) and Zr-1.2Sn-0.4Fe (EZ-1) that served the base of the multicomponent alloy before and after testing in the experimental reactor MIR, as well as Zr-Sn-Nb-Fe and Zr-2.5Nb (E125) alloy samples cut out of pressure tubes that operated for different periods of time (up to 15.5 years) within RBMK process channels. The initial state of the items and reactor testing conditions are given in Table 1. Before irradiation, all tubes were as cold worked and annealed, fully or partially recrystallized. The structure of E125 pressure tubes was also studied in the quenched (870°C in water) and heat-treated condition (TMT-1). The cladding tubes were irradiated as stress-strained in the transverse direction to 4%.

Studies of Structure

The microstructure, phase, and chemical compositions were investigated with a transmission electron microscope (TEM) using electron diffraction (EM-301G, JEM 2000FX), energy dispersive X-ray microanalyses (EDX) (LINK Systems 860, KEVEX-700), as well as X-ray diffraction analysis (XRD) with a diffractometer ADP-2 with monochromatized CuK$_\alpha$ radiation. Foils for TEM studies were prepared in a facility "Struers" by electropolishing disks 3 mm in diameter and 0.1 mm thick with a solution 90% CH$_3$COOH + 10% HClO$_4$ (T = 13°C) and 80% C$_2$H$_5$OH + 20% HClO$_4$ (T = 30°C). The disks were cut out of cladding tubes in the plane of a tube surface and pressure tubes in the cross section. The chemical composition of secondary phase precipitates was determined using foils and carbon replicas with extracted particles. The mean α-grain size, the mean size, and concentration of precipitates and dislocation loops were evaluated from electron-microscopic patterns taking into account the studied foil thickness. The dislocation density was calculated with the formula:

$$\rho = 2L/Nt$$

where

TABLE 1—Materials studied and irradiation conditions.

Alloy	Item	Initial State	Recrystallization Grade, %	Reactor	Temperature, °C	Neutron Fluence ($E \geq 0.1$ MeV), n/m^2, 10^{25}
E110	Cladding, 9.15 by 0.7 mm	C. W. + 580°C, 2h	100	MIR	350	5
EZ1	Cladding, 9.15 by 0.7 mm	C. W. + 580°C, 2h	100	MIR	350	5
E635	Cladding, 9.15 by 0.7 mm	C. W. + 580°C, 2h	100	MIR	350	5
E635	Pressure tube, 88 by 4 mm	C. W. + 550°C, 5h	60–70	RBMK-1500	315	10
E125,A	Cladding, 9.15 by 0.7 mm	C. W. + 540°C, 5h	75	MIR	350	5
E125,A	Pressure tube, 88 by 4 mm	C. W. + 540°C, 5h	75	RBMK-1000	305	10, 15.5 years
E125,B	Pressure tube, 88 by 4 mm	Quenching + 515°C, 24 h (TMT-1)	15	RBMK-1500	315	7

L = a secant circle length,
N = a number of intersections with dislocations, and
t = a foil thickness.

Burgers vector of dislocations ($\bar{b}$) was found from the condition of dislocation image extinction: $\bar{g} \times \bar{b} = 0$, where $\bar{g}$ is a vector of a strongly acting reflection. The X-ray diffraction data were used to assess the strains, domain sizes, as well as $\bar{a}$ and $\bar{c}$ dislocation densities. Strains (ε) and effective domain size (D) were determined from the systems of planes of basal, prism, and pyramidal types. The results of the harmonic analysis were used to determine the density of dislocations containing $\bar{a}$ and $\bar{c}$ components with the formulae [5]:

$$\rho_a = \frac{K_a \, \varepsilon_a^2}{b_a^2} \frac{1}{\ln(D_a/2r_0)}, \, \rho_c = \frac{K_c \, \varepsilon_c^2}{b_c^2} \frac{1}{\ln(D_c/2r_0)}$$

where $K_a = 52.1$, $K_c = 26.1$, $b_a = \frac{1}{3}\langle 11\bar{2}0\rangle$, $b_c = [0001]$, $r_0 = 1$ nm, ε_a, D_a are the strains and domain sizes calculated from the prism planes, and ε_c, D_c are the strains and domain sizes calculated from the basal planes.

Results

E110 Alloy [Zr-1Nb]

The structure of unirradiated E110 alloy cladding contains a secondary phase as globular β-Nb precipitates having the size of about 0.05 μm, the BCC lattice with the parameter $a = 0.328$ nm (the content of niobium up to 90%) in the α-Zr matrix (0.6% niobium) Fig. 1A). The $\bar{a}$ and $\bar{c}$ type dislocation density is not high in the above alloy. The irradiation to a neutron fluence of 5×10^{25} n/m² ($E > 0.1$ MeV) at $T = 350°C$ results in an alteration of the dislocation structure, specifically in irradiation-induced clusters and dislocation loops. The analysis of the distribution of these defects revealed that they are ordered in the direction parallel to the basal plane (0001) (Fig. 1B). Using the condition of the invisibility of dislocations $\bar{g} \times \bar{b} = 0$, the Burgers vector of dislocation loops was found to be ($\bar{b} = \frac{1}{3}\langle 11\bar{2}0\rangle$). The nature of the loops was not established due to their small size (~10 nm); however, in the same row both interstitial and vacancy loops are likely to be available lying on prism planes {10$\bar{1}$0} and {11$\bar{2}$0} [9]. Under the conditions of a strongly acting reflection, $\bar{g} = 0002$, when $\bar{a}$-dislocations are invisible, $\bar{c}$-component dislocations are detected in the edge and screw orientations (Fig. 1C). The dislocation segment bends indicate the climbing to the basal planes of the dislocations that were available prior to irradiation. In this instance the density of the $\bar{c}$-dislocations is even higher than that of the $\bar{a}$-dislocations in network. Under the same reflections, fine defects were identified as dark bands ~6 nm (Fig. 1D). The preliminary analysis of their contrast shows that they are likely to be vacancy loops lying on basal planes. The irradiation did not induce any noticeable changes in the phase composition. The mean β-Nb precipitate size became slightly larger, the precipitate concentration was reduced, and no amorphization of the precipitates was observed. The quantitative results on the microstructure and composition are tabulated in Table 2.

EZ-1 Alloy [Zr-1.2Sn-0.4Fe]

In this alloy there is no niobium Zr-Fe phase. Precipitates are observed within coarse recrystallized grains and along their boundaries (Fig. 2A). The total dislocation density is low. It is established by electron diffraction and X-ray microanalysis that the precipitated

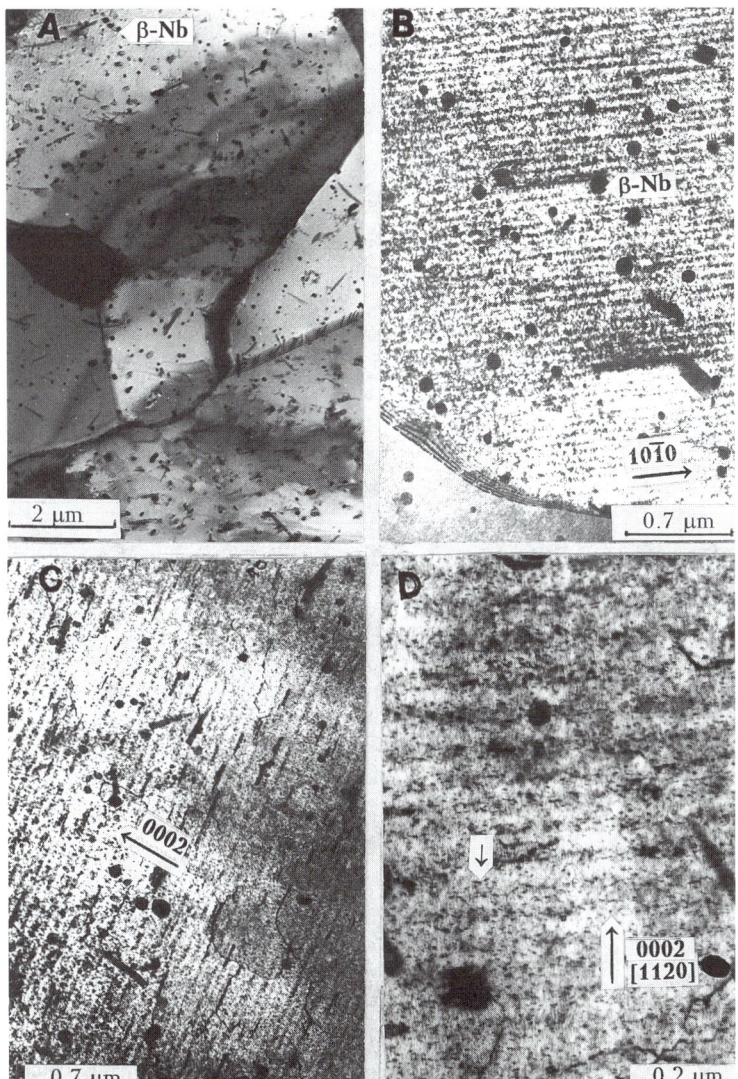

FIG. 1—*Microstructure of recrystallized Zr-1Nb alloy:* (A) *initial and* (B) *irradiated to a fluence of* 5×10^{25} $n \cdot m^{-2}$ *at* 350°C: *grain structure, β-Nb precipitates,* $\vec{a}$-*dislocation loop ordering;* (C) $\vec{c}$-*dislocations, climbing onto basal planes;* (D) *fine defects, possibly* $\vec{c}$-*dislocation loops.*

phase has an orthorhombic crystal lattice with the parameters $a = 0.33$ nm, $b = 0.88$ nm, and $c = 1.10$ nm and contains about 23% (at%) Fe; it was identified as Zr_3Fe. Finer precipitates, Zr_2Fe, having a tetragonal lattice ($a = 0.58$ nm, $c = 0.64$ nm) are observed much less frequently. The total precipitate concentration is low, namely $\sim 3 \times 10^{18}$ m^{-3}. The irradiation of the alloy to a fluence of 5×10^{25} n/m² ($E > 0.1$ MeV) at $T = 350$°C did not effect any prominent changes in its phase composition; no precipitate amorphization was observed (Fig. 2B).

TABLE 2—*Main structure features of irradiated E110, EZ-1, E635 alloys.*

Alloy	Fluence, n/m^2, 10^{25}	Mean Grain size, μm	Precipitates, mean concentration size		Dislocation, Density (TEM) m^{-2}, 10^{14}		Dislocation loops, mean concentration size and type		
			μm	m^{-3}, 10^{20}	$\bar{a}$	$\bar{c}$	nm	m^{-3}, 10^{22}	type
E110	0	4	0.05	2.5	0.2	0.2	...	...	...
	5	8	0.07	1.0	0.1	0.4	9	3	$\bar{a}$
							6	4	$\bar{c}$
EZ-1	0	6	0.6	0.03	0.4	0.4	...	...	...
	5	10	0.6	0.02	0.1	2.0	20	2	$\bar{a}$
E635, recrystallized	0	3.5	0.12	0.5	0.2	0.1	...	...	...
	5	8	0.20	0.4	0.1	0.1	10	5.0	$\bar{a}$
E635, partially recrystalized	0	2.0	0.20	0.15	5.0^a 1.0^b	1.0^a 0.8^b	...	...	...
	10	3.0	0.15	0.1	1.5^a	0.6^a 0.8^b	12	1.4	$\bar{a}$

^{a} TEM measurements.
^{b} XRD measurements.

The irradiation-induced dislocation loops are of a larger size than in all other alloys (Figs. 2C,D). Both interstitial and vacancy loops with $b = \frac{1}{3} \langle 11\bar{2}0 \rangle$ are observed. The $\bar{c}$-dislocation density is increased under irradiation to reach $\sim 2 \times 10^{14}$ m^{-2}. The analysis of the dislocation structure at $\bar{g} = 0002$ reveals that some $\bar{c}$-dislocations nucleated in the process of irradiation and some transformed from the initial ones by climbing to basal planes (Fig. 2E). Faulted basal plane loops visible with diffracting vector $g = 10\bar{1}0$ were detected (Fig. 2F).

E635 Alloy [Zr-1.2Sn-0.4Fe-1Nb]

Structure studies were carried out using recrystallization annealed (fuel rod claddings) and partially recrystallized alloys (pressure tubes). The results of the studies are listed in Table 2. Both recrystallization annealed and partially recrystallized alloys contained intermetallic Zr-Nb-Fe precipitates with the mean size of ~ 0.12 μm and a concentration of $< 5 \times 10^{19}$ m^{-3} (Fig. 3A). The data of the electron microdiffraction (Fig. 3B) and X-ray microanalysis of particles in a thin foil and those extracted on a carbon replica (Fig. 3C,D) indicate the likely availability of several phases having different crystal lattices. The chemical composition of the particles varies in the following ranges: Zr(30–40)-Nb(30–40)-Fe(15–30) (at%). The calculation of the electron diffraction patterns gives the interplanar spacings and the lattice parameters of the orthorhombic phase of the Zr$_3$Fe type, hexagonal of the ZrFe$_2$ type ($a = 0.51$ to 0.53 nm, $c = 0.83$ to 0.87 nm), and the tetragonal one of the Zr$_2$Fe type in which some atoms, most likely those of zirconium, are substituted by niobium atoms.

The Nb/Fe ratio as found by microanalysis is conventionally equal to ~ 1.5 to 2.0, while the ratio between the hexagonal ZrFe$_2$-type lattice parameters $c/a = 1.63$ is typical of the Laves phase having characteristic twins (Fig. 3B). The positions of niobium atoms within lattices have not yet been determined; however, a series of electron diffraction patterns taken at different tilt angles suggest the likely existence of a hexagonal phase of the Zr(Nb,Fe)$_2$ type. It was shown before [6] that ZrFe$_2$ phases with the cubic lattice ($a = 0.71$ nm) and orthorhombic Zr$_3$Fe could

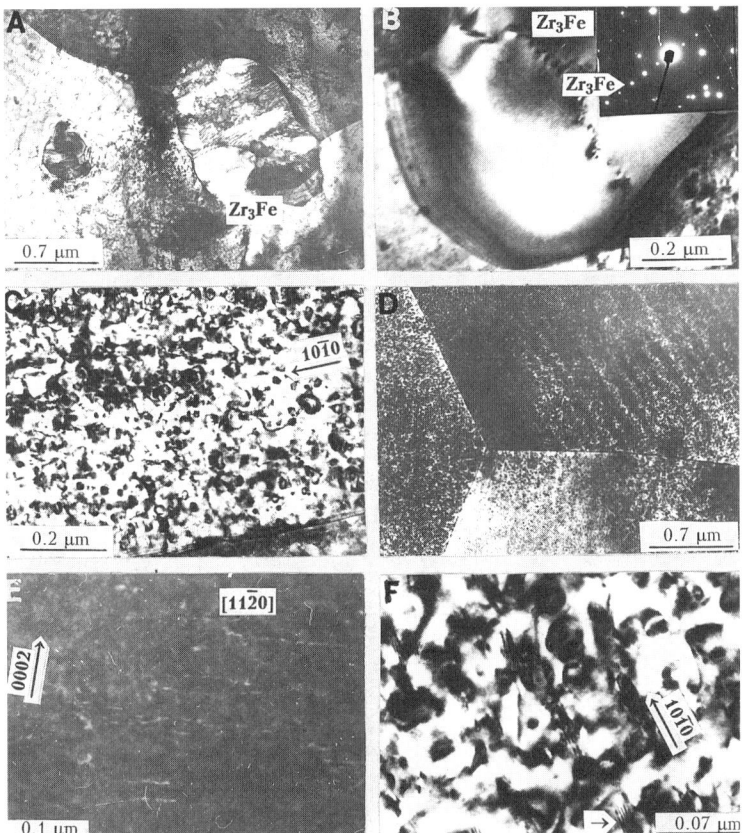

FIG. 2—*Microstructure of recrystallized Zr-1.2Sn-0.4Fe alloy: (A) initial and irradiated to a fluence of 5×10^{25} n · m^{-2} at 350°C: (B) Zr_3Fe precipitate without traces of amorphization; (C) $\vec{a}$-dislocation loops at $\vec{g} = \bar{1}010$ (dark field); (D) grain boundaries, $\vec{a}$-dislocation loop ordering; (E) $\vec{c}$-dislocations with a 0002 diffraction vector; (F) $\vec{c}$ dislocations with stacking faults at $\vec{g} = \bar{1}010$.*

be stable in the Zr-Fe alloy. The availability of niobium and the Nb/Fe ratio of the precipitates can significantly influence not only the lattice parameters but also its type. Hence, the phases have not been identified unambiguously in this alloy. The total dislocation density is low in the recrystallized alloy; it is much higher in the partially recrystallized one, specifically, 6×10^{14} m^{-2}; the $\bar{c}$-dislocation density being no higher than 1×10^{14} m^{-2}; according to the refined data of the X-ray diffraction analysis, it is even lower than 0.1×10^{14} m^{-2} in recrystallized alloy. The structure of the irradiated tubes of both the types was fully recrystallized (Fig. 4A). The secondary phase particle concentration was slightly reduced by irradiation. The particle amorphization was not detected, although characteristic stacking faults vanished from most particles. In individual instances, a partial amorphization of intermetallic particles retaining their crystalline cores was observed (Fig. 4B). The microanalysis of the matrix and precipitate compositions showed that it was difficult to get the absolute values of chemical element contents by studying foils; however, based on the comparison data, one can perceive the element redistribution both within the matrix and the precipitates. The precipitates are depleted in iron, which

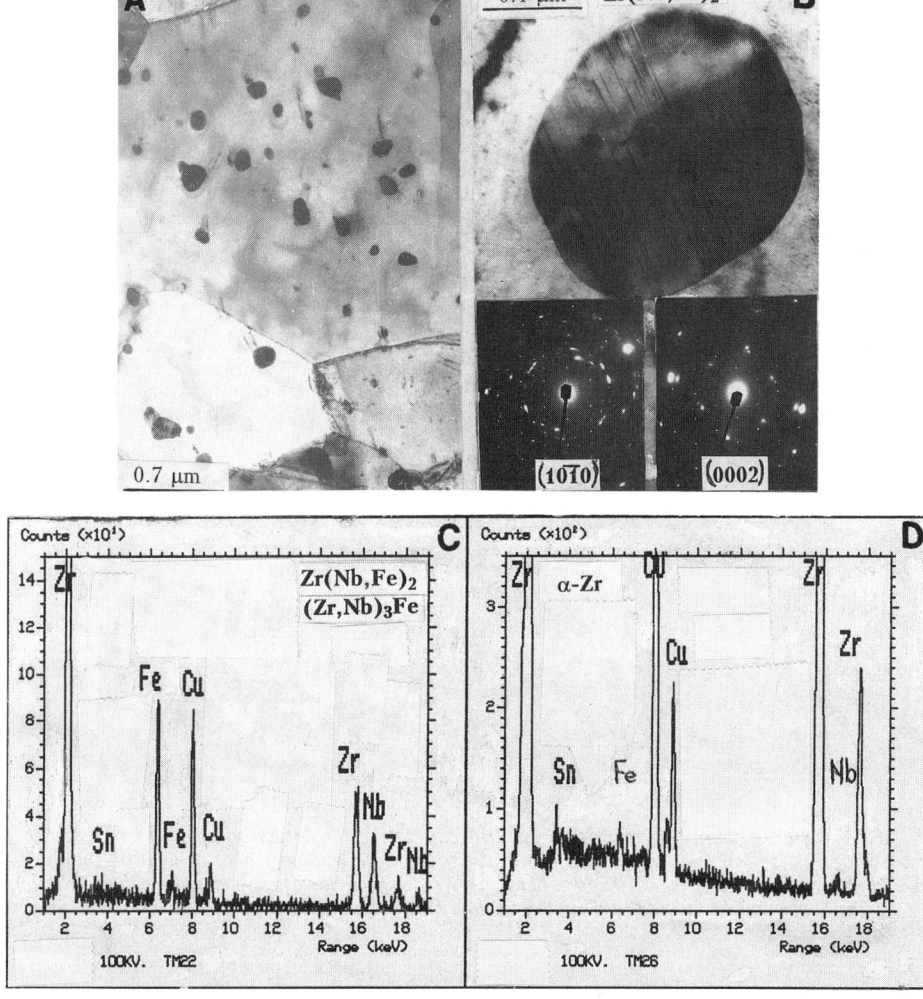

FIG. 3—*Microstructure of unirradiated recrystallized Zr-1.2Sn-1Nb-0.4Fe alloy: (A) grain structure and distribution of Zr-Nb-Fe secondary phase precipitates; (B) precipitate with stacking faults and crystal structure (electron diffraction patterns); (C) X-ray microanalysis spectra showing distribution of elements within a particle extracted on a carbon replica and within a foil on 15 by 15-μm area; (D) within a foil in 15 by 15-μm area.*

results in its content of the solid solution increasing to ~0.2%. A change in particle contrast and their ability for electropolishing with foil thinning is also evidence of this redistribution. An irradiation-effected alteration of the dislocation structure involves a reduction in the $\bar{a}$-dislocation density in fully and partially recrystallized materials and an appearance of dislocation loops with Burgers vector $b = \frac{1}{3}\langle 11\bar{2}0\rangle$ lying on prism planes. This is corroborated by the X-ray diffraction data, which indicate an increase of strains in $\{10\bar{1}0\}$ and $\{10\bar{1}1\}$ planes. Analysis of the dislocation loop distribution revealed that they are ordered parallel to the basal plane (0001) (Fig. 4C). The phenomenon is characteristic of high fluences and irradiation under

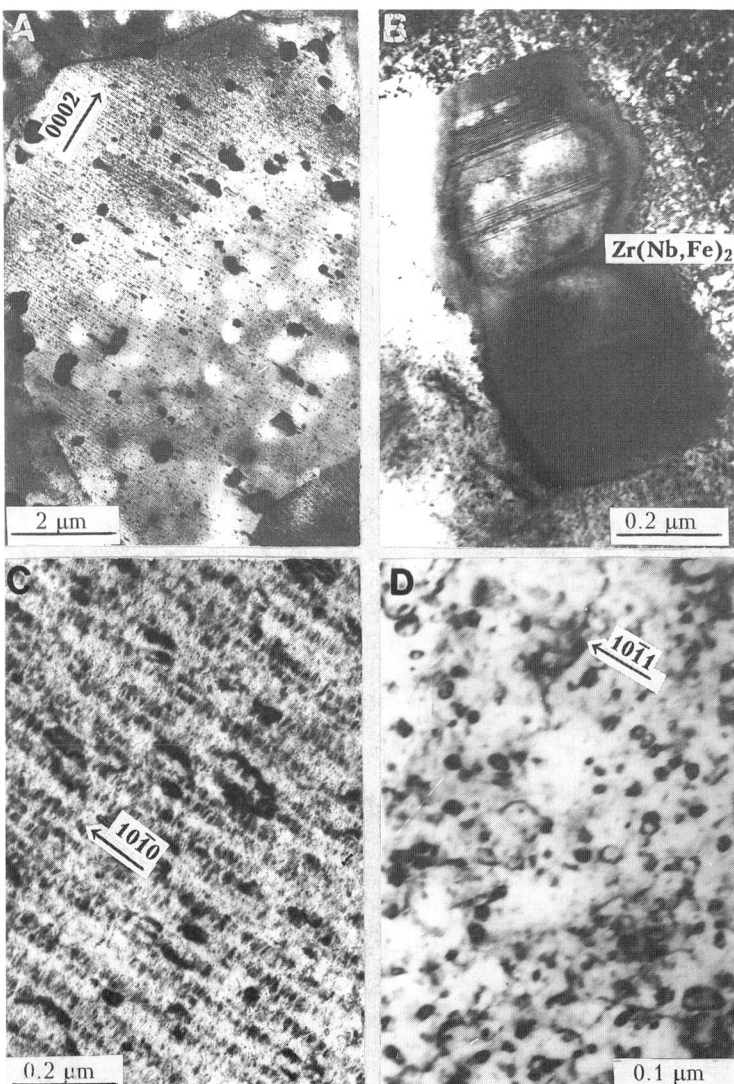

FIG. 4—*Microstructure of recrystallized Zr-1.2Sn-1Nb-0.4Fe alloy irradiated to a fluence of 5 × 10^{25} n · m^{-2} at 350°C: (A) grain structure and intermetallic particles distribution; (B) a rare event of surface amorphization of an intermetallic particle-retaining crystalline core with stacking faults; (C) $\vec{a}$ dislocation loop ordering, $\vec{g} = 10\bar{1}0$; (D) $\vec{a}$-dislocation loops, $\vec{g} = 10\bar{1}1$.*

stress conditions. The $\bar{c}$-dislocation density remained at the level of that under unirradiated condition, i.e., less than 1×10^{14} m^{-2} (TEM data).

After irradiation, the lattice parameter a was slightly decreased, while the parameter c did not essentially change, most likely because of the higher niobium and iron contents of the solid solution. Increased sizes of domains in planes (0001) and a decrease in strains can be related to a reduction in the $\bar{c}$-dislocation density and the likely splitting of $\bar{c}$-dislocations into partial ones.

E125 Alloy [Zr-2.5Nb]

Prior to irradiation, the studied tubes of both the types (cladding and pressure ones) were partially recrystallized. The fraction of recrystallized grains varies with the fabrication process. For State A (Table 3), the fraction of recrystallized areas is ~75%, and for State B it is less than 15% (Figs. 5A,B,D). Non-recrystallized areas are described by an intensive grain fragmentation, low-angle boundaries, and a high density of dislocations within walls. In both states, the main fraction of dislocations is made up by $\bar{a}$-type dislocations; the density of $\bar{c}$-type dislocations does not exceed 1.0×10^{14} m^{-2}.

The phase composition of the unirradiated alloy structure consists of the α-Zr matrix with β-Nb precipitates and residual β-Zr phase. The β-Nb particles are distributed mainly along the grain boundaries, forming stringer pile-ups at the sites of the decomposed β-Zr phase and the former boundaries (Fig. 5C). In State A of the tube structure, the particle-size range is rather wide, specifically from 0.01 to 0.20 μm. In State B of the tube, the particles are finer and their mean size is ~0.03 μm. In the latter case, the higher concentration of precipitates is a result of the operations of quenching and low-temperature anneal. The chemical composition of the α-solid solution, grain boundaries, and alloy was analyzed in an area of 0.2 by 0.1 mm using foils; to determine the composition of precipitates, β-Nb particles extracted onto a carbon replica were used (Fig. 6). The analysis of irradiated samples took into account not only the contribution made by the background, but also the activity induced by niobium isotopes. The investigations showed that the niobium content of the recrystallized α-Zr grains was lower compared to that of non-recrystallized areas in both states of the alloy. In the alloy of State B, the grains of both the types contain more niobium than in the alloy of State A, which is likely to be explained by a high supersaturation as a result of quenching. The composition of β-Nb precipitates is as much as 80% (at%) niobium in unirradiated tubes in State A. The niobium content of coarse β-Zr precipitates is in excess of 9% (at%). The results of the microdiffraction and X-ray diffraction analyses agree well with the data of the microanalysis of the precipitate composition and corroborate the presence of the β-Nb and β-Zr phases having the BCC lattice as well as α-Zr with different niobium contents of solid solutions.

TABLE 3—*Main structure features of irradiated E125 alloy.*

Condition Fluence, n/m^2, 10^{25}		Precipitates, mean concentration size		Composition, %Nb	Dislocation Density, m^{-2}, 10^{14}		Dislocation Loops, mean concentration size and type		
		μm	m^{-3}, 10^{20}		$\bar{a}$	$\bar{c}$	nm	m^{-3}, 10^{22}	Type
E125, A, $T_{ir.}$ = 350°C	0	0.04	7.0	79^c	4.0^a	0.5^a	...	...	...
				83^b	1.0^b	0.9^b			
	5	0.16	0.2	80^c	0.1^a	1.0^a	11	5	$\bar{a}$
				...	...	...	10	4	$\bar{c}$
E125, A, $T_{ir.}$ = 305°C	7	0.05	1.0	75^c	1.7^a	0.5^a	10	1.5	$\bar{a}$
				85^b	...	0.6^b			
	10	0.05	1.0	81^c	1.5^a	0.5^a	14	2.0	$\bar{a}$
				90^b	...	0.4^b			
E125, B, $T_{ir.}$ = 315°C	0	0.03	3.0	79^c	6.0^a	0.4^a	...	...	...
				90^b	2.0^b	0.9^b			
	10	0.04	0.8	80^c	...	0.5^a	10	1.5	$\bar{a}$

^{a} TEM measurements.
^{b} XRD measurements.
^{c} EDX measurements.

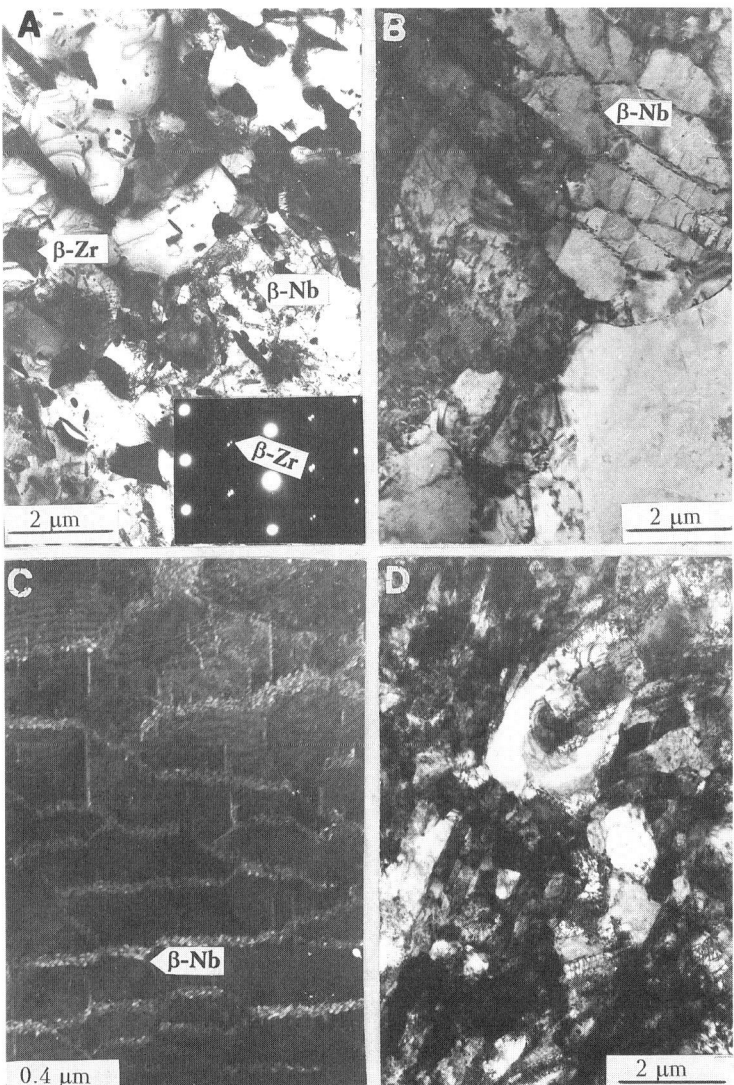

FIG. 5—*Microstructure of partially recrystallized unirradiated Zr-2.5Nb alloys: (A) cold worked and annealed (State A), β-Nb and β-Zr precipitates; (B) distribution of fine dispersive β-Nb precipitates along non-recrystallized grains; (C) distribution of β-Nb precipitates, dark field; (D) quenched and stress-relieved condition (State B).*

The alloy irradiated in different structure states to fluences of 0.5 to 1.0×10^{26} n/m² ($E >$ 0.1 MeV) at temperatures of 305 to 350°C has different grain and dislocation structures as well as phase and chemical compositions (Figs. 7A,B,C). The alloy irradiated to a fluence of 5×10^{25} n/m² at $T = 350$°C experienced a full recrystallization and β-Zr decomposition to form β-Nb precipitates (Fig. 7C). The mean size of precipitates both newly formed and grown from initial ones and coagulated is 0.16 μm, while the grain size is increased from 2 to 4 μm to 5

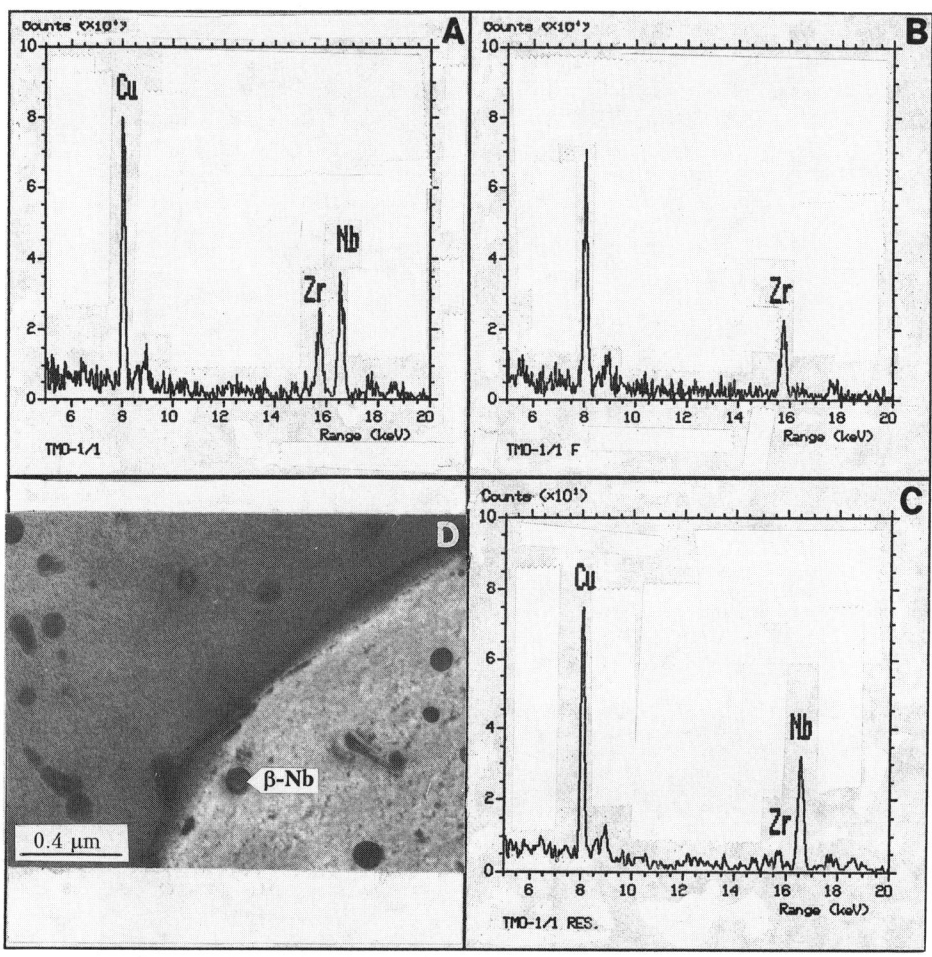

FIG. 6—*X-ray microanalysis spectra of β-Nb precipitates in unirradiated Zr-2.5Nb alloy. Particles extracted on a carbon replica:* (A) *not processed spectrum;* (B) *background spectrum;* (C) *processed spectrum;* (D) *particles.*

to 6 μm. No amorphization of β-Nb precipitates was detected (Fig. 7D). The lower temperature (305°C) irradiation caused little recrystallization of the alloy in either state. However, in the irradiation–thermal anneal process, stress relaxation takes place, the grain boundaries are more clearly defined, and new grains nucleate in non-recrystallized areas. The phase compositions of the alloy irradiated in the A and B states are close (α-Zr + β-Nb), and the mean size of β-Nb precipitates is increased, which is accompanied by an increase in the volume fraction of β-Nb. The niobium volume fraction of the precipitates and the matrix calculated from the TEM results agrees adequately with the data of the X-ray diffraction and microanalysis. The niobium content of β-Nb is slightly increased with the fluence (Fig. 8, Table 3).

The dislocation structure is subject to more substantial changes that are related to a reduction in the $\bar{a}$-dislocation density, nucleation, and growth of irradiation-induced defects as $\bar{a}$-type dislocation loops at the concentration above 2×10^{22} m^{-3} (Fig. 9A). Loop alignments to form

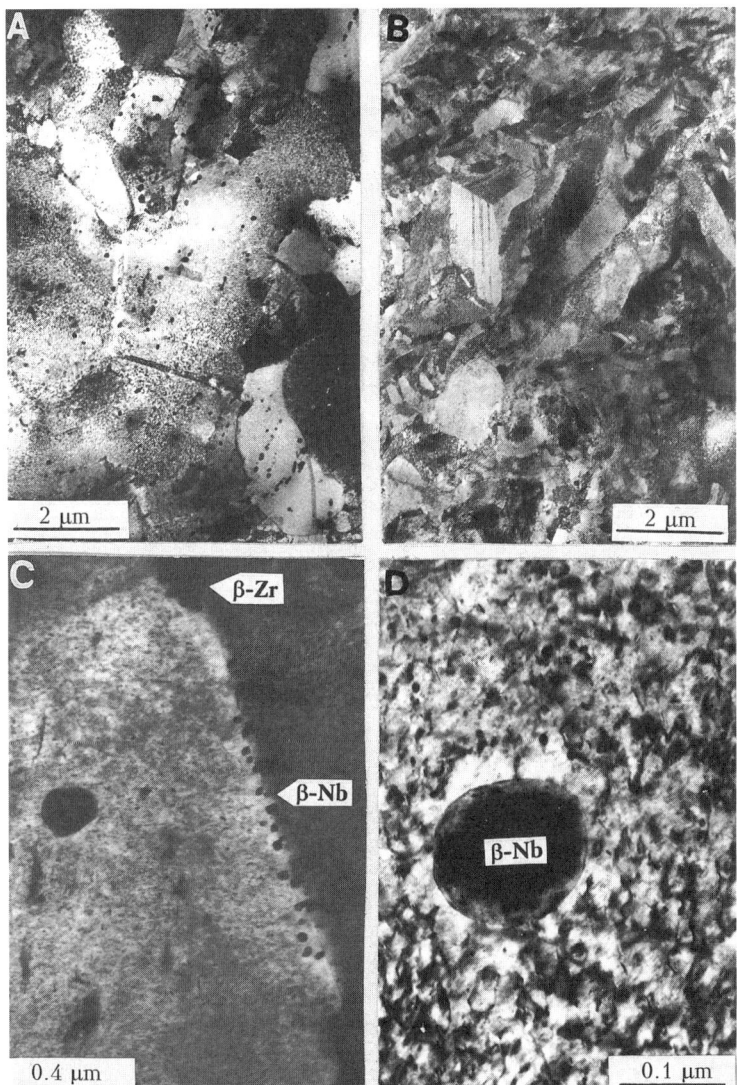

FIG. 7—*Microstructure of irradiated Zr-2.5Nb alloy:* (A) *at 305°C (RBMK-1000), State A;* (B) *at 315°C (RBMK-1500), State B;* (C) *at 350°C (5 × 10^{25} n · m^{-2}, MIR), State A, β-Zr decomposition into β-Nb chains;* (D) *β-niobium precipitation in alloys.*

rows having the direction parallel to (0001) are more clearly revealed in the alloy irradiated under stress conditions. Burgers vector of the loops found by the $\bar{g} \times \bar{b} = 0$ method was established to be $b = \frac{1}{3} \langle 11\bar{2}0 \rangle$. As in the Zr-1Nb alloy at the foil orientation [10$\bar{1}$0], defects in basal planes were detected having the size of ~10 nm; they are likely to be vacancy $\bar{c}$-loops (Fig. 9B). The mean size of the dislocation loops in the rows and their spacing depend on a number of factors that cover the alloy heat treatment conditions, irradiation temperature, and fluence. The common regularities consist in the fact that the mean loop size grows with the

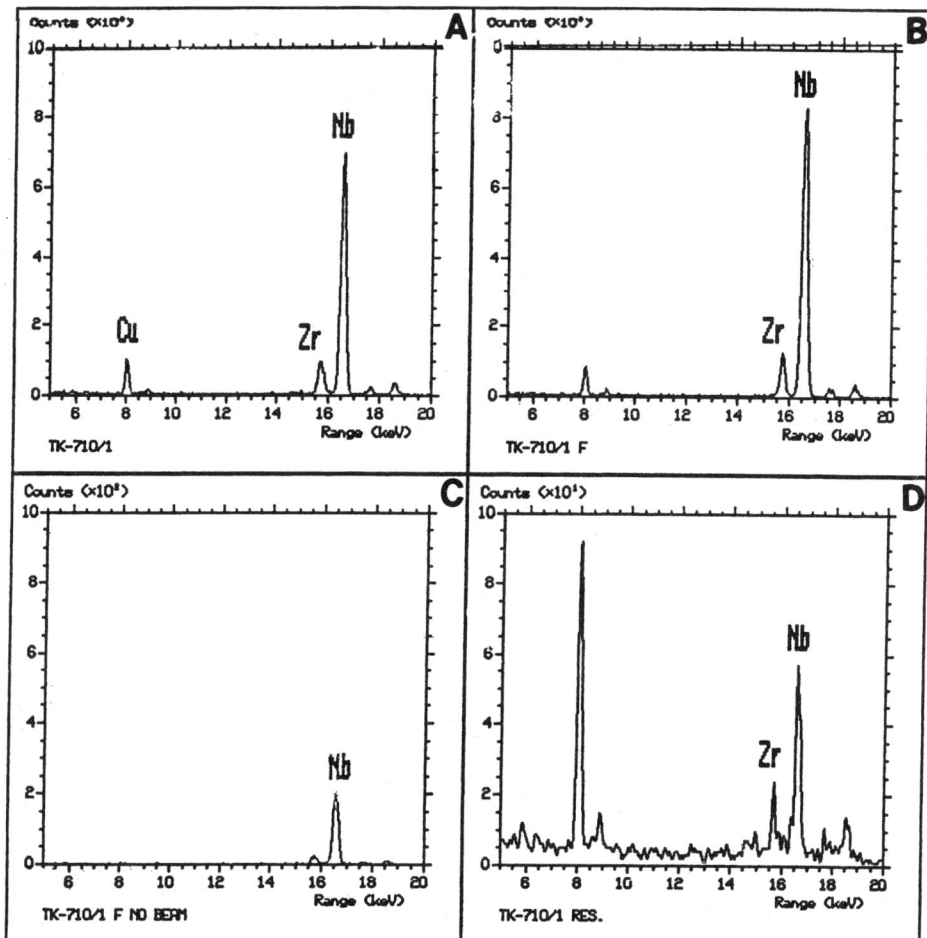

FIG. 8—*X-ray microanalysis spectra of β-Nb precipitates in irradiated Zr-2.5Nb alloy (A) with due account for contribution made by scattered radiation (B), induced activity of "niobium" isotopes (C), (D) processed.*

fluence and temperature and is reduced with an increase of the supersaturation of the α-solid solution. The $\bar{c}$-dislocation densities determined by the TEM method under conditions of the acting reflection $\bar{g} = 0002$ and the X-ray diffraction analysis are in a good agreement and do not exceed 1×10^{14} m^{-2} (Figs. 9C,D). Under the above diffraction conditions, non-basal dislocations with the $\bar{c}$-component were also revealed climbing to the basal planes. The pile-up of the $\bar{c}$-dislocations was detected near the grain boundaries and precipitate aggregates (Fig. 9C). This can be the result of stresses produced by an anisotropic growth of adjacent grains or the growth of precipitates.

Discussion of Results

From the comparison between the irradiation-induced growth (Fig. 10) versus microstructure relationships for binary and multicomponent zirconium alloys, it follows that the pre-transition

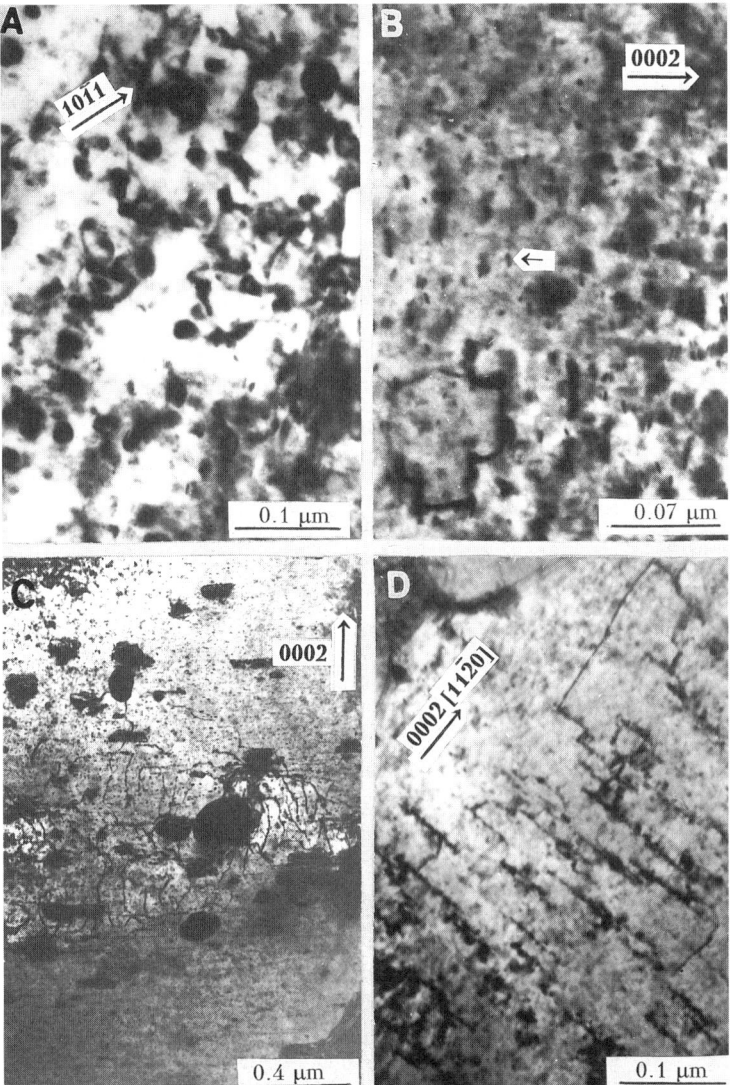

FIG. 9—*Dislocation structure of irradiated Zr-2.5Nb:* (A) $\bar{a}$-*dislocation loops,* $\bar{g} = 10\bar{1}1$; (B) *structure defects at 350°C under stresses likely to be basal dislocation loops;* (C) *image of* $\bar{c}$-*dislocations on basal and non-basal planes,* $\bar{g} = 0002$; (D) $\bar{c}$-*dislocations revealed at* $\bar{g} = 0002$, *basal loops (irradiation at 305°C).*

growth is related to interstitial and vacancy dislocation loop ($\bar{b} = \frac{1}{3} \langle 11\bar{2}0 \rangle$) nucleation and evolution in prism planes. The accelerated growth stage following high fluences is related to the formation of vacancy dislocation loops with $\bar{b} = \frac{1}{6} \langle 20\bar{2}3 \rangle$ in basal planes and an increase in the $\bar{c}$-dislocation density [7]. If one takes into account the tube texture, it becomes clear that this results in an axial elongation and a tangential reduction. In this work, the alloys were studied that did not experience an accelerated growth ($\Phi < 1 \times 10^{26}$ n/m^2); therefore, it was only feasible to assess the tendency toward a dislocation structure evolution. The dislocation

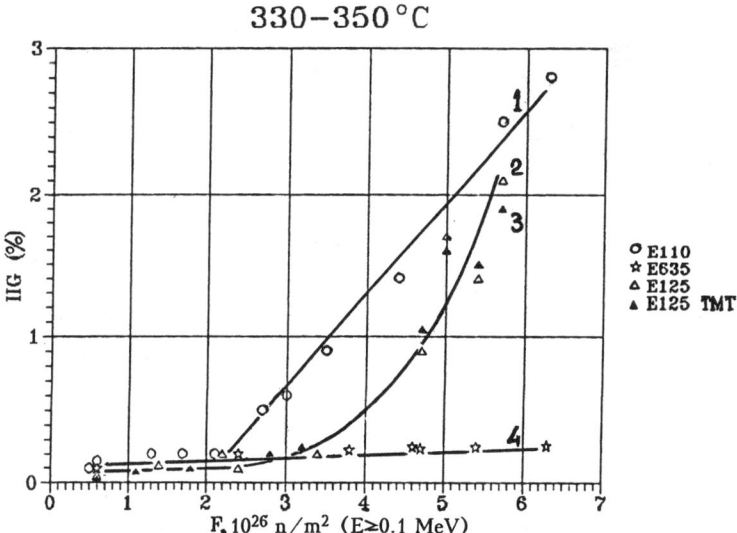

FIG. 10—*Irradiation-induced growth versus neutron fluence for different alloys: 1 = E110; 2 = E125; 3 = E125(TMT); 4 = E635.*

structure and phase composition that are substantially different in unirradiated E110 and E125 alloys were similar in many respects after irradiation. As a result of long-term irradiation and thermal effects, the β-Zr phase decomposed to form a disperse β-Nb phase, the alloy being transformed to a more equilibrium state. The initial fine β-Nb precipitates became coarser, and their volume fraction increased.

The analysis of the results shows that the recrystallization grade of the alloy is much dependent on the irradiation temperature and composition of the alloy. The E125 alloy irradiated at $T = 350°C$ was, e.g., fully recrystallized, while no recrystallization is observed upon irradiation at $T = 305$ to $315°C$ even to higher fluences. Under similar conditions, the E635 alloy was fully recrystallized. The temperature of the multicomponent alloy recrystallization is slightly lower than that of the E125 alloy; moreover, the irradiation-stimulated increase of a diffusion mobility of atoms, particularly iron ones, accelerates recrystallization processes. The results of the dislocation structure studies were assessed based on the influence exerted by the composition and heat treatment of the alloys on the irradiation-effected evolution of the $\bar{c}$-component dislocations. An attempt was made to relate the irradiation resistance of secondary phase precipitates and a solid solution composition to their influence on the dislocation structure evolution and, hence, on the irradiation-induced growth.

The specific feature of Zr-Nb alloys as irradiated under conditions of stresses is defects appearing in basal planes at the acting reflections of the type $\bar{g} = 0002$. They are aligned parallel to (0001) or at an angle of $<20°C$ and have a dislocation loop contrast in "profile" when imaged edge on with a 0002 diffraction vector. Similar features observed in the Zr-2.5Nb alloy structure [8] were identified to be niobium-enriched precipitates. As distinct from Ref 8, in this work the solid solution is not supersaturated since niobium for the most part precipitated from the solution as β-Nb particles. It is counted that precipitates in the E110 and E125 alloys contain about 80% of the total niobium content of the alloys. In the previous work [9] on the E110 alloy also irradiated under conditions of strain, it was shown that until the fluence of 1

$\times 10^{25}$ n/m^2 ($E > 0.1$ MeV) is attained, only mixed dislocation loops of the $\bar{a}$-type were formed, while at higher fluences in the temperature range of 290 to 400°C, $\bar{c}$-type loop lattices with $b = \frac{1}{6}\langle 20\bar{2}3\rangle$ were observed, 70% being vacancy loops of 10 to 15 nm. However, we failed to observe this type of defect in the EZ-1 and E635 alloys irradiated under the same conditions.

In all the alloys studied, $\bar{c}$-type dislocations were observed on basal and non-basal planes. Their density increases with the fluence and depends on the initial density. Of all the recrystallized alloys investigated, the highest density of $\bar{c}$-dislocations was observed in E125 and EZ-1 alloys (0.5 to 2.0 $\times$ 10^{14} m^{-2}); the lowest $\bar{c}$-dislocation density was demonstrated by the E635 alloy ($\sim$0.1 $\times$ 10^{14} m^{-2}). Many of the $\bar{c}$-dislocations are dislocations of the initial structure that climbed onto basal planes. The latter process depends on the $\bar{c}$-type dislocation loop nucleation and is much facilitated if $\bar{c}$-dislocations are available in the unirradiated alloy.

The influence exerted by the alloying elements in solution on the defect structure evolution under irradiation is far from clear. It is believed [10] that as a result of an anisotropic diffusion in α-Zr, sinks for vacancies and interstitials are separated on basal and prism planes, respectively, to form dislocation loops under specific conditions. In the investigated alloys, niobium, tin, and iron are in solid solution and can serve as traps for vacancies, thus reducing their flow to sinks in the form of $\bar{c}$-dislocations or grain boundaries specifically oriented. This is likely to be the cause of the low irradiation-induced growth experienced by the E635 alloy. The experimental data attest to the small number of $\bar{c}$-dislocations in the E635 alloy both fully and partially recrystallized. The joint alloying with iron, tin, and niobium substantially lowers the alloy susceptibility for irradiation-induced growth at high fluences, at least, up to $\sim$9 $\times$ 10^{26} n/m^2 [11]. Iron atoms in zirconium have much more mobility than zirconium self-diffusion at $T = 250$ to 400°C; therefore, while moving together with vacancies, they are capable of enhancing the irradiation-induced growth [19]. However, atoms of niobium and tin in a solid solution act as vacancy traps and can offset the effect of iron. This is likely to depend on the proportions of iron, tin, and niobium in solid solution and the irradiation resistance of secondary phase precipitates. The maximum solubility of iron in α-zirconium is 0.012% and that of niobium is 0.6% [6]; therefore, they precipitate from α-Zr as intermetallic particles having a complex composition. Intermetallic precipitates were also observed in zirconium alloys containing niobium and iron [12–14]. The authors failed to fully identify their structure and assess their stability under neutron irradiation.

The role of the irradiation stability of precipitates in the evolution of $\bar{c}$-dislocations was adequately discussed [4,15,16]. Amorphization, dissolution, and changes in the composition of Zr(Fe,Cr)$_2$ and Zr$_2$(Fe,Ni) particles are observed in Zircaloy-type alloys [20]. The irradiation effected an increase of the iron concentration in α-Zr and hence an enhancement of the mobility of iron atom-vacancy complexes and a nucleation of basal $\bar{c}$-vacancy loops. Amorphization of Zr-Fe or Zr-Nb-Fe containing particles was not essentially detected in E635 or EZ-1 alloys. The more so, it was not observed in binary Zr-Nb alloys. However, from particle-size-distribution histograms, we found out that the sizes of both β-Nb and Zr-Nb-Fe precipitates become larger. So far it is difficult to unambiguously state what this is related to since from the theory of phase stability under irradiation it follows that a dissolution of fine particles, a growth of coarse ones, as well as a growth of particles as a result of coagulation under the action of irradiation and thermal processes are feasible [17]. An increase of the particle sizes due to the dissolution of fine ones was also noted in Zircaloys protons irradiated at 350°C [18]. Nonetheless, under irradiation, the composition of Zr-Nb-Fe precipitates is changed to reduce the amount of iron, its content in α-Zr being increased. But the content of iron in the solid solution is likely to be too low compared to niobium or tin to bring about an accelerated irradiation-induced growth. Changes in the composition of Zr-Nb-Fe containing precipitates to form iron-depleted phases of β-Nb and (Zr,Nb)$_{3-4}$Fe types have been observed already in the multicomponent alloy irradiated to a neutron fluence of 4.1×10^{26} n/m² [1]. Irradiation to moderate fluences ($\sim 5 \times 10^{25}$ n/m²) does not produce prominent changes in the phase composition, which can be due to a wide range of a ternary intermetallic compound existence.

Conclusion

The previously reported [1] data on the neutron irradiation-induced evolution of the multicomponent alloy structure are corroborated. The absence of substantial amounts of $\bar{c}$-dislocations and dislocation loops of the $\bar{c}$-type from the alloy structure before and after irradiation is likely to be the main factor that imparts the high resistance to irradiation-induced creep and growth to the alloy [21]. The radiation-thermal anneal under reactor conditions readily recrystallizes the alloy; the relatively high irradiation stability of $(Zr, Nb)_3Fe$ and/or $Zr(Nb, Fe)_2$ precipitates can make an extra contribution to provide the high resistance of the alloy to dimensional changes. In the structures of the E110 and EZ-1 (that can be considered constituents of the E635 alloy) irradiated to a fluence of $\leq 5 \times 10^{25}$ n/m² ($E \geq 0.1$ MeV) at 350°C, the density of $\bar{c}$-type dislocations is substantially higher and reaches 2×10^{14} m^{-2}. The $\bar{c}$-dislocation density after irradiation of E125 alloy is similar or less in comparison to those of the E110 and EZ-1 alloys. The initial structure of E125 tubes and the time of reactor operation during up to 15.5 years little affect the density of $\bar{c}$-dislocations. All the materials studied did not show any significant evidence of amorphization of secondary phase particles.

Acknowledgments

The authors would like to thank V. L. Panchenko, V. E. Kalachikov, Yu. G. Kostyuk, and V. M. Zyablov for their technical assistance in obtaining TEM, EDX, and XRD measurements used in this paper, and V. B. Kozlyaeva, E. G. Afonina, and O. A. Simonova for translation, typing, and layout.

References

[1] Nikulina, A. V., Markelov, V. A., Peregud, M. M., Voevodin, V. N., Panchenko, V. L., and Kobylyansky, G. P., "Irradiation-Induced Microstructural Changes in Zr-1 Sn-1 Nb-0.4 Fe," *Proceedings, Third International Conference on the Evolution of Microstructure in Metals During Irradiation, Clevelands House, Muskoka, Ontario, Canada, 28 Sept. 2 Oct. 1992*, to be published in *Journal of Nuclear Materials*.

[2] Buckley, S. N., "Properties of Reactor Materials and the Effects of Radiation Damage," Butterworths, London, 1964, p. 443.

[3] Griffiths, M., "A Review of Microstructure Evolution in Zirconium Alloys during Irradiation," *Journal of Nuclear Materials*, Vol. 159, 1988, p. 190.

[4] Griffiths, M., "Evolution of Microstructure in hcp Metals During Irradiation," *Journal of Nuclear Materials*, Vol. 205, 1993, p. 225.

[5] Griffiths, M., Winegar, I. E., Mecke, I. F., and Holt, R. A., "Determination of Dislocation Densities in Hexagonal Close-Packed Metals Using X-Ray Diffraction and Transmission Electron Microscopy," *Advances in X-Ray Analysis*, Vol. 35, Plenum Press, New York, 1992, p. 593.

[6] Aubertin, F., Gonser, U., Campbell, S. I., and Wagner, H.-G., "An Appraisal of the Phases of the Zirconium-Iron System" *Zietschrift Metallkunde*, Vol. 76, No. 4, 1985, p. 237.

[7] Holt, R. A. and Gilbert, R. W., *Journal of Nuclear Materials*, Vol. 137, 1983, p. 127.

[8] Coleman, C. E., Gilbert, R. W., Carpenter, G. I. C., and Weatherly, G. C., "Precipitation in Zr-2.5 Nb During Neutron Irradiation," *Phase Stability Under Irradiation*, Proceedings of the Fall Meeting of AIME, Pittsburgh, PA, October 1980.

[9] Gurovich, B. A., Astrahantsev, M. S., Platonov, P. A., and Yelesin, L. A., *Fizika Metallov i Metallovedenie*, Vol. 61, 1986, p. 922.

[10] Holt, R. A., Woo, C. H., and Chow, C. K., "Production Bias—A Potential Driving Force for Irradiation Growth," *Journal of Nuclear Materials*, Vol. 205, 1993, p. 293.

[11] Nikulina, A. V., Markelov, V. A., et al., "Zirconium Alloy E635 as a Material for Fuel Rod Cladding and Other Components of VVER and RBMK Cores," presented at this conference.
[12] Northwood, D. O., Meng-Burany, X., and Warr, B. D., "Microstructure of Zr-2.5 Nb Alloy Pressure Tubing," *Zirconium in the Nuclear Industry: Ninth International Symposium, ASTM STP 1132,* American Society for Testing and Materials, West Conshohocken, PA, 1991, p. 156.
[13] Kruger, R. M. and Adamson, R. B., "Precipitate Behaviour in Zirconium-Based Alloys in BWRs," *Journal of Nuclear Materials,* Vol. 205, 1993, p. 242.
[14] Sabol, G. P., Schoenberger, G., and Balfour, M. G., "Improved PWR Fuel Cladding," *Proceedings,* IAEA Technical Committee Meeting on Materials for Advanced Water Cooled Reactors, IAEA, Vienna, IAEA-TECDOC-665, 1992, p. 122.
[15] Griffiths, M., "HVEM Study of the Effects of Alloying Elements and Impurities on Radiation Damage in Zr-Alloys," *Journal of Nuclear Materials,* Vol. 205, 1993, p. 273.
[16] Griffiths, M., Gilbert, R. W., and Carpenter, G. I. C., "Phase Instability Decomposition and Redistribution of Intermetallic Precipitates in Zircaloy-2 and -4 during Neutron Irradiation," *Journal of Nuclear Materials,* Vol. 150, 1987, p. 53.
[17] Russel, K., "The Theory of Phase Stability under Irradiation," *Journal of Nuclear Materials,* Vol. 83, 1979, p. 176.
[18] Lee, Y. S., Huang, K. Y., Huang, C. Y., Kai, I. I., and Hsieh, W. F., "Effects of Proton Irradiation on the Microstructural Evolution and Uniform Corrosion Resistance of Zircaloys," *Journal of Nuclear Materials,* Vol. 205, 1993, p. 476.
[19] Hood, G. M., et. al., *Journal of Nuclear Materials,* Vol. 185, 1991, p. 174.
[20] Motta, A. T., Lefebvre F., and Lemaignan, C., "Amorphization of Precipitates in Zircaloy Under Neutron and Charged Particle Irradiation," *Zirconium in the Nuclear Industry: Ninth International Symposium, ASTM STP 1132,* American Society for Testing and Materials, West Conshohocken, PA, 1991, pp. 718–739.
[21] Kobylyansky, G. P., Shamardin, V. K., Ostrovsky, Z. E., Raetsky, V. M., Nikulina, A. V., Peregud, M. M., and Grigiryev, V. M., "Radiation from Change of Zr Alloy Cladding and Channel Tubes at High Neutron Fluences," *Proceedings,* International Conference on Radiation Materials Science, Alushta, 22–25 May 1990, Vol. 4, Kharkov, Russia, 1990, pp. 64–72.

DISCUSSION

A. T. Motta[1] (written discussion)—Could the fact that you observe amorphization of $(Zr,Nb)_2$ Fe Laves phase only infrequently be related to the relatively high temperature in your experiments? It seems that there is a difference in critical temperature for amorphization under ion and neutron irradiation for all compounds studied except for $Zr(Cr,Fe)_2$. The critical temperatures under ion irradiation of Zr_3Fe, $Zr_2(Ni,Fe)$ and $Zr(Cr,Fe)_2$ are all approximately 600 to 650 K. Under neutron irradiation, only $Zr(Cr,Fe)_2$ amorphousizes near that temperature. Have you any comment on this difference?

V. N. Shishov et al. (authors' closure)—Yes, we observe some Fe depletion of particles, but without essential amorphization. Possibly this is related to the high temperatures in our experiments (350°C) and to moderate fluence (less than 10^{26} n/m²). It seems also that $Zr(Nb, Fe)_2$ precipitates have higher irradiation resistance than $Zr(Cr, Fe)_2$ in this temperature range.

H. M. Chung[2] (written discussion)—Number density of $\langle c \rangle$-type dislocation loops was quantified by dark-field imaging of (0002) reflections of the Zr matrix. This method may produce misleadingly higher density because reflections of several oxides and hydrides (some artifacts, some real precipitates contained in bulk) are usually superposed on (0002) Zr reflections.

V. N. Shishov et al. (authors' closure)—$\langle c \rangle$-component dislocations and loops density were estimated not only by dark-field imaging of (0002) reflections of Zr matrix, but, in the main, by bright-field imaging with the diffracting vector **g** = 0002, showing $\langle c \rangle$-type contrast. Oxides and hydrides were excluded by defect examination after specimen tilting.

---

[1] Pennsylvania State University, University Park, PA.
[2] Argonne National Laboratory, Argonne, IL.

R. A. Holt,[1] A. R. Causey,[1] N. Christodoulou,[1] M. Griffiths,[1] E. T. C. Ho,[2] and C. H. Woo[3]

Non-Linear Irradiation Growth of Cold-Worked Zircaloy-2

REFERENCE: Holt, R. A., Causey, A. R., Christodoulon, N., Griffiths, M., Ho, E. T. C. and Woo, C. H., "Non-Linear Irradiation Growth of Cold-Worked Zircaloy-2," *Zirconium in the Nuclear Industry: Eleventh International Symposium, ASTM STP 1295*, E. R. Bradley and G. P. Sabol, Eds., American Society for Testing and Materials, 1996, pp. 623–637.

ABSTRACT: Accelerating irradiation growth has been reported for several zirconium alloys with a range of metallurgical states during high-temperature tests in fast-breeder reactors (673 to 723 K) for annealed Zircaloys in thermal test reactors at power reactor temperatures (523 to 623 K) and in power reactor core components fabricated from annealed or recrystallized Zircaloy. In the latter case, there was a transition from low to high irradiation growth rates at moderate fluences (about 3×10^{25} n/m^2, $E > 1$ MeV, at 580 K) related to the nucleation and growth of basal plane c-component loops.

It was recently reported that the elongation rate of cold-worked Zircaloy-2 pressure tubes also accelerates with increased fluence at about 550 K. This gradual acceleration coincides with a gradual increase in the density of c-component dislocations as a result of the helical climb of those dislocation segments with a predominantly screw character.

New data showing the acceleration of growth of cold-worked Zircaloy-2 at 550 K are presented, and a model is presented for the irradiation growth behavior of cold-worked Zircaloy-2 based on the evolution of the sink strengths of c-component dislocations (net vacancy sinks) and a-type dislocations (net interstitial sinks) as a function of fast fluence. According to the model, the growth rate reaches a maximum when the sink strength of c-component dislocations is approximately equal to that of the a-type dislocations, and hence the growth rate tends to saturate with increasing fluence.

KEYWORDS: zirconium alloys, pressure tubes, irradiation growth, Zircaloy, pressure tubes, growth breakaway

The general characteristics of irradiation growth of annealed Zircaloy-2 and -4 are well documented. In the early stages of growth, the rate of growth is rapid, but then tends to saturate at strains of about 0.1% or less [1,2]. In the early stages, growth is generally positive in directions containing a small fraction of basal plane normals and negative in directions containing a large fraction of basal plane normals [2,3], but no clear relationship to crystallographic texture has been established.

After some threshold fast neutron fluence, the growth rate tends to accelerate [4–6]. The fluence at the onset of acceleration decreases with increasing temperature [7,8] and, at high temperatures (>650 K), the threshold fluence may be negligible. It is not clear that acceleration occurs at low temperatures (330 to 350 K). Once acceleration occurs, the growth roughly correlates with the quantity $(1 - 3F)$, where F is the resolved fraction of basal plane normals in the direction of measurement [7,9]. This ideal relationship corresponds to simple averaging

---

[1] Atomic Energy of Canada Ltd., Chalk River Laboratories, Chalk River, Ontario, Canada.
[2] Ontario Hydro Technologies, 800 Kipling Avenue, Toronto, Canada.
[3] Atomic Energy of Canada Ltd., Whiteshell Laboratories, Pinawa, Manitoba, Canada.

of the shape change of all the grains in a polycrystal when there is expansion along the **a**-axis and contraction along the **c**-axis, but is modified by grain-to-grain interaction stresses [10] or if the grain shape is anisotropic [11].

During the transient stage of growth of annealed Zircaloy, the radiation-induced microstructure comprises prismatic loops on prism planes with the Burgers vector, $\mathbf{b} = \frac{1}{3} <11\bar{2}0>$ [12]. The loop character is mixed, some being vacancy and some interstitial. The acceleration coincides with the appearance in the microstructure of **c**-component defects lying on the basal planes [13]. These were shown to be faulted vacancy loops with $\mathbf{b} = \frac{1}{6} <20\bar{2}3>$ [14], and their appearance is generally accompanied by the radiation-induced redistribution of β stabilizing elements (Fe,Cr,Ni) [15]. The observation of **c**-component defects accompanying the acceleration has subsequently been confirmed for β- as well as α-annealed Zircaloy [6,16].

The very early stage of irradiation growth in cold-worked and stress-relieved Zircaloy-2 and -4 is similar to that of annealed material, but instead of tending to saturate, the growth continues at a steady rate [2,8] to fairly high fluences. The $(1 - 3F)$ relationship is approximately obeyed in the steady state growth regime [17,18]. In materials cold-worked without stress relief, the early stages of growth are dominated, especially at low temperatures, by the relaxation, due to anisotropic creep, of residual intergranular constraints [10,19].

We have recently reported that the long-term growth rate of cold-worked Zircaloy-2 pressure tube material accelerates gradually with fast fluence, as did the elongation rate of the pressure tubes in-service in the steam-generating heavy water reactor (SGHWR) at Winfrith (UK) [20]. The acceleration of the elongation rate of the pressure tubes corresponds to an increase in the density of **c**-component dislocations with fast fluence, mainly by helical climb of existing dislocation segments (from cold work) of predominantly screw character threading the basal plane.

The acceleration of growth of cold-worked and stress-relieved Zircaloy is not a universal observation. Garzarolli et al. [6] indicate a tendency for the growth rate of some lots of Zircaloy-4 fuel cladding to decrease slightly with fluence up to 7×10^{25} n/m^2, $E > 0.82$ MeV, at 573 K.

In this paper, we present further irradiation growth data illustrating the acceleration of irradiation growth of Zircaloy-2 pressure tube material and new microstructural observations. We discuss the relationship of the acceleration in growth to the acceleration in the pressure tube elongation, and we present a theoretical model relating the acceleration to the increased density of **c**-component dislocations.

Experimental

A total of seven pressure tubes manufactured for the Winfrith SGHWR by Accles and Pollock were employed in this study. The "Winfrith" tubes were manufactured by hot-extrusion, followed by cold drawing 15% with an intermediate anneal and further 30% cold drawing. The average grain size was ~11 by 14 by 28 μm in the radial, circumferential, and axial directions of the tube, respectively [21]. One tube of a second type was employed, manufactured by Chase and used by Murgatroyd and Rogerson [3,8,20,22] for irradiation growth experiments in DIDO. The "Chase" tube was fabricated by hot extrusion followed by cold drawing 25% and was reported to have fine equiaxed grains 5 to 8 μm in diameter [22].

Two specimens 1.5 by 38.1 by 6.4 mm were manufactured with their long axis in the axial direction from the Chase Tube 002. The specimens were irradiated in the OSIRIS reactor at Saclay, France in a flux of $\sim 1.8 \times 10^{18}$ n/m^2/s, $E > 1$ MeV to fluences of 9.8 and 10.8 × 10^{25} n/m^2, $E > 1$ MeV. The experimental details were described earlier [23]. The nominal irradiation temperature was 553 K, while actual irradiation temperatures for the two specimens, GVU and GVR, were 550 and 561 K, respectively.

The specimens were removed from the reactor five times, and length measurements were made in a hot cell using a computer-controlled comparator with two linearly variable differential transformers with an estimated accuracy of ±0.5 μm. The fast fluences for the first four measurement campaigns were determined by the activation of Cu wire monitors included in the insert. The fluence for the last campaign was calculated from the reactor operating history.

X-ray line broadening, texture, and transmission electron microscopy (TEM) data were obtained from:

1. Three positions in the Winfrith Tube 057 removed from service in Winfrith SGHWR Channel N13.
2. Unirradiated offcuts of four Winfrith tubes (no unirradiated material from 057 was available).
3. Unirradiated offcuts of two Winfrith tubes for which elongation data were reported previously, 009 and 113 (Channels N11 and N17).
4. An unirradiated section of Chase Tube 002.

Because the coolant in SGHWR is boiling over most of the length of the pressure tube, the operating temperatures at all three locations in Tube 057 were similar (estimated mid-wall temperatures of 558 to 565 K) while the fast fluence varied.

Complete details of the experimental procedures were given previously [20,24]. X-ray diffraction specimens about 10 by 10 by 0.5 mm were prepared by cutting slices from the tube perpendicular to each of the three principle axes with a low-speed diamond wheel. The specimens were chemically polished to remove at least 0.025 mm, and the shapes of the second and third order $\{10\bar{1}0\}$ and first and second order (0002) lines were measured in a Rigaku diffractometer with a Rotoflex rotating anode generator. Shapes of the first and second order $\{11\bar{2}0\}$ lines were measured using a Siemens Type F Diffractometer and a Crystalloflex 4 generator. Basal pole figures were determined using the three orthogonal specimens and a Siemens texture goniometer with the Crystalloflex 4 generator. Cu K_α radiation was used in all cases. Dislocation densities were estimated from the line broadening data [24,25].

Specimens for TEM were prepared by punching 3-mm-diameter disks from 0.1-mm- thick slices and jet electropolishing with a solution of 10% perchloric acid in methonol at about 230 K. TEM analysis was performed using a Philips CM 30 (300 kV).

Results

The irradiation growth data for Chase Tube 002 are shown in Fig. 1. There is an initial transient strain; then the rate remains approximately constant until about 4×10^{25} n/m^2, $E >$ 1 MeV, then accelerates. The rate of acceleration is slightly larger for Specimen GVU than for Specimen GVR. Both the rate of growth and the rate of acceleration are larger than those found by Rogerson (included in the diagram[4]). The discrepancy in the rate of growth noted previously [23] is consistent with results for cold-worked Zr-2.5Nb pressure tube material irradiated in both reactors and appears to represent a real systematic difference between DIDO and OSIRIS. We find the data from OSIRIS to be consistent with low fluence results from NRU and from specimens irradiated in CANDU power reactors; however, the results from DIDO are reasonably consistant with the other reactors only if the 24% increase in the DIDO fluences[4] is not applied [27,28].

---

[4] The growth data originally reported by Rogerson [8] for tests in DIDO were based on fast fluences determined using iron wire activation monitors assuming a cross section based on a fission flux spectrum. Subsequent analysis by the AEA Technologies staff [26] using the actual DIDO flux spectrum resulted in an increase in the estimated fluences of 24%. The increased fluences are used here as in two previous reports [20,23].

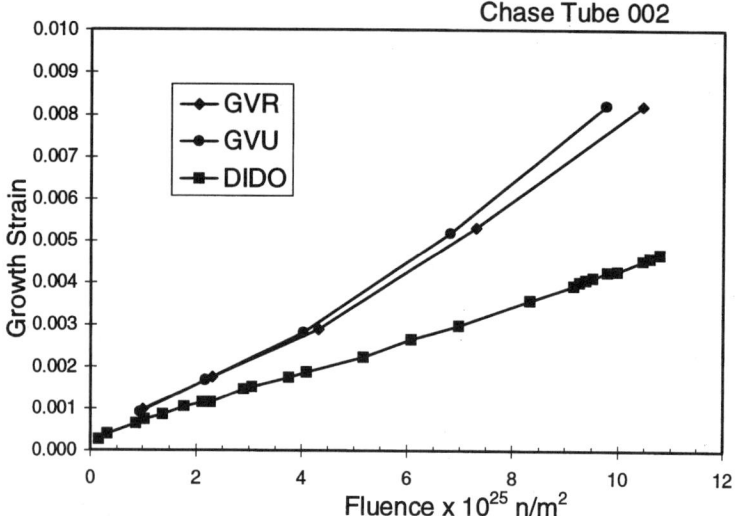

FIG. 1—*Irradiation growth of Chase Tube 002 irradiated in the OSIRIS (GVR and GVU) and DIDO reactors.*

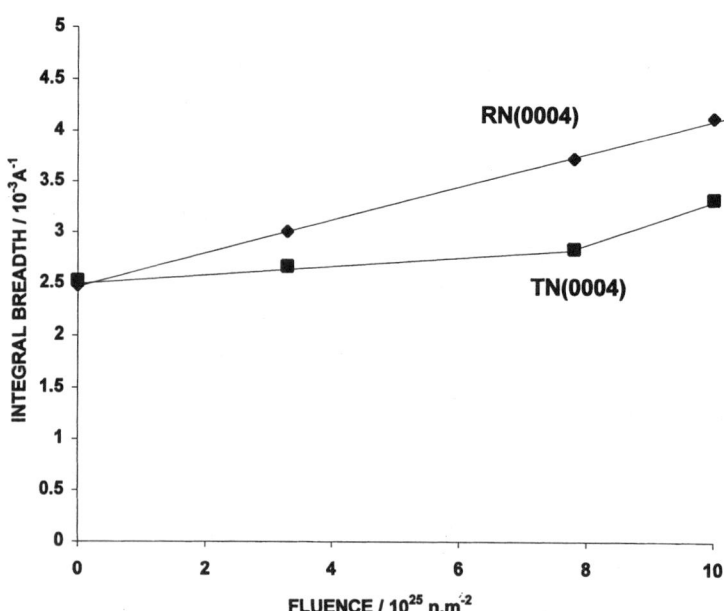

FIG. 2—*Basal plane line broadening as a function of fluence for Winfrith Tube 057.*

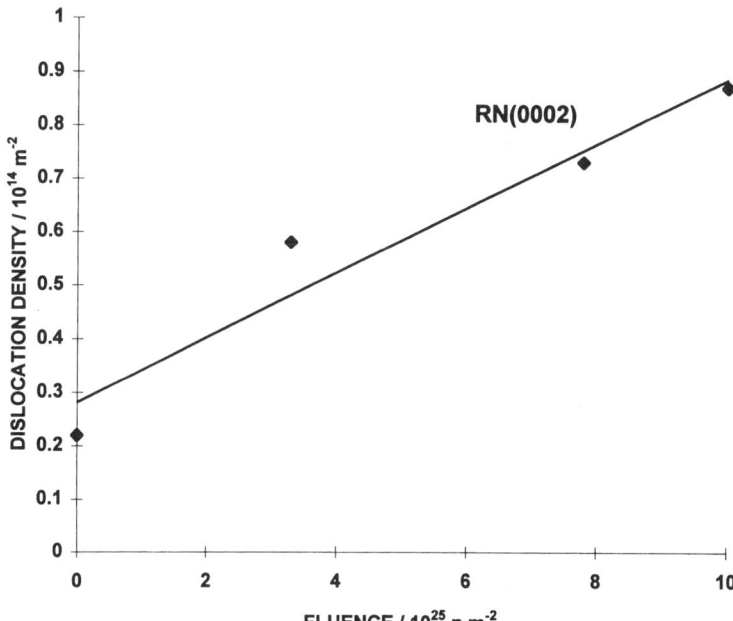

FIG. 3—*Calculated c-component dislocation density as a function of fast neutron fluence for Winfrith Tube 057.*

The X-ray line broadening data for the irradiated SGHWR pressure tube 057 (Channel N13) are given in Figs. 2 to 4. Figure 2 shows the integral breadth of the (0004) reflection for crystals with their **c**-axes in the radial direction and in the transverse direction of the tube (there are very few grains with their **c**-axes in the axial direction of the tube, Table 1). The breadth of the X-ray line, corresponding to an increase in the density of dislocations with a **c**-component Burgers vector, increases continuously with fast fluence for grains with radial **c**-axes. The breadth for grains with transverse **c**-axes increases slowly up to $\sim 8 \times 10^{25}$ n/m², $E > 1$ MeV, then increases more rapidly.

Figure 3 shows the increase in dislocation density with fluence for grains with radial **c**-axes calculated from the shapes of the (0002) and (0004) lines, assuming that all the line broadening is from the strain fields of dislocations and that the original dislocations have a Burgers vector of $\frac{1}{3}<11\bar{2}3>$, while the increase generated by radiation is from dislocations with a Burgers vector of $\frac{1}{6}<20\bar{2}3>$. The intensity of the (0002) lines is not sufficient to estimate the dislocation densities for grains with transverse **c**-axes.

Figure 4 shows the increase in line breadth of the $\{30\bar{3}0\}$ peaks with fluence for grains with $<10\bar{1}0>$ in the axial direction of the tube—a large majority of grains—and the density of dislocations with an **a**-type Burgers vector calculated from the shapes of the $\{20\bar{2}0\}$ and $\{30\bar{3}0\}$ peaks for these grains.

Figure 5 shows electron micrographs of unirradiated offcuts of Winfrith tubes showing a large grain-to-grain variation in the character of the **c**-component dislocation substructure. The density and orientation (screw or edge character) of the **c**-component dislocations varies considerably from grain to grain.

The basal pole figures for Winfrith pressure Tubes 009 and 113 are shown in Figs. 6a and

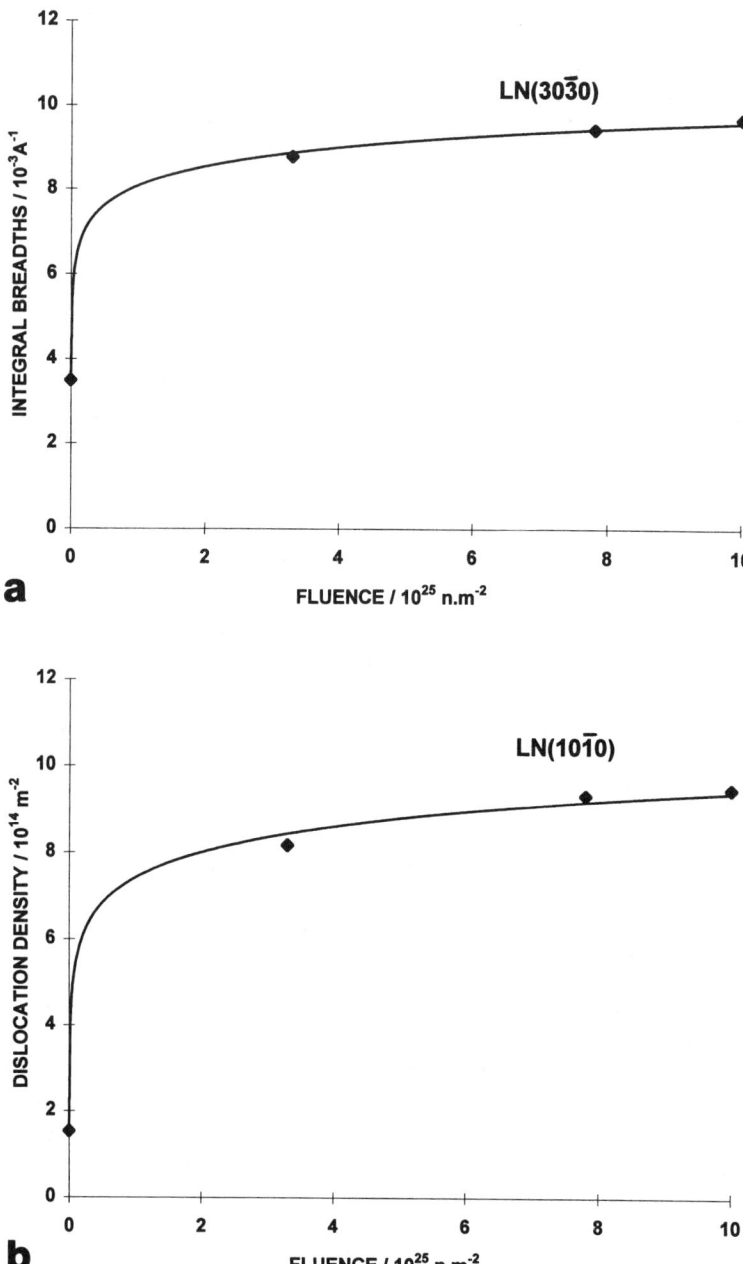

FIG. 4—(a) *Prism plane line broadening and* (b) *a-type dislocation density as a function of fast neutron fluence for Winfrith Tube 057.*

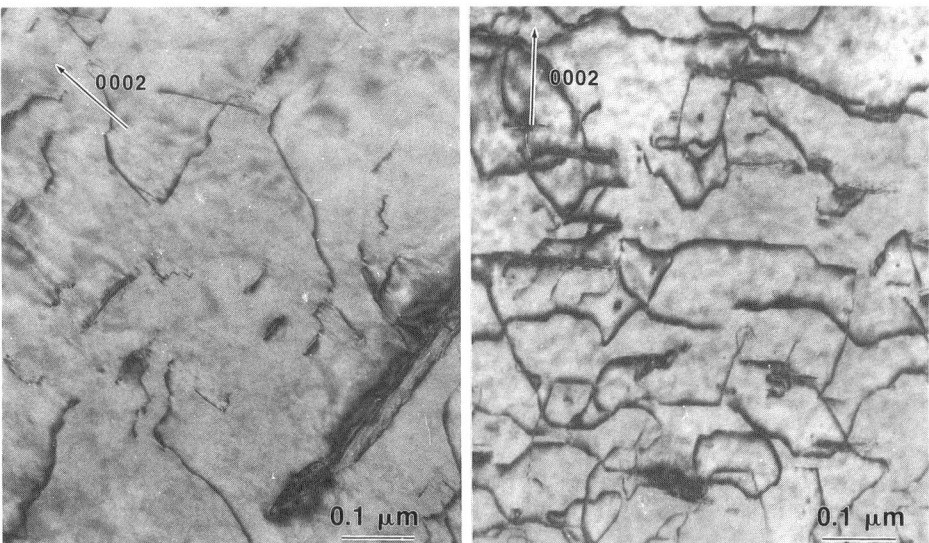

FIG. 5—*Micrographs showing variability in c-component network dislocation structure in a sample of unirradiated cold-worked Winfrith pressure tubing.*

6b and that for Tube 002 is shown in Fig. 6c. Table 1 shows the resolved fractions of basal plane normals, the integral breadths of the (0004) and {30$\bar{3}$0} lines, and the calculated dislocation densities for the unirradiated offcuts of the three tubes.

Table 2 gives the dislocation densities for Winfrith Tube 057 as a function of fluence for grams with their c-axes in the radial direction of the tube and for grains with <10$\bar{1}$0> along the tube axis. The following assumptions are made on the basis of TEM observations:

1. The initial c-component network dislocation structure consists entirely of dislocations with **b** = ⅓<11$\bar{2}$3>.
2. The c-component dislocations fromed during irradiation have **b** = ⅙<20$\bar{2}$3>.
3. The c-component dislocations above contribute to the **a** line broadening assuming that their **a**-components are ⅓<11$\bar{2}$0> and ⅓<10$\bar{1}$0>, respectively.

Discussion

The increase of the irradiation growth rate of the specimens irradiated in OSIRIS is much more pronounced than for the specimen irradiated in DIDO, approaching that of the elongation of pressure tubes in Winfrith. This is reflected in the parameters of quadratic curves fitted to the data (Table 3). [Note that a quadratic curve does not necessarily represent the true behavior, see below, but is chosen for simplicity to illustrate that the rate changes.]

Previous analysis of the deformation behavior of the Winfrith SGHWR pressure tubes [21] indicated that the elongation is largely due to irradiation growth, but with a significant negative contribution from irradiation creep (because the basal plane normals are predominantly near the radial direction). More recently, reactor creep tests at 550 K on small Zr-2.5Nb tubes with crystallographic textures similar to that of the Winfrith tubes [28] indicate that the negative axial creep component is about 10% of the diametral creep rate, or, in the case of the Winfrith

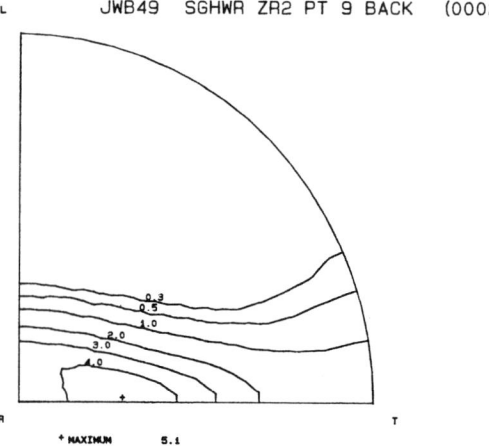

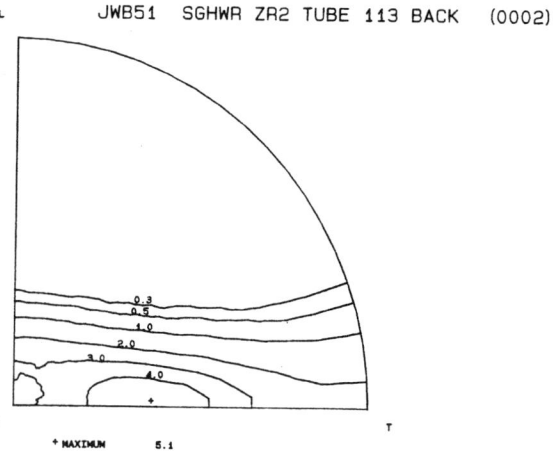

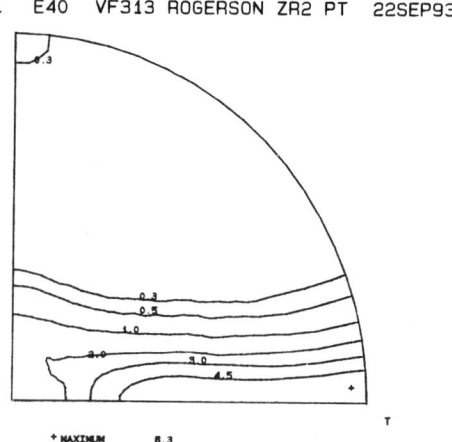

FIG. 6—*Basal pole figures for Winfrith Tubes:* (a) *009,* (b) *113, and* (c) *Chase Tube 002.*

TABLE 1—*Resolved fractions of basal pane normals, F, dislocation densities, ρ ($\times 10^{-4}$ m^{-2}) and integral line breadths, β ($\times 10^{-3}$ $Å^{-1}$), for three unirradiated Zircaloy-2 pressure tubes.*

Tube No.	F_R	F_T	F_L	$\rho_{\{10\bar{1}0\}}$	$\rho_{\{0002\}}$	$\beta_{\{30\bar{3}0\}}$	$\beta_{\{0004\}}$
009	0.61	0.32	0.07	1.37	0.23	3.53	2.32
113	0.58	0.35	0.07	1.48	0.30	3.57	2.38
002	0.45	0.49	0.06	1.98	0.28	4.13	2.67

tubes, ~1.1 × 10^{-29} m^2/n, $E > 1$ MeV. Thus, the initial elongation rates of the Winfrith pressure tubes can be interpreted as representing growth rates of ~5.2 and ~6.4 × 10^{-29} m^2/n, $E > 1$ MeV. These are reasonably consistent with the growth rates of the growth specimens in OSIRIS, especially considering that Tubes 009 (Channel N11) and 113 (Channel N17) had the highest elongation rates of any of the tubes in Winfrith and that they are metallurgically different from Tube 002 (see Table 1). Any tendency of the creep rate to saturate would result in an acceleration of the elongation rate, but the upper limit would be an increase of ~20 to 25% even if creep were to stop altogether. The quadratic fits to the elongation data indicate a much larger increases in elongation rate over the fluence range of the data, >100%.

Previous work [13–15,20] has shown that increases in the irradiation growth rate of zirconium alloys with increasing fast neutron fluence correlate clearly with the appearance and multiplication of **c**-component dislocations in the microstructure. In the case of materials like the Winfrith and Chase tubes, with a pre-existing **c**-component dislocation substructure, the multiplication occurs primarily by helical climb (in the form of loops spreading out on the basal plane) of dislocation segments of predominantly screw character threading the basal plane (Fig. 7). The contrast from the basal plane segments appears very similar to the faulted loops formed in annealed material during irradiation, and these segments were assumed to be faulted loops for the purposes of calculating dislocation densities from the line-broadening data. In the case of cold-worked material, there is no clear correlation with the presence of solutes.

The new X-ray line broadening data, presented in Fig. 2, show that the multiplication of the **c**-component dislocations depends on crystal orientation with respect to the tube axes. Grains with their **c**-axis in the radial direction experience a continuous increase in the breadth of the peaks with fast fluence. Grains with their **c**-axes in the transverse direction show only a gradual increase in the breadth of the (0004) X-ray peak up to 8 × 10^{25} n/m^2, $E > 1$ MeV, followed by a more rapid increase. A variation in the density of **c**-component dislocations with crystal orientation was previously observed by TEM for a cold-worked Zircaloy-2 pressure tube irradiated to 7.2 × 10^{25} n/m^2, $E > 1$ MeV, in Pickering-2 [20], but the relationship to the tube axes was not determined.

TABLE 2—*Dislocation densities ($\times 10^{14}$ m^{-2}) for Winfrith tube 057 after service in Channel N13 and average data for four unirradiated offcuts. The "ratio (c/a)" is the ratio of the "total c-component" dislocation density to the "actual a-type" dislocation density.*

Burgers Vector	Unirradiated	3.3 × 10^{25} n/m^2	7.8 × 10^{25} n/m^2	10.0 × 10^{25} n/m^2
⅓<11$\bar{2}$3>	0.25	0.25	0.25	0.25
⅙<20$\bar{2}$3>	0	1.36	1.92	2.48
Total **c**-component	0.25	1.61	2.16	2.73
Total **a**-component	1.51	8.1	9.3	9.5
Actual **a**-type	1.26	7.3	8.3	8.4
Ratio, c/a	0.19	0.22	0.26	0.32

TABLE 3—*Parameters of quadratic curves fitted to growth and elongation data (Fig. 1 and Ref 20).*

Designation	Reactor	Intercept	Initial Slope, $\times 10^{-29}$ (m^2/n)	Quadratic Term, $\times 10^{-55}$ (m^2/n)^2
L2D	DIDO	4.5×10^{-4}	3.2	0.7
GVR	OSIRIS	5.8×10^{-4}	4.2	2.9
GVU	OSIRIS	6.9×10^{-4}	4.1	3.9
N11	SGHWR	2.1×10^{-4}	4.1	5.4
N17	SGHWR	0.7×10^{-4}	5.3	3.7

These results suggest that nucleation of the loops on the screw segments is easier in the grains with radial c-axes than for grains with transverse c-axes. Possibly, stress influences the ease of nucleation, the tubes being internally pressurized with a tangential tensile stress of 83 MPa and an axial tensile stress of 42 MPa. This stress system favors the nucleation of basal plane vacancy defects as distinct loops on the basal plane or on dislocations (i.e., helical climb) in grains with radial c-axes and inhibits nucleation of basal plane vacancy loops on dislocations in grains with tangential c-axes. A different, fast-fluence-dependent factor appears necessary to influence the evolution of the c-component dislocation structure in the latter case. Possibly, the irradiation-induced dissolution of precipitates plays a role in enhancing basal plane vacancy loop formation in grains with transverse c-axes [20].

On the basis of the above observations, one might expect that the rate of acceleration would increase with the proportion of grains with their c-axes near the radial direction. However, one unresolved question is the reason for the difference in the growth rate between the specimens of Chase Tube 002 in DIDO and in OSIRIS. Even if the fast fluence scales are rationalized to make the growth curves coincide over most of their length (e.g., by reducing the DIDO fluences

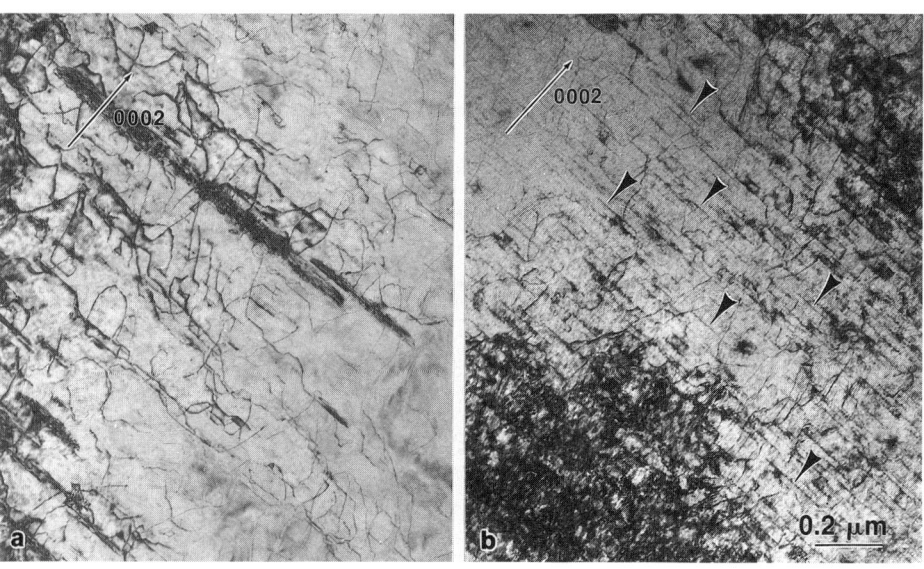

FIG. 7—*Comparison of c-type dislocation structures in Zircaloy-2 (a) before and (b) after irradiation to a fluence of 10.0×10^{25} n.m^{-2} at 553 K. Edge segments (arrowed) are produced during irradiation.*

by a factor of ~1.5), the acceleration rate in DIDO is still a factor of 3 to 4 lower than in OSIRIS.

Theory

Early theoretical models predicting pressure tube elongation [29,30] anticipated possible increases in the network dislocation density with fluence either due to loop growth or irradiation growth-induced plastic strain, but did not forcsee the role of **c**-component dislocations. Woo and Gösele [31] recognized the need for a new driving force for growth to explain the correlation of high growth rates with high **c**-defect densities [13]. The model, based on the diffusional anisotropy difference (DAD) between vacancies and self-interstitial atoms, was able to explain several peculiarities of the growth of zirconium alloys, including the increase in growth rate with **c**-component dislocation density in annealed Zircaloy [32,33].

Here we present a simplified model based on a rate theory formulation for DAD-driven growth of a single crystal to illustrate the effects of competition for point defects by two types of sinks (in this case **a**-type and **c**-component dislocations). A full accounting of the more extensive theoretical treatment is given in Refs 31 and 32.

Denoting the sink strength as $z\rho$, where ρ is the total dislocation density, the relative sink strengths are:

1. For the **a**-type dislocations: $z\rho_a$ for vacancies, and $z(1 + p)\rho_a$ for interstitials.
2. For the **c**-component dislocations: $z\rho_c$ for vacancies, and $z\rho_c$ for interstitials.

Here $z(1 + p)$ is the bias factor and is a function of $\{(D_c/D_a)_{int}/(D_c/D_a)_{vac}\}^{1/6}$, where $(D_c/D_a)_m$ is the ratio of the diffusion coefficients in the **c** and **a** directions for interstitials ($m =$ int) and for vacancies ($m =$ vac) [31].

For vacancies, the total sink strength is

$$k_v^2 = z(\rho_a + \rho_c) \tag{1}$$

and for interstitials

$$k_i^2 = z(\rho_a[1 + p] + \rho_c) \tag{2}$$

The fluxes of interstitials and vacancies to the sinks are:

$$D_i C_i = \frac{G}{z(\rho_a[1 + p] + \rho_c)} \tag{3}$$

and

$$D_v C_v = \frac{G}{z(\rho_a + \rho_c)} \tag{4}$$

where G is the production rate of point defects. The net flux of interstitials to **c**-dislocations is:

$$J = z\rho_c(D_i C_i - D_v C_v) = \frac{-pG\rho_a\rho_c}{(\rho_a + \rho_c)(\rho_a[1 + p] + \rho_c)} \tag{5}$$

or in terms of the ratio $\gamma = \rho_c/\rho_a$

$$-\frac{J}{G} = \frac{p\gamma}{\gamma^2 + (2 + p)\gamma + 1 + p} \qquad (6)$$

The ratio $-J/G$ in Eq 6 is proportional to the growth rate of the single crystal. The last expression is plotted in Fig. 8 as a function of γ for different values of p. The important points to note from this plot are:

1. The growth rate increases with γ for $\gamma < 1$. Note that for the materials and fluence range here, $\gamma < 1$, has values corresponding to the steeply rising part of the curves in Fig. 8 and increases with fluence (see Table 2). This is consistent with the observations of accelerating growth and elongation.
2. The growth rate is a maximum for $\gamma = 1$. This implies that, even if γ continues to increase indefinitely, the growth rate will eventually saturate and possibly even start to decrease. For materials with an initial value of γ near or greater than 1, the growth rate would be expected to decrease with increasing density of **c**-component dislocations.

Conclusions

1. The previously reported acceleration of growth of cold-worked and stress-relieved Zircaloy-2 pressure tube material [20] has been confirmed.
2. The new growth results are consistent with the observed acceleration of elongation rates of Zircaloy-2 pressure tubes in the SGHWR at Winfrith.
3. The acceleration coincides with an increasing density of **c**-component dislocations, and, based on X-ray line broadening measurements, an increase in γ, the ratio of the density of **c**-component dislocations to the density of **a**-type dislocations.
4. The increase in the **c**-component dislocation density is less for crystals with their **c**-axes in the transverse direction than for crystals with their **c**-axis in the radial direction.

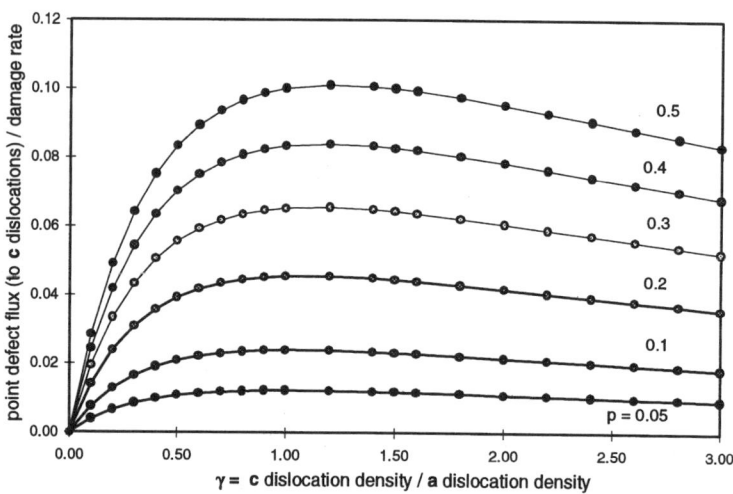

FIG. 8—*Dependence of the flux of vacancies to **c**-component dislocations on the ratio of dislocation densities, γ, for different values of the bias, p.*

5. A simple model for the growth rate as a function of the density of c-component and a-type dislocations shows that the growth rate increases with γ for $\gamma < 1$ is a maximum for $\gamma = 1$ and decreases for $\gamma > 1$. This implies that the growth rate will eventually saturate in materials with $\gamma < 1$ even if γ increases indefinitely.

Acknowledgments

This work was funded by the CANDU Owners Group (COG), Working Party 32. We thank M. A. Miller, I. R. Emmerton, F. J. Butcher, J. F. Mecke, and J. E. Winegar and the staff of CEN Saclay for technical assistance and B. E. Sheldon for useful discussions.

References

[1] Harbottle, J. E. in *Irradiation Effects on Structural Alloys for Nuclear Applications, ASTM STP 484*, American Society for Testing and Materials, West Conshohocken, PA, 1970, p. 287.
[2] Adamson, R. B. in *Zirconium in the Nuclear Industry (Third Conference), ASTM STP 633*, A. L. Lowe, Jr. and G. W. Parry, Eds., American Society for Testing and Materials, West Conshohocken, PA, 1977, pp. 326–343.
[3] Murgatroyd, R. A. and Rogerson, A. in *Journal of Nuclear Materials*, Vol. 79, 1979, p. 302.
[4] Melehan, J. B., DeStefano, J., Balfour, M. G., and Cerni, S. in *Proceedings*, Topical Meeting on Light Water Reactor Fuel Performance, American Nuclear Society, La Grange Park, IL, 1979, pp. 29–38.
[5] Murgatroyd, R. A. and Rogerson, R. in *Journal of Nuclear Materials*, Vol. 113, 1983, p. 256.
[6] Garzarolli, F., Dewes, P., Maussner, G., and Basso, H.-H. in *Zirconium in the Nuclear Industry (Eighth Symposium), ASTM STP 1023*, L. F. P. Van Swan and C. M. Eucken, Eds., American Society for Testing and Materials, West Conshohocken, PA, 1989, pp. 641–657.
[7] Fidleris, V., Tucker, R. P., and Adamson, R. B. in *Zirconium in the Nuclear Industry (Seventh International Symposium), ASTM STP 939*, R. B. Adamson and L. F. P. Van Swan, Eds., American Society for Testing and Materials, West Conshohocken, PA, 1987, pp. 49–85.
[8] Rogerson, A. in *Journal of Nuclear Materials*, Vol. 159, 1988, pp. 43–61.
[9] Tucker, R. P., Fidleris, V., and Adamson, R. B. in *Zirconium in the Nuclear Industry (Sixth International Symposium), ASTM STP 824*, D. G. Franklin and R. B. Adamson, Eds., American Society for Testing and Materials, West Conshohocken, PA, 1984, pp. 427–451.
[10] Causey, A. R., Woo, C. H., and Holt, R. A. in *Journal of Nuclear Materials*, Vol. 159, 1980, pp. 225–236.
[11] Holt, R. A. and Ibrahim, E. F. in *Acta Metallurgica*, Vol. 27, 1979, pp. 1319–1328.
[12] Northwood, D. O., Blake, R. G., Jostsons, A., Madden, P. K., Faulkner, D., Bell, W., and Adamson, R. B. in *Journal of Nuclear Materials*, Vol. 79, 1979, pp. 379–394.
[13] Holt, R. A. and Gilbert, R. W. in *Journal of Nuclear Materials*, Vol. 137, 1986, pp. 185–189.
[14] Griffiths, M. and Gilbert, R. W. in *Journal of Nuclear Materials*, Vol. 150, 1987, p. 169.
[15] Griffiths, M., Gilbert, R. W., and Fidleris, V. in *Zirconium in the Nuclear Industry (Eighth Symposium), ASTM STP 1023*, L. F. P. Van Swan and C. M. Eucken, Eds., American Society for Testing and Materials, West Conshohocken, PA, 1989, pp. 658–677.
[16] Gilbon, D. and Simonon, C. in *Zirconium in the Nuclear Industry, ASTM STP 1245*, A. M. Garde and E. R. Bradley, Eds., American Society for Testing and Materials, West Conshohocken, PA, 1994, pp. 521–548.
[17] Harbottle, J. E. in *Philosophical Magazine*, Vol. 13, 1978, p. 49.
[18] Murgatroyd, R. A. and Rogerson, A., *Dimensional Stability and Mechanical Behaviour of Irradiated Metals and Alloys*, British Nuclear Energy Society, London, Vol. 2, 1984, pp. 93–96.
[19] Holt, R. A., Holden, T. M., Causey, A. R., and Fidleris, V. in *Proceedings*, Tenth Risø International Symposium on Metallurgy and Materials Science: Materials Architecture, J. B. Bilde-Sørensen et al., Eds., Risø National Laboratory, Roskilde, Denmark, 1989, pp. 383–389.
[20] Griffiths, M., Holt, R. A., and Rogerson, A. in *Journal of Nuclear Materials*, Vol. 225, 1995, p. 245.
[21] Holt, R. A., *Journal of Nuclear Materials*, Vol. 82, 1979, pp. 419–429.
[22] Murgatroyd, R. A. and Rogerson, A. in *Zirconium in the Nuclear Industry, ASTM STP 681*, J. H. Schemel and T. P. Papazoglou, Eds., American Society for Testing and Materials, West Conshohocken, PA, 1979, pp. 213–226.
[23] Fleck, R. G., Elder, J. E, Causey, A. R., and Holt, R. A. in *Zirconium in the Nuclear Industry, ASTM*

STP 1245, A. M. Garde and E. R. Bradley, Eds., American Society for Testing and Materials, West Conshohocken, PA, 1994, pp. 168–182.

[24] Griffiths, M., Winegar, J. E., Mecke, J. F., and Holt, R. A., "Determination of Dislocation Densities in HCP Metals using XRD and TEM," *Advances in x-ray Analysis*, C.15. Barrett et al., eds., Plenum Press, New York, 1991, p. 593.

[25] Holt, R. A. in *Journal of Nuclear Materials*, Vol. 59, 1976, pp. 234–242.

[26] Sheldon, B. E., unpublished work, AEA Technologies, 1990.

[27] Gendron, T. S., Holt, R. A., and Fleck, R. G., Atomic Energy of Canada Ltd. Report AECL-10632, 1992.

[28] Causey, A. R., Elder, J. E., Holt, R. A., and Fleck, R. G. in *Zirconium in the Nuclear Industry, ASTM STP 1245*, A. M. Garde and E. R. Bradley, Eds., American Society for Testing and Materials, West Conshohocken, PA, 1994, pp. 202–219.

[29] Holt, R. A. in *Journal of Nuclear Materials*, Vol. 90, 1980, pp. 193–204.

[30] Willard, H. J. in *Zirconium in the Nuclear Industry, ASTM STP 824*, D. G. Franklin and R. B. Adamson, Eds., American Society for Testing and Materials, West Conshohocken, PA, 1984, pp. 452–480.

[31] Woo, C. H. and Gösele U. in *Journal of Nuclear Materials*, Vol. 119, 1983, p. 219.

[32] Woo, C. H. in *Journal of Nuclear Materials*, Vol. 159, p. 237, 1988, pp. 237–256.

[33] Holt, R. A. in *Journal of Nuclear Materials*, Vol. 159, p. 310, 1988, pp. 310–338.

DISCUSSION

J. Harbottle[1] (written discussion)—Does your model predict or assume that volume is conserved during irradiation growth?

R. A. Holt et al. (authors' closure)—The model assumes, in accordance with experimental observation for high fluence, that volume is conserved during growth.

S. Yagnik[2] (written discussion)—Would your model (DAD) also predict non-linear growth in CWSRA Zr-4 at higher temperatures (600 to 625 K) of irradiation?

R. A. Holt et al. (authors's closure)—The model predicts non-linear growth under any circumstances where the ratio of the sink strength of **c**-component dislocations to that of **a**-type dislocations is changing. This could result in either an acceleration or a deceleration of the growth rate (see Fig. 8 of the paper), although the change in rate would be very slow when the two sink strengths are nearly equal.

A. T. Motta[3] (written discussion)—Does the fact that you have used DAD to explain your results mean that one does not need the production bias model to rationalize the irradiation growth data?

R. A. Holt et al. (authors' closure)—One needs the production bias model or at least some modification to the rate theory model used in our paper to rationalize irradiation growth data at temperatures above about 620 K, where the growth rate increases rapidly with temperature, and below about 400 K, where intrinsic vacancy mobility is too low to account for much growth.

E. Kohn[4] (written discussion)—Is there experimental support from diffusion studies relating to the differential diffusion rates of vacancies and interstitials in the *a* and *c* directions of the zirconium lattice?

R. A. Holt et al. (authors' closure)—The work of Hood et al. (*Journal of Nuclear Materials*, Vol. 223, 1995, pp. 122–125) has shown that self- and substitution diffusion in α-zirconium are anisotropic, illustrating that vacancy migration is anisotropic. I'm not aware of any experimental evidence that self-interstitial diffusion is anisotropic; however, there is evidence that interstitially diffusing solutes diffuse anisotropically (e.g., Hood, G. M., *Journal of Nuclear Materials*, Vol. 159, 1988, pp. 149–175). Theoretical studies of point defects in Zr (e.g., Bacon, D. J., *Journal of Nuclear Materials*, Vol. 159, 1988, pp. 176–189) predict anisotropic diffusion of self-interstitials.

---

[1] The S. M. Stoller Corp., Erlangen, Germany.
[2] Electric Power Research Institute, Palo Alto, CA.
[3] Pennsylvania State University, University Park, PA.
[4] Ontario Hydro Technologies, Toronto, Ontario, Canada.

Y. de Carlan,[1] C. Regnard,[1] M. Griffiths,[2] D. Gilbon,[3] and C. Lemaignan[1]

Influence of Iron in the Nucleation of ⟨c⟩ Component Dislocation Loops in Irradiated Zircaloy-4

REFERENCE: de Carlan, Y., Regnard, C., Griffiths, M., Gilbon, D., and Lemaignan, C., "**Influence of Iron in the Nucleation of ⟨c⟩ Component Dislocation Loops in Irradiated Zircaloy-4,**" *Zirconium in the Nuclear Industry: Eleventh International Symposium, ASTM STP 1295*, E. R. Bradley and G. P. Sabol, Eds., American Society for Testing and Materials, 1996, pp. 638–653.

ABSTRACT: Under irradiation, the acceleration of growth of nuclear reactor components made of Zircaloy is clearly correlated to the presence of ⟨c⟩ component dislocation loops. In the early stages, these loops appear to be essentially located close to the intermetallic precipitates. At higher doses, when ⟨c⟩ component dislocation loops are observed all over the microstructure, analysis of the Zr matrix reveals an homogeneous iron content due to the dissolution of the precipitates. Thus, iron may play a significant role in the nucleation of ⟨c⟩ component dislocation loops.
  A specific study has been performed on Zircaloy thin foils implanted with increasing amounts of iron. The conditions of nucleation and growth of these ⟨c⟩ component dislocation loops have been followed in a 1-MeV transmission electron microscope. The dose level for ⟨c⟩ component dislocation loop nucleation is weakly Fe-content dependent. However, their size and density increase with increasing iron implantation. These results are compared to direct observations made on irradiated reactor components. The role of iron on accelerated growth is then discussed.

KEYWORDS: Zr alloys, irradiation growth, ⟨c⟩ dislocation loops, Fe supersaturation

The change in geometry due to irradiation of Zr alloys (creep and growth) is of great concern for the design and operation of power plants. Indeed, structural components or fuel cladding tubes in many water-cooled nuclear reactors deform as the irradiation proceeds. It is important for design purposes to be able to predict the deformation behavior over the lifetime of the reactor.

The deformation behavior of textured, polycrystalline Zr-alloy components is dependent on the combined interaction of all the grains and is therefore a function of the single-crystal deformation behavior of individual grains combined in a polycrystalline agglomerate [*1,2*]. One component of this deformation is the single-crystal growth behavior, which is the subject of this paper.

The mechanism of irradiation growth in a single grain of Zr is still under discussion. It is based on a number of factors, some of which are an intrinsic property of the material, such as the anisotropic diffusion of point defects, and some of which are dependent on the microstruc-

---

[1] CEA Grenoble, DTP SECC, 17 rue des Martyrs, 38054 Grenoble Cedex 9, France.
[2] AECL, Chalk River Laboratory, Ontario, Canada.
[3] CEA Saclay, DECM SRMA, 91191 Gif sur Yvette Cedex, France.

ture, for example, network dislocations, grain size and shape, and the formation of various types of dislocation loops.

In all the Zr alloys, a neutron irradiation induces interstitial and vacancy $\langle a \rangle$-type loops. In addition, at high fluences and mostly in the case of Zircaloy, accelerated growth is observed in conjunction with the occurrence of $\langle c \rangle$-type loops. In this case, $\langle c \rangle$-type loops refer to those dislocation loops having a component of their Burger's vector in the [0001] direction. For Zircaloy-4 irradiated at about 573 K, these $\langle c \rangle$-type loops are typically vacancy in nature and are first observed to nucleate close to $Zr(Cr,Fe)_2$ intermetallic precipitates. An acceleration of the growth rate is observed at the same time as the $\langle c \rangle$-type loops are observed. In addition, the Fe and Cr content within the matrix around the precipitates is increasing as the irradiation progresses, and there is also an amorphous transformation of the precipitates [3].

There may be some influence of the amorphous transformation on $\langle c \rangle$-loop nucleation; however, since the amorphous transformation of the precipitates is combined with the release of Fe (and Cr) from the precipitates, the change in impurity concentration may be a more important factor, especially as impurities are linked with $\langle c \rangle$-type loop formation in other Zr alloys and at other irradiation temperatures [4,5].

An experimental program has been undertaken to analyze the effect of Fe in solution in the Zr matrix on the formation of $\langle c \rangle$-type loops. The main frame of this program consists of two parts: (1) analyze the Fe content in the surroundings of amorphizing precipitates and relate the Fe concentration in the matrix with the formation of the $\langle c \rangle$-type loops; (2) analyze the irradiation behavior, simulated with electrons, of a Zircaloy matrix previously doped with Fe at low temperature.

Testing Procedure

The first part of the study was performed on neutron-irradiated fuel cladding. With the aim of finding a correlation between Fe content and the presence of $\langle c \rangle$-type loops, examinations of the Fe content in the surroundings of the amorphous precipitates were performed on RX Zry-4 guide tubes and cladding irradiated in a PWR for up to three cycles. The thin foils were fabricated using standard electrochemical thinning procedures in hot cells. X-ray analytical TEM observations were performed with a beam size of 6 nm, and the minimum detectable level was found to be 0.15 at% Fe in the Zr matrix.

The second phase was based on the analysis of the behavior, under irradiation, of a high-Fe-content Zr matrix. Due to the very low solubility of iron in the Zr matrix [6,7], it is not possible to simulate the release of Fe from the precipitates thermally. Thus, Fe supersaturation in the matrix was obtained using implantation directly on the foils. The thin foils were prepared out of a thick recrystallized Zry-4 plate having a rather large grain size (12 to 16 μm). The orientation of the foil was chosen to obtain easily the electron diffraction condition $g = 0002$. The Fe doping was performed after preparation of the thin foils, which were cooled to liquid nitrogen on the irradiation stage of the implanter. The implantations were then performed with 110, 180, and 390-keV Fe ion beams to concentration levels up to 1.8 at% as computed using the TRIM code. Typical profiles after Fe implantation are shown in Fig. 1.

After the implantations, electron irradiations and observations were performed in a high-voltage electron microscope (HVEM) operating at 1 MeV using a heating stage to control the temperature of the foil during irradiation within 5 K. The thicknesses of the electron-irradiated foils were in the range of 300 to 450 nm. The electron beam intensities were the same for all the experiments, leading to a damage creation rate of 25 dpa $\cdot$ h^{-1}. The correspondence between electron beam intensity and damage rate is 12.6 dpa per each 10^{20} e$^-\cdot$ cm$^{-2}\cdot$ s^{-1}. In specific cases, additional observations were performed after the end of the irradiation on an advanced

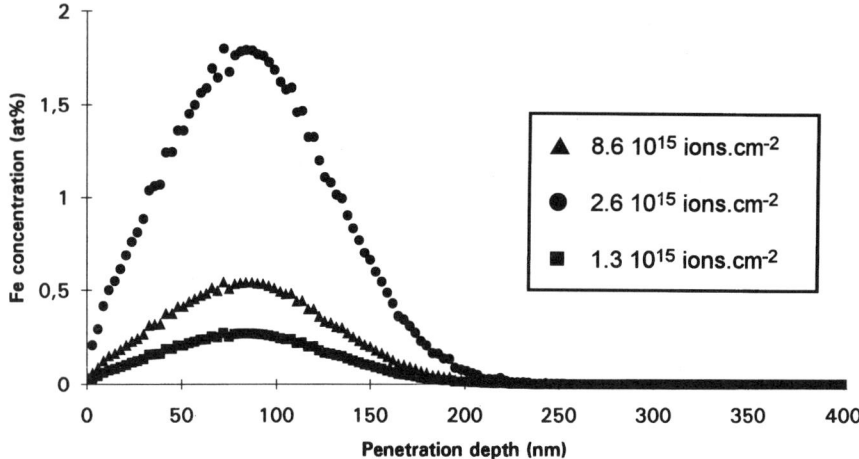
FIG. 1—*Implantation profile of 180-keV Fe ions in the thin foil.*

TEM with analytical capability at a spatial resolution of 6 nm (beam size). Table 1 gives the different conditions examined in the course of the electron irradiation work.

The damage induced by the implantations was high (about 55 dpa for 1.8 at%); thus, a series of implantations with Zr ions was also performed to observe any effects induced by the ion irradiation damage itself.

Results

For all the damage-producing irradiations performed on any Zr alloy, $\langle a \rangle$-type loops nucleate at the beginning of the irradiation, as observed in many earlier studies reviewed, e.g., in Refs 8 to 11. These loops are believed to contribute to the initial stage of growth along with creep recovery of the internal stresses and strain induced by the climb of the dislocations produced during the thermomechanical processing. These $\langle a \rangle$-type loops have not been analyzed in detail during the course of this work. Since they are very numerous and dense, their presence restricts the observation of the $\langle c \rangle$-type loops, except when using $g = 0002$ diffraction conditions. The $\langle a \rangle$-type loops will not be described further in the course of this paper, and we will focus only on the $\langle c \rangle$-type loops.

No specific determination of the components of their Burger's vector has been performed in

TABLE 1—*Test matrix for electron irradiation experiments (dose in dpa, DME = detailed microanalysis examination).*

Temperature, K	523	573	623
Standard Zircaloy (reference)	79	78	88
0.3 at% Fe			85
0.6 at% Fe	70		40
0.8 at% Fe	74 DME	60	
1.8 at% Fe		44	100 DME
Zr implantation (55 dpa)		53	82

this study, but their characteristics are consistent with earlier loop characterization that showed the loops to be $\frac{1}{6} \langle 20\bar{2}3 \rangle$ [3].

Neutron Irradiation

The general trend for $\langle c \rangle$-loops around amorphizing precipitates is shown in Fig. 2. Several features are noted: (1) no $\langle c \rangle$-type loops were observed prior to the amorphous transformation of the precipitates; (2) a few $\langle c \rangle$-type loops were first observed in the vicinity of the precipitates at the same time as the structural and microchemical changes within the precipitates; (3) the $\langle c \rangle$-type loops were not distributed uniformly around the precipitates but were much more dense in the sides of the precipitate parallel with the (0001) plane; (4) as the irradiation fluence was increased, the density of the $\langle c \rangle$-type loops increased ($>10^{26}$ n · m^{-2}) and the $\langle c \rangle$-type loops were spread increasingly more homogeneously throughout the matrix.

Using other diffraction contrast, it was possible to observe the faulted nature of these loops [3]. In the case of the non-uniform distribution of the loops (see Fig. 2), a detailed X-ray analysis mapping has shown an increase in Fe in the matrix up to 1.5 at% at a distance of 30 nm from the precipitate with no local variation with respect to the crystallographic directions of the matrix, i.e., with a spherical symmetry for the enrichment of Fe in the matrix around the precipitates. Mapping for Cr reveals some asymmetry with Cr content increasing in directions parallel with the basal plane around the precipitates, and post-irradiation annealing shows increased precipitate formation within these regions of high Cr content [4].

The local concentration of Fe with respect to the $\langle c \rangle$-type loop locations was analyzed using an analytical TEM, and no specific enrichment of Fe was found when the measurements were performed over such a loop. Due to the radioactivity of the neutron-irradiated foil, the minimum detectable was then close to 0.3 at% Fe.

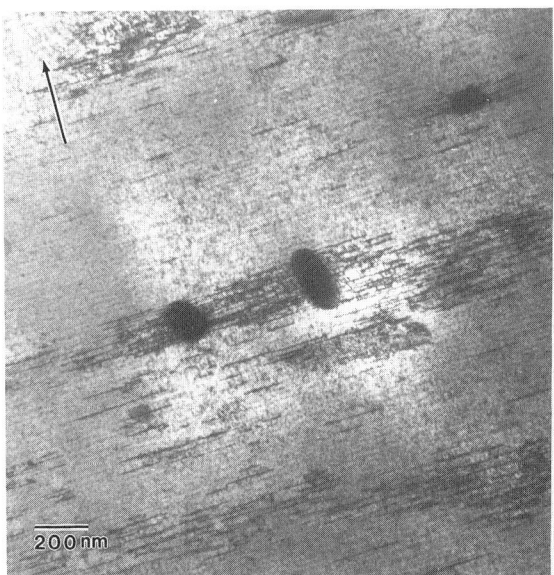

FIG. 2—$\langle c \rangle$-*type loops in the vicinity of precipitates under amorphous transformation, 585 K, 10 dpa, diffracting vector* $\mathbf{g} = 0002$.

Electron Irradiation

Electron Irradiation at 523 K—Electron irradiation of archive (unimplanted) Zircaloy material at 523 K shows the development of a band contrast due to planar arrays of very small cavities after a damage of 6 to 10 dpa. At a total dose of 70 dpa, these cavities have a mean diameter of about 8 to 10 nm and are aligned as (0001) planes 40 to 45 nm apart (Fig. 3). Very few ⟨c⟩-type dislocation loops were observed in the whole irradiation area. The presence under electron irradiation of these cavities could be connected to a similar occurrence in neutron irradiation of high-purity Zr samples [*12*].

For the 0.6 and 0.8 at% Fe-doped samples irradiated at 523 K, no cavities were observed, even at the maximum fluence of 80 dpa. The effect of Fe doping is to allow the nucleation and growth of ⟨c⟩-type loops. The suppression of cavity nucleation in this case may be related to the nucleation and growth of ⟨c⟩-type loops that act as alternative vacancy sinks. The development of these loops is observed as the irradiation proceeds (Fig. 4). They form a regular array of lines perpendicular to the [0001] direction. Their density was measured using a linear intercept method using a grid of parallel lines drawn in the [0001] direction. The density of the loops is plotted as a function of dose in Fig. 5. The nucleation occurs after a threshold dose of 8 to 10 dpa, and the loop densities increase until a saturation occurs above 35 to 40 dpa.

After the electron irradiation of 74 dpa, the foil implanted with 0.8 at% Fe was transferred to the analytical TEM to determine the distribution of Fe with respect to the ⟨c⟩-type loops. Using a large beam size, the expected average concentration due to implantation was confirmed (0.84 at% of Fe measured). With the minimal beam size (6 nm), the concentration of Fe in the areas between the ⟨c⟩-loops was found to be below the detection level (<0.1 at%), while the average concentration of Fe when the electron beam was focused on a zone embracing a ⟨c⟩-loop was 2.0 ± 0.4 at% Fe (Fig. 6). No similar local changes were noticeable for Sn, the other main constituent of the α phase.

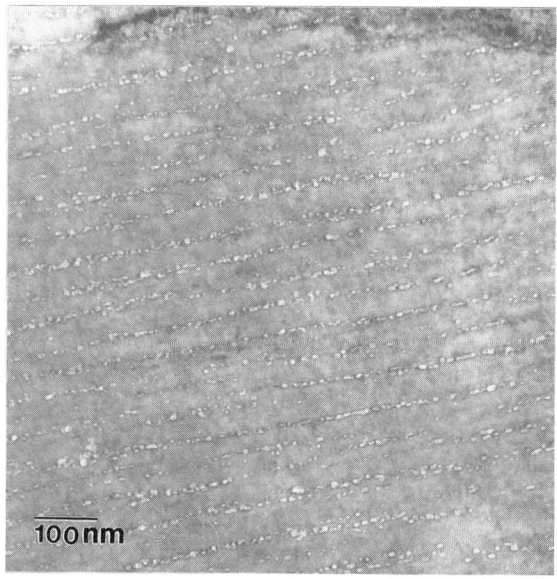

FIG. 3—*Formation of bands of fine cavities after electron irradiation of archive material (70 dpa, 523 K).*

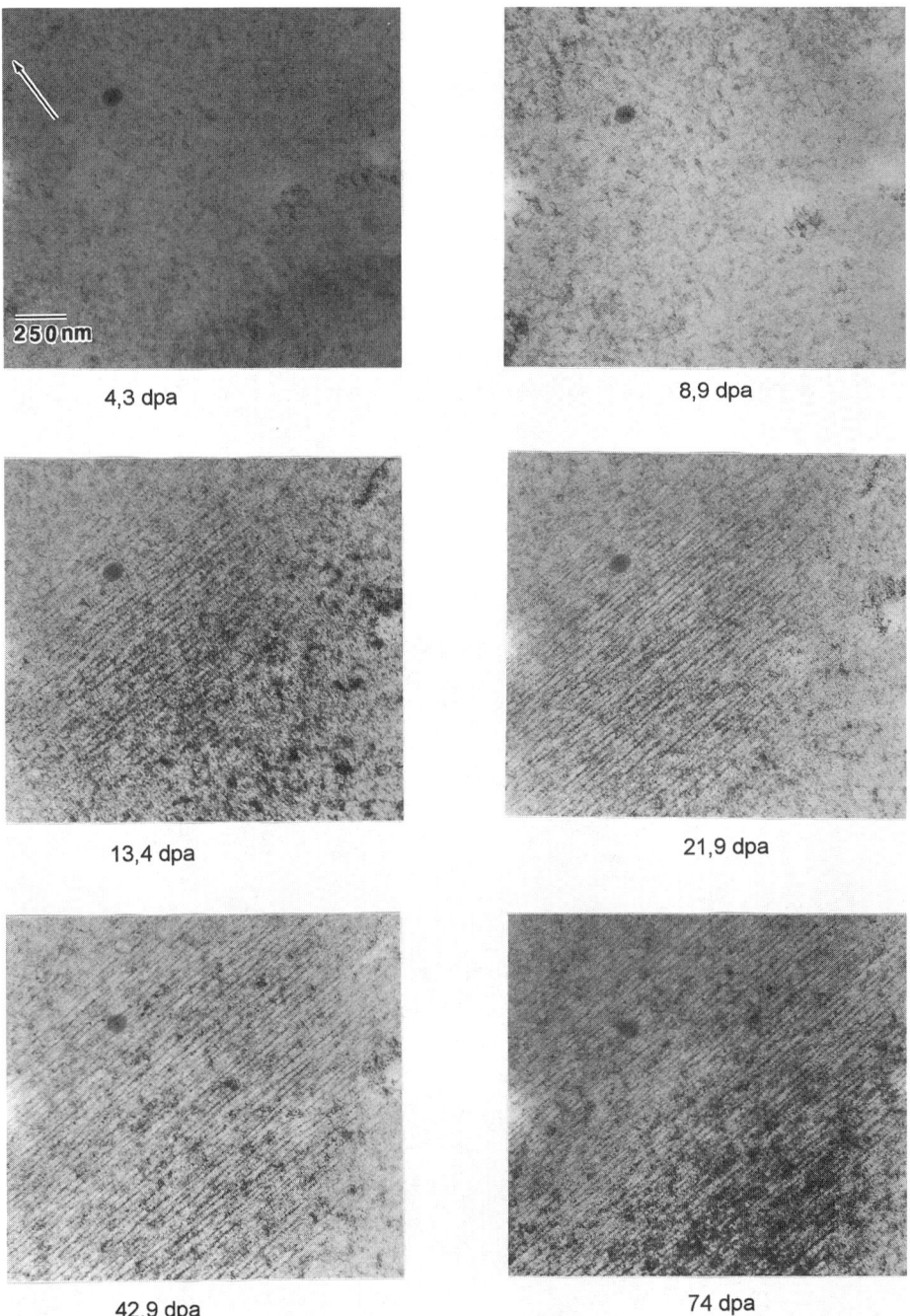

FIG. 4—*Formation of ⟨c⟩-type loop during e^- irradiation at 523 K of 0.8 at% Fe sample (**g** = 0002).*

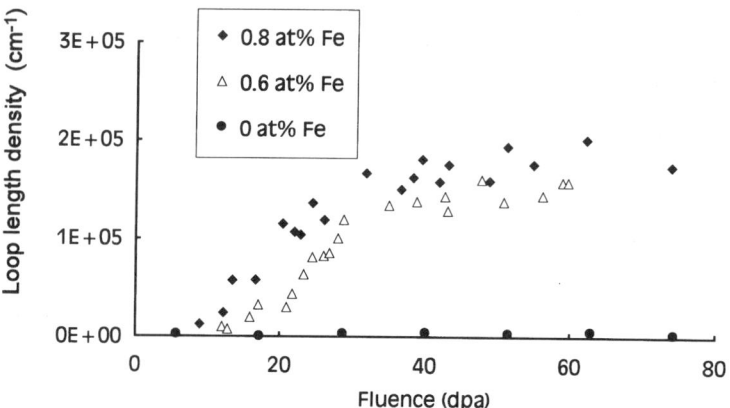

FIG. 5—*Density of $\langle c \rangle$-type loops as a function of irradiation damage and Fe doping (523 K).*

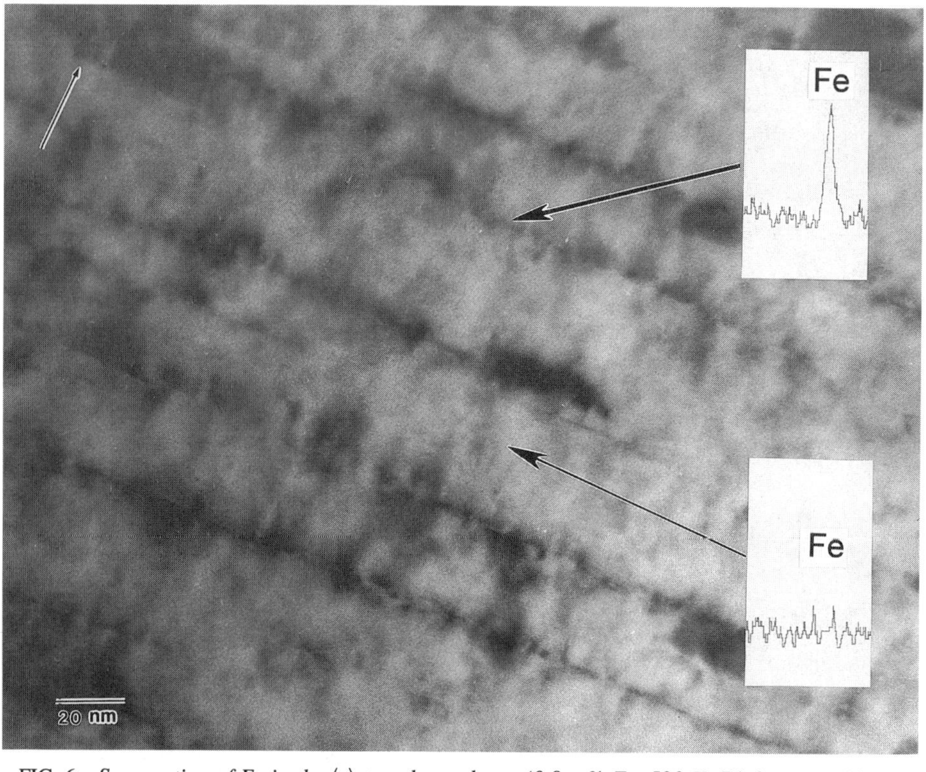

FIG. 6—*Segregation of Fe in the $\langle c \rangle$-type loop planes (0.8 at% Fe, 523 K, 74 dpa, **g** = 0002).*

Electron Irradiation at 573 K—During electron irradiation at 573 K, the general trend was similar to that observed at 523 K; however, ⟨c⟩-type loop production was more prevalent in the undoped sample at this higher temperature. The archive Zircaloy matrix showed the formation of cavities after 5 to 8 dpa and ⟨c⟩-loops were also found, as shown in Fig. 7. The density of the ⟨c⟩-type loops was low compared to that from Fe-doped materials. In these cases, the ⟨c⟩-type loops were distributed in a less regular manner and appear to be less planar. Their densities increased linearly with irradiation, and no saturation was found.

For the Fe-doped samples, no cavities were found, a result consistent with the lower-temperature irradiations. Numerous ⟨c⟩-loops occurred after a short incubation period of 2 to 5 dpa, and the saturation density, occurring above 40 dpa for the 0.8 at% Fe, was similar to that obtained at 523 K, as shown in Fig. 8. The saturation density for the 1.8 at% sample was probably just obtained at 40 dpa when the irradiation was discontinued.

For the sample implanted with Zr only at a damage dose equal to the one received by the 1.8 at% Fe, the microstructure during electron irradiation was very similar to the reference material: a high concentration of cavities was observed, and very few ⟨c⟩-loops were present for a higher incubation time.

Electron Irradiation at 623 K—At 623 K, no cavities were found anymore in the undoped material. Instead, ⟨c⟩-loops were observed nucleating inhomogeneously in higher densities than at 573 K. Irradiations at different points in the thin foil showed that slight changes in loops density were found, depending of the local foil thickness.

For the 0.3 and 0.6 at% Fe doping, the ⟨c⟩-type loop density was comparable with the reference material irradiated to the same dose. There appeared to be a slightly higher ⟨c⟩-type loop density, and the incubation dose was below 5 dpa.

For the sample doped with 1.8 at% Fe, some ⟨c⟩-type line contrast was observed, but the thickness of the lines increased during irradiation, being completed at a dose below 20 dpa. A detailed analysis of this high-Fe, high-temperature irradiation sample showed that the banded contrasts remain thick at high magnification and were no longer due to ⟨c⟩-type loops but rather to the nucleation of very fine Fe-rich precipitates (10 by 150 nm). The common diffraction of these precipitates in a dark field imaging indicates an epitaxy of these precipitates to the Zircaloy matrix (Fig. 9).

The Zr-implanted foil contained large number of small cavities after electron irradiation at 623 K, but they were less numerous than the reference material at 523 K.

Discussion

The state of the Fe is obviously important with respect to microstructural evolution. As the Fe is essentially insoluble in α-Zr, it is normally contained within intermetallic particles. The effect of neutron irradiation at temperatures of about 573 K and Fe-ion implantation at low temperature is to produce a dispersed distribution of Fe throughout the matrix. The Fe may be in some form of supersaturated solid solution or in some other finely divided state. With no implantation of Fe, few, if any, ⟨c⟩-type loops are produced during electron irradiation at 523 K. Their formation is enhanced at higher temperatures, but their number density is always much lower than in Fe-implanted material.

Both ⟨c⟩-type loop formation and Fe segregation are observed only within the irradiated area in the materials tested. Precipitation of Fe can be induced thermally, but it is necessary to heat the material for at least 30 h at 673 K, and, in this case, the precipitates are small (<30 nm in diameter) and are distributed uniformly. Irrespective of the state of the as-implanted Fe, dynamically the Fe has to be available in some form of solid solution in the matrix in order to interact with vacancy point defects induced by irradiation and then segregate to the ⟨c⟩-type

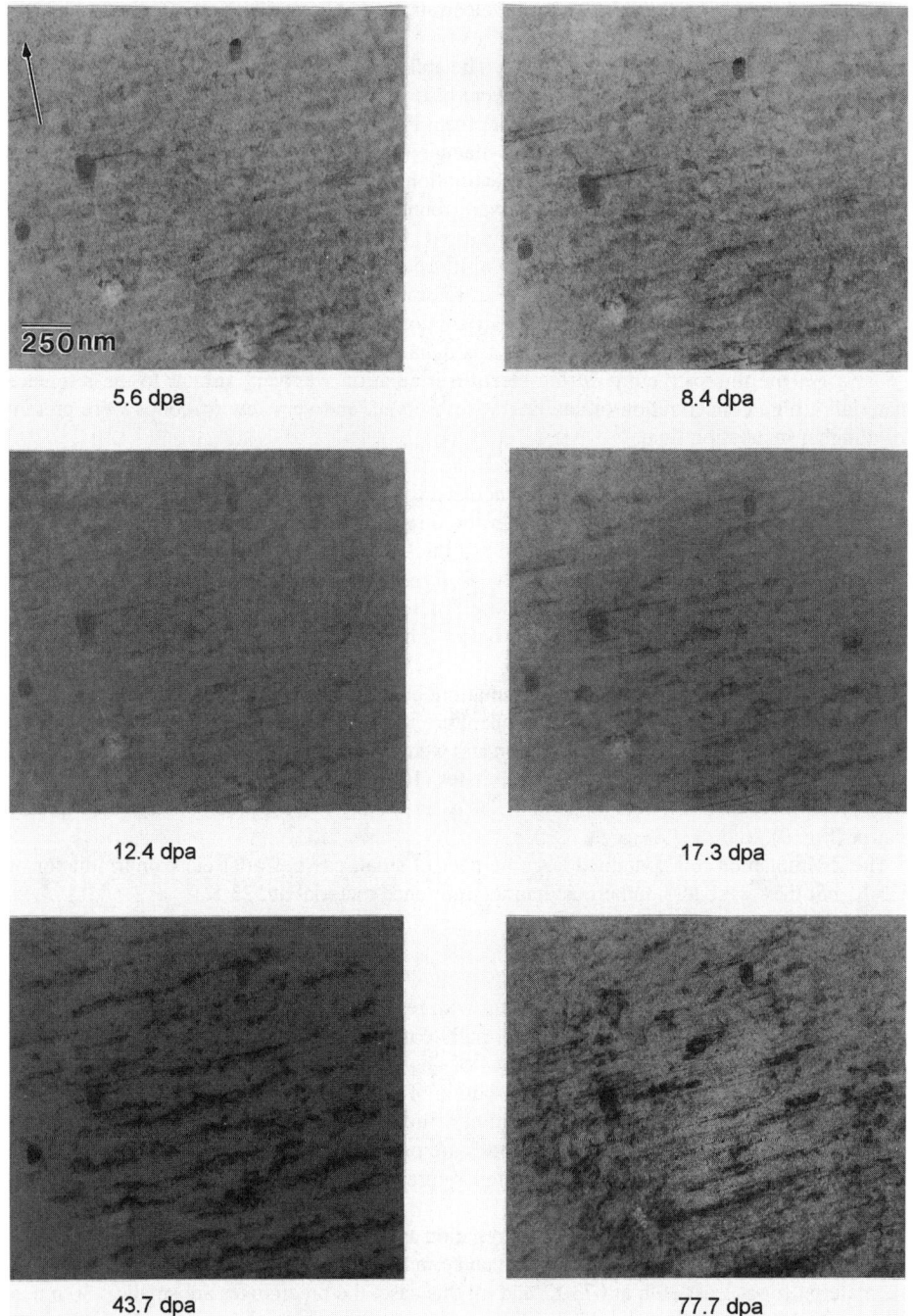

FIG. 7—*Formation of ⟨c⟩-type loop during e^- irradiation at 573 K of undoped sample (**g** = 0002).*

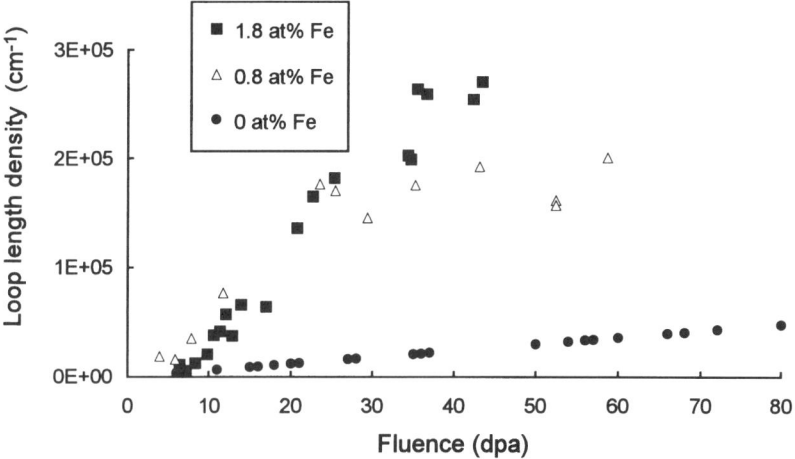

FIG. 8—*Density of $\langle c \rangle$-type loops as a function of irradiation damage and Fe content (573 K).*

vacancy loops. One can conclude that the Fe contributes to the stabilization of the $\langle c \rangle$-type loops.

Although obscured at 623 K, the segregation of Fe is very apparent at the $\langle c \rangle$ type loops analyzed following lower temperature irradiation, 523 and 573 K. For the local analysis performed, there is a "beam size dilution factor," d_f, that reduces the local composition during the X-ray analysis. For a beam diameter, D_b, and an observation of atomic rows perpendicular to the [0001] direction, the value of d_f is:

FIG. 9—*Formation of Fe-rich precipitates after e^- irradiation at 623 K (1.8 at% Fe, 20 dpa, dark field showing the epitaxy of the precipitates).*

$$d_f = \frac{4 \cdot D_b \cdot \frac{c}{2}}{\pi \cdot (D_b)^2} = \frac{2 \cdot c}{\pi \cdot D_b}$$

where c is the modulus of the [0001] vector (0.5148 nm).

For the typical beam size used (6 nm), the dilution factor is 1/18. Thus, if we consider that the iron is located in one plane parallel to the basal planes of the Zr, the measured concentrations of Fe in the analyzed volume correspond to a segregation of iron of 0.65 to 1 monolayer. A more precise analysis following the Goldstein procedure with considerations for the beam lateral straggling in our experimental conditions would have increased this value by a factor of 1.5 at most [13].

It is well known from Suzuki [14] that a segregation of substitutional alloying elements can occur at stacking faults, reducing their energies. Recent measurements of this effect have been made showing both segregation in the stacking fault and stacking fault energy decreases [15]. However, stacking faults are not present in Zr alloys without irradiation, and, thus, during heat treatment without irradiation, no thermal segregation can be obtained, but precipitation. One could consider a substitution-type segregation on some conceivable irradiation-induced stacking faults by a type of Suzuki effect. However, specific crystallographic considerations allowed us to propose an ordered structure of a planar segregation in a 2-D precipitation of the Fe atoms linked with vacancies.

Zirconium and iron have a large affinity for each other and react with the formation of a series of phases. In the Zr-rich side, Zr_3Fe and Zr_2Fe have to be considered. Between them, Zr_2Fe is the most stable. Thus, the spatial arrangement of Fe and Zr in Zr_2Fe corresponds to the lowest interaction energy between these two species. A stacking geometry similar to it would minimize the free energy of the system. This compound crystallizes in an tetragonal C16 crystal structure (Al_2Cu type) of parameters $a = 0.6385$ and $c = 0.5596$ nm [16]. It has a free energy of formation of $\Delta G = -27$ kJ.mol^{-1} [17]. The structure has a specific feature: it consists of a series of layers of Zr and Fe in the {110} planes, which have the following characteristics (Fig. 10): The Zr planes consist of almost perfect hexagonal-packed planes of Zr-Zr distances very close to the ones of the basal plane of the pure Zr, with an extra Zr atom off-plane. In the Fe planes, the Fe atoms form a rectangular array that, with minor translations, corresponds to a 2/3 filling of a basal plane.

A very good matching of the phases occurs with the following epitaxy relationships: [001] Zr_2Fe // [$\bar{1}100$] Zr ($\Delta \approx 0.05\%$) and [110] Zr_2Fe // [$11\bar{2}0$] Zr ($\Delta \approx 3.35\%$). In addition, if one considers the sequence Zr, Fe, Zr, starting from a basal plane of hcp Zr, the second Zr layer in Zr_2Fe is shifted in the Zr reference system by a vector 1/3 [$\bar{1}100$], one of the components of the shear in the 1/6 [$20\bar{2}3$] type loops.

A possible behavior for the Fe in those $\langle c \rangle$-loops could be that the Fe atoms diffuse as interstitials and interact with mobile vacancies to become substitutional atoms. Being in a favorable configuration, several of these substitutional atoms will cluster in a basal plane of Zr and start to form, with additions of vacancies, a 2-D layer of the Zr_2Fe structure, as a {110} plane. It can grow laterally very easily because the energy balance is in favor of this structure, but no extension occurs in the perpendicular direction due to the large misfit between the two structures in the [0001] direction of the Zr ($\Delta \approx 14\%$), which induces a large strain energy at the beginning of the second Fe layer.

A similar crystallographic analysis could be performed using $ZrFe_2$ as a reference since it is the most stable intermetallic phase between Zr and Fe. This fcc compound has pseudo-hexagonal layers of Zr when projected in the $\langle 111 \rangle$ direction, but with a much less precise epitaxy on Zr [0001] planes ($\Delta \approx 7$ to 8%) and a surface concentration of Fe larger than a monolayer.

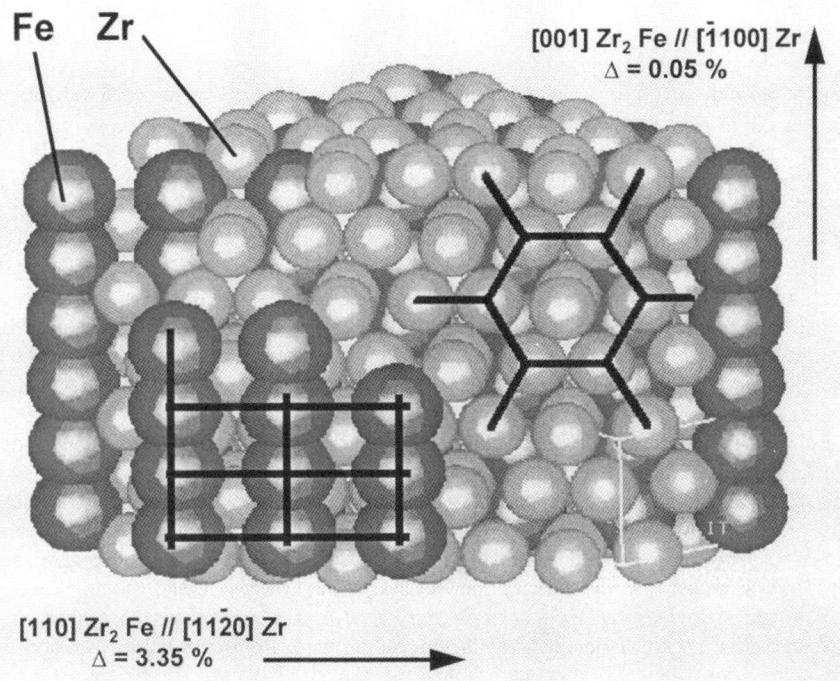

FIG. 10—*Zr and Fe layers in the Zr$_2$Fe structure (projected along the [110] direction) showing the similarity between the Zr basal plane and the [110] plane in Zr$_2$Fe.*

The real 2-D layer structure should have a configuration in between, with a layer of Fe having a partial coverage between two basal planes of Zr.

Thus the ⟨c⟩-loops obtained in Fe-doped materials could be considered as a 2-D precipitation of Fe in the basal plane of Zr, one basal plane being substituted by a partial monolayer of Fe, leading to a shorter interatomic distance across this plane. Since some relaxation also occurs in the basal plane, in order to be closer to the hcp stacking sequence, the exact component of the Burger's vector of this planar defect cannot be precisely determined, but can easily be consistent with the extinction conditions observed.

In addition, it seems that this segregation mechanism could not occur in the other directions of the Zr matrix. Indeed, there is no similar possibility to match Zr$_2$Fe to Zr on the prismatic planes of the Zr. Consequently, no segregation of Fe on the ⟨a⟩-loops is to be expected during irradiation of Zr alloys, with Fe available as interstitial solution in the matrix. In addition, the higher energy of the basal ⟨c⟩ loops with respect to the prismatic ⟨a⟩ loops is balanced by the energy released occurring during the formation of this 2-D, Zr$_2$Fe compound. Indeed, the surface energy due to this planar precipitation, $\Delta\gamma_s$, can be computed according to the following expression:

$$\Delta\gamma_s = \frac{4 \cdot \Delta G}{\sqrt{2} \cdot a.c.N}$$

where a and c are the unit cell parameters, and N is Avogadro's number. It corresponds to $\Delta\gamma_s = -0.355$ J · m^{-2}, a quantity equivalent to the one required for ⟨c⟩ loops to be more stable [8].

The saturation observed in the $\langle c \rangle$-loop density during electron irradiation could be explained by the end of the 2-D precipitation of the supersaturated Fe solution. For a coverage of Fe in the basal plane of ⅔ and a concentration of Fe c_{Fe}, a total precipitation of Fe would lead to a density of those defects of one out of $2/(3 \cdot c_{Fe})$ basal planes. The experimental distance between the $\langle c \rangle$-type loops at saturation for 0.8 at% (26 nm) is in good agreement with the corresponding computed value (21.5 nm).

This ⅔ coverage of one basal plane leads us to consider that those defects are strong sinks for vacancies; indeed, one needs to remove every third atom on those planes. Without those defects, such sinks do not exist anymore, and this explains why cavities are found in Fe-free materials.

For the irradiation performed at higher temperature, some precipitation is observed that reduces the $\langle c \rangle$-loop density. The reason for the formation of 3-D precipitates during higher temperature treatments, outside of irradiation, is the homogenous nucleation of a Zr-Fe compound. In that case, no 2-D precipitation could occur, probably due to the fact that when interstitial Fe atoms aggregate in the absence of vacancies, they form a layer of interstitials and, right at the beginning, the high strain energy in the [0001] direction is present, requiring the formation of a 3-D nucleus that grows afterwards with the same epitaxy.

The results show that increased Fe in the matrix enhances $\langle c \rangle$-type loop formation in Zircaloy-4 during electron irradiation. It is likely that the nucleation step of the $\langle c \rangle$-type loop formation could correspond to the same mechanism in neutron-irradiated materials. In this case, both Fe and Cr are dispersed into the matrix, and the localization of the slower moving Cr may be important when considering the distribution of $\langle c \rangle$-type loops close to the intermetallic particles. The local concentration of Fe and Cr will be proportional to their respective diffusion coefficients. Their anisotropic behavior contributes to an azimuthal concentration gradient in the matrix in the vicinity of the precipitates for short times, when the diffusion distances remain small compared to the interprecipitate distances.

One alternative explanation for the $\langle c \rangle$-type loop stability in the vicinity of the Zr(Cr,Fe)$_2$ precipitates in neutron-irradiated material at 573 K could have been that their nucleation is favored by the stress state induced by the expansion of the precipitates during the amorphous transformation (typically 1 to 2% during liquid-quenching amorphous transformation, but possibly negative in our case due to Fe and Cr depletion). However, the inhomogeneous distribution of $\langle c \rangle$-type loops close to the Zr(Cr,Fe)$_2$ precipitates is not observed during low-temperature irradiations (about 350 K) even though the precipitates are rendered amorphous. Also, $\langle c \rangle$-type loop clustering close to the Zr(Cr,Fe)$_2$ precipitates is observed at temperatures about 673 K. In this latter case, the precipitates remain crystalline while the matrix is enriched to a high concentration of dispersed Fe and Cr [18]. In addition, no bending contrast was observed in the vicinity of the amorphizing precipitates that could support the existence a stress gradient around them.

During long-term in-reactor irradiation, when those $\langle c \rangle$-loops have nucleated and have grown large enough in size, the ballistic mixing, at a rate of a few dpa per year, will remove the Fe far away from their 2-D layer positions. This allows the formation of local, stable missing layers of atoms in the basal planes, which would be the nuclei of intrinsic stacking faults, while the Fe released at long distances in the matrix is available for the nucleation of new 2-D precipitation defects. A continuous process of nucleation of true $\langle c \rangle$-loops can be conceived in which the Fe plays a catalytic role. If this mechanism is correct, this would lead to the absence of saturation of the accelerated growth induced by the $\langle c \rangle$-type loops.

Conclusions

A specific study has been performed on Zircaloy thin foils implanted with increasing amounts of iron. The conditions of nucleation and growth of these $\langle c \rangle$-type dislocation loops have been followed in a 1-MeV TEM. The dose level for ⟨c⟩-type loop nucleation was found to be weakly Fe-content dependant. However, their size and density increased with increasing iron implantation.

For the electron irradiation at moderate temperature of Fe-doped material, a strong segregation of Fe was measured in the plane of the ⟨c⟩-type loops corresponding to a fraction of a monolayer, while this was not the case in the neutron irradiated Zry-4.

A mechanism of planar Fe precipitation is proposed that allows the nucleation of ⟨c⟩-type loops with the assistance of a supersaturation of Fe released in the matrix during irradiation. During neutron irradiation, this mechanism is expected to act as the initiation step of the ⟨c⟩-type loop formation. Thus, the density of these ⟨c⟩-type loops, strongly dependent on their nucleation rates, could be controlled by the concentration of impurities, such as Fe, in the matrix.

Acknowledgments

This work would have not been completed without the efficient contribution of M. Dupuis (CEA Grenoble) for fine analytical TEM work and the continuous interest and support of Framatome and EDF. Fe implantations were performed with the help of E. Ligeon (CEA Grenoble) and Mrs O. Kaïtsasov (IN2P3). The authors thank them gratefully.

References

[1] Griffiths, M., Gilbert, R. W., and Fidleris, V., "Accelerated Irradiation Growth of Zr Alloys," *Zirconium in the Nuclear Industry: Eighth Symposium, ASTM STP 1023,* American Society for Testing and Materials, West Conshohocken, PA, 1989, p. 658.
[2] Holt, R. and Fleck, R. G., "The Contribution of Irradiation Growth to Pressure Tube Deformation," *Zirconium in the Nuclear Industry: Ninth Symposium, ASTM STP 1132,* American Society for Testing and Materials, West Conshohocken, PA, pp. 218–229.
[3] Griffiths, M. and Gilbert, R. W. in *Journal of Nuclear Materials,* Vol. 150, 1987, pp. 169–181.
[4] Griffiths, M., Holt, R. A., and Rogerson, A. in *Journal of Nuclear Materials,* Vol. 225, 1995, pp. 245–258.
[5] Griffiths, M., Gilbon, D., Regnard, C., and Lemaignan, C. in *Journal of Nuclear Materials,* Vol. 205, 1993, pp. 273–283.
[6] Charquet, D. et al. in *Zirconium in the Nuclear Industry: Eighth Symposium, ASTM STP 1023,* American Society for Testing and Materials, West Conshohocken, PA, 1989, pp. 405–422.
[7] Zou et al. in *Journal of Nuclear Materials,* Vol. 210, 1994, pp. 239–243.
[8] Griffiths, M. in *Journal of Nuclear Materials,* Vol. 205, 1993, pp. 225–241.
[9] Northwood, D. O. et al. in *Journal of Nuclear Materials,* Vol. 79, 1979, pp. 379–394.
[10] Hellio, C., de Novion, C. H., and Boulanger, L. in *Journal of Nuclear Materials,* Vol. 159, 1988, pp. 368–378.
[11] Griffiths, M. in *Journal of Nuclear Materials,* Vol. 159, 1988, pp. 190–218.
[12] Griffiths, M., Gilbert, R. W., Fidleris, V., Tucker, R. P., and Adamson R. in *Journal of Nuclear Materials,* Vol. 150, 1987, pp. 159–168.
[13] Williams, D., "Practical Analytical Electron Microscopy in Material Science," *Verlag Chemie Inter.,* 1984 p. 82.
[14] Suzuki, H., *J. Phys. Soc. Jap.,* Vol. 17, 1962, p. 322.
[15] Coujou, A. and Coulomb, P., *Scripta Metallurgica,* Vol. 22, 1988, pp. 1841–1846.
[16] Nevitt, M. V. and Koch, C. C. in *Intermetallic Compounds. Principles and Practice,* J. H. Westbrook and R. L. Fleischer, Eds., John Wiley and Sons, New York, 1995, pp. 385–400.
[17] Rodriguez, C. et al. in *Journal of Materials Science,* Vol. 30, 1995, pp. 196–200.
[18] Gilbon, D. and Simonot, C. in *Zirconium in the Nuclear Industry: Tenth Volume, ASTM STP 1245,* American Society for Testing and Materials, West Conshohocken, PA, 1994, pp. 521–548.

DISCUSSION

H. M. Chung[1] (written discussion)—Your proposed mechanism for nucleation of $\langle c \rangle$-type dislocation loops is intriguing. However, it appears to be inconsistent with the following observations: (1) for a similar fluence level, density of $\langle c \rangle$-type loops in Zircaloy-2 and -4, Zr-1Nb, Zr-2.5Nb seems to be similar despite the large difference in Fe levels in solution in the matrix; (2) some Fe should be in solution (in equilibrium with Fe in Zr-based intermetallics) even in the beginning stage of irradiation, yet $\langle c \rangle$ loops are observed only after accumulation of the relatively high threshold fluence; and (3) Zircaloys with small intermetallics do not exhibit $\langle c \rangle$ loops at low fluence despite the fact that the intermetallic precipitates are amorphousized and Fe atoms are released at low fluence.

C. Lemaignan et al. (authors' closure)—For Point 1, you have to consider that the density of $\langle c \rangle$ loops is not a measure of the nucleation rate, but a measure of the growth rate since, in reactor, most of the loops grow when they are large and stable and do not contain iron anymore. Points 2 and 3 refer to the same question (i.e., the existence of an incubation dose). The exact mechanism for this is probably linked to the saturation density of the $\langle a \rangle$ loops before having a higher density of vacancies and $\langle c \rangle$ loop nucleation. We didn't measure that quantity and cannot be more precise. However, in the Fe implantation followed by electron irradiation, which we have presented, an incubation dose is also found. So I don't see any inconsistency, but I agree that the model may be further developed.

A. T. Motta[2] (written discussion)—(1) For your thermal irradiation segregation experiment, if there were no c-component dislocations, how is the experiment relevant to testing your idea? (2) Are you proposing that a 2Fe-vacancy complex migrates to the loops? If not, how does the Fe arrive at the c-loop sinks? Is it irradiation-induced segregation?

C. Lemaignan et al. (author's closure)—(1) Our experiments show that $\langle c \rangle$ loop formation needs both Fe in solution in the matrix and irradiation point defects. An experiment on thermal Fe segregation on cladding with $\langle c \rangle$ loops already present (i.e., at high BU) is foreseen to confirm the absence of pure Suzuki segregation. (2) We consider that such a defect will not be mobile, and that an individual Fe atom or vacancy will diffuse and merge on this initial cluster to nucleate the 2D precipitation. It is not an irradiation-induced segregation since the $\langle c \rangle$ loops are not present before the process, but more likely some kind of irradiation-enhanced precipitation.

A. Strasser[3] (written discussion)—Please comment on the effect of the presence or lack of a- and c-loops, thereby Fe distribution, on performance-related properties such as mechanical properties, growth, corrosion resistance, etc. You already noted that there will be an effect on accelerated growth. Could you indicate why?

C. Lemaignan et al. (authors' closure)—I don't think that we could expect any effect of the presence of $\langle a \rangle$ and $\langle c \rangle$ loops on corrosion except through the change in Fe content in the matrix as explained in our paper on corrosion (X. Iltis, F. Lefebvre, and C. Lemaignan). With respect to mechanical properties, the main reason for the increase in strength seems to be the high density of small $\langle a \rangle$ loops. We expect only an effect of Fe in solution during irradiation on the formation of $\langle c \rangle$ loops and therefore on accelerated growth.

---

[1] Argonne National Laboratory, Argonne, IL.
[2] Pennsylvania State University, University Park, PA.
[3] The S. M. Stoller corporation, Pleasantville, NY.

B. Cheng[4] (written discussion)—Your work suggests that dissolution of Fe induces formation of ⟨c⟩ dislocation. How does one imply your results to the performance of LWR components in light of the industry trend of increasing Fe content in general? Do you anticipate an effect of ppt size on ⟨c⟩ product of growth?

C. Lemaignan et al. (authors' closure)—From our work, we could expect that the critical point is not the total amount of Fe, but the Fe solution in the matrix during irradiation. Therefore, the stability of the precipitates under irradiation may be of great importance. A higher tendency for resolution of Fe under irradiation may lead to higher tendency toward accelerated growth. In addition, lower precipitate size will enhance the re-solution kinetics (at least for $Zr(Fe,Cr)_2$).

---

[4] Electric Power Research Institute, Palo Alto, CA.

Effects of Processing on Structures and Properties

R. Choubey,[1] S. A. Aldridge,[2] J. R. Theaker,[3] C. D. Cann,[1] and C. E. Coleman[3]

Effects of Extrusion-Billet Preheating on the Microstructure and Properties of Zr-2.5Nb Pressure Tube Materials

REFERENCE: Choubey, R., Aldridge, S. A., Theaker, J. R., Cann, C. D., and Coleman, C. E., "**Effects of Extrusion-Billet Preheating on the Microstructure and Properties of Zr-2.5Nb Pressure Tube Materials,**" *Zirconium in the Nuclear Industry: Eleventh International Symposium, ASTM STP 1295,* E. R. Bradley and G. P. Sabol, Eds., American Society for Testing and Materials, 1996, pp. 657–675.

ABSTRACT: The effects of extrusion temperature and pre-heat soak time for billets on the mechanical properties of Zr-2.5Nb pressure tubes for CANDU[4] reactors have been examined. The β-quenched billets from a quadruple-melted ingot containing approximately 1200 ppm of oxygen were extruded at 780, 815, and 850°C with pre-heat soak times of 15 to 300 min. The extruded hollows were finished by cold drawing (with a 28% reduction in area) and then stress relieving at 400°C. The α-phase grain structure, tensile strength, and fracture toughness properties were found to vary with the pre-heat temperature and soak time. All the materials were tough because embrittling impurities were absent. The tubes with 780°C preheat had a very fine and uniform α-grain structure, giving high strength and toughness at all soak times. The opposite was true for the 850°C soaks; the grain structure was coarse and inhomogeneous and the materials tended to be less strong and less tough. The tubes with the 815°C soaks showed intermediate values of strength and toughness. These variations in mechanical properties are discussed in terms of α-grain refinement and oxygen enrichment.

KEYWORDS: CANDU reactor, Zr-2.5Nb, pressure tube, extrusion, billet, pre-heat temperature, soak time, microstructure, $(\alpha+\beta)$-phase, oxygen partitioning, SIMS microanalysis, tensile strength, fracture toughness

To minimize material variability and develop optimum mechanical properties, the variables in each step of the production of Zr-2.5Nb pressure tubes for CANDU reactors were examined. Extrusion of hollow billets is the last hot-working step; it determines the α/β-phase microstructure and is mainly responsible for the texture of the α-phase grains in the finished tube [1–4]. During extrusion, the geometric variables (extrusion ratio and die angle) are fixed. The other variables, i.e., rate of deformation (ram speed) and starting microstructure, starting temperature of extrusion, and soaking time for the billet before extrusion, are controlled.

This paper focuses on the latter two variables. In a previous examination of pre-heating variables [5], we observed large variations in microstructure, but little variation in the fracture toughness, probably because of a high concentration of chlorine in these tubes. For the current

---

[1] Research scientist and senior research scientist, respectively, Atomic Energy of Canada Limited, Whiteshell Laboratories, Pinawa, Manitoba, Canada, ROE 1LO.
[2] President, Nu-Tech Precision Metals Inc., 460 McCartney St., Arnprior, Ontario, Canada K7S 3H2.
[3] Research scientist and manager, respectively, Fuel Channel Components Branch, Atomic Energy of Canada Limited, Chalk River Laboratories, Chalk River, Ontario, Canada K0J 1J0.
[4] CANDU® is a registered trademark of Atomic Energy of Canada Limited (AECL).

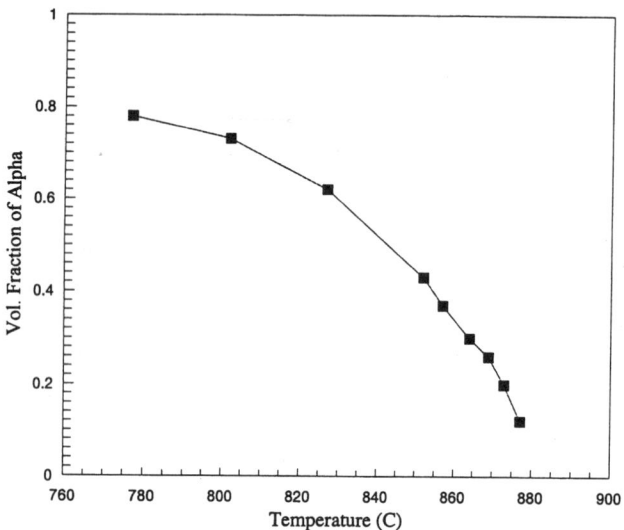

FIG. 1—*Calculated volume fractions of α at different temperatures in the α+β temperature range for Zr-2.5Nb pressure tube alloy.*

experiments, quadruple-melted material was used to eliminate the deleterious effect of chlorine [5,6]. Also, by taking all the billets from a single ingot, the compositional variability was eliminated. The objective of this study was to evaluate whether the current practice of billet pre-heating at 815°C for 15 min can be modified to produce pressure tubes with improved attributes and reduced variability.

After forging and machining, each Zr-2.5Nb billet is given a homogenizing and α-phase grain-refining heat treatment by heating into the β-phase region and then quenching in water [5,7,8]. Before extrusion, the homogenized billet material is heated into the (α + β)-phase region. As this pre-heat temperature is increased, the volume fraction of the α-phase, V_α, decreases rapidly, as shown in Fig. 1. The volume fractions were obtained from a pseudo-binary Zr-Nb (~1200 ppm O) diagram [9]. With time at temperature, the primary α-grains grow, and oxygen, being a strong α-stabilizer, partitions preferentially into the remaining α-phase grains.

After cooling from the extrusion temperature, the microstructure consists of a mixture of primary α-phase enriched in oxygen, transformed α-phase (from prior β-phase) that is oxygen depleted [10]. These α-grains are surrounded by thin ligaments of β-phase. The focus of the present work was to examine a range of V_α between 0.8 and 0.4 in which half-length billets were extruded using pre-heat temperatures in the range of 780 to 850°C, straddling that of current fabrication practice. The pre-heat-soak time was varied between 15 to 300 min. The extruded hollows were then cold drawn with about 28% reduction in area and stress relieved by autoclaving at 400°C for 24 h. In this paper, we present results of tension and fracture toughness tests of the finished tubes and microstructural examinations of the billet materials, the extruded hollows, and the finished tubes.

Test Materials and Procedures

Ten half-length Zr-2.5Nb pressure tubes of standard diameter and wall thickness (~3 m long, 104 mm I.D., and 4.2 mm thick) were produced at Nu-Tech by the hot extrusion of hollow

Consumable-Electrode Vacuum-Arc Melting
(Quadruple-Melted Ingot)
↓
Press Forging after Heating
at a Fully-β Phase Temperature
↓
Rotary Forging after Reheating
at an (α+β)-Temperature
↓
Machining of Hollow Billet (Trepanning)
↓
β Quenching of Billet
(from 1015°C)
↓
Billet-Preheating at an (α+β)-Temperature
for Different Soak Times
↓
Hot Extrusion
(10.5:1 extrusion ratio)
↓
Cold Drawing
(~ 28 % red. in area)
↓
Autoclaving/Stress-Relieving
(at 400°C for 24 h)

FIG. 2—*Manufacturing process flow outline for Zr-2.5Nb pressure tubes.*

billets (~250 mm long, ~45 mm thick) using a variety of billet pre-heat treatments. The manufacturing processes, from ingot melting to stress relieving of finished pressure tubes, are given in the process flow outline in Fig. 2. The billets were manufactured by Teledyne Wah Chang Albany (TWCA) using a consumable vacuum-arc quadruple-melting (Q-M) practice. The billets for these tubes were taken from a single forged log from a one-third section of the ingot. This was to ensure that the starting billet material for all the tubes had the same metallurgical history of ingot melting and processing (i.e., press- and rotary-forging), including the final β-quenching treatment of the billets. The material composition, including the concentrations of major impurity elements, is given in Table 1. The cold working and stress relieving for these tubes followed normal production practice.

Table 2 shows the billet preheating temperatures, T, and soak times, t, used in the production of the tubes for this study. The copper and steel-clad billets were heated in air to the soak temperature and held for the desired time in a "walking beam" electric furnace. The time to reach the soak temperature was approximately 150 min. The billets were extruded in a 2500-tonne horizontal press at an extrusion ratio of 10.5:1 and ram speed of 0.01 m s^{-1}. The extrusion hollows were finally processed to finished half-length tubes. As the heated clad-billets are transferred quickly from the furnace to the extrusion press in a fraction of a minute, little temperature drop is expected before the extrusion. However, judging from previous experience about the front-to-back end variations in the strength and microstructure of the tubes, a small temperature drop occurs due to the net heat loss through the toolings even though the toolings are pre-heated and adiabatic heating occurs during the extrusion.

The microstructural evolution during fabrication was evaluated by examining samples from the following production stages: the quenched billet, simulated pre-heated and soaked billet material, as-extruded hollows, and finished tubes. For the finished tube, the microstructures were examined at both the front and back ends of the tube. The techniques used were X-ray

TABLE 1—*Chemical composition of quadruple-melted Zr-2.5Nb ingot material (TWCA—Heat No. 233074Q).*

Alloying Element	Wt%
Nb	2.6
O	0.12
Zr	Balance

Impurity Element	ppm, wt	Impurity Element	ppm, wt
Al	39	Mn	<25
B	<0.25	Mo	<25
C	111	N	28
Cd	<0.25	Ni	<35
Cl	<0.2[a]	P	6
Co	<10	Pb	<25
Cr	<100	Si	27
Cu	<25	Sn	<25
Fe	458	Ta	<100
H	<5	U	1.1
Hf	38	V	<25
Mg	<10	W	<25

[a] Measured using glow discharge mass spectroscopy (GDMS).

diffraction (XRD), light metallography (LM), transmission electron microscopy (TEM), scanning electron microscopy (SEM), and secondary ion mass spectroscopy (SIMS).

The tensile strength in the axial direction of the tube was measured according to ASTM Test Methods for Elevated Temperature Tension Tests for Metallic Materials (E 21) at a test temperature of 300°C and a strain rate of 1×10^{-4} s^{-1}. This test temperature was chosen because it is the highest reactor-operating temperature. The results presented are averages of triplicate tests on material from the front end of each tube. This location represents the tube's weakest material. The fracture toughness (dJ/da and J (max load) derived from J-R curves [11]) was measured in triplicate tests at 250°C on material from the back end of the half-length tubes,

TABLE 2—*The pre-heat temperature and soak times for Zr-2.5Nb pressure tubes.*

Tube Number	Pre-Heat Temperature, °C	Soak Time, min
RX080	780	15
RX081	780	60
RX082	780	300
RX083	815	15
RX084	815	60
RX085	815	300
RX086	850	15
RX087	850	60
RX088	850	300

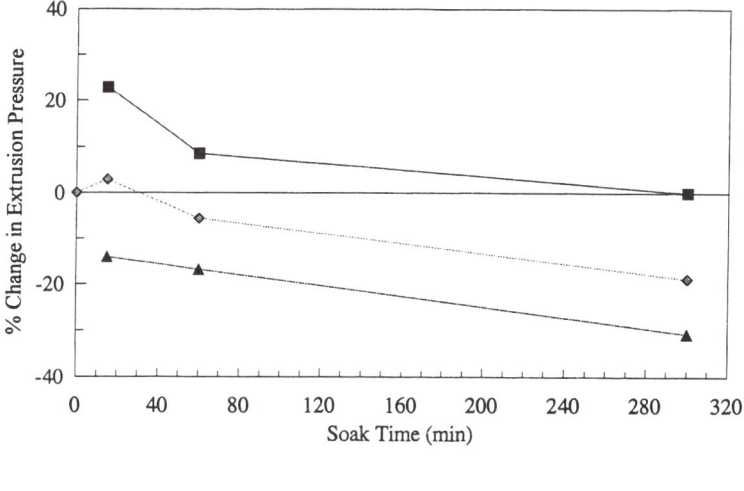

FIG. 3—*Relative (%) changes in the extrusion pressure for the various pre-heat temperatures and soak times.*

with the crack propagating in the axial direction. These conditions represent the least tough material location at the lowest CANDU reactor operating temperature [6].

Results

Extrusion

Figure 3 shows the % change in steady-state extrusion pressures for the different pre-heat temperatures and soak times relative to the extrusion pressure for 0-min soak at 815°C. The extrusion pressure, P, was calculated using the formula:

$$P = K \cdot A \cdot \ln R$$

where K is the extrusion constant, R is the extrusion ratio, A/a, and A and a are the cross-sectional areas of the billet and of the extruded tube, respectively. From Fig. 3, it is seen that the pressure required for extrusion decreases as the soak temperature and time increase.

Tensile and Fracture Toughness Properties

The tensile properties are presented in Fig. 4. The results demonstrate that the strength and ductility are influenced both by soak temperature and time, with the greatest changes occurring in the first 60 min. The tubes made after a pre-heating of 780°C had the highest strength and ductility and the lowest variation with soak time.

The tubes showed similar ranking in fracture toughness, Fig. 5, as with the strength and ductility, with the 780°C pre-heat tubes having the highest fracture toughness. All tubes were tough with values of dJ/da greater than 300 MPa. The J-values at maximum load, J_{ml}, for the 780°C pre-heat tube varied over a narrow range, and, with the exception of one test point, confirmed the effect of pre-heating temperature on toughness.

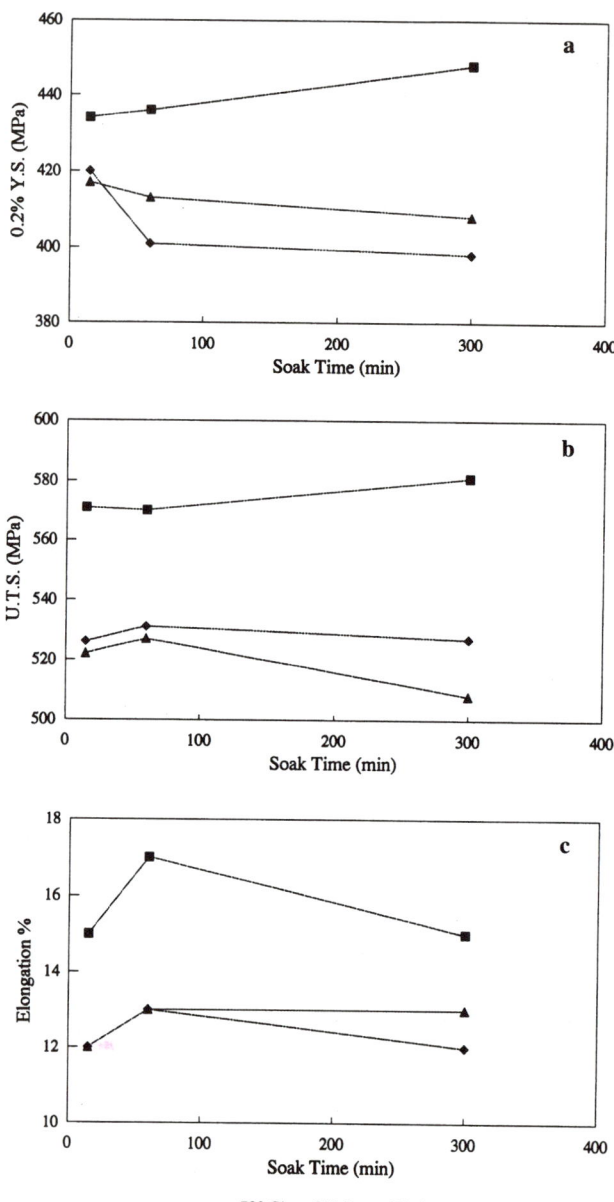

■ 780 C ◆ 815 C ▲ 850 C

FIG. 4—*Effects of billet pre-heat temperature and soak time on the longitudinal tensile properties of Zr-2.5Nb pressure tubes (front ends) tested at 300°C: (a) yield stress, (b) UTS, and (c) elongation %.*

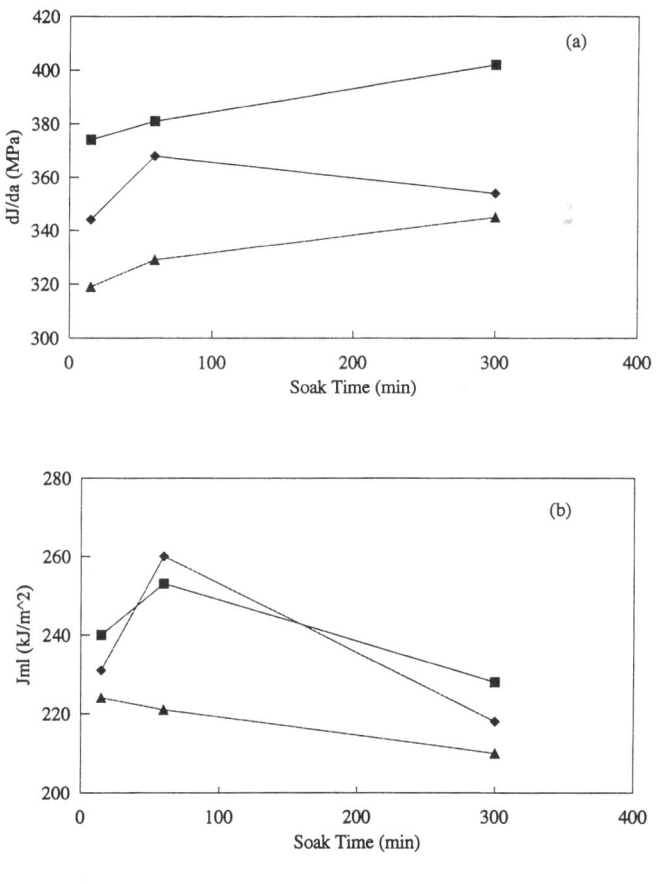

FIG. 5—*Effects of billet pre-heat temperature and soak time on the fracture toughness properties of Zr-2.5Nb pressure tubes (back ends) tested at 250°C.*

Microstructure Examination

The results of the microstructural evaluation are summarized in Tables 3 and 4 with typical examples showing variabilities in microstructures in Figs. 6 to 9. For the short soak times at the lower temperatures, the microstructure of the as-extruded material consisted of a uniform collection of thin, elongated α-phase grains surrounded by a filament of β-phase (Fig. 8a). As the soak temperature increased, the microstructure became increasingly coarse and inhomogeneous and contained regions of a Widmanstätten structure (Fig. 8b). The primary α-grain size appeared to be slightly coarser at the back end than at the front end, which could be attributed to the temperature difference.

SIMS Microanalysis for Oxygen

SIMS microanalysis was conducted to determine the extent of oxygen partitioning between the primary α-phase and the transformed α-phase (prior β-phase) in the as-extruded hollows.

TABLE 3—*Grain thicknesses of α-grains in the extruded hollows.*

Pre-Heat Temperature, °C	Soak Time, min	Mean Grain Thickness, μm	Standard Deviation
780	15	0.33	0.13
780	300	0.33	0.13
815	0	0.33	0.14
815	60	0.44	0.20
815	300	0.47	0.23
850	15	0.54	0.30
850	300	0.31^a	0.13

[a] Very inhomogeneous grain microstructure with different grain morphologies.

The oxygen-rich regions in the SIMS images were identified with the primary α-phase grains and the oxygen-deficient region with the transformed α-phase [10]. The oxygen-concentration ratio for the primary α and prior β region, C_α/C_β, was used as a measure of oxygen enrichment of the primary α. Because the spatial resolution of the SIMS is about 1 μm, an overlap occurs between the measurement regions of primary α-phase and transformed α-phase regions in the materials with smaller grain widths, i.e., those soaked at 780°C and the shorter times at 815 and 850°C. This overlap results in an underestimation of the partitioning in these materials. At the longer times at 815 and 850°C, where overlap was less because of the larger grains, the C_α/C_β values were found to increase from 3 in the 815°C pre-heat tubes to 5 in the 850°C pre-heat tubes, which is consistent with previous observations [10].

XRD Texture and Dislocation Density Measurements

The basal-pole components and dislocation densities (from X-ray line broadening [12]) were determined on the front ends of all the finished pressure tubes, and the results are summarized in Table 5. In general, the textures are similar to those previously observed in regular production tubes [1,2,4] with indications that, as the pre-heat temperature increases, the radial component decreases and the longitudinal component increases. With a pre-heat temperature of 850°C and increasing soak time, the transverse component decreases as the longitudinal component increases.

The effects of the pre-heat temperature and soak time on dislocation density are also shown in Table 5. The c-component dislocation density doubles when the pre-heat temperature and soak time are increased from 780°C/15 min to 850°C/300 min. However, the a-component dislocation density remained unchanged with soak conditions.

Discussion

Current pressure tubes are extruded at 815°C with a 15-min soak time, but early pressure tubes did not have any specification requirements for extrusion, which resulted in a variation of both temperature (up to 850°C [1,2]) and pre-heat soak time. The results of this program show that the properties of Zr-2.5Nb pressure tubes are influenced by both the extrusion temperature and the pre-heat soak time, which must have contributed to the variation of properties measured on the early pressure tubes. We have also shown that the properties of Zr-2.5Nb pressure tubes can be improved by a reduction of the pre-heat temperature.

TABLE 4—*Results of microstructural evolution in the Zr-2.5Nb billet before and after pre-heat and in the extruded tube hollows.*

Source	General Description of Microstructure	Pre-Heat Temperature, °C	Specific Variation in Microstructure	Figure References
Quenched hollow billet	Clearly delineated prior β-phase grain boundaries. Martensitic α' at the outside and inside surfaces changing to. Widmanstätten α composed of parallel plates at the prior β-grain boundaries and a basketweave structure inside the prior β-grains.	1015	The prior β-grain size varies from 2.5 mm diameter at the outer and inner surfaces to 1.7 mm diameter in the center section.	Fig. 6
Simulated, pre-heated hollow billet	Mixture of α-phase islands in the prior β-phase matrix transformed to martensitic α'-phase	780	Martensitic α'-islands (transformed β-phase) in the α-phase matrix. The islands in the sample soaked for 15 min were 0.1 to 0.5 μm wide and in the sample soaked for 300 min were 0.2 to 1.0 μm wide.	Fig. 7a
		850	α-platelets in martensite (transformed β-phase matrix) with a marked coarsening of the structure as the soak time increased.	Fig. 7b
		815	An intermediate structure between the above.	
As extruded tube	Elongated α-grains surrounded by thin filaments of β-phase.	780	Homogeneous, small, uniformly narrow grains with no change of thickness as the soak time is increased.	Fig. 8a, Table 3
		815	Generally similar to the 780°C structure with a small increase in grain thickness, but at longer soaking times there are larger primary α-regions and some areas of Widmanstätten α.	Table 3
		850	In-homogeneous structure with the primary α-phase separated by regions of Widmanstätten α-phase. At the 300-min soak, the microstructure was the same as that of the 15-min soak but more variable consisting of a mixture of: a. elongated primary α-grains b. coarse, elongated regions of Widmanstätten α-platelets c. fine, equiaxed, transformed α-grains.	Figs. 8b, 9b, and 9c, Table 3

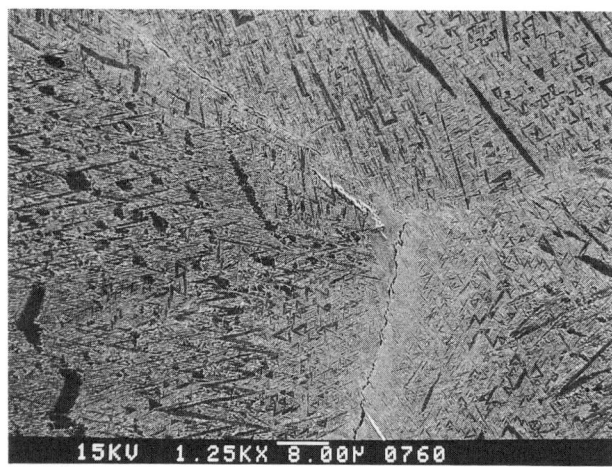

FIG. 6—*An SEM micrograph of billet after water quenching from 1015°C showing the martensitic α'-structure and parallel-plate α'-structure in the region of intersection of three prior β-grain boundaries and some very coarse and very fine martensitic α'-structure inside the grain.*

Strength

The results of the tension tests indicate that, of the three pre-heating temperatures examined, 780°C produces tubes with the best combination of strength and ductility and the least sensitivity to soaking time. The observed microstructural factors, produced by the various pre-heating treatments that may control strength, are:

1. Oxygen distribution.
2. Crystallographic texture.
3. Dislocation density.
4. Grain size and distribution.

Effect of Oxygen

The addition of oxygen to Zr-Nb alloys strengthens the alloys [13]. Partitioning of the oxygen from the β-phase to the primary-α grains during the pre-heating before extrusion results in greater strengthening of these grains while reducing the strength of the α-phase formed from transformation of the β-phase during cooling. Calculations show that this partitioning may result in a change in strength of the alloy depending upon the microstructure (see the Appendix). If both α-phases have the same grain morphology of ribbons running in the longitudinal direction [14], the rule of mixtures predicts that oxygen partitioning between the two α-phases will have no effect on the tensile strength of the alloy.

For our materials, the assumption of both phases as continuous ribbons is most valid for the 780°C extrusion materials based on the observed microstructures. At the higher pre-heating temperatures, the primary α-phase grains are larger, less continuous, and represent a smaller total volume fraction. If the microstructure at the higher pre-heating temperatures is modelled as isolated inclusions in a continuous matrix, the strength of the material is determined primarily by the strength of the α-phase formed from transformation of the β-phase matrix. Calculations for this α-phase after a 850°C pre-heat found that its ultimate tensile strength will be reduced

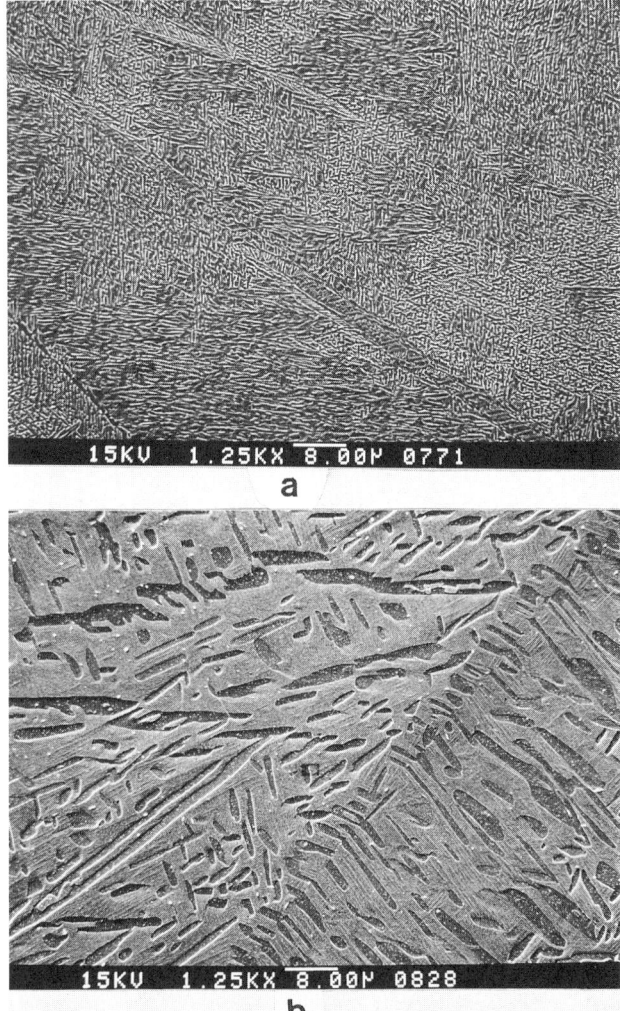

FIG. 7—*SEM micrographs showing the microstructures of β-quenched billet after preheating under two extreme temperatures and soak times: (a) 780°C/15 min, showing residual β-phase particles (light phase) in α-matrix (dark phase) and (b) 850°C/300 min, showing the coarse α-plates (dark phase) oriented along certain crystallographic directions of the matrix β-phase (light phase).*

by about 20 MPa due to oxygen depletion from the partitioning. If the strengthening mechanisms in the alloys are additive, oxygen partitioning can account for about 40% of the difference between the UTS of the material with a pre-heat of 780°C and that which received a pre-heat of 850°C for 300 min.

Effect of Texture

Zirconium alloys are stronger in the direction normal to the basal plane than the prism plane of the hexagonal close-packed (hcp) structure as shown by the higher strengths in the transverse

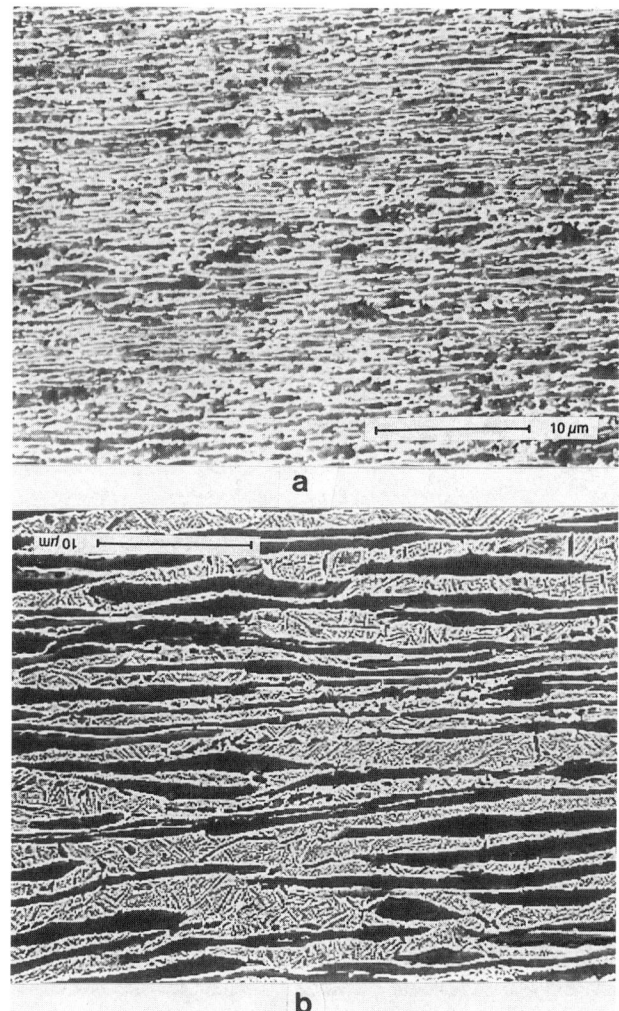

FIG. 8—*SEM micrographs of extruded tube hollows under two extreme preheat temperatures and soak times: (a) 780°C/15 min, showing fine and uniform elongated α-grains (dark phase) with a thin β-filament at grain boundaries and (b) 850°C/300 min, showing regions of Widmanstätten α-plate structure between the elongated primary α-grains (dark phase).*

direction (large basal pole component) compared with the longitudinal direction (small basal pole component) in pressure tubes. When the β-phase transforms to α on the pre-existing primary α-grains, the texture is defined by the pre-existing α and there is no increase in the longitudinal basal pole component. When the volume fraction of β is large and the spacing of the primary α is large, the β transforms to α in a Widmanstätten form within the β-matrix. The texture of this α has a very strong basal pole component in the longitudinal direction. An examination of Table 5 shows that the basal pole fraction is increasing in the longitudinal direction with increasing soak temperature. However, Fig. 4 shows that the tensile strengths are decreasing in the longitudinal direction with increasing temperature. Thus, the predicted

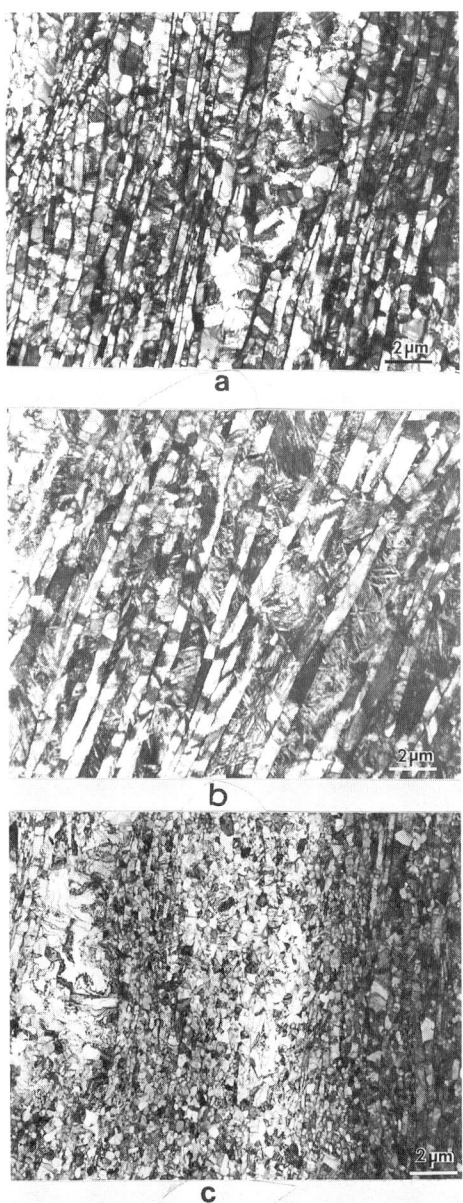

FIG. 9—*TEM micrographs showing the variety of microstructures in pressure tube hollows extruded:* (a) *at 815°C after soaking for 60 min,* (b) *at 850°C for 300 min, and* (c) *850°C for 300 min. The predominant microstructure consists of α-grains (light phase) surrounded by thin, grain boundary β-phase (dark phase) elongated in the extrusion direction. Also visible are regions of Widmanstätten α-plates in* (b) *and some equiaxed, very fine α-grains in* (c).

TABLE 5—*Effects of billet-preheating on basal-pole texture and dislocation density in Zr-2.5Nb pressure tubes.*

Tube I.D.	Preheat Temperature, °C	Soak Time, min	Basal-Pole, Radial Fraction	Basal-Pole, Trans. Fraction	Basal-Pole, Long. Fraction	a-Comp. Disloc. Density, $\times 10^{14}$ m^{-2}	c-Comp. Disloc. Density, $\times 10^{14}$ m^{-2}
RX080	780	15	0.414	0.528	0.058	3.0	0.4
RX081	780	60	0.392	0.542	0.066	3.0	0.5
RX082	780	300	0.365	0.577	0.057	3.3	0.8
RX083	815	15	0.325	0.600	0.074	2.9	0.5
RX084	815	60	0.329	0.595	0.075	3.0	0.6
RX085	815	300	0.338	0.591	0.070	2.9	0.8
RX086	850	15	0.307	0.600	0.092	3.4	0.8
RX087	850	60	0.321	0.561	0.116	3.3	0.8
RX088	850	300	0.306	0.531	0.161	3.3	0.9

effect of texture is opposite to the observed tensile strength behavior, i.e., texture changes do not explain the observed strength behavior.

Effect of Dislocation Density

Increasing the density of dislocations by cold working generally results in stronger materials as the strength is proportional to the square root of the dislocation density. As shown in the data presented in Table 5, the total dislocation density (sum of a- and c-component dislocations) appears to increase slightly with pre-heat temperature. The reason for this increase is not clear. The predicted increase in strength due to this increase is opposite to the observed decrease in strength with increasing pre-heat temperature, i.e., the changes in dislocation density do not explain the observed behavior. The effect of dislocation density appears to be compensated by other factors such as the coarser grain thickness, greater partitioning of the alloying elements O and Nb into the α- and β-phases, respectively, and larger volume fraction of the prior β-phase, which is transformed during cooling to oxygen-deficient α-grains.

Effect of Grain Thickness

As a rule, the strength of a polycrystalline material is inversely proportional to the square root of the average grain diameter. Therefore, an increase in the average grain diameter would be predicted to result in a decrease in the strength. A comparison of Table 3 with Fig. 4 shows that there is a general correlation between increasing grain thickness and decreasing strength with the exception of the tube soaked for 300 min at 850°C; its low strength may be associated with the large fraction of Widmanstätten plates and the general microstructural inhomogeneity of this material.

Fracture Toughness

The material used to make these tubes contained little chlorine, carbon, and phosphorus; thus, few sites for void initiation during fracture existed. Consequently, all the fracture toughness values were high. The observed variations (Fig. 5) are attributed to a combination of oxygen partitioning and the grain structure. The tubes made with an extrusion pre-heat temperature of 780°C contained a high volume fraction of primary α-grains that were uniformly thin. This material was tough because it had little spatial variation in its microstructure or microchemistry. As the pre-heat temperature was increased, the grain structure became more inhomogeneous, and although the volume fraction of oxygen-enriched α-phase was smaller, it contained a higher concentration of oxygen than the material pre-heated to 780°C. The highly oxygenated primary α-phase would be less tough but stronger than the surrounding matrix of transformed β-phase and would provide potential crack initiation sites by acting like semi-brittle particles. Thus, this material would be less tough than the more ductile and homogeneous material produced by the lower pre-heat temperature.

Microstructural Evolution

The current ingot-to-billet manufacturing process follows the route, press forging → rotary forging → β quenching (of hollow billets) [5]. The β-quenched, quadruple-melted material shows large prior β-phase grains with a fine, parallel α-plate structure near the grain boundaries and a mixture of some very coarse ($\sim$5 μm width) and very fine, acicular (martensitic) α'-plate structure (<1 μm width) within the grains. Such microstructures with large prior β-phase grains are attributed to the low concentrations of grain-refining/nucleating elements, carbon (100 ppm maximum) and chlorine (<0.5 ppm) reduced specifically to improve the fracture toughness [5,6]. The α'-plates are oriented along the three crystallographic {111} planes of the parent body-centered-cubic (bcc) lattice of the β-phase, and these plates have an internally twinned sub-structure [15].

When the billet material is pre-heated, the α'-plates, being thermally unstable and supersaturated in Nb, transform to α with precipitation of β-Nb, enriched in Nb, along the grain and twin boundaries of these plates [16,17], effectively pinning the new grain boundaries. As these precipitation processes are diffusion-controlled, their kinetics depend on the billet pre-heating temperature-time parameter. As the temperature increases above the monotectoid temperature, the β-Nb transform to β-Zr by dissolving in the adjacent α-phase to form a continuous boundary network restricting grain growth. As the temperature increases through 750 to 850°C and with longer soaking times, coarsening and spheroidizing of the α-plate structure by Ostwald ripening occurs (compare Fig. 7a with Fig. 7b). During extrusion, the primary α- and β-phases are elongated in the extrusion direction. During cooling, the β-phase transforms to α, with the existing primary α providing nucleation sites and a very thin layer of residual β. This microstructure was found only in tubes with a 780°C pre-heat (all soak times) or 815°C pre-heat (short soak time).

A very different grain structure is developed in tubes extruded under a higher temperature or longer pre-heat time, i.e., at 850 or 815°C (long soak times). In these cases, the billet material, just before extrusion, consists of coarse, oxygen-enriched primary α-phase grains dispersed in the β-phase matrix. The β-phase matrix has a body-centered-cubic (bcc) structure and is softer than the oxygen-enriched α-plates. During extrusion, the softer β-phase matrix flows around the harder (oxygen-strengthened) α-phase plates. Because of the large spacing between the α-phase plates, the hot-worked β-phase matrix transforms predominantly into a Widmastätten α-plate structure. This transformation happens because the thin extruded tube cools at a rate such that the residual β-phase does not have enough time to transform fully at the widely spaced existing primary α-phase grains. In some regions with large primary α-grains, the hot-worked and dynamically recovered sub-grain structure probably leads to a very fine α-phase grain structure during cooling.

During the pre-heating for extrusion, the oxygen partitions from the β- to the α-phase. A ratio of 5 for the concentrations in these two phases was found for the 850°C/300-min pre-heat. This is approaching the value of 6.1 measured in Zr-2.5Nb after soaking at 870°C for a long period to achieve a grain size of a few microns [10] and confirms that significant partitioning is occurring. The present data are also consistent with previous work on Zr-oxygen and Zircaloy-oxygen alloys that showed the ratio of oxygen in primary α-phase grains to that in the prior β-phase can reach up to about 9 [9,18].

Extrusion Pressure

The extrusion constant, K, for the alloy is controlled by the flow stress of each phase, α and β, and their volume fractions at the given temperature. The flow stress for the bcc β-phase is significantly lower than for the hcp α-phase material, which results in a very rapid decrease in the extrusion pressure as the temperature increases within the two-phase field. The relative 22% increase in the extrusion pressure, P, for a pre-heat temperature reduction from 815 to 780°C (with 15-min soak) agrees with the 24% increase in the flow stress measured experimentally in high-temperature compression testing of a β-quenched Zr-2.5Nb material [19]. Similarly, the 19% reduction in P for a pre-heat temperature increase from 815 to 850°C agrees with the 20% reduction in the experimental flow stress. The experimental values were obtained at a true strain rate of 6.5×10^{-2} s^{-1}. The decrease in extrusion pressure with increasing pre-heat soak time has been known for many years by the fabricators and has been attributed to the grain coarsening observed here, but oxygen partitioning may also play a role in decreasing the extrusion pressure.

Implication

For Zr-2.5Nb pressure tubes of CANDU reactors made from quadruple-melted ingots, changing the extrusion temperature from 815 to 780°C would result in a more uniform grain microstructure, an increase in tensile strength and ductility, maintenance of high fracture toughness, and a reduction in sensitivity of these properties to pre-heat soak time.

With an ultimate tensile strength of 570 MPa for the tubes extruded at 780°C, there exists a margin of 90 MPa above the minimum value specified for pressure tubes. In-reactor deformation may be reduced by an intermediate stress relief that lowers the density of dislocations introduced by the cold drawing process. A consistent margin of 90 MPa would allow us to introduce this beneficial intermediate stress relief into the pressure tube fabrication cycle of Zr-2.5Nb pressure tubes. Since, in practice, there will always be some small variations in soak time and temperature, the reduction in sensitivity to pre-heat effects will permit more flexibility for the fabricator and will decrease the dispersion in the mechanical properties from tube to tube.

Conclusions

This study has shown that the extrusion pre-heat conditions have a significant effect on the tensile strength and facture toughness properties of quadruple-melted Zr-2.5Nb pressure tubes for CANDU reactors. In particular, it was found that:

1. Lowering the extrusion temperature to 780°C from the current temperature of 815°C results in a pressure tube material with a fine and more uniform grain microstructure with consistently higher tensile strength, ductility, and fracture toughness with reduced sensitivity of these properties to the soak time at the pre-heat temperature.

2. Increasing the extrusion temperature to 850°C from the current temperature of 815°C results in a tube with a coarse and inhomogeneous microstructure and reduced tensile strength and ductility.

3. The changes in mechanical properties introduced by varying the extrusion pre-heat treatment are controlled by the grain size and distribution together with a contribution from oxygen partitioning at high pre-heat temperatures.

4. The fracture toughness of all the materials is high, with a lessening in toughness caused by the development of an inhomogeneous microstructure and oxygen microchemistry as the pre-heat temperature is increased.

Acknowledgments

The authors would like to thank: J. E. Winegar, J. H. van der Kuur, J. Coleman, R. C. Styles, L. Brown, R. Kaatz, and A. H. Jarvis of AECL; V. Chartrand and J. A. Jackman of CANMET for experimental assistance; Teledyne Wah Chang Albany and Nu-Tech Precision Metals Inc. for providing the material and fabrication facilities. We are also grateful to B. A. Cheadle, E. G. Price, R. Dutton, and M. P. Puls for their helpful advice and encouragement throughout this program. Funding for this work was provided by the CANDU Owners Group through Working Party 33, Work Package 6597.

APPENDIX

Contribution of Oxygen Partitioning to Strength

Rule of Mixtures

In the rule of mixtures, the total ultimate tensile strength is the sum of the products of the volume fractions of the primary α, V_α, and the α from the transformed β-phase, $1 - V_\alpha$, times the strength of these phases, taking into account oxygen partitioning. Then the total ultimate tensile strength, σ_T, can be written as

$$\sigma_T = V_\alpha \sigma_\alpha + (1 - V_\alpha)\sigma_\beta \qquad (1)$$

where σ_α and σ_β are the ultimate tensile strengths of the primary α and the α from the transformation of the β-phase, respectively. In this calculation, the effect of the untransformed β-phase present has been neglected. The effect of oxygen on strengthening is linear [13] so that σ for each phase can be expressed in the form

$$\sigma = Ca + b \qquad (2)$$

where C is the oxygen concentration in that phase, and a and b are constants. (The assumption of a linear relationship between oxygen concentration and strength is based on data [13] ranging from 400 to 2500 ppm oxygen. Since the maximum oxygen concentration of the primary α for the extrusion temperatures used here is 2700 ppm, the linear relationship should be a good approximation.) These concentrations can be found from the expression for the oxygen partitioning

$$C_i = V_\alpha C_\alpha + (1 - V_\alpha)C_\beta \qquad (3)$$

where C_i, C_α, and C_β are the total oxygen concentrations and the oxygen concentrations in the primary α and the α from the transformed β, respectively. For a partitioning ratio of P, when Eqs 1 and 2 are combined, the expression for σ_T becomes

$$\sigma_T = [V_\alpha(P - 1)C_\beta + C_\beta]a + b \qquad (4)$$

TABLE 6—*Calculated oxygen concentrations and strengths of the primary α- and transformed β-phases of annealed Zr-2.5Nb [13] as a function of extrusion temperature for an oxygen partitioning ratio of 6 at a temperature of 300°C.*

Extrusion Temperature, °C	Volume Fraction Primary α	Oxygen Concentration, wt ppm		Ultimate Tensile Strength, MPa	
		Primary α	Transformed β	Primary α	Transformed β
780	0.78	1470	245	315	282
815	0.66	1675	280	320	283
850	0.45	2215	370	335	285

Combining Eqs 3 and 4 then gives

$$\sigma_T = C_i a + b$$

Thus, oxygen partitioning between the two α-phases is predicted to have no effect on the ultimate tensile strength of the tubes.

Inclusions in a Matrix

The equilibrium oxygen concentrations in the primary α- and transformed β-phases were calculated from Eq 3 and the relation $C_\alpha = PC_\beta$ for an oxygen-partitioning ratio of $P = 6$ and the volume fractions of the phases at the different temperatures as given in Fig. 1. An initial bulk oxygen concentration of 1200 ppm was assumed. The predicted ultimate tensile strengths of these phases were then obtained from the measured effect of oxygen on the UTS [13]. The results of these calculations are given in Table 6. For comparison, the ultimate tensile strength of the alloy in the absence of partitioning was calculated to be 307 MPa [13].

From Table 6 it can be seen that, after a pre-heat of 850°C, the strength of the transformed β is about 20 MPa less than the strength in the absence of partitioning. If the strength of the alloy is primarily determined by the transformed β-matrix, since the volume fraction of the primary α-phase is small and present as isolated grains (inclusions) after this pre-heat, oxygen partitioning is predicted to result in a decrease in the ultimate tensile strength of the alloy in this case.

References

[1] Cheadle, B. A., Aldridge, S. A., and Ells, C. E., "Development of Texture and Structure in Zr-2.5Nb Extruded Tubes," *Canadian Metallurgical Quarterly*, Vol. 11, No. 1, 1972, pp. 121–127.

[2] Holt, R. A. and Aldridge, S. A., "Effect of Extrusion Variables on Crystallographic Texture of Zr-2.5 wt% Nb," *Journal of Nuclear Materials*, Vol. 135, 1985, pp. 246–259.

[3] Cheadle, B. A., Coleman, C. E., and Licht, H., "CANDU-PHW Pressure Tubes: Their Manufacture, Inspection and Properties," *Nuclear Technology*, Vol. 57, 1982, pp. 413–425.

[4] Fleck, R. G., Price, E. G., and Cheadle, B. A., "Pressure Tube Development for CANDU Reactors," *Zirconium in the Nuclear Industry: Tenth International Symposium, ASTM STP 824*, D. G. Franklin and R. B. Adamson, Eds., American Society for Testing and Materials, West Conshohocken, PA, 1984, pp. 88–105.

[5] Theaker, J. R., Choubey, R., Moan, G. D., Aldridge, S. A., Davis, L., Graham, R. A., and Coleman, C. E., "Fabrication of Zr-2.5Nb Pressure Tubes to Minimize the Harmful Effects of Trace Elements," *Zirconium in the Nuclear Industry: Tenth International Symposium, ASTM STP 1245*, ASTM, West Conshohocken, PA, 1994, pp. 221–242.

[6] Davies, P. H., Hobsons, R. R., Griffiths, M., and Chow, C. K., "Correlation between Irradiated and Unirradiated Fracture Toughness of Zr-2.5Nb Pressure Tubes," *Zirconium in the Nuclear Industry:*

Tenth International Symposium, ASTM STP 1245, ASTM, West Conshohocken, PA, 1994, pp. 135–167.
[7] Cheadle, B. A. and Condie, R. M., "Effect of Forging Practice on the Microstructure of Zr-2.5Nb Pressure Tube Billets," Report No. CRNL-1333, Atomic Energy of Canada Ltd., Chalk River Laboratories, January 1977.
[8] Condie, R. M. and Cheadle, B. A., "The Effect of β Quenching Heat Treatments on the Microstructure of Zr-2.5 wt% Nb Extrusion Billets," Report No. CRNL-2090, Atomic Energy of Canada Ltd., Chalk River Laboratories, November 1982.
[9] Bethune, I. T. and Williams, C. D., "The $\alpha/(\alpha+\beta)$ Boundary in the Zr-2.5 Nb System," *Journal of Nuclear Materials,* Vol. 29, 1969, pp. 129–132.
[10] Choubey, R. and Jackman, J. A., "Microsegregation of Oxygen in Zr-2.5Nb Alloy Materials," *Metallurgical and Materials Transactions A,* Vol. 27A, 1996, pp. 431–440.
[11] Simpson, L. A., Chow, C. K., and Davies, P. H., "Standard Test Method for Fracture Toughness of CANDU Pressure Tubes," Report No. COG-89-110-1, Atomic Energy of Canada Ltd., Whiteshell Laboratories, 1989 September.
[12] Griffiths, M., Winegar, J. E., Mecke, J. F., and Holt, R. A., "Determination of Dislocation Densities in HCP Metals Using X-ray Diffraction and Transmission Electron Microscopy," *Advances in X-ray Analysis,* Vol. 35, 1992, pp. 593–599.
[13] Winton, J. and Murgatroyd, R. A., "The Effect of Variations in Composition and Heat Treatment on the Properties of Zr-Nb Alloys," *Electrochemical Technology,* Vol. 4, 1966, pp. 358–365.
[14] Stoloff, N. S., "Composite Strengthening," *Alloy and Microstructure Design,* J. K. Tein and G. S. Ansell, Eds., Academic Press, New York, 1976, pp. 83–84.
[15] Srivastava, D., Madangopal, K., Banerjee, S., and Ranganathan, S., "Self Accommodation Morphology of Martensite Variants in Zr-2.5 wt% Nb Alloy," *Acta Metallurgica et Materialia,* Vol. 41, 1993, pp. 3445–3454.
[16] Ells, C. E. and Cheadle, B. A., "Ageing and Recovery in Cold Rolled Zr-2.5 wt% Nb Alloy," *Journal of Nuclear Materials,* Vol. 23, 1967, pp. 257–269.
[17] Kishore, R., Singh, R. N., Dey, C. K., and Sinha, T. K., "Age Hardening of Cold-Worked Zr-2.5wt% Nb Pressure Tube Alloy," *Journal of Nuclear Materials,* Vol. 187, 1992, pp. 70–73.
[18] Garde, A. M., Chung, H. M., and Kassner, T. F., *Acta Metallurgica,* Vol. 26, 1978, pp. 153–166.
[19] Choubey, R. and Jonas, J. J., "High-temperature Flow and Aging behaviour of Oxygen-enriched Zircaloy-2 and Zr-2.5Nb," *Metal Science,* Vol. 15, 1981, pp. 1–9.

C. David Williams,[1] Mick O. Marlowe,[1] Ronald B. Adamson,[2] Steven B. Wisner,[2] Robert A. Rand,[1] and J. Sam Armijo[1]

Zircaloy-2 Lined Zirconium Barrier Fuel Cladding

REFERENCE: Williams, C. D., Marlowe, M. O., Adamson, R. B., Wisner, S. B., Rand, R. A., and Armijo, J. S., **"Zircaloy-2 Lined Zirconium Barrier Fuel Cladding,"** *Zirconium in the Nuclear Industry: Eleventh International Symposium, ASTM STP 1295,* E. R. Bradley and G. P. Sabol, Eds., American Society for Testing and Materials, 1996, pp. 676–694.

ABSTRACT: The introduction of Zr-lined "barrier" fuel clad tubing by GE in the early 1980s to counter the pellet-clad interaction (PCI) failure mechanism in fuel for boiling water reactors provided a major improvement in fuel reliability and operational flexibility. While the frequency of fuel failures has been substantially reduced in the past decade, an increased tendency has been observed for failed fuel rods to exhibit post-failure degradation in the form of longer cracks that allow release of radioactive off-gas and contamination of the reactor coolant circuit with tramp fuel material. One factor involved in this degradation is hydriding of the cladding at a location remote from the initial perforation of the fuel rod. This local hydriding can lead to secondary crack initiation in the cladding when stressed by the fuel expansion accompanying a power increase.

A modification of Zr-lined "barrier" fuel clad tubing has been developed to retard post-failure local hydriding while retaining the proven PCI resistance of the high-purity sponge Zr barrier. By adding a thin inner layer of corrosion-resistant Zircaloy-2 bonded to the inner surface of the Zr-barrier tube, the resistance to internal corrosion and hydrogen generation in a perforated fuel cladding tube is made equivalent to that of an all-Zircaloy-2 tube. Tests show that the PCI mitigating capability of the Zr barrier is not compromised by this inner Zircaloy-2 liner. Materials considerations and manufacturing technology used to integrate this optional inner liner with other Zr barrier tubing properties and performance requirements are discussed with a summary of testing experience.

KEYWORDS: nuclear fuel cladding, zirconium barrier, inner Zircaloy liner, pellet clad interaction, post-failure degradation, corrosion resistance, hydriding, manufacturing processes, in-reactor fuel performance, zirconium alloys, nuclear applications, radiation effects, crack propagation

1. Background

1.1 Zr Barrier Implementation

The development and introduction of Zr-lined barrier cladding have been reviewed recently by Armijo et al [*1*]. This barrier cladding has effectively mitigated the pellet-cladding-interaction (PCI) failure mechanism and provided reactor operators with a high degree of flexibility in power management [*2–4*]. Since commercial introduction in 1981, approximately 3 million fuel rods having Zr-lined barrier cladding have been installed in BWRs around the world.

The broad implementation of barrier cladding through the mid-1980s contributed to a continuous reduction in fuel failure rate for BWRs. However, through the late 1980s and early

General Electric Nuclear Energy, [1]Wilmington, N.C. and [2]Pleasanton, CA.

1990s, there has been an increasing trend for the gaseous fission product release levels associated with individual fuel rod failures to be higher than was generally observed with the failure of earlier fuel. This tendency for substantial degradation of both barrier and non-barrier fuel rods after a perforation has occurred has been the subject of recent reports and discussion [5–9].

GE examination of failed fuel rods [8] has confirmed that the high "off-gas" rates coincide with the formation of brittle cracks extending for lengths of 15 to 330 cm (6 to 130 in.) along the cladding tube, these cracks not generally being associated with hydrides. The primary driving forces for initiation and propagation of these cracks are stresses generated in the cladding by fuel pellet expansion during rod power increases. Examinations of degraded rods showed that the long cracks were sometimes extensions of the initial or primary perforation of the cladding tube; in many cases, however, the long cracks were initiated in regions of localized hydrides formed at a location on the rod well separated from the primary failure.

Davies [10] has explained the chemistry and kinetics of secondary cladding hydriding after initial perforation in terms of the oxygen/hydrogen ratio that is developed at the cladding inner surface as a function of axial position inside the rod after initial perforation. In short, hydrogen-rich regions are developed some distance away from the source of steam ingress, and these become sites of local hydride formation in the cladding wall.

Early assessments [11–12] of the apparently new tendency for fuel degradation included the possibility that the problem was related to lower oxidation resistance of the inner Zr layer in Zr barrier cladding compared with conventional non-barrier cladding. The assessments suggested that the observed degradation was caused by hydrogen-related embrittlement of the Zircaloy-2 cladding promoted by the low oxidation resistance of the Zr barrier material and could thus be an inherent post-failure characteristic limitation of Zr barrier cladding. Data now available [5–9], however, show that degradation via growth of long cracks occurs in both Zr-barrier and modern non-barrier fuel rods [3–4].

GE's examination of hydride distributions in degraded rods confirms the pattern of local hydrides as common initiation sites for this severe crack growth [8]. The examinations have also shown that there is no general evidence of very extensive Zr barrier oxidation in the crack region in degraded barrier cladding. Nonetheless, it is clear that, under some conditions of internal oxidation, hydrogen may be generated more rapidly in a Zr-barrier-clad fuel rod than in a Zircaloy-2-clad rod because of the relatively low oxidation resistance of the Zr barrier compared with that of Zircaloy-2. In this way, the Zr barrier could contribute to a reduction of the operating period between the primary and secondary failures, i.e., the period before onset of degradation.

Therefore, as an option to provide protection to the Zr barrier under conditions of internal oxidation, GE has developed and patented [13] a Zircaloy-2 lined Zr-barrier cladding (introduced as TRICLAD™ tubing) using a multi-layer cladding concept tested in the early 1970s in combination with the extensive experience accumulated in tubeshell and tube manufacture for the Zr-lined barrier cladding.

1.2 Related Zircaloy-2 Microstructural Effects and Crack Propagation Resistance

A conclusion from GE's analysis [8] of degradation in BWR fuel rods was that the tendency for formation of long secondary cracks has been higher in cladding manufactured in the period since significant process changes were made throughout the BWR fuel industry to ensure high resistance to nodular corrosion. Important among these changes was a marked refinement of the size distribution of intermetallic, second-phase particles [14–15]. In the case of Zircaloy-2 cladding for GE BWR applications, the median diameter of second-phase particles (SPPs) in the tube wall—measured using transmission electron metallography on foils prepared from the tube mid-wall region—was reduced from near 0.15 μm to less than 0.05 μm [17–18]. This refinement was accomplished by methods including higher average concentrations and tighter control of alloying elements (notably Fe and Ni), improved beta-quenching during the ingot-to-tubeshell process sequence, limiting of total thermal exposure after the beta quench, and use of additional heat treatments late in the cladding tube manufacturing sequence.

Dissolution of second-phase particles occurs in the neutron fluence levels to which BWR cladding operates [19–21]. The rate of dissolution, and the resultant increase in concentration of the alloying elements in the Zircaloy-2 matrix, are relatively high for the small SPPs characteristic of Zircaloy-2 processed for high resistance to nodular corrosion in BWR conditions [22].

The transfer of the alloying elements from intermetallic particles into the Zircaloy-2 matrix is known to influence corrosion resistance and is expected to have significant impact on some aspects of mechanical behavior [14–22]. GE's analysis of the degradation phenomenon has lead to the conclusion that this impact includes a reduction of the resistance of irradiated Zircaloy-2 to crack propagation. It is hypothesised that the smaller SPPs characteristic of BWR cladding material in use in the late 1980s and early 1990s promoted enhanced dissolution in the neutron flux, contributing to reduced resistance to crack propagation.

Therefore, GE has implemented modifications to the processing of the Zircaloy-2 used in the parent cladding tube [16–18]. The manufacturing sequence used by tubeshell suppliers, shown schematically in Section 3 below, has been modified to increase the SPP size in the Zircaloy-2 prior to extrusion by reducing the cooling rate in the beta quench and increasing the allowable ΣA factor by approximately 10^3. Zircaloy-2 tubeshells and the resultant fuel clad tubing manufactured in this way exhibit a mid-wall median SPP diameter in the range observed in tubeshells and BWR tubing manufactured in the early 1980s. The adjustment of SPP size distribution in the parent Zircaloy-2—to promote crack growth resistance in the parent tube in the irradiated condition—is complemented by an outer-surface heat treatment carried out at an intermediate tube size during the tube reduction process. This ensures SPP refinement in the outer surface layer as required for nodular corrosion resistance of fuel cladding in the BWR environment [14–15], consistent with GE's practice since 1984.

The implementation of the optional inner Zircaloy-2 liner for this Zr barrier tube, the primary subject of this paper, has been carried out in parallel to these metallurgical process modifications.

In this way, a cladding tube may be provided for BWR service with up to four functional layers, shown schematically in Fig. 1: (1) an outer surface layer of Zircaloy-2 processed to

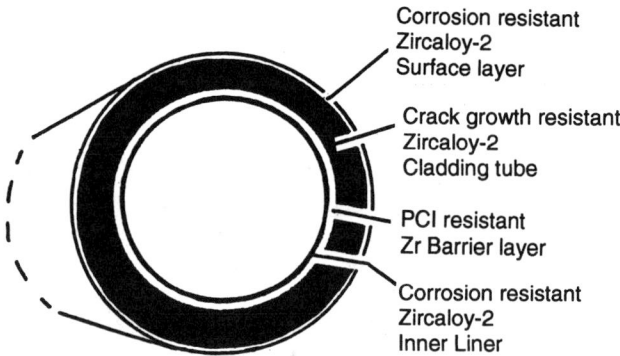

FIG. 1—*Zircaloy-2-lined Zr-barrier fuel cladding tube.*

have high resistance to nodular corrosion, (2) a parent Zircaloy-2 tube having a microstructure to resist extensive crack growth after irradiation in the event of rod perforation, (3) a Zr barrier layer to blunt PCI cracks nucleated at the inner surface, and (4) an optional corrosion-resistant inner surface layer of Zircaloy-2 to slow oxidation and hydrogen generation and delay local hydride formation in the case of rod perforation.

2. Design Approach for Zircaloy-2-Lined Zr Barrier Tubing

The specifications for dimensions, alloy chemistry, and process for each of the layers in this tubing are defined to ensure that: (1) the parent cladding tube meets mechanical properties and corrosion resistance requirements; (2) the PCI resistance of the proven Zr barrier technology is maintained; and (3) in the event of a loss of hermeticity of the fuel rod, the capability of the inner Zircaloy-2 layer to withstand oxidation will meet operating requirements.

2.1 Crack Initiation

The presence of a Zircaloy-2 layer at the inner surface of the tube is expected to result in the formation of stress corrosion cracks in this layer under the conditions of fuel power increases (fuel swelling and fission product release), known to cause PCI crack initiation in non-Zr-barrier fuel cladding. Such cracks could extend through the inner Zircaloy-2 liner and propagate into the Zr barrier layer. The ability of the soft Zr barrier layer to arrest cracks initiated at the inner surface of the Zr barrier is the major factor in the outstanding PCI resistance of barrier cladding. To preserve this PCI resistance, the thickness of the inner Zircaloy-2 liner in the modified Zr barrier tube is selected so that PCI cracks able to propagate through this inner liner will be less than the minimum length capable of driving the crack through the Zr barrier layer under rod operating conditions.

It is known that fission product absorption occurs at the inner surface of the conventional Zr barrier during fuel operation, creating a surface layer within which microcracks are formed under conditions of fuel expansion and fission product release. These microcracks do not propagate through the Zr barrier layer. Ohara et al. [23] have shown that microcracks approximately 10 μm (0.0004 in.) deep are present at the inner surface of Zr barrier layers in fuel cladding after high exposure (fuel exposure up to 49 GWd/t), and that an increased hardness can be measured in this region. Addition of a Zircaloy-2 inner layer will provide protection to the Zr barrier layer relative to this fission product effect, conserving the ductility of the Zr layer for the purpose of PCI resistance.

2.2 Corrosion Behavior of the Inner Zircaloy-2 Liner

For economical and efficient reactor operation, it is desirable that, in the event of a rod perforation detected by an increase in radioactive off-gas level, the fuel rod can be retained in operation until the end of the cycle in which the primary failure occurs. The increasing use of longer reactor cycles requires that the protective lifetime of the inner Zircaloy-2 liner with respect to corrosion by an aqueous environment inside the fuel rod (loss of hermeticity) be not less than around 24 months. In defining the basic requirements of a cladding tube having an inner layer of Zircaloy-2 to provide corrosion protection for the Zr barrier layer, it is possible to identify from available data the minimum thickness of the Zircaloy-2 layer that will provide this protection.

For an operating BWR fuel rod, the cladding inner surface temperature will be close to 350°C, and the average temperature of the aqueous mixture in the pellet cladding gap after a rod perforation will be near 400°C. From measurements of oxidation rates in laboratory tests at 350 and 400°C steam at 1000 psi (6.8 MPa) and from oxide thickness measurements from both sound and failed BWR rods, it is concluded that an inner Zircaloy-2 layer of minimum thickness near ~10 μm (0.0004 in.) can provide effective protection of the Zr barrier layer for an operating period of approximately 24 months. The maximum thickness for the inner Zircaloy-2 liner will be subject to considerations of conservation of the integrated mechanical properties of the final tube wall and of the need to limit—as noted in 2.1 above—the length of PCI cracks that can be initiated in, and propagate through, this liner.

3. Manufacture of Triclad® Tubing

The manufacture of Zircaloy-2-lined Zr barrier cladding is carried out by a sequence of operations similar to those successfully used commercially since 1981 when fuel bundles were built for the first large-scale demonstration of Zr barrier cladding [24]. Co-extrusion in the alpha phase, capable of providing a sound metallurgical bond between Zr and Zircaloy-2, is used to "sandwich" the Zr barrier between the parent Zircaloy-2 tube and the thinner Zircaloy-2 layer at the inner surface. The extrusion is then reduced to near the required final dimensions using conventional pilger mill technology for finishing by surface treatment methods.

The manufacturing sequences used for the components of the extrusion billet and for the completion of the tube-reduced extrusion (TREX) are shown schematically in Fig. 2. The sequence used for the parent Zircaloy-2 component of the tubeshell includes a thermal process cycle including a beta quench to ensure relatively larger SPPs in the Zircaloy-2 tube, in contrast

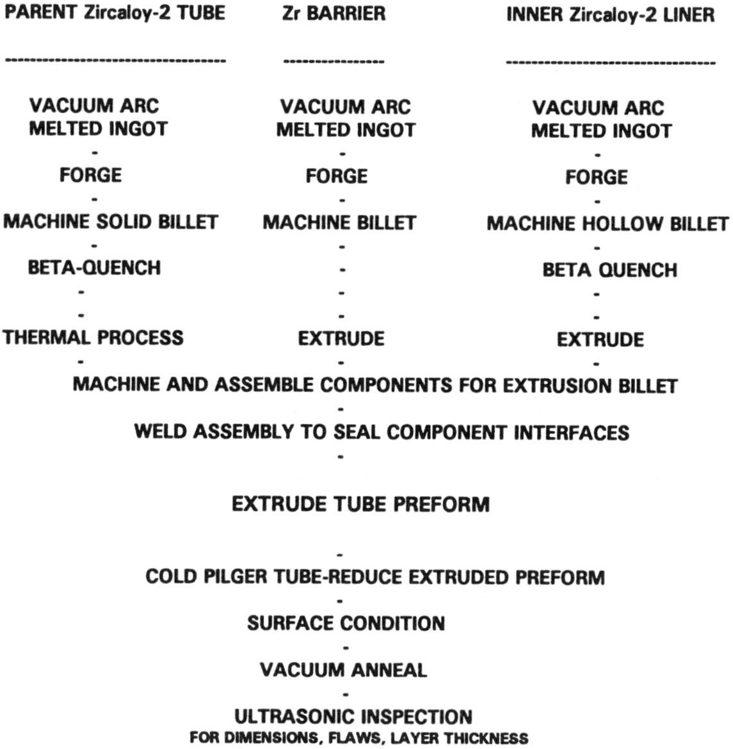

PARENT Zircaloy-2 TUBE	Zr BARRIER	INNER Zircaloy-2 LINER
VACUUM ARC MELTED INGOT	VACUUM ARC MELTED INGOT	VACUUM ARC MELTED INGOT
FORGE	FORGE	FORGE
MACHINE SOLID BILLET	MACHINE BILLET	MACHINE HOLLOW BILLET
BETA-QUENCH	-	BETA QUENCH
THERMAL PROCESS	EXTRUDE	EXTRUDE

MACHINE AND ASSEMBLE COMPONENTS FOR EXTRUSION BILLET

WELD ASSEMBLY TO SEAL COMPONENT INTERFACES

EXTRUDE TUBE PREFORM

COLD PILGER TUBE-REDUCE EXTRUDED PREFORM

SURFACE CONDITION

VACUUM ANNEAL

ULTRASONIC INSPECTION
FOR DIMENSIONS, FLAWS, LAYER THICKNESS

FIG. 2—*Process for manufacture of extruded tubeshells for Zircaloy-2-lined Zr barrier tubing.*

to the sequence used for the inner Zircaloy-2 liner component, which is controlled to provide a smaller-size distribution of SPPs and ensure high corrosion resistance.

As for the established Zr barrier tubing, the steps taken to ensure geometrical and cleanliness control of the components and assembly of the extrusion billet are critically important for the successful production of the tubeshells and of the subsequent tubing.

Figure 2 shows that the final control step carried out by the tubeshell supplier is inspection. Conventional ultrasonic inspection is used to verify that overall TREX dimensional requirements are met; that inner and outer surfaces meet integrity requirements; and that bonding between the parent tube, the Zr barrier, and the inner liner meets requirements (the sensitivity of ultrasonic inspection to debonding at such interfaces has been well established in the manufacturing experience with Zr barrier tubeshells). Ultrasonic inspection is also used to measure the thickness of the Zr barrier and inner Zircaloy-2 liner in the TREX. The accuracy of this ultrasonic measurement has been verified using metallographic data obtained from destructive sampling along the length of individual tubeshells. The liner thickness measurements at this intermediate stage verify uniformity and acceptability of the TREX for the subsequent tube manufacturing operations.

The process for conversion of the Zircaloy-2-lined Zr barrier TREX to near-final-sized tubing is carried out in the tube reduction sequence shown in Fig. 3. The GE manufacturing process for fuel clad tubing includes an induction heat treatment of the outer layer of the parent Zircaloy-2 tube to ensure corrosion resistance in the BWR environment [5].

During tube reduction of this Zircaloy-2-lined Zr barrier tubing, the inner surface responds to parameters including pilger mill feed rate, stroke rate, reduction imposed per pilger step, and surface lubrication in a manner similar to that experienced on Zircaloy tubes in general [25]. Measurement of the variability in thickness of the as-reduced inner Zircaloy-2 liner has shown a standard deviation of approximately 6% of the average liner thickness.

The process used for manufacture of the finished Zircaloy-2-lined Zr barrier tubing accommodates the potential for variabilities in surface characteristics after tube reduction with a post-reduction surface removal. A combination of chemical and mechanical surface conditioning provides for control of the tube wall thickness and the inner liner thickness and also ensures that both outer and inner surfaces meet requirements for integrity and smoothness.

COLD PILGER TUBE REDUCTION

RECRYSTALLIZATION ANNEAL

INDUCTION HEAT-TREATMENT OF THE OUTER SURFACE LAYER

COLD PILGER TUBE REDUCTION

RECRYSTALLIZATION ANNEAL

COLD PILGER TUBE REDUCTION

RECRYSTALLIZATION ANNEAL

CHEMICAL/MECHANICAL FINISHING PROCESS COMBINATION

ULTRASONIC INSPECTION

CUT-TO-LENGTH / END FACE FOR WELDING

FINAL VISUAL INSPECTION

FIG. 3—*Process for manufacture of Zircaloy-2-lined Zr-barrier fuel cladding.*

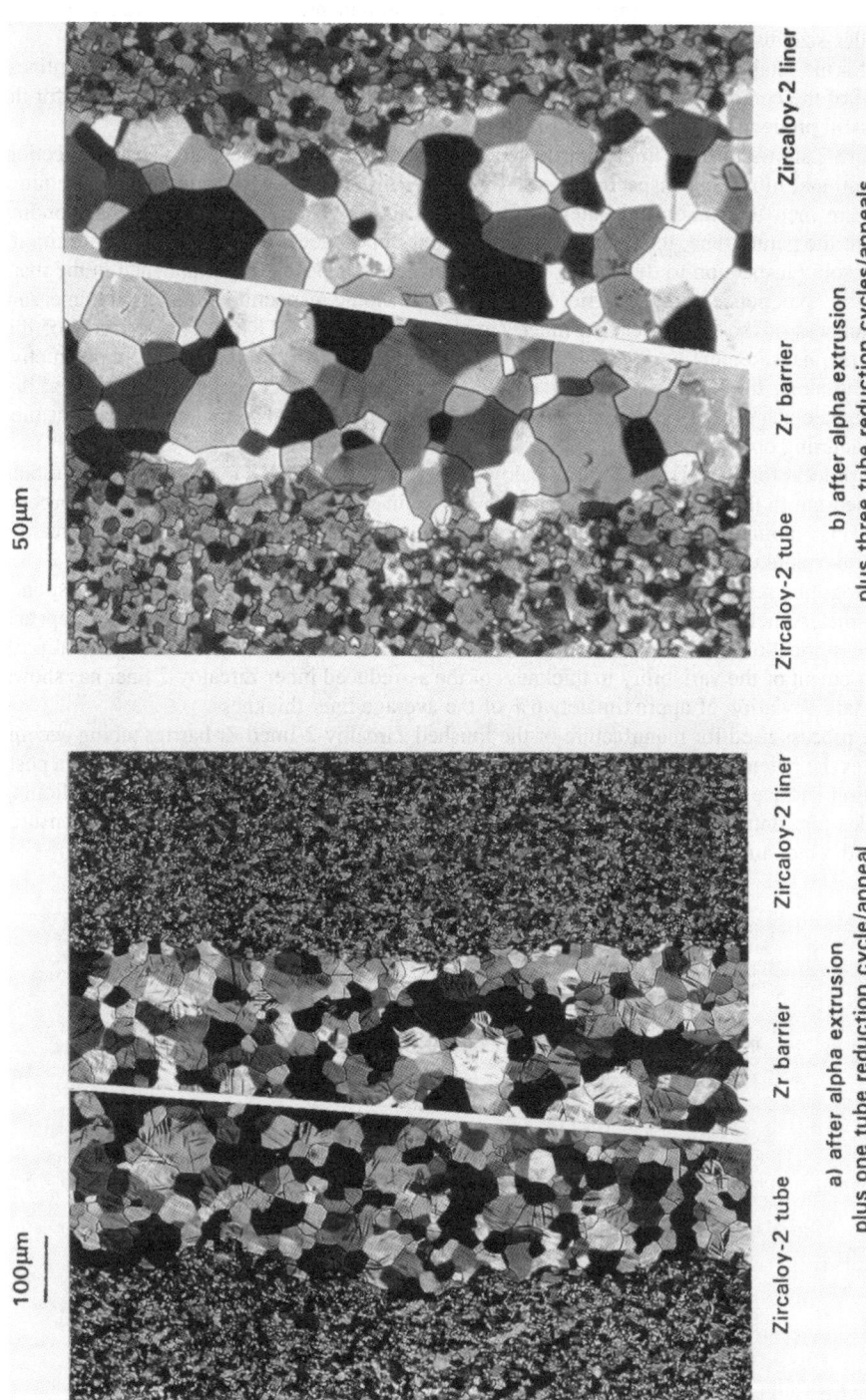

FIG. 4—*Interface microstructures at boundaries between Zircaloy-2 tube, Zr barrier, and inner Zircaloy-2 liner.*

Ultrasonic inspection of the finished tubes is used to verify that the tubes meet design requirements for overall dimensions and surface quality. The verification of inner Zircaloy-2 layer thickness in the finished tubes is done in the absence of a suitable non-destructive inspection method by metallographic measurements in accordance with a sampling plan qualified by data gathered from samples taken at intervals along the length of finished tubes.

An important feature of the product tubing is the integrity of the metallurgical bonding between the Zr barrier and both the parent Zircaloy-2 tube and the inner Zircaloy-2 liner. The ultrasonic inspection carried out at the TREX stage is sensitive to the presence of debonding between layers in the tubes. The extensive manufacturing experience gained with Zr barrier TREX has established high confidence in the overall process capability to provide fuel clad tubing with fully bonded Zircaloy-2 and Zr layers. Data generated in the manufacture of Zircaloy-2-lined Zr barrier tubing shows that this capability is extended to this modification of the Zr barrier tube. Figure 4 illustrates typical interface morphology, shown at the two Zircaloy-2-to-Zr interfaces for the tube reduced extrusion (TREX) and intermediate tube sizes after the reduction and anneal cycles at each size. It is noted that, as in the case of the conventional Zr-barrier tube, these interfaces deviate on a microscopic scale from a smooth cylindrical form, a result attributed to the flow of material during the extrusion when bonding is established, combined with the incremental nature of the deformation, both axial and azimuthal, incorporated in the pilger reduction process.

4. PCI Resistance and Corrosion Behavior of Zircaloy-2-Lined Zr Barrier Tubing

4.1 PCI Resistance

The ability of Zircaloy-2-lined Zr barrier tubing to resist stress corrosion cracking has been compared to that of Zr barrier cladding and non-barrier cladding in laboratory tests. The configuration used for this test by GE has been described previously [26] and is shown in Fig. 5. An expanding mandrel is used to apply a localized circumferential stress distribution similar to that which is generated over cracks in expanding fuel pellets during and following a power increase.

The sample, taken from as-cold reduced tubing to simulate the hardened state obtained by irradiation, is exposed to an iodine environment at 315°C. The load is transmitted to the tube sample via an alumina annulus that simulates the fuel pellet surface; the alumina annulus is around tungsten carbide segments in contact with a soft zirconium cylinder that is compressed by the testing machine. Outer surface diametral strain is measured by a caliper extensometer in contact with opposite points on the sample at locations corresponding to the midpoint be-

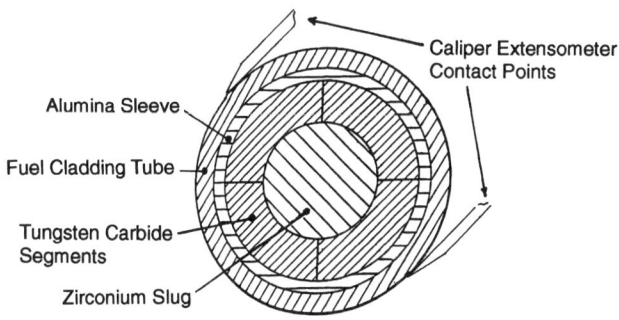

FIG. 5—*Expanding Mandrel configuration for PCI simulation tests.*

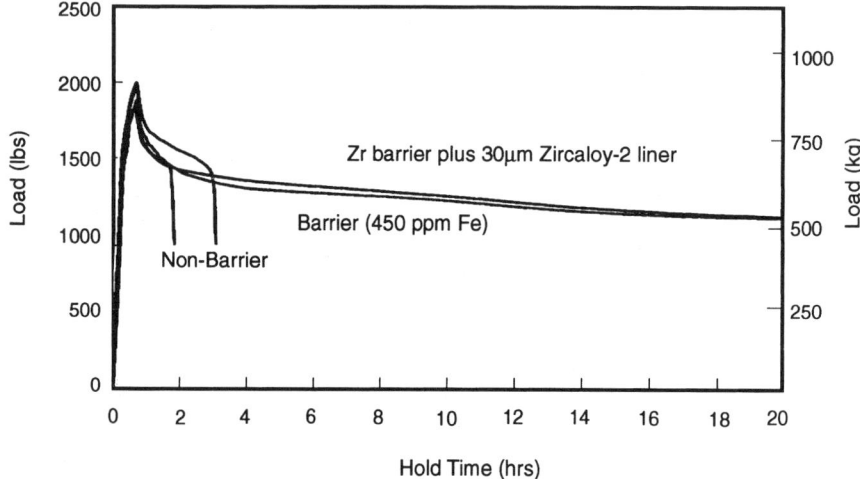

FIG. 6—*Simulated PCI behavior for non-barrier, Zr-barrier, and Zircaloy-2 (0.0012 in. thickness)-lined Zr-barrier cladding. Samples were tested in the ~70% cold-worked condition. Simulated ramp cladding strain was 4% in a 40-Pa iodine atmosphere at 315°C.*

tween gaps in the two opposing tungsten carbide segments. The sample is heated using an induction coil, and iodine at 40 Pa is directed through the segment gaps. To simulate the effects of a power increase in a fuel rod, the load is applied rapidly and constant deflection maintained as the load decays.

Figure 6 shows the marked difference in load-time curves for samples of non-barrier and Zr-barrier tubing. Failure due to initiation and propagation of stress-corrosion cracks from the point of localized loading over spaces between segments occurs in less than 4 h for the non-barrier cladding. No failure occurred in Zr barrier tubing during tests of up to 60 h in duration. Tests on samples of Zircaloy-2-lined Zr barrier tubing under similar loading conditions have shown that, while stress corrosion cracks are nucleated in and do penetrate the inner Zircaloy-2 liner as expected, they do not propagate through the Zr barrier layer. Figure 7 shows the stable non-propagating state of cracks formed in the inner Zircaloy-2 liner over test periods of up to 60 h.

Figure 8 shows data from tests carried out on samples cut from cold-reduced tubing having inner Zircaloy-2 layers of three average thicknesses covering the range from 15 μm (0.0006 in.) to 50 μm (0.002 in.). This represents a range of thickness of interest for this inner liner based on considerations referred to in Section 2. These samples performed in a similar manner, each inner liner showing stress corrosion cracks reaching the Zr barrier, and the cracks in each case not propagating through the Zr barrier layer.

These data indicate that a Zircaloy-2-lined Zr barrier tube with nominal inner liner thickness in the range 15 to 50 μm is viable as an alternative to a Zr barrier tube for mitigating PCI failure.

4.2 Corrosion Behavior

Figure 9 shows the lustrous black oxide layer formed on an inner Zircaloy-2 liner of ~25 μm (~0.001 in.) average thickness subjected to a high-temperature steam corrosion test [in steam at 1750 psi (12 MPa) for 4 h at 410°C, followed by 16 h at 520°C]. The black oxide

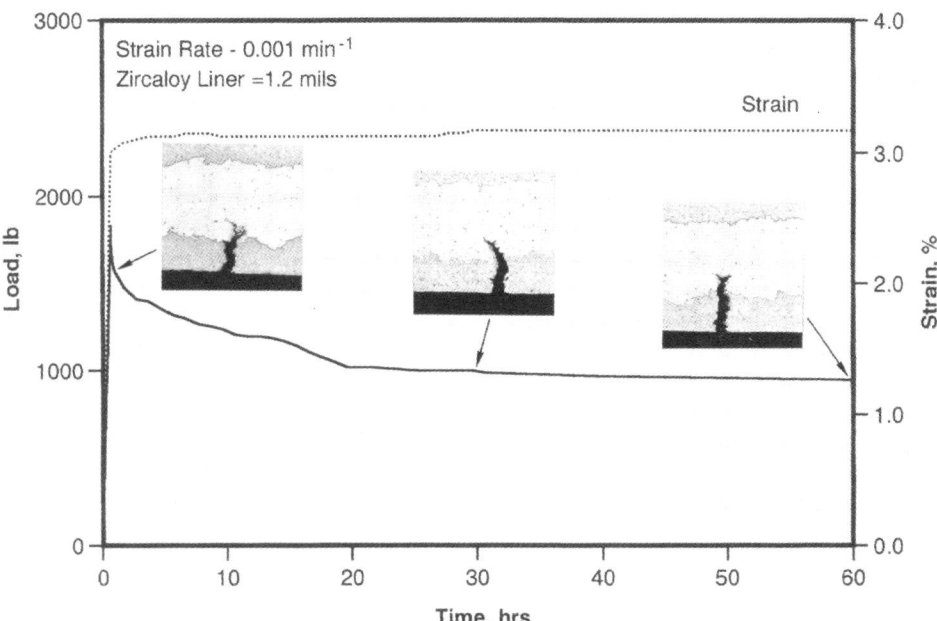

FIG. 7—*Simulated PCI behavior for Zircaloy-2 (0.0012 in. thickness)-lined Zr barrier cladding. Sample was tested in ~70% cold-worked condition. Strain rate was 0.001 min^{-1} in a 40-Pa iodine atmosphere at 315°C. Strain level was maintained through 60 h. Inset micrographs show blunting by the Zr barrier layer of cracks initiated in the inner Zircaloy-2 liner.*

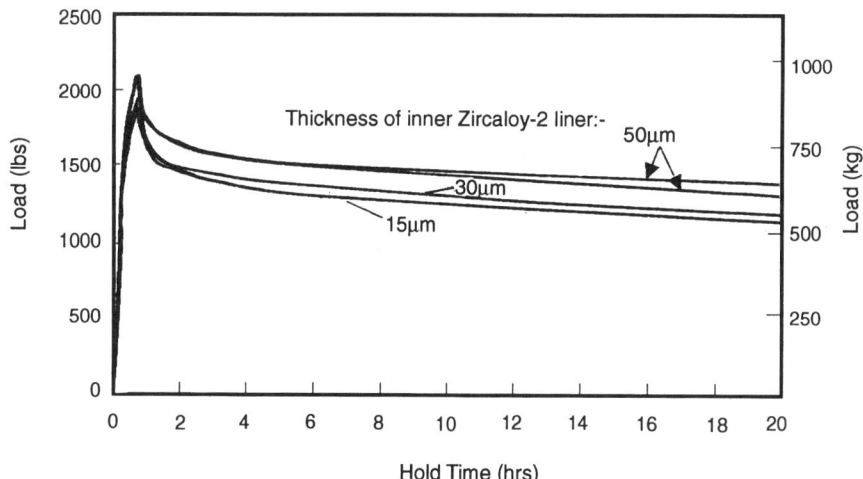

FIG. 8—*Simulated PCI behavior for Zircaloy-2-lined Zr barrier cladding having three thicknesses for the inner Zircaloy-2 liner. Samples were tested in ~70% cold-worked condition. Simulated ramp cladding strain was 4%, in a 40-Pa iodine atmosphere at 315°C.*

FIG. 9—*Inner Zircaloy-2 liner surface oxidized in steam for 4 h at 410°C at 1750 psi (12 MPa) followed by 16 h at 520°C at 1750 psi (12 MPa). Lustrous black oxide is on Zircaloy-2. White oxide is visible at "A" and "B" where Zr barrier layer was exposed at sample ends. Print magnification X3.*

layer formed in this test has a measured thickness of 3 ± 1 μm ($0.000\,12 \pm 0.00004$ in.). It is noted that the flaking white oxide that forms on the Zr barrier layer exposed at the machined ends of the sample under these aggressive test conditions is visible in Fig. 9, highlighting the protective function of the inner Zircaloy-2 liner.

An additional corrosion test that simulates aspects of the behavior of a perforated fuel rod has been used to examine the capability of the inner Zircaloy-2 liner to protect the Zr barrier

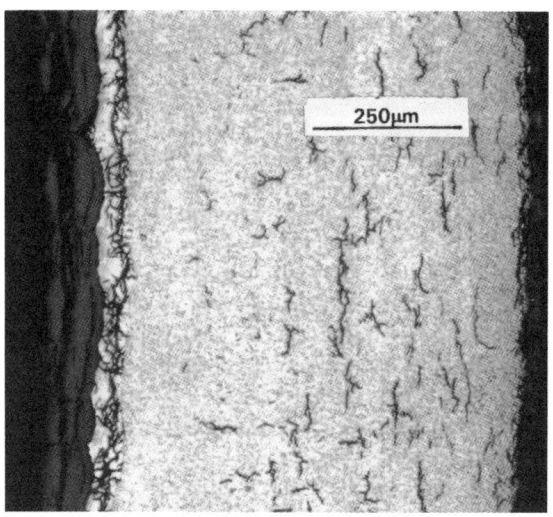

FIG. 10—*Oxidized Zr barrier layer in BWR fuel rod operated four months after clad perforation. Zr barrier was bonded to fuel pellet. The Zircaloy-2 cladding had picked up ~150 ppm of hydrogen.*

from corrosion and hydrogen pickup. Closed-end tubing samples 22.8 cm. (9 in.) in length having an axial pre-machined slit 2.5 cm (1 in.) in length penetrating the wall and containing UO_2 pellets to modify both the chemical and steam transport conditions were subjected to autoclave exposure in 400°C steam at 1000 psi (6.8 MPa) for five days.

As a basis for comparison of performance of unirradiated cladding tube samples, Fig. 10 shows a section through a Zr-barrier fuel rod that operated in a commercial BWR for four months after rod perforation was detected in radioactive off-gas measurements. The hydride distribution evident in the body of the Zircaloy-2 tube represented a hydrogen content of 150 ppm. A region relatively denuded of hydride precipitates is present adjacent to the Zr barrier layer, and it was noted that the Zr barrier layer was oxidized at the inner surface and was bonded to the fuel pellet in contact with it.

The samples shown in Figs. 11, 12, and 13 were cut from tubes subjected to the simulation test described above, the samples being taken at a location 180° around from the axial slit machined in the tube.

Figure 11 shows the result obtained for a Zr-barrier tube sample subjected to the simulation test described above. A hydride distribution similar to that shown in Fig. 10 is evident, with a hydrogen concentration near 300 ppm. Again the Zr barrier layer was oxidized at the inner surface and was bonded to the contacting fuel pellet. Thus the simulation has reproduced the key features seen in the irradiated fuel rod cladding shown in Fig. 10. It is noted that the presence of radially oriented hydrides in this sample, as shown in Fig. 11, is attributed to stresses generated in the tube wall due to closure of the gap between the tube and the pellets by oxidation of the Zr barrier layer during the test.

Figure 12 shows a sample of Zircaloy-2-lined Zr barrier cladding exposed to similar conditions to those experienced by the Zr barrier sample shown in Fig. 11. The inner Zircaloy-2

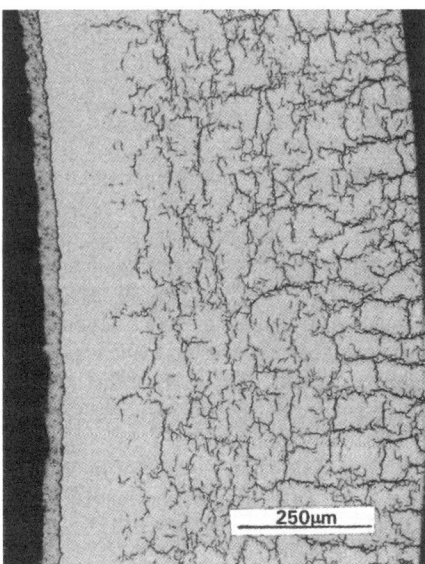

FIG. 11—*Zr-barrier tube sample oxidized in degradation simulation test described in Section 4.2. In steam for five days at 400°C at 1000 psi (6.8 MPa). The Zr barrier was bonded to fuel pellet. The Zircaloy-2 cladding had picked up 300 ppm of hydrogen.*

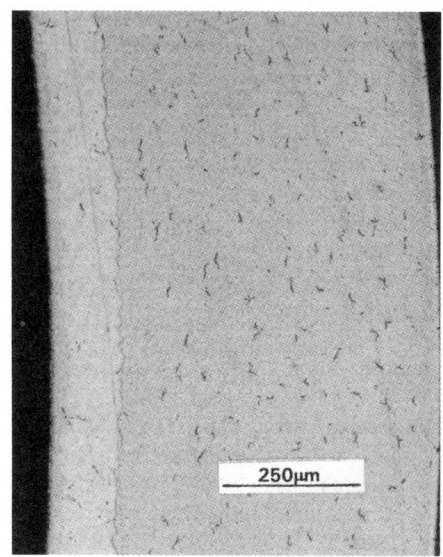

FIG. 12—*Zircaloy-2-lined Zr-barrier cladding sample oxidized in degradation simulation test described in Section 4.2. In steam for five days at 400°C at 1000 psi (6.8 MPa). The inner Zircaloy-2 liner was not bonded to fuel pellet. The Zr barrier was not oxidized. The Zircaloy-2 cladding had picked up 30 ppm hydrogen.*

liner has prevented Zr barrier oxidation and bonding of the fuel pellet to the Zr barrier. Also, the inner Zircaloy-2 liner has restricted the hydrogen pickup to 30 ppm, approximately 10% of the pickup shown by the Zr barrier sample illustrated in Fig. 11.

It is concluded from these results that the test described above provides a useful simulation of effects that may be created in a fuel rod in which a loss of hermeticity has occurred and that the inner Zircaloy-2 liner is capable of providing an effective corrosion protection for the Zr barrier.

To examine the behavior of the inner Zircaloy-2 liner and underlying Zr barrier layer in the event that small perforations occur in the inner Zircaloy-2 liner, samples having defects in the inner Zircaloy-2 liner have been subjected to high-temperature steam testing. Figure 13 shows a metallographic section through a point of Zircaloy-2 inner liner penetration on a sample exposed to steam at 400°C and 1500 psi (10.3 MPa) for 30 days. It is noted that: (a) the oxidation of the Zr barrier layer propagated to a limited extent beyond the dimensions of the initial penetration of the Zircaloy-2 liner; (b) there was no evidence of extensive delamination of the Zircaloy-2 inner liner from the Zr barrier as a result of the localized corrosion and hydrogen pickup; and (c) the oxidation resistance of the Zircaloy-2 body was not diminished in the region directly below the point of Zr barrier oxidation.

It is concluded that the concept of an inner Zircaloy-2 liner for protection of the Zr barrier layer is capable of tolerating local perforation of the inner liner without allowing extensive corrosion of the Zr barrier layer.

5. Summary

This work has shown that the addition of an inner liner of Zircaloy-2 can provide protection for a Zr barrier layer from oxidation and hydrogen pickup. The work has shown that the Zr

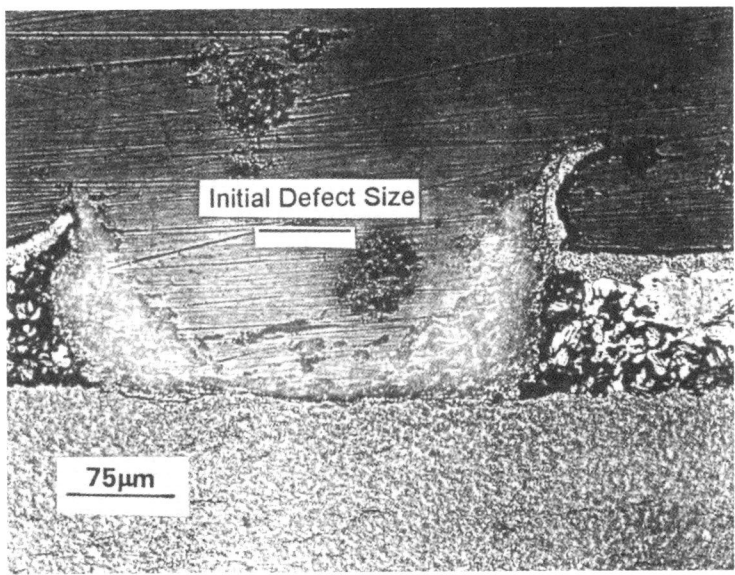

FIG. 13—*Section through corrosion test sample of Zircaloy-2-lined Zr barrier tubing showing oxidation of Zr barrier in region adjacent to defect in inner Zircaloy-2 liner after exposure to steam at 400°C for 30 days at 1500 psi (10.3 MPa).*

barrier layer between the inner Zircaloy-2 liner and the parent Zircaloy-2 tube retains its ability to protect the parent Zircaloy-2 tube from stress corrosion cracks initiated at the cladding inner surface. Technology for manufacture of this three-layer extruded and tube-reduced tube has been developed as an option from the base established with the conventional Zr-barrier tube product. The composite tube concept can be integrated with other metallurgical objectives of tube manufacturing technology to provide the balanced mechanical and corrosion properties required in fuel cladding tubes.

The technology described has been used for commercial manufacture of Zircaloy-2-lined Zr barrier (TRICLAD™) tubing in quantities to support reactor reload assembly targeted for 1995–1996 reactor loading.

Acknowledgments

Many members of GE's Nuclear Fuel Engineering and Manufacturing teams contributed to this work. Contributions from Kobe Steel Company and GE Canada Inc. in the area of tube manufacture are acknowledged. The efforts of M. Perez and his CEZUS colleagues, and—in particular—of T. F. Cook, M. J. Pleinis, and M. V. Morris and their colleagues at Westinghouse/Western Zirconium are acknowledged for collaboration in development and production of TRICLAD™ TREX.

References

[1] Armijo, J. S., Coffin, L. F., and Rosenbaum, H. S., "Development of Zirconium Barrier Fuel Cladding," *Zirconium in the Nuclear Industry (Tenth International Symposium), ASTM STP 1245,* 1994, p. 3.

[2] Yang, R. L., Ozer, O., and Klepfer, H. H., "Fuel Performance Evaluation for EPRI Program Planning," *Proceedings,* Fuel for the 90's, ANS-ENS International Topical Meeting on LWR Fuel Performance, Avignon, France, April 1991, p. 258.
[3] Krebs, W.-D. and Garzarolli, F., "Advances in Nuclear Fuel Design for Improving Economy and Operating Reliability," *Proceedings,* Topfuel'95, 1995, German Nuclear Society KTG, published by Inforum Verlag, Vol. 1, p. 15.
[4] Weidinger, H. G. and Harbottle, J. E., "An Evaluation of LWR Fuel Performance: Current Successes and Problems," *Proceedings,* Topfuel'95, 1995, German Nuclear Society KTG, published by Inforum Verlag, Vol. 1, p. 219.
[5] Potts, G. A. and Proebstle, R. A., "Recent GE BWR Fuel Experience," *Proceedings,* 1994 International Topical Meeting on Light Water Reactor Fuel Performance, American Nuclear Society, West Palm Beach, FL, p. 87.
[6] Harbottle, J. E., Kennard, M. W., Sunderland, D. J., and Strasser, A. A., "The Behavior of Defective BWR Barrier and Non-Barrier Fuel," *Proceedings,* 1994 International Topical Meeting on Light Water Reactor Fuel Performance, American Nuclear Society, West Palm Beach, FL, p. 391.
[7] Vesterlund, G. and Corsetti, L. V. "Recent ABB Fuel Design and Performance Experience," *Proceedings,* 1994 International Topical Meeting on Light Water Reactor Fuel Performance, American Nuclear Society, West Palm Beach, FL, p. 62.
[8] Armijo, J. S., "Performance of Failed BWR Fuel," *Proceedings,* 1994 International Topical Meeting on Light Water Reactor Fuel Performance, American Nuclear Society, West Palm Beach, FL, p. 410.
[9] Lin, K.-F., Chung, C.-S., Yeh, J.-J., Chen, J.-H., Chu, S.-S., and Lin, L.-F., "Investigation on the Post-Defect Deterioration of Non-barrier BWR Failed Fuel Rods," *Proceedings,* 1994 International Topical Meeting on Light Water Reactor Fuel Performance, American Nuclear Society, West Palm Beach, FL, p. 377.
[10] Davies, J. H., "Secondary Damage in LWR Fuel Following PCI Defection: Characteristics and Mechanism," *Proceedings,* IAEA Specialists Meeting, Chalk River, Ontario, Canada, published by IAEA, Vienna, 1979, p. 135.
[11] Jonsson, A., Hallstadius, L., Grapengiesser, B., and Lystell, G., "Failure of Barrier Rod in Oskarshamn 3," *Proceedings,* ANS-ENS International Topical Meeting on LWR Fuel Performance, Avignon, France, 1991, p. 371.
[12] Siebold, A. and Woods, K. N., "BWR Advanced Material," *Proceedings,* 1994 International Topical Meeting on Light Water Reactor Fuel Performance, American Nuclear Society, West Palm Beach, FL, p. 633.
[13] U.S. Patent Number 5,383,228, J. S. Armijo, H. S. Rosenbaum, and C. D. Williams, "Method for Making Fuel Cladding Having Zirconium Barrier Layers and Inner Liners," 17 Jan. 1995.
[14] Cheng, B. and Adamson, R. B., "Mechanistic Study of Zircaloy Nodular Corrosion," *Zirconium in the Nuclear Industry (Seventh International Symposium), ASTM STP 939,* American Society for Testing and Materials, West Conshohocken, PA, 1987, p. 387.
[15] Rudling, P., Vannesjo, K. L., Vesterlund, G., and Massih, A. R., "Influence of Second Phase Particles on Zircaloy Corrosion in BWR Environment," *Zirconium in the Nuclear Industry, ASTM STP 939,* American Society for Testing and Materials, West Conshohocken, PA, 1987, p. 292.
[16] U.S. Patent Number 5,437,747, R. B. Adamson and G. A. Potts, "Method of Fabricating Zircaloy Tubing High Resistance to Crack Propagation," 1 Aug. 1995.
[17] Marlowe, M. O. and Montes, J., "Degradation Resistant Fuel Cladding Materials and Manufacturing," *Proceedings,* Topfuel'95, Wurzburg, Germany, 1995, p. 60.
[18] Marlowe, M. O. Williams, C. D. Adamson, R. B. and Armijo, J. S., "Degradation Resistant Fuel Cladding," *Proceedings,* Jahrestagung Kerntechnik '95, Nurnberg, Germany, 1995, p. 329.
[19] Gilbert, R. W., Griffiths, M., and Carpenter, G. J. C., *Journal of Nuclear Materials,* Vol. 135, 198, p. 265.
[20] Yang, W. J. S., Tucker, R. P., Cheng, B., and Adamson, R. B., *Journal of Nuclear Materials,* 1986, Vol. 138, p. 185.
[21] Kruger, R. M. and Adamson, R. B., *Journal of Nuclear Materials,* Vol. 205, 1993, p. 242.
[22] Huang, P. Y., Mahmood, S. T., and Adamson, R. B., "Effect of Thermomechanical Processing on In-Reactor Corrosion and Post-Irradiation Properties of Zircaloy-2," this publication.
[23] Ohara, H., Irube, M., Futakuchi, M., Nomata, T., and Iwata, S., "Fuel Behavior During Power Ramp Tests," *Proceedings,* 1994 International Topical Meeting on Light Water Reactor Fuel Performance, American Nuclear Society, West Palm Beach, FL, p. 674.
[24] Allen, D. E., Donaghy, R. E., Chambers, R. L., and Mack, R. J., "Fabrication of Demonstration Barrier Fuel," General Electric Topical Report GEAP-22101, Vol. 1, April 1982.
[25] Aubin, J. L., Girard, E., and Montmitonnet, P., "Modeling of Damage in Cold Pilgering," *Zirconium*

in the Nuclear Industry (Tenth International Symposium), ASTM STP 1245, American Society for Testing and Materials, West Conshohocken, PA, 1994, p. 245.

[26] Tomalin, D. S., Adamson, R. B., and Gangloff, R. P., "Performance of Irradiated Copper and Zirconium Barrier-Modified Zircaloy Cladding Under Simulated Pellet Cladding Interaction Conditions," *Zirconium in the Nuclear Industry (Fourth Conference), ASTM STP 681,* American Society for Testing and Materials, West Conshohocken, PA, 1979, pp. 122–144.

DISCUSSION

P. Rudling[1] (written discussion)—In the presentation by V. Grigoriev from Studsvik Material, three different materials (irradiated) with large differences in initial second-phase particle-size distribution, LK0, LK1, and LK2, were investigated. Despite the large difference in initial second-phase size distribution in the irradiated materials, no significant difference in fracture toughness between the materials could be detected (all materials contrived low hydrogen contents, 50 to 100 wppm).

C. D. Williams et al. (authors' closure)—The testing technique and specimen geometry presented by Dr. Grigoriev are very interesting. However, it is not clear at this time how directly the results are related to fracture toughness or to the type of brittle fracture exhibited by "degraded" fuel rods. At the present stage of development, we believe that the test results should be treated with caution, particularly until the test includes propagation measurements of an initially sharp crack. With reference to the microstructure of the materials tested in Dr. Grigoriev's work, no specific information was given in the presentation and specific comments would seem inappropriate. In cases where detailed information is available on initial and irradiation-affected microstructure, we have found that mechanical properties are affected.

J. B. Bai[2] (written discussion)—(1) Our experience with Duplex tubes show that the manufacture residual stresses and consequently the loading rate play an important role on the hydride distribution. How about the inner liner in your case?

C. D. Williams et al. (authors' closure)—The final recrystallization anneal used in the manufacture of GE BWR fuel clad tubing leaves very low residual stress levels in the tubing, including the inner Zircaloy-2 liner. We would not expect a significant related effect on hydride orientation and have seen no evidence of such an effect.

J. B. Bai[2] (written discussion)—(2) On the right picture of your slide showing hydrides distribution in the thickness direction, it seems to me that the concentration of hydrogen would be higher than the 300 ppm as indicated.

C. D. Williams et al. (authors' closure)—The hydrogen concentration in the sample illustrated was estimated to be ~300 ppm based on established correlations between visible hydride distributions and results of chemical analysis. We believe that this is accurate to within ±50 ppm. . . .

Bo Cheng[3] (written discussion)—One of the important features of secondary degradation of fuel rods is hydride damage at the high power location when the primary defect occurs at either end of the rod. In such a case, the high power locations will be exposed to a dry hydrogen environment. So my key question is whether the triclad cladding will have better or worse characteristics in a dry hydrogen environment. Do you have any test results on this issue?

C. D. Williams et al. (authors' closure)—Our expectation is that hydrogen absorption at the inner surface Zircaloy-2 in the TRICLAD® tubing will be similar to that at the inner surface of a non-barrier Zircaloy-2 cladding. Tests are in progress.

Pauline H. Davies[4] (written discussion)—Have any void nucleation studies been conducted

---

[1] ABB Atom, Sweden.
[2] Lab MSS/MAT, Ecole Centrale Paris, France.
[3] Electric Power Research Institute, Palo Alto, CA.
[4] AECL, Chalk River, Ontario, Canada.

DISCUSSION ON ZIRCALOY-2 LINED CLADDING   693

to optimize the size and distribution of particles required for high crack growth resistance in the Zircaloy-2?

C. D. Williams et al. (authors' closure)—No studies of this type have been carried out.

G. Bart[5] (written discussion)—Apart from the differing hydride concentration after autoclave corroding of the earlier barrier cladding versus the new triclad tube, I am surprised about the significant radial orientation of the hydrides in the earlier barrier clad sample. Could you comment on the stress fields that were active during the tests?

C. D. Williams et al. (author's closure)—The radial hydride orientation evident in Fig. 11 was caused by circumferential stresses generated in the cladding due to the closure, as a result of oxidation of the Zr barrier layer during the test, of the relatively small pellet-to-cladding gap existing in this sample before testing.

Brian Cox[6] (written discussion)—Your hypothesis for the effect of precipitate size on the incidence of major degradation in perforated barrier cladding is that the small precipitates dissolve during irradation and thereby reduce the crack growth resistance. Can you explain how this can affect a cracking process which is essentially a fast fracture through hydride or at least an essentially radial crack with no evidence for shear or ductility due to a notched condition supplied by cracked hydride?

C. D. Williams et al. (authors' closure)—The GE view is that the available evidence does not support the contention that the observed crack growth generally occurs through radial hydrides. Extensive crack growth has been observed through regions of cladding having only low-to-moderate hydride concentration (<300 ppm H), with no significant radial hydride component. The ability of cracks to run in a radial direction with no apparent ductile component is, on the present evidence, attributed to the combination of the material properties obtained in current material as a result of fluence effects and the mechanical loading obtained in this fuel cladding under current operating conditions.

E. Kohn[7] (written discussion)—The use of Zircaloy-2 as a barrier will result in a distribution of PCI cracks in this liner. As a result, following a defect of the fuel pin (due to another cause), the defects in the barrier will cause more corrosion through the Zr barrier, leading to fuel pin degradation. Has this been considered?

C. D. Williams et al. (authors' closure)—It is recognized that PCI cracks will form in the Zircaloy-2 inner liner under fuel operating conditions and that the resulting cracks will reduce locally the protective capability of the inner liner. However, our laboratory experiments have shown that local oxidation in a defected Zircaloy-2 liner does not lead to extensive oxidation of the Zr barrier layer or degradation of the main body of the Zircaloy-2 cladding tube. For example, see Fig. 13 in the text.

A. Strasser[8] (written discussion)—Was there a difference in the hydrogen content between the barrier and base Zircaloy-2 in defected barrier and defected TRICLAD samples after the autoclave test, and, if so, what difference?

C. D. Williams et al. (authors' closure)—The tests reported here and illustrated in Figs. 10 to 12 in the paper showed ~300 ppm hydrogen in the Zircaloy-2 region of the Zr-barrier

---

[5] Paul Scherrer Institute, Switzerland.
[6] University of Toronto, Toronto, Ontario, Canada.
[7] Ontario Hydro Technologies, Toronto, Ontario, Canada.
[8] The S. M. Stoller Corp., Pleasantville, NY.

tube sample and only ~30 ppm hydrogen in the corresponding region of the Zircaloy-2-lined Zr barrier tube sample. The hydrogen content of the Zr barrier layer in both of these samples was not discernible from visible hydrides and is unknown, but is undoubtedly relatively lower in the TRICLAD™ tubing sample.

Y. S. Kim[9] (written discussion)—(1) I would like to know the position of the defective cladding where the photo with concentrated hydrides near the outer surface was taken.

C. D. Williams et al. (authors' closure)—The cladding section was taken from a position 180° around the tube from the machined slit, and at the same elevation in the tube.

Y. S. Kim[9] (written discussion)—(2) Can you comment on the tensile stress applied to the two types of cladding—TRICLAD™ and the double layer cladding—probably caused by fuel expansion during the testings?

C. D. Williams et al. (authors' closure)—See response to Question 5 above. The relative lack of radial hydrides in the Zircaloy-2-lined Zr barrier tubing is due to the low hydrogen pickup achieved in this sample. Also, this sample would not have achieved the level of circumferential stress noted in the Zr barrier sample because of the relatively low-level oxidation on the inner surface of the Zircaloy-2-lined tube.

---

[9] Korea Atomic Energy Research Institute, Daejom, Korea.

Sergei A. Nikulin,[1] Vladimir I. Goncharov,[1] Vladimir A. Markelov,[2] and Vyacheslav N. Shishov[2]

Effects of Microstructure on Ductility and Fracture Resistance of Zr-1.2Sn-1Nb-0.4Fe Alloy

REFERENCE: Nikulin, S. A., Goncharov, V. I., Markelov, V. A., and Shishov, V. N., "**Effects of Microstructure on Ductility and Fracture Resistance of Zr-1.2Sn-1Nb-0.4Fe Alloy**," *Zirconium in the Nuclear Industry: Eleventh International Symposium, ASTM STP 1295,* E. R. Bradley and G. P. Sabol, Eds., American Society for Testing and Materials, 1996, pp. 695–709.

ABSTRACT: The effect of microstructure on the ductility and fracture resistance of Zr-1.2Sn-1Nb-(0.2-0.5)Fe alloy has been studied. Different structural states of the alloy were attained by varying the iron content and working/heat treatment schedules, which comprised quenching, cold work, and anneal. The results of the tests for uniaxial tension, impact toughness, and static crack resistance as well as the electron microscope analysis of the microstructure revealed that the main structural factors governing the level of the as-recrystallized alloy ductility and fracture resistance are the sizes and uniformity of distribution of intermetallic particles of different types in the matrix. The highest ductility and impact toughness are reached when fine intermetallic particles from 0.03 to 0.20 μm are distributed uniformly within the structure. The impact toughness and critical crack opening grow linearly with an increase of particle distribution density and a decrease in interparticle spacing. Changes in the alloy microstructure and mechanical properties were investigated upon its anneal after β-quenching. It is demonstrated that the highest values of ductility and impact toughness are reached with the formation of a polygonized matrix structure without intermetallic particle precipitation.

KEYWORDS: Zr-1.2Sn-1Nb-0.4Fe alloy, microstructure, intermetallic particle, ductility, fracture resistance

The multicomponent zirconium alloy Zr-1.2Sn-1Nb-0.4Fe, designed in Russia, is a promising structural material for different demanding components of nuclear reactor cores [1]. The combined alloying of zirconium with tin, niobium, and iron as well as the application of working/heat treatment result in high resistance of the alloy to irradiation-induced creep, growth, and corrosion both in water and in a boiling coolant, which makes it a suitable structural material for different core components. In Russia, the alloy was studied comprehensively as a structural material for pressure tubes of boiling reactors (RBMK), and its superior characteristics were corroborated as applied to those components [1]. However, even at the beginning of the studies it was noted that its ductility and toughness are sensitive to some specific features of the structure, which is not observed for binary zirconium-niobium alloys.

Working/heat treatment of the multicomponent alloy, comprised of quenching and cold

---

[1] Professor and senior scientific officer, respectively, Moscow State Institute of Steel and Alloys, Leninsky av. 4, 117936, Moscow, Russia.
[2] Leading scientific officer, A. A. Bochvar All-Russia Scientific Research Institute of Inorganic Materials, Rogov st. 5, 123060, Moscow, Russia.

TABLE 1—*Microstructures of alloy in states studied.*

No. of State	Type of Treatment	Type of Structure	α-Phase Grain Size, μm	Secondary Phase Characteristics Composition and Size, μm	Secondary Phase Characteristics Distribution within Matrix
1	Cold Work (C/W) + α-anneals, 550 to 630°C	Recrystallized with secondary phase precipitates	Uniaxial grains, 3.9 to 5.0	$Zr(NbFe)_2$ 0.10 to 0.15 with high Fe $ZrFe_3$ type 0.4 to 1.2	Stringers in grains; aggregates close to grain boundaries
2	β-quenching + multiple C/W + α-anneals, 550 to 630°C	Recrystallized with only fine particles of secondary phase precipitates	Uniaxial grains, 4.4 to 4.8	$Zr(NbFe)_2$ 0.09 to 0.10 with high Fe $ZrFe_3$ type < 0.3	Uniformly in grains
3	β-quenching, 2000°C/s	Lath α'-martensite, $β_{Zr}$-interlayers	Lath width 0.3 to 0.6, platelet width, 0.5 to 1.4	β-Zr Interlayer width 0.05 to 0.10	Along lath and platelet boundaries
4	β-quenching, 400°C/s	Coarse platelet α'-martensite; $β_{Zr}$-interlayers	Platelet width, 0.7 to 2.0	β-Zr Interlayer width 0.06 to 0.13	Along platelet boundaries
5	Cooling from β-region, 25°C/s	Coarse platelet α-phase; $β_{Zr}$-interlayers	Platelet width, 0.9 to 2.2	β-Zr Interlayer width 0.08 to 0.19	Along platelet boundaries
6	β-quenching 2000°C/s + α-anneal, 380 to 520°C	Polygonized α-matrix; interlayers of fine dispersed eutectoid	Lath width, 0.3 to 0.6; platelet width, 0.5 to 1.6	β-Zr (eutectoid) interlayer thickness, 0.10 to 0.15	Along platelet boundaries
7	β-quenching 400°C/s + α-anneal, 600 to 650°C	Partially recrystallized; α-phase grains and platelets; stringers of secondary phase particles	Grains 1.5 to 3.0; platelet width 0.9 to 3.5	$Zr(NbFe)_2$, 0.05 to 0.25	Along platelet boundaries
8	Cooling from β-region 25°C/s + α-anneal, 550 to 650°C	Polygonized coarse platelet α-phase; stringers of secondary phase particles	Platelet width, 0.9 to 3.0	$Zr(NbFe)_2$, 0.1 to 0.4	Along platelet boundaries

work combined with annealing, can result in the formation of a non-uniform microstructure that differs in composition, morphology, size, and distribution of secondary phases within the matrix. Intermetallic phase particles, the sizes and the volume fraction of which can vary from 0.03 to 1.0 μm and 3 to 3.5 vol%, respectively, can in many respects affect the alloy ductility and fracture resistance [2,3]. This work was carried out to gain a better understanding of this interrelation.

Experimental

For the studies, use was made of 4-mm-thick strips of Zr alloy containing 1.20 to 1.36 wt% Sn, 0.97 to 1.10 wt% Nb, and 0.20, 0.35, or 0.46 wt% Fe. The strips were manufactured from laboratory-prepared ingots 15 kg in mass and 100 mm in diameter using various workings/heat treatment (States 1 to 8 shown in Table 1). The workings/heat treatments were comprised of heating in the β-region from a temperature of 900 to 1000°C followed by cooling at 25 to 2000°C/s and α-annealing from temperatures of 380 to 650°C prior to cold rolling, and multiple cold rollings with intermediate and final anneals at temperatures from 550 to 630°C. The application of different workings/heat treatments resulted in zirconium matrices with different microstructures and secondary phase particles of various sizes dissimilarly distributed within them. Specimens for tension tests and structure analyses were cut out of differently processed billets.

The microstructures of the alloys were studied using an electron microscope YEM-200CX at an acceleration voltage of 160 kV. Foils for the electron microscope studies were prepared in the standard way, namely, electrolytic polishing in TENUPOL-2 in the following electrolyte, C_2H_5OH 75 mL + C_4H_9OH 125 mL + $HClO_4$ 4 mL at a voltage of 20 V and a temperature of 25 to 35°C.

The grain sizes were measured using an automated analyzer "Epiquant." The quantitative parameters of the secondary phase particles were determined by analyzing ten patterns of the microstructures of each alloy. The mean size of particles of each type was measured, and particle-size distribution histograms were constructed. Using the techniques described in Ref 2, the number of particles was counted per unit area of the projection image of foil, and the volume density of particles, N_v, and the mean interparticle spacing were calculated.

Tension tests were conducted at room temperature using a screw-driven Instron machine at a strain rate of 4×10^{-4} s^{-1}. Specimens 3 mm in diameter and 30 mm in gage length were cut out of strips along the rolling direction.

The fracture resistance was defined by impact toughness, KCV, and critical crack opening, δ_c. The KCV was determined using specimens 4 by 4 by 30 mm having a sharp notch 0.8 mm deep and a radius of 0.1 mm oriented across the rolling direction [3]. The δ_c, similarly to the KCV, was determined for statically bent specimens of the same size and orientation having an electric-spark-initiated fatigue crack [4].

In the tension tests, an acoustic emission (AE) from a piezo transducer placed at the specimen was registered by the technique described in Ref 5 simultaneously with the strain diagram recording. Receiver-amplifier devices processed input signals linearly by a level of ±15 V in the frequency range of 0.01 to 10 MHz; the dynamic range is 72 dB. The level of the amplifier noise brought to the inlet is 10 mV.

The detected signal was registered and processed digitally using a microprocessor analyser of AE signals designed at the Moscow State Institute of Steel and Alloys [5]. The schematic of the measuring unit is illustrated in Fig. 1.

Results and Discussion

Microstructure

The general characteristics of the created microstructures studied in their interrelation to the properties as well as the quantitative parameters of the main elements were tabulated (Table 1). The State 1 structure formed after multiple cold rollings with intermediate and final anneals

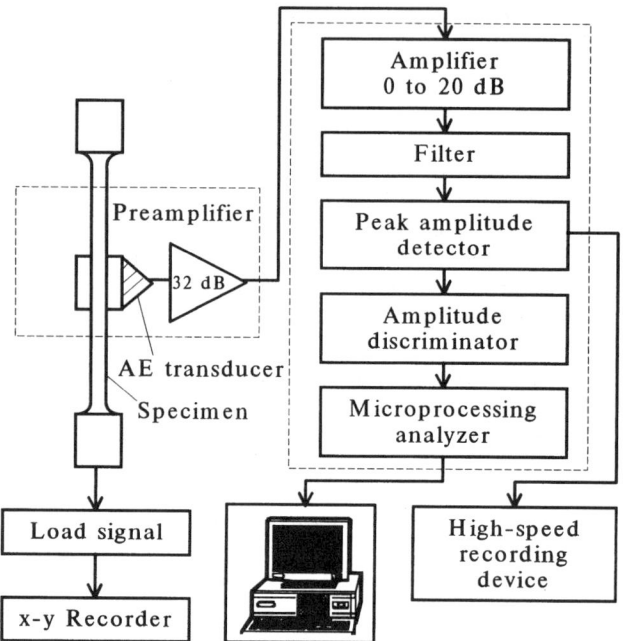

FIG. 1—*Schematic of acoustic-emission facility.*

in the α-phase region was, in essence, a recrystallized phase matrix having uniaxial grains 3.9 to 5.0 μm with stringers of fine particles 0.10 to 0.12 μm that were previously identified by a diffraction analysis to be Zr(NbFe)$_2$ with an Nb:Fe ratio (at%) of ~1.5 [2] (Fig. 2a). These particles are normally observed in this alloy after anneal. The recrystallized grain volume fraction was 80 to 95%. Aggregates of coarse (0.4 to 1.2-μm) particles with high iron, which were previously identified also by a diffraction analysis to be ZrFe$_3$ [2], were arranged closed to the α-grain boundaries. This phase is rarely encountered in commercial alloys of this type.

The alloy in State 2, β-quenched at a cooling rate of 2000°C/s prior to cold rolling and α-annealed, has a similar matrix structure with recrystallized grain sizes 4.4 to 4.8 μm. However, in this state there were only fine Zr(NbFe)$_2$ particles, 0.09 to 0.10 μm, and ZrFe$_3$ particles of <0.3 μm uniformly distributed within the α-matrix grains (Fig. 2b). In State 2, the particle volume density is slightly higher than that of State 1, specifically, N_v = 1.1 to 1.7 × 10^{20} m^{-3} against 0.9 to 1.3 × 10^{20} m^{-3} (see Table 1). The histograms of the intermetallic particle-size distribution in the alloy of States 1 and 2 are given in Fig. 3.

Thus, the alloy processing involving β-quenching prior to cold rollings enhanced the dispersity and uniformity of the intermetallic particle distribution within the matrix through a refinement of coarse particles and elimination of fine Zr(NbFe)$_2$ particle stringers. The higher dispersity and uniformity of the particle distribution within the alloy structure is attained by cooling at higher rates from the β-region so as to study the interrelation between the structure and properties of the alloy as β-quenched and α-annealed States 3 to 8 were created (Table 1).

The alloy as β-quenched at the cooling rate of 2000°C/s (State 3) had for the most part the structure of lath martensite of α'-phase, the lath width being 0.3 to 0.6 μm with individual α'-phase platelets 0.5 to 1.4 μm wide and 0.05 to 0.10-μm-thick $β_{Zr}$-phase interlayers arranged along the lath and platelet boundaries (see Fig. 2c). By decelerating the cooling rate from the

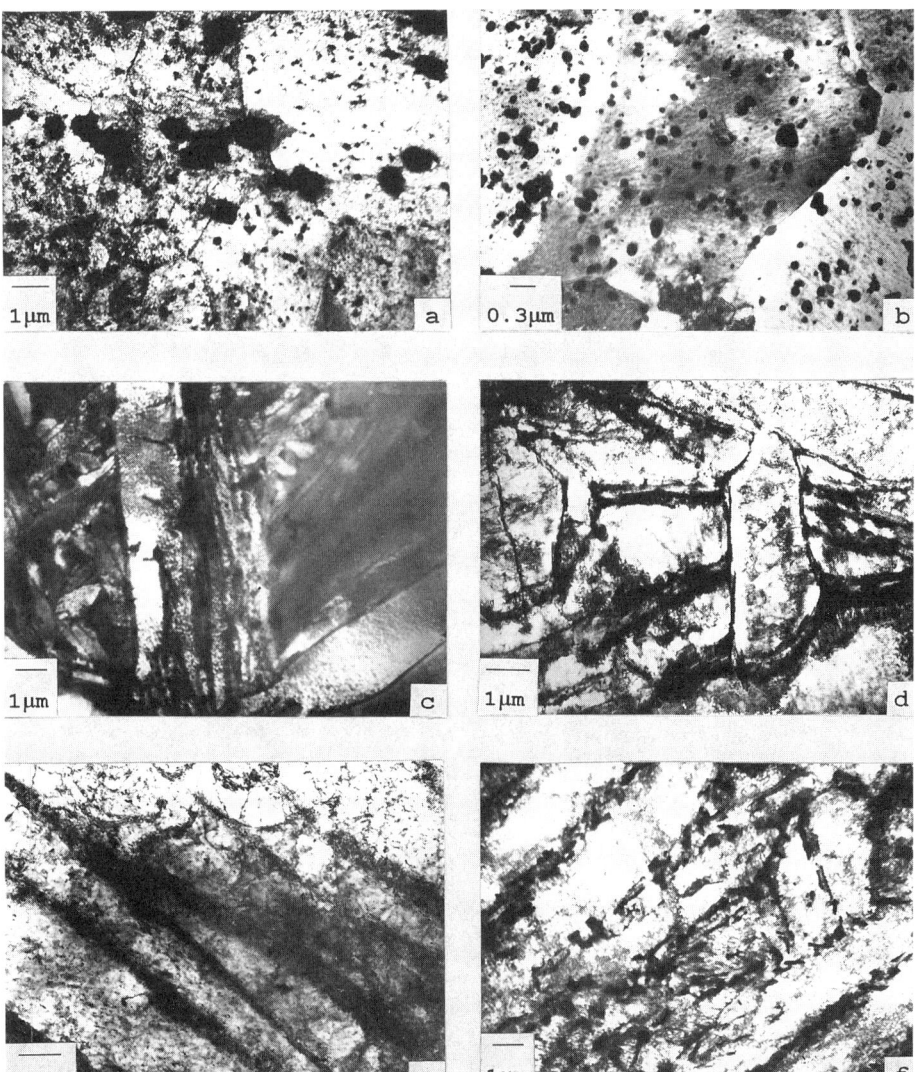

FIG. 2—*Microstructure of Zr-1.2Sn-1Nb-0.4Fe alloy as worked and heat treated: (a) not β-quenched (State 1), (b) β-quenched (State 2); (c) β-quenched (2000°C/s) (State 3), (d) cooled from β-region (25°C/s) (State 5), (e) β-quenched (2000°C/s) and annealed at 380 to 520°C (State 6), (f) cooled from β-region (25°C/s) and annealed at 550 to 650°C (State 8).*

β-region to 400°C/s, State 4 was created, having a much coarser structure of the α'-phase, containing platelets 0.7 to 2.0 μm wide and residual β-phase interlayers 0.06 to 0.13 μm thick. By cooling at a rate of 25°C/s (State 5), a still less uniform coarse platelet structure of the α-matrix containing coarse β-phase interlayers 0.08 to 0.19 μm thick was created (see Fig. 2d). This state is made up of more diffuse and discontinuous interlayers, which indicates the onset of metastable β_{Zr}-phase decomposition.

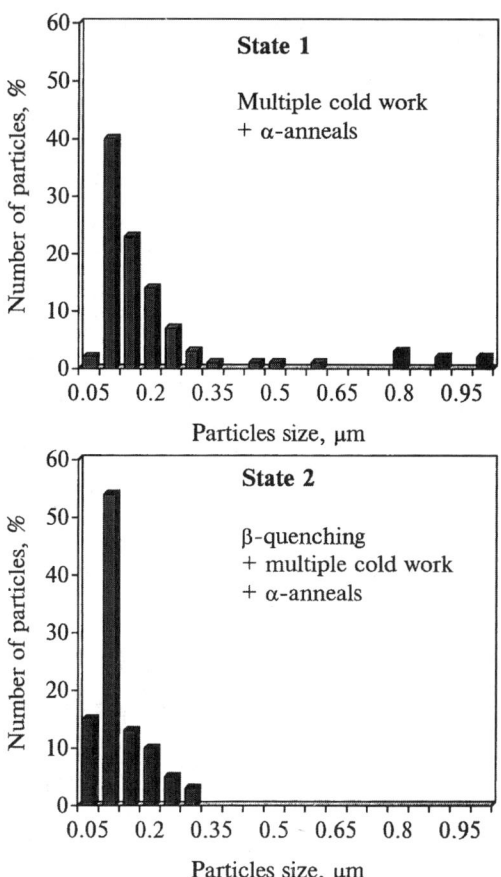

FIG. 3—*Histograms of particle-size distribution in Zr-1.2Sn-1Nb-0.4Fe alloy.*

In the structure of the alloy as annealed after cooling from the β-region, the following processes take place: a decomposition of the supersaturated α'-solid solution, a reversal and recrystallization in the α-matrix, and a eutectoid-type decomposition of β-interlayers followed by intermetallic particle formation. The degree of evolution of those processes is governed by the cooling rate from the β-region and anneal.

By taking into account these circumstances and varying the cooling rate and the schedule of the subsequent anneal, states were prepared having different structures of the α-matrix, and β-interlayers decomposed to different extents (States 6–8) (see Table 1). By annealing from 380 to 520°C after quenching at a cooling rate of 2000°C/s (State 6), a uniform structure without intermetallic particle stringers was prepared. The structure consists of polygonized (fragmented) α-phase laths and interlayers of finely dispersed eutectoid formed at the initial stage of the β-phase decomposition (see Fig. 2e). Other states, 7 and 8 (Table 1), were distinguished by a much less uniform microstructure of the alloy. In State 7, the partially recrystallized α-matrix revealed stringers of particles 0.05 to 0.2 μm in place of the former β-interlayers. In State 8, similar stringers of particles 0.1 to 0.4 μm were observed in the partially polygonized coarse platelet α-matrix (see Fig. 2f).

Tensile Properties and Fracture Resistance

The tensile properties, impact toughness, and critical crack opening of the alloys in the above structure states are compared in Fig. 4. Differences in the microstructures (States 1 and 2) of the alloy subjected to working/heat treatment affect most essentially the fracture resistance. The δ_c and KCV of the alloy not β-quenched and the alloy β-quenched prior to cold rolling and anneals are shown versus the iron content of the alloy (Fig. 5). With the iron content increased from 0.2 to 0.46 wt%, the KCV and δ_c of the alloy not β-quenched become lower, while those of the β-quenched alloy increase with the iron content of the alloy.

The highest increase in the KCV and δ_c effected by an enhanced dispersity and distribution

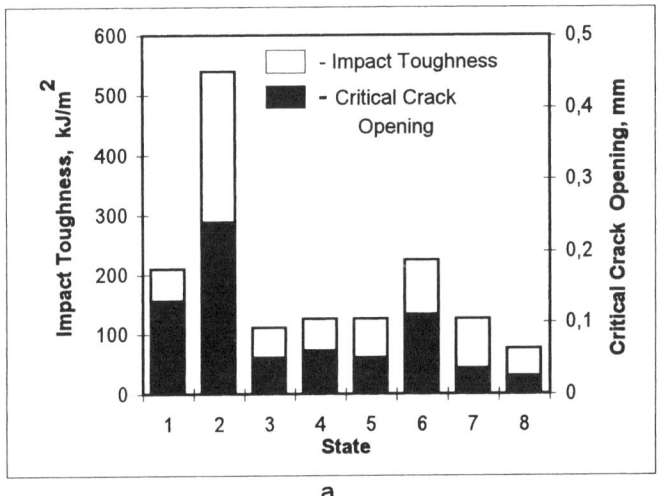

a

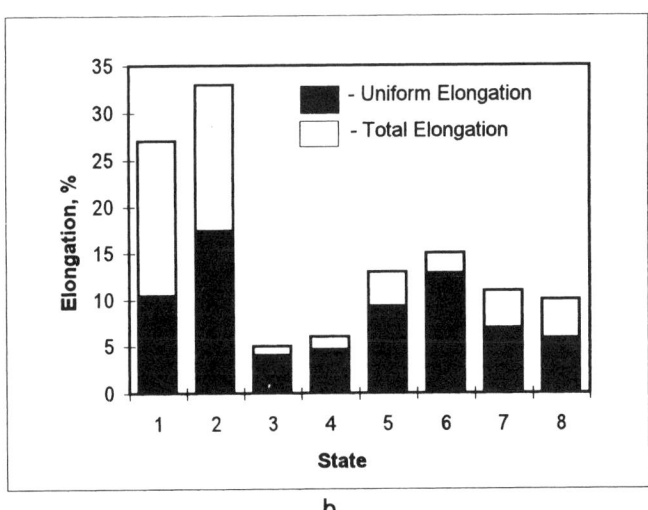

b

FIG. 4—*Impact toughness and critical crack opening* (a), *full and uniform percent strain* (b), *yield strength and ultimate tensile strength* (c) *of Zr-1.2Sn-1Nb-0.4Fe alloy in different structure states.*

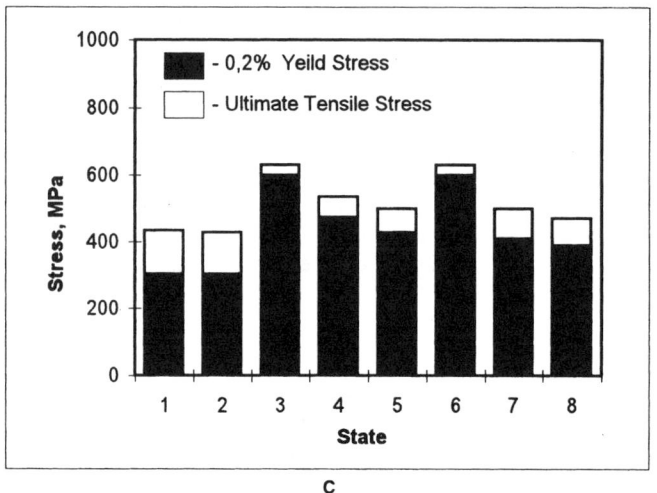

c

FIG. 4—*Continued*

of intermetallic particles in the β-quenched alloy compared to the one worked without β-quenching is observed for the alloy containing 0.46 wt% iron. In this case, the δ_c and KCV values increase by a factor of 1.5 to 1.8 and 2.0 to 2.5, respectively, with the 0.2% yield strength ($\sigma_{0.2}$ = 305 to 330 MPa) and the ultimate tensile strength (σ_u = 420 to 435 MPa) remaining at the same level. On the contrary, at an iron content of 0.20 wt%, the alloy working comprised of β-quenching does not improve its KCV or δ_c compared to the working without β-quenching (see Fig. 5).

The level of the tensile properties of the annealed alloy is determined by the extent of the evolution of the two main competing structural processes: first, the extent of the evolution of the reversal and polygonization processes in the α-matrix that result in its softening and, hence, higher toughness and ductility; second, the extent of the metastable β-phase interlayer decomposition to form intermetallic particles which, on the contrary, results in lower toughness and ductility. The temperature ranges of the above processes during anneal depend on the state of the as-quenched alloy.

β-quenching at a cooling rate of 2000°C/s results in a hardened supersaturated solid solution of α'-phase having a fine, disperse martensite-type structure. This accelerates the processes of reversal and polygonization of the matrix during the subsequent anneal that go ahead of the β_{Zr}-interlayer decomposition.

Cooling at rates of 400 and 25°C/s leads to formation of a less hardened matrix and a partial decomposition of the β_{Zr}-interlayers even during cooling from the β-region. In this case, intermetallic particles precipitate upon anneal at lower temperatures and the precipitation process is superimposed on the loss of strength by the matrix.

The highest full and uniform percent elongation, impact toughness, and critical crack opening are attained by the β-quenched and annealed alloy in State 6, which exhibits a thin platelet (lath) polygonized structure of the α-matrix with fine disperse eutectoid interlayers along the lath boundaries (see Fig. 4). In this state of the alloy, its KCV and δ_c are a factor of 1.5 or 2 higher than in State 3 immediately after β-quenching. Intermetallic particle stringers typical of States 7 and 8 that form upon the β-phase decomposition substantially reduce the ductility and

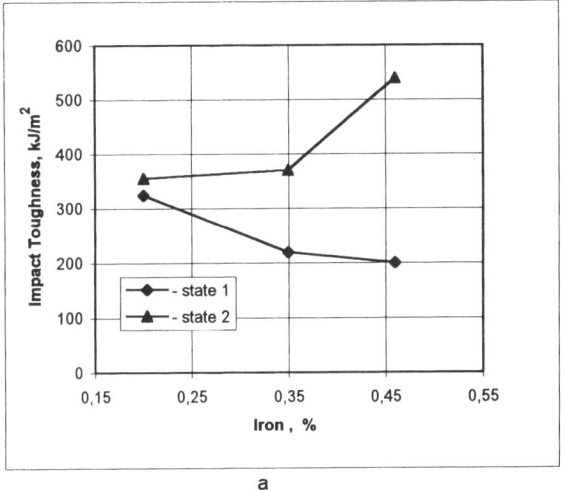

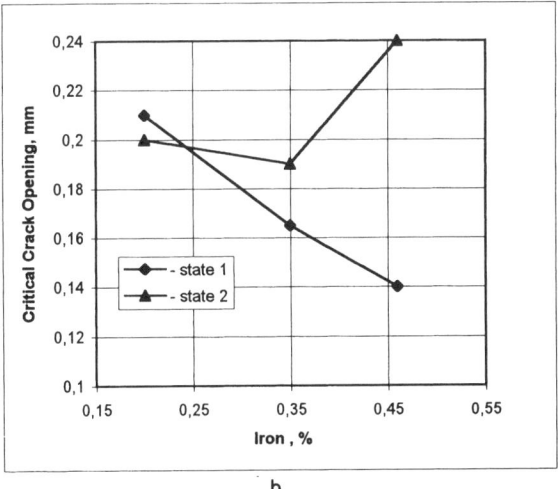

FIG. 5—*Impact toughness* (a) *and critical crack opening* (b) *versus iron content of Zr-1.2Sn-1Nb-0.4Fe alloy.*

toughness of the alloy (Fig. 4). This reduction in ductility and toughness takes place despite the loss of strength by the partially recrystallized matrix (State 7). The β-quenched and annealed alloy has the lowest ductility and toughness in State 8 when a high non-uniformity in the α-phase structure and precipitation of intermetallic particle stringers along the β-phase boundaries are observed (see Fig. 2f).

The changes in the β-quenched and annealed alloy structure affect its tensile strength to a much less extent than its ductility. The $\sigma_{0.2}$ and the σ_u of the alloy in States 6 and 8 remain in the ranges of 390 to 600 MPa and 470 to 630 MPa, respectively. The highest strength of the alloy is observed in State 6 when it also has the highest ductility and toughness (Fig. 4).

Influence of Intermetallic Particles on Ductility and Toughness

The analysis of the influence effected by the Zr-1.2Sn-1Nb-0.4Fe alloy structure on its tensile properties revealed that the main factor that governs its ductility and fracture resistance is intermetallic particles available in the as-recrystallized or polygonized matrix. Coarse intermetallic particles available close to the grain boundaries in the as-worked and heat-treated alloy not subjected to β-quenching prior to cold rolling substantially reduce its ductility and toughness. This fact brought about more detailed studies of the effect of particles on alloy ductility and toughness.

The ductility and toughness of the alloys are known to be defined mostly by their ability for a stable plastic strain [6]. It is possible to assess the ability of a material for a stable flow in the process of deformation by analyzing stress-strain curves recorded in tension tests. This type of analysis showed that for all structure states of alloys, the stress-strain curves in the true stress σ–true strain ε coordinates comply with the Holloman equation [7]

$$\sigma = K \times \varepsilon^n \qquad (1)$$

where n is the coefficient of strain hardening.

For the material flow to be stable in the process of deformation under the action of stress σ and strain ε, the intensity of hardening must be $d\sigma/d\varepsilon > \sigma$ [6]. Then, instability of the flow accompanied by necking sets in at $d\sigma/d\varepsilon = \sigma$, which usually occurs without disturbing the material continuity by the "geometrical loss of strength" during uniform strain ε_u [6]. Equation 1 predicts the loss of flow stability when strain $\varepsilon_u = n$ is reached. Hence, the maximum uniform strain is determined by the intensity of strain hardening, i.e., by the coefficient n. If $\varepsilon_u < n$, the localization of tensile strain sets in earlier than is predicted by Eq 1. The latter may take place, specifically, due to microcracking that occurred during uniform strain and may evidence the reduction in ductility and toughness of a material [8].

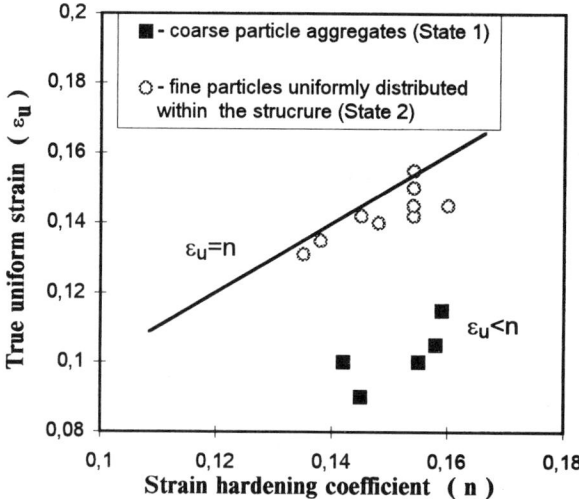

FIG. 6—*Relationship between true uniform strain and coefficient of specimen hardening on tension of Zr-1.2Sn-1Nb-0.4Fe alloy having different structures.*

The true uniform strain, ε_u, versus the coefficient of hardening, n, are summarized for all states (Fig. 6). It can be seen (see Fig. 6) that the relation $\varepsilon_u = n$ is fulfilled only for structure states having a uniform fine intermetallic particle distribution within a matrix that is achieved by β-quenching carried out prior to cold rolling and anneal (State 2, shown in Table 1). For the structure states characterized by aggregates of coarse intermetallic particles or fine particle stringers within a matrix of an alloy not subjected to β-quenching (State 1), the relation $\varepsilon_u < n$ is fulfilled at the same values of the coefficient n, i.e., the strain was localized in the neck before it had to occur due to the geometrical loss of strength at the given coefficient, n.

The measurement of acoustic emission (AE) in the process of tension of alloy specimens having different structures allowed a direct observation of differences in the mechanisms of the loss of stability of the plastic flow during tension. The specific AE and strain diagrams recorded upon tension of alloy specimens having different structures are presented in Fig. 7. The mode of the AE signals generated in the process of tension of specimens with different distributions and sizes of secondary phase particles is quite different. In the AE diagrams

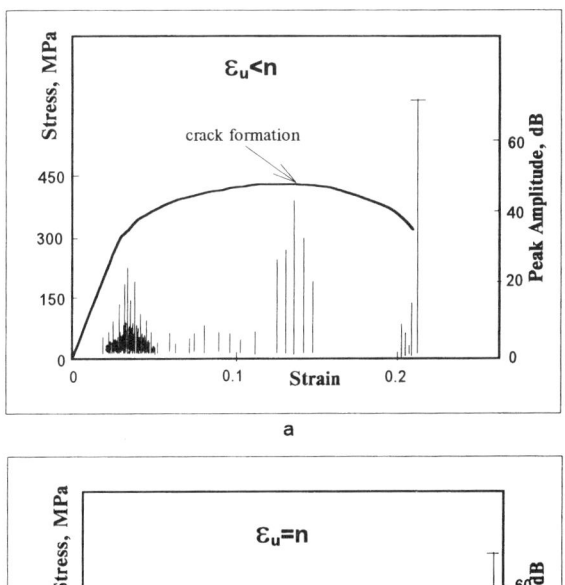

FIG. 7—*Acoustic emission during tension testing of the Zr-1.2Sn-1Nb-0.4Fe alloy having different structures:* (a) *aggregates of coarse particles (State 1 shown in Table 1);* (b) *uniform distribution of fine particles (State 2).*

pertinent to the specimens with aggregates of coarse intermetallic particles (State 1), after the conventional increase in the AE intensity that corresponds to the elastic-plastic transition, from two to ten strong pulses of amplitude of 30 to 45 dB are registered in the uniform strain area. After that, a neck is formed in a specimen.

After such a series of strong AE pulses were registered, testing of the specimens was stopped. Subsequent metallographic analysis of layer after layer of longitudinal sections revealed, in the area of the neck, aggregates or individual microcracks (voids) 0.05 to 0.2 mm, which were the cause of the premature instability of the flow and necking at the strain $\varepsilon_u < n$. In the uniform strain area in the specimens with uniformly distributed fine particles (State 2), no AE signals above the noise level were recorded. In this case, the loss of stability by plastic flow and necking proceed by the conventional mechanism, i.e., due to geometrical loss of strength at a strain $\varepsilon_u = n$.

Thus, joint analysis of the strain and AE diagrams revealed two different mechanisms of the loss of plastic flow stability upon tension of the alloy specimens having different structures. With fine secondary phase particles uniformly distributed in the structure, the flow instability starts at the maximum feasible strain $\varepsilon_u = n$ due to the geometrical loss of strength. With coarse particle aggregates available in the structure at the same coefficient n, the strain localization in the neck starts earlier at $\varepsilon_u < n$ due to the fact that cracks on the aggregates had formed during the uniform strain. Similar fracture mechanisms upon specimen tension were also observed in manganese two-phase steels subjected to intensive hardening ($n = 0.3$ to 0.5) [8] and in Zr-2.5Nb alloy with a partially recrystallized structure [9]. However, in alloys not having high intensity of hardening ($n = 0.13$ to 0.17) and a relatively uniform recrystallized structure, the loss of the flow stability effected by cracks formed on particles was not observed.

With the unvaried texture and, hence, the unvaried coefficient n, the main factor improving ductility and toughness is that element of the structure that defines either the flow stability or the process of void formation at the crack tip in a ductile fracture.

It has been shown that the fracture resistance of Zr-2.5Nb alloy can be enhanced by improving the total and local ductility through the degree of recrystallization of the zirconium matrix [9]. However, it is not sufficient for alloys of the Zr-1.2Sn-1Nb-0.4Fe type; in this case, the sizes and density of intermetallic particle precipitates need to be controlled. Here, as is the case for a ductile fracture, the void formation process is controlled by fine particles up to 0.1 μm in size, while the growth and coalescence of voids are dependent upon the coefficient n and interparticle spacing.

The strain to fracture is related to a volume fraction of particles and voids. Therefore, one usually expects the toughness to increase with a decrease in the volume fraction of particles, which was observed in precipitation-hardened copper alloys with a volume fraction of particles varying from 2 to 20 vol% [10]. Since there is a constant volume fraction of particles and a constant coefficient of hardening, the fracture resistance of multicomponent zirconium alloy depends on the particle distribution density within the matrix and interparticle spacing. Indeed, with a uniform particle distribution, the impact toughness and the critical crack opening in the Zr-1.2Sn-1Nb-0.4Fe alloy grow linearly with an increase in the particle volume density and a decrease in interparticle spacing (Fig. 8). The relationships do not cover only states having aggregates of coarse particles and fine particle stringers, the availability of which in the structure is the main cause of early instability of flow upon strain and a tough crack propagation. Thus, the fracture resistance of the alloy can be improved by increasing the density of the intermetallic particle distribution with their volume fraction remaining almost invariable.

This approach to toughness improvement is feasible only when the volume fraction of the secondary phase particles is relatively low (<10 vol%), i.e., when individual particles behave independently during fracture by the mechanism of void formation and growth.

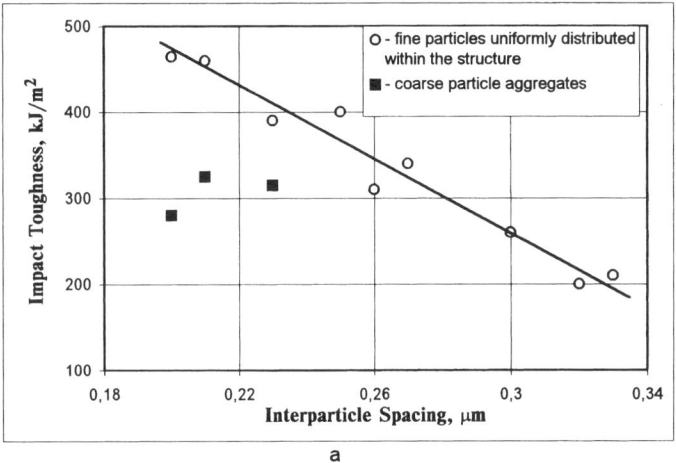

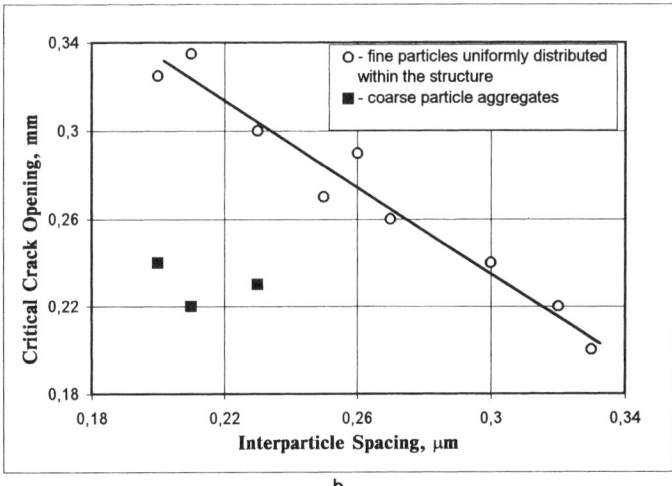

FIG. 8—*Impact toughness* (a) *and critical crack opening* (b) *of Zr-1.2Sn-1Nb-0.4Fe alloy as a function of interparticle spacing.*

Conclusions

The main factor of the structure that governs the ductility and fracture resistance of recrystallized Zr-1.2Sn-1Nb-0.4Fe alloy is the size and mode of intermetallic particle distribution within the matrix. The highest ductility and toughness of the alloy are reached for the structure with a uniform distribution of fine intermetallic particles from 0.03 to 0.20 μm within the matrix. A higher density of their distribution results in improved impact toughness and static crack resistance of the alloy. The structure can be formed through working/heat treatment comprised of β-quenching, cold work, and anneal.

Coarse intermetallic particles from 0.4 to 1.2 μm or fine particle stringers present in the structure lead to substantially reduced ductility and fracture resistance due to formation of microcracks that effect the early loss of plastic flow stability upon strain or propagation of a tough crack. This can induce a reduction in the manufacturing ductility of the alloy having this type of structure.

In the alloy annealed after β-quenching, the highest ductility and toughness are reached if the matrix is polygonized without an intermetallic particle precipitation.

The decomposition of β-interlayers to form intermetallic particle stringers results in reduced ductility and toughness of the alloy despite loss of strength by the zirconium matrix.

Acknowledgments

The authors would like to thank the following individuals who helped carry out these experiments: V. Khanzhin for acoustic emission measurements, A. Melnicov and E. Kurianova for overall technical assistance.

References

[1] Nikulina, A. V., Markelov, V. A., et al., "Zr-1%Nb-1%Sn-0.5%Fe Alloy for Fuel Channel Tubes of RBMK Type Reactors," *Voprosy Atomnoi Nauki i Tekhniki. Ser. Materialovedenie i Novye Materialy*, Vol. 2, No. 36, 1990, pp. 58–66.

[2] Markelov, V. A., Rafikov, V. Z., Nikulin, S. A., and Shishov, V. N., "Changes in Microstructure of Zr-Sn-Nb-Fe Alloy upon Working/Heat Treatment," *Fizika Metallov i Metallovedenie*, Vol. 77, No. 4, April 1994, pp. 70–79.

[3] Markelov, V. A., Nikulin, S. A., Gusev, A. Yu., et al., "Effect of Composition and Working–Heat Treatment with β-Quenching on Structure and Fracture Resistance of Zr-Sn-Nb-Fe Alloy," *Voprosy Atomnoi Nauki i Tekhniki. Ser. Materialovedenie i Novye Materialy*, Vol. 1, No. 48, 1993, pp. 37–43.

[4] Nikulin, S. A., Markelov, V. A., and Fateev, B. M., "Determination of Critical Crack Opening in Small Size Specimens," *Izvestiya Vysshikh Uchebnykh Zavedeniy. Chernaya Metallurgia*, No. 11, 1987, pp. 156–157.

[5] Nikulin, S. A., Shtremel, M. A., Khanzhin, V. G., and Markelov, V. A., "Influence of Hydrides on Ductile Fracture in the Zr-2.5Nb Alloy," *Nuclear Science and Engineering*, No. 115, 1993, pp. 193–204.

[6] Backofen, W. A., "Deformation Processing," *Metallurgia*, Moscow, 1977.

[7] Hollomon, J. H., "Tensile Deformation," *Transactions AIME*, Vol. 162, 1945, pp. 268–290.

[8] Nikulin, S. A., Shtremel, M. A., and Khanzhin, V. G., "On Ductile Fracture of High Manganese Steel upon Tension," *Izvestiya Akademii Nauk SSSR. Metally*, No. 1, 1990, pp. 145–151.

[9] Nikulin, S. A., Markelov, V. A., and Fateev, B. M., "Influence of Structure on Zr-2.5% Nb Alloy Stress-Strain Diagrammes," *Izvestiya Akademii Nauk SSSR. Metally*, No. 3, 1991, pp. 134–139.

[10] Martin, J. W, "Micromechanisms in Particle-Hardened Alloys," *Metallurgia*, Moscow, 1983.

DISCUSSION

P. Bouffioux[1] (written discussion)—Have you investigated the effect of the microstructure, more particularly, the size and distribution of intermetallic particles, on the thermal creep behavior of the material? If yes, have you established any correlation between creep resistance and the ductility?

S. A. Nikulin et al. (authors' closure)—No, we have not investigated the effect size and distribution of intermetallic particles on thermal creep.

J. B. Bai[2] (written discussion)—Is there any difference in AE response for fine particle and coarse particle aggregates?

S. A. Nikulin et al. (authors' closure)—AE measurements were used to analyze the process of a crack initiation on tension of specimens leaving different particle sizes and distributions. For tension of samples having fine particles, no microcracks are formed until a specimen fracture in the neck. There is no AE with a large amplitude of a signal. For alloy specimens containing coarse particle aggregates, AE indicated a crack initiation even at the stage of uniform strain. Initiation of this type of crack is accompanied by AE of a large amplitude. From the peak amplitudes of signals is measured, we determined crack sizes of 20 to 100 μm.

R. B. Adamson[3] (written discussion)—You have shown a very nice correlation between interparticle spacing and toughness. Have you been able to check the relationship with irradiated material, where some particles may have dissolved and thus changed the particle spacing?

S. A. Nikulin et al. (authors' closure)—So far data are not available on the relationship between interparticle spacing and toughness for irradiated material.

---

[1] Electricite de France, Direction des Etudes Rescher, France.
[2] Lab MSS/MAT, Ecole Centrale Paris, France.
[3] GE Nuclear Energy, Pleasantville, NY.

Robert J. Comstock,[1] *Gerald Schoenberger,*[2] *and George P. Sabol*[3]

Influence of Processing Variables and Alloy Chemistry on the Corrosion Behavior of ZIRLO* Nuclear Fuel Cladding

REFERENCE: Comstock, R. J., Schoenberger, G., and Sabol, G. P., **"Influence of Processing Variables and Alloy Chemistry on the Corrosion Behavior of ZIRLO* Nuclear Fuel Cladding,"** *Zirconium in the Nuclear Industry: Eleventh International Symposium, ASTM STP 1295,* E. R. Bradley and G. P. Sabol, Eds., American Society for Testing and Materials, 1996, pp. 710–725.

ABSTRACT: Variations in the thermal heat treatments used during the fabrication of ZIRLO (Zr-1Nb-1Sn-0.1Fe) fuel clad tubing and in ZIRLO alloy chemistry were explored to develop a further understanding of the relationship between processing, microstructure, and cladding corrosion performance. Heat treatment variables included intermediate tube annealing temperatures as well as a beta-phase heat treatment during the latter stages of the tube reduction schedule. Chemistry variables included deviations in niobium and tin content from the nominal composition. The effects of both heat treatment and chemistry on corrosion behavior were assessed by autoclave tests in both pure and lithiated water and high-temperature steam. Analytical electron microscopy demonstrated that the best out-reactor corrosion performance is obtained for microstructures containing a fine distribution of beta-niobium and Zr-Nb-Fe particles. Deviations from this microstructure, such as the presence of beta-zirconium phase, tend to degrade corrosion resistance.

ZIRLO fuel cladding was irradiated in four commercial reactors. In all cases, the microstructure in the cladding included beta-niobium and Zr-Nb-Fe particles. ZIRLO fuel cladding processed with a late-stage beta heat treatment to further refine the second-phase particle size exhibited in-reactor corrosion behavior that was similar to reference ZIRLO cladding. Variations of the in-reactor corrosion behavior of ZIRLO were correlated to tin content, with higher oxide thickness observed in the ZIRLO cladding containing higher tin. The results of these studies indicate that optimum corrosion performance of ZIRLO is achieved by maintaining a uniform distribution of fine second-phase particles and controlled levels of tin.

KEYWORDS: zirconium, zirconium alloys, ZIRLO, fuel cladding, corrosion, in-reactor, autoclave, microstructure

Prior studies on zirconium-niobium (Zr-Nb) alloys [1–4] established a well-documented correlation between microstructural features and in-reactor corrosion characteristics. For optimized corrosion performance, the preferred microstructure included the complete precipitation of the niobium as a very fine, uniform distribution of second-phase beta-niobium particles.

The ZIRLO (Zr-1Nb-1Sn-0.1Fe) fuel cladding manufacturing parameters were developed with these microstructural features as the objective. To achieve a final microstructure containing a fine uniform distribution of second-phase precipitates, the resultant fabrication process was

---

[1] Program manager, Westinghouse Electric Corporation, Science and Technology Center (STC), Pittsburgh, PA 15235.
[2] Principal engineer, Westinghouse Electric Corporation, Specialty Metals Plant, Blairsville, PA 15717.
[3] Manager, Development Programs, Westinghouse Electric Corporation, Commercial Nuclear Fuel Division (CNFD), Pittsburgh, PA 15230.
* ZIRLO is a trademark of Westinghouse Electric Corporation, Pittsburgh, PA.

based on the principles of a quench and temper treatment [5]. An upstream beta-quenching operation was performed prior to extrusion. Subsequent anneals were performed prior to each pilger pass to provide workability for the cold reduction operation but were restricted to low temperatures to minimize particle growth. Both the BR-3 [6] and North Anna [7] demonstration ZIRLO fuel claddings were fabricated via the above process, which successfully produced fine precipitates with an average particle size less than 80 nm. Having achieved the desired microstructures, both materials showed significantly lower in-reactor corrosion than Zircaloy-4 cladding [6,7].

To further develop an understanding of the relationship between microstructure and corrosion performance, a small quantity of ZIRLO cladding was subjected to a late-stage beta-quench prior to the final two pilger operations. This process resulted in a very fine distribution of second-phase particles. Some of the late-stage beta-quenched cladding was fabricated into fuel rods as part of the North Anna demonstration program and was irradiated for two cycles. The results of the program on beta-quenched tubing have not been previously reported and are therefore covered in this paper.

Corrosion performance of ZIRLO cladding manufactured from several production ingots and irradiated in three additional commercial reactors has recently been evaluated. ZIRLO fuel rods continued to maintain lower corrosion levels than similarly irradiated Zircaloy-4 fuel rods but exhibited some variability. Follow-up investigative studies were initiated that focused on alloy chemistry effects and the subsequent impact on corrosion performance. This paper reports the results of that study.

Materials, Test Conditions, and Reactor Parameters

Processing of Tubing

Several variations in both ZIRLO alloy chemistry and thermal heat treatments used during the fabrication of ZIRLO nuclear fuel cladding were examined to identify the effect of these variables on subsequent cladding performance. The chemistry variables included deviations from the nominal ZIRLO composition of 1 wt% niobium and 1 wt% tin. Niobium and tin varied from their nominal compositions by up to 0.3 wt%, while iron was maintained close to its nominal concentration of 0.1 wt%.

The ingot compositions are listed in Table 1 in order of increasing tin content. The ingots included several production ingots along with three small ingots (approximately 90 kg/ingot) that were melted to provide chemistry variations beyond those of the production ingots. Multiple analyses were performed on samples from the production ingots and from final size tubing for the small ingots. The tin in the production ingots ranged from 0.96 to 1.20 wt%, while the niobium remained close to the nominal concentration of 1 wt%. The small ingots extended the tin concentration down to 0.85 wt% and included extremes in niobium from 0.90 to 1.33 wt%. The cladding from the production ingots was made into fuel rods and irradiated in reactors identified in Table 1.

The production ingots were fabricated into fuel cladding via conventional fabrication processes, i.e., via forging into billets, beta-quenching, extruding to tube hollows, and multiple pilgering operations to final size. This process was previously presented [7,8] and is referred to as reference or conventional ZIRLO processing. The small non-production ingots were beta-quenched and then fabricated into fuel cladding via the reference process.

In addition to the above, two lots of cladding were fabricated by alternate process routes. In the first lot, a small quantity of production material was beta-quenched at an intermediate tube size prior to the final two pilger reductions. The beta-quench process was similar to that given to Zircaloy-4 cladding as previously reported [7]. Subsequent fabrication to final size cladding

TABLE 1—Chemical analysis of ZIRLO materials.

		Ingot, percent by weight						
		1[a]	2[b]	3[a]	4[b]	5[b]	6[a]	7[b]
Tin	Average	0.87	0.97	1.08	1.11	1.14	1.17	1.18
	Range	0.85–0.89	0.96–0.98	1.07–1.09	1.07–1.13	1.09–1.20	1.16–1.17	1.14–1.20
Niobium	Average	1.32	1.03	0.91	1.05	1.03	1.32	1.00
	Range	1.30–1.33	1.02–1.04	0.90–0.91	1.00–1.08	0.99–1.06	1.31–1.33	0.99–1.03
Iron	Average	0.10	0.10	0.10	0.11	0.11	0.11	0.12
	Range	0.10–0.10	0.09–0.11	0.10–0.10	0.10–0.13	0.09–0.12	0.10–0.12	0.11–0.12
Zirconium		Balance	Balance	Balance	Balance	Balance	Balance	Balance
Reactor exposure		None	N. Anna, Reactor X	None	Reactor Z	Reactor Y	None	Reactor Z

[a] Non-production ingot.
[b] Production ingot.

was identical to conventionally processed ZIRLO cladding. In the second lot, cladding was fabricated according to the reference process except for the use of a higher annealing temperature (950 K) between pilger reductions. The final anneal remained the same as the reference process.

Autoclave Testing

Long-term autoclave testing of ZIRLO cladding was performed in several different autoclave environments. The environments included pure and lithiated water at 633 K and high-temperature steam at temperatures from 672 to 727 K. Autoclaves were run at saturation pressure for the water tests and 10.3 MPa for the steam tests. Lithium was added as the hydroxide at a lithium concentration of 70 ppm by weight.

The specimens included samples representative of the range of ZIRLO processes and chemistries described above. In several instances, the specimens were from the same or companion tubing lots from which in-reactor data were obtained to provide a one-to-one comparison between out-reactor and in-reactor corrosion behavior.

Multiple specimens (three to five) from each lot were tested in a given corrosion environment. Specimens, typically 2 cm long, were tested in the as-fabricated condition and were prepared by cleaning in Alconox detergent, rinsing in tap water, deionized water, and alcohol, and then blow drying with warm air. Weight gains during autoclave testing were periodically measured.

Reactor Operating Conditions

ZIRLO fuel cladding was irradiated in four commercial reactors identified as North Anna Unit 1, Reactor X, Reactor Y, and Reactor Z. Operating conditions of the reactors are discussed below:

North Anna Unit 1—Demonstration Assembly AM2 operated in North Anna Unit 1, Cycles 7 and 9, with a total accumulated assembly burnup of 45.8 gigawatt days per metric ton of uranium (GWd/MTU). The reactor operated at essentially 100% power throughout Cycle 7, except for an end-of-cycle coastdown to approximately 80% power, for a total exposure of 10 150 h. In Cycle 9, North Anna Unit 1 operated at 100% power for the first 6000 h; this was followed by about 4500 h at 95% power and an extended end-of-cycle coastdown to 40% power over approximately 2850 h. The coolant core outlet temperature at full power was 602 K in Cycle 7 and 600 K in Cycle 9. At full power, the fuel rod average linear powers were 24.2 kW/m in Cycle 7 and 22.5 kW/m in Cycle 9, and the rod burnups at the end of Cycle 9 were between 45.0 and 46.5 GWd/MTU. In both Cycles 7 and 9, the coolant was maintained at a pH of 6.9 at the core average temperature with a lithium concentration near the start of the cycle of approximately 2.5 ppm.

Reactor X—Reactor X is a Westinghouse 3-loop plant with each fuel assembly containing a 17 by 17 array of fuel rods. The ZIRLO demonstration assembly has operated for five cycles for a total exposure of approximately 33 300 effective full power hours (EFPH). The core coolant outlet temperature was 597 K in all cycles. The fuel rod average linear powers ranged from 20.6 to 24.1 kW/m in the first three cycles, dropped to 6.3 kW/m in the fourth cycle, and increased to 13.8 kW/m in the fifth cycle.

Reactor Y—Two demonstration assemblies, each containing a 17 by 17 array of fuel rods, were irradiated in Cycles 4 and 5 of Reactor Y, a Westinghouse 4-loop plant. The cycle lengths were approximately 10 800 and 11 500 EFPH, respectively. Reactor Y operated at essentially 100% power except for end-of-cycle coastdowns in both cycles with a 4.5% uprating in Cycle 5. The coolant core outlet temperature was 600 K in both cycles. The demonstration assemblies operated in symmetric locations in both cycles and accumulated assembly burnups of 22.4

GWd/MTU at the end of their first cycle and 36.3 GWd/MTU at the end of their second cycle. In Cycle 4, the fuel rod average linear powers ranged from 19.5 to 24.1 kW/m, and, in Cycle 5, the fuel rod powers ranged from 11.4 to 17.7 kW/m. Reactor Y operated with modified coolant chemistry in both cycles with the lithium concentration maintained at 2.2 ppm from early in the cycle until the coolant pH at the core average temperature reached 7.4. Operation then continued at 7.4 pH until the end of the cycle.

Reactor Z—Reactor Z, a Westinghouse 17 by 17 three-loop plant, was fueled with ZIRLO clad fuel in Cycles 7 and 8, with cycle lengths of approximately 10 800 and 12 100 EFPH, respectively. Reactor Z operated at essentially 100% power throughout Cycle 7, but experienced approximately 2300 h of partial power operation between 65 and 80% of full power, plus a coastdown, in Cycle 8. The reactor Z coolant core outlet temperature was 601 K at 100% power. The fuel rod average linear powers for the fuel exposed for two cycles ranged from 15.0 to 23.2 kW/m in Cycle 7 and from 17.2 to 20.4 kW/m in Cycle 8. The rod average powers for the fuel initially loaded in Cycle 8 was 23.6 kW/m. In both Cycles 7 and 8, the coolant was maintained at a pH of 6.9 at the core average temperature with a lithium concentration near the start of the cycle of approximately 2.3 ppm.

PIE Data Acquisition

Oxide thickness corrosion measurements were obtained on the fuel rods by the eddy current non-destructive measuring system developed by Westinghouse [7]. The system, utilizing a probe manufactured by the Fischer Tech Corp. of Windsor, Connecticut, operates over a range of 0 to 400 μm and is considered very accurate in the range of 10 to 390 μm. Calibration of the instrument was accomplished by the use of known standards, which requires that the eddy current probe response be linear over the range of that standard. A two-sigma uncertainty factor of ± 1.6 μm for ZIRLO has been statistically determined for the oxide thickness measuring system.

Results and Discussion

Properties of Tubing

Microstructure—Transmission electron microscopy (TEM) was used to characterize the microstructure of selected ZIRLO materials at final size. The goal was to verify that the thermal-mechanical processing of the tubing achieved a uniform distribution of very fine precipitates in the final product.

The microstructure of conventionally processed ZIRLO cladding was previously described [7]. Two types of particles were present: one was a niobium-rich phase identified as beta-niobium, while the second type was a hexagonal Zr-Nb-Fe particle. The average size, including both types of particles, was 79 nm. The late-stage beta quenching was designed to produce a finer particle size in the final cladding. The as-beta-treated microstructure revealed no precipitation of secondary phases in the matrix. However, the subsequent cold pilger reductions and intermediate annealing resulted in fine second-phase particles in the final product with an average particle size of 35 nm. STEM analysis of the particles identified two particle types that were comparable to those found in conventionally processed material. Movement of the beta-phase heat treatment closer to final size resulted in a refinement in particle size as shown in Fig. 1.

The impact of higher intermediate annealing temperatures was evaluated in intermediate size tubing annealed at 950 K. The microstructure, shown in Fig. 2, revealed the presence of beta-zirconium. This was a departure from the desired microstructure obtained in material processed

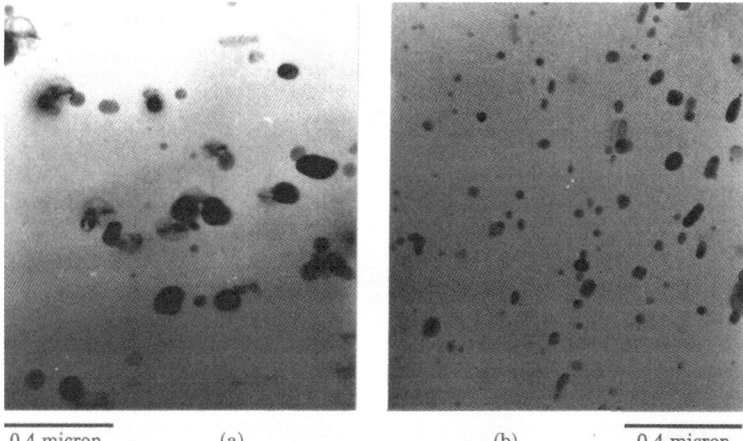

FIG. 1—*Microstructure of ZIRLO cladding processed by* (a) *reference process and* (b) *late-stage beta quenching.*

at the lower annealing temperature used in the reference process [7,8]. This material was not irradiated as the microstructure was sufficiently different from that previously shown to yield good corrosion behavior.

Autoclave Corrosion Tests—Table 2 summarizes the autoclave tests performed on the several ZIRLO cladding lots representative of material irradiated in three of the four nuclear reactors from which in-reactor corrosion data were obtained. No autoclave tests were performed on cladding irradiated in Reactor Z. The out-reactor corrosion behavior of these lots was characterized by the post-transition corrosion rates in pure and lithiated water as well as high-temperature steam. Comparison of the two processes revealed lower corrosion rates for the cladding processed with a late-stage beta quench.

Additional corrosion tests were performed on cladding fabricated from the small nonpro-

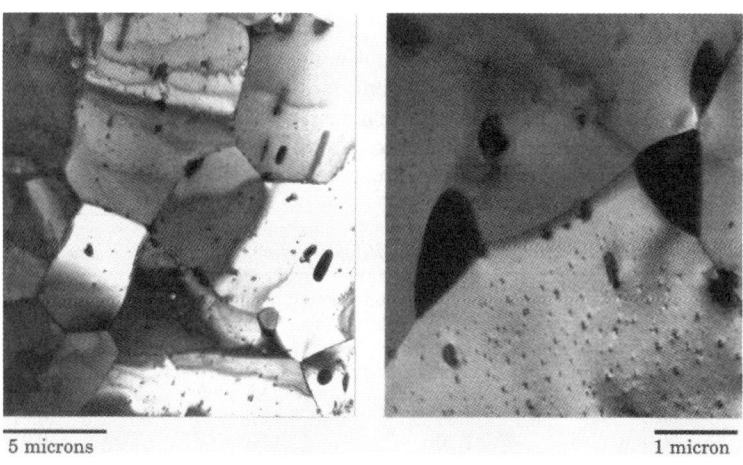

FIG. 2—*Microstructure of an intermediate-size ZIRLO tube annealed at 950 K.*

TABLE 2—*Autoclave corrosion rates of ZIRLO cladding.*

			Post Transition Corrosion Rates, mg/dm^2/d		
Process	Ingot	Reactor	633 K Water	672 K Steam	633 K Water, 70 ppm Li
Reference	2	North Anna, X	0.318–0.371	0.948–1.02	0.549–0.591
	5	Y	0.359–0.379	1.11–1.16	0.605–0.658
Late–stage beta quench	2	North Anna	0.317	0.914	0.467

duction size ingots to provide information regarding the influence of chemistry variations on corrosion behavior. Table 3 summarizes the corrosion rates of the cladding for the four test environments. Also included in the table are corrosion rates from a Zr-1Nb alloy [6]. This provides a reference point for expected corrosion rates in ZIRLO as the tin content approaches zero weight percent.

A final set of corrosion tests were performed on final-size cladding following vacuum recrystallization anneals at temperatures up to 950 K. Corrosion rates are summarized in Table 4. The higher corrosion rates of the cladding annealed at 950 K coincide with the observed formation of beta-zirconium discussed above. These results are further evidence of the importance of maintaining low annealing temperatures during fabrication of ZIRLO cladding.

In-reactor Corrosion Performance

Oxide thickness measurements were performed on eight peripheral ZIRLO fuel rods on North Anna Unit 1 Fuel Assembly AM2 after two cycles of exposure to an average assembly burnup of 45.8 GWd/MTU. Five of the rods were manufactured via the reference process, and three of the rods were manufactured by late-stage beta quenching. Typical oxide film eddy current traces for each material type are shown in Fig. 3; the maximum oxide thickness observed on the individual ZIRLO rods is summarized in Table 5. Both the reference ZIRLO and the beta-quenched ZIRLO exhibited on average approximately 76% lower corrosion than comparably irradiated Zircaloy-4 cladding. The relative performance of Zircaloy-4 versus ZIRLO was consistent with the performance of the previously reported North Anna 2-cycle fuel assembly that had been irradiated to an average burnup of 37.8 GWd/MTU [7]. In this latter assembly, the ZIRLO fuel rods showed 68% lower corrosion than the Zircaloy-4 fuel rods.

Oxide thickness measurements were also performed on ZIRLO fuel rods manufactured via the reference process and irradiated in three commercial reactors. As shown in Table 6, maximum oxide thickness was measured in Reactor X on ZIRLO test rods after the first, second, fourth, and fifth exposure cycles up to an average assembly burnup of 50.0 GWd/MTU. The

TABLE 3—*Autoclave corrosion rates of ZIRLO cladding fabricated from small ingots.*

	Post Transition Corrosion Rates, mg/dm^2/d			
Ingot	633 K Water	700 K Steam	727 K Steam	633 K Water, 70 ppm Li
Zr-1Nb	0.204	1.06	2.83	>10
1	0.337	1.89	4.65	0.578
3	0.418	2.20	4.98	0.776
6	0.384	2.40	5.54	0.675

TABLE 4—*Autoclave corrosion rates of ZIRLO cladding as a function of annealing temperature.*

Annealing Temperature	Post-Transition Corrosion Rates, mg/dm^2/d			
	633 K Water	672 K Steam	727 K Steam	633 K Water, 70 ppm Li
As-pilgered	0.323	1.02	4.34	0.675
869 K	0.382	1.14	4.74	0.684
894 K	0.356	1.47	7.22	0.610
950 K	0.531	2.05	9.09	1.033

maximum oxide thickness on the ZIRLO rods after five cycles of operation was 24 μm, 76% lower than similarly irradiated Zircaloy-4 rods with maximum oxide thickness of 99 μm. The corrosion performance of the ZIRLO fuel rods in Reactor X was consistent with their performance in North Anna.

Two demonstration fuel assemblies containing ZIRLO fuel rods fabricated by the reference process were irradiated in commercial Reactor Y. The purpose of the demonstration program was to verify the improved corrosion performance of ZIRLO fuel rods. Oxide thickness measurements were performed on two 2-cycle fuel assemblies on a total of ten peripheral ZIRLO fuel rods with burnups ranging from 36.9 to 40.2 GWd/MTU. Linear scans were made on the area of the fuel rods between the grids for the entire length of the rods. The maximum oxide thickness measured for each rod is summarized in Table 7. The maximum oxide thickness on 2-cycle Zircaloy-4 clad rods with burnups of about 45 GWd/MTU ranged from 48 to 68 μm.

In Reactor Z, a full region of 64 ZIRLO fuel assemblies was irradiated for two cycles, achieving rod burnups in the range of 38.1 to 42.2 GWd/MTU. Peripheral rod oxide measurements were performed on five of these 2-cycle ZIRLO fuel assemblies along with one 1-cycle ZIRLO fuel assembly. Linear scans were performed on a total of 28 rods along the entire length except for the area of the rods at the grid locations. The maximum measured oxide thickness for the rods is summarized in Table 8 and Fig. 4. A data band representing the maximum measured oxide thickness on the North Anna Unit 1 ZIRLO demonstration rods has been added

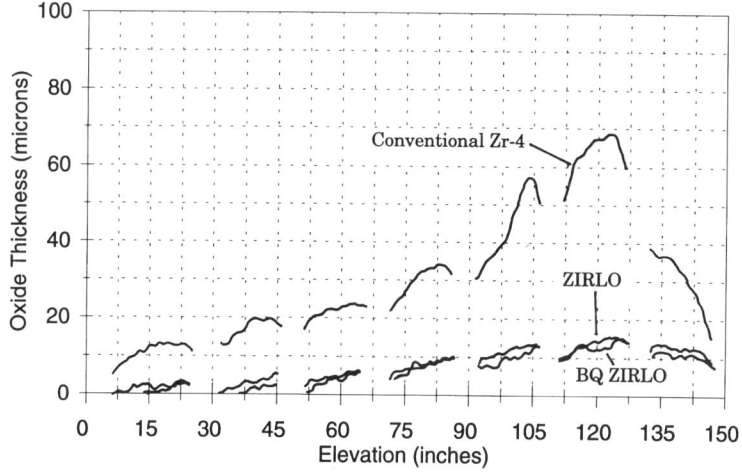

FIG. 3—*North Anna Unit 1 two-cycle oxide film eddy current traces.*

TABLE 5—*ZIRLO fuel rod[a] corrosion data from North Anna Unit 1.*

Process	Rod	Average Peak Oxide, μm
Reference	A01	17
	K17	16
	L17	17
	A01	15
	F01	19
Beta quench	A06	15
	R01	21
	R11	15

[a] Rods from Assembly AM2 with an average burnup of 45.8 GWd/MTU.

to Fig. 4 and shows that for a given rod burnup, the maximum measured oxide thickness of the North Anna rods was lower than the ZIRLO fuel rods irradiated in Reactor Z. As the entire region contained ZIRLO rods, there were no data on equivalent Zircaloy-4 rods for comparison.

Correlation of In-Reactor Corrosion to ZIRLO Processing and Chemistry

The philosophy for achieving optimum corrosion performance in niobium-bearing alloys of zirconium is to attain complete precipitation of fine second-phase particles uniformly distributed in the matrix. For the binary Zr-Nb alloys, the second-phase particles are beta-niobium with an equilibrium concentration of niobium in the matrix [3,4]. Further studies on Zr-Nb and Zr-Nb-Sn alloys revealed good corrosion resistance when the material was processed to achieve an average particle size of approximately 50 nm [5]. This was achieved by low-temperature aging following a beta-phase heat treatment. When the average particle size was coarsened to 200 to 300 nm, a significant degradation in corrosion resistance was observed. Finally, good in-reactor corrosion performance of ZIRLO was observed in both BR-3 and North Anna Unit 1 for cladding with an average reactor particle size less than 80 nm [6,7].

In the current study, in-reactor corrosion data were obtained on cladding processed from the same ingot using two different process routes, the reference ZIRLO process, and the late-stage beta-quench process. In both processes, low-temperature thermal anneals were used following the beta heat treatment to achieve a fine distribution of second-phase particles in the final microstructure. Both processes were successful in achieving the desired microstructure, with average particle sizes ranging from 35 to 79 nm. The autoclave corrosion results showed slightly lower corrosion rates for the late-stage beta-quenched material with the finer particle size.

TABLE 6—*Fuel rod corrosion data from Reactor X.*

Cycles	Assembly Burnup, GWd/MTU	Maximum Oxide, μm	
		ZIRLO	Zircaloy-4
1	12.3	6	...
2	26.2	6	...
4	42.0	12	...
5	50.0	24	99

TABLE 7—*Fuel rod corrosion data from Reactor Y.*

Assembly	Rod	Rod Burnup, GWd/MTU	Maximum Oxide, μm
F59	F01	40.2	32
	J01	39.3	38
	E01	40.0	32
	I01	39.6	40
	N01	36.9	44
F69	I17	39.6	30
	J17	40.0	33
	E17	37.9	36
	H17	39.3	34
	N17	39.4	37

In-reactor corrosion behavior of the two ZIRLO materials was quite similar after two cycles of irradiation in North Anna Unit 1. The late-stage beta-quenched material had been expected to demonstrate slightly lower in-reactor performance than the reference process material based on its finer reactor particle size and better autoclave corrosion behavior. However, the data

TABLE 8—*Fuel rod corrosion data from Reactor Z.*

Ingot	Cycles	Assembly	Rod	Rod Burnup, GWd/MTU	Maximum Oxide, μm
4	1	K07	F01	24.2	17
			G01	24.2	15
			H01	24.1	17
			I01	24.0	18
			J01	24.1	14
			K01	24.2	15
	2	J37	F01	39.9	28
			G01	39.4	25
			H01	38.8	30
			I01	38.4	29
			J01	38.1	26
		J50	A07	42.2	32
			A10	42.2	29
		J51	Q08	42.2	35
			Q10	42.2	36
			Q11	42.2	33
7	2	J37	K01	37.5	28
		J49	G17	42.2	41
			H17	42.2	34
			J17	42.2	36
			K17	42.2	42
		J50	A08	42.2	31
			A11	42.2	35
		J51	Q07	42.2	35
		J52	G01	42.2	34
			H01	42.2	33
			J01	42.2	34
			K01	42.2	32

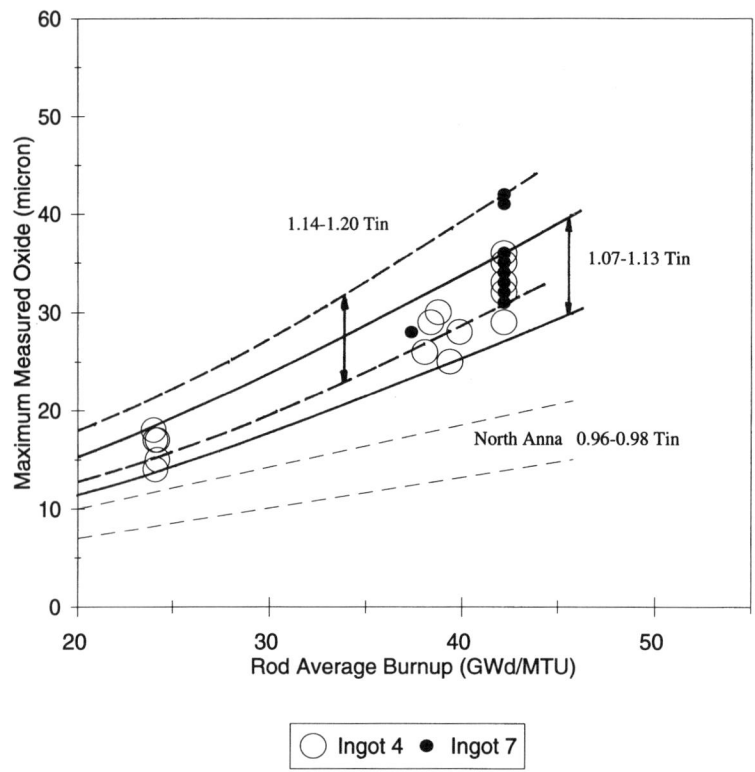

FIG. 4—*Reactor Z corrosion data.*

indicate that refinement of the particle size from 79 to 35 nm does not result in significant corrosion advantage for the burnups attained.

These results are in direct contrast to the behavior of late-stage beta-quenched Zircaloy-4 [7,9–11]. Zircaloy-4 processed in this manner with a corresponding low A-parameter [12] exhibits higher out-reactor corrosion at low temperatures (e.g., 633 and 672 K) and higher in-reactor corrosion than conventionally processed Zircaloy-4. Such corrosion degradation in both autoclave and in-reactor testing was not observed in ZIRLO processed with a late-stage beta quench, suggesting that ZIRLO does not exhibit a similar sensitivity to A-parameter effects.

After completion of the North Anna program, additional ZIRLO cladding was fabricated from new production ingots and irradiated in three additional commercial reactors identified as X, Y, and Z. The microstructure of the cladding was similar to that irradiated in North Anna, i.e., a uniform distribution of second-phase particles with an average diameter less than 80 nm. Autoclave test results on Reactor X and Y cladding indicated a relatively consistent behavior with the North Anna cladding (see Table 2). However, the measured in-reactor peak oxide thicknesses measured in Reactor Y and Reactor Z rods were higher than expected when compared to the in-reactor results from North Anna and Reactor X. Since these rods were fabricated from different ingots than that used for North Anna and Reactor X, a review of ingot chemistries was performed focusing on significant ingot differences. Niobium concentrations in the North Anna/X ingot ranged from 1.02 to 1.04 wt%, which fell within the range of niobium (0.99 to 1.08 wt%) of the Y/Z ingots. However, tin concentrations in the North Anna/X ingot (0.96 to 0.99 wt%) were lower than the tin range of 1.07 to 1.20 wt% in the Y/Z ingots, suggesting the

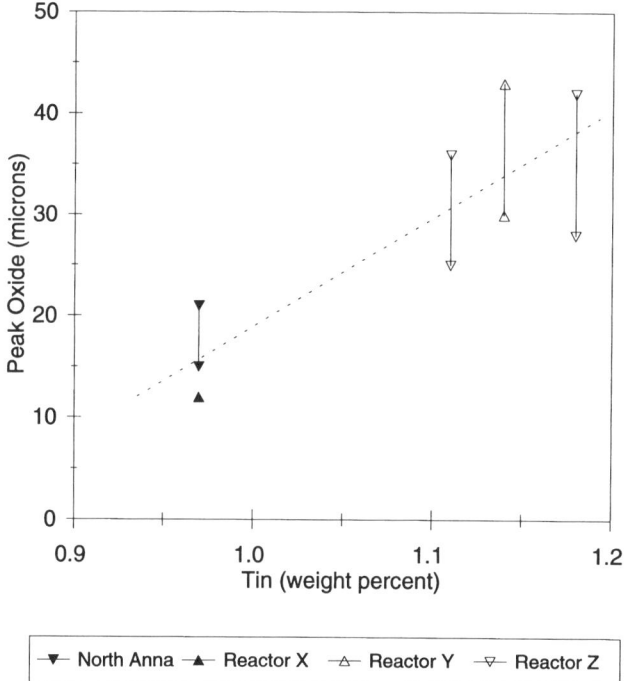

FIG. 5—*Peak oxide measurements in irradiated ZIRLO fuel rods as a function of tin content.*

potential role of tin in in-reactor ZIRLO corrosion performance. Figure 5 shows the maximum oxide thickness measured on rods with burnups in the range of 40 GWd/MTU as a function of tin concentration for all four commercial reactors. The corrosion data within Reactor Z were also reviewed by separating the rods from Ingot 4 with a tin level of 1.07 to 1.13 wt% from the rods from Ingot 7 with a tin level of 1.14 to 1.20 wt%. Figure 4 shows an overlap in the maximum measured oxide thicknesses, but suggestive of higher corrosion with higher tin content.

The implied effect of tin is consistent with recent efforts that demonstrate lower corrosion with decreased tin levels [7,13–16]. Lower in-reactor corrosion was observed in Zircaloy-4 as a result of lowering tin from about 1.5 to 1.3 wt% [7,16]. In addition to the effect on Zircaloy-4, Isobe and Matsuo [14] reported decreased corrosion weight gains in Zr-1Nb-XSn-0.2Fe-0.1Cr alloys as tin concentration decreased from 1.5 to 0.5 wt%. Results from the small non-production ingots melted to explore the effect of ZIRLO chemistry variations on corrosion are supportive of a tin effect. A plot of corrosion rates given in Table 3 as a function of the tin show lower corrosion as tin concentration decreases (Fig. 6). Rates appear to decrease down to zero weight percent tin as shown by the Zr-1Nb alloy. However, prior corrosion data on this alloy (Zr-1Nb) along with its poor corrosion resistance in lithiated environments [7] indicate that tin is an important alloy addition in maintaining good in-reactor performance.

The observed tin effect in ZIRLO is similar to that observed and well-documented for Zircaloy-4, where a reduced tin content provides the basis for the lower in-reactor corrosion of improved Zircaloy-4 over conventional Zircaloy-4 [7,16]. Revised target levels for tin content in ZIRLO materials have been similarly adjusted to levels consistent with, or lower than, cladding irradiated in North Anna Unit 1 and Reactor X. This is expected to provide consistent in-reactor ZIRLO corrosion performance that is significantly lower than Zircaloy-4.

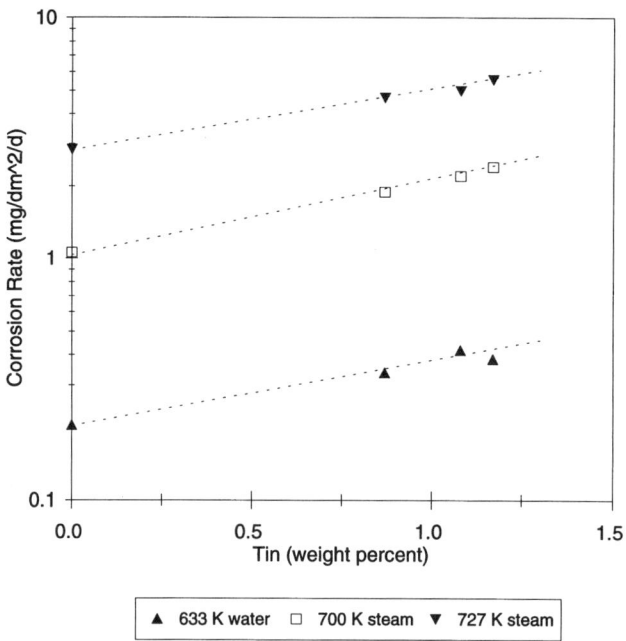

FIG. 6—*Autoclave corrosion rates as a function of tin concentration.*

Conclusions

1. Microstructural characterization of ZIRLO cladding specimens manufactured by maintaining low-temperature processing subsequent to the beta-quenching operation produces a desirable microstructure containing a fine dispersion of second-phase particles. Higher-temperature processing results in undesirable microstructural features containing areas of residual beta-zirconium.

2. A demonstration fuel assembly containing ZIRLO cladding fabricated via both conventional processing and late-stage beta-quenching was irradiated in North Anna Unit 1 reactor for two cycles to an average assembly burnup of 45.8 GWd/MTU. Both ZIRLO cladding types exhibited comparable corrosion performance with an average peak oxide thickness of 15 to 21 μm, an average 76% lower than comparably irradiated Zircaloy-4 rods.

3. Irradiation programs were performed in three other commercial reactors containing additional variants of ZIRLO processing and ingot chemistry. Results were consistent with those from North Anna when tin concentrations were maintained at comparable levels. Reduced corrosion benefits were observed with an increase in tin content.

4. Optimum corrosion performance of ZIRLO is achieved by a maintaining a uniform distribution of fine second-phase particles and controlled levels of tin.

References

[1] LeSurf, J. E., "The Corrosion Behavior of 2.5Nb Zirconium Alloy," *Application-Related Phenomena for Zirconium and Its Alloys, ASTM STP 458,* American Society for Testing and Materials, West Conshohocken, PA, 1969, pp. 286–300.
[2] Urbanic, V. F., LeSurf, J. E., and Johnson, A. B., Jr., "Effect of Ageing and Irradiation on the Corrosion of Zr-2.5 Wt% Nb," *Corrosion,* Vol. 31, 1975, p. 15.

[3] Urbanic, V. F., Warr, B. D., Manolescu, A., Chow, C. K., and Shanahan, M. W., "Oxidation and Deuterium Uptake of Zr-2.5Nb Pressure Tubes in CANDU-PHW Reactors," *Zirconium in the Nuclear Industry: Eighth International Symposium, ASTM STP 1023*, L. F. P. Van Swam and C. M. Eucken, Eds., American Society for Testing and Materials, West Conshohocken, PA, 1989, pp. 20–34.
[4] Urbanic, V. F. and Gilbert, R. W., "Effect of Microstructure on the Corrosion of Zr-2.5Nb Alloy," *Proceedings*, IAEA Technical Committee Meeting on Fundamental Aspects of Corrosion on Zirconium Base Alloys in Water Reactor Environments, International Atomic Energy Agency, Vienna, IWGFPT/34, 1990, pp. 262–272.
[5] Sabol, G. P., Schoenberger, G., and Balfour, M. G., "Improved PWR Fuel Cladding," *Proceedings*, IAEA Technical Committee Meeting on Materials for Advanced Water Cooled Reactors, International Atomic Energy Agency, Vienna, IAEA-TECDOC-665, 1992, p. 122.
[6] Sabol, G. P., Kilp, G. R., Balfour, M. G., and Roberts, E., "Development of a Cladding Alloy for High Burnup," *Zirconium in the Nuclear Industry: Eighth International Symposium, ASTM STP 1023*, L. F. P. Van Swam and C. M. Eucken, Eds., American Society for Testing and Materials, West Conshohocken, PA, 1989, pp. 227–244.
[7] Sabol, G. P., Comstock, R. J., Weiner, R. A., Larouere, P., and Stanutz, R. N., "In-Reactor Corrosion Performance of ZIRLO and Zircaloy-4," *Zirconium in the Nuclear Industry: Tenth International Symposium, ASTM STP 1245*, A. M. Garde and E. R. Bradley, Eds., American Society for Testing and Materials, West Conshohocken, PA, 1994, pp. 724–744.
[8] Sabol, G. P. and McDonald, S. G., "Process for Fabricating a Zirconium-Niobium Alloy and Articles Resulting Therefrom," U. S. Patent 4,649,023, March 1987.
[9] Kilp, G. R., Thornburg, D. R., and Comstock, R. J., "Improvement in Zirconium Alloy Corrosion Resistance," *Proceedings*, IAEA Technical Committee Meeting on Fundamental Aspects of Corrosion on Zirconium Base Alloys in Water Reactor Environments, International Atomic Energy Agency, Vienna, IWGFPT/34, 1990, pp. 145–157.
[10] Garzarolli, F., Steinberg, E., and Weidinger, H. G., "Microstructure and Corrosion Studies for Optimized PWR and BWR Zircaloy Cladding," *Zirconium in the Nuclear Industry: Eighth International Symposium, ASTM STP 1023*, L. F. P. Van Swam and C. M. Eucken, Eds., American Society for Testing and Materials, West Conshohocken, PA, 1989, pp. 202–212.
[11] Charquet, D., Steinberg, E., and Millet, Y., "Influences of Variations in Early Fabrication Steps on Corrosion, Mechanical Properties, and Structure of Zircaloy-4 Products," *Zirconium in the Nuclear Industry: Seventh International Symposium, ASTM STP 939*, R. B. Adamson and L. F. P. Van Swam, Eds., American Society for Testing and Materials, West Conshohocken, PA, 1987, pp. 431–447.
[12] Steinberg, E., Weidinger, H. G., and Schaa, A., "Analytical Approaches and Experimental Verification to Describe the Influence of Cold Work and Heat Treatment on the Mechanical Properties of Zircaloy-4 Cladding Tubes," *Zirconium in the Nuclear Industry: Sixth International Symposium, ASTM STP 824*, D. G. Franklin and R. B. Adamson, Eds., American Society for Testing and Materials, West Conshohocken, PA, 1984, pp. 106–122.
[13] Eucken, C. M., Finden, P. T., Trapp-Pritsching, S., and Weidinger, H. G., "Influence of Chemical Composition on Uniform Corrosion of Zirconium-Base Alloys in Autoclave Tests," *Zirconium in the Nuclear Industry: Eighth International Symposium, ASTM STP 1023*, L. F. P. Van Swam and C. M. Eucken, Eds., American Society for Testing and Materials, West Conshohocken, PA, 1989, pp. 113–127.
[14] Isobe, T. and Matsuo, Y. "Development of Highly Corrosion Resistant Zirconium-Base Alloys," *Zirconium in the Nuclear Industry: Ninth International Symposium, ASTM STP 1132*, C. M. Eucken and A. M. Garde, Eds., American Society for Testing and Materials, West Conshohocken, PA, 1991, pp. 346–367.
[15] Harada, M., Kimpara, M., and Abe, K., "Effect of Alloying Elements on Uniform Corrosion Resistance of Zirconium-Based Alloys in 360°C Water and 400°C Steam," *Zirconium in the Nuclear Industry: Ninth International Symposium, ASTM STP 1132*, C. M. Eucken and A. M. Garde, Eds., American Society for Testing and Materials, West Conshohocken, PA, 1991, pp. 368–391.
[16] Garde, A. M., Pati, S. R., Krammen, M. A., Smith, G. P., and Endter, R. K., "Corrosion Behavior of Zircaloy-4 Cladding with Varying Tin Content in High-Temperature Pressurized Water Reactors," *Zirconium in the Nuclear Industry: Tenth International Symposium, ASTM STP 1245*, A. M. Garde and E. R. Bradley, Eds., American Society for Testing and Materials, West Conshohocken, PA, 1994, pp. 760–778.

DISCUSSION

J. Harbottle[1] (written discussion)—You showed the significant effect of tin on the corrosion behavior of ZIRLO. Assuming that tin also changes the creep properties as in Zircaloy, how do you compensate for this in the composition and/or processing of ZIRLO?

R. J. Comstock et al. (authors' closure)—In-reactor creep of ZIRLO cladding was previously reported for the North Anna cladding and shown to be about 80% of the creep of Zircaloy-4 [7]. Tin in this material was 0.97 wt%, which is close to the nominal ZIRLO concentration of 1 wt%. Since the creep behavior of the cladding is acceptable, no changes in processing or chemistry were required to refine the creep properties of ZIRLO cladding. The presence of niobium in ZIRLO offsets the potential decrease in creep strength one might expect due to the lower tin in ZIRLO than in Zircaloy-4.

B. Cox[2] (written discussion)—(1) ZIRLO appears to be relatively insensitive to accelerated corrosion in LiOH at Li concentrations well above 70 ppm. Did you test these ZIRLO batches in more concentrated LiOH solutions, and, if so, did they all maintain this resistance in degradation? (2) In comparing the in-reactor behavior, were the conditions of water chemistry, thermal hydraulics, and heat rating in Reactors Y and Z the same as North Anna?

R. J. Comstock et al. (authors' closure)—(1) These specific lots of ZIRLO cladding were not tested in lithium concentrations greater than 70 ppm. However, ZIRLO cladding does exhibit accelerated corrosion in very high lithium concentrations. While remaining relatively insensitive to accelerated corrosion at lithium concentrations of 70 ppm, the BR-3 ZIRLO cladding exhibited accelerated corrosion when tested in 589 K water containing 700 ppm Li [6]. Such behavior would be expected for ZIRLO cladding lots processed similar to those discussed in this paper. (2) The thermal hydraulics and heat ratings of Reactors Y and Z were similar to North Anna. The water chemistry in Reactor Z and North Anna was a coordinated Li/B control maintaining a coolant $pH_{(T)}$ of 6.9. Reactor Y operated with a modified coolant chemistry with a lithium concentration of 2.2 ppm until the coolant reached a $pH_{(T)}$ of 7.4, which was then maintained to the end of the cycle.

J. Thomazet[3] (written discussion)—What are the effects of the microstructure and the chemistry on the hydrogen pickup?

V. Urbanic (written discussion)—Did you measure the hydrogen pickup associated with the autoclave tests and in the irradiated cladding and, if so, does the hydrogen pickup follow the same trends as the corrosion response or do the percentage uptakes vary?

R. J. Comstock et al. (authors' closure to both of the above questions)—In assessment of the chemistry effects, the hydrogen pickup in the autoclaved specimens showed similar percentage uptakes over the range of chemistries evaluated. Departures from the target ZIRLO microstructure were not explored relative to their impact on hydrogen uptake. Hydrogen pickup in irradiated ZIRLO with nominally 1% Sn content has been reported [6]. Additional measurements of hydrogen in irradiated ZIRLO cladding with higher Sn levels will be performed in the future.

---

[1] The S. M. Stoller Corp., Erlangen, Germany.
[2] University of Toronto, Toronto, Ontario, Canada.
[3] FRAMATOME, Lyons, France.

Dr. B. Warr[4] *(written discussion)*—Did you look at the effect of Nb or Fe content on in-reactor corrosion performance?

R. J. Comstock et al. (authors' closure)—The ZIRLO rods that have been subjected to in-reactor corrosion measurements were all fabricated close to the nominal niobium content of 1.0% and the nominal iron content of 0.1%. However, autoclave tests of ZIRLO samples with niobium contents ranging from ~0.8% to 1.2% (iron was maintained at nominal) did not reveal any significant difference in corrosion behavior. It is therefore expected that in-reactor corrosion behavior of ZIRLO would similarly show no niobium effects within the compositional range established for ZIRLO.

Y. S. Kim[5] *(written discussion)*—I would like you to comment on why ZIRLO with no Sn showed accelerated corrosion in the lithiated condition compared to ZIRLO with the addition of Sn.

R. J. Comstock et al. (authors' closure)—Removal of both tin and iron from ZIRLO results in a Zr-1Nb binary alloy that exhibits accelerated corrosion in lithiated environments. The processing of both ZIRLO and Zr-1Nb were similar and resulted in the development of a uniform distribution of small second-phase particles in the matrix. The major difference in the microstructure of the materials is the presence of Zr-Nb-Fe particles and tin in ZIRLO. We do not understand why the two alloys perform differently in lithium but believe that the presence of tin in the matrix plays an important role. Further work is required to understand the role of tin during oxide formation and growth in lithiated water.

---

[4] Ontario Hydro, Toronto, Ontario, Canada.
[5] Korea Atomic Energy Research Institute, Daejom, Korea.

Peter Y. Huang,[1] *S. Tahir Mahmood*,[1] *and Ronald B. Adamson*[2]

Effects of Thermomechanical Processing on In-Reactor Corrosion and Post-Irradiation Mechanical Properties of Zircaloy-2

REFERENCE: Huang, P. Y., Mahmood, S. T., and Adamson, R. B., "**Effects of Thermomechanical Processing on In-Reactor Corrosion and Post-Irradiation Mechanical Properties of Zircaloy-2,**" *Zirconium in the Nuclear Industry: Eleventh International Symposium, ASTM STP 1295*, E. R. Bradley and G. P. Sabol, Eds., American Society for Testing and Materials, 1996, pp. 726–757.

ABSTRACT: Interest continues to be high in the effects of irradiation on the microstructure and microchemistry of Zircaloy and on the resultant effects on performance of Zircaloy components in-reactor. In this study, we have investigated the behavior of material prepared to have excellent resistance to nodular corrosion in a BWR. Coupon specimens of Zircaloy-2 were prepared having a range of critical microstructural parameters (precipitate size and distribution) by varying the thermomechanical processing history (quench rate and cumulative annealing parameter). Irradiation was conducted in a BWR at 561 K to fluences between 1.3 and 8.5×10^{25} n/m^2 ($E > 1$ MeV).

As expected, nodular corrosion was absent in these materials, and, at low fluence, corrosion and hydriding of all materials was very low. At the highest fluences, patch-type uniform corrosion developed, with oxide thickness increasing with decreasing initial precipitate size. Detailed STEM investigation revealed that precipitates dissolved and decreased in size continuously until at the highest fluence no precipitates remained for material having the smallest initial precipitate size. Post-irradiation mechanical property tests showed strength somewhat higher than expected.

The results indicate that corrosion resistance and mechanical properties change as the microstructure evolves during irradiation. However, at the highest fluence tested, strength, ductility, and corrosion resistance ensure excellent performance of BWR components.

KEYWORDS: Zircaloy-2, BWR, thermomechanical processing, intermetallic precipitates, precipitate dissolution, irradiation, corrosion resistance, tensile properties, irradiation hardening

In 1985, at the Seventh International Symposium of the current series, a group of papers was presented that described effects of microstructure on nodular corrosion of Zircaloy. The interest in nodular corrosion, which had always been present in BWRs and which had by itself not resulted in fuel performance problems, arose for two primary reasons:

1. Heavy nodular corrosion combined with certain soluble chemical impurities (primarily copper) had caused fuel failures several years earlier [1].
2. Corrosion requirements became more stringent as utilities moved toward higher fuel burnups and residence times.

It was shown in that symposium that corrosion could be related to details of the microstructure (particularly characteristics of the second phase precipitates) found in either Zircaloy-2

---

[1] Senior engineer and [2] manager, Materials Technology, GE Nuclear Energy, Vallecitos Nuclear Center, Pleasanton, CA.

and -4) [2–6] and to the microchemistry (particularly the amounts of solute Fe and Ni found in the matrix [2,6]). In addition, an engineering approach that aided in keeping track of the metallurgical state during fabrication was proposed known as the cumulative annealing parameter (ΣA_i) [7]. Since that time, BWR fuel suppliers have moved to small mean precipitate sizes (0.03 to 0.10 μm) and to heat treatments that maximized solute contents and minimized precipitate size. Interestingly, at the same time, PWR fuel suppliers have moved to larger precipitate sizes, a consequence of the specific mechanisms associated with uniform rather than nodular corrosion that occurs in the low-oxygen environment of a PWR. Although there is still considerable room for improvement in our understanding of corrosion mechanisms, the ideas presented in 1985 provided a framework on which to base decisions involving fabrication variables.

As access to high-fluence material increased in the mid-1980s, sophisticated analytical techniques revealed that the effects of neutron irradiation on the microstructures were more complicated than previously thought. The early work of Bell et al. [8] indicated that small precipitates in the weld structures of Zircaloy-4 were not stable and, in fact, dissolved during irradiation. Subsequent work, led by studies at Vallecitos [6,9–13], Chalk River [14,15], and NFD [16,17] showed that at BWR temperatures (~573 K) precipitates steadily lost their Fe, Cr, and Ni to the matrix as a function of fluence. The effects on corrosion appear to be as predicted [6], with nodular corrosion decreasing or eliminated and uniform corrosion increasing as the matrix becomes supersaturated with solutes under the influence of the neutron flux [16,18]. In addition, even in unirradiated material it has been shown that small increases in the matrix concentration of the alloying elements can greatly reduce nodular corrosion in Zircaloy-2 [19]. Most recent work has indicated that for high fluences there may be a range of precipitate sizes that result in an optimum balance of uniform and nodular corrosion in BWRs [20].

In the past, most detailed studies of the effects of irradiation on corrosion and mechanical properties have been conducted on materials having relatively large mean precipitate sizes in the range of 0.2 to 0.5 μm. With increasing fluence, the mechanical properties of large precipitate materials appear to saturate after less than 2×10^{25} n/m² ($E > 1$ MeV) [21]. In PWRs, the corrosion rate of Zircaloy-4 accelerates at fluences above about 5×10^{25} n/m² ($E > 1$ MeV), the phenomenon being a complicated function of microstructural changes, metal/oxide temperatures, and water chemistry. In a BWR, the corrosion rate also appears to increase to some extent at high fluences [20] for all precipitate sizes. For the large precipitate material, the acceleration takes the form of local patches of thicker uniform oxide (30 to 40 μm rather than the average <10 μm), but only at very high fluence, about 1×10^{26} n/m² [22]. The current study examines in detail the effects of initial Zircaloy-2 microstructures on in-reactor corrosion and post-irradiation mechanical properties for material processed to have "small" precipitates and excellent resistance to nodular corrosion. The evolution of microstructure and microchemistry during irradiation is related to the change in properties observed.

Experimetal Procedure

Materials and Processing

One 19-mm (¾-in.)-thick slab of Zircaloy-2 plate was used to investigate the effects of thermomechanical processes on corrosion behavior. The composition of the plate is listed in Table 1. The thermomechanical processes are β (1000°C) water quench at one of three different thicknesses followed by either five, three, or one cold work and annealing steps to a common final thickness. The details of sample preparation and the corresponding ΣA_i values are shown in Table 2.

Optical metallography shows that the specimens are recrystallized, although in Specimen 1

TABLE 1—*19-mm (¾-in.) Zircaloy-2 plate (1986 TWCA)*.

Elements	Sn	Fe	Cr	Ni	O	C	Zr
wt%	1.48	0.14	0.1	0.06	0.127	0.016	Balance

a few grains are not fully recrystallized, with some remnants of the β-quench remaining. 2R and 3R specimens have equiaxed grains.

STEM Examination

The microstructure was examined by scanning transmission electron microscopy (STEM). Standard STEM 3-mm disks were obtained by grinding coupons from one side, then mechanically punching disks from the foil. Electrochemical thinning was done in a dual-jet polishing unit using a solution of 50-mL perchloric acid and 450-mL ethanol maintained at 20 mA, ~30 V, and -30 to -35°C. The specimens were examined using a JEOL 100-CX STEM at the Vallecitos Nuclear Center (VNC). The intermetallic precipitates were identified by selected area diffraction (SAD) and chemical analysis (EDS). Precipitate diameters were determined by scanning charged-couple device (CCD) images at 29 000 magnification that measured the precipitates using a computer-imaging analysis program.

In most cases, STEM data were taken from the surface areas of the specimen (within 75 μm of the surface). Data were also gathered from the specimen mid-walls. The measured parameters are

$$X_{Fe} = \frac{Fe}{Fe + Cr}, \quad Y_{Fe} = \frac{Fe}{Fe + Ni} \tag{1}$$

$$\overline{X}_{Fe} = \frac{(X_{Fe,initial})(V_{crystalline}) + (X_{Fe,min})(V_{amorphous})}{(V_{crystalline}) + (V_{amorphous})} \tag{2}$$

Texture Determination

Crystallographic textures of the archive materials were characterized using the standard back reflection X-ray diffraction technique. Small pieces cut from each of the sheet materials were assembled to make three orthogonal specimens with surface normals along the rolling (RD), transverse (TD), and through-thickness (ND) directions of each of the sheets. After mounting, polishing, and chemical etching of the specimens, Cr K_α X-rays were used to collect diffraction data from basal (0002) and prismatic ($10\overline{1}0$) planes of each of the specimens. These data were then used to construct complete basal and prismatic pole figures and to calculate basal and prismatic plane f-factors along ND, RD, and TD of the materials. The measured f-factors are

TABLE 2—*Processing schedules*.

Thickness	19 mm (0.75 in.)	9.5 mm (0.375 in.)	4.8 mm (0.190 in.)	3.3 mm (0.130 in.)	2.5 mm (0.100 in.)	0.8 mm (0.030 in.)	$\Sigma A = \Sigma t \cdot \exp(-40\,000/T)$
3R	βWQ	893 K/2h	783 K/2h	783 K/2h	893 K/2h	849 K/2h	1.45E-19
2R	...	βWQ	783 K/2h	...	893 K/2h	849 K/2h	7.40E-20
1R	...	...	...	...	βWQ	849 K/2h	3.46E-21

TABLE 3—*Crystallographic texture orientation parameters for archive materials.*

Material	(0002) Plane				($10\bar{1}0$) Plane			
	f_{ND}	f_{RD}	f_{TD}	Σf	f_{ND}	f_{RD}	f_{TD}	Σf
1R	0.671	0.082	0.247	1.000	0.166	0.458	0.374	0.998
2R	0.736	0.051	0.210	0.997	0.134	0.468	0.394	0.996
3R	0.748	0.060	0.192	1.000	0.127	0.474	0.400	1.001

given in Table 3. All three materials exhibit texture typical of recrystallized Zircaloy-2, with most of the basal poles oriented close to the ND and prismatic poles close to the RD and TD of the materials. A comparison of these numbers shows that while Materials 2R and 3R have similar texture, Material 1R has about 6 to 7% less basal poles oriented along the ND and, therefore, more along RD and TD. This difference in texture is also reflected by differences in mechanical properties discussed later.

Mechanical Property Measurements

Mini-tensile specimens with a 3.8 by 1.27 by 0.76-mm gage section were machined from archive and irradiated coupons using electro-discharge machining. The tensile axis of the specimens is in the rolling (or longitudinal) direction. The specimens were tensile tested at 288°C in a three-zone furnace in air at a strain rate of 8.3×10^{-4} per second. The test temperature was monitored by a thermocouple welded to the grips near the specimen. The test temperature was maintained within ±3°C. The load on the specimen was measured using a 106.5-N (1000-lb) load cell, while the specimen extension was determined from the preset strain rate and chart speed of the strip chart recorder. Two tests were conducted on each material. The load-displacement data collected on a strip chart recorder were digitized to construct engineering stress-strain curves and to determine the 0.2% offset yield strength (YS), ultimate tensile strength (UTS), uniform elongation (UE), and total elongation (TE).

Knoop microhardness measurements were made on one of the shoulder regions of the tensile specimens using a 300-g load and a 15-s hold time. The Knoop indentation was aligned with the long axis transverse to the rolling direction of the specimen.

Autoclave tests

Ex-reactor autoclave tests were conducted to study the nodular and uniform corrosion behavior of these samples. Before testing, samples were etched in a solution containing 5 vol% concentration HF, 45 vol% concentrated HNO_3, and the remainder distilled water.

Nodular corrosion testing was performed using the two-step (MAT) test (683 K/4 h + 793 K/16 h at 12 MPa (1750 psi) and uniform corrosion testing using 673-K, 10.3-MPa (1500-psi) steam.

Hydrogen Pickup

Sections of corrosion coupons were used for hydrogen determination using the tin flux fusion technique (LECO RH-2). The sample was ultrasonically cleaned in acetone, air dried at 303 K, and weighed. A single-use graphite crucible was loaded with 3 g of tin metal and outgassed for 4 min in the induction furnace at a current of 0.72 A. After cooling for 2.5 min, the furnace was opened and the sample was added to the crucible. The induction furnace was then heated in a two-step process, first at a current of 0.25 A and then at a current of 0.72 A. The hydrogen evolved during heating is swept by a high-purity nitrogen gas stream to a thermal conductivity cell. The thermal conductivity of the gas stream is related to the amount of hydrogen present.

The two-step process produces two-time resolved peaks indicative of the hydrogen concentration in the oxide layer and in the metal. As determined by Reager [23], the first peak occurs at relatively low temperature ($\leq$873 K) and is associated with hydrogen evolved from the oxide layer. The second peak occurs at relatively high temperatures and is associated with hydrogen evolved from the metal, or bulk, of the sample. The hydrogen content of the oxide is reported in ppm as if the oxide had the same weight as the metal. (This is only a convenience since the oxide weight at the time of hydrogen analysis is not known as a result of post-weighing spalling due to handling.)

Irradiation Conditions

Coupons were irradiated in a dummy neutron source holder (DNSH) in a commercial BWR. Coupons were in contact with non-boiling coolant water at 561 K. The fast flux was near 5×10^{17} n/m^2/s ($E > 1$ MeV). Three DNSHs were irradiated: for one cycle (284 days), two cycles (543 days), and four cycles (1515 days). Fast fluences were 1.3, 2.5, and 8.5×10^{25} n/m^2, as measured experimentally using flux monitors.

Results

Nodular Corrosion—Laboratory

For unirradiated archives, a few white nodules were found on the edges of the 3R and 2R specimens, but not on the 1R. Nodules were not observed on the main surfaces (rolling planes) of any specimen. Weight gain decreased with decreasing thickness at the quenching stage (increasing quench rate) or numbers of cold work passes to final thickness, although the differences were small and can be attributed to edge effects. For 3R, 2R, and 1R specimens, the weight gains were 60, 50, and 50 mg/dm^2, respectively.

Uniform Corrosion—Laboratory

Figure 1 shows the weight gain for unirradiated archives for up to 150 days. Specimens had similar weight gains (~100 mg/dm^2) for this relatively short testing time.

In-Reactor Corrosion

Figure 2 gives the corrosion weight gain data versus accumulated fast neutron fluence ($E > 1$ MeV). At fluences of 1.3 and 2.5×10^{25} n/m^2, there are no differences between specimens, the weight gain being near 50 mg/dm^2. However, at the high neutron fluence (~8.5×10^{25} n/m^2), the weight gain increases significantly and there is a clear separation between specimens having a different thermomechanical history. Namely, the 1R specimens have exhibited the highest weight gains, while the 3R specimens have the lowest weight gains.

Hydrogen Analysis

The hydrogen analysis results are summarized in Table 4. The measured hydrogen contents and in-reactor corrosion weight gains were used to calculate the percent theoretical hydrogen pickup according to the oxidation reaction:

$$Zr + 2H_2O \Rightarrow ZrO_2 + 2H_2 \quad (3)$$

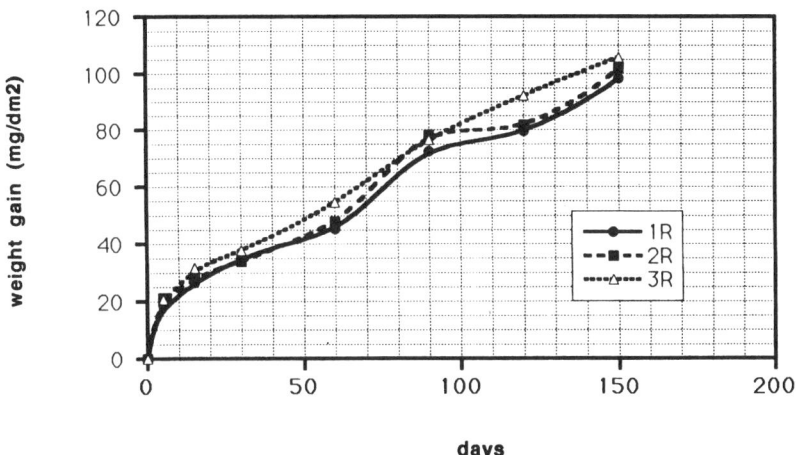

FIG. 1—*Archive materials tested in 10.3-MPa (1500-psi) steam at 673 K for 150 days.*

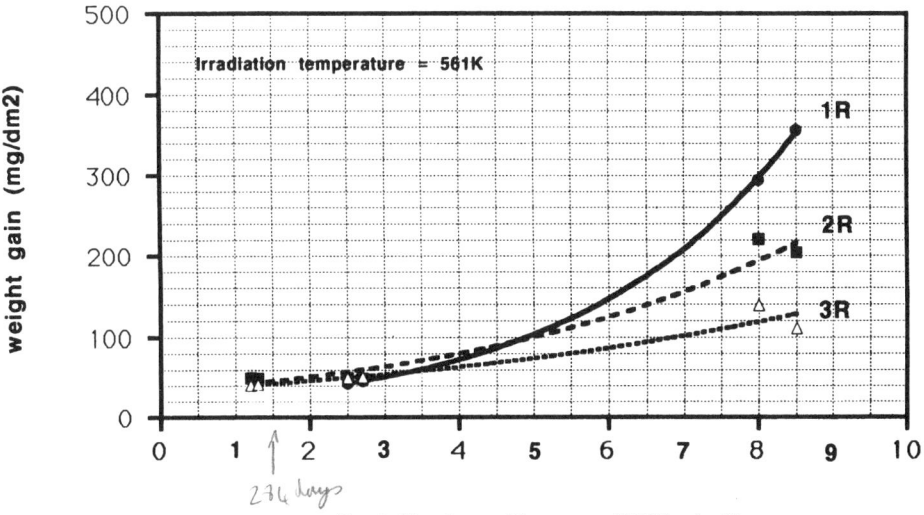

FIG. 2—*Weight gain as a function of fluence at an irradiation temperature of 561 K.*

TABLE 4—*Corrosion weight gain and % hydrogen pickup of two and four-cycle irradiated mini-tensile specimens.*

Specimen	Fluence, 10^{25} n/m^2	Weight Gain, mg/dm^2	Hydrogen Content, ppm		Hydrogen Pickup, % (in metal)
			Oxide	Metal	
3R	2.5	40	13	11	2
2R	2.5	53	15	14	3
1R	2.5	45	15	11	2
3R	8.5	112	13	147	26
2R	8.5	204	52	365	35
1R	8.5	294	60	370	25

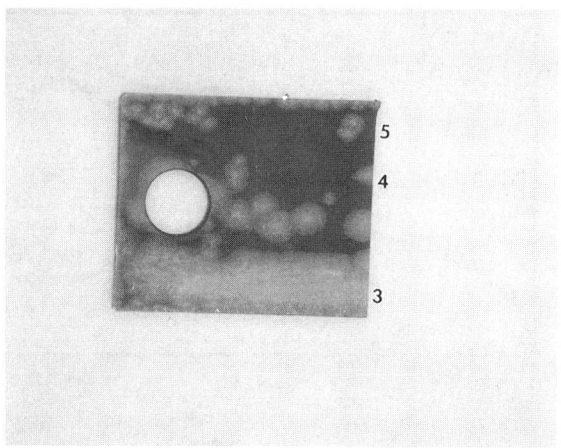

FIG. 3—*Surface appearance of 3R materials (112 mg/dm²).*

The % hydrogen pickup increased from about 2% after two irradiation cycles to about 30% after four irradiation cycles as shown in Table 4. The hydrogen content of the unirradiated archives was about 7 ppm. This value was subtracted from the metal value for pickup fraction calculations.

Post-Irradiation Exam (PIE)

Visual and Metallographic Exam—Visual exams of one and two-cycle specimens reveal no distinguishing features, i.e., no nodules or corrosion patches. Cross-sectional metallography showed uniform oxide a few microns thick. Visual exams of four-cycle specimens revealed non-homogeneous corrosion on all specimens. Figure 3 shows a photograph of one side of half of the 3R specimen (weight gain $\sim$ 112 mg/dm²). The surface consists of various features, e.g., scattered "patch oxides," white strips of oxide near the edges of specimen, and grayish areas. Because of the inhomogeneous surface appearance, careful measurement of the oxide thickness

FIG. 4a—*Oxide morphology of 3R materials at Location 3 in Fig. 3.*

FIG. 4b—*Oxide morphology of 3R materials at Location 4 in Fig. 3.*

was performed on specific features. Examples are given in Figs. 4a–c. The maximum oxide thickness is ~22.9 μm (~0.9 mil) measured at Location 3 in Fig. 3 (Fig. 4a), while the minimum oxide thickness is ~1.5 μm (0.06 mil) measured at Location 5 in Fig. 3 (Fig. 4c) that corresponds to the grayish area. The average oxide thickness of the 3R specimen measured at a total of 10 locations is ~7 μm (0.28 mil). It should be noted that the average oxide thickness was obtained from various surface features and does not represent a "true" average thickness. The hydride distribution was uniform (Fig. 5).

Figure 6 shows the visual appearance of the 2R specimen (weight gain ~ 204 mg/dm^2). The surface is more homogeneous than 3R, with many small, closely spaced "white dots" distributed on the entire surface. Figure 7a shows typical oxide morphology. The appearance of the "white dots" does not resemble the typical nodular oxide observed on the fuel rods in commercial BWRs. The maximum oxide thickness is ~27.9 μm (~1.1 mil). The minimum is ~5.1 μm (0.2 mil). The "average" oxide thickness of the 2R specimen was measured at a total of 10 locations and was ~18.3 μm (0.73 mil). Figure 7b shows uniform distribution of hydride in the 2R specimen.

FIG. 4c—*Oxide morphology of 3R materials at Location 5 in Fig. 3.*

FIG. 5—*Hydride distribution in 3R materials.*

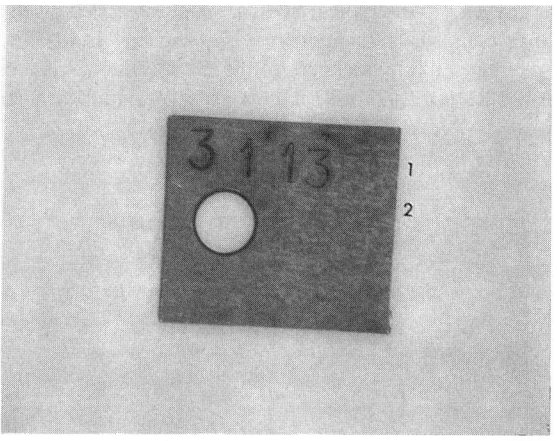

FIG. 6—*Surface appearance of 2R materials (204 mg/dm^2).*

FIG. 7a—*Oxide morphology of 2R materials at Location 2 in Fig. 6.*

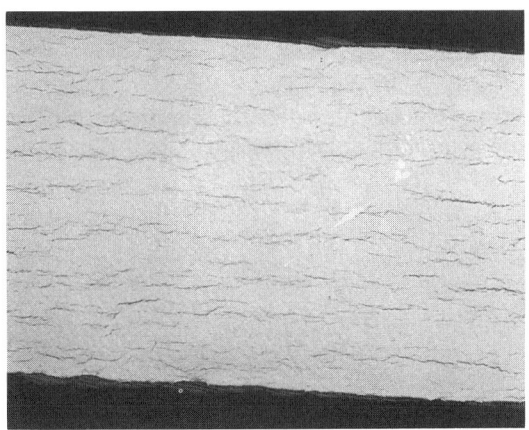

FIG. 7b—*Hydride distribution in 2R materials.*

FIG. 8a—*Surface appearance of front side of 1R material (293 mg/dm^2).*

Figure 8a shows the 1R specimen (weight gain ~293.8 mg/dm^2). The front surface has a grayish appearance, while the other side consists of scattered and coalesced "patch oxides." Figure 9 shows the oxide morphology of the 1R specimen at the various locations as indicated in Fig. 8a. On one side of the surface (Fig. 9), the oxide morphology appears to be wavy and uniform. However, on the other side (Fig. 8b) where "patch oxide" is seen, only uniform oxide morphology is observed (Fig. 10). The maximum oxide thickness measured is ~43.2 µm (~1.7 mil), and the minimum oxide thickness measured (at Location 4) is ~2.5 µm (0.1 mil). The "average" oxide thickness of the 1R specimen measured at a total of ten locations is ~20.3 µm (0.8 mil). Figure 11 shows uniform hydride distribution in the 1R specimen.

STEM Microstructural Examination

Unirradiated Archives—For the 1R and 2R specimens, the intermetallic precipitates are formed along the phase boundaries of lath structures (Figs. 12a and 12b), while for the 3R

FIG. 8b—*Surface appearance of back side of 1R material (293 mg/dm^2).*

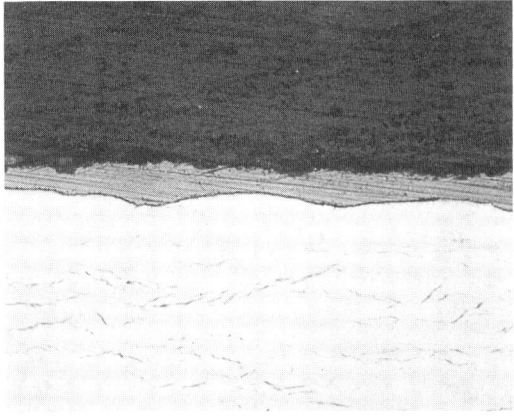

FIG. 9a—*Oxide morphology of 1R material at Location 1 in Fig. 8a.*

FIG. 9b—*Oxide morphology of 1R material at Location 2 in Fig. 8a.*

specimen, they are more uniformly distributed in the matrix (Fig. 12c). For the archive 2R and 3R specimens, completely recrystallized grains with few dislocations are observed. The 1R specimen exhibits a few partially recrystallized grains with a higher density of dislocations. The precipitate size distribution of these specimens is given in Figs. 13a–c. It should be noted that these precipitate size distributions were determined near the outer surface of samples. Examination near the midplane revealed a larger median precipitate size than at the surface for the 3R specimens, but not for the 2R and 1R specimens. The relatively larger thickness of 3R at the quenching step probably produced a through-thickness quench rate gradient that resulted in a corresponding precipitate size gradient. It is noted that we prefer to quote median precipitate size as a figure of merit. The size distributions are generally log normal, and the median is slightly smaller than the mean.

Low-Fluence ($\sim 2.5 \times 10^{25}$ n/m^2) Two-Cycle Irradiation—Irradiation for two cycles induced several microstructural changes for these specimens. In all materials, a high density of

FIG. 10a—*Oxide morphology of 1R material at Location 3 in Fig. 8b.*

FIG. 10b—*Oxide morphology of 1R material at Location 4 in Fig. 8b.*

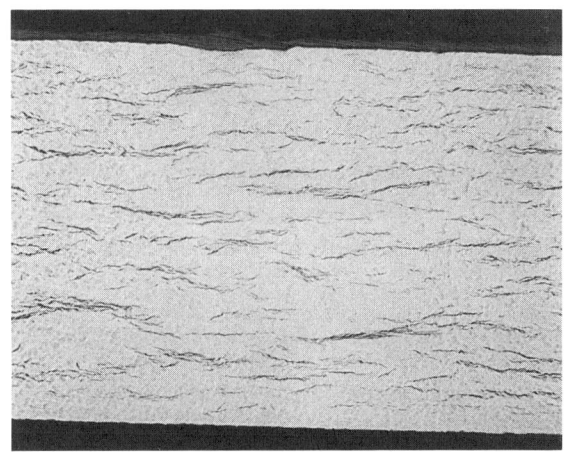

FIG. 11—*Hydride distribution in 1R material.*

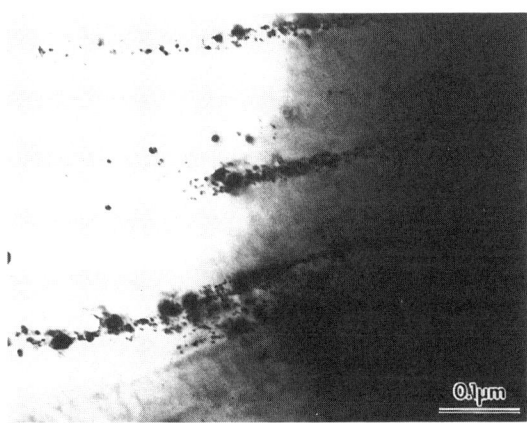

FIG. 12a—*STEM micrograph of archive 1R material.*

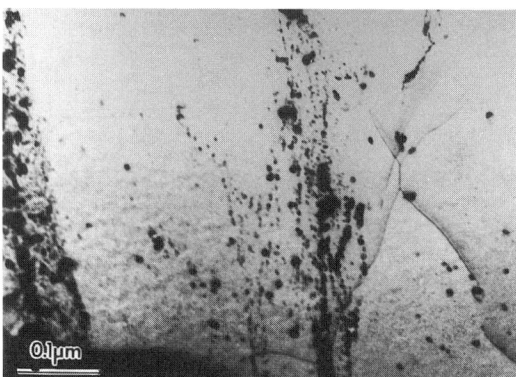

FIG. 12b—*STEM micrograph of archive 2R material.*

"black dot" damage (small $<a>$-loops) was detected. A low density of $\frac{1}{2}\langle 0001 \rangle$ vacancy type $\langle c \rangle$-component dislocations was also observed. Irradiation caused changes in precipitate size distribution, composition, and morphology (Fig. 14). Partial and complete amorphization were seen in large and small $Zr(Fe,Cr)_2$ precipitates, respectively. The width of the amorphous layer was measured to be around 0.025 μm. The reported linear correlation of amorphous layer width with fluence of 0.01 μm per 10^{25} n/m² ($E > 1$ MeV) [24] corresponds to the measured fluence of 2.5×10^{25} n/m². The average (defined in Eq 2) decreased from 0.45 to 0.32, indicating that some Fe dissolution had occurred. For $Zr_2(Fe,Ni)$ precipitates, as expected, no amorphization was observed. Furthermore, Y_{Fe} (i.e., Fe/(Fe+Ni) ratio) remained the same as for unirradiated samples. This implies that either there is no dissolution occurring or the dissolution rate of Fe and Ni is similar.

The precipitate size for those specimens subjected to the two-cycle irradiation changed relative to the unirradiated material. For the 1R specimens, the median decreased while, for both the 2R and 3R specimens, the median increased as compared to the archive materials (Table 5). Furthermore, the number density in all specimens decreased, indicating that complete dissolution (shrinkage) of small precipitates occurred. The dissolution of the precipitates, in turn, implies supersaturation of the Zr-matrix with Fe and Cr. No re-precipitation was observed in either the matrix or grain boundaries.

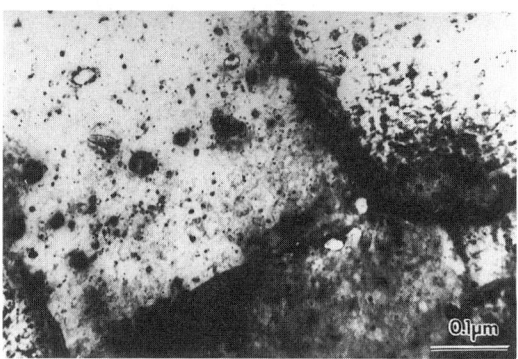

FIG. 12c—*STEM micrograph of archive 3R material.*

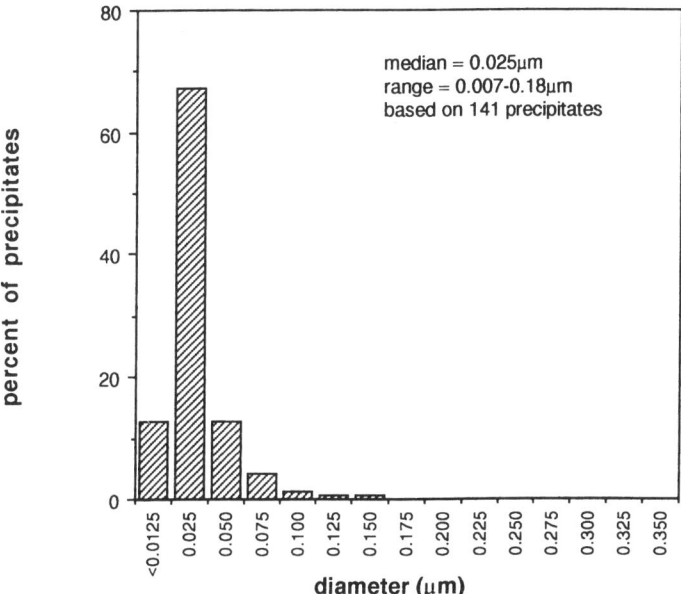

FIG. 13a—*Precipitate size distribution of archive 1R material.*

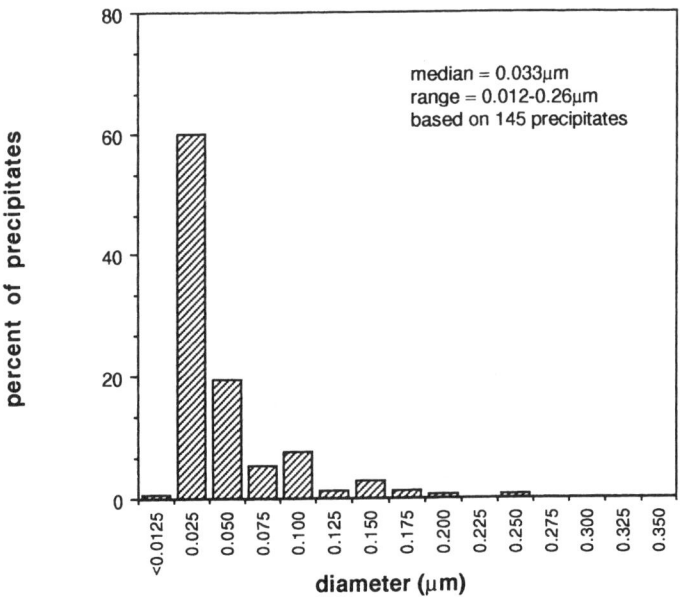

FIG. 13b—*Precipitate size distribution of archive 2R material.*

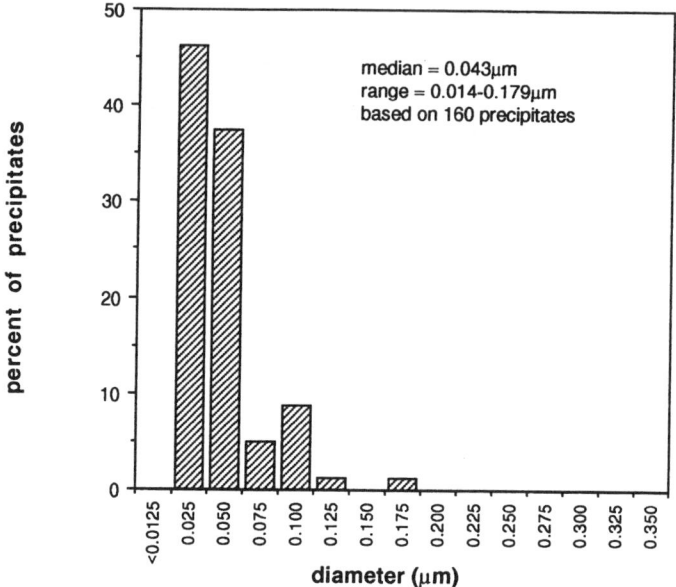

FIG. 13c—*Precipitate size distribution of archive 3R material.*

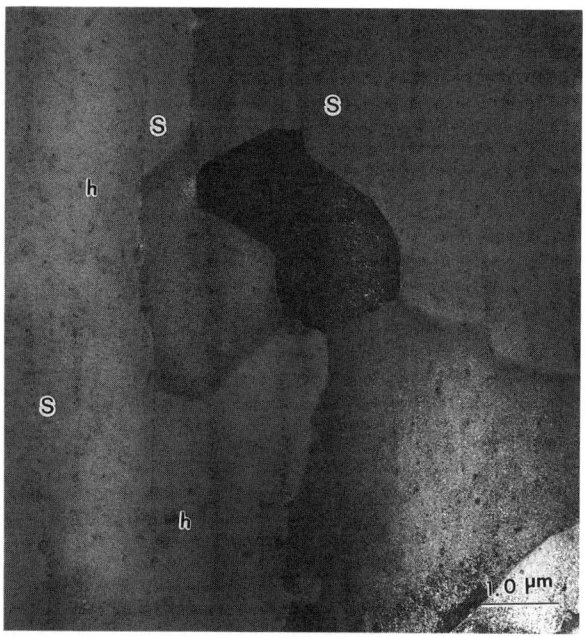

FIG. 14a—*STEM micrograph of two-cycle 1R material: "h" denotes hydrides and "s" denotes stringers.*

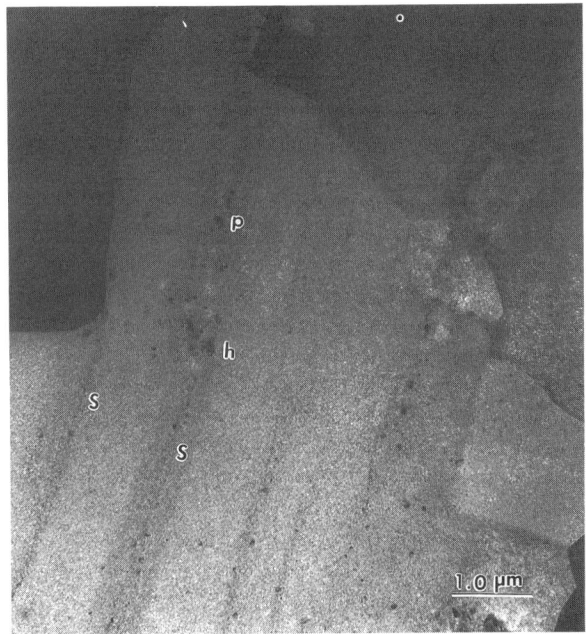

FIG. 14b—*STEM micrograph of two-cycle 2R material: "P" denotes precipitate, "h" denotes hydrides, and "s" denotes stringers.*

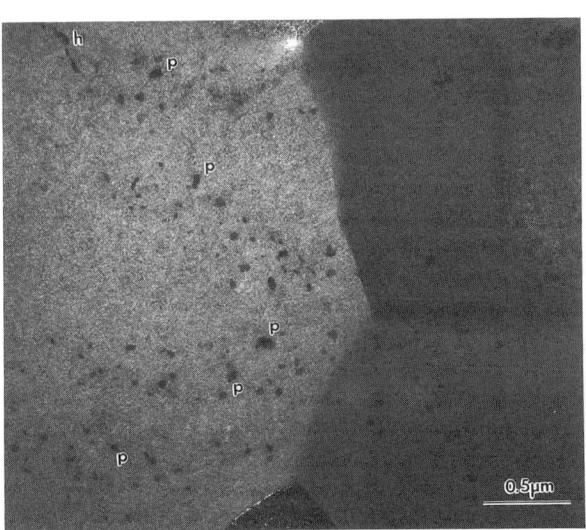

FIG. 14c—*STEM micrograph of two-cycle 3R material: "P" denotes precipitate, "h" denotes hydrides.*

TABLE 5—*STEM characterization on intermetallic precipitates of archive and irradiated Zircaloy-2 specimens.*

Fluence, 10^{25} n/m^2	Sample	Median of Size Distribution, (μm)	$\overline{X}_{Fe}$	Y_{Fe}	Number Density of Precipitates, 10^7/cm^2
0	1R	0.025	0.45±0.02	0.53±0.03	3.6
	2R	0.032	0.46±0.04	0.55±0.04	2.1
	3R	0.042	0.44±0.03	0.52±0.03	1.6
2.5	1R	0.016	0.32±0.04	0.48±0.05	1.9
	2R	0.063	0.33±0.04	0.49±0.04	1.3
	3R	0.082	0.32±0.03	0.48±0.04	1.3
8.5	1R	~0	...	...	~0
	2R	0.038	0.17±0.06	n/a	0.1
	3R	0.112	0.17±0.05	0.40±0.08	0.5

$$Y_{Fe} = \frac{Fe}{Fe + Ni}$$

$$\overline{X}_{Fe} = \frac{(X_{Fe,initial})(V_{crystalline}) + (X_{Fe,min})(V_{amorphous})}{V_{crystalline} + V_{amorphous}}$$

It is noted that for the 1R and 2R specimens, "dark" bands with a width of ~0.5 μm were observed along the precipitate stringers (Figs. 14a–b). This "dark" band was not observed in archive materials, indicating that the formation is probably due to irradiation. Microanalysis in these regions measured high Fe concentrations (2000 wppm) but no Cr. However, the spatial resolution and sensitivity of the technique does not entirely rule out the possibility that small Fe-rich precipitates are included in the measurement. The band width does appear to compare favorably to the expected diffusion distance of Fe based on the diffusivity of Fe during irradiation at BWR temperature suggested by Motta et al. [25]. Microdiffraction taken from this region revealed that the lattice parameter has been increased by ~2%, i.e., the c/a ratio is ~1.63 rather than c/a ~ 1.59 observed from the matrix.

High-Fluence (~8.5 × 10^{25} n/m^2) Four-Cycle Irradiation—High-fluence irradiation of these specimens produced even more distinct microstructural changes. Changes in the shape of the grains indicated that grain boundary migration had occurred; for example, in 2R and 3R materials, grain boundary migrations were observed (Figs. 15a–b). For the 1R material, which possessed some partially recrystallized grains in the unirradiated condition, subgrain formation and grain boundary migration were also observed (Fig. 15c). Furthermore, for all the four-cycle irradiation materials, a high density of pure $\langle c \rangle$-dislocations was detected (i.e., ½[0001] type). This is consistent with previous results [26] for materials irradiated at high fluence near 573 K.

For the 1R material, complete dissolution of the precipitates (both $Zr_2(Fe,Ni)$ and $Zr(Fe,Cr)_2$ types) occurred (Fig. 16). For the 2R material, a trace of pre-existing stringers of precipitates was detected, but most of the precipitates were also dissolved (Fig. 15b). For the 3R materials, many precipitates were still observable; however, all $Zr(Fe,Cr)_2$ type of precipitates underwent complete amorphization with the average $X_{Fe} = 0.17$. On the other hand, "disintegration" of $Zr_2(Fe,Ni)$ precipitates occurred (Fig. 17) in accordance with the observation reported by Etoh et al. [27]. Furthermore, Y_{Fe} decreased only slightly to 0.4 ± 0.06, indicating that the dissolution rate of Fe and Ni was nearly the same.

The number density of precipitates for both 2R and 3R materials drastically decreased. A large increase of the median of the precipitate size distribution in these materials was also

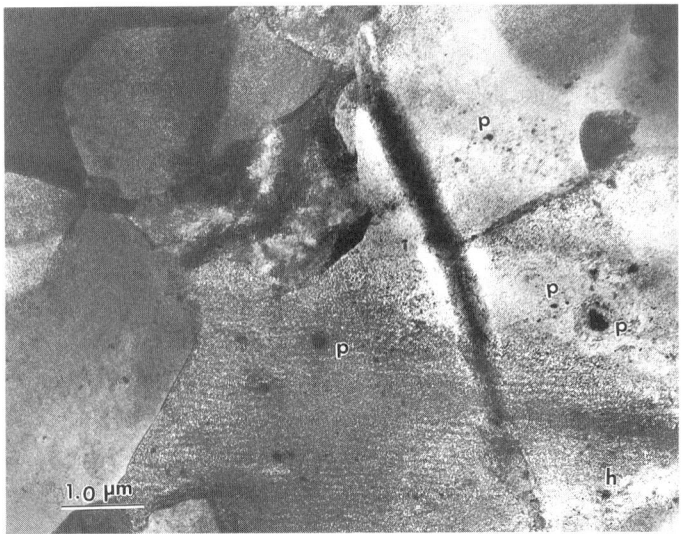

FIG. 15a—*STEM micrograph of four-cycle 2R material: "h" denotes hydrides, "p" denotes precipitate.*

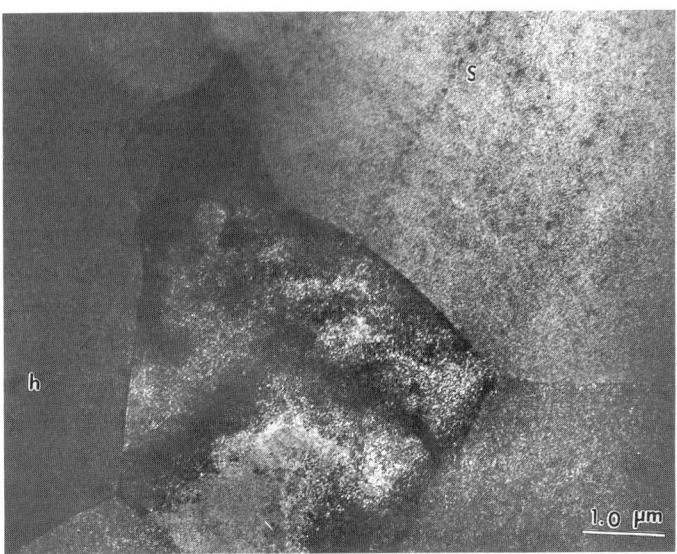

FIG. 15b—*STEM micrograph of four-cycle 3R material: "h" denotes hydrides, and "s" denotes stringers.*

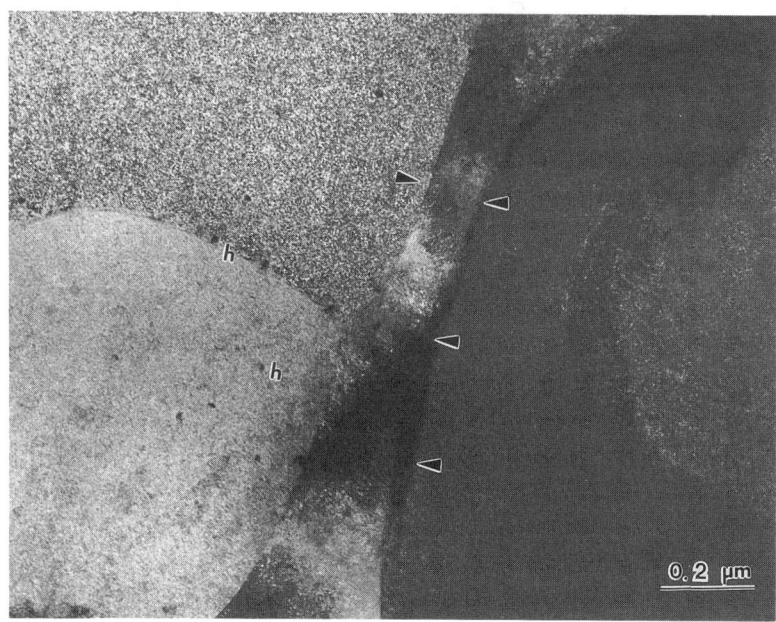

FIG. 15c—*STEM micrograph of four-cycle 1R material: "h" denotes hydrides, while arrows refer to migrating grain boundaries.*

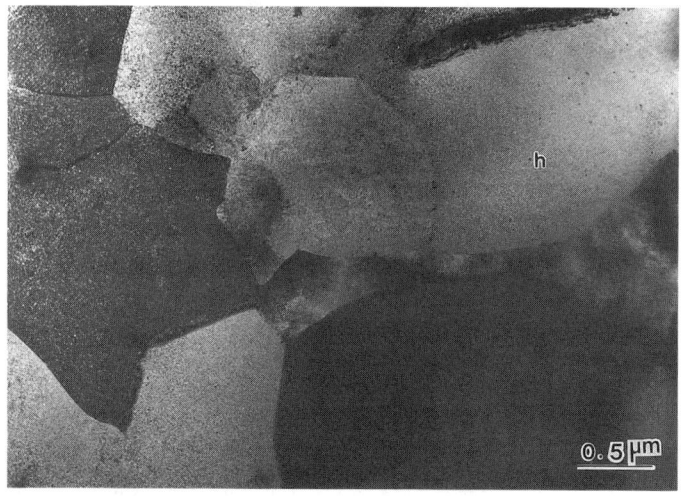

FIG. 16—*STEM micrograph of four-cycle 1R material: "h" denotes hydrides.*

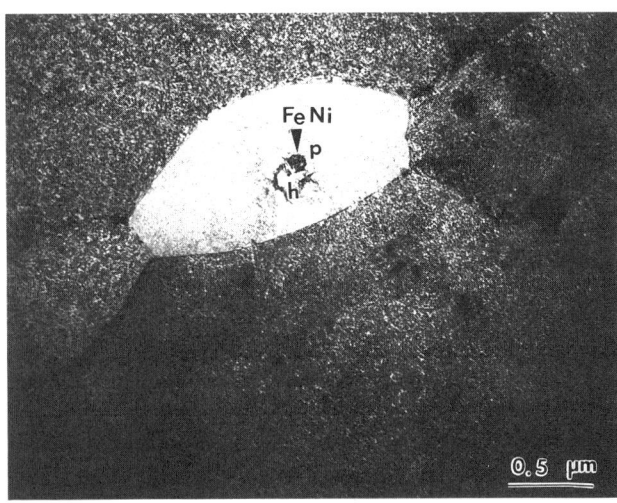

FIG. 17—*Zr-Fe-Ni precipitate observed in 3R material after four cycles of irradiation.*

observed (Table 5). These changes are attributed to the precipitate dissolution, with only large precipitates remaining. The complete dissolution for the 1R sample and near complete dissolution of the precipitates for the 2R and 3R samples indicate that the zirconium matrix is supersaturated with a large amount of Fe, Ni, and Cr. Some reprecipitation at grain boundaries was observed, the precipitates being of the Zr_3Fe type. Figure 18 illustrates the change of precipitate distribution as a function of fluence for Material 3R.

It is of interest to note that the "dark" bands observed in the two-cycle 1R and 2R materials were no longer observable after four cycles of irradiation, indicating that the alloying elements were more uniformly distributed.

Irradiation Growth

In a separate irradiation program, irradiation growth was determined for materials in this study. Specimens were irradiated in an inert-gas (He), gamma-heated capsule in the advanced test reactor (ATR). The capsule was designed to operate at 573 K, but the exact temperature has not yet been determined. However, STEM examination of Material 3R after a fluence of 6.5×10^{25} n/m^2 ($E > 1$ MeV) (measured experimentally) reveals a microstructure typical of irradiation near 573 K, that is, black spot damage, absence of the original smallest precipitates, amorphous $Zr(Fe,Cr)_2$ precipitates, and $<c>$-component dislocations. The amorphous layer widths were about 0.06 μm. Although not yet confirmed by STEM, it is expected that precipitates in 2R and 1R would be nearly completely dissolved. Growth data are given in Table 6. Growth of the materials is as expected [28]. For the recrystallized materials, the "breakaway" growth regime has not yet been reached, and the growth magnitudes for all specimens are about the same. Adjusted for texture differences, the growth of Material 1R is slightly higher than the others, which may reflect the observation that 1R is not quite fully recrystallized. The stress-relieved Material 2S (data included here to illustrate the differences between fully recrystallized and stress-relieved materials) has a substantially higher growth and growth rate compared to the recrystallized Specimen 2R, which again is consistent with expected behavior.

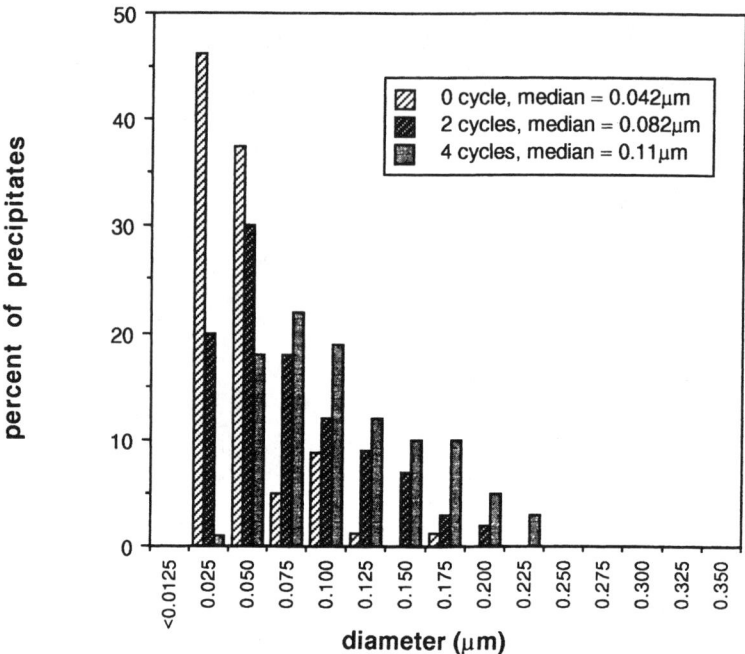

FIG. 18—*Change of the precipitate size distribution of 3R material as a function of neutron fluence.*

Mechanical Properties

The mechanical properties of the archive and irradiated materials after four cycles in-reactor obtained from testing of mini-tensile specimens (0.76-mm gage length) are given in Table 7. Stress-strain curves are given in Fig. 19. Typical irradiation effects are observed, that is, an increase in strength and decrease in ductility compared to unirradiated material. Substantial work softening occurs at the point of maximum load where dislocation channeling concentrates the deformation in one main sheer band. Fractographic analysis revealed completely ductile fracture surfaces. Knoop microhardness measurement results on archive and irradiated materials are also given in Table 7. These measurements also show significant radiation hardening. The properties were roughly the same for all three materials.

TABLE 6—*Irradiation growth at about 573 K as a function of fluence for recrystallized Materials 1R, 2R, 3R and stress-relieved Material 2S.*

Material Specimen	Texture, f_{RD}	Growth, %	
		4.0×10^{25} n/m^2	6.5×10^{25} n/m^2
1R	0.08	0.10	0.11
2R	0.05	0.09	0.10
3R	0.06	0.09	0.09
2S[a]	0.05	0.15	0.30

[a] Stress relieved.

TABLE 7—*Tensile properties (at 561 K) of archive and irradiated Zircaloy-2 specimens.*

Fluence, n/m^2	Material		
	1R	2R	3R
	YS, MPa		
0	249	220	196
1.3×10^{25}	...	657	601
2.5×10^{25}	607	656	643
8.5×10^{25}	666	689	689
	UTS, MPa		
0	323	252	256
1.3×10^{25}	...	668	622
2.5×10^{25}	696	683	654
8.5×10^{25}	792	782	728
	UE, %		
0	13.0	8.0	14.0
1.3×10^{25}	...	0.6	0.6
2.5×10^{25}	1.9	1.1	0.4
8.5×10^{25}	1.6	1.4	1.4
	TE, %		
0	39.0	39.0	45.0
1.3×10^{25}	...	18.3	18.0
2.5×10^{25}	16.4	16.9	14.8
8.5×10^{25}	12.1	8.0	11.6

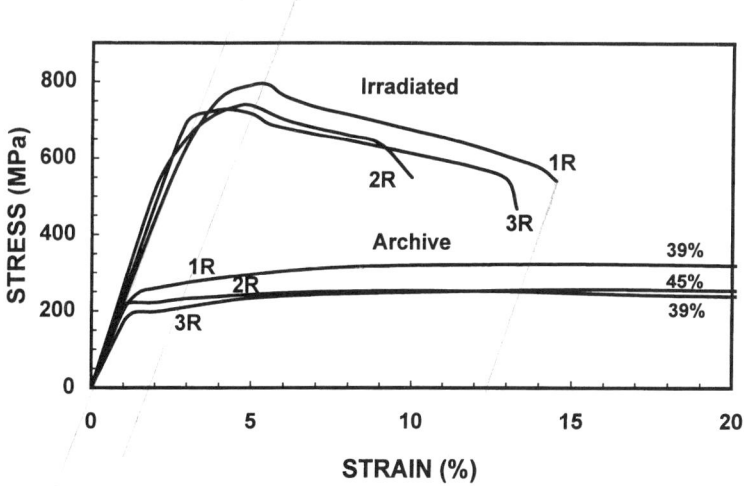

FIG. 19—*Stress-strain curves for archive and irradiated specimens at 8.5 $\times$ 10^{25} n/m^2.*

Discussion

The rate of precipitate dissolution, the type of "solute" dissolved, and the distribution of the dissolved "solute" are important factors to the corrosion resistance and mechanical behavior observed in these thermomechanically processed materials.

Precipitate Dissolution

It is clear that at 573 K in a BWR the major precipitates undergo irradiation-induced dissolution. We find no firm evidence of precipitate coarsening. Logically, small precipitates disappear first, causing a decrease in the average or median size. Eventually, however, at a fluence that depends on the initial size distribution, the as-irradiated median exceeds the original value as the smaller precipitates disappear. Our data confirm the following scenarios. For the Zr $(Fe,Cr)_2$ laves phase precipitates, the process occurs in two simultaneous steps: (1) an amorphous transformation accompanied by a loss of Fe to the matrix, and (2) an anisotropic [9] dissolution with Cr and Fe loss to the matrix. The $Zr_2(Fe-Ni)$ precipitates do not become amorphous but dissolve continuously in a non-uniform manner [27].

A model has been developed that calculates the amount of Fe, Cr, and Ni in solution as a function of fluence [29]. The model, for instance, calculates values of Fe content in the matrix of about 1400, 1200, and 1100 ppm for Specimens 1R, 2R, 3R, respectively, for a fluence of 8.5×10^{25} n/m². Correspondingly, smaller values are calculated at lower fluences, with the rates of Fe buildup being dependent on the initial size distribution of the precipitate. There appears to be a strong correlation between buildup of solute in the matrix with corrosion and, perhaps, also for mechanical properties. Details of the model are beyond the scope of this paper and will be published later.

Corrosion

The corrosion behavior observed in this study provides confirmation of earlier work on nodular and uniform corrosion processes [6,18]. For material with small precipitate size (and, almost coincidentally, low ΣA_i values), nodular corrosion is virtually eliminated. At low fluences, where partially amorphous precipitates co-exist with relatively low values of dissolution-produced alloying element supersaturation, corrosion is low and equivalent for all three materials. Apparently, only a small supersaturation is required to suppress nodules in-reactor, as was observed for unirradiated material [19] where an increase in matrix Fe concentration from 37 to 112 wppm was sufficient to greatly reduce nodular corrosion. As fluence increased, the microchemistry and microstructure continually evolved: all precipitates shrunk, smaller precipitates disappeared, a lower density of large precipitates remained, and the matrix became increasingly supersaturated with solute. At that point, the "semiconducting theories" of Zircaloy corrosion [6,30–32] and some experimental results [6,18] predict an increasing rate of uniform corrosion. This was observed here, with patches of uniform oxide forming, as previously observed in BWRs [22] and PWRs [33]. The rate of acceleration of the uniform corrosion is greatest for the highest level of solute supersaturation and for the most uniform distribution of alloying elements (Specimen 1R)(Fig. 2).

It is noted that the uniform corrosion hypothesis of Hutchinson and Lehtinen [34] suggests that there is an optimum size and distribution of precipitates that results in optimum resistance (or acceleration) of uniform corrosion. With the continually changing precipitate size distribution that occurs at BWR temperatures, this optimum distribution probably exists at one point in time, perhaps in the current case at low fluence. However, the "worst" case for uniform corrosion by this hypothesis is either to have no precipitates at all or to have a low density of large precipitates. This is exactly the situation at high fluence in the current case, where corrosion acceleration is observed.

A surprising result in this study is the exceedingly low value of hydrogen pickup fraction that occurs at low fluences. We suspect that hydrogen pickup is strongly affected by the ratios of Fe, Cr, and Ni in the zirconium matrix as well as in the precipitates. These ratios, of course, are continually changing under irradiation conditions. At low fluences in materials in this study, there is relatively more Fe in the matrix than Ni and, particularly, Cr due to particulars of the dissolution process. This fact may provide some insight into hydrogen pickup mechanisms and is being persued. At high fluence, the elemental ratios in the matrix are about the same as in the initial precipitates, and the pickup fractions, although on the high side, are fairly normal [35]. The oxide thickness and, perhaps, porosity is also different at low and high fluences. This may also affect the hydrogen pickup fractions [36].

It is clear that the corrosion rate accelerates at higher fluences as the microchemistry is severely changed by precipitate dissolution. Since in this experiment there is no heat flux, increased metal/oxide temperatures are highly unlikely. It is also clear that the acceleration is greater for materials having smaller initial precipitates and that the precipitate distribution is also important. However, for materials like Specimens 2R and 3R, which are similar in precipitate characteristics to two types of GE BWR fuel cladding, the acceleration is not great and the corrosion magnitude is relatively small. For both the specimens in this study and current GE cladding materials [37], the amount of corrosion is considerably less than is predicted by the data of Garzarolli et al. [20], as shown in Fig. 20. In all cases, the corrosion and hydriding levels are moderate and would not be expected to be the source of any performance problem of components in a BWR.

Tensile Properties

In the past, literature reports of the effect of irradiation on tensile properties of Zircaloy have not emphasized precipitate size effects consistent with the tacit acknowledgement that precip-

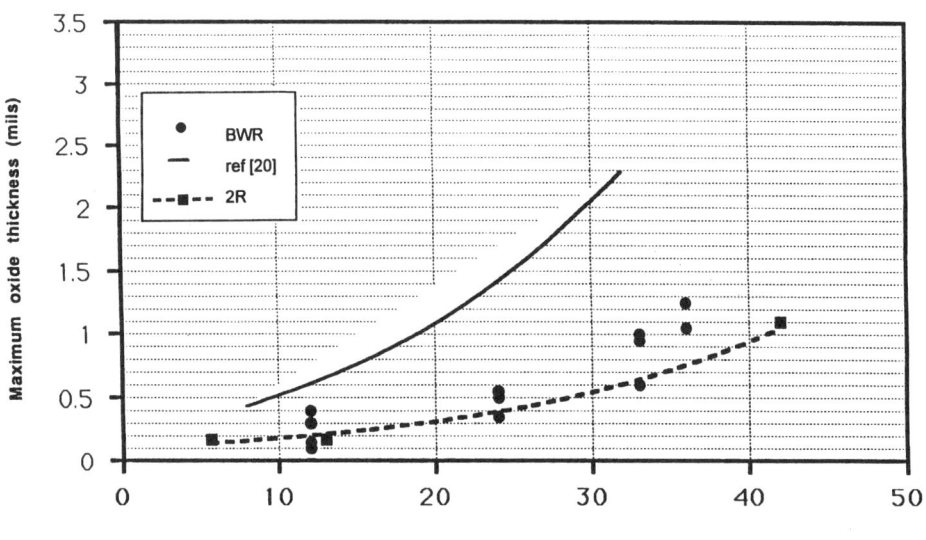

FIG. 20—*Maximum oxide thickness as a function of burnup for Zircaloy-2 in BWRs.*

itation hardening was not a factor. Saturation of strength with fluence is the expected behavior. Irradiation hardening is associated primarily with the creation of "black spot" (small (<10 nm) ⟨a⟩-dislocation loops) damage. Confirming data indicated that thermal annealing that removed the ⟨a⟩-loops also removed the irradiation hardening; even though for high-fluence material, all of the irradiation-induced ⟨c⟩-component dislocations remained in the structure [12,38].

The data reported here are not sufficient to correlate mechanical properties with initial precipitate size or with precipitate dissolution. Comparison with other data generated at Vallecitos, however, indicates that the as-irradiated strength of the test materials is higher than expected. Comparative tensile tests were conducted on a large precipitate (range 0.1 to 0.8 μm) material that has been well characterized in other programs [9–12] and has a hydrogen content similar to Specimen 3R. Our model [29] calculates a relatively small concentration of Fe, Cr, and Ni in the matrix of this material at 8.5 × 10^{25} n/m², and the mechanical tests give an ultimate tensile strength of 620 MPa, which is well below values for Specimens 1R, 2R, and 3R. Hardness of these specimens also increased significantly after irradiation. Although speculative at this time, we suggest that a form of solid solution hardening may be occurring at high fluence. For Specimen 1R, in which all precipitates have dissolved, the "solute" concentration of Fe and Cr is 2400 ppm as compared to, for instance, 1300 ppm oxygen, which is a known hardener. Also, the solutes may be interacting with irradiation-produced defects, thus changing their hardening capability. It is also possible that the dissolution processes change the stacking fault energy of the matrix, thus affecting work hardening, dislocation cross slip propensity, and the choice of dislocation channeling planes. It is noted that if the lattice parameter of the matrix is changed (larger c/a ratio), as indicated by measurements in the "dark bands" after two cycles, details of the deformation process may also change.

It is noted that the ductility of these specimens at 561 K, although reduced by irradiation, is still adequate to support good mechanical performance under normal reactor conditions.

Conclusions

For Zircaloy-2 processed to have small precipitates (medians of 0.02, 0.03, 0.04 μm) and irradiated in a BWR at 561 K to fluences between 1.5 to 8.5 × 10^{25} n/m², the following conclusions can be drawn:

1. Nodular corrosion is absent in all materials.
2. Enhanced uniform corrosion develops for fluences greater than 2.5 × 10^{25} n/m².
3. For fluences near 2.5 × 10^{25} n/m², corrosion of all specimens is about the same. Differences in corrosion rate develop only above that fluence level.
4. Hydrogen pickup fraction is very low at low fluence and increases at high fluence.
5. Complete dissolution of precipitates occurs below 8.5 × 10^{25} n/m² for an original precipitate distribution with a median size of 0.02 μm. The fluence for achievement of complete amorphization and dissolution increases with increasing initial precipitate size.
6. Enhanced uniform corrosion at high fluences is attributed to precipitate dissolution and/or increases in matrix levels of solute. The initial and final spatial distributions of precipitates may also influence the acceleration.
7. Mechanical properties at the highest fluence show typical irradiation effects, although the strength is somewhat higher than expected. Ductilities are quite adequate for good mechanical performance at BWR temperatures.
8. Irradiation growth in the pre-breakaway regime is not affected by precipitate size or dissolution.

9. The increase in corrosion at high fluences, although real, would not result in performance problems for fluences in the ranges of this study corresponding to fuel burnups of about 40 MWd/mt.

Acknowledgments

The authors are grateful to the many people at Vallecitos who provided expert help in making this paper possible: Fred Chen, for gathering the data and for useful discussions; Ross Reager for hydrogen analysis; Steve Wisner for mechanical property and irradiation growth data; Dan Lutz for help in many areas; the Hot Cell staff for expert assistance; Bob Blood for help in lab operations and specimen preparation; John Lewis for texture analysis; Jerry Martin for flux analysis; Jim Lakner for metallography; Bob Warner for fractography; Donna Crawford for report preparation; Bo Cheng for early program planning; and Dale Taylor for mechanistic insights.

A major portion of this work was funded in part by Tokyo Electric Power Co., Tohoku Electric Power Co., Inc., Chubu Electric Power Co., Inc., Hokuriku Electric Power Co., Inc., Chugoku Electric Power Co., and the Japan Atomic Power Co., Nippon Nuclear Fuel Development Co., Ltd., Toshiba Corp. and Hitachi Ltd.

The BWR fuel rod corrosion data reported here and in Ref *37* were gathered as part of a program sponsored by the Electrical Power Research Institute, Palo Alto, California.

References

[1] Marlowe, M. O., Armijo, J. S., Cheng, B., and Adamson, R. B., "Nuclear Fuel Cladding Localized Corrosion," *Light Water Reactor Fuel Performance*, Proceedings of American Nuclear Society Topical Meeting, Orlando, FL, 1985, ANS, La Grange Park, IL, pp. 3–73.
[2] Rudling, P., Vannesjo, K. L., Vesterlund, G., and Massih, A. R., "Influence of Second-Phase Particles on Zircaloy Corrosion in BWR Environment," *Zirconium in the Nuclear Industry: Seventh International Symposium, ASTM STP 939*, R. B. Adamson and L. F. P. Van Swam, Eds., American Society for Testing and Materials, West Conshohocken, PA, 1987, pp. 292–306.
[3] Maussner, G., Steinberg, E., and Tenckhoff, E., "Nucleation and Growth of Intermetallic Precipitates in Zircaloy-2 and Zircaloy-4 and Correlation to Nodular Corrosion Behavior," *Zirconium in the Nuclear Industry: Seventh International Symposium, ASTM STP 939*, American Society for Testing and Materials, West Conshohocken, PA, 1987, pp. 307–320.
[4] Andersson, T. and Thorvaldsson, T., "Nodular Corrosion Resistance of Zircaloy-2 in Relation to Second-Phase Particle Distribution," *Zirconium in the Nuclear Industry: Seventh International Symposium, ASTM STP 939*, R. B. Adamson and L. F. P. Van Swan, Eds., American Society for Testing and Materials, West Conshohocken, PA, 1987, pp. 321–337.
[5] Bangaru, N. V., Busch, R. A., and Schemel, J. H., "Effect of Beta Quenching on the Microstructure and Corrosion of Zircaloys," *Zirconium in the Nuclear Industry: Seventh International Symposium, ASTM STP 939*, R. B. Adamson and L. F. P. Van Swam, Eds., American Society for Testing and Materials, West Conshohocken, PA, 1987, pp. 341–363.
[6] Cheng, B. and Adamson, R. B., "Mechanistic Studies of Zircaloy Nodular Corrosion," *Zirconium in the Nuclear Industry: Seventh International Symposium, ASTM STP 939*, American Society for Testing and Materials, West Conshohocken, PA, 1987, pp. 387–416.
[7] Garzarolli, F., Stehle, H., Steinberg, E., and Weidinger, H., "Progress in the Knowledge of Nocular Corrosion," *Zirconium in the Nuclear Industry: Seventh International Symposium, ASTM STP 939*, R. B. Adamson and L. F. P. Van Swam, Eds., American Society for Testing and Materials, West Conshohocken, PA, 1987, pp. 417–430.
[8] Cheng, B., Adamson, R. B., Bell, W. L., and Proebstle, R. A., "Corrosion Performance of Some Zirconium Alloys Irradiated in the Steam Generating Heavy Water Reactor Winfrith," *Proceedings of International Symposium on Environmental Degradation of Materials in Nuclear Power Systems-Water Reactors*, Myrtle Beach, SC, 22–25 Aug. 1983, TMS, Warrendale, PA, pp. 274–285.
[9] Yang, W. J. S. in *Journal of Nuclear Materials*, Vol. 158, 1988, p. 71.

[10] Yang, W. J. S. and Adamson, R. B., "Beta-Quenched Zircaloy-4: Effects of Thermal Aging and Neutron Irradiation," *Zirconium in the Nuclear Industry: Eighth International Symposium, ASTM STP 1023*, L. F. P. Van Swam and C. M. Eucken, Eds., American Society for Testing and Materials, West Conshohocken, PA, 1989, pp. 451–477.
[11] Yang, W. J. S., Tucker, R. P., Cheng, B., and Adamson, R. B. in *Journal of Nuclear Materials*, 1986, Vol. 138, p. 185.
[12] Kruger, R. M., "Precipitate Stability in Zircaloy-2," EPRI NP-6845-D, Electric Power Research Institute, Palo Alto, PA, 1990.
[13] Kruger, R. M. and Adamson, R. B. in *Journal of. Nuclear Materials*, Vol. 205, 1993, p. 242.
[14] Griffiths, M., Gilbert, R. W., and Carpenter, G. J. C. in *Journal of Nuclear Materials*, Vol. 150, 1987, p. 53.
[15] Gilbert, R. W., Giffiths, M., and Carpenter, G. J. C. in *Journal of Nuclear Materials*, Vol. 135, 1985, p. 265.
[16] Etoh, Y., Kikuchi, K., Yasuda, T., Koizumi, S., and Oishi, M. in *Proceedings*, International Topical Meeting on LWR Fuel Performance, Avignon, 1991, American Nuclear Society, La Grange Park, IL, p. 691.
[17] Etoh, Y., Shimada, S., Kikuchi, K., and Kawai, K. in *Journal of Nuclear Science and Technology*, Vol. 29, 1992, pp. 1173–1183.
[18] Cheng, B.-C., Kruger, R. M., and Adamson, R. B., "Corrosion Behavior of Irradiated Zircaloy," *Zirconium in the Nuclear Industry: Tenth International Symposium, ASTM STP 1245*, A. M. Garde and E. R. Bradley, Eds., American Society for Testing and Materials, West Conshohocken, PA, 1994, pp. 400–418.
[19] Kruger, R. M., Adamson, R. B., and Brenner, S. S. in *Journal of Nuclear Materials*, Vol. 189, 1992, p. 193.
[20] Garzarolli, F., Schumann, R., and Steinberg, E., "Corrosion Optimized Zircaloy for Boiling Water Reactor (BWR) Fuel Elements," *Zirconium in the Nuclear Industry: Tenth International Symposium, ASTM STP 1245*, A. M. Garde and E. R. Bradley, Eds., American Society for Testing and Materials, West Conshohocken, PA, 1994, pp. 709–723.
[21] Garde, A. M., "Effects of Irradiation and Hydriding on the Mechanical Properties of Zircaloy-4 at High Fluence," *Zirconium in the Nuclear Industry: Eighth International Symposium, ASTM STP 1023*, L. F. P. Van Swam and C. M. Eucken, Eds., American Society for Testing and Materials, West Conshohocken, PA, 1989, pp. 548–569.
[22] Baumgartner, J. A., "BWR Fuel Bundle Extended Burnup Final Report," DOE/ET/34031-18, UC-78, U.S. Department of Energy, December 1984.
[23] Reager, R. D., to be published.
[24] Griffiths, M., Gilbert, R. W., and Carpenter, G. J. C. in *Journal of Nuclear Materials*, Vol. 150, 1987, p. 53.
[25] Motta, A. T. and Lemaignan, C. in *Journal of Nuclear Materials*, Vol. 195, 1992, pp. 277–285.
[26] Holt, R. A. and Gilbert, R. W., "<c> Component Dislocations in Neutron Irradiated Zircaloy-2," *Journal of Nuclear Materials*, Vol. 116, 1983, p. 127.
[27] Etoh, Y. and Shimada, S. in *Journal of Nuclear Materials*, Vol. 200, 1993, p. 59.
[28] Fidleris, V., Tucker, R. P., and Adamson, R. B., "An Overview of Microstructural and Experimental Factors That Affect the Irradiation Growth Behavior of Zirconium Alloys," *Zirconium in the Nuclear Industry, Seventh International Symposium, ASTM STP 939*, R. B. Adamson and L. F. P. Van Swam, Eds., American Society for Testing and Materials, West Conshohocken, PA, 1987, pp. 49–85.
[29] Edsinger, K. W., Kruger, R. M., and Huang, P. Y., to be published.
[30] Taylor, D. F., "An Oxide-semiconductance Model of Nodular Corrosion and Its Application to Zirconium Alloy Development," *Journal of Nuclear Materials*, Vol. 184, 1991, pp. 65–77.
[31] Inagaki, M., Kanno, M., and Maki, H., "Effect of Alloying Elements in Zircaloy on Photo-Electrochemical Characteristics of Zirconium Oxide Films," *Zirconium in the Nuclear Industry: Ninth International Symposium, ASTM STP 1132*, C. M. Eucken and A. M. Garde, Eds., American Society for Testing and Materials, West Conshohocken, PA, 1991, pp. 437–460.
[32] Harding, J. H., "The Effect of Alloying Elements on Zircaloy Corrosion," *Journal of Nuclear Materials*, Vol. 202, 1993, pp. 216–221.
[33] Van Swam, L. F. P. and Shann, S. H., "The Corrosion of Zircaloy-4 Fuel Cladding in Pressurized Water Reactors," *Zirconium in the Nuclear Industry: Ninth International Symposium, ASTM STP 1132*, C. M. Eucken and A. M. Garde, Eds., American Society for Testing and Materials, West Conshohocken, PA, 1991, pp. 758–781.
[34] Hutchinson, B. and Lehtinen, B., "A Theory of the Resistance of Zircaloy to Uniform Corrosion," *Journal of Nuclear Materials*, Vol. 217, 1994, pp. 243–249.
[35] Cheng, B., Adamson, R. B., Machiels, A. J., and O'Boyle, D., "Effect of Hydrogen Water Chemistry

on Fuel Performance at Dresden-2," *International Topical Meeting on LWR Fuel Performance, Proceedings of the American Nuclear Society Topical Meeting*, Avignon, France, 1991, American Nuclear Society, La Grange Park, IL, p. 664.

[36] Charquet. D., Rudling, P., Mikes-Lindback, M., and Barberis, P., "Hydrogen Absorption Kinetics During Zircaloy Oxidation in Steam," *Zirconium in the Nuclear Industry: Tenth International Symposium, ASTM STP 1245*, A. M. Garde and E. R. Bradley, Eds., American Society for Testing and Materials, West Conshohocken, PA, 1994, pp. 80–97.

[37] Levin, H. A. and Garcia, S. E., "BWR Fuel Experience with Zinc Injection," *NACE 7th International Symposium on Environmental Degradation of Materials in Nuclear Power Systems-Water Reactors*, Breckenridge, CO, 6–10 Aug. 1995, Paper 63, TMS, Warrendale, PA.

[38] Adamson, R. B. and Bell, W. L., "Effects of Neutron Irradiation and Oxygen Content on the Microstructure and Mechanical Properties of Zircaloy," *Microstructure and Mechanical Behaviour of Materials*, X'ian, China, Vol. 1, 21–24 Oct. 1985, Engineering Materials Advisory Services, Ltd., Warley, UK.

DISCUSSION

B. Cheng[1] (written discussion)—(1) With hydrogen pickup fraction of 25 to 35%, you will reach 600 ppm H_2 at ≤ 60 μm oxide. How can fuel with burnups exceeding 50 GWd/MT be made with the current cladding? (2) You showed that the ductility of the 1R material is about the same as the 3R material even when all ppts in the 1R material were all dissolved. Don't you anticipate a decrease in ductility as particles are dissolved in the 1R material?

R. B. Adamson et al. (authors' closure)—(1) Figure 20 gives data for current GE BWR Fuel to 37 GWd/MT. It appears that this fuel would reach 50 GWd/MT with maximum oxide thickness below 50 μm (which implies an average thickness below that value). We anticipate no (nor have we observed) performance problems whatsoever with such cladding. Development programs now in progress (of which the experiments reported here are a small part) are aimed at understanding the metallurgical parameters needed to allow fuel to reach even higher burnups. (2) We do anticipate some reduction of conventional ductility parameters as the precipitates dissolve and the matrix becomes supersaturated with alloying elements. The level of supersaturation is very high for all three materials (see text) after four cycles of irradiation, so we don't expect much difference in tensile ductility for these three small precipitate materials.

B. Cox[2] (written discussion)—The radiation chemistry inside the "thimble tubes" in which these specimens were irradiated must be quite different from that seen by fuel cladding. Did you have control specimens known to suffer nodular corrosion along with your test specimens in order to demonstrate that typical nodular corrosion could be observed under these water chemistry conditions?

R. B. Adamson et al. (authors' closure)—Nodular corrosion was observed on companion specimens. Some of the corrosion results are given in another paper at this meeting, "Development of New Zr Alloys for a BWR," by Y. Etoh et al.

A. Garde[3] (written discussion)—(1) Did you observe a minimum uniform elongation at intermediate fluence levels? (2) With reference to the IR and 3R irradiated specimens showing different uniform and total elongations, was there a difference in dislocation channeling (parallel channels or intersecting channels) or surface deformation bands between 1R and 3R tensile specimens?

R. B. Adamson et al. (authors' closure)—(1) For this paper, we have only high fluence tensile properties to report. Data at lower fluences will be available to report soon. (2) We have not yet conducted TEM exams to characterize the dislocation channeling phenomena. Optical microscopy shows similar deformation band characteristics for the different materials. A single, sharp deformation band and fracture predominate.

B. Warr[4] (written discussion)—(1) Could % hydrogen uptake be much higher than your average values of ~ 35%. (2) Can you comment on why there is no correlation between % hydrogen pickup and oxide thickness? (3) The carbon dating technique (in an earlier paper) may be used to infer corrosion kinetics.

[1] Electric Power Research Institute, Polo Alto, CA.
[2] University of Toronto, Toronto, Ontario, Canada.
[3] ABB Combustion Engineering Nuclear Operations, Windsor, CT.
[4] Ontario Hydro, Toronto, Ontario, Canada.

R. B. Adamson et al. (authors' closure)—(1) We have calculated percent hydrogen uptake only after two cycles and four cycles of irradiation. It is possible to speculate that the uptake increases dramatically only at the highest fluences, for instance between three and four cycles, but we have no data to confirm or refute such speculation. (2) Our data say that for thin oxides (~2 μm) the pickup fraction is small (~2%) and that for thicker oxides (28 to 43 μm maximum) it is larger (25 to 35%). However, we suspect that changes in microchemistry and microstructure due to irradiation are as important as the oxide thickness in determining the pickup kinetics.

V. Urbanic[5] (written discussion)—Can you speculate as to why the percent hydrogen uptake increased after the four-cycle irradiation and also why there was little variation in the percentage uptake for the three material types despite differences in microstructure with respect to the presence or absence of precipitates after the fourth cycle?

R. B. Adamson et al. (authors' closure)—As noted in the text, our calculations indicate that the concentration of Fe in the matrix is very high for all three materials after four cycles. The percentage hydrogen uptake is also on the high side for all three materials, with 2R being only a little higher than the others. At this point, we do not expect the percentage uptake to depend only on the "dissolved" Fe content, but we are not yet prepared to speculate further. Hopefully, further experiments and analysis in progress will help clear up this interesting point.

M. Griffiths[6] (written discussion)—You showed mechanical property data for 0 and 8.5×10^{25}nm^{-2}. Have you any more information to show how these properties change as a function of fluence?

R. B. Adamson et al. (authors' closure)—We will soon be able to report additional fluence versus property data, but unfortunately it is not yet ready.

A. T. Motta[7] (written discussion)—(1) Do you believe that the lack of irradiation growth enhancement with fluence in your samples is due to absence of $\langle c \rangle$ component dislocations? (2) Would you care to speculate on the mechanism for the high Fe supersaturation in the matrix during irradiation? Given that the maximum concentration of point defects is $\sim 10^{-4}$, do you think this Fe is associated with dislocation and other extended defects?

R. B. Adamson et al. (authors' closure)—(1) To date we have examined only one growth specimen by TEM, Specimen 3R. In this specimen, a "normal" density of $\langle c \rangle$- type dislocations was observed. (2) At this point, we can only speculate; however, we agree that the high Fe (and Cr and Ni) supersaturation may be due to interaction with both line and point defects. We have previously noted that there appears to be an interaction with $\langle c \rangle$-type line dislocations, but that no such interaction with $\langle a \rangle$-type line dislocation was observed. ("Effect of Prior Deformation on the Stability of the Intermetallic Precipitate in Zircaloy-2" by P. Y. Huang and R. B. Adamson, *Proceedings*, Materials Research Society Symposium on Microstructure of Irradiated Materials, Boston, January 1994.)

---

[5] Atomic Energy of Canada, Ltd., Chalk River, Ontario, Canada.
[6] Atomic Energy of Canada, Ltd., Chalk River, Ontario, Canada.
[7] Pennsylvania State University, University Park, PA.

P. H. Kreyns,[1] W. F. Bourgeois,[1] C. J. White,[2] P. L. Charpentier,[1] B. F. Kammenzind,[1] and D. G. Franklin[1]

Embrittlement of Reactor Core Materials

REFERENCE: Kreyns, P. H., Bourgeois, W. F., White, C. J., Charpentier, P. L., Kammenzind, B. F., and Franklin, D. G., "**Embrittlement of Reactor Core Materials,**" *Zirconium in the Nuclear Industry: Eleventh International Symposium, ASTM STP 1295*, E. R. Bradley and G. P. Sabol, Eds., American Society for Testing and Materials, 1996, pp. 758–782.

ABSTRACT: Over a core lifetime, the reactor materials Zircaloy-2, Zircaloy-4, and hafnium may become embrittled due to the absorption of corrosion-generated hydrogen and neutron irradiation damage. Results are presented on the effects of fast fluence on the fracture toughness of: (1) wrought Zircaloy-2, Zircaloy-4, and hafnium; (2) Zircaloy-4 to hafnium butt welds; and (3) hydrogen-precharged beta-treated and weld-metal Zircaloy-4 for fluences up to a maximum of approximately 150×10^{24} n/m^2 (>1 MeV). While Zircaloy-4 did not exhibit a decrement in K_{IC} due to irradiation, hafnium and butt welds between hafnium and Zircaloy-4 are susceptible to embrittlement with irradiation. The embrittlement can be attributed to irradiation strengthening, which promotes void formation in the high-strain crack-tip region, and, in part, to the lower chemical potential of hydrogen in Zircaloy-4 compared to hafnium, which causes hydrogen to drift over time from the hafnium end toward the Zircaloy-4 end and to precipitate at the interface between the weld and base-metal interface. Neutron radiation apparently affects the fracture toughness of Zircaloy-2, Zircaloy-4, and hafnium in different ways. Possible explanations for these differences are suggested. It was found that Zircaloy-4 is preferred over Zircaloy-2 in hafnium-to-Zircaloy butt-weld applications due to absence of a radiation-induced reduction in K_{IC} plus its lower hydrogen absorption characteristics compared with Zircaloy-2.

KEYWORDS: Zircaloys, hafnium, fracture toughness, irradiation effects, hydrides

Reactor core components, especially weldments, often contain regions of local stress concentration such as notches and small linear separations. Also, the effect of localized regions of high hydrogen concentration on the potential embrittlement of wrought or weld-metal Zircaloy-4 must be considered. The structural analysis of these components utilizes linear elastic fracture mechanics (LEFM). Therefore, it is important to have available the plane-strain fracture toughness (K_{IC}) of PWR-type cladding and control rod materials as a function of fast neutron fluence, temperature, hydrogen concentration, and hydride morphology. A previous study [1] reported K_{IC} results for Zircaloy-4 containing up to 250 ppm hydrogen and irradiated to 20×10^{24} n/m^2. The maximum test temperature was 316°C. The present work extends the fluence and hydrogen ranges to 150×10^{24} n/m^2 and 4000 ppm, respectively. In addition, results are provided for irradiated Zircaloy-2, hafnium, and Zircaloy-4-to-hafnium butt welds. Particular emphasis has been placed on the effect of hydride morphology in the Zircaloy-4 and the hydride redistribution within the butt weld on the toughness of reactor core components over their lifetime. These factors are of critical importance in developing a useful K_{IC} database.

---

[1] Consultant, principal engineer, senior engineer, manager of Structural Materials Performance, and manager of PWR Materials Technology, respectively, Bettis Atomic Power Laboratory, Westinghouse Electric Corporation, West Mifflin, PA 15122.

[2] Manager of Clad Test and Analysis, Knolls Atomic Power Laboratory, Martin Marietta Corporation, Schenectady, NY 12301.

Experimental Details

Materials and Specimens

Fracture toughness tests were conducted on the following materials: alpha-annealed Zircaloy-2, beta-treated and weld-metal Zircaloy-4, alpha-annealed hafnium, and Zircaloy-4-to-hafnium butt welds. All materials were reactor grade with nominal compositions shown in Table 1. The tin content of the Zircaloy-2 and Zircaloy-4 was 1.50 wt%. Modified compact tension (CT) specimens (Fig. 1a) were machined from various source pieces. Alpha-annealed specimens [(S-T) orientation per ASTM Terminology Relating to Fracture Testing (E 616)] were machined from a wrought plate and were fatigue precracked prior to irradiation. Also, the beta-treated and weld-metal Zircaloy-4 specimens were fatigue precracked and hydrided prior to irradiation. Figure 1b shows the typical $\beta \rightarrow \alpha$ Widmanstätten or basketweave microstructure within the prior-beta-phase grain boundaries. This microstructure was typical of all the beta-quenched and weld-metal Zircaloy specimens. The Zircaloy-hafnium butt weld was made by a single-pass tungsten-inert-gas (TIG) autogenous weld with 100% penetration. Hafnium and Zircaloy-4-to-hafnium butt-weld specimens were machined and fatigue precracked after irradiation. Specimen orientations relative to the weld region are shown in Fig. 2. The specimen notch was located along the centerline of the butt weld. The initial hydrogen concentrations in the hafnium and Zircaloy-4 were both $\sim$ 10 ppm. The hydrogen concentration at the centerline of the butt weld remained essentially unchanged after irradiation.

Hydrogen Precharging and Hydride Morphology

Two methods were used to hydrogen precharge the Zircaloy-4 specimens. One method equilibrated the specimens with a predetermined amount of zirconium hydride powder in evacuated Vycor bulbs at 565 or 649°C. The second method utilized a Sievert's apparatus in which the desired amount of hydrogen in the specimens was established by maintaining the appropriate partial pressure of hydrogen gas at 527°C.

The hydride morphology of the precharged specimens depended on the hydrogen content and the hydriding temperature and was consistent with the known equilibrium and kinetic characteristics of the binary zirconium-hydrogen system (see Fig. 3). On cooling specimens that had been hydrided in the alpha phase, hydride platelets precipitated on crystallographic habit planes within the alpha platelets (see Fig. 3a). This was typical of specimens containing up to approximately 600 ppm and hydrided at either 649 or 527°C. For higher hydrogen levels, the resulting hydride morphology depended on whether the hydriding temperature was above or below the Zr-H eutectoid temperature (550°C). At 649°C, the alpha and beta zirconium phases are present for hydrogen concentrations > 600 ppm. A study [2] of the nucleation of the beta phase in an alpha platelet-type structure showed that in this two-phase field, the beta phase forms at the alpha/alpha platelet boundaries where the beta-stabilizing elements iron and chromium are also located. Because the solubility of hydrogen is much greater in the beta phase, these interfaces become rich in hydrogen and on cooling become preferred nucleation

TABLE 1—*Chemical compositions of Zircaloy-2, Zircaloy-4, and hafnium (wt%).*

Alloy	Zr	Hf	Sn	Fe	Cr	Ni	O
Zircaloy-2	Bal.	0.01	1.5	0.13	0.10	0.06	0.17
Zircaloy-4	Bal.	0.01	1.5	0.21	0.10	0.007	0.12
Hafnium	3.0	Bal.	...	0.04	0.01	0.01	0.04

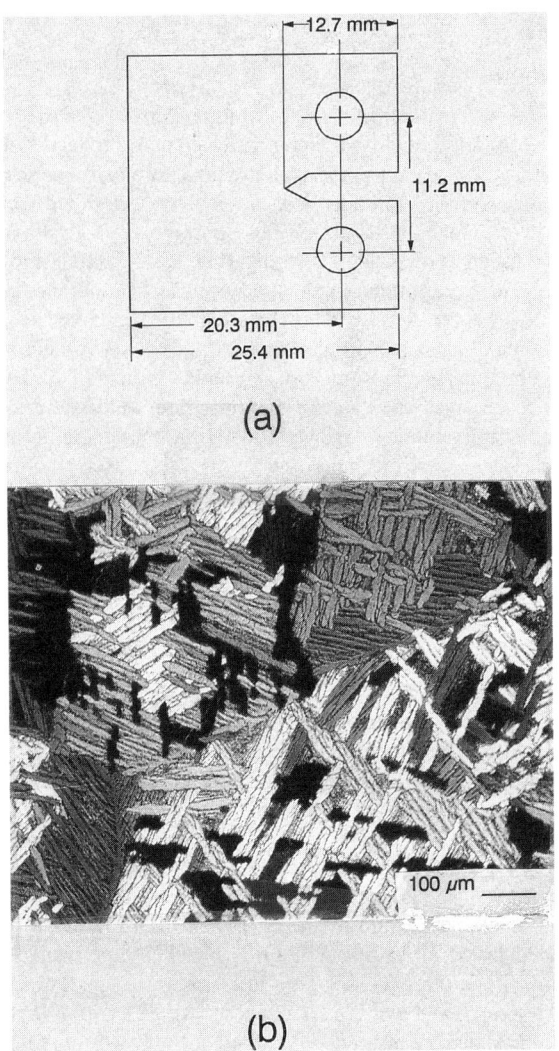

FIG. 1—(a) *Compact-tension specimen: thickness varied between 2 and 10 mm;* (b) *Basketweave microstructure typical of beta-quenched Zircaloy-4 specimens (10 mm) and weld-metal Zircaloy-4 specimens (2 to 10 mm).*

sites for hydride precipitation. Figure 3b shows hydride at the alpha/alpha interface in a specimen containing 1480 ppm hydrogen and hydrided at 649°C. At 527°C, the phases present were alpha zirconium and zirconium hydride. In the Sievert's apparatus, the hydrogen concentration builds up slowly. When the solubility of hydrogen exceeds the solubility limit, zirconium hydride precipitates at the prior-beta-phase grain boundaries, while that of hydrogen in solution precipitates within the alpha platelets on cooling (see Fig. 3c). In this study, the hydride morphology is described relative to the prior-beta-phase grain boundaries. That typical of Figs. 3a and 3b is classified as *transgranular* and that typical of Fig. 3c as *intergranular*.

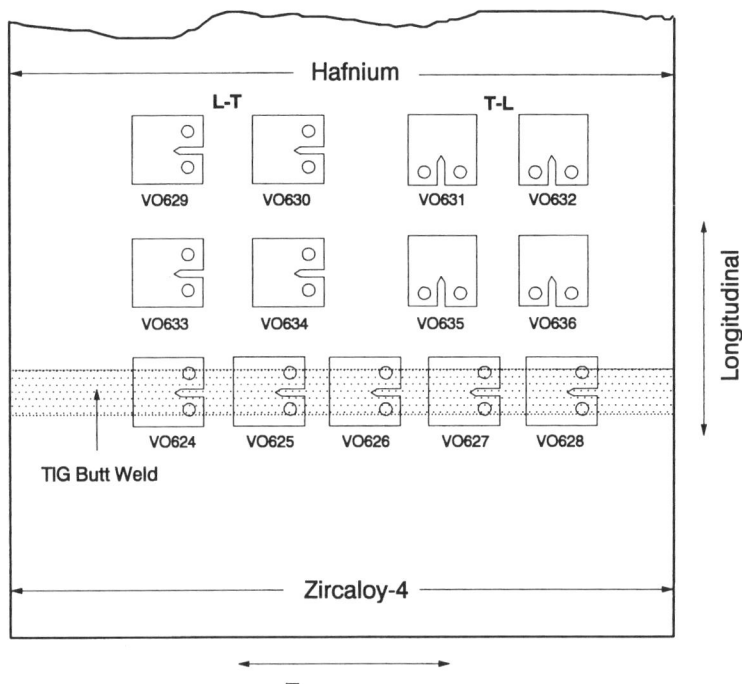

FIG. 2—*Locations and orientation of compact-tension specimens machined from Zircaloy-to-hafnium butt weld and from wrought hafnium plate stock.*

Irradiation Histories

Compact-tension specimens of nonhydrided Zircaloy-2 and hydrogen-precharged Zircaloy-4 were irradiated in deionized pressurized water (DPW) at 260 ± 5°C in an average fast neutron flux of 10^{18} n/m²-s. Fast fluences ranged between approximately 10×10^{24} n/m² and 150×10^{24} n/m² ($E > 1$ MeV). Compact-tension specimens of hafnium and hafnium-to-Zircaloy-4 butt welds were machined from material that had been irradiated in deionized pressurized water (DPW) at 285 ± 5°C in an average fast neutron flux of 10^{17} n/m²-s to a fast fluence of approximately 50×10^{24} n/m².

Fracture Toughness Testing

Constant-displacement-rate plane-strain fracture toughness K_{IC} tests on Zircaloy-2 and Zircaloy-4 were conducted in accordance with ASTM Test Method for Plane-Strain Fracture Toughness of Metallic Materials (E 399). Several specimens tested at elevated temperature did not meet all of the criteria for valid plane-strain fracture toughness and are identified in the data presentation.

Because of the low yield strength of hafnium in both the nonirradiated and irradiated condition and the limitation on the butt-weld thickness to 5 mm, the hafnium and hafnium-to-Zircaloy-4 butt weld specimens were J_{IC} tested in accordance with ASTM Test Method for J_{IC}, a Measure of Fracture Toughness (E 813-87) in an attempt to obtain valid fracture toughness

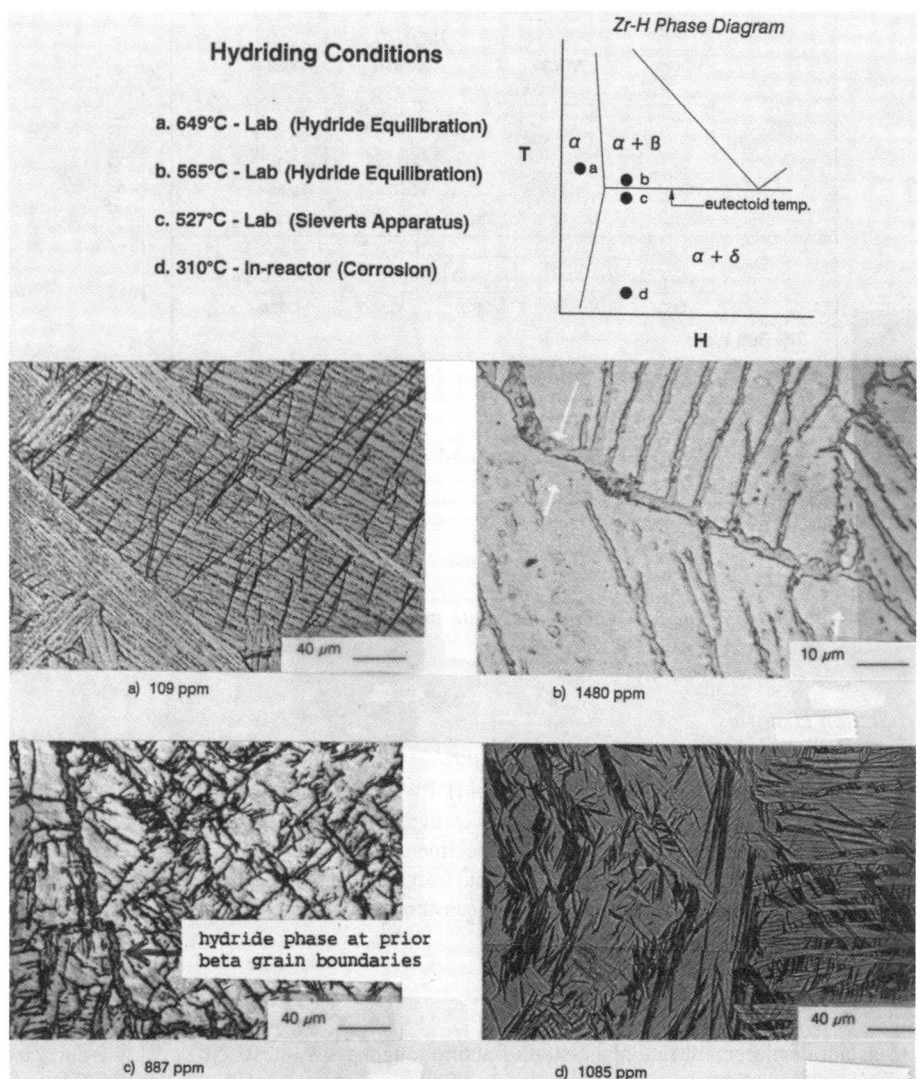

FIG. 3—*Hydride morphologies in beta-quenched Zircaloy-4 as related to the Zr-H equilibrium phase diagram shown in figure: (a) hydrided in single-phase alpha region; (b) hydrided above eutectoid temperature in the two-phase alpha+beta region; (c) hydrided just below eutectoid temperature in the alpha+hydride region; (d) hydride in-reactor via hot-water corrosion.*

TABLE 2—*Yield strengths used in evaluating fracture toughness tests.*

Materials	Condition[a]	Texture,[b] (Kearns's f-factor)	Temperature, °C	Yield Strength, MPa	
				Nonirradiated	Irradiated
Zircaloy-2	Wrought	0.07	24	410	665
Zircaloy-4	Weld	0.33	24	438	775
Hafnium (T-L)	Wrought	0.36	24	377	655[c]
Hafnium (L-T)	Wrought	0.07	24	222	655
Hf-to-Zircaloy butt weld	Weld	0.33	24	414	655[c]
Zircaloy-2	Wrought	0.07	149	290	546
Zircaloy-4	Weld	0.33	149	299	637
Hafnium (T-L)	Wrought	0.36	149	302	580[c]
Hafnium (L-T)	Wrought	0.07	149	147	580
Hf-to-Zircaloy butt weld	Weld	0.33	149	339	580[c]

[a] Cooling rates for beta-quenched and weld-metal specimens were generally less than 30°C/s.
[b] Crystallographic texture in direction parallel to specimen axis in the direction of tension test.
[c] Conservatively assumed post-irradiation strength to be the same as that of hafnium in longitudinal direction.

values. However, due to the tendency of hafnium and zirconium-hafnium alloy welds to exhibit successive *pop-ins*, no valid J_{IC} tests were obtained. Alternatively, the J_{IC} load-displacement curves were analyzed per E 399 for plane-strain fracture toughness. Table 2 summarizes the yield strength values used to evaluate the validity of the K_{IC} fracture toughness tests.

Results and Discussion

Hydrogen Precharged Zircaloy-4

Table 3 summarizes the plane-strain fracture toughness data obtained for beta-treated and weld-metal Zircaloy-4. The bulk of the specimens tested had a transgranular hydride morphology and exhibited a transgranular fracture relative to the prior-beta-phase grain boundaries. The remaining specimens had an intergranular hydride morphology and fractured along the prior-beta-phase grain boundaries.

Figure 4 shows that increasing the hydrogen concentration of the transgranular hydride specimens at room temperature results in an exponential decrease in K_{IC}. Specimens with an intergranular hydride morphology had K_{IC} values ~ 9 MPa$\sqrt{m}$ less than those with a transgranular hydride distribution. Both nonirradiated and postirradiated specimens are included in Fig. 4. Examination of the data in Table 3*a* shows no detectable effect of irradiation on the fracture toughness of these Zircaloy-4 specimens.

Figure 5 shows, in general, that increasing the test temperature leads to higher fracture toughness of beta-treated and weld-metal Zircaloy-4. Figure 6 shows three subsets of these data, namely, 40 to 46 ppm hydrogen, 240 to 267 ppm hydrogen, and 441 to 559 ppm hydrogen. This temperature behavior is consistent with the results reported for Zr-2.5Nb [5] and Zircaloy-2 [6], which showed an increase in K_{IC} with increasing temperature eventually resulting in a definite transition from ductile to brittle behavior at about 250°C. This transition is thought to be due to the decreased strength of the matrix and to some extent the increased ductility of the hydride phase at higher temperatures. Higher test temperatures are needed to confirm this behavior for Zircaloy-4. The results in Fig. 5 suggest that the temperature dependency is independent of fluence levels. Figure 6 also includes the data for nonhydrided, highly irradiated

TABLE 3a—*Fracture toughness of beta-treated and weld-metal Zircaloy-4: room temperature–transgranular fractures.*

H, μg/g	Fluence, 10^{24} n/m^2	K_{IC}, MPa$\sqrt{m}$	P_{max}/P_q
10	10.4	43.9	1.00
46	10.3	33.3	1.13
240	47.0	27.6	1.02
240	47.0	28.5	1.02
240	47.0	28.2	1.06
240	47.0	31.1	1.05
240	47.0	25.6	1.09
240	47.0	27.7	1.04
246	0.0	26.7	1.11
247	28.8	22.4	1.00
250	10.1	26.7	1.00
250	20.4	27.1	1.00
250	0.0	23.5	1.00
250	0.0	26.0	1.00
262	10.3	26.2	1.00
267	0.0	25.1	1.04
296	0.0	24.5	1.14
300	0.0	24.2	1.13
400	0.0	21.8	1.17
401	0.0	20.1	1.14
405	29.9	21.1	1.07
411	0.0	20.1	1.10
462	28.3	21.5	1.01
465	28.2	20.0	1.06
483	12.8	18.4	1.05
498	27.8	22.0	1.04
500	28.9	19.4	1.10
500	0.0	16.5	1.00
500	0.0	19.1	1.04
500	0.0	18.2	1.01
521	0.0	20.6	1.09
526	0.0	20.2	1.02
531	28.5	19.1	1.00
559	0.0	20.9	1.05
579	0.0	18.3	1.06
1200	0.0	18.2	1.00
1275	0.0	15.1	1.03
1278	0.0	16.6	1.13
1449	29.9	16.2	1.09
1558	30.8	16.8	1.09
1647	12.8	13.5	1.10

NOTE:
P_{MAX} = maximum load achieved during the test.
P_Q = load associated with the intersection of the 5% secant line drawn through the origin with the initial linear load-displacement curve.

TABLE 3b—*Fracture toughness of beta-treated and weld-metal Zircaloy-4: room temperature–intergranular fractures.*

H, μg/g	Fluence, 10^{24} n/m^2	K_{IC}, MPa$\sqrt{m}$	P_{max}/P_q
668	0.0	14.8	1.02
707	0.0	12.4	1.13
950	0.0	13.1	1.04
1012	28.6	15.8	1.00
1485	28.6	13.0	1.11
1830	0.0	11.2	1.11
2225	0.0	17.8	1.00
2540	0.0	15.1	1.00
4000	0.0	7.4	1.13

TABLE 3c—*Fracture toughness of beta-treated and weld-metal Zircaloy-4: elevated temperature–transgranular fractures.*

Temperature, °C	H, μg/g	Fluence, 10^{24} n/m^2	K_{IC}, MPa$\sqrt{m}$	P_{max}/P_q
149	240	53.0	34.5	1.01
149	253	10.3	31.2	1.00
149	253	29.3	29.3	1.00
149	453	26.6	21.5	1.02
149	494	28.6	21.7	1.00
149	505	28.5	22.4	1.00
177	446	64.0	20.7	1.15
177	1337	64.0	21.3	1.04
204	489	12.8	24.5	1.00
204	586	12.8	23.0	1.00
204	1606	12.8	13.3	1.14
260	237	14.0	46.9	1.16
260	441	14.0	33.9	1.00
260	446	14.0	27.1	1.00

TABLE 3d—*Fracture toughness of beta-treated and weld-metal Zircaloy-4: elevated temperature–intergranular fractures.*

Temperature, °C	H, μg/g	Fluence, 10^{24} n/m^2	K_{IC}, MPa$\sqrt{m}$	P_{max}/P_q
149	1620	27.7	13.8	1.06
149	1825	27.7	11.3	1.10
149	2197	27.7	14.6	1.00
149	3192	27.7	12.3	1.04

TABLE 3e—*Fracture toughness of alpha-annealed Zircaloy-4: transgranular fractures.*

Temperature, °C	H, μg/g	Fluence, 10^{24} n/m^2	K_{IC}, MPa$\sqrt{m}$	P_{max}/P_q
29	40	150	37.2	1.17
29	40	150	35.8	1.15
29	40	150	46.3	1.27
149	40	150	54.5	1.19
204	40	150	61.2	1.14
204	40	150	57.3	1.12
204	40	150	48.7	1.17
260	40	150	56.1	1.18

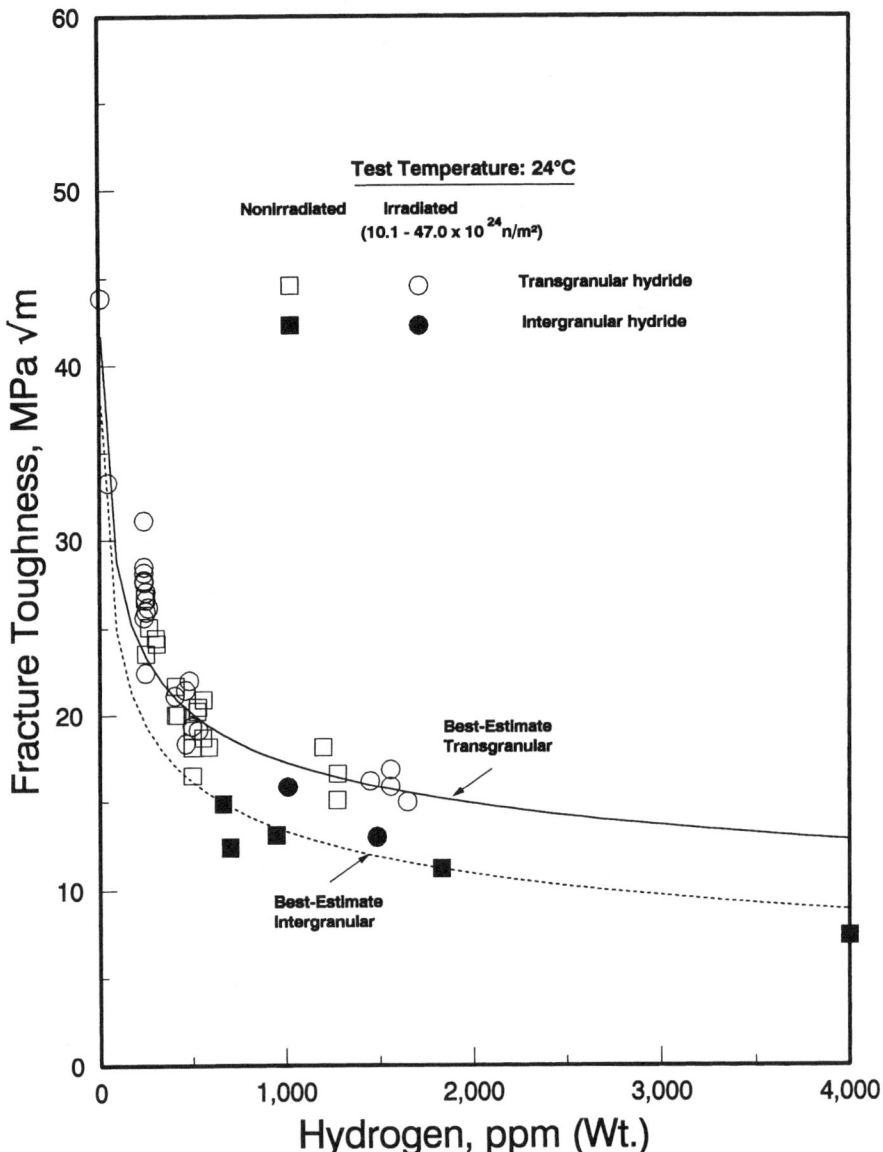

FIG. 4—*Room temperature plane-strain fracture toughness of nonirradiated and irradiated beta-treated and weld-metal Zircaloy-4.*

alpha-annealed Zircaloy-4. The fracture toughness gradually increases with increasing test temperature. In Fig. 6, the datum point for beta-treated Zircaloy-4 (46 ppm H) irradiated to 10.3×10^{24} n/m² (■) and tested at room temperature lies about 7 MPa$\sqrt{m}$ below the results for alpha-annealed Zircaloy-4 (40 ppm H) irradiated to 150×10^{24} n/m².

The observations that the fracture toughness of Zircaloy-4 decreases with hydrogen content

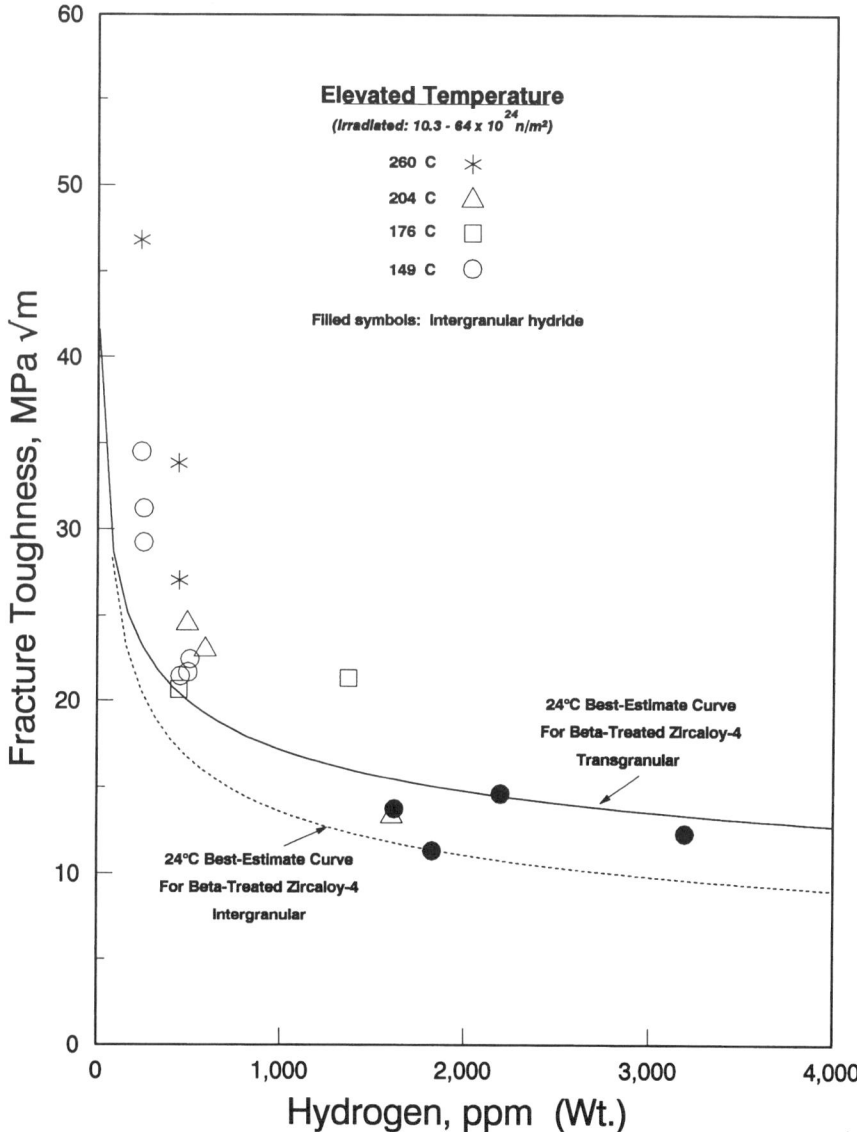

FIG. 5—*Elevated temperature plane-strain fracture toughness of irradiated beta-treated Zircaloy-4.*

and that specimens with the intergranular hydride morphology have lower toughness than those with the transgranular hydride morphology is consistent with previously reported work on zirconium-based alloys [5]. It is therefore important that the hydride morphology of precharged specimens used to predict lifetime embrittlement have a hydride morphology similar to that developed under representative conditions. Figure 3*d* shows the hydride morphology developed due to the absorption of corrosion-generated hydrogen during in-reactor exposure to high-

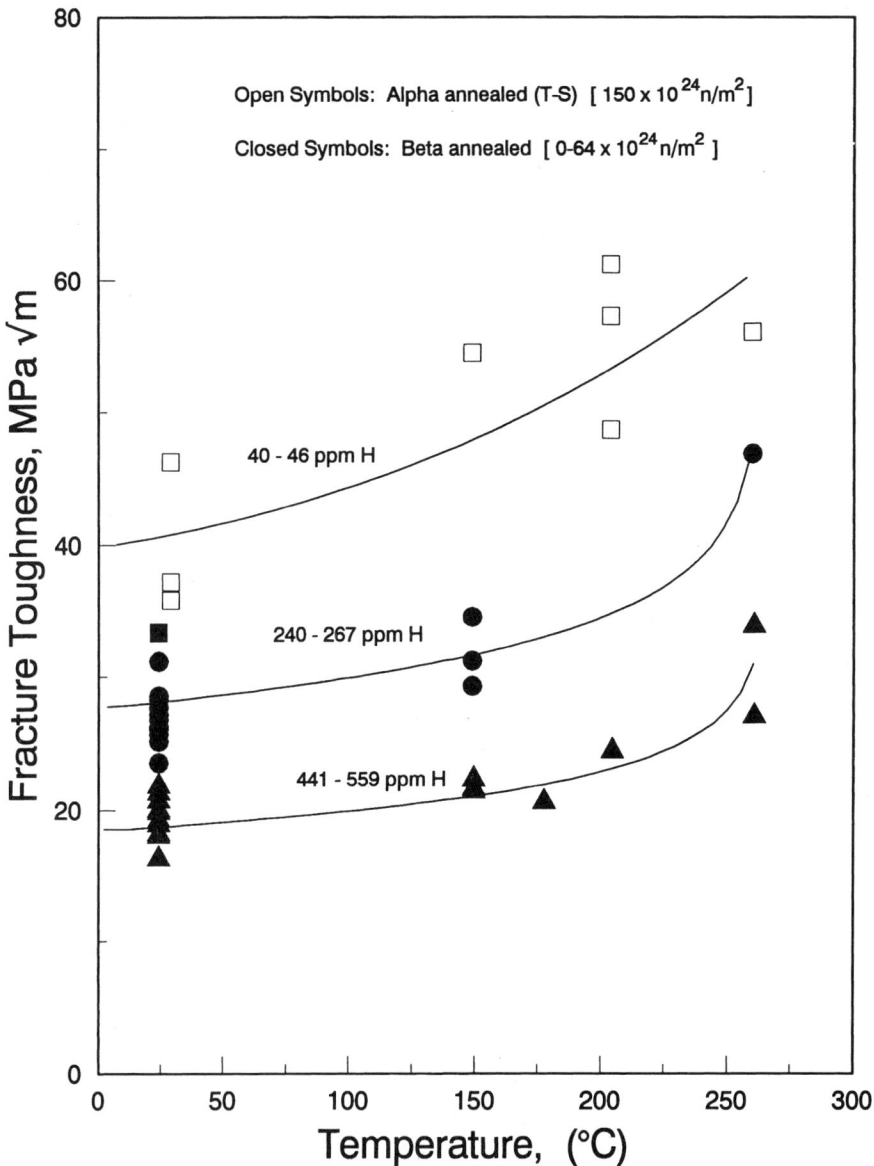

FIG. 6—*Effect of temperature on the fracture toughness of Zircaloy-4 for various hydrogen levels.*

temperature water. This hydride morphology is transgranular. From an unstable crack propagation viewpoint, it is equivalent to that in the transgranular specimens reported here.

Irradiated Nonhydrided Zircaloy-2

One nonirradiated and two irradiated specimens of nonhydrided alpha-annealed Zircaloy-2 with an S-T orientation were tested at room temperature. As shown in Fig. 7, these results are

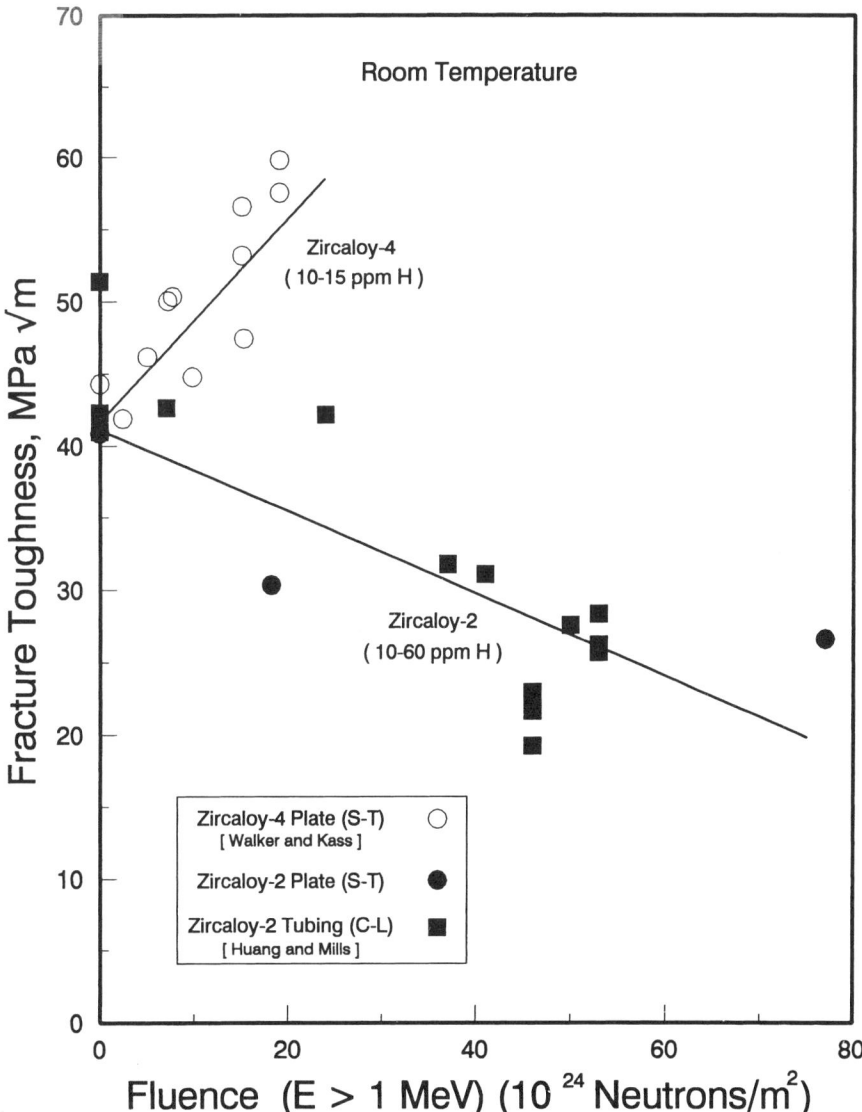

FIG. 7—*Comparison of the effects of irradiation on the fracture toughness of alpha-annealed Zircaloy-4 and Zircaloy-2.*

consistent with those reported for Zircaloy-2 tubing [3,4], namely, increasing the fluence decreases K_{IC}.

The decrease in fracture toughness of Zircaloy-2 with fast fluence is in contrast with previously reported results on Zircaloy-4 [1]. As shown in Fig. 7, the fracture toughness of alpha-annealed Zircaloy-4 increased with fluence, while that of alpha-annealed Zircaloy-2 decreased. For Zircaloy-4 weld metal (Fig. 4), no increase in toughness was seen for hydrogen concentrations above ~300 ppm hydrogen. Note that the Zircaloy-2 pressure-tube specimens were machined in the C-L orientation, which means that the specimens were stressed in the circumferential direction. According to Huang and Mills [3], the hydride normals were generally in the radial direction; hence, the hydride phase did not have a significant effect on the fracture toughness of these specimens. However, the difference in the behavior of these similar alloys may be explained in terms of their intermetallic precipitates. The fracture toughness of non-hydrided zirconium alloys is determined by the nucleation and growth of voids and the linking up of these voids, or coalescence, by ductile rupture. Common sites for void nucleation are second-phase intermetallic precipitates. While the important properties of strength and strain-rate sensitivity of the Zircaloys are expected to be essentially the same, one significant difference is the composition of their principal intermetallic precipitates and the manner in which these precipitates are affected by irradiation. For Zircaloy-4, the principal precipitate is $Zr(Fe,Cr)_2$. During irradiation at temperatures of $\sim 260°C$, the $Zr(Fe,Cr)_2$ Laves phase both rejects iron and undergoes a crystalline-to-amorphous transformation. The reaction begins at the outer surface of the precipitate diameter and proceeds inward with increasing fluence [7]. It is suggested that this diffuse particle-matrix interface has a higher effective cohesive strength and is more resistant to particle-matrix decohesion. Void nucleation is forestalled, which results in higher loads, which lead to an apparent higher toughness. In Zircaloy-2, the principal precipitate is $Zr_2(Fe,Ni)$. At test temperatures of 260°C, this precipitate does not undergo amorphization [8]. Furthermore, within the precipitate, helium is generated by the Ni (n,α) transmutation process. Although classic helium embrittlement typically occurs in the high-temperature regime, in this case the presence of helium atoms at the interface may decrease its cohesive strength. The radiation-induced increase in yield strength (and hydrostatic tensile stress), the reduced strain-to-fracture, and the potentially weakening of the particle-matrix interface by helium generation are all contributing, presumably, to the irradiation embrittlement of Zircaloy-2.

Irradiated Hafnium and Hafnium-to-Zircaloy-4 Butt Welds

The post-irradiation fracture toughness results at room temperature and 149°C, along with the results from non-irradiated room temperature controls, are given in Table 4. None of the fracture toughness values was valid per E 399 due principally to insufficient specimen thickness (5 mm). Examples of the *load versus crack-mouth-opening displacement* curves at room temperature for the irradiated butt weld, wrought hafnium (T-L), and wrought hafnium (L-T) are shown in Fig. 8. The specimen strength ratios, R_{SC}, are listed in Table 4 and are compared graphically in Fig. 9. The butt welds tended to exhibit the least toughness, while wrought hafnium in the L-T orientation had higher toughness than the wrought hafnium in the T-L orientation.

The fracture toughness (K_Q) of irradiated hafnium-to-Zircaloy-4 butt welds at room temperature ranged between 33 and 37 MPa$\sqrt{m}$. Because the nonirradiated room temperature yield strength was too low to yield a meaningful K_Q value, one butt weld specimen was tested at $-196°C$ in liquid nitrogen to increase its yield strength. The measured plane-strain toughness value was 45.3 MPa$\sqrt{m}$ and approximately 40% greater than its post-irradiation value at room temperature. It was concluded that the apparent lower K_Q and R_{SC} values of the irradiated wrought hafnium and butt weld specimens compared with their companion nonirradiated specimens indicated that there was a real decrease in their fracture toughness after irradiation. The reason for this decrease in K_Q was the irradiation-induced increase in yield strength that promoted more plane-strain loading and higher hydrostatic tensile stresses that enhanced void nucleation and growth.

Fractography—Unlike zirconium, in which no cleavage has been observed, the fracture of hafnium can occur by transgranular cleavage if the yield strength is sufficiently high [9]. Figure

TABLE 4—*Irradiated hafnium fracture toughness results.*[a]

ID	Alloy	Condition	Oriented	Temperature, °C	Fluence, 10^{24} n/m^2	K_Q,[a] MPa$\sqrt{\text{m}}$	P_{max}/P_Q	R_{SC}[b]
HFZR1	Hf/Zr-4	Butt weld	Parallel	RT	0	28.4[f]	2.15	2.15
HFTL1	Hf	Wrought	T-L[c]	RT	0	23.8[f]	2.40	1.97
HFLT1	Hf	Wrought	L-T[d]	RT	0	18.4[f]	2.22	2.70
V0624	Hf/Zr-4	Butt weld	Parallel	RT	55.0	33.4[c]	1.00	0.99
V0626	Hf/Zr-4	Butt weld	Parallel	RT	55.0	37.4[c]	1.05	0.79
V0628	Hf/Zr-4	Butt weld	Parallel	RT	55.0	33.0[c]	1.04	0.72
V0632	Hf	Wrought	T-L	RT	49.5	39.4[c]	1.08	0.89
V0635	Hf	Wrought	T-L	RT	49.5	41.9[c]	1.05	0.90
V0630	Hf	Wrought	L-T	RT	49.5	50.6[f]	1.07	1.15
V0633	Hf	Wrought	L-T	RT	49.5	44.2[f]	1.36	1.30
V0625	Hf/Zr-4	Butt weld	Parallel	149	55.0	42.1[f]	1.21	1.15
V0627	Hf/Zr-4	Butt weld	Parallel	149	55.0	39.6[f]	1.23	1.16
V0631	Hf	Wrought	T-L	149	49.5	29.4[c]	1.71	1.10
V0636	Hf	Wrought	T-L	149	49.5	40.1[f]	1.22	1.14
V0629	Hf	Wrought	L-T	149	49.5	42.5[f]	1.35	1.31
V0634	Hf	Wrought	L-T	149	49.5	46.7[f]	1.23	1.33

[a] K_Q defined as K_i at 5% secant offset (P_Q).
[b]
$$\text{Specimen Strength Ratio} = \frac{\text{Nominal crack tip stress at } P_{MAX}}{\text{Yield strength}} = \frac{2P_{MAX}(2W + a)}{B(W - a)^2 \sigma_{YS}} = \frac{2P_{MAX}}{P_{\text{LIMIT LOAD}}}$$

[c] Specimen loaded in the transverse direction, cracking in the longitudinal direction.
[d] Specimen loaded in the longitudinal direction, cracking in the transverse direction.
[e] Thickness requirement not met.
[f] Thickness, flaw, and ligament size requirements not met.

10 compares the scanning electron micrographs (SEM) on fractures in nonirradiated Zircaloy-4, hafnium, and a hafnium-to-Zircaloy-4 butt weld tested in liquid nitrogen at −196°C. Large areas of cleavage fracture can be seen in the hafnium and hafnium-Zircaloy-4 butt weld; no cleavage was observed in the Zircaloy-4 specimen. At room temperature, where the yield strengths are lower, cleavage is replaced by a mixture of coarse and fine dimpled fractures indicative of microvoid coalescence. Figure 11 shows the fracture surfaces in wrought hafnium and a hafnium-to-Zircaloy-4 butt weld tested at 24°C before and after irradiation. In each case, after irradiation the area fraction of very coarse dimples decreased and the area fraction of fine dimpling increased. This was more prominent in the butt weld. Thus, the effect of irradiation hardening on the fracture mechanism is to increase the local crack-tip stresses, thereby facilitating the formation of voids, which reduces the fracture toughness. Note, however, that even though the yield strengths of the irradiated butt welds tested at 24°C and that of the nonirradiated butt welds tested at −196°C are quite similar, being 655 and 662 MPa, respectively, the cleavage seen in the nonirradiated specimen at −196°C was not evident at 24°C after irradiation. Evidently, radiation hardening raises the stresses required for cleavage, e.g., twinning stress, and yielding proportionately. As a result, there was no apparent increase in the transition temperature from ductile-to-cleavage fracture after irradiation.

Hydrogen Redistribution—Metallographic examination of butt-weld joints consistently showed the hydrogen concentration in the Zircaloy-4 to be significantly greater than that of the hafnium. Two factors contribute to this result. First, the Zircaloy-4 corrodes faster and absorbs more hydrogen. Second, the chemical potentials of both the hydride and hydrogen dissolving

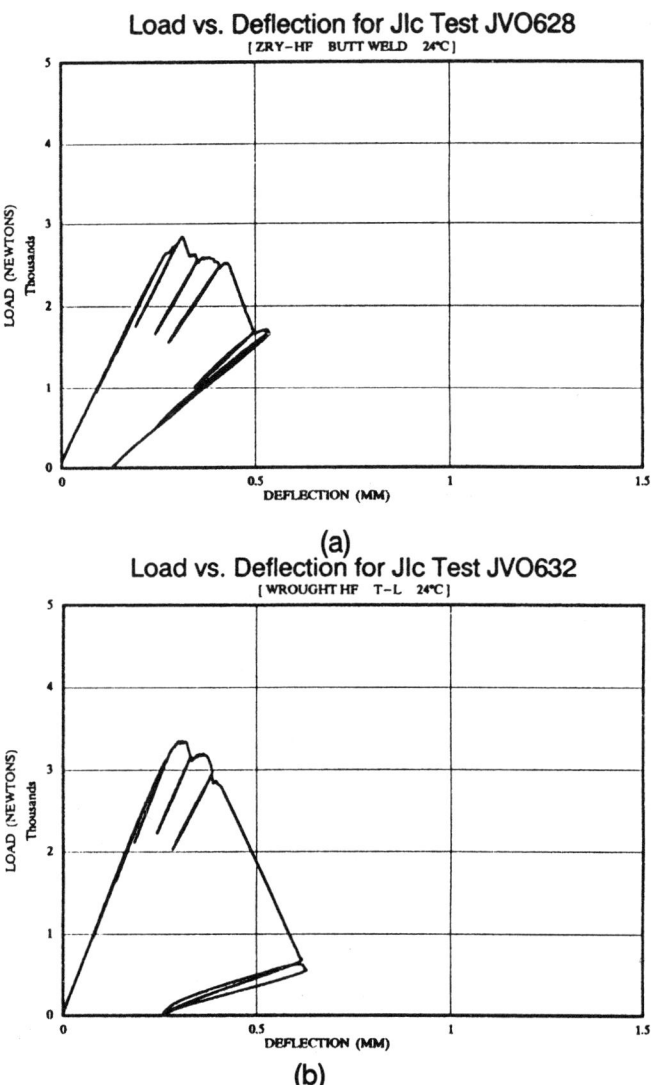

FIG. 8—J_{IC} load-displacement curves for irradiated hafnium and hafnium-to-Zircaloy-4 butt weld at 24°C: (a) butt weld—V0628; (b) wrought (T-L)—JV0632; and (c) wrought (L-T)—JV0630.

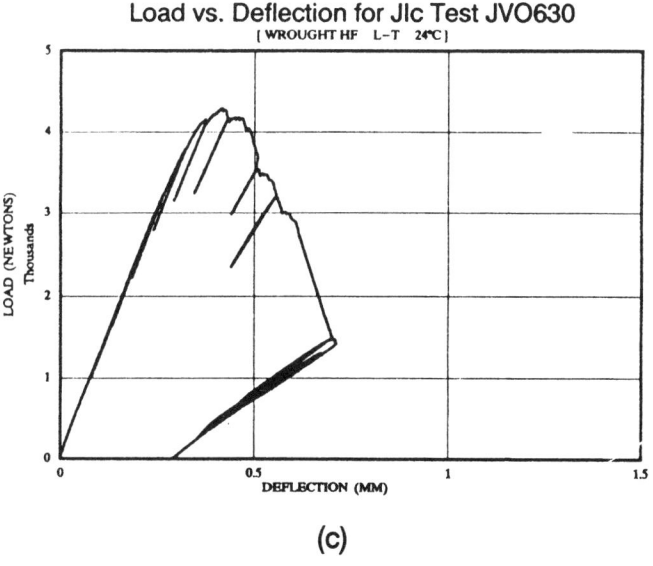

(c)

FIG. 8—*Continued*

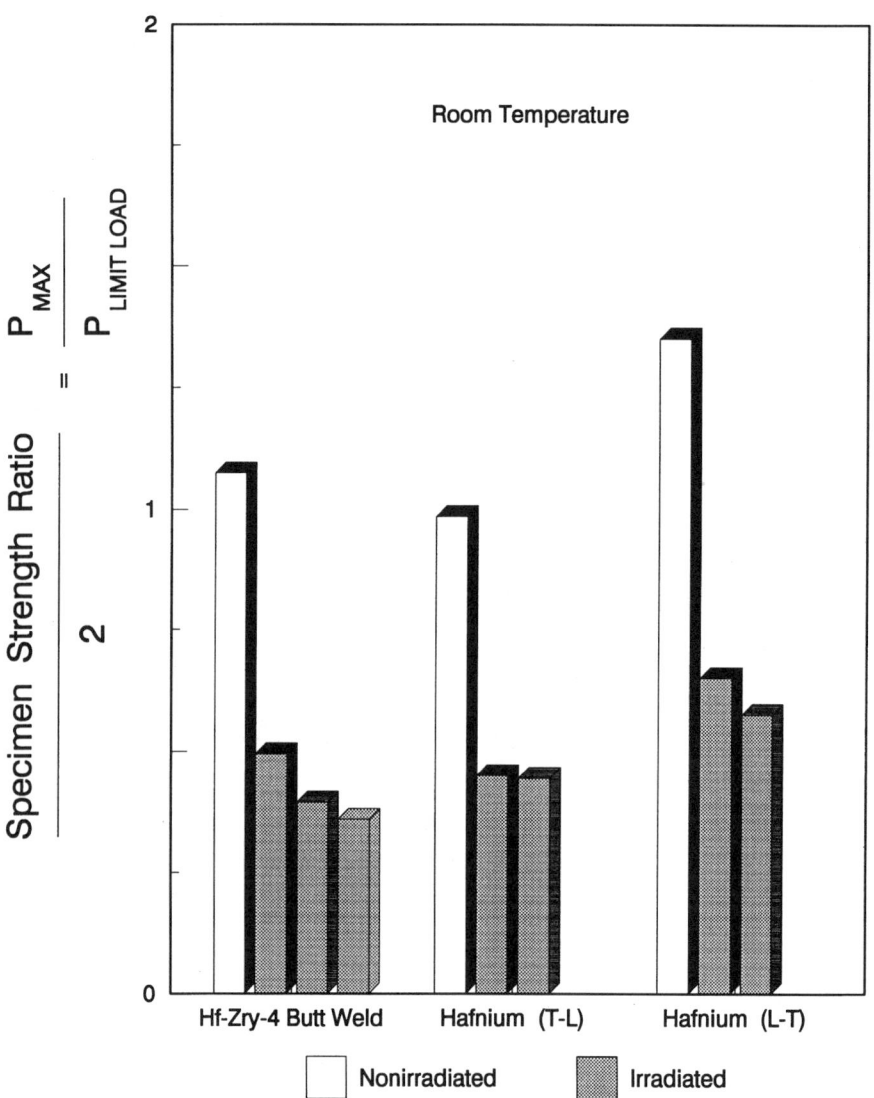

FIG. 9—*Neutron-radiation embrittlement of hafnium and hafnium to Zircaloy 4 butt welds.*

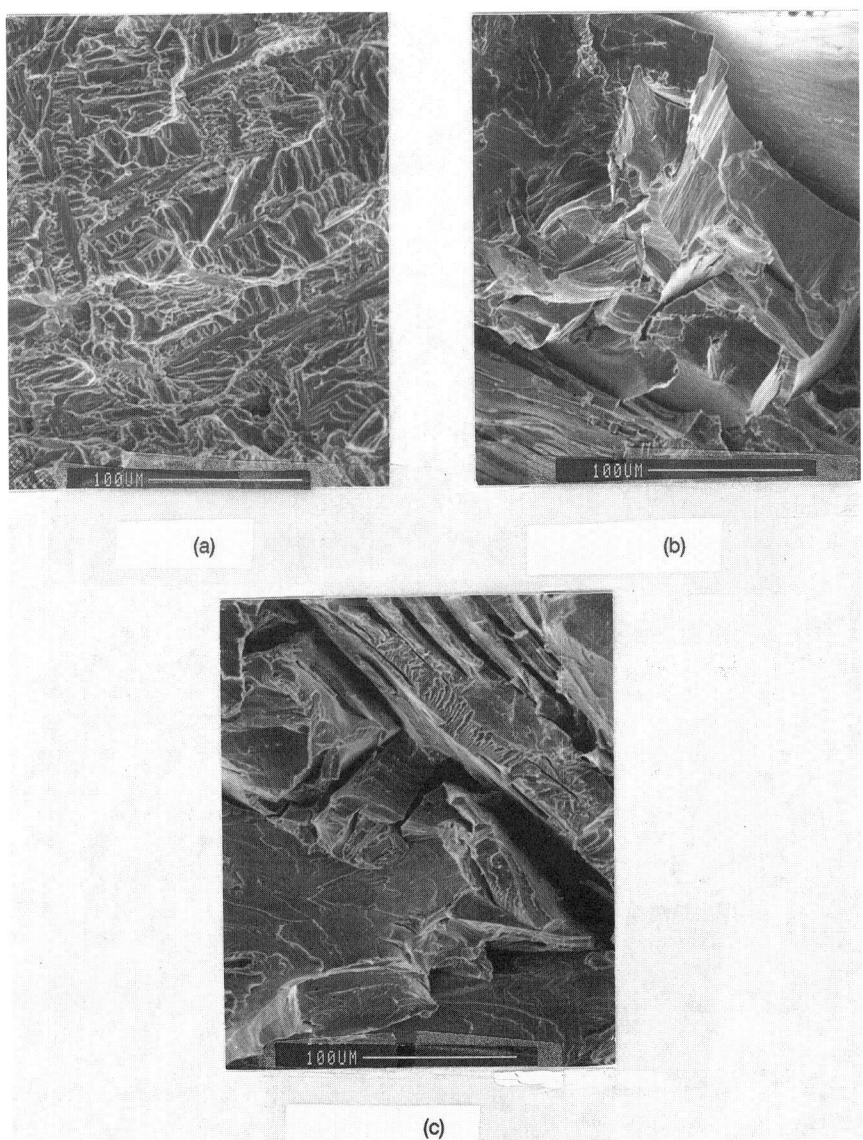

FIG. 10—*Scanning electron micrographs of fracture surfaces of nonirradiated* (a) *Zircaloy-4,* (b) *hafnium, and* (c) *a hafnium-to-Zircaloy-4 butt weld tested in liquid nitrogen.*

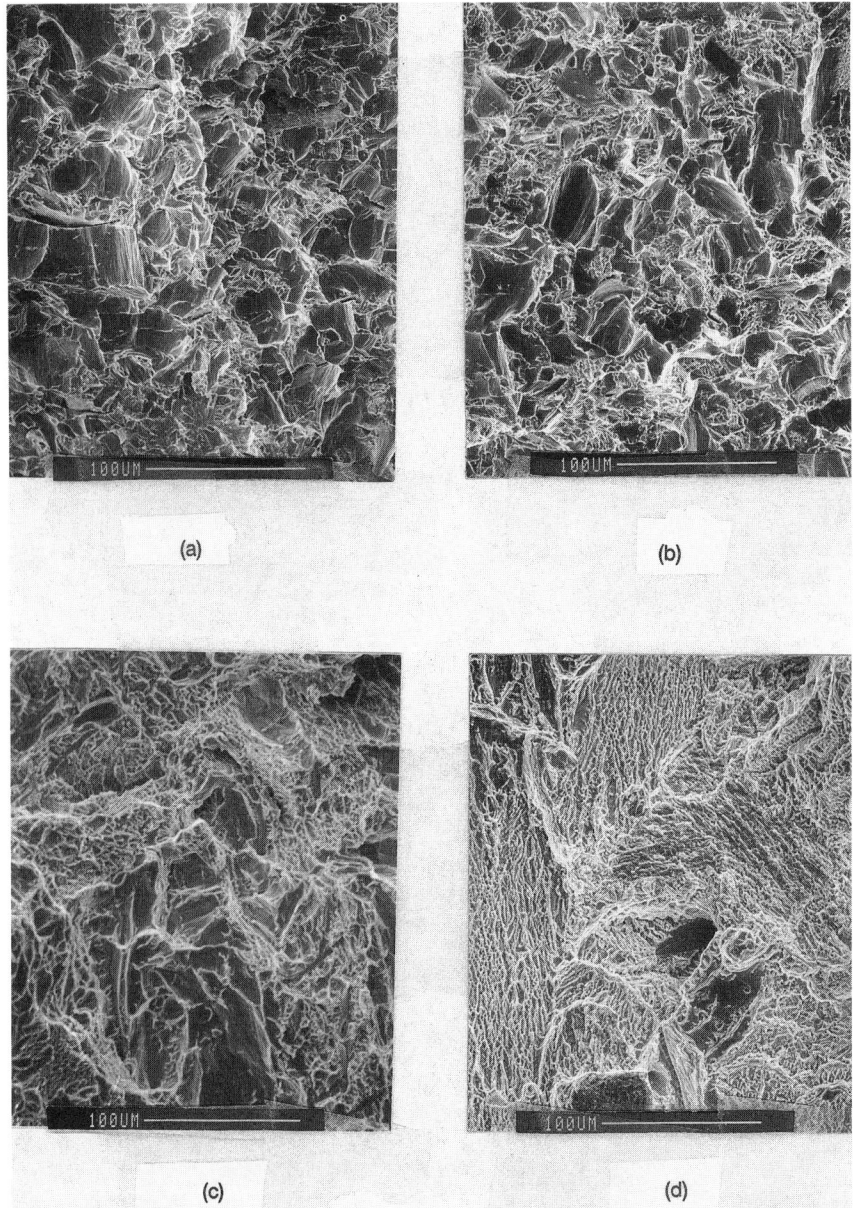

FIG. 11—*Scanning electron micrographs showing the effect of irradiation on fracture morphology in hafnium and hafnium-to-Zircaloy-4 butt weld tested at room temperature: (a) nonirradiated wrought hafnium (T-L); (b) irradiated wrought hafnium (T-L); (c) nonirradiated hafnium-to-Zircaloy-4 butt weld; (d) irradiated hafnium-to-Zircaloy-4 butt weld.*

in the solid solution are less in zirconium than in hafnium [10]. It is expected that this latter factor keeps the absorbed hydrogen within the Zircaloy and makes the Zircaloy a sink for hydrogen present in the hafnium.

Figure 12 shows the results of hydrogen concentration measurements made along the length of the irradiated hafnium-Zircaloy-4 butt weld. The hydrogen concentration in the Zircaloy-4 increased from an initial concentration of ~10 to ~42 ppm, all of which is soluble at the irradiation temperature of 285°C. In comparison, the hydrogen concentration in the hafnium decreased from its starting hydrogen concentration of ~10 to ~ 2 ppm. The hydrogen concentration in the Zircaloy can be accounted for by its initial hydrogen concentration plus the

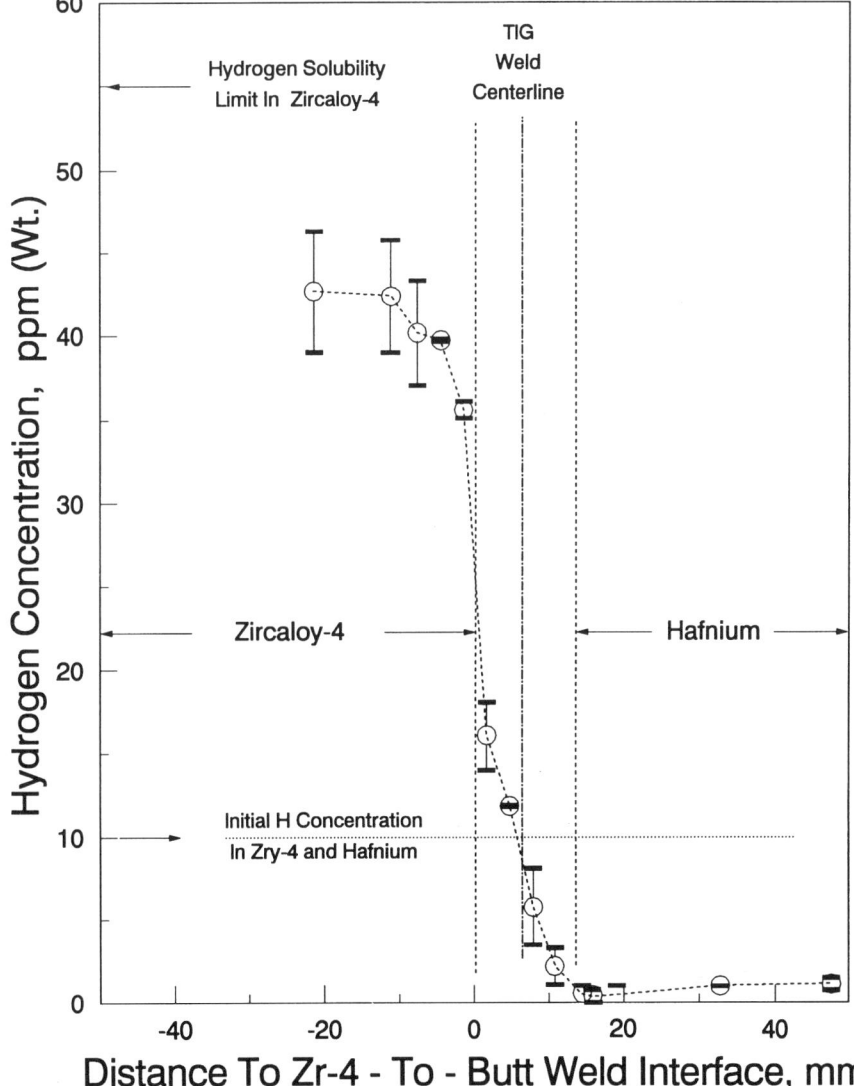

FIG. 12—*Hydrogen concentration profile through hafnium-Zircaloy-4 weld interface.*

hydrogen picked up due to corrosion and the ingress of hydrogen from the hafnium side of the weld. The hydrogen concentration in the butt weld ranges from about 2 ppm at the hafnium interface to about 30 ppm at the Zircaloy-4 interface. In this case, the hydrogen pickup in the butt weld was small and likely did not contribute significantly to its embrittlement.

However, if the hydrogen solubility limit in the Zircaloy was exceeded, further *ingress of hydrogen from the hafnium side* would result in precipitation of hydride at the Zircaloy side of the Zircaloy-butt weld interface. Figure 13 shows an example where, due to greater corrosion and a higher hydrogen pickup ratio in Zircaloy-2, the hydrogen solubility limit in the Zircaloy-2 was exceeded before all of the hydrogen in the hafnium diffused across the weld joint. As a

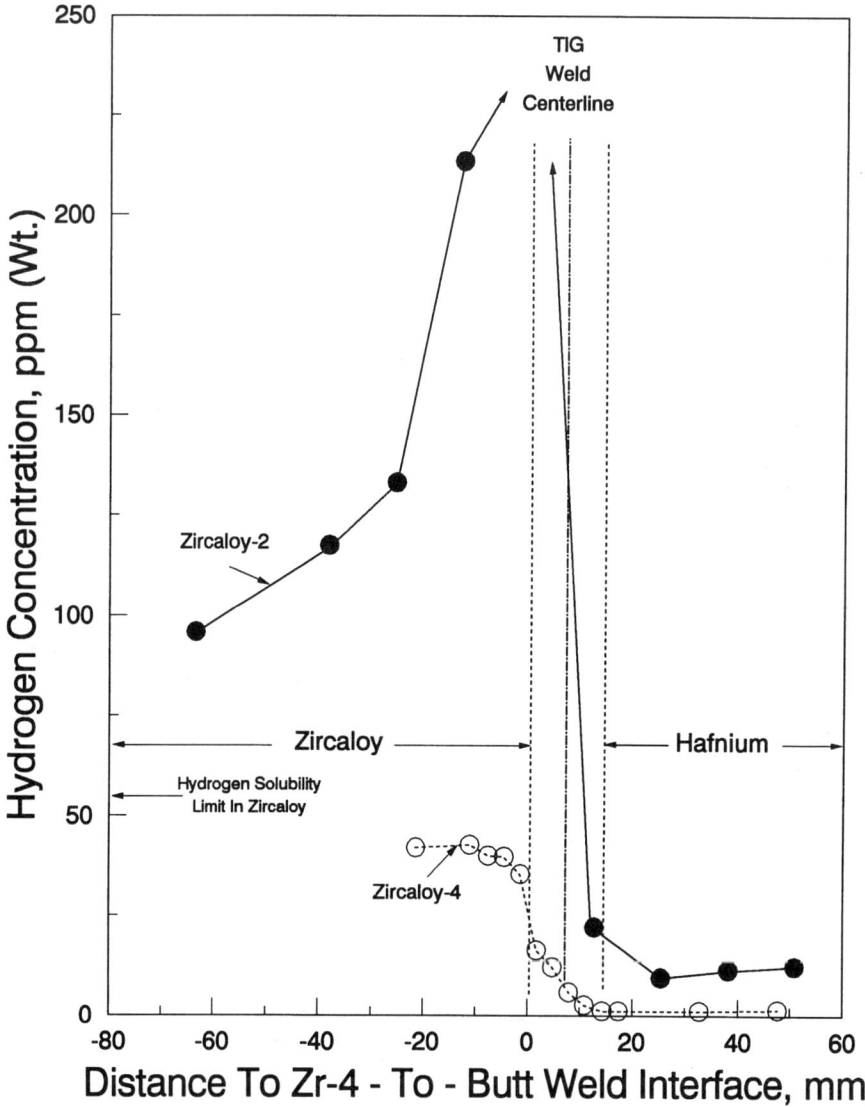

FIG. 13—*Hydrogen concentration profile through hafnium-Zircaloy-2 weld interface.*

consequence, a large amount of hydride, over 200 ppm, precipitated at the Zircaloy-butt weld interface, thereby creating a highly embrittled zone parallel to the weld joint. In this case, the fracture toughness of the weld interface would be expected to be significantly less than the values reported.

As stated above, hydrogen will not accumulate in the Zircaloy at the interface region until the corrosion of the Zircaloy has increased the overall hydrogen concentration of the Zircaloy to above the solubility limit. Until that time, solid solution gradients can exist in the Zircaloy, which will remove the hydrogen from the interface and allow essentially the entire Zircaloy extension to act as a sink for hydrogen. Once the corrosion of the Zircaloy has increased the hydrogen concentration above the solubility limit, solid solution concentration gradients can no longer exist to support a net flux of hydrogen away from the interface. For this condition, the buildup of hydrogen within the first 1.25 cm of the interface can be approximated using the simple model shown in Fig. 14 and the following expression:

$$C_{\text{Interface}} = C_{\text{sol}} + \frac{3.2}{\sqrt{\pi}} C_{0\text{Hf}} \sqrt{D_{\text{Hf}}} \sqrt{t - t_{\text{sol}}} \text{ (for } t < t_{\text{sol}}\text{)} \tag{1}$$

where

$C_{\text{Interface}}$ = average hydrogen concentration in the first 1.25 cm of the Zircaloy (ppm),
C_{sol} = solubility of hydrogen in Zircaloy at operating temperature (ppm),
$C_{0(\text{Hf})}$ = initial concentration of hydrogen in hafnium (ppm),
D_{Hf} = diffusion coefficient of hydrogen in hafnium (ppm),
t = time (seconds), and
t_{sol} = time to pickup equilibrium concentration of hydrogen in Zircaloy through corrosion of Zircaloy (seconds).

From inspection of this equation, one can see that the lower the initial hydrogen concentration in the hafnium and the lower the corrosion rate of the Zircaloy, the less will be the buildup of hydrides at the weld interface.

Optimizing Lifetime Integrity of Butt Welds—There are several approaches that can be taken to minimize the hydride embrittlement of butt welds during service. The first approach is to begin with a starting hydrogen concentration in the hafnium and the Zircaloy material as low as possible. The second approach is to select Zircaloy-4 over Zircaloy-2 because of its inherent higher irradiated fracture toughness and its lower hydrogen uptake compared with Zircaloy-2. The third approach is to use Zircaloy-4 optimized for corrosion resistance. These approaches are in the direction of having essentially all of the hydrogen in the hafnium "drained" into the Zircaloy-4 before the hydrogen concentration in the Zircaloy-4 reaches it solubility limit, thereby avoiding local hydrogen peaking at the weld interface.

Conclusions

1. Increasing the hydrogen content of Zircaloy-4 results in an exponential decrease in K_{IC}.
2. In hydrogen precharged Zircaloy-4 with a Widmanstätten microstructure, the fracture toughness of specimens with a transgranular hydride morphology is higher than those with an intergranular hydride morphology.
3. The hydride morphology developed in-reactor in a Widmanstätten Zircaloy-4 micro-structure, as the result of the absorption of corrosion-generated hydrogen, is transgranular and effectively equivalent to that of the transgranular-hydride specimens reported here.

Zircaloy-to-hafnium couple of cross-sectional area A

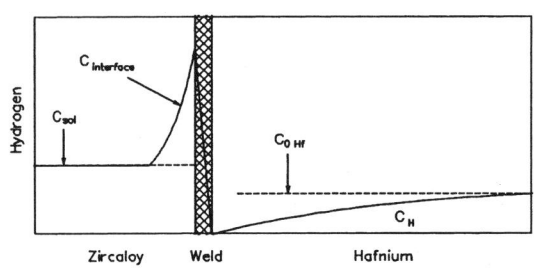

$x < 0 \qquad \uparrow x = 0 \qquad x > 0$

$$C_{Interface} = C_{sol} + \frac{3.2}{\sqrt{\pi}} C_{0_{Hf}} \sqrt{D_{Hf}} \sqrt{t - t_{sol}} \qquad (\text{for } t > t_{sol})$$

Model conservatively assumes hydrogen diffusion out of hafnium, of density ρ_{Hf}, does not occur until $t \geq t_{sol}$

For Boundary Conditions:
@ $x = 0 \quad C_H = 0$
@ $t = 0 \quad C_H = C_{0(Hf)}$
@ $x = \infty \quad C_H = C_{0(Hf)}$

$$C_H = C_{0_{Hf}} \, \text{erf} \frac{x}{2\sqrt{Dt}} \; ; \quad J_{x=0} = -D_{Hf} \frac{\delta C}{\delta x}\bigg|_{x=0} = \frac{-D_{Hf} C_{0_{Hf}}}{\sqrt{\pi D t}}$$

Grams of hydrogen, M_H, flowing into Zircaloy once solubility is exceeded[1]:

$$M_H = \int_{t_{Sol}}^{t} A D_{Hf} \frac{(C_{0_{Hf}} \rho_{Hf})}{\sqrt{\pi D_{Hf} t}} \times 10^{-6} \, dt = \frac{2}{\sqrt{\pi}} D_{Hf}^{\frac{1}{2}} C_{0_{Hf}} \rho_{Hf} A \, (t^{\frac{1}{2}} - t_{Sol}^{\frac{1}{2}}) \times 10^{-6}$$

---

[1] Terminology and appropriate units are provided in text.

FIG. 14—*Model for buildup of zirconium hydride at Zircaloy-butt weld interface.*

4. The fracture toughness of alpha-annealed Zircaloy-4 increases with irradiation, while that of alpha-annealed Zircaloy-2 decreases with irradiation at 260°C. The difference is attributed to the way irradiation affects their second-phase intermetallic precipitates. It is speculated that the obliteration of the particle-matrix interface in irradiated Zircaloy-4 leads to a difficult-void-nucleation microstructure, and this is reflected in increased postirradiation toughness.

5. For hydrogen concentrations less than 10 ppm, the postirradiation fracture toughness of the hafnium-to-Zircaloy-4 butt weld is too high to yield a valid K_{IC} per ASTM E 399 in specimens tested in this program. The estimated room-temperature fracture toughness $K_{IC}^{est.}$ at 55×10^{24} n/m^2 is 33 MPa$\sqrt{m}$.

6. Zircaloy getters hydrogen from hafnium, resulting in hydrogen accumulation in the Zircaloy side of a hafnium-Zircaloy butt weld.

7. Hydrogen will not accumulate at the butt weld interface until the Zircaloy corrosion plus hydrogen ingress from the hafnium have increased the overall hydrogen concentration in the Zircaloy above the solubility limit. Any subsequent ingress of hydrogen from the hafnium, either from its initial hydrogen inventory or from corrosion, will precipitate near the Zircaloy-weld interface.

Acknowledgment

This work was supported by DOE contract DE-AC11-93PN38195. Also, the authors wish to thank Dr. W. J. Mills for his helpful comments and suggestions on the manuscript.

References

[1] Walker, T. J. and Kass, J. N., "Variation of Zircaloy Fracture Toughness in Irradiation," *Zirconium in Nuclear Applications, ASTM STP 551*, J. H. Schemel and H. S. Rosenbaum, Eds., American Society for Testing and Materials, West Conshohocken, PA, 1974, pp. 328–354.
[2] Woo, O. T. and Tangri, K. "Transformation Characteristics of Rapidly Heated and Quenched Zircaloy-4 Oxygen Alloys," *Journal of Nuclear Materials*, Vol. 79, 1979, pp. 82–94.
[3] Huang, F. H. and Mills, W. J., "Fracture and Tensile Properties of Irradiated Zircaloy-2 Pressure Tubes," *Nuclear Technology*, Vol. 102, 1993, pp. 367–375.
[4] Huang, F. H., "Brittle-Fracture Potential of Irradiated Zircaloy-2 Pressure Tubes," *Journal of Nuclear Materials*, Vol. 207, 1993, pp. 103–115.
[5] Simpson, L. A. and Chow, C. K., "Effect of Metallurgical Variables and Temperature on the Fracture Toughness of Zirconium Alloy Pressure Tubes," *Zirconium in the Nuclear Industry: 7th International Symposium, ASTM STP 939*, R. B. Adamson and L. F. P. van Swam, Eds., American Society for Testing and Materials, West Conshohocken, PA, 1987, pp. 579–596.
[6] Davies, P. H. and Stearns, C. P., "Fracture Toughness Testing of Zircaloy-2 Pressure Tube Material with Radial Hydrides Using Direct-Current Potential Drop," *Fracture Mechanics, ASTM STP 905*, J. H. Underwood et al., Eds., American Society for Testing and Materials, West Conshohocken, PA, 1986, pp. 379–400.
[7] Yang, W. J. S., Tucker, R. P., Cheng, B., and Adamson, R. B., "Precipitates in Zircaloy: Identification and the Effects of Irradiation and Thermal Treatment," *Journal of Nuclear Materials*, Vol. 138, 1986, pp. 185–195.
[8] Griffiths, M., Gilbert, R. W., and Carpenter, G. J. C., "Phase Instability, Decomposition and Redistribution of Intermetallic Precipitates in Zircaloy-2 and -4 During Neutron Irradiation," *Journal of Nuclear Materials*, Vol. 150, 1987, pp. 53–66.
[9] Straw, R. L., "Effective of Hydrogen and Oxygen on the Deformation and Fracture of Reactor Grade Hafnium," Ph.D. dissertation, Rensselaer Polytechnical Institute, 1971.
[10] Beck, R. L. and Mueller, W. M., "Zirconium Hydrides and Hafnium Hydrides," *Metal Hydrides*, W. M. Mueller, J. P. Blackledge, and G. G. Libowitz, Eds., Academic Press, Inc., New York, 1968, pp. 241–335.

DISCUSSION

P. H. Davies[1] (written discussion)—Please clarify the cooling rate used in the experiments showing the difference in the effect of fast neutron fluence on the toughness of Zr-4 and Zr-2, i.e., would the hydride morphology have been similar in both cases?

P. H. Kreyns et al. (authors' closure)—The Zircaloy-2 and Zircaloy-4 specimens were machined from wrought plate stock and irradiated in the nonhydrided condition. Their cooling rates from the reactor operating temperature were comparable. In each case, the small amount of hydride picked up in-reactor due to corrosion had a similar morphology.

B. Cox[2] (written discussion)—Your fractographs of Zircaloy and hafnium at 77 K and room temperature did not reveal enough information of hydrogen content and hydride morphology to allow a judgment of whether any of the fractographic features could have been fractured hydrides. Since fractured hydrides commonly appear in mechanical fracture surfaces, the inclusion of both H content and morphology would be useful.

P. H. Kreyns et al. (authors' closure)—The initial hydrogen concentration of the nonhydrided, nonirradiated Zircaloy-4, hafnium, and Zircaloy-4 hafnium butt welds was nominally 10 ppm. After irradiation, the centerline of the butt weld was still approximately 10 ppm hydrogen, while the hydrogen concentration of the wrought hafnium was reduced to about 1 to 2 ppm (see Fig. 12). The fractography consisted either of ductile dimpling and metal cleavage at 77 K or entirely of ductile dimpling at room temperature.

---

[1] Atomic Energy of Canada, Ltd., Chalk River, Ontario, Canada.
[2] University of Toronto, Toronto, Ontario, Canada.

In-Reactor Behavior

Antonina V. Nikulina,[1] Vladimir A. Markelov,[1] Mikhail M. Peregud,[1] Yury K. Bibilashvili,[1] Vladimir A. Kotrekhov,[2] Anatoly F. Lositsky,[2] Nikolay V. Kuzmenko,[2] Yuriy P. Shevnin,[2] Valentin K. Shamardin,[3] Gennady P. Kobylyansky,[3] Andrey E. Novoselov[3]

Zirconium Alloy E635 as a Material for Fuel Rod Cladding and Other Components of VVER and RBMK Cores

REFERENCE: Nikulina, A. V., Markelov, V. A., Peregud, M. M., Bibilashvili, Y. K., Kotrekhov, V. A., Lositsky, A. F., Kuzmenko, N. V., Shevnin, Y. P., Shamardin, V. K., Kobylyansky, G. P., and Novoselov, A. E., **"Zirconium Alloy E635 as a Material for Fuel Rod Cladding and Other Components of VVER and RBMK Cores,"** *Zirconium in the Nuclear Industry: Eleventh International Symposium, ASTM STP 1295*, E. R. Bradley and G. P. Sabol, Eds., American Society for Testing and Materials, 1996, pp. 785–804.

ABSTRACT: Data are given on Zr alloy E635 (Zr-1.2Sn-1Nb-0.4Fe), developed in Russia as a fuel rod cladding and other component material for use in cores of VVER and RBMK types. The alloy is much superior to binary alloys with 1.0 and 2.5% Nb and Zircaloys in terms of its resistance to irradiation-induced creep and growth and nodular corrosion. The creep rate of the alloy is slightly dependent on irradiation temperature, stress, neutron fluence, and neutron density. The alloy is subject to substantial irradiation hardening while retaining its high-percent elongation. Corrosion, creep, and growth resistances are slightly dependent on the structure of components (alloy, final product). Based on the previously studied influence of impurities, structure, heat treatment, and working schedules, the technological processes were designed and mastered commercially for fabrication of tubes, bars, strips, and fuel rod claddings from this alloy. Components are produced commercially. Fuel assemblies with fuel rods clad in the E635 alloy were successfully tested in the RBMK reactor at the Leningrad NPP as well as in experimental reactors under VVER-1000 conditions. Today, the E635 alloy is recommended as a promising material for use in cores of VVER-1000 and VVER of new generations as well as RBMK-type reactors having a longer fuel cycle.

KEYWORDS: E635, E110, Zircaloy-4, ZIRLO, in-reactor, creep, growth, fuel cladding, corrosion

Fuel rods clad in E110 alloy (Zr-1Nb) are in successful operation in VVER and RBMK. The chosen alloy composition and the process of tube fabrication thereof promoted the reliable operation of the fuel rods during a three-year cycle in VVER-1000 and RBMKs and a four-year cycle in VVER-440 reactors. Improvements of the fabrication processes are underway to

---
[1] Leader of laboratory, leading scientific officer, senior scientific officer, deputy director, respectively, A. A. Bochvar All-Russia Scientific Research Institute of Inorganic Materials, Rogov st. 5, 123060, Moscow, Russia.
[2] Chief engineer, chief production engineer, deputy leader of laboratory, group manager, respectively, PU Chepetsky Mechanical Plant, 427600, Glasov, Republic of Udmurtia.
[3] Leader of laboratory, senior scientific officer, deputy leader of laboratory, respectively, V. I. Lenin Scientific Research Institute of Atomic Reactors, 433510, Dimitrovgrad 10, Ulyanov region, Russia.

further enhance the operational reliability of E110 alloy tubes, thus promoting longer-term (five or six-year) cycles in VVERs with the fuel burnup extended to 65 MW day/kg U. However, some limitations are inherent in the performance of this alloy that make one uncertain of its ability to ensure targeted service conditions (fuel burnup, cycle time). Its limitations are:

1. A transition in the irradiation-induced growth kinetics at the neutron fluence of $\sim 2 \times 10^{22}$ n/cm^2 ($E \geq 0.1$ MeV) followed by accelerated growth.
2. High irradiation-induced creep.
3. Degraded corrosion resistance at the oxygen content of the water coolant above 0.02 ppm and upon boiling.

Late in the 1960s in Russia a new alloy was designed that in addition to niobium (0.9 to 1.1%) contains 1 to 1.4% tin and 0.3 to 0.5% iron; it is devoid of the above-mentioned limitations [1]. In 1971, at the Fourth International Conference, this alloy was reported to be corrosion resistant in a water-coolant environment [2]. In Russia, this alloy was called E635. For more than 20 years experimental data have been collected on this alloy in Russia that corroborate the significant superiority of some of its characteristics over binary niobium-containing alloys and Zircaloys [3,4]. The advantages of this alloy were also reported by L. Kastaldelli and L. Lunde in 1980 at the Fifth International Conference on zirconium [5]. In compliance with Russian programs aimed at improving the VVERs and designing a new generation of units of this type meant for extended burnup of fuel and longer-term service, zirconium materials were needed that had a much greater level of performance. This gave rise to further improvement of commercial E110 alloy tubes and a more comprehensive study of the E635 alloy. The properties of the latter led to its recommendation as a structural material for fuel claddings, guide tubes, and spacer grids. This paper presents data never before given on the E635 alloy that describe its performance, the extent to which it has been studied, and the state of the art of the production of fuel assembly (FA) components for water-cooled reactors. We have previously reported [6] the use of this alloy as a material for RBMK pressure tubes and its advantages over Zr-2.5Nb. However, because of the inadequately elaborated process of the fabrication of these large-scale tubes (8000 mm in length, 88 mm in diameter, 4.0 mm in wall thickness), the fracture toughness margin proved to be lower. Unfortunately, the pressure tube related activities were essentially haltered due to the decision to replace fuel channels in RMBK reactors after 15 to 16 years of operation. Zr-2.5Nb pressure tubes operate reliably during that period. However, investigations aimed at improving the process of fabricating the large-scale E635 tubes were continued, and nowadays beneficial techniques are available that will permit elimination of the earlier revealed defects in the tubes.

Materials, Methods, and Test Conditions

Materials

The results discussed in this paper refer to samples cut out of cladding tubes 9.15 mm in outer diameter and 0.65 mm in wall thickness and 13.65 in outer diameter and 0.9 mm in wall thickness, bars 10 mm in diameter, and plates laboratory tested before and after neutron irradiation, as well as to fuel claddings tested in research and commercial reactors. The typical composition of the E635 alloy is shown in Table 1. The samples and cladding tubes examined were commercially fabricated from commercial billets and had a condition close to recrystallyzed. The recrystallization grade is ~95%. Our experience in operating zirconium materials, our studies of their properties versus their structure, the effect of the neutron field, and the requirements placed by designers corroborated the advisability of having the components op-

TABLE 1—*Chemical composition of ingots.*

Alloying Elements and Major Impurities	Typical Content, Mass Fraction, %	
	E635	E110
Niobium	0.95–1.05	0.95–1.05
Tin	1.20–1.30	...
Iron	0.34–0.40	0.006–0.012
Oxygen, ppm	500–700	500–700
Carbon, ppm	50–100	50–100
Silicon, ppm	50–100	50–100
Nitrogen, ppm	30–60	30–60

erated in VVER and RBMK reactors in a condition close to equilibrium (phase composition, recrystallization grade). The technological process of E635 alloy tube fabrication is similar to that used for the E110 alloy and involves the following major stages:

1. Vacuum-arc furnace production of a 1.2-ton ingot from a mixture of an electrolytic powder and iodide zirconium bars in the ratio ~70:30.
2. β-forging (~1070 to 900°C).
3. Water quenching of forged alloy billets from the temperature of heating in the β-region (~1050°C).
4. α-pressing (600 to 650°C).
5. Cold rolling in four or five stages with intermediate α-anneals (560 to 620°C).
6. Final recrystallization anneal in the α-region (560 to 620°C).
7. Straightening, grinding, chemical pickling.

As a rule, all the tests were carried out simultaneously and in comparison to the commercial E110 alloy (Table 1), which was in a similar condition, and in some instances also with stress-relieved Zircaloy-4 manufactured by foreign companies. The results presented in the paper cover a large quantity of samples manufactured in different time periods.

Test and Study Methods

Corrosion and tensile property tests of samples and investigations of unirradiated sample microstructures were performed using essentially ASTM methods. The autoclave corrosion tests were carried out in deionized water at 320 to 360°C as well as in steam at 400 to 500°C. The tests were also conducted with oxygen, alkalies, or acids added to the water.

The microstructure was examined using optical as well as transmission electron microscopes to determine the size, distribution, and type of secondary phase particles. The phase composition was determined by the energy dispersive analysis using LINK 860.

The effect of neutron irradiation on the short-term properties was studied using annular samples 3 mm high cut from fuel rod claddings. The creep and stress corrosion cracking (SCC) tests without and under irradiation were carried out using internally and sometimes externally pressurized tubular samples. The accuracy of the strain measurement was 0.05%. The irradiation-induced growth was studied in samples cut out of tubes and bars in the axial and tangential directions. The accuracy of the strain measurement was 0.003%. The samples were irradiated mainly in the experimental high flux reactor BOR-60. The condition of the tested fuel rod claddings was studied visually in hot cells based on their appearances and also using samples

TABLE 2—*Microstructure characteristics of E635 alloy cladding tubes.*

Characteristic	Mean Value Range
Grain number	12.2–12.7
Average grain size, μm	3.9–4.7
Average particle size, Å	900–1000
Particle density, cm^{-3}	$(2-4)10^{13}$
Particle composition, at%	(30–45)%Zr-(15–40)%Nb-(10–40)%Fe
Type of particles	Zr(Nb,Fe)$_2$, (Zr,Nb)$_2$Fe, (Zr,Nb)$_3$Fe
Texture parameters	$f_r = 0.59$–0.65, $f_t = 0.28$–0.32, $f_l = 0.06$–0.10

of different fuel rod claddings cut from areas located at different heights. Also investigated were the oxide film thicknesses and modes of hydride precipitation; spectral-isotopic analysis for the content of hydrogen was conducted; tensile properties were studied at different temperatures. Examinations of the irradiated sample and tube microstructures are dealt with in a special paper [7].

Results of Investigations and Tests

Tube Structure

The tube microstructure (Table 2) consists of α-grains and secondary phase particles at a distribution density of 2 to 4×10^{13} cm^{-3} (Fig. 1). All particles contain zirconium, niobium, and iron. Based on the results of the X-ray diffraction microanalysis and energy dispersive analysis, the E635 alloy comprises particles of three types. For the most part, the identified particles have a hexagonal lattice with $a = 0.51$ nm and $c = 0.83$ nm, which was assumed by us to correspond to Zr(Nb,Fe)$_2$. The other particles were identified as (Zr,Nb)$_2$Fe having a tetragonal lattice with $a = 0.636$ nm and $c = 0.58$ nm and (Zr,Nb)$_3$Fe having an orthorhombic lattice with $a = 0.88$ nm, $b = 0.33$ nm, and $c = 1.10$ nm. The tubes are characterized by the preferred radial basal pole texture.

Autoclave Corrosion

The autoclave corrosion of the E635 alloy at a water temperature of 320 to 360°C and pressure up to 18.6 MPa always forms dark oxide films firmly adhering to the metal with an insignificant pickup of hydrogen even if significant amounts of oxygen and additives of alkalies and acids are available (Tables 3, 4). Under the conditions of ammonia-potassium water chemistry with boron control (the VVER water chemistry), corrosion of the E635 alloy is similar to that of the E110 alloy. In lithium-containing water, the behavior of the E635 alloy is much superior to that of the E110 alloy or Zircaloy-4; moreover, compared to the data of Westinghouse on ZIRLO[4] [8,9], there is some advantage over the ZIRLO alloy (Fig. 2). In deionized water at 350 and 360°C, the corrosion of the E635 alloy is attended by a higher weight gain compared to that of the E110 alloy and comparable to that of ZIRLO (Fig. 3). In steam at 400°C, the E635 alloy is inferior to the E110 alloy or Zircaloy-4 and comparable to ZIRLO (Fig. 4). Similarly to the E110 alloy after 45-day testing in steam at 500°C, the E635 alloy is much superior to Zircaloy-4 even after 24-h testing resulted in a severely corrosion-damaged

---

[4] ZIRLO is a registered trademark of Westinghouse Electric Corporation, Pittsburgh, PA.

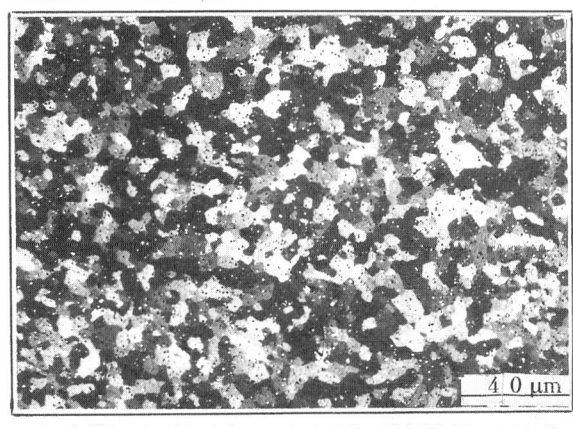

FIG. 1—*Microstructure of E635 alloy cladding tubes:* (a) *polarized light,* ×500; (b) *TEM,* ×10 000.

TABLE 3—*E635 alloy corrosion in water containing oxygen (350°C, 16.8 MPa, 180 days).*

Oxygen Content of Water, ppm	Weight Gain, mg/dm²	Hydrogen Content	
		Mass %	Exper./Theor.%
≤0.045	38.0	0.004	38
0.2-0.9	45.0	0.005	41
5.0-7.0	58.0	0.004	25

TABLE 4—*E635 alloy corrosion in water with alkali and acid additives (320°C, 11.5 MPa, 120 days).*

Additives	Additive Concentration		Weight Gain, mg/dm²	Hydrogen Content	
	pH	mg/dm³		Mass. %	Exper./Theor.%
Deionized water	5.6	...	25.0	0.003	44
LiOH	7.0	0.0007	18.0	<0.002	<41
LiOH	9.0	0.07	16.0	<0.002	<45
NH_4OH	7.0	0.004	16.0-18.0	<0.002	<44-41
NH_4OH	9.0	0.35	16.0-18.0	<0.002	<44-41
NH_4OH	11.0	35	16.0-18.0	<0.002	<44-41
KOH	9.0	0.39	14.0	<0.002	<53
NaOH	9.0	0.23	15.0	<0.002	<49
H_2SO_4	5.0	0.0001	23.0	0.003	48

surface that was, hence, much deformed (Fig. 5). The films on both the E635 and E110 alloys remained dark and smooth.

Tensile Properties

The tensile properties of the tubes were determined by uniaxial tension and burst tests (Table 5). In terms of strength, E635 alloy tubes are superior to E110 tubes, both having a similarly high ductility typical of the recrystallized condition. These properties to a greater degree comply with the requirements of VVER and RBMK fuel rod designers.

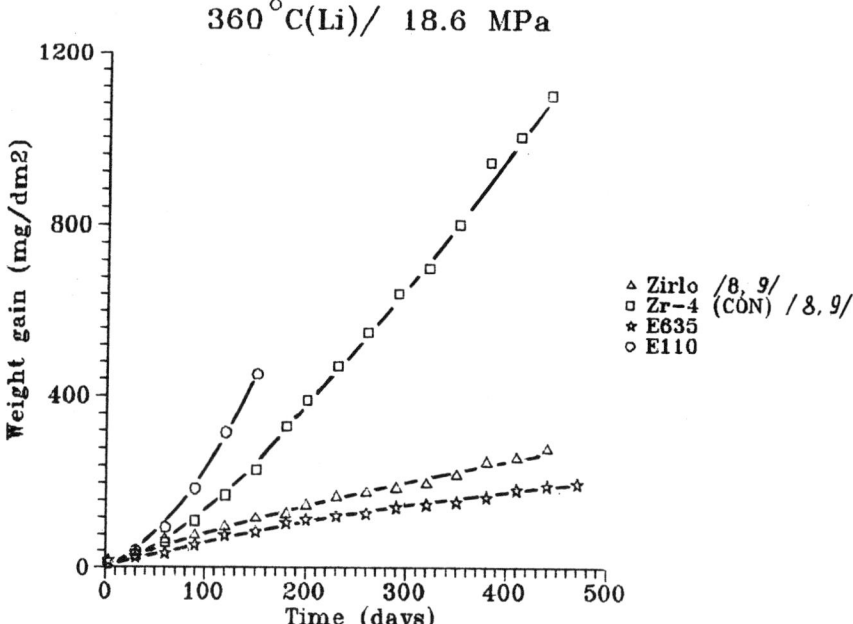

FIG. 2—*Autoclave corrosion of alloys in water containing 70 ppm lithium at 360°C, 18.6 MPa.*

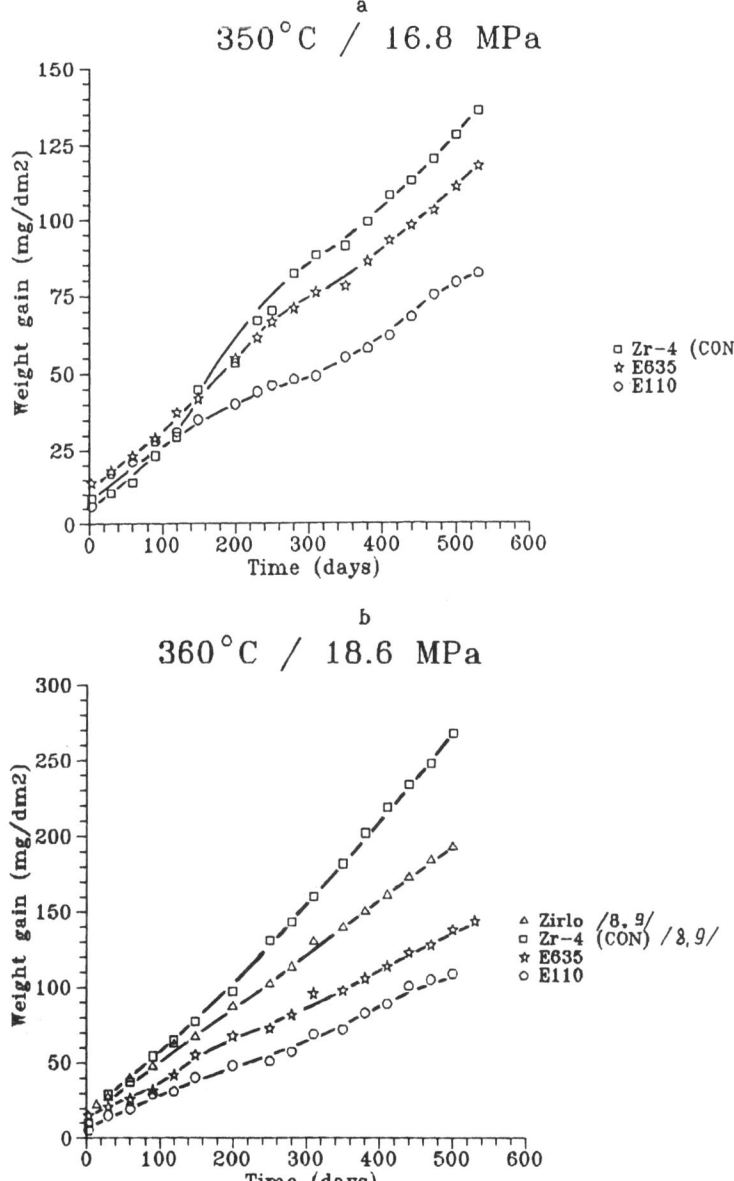

FIG. 3—*Autoclave corrosion of alloys in deionized water:* (a) *at 350°C, 16.8 MPa;* (b) *at 360°C, 18.6 MPa.*

Creep

The E635 alloy and items thereof have a high creep resistance that is slightly dependent on temperature, stress, and time of testing typical of VVER and RBMK type reactors (Figs. 6, 7). The creep of the E635 alloy is little influenced by the neutron field (fluence, neutron density)

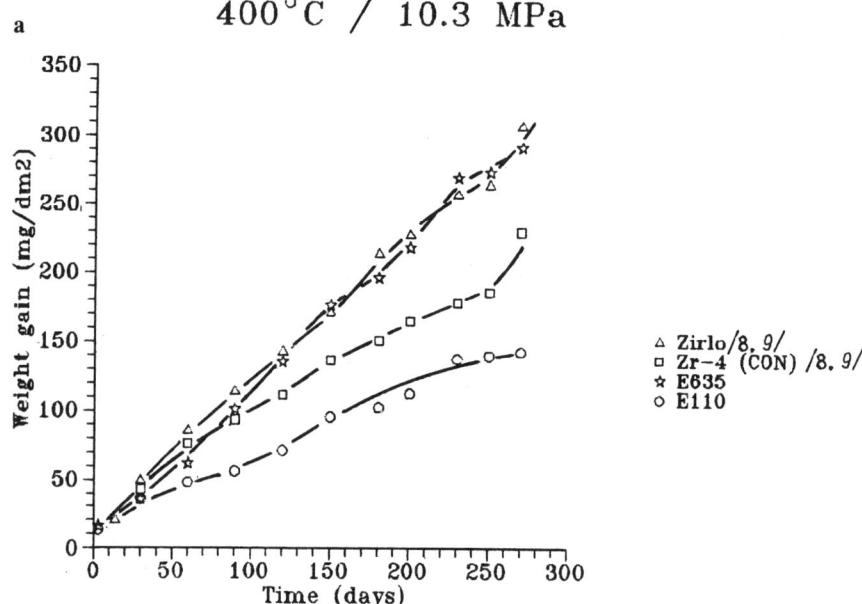

FIG. 4—*Autoclave corrosion of alloys in steam at 400°C, 10.3 MPa.*

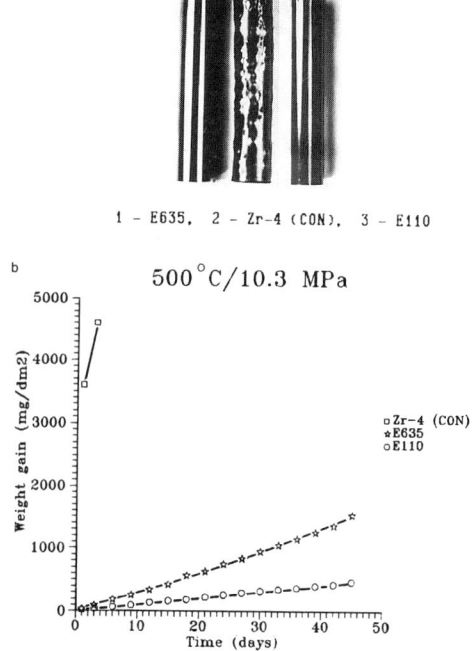

FIG. 5—*Autoclave corrosion of alloys in steam at 500°C, 10.3 MPa: (a) appearance of samples as tested for 72 h; (b) weight gain of samples versus testing time.*

TABLE 5—*Properties of tubing.*

Property	E635	E110
	Room Temperature	
Tension Test Properties		
Yield Strength (0.2% offset), MPa	300	209
Tensile strength, MPa	480	380
Elongation, %	33	35
Burst Test Properties		
Ultimate hoop strength, MPa	600	525
Total Circumferential Elongation, %	45	51
Contractile Strain Ratio	2.3	2.2
	300°C	
Tension Test Properties		
Yield Strength (0.2 Offset), MPa	160	112
Tensile Strength, MPa	280	209
Elongation, %	38	38
Burst Test Properties		
Ultimate Hoop Strength, MPa	340	285
Total Circumferential Elongation, %	50	51
	400°C	
Tension Test Properties		
Yield Strength (0.2 Offset), MPa	150	98
Tensile Strength, MPa	250	196
Elongation, %	35	38

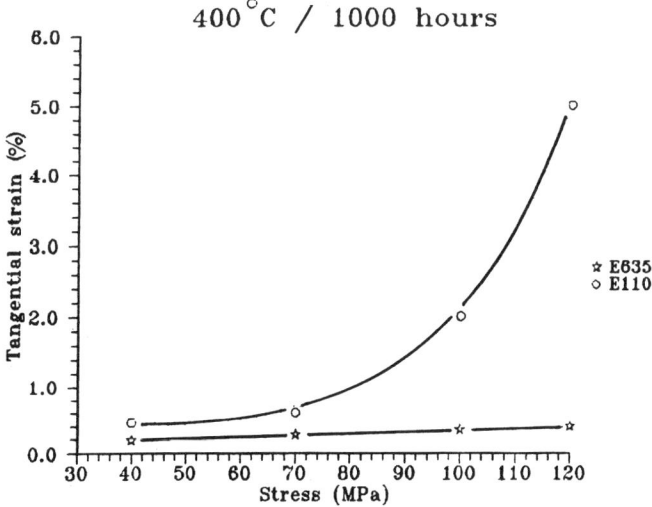

FIG. 6—*Effect of stress on tangential strain of cladding tubes at 400°C, 1000 h.*

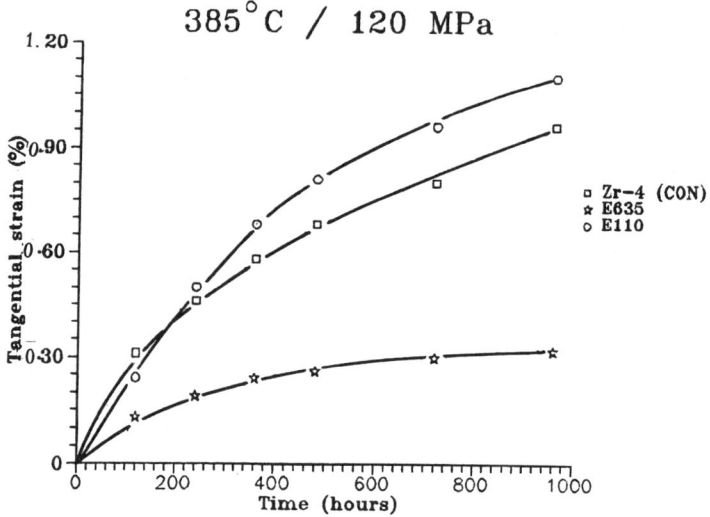

FIG. 7—*Effect of testing time on tangential strain of cladding tubes at 385°C, 120 MPa.*

(Figs. 8–10). The high creep resistance of the E635 alloy compared to that of other alloys has been corroborated by reactor-operated fuel rods and pressure tubes.

Irradiation-Induced Growth

The E635 alloy is subject to a slight irradiation-induced growth. The axial and tangential deformation of tubes due to the irradiation-effected growth is most insignificant. The alloy does not reveal a transition in growth kinetics encountered in Zr-Nb alloys and Zircaloys (Fig. 11). The irradiation-induced growth of this alloy remains quite insignificant with irradiation temperature varying between 240 to 380°C (Fig. 12).

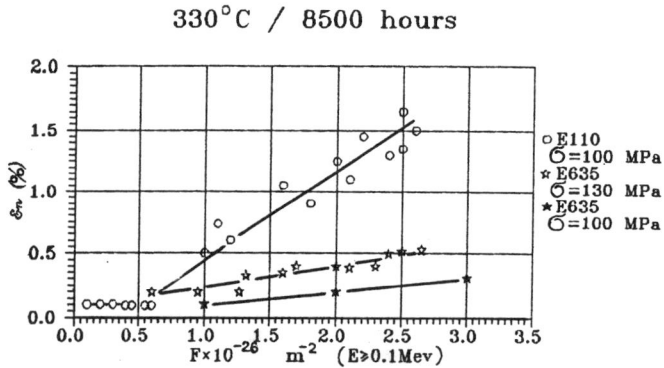

FIG. 8—*Influence of neutron fluence on creep strain of tubes at 330°C, 8500 h.*

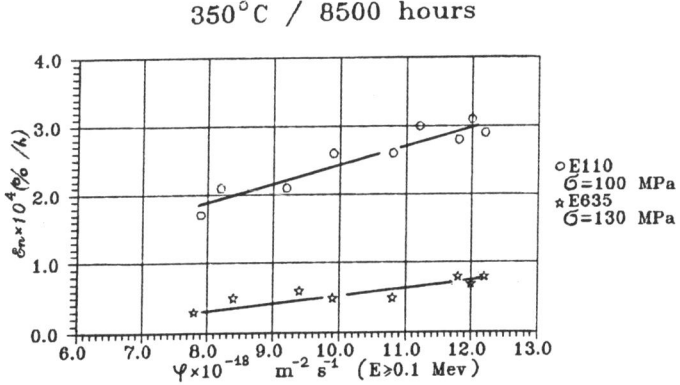

FIG. 9—*Influence of neutron flux density on creep rate of tubes at 350°C.*

Stress Corrosion Cracking in Iodine

Tests of unirradiated tubular samples and those under irradiation internally pressurized with argon containing iodine vapors at a concentration of ~13 mg/cm^2 to reach a specified pressure effecting different stresses in a tube wall indicated that the E635 alloy tubes have a much higher resistance to SCC (Fig. 13). Tests using different techniques are underway.

E635 Alloy Claddings

Rather a large number of fuel rods clad in the E635 alloy have been tested under VVER and RBMK conditions in experimental reactor loops and a commercial quality reactor. The fuel rods were tested during different time periods starting in 1970. In the loops of experimental reactors MR and MIR under RBMK and VVER conditions, pilot fuel rods were tested. They were 250 to 1000 mm long and clad in 9.15 by 0.65 mm (VVER fuel rods) and 13.65 by 0.9 mm (RBMK fuel rods) tubes. Altogether, approximately 100 fuel rods were tested. 1404 full-scale fuel rods (4000 mm long) were tested in the boiling RBMK-1000 at the Leningrad NPP.

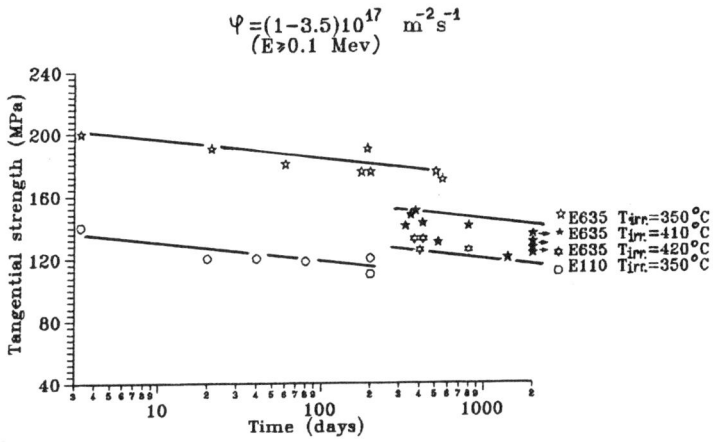

FIG. 10—*External pressure effected collapse of tubes.*

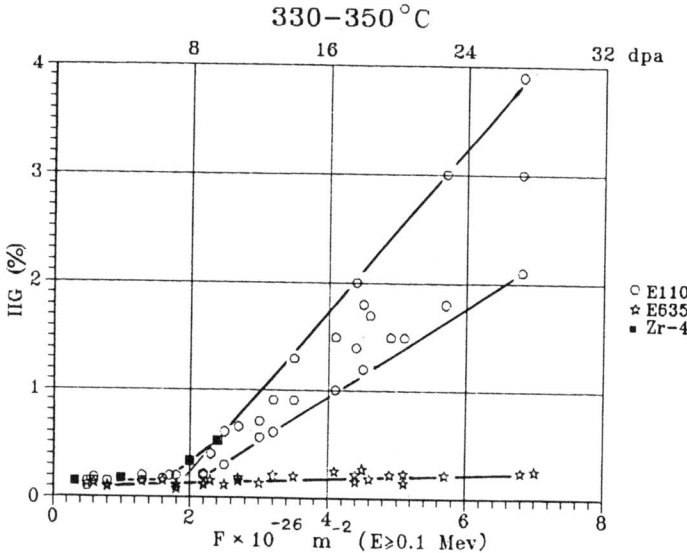

FIG. 11—*Irradiation-induced growth (IIG) of tubes versus neutron fluence at 330 to 350°C.*

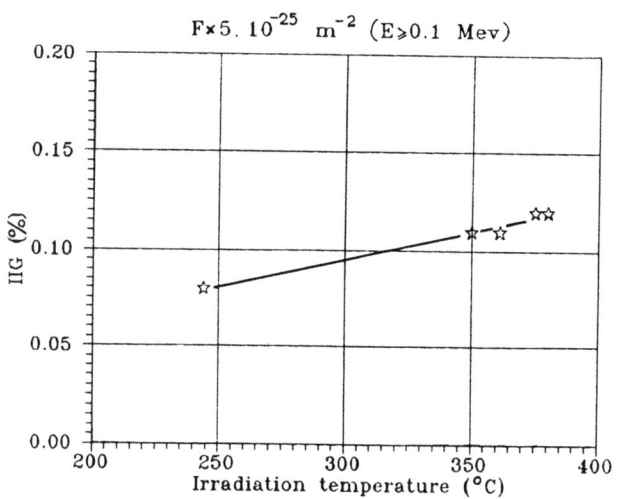

FIG. 12—*Influence of irradiation temperature on irradiation-induced growth of E635 alloy (neutron fluence of 5×10^{25} m^{-2} $E \geq 0.1$ MeV).*

The fuel rods have fully achieved their targeted lifetime (maximum burnup of 27.0 MW·day/kg U); the cladding maintained its good condition and, according to expert opinion, could operate longer [10]. Not one of the fuel rods tested in the experimental or commercial reactors had a fractured E635 alloy cladding. The typical water chemistry of VVER and RBMK is presented in (Table 6).

Corrosion under Boiling Conditions--The corrosion behavior of the E635 alloy cladding under boiling conditions typical of RBMK reactors is given as an oxide film thickness versus

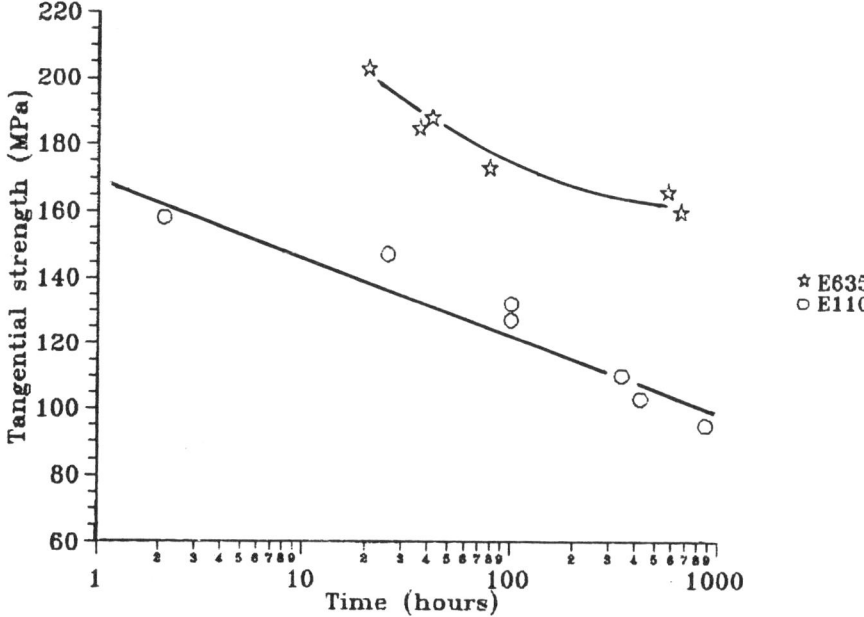

FIG. 13—*Long-term strength of cladding tubes at 400°C under conditions of irradiation $\vartheta \sim 3.5 \times 10^{16}$ m^{-2} s^{-1}, $E \geq 0.1$ MeV and SCC at iodine vapor concentration of 13 mg/cm².*

TABLE 6—*Service conditions in commercial VVER-440, VVER-1000, RBMK-1000.*

	VVER-440	VVER-1000	RBMK-1000
Water Chemistry			
pH (25°C)	≥6	5.7-10.2	6.5-8.0
Chlorides and fluorides ($Cl^- + F^-$), μg/kg	≤100	≤100	≤100
Iron (Fe), μg/kg	≤200	≤200	≤50
Copper (Cu), μg/kg	...	≤20	≤20
Oxygen (O), μg/kg	≤10	≤5	...
Hydrogen (H), mg/kg	2.7-5.4	2.7-5.4	...
Potassium + lithium + sodium ($K^+ + Li^+ + Na^+$), mg·eq/kg	...	0.05-0.45	...
Potassium (K), mg/kg	2.0-16.5	...	...
Ammonia (NH_3), mg/kg	≥5.0	≥5.0	...
Boric acid (H_3BO_3), g/kg	0-8.0	0-13.5	...
Silicic acid (SiO_2 nH_2O), μg/kg	...	...	≤1000
T/H Conditions			
Coolant temperature, °C			
Inlet	268	290	270
Outlet	290	320	289
Coolant pressure, MPa	12.5	16.0	8.0
Liner heat generation rate, kW/m	12.7	16.7	21.0

time of testing, fuel burnup, and heat flux (Fig. 14). After a ~3 year operation at the maximum fuel burnup of <60 MW·day/kg U and heat fluxes up to 170 W/cm^2, the fuel claddings were covered with uniform oxide films not more than 30 μm thick at an hydrogen content <0.01 to 0.02% mass (%exper./theor. < 11 to 22). Corrosion dependence upon test time, burnup, and heat flux is slight. As distinct from the E110 alloy, no evidence of nodular corrosion and essentially no traces of fretting corrosion at the point of contact with steel spacer grids were detected.

Corrosion under VVER Conditions—Properties of the VVER fuel E635 claddings were studied after testing in experimental reactor loops. Today tests of this kind are underway in the MIR reactor, and E635 alloy clad fuel rod assemblies are to be loaded into commercial reactors VVER-1000. In compliance with specifications, the oxygen concentration of the VVER coolant in the range of 0.005 to 0.010 ppm is tolerable (Table 6). At this oxygen content of water, the corrosion experienced by E635 alloy claddings is similar to that of E110 alloy claddings, i.e., dark oxide films and a low hydrogen pickup are encountered, but the oxide films are slightly thicker (a factor of 1.5). However, it should be kept in mind that the corrosion behavior of the E110 alloy is unique, namely, with no oxygen and no boiling of a coolant the oxide films developed on claddings after three to five-year testing are very fine, less than 10 μm [11]. At a higher oxygen concentration of water or on short-term boiling of a coolant at a fuel cladding

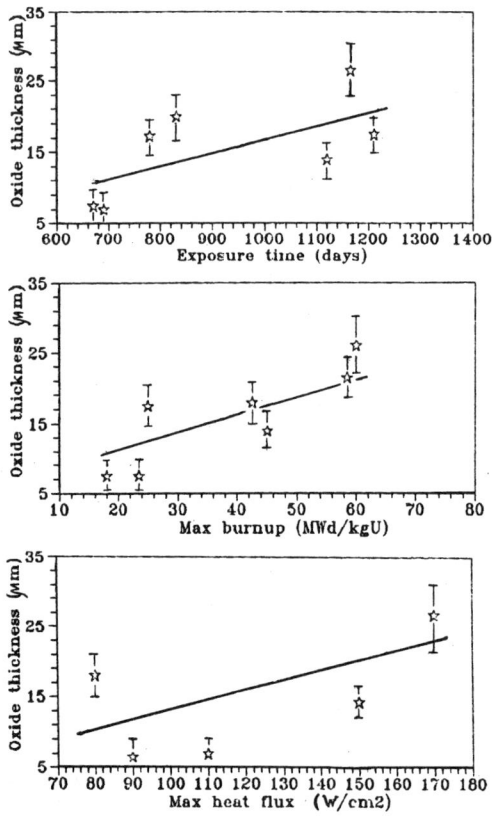

FIG. 14—*Dependence of E635 alloy corrosion on boiling RBMK conditions.*

surface, the E635 alloy demonstrates a higher corrosion resistance compared to that of the E110 alloy on the cladding, which was covered with a very fine oxide film upon which there appear centers of nodular corrosion (Fig. 15). Similarly to RBMK fuel rods under VVER conditions at the maximum fuel burnup of 45 MW·day/kg U and heat fluxes at the cladding < 200 W/cm^2 after ~650 days, the corrosion of E635 alloy cladding is slightly dependent upon variations in the above parameters (Fig. 16). In all instances, even at a high oxygen content of water (<0.02 ppm), the oxide film thickness was less than 30 μm and the hydrogen content of the cladding was not more than 0.01 to 0.02% mass.

Mechanical Properties of Cladding—The mechanical properties of fuel claddings were tested under VVER and RBMK conditions (Fig. 17). The alloy is subject to significant irradiation-induced hardening while retaining high ductility; specifically, on the average, δ_{total} is ~10 to 11% and $\delta_{uniform}$ is 4 to 5% at room temperature.

Production State of the Art

With regard to the E365 alloy, all the technological processes used to fabricate a wide range of products, namely tubes, sheets, and bars, have been developed and mastered commercially. The Russian plant (in Glasov) has mastered the production in compliance with ASTM specifications and is delivering E635 tubes for claddings, guide tubes, and spacer grids. Underway are improvements of the alloy and the technological processes as applied to the fabrication of semi-final and final products to carry out to perfection the quality and properties of products and to enhance production efficiency. The E635 alloy has a potential for further evolution to

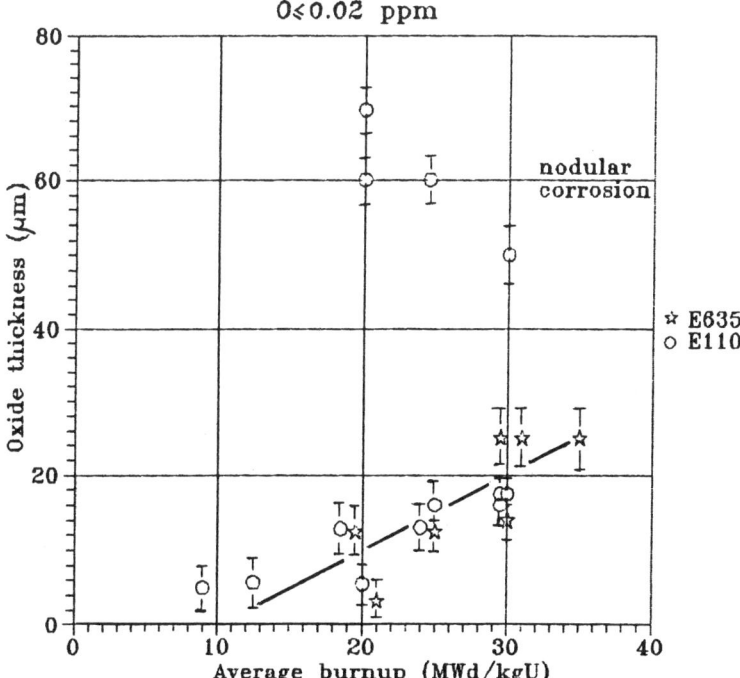

FIG. 15—*Corrosion of VVER-type fuel cladding in experimental reactor MIR.*

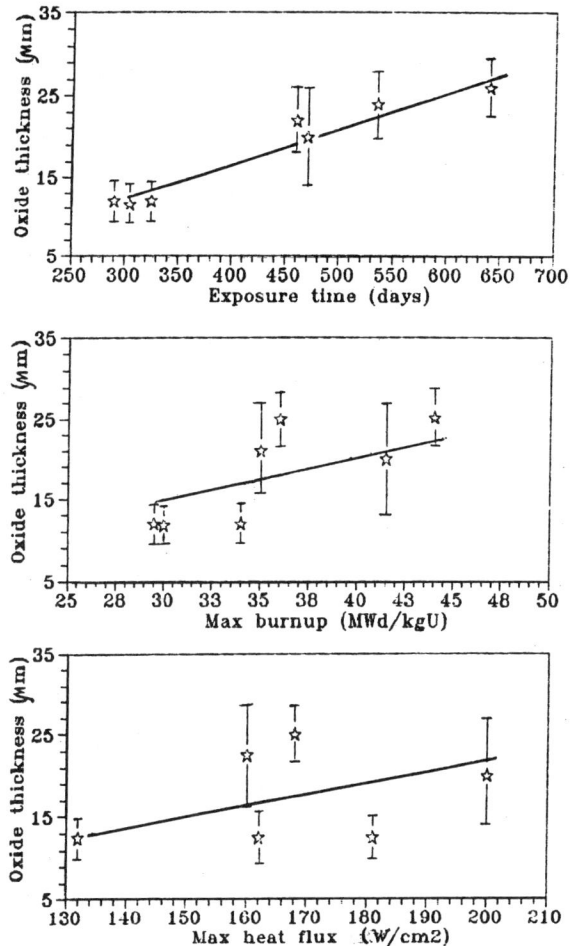

FIG. 16—*Dependence of E635 alloy corrosion on VVER conditions.*

comply with the tightening requirements of designers in terms of its higher resistance to corrosion, irradiation-induced creep, growth, and cracking.

Discussion of Results

Studies of E635 alloy characteristics demonstrate that as to the complexity of its properties this alloy largely provides for the irradiation stability of the components of the VVER and RBMK cores. As our investigations show, the higher alloy properties are promoted by alloying that is close, in our opinion, to the optimum. Niobium (1.0% mass) and iron (0.3 to 0.4% mass) introduced into the alloy allowed for bonding of those elements to form rather stable compounds [7]. As a result of the niobium-iron ratio specific for the E635 alloy and mastering of the technological process at all manufacturing stages, the structure of the E635 alloy consists of an α-solid solution and particles comprising three elements, i.e., zirconium, niobium, and iron. The β_{Nb}-phase is not available in thin-walled E635 tubing. This is its distinction from ZIRLO

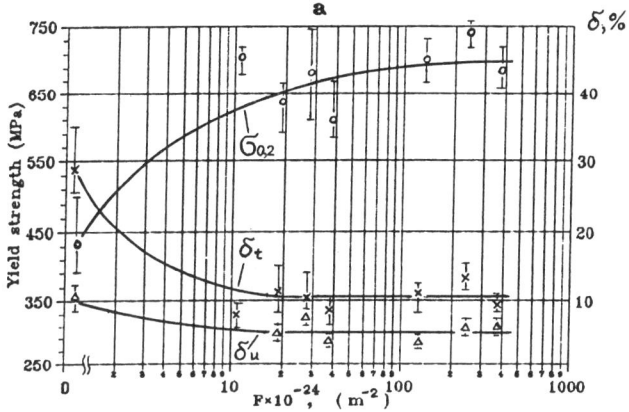

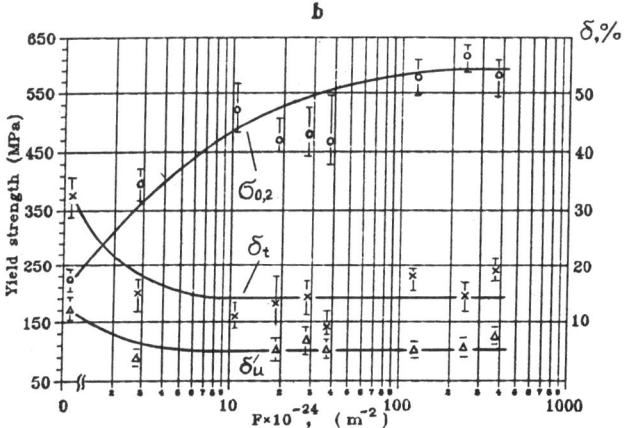

FIG. 17—*Mechanical properties of E635 alloy annular specimens cut out of fuel claddings:* (a) *tested at 20°C;* (b) *tested at 350°C.*

[8,9]. The β_{Nb}-phase was identified in individual items including those manufactured on a laboratory scale at earlier stages of the technological process, and this was reported previously [12]. Niobium and iron, which are available only in the "as bonded" condition, made it possible, in our opinion, to avoid diffusion-induced changes of phases containing those elements as individual ones, thus ensuring improved irradiation stability of the structure and the lower dependence of alloy corrosion on heat treatment and working compared to Zr-1 Nb alloys.

The high resistance of the E635 alloy to irradiation-induced creep and growth, which is much superior to those of E110 and Zircaloys, can in many respects facilitate the resolution of the problems relevant to the cladding-fuel interaction, fission products included. It is clear to us now that, distinct from other alloys, the E635 dimensional stability, particularly under irradiation, is to a larger extent explained by the essential freedom from the $\langle c \rangle$-dislocation component in its structure. This interesting phenomenon, reported earlier by us [13], was corroborated by our subsequent investigations [7]. More comprehensive studies are needed.

Analysis of the data available on the E635 alloy reveals that its substantial advantage over binary niobium-containing alloys and Zircaloys is to show up at higher fluences and exposures due to the higher resistance and stability of its structure under irradiation. Comparison between the kinetics of alloy corrosion in autoclaves and in-pile points to good agreement between the results of corrosion testing in autoclaves at water temperature of 350°C and under RBMK conditions where the temperature of the cladding outer surface is ~305°C.

The corrosion of E635 alloy cladding under VVER-1000 conditions correlates well with corrosion observed in autoclave tests at the temperature of lithium containing water of ~360°C. The temperature of the fuel rod cladding outer surface is ~350°C in this reactor. Since the in-pile tests of the E635 cladding tubes did not reveal nodular corrosion, it is difficult to interpret the results of the autoclave tests in steam at 400 and 500°C. These results most likely describe E635 alloy behavior in inadvertent situations when local overheating of the cladding outer surface is possible. In a sense, E635 alloy corrosion can be accounted for in terms of an evolution of the phase composition of growing oxide films [12,14]. The beneficial part played by metallic phases available in oxide films, which is the case for alloys of the E635 type, has a softening effect on an oxide film, which promotes stress relaxation in zirconium dioxide without its rupture. According to Ref 14, the resultant corrosion resistance of the alloy is governed by the competing effects of oxide film softening under the action of metallic phases and their heterogenization by non-metalllic ones. Lithium available throughout an oxide film on Zircaloy, noted in Ref 15, can promote oxide film heterogenization, thus enhancing corrosion. The data available do not allow for an adequate explanation of the mechanism of the E635 alloy corrosion under the conditions studied.

To gain a scientific understanding of the revealed characteristics of the E635 alloy, the relevant issues need to be further studied. This will allow for a more credible prediction of the service reliability of zirconium components. We are continuing our efforts to resolve this problem and to further improve the quality of E635 alloy products.

Conclusion

Zirconium alloying with tin, niobium, and iron in the quantity of 1.0 to 1.4, 0.9 to 1.1, and 0.3 to 0.5% mass, respectively, and worked out technological processes resulted in the creation of a corrosion-resistant and highly irradiation-resistant zirconium alloy, E635. As far as properties needed for long-term operation of zirconium components in water-cooled reactor cores are concerned, the E635 alloy is superior to known commercial zirconium alloys. The alloy can be recommended for use as a material for fuel claddings, guide tubes, spacer grids, and other core components in water-cooled reactors of VVER, RBMK, PWR, and BWR types.

References

[1] Zaimovsky, A. S., Nikulina, A. V., et al. "Zr-Base Alloy," A. C. No. 64815, *Bulletin of Inventions*, No. 28, 1992.

[2] Amaev, A. D., Anisimova, I. A., Nikulina, A. V., et al., "Corrosion Behavior of Zr Alloys in Boiling Water under Irradiation," *Proceedings*, Fourth International Conference on the Peaceful Uses of Atomic Energy, Geneva, Vol. 10, 1971, p. 537.

[3] Zaimovsky, A. S., Nikulina, A. V., et al., "Influence of Zr-Alloy Chemical Composition and Microstructure on Its Creep Resistance," *Proceedings*, Reactor Science Conference, 1978, Vol. 6, p. 52.

[4] Kobylyansky, G. P., Shamardin, V. K., and Stukalov, A. I., "Irradiation Induced Growth of Zr-Alloy Tubes at High Neutron Fluences," *Issues of Atomic Science and Engineering, Series: Physics of Irradiation Induced Damages and Irradiation Material Science*, 1990, Vol. 2, No. 53, pp. 32–34.

[5] Castaldelli, L., Fizzotti, C., and Lunde, L., "Long-Term Test Results of Promising New Zirconium

Alloys," *Zirconium in the Nuclear Industry: 5th Conference, ASTM STP 754*, American Society for Testing and Materials, West Conshohocken, PA, 1982, pp. 105–125.

[6] Nikulina, A. V., Markelov, V. A., Gusev, A. Y., et al, "Zr-1 Sn-1 Nb-0.5-Fe as a Material for Pressure Tubes of RBMK Type," *Issues of Atomic Science and Engineering, Series: Materials Science and New Materials*, 1990, Vol. 2, No. 36, pp. 58–66.

[7] Shishov, V. N., Nikulina, A. V., Markelov, V. A., et al., "Influence of Neutron Irradiation on Dislocation Structure and Phase Composition of Zr-Base Alloys," this publication.

[8] Sabol, G. P., Kilp, G. P., Balfour, M. G., and Roberts, E., "Development of a Cladding Alloy for High Burnup," *Zirconium in the Nuclear Industry: Eighth International Symposium, ASTM STP 1132*, American Society for Testing and Materials, West Conshohocken, PA, 1988.

[9] Sabol, G. P., Comstock, R. J., Weiner, R. A., Larouere, P., and Stanutz, R. N., "In-Reactor Corrosion Performance of ZIRLO^{tm} and Zircaloy-4," *Proceedings*, Zirconium in the Nuclear Industry: Tenth International Symposium, *ASTM STP 1245*, American Society for Testing and Materials, West Conshohocken, PA, 1993.

[10] Shamardin, V. K., Pokrovsky, A. S., Kobylyansky, G. P., Stupina, L. N., and Maershina, G. I., "Results of Investigations of Life-Time Operated RBMK Fuel Rods Clad in Zr-Nb, Zr-Nb-Sn-Fe and Zr-Sn-Fe Alloys," Preprint, NIIAR (Research Institute of Nuclear Reactors), Vol. 8, No. 654, 1985, Dimitrovgrad.

[11] Smirnov, A. V., Kuzmin, V. I., and Dubrovin, K. P., "VVER-1000 and VVER-440 Fuel Operation Experience," *Proceedings*, International Topical Meeting on LWR Fuel Performance, Florida, 16–19 April 1994, pp. 31–44.

[12] Glazkov, A. G., Grigor'ev, V. M., Kon'kov, V. F., Moinov, A. S., Nikulina, A. V., and Sidorenko, V. I., "Corrosion Behaviour of Zr-Nb-Sn-Fe Alloy," *Proceedings*, IAEA Technical Committee Meeting on Fundamental Aspects of Corrosion on Zirconium Base Alloys in Water Reactor Environments, IAEA, Vienna, 1990, pp. 158–164.

[13] Nikulina, A. V., Markelov, V. A., Panchenko, V. L., et al., "Irradiation-Induced Changes in Zr-1Sn-1Nb-0.4Fe Alloy Structure," *Proceedings*, International Conference on Evaluation of Metal Microstructure under Irradiation, Canada, Muskoqee, 29 Sept.–2 Oct. 1992.

[14] Khaikovsky, A. A. and Abramtsev, V. N., "Interrelation between Kinetics of Zr-Alloy Corrosion and Oxide Film Composition," *Proceedings*, International Conference on Reactor Materials Science, Alushta, 1978, Vol. 5, pp. 216–254.

[15] Ramasubramanian, N., et al., "Lithium Uptake and the Accelerated Corrosion of Zirconium Alloys," *Zirconium in the Nuclear Industry: Eighth International Symposium, ASTM STP 1132*, American Society for Testing and Materials, West Conshohocken, PA, 1988.

DISCUSSION

R. B. Adamson[1] (written discussion)—Could you tell us the final surface finishing process of your RBMK fuel cladding? For example, etching, autoclave, or grinding?

A. V. Nikulina et al. (authors' closure)—In this work, we used RBMK fuel cladding with final surface finishing in autoclaves at 300°C, 100 h.

G. P. Sabol[2] (written discussion)—Both the E635 and Zirlo display low irradiation growth. Furthermore, Shishov has shown that the formation of $\langle c \rangle$ dislocations is retarded in Alloy 635. What factors inhibit the formation of $\langle c \rangle$ dislocations and, hence, irradiation growth?

A. V. Nikulina et al. (authors' closure)—Today we cannot unambiguously say what factors inhibit the formation of $\langle c \rangle$ dislocations in E635 alloy. It is likely that the important part is played by the favorable combination of alloying elements that, in our opinion, is close to the optimal one. We also continue our investigations to get an answer to the question put by you. And we hopefully think that in future this behavior of the alloys will find its explanation.

A. Strasser[3] (written discussion)—What are the advantage of the high Fe content (0.3 to 0.5%) of Alloy 635 compared to the lower Fe content (0.1%) of Zirlo?

A. V. Nilulina et al. (authors' closure)—The higher iron content of E635 alloy makes it possible to find almost all Nb available to form rather stable ternary compounds $Zr(Nb,Fe)_2$ having the hexagonal lattice and to avoid the formation of β-Nb observed in Zirlo. In our opinion, this fact provides for the higher stability of E635 alloy structure and properties under irradiation.

---

[1] GE Nuclear Energy, Pleasanton, CA.
[2] Westinghouse NMO, Energy Center, Pittsburgh, PA.
[3] The S. M. Stoller Corp., Pleasantville, NY.

Otto A. Besch,[1] Suresh K. Yagnik,[2] Keith N. Woods,[3] Craig M. Eucken,[4] and E. Ross Bradley[5]

Corrosion Behavior of Duplex and Reference Cladding in NPP Grohnde

REFERENCE: Besch, O. A., Yagnik, S. K., Woods, K. N., Eucken, C. M., and Bradley, E. R., "**Corrosion Behavior of Duplex and Reference Cladding in NPP Grohnde,**" *Zirconium in the Nuclear Industry: Eleventh International Symposium, ASTM STP 1295*, E. R. Bradley and G. P. Sabol, Eds., American Society for Testing and Materials, 1996, pp. 805–824.

ABSTRACT: The Nuclear Fuel Industry Research (NFIR) Group undertook a lead test assembly (LTA) program in NPP Grohnde PWR in Germany to assess the corrosion performance of duplex and reference (i.e., non-duplex) cladding. Two identical 16 by 16 LTAs, each containing 32 peripheral test rods, completed four reactor cycles, reaching a peak rod burnup of 46 MWd/kgU. The results from poolside examinations performed at the end of each cycle, together with power histories and coolant chemistry, are reported.

Five different cladding materials [three types of duplex (D1–D3) and two types of reference (R1, R2)] were characterized during fabrication. The corrosion performance of the cladding materials was tracked in long-term tests (~900 days) in high-pressure, high-temperature autoclaves. The relative ranking of corrosion behavior in such tests corresponded well with the in-reactor corrosion performance. The extent and distribution of hydriding in duplex and reference specimens during the autoclave testing has been characterized.

The in-reactor corrosion data indicate that the low-tin (1.3% Sn) Zircaloy-4 reference cladding, R2, had an improved corrosion resistance compared to high-tin (1.5% Sn) Zircaloy-4 reference cladding, R1. Two types of duplex cladding, D1 (Zr-2.5% Nb) and D2 (Zr-0.4% Fe-0.5% Sn), showed an even further improvement in corrosion resistance compared to R2 cladding. The third duplex cladding, D3 (Zr-4 + 1.0% Nb), had significantly less corrosion resistance, which was inferior to R1. The in-reactor and out-reactor corrosion performances have been ranked.

KEYWORDS: corrosion, zirconium alloys, fuel cladding, in-reactor performance, alloy development, autoclave, water, steam, hydrides, microstructure, nuclear applications, radiation effects

The current trend in the nuclear industry to extend fuel burnup is driven by economic incentives related to fuel-cycle costs, spent-fuel storage requirements, and improved uranium utilization. In PWRs, the trend is often accompanied by new operating regimes, including higher pH and higher temperatures for the primary coolant. Under these operating regimes, the waterside corrosion of Zircaloy-4 could be a performance-limiting factor, especially for high-temperature plants. To address this concern and to limit general in-reactor cladding corrosion, the fuel industry has developed a number of material solutions ranging from the optimization of chemical composition and microstructure of the cladding to the introduction of new cladding alloys.

---

[1] Senior nuclear fuel engineer, PreussenElektra AG, Tresckowstrasse 3, 30457 Hannover (Germany).
[2] NIFR program manager, Electric Power Research Institute, Nuclear Power Group, Palo Alto, CA 94303.
[3] Technology development manager, Siemens Power Corporation, Richland, WA 99352.
[4] Project leader, Teledyne Wah Chang, Albany, OR 97321.
[5] Senior development metallurgist, Sandvik Special Metals Corp., Kennewick, WA 99336.

The Nuclear Fuel Industry Research (NFIR) Group undertook a lead test assembly (LTA) program in NPP Grohnde PWR in Germany to compare the corrosion performance of other zirconium-based alloys with reference Zircaloy-4 cladding. The concept was to produce a new zirconium-based alloy at the outer surface of normal Zircaloy-4 cladding to facilitate the irradiation of the former alloy in a commercial PWR. This concept was modified to the use of duplex cladding, which is comprised of two layers of distinct compositions. The inner 90% of the cladding wall is Zircaloy-4; but the outer 10% of thickness consists of an alloy of entirely different composition. By fabricating the outer 10% of thickness of a more corrosion-resistant alloy than Zircaloy-4 while retaining the 90% inner layer as Zircaloy-4 for its proven strength and neutron economy, the duplex cladding offers a cladding design alternative for PWR application.

The duplex and reference (i.e., non-duplex) cladding was manufactured by Sandvik Special Metals (SSM) from Teledyne Wah Chang (TWC) TREXs. Siemens Power Corporation (SPC) supplied two identical LTAs, each containing 32 peripheral test rods. These LTAs, introduced in the reactor in 1989, have since completed four reactor cycles, reaching a peak rod burnup of 46 MWd/kgU.

This paper reviews fabrication processes as well as pre-characterization of all cladding materials in terms of chemical composition, microstructure, and mechanical properties. Next, out-reactor corrosion performance of the cladding materials in long-term tests in high-pressure, high-temperature autoclaves is reported. The results from poolside examinations performed at the end of each reactor cycle, power histories, and coolant chemistry are also discussed. Finally, the in-reactor and out-reactor corrosion performances of the alloys have been ranked.

Fabrication and Pre-Characterization

Chemical Compositions

The nominal alloy compositions for the outer layers in three of the duplex claddings were as follows: D1 (Zr-2.5% Nb), D2 (Zr-0.4% Fe-0.5% Sn), and D3 (Zr-4 + 1.0% Nb). The base layer in all three duplex variants was high-tin (~1.5% Sn) Zircaloy-4. The two reference cladding variants R1 and R2 were nominally high-tin and low-tin Zircaloy-4, respectively.

The results from ingot chemical analyses are summarized in Table 1, where the listed values represent from two to six axial locations depending on the ingot size. Impurity levels for elements not listed in Table 1 were all within ASTM limits for zirconium alloys given in ASTM Specification for Zirconium and Zirconium Alloy Ingots for Nuclear Application (B 350).

Tubeshell Fabrication

The inner Zircaloy-4 material for the duplex tubes was taken from a 685-mm-diameter, triple-melted ingot. This ingot was heated to 1323 K and press forged to 240 mm. The forged pieces

TABLE 1—*Alloy composition.*

Cladding Variant	Sn	Fe	Cr	Nb	O
D1	...	...	...	2.57–2.61	0.102–0.114
D2	0.47–0.50	0.40–0.42	...	...	0.102–0.105
D3	1.43–1.46	0.2	0.1	1.0	0.097–0.098
Base Zr-4	1.57–1.59	0.20–0.22	0.10–0.11	...	0.121–0.136
R1	1.54–1.59	0.19–0.21	0.10–0.11	...	0.123–0.133
R2	1.28–1.31	0.22–0.23	0.11–0.12	...	0.125–0.133

were beta quenched at this size and reheated to 998 K for a minimum of 2 h and then rotary forged to 150 mm in diameter.

The outer layer material for D1 originated from a large (575-mm-diameter) double-melted ingot that had been heated to 1303 K and press forged to 240 mm in diameter. It was reheated to 1303 K for 20 min and water quenched. Subsequently, the material was heated to 1073 K and rotary forged to a size of ~210 mm in diameter. One small (200 mm in diameter) ingot each for D2 and D3 outer layer materials were specially double vacuum arc melted, reheated to 1323 K for 30 min, and press forged to produce a final billet size ~160 mm in diameter. The as-forged billets were reheated to 1323 K for 15 min and quenched in water.

Finally, an annular billet of 153 mm OD by 141 mm ID was machined from the as-forged (D1) or as-quenched (D2 and D3) pieces of all three outer layer materials, D1-D3.

The inner and outer components were machined, cleaned by pickling in HF-HNO$_3$ acid, assembled, and electron-beam welded (Fig. 1) at the ends to seal the annulus. Billets were heated to 923 K for ~5 min and extruded to form 63.5 mm OD by 10.9-mm wall-thickness tubeshells. Subsequently, the D1 and D3 tubeshells were vacuum annealed at 853 K for 2 h, while D2 was vacuum annealed at 893 K for 2 h.

The reference Zircaloy-4 tubeshells were manufactured from triple-melted ingots by standard fabrication processes.

Cladding Fabrication

Final cladding tubes of all five variants were produced by cold pilgering and vacuum annealing several times to obtain final dimensions. All variants were delivered in the stress-relieved condition with abraded OD and etched ID surfaces. The final anneal conditions and accumulated anneal parameters ($Q/R = 40\ 000$ K) for the five variants are summarized in Table

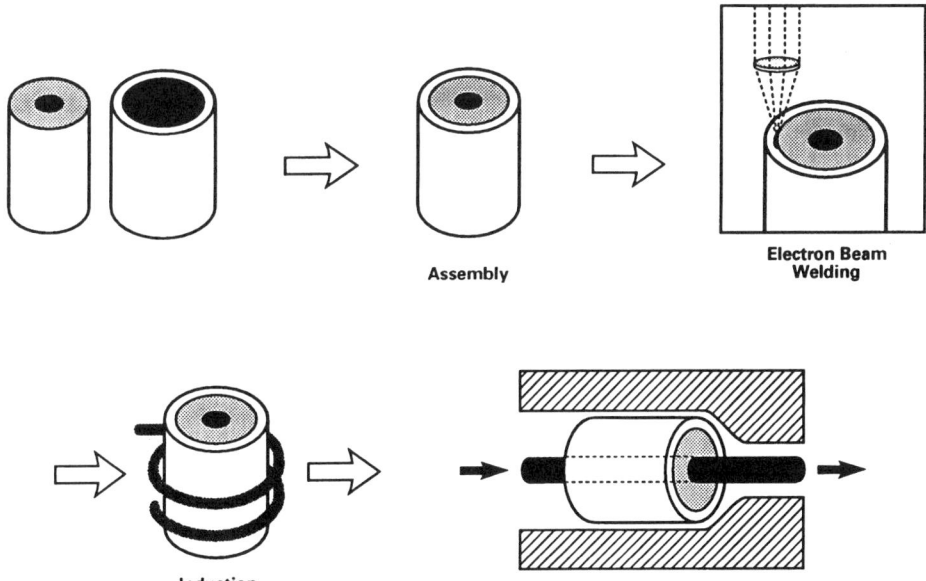

FIG. 1—*Duplex cladding fabrication process.*

TABLE 2—*Fabrication summary.*

Cladding Variant	Cumulative Anneal Parameter, h	Final Anneal Conditions, K/h
D1	6.7E-17	764/4
D2	5.0E-18	764/4
D3	8.5E-20	764/4
R1	1.5E-17	764/4
R2	1.5E-17	764/4

2. The cumulative anneal parameter (ΣA) values for the duplex cladding variants in Table 2 pertain to the outer layers.

Mechanical Properties

Tensile and closed-end burst tests were carried out on unirradiated cladding specimens at both room temperature and 655 K. As summarized in Table 3, the results from all five cladding variants fall within the same scatter band with respect to both strength and ductility. The strength of D2 cladding is somewhat lower than those containing niobium in the outer layer, but all variants were well above the design specification for strength and ductility.

Microstructure

The microstructures of all cladding variants were examined by optical metallography, scanning electron microscopy (SEM), and transmission electron micrography (TEM). Typical mi-

TABLE 3—*Summary of mechanical properties*

	655-K Tensile Properties			655-K Burst Properties	
Cladding Variant	Yield Strength, MPa	Ultimate Tensile Strength, MPa	Total Elongation, %	Ultimate Hoop Strength, MPa	Total Circular Elongation, %
D1	336	437	27	504	16
D2	314	406	26	466	18
D3	364	460	26	524	14
R1	349	445	24	495	18
R2	320	424	24	478	14

	RT Tensile Properties			RT Burst Properties	
Cladding Variant	Yield Strength, MPa	Ultimate Tensile Strength, MPa	Total Elongation, %	Ultimate Hoop Strength, MPa	Total Circular Elongation, %
D1	566	752	22	907	31
D2	542	731	21	883	27
D3	587	766	21	925	27
R1	580	773	22	921	22
R2	576	752	22	894	24

crographs showing the basic microstructure are presented in Fig. 2 where the micrographs for the duplex cladding illustrate the Zircaloy-4/outer layer interface. Some general features are:

- The outer layer in D1 has a finer grain structure than the Zircaloy-4 base alloy.
- The outer layer in D2 has equiaxed grains, indicating that the material is fully recrystallized.
- The outer layer in D3 has a microstructure very similar to base Zircaloy-4 material, making it difficult to discern the interface.

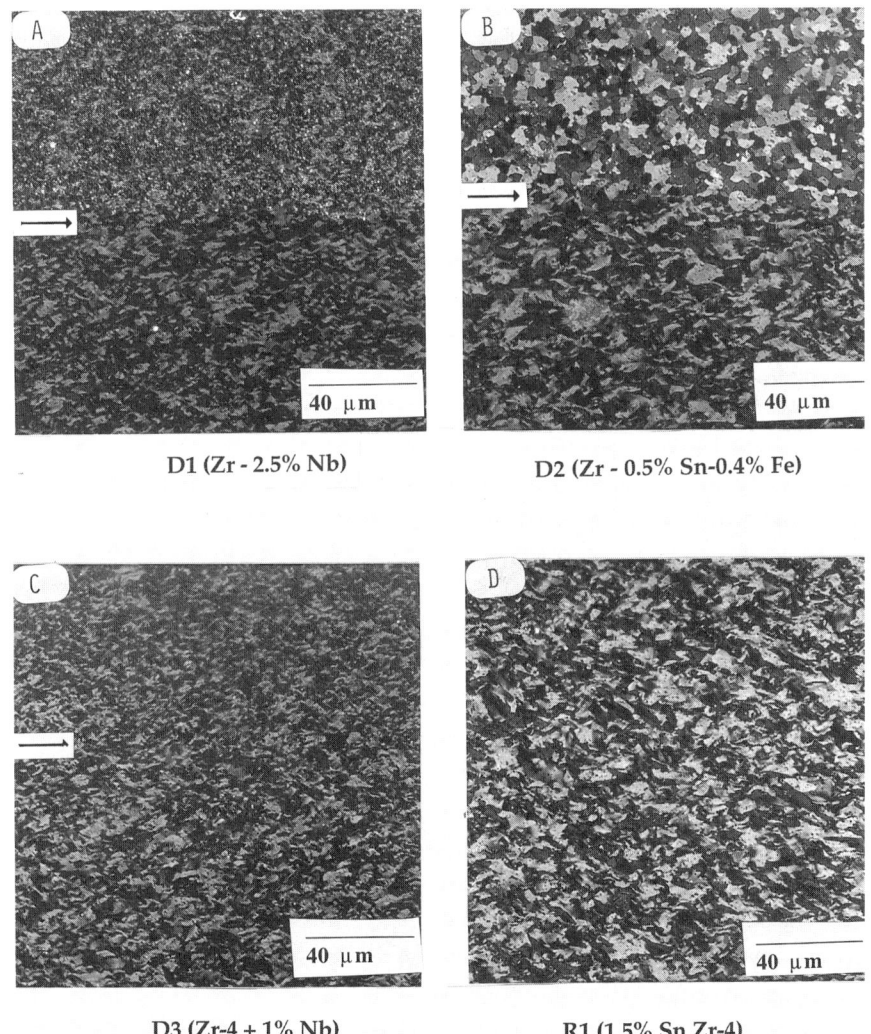

FIG. 2—*Optical micrographs of as-fabricated cladding variants [X575]. The outer layer is at the top of the micrographs for duplex variants, with the arrow marking the interface. (Figure reduced one third for publication.)*

TABLE 4—*Microstructure and intermetallic particle size measurements.*

Cladding Variant	Degree of Recrystallization		Average Particle Diameter, μm	
	Zr-4 Base, %	OD Clad, %	SEM	STEM[a]
D1	22	30	0.12	0.07
D2	9	100	0.35	0.25
D3	Not Measured	2	0.16	0.09
R1	2	...	0.23	0.17
R2	10	...	0.22	0.18

[a] Data provided by Institute of Nuclear Energy Research, Taiwan.

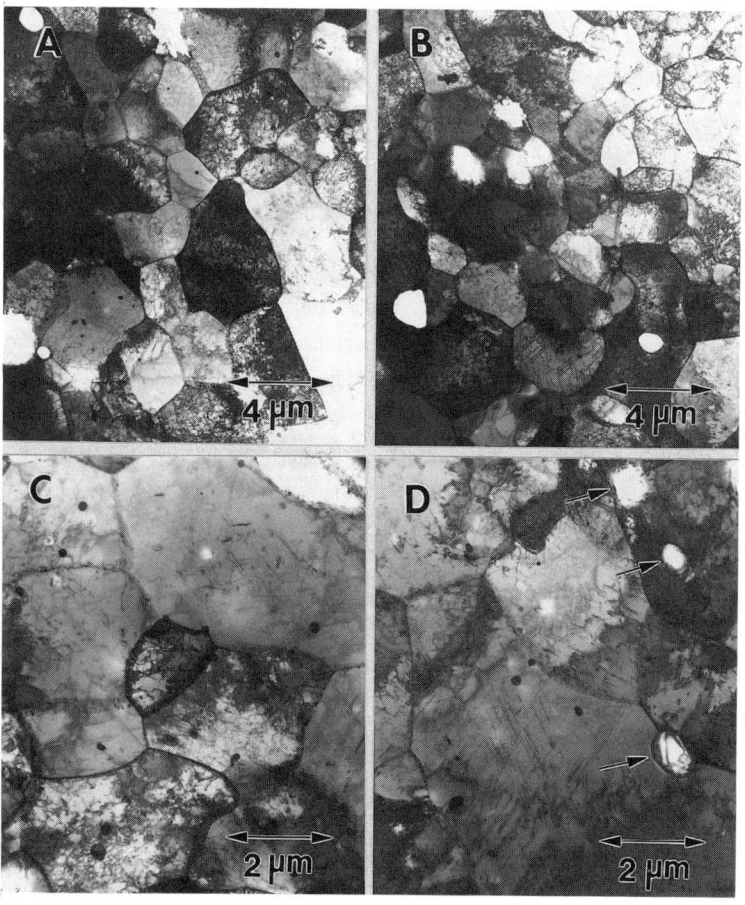

FIG. 3—*TEM micrographs showing the outer layer microstructure in D2 (Zr-0.5% Sn-0.4% Fe).*

- The microstructures of the base Zircaloy-4 and the reference claddings R1 and R2 are similar to that in typical stress-relieved fuel cladding.

The degree of recrystallization was estimated from TEM based on two or three thin foil specimens. A point-counting technique was used to measure the percent recrystallization on 15 micrographs, and the results are summarized in Table 4. This analysis verified that the outer layer in D2 is fully recrystallized (see Fig. 3), while the degree of recrystallization in other outer alloys, reference claddings, and Zircaloy-4 base alloy varied between 2 to 30%. Small differences within this range may not be significant considering the extremely small volume of material examined by this technique.

Particle (second-phase precipitate) size determinations from SEM and TEM are also summarized in Table 4. Particle size in the two niobium-containing duplex alloys is significantly smaller than the Zircaloy-4 reference materials. Further, the particle size measured by SEM is consistently larger than measured by TEM. This can be attributed to a combination of relatively poor detection of small particles in SEM and the difficulty in preparing and examining thin TEM specimens with large precipitates. The true precipitate size is expected to be between the two measured values.

Figures 4 and 5 show bright field TEM images of duplex alloy D1. Compared to the CANDU pressure tube alloy [1] of the same composition, the microstructure of D1 shows spherical and elongated beta niobium precipitates (Fig. 4) and some non-uniform alpha grain growth (Fig. 5). The precipitates are rich in Nb and have a bcc structure.

The outer layer alloy in D2 contains mostly orthorhombic Zr_3Fe particles. Other types of precipitates Zr_2Fe (bct) and $Zr(Fe,Cr)_2$ were also found in D2 but were less abundant. By contrast, the outer layer alloy in D3 contains hexagonal $(Zr_{1-x}Nb_x)(Fe_{1-y}Cr_y)_2$ type precipitates. Such precipitates are expected with niobium addition to Zircaloy-4 composition and have been reported previously, for example, by Miyake and Gotoh [2]. On the other hand, precipitates of this type have not been found in ZIRLO [3].

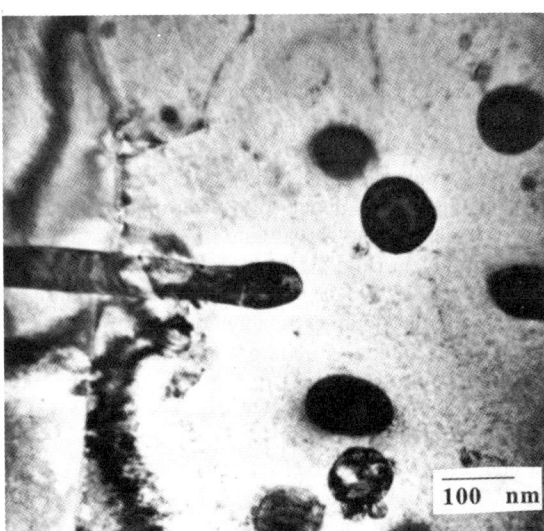

FIG. 4—*Bright field TEM image showing precipitates in cladding variant D1 (Zr-2.5% Nb) [X150 000]. (Figure reduced one third for publication.)*

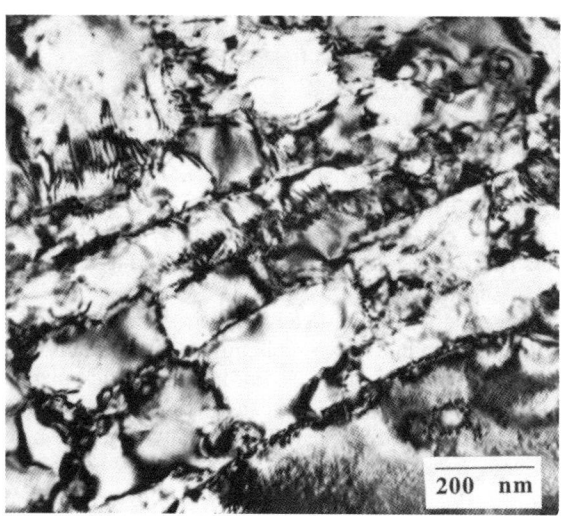

FIG. 5—*Bright field TEM image showing recovered and recrystallized regions in variant D1 (Zr-2.5% Nb) [X100 000]. (Figure reduced one third for publication.)*

Out-Reactor Corrosion and Hydrogen Pickup

Tubular samples of all cladding variants were autoclave tested in 633 K water and 673 K steam environments for exposures times up to 924 and 787 days, respectively. Since the sample ends were not closed, the inner and the outer surfaces of the sample corroded at different rates, especially for the duplex tubing. For samples autoclaved in 633 K water, metallographic sections were prepared from samples at different exposure times, and the OD oxide thickness was measured. For samples autoclaved in 673 K steam, eddy current (EC) measurements of oxide thickness were made at different time intervals, and metallographic sections were prepared only at the conclusion of the test. The autoclave corrosion results for 633 and 673 K conditions are shown graphically in Figs. 6 and 7, respectively.

As shown in Fig. 6, in the 633 K water environment, all variants generally showed a linear increase in oxide thickness with time. Regression analysis confirmed the linear growth rates, with correlation coefficients better than 0.77 (Table 5). The oxide thickness data for duplex variants D1 and D2 are comparable and significantly lower than for the remaining three variants. The low-tin reference cladding R2 has about 20% lower oxide thickness than R1 or D3 after 924 days exposure, and the steady state growth rate is about 30% lower than for R1 or D3.

In the 673 K steam environment (Fig. 7), the high-tin reference cladding R1 deviated from linear corrosion kinetics after about 500 days exposure. Optical metallography on this cladding variant at the end of the test indicated an extremely non-uniform (30 to 150-μm) oxide layer, while the other cladding variants showed quite uniform oxide thicknesses. It is believed that the non-linearity of the corrosion data for R1 in Fig. 7 is consistent with the observed non-uniform oxide growth.

The relative ranking of the cladding variants is different between the two autoclave conditions. The D3 duplex and the R1 reference cladding performed poorly in both environments. The major difference in the 673 K autoclave data is the relative position of R2 reference cladding compared to the D1 and D2 duplex variants. The R2 reference cladding is no longer much worse than the D1 and D2 variants, but exhibits an oxide thickness midway between these two duplex variants.

At the conclusion of autoclave testing, the hydrogen uptake was measured by hot extraction,

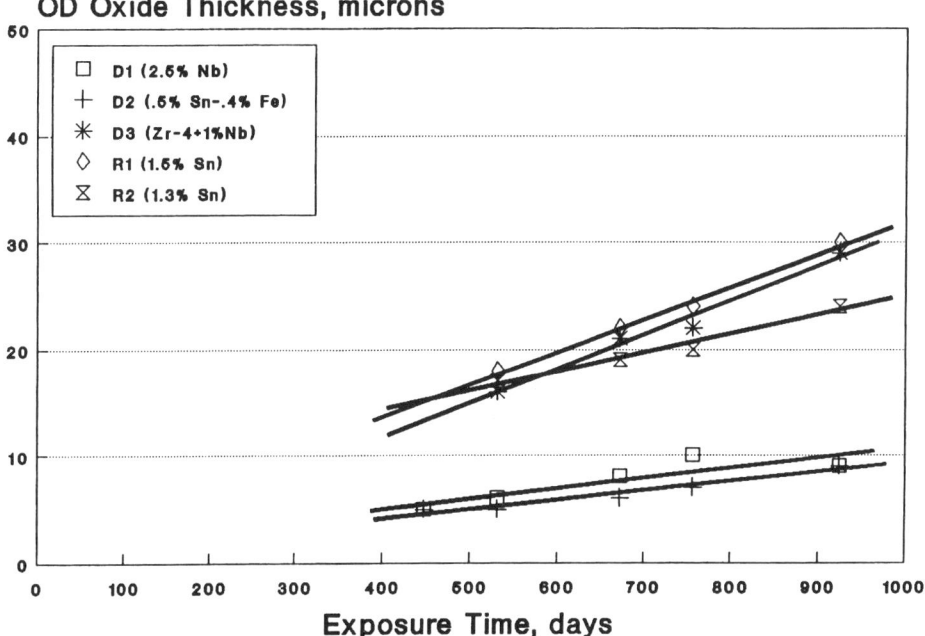

FIG. 6—*Oxide thickness by metallography as a function of exposure time in the 633-K water environment.*

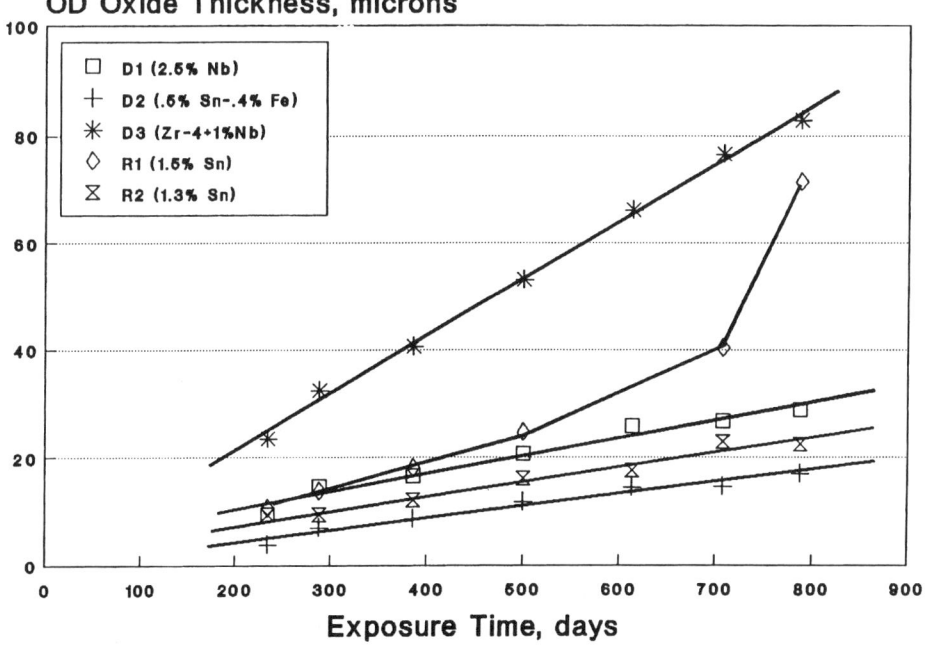

FIG. 7—*Oxide thickness by eddy current as a function of exposure time in the 673-K steam environment.*

TABLE 5—*Summary of oxide growth regression analysis.*

	360°C Water Autoclave					400°C Steam Autoclave				
	D1	D2	D3	R1	R2	D1	D2	D3	R1[a]	R2
Exposure, days	924	924	924	924	924	787	787	787	787	787
Oxide thickness										
Eddy Current μm						29	17	83	71	23
Metallography μm	9	9	29	30	24	25	13	79	...[b]	20
Regression analysis										
Intercept	1.10	0.62	−1.50	1.60	7.20	3.20	0.16	−0.16	−1.20	1.60
Slope	0.01	0.01	0.03	0.03	0.02	0.03	0.02	0.11	0.05	0.03
Correlation Coefficients R^2	0.77	0.94	0.99	0.99	0.97	0.96	0.97	1.00	0.99	0.98

[a] Regression Analysis 235 to 500 days' exposure.
[b] Nonuniform Oxide (30 to 150 μm).

and microstructures of selected corroded samples were examined by optical microscopy and SEM. Total hydrogen content (pickup from both OD and ID surfaces) as a function of weight gain at 633 and 673 K are shown in Figs. 8 and 9, respectively. Excluding the niobium-containing duplex alloys, a good correlation exists between weight gain and hydrogen concentration at both these temperatures. Linear regression of data for non-niobium-containing variants in Figs. 8 and 9 gave correlation coefficients of 0.77 and 0.96, respectively. The niobium-containing samples exhibit a relatively lower hydrogen pickup at equivalent weight gains.

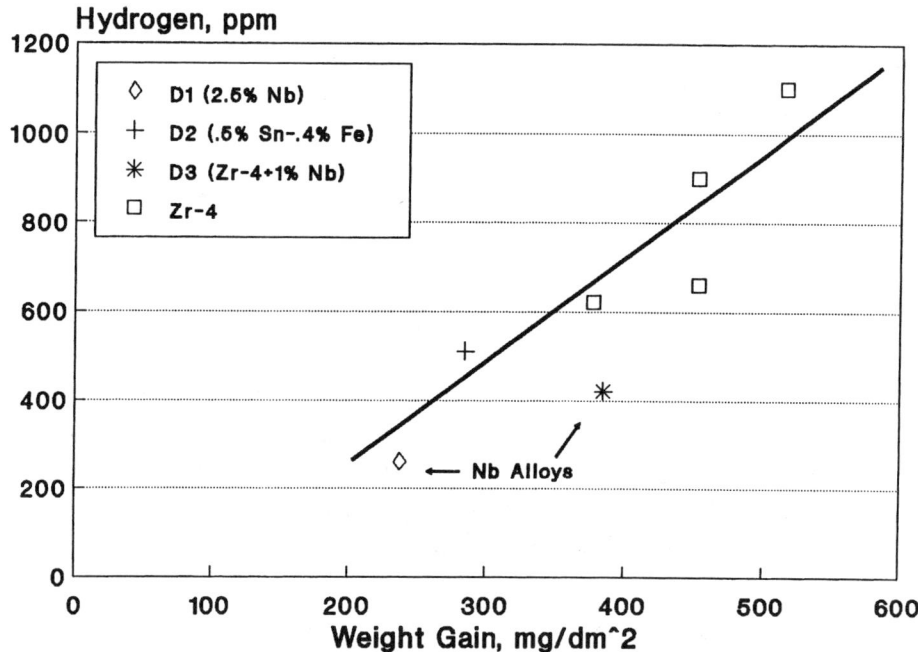

FIG. 8—*Hydrogen concentration as a function of weight gain for cladding variants after long-term autoclave tests in the 633-K water environment.*

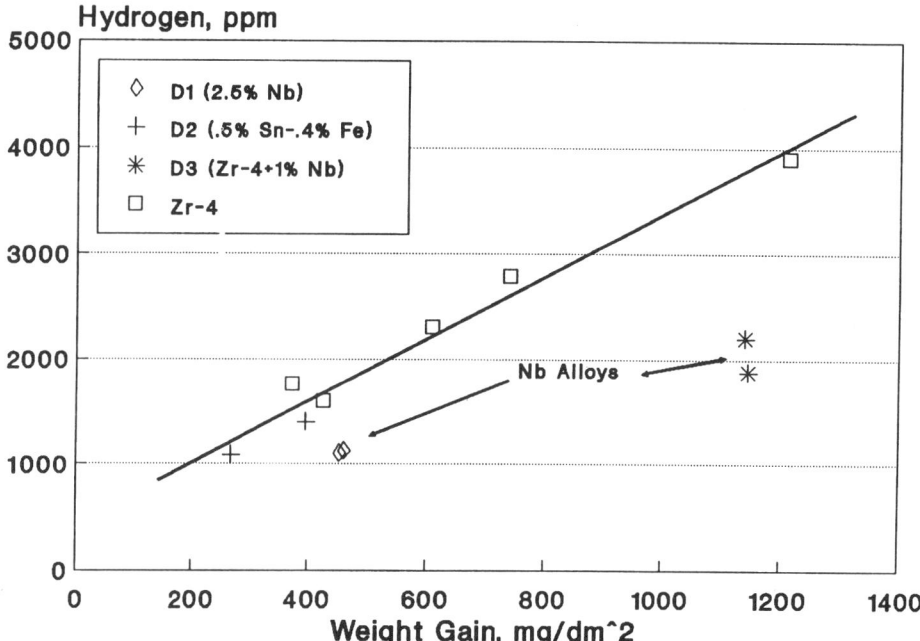

FIG. 9—*Hydrogen concentrations as a function of weight gain for cladding variants after long-term autoclave tests in the 673-K steam environment.*

Microstructural examination of the duplex variants after long-term autoclave testing revealed an increased uniform concentration of hydrides in the outer layers of D1 and D2 as illustrated in Fig. 10. The hydride distribution in the base Zircaloy-4 layer in Fig. 10 is typical of that seen in autoclave corroded tubular samples except for a hydride-denuded zone close to its interface with the outer layer. SEM examination confirmed the elevated concentration of hydride precipitates throughout the thickness of the outer layers with no evidence of any loss in mechanical bonding at the interface. These features are illustrated in Fig. 11 for the D1 duplex variant and Fig. 12 for the D2 duplex variant. The hydride distribution shown in Figs. 10 to 12 resulted from corrosion of open-ended specimens in static autoclaves, with relatively high dissolved hydrogen and hydrogen over pressure. Such conditions are not typical of in-reactor hydriding.

In-Reactor Program

The nominal thermal hydraulic parameters of NPP Grohnde PWR are listed in Table 6. The coolant temperature is relatively high and is representative of many other modern plants that have been categorized as "hot" PWRs. The plant operated for all four cycles with a nominal pH of 7.3. The maximum lithium hydroxide concentration at beginning-of-cycle was 2.0 ppm Li.

The cladding variants were distributed around the periphery of two identically configured assemblies (FA 436 and FA 437). Each assembly contained a total of 32 fuel rods with the five cladding variants, as shown in Fig. 13. The assembly design included nine structural spacers and four intermediate flow mixers, as shown in Fig. 14.

During each of the four irradiation cycles, the two assemblies were in symmetrical core

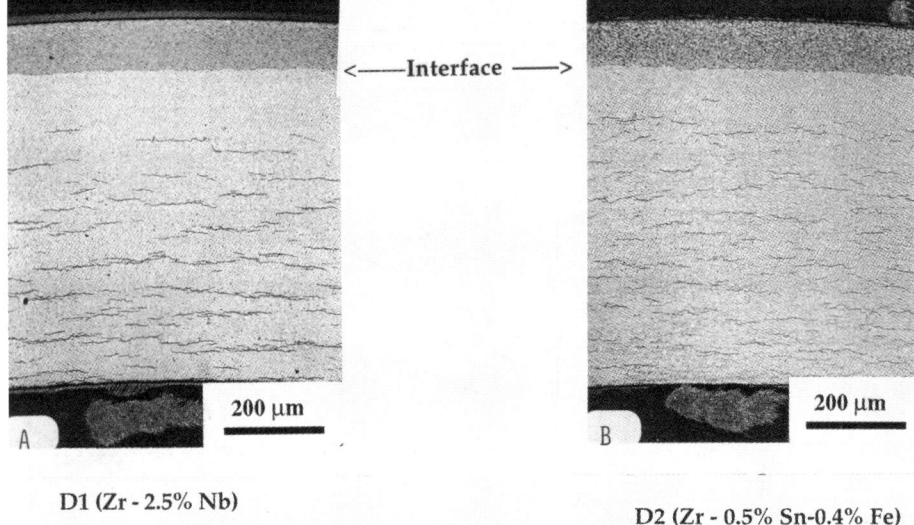

FIG. 10—*Optical micrographs showing the hydride distribution in duplex variants D1 (Zr-2.5% Nb) and D2 (Zr-0.5% Sn-0.4% Fe) after long-term autoclave tests in the 673-K steam environment.*

positions, as shown in Fig. 15. The individual rod average burnup values after four irradiation cycles ranged from 39.4 to 45.4 MWd/KgU.

During refueling outages after each irradiation cycle, individual fuel rods containing the cladding variants were removed from the two fuel assemblies and non-destructively examined at poolside.

Since the primary objective of the irradiation program was to determine the relative corrosion resistance of the cladding variants, detailed oxide thickness measurements were made on fuel rods representing each of the five cladding variants. Helical oxide thickness traces were made along the full length of the rods using an eddy current probe technique (the so-called "lift-off" method). Representative traces for each of the five cladding variants are shown in Fig. 16.

The peak oxide thickness, which typically occurred between Spacer 7 and Spacer 8 (see Fig. 14), was used to characterize the corrosion behavior of each measured fuel rod. A summary of the peak oxide thickness versus rod average burnup is shown in Fig. 17 for all rods measured during the four poolside exams that were part of this irradiation program. The highest corrosion occurred in cladding type D3 (100 μm peak oxide), and the lowest corrosion occurred in cladding type D1 (12 μm peak oxide). The relative corrosion performance among the five cladding types was: D3 (highest corrosion), R1, R2, D2, D1 (lowest corrosion).

Several observations regarding the relative corrosion behavior can be made. The lower tin concentration in the R2 cladding (1.3% Sn) compared to the R1 cladding (1.5% Sn) significantly improved the corrosion resistance of the reference cladding. This improved corrosion resistance is consistent with the industry trend toward "low tin" Zircaloy-4 cladding for PWR applications.

The D2 cladding (Zr-0.5% Sn-0.4% Fe), with a much lower tin concentration and higher iron concentration, exhibited an even further improvement in corrosion resistance and was almost equivalent to the niobium-bearing variant D1 (Zr-2.5% Nb). The other niobium-bearing cladding variant D3 (Zr-4 + 1% Nb) exhibited the worst corrosion resistance. This result is somewhat surprising in light of the good corrosion resistance reported for ZIRLO [*3*] since the

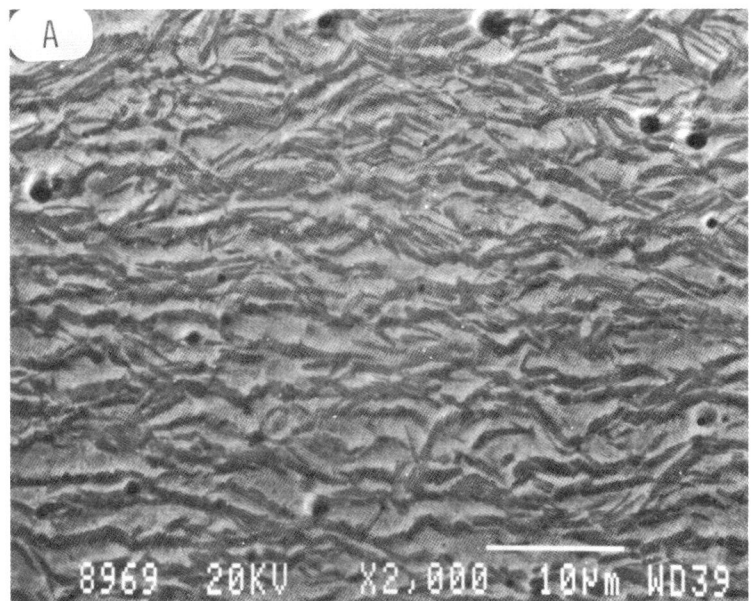

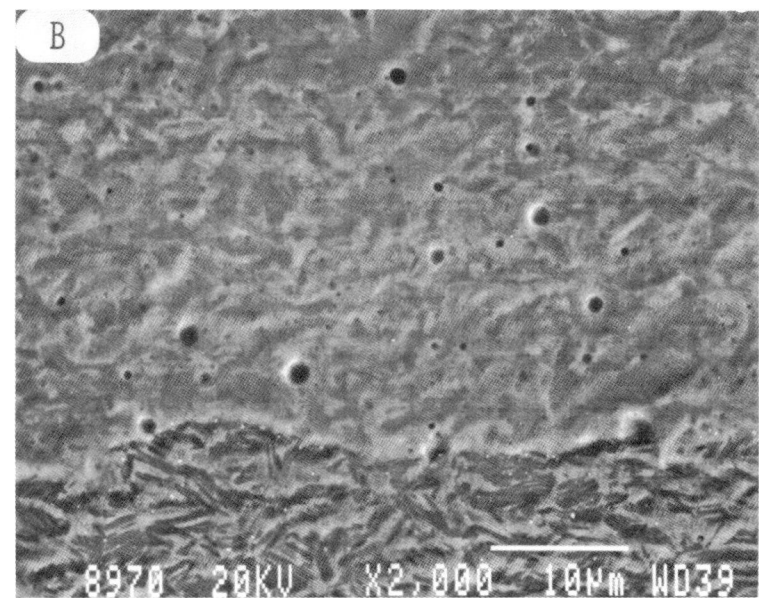

Zircaloy-4

Interface

Zr-2.5% Nb

FIG. 11—*SEM micrographs showing the hydride distribution for cladding variant D1 (Zr-2.5% Nb) in (A) the outer layer and (B) the interface region between the outer layer and the base layer.*

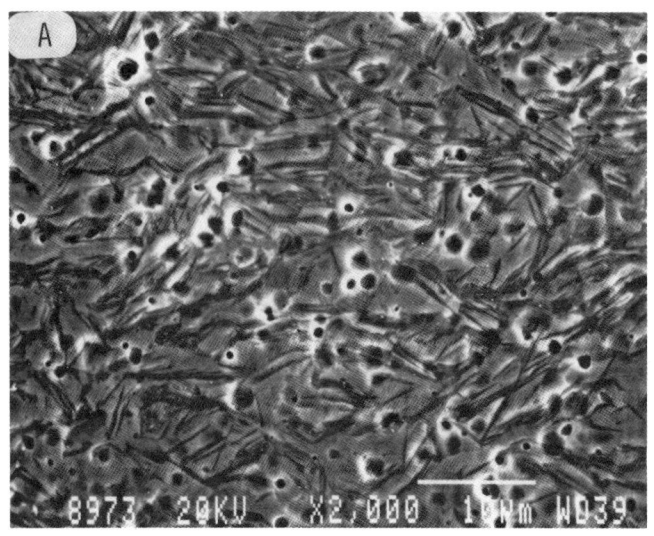
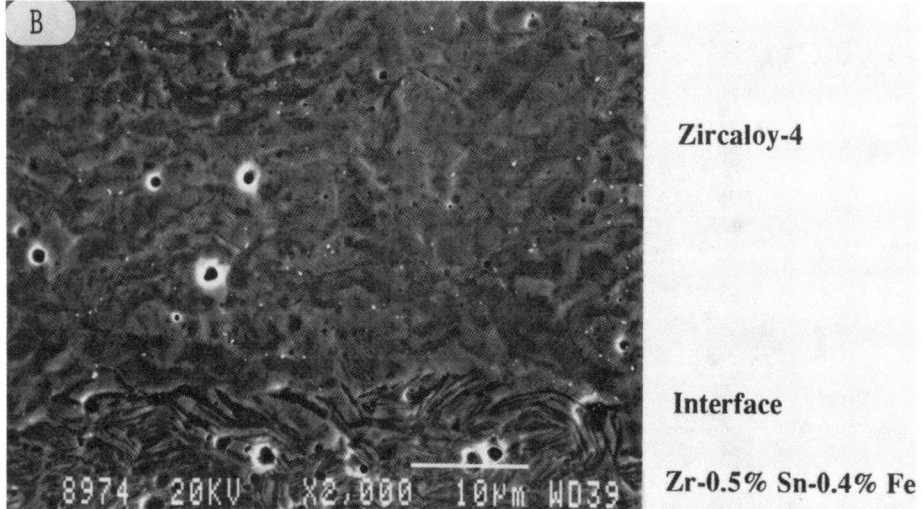

FIG. 12—*SEM micrographs showing the hydride distribution for cladding variant D2 (Zr-0.5% Sn-0.4% Fe) in (A) the outer layer and (B) the interface region between the outer layer and the base layer.*

TABLE 6—*Design thermal hydraulic parameters.*

	Original Power	Uprated Power[a]
Nominal thermal power	3765 MW	3850 MW
Number of coolant pumps	4	4
Coolant flow rate (including bypass)	27.12 m^3/s	27.12 m^3/s
Core bypass percentage	6%	6%
Fraction of heat produced in fuel pin	98%	98%
Average linear heat generation rate	207.0 W/cm	212 W/cm
Primary coolant pressure	158 bar absolute	158 bar absolute
Coolant inlet temperature	291.5°C	292.1°C
Average coolant temperature	308.7°C	309.6°C
Coolant outlet temperature	325.9°C	327.2°C

[a] The nominal core power was uprated in January 1990.

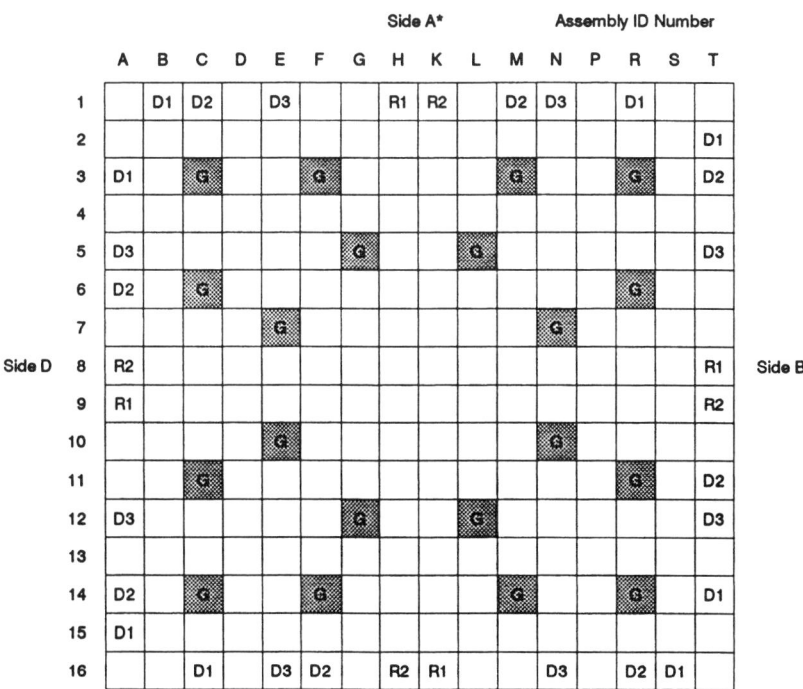

* = Spacer Serial Numbers are on Side A
G = Guide Tube
D1, D2, D3 = Duplex Clad Test Rods
R1, R2 = Reference Clad Test Rods

FIG. 13—*Fuel assembly map.*

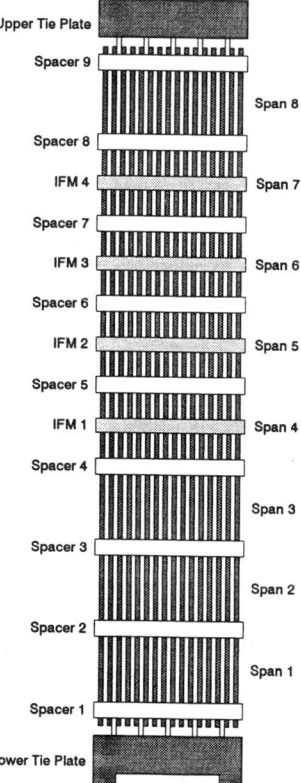

FIG. 14—*Fuel assembly spacer configuration.*

composition of Variant D3 is quite similar to ZIRLO. Apparently, the presence of slightly higher levels of iron, tin, and chromium in D3 and possible differences in processing are responsible for the poor corrosion resistance of D3.

The relative in-reactor corrosion behavior of the five cladding variants is generally consistent with the 633-K water and the 673-K steam autoclave behavior, as previously shown in Figs. 6 and 7. The 633 K water test does not correctly predict the distinction between the in-reactor corrosion behavior of R1 versus D3 and the D1 versus D2 cladding types. The 673 K steam test has a better correlation with the in-reactor corrosion behavior but overpredicts the relative in-reactor corrosion of the D1 cladding variant.

Conclusions

An LTA program involving comprehensively pre-characterized duplex and reference non-duplex cladding has been completed in a high-temperature PWR. The four-cycle program reached a peak rod burnup of 46 MWd/kgU.

The in-reactor corrosion data corroborate previous experience of improvement in corrosion resistance of low-tin reference cladding compared to high-tin reference cladding. The corrosion of duplex cladding, D1 (Zr-2.5% Nb) and D2 (Zr-0.4% Fe-0.5% Sn), shows an improvement over the low-tin reference cladding R2. The third duplex cladding, D3 (Zr-4 + 1% Nb), had significantly less corrosion resistance, which was inferior to high-tin reference cladding R1.

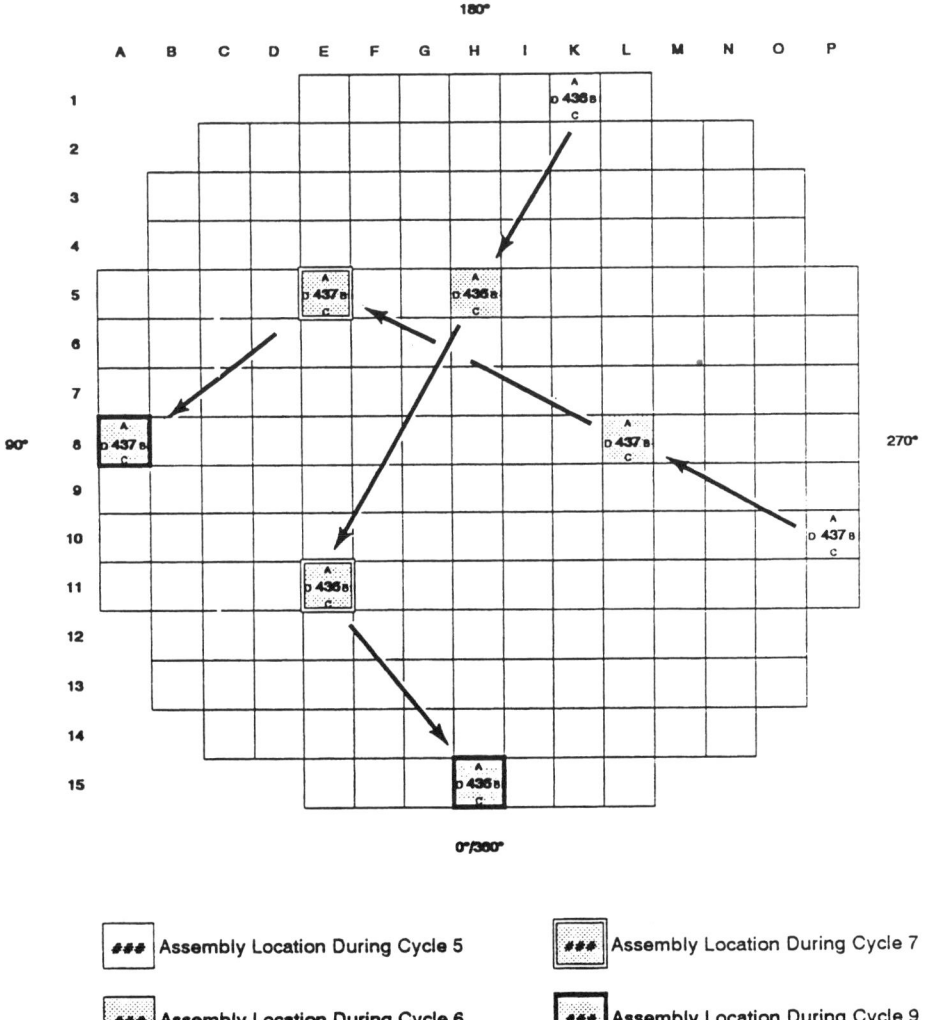

FIG. 15—*Fuel assembly locations in the core during irradiation cycles.*

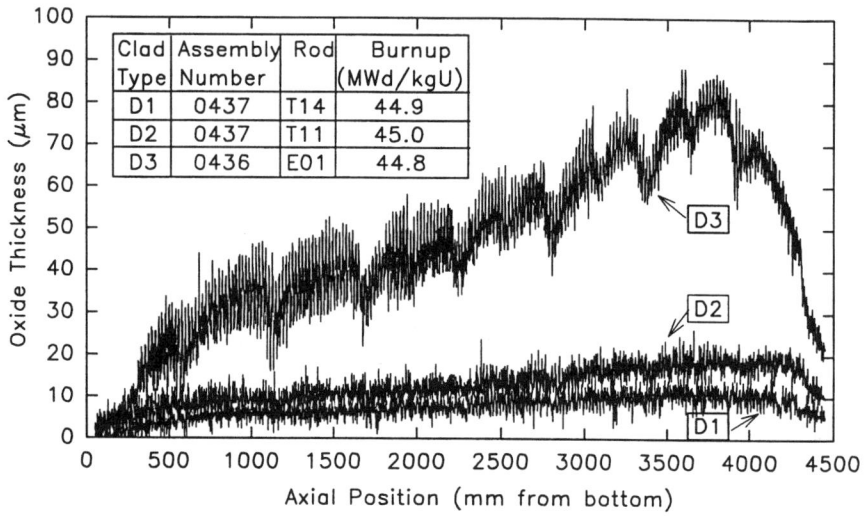
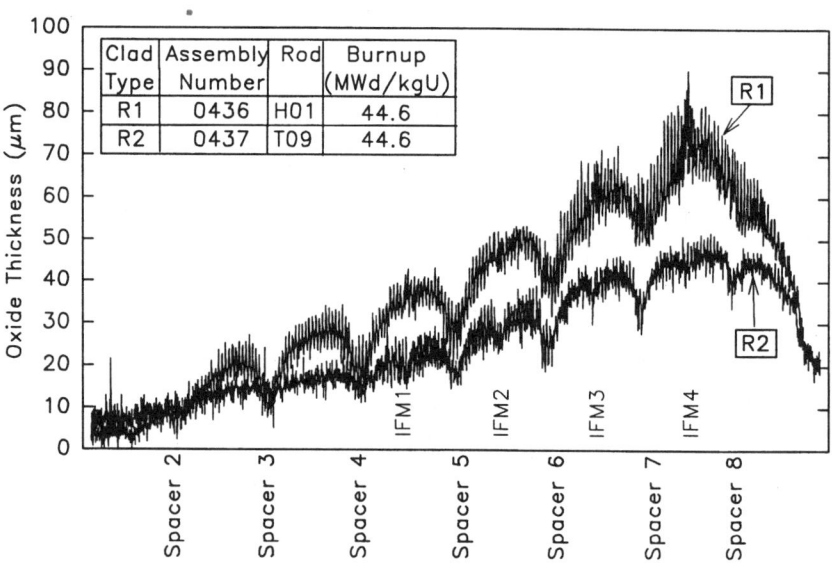

FIG. 16—*Four-cycle axial oxide thickness profiles for duplex and reference cladding.*

The relative ranking of the out-reactor corrosion in 633 K water better correlated to the in-reactor ranking, but the 673 K steam corrosion behavior did not replicate the good in-reactor corrosion behavior of D1. In addition, basic corrosion rates of various duplex alloys can be discerned from the autoclave and reactor data. Although the in-reactor hydrogen pickup is expected to be different compared to the out-reactor tests, no evidence of segregation of hydrides at the duplex interface was observed even after long-term out-reactor exposures in 633 K water or 673 K steam environments.

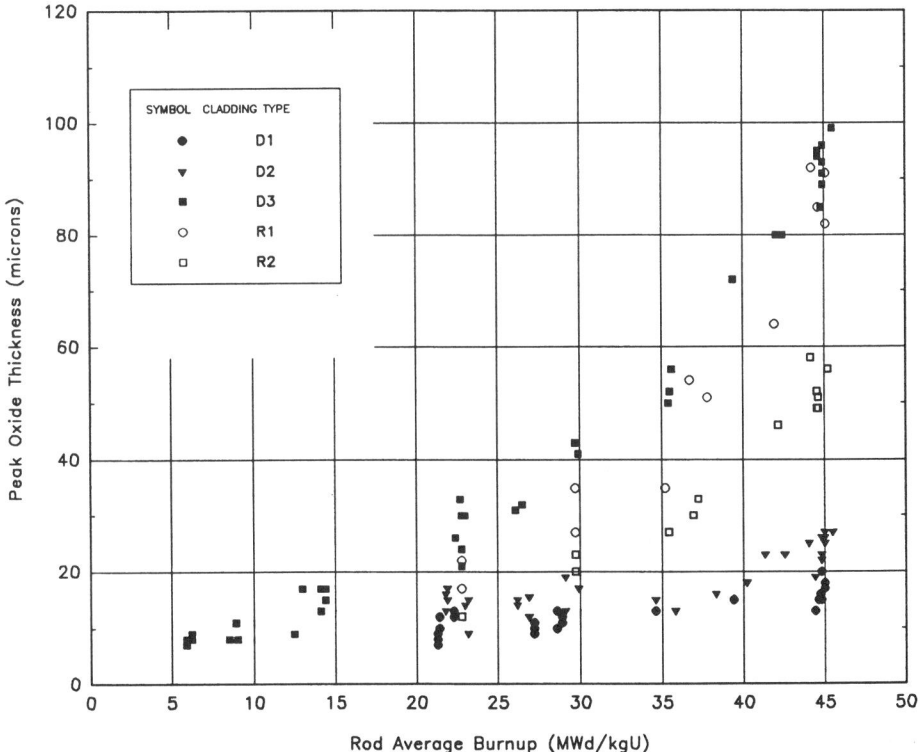

FIG. 17—*Peak oxide thickness of each measured rod versus rod average burnup.*

References

[1] Perovic, A., Pervoic, V., Weatherly, G. C., Purdy, G. R., and Fleck, R. G., "A Study of the Distribution of Nb and Fe in Two-Phase Zr-2.5 wt% Nb Alloys," *Journal of Nuclear Materials,* Vol. 199, 1993, pp. 102–111.
[2] Miyake, C. and Gotoh, K., "Magnetic Study of Zircaloy, II Zircaloy with Additive of Niobium," *Journal of Nuclear Materials,* Vol. 184, 1991, pp. 212–220.
[3] Sabol, G. P., Comstock, R. J., Weiner, R. A., Larouere, P., and Staunz, R. N., "In-Reactor Corrosion Performance of ZIRLO and Zircaloy-4," *Zirconium in the Nuclear Industry: Tenth International Symposium, ASTM STP 1245,* A. M. Garde and E. R. Bradley, Eds., American Society of Testing and Materials, West Conshohocken, PA, 1994, pp. 724–744.

Acknowledgments

This work was sponsored by the EPRI-managed NFIR Group. The collaboration among PE, SPC, SSM, and TWC within NFIR auspices made the Grohnde lead assembly program possible. The authors are grateful to the members of the NFIR Steering Committee for their technical guidance and the permission to publish the results. Some of the TEM pre-characterization results used in the paper were contributed by Institute of Nuclear Energy Research (INER), Taiwan.

DISCUSSION

P. Beslu[1] *(written discussion)*—You presented test results from autoclaves, and you concluded that tests at 635 K gave the same ranking for corrosion as the reactor observations. Do you have results from tests at 70 ppm of lithium? What was the coolant chemistry in the Grohnde reactor? Was it a high lithium operation?

S. K. Yagnik et al. (authors' closure)—No autoclave tests were performed under lithiated water. Grohnde reactor primary water chemistry was 2.0 to 0.6 ppm Li with boric acid additions to keep pH = 7.3 throughout the annual cycles.

V. Urbanic[2] *(written discussion)*—With respect to hydride distribution, was the denuded zone only observed for the Zr-2.5% Nb (D1) outer layer or was it observed in the other duplex claddings as well?

S. K. Yagnik et al. (authors' closure)—Hydride denuded zone was clearly observed in both D1 and D2. In D3, such a zone was either absent or undetectable at the magnifications used.

C. Lemaignan[3] *(written discussion)*—Is there any difference in oxygen content in the external and bulk alloys for the duplex? Could that explain hydrogen segregation due to change of H solubility with alloy chemistry?

S. K. Yagnik et al. (authors' closure)—There is indeed a small difference in oxygen content between the outer layer and the base alloy as shown in Table 1. The difference ranged from 200 ppm for D1 to 310 ppm for D3. Since D3 exhibited the least hydrogen segregation of the three duplex alloys, the oxygen content does not seem to explain the hydrogen segregation. On the other hand, the tin content difference between D1 or D2 and the base alloy is much larger than that in D3.

---

[1] CEA, Grenoble, France.
[2] Atomic Energy of Canada, Ltd., Chalk River, Ontario, Canada.
[3] CEA, Grenoble, France.

Yoshinori Etoh,[1] Sachio Shimada,[1] Takayoshi Yasuda,[2] Tadahiro Ikeda,[3] Ronald B. Adamson,[4] J.-S. Fred Chen,[4] Yoshiaki Ishii,[5] and Kazuhiro Takei[5]

Development of New Zirconium Alloys for a BWR

REFERENCE: Etoh, Y., Shimada, S., Yasuda, T., Ikeda, T., Adamson, R. B., Chen, J.-S. F., Ishii, Y., and Takei, K., **"Development of New Zirconium Alloys for a BWR,"** *Zirconium in the Nuclear Industry: Eleventh International Symposium, ASTM STP 1295,* E. R. Bradley and G. P. Sabol, Eds., American Society for Testing and Materials, 1996, pp. 825-849.

ABSTRACT: Specimens for irradiation testing in a commercial BWR were prepared from 24 kinds of Zr alloys. The corrosion performance of these specimens was investigated after irradiation for up to four cycles. Two cycles of irradiation were not always sufficient to identify the lowest corrosion alloys. Marked acceleration of corrosion occurred for many alloys between one and four cycles. The onset of mildly accelerated oxidation was observed for standard Zircaloy-2 at four cycles of irradiation. On the other hand, for Zircaloy-like alloys, increasing Fe and Ni contents or decreasing the Sn content promoted a saturation oxidation between two and four cycles of irradiation. The effects of alloying elements on corrosion behavior were evaluated by both in-pile and out-of-pile corrosion tests. The alloying elements Fe, Cr, and Ni, which have smaller valences than Zr, improved the in-pile corrosion resistance of Zr alloys, while the alloying elements Nb, Mo, and Te, which have larger valences than Zr, were responsible for increased weight gain during the irradiation test in the BWR.

KEYWORDS: Zr alloys, BWR, corrosion, accelerated corrosion, valence, nodular corrosion, hydrogen pickup, high burnup

In the past, nodular corrosion has appeared on the surfaces of Zircaloys irradiated in BWRs as fuel cladding tubes and assembly components [1–5]. Nodular corrosion did not affect the performance of the fuel claddings and assembly components, and it showed a saturation tendency for maximum oxide thickness as a function of burnup [2,4,5]. The small size and high density of second phase particles in Zircaloy were found to improve the nodular corrosion resistance [6–7]. This microstructure was achieved by rapid quenching from the beta or the alpha + beta phase. Current Zircaloy products for BWRs have very small particles and show good nodular corrosion resistance. Recently, it was pointed out that at high burnup an accelerated corrosion occurs in Zircaloy with very fine particles and very high resistance against nodular corrosion [8]. Accelerated corrosion is very important for the performance of high burnup fuels.

In this study, many new Zr alloys developed to improve corrosion performance of fuel

---

[1] Researcher and senior researcher, respectively, Nippon Nuclear Fuel Development Co., Ltd., 2163, Narita-cho, Oarai-machi, Higashi-Ibaraki-gun, Ibaraki-ken, 311-13, Japan.
[2] Senior engineer, Hitachi Ltd., 1-1, Saiwai-cho 3-chome, Hitachi-shi, Ibaraki-ken, 317, Japan.
[3] Senior specialist, Toshiba Corp., 8, Shinsugita-cho, Isogo-ku, Yokohama-shi, Kanagawa-ken, 235, Japan.
[4] Manager and senior engineer, respectively, GE Nuclear Energy, P.O. Box 460, Pleasanton, CA 94566.
[5] Staff researcher and researcher, respectively, Tokyo Electric Power Co., 4-1, Egasaki-cho, Tsurumi-ku, Yokohama-shi, Kanagawa-ken, 230, Japan.

cladding tubes and assembly components for high burnup fuel were irradiated in a commercial BWR. These new Zr alloys were expected to suppress nodular corrosion at the initial stage of irradiation and to delay or to prevent the onset of accelerated corrosion at high burnup. Based on the post-irradiation examination data on these irradiated materials, promising candidates were selected for high burnup fuels in BWRs. Additionally, considerations were given to what kind of alloying elements improved the corrosion performance of Zr alloys in both a BWR environment and out-of-pile corrosion tests, how corrosion behaviors correlated between in-pile and out-of-pile corrosion tests, and how irradiation time and fast neutron flux affected the corrosion behavior of Zr alloys.

Experimental Procedures

Materials

Specimens are listed in Tables 1 and 2. Alloy 1 is standard Zircaloy-2 (Zry-2). Alloys 2 to 10 are variants of Zry-2. Alloy 11 is Zr-2.5Nb. Alloys 12 to 16 are Zr-Sn-Nb-Mo alloys; Alloy 16 is a modification of EXCEL [9]. Alloys 17 to 19 are Zr-Sn-Nb. Alloy 20 is Zr-Sn-Te, and Alloy 21 is Zr-Bi-Nb. Alloys 22 to 24 are Zry-like alloys with no Cr. There are two types of specimens, T series and I series. All T series specimens are sheet specimens, and their fabrication processes were basically the same except for the high-strength alloys, which contain high contents of Sn, Nb, Mo, or Te. I series specimens have almost the same chemical compositions as T series specimens, except for I5 and I7. There are some tube specimens in the I

TABLE 1—*T series specimens.*

No.	Composition, wt%						Impurities, ppm				Process
	Sn	Nb	Fe	Cr	Ni	Others	Si	N	H	C	
T1	1.61	...	0.14	0.11	0.06	...	99	21	<5	155	A
T2	2.95	...	0.16	0.10	0.05	...	84	12	<5	50	C
T3	1.04	...	0.16	0.10	0.06	...	89	21	<5	85	A
T4	1.46	...	0.26	0.10	0.05	...	98	26	<5	170	A
T5	1.43	...	0.26	0.10	0.10	...	103	22	<5	150	A
T6	1.52	...	0.30	0.10	0.15	...	105	35	<5	175	A
T7	1.37	...	0.16	0.09	0.05	C < 50ppm	93	22	<5	35	A
T8	1.47	0.22	0.15	0.11	0.05	...	96	17	<5	165	A
T9	1.34	0.50	0.16	0.09	0.05	...	91	20	<5	150	A
T10	1.48	...	0.16	0.10	0.05	0.46 Mo	108	24	<5	145	C
T11	...	2.60	...	...	...	...	50	16	<5	130	D
T12	1.00	1.00	...	...	...	0.50 Mo	100	22	<5	120	E
T13	1.00	1.50	...	...	...	0.50 Mo	99	18	<5	65	E
T14	0.96	2.02	...	...	...	0.19 Mo	103	27	<5	160	E
T15	1.00	1.90	...	...	...	0.50 Mo	115	15	<5	100	E
T16	1.24	0.30	...	...	...	0.29 Mo	105	17	<5	150	A
T17	1.00	1.01	0.21	...	...	...	96	22	<5	165	A
T18	1.04	0.58	0.21	...	0.05	...	87	17	<5	80	A
T19	1.40	0.38	...	...	...	0.18 Te	94	20	7	150	A
T20	1.15	...	...	...	...	0.62 Te	92	20	7	155	B
T21	...	0.50	...	...	...	1.00 Bi	78	30	<5	85	A
T22	1.48	...	0.17	...	0.17	...	91	23	5	50	A
T23	0.49	...	0.18	...	0.11	...	94	23	9	170	A
T24	0.50	...	0.18	...	0.13	0.10 Si	...	26	<5	65	A

TABLE 2—*I series specimens.*

No.	Composition, wt%						Process
	Sn	Nb	Fe	Cr	Ni	Others	
I1	1.50	...	0.15	0.10	0.05	...	M
I2	3.29	...	0.15	0.09	0.05	...	I
I3	1.28	...	0.17	0.10	0.07	...	N
I4	1.45	...	0.24	0.10	0.05	...	F
I5	1.50	...	0.12	0.10	0.09	...	F
I6	1.50	...	0.24	0.10	0.09	...	F
I7	1.23	...	0.17	0.05	0.07	...	N
I8	1.50	0.20	0.16	0.09	0.07	...	...[a]
I9	1.59	0.51	0.15	0.09	0.05	...	H
I10	1.58	...	0.15	0.10	0.05	0.50 Mo	H
I11	...	2.50	...	...	...	...	L
I12	1.05	1.00	...	...	...	0.44 Mo	G
I13	1.00	1.60	...	...	...	0.40 Mo	G
I14	1.10	2.00	...	...	...	0.23 Mo	G
I15	1.12	2.03	...	...	...	0.46 Mo	G
I16	1.50	0.30	...	...	...	0.30 Mo	J
I17	1.00	1.00	0.20	...	...	...	L
I18	1.00	0.60	0.20	...	0.05	...	L
I19	1.50	0.40	...	...	...	0.20 Te	K
I20	1.40	...	...	...	...	0.60 Te	K
I21	...	0.50	...	...	...	1.20 Bi	K
I22	1.50	...	0.17	...	0.17	...	J
I23	0.52	...	0.17	...	0.12	...	H
I24	0.52	...	0.18	...	0.11	0.10 Si	H

[a] Fabrication process was not clear.
NOTE: Impurities were not analyzed for I-series specimens.

series; I1, I4, I5, and I6 are tube shell heat-treated cladding tubes fabricated by a conventional process in Japan, and I8, I16, and I22 lack the tube shell heat treatment. The tube shell heat treatment is the solution heat treatment in the alpha + beta phase followed by quenching for the outer surface region of a tube shell. Fabrication processes are summarized in Table 3.

Out-of-Pile Corrosion Tests

Four types of out-of-pile corrosion tests were performed using T series specimens. Two were long-term, uniform corrosion tests in 589 K, 7.6 MPa water and 673 K, 10.3 MPa steam. The others were nodular corrosion tests, i.e., a MAT test (683 K/4 h + 793 K/16 h in 12.5 MPa steam) and an 803 K steam test up to five days. Feed water to an autoclave was degassed by the bubbling of Ar gas and deionized by ion exchange resin. At intervals during corrosion tests, visual appearances of specimens were inspected and weight gains were measured. After corrosion tests, oxide thicknesses were measured by metallography to evaluate corrosion behaviors of Zr alloys.

Irradiation Tests

Both T and I series specimens were irradiated in dummy neutron source holders in a commercial BWR. Irradiation temperature was the same as coolant temperature. T series specimens

TABLE 3—The fabrication process of specimens.

Process	A	B	C	D	E
Ingot size[a]			200 φ × 600, 43 kg		
Billet quench[a]			1303 K × 0.5 h, 184 φ × 35		
Extrusion			908 K × 0.2 h		
Anneal	773 K × 1 h	773 K × 1 h	773 K × 1 h	973 K × 1 h	973 K × 1 h
Number of cold reduction passes	3	3	4	4	4
Intermediate anneals	839 K × 0.1 h	839 K × 0.1 h	839 K × 0.1 h	977 K × 0.11 h	977 K × 0.11 h
In-process heat treatment	None	None	None	α + β quench	α + β quench
Final anneal	977 K × 0.05 h	1005 K × 0.05 h	1005 K × 0.05 h	773 K × 1 h	773 K × 1 h
Form	Sheet	Sheet	Sheet	Sheet	Sheet

Process	F	G	H	I	J
Ingot size[a]	370 φ × 600, 450 kg	280 φ, 270 kg	230 × 32 × 16ℓ, 0.7 kg	230 × 32 × 16ℓ, 0.7 kg	260 φ × 1800, 680 kg
Billet quench[a]	1313 K × 1.0 h, 180 φ solid	1143 K × 1.0 h	1289 K × 0.5 h	1289 K × 0.5 h	1283 K × 1.0 h
Extrusion	923 K × 0.2 h	None	1023 K × 0.5 h	1023 K × 0.5 h	973 K × 0.2 h
Anneal	None	None	None	923 K × 0.5 h	893–923 K × 1.0 h

Tube shell heat treatment	63.5 φ × 11t, 1203 K, 50 K/s	None	None	None	None
Number of cold reduction passes	3	1	1	1	3
Intermediate anneals	873 K × 1.25 h	None	None	None	898 K × 1.25 h
Final anneal	850 K × 2.5 h	773 K × 24 h	950 K × 2.5 h	950 K × 2.5 h	849 K × 2.5 h
Form	Tube	Sheet	Sheet	Sheet	Tube

Process	K	L	M	N
Ingot size[a]	260 φ × 1800, 680 kg	200 φ	Production ingot	75 φ
Billet quench	1283 K × 1.0 h	1283 K × 1.0 h	uncertain	1283 K × 0.2 h
Extrusion	973 K × 0.2 h	953–963 K × 0.2 h	uncertain	None
Tube shell heat treatment	None	None	Yes	None
Anneal	893–923 K × 1.0 h	893–923 K × 1.0 h	893–923 K × 1.0 h	None
Number of cold reduction passes	3	3	3	1
Intermediate anneals	893 K × 1.25 h	893 K × 1.25 h	893 K × 2.5 h	None
Final anneal	849 K × 2.5 h	849 K × 2.5 h	849 K × 2.5 h	950 K × 2.5 h
Form	Sheet	Sheet	Tube	Sheet

[a] Numbers without dimension are in mm.

830 ZIRCONIUM IN THE NUCLEAR INDUSTRY: ELEVENTH SYMPOSIUM

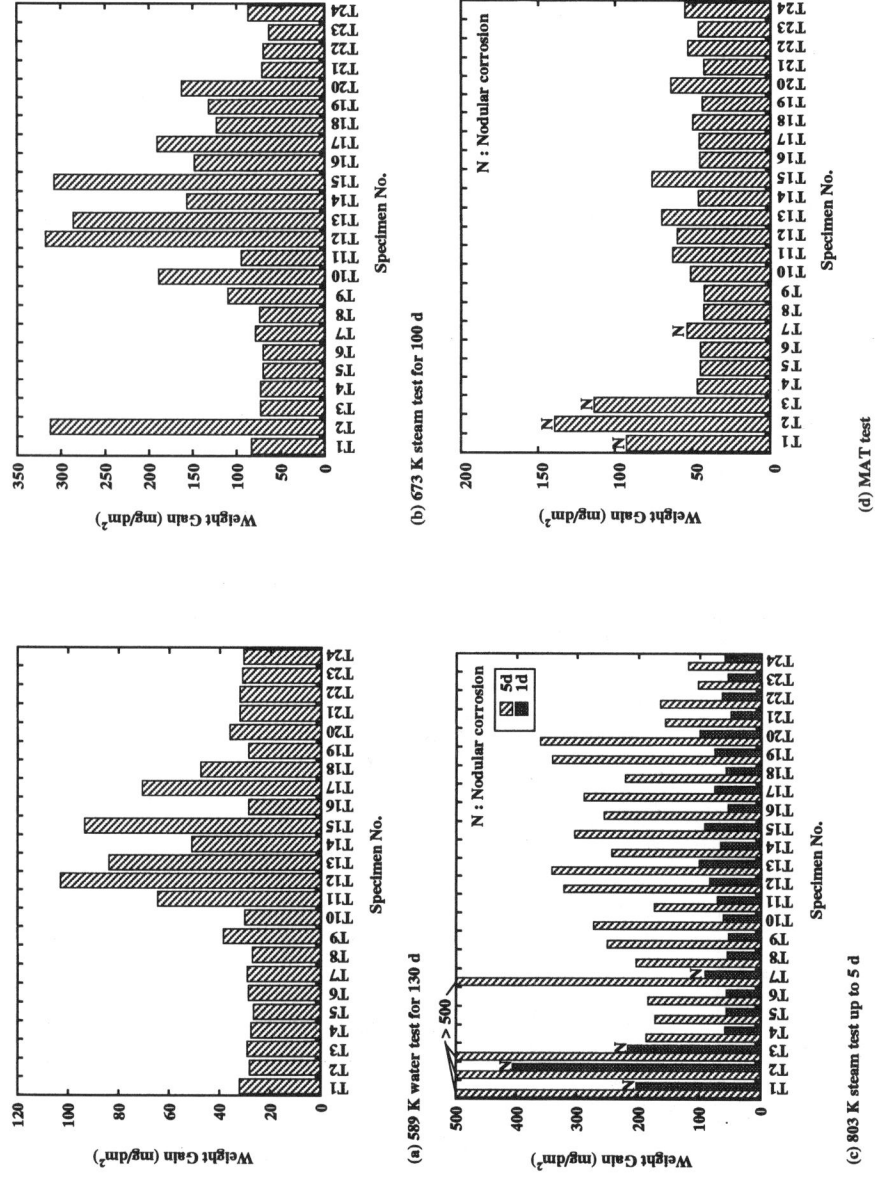

FIG. 1—*Comparisons of weight gain data from out-of-pile corrosion tests using T series specimens.*

were irradiated for two cycles (777 days), and their fast neutron fluences exceeded 2.5 to 3.9×10^{25} n/m² ($E > 1$ MeV). I series specimens were placed into three dummy neutron source holders. The first one was irradiated for one cycle (284 days), the second, for two cycles (543 days), and the last, for four cycles (1515 days). Fast neutron fluences of I series specimens exceeded 0.8 to 8.5×10^{25} n/m².

Post-Irradiation Examinations

To evaluate the in-pile corrosion behavior of Zr alloys, visual appearances were inspected and weight gains were measured for all irradiated specimens. Two or three specimens were irradiated for each alloy. The scattering of weight gain data was less than 30% of average values. The morphology of the oxide layers was inspected, and oxide thicknesses were measured on cross-sectional metallographs of some irradiated specimens. Hydride distributions were also observed on the cross-sectional metallographs. Hydrogen contents in the specimens were measured by gas chromatography.

Results

Out-of-Pile Corrosion Tests

Weight gain data of out-of-pile corrosion tests are compared among T series specimens in Fig. 1. All specimens tested in 589 K water up to 130 days show only shiny, black oxide. Alloys containing a high Nb concentration (T11 to T15, and T17) show high weight gains in the 589 K water test up to 130 days, as shown in Fig. 1a. Weight gains are about 30 mg/dm² among the other alloys. The latter alloys do not show any transition in oxidation rate up to 130 days. Figure 2 shows the effect of Nb content in Zr alloys on weight gain data in the 589 K water test up to 130 days. There appears to be a correlation between the Nb content and the

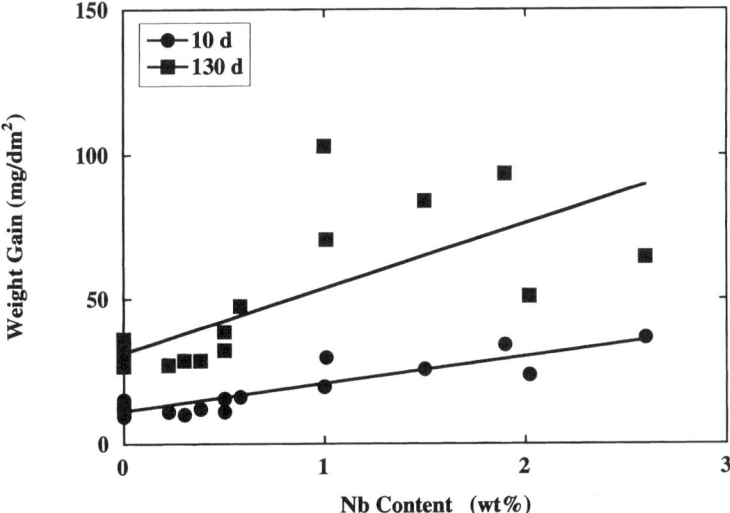

FIG. 2—*Nb content dependence of weight gain data from a 589 K water test using T series specimens.*

corrosion of Zr alloys in 589 K water, especially early in the corrosion test. The correlation coefficients from linear regression analyses were 0.911 for 10 days and 0.750 for 130 days.

All specimens tested in 673 K steam up to 100 days show black or dark oxide. The steam test results at 673 K, shown in Fig. 1b, demonstrate a similar tendency to those of the 589 K water test. Higher weight gains are observed for alloys containing a high Nb content, except for Alloy T11. In addition, Mo or Te added alloys (T10, T16, T19, and T20) and a high Sn content alloy (T2) indicate higher weight gains than the other alloys. Weight gains of other alloys are 70 to 80 mg/dm^2, and these alloys do not show any transition in oxidation rate up to 100 days.

Nodular corrosion appears on four alloys (T1 to T3, and T7) in the 803 K steam test and MAT test. The other alloys show only uniform corrosion. Therefore, the latter alloys have good nodular corrosion resistance for out-of-pile tests. Alloys susceptible to nodular corrosion are Zry-2 and modified Zry-2 compositions containing Sn and lower intermetallic concentrations. Nodular corrosion grows very rapidly for the out-of-pile tests, and spalling of oxide occurs after five days in 803 K steam. This is different from that of in-pile nodular corrosion, which shows a decrease in growth rate [2,4,5]. The steam test at 803 K up to five days can be interpreted as an accelerated uniform corrosion test for alloys not susceptible to nodular corrosion. Figure 1c shows that the weight gain data up to five days without nodular corrosion are similar to those from the 673 K steam test up to 100 days. Namely, Nb, Mo, and Te additions increase weight gain, low Sn content decreases it, and alloy T11 shows a relatively lower weight gain.

Two-Cycle Irradiation Test of T Series Specimens

Weight gain data of the two-cycle irradiation test (777 days) for T series specimens are shown in Fig. 3. A letter code is used to report visual inspection results along with the weight gain data in this figure. Many alloys are subject to nodular corrosion. Some of them do not show nodular corrosion in the out-of-pile nodular corrosion tests. Conversely, Alloy T3, having extensive nodular corrosion in the out-of-pile nodular corrosion tests, does not reveal nodular corrosion in the in-pile tests. Evidently, the development of nodular corrosion is more dependent on Sn concentration for in-pile exposures, however more dependent on intermetallic concentrations for out-of-pile tests. The more general conclusion is that in-pile nodular corrosion is difficult to predict on only out-of-pile tests. The weight gain data indicate Zr alloys containing Nb, Mo, and/or Te experience more extensive in-pile corrosion than Zr alloys without these elements.

To investigate the effects of alloying elements on corrosion behavior, bi-variant plots are used for alloys having similar chemical compositions. Weight gain data of the in-pile test, MAT test, and 673 K steam test as a function of Fe and Ni, Sn, and Nb are shown in Figs. 4a, 4b, and 4c, respectively.

Figure 4a shows the effect of Fe and Ni content on weight gains of Zry-2 and modified Zry-2, Specimens T4, T5, T6, and T22. These alloys have almost the same content of Sn, while the Cr content of T22 is different from the other alloys. The x-axis is Fe + 2Ni content and is chosen based on a hypothesis that the valence difference between Zr and the alloying elements is the dominant factor influencing the corrosion performance, as discussed later. Out-of-pile tests show that increasing the Fe and Ni content improves corrosion resistance against both nodular and uniform corrosion regardless of Cr addition or absence. For the in-pile test, one alloy (T22) shows a high weight gain while the other alloys have similar but lower weight gains. These results suggest that Cr does not affect the corrosion performance in out-of-pile tests, but that its addition decreases weight gains in the in-pile tests.

Figure 4b shows the dependence of weight gain on Sn content. Weight gain data from Alloys

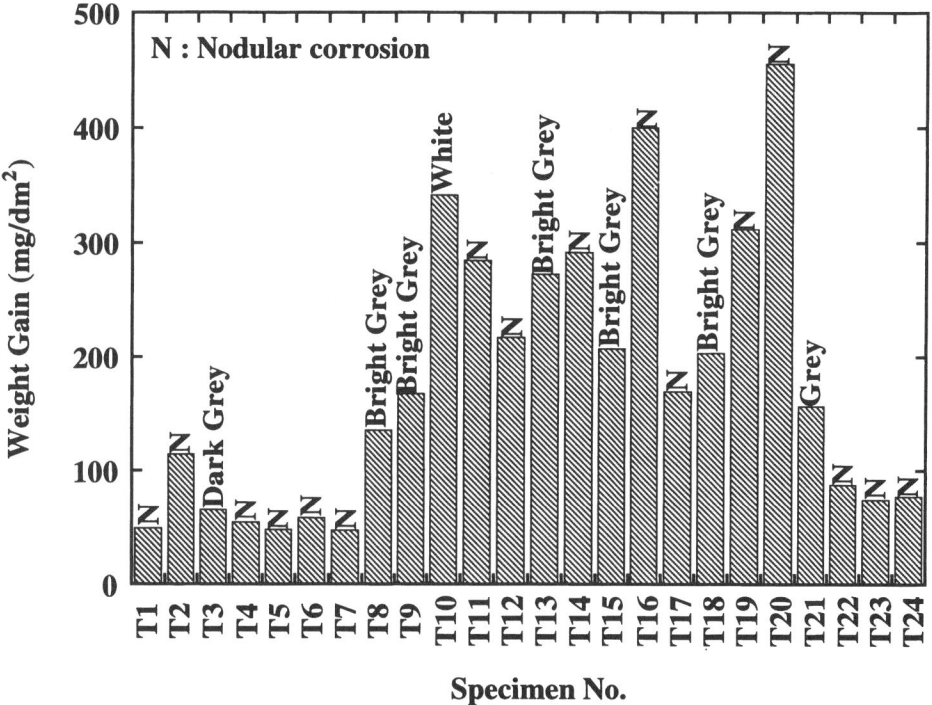

FIG. 3—*Comparison of weight gain data from an in-pile corrosion test for two cycles using T series specimens.*

T1, T2, T3, and T7 are plotted in this figure. These alloys have almost the same contents of Fe, Cr, and Ni. For the 673 K steam test, a decrease of Sn content clearly improves the corrosion resistance. MAT test results show that nodular corrosion appears on all four of these alloys. Then nodular corrosion susceptibility apparently is not dependent on Sn content for the out-of-pile test. Alloy T7, which reduces C content less than 50 ppm, shows smaller weight gain than the other alloys. It is not clear whether this effect is significant, or whether it is due to a decrease of C content. In-pile test results indicate that high Sn content (Alloy T2) increases weight gain, but, in the range from 1.0 to 1.6 wt% Sn, it is difficult to observe any significant difference among specimens irradiated for only two cycles.

The Nb-addition effect on Zry-2 is shown in Fig. 4c. While the Nb addition suppresses nodular corrosion, high Nb content increases uniform corrosion in out-of-pile tests. In-pile test results indicate that Nb-added Zry-2 is effective to reduce nodular corrosion, but has a greater amount of uniform corrosion. Nb addition accelerates uniform corrosion for both in-pile and out-of-pile corrosion.

Four-Cycle Irradiation Test of I Series Specimens

Photos showing specimen appearances are reproduced in Fig. 5 for selected I series specimens after one-, two-, and four-cycle BWR irradiations. None of the specimens demonstrate nodular corrosion to the unaided eye after one-cycle irradiation, but at higher magnifications nodular corrosion is observed on Specimens I3, I4, I8, I9, I14, and I16. After four-cycle irra-

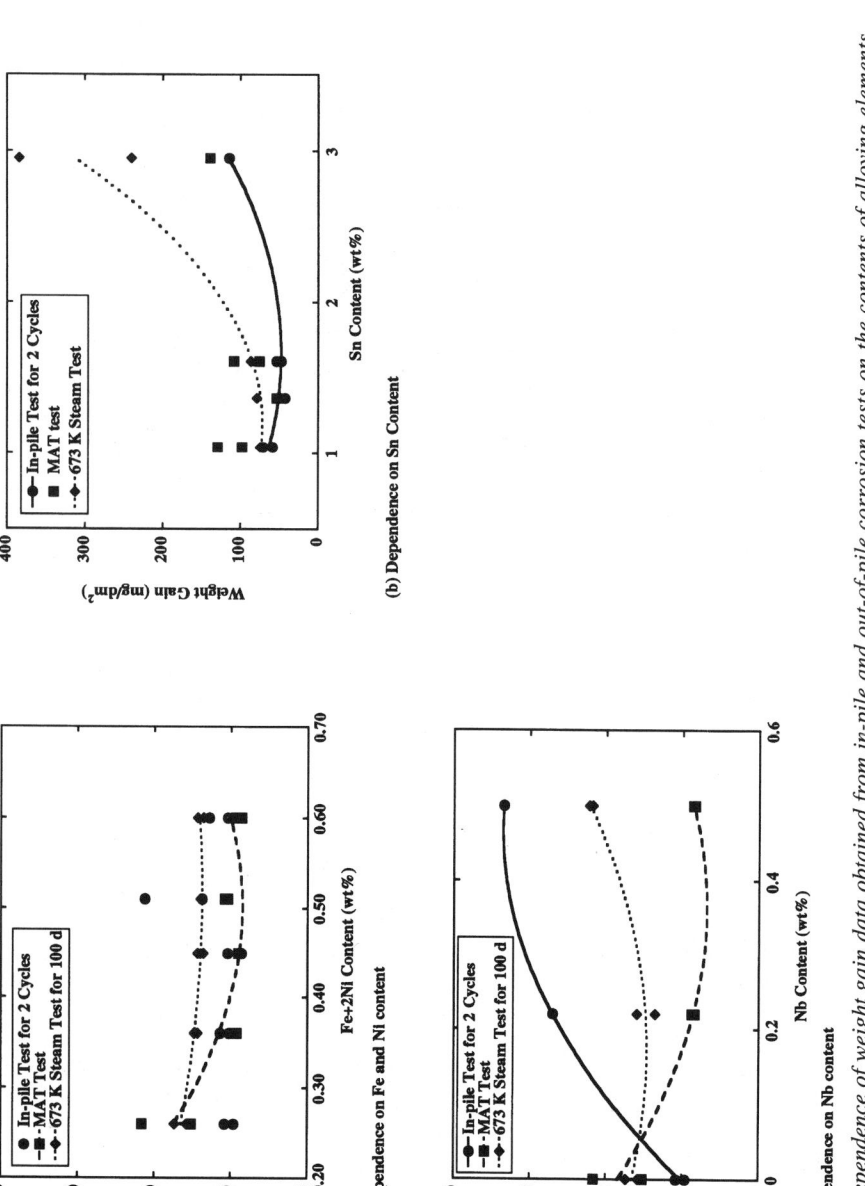

FIG. 4—*Dependence of weight gain data obtained from in-pile and out-of-pile corrosion tests on the contents of alloying elements.*

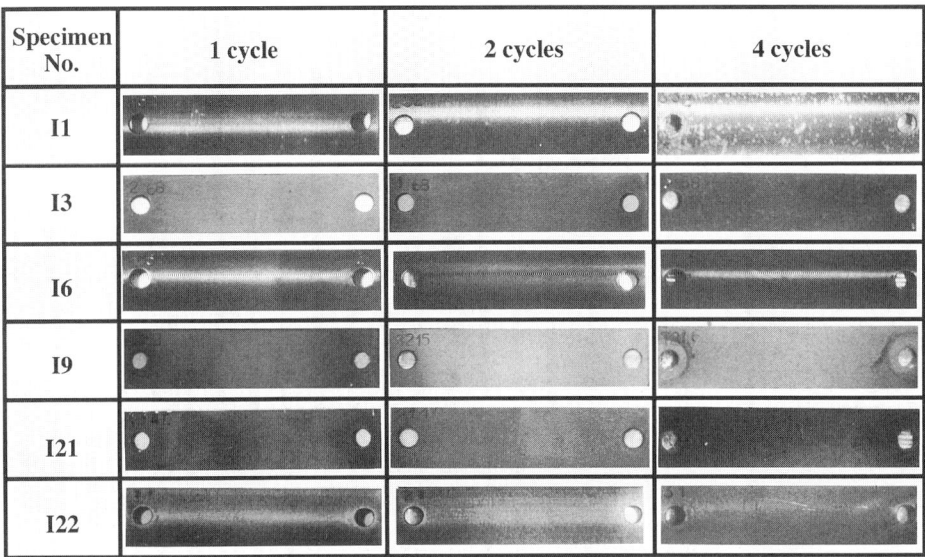

FIG. 5—*Typical visual appearances of I series specimens after in-pile tests.*

diation, localized areas of white oxide are observed on the surfaces of some alloys, such as I1, I3, I5, I7, I11, and I19. This disk-shaped, white oxide differs from nodular corrosion appearing at the initial stage of irradiation for other specimens. Disk-shaped, white oxide does not show any wavy metal/oxide interface like a coalescence of nodules, but shows a uniform corrosion in cross-sectional metallographs. Alloys I5 and I6 show black, shiny oxide; however, the higher fluence specimens show white, disk-shaped oxides a few millimetres in diameter. The other alloys were covered with white or grey uniform oxide.

Weight gain data are compared among I series specimens irradiated up to four cycles in Fig. 6. Significant differences in weight gains are not observed between Zry-2 and the other alloys except for alloys containing a high Nb concentration (Specimens I11 to I15, I18, and I20). Marked acceleration of corrosion occurs for many alloys after two-cycle irradiation. This acceleration is clearly observed for Nb, Mo, and/or Te added alloys (I8 to I21) and Alloys I23 and I24, which are modified Zry-2 specimens and contains no Cr. Zry-2 (I1) also shows slight acceleration.

Irradiation time dependencies on weight gain data are shown in Fig. 7 for some alloys. Specimen I1 (Zry-2) demonstrates typical pre-transition and transition behavior, and post-transition is not reached after four cycles. Nb-added alloys, such as I8 and I21, show clear transition during three and four cycles. Alloy I22 has a linear corrosion rate from the second to fourth cycles. Alloys I3 and I6, containing either lower Sn concentration or higher intermetallics element concentrations, respectively, show strong saturation tendencies up to four cycles. These alloys are expected to be promising materials for high burnup fuels.

Thickness of oxide layers and hydride distributions are shown in Fig. 8 using cross-sectional metallography for three specimens irradiated for four cycles. Uniform corrosion is observed for all specimens. Average oxide thicknesses, determined from metallographs dividing the cross-sectional area of the oxide layer by the length of the oxide/metal interface, are shown in Fig. 9a as a function of fast neutron fluence. The average oxide thickness of Alloy I6 does not

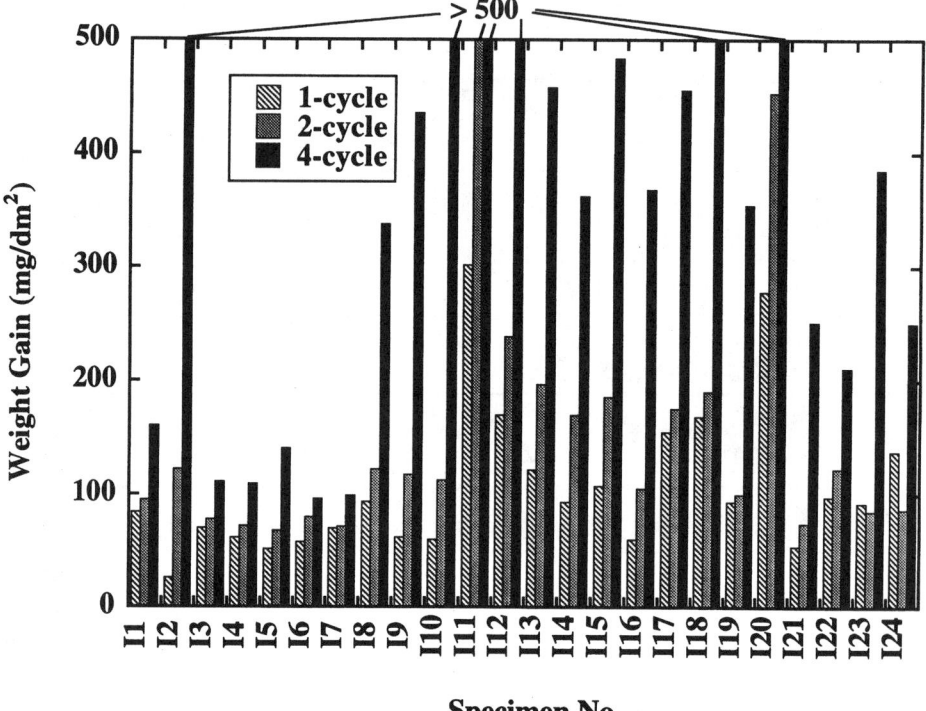

FIG. 6—*Comparison of weight gain data from in-pile corrosion tests up to four cycles using I series specimens.*

increase visibly from 1 to 8 × 10^{25} n/m². Alloy I6 is a tube shell heat-treated specimen, and the average oxide thickness was measured on the outer surfaces. Corrosion performance of the outer surface is important for the fuel cladding; thus, Fig. 9a suggests that Alloy I6 is suitable for high burnup fuel.

Hydrogen contents of some specimens were measured by gas chromatography or estimated from hydride distributions in micrographs. Hydrogen contents by both methods show good agreement. Heavily oxidized alloys, for example Alloy I9 (~500 ppm), show higher hydrogen pickup than Zry-2 (~150 ppm). The hydrogen content of Alloy I6 (~150 ppm) is comparable to that with Zry-2 (I1). Figure 9b shows the dependence of hydrogen content on the fast neutron fluence of Alloys I1 and I4 to I6. Hydrogen content strongly saturates as a function of fast neutron fluence for Alloy I6.

Zry-2 has thicker oxide than Alloy I6; therefore, the hydrogen pickup fraction of Alloy I6 is higher than Zry-2. Hydrogen pickup fractions are compared in Fig. 9c as a function of fast neutron fluence. After two cycles of irradiation, hydrogen pickup fractions among Zry-like alloys show only a small difference; however, the difference becomes clear after four cycles of irradiation. The effects of Fe and/or Ni contents on the hydrogen pickup fraction were investigated using four-cycle irradiation data. The results and correlation coefficients are shown in Fig. 10. Hydrogen pickup fractions are the average values of two specimens. These figures indicate that Ni content or the Ni/Fe ratio strongly affect the hydrogen pickup fraction of Zry-like alloys.

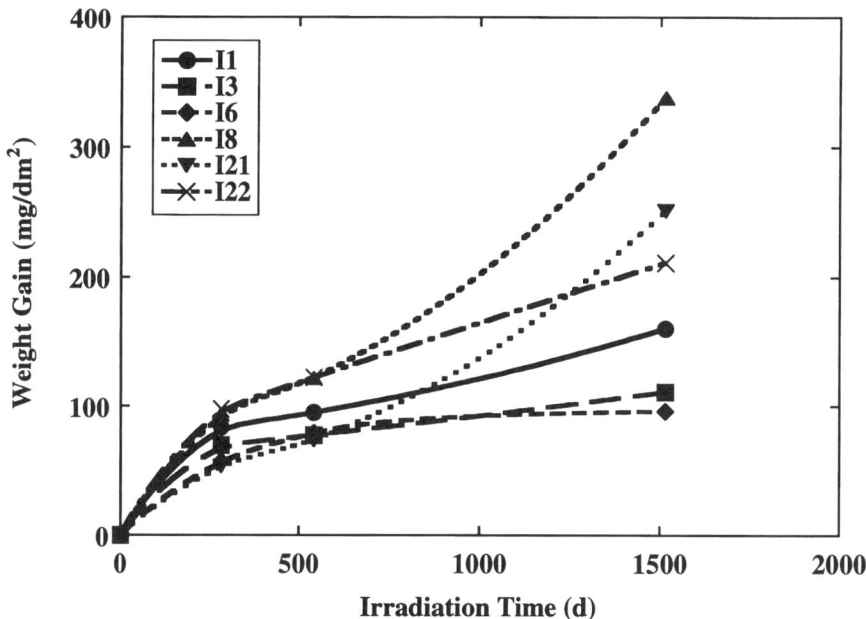

FIG. 7—*Irradiation time dependence of weight gain data for typical I series specimens up to a four-cycle irradiation test.*

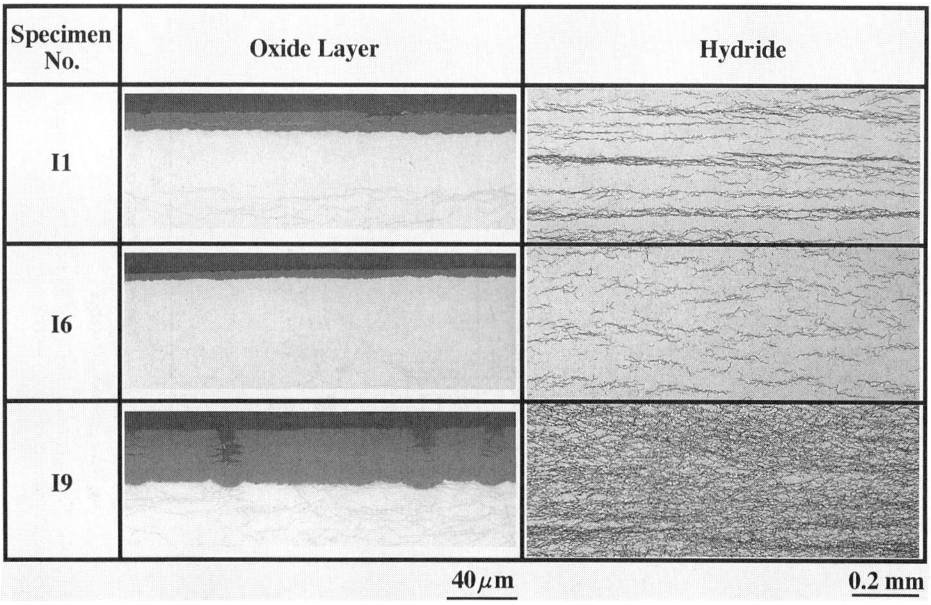

FIG. 8—*Typical metallographs of I series specimens irradiated for four cycles.*

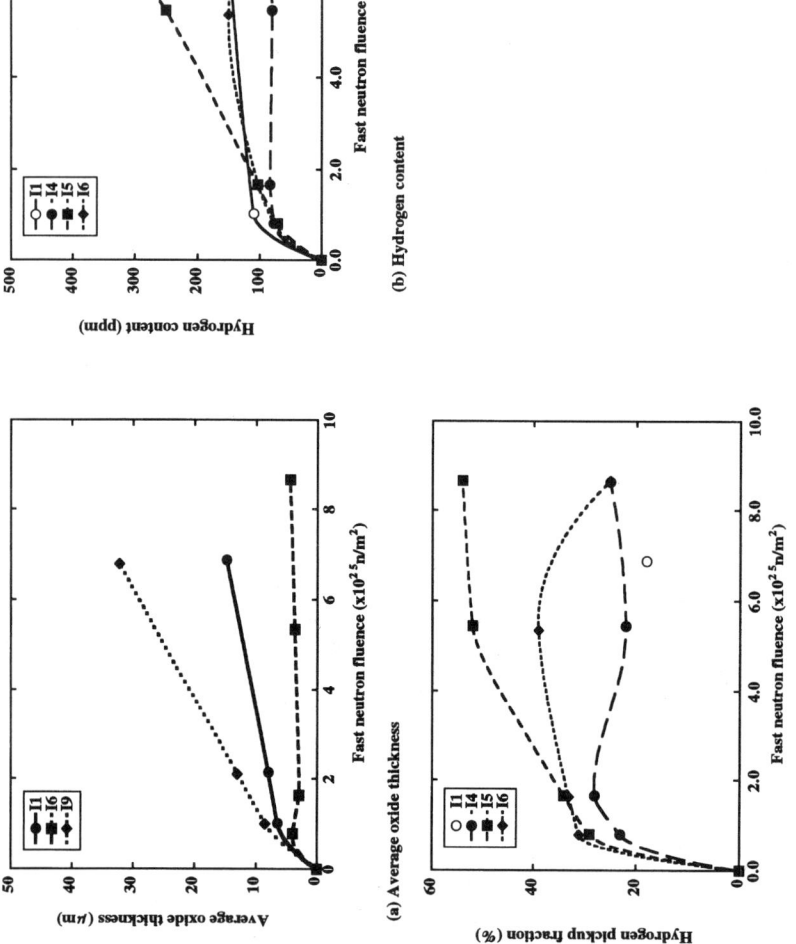

FIG. 9—*Fast neutron fluence dependence of average oxide thicknesses, hydrogen contents, and hydrogen pickup fractions for typical I series specimens irradiated up to four cycles.*

ETOH ET AL. ON NEW ZIRCONIUM ALLOYS 839

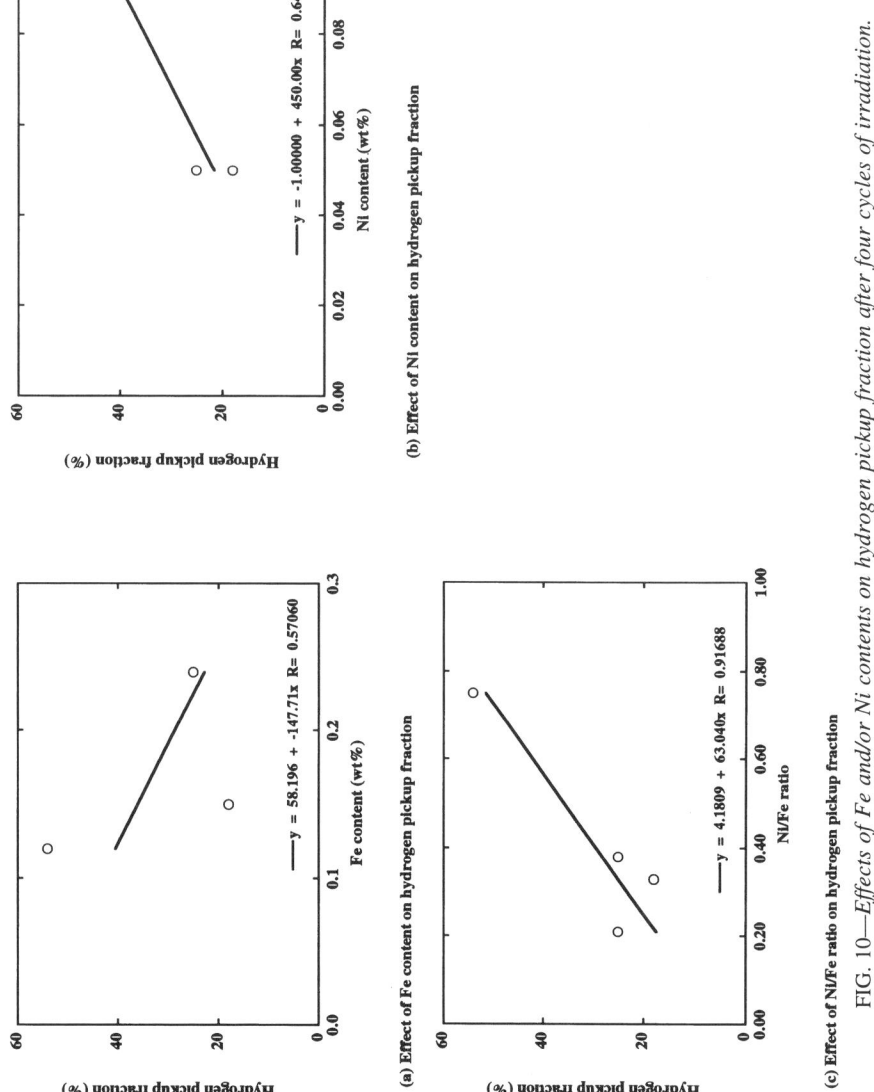

FIG. 10—*Effects of Fe and/or Ni contents on hydrogen pickup fraction after four cycles of irradiation.*

Effect of Fast Neutron Flux on In-Pile Corrosion

A few specimens of the same alloy were irradiated in the dummy neutron source holders and exposed at different fast neutron flux positions. It is possible to evaluate the effects of fast neutron flux and irradiation time on the corrosion behavior of Zr alloys, separately, by multiple regression analysis. The equation used for multiple regression analysis is

$$W.G. = a\phi^b t^c \tag{1}$$

where $W.G.$ is weight gain, ϕ is fast neutron flux, t is irradiation time, and a, b, and c are constants.

Figure 11 shows the fast neutron fluence dependence of weight gains of three alloys for each irradiation position. For Alloy I3, three specimens were irradiated at different positions for each cycle. Fluence dependence of weight gains are shown in Fig. 11a. The weight gain of Alloy I3 depends on fast neutron fluence. Multiple regression analysis gives values $b = 0.21 \pm 0.14$ and $c = 0.25 \pm 0.04$. Thus, Alloy I3 has a rate constant, c, below that for cubic oxidation ($c = 0.33$) and is relatively insensitive to fast neutron flux as implied by the large error in the fitted value.

Figure 11b shows the case for Alloy I6. The rate constant is similar to Alloy I3; however, the sensitivity to fast neutron flux is slightly higher. Multiple regression analysis gives $b = 0.69 \pm 0.15$ and $c = 0.20 \pm 0.06$. Figure 11c is for Alloy I9, which shows almost no dependence on fast neutron flux or negative effect and near linear oxidation kinetics. Multiple regression analysis gives $b = -0.42 \pm 0.30$ and $c = 1.23 \pm 0.07$.

Discussion

Comparison of Corrosion Behaviors of Zr Alloys

From the in-pile corrosion tests of I series specimens, it is clear that transition occurred for many Zr alloys between one and four cycles. Even Zry-2 showed a mild transition behavior. On the other hand, for Zircaloy-like alloys, increasing Fe and Ni contents or decreasing Sn content delayed transition beyond four cycles. This result suggested that these modified alloys of Zry-2 were promising for high burnup fuels regarding corrosion behavior.

Hydrogen pickup is an important behavior in high burnup fuels. Increasing Ni content can accelerate hydrogen pickup. The hydrogen pickup fraction depends on Ni content or the Ni/Fe ratio and, for example, Alloy I6 has a higher Ni content than Zry-2. As a result, Alloy I6 showed a higher hydrogen pickup fraction than Zry-2; however, the hydrogen content of Alloy I6 was comparable to that of Zry-2 after the four-cycle irradiation because of lower oxidation than Zry-2. Therefore, hydrogen pickup is not a critical problem for Alloy I6.

Based on the results of this study, it could be concluded that modified alloys of Zry-2, such as increasing Fe and Ni content or decreasing Sn content, were the most promising alloys for high burnup fuels in BWRs. The same result was reported for out-of-pile corrosion tests by R. A. Graham and C. M. Eucken [10]. It should be noted that Ni decreases oxidation but increases the hydrogen pickup fraction, so that Ni content should be determined to optimize both oxidation and hydrogen pickup. Since the concepts of high Fe and Ni contents and low Sn content are compatible, the combination of the above two concepts offers interesting possibilities.

Correlation Between In-Pile and Out-of-Pile Corrosion

T series specimens were tested under both in-pile and out-of-pile conditions. Investigating the correlations between the results from in-pile and out-of-pile corrosion tests offers the chance

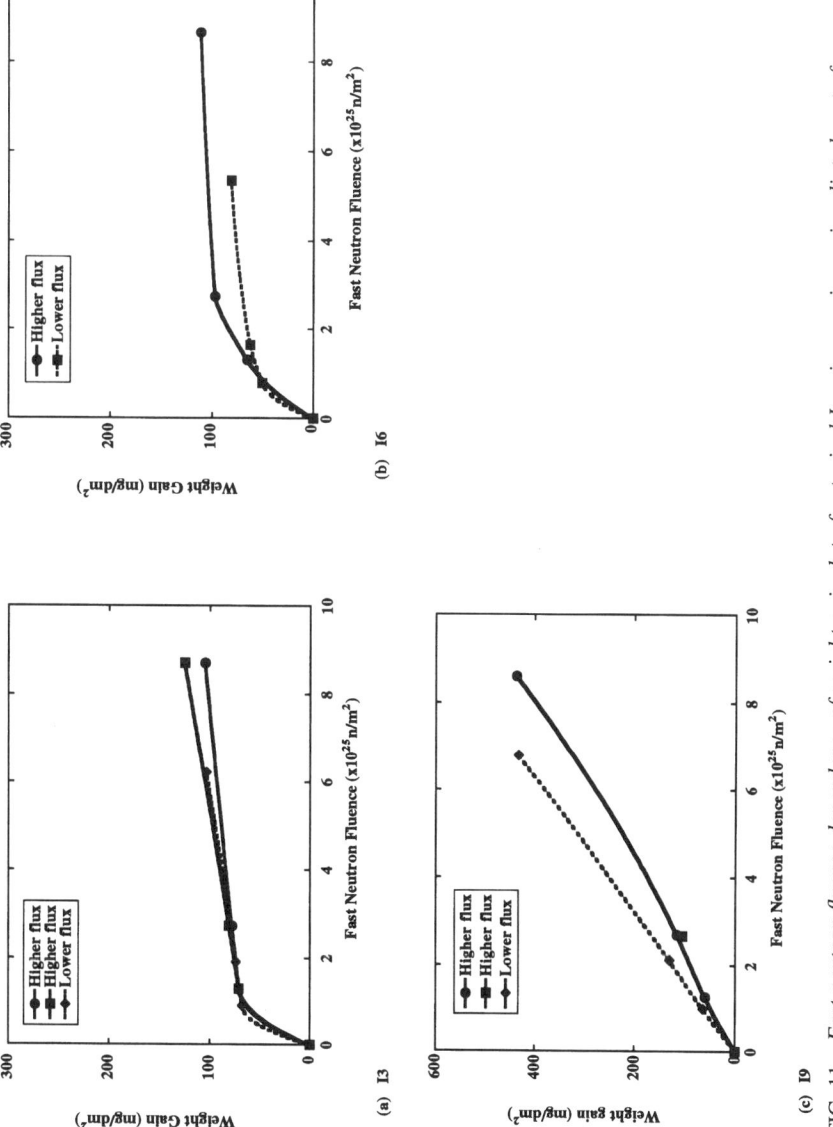

FIG. 11—*Fast neutron fluence dependence of weight gain data for typical I series specimens irradiated up to four cycles.*

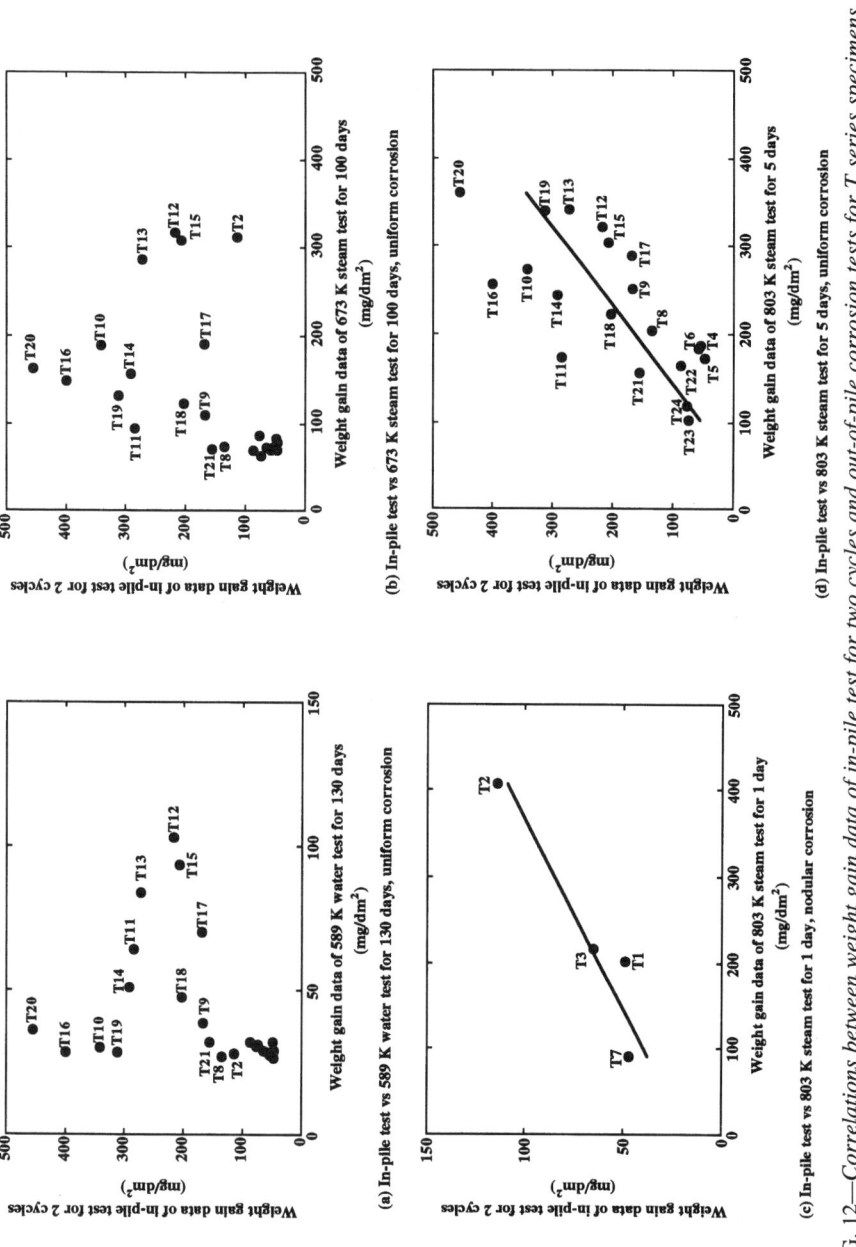

FIG. 12—*Correlations between weight gain data of in-pile test for two cycles and out-of-pile corrosion tests for T series specimens.*

to clarify their difference. Figure 12 shows the correlations between weight gain data obtained from the two-cycle irradiation test and the four types of out-of-pile corrosion tests. Figure 12*a* is for the 589 K water test up to 130 days. Under this test condition, some alloys show high weight gains. These alloys contain high Nb concentrations, i.e., more than 0.5 wt%, and show high weight gains for the in-pile test. Weight gains of four alloys, such as T10, T16, T19, and T20, are low for the 589 K water test, while they are highest for the in-pile test. The corrosion performances of these alloys cannot be simulated by the 589 K water test up to 130 days. These alloys contain Mo or Te, and their Nb contents are less than 0.4 wt%. These results suggest that the 589 K water test up to 130 days may be used to predict the behavior of Nb, but not Mo or Te effects for in-pile corrosion.

Results of the steam test at 673 K up to 100 days correlate a little better than the 589 K water test results, as shown in Fig. 12*b*. The high weight gains of alloys containing Mo or Te are reproduced by the 673 K steam test, while alloys containing high concentrations of Nb and Mo (T12, T13, and T15) or Sn (T2) show higher weight gains than the other alloys. Alloys T11 and T14 show high weight gains for the 589 K water test; however, their weight gains agree with the other alloys for the 673 K steam test. Alloys T11 and T14 contain an impurity level and 0.19 wt% Mo. These amounts are smaller than that of Alloys T12, T13, and T15. These results suggest that effects of Mo, Te, and Sn can be demonstrated for the 673 K steam test, while the effect of Nb cannot.

Out-of-pile nodular corrosion tests show that only four alloys are susceptible for nodular corrosion. The results from the MAT test and 803 K steam test for one day are very similar, and these tests show good correlation with the in-pile test for alloys, which shows nodular corrosion in the out-of-pile test, as shown in Fig. 12*c*. The other alloys show only uniform corrosion in the 803 K steam test for five days, and their weight gains have a good correlation with the in-pile test, as shown in Fig. 12*d*. For nodular corrosion, the 803 K steam test for one day has a good correlation with the in-pile test, and for uniform corrosion, the 803 K steam test for five days shows the best correlation among the three out-of-pile corrosion tests in this study.

Effects of Valence of Alloying Elements on Corrosion

Based on the results obtained in this study, increasing Fe and Ni content and decreasing Sn content improved corrosion performance of Zr alloys under BWR conditions, while Nb, Mo, and Te addition increased the corrosion of Zr alloys. Except for Sn, elements that improved corrosion resistance have a different valence than Zr. The effect of valence on the corrosion behavior of Zr alloys was investigated from a macroscopic viewpoint. Figure 13 shows the dependence of weight gains of T and I series specimens on the valence of the alloying element. Oxidation states used in Fig. 12 are Ni^{2+}, Cr^{3+}, Fe^{3+}, Bi^{3+}, Zr^{4+}, Sn^{4+}, Si^{4+}, Nb^{5+}, Mo^{6+}, and Te^{6+}. The horizontal axis indicates the total valence difference with Zr, i.e., $\Sigma \Delta Vi \times Ci$, where ΔVi is the valence difference between alloying element i and Zr, and Ci is added concentration of the alloying element i. Figure 13 shows a large scattering of data because other major effects on corrosion performance (i.e., Sn content, microstructures, distributions of alloying elements, changes of microstructures and alloying element distributions during corrosion tests and irradiation, ionic sizes of alloying elements) are not taken into account. In spite of this, dependence of weight gain data on $\Sigma \Delta Vi \times Ci$ is found for all of the corrosion tests in this study. This suggests that the valence of alloying elements is a factor influencing the corrosion performance of Zr alloys. Lower valence elements seem to improve corrosion resistance, while higher valence elements degrade corrosion resistance for both in-pile and out-of-pile corrosion tests. The valence effect on nodular corrosion was discussed by Taylor et al.

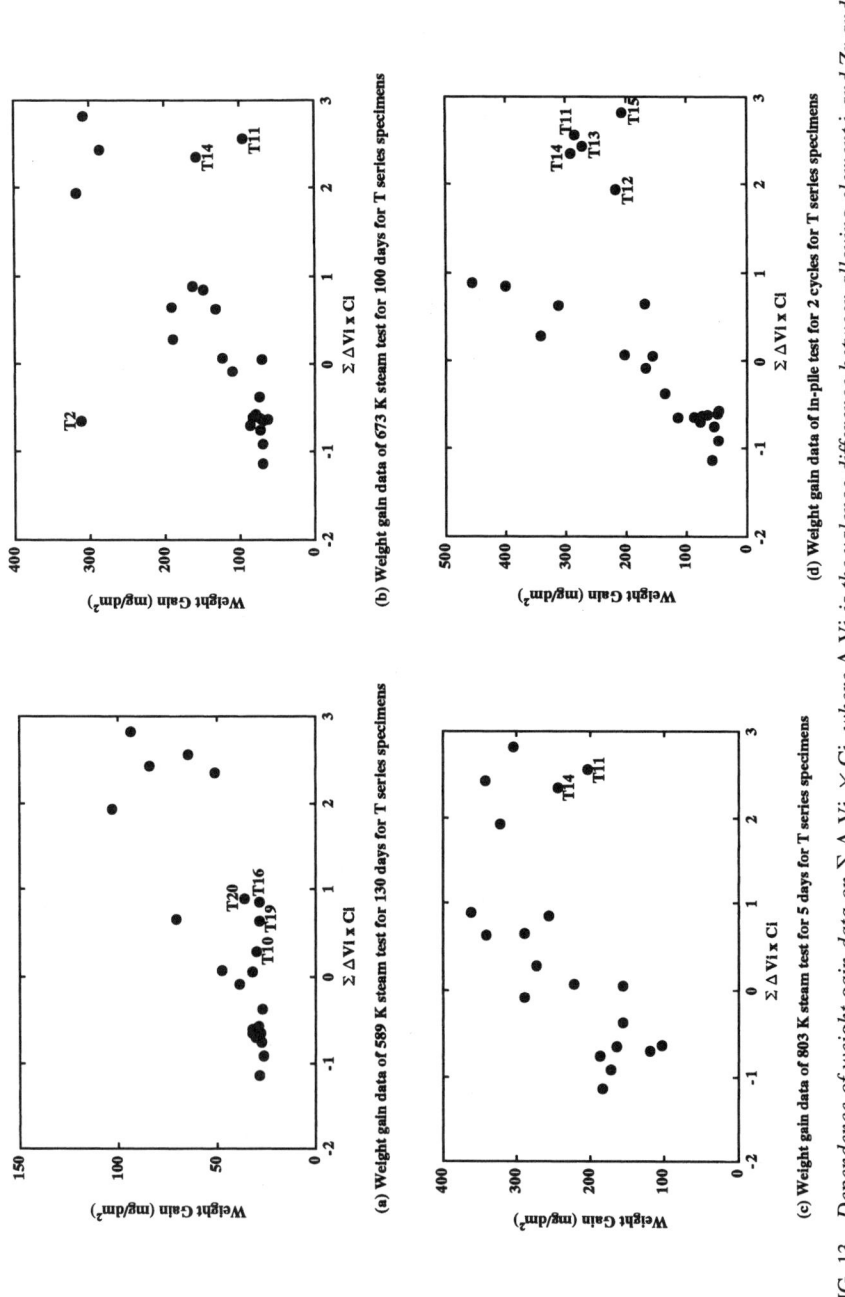

FIG. 13—*Dependence of weight gain data on $\Sigma \Delta V_i \times C_i$, where ΔV_i is the valence difference between alloying element i and Zr and C_i is the added concentration of alloying element i in at%.*

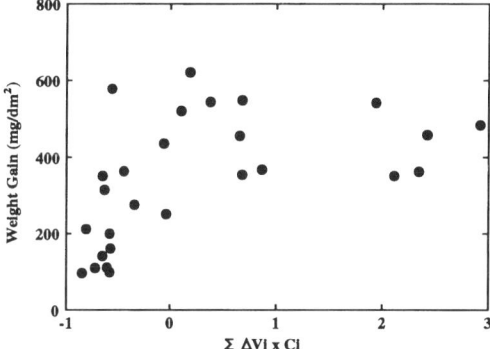

(e) Weight gain data of in-pile test for 4 cycles for I series specimens

FIG. 13—Continued

[11,12] and Harding [13] for the 783 K steam test. They conclude that aliovalent oxidation states of alloying elements can protect Zr alloys from nodular corrosion. Results from this study are mainly for uniform corrosion, and different effects of valence are observed.

There are two ways to improve the corrosion resistance of Zr alloys. One is to elongate the lifetime of protective, dense oxide films, i.e., to prevent or delay a breakaway of oxidation kinetics. Another is to suppress the transportation of oxidizing species, protons, or electrons through protective, dense oxide films. Therefore, two mechanisms have been considered for the valence effect on the corrosion behavior of Zr alloys. When a lower-valency metal is substituted for Zr in ZrO_2, oxygen vacancies increase to compensate charge balance. In the first mechanism, an increase of oxygen vacancies stabilizes the protective oxide film in a water or steam environment, while near-stoichiometric ZrO_2 film is unstable and transforms to granular oxide [3]. In the second mechanism, oxygen vacancies act as holes decreasing the conductivity of electrons [14]. If the counter diffusion of electrons against O^{2-} is not enough, a potential barrier is formed for O^{2-} diffusion from the oxide surface to the oxide/metal interface, so that the growth rate of oxide decreases.

Corrosion resistance is not controlled by only the valence effect. Alloying elements also affect the strength of the metal. High strength of the metal produces high compressive stresses in the oxide due to the volume difference between the metal and oxide and the succeeding early breakaway [15–18]. Bryner [19] and Dollins and Jursich [20] proposed the oxidation model in which oxygen diffusivity in the oxide depended on the compressive stresses. Therefore, compressive stresses in the oxide are also considered to be an important factor influencing corrosion behavior of Zr alloys.

Behaviors of Alloying Elements During Corrosion

Fe and Ni improved the corrosion resistance of Zr alloys for both in-pile and out-of-pile tests. These elements have very small solubility limits in α-Zr [21] or α–Zr-Sn matrix [22]. The majority of added Fe and Ni atoms form bct-Zr_2(Fe,Ni) intermetallic precipitates [23]. This type of intermetallic precipitate dissolves easily in the oxide [22], and a part of the added Fe and Ni might be substituted for Zr in the oxide matrix. Due to the valence effect, Fe and Ni would be effective for both in-pile and out-of-pile corrosion.

Cr does not influence the corrosion resistance of Zr alloys for out-of-pile tests, while Cr improves it for in-pile tests. Cr has very small solubility limits similar to Fe and Ni [21,22]. The majority of the added Cr atoms form hcp-Zr(Fe,Cr)$_2$ intermetallic precipitates [23]. This type of intermetallic precipitate shows a slower dissolution rate in the oxide than bct-Zr$_2$(Fe,Ni) [24], while Fe diffuses from hcp-Zr(Fe,Cr)$_2$ to the metal matrix during neutron irradiation under BWR conditions [25]. Therefore, diffused Fe during irradiation might occupy the substitutional site for Zr in oxide and enhance corrosion resistance. Cr itself has a minor effect on corrosion due to slower diffusion into the matrix during irradiation [26]. Therefore, Cr addition is effective only for an in-pile test where Cr might control the distribution of Fe released under neutron irradiation.

An increase of added Sn content deteriorated the corrosion resistance of Zr alloys for both in-pile and out-of-pile tests, yet the valence of Sn is the same as that of Zr. This apparent anomaly can be explained by the fact that Sn dissolves in a α-Zr matrix and above certain concentration disrupts the formation of protective oxide [27]. The reason for the Sn effect could be because the Sn addition increased the strength of Zr alloys and oxide films on the surface suffered high compressive stresses [15–18]. On the other hand, the diffusion coefficients of Fe and Cr in Zr and Zry-2 are very different. Diffusivity of Fe at 1003 K in α-Zr is about five orders of magnitude greater than that in Zry-2 [28]. It could be expected that Sn content limits the Fe and Ni diffusion in both the metal and oxide during oxidation and/or irradiation and suppresses lower valence effect.

Nb is partially soluble in the α-Zr matrix. Due to the valence effect and solubility, Nb accelerates the pre-transition corrosion rate in the 589 K water test for ten days, as shown in Fig. 2. This result suggests that Nb primarily affects corrosion pretransition behavior and that alloying elements such as Fe, Cr, Ni, Mo, and Te play a lesser role in corrosion behavior. During oxidation and irradiation, these elements gradually increase their dissolved concentrations in the oxide matrix and contribute to improved or deteriorated corrosion resistance.

Conclusions

Based on the above results and discussions, the following conclusions were obtained:

1. Two cycles of irradiation were not always sufficient to rank longer-term corrosion behavior. The transition to accelerated corrosion occurred for many alloys between one and four cycles.

2. The onset of transition was observed for standard Zircaloy-2 at four cycles of irradiation. Zircaloy-like alloys containing increased Fe and Ni contents or decreased Sn content delayed the onset of transition over four cycles of irradiation.

3. Removal of Cr from Zircaloy-2 did not affect the corrosion behavior for the out-of-pile tests; however, it increased weight gains in the in-pile tests.

4. Alloying elements, such as Nb, Mo, and Te, which have larger valences than Zr, increased the weight gain during the irradiation test in the BWR.

5. Alloying elements Fe, Cr, and Ni, which have smaller valences than Zr, improved the in-pile corrosion resistance of Zr alloys.

6. The 803 K steam test showed the best correlation with in-pile and out-of-pile tests for both nodular and uniform corrosion.

7. The corrosion rate of modified Zircaloy-2 alloys increases with increasing neutron flux, while the corrosion rates of alloys with Nb additions either decreased or remained unchanged with increasing neutron flux.

8. Although the alloys with increasing Fe and Ni contents in Zircaloy-2 showed a higher percent hydrogen pickup, the amount of hydrogen pickup was almost the same as for Zircaloy-2.

Acknowledgments

This work is a cooperative study between Tokyo Electric Power Co., Tohoku Electric Power Co., Chubu Electric Power Co., Hokuriku Electric Power Co., Chugoku Electric Power Co., Japan Atomic Power Co., Hitachi Ltd., Toshiba Corp., General Electric Company, and Nippon Nuclear Fuel Development Co., Ltd. The authors acknowledge these member organizations for their roles in the study.

References

[1] Garzarolli, F., et al., "Waterside-Corrosion and Growth of Zry-Clad Fuel Rods Under the Aspect of High Burnup," *Proceedings,* IAEA Specialists' Meeting on High Burnup in Power Reactor Fuel, Mol, Yugoslavia, 1981.
[2] Stehle, H. and Holtzer, R., "KWU LWR Fuel Experience, New Design and Development Activities," *Proceedings,* BNES Conference on Nuclear Fuel Performance, Stratford-on-Avon, England, 1985.
[3] Cheng, B. and Adamson, R. B., "Mechanistic Studies of Zircaloy Nodular Corrosion," *Zirconium in the Nuclear Industry: Seventh International Symposium, ASTM STP 939,* American Society for Testing and Materials, West Conshohocken, PA, 1987, p. 387.
[4] Mishima, Y. and Aoki, T., "Proving Test on the Reliability of BWR 8×8 Fuel Assemblies in Japan," *Proceedings,* International Symposium on Improvements in Water Reactor Fuel Technology and Utilization, Paper IAEA-SM-288/58, Stockholm, 1986.
[5] Etoh, Y. in *Journal of Nuclear Science and Technology,* Vol. 26, No. 8, 1989, p. 752.
[6] Rudling, P., Vannesjo, K. L., Vesterlund, G., and Massih, A. R., "Influence of Second Phase Particles on Zircaloy Corrosion in BWR Environment," *Zirconium in the Nuclear Industry: Seventh International Symposium, ASTM STP 939,* American Society for Testing and Materials, West Conshohocken, PA, 1987, p. 292.
[7] Andersson, T. and Thorvaldsson, T., "Nodular Corrosion Resistance of Zircaloy-2 in Relation to Second-Phase Particle Distribution," *Zirconium in the Nuclear Industry: Seventh International Symposium, ASTM STP 939,* American Society for Testing and Materials, West Conshohocken, PA, 1987, p. 321.
[8] Garzarolli, F., Schumann, R., and Steinberg, E., "Corrosion Optimized Zircaloy for Boiling Water Reactor (BWR) Fuel Elements," *Zirconium in the Nuclear Industry: Tenth International Symposium, ASTM STP 1245,* American Society for Testing and Materials, West Conshohocken, PA, 1994, p. 709.
[9] Tucker, R. P., Fidleris, V., and Adamson, R. B., "High-Fluence Irradiation Growth of Zirconium Alloys at 644 to 725 K," *Zirconium in the Nuclear Industry: Sixth International Symposium, ASTM STP 824,* American Society for Testing and Materials, West Conshohocken, PA, 1984, p. 427.
[10] Graham, R. A., and Eucken, C. M., "Controlled Composition Zircaloy-2 Uniform Corrosion Resistance," *Zirconium in the Nuclear Industry: Ninth International Symposium, ASTM STP 1132,* American Society for Testing and Materials, West Conshohocken, PA, 1991, p. 279.
[11] Taylor D. F., Cheng, B., and Adamson R. B., "Nodular Corrosion Mechanisms and Their Application to Alloy Development," *Proceedings,* IAEA Technical Committee Meeting on Fundamental Aspects of Corrosion of Zirconium-based Alloys in Water Reactor Environments, Portland, 11–15 Sept. 1989.
[12] Taylor D. F. in *Journal of Nuclear Materials,* Vol. 184, 1991, p. 65.
[13] Harding, J. H. in *Journal of Nuclear Materials,* Vol. 202, 1993, p. 216.
[14] Inagaki, M., Kanno, M., and Maki, H., "Effect of Alloying Elements in Zircaloy on Photo-Electrochemical Characteristics of Zirconium Oxide Films," *Zirconium in the Nuclear Industry: Ninth International Symposium, ASTM STP 1132,* American Society for Testing and Materials, West Conshohocken, PA, 1991, p. 437.
[15] Bradhurst, D. H. and Heuer, P. M. in *Journal of Nuclear Materials,* Vol. 37, 1970, p. 35.
[16] Evans, H. E. in *Corrosion Science,* Vol. 23, 1983, p. 495.
[17] Srolovitz, D. J. and Anderson, M. P. in *Acta Metallurgica,* Vol. 32, 1984, p. 1089.
[18] Evans, H. E. in *Materials Science and Engineering,* Vol. A 120, 1989, p. 139.
[19] Bryner, J. S. in *Journal of Nuclear Materials,* Vol. 82, 1979, p. 84.
[20] Dollins, C. C. and Jursich, M. in *Journal of Nuclear Materials,* Vol. 113, 1983, p. 19.
[21] Hansen, M. and Anderko, K., *Constitution of Binary Alloys,* McGraw-Hill, New York, 1958.
[22] Charquet, D., Hahn, R., Ortlieb, E., Gros, J. P., and Wadier, J. F., "Solubility Limits and Formation of Intermetallic Precipitates in ZrSnFeCr," *Zirconium in the Nuclear Industry: Eighth International*

Symposium, ASTM STP 1023, American Society for Testing and Materials, West Conshohocken, PA, 1989, p. 405.
[23] Rao, P., et al., General Electric Company Report 76 CRD, 1977, p. 183.
[24] Kubo, T. and Uno, M., "Precipitate Behavior in Zircaloy-2 Oxide Films and Its Relevance to Corrosion Resistance," *Zirconium in the Nuclear Industry: Ninth International Symposium, ASTM STP 1132,* American Society for Testing and Materials, West Conshohocken, PA, 1991, p. 476.
[25] Etoh, Y. and Shimada, S. in *Journal of Nuclear Materials,* Vol. 200, 1993, p. 59.
[26] Etoh, Y. and Shimada, S. in *Journal of Nuclear Science and Technology,* Vol. 29, No. 4, 1992, p. 358.
[27] Kass, S., "The Development of the Zircaloys," *Corrosion of Zirconium Alloys, ASTM STP 368,* American Society for Testing and Materials, West Conshohocken, PA, 1964, p. 3.
[28] Pande, B. M., Naik, M. C., and Agarwara, R. P. in *Journal of Nuclear Materials,* Vol. 28, 1968, p. 324.

DISCUSSION

A. T. Motta[1] (written discussion)—(1) Do you attribute the valence effect to the availability of electrons in the oxide layer affecting transport properties? (2) Have you measured the irradiation growth of the NOCR alloys?

Y. Etoh et al. (authors' closure)—(1) Yes, we do. And we are thinking of another effect of lower valence on corrosion, which is the stabilization of protective oxide film. (2) We are now conducting an irradiation test of some alloys to measure the irradiation growth. The NOCR alloys are included in this test.

B. Cheng[2] (written discussion)—You suggest a higher corrosion rate of the Nb alloy in terms of the valence state. But, another factor in corrosion of zirconium alloys is the O_2 content in the coolant. This is an important factor in explaining the difference of Nb-containing alloys in BWR and PWRs.

Y. Etoh et al. (authors' closure)—This paper shows that Nb addition accelerates the corrosion rate of Zr alloys in BWR conditions. But Nb addition improves corrosion resistance in PWR. I think this difference is derived mainly from the O_2 content in the coolant. Different O_2 content produces different distribution of oxygen potential across the oxide and causes different valence states of Nb in the oxide. We can guess that Nb would be oxidized to higher valence in BWR than in PWR.

B. Cox[3] (written discussion)—The correlations between valence of alloying additions and corrosion resistance in BWR conditions have always been difficult to understand because both from theory and observation these elements (whether of lower or higher valence than Zr) tend to be present in the oxide near the oxide metal interface in low or zero valence states. Thus, another explanation needs to be sought and perhaps their effect on oxide conductivity is the direction to consider.

Y. Etoh et al. (authors' closure)—We suppose that there is an oxygen potential distribution across the oxide; therefore, the valence state of multi-valent elements varies from zero to the maximum valence state with increasing distance from the metal/oxide interface. But the thickness of oxide containing lower valent elements might be different between the elements of lower and higher valence than Zr. We suppose that this thickness difference is the reason for the valence effect on corrosion. Lower valence than Zr would prohibit electron transportation and stabilize the protective oxide.

---

[1] Pennsylvania State University, University Park, PA.
[2] Electric Power Research Institute, Palo Alto, CA.
[3] University of Toronto, Toronto, Ontario, Canada.

F. Garzarolli,[1] Y. Broy,[1] and R. A. Busch[2]

Comparison of the Long-Time Corrosion Behavior of Certain Zr Alloys in PWR, BWR, and Laboratory Tests

REFERENCE: Garzarolli, F., Broy, Y., and Busch, R. A., "**Comparison of the Long-Time Corrosion Behavior of Certain Zr Alloys in PWR, BWR, and Laboratory Tests,**" *Zirconium in the Nuclear Industry: Eleventh International Symposium, ASTM STP 1295,* E. R. Bradley and G. P. Sabol, Eds., American Society for Testing and Materials, 1996, pp. 850–864.

ABSTRACT: Laboratory corrosion tests have always been an important tool for Zr alloy development and optimization. However, it must be known whether a test is representative for the application in-reactor. To shed more light on this question, coupons of several Zr alloys were exposed under isothermal conditions in all or most of the following environments:

In-Reactor:

(1) PWR core at 300 to 340°C up to six years.
(2) BWR core with a low sensitivity to nodular corrosion up to four years.
(3) BWR core with a high sensitivity to nodular corrosion up to two years.

Ex-Reactor (in Autoclave):

(1) 350°C/pressurized water up to three years.
(2) 400°C/100-bar steam up to two years.
(3) 350°C/0.01 M LiOH water up to two years.
(4) 500 to 515°C/high-pressure steam 16 to 24 h.

In addition, the material condition of several of the examined Zr alloys was varied over a wide range.

For evaluation of the in-PWR tests and for comparison of out-of-pile and in-pile tests, the different temperatures and times were normalized to a temperature-independent "normalized time" by assuming an activation temperature (Q/R) of 14 200 K. Comparison of in-PWR and out-of-pile corrosion behavior of Zircaloy shows that corrosion deviates to higher values in PWR if a weight gain of about 50 mg/dm^2 is exceeded. In the case of the Zr2.5Nb alloy, a slight deviation of corrosion as compared to laboratory results starts in PWR only above a weight gain of 100 mg/dm^2. In BWR, corrosion of Zircaloy is enhanced early in time if compared with out-of-pile. Zr2.5Nb exhibits higher corrosion results in BWR than Zircaloy-4.

Alloying chemistry and material condition affect corrosion of Zr alloys. However, several of the material parameters have shown a different ranking in the different environments. Nevertheless, several material parameters influencing in-reactor corrosion like the second phase particle (SPP) size or in-PWR behavior as the Sn and Fe content can be optimized by out-of-pile corrosion tests.

KEYWORDS: zirconium alloys, Zircaloy, corrosion, in-PWR corrosion, in-BWR corrosion, out-of-pile corrosion, neutron irradiation, radiation effects, nuclear application

In light water reactor (LWR) fuel assemblies, Zr alloys are used for fuel rod claddings and structural components in the high neutron flux range. In PWR, a uniform oxide layer is formed

---

[1] Siemens AG, Power Generation Group (KWU), 91050 Erlangen, Germany.
[2] Siemens Power Corporation, Nuclear Division, Richland, Washington.

that increases with increasing operating temperature and time. In BWR, which operates at a lower peak operating temperature, a nodular oxide is formed besides a thin uniform oxide layer. The corrosion behavior of the Zr alloys is an important parameter for optimizing fuel assembly economy. It is an everlasting attempt to optimize the corrosion behavior of the Zr alloys by using simple tests that can be performed in the laboratory. Out-of-pile long-time corrosion tests in pressurized water at 350 to 370°C or 100-bar steam at 400 to 420°C are being used to study uniform corrosion and short-time corrosion tests at 500 to 550°C in high-pressure steam to study nodular corrosion [1–4]. Furthermore, tests in diluted LiOH have been performed to study the potential effects of a concentration of Li at the surface of PWR fuel rods. Certainly, such rather simple tests can only be used for material optimization if the correctness of the test answer for the application is validated.

In-reactor corrosion data gained by examinations of standard or experimental fuel assemblies [3–6] are the most relevant basis for material optimization. However, it is almost impossible to study all variants to be considered for material optimization with experimental fuel rods in a power reactor. Furthermore, such tests have to be done in both systems because irradiation effects on corrosion behavior are principally different in the hydrogenated PWR and the oxygenated BWR environments. Many attempts were made to compare the corrosion in different environments for a validation of the out-of-pile tests, but mostly the available in-reactor and out-of-pile data are from different material lots. Therefore, a parameter study of the different influencing factors was performed with a consistent set of material samples in all the different environments of interest. Several archive lots and experimental Zr alloy melts were selected. The material conditions of several of these alloys were varied over a wide range. For this study, mini-coupons have been used that also allow a quite large number of test variants in reactor.

Materials and Experimental Methods

The material coupons that have been studied in PWR, BWR, and laboratory tests are from different Zircaloy-2 and Zircaloy-4 tubing lots and different small melts with large variations of the alloy composition (inside and outside the ASTM range for Zircaloy-2 and -4). The chemical composition of these alloys is given in Table 1. The claddings were all fabricated according to standard routines. The small melts were fabricated to strips using a process as close as possible to the fabrication process of cladding tubes. The accumulated annealing parameters of these material coupons were in the range used for PWR and BWR cladding tubes and were varied to higher and lower values. The materials can be subdivided in such with low process temperatures (LTP) with accumulated annealing parameters below <2E-18 h and such with high process temperatures (HTP) with accumulated annealing parameters between 4E-18 and 1E-16 h (using a Q/R value of 40 000 K). The materials received as final annealing are either a recrystallization anneal or a stress relive anneal. In several cases, both treatments were used in parallel for the same materials. Certain coupons were quenched from the β-temperature with different controlled quenching rates and afterwards annealed between 400 and 750°C.

TABLE 1—*Chemical composition of materials.*

Material	Melt	Sn, %	Nb, %	Fe, %	Cr, %	Ni, %
Zry-2 BWR cladding	Standard	1.3–1.6	...	0.12–0.16	0.09–0.12	0.04–0.06
Zry-4 PWR cladding	Standard	1.3–1.6	...	0.19–0.23	0.09–0.11	...
Zr2.5Nb cladding	Standard		2.5–2.7			
Zry-4 variants	Small	1.1–1.8	...	0.15–024	0.06–0.12	...
Alternate Zr alloys	Small	0–1.6	...	0.05–0.74	0.01–1.05	0–0.06

For this study, mini-coupons (35 by 7 by 0.5 mm) were used. The coupon numbers were engraved before the final surface treatment. Some of the coupons were sand-blasted, and all were finally pickled. Coupons of different materials were exposed under isothermal conditions in all or most of the following environments:

In-Reactor

(1) PWR core at 300 to 340°C up to six years.
(2) BWR core with a low sensitivity to nodular corrosion up to four years.
(3) BWR core with a high sensitivity to nodular corrosion up to two years.

Ex-Reactor (in Autoclave)

(1) 350°C/pressurized water up to three years.
(2) 400°C/100 bar steam up to two years.
(3) 350°C/0.01 M LiOH-water up to two years.
(4) 500 to 515°C/high-pressure steam 16 to 24 h.

The out-of-pile corrosion tests in 350°C water were performed in 5 and 10-L stainless steel autoclaves in deionized water or in water with lithium additions under static conditions. The weight gain of the coupons was measured in intervals of 7 to 30 days. Prior to each corrosion run, the autoclave was purged with argon gas, evacuated, and a pressure of 0.25 bar hydrogen and 5 to 6 bar argon was applied. The corrosion tests in 400 and 500°C steam were performed at 105 and 125 bar in a refreshed-type autoclave. This latter test is described in more detail in Ref *3*. The corrosion rate was determined by weight gain measurements and in some cases also by infrared spectroscopy.

The in-reactor tests were performed in one PWR and two BWRs (A and B). BWR-A has a low and BWR-B has a high sensitivity to nodular corrosion. The PWR and BWR-A corrosion coupons were inserted in segmented water test rods (WTRs) replacing fuel rods. For the cladding tubes of the individual segments of the WTRs, different Zircaloys and Zr alloys were used. This technique is described in more detail in Ref *6*. In BWR-B, the corrosion coupons are arranged in sample containers. Figure 1 shows a section of one of these containers, which are situated in the central water channel of a Siemens ATRIUM TM 9 type fuel assembly (9 by 9 lattice). The interior of these WTRs and containers is open to the coolant. The values of coolant temperature, number of cycles, and an equivalent burnup are listed in Table 2.

The different materials used for the cladding tubes of the PWR and BWR-A segments were measured after each reactor cycle by a EC testing method for oxide layer thickness. The corrosion coupons from PWR were examined in hot cells after two, four, and six cycles by measuring the weight gain. Examination, including visual inspection of corrosion coupons from BWR-A, was done after one, two, and four cycles. From BWR-B, weight gain data and visual inspection results are available for two cycles up to now. Irradiation in BWR-B is continuing up to six cycles.

Results and Discussion

Effect of Irradiation in BWR and PWR Environment

The effect of irradiation in PWR was deduced by comparing the corrosion results of in-PWR segments and coupons with the out-of-pile corrosion behavior of reference coupons in deionized 350°C water. For this comparison, a new formalism was used to normalize the temperature of the different axial positions and operating periods in reactor (284 to 340°C, see Table 2) and

GARZAROLLI ET AL. ON CORROSION BEHAVIOR OF ALLOYS 853

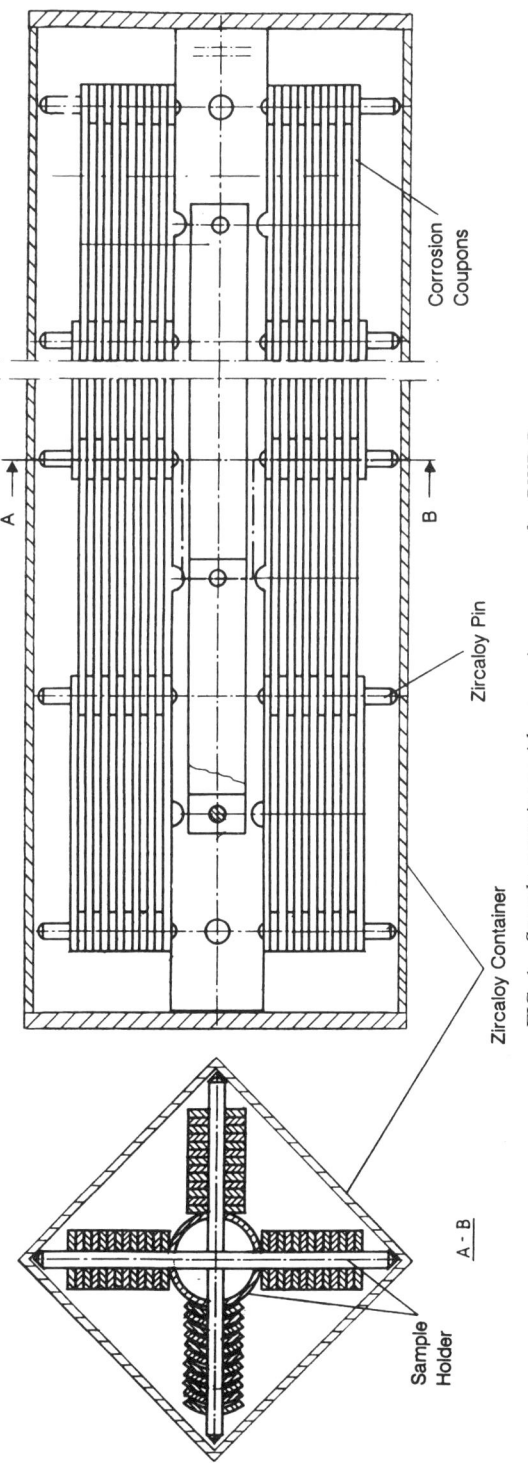

FIG. 1—*Sample container with corrosion coupons for BWR-B.*

TABLE 2—*Irradiation data of coupons in PWR and BWR environment.*

Reactor Type	Cycles	Equivalent Burnup, MWd/kgU	Coolant Temperature, °C
PWR	6	70	295–340
BWR A	4	30.2	274–286
BWR B	2	15.5	274–286

out of pile (350°C). For normalization of the temperature differences and variations, a normalized time was calculated for each sample considering the actual exposure time, t_i, and temperature, T_i, of the individual exposure steps and an activation temperature (Q/R) of 14 200 K by the following equation:

$$\text{Normalized time} = \Sigma\ t_i * \exp(Q/(R * T_i))$$

The value 14 200 K for Q/R was used accordingly to previous studies [7;]; however, best fit analysis of the PWR data from different Zry-4 and Zr2.5Nb claddings after four cycles revealed Q/R values between 12 200 and 15 300 K for the different materials and agree well with the selected number. Figure 2 shows the measured weight gains of PWR mini-coupons and the calculated weight gains from measured oxide layer thickness of PWR segment cladding tubes for a certain Zircaloy-4 material lot versus the normalized time. In this figure, the data from coupons exposed out of pile at 350°C in deionized water and in water with LiOH are also included. All in-PWR data up to five cycles lay in a small scatter band confirming the validity of the normalized time concept. Only the six-cycle data exhibit a deviation indicating a change of the corrosion kinetics. During the sixth cycle, corrosion obviously increases, especially at the lower temperatures (smaller normalized times at same actual time), and the normalized time

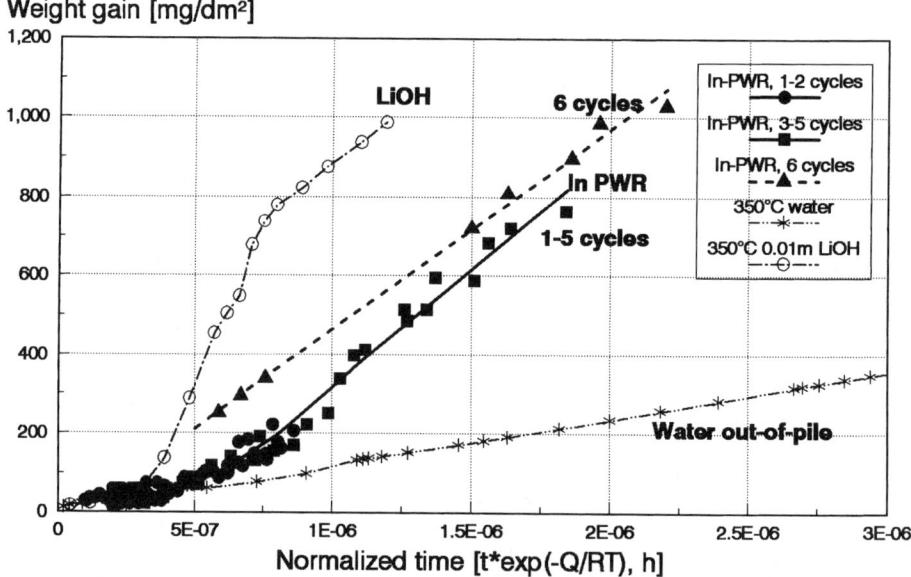

FIG. 2—*Corrosion of Zircaloy-4 in PWR in deionized water and in LiOH.*

concept is no longer applicable. This late acceleration is probably due to irradiation-induced damage, irradiation-induced precipitates, or irradiation-reduced changes of the precipitates existing already before irradiation and has also been seen in other lots. A comparison of the in-PWR data up with the out-of-pile behavior in deionized water shows the same corrosion behavior up to a weight gain of about 50 mg/dm^2 (3.4 μm oxide layer thickness). Afterwards, corrosion is increased by a factor of about 5. This can be seen more accurately in the detailed view in Fig. 3. Scattering of the EC data from segments can be explained by the low accuracy of this technique. However, the values are confirmed by the weight gain data from mini-coupons. In PWR, corrosion is fully identical to out-of-pile corrosion up to a weight gain of about 50 mg/dm^2. Furthermore, Fig. 2 shows that the start of the corrosion increase in PWR does not occur at the classical rate transition generally observed at about 30 mg/dm^2 but at a larger weight gain value. Figures 2 and 3 reveal also that 0.01 M LiOH increases corrosion even more than irradiation in PWR.

Figure 4 shows a similar comparison between in-PWR and out-of-pile corrosion for the binary Zr2.5Nb alloy. This alloy exhibits increased corrosion in PWR only at a weight gain of about 100 mg/dm^2 if compared with out-of-pile behavior in deionized water. This is, compared to Zircaloy-4, a two times larger weight gain. Above 100 mg/dm^2, corrosion of Zr2.5Nb increases by a factor of 3.5, which is smaller as in the case of Zircaloy-4. From Fig. 4 the high sensitivity of Zr2.5Nb to LiOH can be seen.

In BWR-A, corrosion of Zircaloy is increased early in time if compared with out-of-pile behavior (Fig. 5). As mentioned before, BWR-A has a low sensitivity to nodular corrosion, and all standard Zircaloy-2 have shown almost exclusively only uniform corrosion. Compared with expectations from out of pile, the uniform corrosion is increased in BWR-A at the early beginning, but the irradiation effect decreases at higher exposure times. In contrast to PWR, Zr2.5Nb behaves significantly worse than Zircaloy in BWR. In the segmented water rods exposed in BWR-A, there were two identical segments containing Zr2.5Nb samples and tubes.

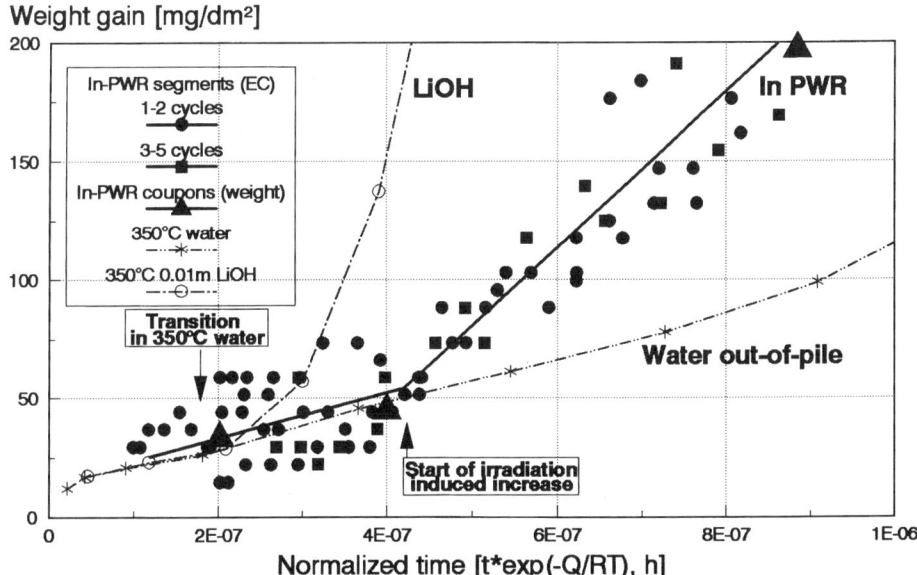

FIG. 3—*Corrosion of Zircaloy-4 in PWR in deionized water and in LiOH. Detailed view of initial range.*

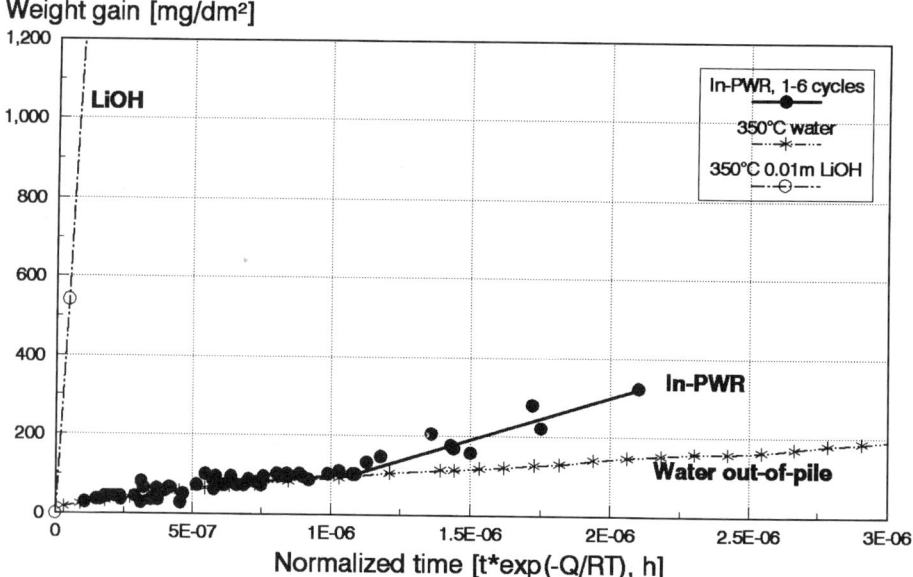

FIG. 4—*Corrosion of Zr2.5Nb in PWR in deionized water and in LiOH.*

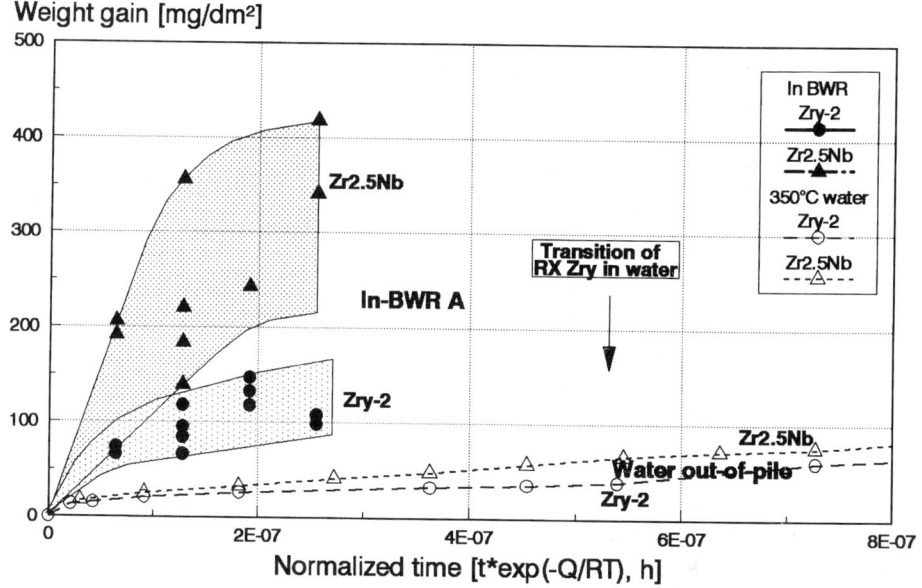

FIG. 5—*Corrosion of Zircaloy and Zr2.5Nb in BWR and out of pile in water.*

One of these segments was positioned at the lower end, where neutron flux is low, and the other was in a high flux position. The low-flux samples exhibited much larger weight gains than the high-flux samples shown in Fig. 5. A beneficial effect of irradiation on corrosion of Zr-Nb alloys was deduced before from experiments in the SGHWR-Winfrith [8]. The poor behavior of Zr2.5Nb is probably due mainly to the oxygen content of the coolant and not due to irradiation. The large effect of oxygen content of the water on corrosion of Zr2.5Nb was also noticed in one out-of-pile test run in 400°C steam. The 400°C steam tests were normally performed under oxygen-free condition. However, during one test run, hydrogen peroxide was added inadvertently. During this period, the Zr2.5Nb showed much higher weight gains than in the following repeated tests.

Influence of Material Parameters

In the following, only those parameters are discussed that showed significant effects on corrosion in the different environments. As far as the material condition is concerned, the size of the second-phase particles (SPP) has the largest influence. The size of the SPP in the material depends on the quenching rate from the β-temperature range (>950°C, where Fe, Cr, and Ni are soluble) and on the subsequent process and annealing temperatures in the α-temperature range (<800°C, where Fe, Cr, and Ni are insoluble). The effect of the process temperatures and times can be normalized by the accumulated annealing parameter ($\Sigma A_i = \Sigma t_i \cdot \exp(-Q/RT_i)$ with t_i and T_i being the time (h) and temperature (K) of the Annealing Steps i after β-quenching and $Q/R = 40\,000$ K). The Zircaloy-2 and -4 mini-coupons with a ΣA_i below 1E-19 to 1E-18 h showed high uniform corrosion (in PWR and out of pile in 350°C water, in 400°C steam) but low nodular corrosion (in BWR-B with a high tendency to nodular corrosion and in 500 to 515°C tests). The dependencies found with the mini-coupons in these environments are very similar, as has been reported before from fuel rods and other out-of-pile tests [3,4,6]. In BWR-A (with a low tendency to nodular corrosion), a variation of ΣA_i exhibits only a small effect on corrosion of Zircaloy-2 and -4. The influence of ΣA_i on corrosion of Zircaloy-2 and -4 in 350°C 0.01 M LiOH was rather small as compared to the other environments. In difference to Zircaloy-2 and -4, alternative Zr-SnFeCr alloys with Fe, Ni, and Cr outside the ASTM specification showed different ΣA_i dependencies than Zircaloy-2 and -4. This is reported in detail in an other paper [9].

One other parameter considered to be important for corrosion due to its effect on size and the distribution of second-phase particles is the quenching rate from the β-solution anneal that is always applied in the tube and strip fabrication process today. To get some insight into this parameter, several tube sections from a Zircaloy-4 lot were quenched under controlled conditions at varying rates. A part of these quenched sections was then annealed for 8 h at 750°C ($\Sigma A_i = 8.4$ E-17 h) to get a large SPP size. Figure 6 shows a large effect of the quenching rate if tested without any annealing in BWR-A and in 500°C steam. Here the quench rate dependency probably reflects mostly the SPP size, which is small only in the case of a fast quenching rate. The coupons that were quenched fastest showed low corrosion only after two cycles in BWR. After four cycles, increased uniform corrosion is already seen. The high sensitivity of Zircaloy with very fine intermetallics to increased uniform corrosion at long exposure times was already discussed in detail in Ref 6. For the annealed samples where only the distribution of SPP is expected to vary with the quenching rate, the figure reveals a small influence on uniform corrosion in PWR and out of pile and a moderate on in-BWR-A corrosion. In cladding tubes, where quite large hot and cold deformations are applied after the β-quenching, the SPP distribution becomes more uniform and less dependent on the β-quench rate. Therefore, the β-quenching rate is not considered to have a large influence for corrosion if a certain size required for good corrosion resistance is achieved by the following annealing treatment as usual for

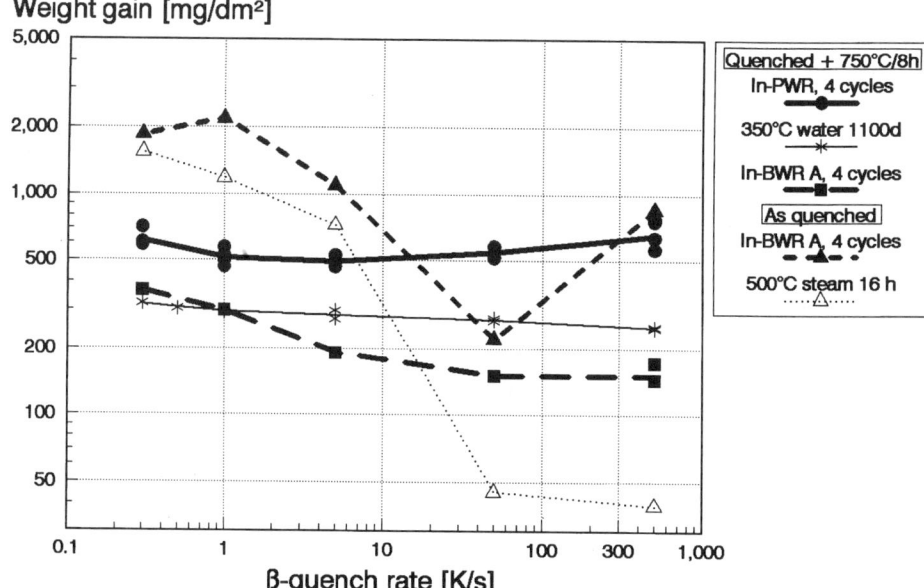

FIG. 6—*Effect of β-quench rate on corrosion of Zircaloy-4 in PWR, in BWR, and out of pile.*

PWR cladding. However, a high quenching rate is certainly advisable for a controlled adjustment of a medium SPP size as necessary for optimized BWR cladding.

Another parameter influencing out-of-pile corrosion is the degree of recrystallization. However, in reactor no obvious effect of the degree of recrystallization has been found. Some results that show this will be discussed later.

As far as alloying chemistry is concerned, Sn, Nb, and Fe were found to have large effects on corrosion, as can be seen from Figs. 7 to 10. Sn exhibited the most different effects in the different environments. Increasing Sn contents reduces nodular corrosion in BWR (Fig. 7) and corrosion in LiOH (Fig. 8) but increases nodular corrosion in 500°C steam and uniform corrosion in PWR (Fig. 7) in 350°C neutral water (Fig. 8) and in 400°C steam. In Fig. 7, this different effect is shown for in reactor and in Fig. 8 for 350°C out-of-pile tests in deionized water and diluted LiOH. The in-PWR data are from coupons with a normalized time of 1E-6 h. The BWR data scatter quite noticeably, depending on the sensitivity of the reactor to nodular corrosion (B is much more sensitive than A) and the fabrication temperatures. Low-temperature-processed (LTP) coupons with small SPPs show lower weight gains in BWR than high-temperature-processed (HTP) ones with large SPPs, as discussed before. Independent of the SPP size, all BWR coupons with Sn contents below 1% exhibit a high weight gain independent also of the tendency of the BWR to nodular corrosion. In addition to the effect of Sn, from Fig. 8 it can also be seen that fully recrystallized material corrodes slower out of pile in 350°C water and LiOH than stress relieved (SR) cladding. As already mentioned, this effect of the degree of recrystallization does not exist in PWR as is obvious from Fig. 7.

The effect of the alloying element Fe is also not the same in the different environments. Uniform corrosion in PWR in 350°C neutral water and 0.01 M LiOH and 400°C is low when the Fe content is above the ASTM range for Zircaloy-4. Figure 9 shows in-reactor data on the Fe effect for different ZrSnFeCr alloys. The four-cycle PWR data represent a normalized time of 1.5E-6 h. Whereas the PWR data reveal an obviously beneficial effect of Fe, the two-cycle

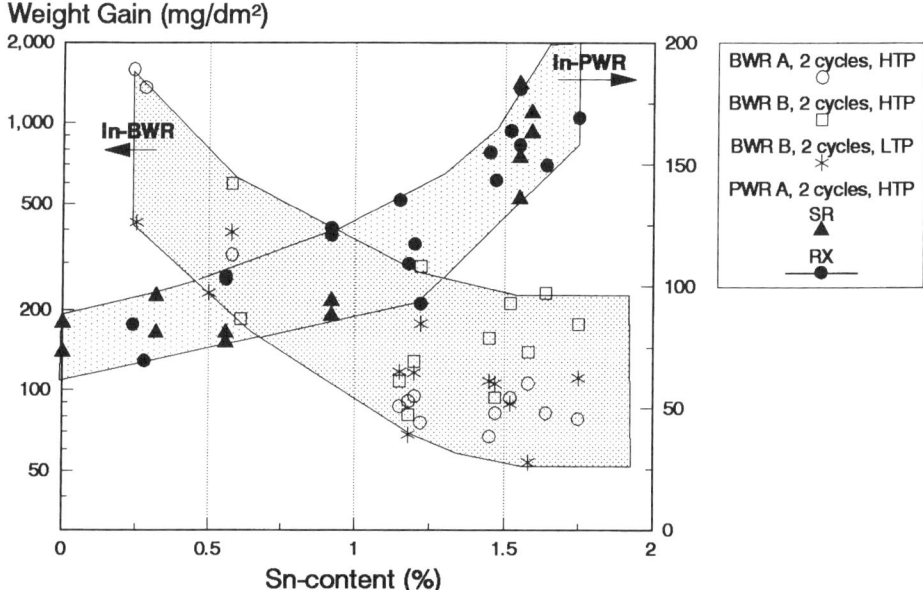

FIG. 7—*Influence of tin on in-reactor corrosion of Zr alloys fabricated at high temperatures (HTP) or low temperatures (LTP).*

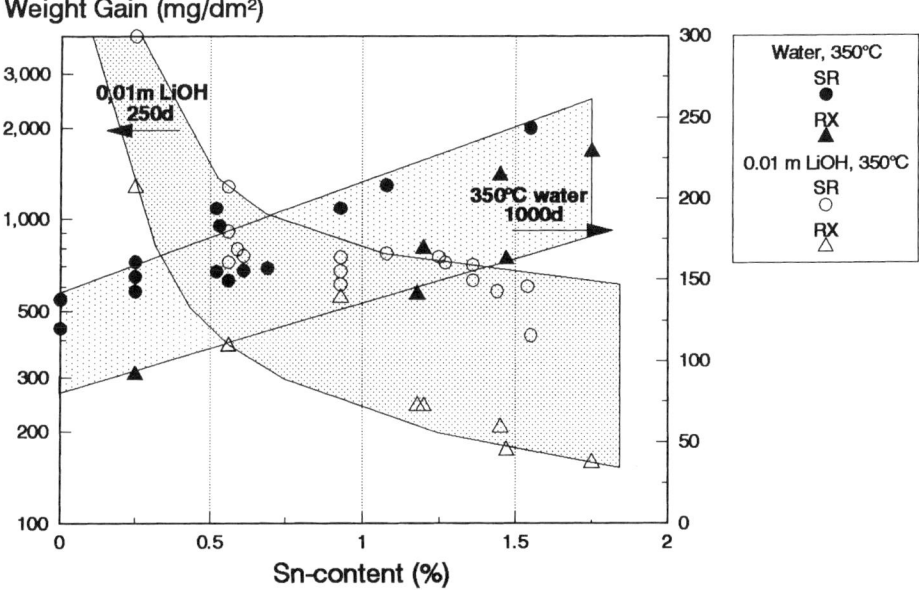

FIG. 8—*Influence of tin on out-of-pile corrosion of Zr alloys.*

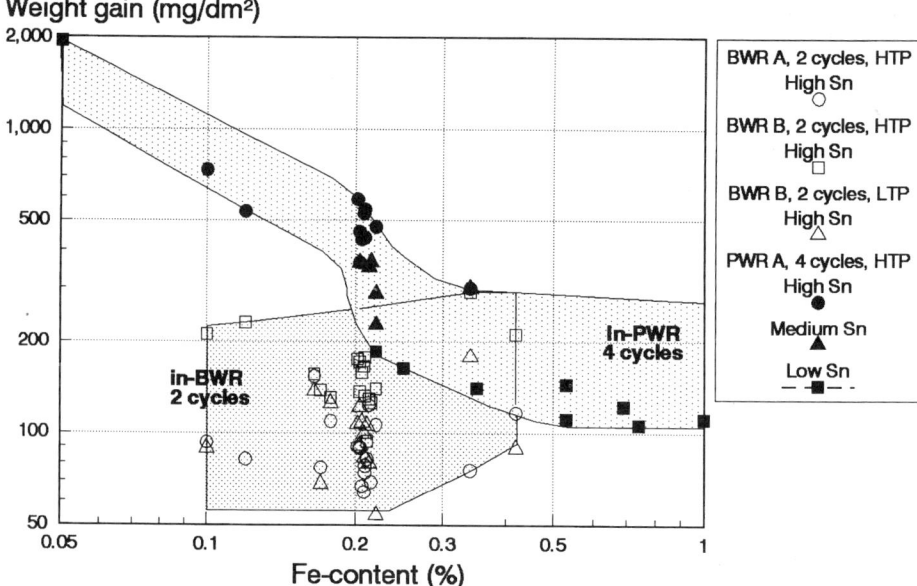

FIG. 9—*Influence of iron on in-reactor corrosion of Zr alloys fabricated at high temperatures (HTP) or low temperatures (LTP).*

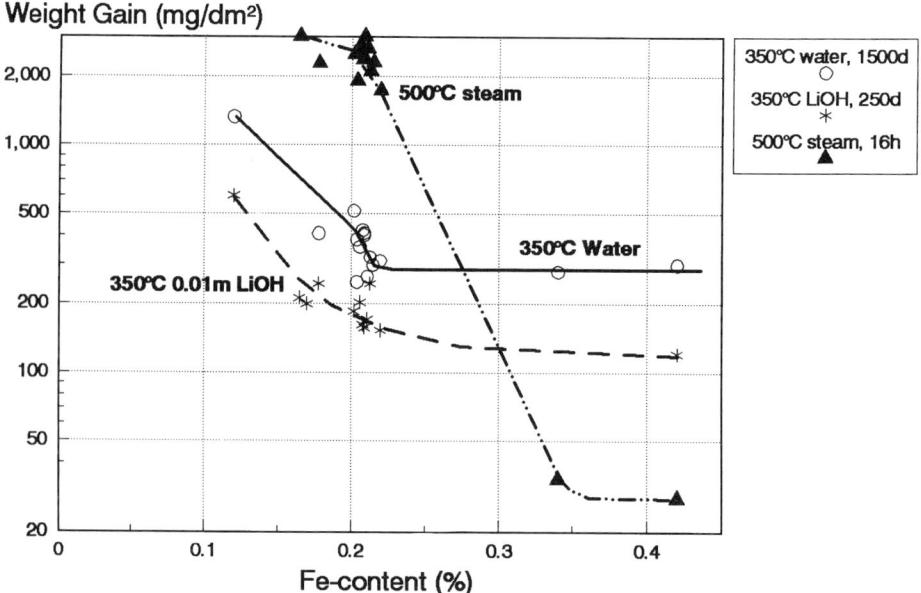

FIG. 10—*Influence of iron on out-of-pile corrosion of Zr alloys fabricated at high temperatures (HTP).*

BWR data exhibit scatter due to a different degree of nodular corrosion and no or only a small effect of Fe. Out-of-pile Fe was found to be beneficial under all test conditions. The data shown in Fig. 10 are from coupons with a Sn content in the upper ASTM range. The effect of Fe is more pronounced for materials with Sn contents below the ASTM range. Such data were reported by Ref *10*.

Knowing these dependencies, material development for cladding and structural materials can be performed in an efficient way. Several material parameters influencing in-reactor corrosion, like the SPP size or the Sn and Fe content for PWR application, can be optimized by out-of-pile corrosion tests.

Conclusions

1. In PWR, corrosion of Zircaloy-4 compared with out-of-pile corrosion in pressurized water deviates from laboratory results above about 50 mg/dm^2. At very high fluences, a further change of corrosion kinetics occurs.

2. In-BWR corrosion of Zircaloy-2/4 is increased from the early beginning as compared to out-of-pile tests.

3. Zr2.5Nb shows less effect of irradiation in PWR than Zircaloy. In-BWR Zr2.5Nb corrodes faster than Zircaloy-2 and -4.

4. A beneficial effect of a low accumulated A-parameter on nodular corrosion was seen in BWR-B, which has a high tendency to nodular corrosion but not in BWR-A.

5. The rate of β-quenching is only important if a low to median SPP size is required as in BWR.

6. The degree of recrystallization affects corrosion significantly only out of pile.

7. Sn exhibits a different effect in the various environments. Zr alloys with low Sn exhibit high corrosion in BWR and in LiOH and low corrosion in all other environments.

8. Fe shows a beneficial effect on corrosion in PWR, in 350°C water, in 400 to 500°C high-pressure steam, and in LiOH, and very little or no effect on nodular corrosion in BWR after two cycles.

9. Thus, out-of-pile tests give relevant information for most of the material parameters for PWR application. For corrosion behavior in BWR, out-of-pile tests are only meaningful as regards the size and distribution of precipitates.

References

[1] Weidinger, H. G., Garzarolli, F., Eucken, C. M., and Baroch, E. F., "Effect of Chemistry on Elevated Temperature Nodular Corrosion," *Zirconium in the Nuclear Industry: Seventh International Symposium, ASTM STP 939*, American Society for Testing and Materials, West Conshohocken, PA, 1987, pp. 364–386.

[2] Eucken, C. M., Finden, P. T., Trapp-Pritsching, S., and Weidinger, H. G., "Influence of Chemical Composition on Uniform Corrosion of Zirconium Base Alloys in Autoclave Tests," *Zirconium in the Nuclear Industry: Eighth International Symposium, ASTM STP 1023*, American Society for Testing and Materials, West Conshohocken, PA, 1989, pp. 113–127.

[3] Garzarolli, F., Stehle, H., Steinberg, E., and Weidinger, H. G., "Progress in the Knowledge of Nodular Corrosion," *Zirconium in the Nuclear Industry: Seventh International Symposium, ASTM STP 939*, American Society for Testing and Materials, West Conshohocken, PA, 1987, pp. 364–386.

[4] Garzarolli, F., Steinberg, E., and Weidinger, H. G., "Microstructure and Corrosion Studies for Optimized PWR and BWR Zircaloy Cladding," *Zirconium in the Nuclear Industry: Eighth International Symposium, ASTM STP 1023*, American Society for Testing and Materials, West Conshohocken, PA, 1989, p. 202.

[5] Fuchs, H. P. et al., "Cladding and Structural Material Development for the Advanced Siemens PWR Fuel FOCUS," *Proceedings*, ANS-ENS Internal Topical Meeting on LWR Fuel Performance, Avignon, France, 1991, p. 682.

[6] Garzarolli, F., Schumann, R., and Steinberg, E., "Corrosion Optimized Zircaloy for BWR Fuel Elements," *Zirconium in the Nuclear Industry: Tenth International Symposium, ASTM STP 1245*, American Society for Testing and Materials, West Conshohocken, PA, 1994, p. 709.

[7] Garzarolli, F., Bodmer, R. P., Stehle, H., and Trapp-Pritsching, S., "Progress in Understanding PWR Fuel Rod Waterside Corrosion," *Proceedings*, ANS Topical Meeting on LWR Fuel Performance, Orlando, Vol. I, 1985, pp. 3–55.

[8] Cheng, B. et al.. "Corrosion Performance of Some Zirconium Alloys Irradiated in the Steam Generating Heavy Water Reactor Winfrith," *Proceedings*, International Symposium on Environmental Degradation of Materials in Nuclear Power Systems-Water Reactors, Myrtle Beach, SC, 1983, NACE, p. 274.

[9] Ruhmann, H., Manzel, R., and Charquet, D., "In BWR and Out-of-Pile Nodular Corrosion Behavior of Zry 2/4 Type Melts with Varying Fe, Cr, and Ni Contents and Varying Process History," *Proceedings*, Eleventh ASTM Symposium on Zirconium in the Nuclear Industry, 11–14 Sept. 1995.

[10] Seibold, A., Garzarolli, F., and Steinberg, E., "Optimized Zry-4 with Enhanced Fe and Cr Content and DUPLEX Cladding: the Answer to Corrosion on PWR," *Proceedings*, International KTG/ENS Topical Meeting on Nuclear Fuel, TOPFUEL 95, Würzburg, Germany, 12–15 March 1995, Vol. II, p. 117.

DISCUSSION

P. Billot[1] *(written discussion)*—Do you have an idea of the mechanism that is at the origin of the experimental fact which consists in: improving corrosion resistance when lowering Sn in the absence of lithium, and in having the reverse effect in the presence of lithium (70 ppm), whatever the materials, like Zircaloy or Zr, Nb, Sn alloys.

F. Garzarolli et al. (author's closure)—Sn obviously improves the resistance against several species that can degrade corrosion resistance. This positive effect is definitely known for N in metal and LiOH or H_2S in water. Corrosion degradation exhibited in all examined cases an equiaxed crystal structure in the oxide layer, whereas normal corrosion layers always showed a columnar structure. Thus, it can be speculated that Sn affects the crystallization at the metal oxide interface. The negative effect of Sn appears only in case of normal corrosion without Li. This negative effect could be due either to an increase of the operating stresses in the oxide layer as a consequence of metal hardening or due to an effect on oxygen ion mobility. The studies [1,2] tend rather to the second possibility. Certainly more work is necessary to really understand the mechanism behind the dual role of Sn. The reverse effect of Sn in Li-containing and Li-free environments was not seen only in Zr-Sn-TM alloys (TM = transition metals) but also in Zr-Nb-Sn-(TM) alloys.

[1] Beie, H. J. et al., "Examinations of the Corrosion Mechanism of Zirconium Alloys," *Zirconium in the Nuclear Industry: Tenth International Symposium, ASTM STP 1245,* 1994, p. 615.
[2] Göhr, H. et al., "Long Term In Situ Corrosion Investigation of Zr Alloys in Simulated PWR Environment by Electrochemical Measurements," *Proceedings,* Eleventh Symposium on Zirconium in Nuclear Industry, Garmisch-Partenkirchen, 11–14 Sept. 1995.

B. Cheng[2] *(written discussion)*—Please clarify your concluding remark that decreasing the Sn content will increase corrosion in BWRs. This is not consistent with the industry trend of reducing the Sn content for ~1.55% Sn to ~1.3% Sn for both BWRs 04 and PWRs.

F. Garzarolli et al. (author's closure)—The positive effect of Sn on nodular corrosion in BWR is strong only below the ASTM specification range for Zircaloy-2 and -4, below 1.2% and weak within this range. The effect on uniform corrosion in a PWR, on the other hand, is also strong within the ASTM specification range for Zircaloy-4. Thus, lowering Sn to the lower ASTM specification range will reduce normal corrosion in PWR. However, for in-BWR application, no real benefit can be deduced from in-reactor experiments for such a step.

P. Rudling[3] *(written discussion)*—We have seen that adding Fe to Zr-2 does not affect the corrosion rate after two cycles, but there is a significant decrease in corrosion rate at high burnup. Please comment.

F. Garzarolli et al. (author's closure)—The finding that corrosion in BWR is not much affected by Fe after two cycles is fully consistent with the conclusions from this study. The significant decrease of in-BWR corrosion after four cycles with increasing Fe, you have found, is probably characteristic only for a certain material condition. In any case, we did not see a

---

[1] Commissariat a l'Energie Atomique, St. Paul lez Durance, France.
[2] Electric Power Research Institute, Palo Alto, CA.
[3] ABB Atom, Sweden.

significant effect of Fe for material with relatively large intermetallics (high A parameter) in BWR-A.

Brian Cox[4] (written discussion)—Have you considered whether the poor correlaton between your 0.01 M LiOH tests and PWR behavior might arise because you have ignored the major effect of boric acid on corrosion in LiOH. Boric acid is present most of the time in PWRs, and LiOH concentration effects in isothermal specimens are readily observed (both in laboratory tests and in-pile) in crevices and should be equally apparent in thick, porous ZrO_2 films.

F. Garzarolli et al. (Author's closure)—Measurements of the Li content in the oxide of PWR fuel rods have shown high concentrations indicative for a Li acceleration of corrosion only at high-oxide layer thickness and high heat fluxes. Thus, we do not believe that the corrosion of isothermal samples in a PWR with maximum 2 ppm Li could be affected by Li at all. Several of the materials discussed in this paper were, in addition, tested in mini autoclaves under different LiOH and boric acid contents (70, 300, and 700 ppm Li and 0, 100, 1000, and 8000 ppm B). These tests have shown after some exposure time the same material ranking for all environments. From these tests it can be definitely concluded that Sn reduces the accelerated corrosion under all Li+B environments, although the onset of accelerated corrosion occurs at least in less aggressive combinations first in high-Sn alloys. In PWR and out of pile in water and steam without any Li, Sn always increases corrosion.

---

[4] University of Toronto.

H. Ruhmann,[1] R. Manzel,[1] H.-J. Sell,[1] and D. Charquet[2]

In-BWR and Out-of-Pile Nodular Corrosion Behavior of Zry-2/4 Type Melts with Varying Fe, Cr, and Ni Content and Varying Process History

REFERENCE: Ruhmann, H., Manzel, R., Sell, H.-J., and Charquet, D., **"In-BWR and Out-of-Pile Nodular Corrosion Behavior of Zry-2/4 Type Melts with Varying Fe, Cr, and Ni Content and Varying Process History,"** *Zirconium in the Nuclear Industry: Eleventh International Symposium, ASTM STP 1295,* E. R. Bradley and G. P. Sabol, Eds., American Society for Testing and Materials, 1996, pp. 865–883.

ABSTRACT: Zircaloy-based materials with constant tin content, a constant sum of Fe and Cr content, and different Fe/Cr ratios were manufactured from small ingots (6 kg) by forging, β-quenching, hot rolling, and cold rolling with two different annealing sequences resulting in two accumulated annealing parameters (A-parameter $A = 0.1$ E-18 and $A = 10$ E-18 h). To study the effect of Ni, chromium was substituted by nickel in one alloy. As a reference material standard, ASTM Zry-2 was manufactured in a similar way.

These materials were examined for precipitate size by TEM and for their out-of-pile corrosion behavior in 400°C steam up to 319 days and in 500°C steam up to 24 h in a static and in a refreshed autoclave. Furthermore, samples made from these alloys were irradiated in a commercial BWR for two years.

By comparing results from different out-of-pile tests to the data from in-pile irradiation, the following conclusions can be drawn:

1. The A-parameter influence with respect to corrosion is the dominant factor for out-of-pile and in-pile corrosion behavior.
2. The (Fe+Ni)/Cr ratio influences the corrosion resistance, especially at low values below about 5. The effect is found to be strong for low A-parameter materials under irradiation of two cycles comparable to the behavior found for the high A-parameter materials in out-of-pile tests. The nodular corrosion test reproduces the in-pile behavior found for the high A-parameter materials.
3. Nickel substituted for Cr reduces the effect of the A-parameter on corrosion.
4. Sizes of intermetallic precipitates depend on chemical composition and structure. Significant numbers of Zr_3Fe particles are found only at Fe/Cr ratios above 5. These particles are significantly larger than the Laves phases.
5. The influence of the chemical composition of the precipitates to the corrosion behavior cannot be finally evaluated on the two-cycle irradiated samples. Further information is expected when the samples reach higher in-pile exposure.

KEYWORDS: Zircaloy, nodular corrosion, in-pile corrosion, BWR cladding materials, annealing parameter

A continuous effort is being undertaken worldwide on the optimization of Zirconium alloys with respect to corrosion for application in pressurized or boiling nuclear reactors. The incentive

---

[1] Siemens AG, Power Generation Group (KWU), D-91050 Erlangen, Germany.
[2] CEZUS CRU, Cedex 73403, Ugine, France.

for this development is the substantial reduction of fuel cycle costs from the extended lifetime of corrosion-optimized fuel assemblies and high burnup.

The in-pile and out-of-pile corrosion behavior of zirconium-based materials is governed by a set of non-independent parameters. One of the most influencial is the chemical composition of the alloys; another important parameter is thermo-mechanical treatment during the manufacturing sequence. The most promising results are to be expected if only one parameter is varied while keeping the others as constant as possible. The disadvantage of such an ideal procedure is the high number of test samples necessary if a high number of non-independent influencial parameters exist. Another consequence of a high number of test samples are long experimentation times and costs.

In this experimental study, a nearly ideal approach was realized. A low number of test samples was obtained by varying alloy composition only (in principle) by one pair of elements (iron and chromium). It is known that the thermo-mechanical treatment, expressed as the A-parameter [1], is important for corrosion resistance, which is governed by the size and distribution of intermetallic precipitates. In order to see the effect of the annealing parameter, the alloys were manufactured in a similar sequence. Only one annealing step was varied to obtain two different values of the A-parameter ($A = 0.1$ E-18 h and $A = 10$ E-18 h) for the specimens. Both sets were investigated thoroughly in the laboratory and irradiated in an, with respect to nodular corrosion, aggressive boiling water reactor for two cycles up to the present time. The results presented in this paper, therefore, represent an intermediate report as far as in-pile behavior is concerned.

Experimental Methods

Materials

A set of ten different alloys was melted for this study. With respect to the variation in chemical composition, the following guidelines were followed (specimen numbers are given in brackets):

1. Keep Sn content constant (range covered is 1.3 to 1.8%).
2. Keep the sum of Fe and Cr content constant (approximately 0.3%) (No. 1 to No. 8).
3. Vary the ratio between Fe (+Ni) and Cr from 2 to 52 (No. 1 to No. 8).
4. Substitute in one Zry-2 type alloy chromium by nickel with a high value outside of the ASTM specification for Zry-2 (No. 9).
5. Compare to standard Zry-2 as a reference material (No. 10).

The exact chemical composition obtained by the small-scale ingot melting of the alloys is given in Table 1. The impurities analyzed for the different alloys show no significant differences, as can be seen from Table 2. Note that the silicon value measured (<20 ppm) is much lower than for standard materials. The variation range of tin and oxygen was a consequence of the small-scale ingot melting process.

From all melts, sheets were manufactured applying a similar manufacturing sequence of forging, β-quench, and hot and cold rolling as shown in Table 3, where the detailed schedule of sample fabrication is outlined. The two temperatures selected for intermediate vacuum anneal were 500 and 750°C, respectively, at a constant annealing time. This procedure resulted in two different accumulated annealing Parameters A, respectively, of 0.1 E-18 h (denoted as "A") and 10 E-18 h (denoted as "B") for the two fabrication sequences.

From the recrystallized (final anneal 620°C, 1 h) sheets, 1-mm-thick, 35 by 10-mm samples were machined, drilled, numbered, and pickled in hydrofluoric acid. Multiple sets of the 2 ×

TABLE 1—*Chemical composition of the melted alloys.*

Alloy	O, %	Ni, %	Sn, %	Fe, %	Cr, %	Fe+Cr, %	(Fe+Ni)/Cr
1	0.09	...	1.3	0.26	0.005	0.27	52
2	0.08	...	1.5	0.3	0.05	0.35	6.0
3	0.08	...	1.4	0.26	0.06	0.32	4.3
4	0.09	...	1.5	0.24	0.07	0.31	3.4
5	0.08	...	1.6	0.22	0.1	0.32	2.2
6	0.07	...	1.6	0.27	0.07	0.33	3.9
7	0.09	...	1.6	0.25	0.06	0.31	4.2
8	0.07	...	1.8	0.27	0.07	0.34	3.9
9	0.09	0.2	1.5	0.18	0.004	...	93
10	0.09	0.06	1.3	0.18	0.09	...	3

10 different specimens were used for different investigations in the laboratory and for reactor irradiation.

Experimental Procedures

The following investigations were performed with the non-irradiated specimen:

1. Microstructure evaluation: metallographic examination by optical microscopy and transmission electron microscopy (TEM) with respect to size and chemical composition of the intermetallic particles.

TABLE 2—*Impurity analysis of the investigated alloys.*

	C	N	H	Cu	Hf	Si
1	85-66 89-87	16 13	17-16-26	18-18-17	58-58-58	<10-<10-<10
2	78-79 68-70	12 12	17-15-25	18-19-18	59-59-58	10-11-11
3	91-83 91-93	12 15	12-13-13	19-20-20	59-59-59	15-16-17
4	70-73 66-71	13 <10	12-11-15	16-16-17	60-59-59	13-11-12
5	71-87 84-63	16 16	14-16-15	25-25-18	54-55-55	16-14-13
6	75-74 82-71	15 17	13-13-15	17-17-17	53-53-53	11-11-11
7	76-72 83-84	21 19	13-14-15	17-17-18	53-53-52	10-11-10
8	64-110 64-73	15 15	17-15-17	17-17-17	53-53-52	12-11-13
9	82-91 85-91	18 16	17-16-15	20-20-20	51-52-52	<10-<10-<10
10	107-102 109-100	15 15	14-13-15	19-19-19	54-54-54	10-<10-<10

C and N = analysis of plate.
H,Cu,Hf,Si = analysis of final sheet.

TABLE 3—*Fabrication sequence for the specimen.*

—Compact 50 × 50 × 500 mm
—Double VAR: Ingot ϕ 130 × 60 mm, 5.7 Kg
—Forging 1030°C → ⌀ 50 mm
  + 30 min at 1030°C → ⌀ 60 × 20 mm (L = 550–600 mm)
—Machining. Thickness = 15 mm
—Water quench = 1030°C-30 min
—Hot Rolling: 650°C-20 min → 5.8 mm
—Descaling. Pickling
—Vacuum Annealing. Argon cooling (V.A.-A.C.)
  Variant 1 (A) = 500°C-1 h
  Variant 2 (B) = 750°C-1 h
—Same Fabrication Schedule for A and B:
  + Cold Rolling → 3.5 mm
  + V.A.-A.C. 600°C 1 h
  + Cold Rolling → 1.8 mm
  + V.A.-A.C. 600°C 1 h
  + Cold Rolling → 1 mm
  + V.A.-A.C. 620°C 1 h

A-parameter: Variant 1 (A) = 1.06×10^{-19}
Variant 2 (B) = 1.05×10^{-17}

2. Investigation of the nodular corrosion resistance in static and refreshed autoclaves at 500°C, and 520°C, 12.5 MPa for 24 h.
3. Investigation of the uniform corrosion resistance in a static autoclave at 400°C, 10.3 MPa test with exposure times up to 319 days.
4. Determination of the hydrogen uptake (pickup fraction) from samples at comparable weight gains from the nodular and uniform corrosion test.

Four equivalent sets of samples were prepared for reactor irradiation. For that purpose, advantage has been taken of the special design of Siemens ATRIUM fuel assemblies. These assemblies are equipped with a 5 by 5-cm water channel as shown in Fig. 1. The samples are positioned on pins of Zry in a sample container located within the water channel. The coolant flow has direct access to the sample stacks, heating them up isothermally to temperatures between 274 and 286°C. One set of the samples has been removed after two cycles of irradiation, reaching a maximum equivalent burnup of 15.5 MWd/kgU. The samples have been transferred to hot cells, where optical examinations and weight gain measurements were performed.

Experimental Results and Discussion

Microstructural Investigations

Grain Size—The sheets were examined by optical metallography and transmission electron microscopy at the final dimension. Generally, differences in the grain size were measured only for samples with different A-parameters. Higher grain size is correlated to high A-parameter. The influence on the chemical composition is not as pronounced, as can be seen from Fig. 2, where the mean intercept grain size is plotted versus the (Fe+Ni)/Cr ratio. The mean values are affected by grain inhomogeneity observed in the sheets. This may be influenced by the small-scale manufacturing sequence. Obviously, substituting Cr for Ni (Sample No. 9) docs not significantly influence the grain size in comparison to the standard material Zry-2 (Sample

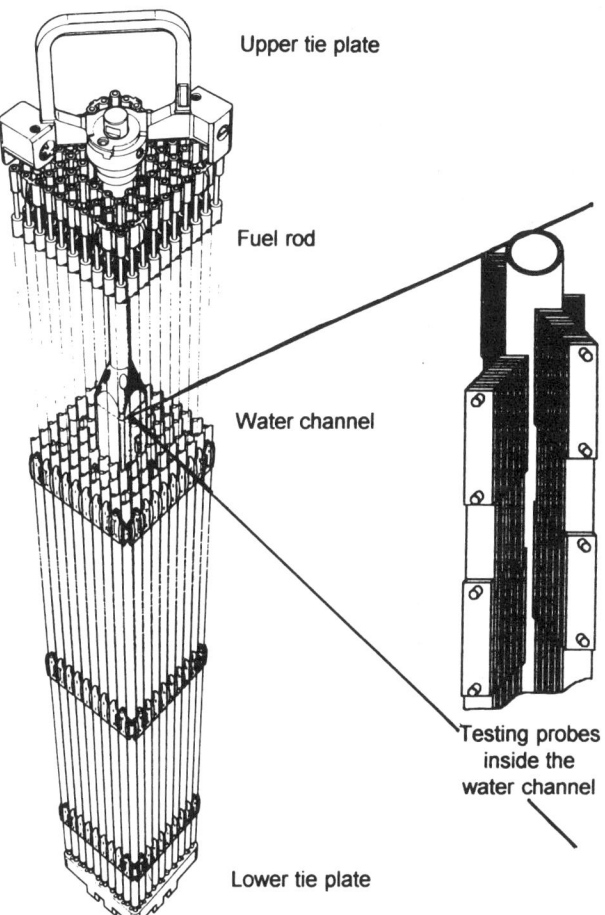

FIG. 1—*Experimental arrangement for in-BWR irradiation of specimens inside the water channel of an Atrium Siemens fuel assembly for boiling water reactors.*

No. 10), indicated with circles in the diagram. It is concluded that grain size of the materials is mainly sensitive to fabrication parameters and only to a minor extent to the chemistry ratio of Fe, Cr, and Ni in the alloy.

Precipitates—The size and chemical composition of the intermetallic particles in the alloys have been determined by transmission electron microscopy (TEM). The results of this investigation are compiled in Table 4 with respect to structure and chemical composition. In Table 5, the measured particle sizes are compiled. TEM structure analysis revealed two different types of precipitates in the alloys without Ni (Sample No. 1 to No. 8): Precipitates with the composition Zr_3Fe and the classical Laves phase $Zr(Fe,Cr)_2$. Only in alloy No. 1 was an additional, not-as-well defined hexagonal Phase X detected with lattice parameters $a = 0.82$ nm and $c = 0.347$ nm. As shown in Table 5, the size of the precipitates depends on the annealing parameter and on the structure of the intermetallic compound. The Zr_3Fe particles are in all investigated alloys significantly bigger than the Laves phases. This points to different growth mechanisms for different intermetallic phases, as has been pointed out in Ref 2.

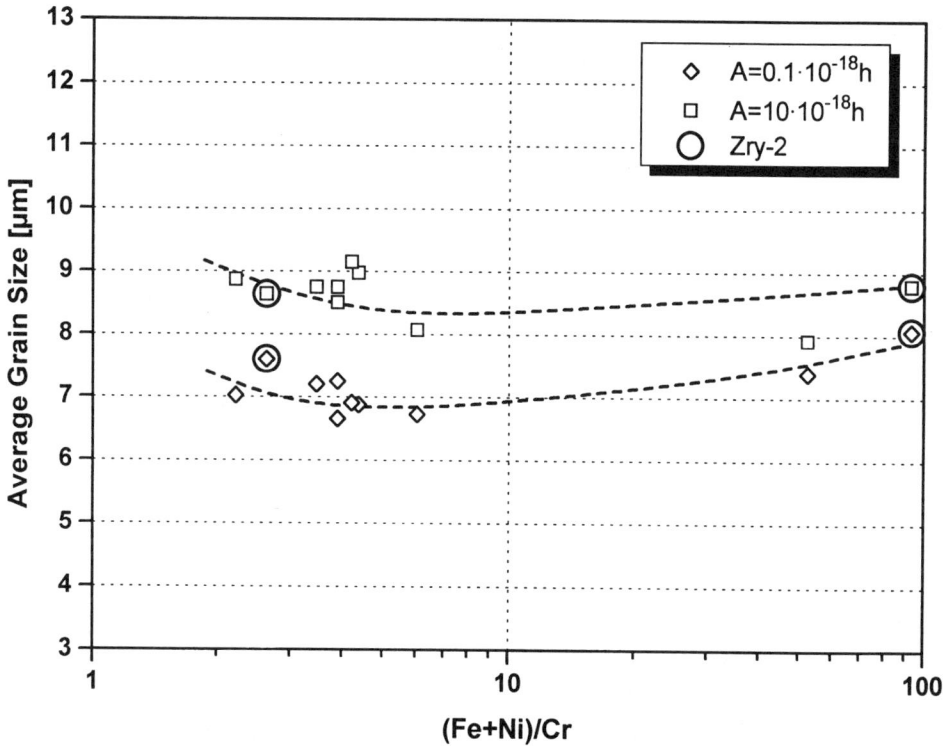

FIG. 2—Grain size as a function of the (Fe+Ni)/Cr ratio.

TABLE 4—Chemical composition of precipitates found in the investigated alloys by TEM.

Alloy	Fe/Cr in the Alloy	A Low A-Parameter		B High A-Parameter	
		SPP	Fe/Cr in $Zr(CrFe)_2$	SPP	Fe/Cr in $Zr(CrFe)_2$
1	57	Zr_3Fe $Zr(Cr,Fe)_2$ x		Zr_3Fe $Zr(Cr,Fe)_2$ x	
2	6.3	Zr_3Fe $Zr(Cr,Fe)_2$	3.9	Zr_3Fe $Zr(Cr,Fe)_2$	3.9
3	4.5	Zr_3Fe $Zr(CrFe)_2$	3.5	Zr_3Fe $Zr(Cr,Fe)_2$	3
4	3.4	Zr_3Fe $Zr(Cr,Fe)_2$	3.3	Zr_3Fe $Zr(Cr,Fe)_2$	3
5	2.1	Zr_3Fe $Zr(Cr,Fe)_2$	2.2	Zr_3Fe $Zr(Cr,Fe)_2$	2

NOTE: x = Zr, Fe, Cr, Sn; hex a = 0.82 nm; c = 0.347 nm.

TABLE 5—*Size of intermetallic particles in the alloy with different Fe/Cr ratios.*

	Low A-parameter	High A-parameter
Zr_3Fe^a	0.1-0.6 μm	0.5-2 μm
$Zr(Cr,Fe)_2$	0.04-0.1 μm	0.07-0.5 μm

[a] Decreasing amount from 1 to 5 (with decreasing Fe/Cr ratio).

The relative amount of the large-sized Zr_3Fe particles depends on the Fe+Ni/Cr-ratio [3,4]. It decreases with decreasing Fe/Cr ratio, and mainly Laves phases exist in an Fe/Cr ratio below 2. If occasionally at a low Fe/Cr ratio (Zry-4 type specimen No. 4) other phases besides the Laves phases could be detected, they were found only as Zr_3Fe and not as Zr_2Fe particles. For a Fe/Cr ratio below a value of 4, the existence of Zr_2Fe would have been expected [5]. Possible reasons for the absence of Zr_2Fe may be found in the fabrication sequence: (1) a high quenching rate during beta quench (resulting from the geometry of the forged plate), (2) the sequence of three annealing steps with rather low temperatures (600, 600, and 620°C—refer to Table 3) [6], and the very low silicon content.

Out-of-Pile Corrosion Behavior

The materials have been exposed to autoclave steam corrosion testing at three different test temperatures. The sensitivity to nodular corrosion was tested at 500 and 520°C in a static and a refreshed autoclave system for 16 h at temperature. Testing performed at 520°C for 16 h has to be considered as very aggressive with respect to nodular corrosion.

To investigate the resistance to uniform corrosion, a static autoclave test was performed at 400°C. For this test, exposure times up to 319 days were obtained.

The results of all corrosion tests performed are compiled in Table 6. The weight gains are, in most cases, mean values from at least two samples. In order to allow a direct comparison of the test sensitivity with respect to the alloy composition for every test, relative weight gains are calculated by normalizing the weight gain to the reference Zry-2 (Sample No. 10). From the graphical presentation of these values, given in Fig. 3, we conclude different sensitivities of the different test conditions for alloys with different composition and with different A-parameters. For low A-parameter material (upper diagram), the most sensitive test with respect to chemistry variation ((Fe+Cr)/Ni-ratio) is a 520°C corrosion test performed in a refreshed autoclave followed by a refreshed test at 500°C. The most aggressive test for high A-parameter materials was found to be the static 500°C test. On the other side, this test condition was found to show the lowest sensitivity with respect to the alloy composition for low A-parameter materials. For high A-parameter materials, a low sensitivity was found for the 520°C refreshed corrosion test.

For material optimization in an out-of-pile test, a high sensitivity is demanded, meaning that small variations in alloy composition induce big differences in the test result. A high sensitivity is connected with high aggressiveness, meaning the corrosion effect is much more exaggerated in the test than under real operation conditions. The higher temperature in the laboratory test (e.g., 500 or 520°C out-of-pile in comparison to 290°C in-pile in isothermal tests) is the most responsible parameter. Out-of-pile tests allow study of parameters under conditions that magnify their effect. The isothermal in-pile test, as can be seen from the diagrams, is less informative in this respect. The conclusion from this is not to rely exclusively on one test temperature and test condition for out-of-pile optimization, especially of alloys with different A-parameters.

TABLE 6—In- and out-of-pile corrosion results.

	Chemical Composition A: Annealing parameter $A = 0.1$ E-18 h B: Annealing parameter $A = 10$ E-18 h							In Pile after 2 Cycles Irradiation				Corrosion Out of pile									
													500 °C refreshed		500°C static		520°C refreshed		400 °C static WG(319 d)		Lin. rate
	Sn, %	Fe, %	Cr, %	O, ppm	Ni, %	Fe+ Ni/Cr	WG mg/dm²	Rela- tive	Nod. Cov. %	Rela- tive	WG mg/dm²	Rela- tive	WG mg/dm²	Rela- tive	WG mg/dm²	Rela- tive	mg/dm²	Rela- tive	mg/dm²
1A	1.3	0.26	0.005	908	0.002	52	107	1.4	35	1.8	38	1.3	45	1.0	562	7.8	141	1.2	0.37
2A	1.5	0.30	0.05	808	0.002	6	115	1.5	55	2.8	29	1.0	46	1.1	466	6.5	124	1.0	0.28
3A	1.4	0.26	0.06	825	0.002	4	75	1.0	34	1.7	51	1.7	53	1.2	567	7.9	151	1.2	0.36
4A	1.5	0.24	0.07	883	0.002	3	73	1.0	18	0.9	80	2.7	55	1.3	700	9.7	369	3.0	0.46
5A	1.6	0.22	0.10	850	0.002	2	60	0.8	15	0.8	161	5.4	77	1.8	872	12.1	...	...	8.50
6A	1.6	0.27	0.07	683	0.002	4	79	1.0	32	1.6	93	3.1	53	1.2	1196	16.6	132	1.1	0.29
7A	1.6	0.25	0.06	908	0.002	4	113	1.5	18	0.9	108	3.6	64	1.5	708	9.8	311	2.6	0.56
8A	1.8	0.27	0.07	680	0.002	4	61	0.8	16	0.8	42	1.4	55	1.3	809	11.2	192	1.6	0.49
9A	1.5	0.18	0.004	925	0.1929	93	52	0.7	14	0.7	27	0.9	43	1.0	62	0.9	137	1.1	0.28
10A	1.3	0.18	0.09	870	0.0571	3	76	1.0	20	1.0	30	1.0	43	1.0	72	1.0	121	1.0	0.28
1B	1.3	0.26	0.005	908	0.002	52	137	0.9	60	1.0	684	17.5	337	6.4	1460	2.1	120	1.0	0.16
2B	1.5	0.30	0.05	808	0.002	6	101	0.6	34	0.6	164	4.2	582	11.0	2813	4.0	120	1.0	0.18
3B	1.4	0.26	0.06	825	0.002	4	121	0.8	15	0.3	857	22.0	2002	37.8	3992	5.7	136	1.2	0.24
4B	1.5	0.24	0.07	883	0.002	3	124	0.8	70	1.2	1256	32.2	1742	32.9	4375	6.2	121	1.0	0.22
5B	1.6	0.22	0.10	850	0.002	2	194	1.2	78	1.3	1194	30.6	1348	25.4	3987	5.7	145	1.2	0.16
6B	1.6	0.27	0.07	683	0.002	4	157	1.0	51	0.9	901	23.1	1938	36.6	3861	5.5	123	1.0	0.20
7B	1.6	0.25	0.06	908	0.002	4	85	0.5	19	0.3	1291	33.1	2978	56.2	4323	6.1	236	2.0	0.61
8B	1.8	0.27	0.07	680	0.002	4	150	1.0	54	0.9	1318	33.8	2160	40.8	4049	5.7	143	1.2	0.25
9B	1.5	0.18	0.004	925	0.1929	93	136	0.9	20	0.3	33	0.8	43	0.8	71	0.1	137	1.2	0.26
10B	1.3	0.18	0.09	870	0.0571	3	157	1.0	58	1.0	39	1.0	53	1.0	705	1.0	118	1.0	0.18

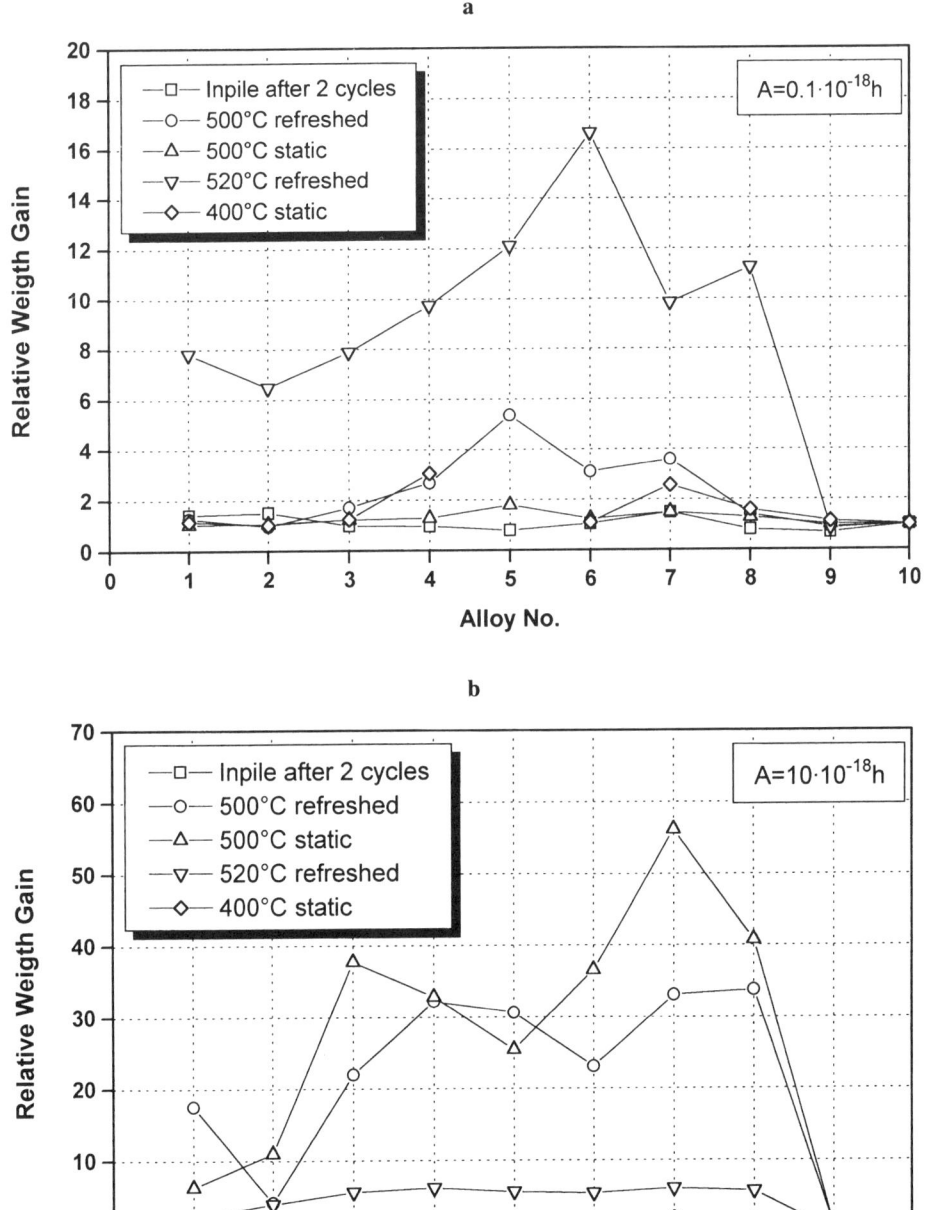

FIG. 3—*Sensitivity of the different corrosion tests performed for samples with low and high A-parameters.*

Results from the measured weight gains versus the alloy composition variation, the (Fe+Ni)/Cr ratio, are presented in the following paragraph. As an outstanding feature of all the experimental results summarized in Fig. 4 (520°C refreshed test), Fig. 5 (500°C static and refreshed tests), and Fig. 6 (400°C static test), a clear separation for materials with different fabrication routines is observed. Low A-parameter-fabricated materials show the lower weight gains in comparison to samples with high A-parameters in nodular tests. Just the opposite is true for the nodular test at 400°C, as shown in Fig. 6. Here the lower weight gains (upper diagram) are found for the high A-parameter materials. For the late linear corrosion rates, as plotted in the lower diagram, similar behavior is found in principle.

The correlation between corrosion (weight gain) and alloy composition (Fe/Cr ratio) shows independently from the test temperatures a decrease with increasing (Fe+Ni)/Cr ratio, more pronounced for the high A-parameter materials and very steep for ratios above 5. Only in the uniform test at 400°C and only if corrosion rates are considered (Fig. 6 lower, diagram) is higher corrosion resistance observed with decreasing (Fe+Ni)/Cr ratio. These findings agree well with results from the literature, where an optimum for corrosion resistance was reported for a Fe/Cr ratio of about 2 [7].

Nickel has a significant effect on the corrosion behavior. The influence of adding nickel instead of chromium to Zry-2 can be seen if the weight gain data of Specimen 9 and Specimen 10 are compared (these specimens are indicated in the plots of Figs. 4 to 6 by circles). Regardless of the A-parameter, the high Ni alloy shows high corrosion resistance in all 500 and 520°C nodular tests. Even a better corrosion resistance than measured for the reference material Zry-2 is obtained in the nodular tests for the high A-parameter fabricated specimen. Similar influences were reported earlier for the resistance to nodular corrosion [8].

Out-of-Pile Hydrogen Uptake

The influence of the alloy composition to hydrogen uptake was investigated for materials with different A-parameters on specimens with comparable weight gains. The hydrogen content

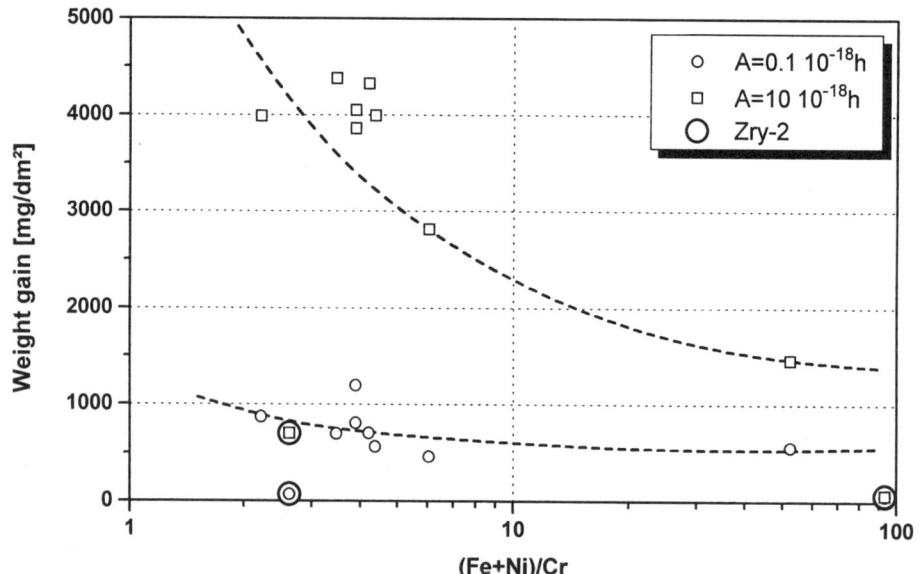

FIG. 4—*Corrosion results from a 520°C, 12.5 MPa refreshed test.*

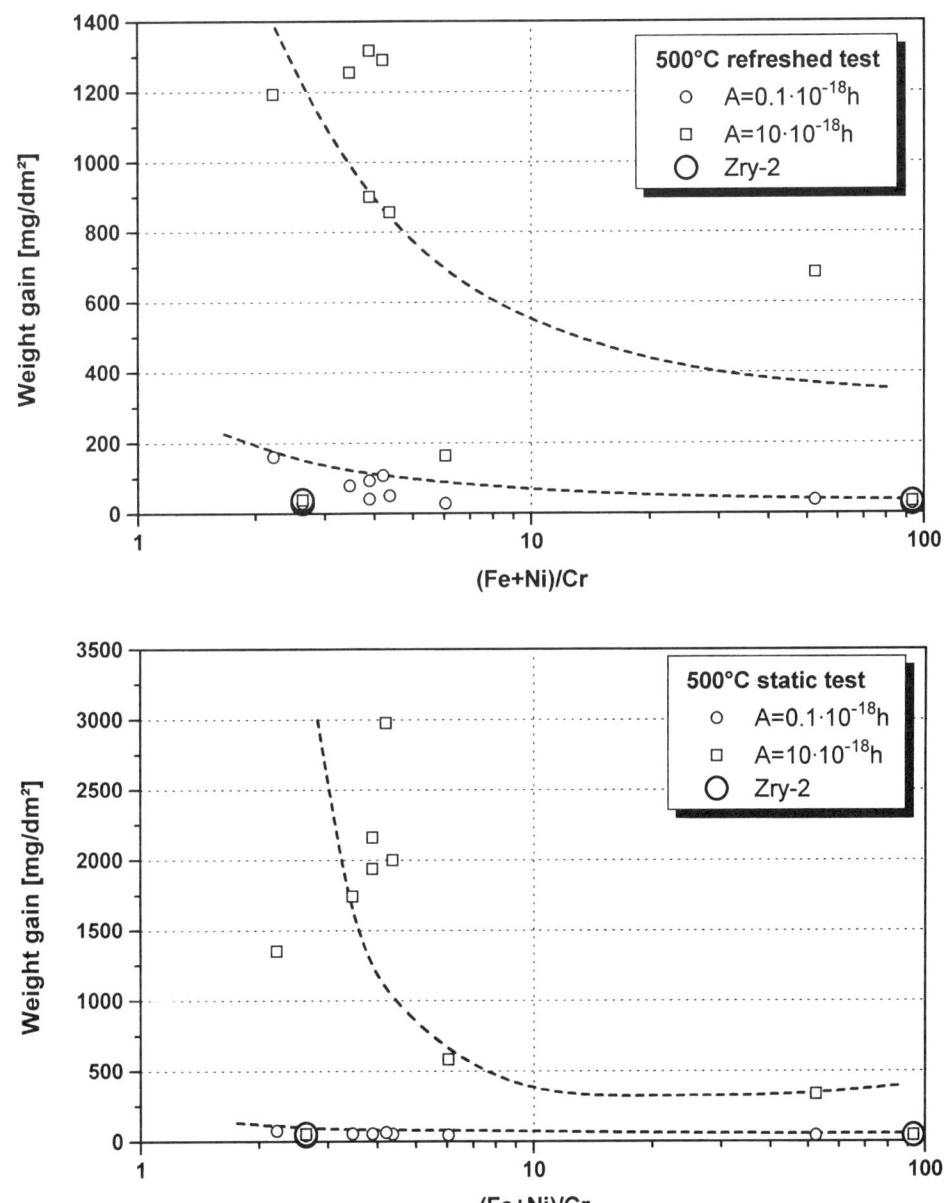

FIG. 5—*Corrosion results from 500°C steam testing under static and refreshed conditions. Exposure for 16 h at temperature.*

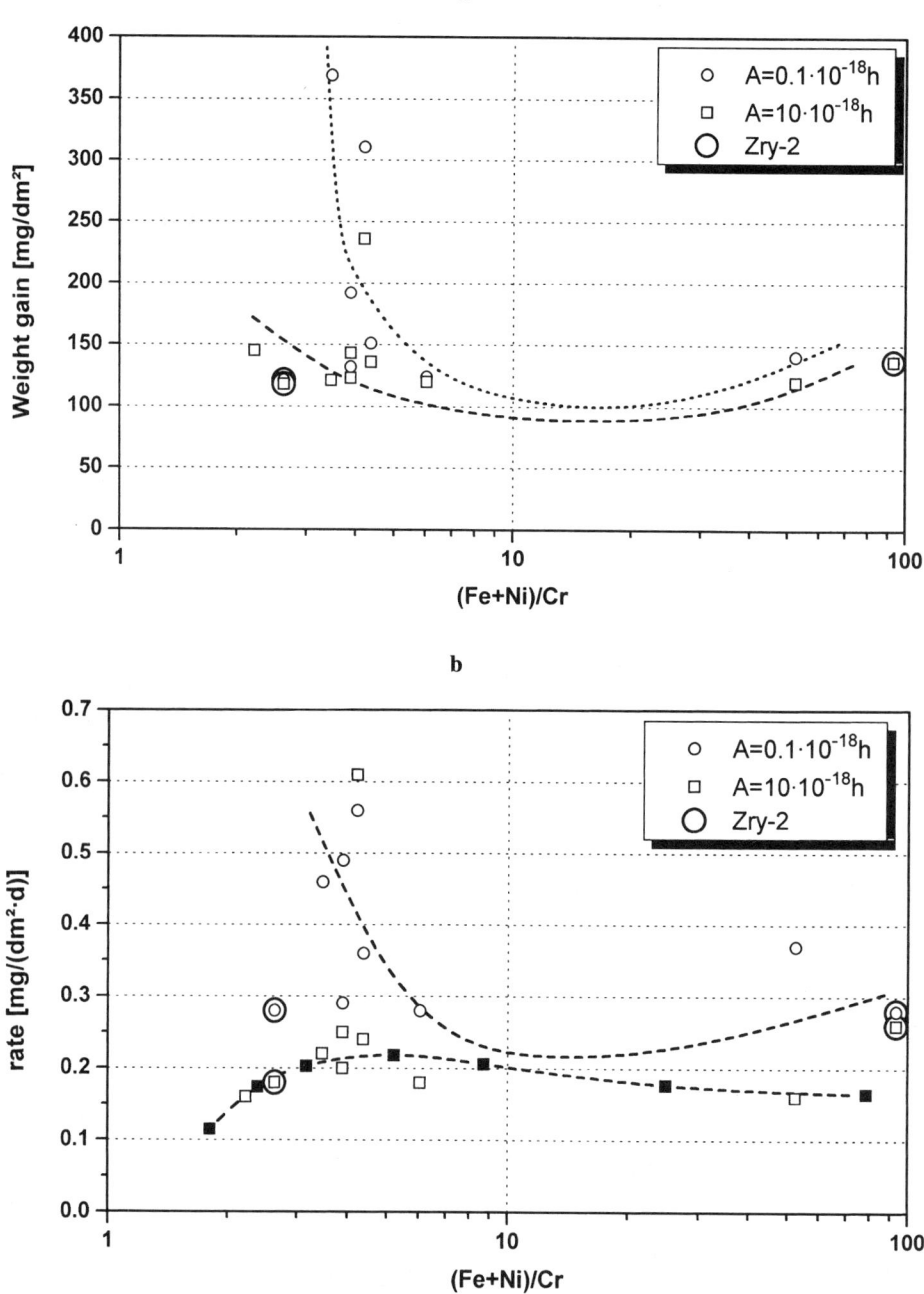

FIG. 6—*Corrosion results from a 400°C, 10.5 MPa static steam test.*

TABLE 7—*Hydrogen pickup values determined at comparable weight gains for low and high A-parameter materials.*

Alloy No.	Ni, %	500°C Standard Test $A = 0.1$ E-18 h %	400°C Standard Test $A = 10$ E-18 h %
1	0.002	28	46
2	0.002	23	38
3	0.002	24	39
4	0.002	26	39
6	0.002	22	44
9	0.1929	42	65
10	0.0571	30	53

was determined by hot extraction for the low A-parameter samples corroded in 500°C steam and for a set of corroded samples with high A-parameters originating from the 400°C test. Relative hydrogen pickup values were calculated and are given in Table 7. No strong influence can be concluded from the values as a function of the chromium range covered by the alloys, but a strong dependence is found for nickel-containing alloys. In Figure 7, the hydrogen pickup is plotted as a function of the Ni content. As can be seen from this plot, an increase of the hydrogen pickup fraction is observed with increasing nickel content for both A-parameters. This effect is one of the main restrictions for applying Ni-containing alloys under conditions that are sensitive to extended hydrogen uptake.

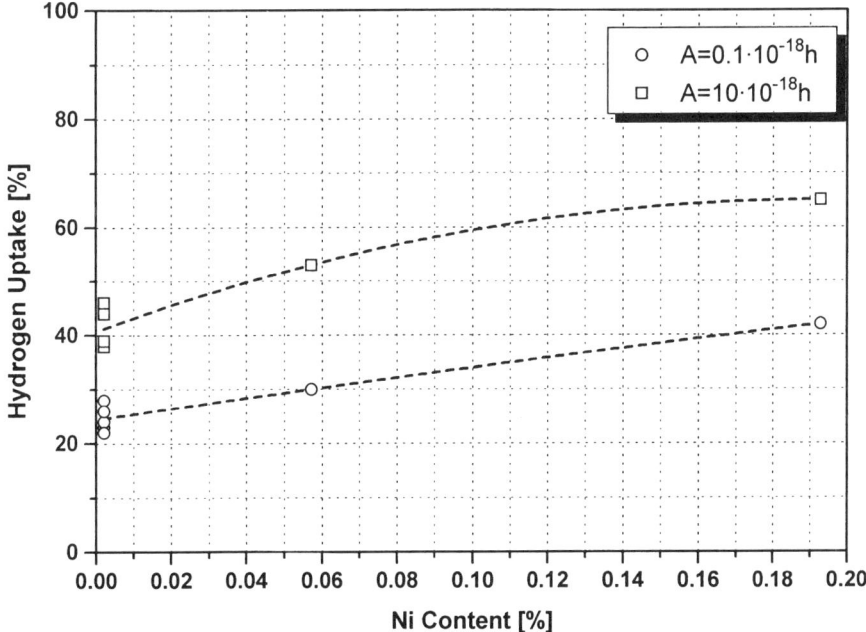

FIG. 7—*Hydrogen uptake for materials with low and high A-parameters determined from a specimen corroded in steam tests.*

In-Pile Corrosion Behavior

After irradiation of two cycles, the specimens were taken out of the rector and transported to the hot cells. There the examination started with visual inspection. In Fig. 8, typical examples are documented. For each specimen, the front and back sides are shown. Obviously, corrosion appears very different on both surfaces. None of the specimens was totally free from nodular corrosion. The appearance of the nodules varies from very fine dense layer forming nodules to singular big ones, sometimes isolated and sometimes consisting of more or less continuous light layers on a black or grayish background. In some cases, surface regions totally free from nodular oxide are observed, and on some specimens excessive corrosion at the edges or near stamped sample identification numbers is visible. All these effects, of course, influence the interpretation of weight gain data determined for the different irradiated specimens. Therefore, in addition to the weight gains, the percentage of the sample area covered by nodular oxide has been determined by automatic image analysis on the basis of photographs of the specimens. The results of this determination together with measured weight gains are listed in Table 6 and

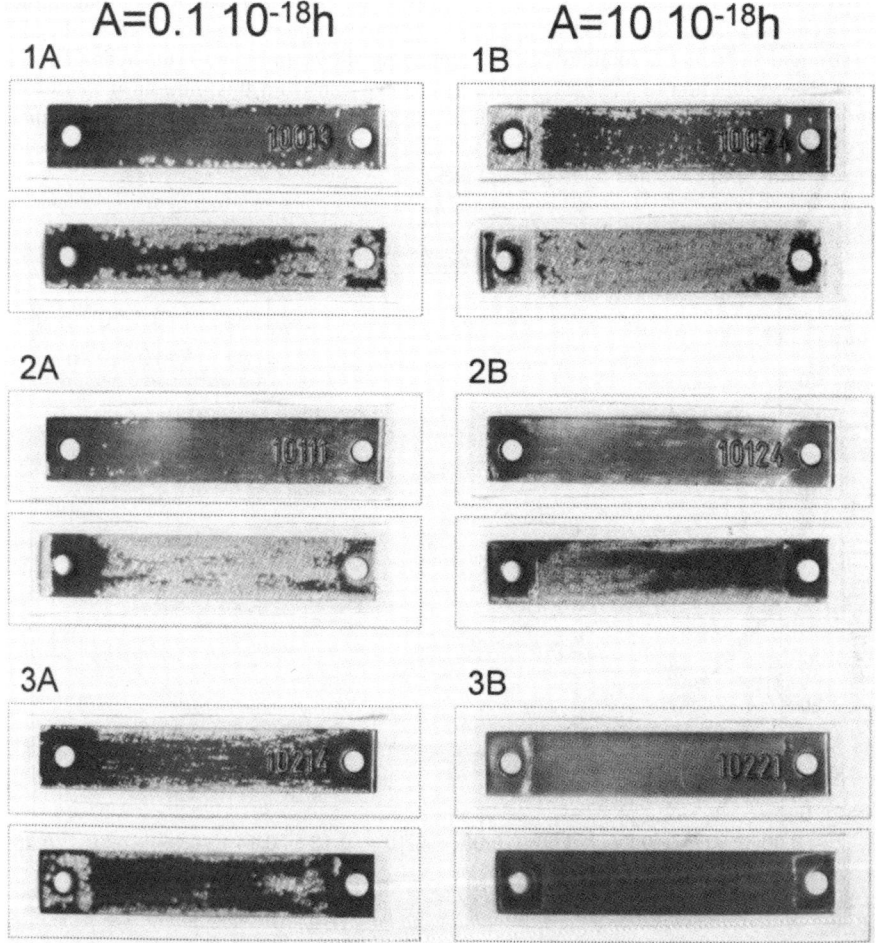

FIG. 8—*Examples of the coupons exposed to BWR irradiation for two cycles.*

plotted in Fig. 9. A reasonable correlation between weight gain data and percentage of nodular oxide coverage can be seen. Due to the fact that deviations from the mean occur statistically in both directions, the correlation is considered valid and expressed by the solid line. No differences are obvious, as expected for this correlation, for specimens with different manufacturing sequences.

Weight gains and percentages of nodular coverage are plotted versus the (Fe+Ni)/Cr ratio in Fig. 10. As seen in out-of-pile tests, different dependencies for low and high A-parameter specimens are found. For (Fe+Ni)/Cr values below 4 to 5, low A-parameter specimens show significant lower weight gains or nodule coverage than high A-parameter materials. This is in agreement with the out-of-pile results from the nodular tests. But, in contrast to the out-of-pile findings in the nodular test, the dependence on the (Fe+Ni)/Cr ratio for the low A-parameter specimen goes in an opposite direction. The weight gains increase with increasing transition metal ratio very similar as in the 400°C laboratory test.

The influence of nickel is comparable to the findings from out-of-pile testing. As can be seen from the values plotted in Fig. 10, nickel contents above the ASTM range for Zry-2 are beneficial with respect to nodular corrosion, in a way reducing the strong influence of the A-parameter.

Summary and Conclusions

The corrosion of Zircaloys is known to be dependent on the size distribution of second-phase particles [9]. Observations from out-of-pile and in-pile behavior show the strong dependence between precipitate size and corrosion resistance. Materials with small precipitates, cor-

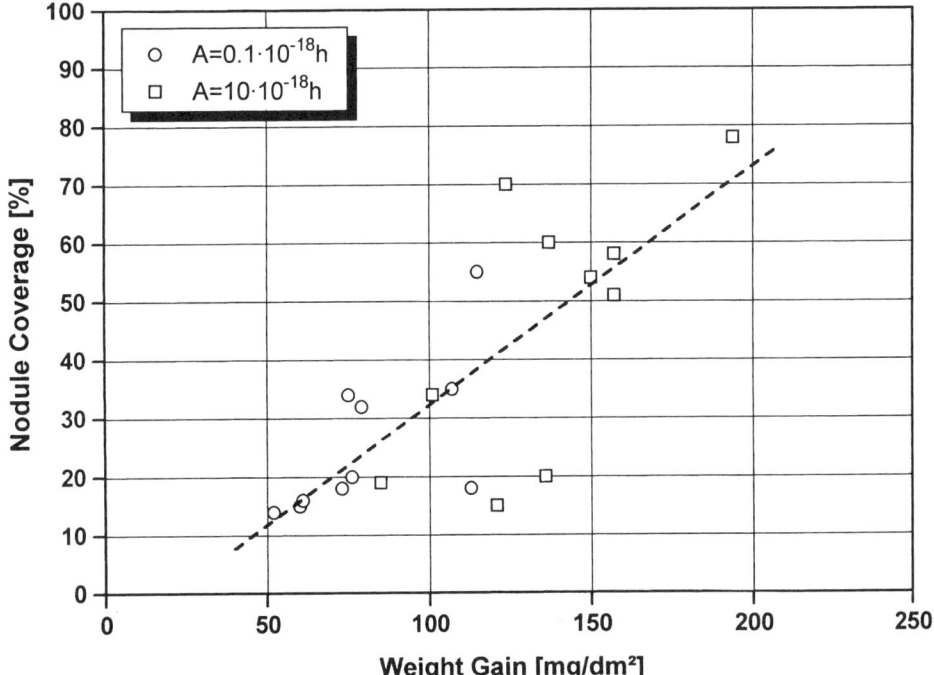

FIG. 9—*Correlation between weight gain and nodule coverage measured on samples exposed to a boiling water reactor.*

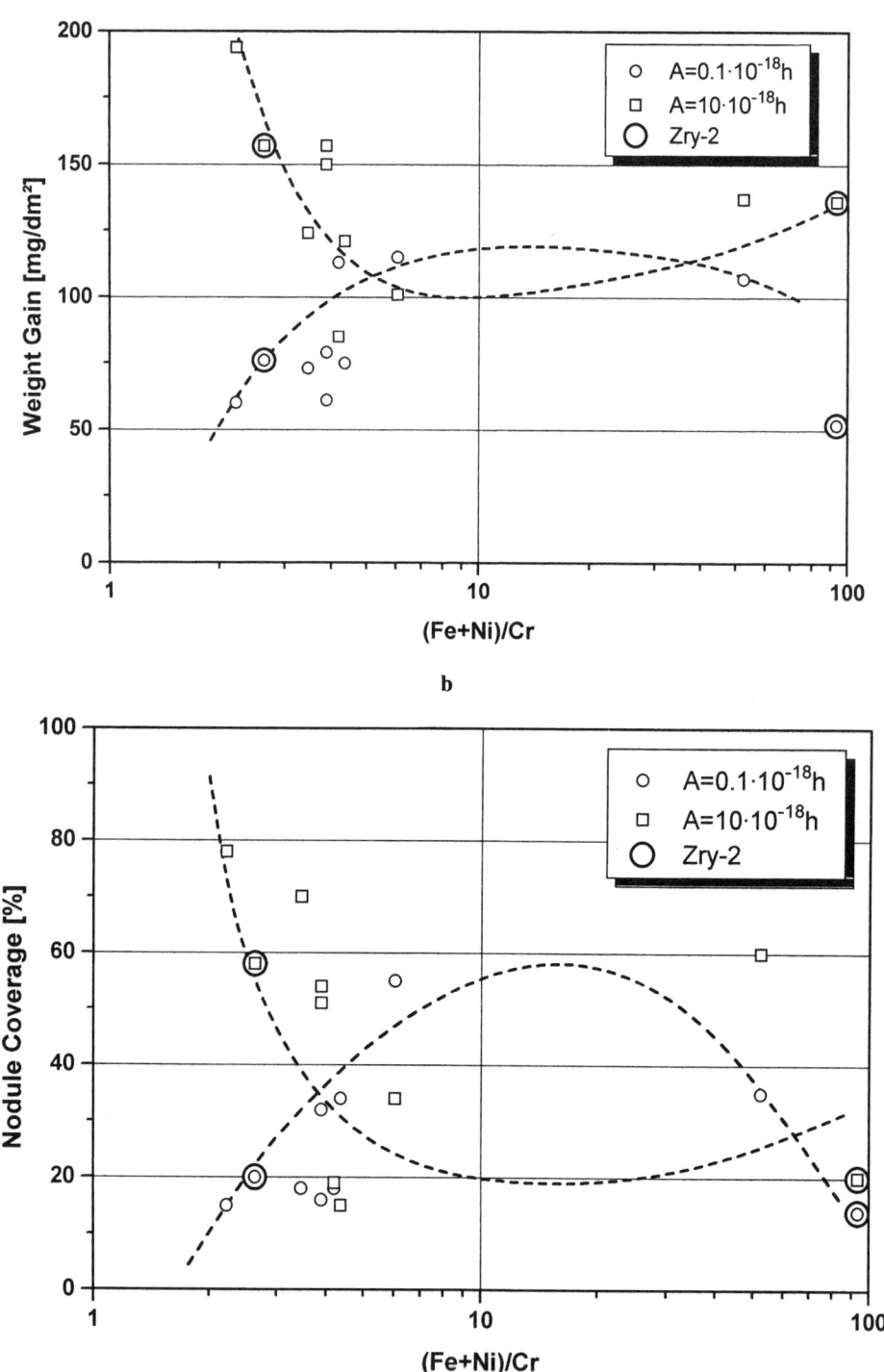

FIG. 10—*In-pile results after two-cycle irradiation. Weight gain and nodule coverage versus (Fe+Ni)/Cr content.*

related by low A-parameters, behave well under in-BWR conditions. Materials with larger mean precipitate sizes, correlated to A-parameters above 10 E-18, show good corrosion resistance in pressurized water reactors. As a result of this study, it has to be mentioned that those findings are obviously only strongly valid for zirconium-based alloy compositions around the ASTM specified ranges, i.e., with a Fe/Cr ratio of about 2. If this ratio is different, significant deviations are observed when comparing autoclave and in-pile behavior. This effect has to be related, on the one hand, to the different types of precipitates occurring in alloys with different Fe/Cr ratios and, on the other hand, to precipitate size distribution established by the thermo-mechanical treatment during fabrication.

The TEM results presented show for a Fe/Cr ratio larger than about 4 the existence of large (in comparison to the Laves phase particles) Zr_3Fe particle predominates for the high A-parameter materials. For those materials, the A-parameter is neither important for corrosion in a 400°C test nor under nodular test conditions. The in-pile result confirms these laboratory findings.

Different findings have to be stated for alloys with a lower Fe/Cr ratio. In these materials, TEM revealed the existence of mainly Laves phase precipitates ($Zr(Fe,Cr)_2$) with different sizes depending on the heat treatment during fabrication (A-parameter). Now small mean sizes (low A-parameter materials) are correlated with poor corrosion resistance in the 400°C test but high corrosion resistance in 500/520°C steam tests. The BWR results show beneficial effects with decreasing Fe/Cr ratios only for low A-parameter materials.

If nickel is added, the influence of the precipitate size is reduced as concluded from the results found for all test conditions. In-pile results show a similar trend, but not to the same extent.

The results show that for material optimization not only the amount of alloying elements but also the size and composition of the precipitated elements are important factors of corrosion behavior.

The following conclusions can be drawn from this study:

1. The A-parameter influence with respect to corrosion is the dominant factor for out-of-pile and in-pile corrosion behavior.

2. The (Fe+Ni)/Cr ratio influences corrosion resistance, especially at low values below about 5. The effect is found to be strong for low A-parameter materials under irradiation of two cycles comparable to the behavior found for the high A-parameter materials in out-of-pile tests. The nodular corrosion test reproduces the in-pile behavior found for high A-parameter materials.

3. Nickel substituting for Cr reduces the effect of the A-parameter on corrosion.

4. Sizes of intermetallic precipitates depend on chemical composition and structure. Significant numbers of Zr_3Fe particles are found only at Fe/Cr-ratios above 5. These particles are significantly larger than the Laves phases.

References

[1] Steinberg, E., Weidinger, G., and Schaa, A. in *Zirconium in the Nuclear Industry: Sixth International Symposium, ASTM STP 824*, American Society for Testing and Materials, West Conshohocken, PA, 1984, pp. 106–122.
[2] Garzarolli, F., Goll, W., Seibold, A., and Ray, I., "Effect of In-PWR Irradiation on Size, Structure, and Composition of Intermetallic Precipitates of Zr Alloys," this publication.
[3] Charquet, D. et al., "Solubility Limits and Formation of Intermetallic Precipitates in ZrSnFeCr Alloys," *Zirconium in the Nuclear Industry: Eighth International Symposium, ASTM STP 1023*, American Society for Testing and Materials, West Conshohocken, PA, 1988, pp. 405–422.
[4] Mauβner, G., Ortlieb, E., and Weidinger, H.-G., *Basic Properties of Zirconium Alloys with Respect to Mechanical and Corrosion Behavior*, BNES, London 1987, pp. 49–55.
[5] Charquet, D., "Influence of Precipitates on the Corrosion of Zry-4 in 400°C Steam," *Journal of Nuclear Materials*, Vol. 211, 1994, pp. 259–261.

[6] Bangaru, N. V., Busch, R. A., and Schemel, J. H., "Effect of Beta Quenching on the Micro Structure and Corrosion of Zircaloys," *Zirconium in the Nuclear Industry: Ninth International Symposium, ASTM STP 939*, American Society for Testing and Materials, West Conshohocken, PA, 1987, pp. 341–363.

[7] Eucken, C. M., Finden, P. T., Trapp-Pritsching, S., and Weidinger, H.-G., "Influence of Chemical Composition on Uniform Corrosion of Zirconium-Based Alloys in Autoclave Tests," *Zirconium in the Nuclear Industry: Eighth International Symposium, ASTM STP 1023*, American Society for Testing and Materials, West Conshohocken, PA, pp. 113–127.

[8] Weidinger, H.-G., Garzarolli, F., Eucken, C. M., and Baroch, E. F., "Effect of Chemistry on Elevated Temperature Nodular Corrosion," *Zirconium in the Nuclear Industry: Ninth International Symposium, ASTM STP 939*, American Society for Testing and Materials, West Conshohocken, PA, 1987, pp. 364–386.

[9] Garzarolli, F., Steinberg, E., and Weidinger, H.-G., "Microstructure and Corrosion Studies for Optimized PWR and BWR Zircaloy Cladding," *Zirconium in the Nuclear Industry: Eighth International Symposium, ASTM STP 1023*, American Society for Testing and Materials, West Conshohocken, PA, 1989, pp. 202–212.

DISCUSSION

P. Rudling[1] *(written discussion)*—By adding Ni to Zr-2, hydrogen pickup is increased. How much of this increase is due to the inherent effect of Ni and how much in inherently due to the decrease in corrosion rate? (D. Charquet showed at the ASTM conference in Baltimore that decreasing the corrosion rate inherently increases the hydrogen pickup.)

Dr. H. J. Sell et al. (authors' closure)—The decrease of the corrosion rate inherently increases the hydrogen pickup, mainly when the process route is concerned. When the composition is concerned, the influence on corrosion and hydriding rate depends on the elements: (1) For example, for ZrFeV alloys, we observe by comparison to Zry-4 a decrease of both corrosion and hydriding rate (D. Charquet, International Topical Meeting on LWR Fuel Performance, Avignon, 21–24 April 1991); (2) In Zry-2, silicon does not change the corrosion rate, but decreases the hydrogen pickup (J. Delafosse, Marianské Lazne, 1–3 October 1968). So, in our alloys we found no clear dependence between corrosion rate and Ni content. Ni-containing materials do not corrode much differently than Ni-free alloys at 400°C steam (see Fig. 6), but have an influence on hydrogen pickup (see Fig. 7). We do not exclude that this influence of the Ni on the hydrogen pickup is exacerbated by the low silicon contents of our alloys.

---

[1] ABB Atom, Sweden.

C. E. Coleman,[1] B. A. Cheadle,[1] C. D. Cann,[2] and J. R. Theaker[1]

Development of Pressure Tubes with Service Life Greater Than 30 Years

REFERENCE: Coleman, C. E., Cheadle, B. A., Cann, C. D., and Theaker, J. R., "**Development of Pressure Tubes with Service Life Greater Than 30 Years,**" *Zirconium in the Nuclear Industry: Eleventh International Symposium, ASTM STP 1295,* E. R. Bradley and G. P. Sabol, Eds., American Society for Testing and Materials, 1996, pp. 884–898.

ABSTRACT: The steps required for the development of a Zr-2.5Nb pressure tube with a 30-year service life are outlined. The life-limiting factors of fracture and deformation are described, and progress towards closing the gap between the properties required to meet the target lifetime and the behavior of current materials is summarized. Fracture is avoided by eliminating hydrides at reactor operating temperatures, by controlling the initial hydrogen concentration and minimizing deuterium ingress from corrosion, and maintaining high toughness. Tube elongation has been reduced by microstructural modification, but transverse deformation remains a challenge.

KEYWORDS: Zr-2.5Nb, pressure tube, service life, delayed hydride cracking, fracture toughness, hydrogen, deformation

In the CANDU, RBMK, and Fugen reactors, the pressure tubes are the pressure vessel. These tubes are long zirconium alloy tubes that pass through the center of the reactor core and therefore are subjected to neutron irradiation, high temperatures and pressures, and corrosion from the heat-transport fluid. The challenge is for the tubes to remain fit for service for an economic lifetime. For CANDU, this is at least 30 years at a 90% capacity factor. Worldwide experience (Table 1) since the late 1950s has indicated that this performance goal is achievable for some reactors but has shown the type and source of problems that have been and have to be solved (Table 2).

In early pressure tube reactors, problems arose as a result of design, fabrication, installation, and reactor operation. These problems have been resolved, and we are now focusing on improvements to the metallurgy of the pressure tubes and the 403 stainless steel end fittings used to link the pressure tubes to the remainder of the heat-transport system via mechanical joints called rolled joints. In this paper, the current lifetime limitations will be outlined, target properties will be defined to provide margins against failure, methods and progress towards achieving these targets will be described, and implementation strategies will be discussed.

Lifetime Limitations of CANDU Pressure Tubes

When the fuel channels were designed for the first pressure tube reactors, the following properties of zirconium alloys that could affect the performance of pressure tubes were not appreciated.

---

[1] Branch manager, division director, and scientist engineer, respectively, AECL, Chalk River Laboratories, Chalk River, Ontario, Canada, K0J 1J0.
[2] Senior scientist engineer, AECL, Whiteshell Laboratories, Pinawa, Manitoba, Canada, R0E 1L0.

TABLE 1—Examples of pressure tube reactors.

Reactor	Description					Lifetime of Pressure Tubes		Comments
	Configurations and Moderator	P/T Material	Hoop Stress, MPa	Outlet Temperature, °C	Fast Neutron Flux, 10^{17} n/m^2·s	Start-up Date	Years (Power Days) as of 1995	
Fugen	Vertical D$_2$O	HT Zr-2.5Nb	96	280	2.7	1979	16 (4050)	Still operating.
Leningrad (RBMK)	Vertical Graphite	Ann Zr-2.5Nb	86	288	1.7	1973	16 (4930)	Still operating. Retubed because of interference between pressure tube and graphite moderator.
N-reactor	Horizontal Graphite	CW Zircaloy-2	54	282	2.0	1963	25 (4242)	Shut down because product no longer required.
SGHWR	Vertical D$_2$O	CW Zircaloy-2	87	282	2.6	1967	24 (5080)	Shutdown because of high inspection costs.
CANDU NPD	Horizontal D$_2$O	CW Zircaloy-2	80	277	1.0	1962	25 (6540)	Shutdown because leak-before-break could not be assured.
CANDU Bruce-3	Horizontal D$_2$O	CW Zr-2.5Nb	135	308	3.7	1978	17 (4725)	Still operating.
CANDU Point Lepreau	Horizontal D$_2$O	CW Zr-2.5Nb	135	312	3.7	1983	12 (4155)	Still operating.

NOTE: HT = heat treated, Ann = annealed, CW = cold worked.

TABLE 2—*Solutions to problems with pressure tubes.*

Problem	Solutions
Fabrication residual stresses leading to DHC in RBMK reactors.	Stress relieve before installation.
Installation residual stresses leading to DHC in rolled joints in CANDU reactors.	In-situ stress relief. Revise operating procedures. Redesign rolled joints.
Fabrication flaws leading to DHC in tubes.	Revise melting and billet inspection practice.
High hydrogen concentration leading to intolerable fracture properties.	Change alloy from Zircaloy-2 to Zr-2.5Nb.
Formation of hydride blisters leading to DHC at temperature gradients caused by contact between hot pressure tube and cold calandria tube.	Eliminate possibility of temperature gradients by redesigning spacers.
Elongation resulting in insufficient bearing allowance.	Redesign bearings on end fittings. Modify microstructure to reduce axial growth.

1. Irradiation enhancement and anisotropy of creep.
2. Shape change in a neutron flux with no applied stress (growth).
3. Hydrogen ingress from corrosion.
4. Delayed hydride cracking (DHC).
5. Formation of hydride blisters.
6. Reduction in fracture toughness by neutron irradiation.

The current CANDU pressure tubes have been improved considerably from the ones installed in the early reactors. The alloy was changed from Zircaloy-2 to Zr-2.5Nb, and many changes have been made to their fabrication to improve their properties. Our current goal is to produce tubes that will have a service life greater than 30 years with the reactor operated at a capacity factor of 90%. Our extensive knowledge of the properties of our pressure tubes and the models we have developed for their behavior have enabled us to assess the condition and properties of the current tubes after 30 years of service (Table 3). Pressure tubes have reached the end of their service life when they can no longer perform their design function.

The most important factors are changes in dimensions and an increase in the probability of fracture. The concerns for deformation are elongation and increase in diameter. Elongation may lead to problems for the support at the ends of the fuel channels. Diametral expansion causes flow bypass around the fuel bundles, which may reduce the margin between normal operating power and the critical channel power that could result in derating of reactors. The probability of fracture and the need to involve leak-before-break (LBB) depend on flaw development, hydrogen concentration, and fracture toughness. (For the pressure tubes of CANDU reactors, LBB is assured if a crack penetrating the tube wall is less than the critical crack length (CCL) for unstable propagation, water leakage is detected, and action is taken before CCL is exceeded. The CCL is estimated from dynamic or static pressure tests on full sections of pressure tube or by matching the *J-R* curve from small specimens to the crack-driving force curve for an axial crack in a tube [1].) In CANDU reactors, hydrogen has been associated with all the fractures, with the problem being exacerbated by neutron irradiation (Table 2).

To control fracture, a defence-in-depth approach is used; crack initiation is prevented, and LBB is demonstrated by showing that adequate time is available to detect a through-wall crack by its leakage. We also need to build in adequate margins on the life-limiting processes. Of

TABLE 3—*End-of-life properties of pressure tubes. Assumes 30 years at 90% capacity factor.*

Property	Current Tubes		Target for Future Tubes	Comments
	Pre-1993 Specification	Post-1993 Specification		
Hydrides present?				
Outlet end	Yes	No	No	Initial concentration at limit of specification. Current maximum ingress rate 0.015 at %/year (3 ppm D/year). Assumes rate of pickup constant for 30 years.
Inlet end	Yes	Yes	No	Initial concentration at limit of specification. Current maximum ingress rate 0.005 at %/year (1 ppm D/year).
Region of high tensile stress at rolled joints	Yes	Yes	No	Little further decrease with irradiation.
K_{IH}, MPa$\sqrt{m}$	4.5	4.5	>10	
DHCV at 250°C, m/s	$<2 \times 10^{-7}$	$<2 \times 10^{-7}$	$<7 \times 10^{-8}$	
CCL at 250°C and 10 MPa, mm	42	>80	>80	Data from burst tests. See Davies and Shewfelt, this conference.
Diameter change, %	5	5	<3	Linear behavior, no rupture.
Elongation, mm	230	230	<100	Current value based on increasing rate.

the various mechanisms of crack initiation and propagation, the most likely is DHC. To prevent DHC, the following risk factors must be absent:

1. *Hydrides*—the hydrogen concentration must be less than the terminal solid solubility limit (TSS) so that brittle hydrides do not form.
2. *Tensile stress*—the total applied tensile stress and its amplification by flaws must be less than the critical value to fracture hydrides.
3. *Time*—the duration of exposure to the first two risk factors must be minimized.

The first line of defence is to eliminate hydrides. Thus, the total hydrogen concentration, consisting of hydrogen present after fabrication and deuterium absorbed from corrosion, should not exceed 0.3 at% anywhere in the pressure tube during operation for 30 years. Hydrides are always present at room temperature, and reactor operation would be simplified if crack initiation was minimized at low temperatures. Our target is for the minimum value of K_{IH}, the threshold stress intensity factor for DHC, to be 10 MPa$\sqrt{m}$. With current rolled joints, where the maximum tensile stress, σ, is 200 MPa, a flaw with a depth, a, of 1 mm would be required before cracking would start based on

$$a = 0.4(K_{IH}/\sigma)^2 \tag{1}$$

Such a flaw is easily detected by modern NDE techniques. K_{IH} is chosen to represent crack initiation because it is straightforward and quick to measure, whereas DHC initiation from rounded flaws is difficult to measure because of scatter and because it takes a long time. The principle is that if K_{IH} is high, then crack initiation is difficult.

If a crack does develop and it is not detected by NDE, it should manifest itself as water leakage detected in the annulus gas system surrounding the pressure tube. The principle of LBB, as practiced in CANDU reactors, requires that the time *available* to detect moisture from a leaking crack before it becomes unstable, t_a, is much greater than the time *required* to detect the leak, t_r [2]. To illustrate the idea, let us assume a crack propagates through the tube wall thickness, W, and has a length, L, at leakage. If the crack then grows in both directions.

$$t_a = (C-L)/2V \tag{2}$$

where

C = the minimum length of an unstable crack, CCL, and
V = the crack velocity of DHC along the tube.

Using typical bounding values of $C = 42$ mm (at 10 MPa), $L = 4W = 16$ mm and $V = 2 \times 10^{-7}$ m/s (at 250°C), $t_a < 20$ h. Since t_r is about 10 h, the safety margin or confidence in LBB, t_a/t_r, is about 2. If C can be increased to 80 mm and V is reduced to 7×10^{-8} m/s, t_a will be increased to about 120 h, thus increasing the confidence factor for LBB to 12. From this information, we have defined the values of the properties that will be required to achieve the target lifetime (Table 3).

Achievement of the Target Properties

A Hydride-Free Tube

The potential methods for maintaining the pressure tubes hydride free are:

1. Minimize the initial hydrogen concentration.
2. Minimize ingress during operation.
3. Move the hydrogen to innocuous locations.

Initial Hydrogen Concentration—Up until 1993, the specification allowed a finished pressure tube to contain up to 0.23 at% (25 ppm) of hydrogen. In practice, tubes contained an average of 0.095 at% (10.3 ppm) hydrogen, but values as high as 0.16 at% (18 ppm) were observed. With the latter concentration, hydrides would be present at 217°C [*3*]. If deuterium is picked up at 0.005 at%/year (1 ppm/year) at the inlet end of a channel (250°C), hydrides would be present after 26 years of operation in a tube with the high initial hydrogen concentration. Thus, there was a large incentive to reduce the initial hydrogen concentration.

In cooperation with the manufacturers, we identified at what stages and locations the hydrogen enters the zirconium during tube fabrication and used this knowledge in its reduction. The implementation of these improvements [*4*] led to a reduction to 0.05 at% (5 ppm) in the initial hydrogen allowed by the specification and is illustrated for recent production in Fig. 1.

Since there is extra deuterium ingress at rolled joints, hydrides form early at the end of the pressure tube, where it is supported by the end fitting. With time, the front of hydrides moves inboard and eventually passes the location where the tensile stresses are high. The aim is to postpone when the hydride front reaches this location. Using the maximum ingress rate of the inlet end of a Bruce-type reactor and the maximum pick-up rate of deuterium from corrosion (0.005 at%/year), the predicted position of the hydride front as a function of time (Fig. 2) shows that a tube with an initial hydrogen concentration of 0.15 at% had hydrides present at the critical location after 15 years, but this time was increased to 25 years when the initial hydrogen concentration was reduced to 0.05 at%.

Ingress During Operation—Two approaches to reducing ingress during reactor operation are to modify the surface and to control the composition of the alloy. Several surface modifications have been studied, and shot peening has been selected as the most promising technique. Shot-peening the surface obliterates the original microstructure with a heavily cold-worked

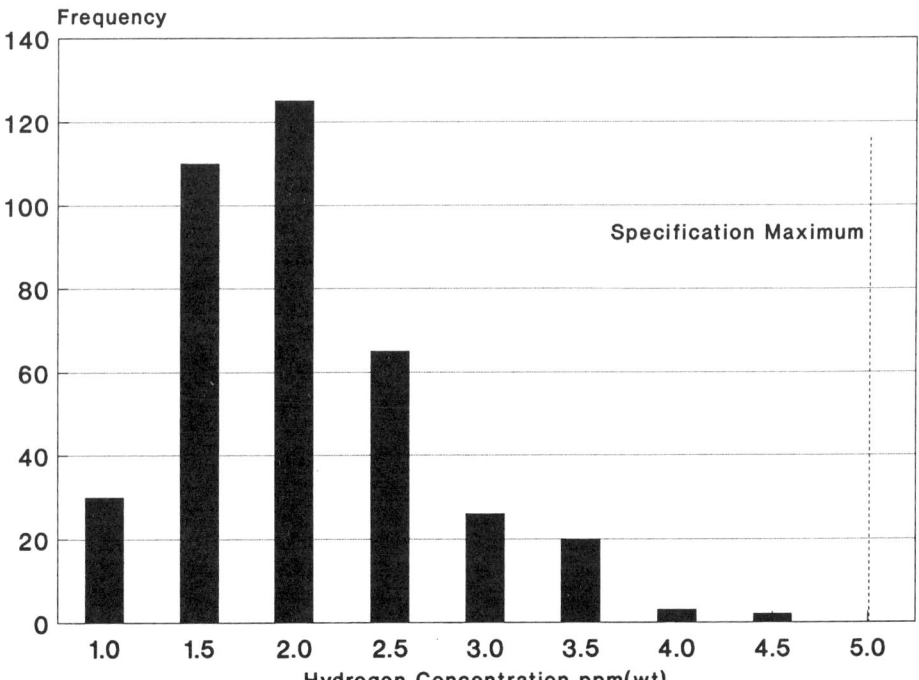

FIG. 1—*Histogram of hydrogen concentration in recently produced Zr-2.5Nb pressure tubes.*

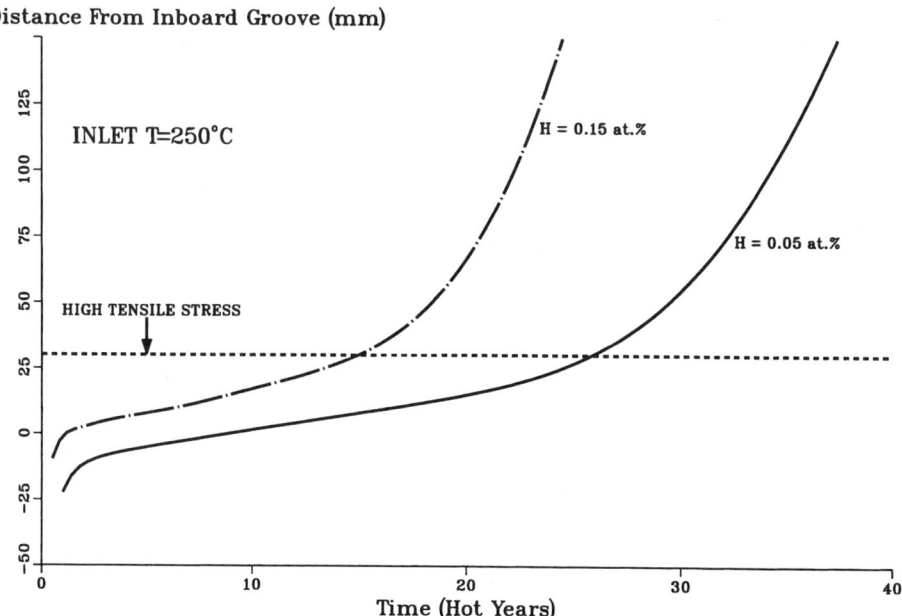

FIG. 2—*The effect of lowering the initial hydrogen concentration on the position of the hydride front at the ends of pressure tubes as a function of time. The position is referenced to the inner of the three grooves in the stainless steel end fitting. The position of the region of highest tensile stress is assumed to be 80 mm from the end of the pressure tube.*

layer that may be recrystallized with subsequent heat treatments to produce a microstructure almost optimum for corrosion resistance with a reduced concentration of niobium in the α-phase and a uniform dispersion of small β-Nb precipitates.

The results of out-reactor tests are very promising [5]: in either the shot-peened or shot-peened and heat-treated conditions, both oxidation and deuterium pick-up are approximately halved compared with as-received material. If this behavior can be maintained in-reactor, we can have high confidence in having hydride-free pressure tubes for 30 years. Using the rolled joint to illustrate the improvement (Fig. 3), we find that hydrides are predicted to be absent at the critical position for over 40 years. The effects of the shot-peening penetrate about 200 μm, about 5% of the tube wall thickness; thus, the remainder of the tube is unchanged. A potential bonus of shot-peening and heat-treatment is a much improved resistance to fretting wear.

Laser glazing the surface to produce martensite [6] and coating the pressure tube with a thin, dense layer of pure zirconia or a mixture of zirconia and niobia have also been investigated [7]. However, the laser process is complicated and needs great care with atmospheric control during melting, and the efficacy of zirconia coatings in reducing deuterium ingress in-reactor and their practical applicability to full-size tubes has yet to be determined.

It is unlikely that our current composition of Zr-2.5Nb is optimum for corrosion and deuterium ingress. Small additions of single elements to Zr-2.5Nb have been made [8], and the results to date show that Ni, Mn, and Ti should be avoided while Cr, Fe, Mo, and Si are beneficial. Batches of Zr-2.5Nb-400 ppm Cr, -1100 ppm Mo, and -Fe/Sn/Cr have been made commercially and are showing promise in out-reactor tests.

The extra deuterium absorbed at the ends of the pressure tubes is caused by galvanic and crevice corrosion between the pressure tube and end fitting. A chromium layer between the

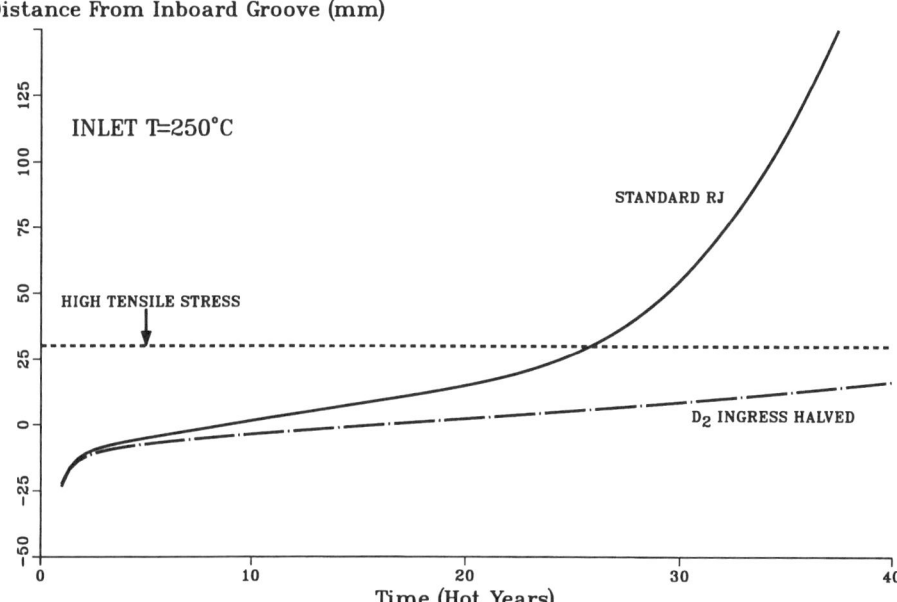

FIG. 3—*The effect of halving the deuterium pick-up rate, for example, by shot peening, on the position of the hydride front at the ends of pressure tubes as a function of time.*

two components reduces the ingress through minimizing galvanic effects and entry of deuterium from the steel into the zirconium [9]. Many joints with the chromium layer have been made and are being tested in a D_2O loop. Results indicate a reduction of the extra deuterium ingress by between 50 and 90%. The illustrative calculation at the rolled joint shows that a reduction of deuterium ingress by 50% would provide an increase of four years before hydrides are present at the critical point, while a reduction of 80% would provide a margin of over 30 years on the target operating lifetime of 30 years.

Relocating the Hydrogen—The two methods being studied are gettering and ejection.

1. *Gettering*—The principle depends on placing a material, such as yttrium, that forms a more stable hydride than zirconium into an innocuous position, where it can getter the hydrogen and minimize the hydrogen concentration in the tube [10,11]. One application is at the ends of the pressure tube, where the yttrium can protect the rolled joints. An experiment to demonstrate the efficacy of the getter is in progress. After 200 days of testing, the hydrogen distribution in the pressure tube at the rolled joint is following the prediction (Fig. 4). If this behavior continued for 30 years, calculations with ingress and gettering show that no hydrides would be present anywhere in the pressure tube, even in the rolled joint, during reactor operation.
2. *Ejection*—Palladium can act as a window for hydrogen and as a catalyst for the reaction between hydrogen and oxygen to form water. If a layer of palladium were placed on the outside of the pressure tube just inboard of the rolled joint and the annulus gas was oxidizing, the hydrogen in the pressure tube would tend to flow through the palladium, react with the oxygen, and the resultant water could be swept out by purging the gas annulus. A laboratory experiment has demonstrated the principle. A piece of Zr-2.5Nb

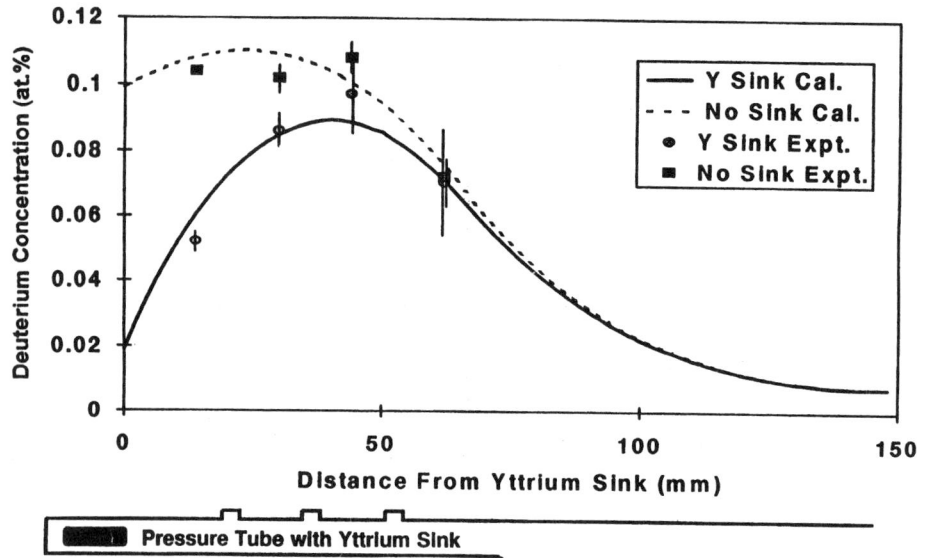

FIG. 4—*Deuterium distribution in a standard pressure tube at the rolled joint compared with one protected by an yttrium sink after 200 days at 300°C in a heavy water loop.*

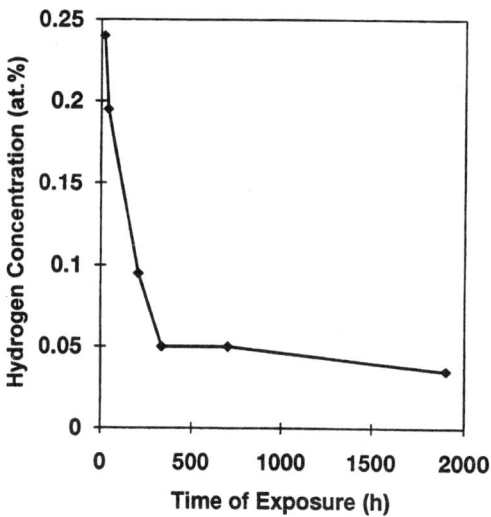

FIG. 5—*Decrease in hydrogen concentration with time in Zr-2.5Nb coupons coated with palladium exposed to $CO_2 + 5\%O_2$ at 300°C.*

pressure tube, containing 0.24 at% hydrogen, was coated with a 4-μm-thick layer of palladium, placed in a simulated annulus gas of CO_2 + 5%O_2 at 300°C, and sampled at intervals. The hydrogen concentration in the Zr-2.5Nb gradually decreased to 0.04 at% (Fig. 5). If this ejection process were to continue for 30 years, hydrides are calculated to be present only at the ends of the pressure tube where the residual stresses are compressive; thus, the rolled joint would be protected.

Reducing Susceptibility to DHC

The reference lower bound for K_{IH} is 4.5 MPa$\sqrt{m}$. Although the theory of K_{IH} is insufficiently well advanced to provide a specific guide to the development of a crack-resistant microstructure, one possibility is to modify the crystallographic texture [12,13]. In current pressure tubes, the basal plane normals of the zirconium crystals are aligned in the transverse direction. In as-fabricated tubes, hydrides have their plate normals in the radial direction (circumferential hydrides). If there is a large tensile stress parallel to the basal plane normals, the hydride platelets precipitate with their normals parallel to the stress. Thus, in a tube, a hoop tensile stress may produce such radial hydrides. This orientation of hydrides leads to low resistance to crack initiation. If the production of such hydrides is difficult, crack initiation is difficult. Thus, if the basal plane normals are oriented away from the stressing direction, cracking may be minimized. Support for this idea comes from experimental results on many different materials (Fig. 6), where K_{IH} is plotted as a function of the resolved fraction of basal plane normals in the stressing direction [12–17].

A strong transverse texture is developed by the extrusion process, and it has been very difficult to rotate these crystals towards the radial direction by cold-working operations. An alternative is to cold work immediately after forging and eliminate extrusion. Short sections of tubes were produced by these routes, and their crystallographic textures had the basal planes concentrated more towards the radial direction. However, the results of preliminary tests, using the load reduction method to measure K_{IH} [14], show little benefit from the change in texture—compare Points A and B in Fig. 6.

These results show that the simple picture is inadequate to describe all the behavior—we need a better way to represent the features of texture critical to DHC and to take account of the change in strength with the change in texture. The good result is that the average values of K_{IH} of some material with a strong transverse texture are acceptably higher (B on Fig. 6). Thus, we need to understand what other microstructural features are important so that we can continue to exploit them. The added importance of this result is that, for low transverse creep, a strong transverse texture is desirable; thus, we will gain a large advantage if we can attain high resistance to DHC without modifying the current texture.

Increasing Fracture Toughness

If a crack does initiate and propagate, we need high-fracture toughness to support the case for LBB. The fracture toughness of pressure tubes removed from power reactors is very variable. The conclusion from programs to understand this variability were that for high toughness and resistance to degradation by irradiation we should minimize the concentration of the trace elements chlorine and phosphorus and control the concentration of carbon [4,18,19]. Chlorine, present as a residue from the Kroll process, is kept to acceptable concentrations by evaporation during four melts. (The previous process used only two melts, although some tubes were made from ingots composed of recycled material and were effectively quadruple melted.) Phosphorus is present in the starting zircon sand, and carbon is introduced as a by-product of the sponge manufacturing process and from contaminated, recycled material.

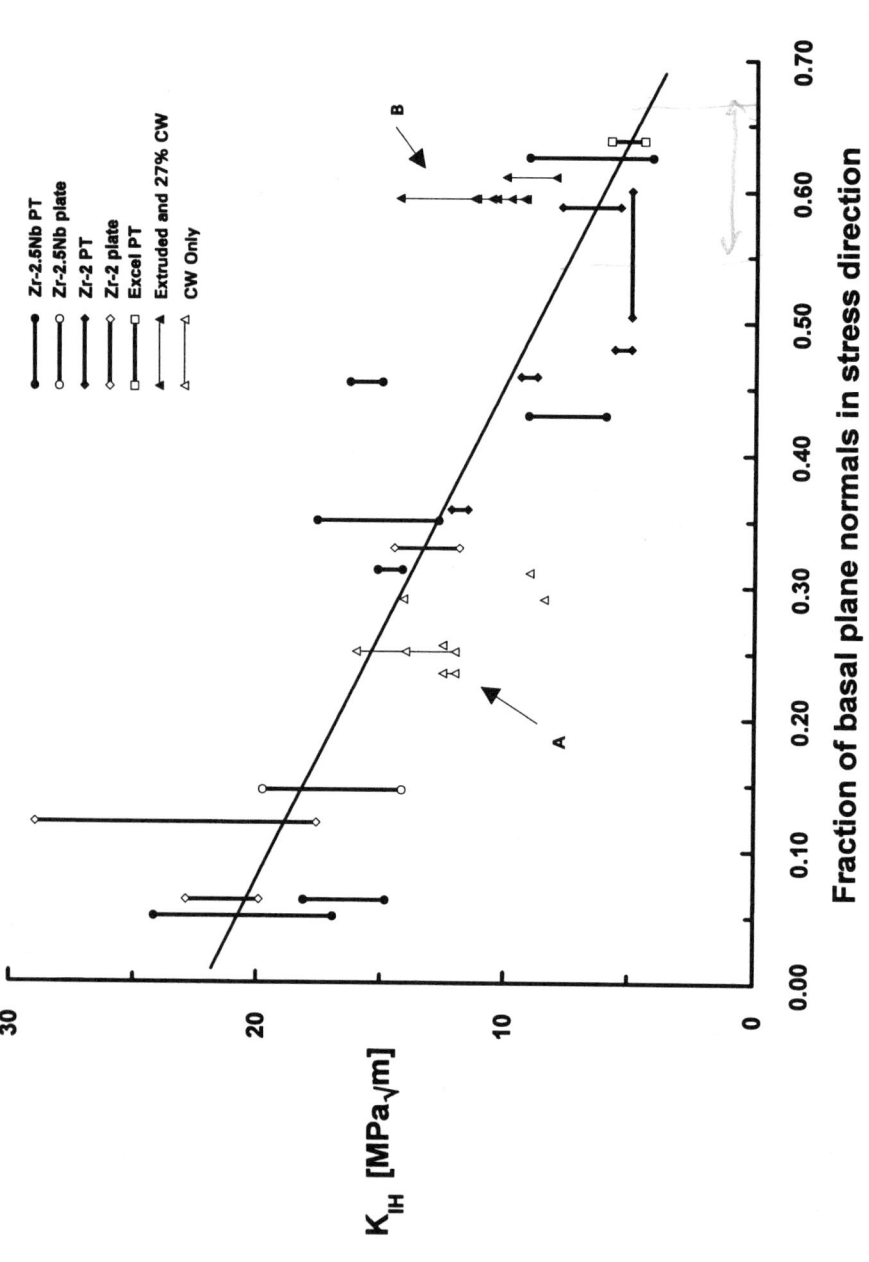

FIG. 6—*Effect of crystallographic texture on threshold for DHC, K_{IH}, as a function of resolved fraction of basal plane normals in the direction of tensile stress, based on new data and data from the literature [12–17]. The line through the data indicate the trend. Points at A are from tests on tubes made by roll extrusion and at B on tubes made from the same ingot but by the standard process.*

Modification to the sponge manufacturing cycle and control of recycled material minimize the phosphorus concentration and maintain the carbon concentration at innocuous values. The success of the transfer of these developments to tube production is illustrated by comparing the distribution of fracture toughness, characterized by dJ/da, for tubes made from double-melted material with that for the latest tubes made from selected sponge that was quadruple melted (Fig. 7). This increased toughness will translate into greater critical crack lengths even after much irradiation [19,20] and therefore contribute to improved confidence in LBB.

Deformation

Progress has been made towards reducing elongation in-reactor. A modified fabrication route [21] that decreases the dislocation density but maintains the fine, elongated grain structure [22] by allowing a heat treatment at 500°C for 6 h has been shown to reverse the direction of irradiation growth in the longitudinal direction. The total in-reactor deformation is usually taken as the sum of irradiation creep and irradiation growth; thus, the reversal of growth implies a reduction in elongation. This reduction has been demonstrated in tubes installed in Bruce Unit 8, which are elongating about 30% slower than standard cold-worked tubes in neighboring channels.

Evaluation of the probable effects of microstructural features that could reduce the diametral expansion without increasing the elongation has indicated the desirability of retaining the strong transverse texture and reducing the thickness of the plate-like α-grains sufficiently for their

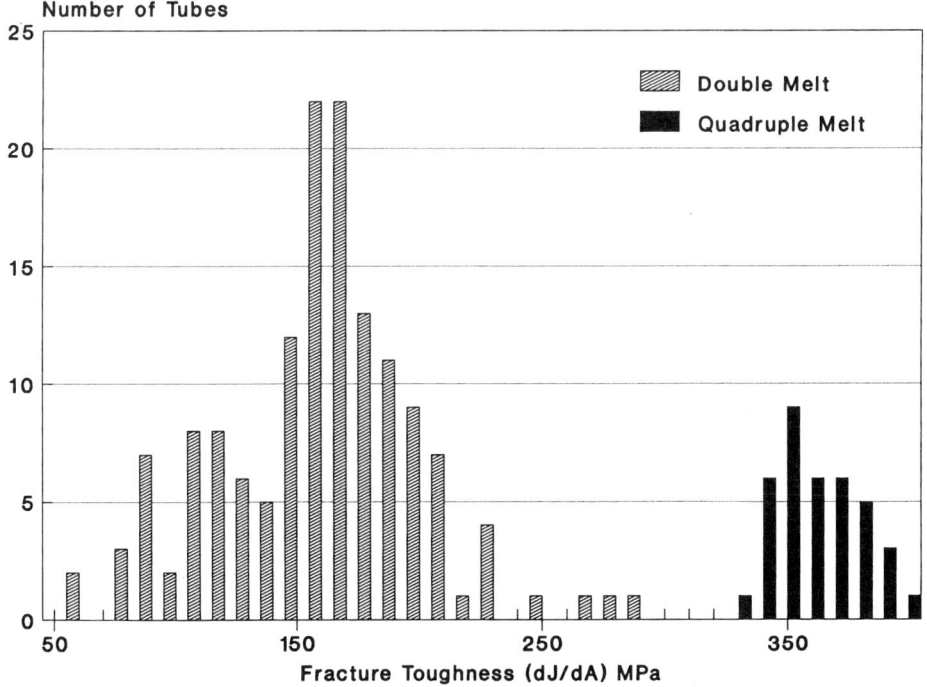

FIG. 7—*Histogram of fracture toughness*, dJ/da, *at 250°C for tubes made to pre- and post-1993 specifications.*

boundaries to dominate as sinks for mobile point defects, the source of irradiation-induced deformation. This will require a modification to the extrusion process or higher cold reduction of the wall thickness than the current practice, followed by stress relieving to reduce the dislocation density without recrystallizing the α-grains, or both, to achieve the required microstructure.

Implementation

For the successful implementation of an improvement, the reactor designers and utilities must first establish the needs and set the technical goals, the researchers provide the ideas, understanding, and evaluation, the manufacturers translate the laboratory results into practical application by defining what is possible and cost effective, and the utilities further assist by using the power reactors as the ultimate proving ground.

Once the principle of an improvement has been shown to work in the laboratory, we must demonstrate that:

1. Full-size tubes, made by an economical route, meet or exceed the target properties in Table 3.
2. The benefit is retained for the target life of the component. This criterion is one of our greatest challenges because it is impractical to test for 30 years. We thus have to rely on short-term tests, often on specimens rather than components, and assess tube behavior using our models, which are based on theories, experiments, and power reactor experience. For example, full-size chromium-plated rolled joints are being tested in an out-reactor heavy water loop since in power reactors the rolled joints are not in a high-neutron flux. Accelerated tests using high-neutron fluxes are needed for the remainder of the pressure tube, but these tests take many years to reach the equivalent of end-of-life fluences. For example, small-fracture specimens have now been irradiated in the high-flux test reactor OSIRIS to a higher fluence than pressure tubes have received in CANDU reactors.
3. The benefit is reproduced in each component. To achieve this criterion, we are studying the causes of variability. For example, specifying the preheat conditions used for extrusion is expected to both reduce the variability and improve the mechanical properties [23].
4. By solving one problem, another has not been created. For example, changing the crystallographic texture from strong circumferential to strong radial distribution of basal poles could increase the resistance to DHC but would increase the rate of diametral expansion.

Conclusion

CANDU Zr-2.5Nb pressure tubes will have a lifetime of more than 30 years when protected from fracture and excess deformation. Fracture is avoided by preventing hydrides from forming during reactor operation and by using tough materials. Progress has been made in reducing elongation, but transverse creep has not been decreased.

Acknowledgments

We would like to thank the many people who are contributing to this development program, the working parties of the CANDU Owners Group, especially Working Party 33, the manufacturers, and the utilities. We are grateful to A. A. Bahurmuz for the calculations of hydrogen distributions.

References

[1] Simpson, L. A., "The Application of Ductile Fracture Analysis to Predictions of Pressure Tube Failure," AECL-6805, report of the Atomic Energy of Canada Limited, Pinawa, Manitoba, Canada, 1981.
[2] Moan, G. D., Coleman, C. E., Price, E. G., Rodgers, D. K., and Sagat, S., "Leak-Before-Break in the Pressure Tubes of CANDU Reactors," *International Journal of Pressure Vessels and Piping*, Vol. 43, 1990, pp. 1–21.
[3] Kearns, J. J., "Terminal Solubility and Partitioning of Hydrogen in α-Phase of Zirconium, Zircaloy-2 and Zircaloy-4," *Journal of Nuclear Materials*, Vol. 22, 1967, pp. 292–303.
[4] Theaker, J. R., Choubey, R., Moan, G. D., Aldridge, S. A., Davis, L., Graham, R. A., and Coleman, C. E., "Fabrication of Zr-2.5 Nb Pressure Tubes to Minimise the Harmful Effects of Trace Elements," *Zirconium in the Nuclear Industry: Tenth International Symposium, ASTM STP 1245*, A. M. Garde and E. R. Bradley, Eds., American Society for Testing and Materials, West Conshohocken, PA, 1994, pp. 221–242.
[5] Amouzouvi, K. F., Clegg, L. J., Styles, R. C., and Winegar, J. E., "Effect of Shot Peening and Post-Peening Heat Treatments on the Microstructure, the Residual Stress and Hardness, Corrosion and Deuterium Uptake Resistance of Zr-2.5 Nb Pressure Tube Material," *Computer Methods and Experimental Measurements for Surface Treatment Effects*, M. H. Aliabadi and C. A. Bzebbia, Eds., Computational Publications, Southampton, 1993.
[6] Amouzouvi, K. F., Clegg, L. G., and Styles, R. C., "Surface Modification of Zirconium Alloys by Laser Glazing," *Proceedings*, Conference on Surface Engineering, Toronto, 1990, S. A. Meguid, Ed., Elsevier Applied Science, New York, 1990, pp. 270–279.
[7] Ploc, R. A., Amouzouvi, K. F., and Turner, C. W., "The Reduction of Corrosion in Zr-2.5 Nb," *Proceedings*, 24th Annual Convention of the International Metallographic Society, Monterey, CA, July 1991, IMS, Columbus, OH.
[8] Ploc, R. A., "The Influence of Impurities in Zr-2.5 Nb on Oxygen and Deuterium Pickup at 573 K in D_2O," 3^e *Colloque International sur la Corrosion et la Protection des Materiaux a Haute Temperature*, Les Embiez, France, May 1992. (Also available as Report AECL-10587, Atomic Energy of Canada Limited, Chalk River, Ontario, Canada.)
[9] White, A. J., Urbanic, V. F., Bahurmuz, A. A., Clendening, W. R., Joynes, R., McDougall, G. M., Skinner, B. C., and Venkatapathi, S., "Plating End Fittings to Reduce Hydrogen Ingress at Rolled Joints in CANDU Reactors," *Proceedings*, International Conference on Expanded and Rolled Joint Technology, Canadian Nuclear Society, Toronto, Ontario, Canada, September 1993, pp. G47–G55.
[10] Cann, C. D., Bahurmuz, A. A., Sexton, E. E., DeGregorio, R., Grant, I., Inglis, I., Murphy, E. V., and Natesan, M., "Removal of Hydrogen from Rolled Joints in CANDU Reactors by Yttrium Sinks," *Proceedings*, International Conference on Expanded and Rolled Joint Technology, Canadian Nuclear Society, Toronto, Ontario, Canada, September 1993, pp. H15–H24.
[11] Spalthoff, W. and Wilhelm, H., "The Use of Hydrogen Getters for Prevention of Hydrogen Embrittlement in Zirconium-Alloy Fuel Cans," *Zirconium in the Nuclear Industry, ASTM STP 458*, E. F. Baroch, Ed., American Society for Testing and Materials, West Conshohocken, PA, 1969, pp. 338–344.
[12] Coleman, C. E., "Effect of Texture on Hydride Reorientation and Delayed Hydrogen Cracking in Cold-Worked Zr-2.5 Nb," *Zirconium in the Nuclear Industry, ASTM STP 754*, D. G. Franklin, Ed., American Society for Testing and Materials, West Conshohocken, PA, 1982, pp. 393–411.
[13] Coleman, C. E., Sagat, S., and Amouzouvi, K. F., "Control of Microstructure to Increase the Tolerance of Zirconium Alloys to Hydride Cracking," AECL-9524, Atomic Energy of Canada Ltd., Chalk River Laboratories, Chalk River, Ontario, Canada, December 1987.
[14] Sagat, S., Coleman, C. E., Griffiths, M., and Wilkins, B. J. S., "The Effect of Fluence and Irradiation Temperature on Delayed Hydride Cracking in Zr-2.5Nb," *Zirconium in the Nuclear Industry, ASTM STP 1245*, A. M. Garde and E. R. Bradley, Eds., American Society for Testing and Materials, West Conshohocken, PA, 1994, pp. 35–61.
[15] Chow, C. K., Coleman, C. E., Koike, M. H., Causey, A. R., Ells, C. E., Hosbons, R. R., Sagat, S., Urbanic, V. F., and Rodgers, D. K., "Properties of an Irradiated Heat-treated Zr-2.5Nb Pressure Tube Removed from the NPD Reactor," this conference.
[16] Mills, J. W. and Huang, F. H., "Delayed Hydride Cracking Behavior for Zircaloy-2 Plate," *Engineering Fracture Mechanics*, Vol. 39, 1991, pp. 241–257.
[17] Huang, F. H. and Mills, W. J., "Delayed Hydride Cracking Behavior for Zircaloy-2 Tubing," *Metallurgical Transactions*, Vol. 22A, 1991, pp. 2049–2060.
[18] Aitchison, I. and Davies, P. H., "Role of Microsegregation in Fracture of Cold-Worked Zr-2.5 Nb Pressure Tubes," *Journal of Nuclear Materials*, Vol. 203, 1993, pp. 206–220.

[19] Davies, P. H., Hosbons, R. R., Griffiths, M., and Chow, C. K., "Correlation Between Irradiated and Unirradiated Fracture Toughness of Zr-2.5 Nb Pressure Tubes," *Zirconium in the Nuclear Industry, ASTM STP 1245,* A. M. Garde and E. R. Bradley, Eds., American Society for Testing and Materials, West Conshohocken, PA, 1994, pp. 135–167.

[20] Davies, P. H. and Shewfelt, R. S. W., "Link between Results of Small and Large-Scale Toughness Tests on Irradiated Zr-2.5Nb Pressure Tube Material," this conference.

[21] Fleck, R. G., Price, E. G., and Cheadle, B. A., "Pressure Tube Development for CANDU Reactors," *Zirconium in the Nuclear Industry, ASTM STP 824,* D. G. Franklin and R. B. Adamson, Eds., American Society for Testing and Materials, West Conshohocken, PA, 1984, pp. 88–105.

[22] Holt, R. A., "Mechanisms of Irradiation Growth of Alpha-Zirconium Alloys," *Journal of Nuclear Materials,* Vol. 159, 1988, pp. 310–338.

[23] Choubey, R., Aldridge, S. A., Theaker, J. R., Cann, C. D., and Coleman, C. E., "Effects of Extrusion-Billet Pre-Heating on the Microstructure and Properties of Zr-2.5 Nb Pressure Tube Materials," this conference.

Author Index

A

Adamson, Ronald B., 676, 726, 825
Ahlberg, Elisabet, 55
Aldridge, S. A., 657
Anada, Hiroyuki, 35, 74
Anderson, Thomas, 448
Armijo, J. Sam, 676
Averin, Sergey A., 603

B

Badie, N., 518
Bart, Gerhard, 218
Besch, Otto A., 805
Bibilashvili, Yury K., 785
Billot, Philippe, 94
Blank, Hubert, 218
Blat, Martine, 319
Bouffioux, Pol, 373
Bourgeois, W. F., 758
Bradley, E. Ross, 805
Broy, Y., 850
Busch, R. A., 850

C

Cann, C. D., 657, 884
Causey, A. R., 469, 518, 623
Charpentier, P. L., 758
Charquet, D., 865
Cheadle, B. A., 884
Cheng, B., 137
Chen, J.-S. Fred, 825
Choubey, R., 657
Chow, C. K., 469
Christodoulou, N., 518, 623
Coleman, C. E., 469, 657, 884
Comstock, Robert J., 710
Cox, Brian, 114

D

Davies, Pauline H., 492
de Carlan, Y., 628

Delobelle, Patrick, 373
Duffin, Walter J., 338

E

Efsing, P., 394
Ells, C. E., 469
Etoh, Yoshinori, 825
Eucken, Craig M., 805

F

Faldowski, Joseph A., 557
Franklin, David G., 338, 758
Furuya, Takemi, 163

G

Garde, A. M., 407
Garzarolli, Friedrich, 12, 181, 218, 541, 850
Gebhardt, Olaf, 218
Geyer, Philippe, 373
Gilbon, D., 628
Gilmore, P. M., 137
Godlewski, Joel, 94
Gohr, H., 181
Goll, W., 541
Goncharov, Vladimir I., 695
Graham, Ronald A., 74
Griffiths, M., 580, 623, 628
Grigoriev, V., 431

H

Hagi, Shigeki, 74
Harlow, Dr. John L., 295
Herb, Brett J., 74
Hermann, Armin, 218
Ho, E. T. C., 623
Holt, R. A., 518, 623
Hosbons, R. R., 469
Howe, Lawrence M., 557
Huang, Peter Y., 726
Hutchinson, Bevis, 55

I

Ikeda, Tadahiro, 825
Iltis, X., 242
Ishii, Yoshiaki, 825
Isobe, Takeshi, 203
Ita, Yoichi, 163

J

Josefsson, B., 431

K

Kammenzind, Bruce F., 338, 758
Klassen, R. J., 518
Klepfer, H. H., 137
Kobylyansky, Gennady P., 785
Koike, M. H., 469
Kolbenkov, Stanislav A., 603
Kotrehkov, Vladimir A., 785
Kozlov, Alexander V., 603
Kreyns, P. H., 758
Kuroda, Takahiro, 74
Kuzmenko, Nikolay V., 785

L

Lefebvre, Florence, 242
Lehtinen, Borje, 55
Lemaignan, Clément, 242, 628
Le Pichon, Isabelle, 373
Limback, Magnus, 448
Lositsky, Anatoly F., 785

M

Mae, Yoshiharu, 203
Maguire, M. A., 265
Mahmood, S. Tahir, 726
Manzel, R., 865
Markelov, Vladimir A., 603, 695, 785
Marlowe, Mick O., 676
Mecke, J. F., 580
Motta, Arthur T., 557
Murai, Takuya, 203

N

Nikulin, Sergei A., 695
Nikulina, Antonina V., 603, 785
Noel, Didier, 319
Nomoto, Ken-ichi, 74
Novoselov, Andrey E., 603, 785

O

Okamoto, Paul R., 557
Oscarsson, Anders, 55

P

Pêcheur, Dominique, 94
Peregud, Mikhail M., 603, 785
Peters, H. Richard, 295, 338
Pettersson, K., 394
Pirek, R. C., 407

R

Rand, Robert A., 676
Ray, Ian L. F., 218, 541
Regnard, C., 628
Robertson, J. A. L., 3
Robinet, Pascol, 373
Rodgers, D. K., 469
Rosborg, B., 431
Rudling, Peter, 55
Ruhmann, H., 181, 865

S

Sabol, George P., 710
Sagat, S., 469
Sauve, R., 518
Schaller, J., 181
Schoenberger, Gerald, 710
Seibold, A., 541
Sell, H.-J., 865
Shamardin, Valentin K., 785
Shevnin, Yuriy P., 785
Shewfelt, Robert S. W., 492
Shimada, Sachio, 825
Shishov, Vyacheslav N., 603, 695
Smith, G. P., 407
Stehle, H., 12
Steinberg, E., 12

T

Takeda, Kiyoko, 35
Takei, Kazuhiro, 825
Theaker, J. R., 657, 884
Thomazet, Joel, 94
Tome, C. N., 518

U

Ungurelu, Mihaela, 114
Urbanic, V. F., 469

V

Van Der Heide, P. A. W., 265

W

Warr, B. D., 265
White, C. J., 758
Wikmark, Gunnar, 55
Williams, C. David, 676

Winegar, J. E., 580
Wisner, Steven B., 676
Wong, Yin-Mei, 114
Woo, C. H., 518, 623
Woods, Keith N., 805
Wu, Chenguang, 114

Y

Yagnik, Suresh K., 805
Yasuda, Takayoshi, 825

Subject Index

A

Alloy development, 805
Alloying element, redistribution, 242
Alpha-phase
    extrusion-billet preheating effects on microstructure, 657
    niobium, 580
Amorphization, 557
    precipitates, 541
Analytical electron microscopy, zirconium alloys, 163
Anisotropic viscoplastic behavior, Zircaloy cladding tube model, cladding tubes, 373
Annealing
    final, model for effect on in- and out-of-reactor creep behavior, 448
    parameter, nodular corrosion, 865
    temperature, Zircaloys, 74
Anodic polarization curve, zirconium alloys, 163
Anodic protection, precipitates in Zircaloys, 203
Autoclave tests, 319, 710, 805

B

Beta phase, 557
    extrusion-billet preheating effects on microstructure, 657
Beta-solution treatment, Zircaloys, 74
Billet, effects on microstructure and properties, 657
Boiling water reactors
    cladding materials, nodular corrosion, 865
    long-time corrosion behavior, comparison of zirconium alloys, 850
    thermomechanical processing effects, Zircaloys, 726
    Ziracloy lined zirconium barrier fuel cladding, 676
    zirconium alloy development, 825
Boric acid, inhibition of degradation of zirconium alloys, 114
Burnup, high, zirconium alloy development, 825

C

CANDU reactor
    extrusion-billet preheating effects on microstructure and properties, 657
    modelling in-reactor deformation of pressure tubes, 518
Carbon dating, CANDU pressure tube corrosion, 265
Cathodic charging, 319
c dislocation loops, 628
Characterization, CANDU pressure tubes, 265
Charged-particle irradiation, 557
Chromium
    anodic protection, 203
    corrosion behavior and, 865
    zirconium alloy development, 825
Cladding tubes, Zircaloy, 74
    anisotropic viscoplastic behavior model, 373
    effect of final annealing on in- and out-of-reactor creep behavior, 448
    fracture toughness, 431
Composite, Zircaloy ductility, 407
Corrosion
    boric acid inhibition, 114
    CANDU pressure tubes, 265
    dependence on kind, size, and distribution of intermetallic precipitates, 541
    duplex and reference cladding, 805
    hydrogen pickup and redistribution, 338
    in-reactor, Zircaloys, 726
    LiOH acceleration, 114
    long-term in situ investigation, zirconium alloys, 181
    long-time behavior, comparison of zirconium alloys, 850
    modeling, PWR Zircaloy fuel cladding corrosion, 137
    nodular, in-BWR and out-of-pile, 865
    post-transition, 295
    pressurized water reactor Zircaloy fuel cladding, 137
    pretransition rates, 295
    rate, zirconium alloys, hydrogen role, 319
    steady-state, 295
    transition rates, 295

Corrosion—*continued*
  uniform, anodic protection, 203
  Zircaloys, 12, 55
  zirconium alloys, 785
    development, 825
    in-pile grown films, 218
    ZIRLO nuclear fuel cladding, 710
Corrosion resistance
  correlation with electrochemical properties of zirconium alloys, 163
  in-reactor Zircaloys, 726
  precipitates, 203
  Zircaloy lined zirconium barrier fuel cladding, 676
  Zircaloys, 35, 74
Crack growth resistance
  irradiated zirconium alloy pressure tubes, 492
  temperature and yield strength effect, 394
Crack propagation, ziracloy lined zirconium barrier fuel cladding, 676
Creep
  effect of final annealing, 448
  irradiated heat-treated zirconium alloys, 469
  Zircaloys, 12
  zirconium alloys, 785
Crystal structure, oxide on Zircaloys, 35

D

Deformation
  in-reactor, 518
  irradiated heat-treated zirconium alloys, 469
  irradiated zirconium alloy pressure tubes, 492
  service life, 884
Deformation equation, 518
Delayed hydride cracking
  delayed, temperature and yield strength effect on, 394
  irradiated heat-treated zirconium alloys, 469
  service life, 884
Deuterium, irradiated zirconium alloys, 469
Dislocation
  density, irradiation effects, 580, 603
  structure, 518
    effect of neutron irradiation, 603
Ductility
  irradiated Zircaloy, hydride precipitate localization and neutron fluence effects, 407
  microstructure effects, 695

E

E110, 785
E635, 785
EB zirconium, 203

Electrochemical impedance spectroscopy, CANDU pressure tubes, 265
Electrochemical measurements, long-term in situ corrosion investigation, zirconium alloys, 181
Electrochemical properties, correlation with corrosion resistance of zirconium alloys, 163
Electron microscopy, in-pile grown corrosion films, 218
Embrittlement
  reactor core materials, 758
  Zircaloy cladding tubes, 431
Extrusion, effects on microstructure and properties, 657

F

Fatigue, irradiated heat-treated zirconium alloys, 469
Film stress, oxide, 55
Finite element code, 518
Fracture resistance, microstructure effects, 695
Fracture toughness
  extrusion-billet preheating effects, 657
  irradiated heat-treated zirconium alloys, 469
  irradiated zirconium alloy pressure tubes, 492
  irradiation effects, 758
  service life, 884
  Zircaloy cladding tubes, 431
Fuel cladding
  duplex and reference, 805
  in-pile grown corrosion films, 218
  VVER and RBMK cores, 785
  Zircaloy-2 lined zirconium barrier, 676
  ZIRLO, corrosion behavior, 710

G

Gaseous charging, 319
Grains, 603
  interaction stresses, 518
Growth
  breakaway, 623
  zirconium alloys, 785

H

Hafnium, embrittlement, 758
Heat treatment
  effects on
    creep behavior, 448
    ductility and fracture resistance, 695
    ZIRLO corrosion behavior, 710
    zirconium alloy pressure tube properties, 469
High burnup fuel rods, Zircaloys, 137

SUBJECT INDEX   905

High-temperature water, zirconium alloy corrosion, 114
Hydrides
  duplex and reference cladding, 805
  embrittlement, 758
  precipitate, localization effects on irradiated Zircaloy ductility, 407
  precipitation, 338
    at metal/oxide interface, 319
  Zircaloys
    delayed hydride cracking, 394
Hydride volume fraction, Zircaloy ductility, 407
Hydriding
  ductility of Zircaloy, 407
  Zircaloy lined zirconium barrier fuel cladding, 676
Hydrogen
  concentration and service life, 884
  embrittlement, Zircaloy cladding tubes, 431
  pickup and redistribution
    alpha-annealed Zircaloy, 338
    zirconium alloy development, 825
  role on corrosion rate, zirconium alloys, 319
  solubility, 338
  uptake, CANDU pressure tubes, 265
Hydrogen thermal redistribution, effect on PWR Zircaloy fuel cladding corrosion, 137

I

Impedance spectroscopy
  in-pile grown corrosion films, 218
  Zircaloys, 181
    oxide on, 55
Impurities, effect on Zircaloy aqueous corrosion, 295
In-BWR corrosion, zirconium alloy comparison, 850
Infinite velocity, SIMS, 265
Inner Zircaloy liner, 676
In-pile corrosion, Zircaloys, 865
In-PWR corrosion, zirconium alloy comparison, 850
In-reactor
  deformation, 518
  E635 as fuel cladding material, 785
  final annealing effect on creep behavior, 448
  ZIRLO fuel cladding, corrosion behavior, 710
In-reactor corrosion, 12
  Zircaloys, 137
In-reactor creep, 12
In-reactor fuel performance, Zircaloy lined zirconium barrier fuel cladding, 676
in situ measurement, Zircaloy corrosion, 181
Intermetallic precipitates
  anodic protection, 203
  crystalline-to-amorphous transformation, 557
  effects on
    ductility and fracture resistance, 695
    in-PWR irradiation, 541
    PWR Zircaloy fuel cladding corrosion, 137
  irradiation effects, Zircaloys, 726
  structural changes during irradiation, 580
Internal variables model, anisotropic viscoplastic behavior, 373
Iodine stress corrosion, Zircaloys, 12
Iron
  anodic protection, 203
  corrosion behavior and, 865
  effects on
    ductility and fracture resistance, 695
    nucleation of c component dislocation loops, 628
  redistribution, Zircaloy oxide layers, 242
  supersaturation, 628
  zirconium alloy development, 825
Irradiation
  corrosion, duplex and reference cladding, 805
  effects on, 12
    dislocation structure and phase composition, 603
    ductility of Zircaloy, 407
    microstructure evolution, 580
    oxidation, 242
    Zircaloy mechanical properties, 726
    ZIRLO corrosion behavior, 710
  embrittlement, 758
  in-PWR, effect on size, structure, and composition of intermetallic precipitates, 541
  long-time corrosion behavior, comparison of zirconium alloys, 850
  microstructure effects, Zircaloys, 726
  nucleation of c component dislocation loops, 628
  phase transformations, in situ studies, 557
  Zircaloy
    fracture toughness, 431
    lined zirconium barrier fuel cladding, 676
    zirconium alloy pressure tubes, 492
    properties, 469
Irradiation creep, in-reactor deformation, 518
Irradiation growth, 603
  iron effect, 628
  non-linear, cold-worked Zircaloy, 623
  Zircaloys, 12
Irradiation hardening, Zircaloys, 726
Irradiation precipitation, 557

L

Laboratory tests, long-term corrosion behavior, zirconium alloy comparison, 850

Lattice parameters, 580
Lauer phase, 557
Laves phase, 541
Light water reactors, operating conditions, 12
Lithium, effect on oxidation rate, 94
Lithium hydroxide
    degradation of zirconium alloys, 114
    PWR Zircaloy fuel cladding corrosion, 137
Loops, 603

M

Manufacturing processes, Zircaloy lined zirconium barrier fuel cladding, 676
Matrix effects, inside surface oxide, 265
Matsuo creep model, 448
Mechanical properties, Zircaloys, 373
    ductility, 407
    post-irradiation, 726
Metal/oxide examination, 319
Microstructure
    duplex and reference cladding, 805
    effects on
        ductility and fracture resistance, 695
        microstructure and properties, 657
    evolution during irradiation, 557, 580
    extrusion-billet preheating effects, 657
    irradiation effects, Zircaloys, 726
    oxide films, 94
    ZIRLO fuel cladding, corrosion behavior, 710
Modeling
    anisotropic viscoplastic behavior, 373
    in-reactor deformation, 518
Molybdenum, zirconium alloy development, 825
Morphology, oxide on Zircaloys, 35
Multiaxial loadings, Zircaloys, 373

N

Neutron damage, oxide films, 137
Neutron irradiation
    effects on
        dislocation structure and phase composition, 603
        irradiated Zircaloy ductility, 407
        long-time corrosion behavior, comparison of zirconium alloys, 850
        microstructure evolution, 580
        Zircaloys, 12
Nickel
    corrosion behavior and, 865
    zirconium alloy development, 825
Niobium
    in alpha-phase, 580
    effect on ZIRLO corrosion behavior, 710
    zirconium alloy development, 825
Nodular corrosion, zirconium alloys, 163
Nuclear fuel
    case study, 3
    history, 3

O

Omega phase, 557
Out-of-pile corrosion, zirconium alloy comparison, 850
Oxidation
    irradiated heat-treated zirconium alloys, 469
    kinetics, Zircaloys, 242
    rate, lithium effect, 94
    saturation, zirconium alloy development, 825
Oxide
    characteristics, CANDU pressure tubes, 265
    growth rate, zirconium alloys, 55
    microstructure, waterside corrosion of Zircaloy cladding in lithiated environment, 94
    morphology, Zircaloys, 74
    waterside layer thickness, 407
Oxide film, Zircaloys, 35
Oxide layers, Zircaloys
    microstructure evolution and iron redistribution, 242
    thickness, 181
    PWR Zircaloy fuel cladding corrosion, 137
Oxygen partitioning, mechanical properties, 657

P

Pellet clad interaction, 676
Phase composition, effect of neutron irradiation, 603
Phase structure, during irradiation, 580
Phase transformations, under irradiation, zirconium alloys, in situ studies, 557
Pin-loading tension test, 431
Porosity, zirconium alloys, 55
Post-failure degradation, Zircaloy lined zirconium barrier fuel cladding, 676
Precipitates
    dissolution, 726
    intermetallic (*see* Intermetallic precipitates)
    irradiation, 557
        effect on size, distribution, and composition, 541, 603
    nodular corrosion, 865
    Zircaloys, 12
        anodic protection, 203
    zirconium alloys, 163

Pre-heat temperature, effects on microstructure and properties, 657
Pressure tubes, zirconium alloys
  cold-worked, non-linear irradiation growth, 623
  development, service life greater than 30 years, 884
  extrusion-billet preheating effects, 657
  irradiated heat-treated, 469
  modelling in-reactor deformation, 518
Pressurized water reactor
  duplex and reference cladding, corrosion behavior, 805
  in-pile grown corrosion films, 218
  long-time corrosion behavior, comparison of zirconium alloys, 850
  Zircaloys
    aqueous corrosion, 295
    corrosion kinetics, 242
    fuel cladding, 137
    hydrogen pickup and redistribution, 338

R

Radiation (see Irradiation)
Reactor core materials, embrittlement, 758
Recrystallization, effect on creep behavior, 448

S

Scanning electron microscopy
  in-pile grown corrosion films, 218
  irradiated Zircaloys, 726
  oxide on Zircaloys, 55, 74
Science policy, 3
Secondary ion mass spectrometry
  CANDU pressure tubes, 265
  extrusion-billet preheating effects, 657
  oxide films, 94
Self-consistent model, anisotropic deformation, 518
Service life, 30-year, 884
Soak time, effects on microstructure and properties, 657
Steam, corrosion and, 319
Strain rates, 518
Stress intensity factor, delayed hydride cracking, 394

T

Tellurium, zirconium alloy development, 825
Tensile properties, irradiated zirconium alloys, 726
  heat-treated, 469

Tensile strength, extrusion-billet preheating effects, 657
Test method, fracture toughness, 431
Tetragonal phase, 55
Texture, crystallographic, 518
Thorium, effect on Zircaloy aqueous corrosion, 295
Threshold stress intensity, delayed hydride cracking, 394
Tin content
  corrosion of zirconium alloys, 181
  effect on ZIRLO corrosion behavior, 710
  PWR Zircaloy fuel cladding corrosion, 137
Toughness tests, small- and large-scale, irradiated zirconium alloy pressure tubes, 492
Transmission electron microscopy
  in-pile grown corrosion films, 218
  microstructure evolution, 580
  oxide, 74
    films, 94
    microstructure, 35
    layers, Zircaloys, 242

U

Uranium, effect on Zircaloy aqueous corrosion, 295

V

Valences, zirconium alloy development, 825
Voids, formation, 758
Volume-controlled fracture model, 492

X

X-bar zirconium, 203
X-ray diffraction
  microstructure evolution, 580
  oxide on Zircaloys, 55

Y

Yield stress, effect on delayed hydride cracking, 394

Z

Zircaloys
  alpha-annealed, hydrogen pickup and redistribution, 338
  annealing temperature effect on corrosion behavior and oxide microstructure, 74
  aqueous corrosion, trace impurity uranium effect, 295

Zircaloys—*continued*
  behavior and properties in power reactors, 12
  cladding tube
    anisotropic viscoplastic behavior model, 373
    fracture toughness, 431
    VVER and RBMK cores, 785
  cold-worked, non-linear irradiation growth, 623
  hydrided, temperature and yield strength effect on delayed hydride cracking, 394
  in-BWR and out-of-pile nodular corrosion behavior, 865
  in-reactor corrosion and post-irradiation mechanical properties, 726
  irradiated
    ductility, hydride precipitate localization and neutron fluence effects, 407
    iron effect on nucleation of c component dislocation loops, 628
  lined zirconium barrier fuel cladding, 676
  model for effect of final annealing on in- and out-of-reactor creep behavior, 448
  oxide
    film microstructure, waterside corrosion in lithiated environment, 94
    layers, microstructure evolution and iron redistribution, 242
    microstructure, 35
  precipitates, anodic protection, 203
  pressurized water reactor fuel cladding, 137
  reactor core, embrittlement, 758
  tin content and corrosion, 137
  waterside corrosion in lithiated environment, 94
Zirconium alloys
  behavior and properties in power reactors, 12
  boric acid effect, 114
  CANDU pressure tubes, 265
  correlation electrochemical properties and corrosion resistance, 163
  corrosion
    duplex and reference cladding, 805
    effects of processing variables and alloy chemistry, 710
    in-pile grown films, 218
    long-time behavior, 850
    long-term in situ investigation, 181
    rate, hydrogen role, 319
    tin content and, 181
  development
    boiling water reactors, 825
    service life greater than 30 years, 884
  extrusion-billet preheating effects on microstructure and properties, 657
  fuel rod cladding, VVER and RBMK cores, 785
  hydrothermal redeposition, 114
  in-PWR irradiation, effect on size, structure, and composition of intermetallic precipitates, 541
  irradiated heat-treated, pressure tube properties, 469
  LiOH degradation mechanisms, 114
  microstructure effects on ductility and fracture resistance, 695
  neutron irradiation effect on dislocation structure and phase composition, 603
  oxide
    microstructure, 35
    morphology and oxidation rate, 55
  phase transformations under irradiation, in situ studies, 557
  precipitates, anodic protection, 203
  pressure tubes, modelling in-reactor deformation, 518
  Zircaloy lined zirconium barrier fuel cladding, 676
Zirconium barrier, Zircaloy lined, 676
ZIRLO, fuel cladding
  corrosion behavior, 710
  E635 as material, 785